Manual of Montana
Vascular Plants

Mount Jefferson in the Centennial Range
Photograph by Peter Lesica

Manual of Montana Vascular Plants

Peter Lesica

Contributions by Matt Lavin and Peter F. Stickney

Illustrations by Debbie McNiel, Rich Adams, Claire Emery

BRIT
PRESS

2012

ISBN 13-978-1889878-39-3

Editor: Barney Lipscomb
Botanical Research Institute of Texas
1700 University Dr.
Fort Worth, Texas 76107-3400, USA
Production: Peter Lesica
Cover Design: Nancy Seiler (nancyseiler.com)

Manual of Montana Vascular Plants
By Peter Lesica
Contributions by Matt Lavin and Peter F. Stickney
Illustrations by Debbie McNiel, Rich Adams, and Claire Emery

Front cover photographs top to bottom and left to right):
    Arnica cordifolia, Gentiana affinis, Amelanchier alnifolia, Coryphantha vivipara,
Carex glacialis, Eritrichium nanum, Sedum lanceolatum, Anemone patens, Pinus ponderosa, Sedum rho-
danthum, Calochortus macrocarpus (Photos by Peter Lesica)
Back cover and spine photographs
    Geranium viscosissimum, Smilax herbacea (Photos by Peter Lesica)
Maps courtesy of the Montana Natural Resources Information Service
and Montana Natural Heritage Program

Manual of Montana Vascular Plants
is published with the support of:
National Science Foundation, the University of Montana and the U.S. Forest Service, the Bureau of
Land Management, the National Park Foundation, the National Fish and Wildlife Foundation, Westech
Environmental Services, Cinnabar Foundation, and the Montana Native Plant Society

Distribution of copies by:
Botanical Research Institute of Texas
1700 University Dr.
Fort Worth, Texas 76107-3400, USA
Telephone: 1 817 332 4441
Fax: 1 817 332 4112
Website: http://www.brit.org/brit-press/
Email: jbrit@brit.org

BRIT
PRESS

30 June 2012

Printed in the United States of America
Third Printing, 2018

Dedication

*This book is dedicated to The Nature Conservancy Montana and the people
who worked there over the past thirty years and gave me the opportunity to explore the
state from the wetlands of the Missouri Coteau to the sagebrush steppe
of the Beaverhead Country: Bob Keasling, Cindy Williams, Joan Bird, Bee Hall,
Dave Carr, Brian Martin, and Dave Hanna.*

❀❀❀

*The book is also dedicated to my father, George, who first took me camping;
I love it now as much now as I did then.*

Manual of Montana Vascular Plants

*A Book to Promote Botanical and Environmental Education,
Scientific Exploration, and Conservation in Montana.*

*The Montana Native Plant Society has generously provided financial support for the publication
of this book. I would like to thank every member of the Society for their interest and assistance.*

Acknowledgments

The following people gave me advice or identified specimens in their field of expertise: Gerry Allen, Ihsan Al-Shehbaz, Ed Alverson, Wendy Applequist, George Argus, Lois Arnow, Duane Attwood, Gary Baird, Bruce Baldwin, Peter Ball, Harvey Ballard, Ted Barkley, **Rupert Barneby**, Carol Baskin, **Randy Bayer**, John Beaman, Katherine Beamish, David Bouford, T. C. Brayshaw, Ralph Brooks, Luc Brouillet, Steve Brunsfeld, Matt Carlson, **Ken Chambers**, Anita Cholewa, William Cody, **Lincoln Constance**, Dan Crawford, **Arthur Cronquist**, Garrett Crow, Alva Day, **Robert Dorn**, Tim Dickinson, Mark Egger, Reidar Elven, Patrick Elvander, **Barbara Ertter**, Erwin Evert, Carolyn Ferguson, Wayne Ferren, Walt Fertig, **Kanchi Gandhi**, John Gaskin, John Gillett, Joyce Gould, Shirley Graham, Jim Harris, Ron Hartman, Doug Henderson, Fred Hermann, Jim Hickman, Noel Holmgren, Pat Holmgren, Robert Kaul, David Keil, Tass Kelso, Job Kuijt, **Klaus Lackschewitz**, **Matt Lavin**, David Lellinger, Walt Mazior, Joy Mastrogiuseppe, Mildred Mathias, Mark Mayfield, Ronald McGregor, Dale McNeal, Bob Meinke, David Morgan, Nancy Morin, Caleb Morse, Gerry Mulligan, David Murray, Guy Nesom, Steve O'Kane, Dick Olmstead, Gerald Owenby, Alan Prather, Robert Price, Peter Raven, Jim Reveal, **Tony Reznicek**, **Reed Rollins**, Jack Rumely, **Ernie Schuyler**, John Semple, Leila Shultz, Doug Soltis, Rob Soreng, Alan Smith, Rich Spellenberg, Peter Stickney, Janet Sullivan, Jerry Tiehm, Rolla Tryon, Fred Utech, **Herb Wagner**, Warren L. Wagner, Bill Weber, Phillip Wells, Stan Welsh, Michael Windham, Alan Whittemore, Windham, Rich Whitkus, Dieter Wilken, Peter Zika.

Dave Hanna, Drake Barton, Steve Cooper, R. T. Hawke, Scott Mincemoyer, Wayne Phillips, Andrea Pipp, and Norm Weeden provided helpful comments on the manuscript. Michele Disney, Lisa Hensley, Becky Horn, Barney Lipscomb, Kathy Lloyd, Matthew Nordhagen, and David Rockwell helped prepare the manuscript for publication. Nancy Seiler designed the cover, and Ute Langner (Montana Natural Heritage Program) prepared the physiography map.

Funding was provided by the Bureau of Land Management, the U.S. Forest Service, the National Park Foundation, the National Fish and Wildlife Foundation, the Cinnabar Foundation, Westech Environmental Services, and National Science Foundation grant DBI 0447391 to Lila Fishman and Elizabeth Crone.

TABLE OF CONTENTS

Bitterroot (*Lewisia rediviva*), Montana's state wildflower
Painting by Debbie McNiel

Introduction

History of this book

It might be said that I began work on this book in 1972, when I first visited Montana and Glacier National Park. I have spent at least part of nearly every summer since then hiking and learning the flora. I started collecting herbarium specimens in Montana in 1978, and have recently passed collection number 10,000. Thanks to The Nature Conservancy, the Bureau of Land Management, the U.S. Forest Service, the National Park Service, and other land management agencies, I botanized and collected plants throughout Montana. During this time colleagues and I also prepared two monographs on Montana's rare and endangered plants[251,247], and discovered many species new to the state.[227,228,229,237,240,246,249,252,254] In 2002, I published a flora for Glacier National Park.[242]

In 2005, Lila Fishman and Elizabeth Crone were awarded a National Science Foundation (NSF) grant to digitize label information for Montana collections from the University of Montana Herbarium and make it available in an on-line database. I was hired to check determinations and update the nomenclature for the approximately 70,000 in-state specimens at the University of Montana (MONTU). Over the next four years I summarized habitat information and took measurements for each species from these specimens in order to describe morphology. That information forms the basis of this book. I also examined well over half of the Montana specimens at the Montana State University Herbarium (MONT). I recorded habitat and distribution information, but only rarely used MONT specimens for morphological descriptions. Shortly after the NSF project began, the Montana Native Plant Society received grants from the Bureau of Land Management and the U.S. Forest Service to commission illustrations for a new Montana flora.

Comprehensiveness

Nearly all taxa included as members of Montana's flora are supported by specimens in herbaria at the University of Montana, Missoula and/or Montana State University, Bozeman. Literature reports were sometimes given credence, especially if the location of a voucher could be identified. I attempted to verify dubious reports, and in some cases these turned out to be erroneous and are not included. I attempted to mention unverified literature reports, but these species are rarely included in the body of the text. This treatment is exclusive rather than inclusive, as I prefer conservatism to propagating misinformation. Nonetheless, I have attempted to include unverified reports and their sources even though they are not given full treatment.

All vascular plants currently known to occur in Montana are included in this manual's keys and descriptions. Hybrids are usually not included in the body of the text or the keys but may be mentioned, although those thought to be self-perpetuating are usually included.

The range of morphological measurements provided reflects pressed Montana specimens at MONTU and may be quite different from the range of variation found throughout the range of a species. Entire geographic ranges are given for the species as a whole, not the variety or subspecies occurring in Montana. County range maps reflect specimens at MONTU and MONT or credible reports in the literature and refer to the species as a whole. Habitat descriptions refer only to Montana. Approximate habitats and within-state ranges for intraspecific taxa are usually provided in the text. Global distribution information was taken from recent floristic monographs, especially the Flora of North America.[131]

Original illustrations were prepared from pressed specimens collected in Montana that I considered typical for the state. Most common genera are represented by at least one illustration. In general, I chose common rather than rare species for illustration. Plates are of two types: assemblages of three to five full-plant portraits; and genus collages consisting of one portrait of a representative species for that genus and an illustration of "critical morphological characters" for six to ten additional species. The majority of full-plant portraits first appeared in the Flora of Glacier National Park.[242]

Taxonomic philosophy

Providing a definition for "species" is no easy matter. The biological species concept which dictates species be reproductively isolated is hard to apply with plants because many reproduce sexually only rarely, and definitive studies have been done for only a small number of taxonomic groups. I prefer a concept more amenable to field biologists. For the purposes of this book, species are groups of populations that possess at least two covarying morphological characters consistent among them and unique from other groups. I recognize subspecific taxa only if they are either morphologically, ecologically, or geographically distinct in Montana. Differentiation between varieties and subspecies is far from consistent in the taxonomic literature. Rather than imposing my own philosophy in this matter, I have chosen to follow recent taxonomic treatments to avoid cluttering the literature with new and unneeded combinations.

Application of molecular genetics and methods of modern cladistics to plant systematics has resulted in a good deal of recent change in taxonomic nomenclature. Many proposed changes are based on statistically strong data which can't be denied. Unfortunately, some recently proposed name changes are based more on opinion than sound scientific evidence. There may be preliminary evidence suggesting that the traditional scientific names don't accurately reflect evolutionary relationships, but there is often not enough genetic or morphological evidence yet available to determine how the names should be changed to remedy the problem. New Linnaean binomials reflecting inadequate, preliminary evidence will often prove no better than the names they propose to replace. I have attempted to evaluate the recent literature to determine if new taxonomic proposals are justified. There is no way this process can be perfectly objective. In some cases there is a brief discussion of the taxonomic issues involved. I am certain that many of my decisions will prove to be incorrect. In general, I have chosen to err on the side of nomenclatural stability, maintaining names employed by other recent regional manuals unless there is strong evidence otherwise. I attempted to include taxonomic synonyms employed by floras in common use in Montana. [100-104,113,131,152,187] I included common names only when I believe they are truly in vernacular use. Most plant systematists are students of evolution, and having classifications that reflect evolutionary processes is, in the long term, a valuable goal. Unfortunately, in the short term this goal is at odds with the other function of taxonomic nomenclature–stability and standardization. I have done my best to walk the line between these two goals.

How to use this book

The book is organized phylogenetically by family following the treatment of Cronquist[99] and adopted by the Flora of North America Editorial Committee.[131] In a few cases I have adopted changes indicated by the Angiosperm Phylogeny Group;[384] for example, Chenopodiaceae is included in the Amaranthaceae, and the Scrophulariaceae has been recircumscribed. Species within genera and genera within families are arranged alphabetically.

Plants can be identified by means of "dichotomous keys." The first "couplet" (a pair of phrases with identical numbers) consists of two alternate, mutually exclusive statements. The reader must choose the statement that best describes the plant in hand. Following the correct statement is either the name of the family, genus or species the plant belongs to, or the number of another couplet. For example:

1. Plants slender, trailing subshrubs; flowers nodding . . . . *Linnaea*
1. Plants shrubs with erect stems 2

If the first (1) is correct then the plant is in the genus *Linnaea*. If the second (1) is correct then the reader must go to the second couplet (beginning with 2) and again choose the correct statement and continue until a name rather than a number is obtained. The key to the families is used to determine the correct family. Under the description for that family will be a key to the genera that is used to determine the correct genus. Under the genus description will be a key to the species that finally gives the name (scientific binomial) of the plant. Once the name of the plant is obtained, the description for that species in the text should be checked against the plant in hand to make sure it is correctly identified. Characters common to all members of a family or genus are given under the descriptions for that family and genus respectively, and in order to save space are usually not repeated in the species descriptions. A full description requires reading descriptions for family, genus, and species. Botanical terminology is defined in the glossary. Taxonomic references are provided at the end of the description for families and genera.

United States state abbreviations are those used by the U.S. Postal Service. Abbreviations for Canadian provinces are those used by the Canadian Postal Service. These and other abbreviations can be found in the glossary.

Elevational ranges are provided in a qualitative rather than quantitative manner (plains, valleys, montane, subalpine, alpine) as explained under Vegetation Zones. Endemic plants are those with a small global range. I have subjectively chosen to call endemic those taxa with a total range equal to or smaller than one-third the area of Montana, regardless of how much of that range is in Montana.

History of Botanical Exploration in Montana

Many people have collected plant specimens in Montana since the Lewis and Clark expedition. This brief account considers only those collectors who are well-known or who made what I consider significant contributions to the knowledge of botany in the state. Additional accounts of botanical exploration in Montana are provided by Blankenship,[47] Thompson,[393] Knowles and Knowles,[220] Lackschewitz,[226] and Lesica.[242] Specimens from most early Montana collectors are deposited at herbaria outside the state.[47]

Botanical exploration of Montana began early with the Lewis and Clark expedition in 1805-1806. Specimens collected in Montana on the trip west were lost. Approximately 30 specimens from mainly western Montana were collected by **Meriwether Lewis** on the return trip. More than a dozen of these proved to be new to science, and several were named for Lewis by Frederick Pursh. These include *Lewisia rediviva*, *Linum lewisii*, and *Philadelphus lewisii*.

The next significant plant collections from Montana were made by lone explorers. In 1833, **Nathaniel Wyeth**, a fur trader, collected plants on his trip east at the request of Thomas Nuttall. He crossed western Montana from the lower Flathead to the Big Hole Valley. Then after traveling in Idaho and Wyoming, reentered Montana on the Big Horn River and floated down the Yellowstone River. Some of his collections, including *Wyethia amplexicaulis*, were new to science and described by Nuttall. Unfortunately, his notes do not allow accurate location determination for many of his collections. In 1843, **Charles Geyer**, a German botanist, traveled to Wyoming with William Drummond and then went north into Montana. He entered Montana from Henry's Lake, Idaho, and explored the Madison, Beaverhead, Big Hole, Bitterroot, and Clark's Fork valleys. Geyer's notes provide descriptions of the vegetation of western Montana.[140]

Significant plant collections resulted from several organized expeditions in the middle of the 19th Century. **Ferdinand Hayden**, primarily a geologist, collected plants along the Yellowstone and Missouri rivers in eastern Montana in 1854 to 1855, and in 1860 as a member of expeditions to explore the western United States. During this time he discovered *Rorippa calycina*. Hayden also led an expedition to the newly-created Yellowstone National Park in 1871. Some of his plant collections were from adjacent Montana. The Stevens Expedition also explored the northern tier of states from 1853 to 1854, looking for a new railroad route. The physician, **George Suckley**, accompanied the expedition and collected plants in northeastern Montana. Among his discoveries were *Suckleya suckleyana* and *Atriplex suckleyi*. The North American Boundary Commission, a joint British-American operation, was established to mark the international boundary from the Pacific Ocean to the summit of the Rocky Mountains in 1857. In 1858, **David Lyall** was designated surgeon and naturalist to the British contingent of the commission. William Jackson Hooker at Kew Gardens in England was influential in having Lyall appointed to the commission because he was an experienced plant collector. The British contingent of the commission reached the Rockies in the summer of 1861. Lyall collected about 6700 specimens of 1375 species during the four seasons of the survey. Lyall collected plants in both British Columbia and Montana, and it is not always known where his collections were made. He collected several species new to science, many of which are common in Montana, including *Larix lyallii*, *Phacelia lyallii*, and *Penstemon lyallii*.

Towards the end of the 19th Century, Montana was being actively surveyed for mineral wealth and transportation routes. **Frank Tweedy** was a topographer for the U.S. Geological Survey in Montana and adjacent Wyoming in 1884 to 1888, and was the first person to collect extensively in southwest Montana, a part of the state with high endemism. He made the first collections of *Chionophila tweedyi*, *Erigeron tweedyi*, and *Vaccinium globulare*, among others. In 1889 to 1891, Tweedy worked and collected plants in Montana's south-central counties as well. Some of Tweedy's duplicates are deposited at MONT. Frank Tweedy published a small monograph on the flora of Yellowstone National Park.[400] Both **William Canby** and **Frank Lamson Scribner** were members of the Northern Transcontinental Survey and collected plants in western Montana in 1883. Canby's first collections of *Ligusticum canbyi*, *Draba lonchocarpa*, *Synthyris canbyi*, and *Cardamine rupicola* come from the mountains of northwest Montana. Scribner also collected several species new to science from Montana, but his collections of *Agropyron scribneri*, *Melica spectabilis*, and *Calamagrostis montanensis* come from near Helena and Great Falls. **Sereno Watson** of the Gray Herbarium at Harvard University surveyed forests in the headwaters of the Missouri River and Bitterroot Valley in 1880. He also made numerous plant collections, including the type collections of *Astragalus terminalis* in Beaverhead Co. and *Prenanthes sagittata* north of Missoula.

Montana began to have its own resident botanists starting at the end of the 19th Century. The first of these was **Robert S. Williams**, who moved to Montana as a young man in 1879. He soon took up residence in what was to become the city of Great Falls with **Fred Anderson**, an early Montana student of algae. For the next 13 years Williams collected plants in the valleys and mountains around Great Falls. He discovered *Conimitella williamsii* in the Belt Mountains east of Great Falls. After the Great Northern Railroad went over Marias Pass to Columbia Falls in 1891, Williams explored and collected plants west of the Continental Divide in the Flathead Valley and what is now Glacier National Park. He collected *Papaver pygmaeum* and *Carex concinnoides*, both of which were new to science. In 1897 to 1898, he collected plants while helping to survey the eastern boundary of what was to become Glacier National Park. Williams collected mosses during the 20 years he was in Montana, and bryophytes eventually became his specialty at the New York Botanical Garden. The vascular plant collections of R.S. Williams are deposited at MONT and MONTU. Some collections made by Fred W. Anderson in the Great Falls area from 1883 to 1889, are among the oldest at MONTU.

Francis Kelsey was another of Montana's earliest resident botanists. In the spring of 1885, Kelsey and his wife moved to Helena where he became pastor of the Congregational Church. Although Kelsey's primary interest appears to have been mycology, he made over 700 collections of vascular plants.[120] These are mainly from the Helena and Deerlodge valleys and surrounding areas. Six of his vascular plant specimens serve as nomenclatural types for species still recognized today, including *Kelseya uniflora*, *Phlox kelseyi*, *Astragalus atropubescens*, and *Dodecatheon conjugens*. He left Montana in 1892, to eventually become curator of the herbarium at Oberlin College in Ohio. Many of his specimens are deposited at MONT and MONTU.

Joseph Blankenship was Montana's first herbarium curator. He served as professor of Botany at Montana State University (then Montana Agricultural College) from 1898 to 1904. He collected plants throughout the state, but focused primarily on the mountainous portion during this time and for several years after. He collected the type of *Impatiens ecalcarata* in Sanders Co. in 1901. In 1905, Blankenship went to work for the Anaconda Mining Company, helping them defeat farmers of the Deerlodge Valley in a law suit claiming crop damage by Anaconda's smelter. From then on, he resided mainly in California and worked as a consulting plant pathologist for mining companies. Joseph Blankenship published several articles on Montana botany.[46-48] His Montana collections are at MONT and MONTU.

Per Axel Rydberg spent at least part of his summers in Montana between 1895 and 1897. The first two years he was working for the U.S. Division of Agrostology. In 1896, Rydberg was accompanied by **John Flodman**, but the two apparently spent much of their time collecting separately. Flodman explored the Madison, Bridger, Little Belts, and Crazy mountain ranges and collected the type specimens for *Cirsium flodmanii, Antennaria umbrinella, Aster junciformis,* and *Glyceria elata*. By 1897, Rydberg had moved to the New York Botanical Garden. That year he was accompanied to Montana by **Charles Bessey** of the University of Nebraska. The two collected in the mountains and valleys around Bozeman and discovered many species new to science, many of which Rydberg subsequently published in the first flora of Montana.[347] These include *Antennaria anaphaloides, Castilleja pulchella, Cirsium canovirens, Musineon vaginatum,* and *Penstemon aridus*. He subsequently published a flora for the Rocky Mountains.[348] Many of Rydberg and Bessey's duplicates are at MONT.

Morton Elrod was the first professor of Biology at the University of Montana (then Montana State University) from 1897 to 1933, and the founder of the University of Montana Biological Station. Although primarily a zoologist, he collected plants around Missoula, the Flathead Lake area, and Glacier National Park between 1897 and 1928. In 1901, he was accompanied in his collecting trips in the Mission and Swan ranges by **Daniel T. MacDougal**, a botanist from the New York Botanical Garden, and many of MacDougal's duplicates from that summer are housed at MONTU. Elrod also accompanied Marcus Jones on two collecting trips in 1909 to 1910. By the end of the first decade of the Twentieth Century, UM had grown, and it was decided to hire a botanist.

Marcus Jones was a respected western botanist hired by Deerlodge farmers to testify against the Anaconda Company for its alleged smoke damage to their crops. This case was the reason for his first trip to Montana in 1905, and when he first met Morton Elrod and his rival expert witness, Joseph Blankenship. He became friends with Elrod and in 1908, returned to Montana to teach at the UM Biological Station and collect plants for the next three summers. He collected mainly in the Flathead region, including Glacier National Park and Beaverhead Co. In 1909 and 1910, he took long collecting trips in Glacier National Park with Morton Elrod and Joseph Kirkwood. He was the first to collect *Cryptantha sobolifera*, as well as several alpine plants still considered rare today. Many of his Montana duplicates are at MONTU.

The tradition of amateur botanists was kept alive in the middle of the 20th Century by **Pliny Hawkins**, a school teacher, banker, and real estate speculator who lived in Columbus and Absarokee. He collected diligently in the valleys of south-central Montana between 1903 and 1933, and in Glacier National Park in 1917 and 1920. He received an MS in Botany from MSU in 1903, attended the University of Wisconsin in the winters of 1910 to 1922, and received an honorary doctorate from MSU in 1933. Hawkins published a book on the woody plants of Yellowstone National Park in 1924. His collections are at MONT (ca.8,000 specimens) and the Smithsonian (3,000 specimens). **Wilhelm Suksdorf**, a well-known Pacific Northwest botanist, had a nephew north of Livingston and made at least two collecting trips to the area, one in 1916, and a three-month trip in 1921 when he was 71 years old. Many of his Montana duplicates are at MONTU.

Some of the most significant collections housed in the University of Montana Herbarium have been made by its curators. **Joseph Kirkwood** was a professor of Botany at the University of Montana from 1909 to 1928. His collections span almost that entire period. They were all made in western Montana, especially around Missoula, the Bitterroot Valley, and what was to become the Selway-Bitterroot Wilderness. He formally established the UM Herbarium (MONTU) in 1926, and his collections are at MONTU and MONT. **C. Leo Hitchcock**, coauthor of Vascular Plants of the Pacific Northwest[187] and Flora of the Pacific Northwest,[188] also spent time collecting in Montana. After graduate school in St. Louis and a year teaching at Pomona College, Hitchcock became professor of Botany at the University of Montana in Missoula in 1932, and stayed until 1937, when he took a position at the University of Washington. During his time in Montana he collected mainly west of the Continental Divide, especially in the Missoula area and Glacier National Park, but also around Harlowton. He returned to collect along the Front Range of the Rockies in 1948. A large number of his specimens are deposited at MONTU. **The National Youth Administration** (NYA) was created in 1935, during the Great Depression. The "Biological Project" of NYA hired students aged 16-25 to collect plant specimens for museums and universities. Hitchcock supervised botany NYA projects at the University of Montana. During the spring and summer of 1936 and 1937, approximately 70 high school students from across the state were employed to collect and press plants. MONTU has NYA collections from 31 of our 54 counties, with most collections coming from Flathead, Carter, Deerlodge, Custer, Silver Bow, and Wibaux counties.

From 1946 through 1977, the University of Montana herbarium was curated by **LeRoy H. Harvey**. During his first 13 years at UM, Harvey spent summers at the UM Biological Station at Yellow Bay collecting plants throughout nearby Glacier National Park, continuing a tradition that began with Morton Elrod. He collected hundreds of specimens in the Park. In 1954, he published a list of approximately thirty new records for Glacier National Park, most of which he had collected during the forays of the previous seven years.[163] He gave up teaching at the Station after 1963 in order to have his summers free for collecting *Eragrostis* in Mexico and the adjacent U.S., and retired in 1977. **Fred Barkley** came to the University of Montana from Washington University in 1937, and was appointed professor of Botany and Curator of the Herbarium. He collected extensively in western Montana, especially around Missoula, the Flathead Valley, and Glacier National Park from 1937 to 1939.

Several graduate students at the University of Montana made large collections while living in western Montana. **Eugene Addor** collected hundreds of specimens of wind-pollinated plants from the Missoula area in 1957, as part of a project to understand urban, air-borne pollen loads. **Albert Finley** worked on a floristic study of the Front Range of the Rockies near Choteau in 1963 and 1964, and collected several hundred specimens. In 1964, **Richard Pemble** collected extensively in Glacier National Park and the Bitterroot Range while working on his thesis describing alpine phytogeography. **Dennis Woodland** collected extensively in the Cabinet Range of Lincoln Co., as well as around Missoula and Jordan, from 1966 to 1967. **David Clark** collected *Penstemons* from throughout the western U.S., including many from Montana from 1968 through 1970, as part of his dissertation on speciation. In 1980, **John Pierce** collected plants for a floristic study of the Big Hole National Battlefield in Beaverhead Co.

No botanist has contributed more to the knowledge of Montana's flora than **Klaus Lackschewitz**. His collecting for the UM herbarium began in the mid-sixties and continued unabated for the next thirty years. During this time he collected over 12,000 specimens, including 38 published state records and four species new to science: *Physaria humilis, Physaria klausii, Astragalus lackschewitzii*, and *Erigeron lackschewitzii*. Lackschewitz's longest and most concerted floristic research was his study of the Bitterroot and Sapphire ranges and the intervening valleys, ending with publication of Vascular Plants of West-central Montana by the U.S. Forest Service in 1991.[226] Lackschewitz was also active in other parts of the state during this time. He frequently explored the Anaconda Range southeast of the Bitterroot Valley. For many years during the late 1970s and early 1980s, Lackschewitz made annual pack trips into the Bob Marshall Wilderness and the Front Range. Other favorite alpine areas were the Gravelly Range and the Beartooth Mountains. There is no doubt that the high country was Lackschewitz's first love. Nonetheless, he was no stranger to the lower elevation valleys and plains. In 1978, he conducted a floristic survey of the newly created Charles M. Russell National Wildlife Refuge. He also spent part of one summer collecting in Rosebud County in the southeastern part of the state. In the latter half of the 1980s he spent more time in the mesic valleys of Lincoln and Sanders counties in northwest Montana. Lackschewitz's collections are housed at MONTU. Duplicates have been sent to major herbaria around the country. **Jacqueline Cory** collaborated with Lackschewitz by also collecting throughout Ravalli Co. and depositing collections at MONTU.

Frank Rose lived in Missoula and collected and processed seeds and whole plants for drug companies, as well as for the horticulture trade. His collecting took him to eastern as well as western Montana and northern Idaho throughout the 1930s and 40s. While in the field Frank made museum collections in addition to his commercial enterprises. His plant collections are from the Missoula area, but also from the Mission Valley, and the Little Belt, Pryor, and Beartooth ranges. Rose expressed special delight in collecting uncommon plants, travelling countless miles in a World War II surplus jeep. It was Frank Rose who introduced Klaus Lackschewitz to Montana's flora and initiated him into collecting plants for museum specimens. **Wally Albert** lived in or around Missoula all of his life, but started collecting plants in the late 1970s and continued into the 21st Century. He and Klaus Lackschewitz were close friends, and Albert brought in many unusual specimens collected mainly in the Bitterroot Valley. Toward the end of this time he specialized in wetland and aquatic plants, collecting several state records. His collections are at MONTU.

Marie Mooar was a high school biology teacher from Michigan who, at the age of 54, was hired by the University of Montana to collect aquatic and wetland plants in western Montana in the summer from 1965 through 1970. When the money ran out she retired from teaching, moved to Polson, and continued to collect for the next five years. More than 4,100 of her specimens were accessioned into MONTU. Many more were traded to other herbaria.

William E. Booth became professor and curator of the herbarium at Montana State University (then Montana State College) in 1941, after obtaining his doctorate from the University of Kansas. After serving in World War II, Booth began collecting throughout the state in earnest. In addition to collecting in the mountains and valleys around Bozeman, he made collecting trips to many parts of eastern Montana and northwestern Montana, including Glacier National Park. He published the first real flora for Montana in two parts in 1950 and 1959.[51,52] The second volume was coauthored by **J. C. Wright**.

Federal land management agencies played an important role in the botanical exploration of Montana in the latter part of the 20th Century. **Wilfred W. White** was a forester stationed in Missoula. He began his career in 1907, and was Bitterroot National Forest Supervisor from 1909-1921. He made a few plant collections in western Montana during that time. It was not until 1928, a year before he retired from the Forest Service, that his interest in botany really flared. Starting around 1934 through 1947, his interest focused on willows and he collected hundreds of specimens from as far away as Petroleum and Valley cos., culminating in a publication on willows in 1951.[437] White's collections are at MONTU. The U.S. Forest Service had a herbarium in Missoula from 1911 through 2010. **Peter F. Stickney** was curator of this herbarium starting in 1957, presumably the same time he began collecting plants in Montana. His collections span 50 years and were taken throughout western Montana with special focus on woody plants, especially the Ericaceae. In 2010, his herbarium was incorporated into MONTU with many duplicates at MONT. During this same time **Virginia Vincent** made collections from the area around her Forest Service lookout tower in Mineral Co. and deposited them at MONTU.

When the Bureau of Land Management needed a floristic inventory of the Centennial Range, Beartrap Canyon (Madison Co.), and the Humbug Spires (Silver Bow Co.), they hired **Porter Lowry**, who had just finished a M.S. at the University of Illinois. Lowry spent five months in 1979 collecting plants throughout much of southwestern Montana. His duplicate collections are at MONT and MONTU, and he compiled and published an annotated list of 242 species for Beartrap Canyon.[259] An interest in conservation by land management agencies and the public at large sparked renewed interest in plant collecting in the 1980s and 1990s. **J. Stephen Shelly, Maria Mantas, James Vanderhorst, Scott Mincemoyer**, and **Bonnie Heidel** all made numerous collections in Montana and deposited them at MONT and MONTU.

Numerous other well-known botanists have collected plants in Montana. These include **Bassett Maguire, Frederick J. Hermann, A. S. Hitchcock, Rupert Barneby, Reed Rollins**, and **Warren**

H. Wagner. **Arthur Cronquist** collected in the Beartooth Range and around Missoula in 1955, with duplicates at MONTU, and **Robert Dorn** collected plants in southern Beaverhead Co. in 1968, while a student at MSU, and Carbon Co. in 1976, with duplicates at MONT.

Abiotic Description of Montana

Physiography and Geology

Only a brief outline of Montana's physiography and geology is presented here. More detailed accounts can be found in several books and monographs.[10,281,410] Montana encompasses 376,564 square kilometers (145,392 square miles) and is the fourth largest state in the United States. It lies immediately south of the Canadian provinces of British Columbia, Alberta, and Saskatchewan. The northeastern two-thirds and southeastern one-third of the state are the western edge of the Northern Great Plains. This Great Plains section can be divided into the Glaciated Plains north of the glacial terminus, and the Unglaciated Plains in the southeast. The Glaciated Plains support areas of glacial pothole wetlands amongst level to rolling terrain. Much of the southern unglaciated plains is traversed southwest to northeast by low ridges that are foothills of the Big Horn Mountains of Wyoming. Tectonic activity has been uncommon in eastern Montana. Rocks of eastern Montana are almost all sedimentary sandstones, mudstones, and shales with little displacement. The greatest geologic force has been the Pleistocene continental glaciers. They covered the state along a line from near the Yellowstone River at the eastern border to along the Missouri River in central Montana. Terrain on the Glaciated Plains is gentle and rolling with steep breaks along the Missouri River and major tributaries. The surface is often covered with a layer of glacial drift. Marine shales are common and can form very fine-textured soils that shrink and swell with wetting and drying. Soils are more coarse-textured where sandstones predominate. Sandstones of the Fort Union Formation dominate the Unglaciated Plains of southeastern Montana, although marine shales are common south of Ekalaka and north of Baker. Terrain is more rolling than on the Glaciated Plains, and incised drainages between well-defined ridges are more common. Coarse-textured soils predominate in the uplands.

The rise and subsequent erosion of the Rocky Mountains dominates the geology of western Montana. The Continental Divide between the Columbia and Mississippi River drainages crosses the western third of the state from north to south. There are at least 40 named, mainly north-south trending mountain ranges in the state. These range from 2130 to 3960 m (7,000-13,000 ft.) in elevation and may rise 1200 to 1830 m (4,000-6,000 ft.) above the valleys. In general, mountains and valleys occupy higher elevations in the southern part of the state. Some mountain ranges, especially those in the northwest, are sedimentary rocks uplifted along faults. Parent materials in the northwest are predominantly Precambrian Belt Series. Soils in these ranges are derived mainly from argilites, quartzites, and limestones. The former two form near-neutral, sandy or silty soils, while limestone forms basic, calcareous soil. Limestones of the Belt Series are not as calcareous as those of Paleozoic or Mesozoic origin. Granitic basement rock was uplifted along faults to form the Beartooth, Tobacco Root, Madison, and other southwest and central Montana ranges. The overlying sedimentary rocks have mostly been eroded off of these ranges. Several ranges in southwestern Montana, such as the Bitterroot and Pioneer ranges, are granitic batholiths sometimes surrounded by uplifted sedimentary rocks. Smaller granitic intrusions form some ranges in central Montana, such as the Crazy and Judith mountains. These granitic parent materials form sandy, acidic soils. Volcanic activity was also common in southwestern and south-central Montana, forming at least part of the Absaroka and Gallatin ranges. Valleys between mountain ranges are filled to various depths with Tertiary-age alluvium and colluvium. Continental and mainly valley glaciers of the Pleistocene Era carved the mountains and deposited glacial till. Valley soils are mostly silty to loamy; however, valleys that were beneath glacial lakes in the Pleistocene may have very fine-textured soils where glacial silt was deposited.

Climate

Montana is a large state with a great deal of topographic variation, so a variable climate is to be expected. The Continental Divide, the mountainous divide running northwest to southeast between the Columbia and Missouri drainages, is the single most significant determiner of the state's climate. Climate is modified northern-maritime west of the Continental Divide. Most moisture comes from the west. Winters are milder, precipitation is more evenly distributed throughout the year, summers are cooler in general, and winds are lighter than on the east side. West of the Continental Divide in Montana topography consists entirely of mountain ranges with high precipitation, and intervening valleys that can be quite dry

Physiography of Montana

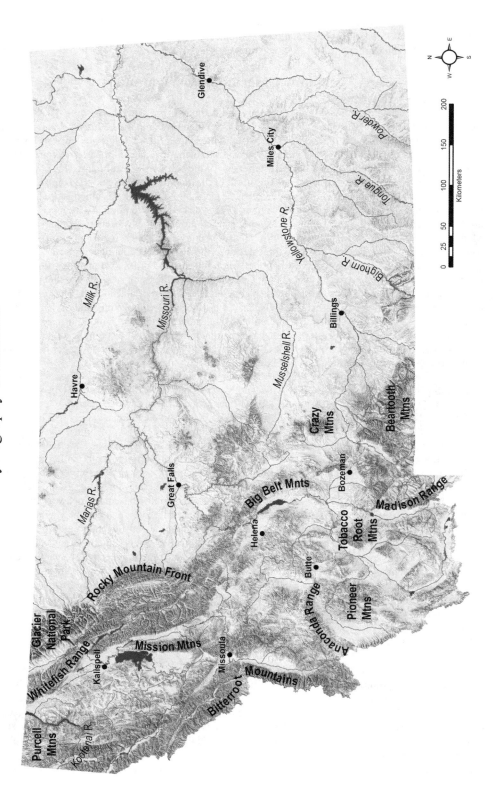

from being in the rain shadow of the north-south trending ranges. The Deer Lodge Valley averages of 280 mm (11.0 in.) of rain annually, while the highest average for an occupied valley in the state is 880 mm (34.7 in.) at Heron in the lower Clark Fork Valley at the western edge of the state. Missoula, the largest city west of the Continental Divide, receives an average of 345 mm (13.6 in.) of precipitation annually with a mean July maximum of 29.2°C (84.6°F) and a mean January minimum of -9.3°C (15.2°F).[436] The Bitterroot Range, about 48 km (30 mi.) southwest of Missoula, receives average annual precipitation of 2030 mm (80 in.).[404] Most precipitation falls between October and June, with May and June being the wettest months. There is more cloudiness in the west in all seasons and humidity is generally higher. Valleys west of the Continental Divide are usually protected from cold arctic air masses moving south from continental Canada by the mountains of the Divide; however, cold air driven by strong east winds will sometimes flush west through mountain passes in the winter. Warmer, less windy winters coupled with cloudier, more humid summers, provide a more temperate climate for vegetation west of the Continental Divide. High-pressure systems developing over the Columbia Plateau in winter cause frequent temperature inversions that trap cold air and fog in the valleys.

Topography east of the Continental Divide is more heterogeneous than on the west side. The northern half is relatively level terrain with some small isolated mountain ranges, while the south has mainly north- to south-trending, high mountain ranges separated by broad, high valleys. Southeastern Montana is primarily low foothill ridges separated by narrow valleys. The climate of eastern Montana, especially the Glaciated Plains north of the Missouri River, is semi-arid continental with hot summers and cold winters. Glasgow, in the Glaciated Plains of northeastern Montana, receives an average of 288 mm (11.3 in.) of precipitation annually, with a mean July maximum of 29.3°C (84.8° F) and a mean January minimum of -16.8°C (1.7° F).[436] Precipitation occurs mainly in spring and summer east of the Divide, and varies widely depending largely upon topographic influences, with May and June the wettest months. In the winter, cold air from Canadian high-pressure systems frequently moves into Montana east of the Divide. Between cold waves there are periods, sometimes longer than 10 days, of mild but often windy weather. These warm, windy winter periods occur almost entirely along the eastern slopes of the Divide and are popularly known as "chinooks." The chinook belt extends from the Browning-Shelby area southeastward to the Yellowstone Valley above Billings. The large winter fluctuations in temperature associated with chinook winds east of the Divide are particularly detrimental to woody plants with living tissue above ground. Summer temperatures in the southwestern and south-central parts of the state are somewhat moderated by higher elevations, and valleys are drier and warmer than in the mountains. Summer precipitation originating in the Gulf of Mexico or Gulf of California is common east of the Divide relative to the west side. The driest area of state is along the Wyoming border south of Billings in the rain shadow of the Beartooth Range, with an average annual precipitation of 167 mm (6.59 in.). Lame Deer, in the hill country of southeastern Montana, receives an average of 394 mm (15.51 in.) of precipitation annually, with a mean July maximum of 31.6°C (88.9°F) and a mean January minimum of -14.5°C (5.9°F).[436] Ennis, in the Madison River Valley of southwestern Montana, receives an average of 302 mm (12.09 in.) of precipitation annually with a mean July maximum of 28.3°C (83.1°F) and a mean January minimum of -10.3°C (13.5°F).[436] Mountains on either side of the Madison Valley receive 1000 to 1270 mm (40 to 50 in.) of precipitation annually.[404]

Vegetation of Montana

Vegetation of Montana is diverse, due primarily to the great topographic relief as well as the size of the state, which provides great variation in environmental factors that affect plant growth and ultimately the vegetation. Climate of the Great Plains portion of the state is relatively homogeneous semi-arid. Temperature and therefore evapotranspiration do increase going from north to south, but there is little change in annual precipitation. Most vegetation patterns on the Great Plains are due to parent material and soils. Climate is much more variable in the mountainous portion of Montana. Temperature generally decreases while precipitation and wind increase with elevation. However, aspect greatly modifies these effects in dissected mountainous terrain. North- and east-facing slopes and valley bottoms are cooler because they receive less direct sunlight and may be wetter when wind redistributes snow to lee slopes. West- and south-facing slopes are warmer and drier because incident radiation is received when ambient air temperature increases over the course of the day.

Merriam recognized that increases in elevation generally result in vegetation changes similar to those found when traveling north in the Northern Hemisphere at the same elevation, and this is the basis of his life zone classification.[276] Five Merriam life zones occur in Montana: Upper Sonoran, Transition, Canadian, Hudsonian, and Arctic-Alpine. I use a five-zone classification based on vegetation that roughly corresponds to the Merriam system. I sometimes differentiate lower and upper subalpine zones that roughly correspond to Merriam's Canadian and Hudsonian zones respectively. The last four zones occur only in the mountainous portion of the state.

This Book	Merriam Zone
1. Plains	Upper Sonoran/Transition
2. Valleys	Upper Sonoran/Transition
3. Montane	Transition/Canadian
4a. Subalpine (lower)	Canadian
4b. Subalpine (upper)	Hudsonian
5. Alpine	Arctic-Alpine

"Plains" refers to the Great Plains portion of the state. Although "plains" and "valleys" are always the lowest elevations in any given area, I differentiate plains from valleys because the Great Plains generally receive more growing-season precipitation than the rain-shadowed valleys in the mountainous portions of the state (valleys), especially in the northwest (see Climate section). Associating plant distribution with vegetation zones is more informative than elevation because vegetation zones occur at higher elevations as one goes south in the state. For example, the alpine zone begins at 2130 m (7,000 ft.) or even lower in Glacier National Park, but is above 2900 m (9,500 ft.) in the area around Yellowstone National Park.

Plains Zone

Riparian and wetland areas occur along rivers and major streams and are dominated by *Populus deltoides* (plains cottonwood).[156] Willows, *Salix exigua, S. bebbiana, S. amygdaloides, S. discolor, S. eriocephala*, as well as *Cornus sericea* (red-osier dogwood) are common understory species in early-seral communities. *Tamarix ramosissima*, an introduced shrub, can be common, especially along rivers regulated by dams. Common ground layer plants in early seral riparian forests include *Phalaris arundinacea* (reed canarygrass), *Carex aquatilis, C. pellita*, and broad-leaved forbs such as *Mentha arvensis*, and *Euthamia occidentalis*. As these forests mature *Fraxinus pennsylvanica* (green ash), *Acer negundo* (boxelder) and *Elaeagnus angustifolia* (Russian olive) may become dominant as cottonwoods succumb to old age. Common understory species of older riparian forests include *Symphoricarpos occidentalis* (western snowberry), *Rosa woodsii* (Wood's rose), and *Shepherdia argentea* (buffaloberry). *Agropyron smithii* (western wheatgrass) is a common native grass, and *Poa pratensis* (Kentucky bluegrass) and *Bromus inermis* (smooth brome) are two introduced grasses that can be abundant. *Artemisia cana* (silver sagebrush) often becomes dominant on old, drier river terraces after cottonwoods have died.

Many ephemeral streams in eastern Montana support woodlands dominated by *Fraxinus pennsylvanica*.[239] These are locally referred to as ash draws or woody draws. *Acer negundo* and *Ulmus americana* (American elm) are other trees that may be common in some stands. Common shrubs include *Prunus virginiana* (chokecherry), *Amelanchier alnifolia* (serviceberry), *Symphoricarpos occidentalis*, and *Prunus americana* (American plum). *Poa pratensis, Carex sprengelii, Galium boreale, Sanicula marilandica*, and *Smilacina stellata* are common herbaceous species. Small-order drainages as well as lee slopes and depressions will often have cover of *Symphoricarpos occidentalis, Rosa woodsii*, and *Elaeagnus commutata* (wolf willow).

Wetlands are uncommon except in the north near the Canadian border where glacial potholes may occur. Vegetation varies within and among these wetlands depending on hydrology (when the soils dries) and degree of salinity.[241,385] Common species include *Alopecurus aequalis, Carex atherodes, C. nebrascensis, Distichlis spicata, Eleocharis palustris, Hordeum jubatum, Juncus balticus* (Baltic rush), *Mentha arvensis, Polygonum amphibium, Puccinellia nuttalliana*, and *Schoenoplectus acutus* (hardstem bulrush).

Grasslands and sagebrush steppe are the most common vegetation types on the Great Plains of Montana. *Agropyron smithii* is the most common and broadly distributed grass species across soil types. *Stipa viridula* (green needlegrass), *Bouteloua gracilis* (blue grama), and *Poa pratensis* are most abundant on silty or clayey soils. Sandier, better-drained soils usually support *Stipa comata* (needle-and-thread) and *Bouteloua gracilis. Agropyron spicatum* (bluebunch wheatgrass) and *Andropogon scoparius* (little bluestem) are common on stony or gravelly soils. Forbs in the Asteraceae and Fabaceae, such as *Aster* spp.,

Solidago spp., *Antennaria* spp., *Astragalus* spp., and *Lupinus* spp. are common. Drought or heavy ungulate grazing results in an increase in *Bouteloua gracilis* and a decline in species of *Stipa*.[423]

Sagebrush steppe most often occurs on clayey, silty, or loamy, relatively deep soils. *Artemisia tridentata* ssp. *wyomingensis* (Wyoming big sagebrush) is the most common shrub, dominating on plains and slopes with heavier soils. *Artemisia cana* (silver sagebrush) is the common species on stream terraces, sandy soils, or in disturbed sites. Herbaceous species in the understory are similar to those of grasslands. The proportions of sagebrush steppe and grassland are determined partially by soils, but also by the frequency of wildfire because *Artemisia tridentata* ssp. *wyomingensis* is killed by fire and recolonizes slowly.[95]

Steppe vegetation dominated by *Sarcobatus vermiculatus* (greasewood), *Atriplex gardneri* (Gardner saltbush), *A. confertifolia* (shadscale), or *A. canescens* (four-wing saltbush) occur on saline soils of badlands, stream terraces, and wetlands.[67] *Rhus aromatica* (skunkbush sumac) can be dominant in stony soils of mid or upper slopes.

Coniferous woodlands are most commonly associated with shallow or stony soils, often on steep, north- or east-facing slopes or hills and breaks with fractured bedrock near the soil surface. The most common trees are *Pinus ponderosa* (ponderosa pine) and *Juniperus scopulorum* (Rocky Mountain juniper); *Pseudotsuga menziesii* (Douglas fir) occurs commonly in the Missouri River Breaks. *Juniperus communis*, *J. horizontalis* (common and horizontal junipers), *Amelanchier alnifolia*, *Prunus virginiana*, and *Symphoricarpos occidentalis* are common shrubs with a ground layer dominated by graminoids such as *Agropyron spicatum*, *Festuca idahoensis* (Idaho fescue), *Andropogon scoparius*, *Oryzopsis micrantha*, and *Carex inops*. In many areas fire suppression has resulted in a dense understory of young pines that facilitate large stand-replacing fires.

Valleys Zone

Intermountain valleys tend to run north-south and vary in elevation from 610 to 915 m (2,000-3,000 ft.) in the northwestern part of the state, to over 1525 m (5,000 ft.) in the southwestern and south-central parts. Valleys tend to be narrower west of the Continental Divide. Grasslands and especially sagebrush steppe dominate most valley bottoms outside of riparian areas. Open forest and savanna are common in the northwestern valleys.

Riparian and wetland areas occur along rivers and major streams. Early-successional forests form on newly-created gravel bars and alluvial surfaces. They are dominated by cottonwoods. *Populus balsamifera* ssp. *trichocarpa* (black cottonwood) is the only common species west of the Continental Divide, while *P. deltoides* becomes more common as one goes east, and *P. angustifolia* (narrow-leaved cottonwood) may dominate in higher southwest valleys. Many species of willow, including *Salix exigua*, *S. bebbiana*, *S. boothii* and *S. eriocephala* establish at the same time under the cottonwoods. Other deciduous shrubs or small trees colonize shortly afterward.[132,156] These include *Acer glabrum* (Rocky Mountain maple), *Alnus incana* (thin-leaf alder), *Crataegus douglasii* (Douglas hawthorn), *Betula occidentalis* (river birch), *Elaeagnus commutata*, and *Prunus virginiana*. *Dryas drummondii*, *Chamerion latifolium*, *C. angustifolia* and *Equisetum* spp. colonize bare gravel. *Pinus ponderosa*, *Juniperus scopulorum*, and *Picea engelmannii* (Engelmann spruce) colonize stream terraces as the cottonwoods mature.

Woodlands dominated by *Populus tremuloides* (quaking aspen) occasionally occur in the valleys but are more common in the montane zone. Small-order drainages as well as lee slopes and depressions will often have cover of *Symphoricarpos occidentalis*, *Rosa woodsii*, *Amelanchier alnifolia*, *Prunus virginiana*, and *Crataegus douglasii*.

Wetlands are more common in the intermountain valleys than on the plains. The most common type of wetland has fresh water and is associated with riparian sloughs, backwaters, or spring creeks. *Deschampsia cespitosa* (tufted hairgrass), *Juncus balticus,* and *Carex nebrascensis* occur in wet meadows often associated with riparian areas. Depressional wetlands are common in areas subject to continental or valley glaciation such as the Blackfoot, Swan, and Mission valleys. *Typha latifolia* (cattail), *Scirpus acutus, Juncus balticus, Eleocharis palustris, Hordeum jubatum, Distichlis spicata, Puccinellia nuttalliana, Carex pellita, C. utriculata, C. vesicaria, C. atherodes, Glyceria borealis*, and *Equisetum fluviatile* are common species.

Grasslands and sagebrush steppe dominate much of the intermountain valleys and foothills, especially on warm or windswept slopes in southwestern and north-central regions. Grasslands generally occur on gravelly to moderately deep soils, while sagebrush steppe is found on deeper or more fine-textured soils. However, fire plays the largest role in determining the presence of sagebrush on the landscape because

Artemisia tridentata is killed by fire, but grasses are usually not. Western wheatgrass, a rhizomatous species, can be common on the nearly level topography of valley bottoms. Otherwise, western Montana grasslands are dominated by bunchgrasses. *Agropyron spicatum* and *Stipa comata* prevail on the most xeric sites with warm aspects and stony soils, while *Festuca idahoensis* becomes more common on deeper soils or cooler slopes. *Festuca campestris* (rough fescue) along with *F. idahoensis* can dominate forb-rich grasslands in the northwestern and north-central parts of the state.

Sagebrush steppe is the most abundant vegetation type in the valleys and foothills. Most of this vegetation is dominated by *Artemisia tridentata* which has two subspecies that occur in different valley habitats.[283,255] Subspecies *wyomingensis* (Wyoming big sagebrush) occurs in the most xeric sites on nearly level valley bottoms or low slopes. Common associated grasses include *Agropyron spicatum*, *Stipa comata*, and *Koeleria macrantha* (junegrass). Subspecies *tridentata* (basin big sagebrush) is less common and usually found in loamy soils of stream terraces in the southwest. *Agropyron smithii* and *Elymus cinereus* (basin wildrye) are common grasses in good-condition stands. Much of this habitat has been lost to hay production.[245]

Coniferous forests are common in the valleys only in the northwestern part of the state. *Pinus ponderosa* and *Pseudotsuga menziesii* can form forests or savannas with an open understory of shrubs such as *Physocarpus malvaceus* (ninebark), *Holodiscus discolor* (oceanspray), *Symphoricarpos albus* (common snowberry), *Purshia tridentata* (bitterbrush), and *Spiraea betulifolia* (birch-leaf spiraea), or simply a ground layer of bunchgrasses such as *Agropyron spicatum*, *Festuca idahoensis*, or *F. campestris*. Forests and savannas are more likely to occur in the valleys on relatively stony or coarse-textured soil.

Montane Zone

This zone encompasses the lowest and warmest portions of mountainous terrain. Grassland and sagebrush vegetation is common but more so east of the Continental Divide. Montane forests and woodlands occur above or intermixed with grasslands and steppe, often on cooler slopes or where snow accumulates. Fire plays an important role in shaping the mosaic of grassland, steppe, and forest. The montane zone occurs between 760 to 1680 m (2,500-5,500 ft.) in the northwest and 1370 to 2130 m (4,500-7,000 ft.) east of the Continental Divide.

Riparian and wetland areas are usually small but more common than in other zones. Riparian areas occur along rivers and major streams and are well-developed only in the lower portion of the montane zone where the gradient is not so steep. They are similar to those described for the valleys. *Populus angustifolia* can be the dominant species along montane rivers and streams in southwestern Montana. The common willows include *Salix bebbiana, S. boothii, S. drummondiana,* and *S. geyeriana. Alnus incana, A. viridis* (green alder), and *Acer glabrum* (Rocky Mountain maple) may occur with the willows. *Picea engelmannii* may invade as soil develops. *Thuja plicata* (western redcedar) is common along streams in the northwestern part of the state.[156]

Wetlands, such as wet meadows, swamps, marshes, and fens also occur in the montane zone. These wetlands are often associated with glacial ponds and lakes and riparian areas. Wet meadows dominated by *Calamagrostis canadensis* often occur in depressions in coniferous forest and along lake margins. Swamps dominated by *Alnus incana, A. viridis,* and willows such as *Salix drummondiana, S. geyeriana, S. bebbiana,* and *S. planifolia* occur most commonly along streams or around beaver impoundments. Marsh vegetation occurs on saturated to flooded mineral soil, often around the shallow margins of lakes or ponds. Marshes are typically dominated by *Carex utriculata, Equisetum fluviatile,* or *Typha latifolia* (cattail). Fen vegetation develops on wet organic soils of glacial depressions or gentle slopes associated with groundwater seepage. Fens are most common west of the Continental Divide, often on nearly level drainage divides. Those developed over calcareous material (rich fens) usually support more species that neutral or acidic (poor) fens.[77] Rich fens are commonly dominated by mosses and sedges such as *Carex limosa, C. simulata, C. lasiocarpa,* and *C. aquatilis. Betula glandulosa* (bog birch) and *Dasiphora fruticosa* (shrubby cinquefoil) are common shrubs.[77,238] *Carex aquatilis, C. utriculata,* and *C. canescens* are common in poor fens, and *Ledum glandulosum* (trapper's tea) is a common shrub.

Grasslands and sagebrush steppe are similar to those of the valleys and are dominated by cool season bunchgrasses.[293] *Agropyron spicatum* is most common on stony soils and warm exposures. *Festuca idahoensis, Bromus carinatus, Stipa nelsonii, S. occidentalis,* and *Danthonia intermedia* are other common grasses. *Festuca campestris* can be common west of the Continental Divide and in north-central mountain ranges. Broad-leaved forbs are diverse, including *Balsamorhiza sagittata, Lupinus* spp., *Senecio* spp., *Astragalus* spp., *Eriogonum* spp., *Potentilla* spp., *Geranium viscosissimum*, and *Geum triflorum*.

Sagebrush steppe is common in the montane zone everywhere except the northwestern counties and east of the Divide along the Front Range. *Artemisia tridentata* ssp. *vasseyana* (mountain big sagebrush) is the dominant shrub, and although it is killed by fire, stands can return to preburn canopy cover after only about three decades.[255] *Ericameria nauseosa* (rubber rabbitbrush) and *Chrysothamnus viscidiflorus* are often common, and *Artemisia tripartita* (three-tip sagebrush) can be common in the southwest. These latter shrubs all resprout to varying degrees after fire. The proportions of grassland and steppe are determined mainly by fire frequency, which was between 25 and 40 years prior to European settlement.[17,194]

Aspen forest is common in the montane zone, especially east of the Divide where *Populus tremuloides* (quaking aspen) groves intermingle with grassland to form extensive parklands.[93,261] Some of these groves are probably in the lower subalpine zone, but they are treated here for convenience. West of the Continental Divide aspen stands are small and usually associated with stream corridors or depressions on slopes otherwise dominated by conifers. Similar sites support aspen east of the Divide, but there stands may be extensive, occupying depressions and cool slopes in the foothills. Aspen forests are leafless until late spring and canopies are not dense, so a good deal of light reaches the ground. As a result, understory vegetation can be luxuriant. Driest stands have *Symphoricarpos* spp., *Amelanchier alnifolia, Berberis repens* (Oregon grape), *Calamagrostis rubescens* (pinegrass), and *Rosa woodsii* in the understory. Stands receiving more moisture or with deeper soil have a ground layer of grasses and tall herbs. Common grasses are *Elymus glaucus, Calamagrostis canadensis, Bromus* spp., and *Poa pratensis. Geranium richardsonii, Osmorhiza occidentalis, Heracleum lanatum,* and *Angelica arguta* are common forbs. *Alnus* spp., *Cornus sericea* (red-osier dogwood), *Prunus virginiana, Amelanchier alnifolia*, and *Rosa woodsii* are common shrubs.

Aspen stands often have conifers, such as *Pinus contorta* (lodgepole pine), *Picea engelmannii*, and *Abies lasiocarpa* (subalpine fir), mixed in. This indicates that many aspen groves are in an early-seral stage eventually leading to spruce-fir forest.[321] Fire is considered essential for maintaining aspen in the Rocky Mountains,[109] but it may not be so important in the aspen parklands of Canada and northwestern Montana.[261] More luxuriant aspen stands demonstrate little tendency to succeed to conifers, probably due to the continuous cover of litter contributed by decomposing forbs.

Coniferous forest of the montane zone occurs on cool slopes of the foothills and on progressively warmer slopes as one proceeds higher. The most common montane forests are dominated by *Pseudotsuga menziesii, Pinus ponderosa*, and *Pinus contorta*; however, *P. ponderosa* does not occur in the mountains of the southwestern ranges.[14] *Larix occidentalis* (western larch) is common west of the Continental Divide, and *Tsuga heterophylla* (western hemlock), *Thuja plicata*, and *Abies grandis* (grand fir) are common in the northwestern mountains. *Pinus monticola* (western white pine) was once common west of the Divide, but has been reduced by timber harvest and blister rust. *Pinus flexilis* (limber pine) is common on limestone east of the Divide. Many species of shrubs occur in the understory. These include *Symphoricarpos albus, Berberis repens, Acer glabrum, Spiraea betulifolia, Rubus parviflorus* (thimbleberry), *Physocarpus malvaceus* (ninebark), *Arctostaphylos uva-ursi* (kinnikinick), *Juniperus communis, Rosa* spp., and *Vaccinium* spp. *Picea engelmannii* and *Abies lasiocarpa* (subalpine fir) are scattered at the upper limits of the montane zone or can be common in lower, cold-air pockets.

The history of presettlement fire frequency and intensity is complex.[129,130] Some of the dominant trees such as *Pinus ponderosa, Pseudotsuga menziesii*, and *Larix occidentalis* have thick, fire-resistant bark, while others do not. In general, montane forests experienced a mixed-severity fire regime; i.e., relatively frequent ground fires that reduced the number of young trees while sparing the older trees with thicker bark, and occasional catastrophic crown fires that killed all or nearly all of the trees. Both fire frequency and severity varied with time and across the landscape. In general, fires were less frequent and more catastrophic on cool slopes with denser trees and a moister understory. Fire suppression during the past century has reduced fire frequency from historic levels and allowed stands to become dense and more prone to crown fire.

Subalpine Zone

Grassland and sagebrush vegetation is confined almost exclusively to the valley and montane zones west of the Continental Divide. However, grasslands and sagebrush steppe can occur up to the lower alpine zone east of the Divide. Dominant subalpine vegetation on both sides of the Divide is forest. The subalpine occurs between 1675 and 2135 m (5,500-7,000 ft.) in the northwest and between 1980 and 2900 m (6,500-9,500 ft.) in the southwestern and south-central mountain ranges.

Grasslands and sagebrush steppe occur mainly on south- or west-facing slopes in the subalpine zone and are similar to those found in the montane zone. *Festuca idahoensis* is the dominant grass; *Agropyron trachycaulum* is also common. The native, high-elevation *Bromus inermis* var. *purpurascens* can be locally common. *Leucopoa kingii* can be a dominant grass in the extreme southern portion of the state. *Geum triflorum*, *Potentilla gracilis*, and *Agoseris glauca* are among the many common forbs. *Artemisia tridentata* ssp. *vasseyana* steppe has a similar ground layer composition. Fire may play a role in maintaining non-forest vegetation at this high elevation, but the deep soils and warm exposures are probably just as important.

Meadows are usually dominated by broad-leaved herbaceous plants and/or sedges. Tall herb meadows usually occur on warm but moist slopes with abundant pocket gopher activity. Common broad-leaved species include *Hackelia micrantha*, *Helianthella uniflora*, *Delphinium* spp., *Angelica* spp., and *Cirsium* spp.

Wet meadows are found on gentle to level terrain with adequate snow cover and may be associated with ponds or streams. Soils are deep with a well-developed organic horizon and remain wet or moist during much of the short growing season. Shortly after snowmelt the early ephemerals, such as *Erythronium grandiflorum*, *Claytonia lanceolata*, *Caltha leptosepala*, and *Ranunculus eschscholtzii* begin flowering. Soon thereafter large, showy wildflowers, such as *Arnica* spp., *Senecio triangularis*, *Erigeron peregrinus*, *Mimulus lewisii*, *Polygonum bistortoides*, *Valeriana sitchensis*, and *Castilleja* spp. begin flowering and continue through the rest of the summer. *Carex paysonis*, *C. epapillosa*, *C. microptera*, *Juncus drummondii*, and *Deschampsia cespitosa* are common graminoids in these habitats.

Warm, moderately steep subalpine slopes, mainly near or west of the Continental Divide, may support extensive meadows dominated by *Xerophyllum tenax* (beargrass). Most of these beargrass meadows are burned-over forests that are regenerating slowly or not at all due to the harsh, snowy environment. Most of the species common in these meadows also occur in open, subalpine forests (see below).

Shrub-dominated wet meadows are found in avalanche chutes that occur on steep, usually warm slopes in the subalpine zone. They are formed by large amounts of snow sliding rapidly downhill, usually in the spring. Trees with rigid trunks are broken off by the moving snow, so the centers of avalanche paths are dominated by herbaceous plants and shrubs with flexible stems. Avalanche chutes are usually associated with long, steep ravines that are moister than the adjacent slopes. Common shrubs include *Alnus viridis*, *Amelanchier alnifolia*, *Spiraea betulifolia*, *Rubus parviflorus*, *Salix scouleriana*, *Sorbus scopulina* (mountain ash), and *Sambucus racemosa* (elderberry). Associated with the shrubs are many tall, luxuriant, moisture-loving, herbaceous plants including *Heracleum lanatum*, *Chamerion angustifolium*, *Urtica dioica*, *Veratrum viride*, *Angelica arguta*, and *Elymus glaucus*. Avalanche chutes merge into subalpine forest at the margins.

Forests are the most common subalpine habitat. The most prevalent trees in subalpine forests are *Abies lasiocarpa*, *Picea engelmannii*, and *Pinus contorta*. *Pinus albicaulis* (whitebark pine) is usually an early-seral component on lower or warmer sites, but can dominate at timberline. West of the Continental Divide open *Larix lyallii* (subalpine larch) forests may occur in the upper subalpine,[16] and *Tsuga mertensiana* (mountain hemlock) is a dominant subalpine tree in scattered locations near the Idaho border.

Warmer and drier subalpine forests occur on warm slopes or well-drained soils. Common understory species include *Vaccinium scoparium* (or *V. myrtillus* in the northwest), *V. globulare* or *V. membranaceum* (huckleberry), *Carex geyeri*, *Calamagrostis rubescens*, and *Arnica cordifolia* or *A. latifolia* in the understory. More mesic sites, especially west of the Continental Divide, support numerous shrubs in the understory including *Menziesia ferruginea*, *Rubus parviflorus*, *Sorbus scopulina*, *Spiraea betulifolia*, *Vaccinium globulare*, *Lonicera utahensis*, and *Alnus viridis*. Common ground layer species include *Arnica cordifolia* or *A. latifolia*, *Chimaphila umbellata*, *Linnaea borealis*, *Clintonia uniflora*, *Tiarella trifoliata*, and *Thalictrum occidentale*. Forests near upper treeline that receive heavy snowfall have open canopies with dense ground layers dominated by *Phyllodoce* spp., *Xerophyllum tenax*, *Erythronium grandiflorum*, and *Luzula hitchcockii*.

Subalpine fir, spruce, and whitebark pine become stunted and dwarfed at upper treeline due to ice-scouring wind or heavy snow accumulations.[18] These "krummholz" woodlands are usually sparse and discontinuous, interspersed with alpine tundra or heath. Tree cover breaks the strong, high-elevation wind, allowing snow accumulation and providing plentiful moisture where there is adequate soil. *Phyllodoce* spp., *Vaccinium scoparium, Luzula hitchcockii, Juncus drummondii, Antennaria lanata, Xerophyllum tenax,* and *Arnica latifolia* are common species in the understory. Krummholz developed on skeletal soils of very wind-exposed sites often has an understory of dwarf shrubs such as *Dryas octopetala, Arctostaphylos uva-ursi, and Dasiphora fruticosa* (shrubby cinquefoil), with many herbaceous species typical of alpine fellfields in the ground layer.

Fire is less frequent in the moister subalpine compared to the montane zone, but is more often catastrophic and stand-replacing. Large areas of lower subalpine forest can be dominated by *Pinus contorta*, a short-lived tree that often requires fire to open cones and release seeds to germinate in recently burned soil.[294] Given enough time, *Picea engelmannii* and *Abies lasiocarpa* can colonize the *Pinus contorta* and eventually come to dominate, but fire intervals are often not long enough to allow this to happen. Unlike the coniferous trees that are killed by fire, nearly all the understory shrubs and herbaceous plants can survive and resprout from stems or roots protected underground. So while fire drastically changes the overstory, the understory recovers quickly and stays much the same. Upper subalpine forests have longer fire-free intervals, but the infrequent fires that do occur promote *Pinus albicaulis* by providing sites for recruitment and reducing competition from *Abies lasiocarpa*.[15] *Pinus albicaulis*, in turn, provides a more benign microclimate for the establishment of *Abies lasiocarpa*.

Disease and insects have also played an important role in shaping subalpine forests. Mountain pine beetles (*Dendroctonus ponderosae*) attack stands of *Pinus contorta* that are 80 or more years old, killing most trees. Fallen and standing-dead lodgepole increase the likelihood of fire. Mountain pine beetles will also attack *Pinus albicaulis*, especially trees weakened by disease or competition from fir trees.[208] A more serious threat to *P. albicaulis* is the introduced fungal disease white pine blister rust (*Cronartium ribicola*). Blister rust has spread rapidly in the Northern Rockies in the past 20 years.[208] As a result, upper subalpine forests show increasing dominance by *Abies lasiocarpa*.

Alpine Zone

Alpine vegetation occurs in the majority of Montana's mountain ranges. There is generally less alpine habitat in the northwestern ranges, either because the mountains are too low or because the terrain is too steep or too snowy for development of alpine vegetation. Extensive areas of alpine habitat occur in southwestern ranges, especially the Absaroka-Beartooth country. The alpine zone is above 1980-2130 m (6,500-7,000 ft.) in the northwest but above 3000 m (9,500 ft.) in the southwest.[25,94,242] Exposure and the redistribution of snow by the wind are the most important determinants of vegetation above treeline. Sparsely vegetated habitats are better thought of as supporting loose assemblages rather than communities.

Wetlands and meadows occupy relatively level terrain and are more common at lower elevations. Soils are moist to wet and usually have a well-developed organic surface layer. Sedges and broad-leaved forbs are usually more common than grasses. Stream terraces and large depressions with late-melting snowbanks above can support alpine swamps dominated by the willows *Salix glauca* and *S. planifolia*. *Deschampsia cespitosa, Carex scopulorum,* and *C. paysonis* are common graminoids in the ground layer.

Other wetland vegetation lacks woody species. Wet meadows occur in depressions or on gentle terrain with adequate snow cover and saturated soil for most of the growing season. *Deschampsia cespitosa, Carex scopulorum, C. paysonis, Juncus drummondii,* and *J. mertensianus* are the common graminoids. Wildflowers such as *Caltha leptosepala, Symphyotrichum foliaceum, Erigeron peregrinus, Pedicularis groenlandica, Trollius albiflorus,* and *Polygonum bistortoides* can create showy displays.

Somewhat drier meadow vegetation can occur on well-drained soils where snow persists late into the season. This snowbed vegetation is less productive because of the shorter growing season. *Carex nigricans, Juncus drummondii,* and *Antennaria lanata* are the dominant species and can form extensive near-monocultures. *Carex paysonis, C. pyrenaica, Sibbaldia procumbens,* and *Erigeron peregrinus* are other common species.

Heath is common on lee slopes or near-level terrain where snow lies late in the growing season. Soils are moist with moderate amounts of organic matter, and most often developed from granite or other crystalline parent materials. The low-shrub heathers, *Phyllodoce empetriformis, P. glanduliflora,* and *Cassiope* spp., are the dominant species along with *Vaccinium scoparium*. Herbaceous species typical of snowbed communities, such as *Juncus drummondii, Antennaria lanata,* and *Sibbaldia procumbens,* occur in the sparse ground layer.

Turf vegetation is dominated primarily by low sedges and broad-leaved forbs and occasionally by grasses. It occurs on somewhat windswept sloping to level terrain, often at very high elevations. Soil development is moderate. Turf grades into fellfield vegetation as wind exposure increases and soil becomes shallower. This is the most abundant alpine vegetation in the southwestern and south-central mountain ranges. Drier turf is dominated by *Carex elynoides*. Other common graminoids include *Carex rupestris* and the alpine form of *Festuca idahoensis*. Low, often cushion-forming forbs are common, including *Oxytropis sericea, Potentilla glaucophylla, P. ovina*, and *Besseya wyomingensis*. Moist turf is dominated by *Carex scirpoidea, C. albonigra, C. phaeocephala, Luzula spicata*, and the alpine form of *Festuca idahoensis*. Broad-leaved forbs, including *Geum rossii, Potentilla glaucophylla, Erigeron simplex, Lupinus argenteus, Lloydia serotina*, and *Solidago multiradiata*, are usually more apparent than the graminoids.

Fellfields are dominated by cushion plants and dwarf shrubs. These communities develop on the most wind-exposed topographic positions such as ridge crests or windward slopes with little soil development. Patterned vegetation, consisting of rows of vegetation that slow the wind and cause some snow deposition, alternate with open ground. Plants are rooting in nearly bare mineral soil without the ameliorating effect of organic matter. Thus, the effect of soil parent material is pronounced.[94,244] *Dryas octopetala* is a dominant subshrub on all soil types and often occurs with *Salix arctica* and dwarfed *Dasiphora fruticosa*. On calcareous soils *Carex rupestris* and *Potentilla ovina* are conspicuous, while *Geum rossii* and *Carex elynoides* are more abundant on crystaline soils. *Minuartia obtusiloba, Silene acaulis, Phlox pulvinata, Senecio canus*, and *Selaginella densa* are common cushion plants on all types of soil.

Talus and scree slopes are common above treeline. Because these slopes are near the angle of repose the surface is always shifting, making it difficult for most plants to take root. However, plants that can grow downslope with the surface rock experience a surprisingly benign environment, especially on warm slopes. Most talus slopes have a surface layer of loose rock over deep, mineral soil. The surface rock allows rain and snowmelt into the soil below but acts as a mulch by preventing evaporation and keeping the soil moist and warm. Plant cover is usually very sparse and the vegetation is better considered an assemblage rather than a community. Nonetheless, several distinctive species, such as *Crepis nana, Stellaria americana, Polemonium viscosum, Claytonia megarhiza, Phacelia sericea, Penstemon* spp., and *Eriogonum* spp. occur most commonly on talus slopes.

Floristics

Floristic biogeography

A primary goal of biogeography, as well as ecology, is understanding the distribution of organisms across landscapes and continents. Methods include molecular genetics to analysis of fossils.[19,304] A common method in plant geography has been floristic analysis: the classification of the flora into groups sharing distinct geographic patterns.[143,196,272] These patterns generally correspond to "floristic provinces," areas with a distinct flora thought to have evolved under relatively static climate and soils over long periods of time.[196] A species whose geographic range largely corresponds to one of these patterns is considered to have affinities with that floristic region and is assumed to have evolved primarily with other species of the same affinity. It is assumed that plants came to be in a given area either by migrating there or evolving in place. Knowing the geographic patterns underlying a local flora provides clues to evolution and important historic migrations.[72,143]

Montana encompasses parts of two floristic provinces, Cordilleran and Great Plains, and is in close proximity to the Boreal Province.[143, 272] Montana has a relatively large flora for a northern continental region due to being at the intersection of these three provinces. Not surprisingly, the largest number of native species has affinities with the North American Cordillera, which stretches from Alaska south to California, Arizona, New Mexico, and to some extent Texas and Mexico. While most of Montana's cordilleran species occupy a large portion of the Cordillera, some are associated with the Pacific Northwest as defined by Hitchcock and Cronquist[188] or the intermountain deserts of the Snake River Plains, Colorado Plateau, and Great Basin (Table 1). Many cordilleran species are regional endemics, having a total range less than one-third the size of Montana (see below). Although at least half of Montana is in the Great Plains Floristic Province, only a small proportion of the state's native species can be unequivocally assigned to a Great Plains affinity (Table 1). Many of the species occurring on Montana's plains have boreal or cordilleran affinities or are too widespread across eastern North America to determine their floristic affinities with reasonable certainty.

Boreal species occur across Canada, and those with a circumboreal range are also found in the boreal regions of Eurasia. Although Montana is south of the Boreal Florsitic Province, species with boreal affinities are more common than Great Plains species (Table 1). Many boreal species occur on the plains as well as in the mountains, especially in the northern half of the state. Plants with arctic affinities are confined to subalpine and alpine habitats in the mountainous portion of the state. Nearly half of these species have a circumpolar distribution, occurring in arctic and alpine regions of Europe and Asia.

There are 100 species of regional endemics (global range of less than 1/3 the area of the state) in Montana. Ninety-four are herbaceous perennials, five are annuals or biennials, and one is a shrub. There are no narrow endemic trees. Nearly two-thirds of Montana's narrow endemics belong to five families: Asteraceae (18), Brassicaceae (17), Fabaceae (11), Plantaginaceae (10), and Apiaceae (9). One-third of these species are in four genera: *Physaria* (9), *Erigeron* (8), *Astragalus* (8), and *Penstemon* (8). The low number of endemics can be attributed to glaciation because much of northern Montana was almost completely covered with ice during the Pleistocene glaciations.[320] Presumably all the vegetation, including narrowly endemic species, was destroyed by the glaciers, and during warmer interglacial periods the barren landscapes were recolonized by more widespread plants that survived south or perhaps north of the glaciers.[108,196]

The occurrence of these regional endemics decreases from southwest to northeast.[251] Thirty-nine and 25 species of narrow endemics are found in the valleys and mountains of the southwestern and northwestern portions of the state respectively, and another 16 species occur in parts of both regions. In addition, there are two "hot spots" of endemism in the state: the northern Bighorn Basin and adjacent valleys south of Billings have ten regional endemics, and the Bitterroot Range of west-central Montana has eight. Two species of narrow endemics are found in the mountains of north-central Montana, and there are none on the northeastern plains. The habitats and elevational ranges of narrow endemics vary. However, only one species occurs on the plains. The majority are found in the montane (59) and/or subalpine (51) zones respectively, while the valleys (26) and alpine zone (32) have fewer. Many endemic species occur in more than one vegetation zone.

Table 1. Biogeographic affinities of Montana's native plant species.

Arctic	**113** (5%)
Circumpolar	53
North American arctic-alpine	60
Boreal	**368** (18%)
Circumboreal	179
North American boreal	189
North American cordilleran	**940** (45%)
Widespread cordilleran	691
Pacific Northwest	107
Intermountain desert	41
Endemic	101
Great Plains	**113** (5%)
Widespread (undetermined)	**548** (26%)

Floristic analysis

More than half of Montana's introduced plants belong to five families: Poaceae (77), Asteraceae (60), Brassicaceae (48), Fabaceae (31), and Caryophyllaceae (29). The largest genera of exotic plants include *Bromus* (9), *Silene* (9), *Centaurea* (8), and *Trifolium* (8). Although similar to natives at the family level, 50% of Montana's introduced plants belong to alien genera. Whereas annuals or biennials are poorly represented in the native flora (15%), they account for half (51%) of the introduced species. Nearly half the introduced species (203) can be found on the plains, and nearly all of them occur in valley habitats (405). Only 13% (55) of Montana's introduced species occur in montane habitats. Only seven introduced species routinely occur in the subalpine, and none in the alpine, a typical pattern for the Northern Rockies.[133]

Table 2. Floristic analysis

Families	128
Total genera	733
Total species	2512
Total taxa (ssp., var.)	2661
Native genera	586
Native species	2082
Introduced genera	248
Introduced species	431

Native species

Largest families

Asteraceae	315
Cyperaceae	163
Poaceae	158
Brassicaceae	99
Fabaceae	98
Rosaceae	85
Ranunculaceae	62
Plantaginaceae	54
Saxifragaceae	50
Apiaceae	49

Largest genera

Carex	121
Astragalus	47
Erigeron	37
Salix	32
Penstemon	30
Senecio	26
Juncus	25
Ranunculus	24
Potentilla	23
Draba	21

Growth forms

Annuals and biennials	318 (15%)
Herbaceous perennials	1554 (75%)
Shrubs and vines	178 (8%)
Trees	32 (2%)

Introduced species

Largest families

Poaceae	77
Asteraceae	61
Brassicaceae	48
Fabaceae	31
Caryophyllaceae	29

Growth forms

Annuals and biennials	218 (51%)
Herbaceous perennials	188 (44%)
Shrubs and vines	17 (4%)
Trees	8 (2%)

Literature Cited

(1) Allen, G.A. 1984. Morphological and cytological variation within the western North American *Aster occidentalis* complex (Asteraceae). Systematic Botany 9:175–191.

(2) Allen, G.A. 2006. *Eucephalus.* Pp. 39–42 *in* FNA Editorial Committee, Flora of North America, Vol. 20, Oxford University Press, New York.

(3) Allen, G.A., M.L. Dean, and K.L. Chambers. 1983 Hybridization studies in the *Aster occidentalis* (Asteraceae) hybrid complex of western North America. Brittonia 35:353–361.

(4) Al-Shehbaz, I.A. 1973. The biosystematics of the genus *Thelypodium* (Cruciferae). Contributions to the Gray Herbarium 204:1–148.

(5) Al-Shehbaz, I.A. 2010a. Brassicaceae. Pp. 224–746 *in* FNA Editorial Committee, Flora of North America, Vol. 7, Oxford University Press, New York.

(6) Al-Shehbaz, I.A. 2010b. *Sandbergia.* Pp. 417–418 *in* FNA Editorial Committee, Flora of North America, Vol. 7, Oxford University Press, New York.

(7) Al-Shehbaz, I.A. and S.L. O'Kane. 2002. *Lesquerella* is united with *Physaria* (Brassicaceae). Novon 12:319–329.

(8) Al-Shehbaz, I.A. and M.A. Belstein. 2010. *Camelina.* Pp. 451–453 *in* FNA Editorial Committee, Flora of North America, Vol. 7, Oxford University Press, New York.

(9) Al-Shehbaz, I.A., M.D. Windham, and R. Elven. 2010. *Draba.* Pp. 269–347 *in* FNA Editorial Committee, Flora of North America, Vol. 7, Oxford University Press, New York.

(10) Alt, D. and D.W. Hyndman. 1986. Roadside geology of Montana. Mountain Press, Missoula, MT.

(11) Alverson, E.R. 1989. A new species of parsley fern, *Cryptogramma* (Adiantaceae) from western North America. American Fern Journal 79:95–102.

(12) Argus, G.W. 2010. *Salix.* Pp. 23–162 *in* FNA Editorial Committee, Flora of North America, Vol. 7, Oxford University Press, New York.

(13) Argus, G.W., J.E. Eckenwalder, and R.W. Kiger. 2010. Salicaceae. Pp. 3–22 *in* FNA Editorial Committee, Flora of North America, Vol. 7, Oxford University Press, New York.

(14) Arno, S.F. 1979. Forest regions of Montana. USDA Forest Service Research Paper INT-218. Ogden, UT.

(15) Arno, S.F. 1986. Whitebark pine cone crops: a diminishing source of wildlife food? Western Journal of Applied Forestry 1:92–94.

(17) Arno, S.F. and G.E. Gruell. 1983. Fire history at the forest-grassland ecotone in southwestern Montana. Journal of Range Management 36:332–336.

(16) Arno, S.F. and J.R. Habeck. 1972. Ecology of alpine larch (*Larix lyallii* Parl.) in the Pacific Northwest. Ecological Monographs 42:417–450.

(18) Arno, S.F. and R.P. Hammerly. 1984. Timberline. Mountain and arctic forest frontiers. The Mountaineers, Seattle, WA.

(19) Axelrod, D.I. and P.H. Raven. 1985. Origins of the cordilleran flora. Journal of Biogeography 12:21–47.

(20) Babcock, E.B. and G.L. Stebbins. 1938. The American species of *Crepis.* Publication of the Carnegie Institute of Washington No. 504.

(21) Backlund, A. and B. Bremer. 1997. Phylogeny of the Asteridae s. str based on *rbc*L sequences, with particular reference to the Dipsacales. Plant Systematics and Evolution 207:225–254.

(22) Bain, J.F., B.S. Tyson, and D.F. Bray. 1997. Variation in pollen wall ultrastructure in New World Senecioneae (Asteraceae) with special reference to *Packera.* Canadian Journal of Botany 75:730–735.

(23) Baird, G.I. 2006. *Agoseris.* Pp. 323–335 *in* FNA Editorial Committee, Flora of North America, Vol. 19, Oxford University Press, New York.

(24) Ball, P.W. and A.A. Reznicek. 2002. *Carex.* Pp. 254–572 *in* FNA Editorial Committee, Flora of North America, Vol. 23, Oxford University Press, New York.

(25) Bamberg, S.A. and J. Major. 1968. Ecology of the vegetation and soils associated with calcareous parent materials in three alpine regions of Montana. Ecological Monographs 38:127–167.

(26) Barkley, T.M. 1962. A revision of *Senecio aureus* L. and allied species. Transactions of the Kansas Academy of Sciences 65:318–408.

(27) Barkley, T.M. 2006. *Senecio.* Pp. 544–570 *in* FNA Editorial Committee, Flora of North America, Vol. 20, Oxford University Press, New York.

(28) Barkley, T.M. and D.F. Murray. 2006. *Tephroseris.* Pp. 615–618 *in* FNA Editorial Committee, Flora of North America, Vol. 20, Oxford University Press, New York.

(29) Barkworth, M.E. and D.R. Dewey. 1985. Genomically based genera in the perennial Triticeae of North America: Identification and membership. American Journal of Botany 72:767–776.

(30) Barkworth, M.E., K.M. Capels, S. Long, and M.B. Peip (eds.). 2003. Magnoliophyta: Commelinidae (in part): Poaceae, part 2. Flora of North America North of Mexico, volume 25. Oxford University Press, New York.

(31) Barkworth, M.E., K.M. Capels, S. Long, and M.B. Peip (eds.). 2007. Magnoliophyta: Commelinidae (in part): Poaceae, part 1. Flora of North America North of Mexico, volume 24. Oxford University Press, New York.

(32) Barkworth, M.E., M.O. Arriaga, J.F. Smith, S.W.L. Jacobs, J. Valdés-Reyna, and B.S. Bushman. 2008. Molecules and morphology in South American Stipeae (Poaceae). Systematic Botany 33:719–731.

(33) Barneby, R.C. 1964. Atlas of North American *Astragalus.* Memoirs of the New York Botanical Garden 13:1–1188.

(34) Barneby, R.C. 1989. Intermountain Flora Vol. 3, Part B. Fabales. Intermountain flora, Vol. 2, part B. The New York Botanical Garden, Bronx, New York.

(35) Bassett, I.J., C.W. Crompton, J. McNeill, and P.M. Taschereau. 1983. The genus *Atriplex* (Chenopodiaceae) in Canada. Agriculture Canada Monograph No. 31, Ottawa.

(36) Bayer, R.J. 1985. Investigations into the evolutionary history of the poyploid complexes in *Antennaria* (Asteraceae: Inuleae) I. The *A. neidioica* complex. Plant Systematics and Evolution 150:143–163.

(37) Bayer, R.J. 1989. Nomenclatural rearrangements in *Antennaria neodioica* and *A. howellii* (Asteraceae: Inulae: Gnaphaliinae). Brittonia 41:396–398.

(38) Bayer, R.J. 1990. Investigations into the evolutionary history of the *Antennaria rosea* (Asteraceae: Inuleae) polyploidy complex. Plant Systematics and Evolution 169:97–110.

(39) Bayer, R.J. 2006. *Antennaria.* Pp. 388–415 *in* FNA Editorial Committee, Flora of North America, Vol. 19, Oxford University Press, New York.

(40) Beaman, J.H. 1957. The systematics and evolution of *Townsendia* (Compositae). Contributions from the Gray Herbarium 183:1–151.

(41) Beetle, A.A. and K.L. Johnson. 1982. Sagebrush in Wyoming. University of Wyoming Agricultural Experiment Station Bulletin 779, Laramie.

(42) Behnke, H.-D. 1997. Sarcobataceae- a new family of Caryophyllales. Taxon 46:495–507.

(43) Benson, L.D. 1948. A treatise on the North American *Ranunculi.* American Midland Naturalist 40:1–261.

(44) Benson, L.D. 1982. Cacti of the United States and Canada. Stanford University Press, Stanford, CA.

(45) Björk, C.R. 2010. *Douglasia conservatorum* (Primulaceae), a new species from Idaho and Montana, U.S.A. Novon:20:9–12.

(46) Blankenship, J.W. 1901. Weeds of Montana. Montana Agricultural Experiment Station Bulletin 30:1–70.

(47) Blankenship, J.W. 1905a. A century of botanical exploration in Montana. Montana Agricultural College Science Studies 1:3–31.

(48) Blankenship, J.W. 1905b. Supplement to the flora of Montana. Montana Agricultural College Science Studies 1:35–109.

(49) Bogler, D.J. 2006a. *Crepis.* Pp. 222–239 *in* FNA Editorial Committee, Flora of North America, Vol. 19, Oxford University Press, New York.

(50) Bogler, D.J. 2006b. *Prenanthes.* Pp. 264–271 *in* FNA Editorial Committee, Flora of North America, Vol. 19, Oxford University Press, New York.

(51) Booth, W.E. 1950. Flora of Montana. Part I, Conifers and Monocots. Montana State College Research Foundation, Bozeman.

(52) Booth, W.E. and J.C. Wright. 1959. Flora of Montana. Part II. Montana State University, Bozeman.

(53) Boraiah, G. and M. Heimburger. 1964. Cytotaxonomic studies on New World *Anemone* (section *Eriocephalus*) with woody rootstocks. Canadian Journal of Botany 42:891–922.

(54) Boufford, D.E. 1997. *Fumaria.* Pp. 356–357 *in* FNA Editorial Committee, Flora of North America, Vol. 3, Oxford University Press, New York.

(55) Brayshaw, T.C. 1985. Pondweeds, and burr-reeds, and their relatives: aquatic families of monocotyledons in British Columbia. British Columbia Provincial Museum Occasional Publication No. 25, Victoria.

(56) Brayshaw, T.C. 1996. Catkin-bearing plants of British Columbia. Royal British Columbia Museum, Victoria.

(57) Bremer, K. and C.J. Humphries. 1993. Generic monograph of the Asteraceae-Anthemideae. Bulletin of the British Museum (Natural History), Botany 23:71–177.

(58) Broich, S.L., 2007. New combinations in North American *Lathyrus* and *Vicia* (Fabaceae: Faboideae: Fabae). Madrono 54:63–71.

(59) Brooks, R.E. and S.E. Clemants. 2000. *Juncus.* Pp. 211–255 *in* FNA Editorial Committee, Flora of North America, Vol. 22, Oxford University Press, New York.

(61) Brouillet, L. 2006a. *Taraxacum.* Pp. 239–252 *in* FNA Editorial Committee, Flora of North America, Vol. 19, Oxford University Press, New York.

(62) Brouillet, L. 2006b. *Matricaria.* 540–542 *in* FNA Editorial Committee, Flora of North America, Vol. 19, Oxford University Press, New York.

(63) Brouillet, L. 2006c. *Eurybia.* Pp. 365–382 *in* FNA Editorial Committee, Flora of North America, Vol. 20, Oxford University Press, New York.

(64) Brouillet, L. and P.E. Elvander. 2009a. *Micranthes.* Pp. 49–70 *in* FNA Editorial Committee, Flora of North America, Vol. 8, Oxford University Press, New York.

(65) Brouillet, L. and P.E. Elvander. 2009b. *Saxifraga.* Pp. 132–146 *in* FNA Editorial Committee, Flora of North America, Vol. 8, Oxford University Press, New York.

(66) Brouillet, L., J.C. Semple, G.A. Allen, K.L. Chambers, and S.D. Sundberg. 2006. *Symphyotrichum.* Pp. 465–539 *in* FNA Editorial Committee, Flora of North America, Vol. 20, Oxford University Press, New York.

(67) Brown, R.W. 1971. Distribution of plant communities in southeastern Montana badlands. American Midland Naturalist 85:458–477.

(68) Brummitt, R.K. 1971. Relationship of *Heracleum lanatum* Michx. of North America to *H. sphondylium* L. of Europe. Rhodora 73:578–584.

(69) Brunsfield, S.J. and F.D. Johnson. 1985. Field guide to the willows of east-central Idaho. University College of Forestry, Wildlife and Range Sciences Bulletin No 39. Moscow.

(70) Brunsfield, S.J. and F.D. Johnson. 1986. Notes on *Betula* ser. *Humiles* (Betulaceae) in Idaho. Madrono 33:147–148.

(71) Brunsfield, S.J. and F.D. Johnson. 1990. Cytological, morphological, ecological, and phenological support for specific status of *Crataegus suksdorfii* (Rosaceae). Madrono 37:274–282.

(72) Cain, S.A. 1944. Foundations of plant geography. Harper Brothers, New York.

(73) Carlquist, S., B.G. Baldwin, and G.D. Carr (eds.). 2003. Tarweeds & Silverswords: Evolution of the Madiinae (Asteraceae). Missouri Botanical Garden Press, St. Louis.

(74) Cantino, P.D. 1982. A monograph of the genus *Physostegia* (Labiatae). Contributions from the Gray Herbarium 211:1–105.

(75) Carolin, R.C. 1987. A review of the family Portulacaceae. Australian Journal of Botany 35:383–412.

(76) Ceska, O. and A. Ceska. 1999. Haloragaceae. Pp. 228–235 in G.W. Douglas, D. Meidinger, and J. Pojar (eds.), Illustrated flora of British Columbia. Vol. 3. British Columbia Ministry of Forests, Vancouver.

(77) Chadde, S. W., J. S. Shelly, R.J. Bursik, R.K. Moseley, A.G. Evenden, M. Mantas, F. Rabe, and B. Heidel. 1998. Peatlands on national forests of the Northern Rocky Mountains: ecology and conservation. USDA Forest Service General Technical Report RMRS-GTR-11, Ogden, UT.

(78) Chambers, K.L. 1993. *Claytonia*. Pp. 898–900 in J. C. Hickman (ed.), The Jepson Manual, higher plants of California. University of California Press, Berkeley.

(79) Chambers, K.L. 2006. *Nothocalais*. Pp. 335–337 *in* FNA Editorial Committee, Flora of North America, Vol. 19, Oxford University Press, New York.

(80) Cherniawsky, D.M. and R.J. Bayer. 1998. Systematics of North American *Petasites* (Asteraceae: Senecioneae) III. A taxonomic revision. Canadian Journal of Botany 76:2061–2075.

(81) Chiapella, J. 2007. A molecular phylogenetic study of *Deschampsia* (Poaceae: Aveneae) inferred from nuclear ITS and plastid trnL sequence data: support for the recognition of *Avenella* and *Vahlodea*. Taxon 56:55–64.

(82) Chiapella, J. and F.O. Zuloaga. 2010. A Revision of *Deschampsia*, *Avenella*, and *Vahlodea* (Poaceae, Poeae, Airinae) in South America. Annals of the Missouri Botanical Garden 97:141–162.

(83) Chinnappa, C.C. and J.K. Morton. 1991. Studies on the *Stellaria longipes* complex (Caryophyllaceae): taxonomy. Rhodora 93:129–135.

(84) Cholewa, A.F., J.J. Pipoly III, and J.M. Ricketson. 2009. Myrsinaceae. Pp. 302–321 *in* FNA Editorial Committee, Flora of North America, Vol. 8, Oxford University Press, New York.

(85) Chuang, T.I. and L.R. Heckard. 1992. A taxonomic revision of *Orthocarpus* (Scrophulariaceae-Tribe *Pediculareae*). Systematic Botany 17:560–582.

(86) Clark, D.V. 1971. Speciation in *Penstemon* (Scrophulariaceae). Ph.D. dissertation, University of Montana, Missoula.

(87) Clausen, J. 1964. Cytotaxonomy and distributional ecology of western North American violets. Madrono 17:173–204.

(88) Clausen, R.T. 1975. *Sedum* of North America north of the Mexican Plateau. Ithaca, N.Y.

(89) Clayton, W.D. and S.A. Renvoize. 1986. Genera Graminum, Kew Bulletin Additional Series XIII, London.

(90) Clemants, S.E. and S.L. Mosyakin. 2003a. *Dysphania*. Pp. 267–275 *in* FNA Editorial Committee, Flora of North America, Vol. 4, Oxford University Press, New York.

(91) Clemants, S.E. and S.L. Mosyakin. 2003b. *Chenopodium*. Pp. 275–299 *in* FNA Editorial Committee, Flora of North America, Vol. 4, Oxford University Press, New York.

(92) Columbus, J.T. 1999. An expanded circumscription of *Bouteloua* (Gramineae: Chloridoideae): new combinations and names. Aliso 18:61–65.

(93) Cooper, S.V., P. Lesica, R.L. DeVelice, and T. McGarvey. 1995. Classification of southwestern Montana plant communities: emphasizing those of Dillon Resource Area, Bureau of Land Management. Montana Natural Heritage Program, Helena.

(94) Cooper, S.V., P. Lesica, and D. Page-Dumroese. 1997. Plant community classification for alpine vegetation on the Beaverhead National Forest, Montana. USDA Forest Service General Technical Report INT-GTR-362, Ogden, UT.

(95) Cooper, S.V., P. Lesica, and G.M. Kudray. 2007. Post-fire recovery of Wyoming big sagebrush shrub steppe in central and southeast Montana. Report to the USDI Bureau of Land Management. Montana Natural Heritage Program, Helena, Montana.

(96) Costea, M., F.J. Tardif, and H.R. Hinds. 2005. *Polygonum*. Pp. 547–571 *in* FNA Editorial Committee, Flora of North America, Vol. 5, Oxford University Press, New York.

(97) Crawford, D.J. 1974. A morphological and chemical study of *Populus acuminata* Rydb. Brittonia 26:74–89.

(98) Cronquist, A. 1947. A revision of the North American species of *Erigeron*, north of Mexico. Brittonia 6:121–302.

(99) Cronquist, A. 1988. The evolution and classification of flowering plants, ed. 2. New York Botanical Garden, Bronx.

(100) Cronquist, A., A.H. Holmgren, N.H. Holmgren, and J.L. Reveal. 1972. Intermountain flora, Vol. 1. Hafner Publishing, New York.

(101) Cronquist, A., A.H. Holmgren, N.H. Holmgren, J.L. Reveal, and P.K. Holmgren. 1977. Intermountain Flora. Vol. 6. Columbia University Press, New York.

(102) Cronquist, A., A.H. Holmgren, N.H. Holmgren, J.L. Reveal, and P.K. Holmgren. 1984. Intermountain Flora, Vol. 4. The New York Botanical Garden, Bronx.

(103) Cronquist, A., A.H. Holmgren, N.H. Holmgren, J.L. Reveal, and P.K. Holmgren. 1994. Intermountain Flora, Vol. 5. The New York Botanical Garden, Bronx.

(104) Cronquist, A., N.H. Holmgren, and P.K. Holmgren. 1997. Intermountain Flora, Vol. 3, part A. The New York Botanical Garden, Bronx.

(105) Crow, G.E. 2005. *Sagina*. Pp. 140–147 275 *in* FNA Editorial Committee, Flora of North America, Vol. 5, Oxford University Press, New York.

(106) Darbyshire, S.J. 1993. Realignment of *Festuca* subgenus *Schedonorus* with the genus *Lolium* (Poaceae). Novon 3:239–243.

(107) Darbyshire, S.J. and L.E. Pavlick. 2007. 14.01. *Festuca* L. Pp. 389–443, In: Barkworth, M.E., K.M. Capels, S. Long, L.K. Anderton, and M.B. Piep. (eds.). Magnoliophyta: Commelinidae (in part): Poaceae, part 1. Flora of North America North of Mexico, Vol.24. Oxford University Press, New York and Oxford.

(108) Davis, M.B. 1981. Quaternary history and the stability of forest communities. Pp. 134–153 in D.C. West et al. (eds.), Forest Succession: concepts and applications. Springer-Verlag, New York.

(109) DeByle, N.V., C.D. Bevins, and W.C. Fischer. 1987. Wildfire occurrence in aspen in the interior western United States. Western Journal of Applied Forestry 2:73–76.

(110) DeSanto, J. 1993. Bitterroot. Lere Press, Babb, MT.

(111) Dickinson, T.A., S. Belaoussoff, R.M. Löve, and M. muniyamma. 1996. North American black-fruited hawthorns. 1. Variation in floral construction, breeding system correlates, and their possible evolutionary significance in *Crataegus* sect. *Douglasii* Louden. Folia Geobotanica & Phytotaxonomica 31:355–371.

(112) Dobes, C.H., T. Mitchell-Olds, and M.A. Koch. 2004. Extensive chloroplast haplotype variation indicates Pleistocene hybridization and radiation of North American *Arabis drummondii, A. x divaricarpa*, and *A. holboellii* (Brassicaceae). Molecular Ecology 13:349–370.

(113) Dorn, R.D. 1984. Vascular plants of Montana. Mountain West Publishing, Cheyenne, WY.

(114) Dorn, R.D. 1997. Rocky Mountain region willow identification field guide. USDA Forest Service, Fort Collins, CO.

(115) Dorn, R.D. 2001. Vascular plants of Wyoming, third edition. Mountain West Publishing, Cheyenne, WY.

(116) Dorn, R.D. 2010. The Genus *Salix* in North America North of Mexico. www.lulu.com/content/8538913

(117) Douglas, G.W., G.B. Straley, D. Meidinger, and J. Pojar (eds.). 1998. Illustrated flora of British Columbia. Vol. 1. British Columbia Ministry of Forests, Vancouver.

(118) Dragon, J.A. and D. S. Barrington. 2009. Systematics of the *Carex aquatilis* and *C. lenticularis* lineages: geographically and ecologically divergent sister clades of *Carex* Section *Phacosystis* (Cyperaceae). American Journal of Botany 96:1896–2009.

(119) Dutton, B.E., C.S. Keener, and B.A. Ford. 1997. *Anemone*. Pp. 139–158 *in* FNA Editorial Committee, Flora of North America, Vol. 3, Oxford University Press, New York.

(120) Elisens, W.J. 1985. The Montana collections of Francis Duncan Kelsey. Brittonia 37:382–391.

(121) Elisens, W.J. and J.G. Packer. 1980. A contribution to the taxonomy of the *Oxytropis campestris* complex in northwestern North America. Canadian Journal of Botany 58:1820–1831.

(122) Elvander, P.E. 1984. The taxonomy of *Saxifraga* (Saxifragaceae) section *Borophila*, subsection *Integrifoliae* in western North America. Systematic Botany Monographs 3:1–44.

(123) Eriksson, T.M.S. Hibbs, A.D. Yoder, C.F. Delwiche, and M.J. Donoghue. 2003. The phylogeny of Rosoideae (Rosaceae) based on sequences of the internal transcribed spacer (ITS) of nuclear rhibosomal DNA and the TRNL/F region of chloroplast DNA. International Journal of Plant Science 164:197–211.

(124) Ertter, B., R. Elven, J.L. Reveal, and D.F. Murray. In ed. *Potentillain* in FNA Editorial Committee, Flora of North America, Vol. 9, Oxford University Press, New York.

(125) Evert, E.F. 2010. Vascular plants of the Greater Yellowstone area. Self-published, Oak Park, IL.

(126) Ewan, J. 1945. A synopsis of the North American species of *Delphinium*. University of Colorado Studies Series D 2:55–244.

(127) Fabijan, D.M., J.G. Packer, and K.E. Denford. 1987. The taxonomy of the *Viola nuttallii* complex. Canadian Journal of Botany 65:2562–2580.

(128) Ferren, W.R. and H.J. Schenk. 2003. *Suaeda*. Pp. 390–398 *in* FNA Editorial Committee, Flora of North America, Vol. 4, Oxford University Press, New York.

(129) Fischer, W.C. and A.F. Bradley. 1987. Fire ecology of western Montana forest habitat types. USDA Forest Service General Technical Report INT-223, Ogden, UT.

(130) Fischer, W.C. and B.D. Clayton. 1983. Fire ecology of Montana forest habitat types east of the Continental Divide. USDA Forest Service General Technical Report INT-141, Ogden, UT.

(131) Flora of North America (FNA) Editorial Committee (eds.) 1993–2010. Flora of North America. Oxford University Press, New York.

(132) Foote, G.G. 1965. Phytosociology of the bottomland hardwood forests in western Montana. M.A. thesis, University of Montana, Missoula.

(133) Forcella, F. and S.J. Harvey. 1983. Eurasian weed infestation in western Montana in relation to vegetation and disturbance. Madrono 30:102–109.

(134) Franzke, A.K. K. Pollmann, and W. Bleeker. 1998. Molecular systematics of *Cardamine* and allied genera (Brassicaceae): Its and non-coding chloroplast DNA. Folia Geobotanica 33:225–240.

(135) Freeman, C.C. and H.R. Hinds. 2005. *Fallopia*. Pp. 541–546 *in* FNA Editorial Committee, Flora of North America, Vol. 3, Oxford University Press, New York.

(136) Fritz-Sheridan, J.K. 1988. Reproductive biology of *Erythronium grandiflorum* varieties *grandiflorum* and *candidum* (Liliaceae). American Journal of Botany 75:1–14.

(137) Furlow, J.J. 1997. Betulaceae. Pp. 507–538 *in* FNA Editorial Committee, Flora of North America, Vol. 3, Oxford University Press, New York.

(138) Galloway, L.A. 1975. Systematics of North American desert species of *Abronia* and *Tripterocalyx*. Brittonia 27:328–347.

(139) Gaskin, J.F. and B.A. Schaal. 2003. Molecular phylogenetic investigation of U.S. Invasive *Tamarix*. Systematic Botany 28:86–95.

(140) Geyer, C.A. 1845–1846. Notes on the vegetation and general character of the Missouri and Oregon territories, made during a botanical journey from the state of Missouri across the South Pass of the Rocky Mountains to the Pacific during years 1843 and 1844. London Journal of Botany 4:479–492; 5:22–41, 198–208, 285–310.

(141) Gillett, J.M. 1957. A revision of the North American species of *Gentianella* Moench. Annals of the Missouri Botanical Garden 44:195–269.

(142) Gillett, J.M. 1963. The gentians of Canada, Alaska and Greenland. Canada Department of Agriculture Publ. 1180.

(143) Gleason, H.A. and A. Cronquist. 1964. The natural geography of plants. Columbia University Press, New York.

(144) Gould, F.W. 1975. The grasses of Texas. Texas A&M University Press, College Station.

(145) Gould, F.W. and R.B. Shaw. 1983. Grass systematics. Texas A&M University Press, College Station.

(146) Grady, B.R. and S.L. O'Kane. 2007. New species and combinations in *Physaria* (Brassicaceae) from Western North America. Novon 17:182–192.

(147) Graham, S.A. 1975. Taxonomy of the Lythraceae in the southeastern United States. Sida 6:80–103.

(148) Grant, A.L. 1924. A monograph of the genus *Mimulus*. Annals of the Missouri Botanical Garden 11:99–388.

(149) Grant, V. 1956. A synopsis of *Ipomopsis*. Aliso 3:351–362.

(150) Grant, V. 2003. Incongruence between cladistic and taxonomic systems. American Journal of Botany 90:1263–1270.

(151) Great Plains Flora Association. 1977. Atlas of the flora of the Great Plains. Iowa State University Press, Ames.

(152) Great Plains Flora Association. 1986. Flora of the Great Plains. University Press of Kansas, Lawrence.

(153) Groff, P.A. 1989. Studies in whole-plant morphology. Ph.D. dissertation (Botany), University of California, Berkeley.

(154) Guppy, G.A. 1978. Species relationships of *Hieracium* (Asteraceae) in British Columbia. Canadian Journal of Botany 56:3008–3019.

(155) Hanks, L.T. and J.K. Small. 1907. Geraniaceae. North American Flora 25:3–24.

(156) Hansen, P., R. Pfister, K. Boggs, B.J. Cook, J. Joy, and D.K. Hinkley. 1995. Classification and management of Montana's riparian and wetland sites. Montana Forest and Conservation Experiment Station, Missoula, MT.

(157) Harris, J.G. 2010. *Braya*. Pp. 546–552 *in* FNA Editorial Committee, Flora of North America, Vol. 7, Oxford University Press, New York.

(158) Hartman, R.L. and R.S. Kirkpatrick. 1986. A new species of *Cymopterus* (Umbelliferae) form northwestern Wyoming. Brittonia 38:420–426.

(159) Hartman, R.L., R.K. Rabeler, and F.H. Utech. 2005a. *Arenaria*. Pp. 51–56 *in* FNA Editorial Committee, Flora of North America, Vol. 5, Oxford University Press, New York.

(160) Hartman, R.L., R.K. Rabeler, and F.H. Utech. 2005b. *Minuartia*. Pp. 116–136 *in* FNA Editorial Committee, Flora of North America, Vol. 5, Oxford University Press, New York.

(161) Hartman, R.L. and R.K. Rabeler. 2005a. *Spergularia*. Pp. 16–23 *in* FNA Editorial Committee, Flora of North America, Vol. 5, Oxford University Press, New York.

(162) Hartman, R.L. and R.K. Rabeler. 2005b. *Pseudostellaria*. Pp. 114–116 *in* FNA Editorial Committee, Flora of North America, Vol. 5, Oxford University Press, New York.

(163) Harvey, L.H. 1954. Additions to the flora of Glacier National Park, Montana. Proceedings of the Montana Academy of Sciences 14:23–25.

(164) Hauffler, C.H., M.D. Windham, F.A. Lang, and S.A. Whitmore. 1993. *Polypodium*. Pp. 315–323 *in* FNA Editorial Committee, Flora of North America, Vol. 2, Oxford University Press, New York.

(165) Hauke, R.L. 1993. Equisetaceae. Horsetail Family. Pp. 76–84 *in* FNA Editorial Committee, Flora of North America, Vol. 2, Oxford University Press, New York.

(166) Hawksworth, F.G. and D. Wiens. 1972. Biology and classification of dwarf mistletoes (*Arceuthobium*). USDA Forest Service Agricultural Handbook 401.

(167) Haynes, R.R. 2000. Rupiaceae. Pp. 75–76 *in* FNA Editorial Committee, Flora of North America, Vol. 22, Oxford University Press, New York.

(168) Haynes, R.R. and C.B. Hellquist. 2000a. Juncaginaceae. Pp. 43–46 *in* FNA Editorial Committee, Flora of North America, Vol. 22, Oxford University Press, New York.

(169) Haynes, R.R. and C.B. Hellquist. 2000b. Potamogetonaceae. Pp. 47–70 *in* FNA Editorial Committee, Flora of North America, Vol. 22, Oxford University Press, New York.

(170) Heckard, L.R. and T.I. Chuang. 1977. Chromosome numbers, polyploidy, and hybridization in *Castilleja* (Scrophulariaceae) of the Great Basin and Rocky Mountains. Brittonia 29:159–172.

(171) Heidel, B. 1996. Noteworthy collections: Montana. Madrono 43:437–440.

(172) Helfgott, D.M. and R.J. Mason-Gamer. 2004. The evolution of North American *Elymus* (Triticeae, Poaceae) allotetraploids: evidence from phosphoenolpyruvate carboxylase gene sequences. Systematic Botany 29:850–861.

(173) Hermann, F.J. 1970. Manual of the carices of the Rocky Mountains and Colorado Basin. USDA Forest Service Agriculture Handbook No. 374, Washington D.C.

(174) Hermann, F.J. 1975. Manual of the rushes (*Juncus* spp.) of the Rocky Mountains and Colorado Basin. USDA Forest Service General Technical Report RM-18, Fort Collins, CO.

(175) Hershkovitz, M.A. 1991. Phylogenetic assessment and revised circumscription of *Cistanthe* (Portulacaceae). Annals of the Missouri Botanical Garden 78:1022–1060.

(176) Hershkovitz, M.A. and S.B. Hogan. 2003. *Lewisia*. Pp. 476–484 *in* FNA Editorial Committee, Flora of North America, Vol. 4, Oxford University Press, New York.

(177) Heywood, V.H., I.B.K. Richardson, N.A. Burgess, et al. 1972. Labiatae. Pp.126–192 *in* Flora Europaea Vol. 3. Cambridge University Press.

(178) Hickman, J.C. (ed.) 1993a. The Jepson Manual, higher plants of California. University of California Press, Berkeley.

(179) Hickman, J.C. 1993b. Polygonaceae. Pp. 854–896 in J. C. Hickman (ed.), The Jepson Manual, higher plants of California. University of California Press, Berkeley.

(180) Higgins, L.C. 1971. A revision of *Cryptantha* subgenus *Oreocarya*. Brigham Young University Science Bulletin Biological Series 13:1–63.

(181) Hinds, H.R. and C.C. Freeman. 2005. *Fagopyrum*. Pp. 572–573 *in* FNA Editorial Committee, Flora of North America, Vol. 5, Oxford University Press, New York.

(182) Hitchcock, A.S. 1951. Manual of the grasses of the United States, 2nd ed. Revised by A. Chase. U.S.D.A. Miscellaneous Publication 200.

(183) Hitchcock, C.L. 1937. A key to the grasses of Montana. John S. Swift Co., Inc. St. Louis.

(184) Hitchcock, C.L. 1941. A revision of the drabas of western North America. University of Washington Publications in Biology Vol. 11, Seattle.

(185) Hitchcock, C.L. 1944. The *Tofieldia glutinosa* complex of western North America. American Midland Naturalist 31:487–498.

(186) Hitchcock, C.L. and B. Maguire. 1947. A revision of the North American species of *Silene*. University of Washington Publications in Biology 13, Seattle.

(187) Hitchcock, C.L., A. Cronquist, M. Owenby, and J. W. Thompson. 1955–1969. Vascular plants of the Pacific Northwest. Parts 1–5. University of Washington Press, Seattle.

(188) Hitchcock, C.L. and A.Conquist. 1973. Flora of the Pacific Northwest. University of Washington Press, Seattle.

(189) Hoch, P. C. 1978. Systematics and evolution of the *Epilobium ciliatum* complex In North America (Onagraceae). Ph.D. dissertation, Washington University, St. Louis, MO.

(190) Hoch, P.C. 1993. *Epilobium*. Pp. 793–798 in J.C. Hickman, The Jepson Manual, higher plants of California. University of California Press, Berkeley.

(191) Hoch, P.C. and P.H. Raven. 1992. *Boisduvalia*, a coma-less *Epilobium* (Onagraceae). Phytologia 73:456–459.

(192) Holmgren, P.K. 1971. A biosystematic study of North American *Thlaspi montanum* and its allies. Memoirs of the New York Botanical Garden 21:1–106.

(193) Holmgren, N.H., P.K. Holmgren, and A. Cronquist. 2005. Intermountain flora, Vol. 2, part B. The New York Botanical Garden, Bronx, New York.

(194) Houston, D.B. 1973. Wildfires in northern Yellowstone National Park. Ecology 54:1111–1117.

(195) Hufford, L. and M. McMahon. 2004. Morphological evolution and systematics of *Synthyris* and *Besseya* (Veronicaceae): A phylogenetic analysis. Systematic Botany 29:716–736.

(196) Hultén, E. 1937. Outline of the history of arctic and boreal biota during the Quaternary Period. Verlag Von Cramer, New York.

(197) Hultén, E. 1962. The circumpolar plants. II Dicotyledons. Almqvist & Wiksell, Stockholm.

(198) Hultén, E. 1968. Flora of Alaska and neighboring territories. Stanford University Press, Stanford, CA.

(199) Hyatt, P.E. 2006. *Sonchus*. Pp. 273–276 *in* FNA Editorial Committee, Flora of North America, Vol. 19, Oxford University Press, New York.

(200) Iltis, H.H. 1965. The genus *Gentianopsis* (Gentianaceae): Transfers and phytogeographic comments. Sida 2:129–154.

(201) Isely, D. 1998. Native and naturalized Leguminosae (Fabaceae) of the United States. Brigham Young University, Provo, UT.

(202) Jacobs, S.W.L. and J. Everett. 2000. Grasses: systematics and evolution. CSRIO Publishing, Melbourne, Australia.

(203) Jacobs, S.W.L. et al. 2000. Relationships within the Stipoid grasses (Gramineae). In Grasses: systematics and evolution, S.W.L. Jacobs and J. Everett (eds.). CSIRO, Melbourne.

(204) Johnson, D.M. 1986. Systematics of the New World species of *Marsilea* (Marsileaceae). Systematic Botany Monographs 11:1–87.

(205) Johnson, L.A., J.L. Schultz, D.E. Soltis, and P.S. Soltis. 1996. Monophyly and generic relationships of Polemoniaceae based on matK sequences. American Journal of Botany 83:1207–1224.

(206) Kallersjo, M., G. Bergqvist, and A.A. Anderberg. 2000. Generic realignment in Primuloid families of the Ericales s.l.: a phylogenetic analysis based on DNA sequences from three chloroplast genes and morphology. American Journal of Botany 87:1325–1341.

(207) Kaul, R.B. 2000. Sparganiaceae. Pp. 270–277 *in* FNA Editorial Committee, Flora of North America, Vol. 22, Oxford University Press, New York.

(208) Keane, R.E., S.F. Arno. 1993. Rapid decline of whitebark pine in western Montana: evidence from 20-year remeasurements. Western Journal of Forestry 8:44–47.

(209) Keck, D.D. 1938. Revision of *Horkelia* and *Ivesia*. Lloydia 1:75–111.

(210) Keil, D.J. 2004. New taxa and new combinations in North American *Cirsium* (Asteraceae: Cardueae). Sida 21:207–219.

(211) Keil, D.J. 2006. *Cirsium*. Pp. 95–164 *in* FNA Editorial Committee, Flora of North America, Vol. 19, Oxford University Press, New York.

(212) Keil, D.J. and J. Ochsmann. 2006. *Centaurea*. Pp. 181–194 *in* FNA Editorial Committee, Flora of North America, Vol. 19, Oxford University Press, New York.

(213) Keil, D.J. and C.E. Turner. 1993. *Cirsium*. Pp. 232–239 in J. C. Hickman (editor), The Jepson Manual, higher plants of California. University of California Press, Berkeley.

(214) Kellogg, E.A. 1989. Comments on genomic genera in the Triticeae (Poaceae). American Journal of Botany 76:796–805.

(215) Kellogg, E.A. 1992. Tools for studying the chloroplast genome in the Triticeae (Gramineae): an EcoRI map, a diagnostic deletion, and support for *Bromus* as an outgroup. American Journal of Botany 79:186–197.

(216) Kellogg, E.A. 2001. Evolutionary history of the grasses. Plant Physiology 125(3):1198–1205.

(217) Kerstetter, T.A. 1994. Taxonomic investigation of *Erigeron lackschewitzii*. M.S. thesis, Montana State University, Bozeman.

(218) Kiger, R.W. and D.F. Murray. 1997. *Papaver.* Pp. 323–334 *in* FNA Editorial Committee, Flora of North America, Vol. 3, Oxford University Press, New York.

(219) King, R.M. and H. Robinson. 1987. The genera of the Eupatorieae (Asteraceae). Monographs in Systematic Botany from the Missouri Botanical Garden 22.

(220) Knowles, C.J. and P.R. Knowles. 1993. A bibliography of literature and papers pertaining to presettlement wildlife and habitat of Montana and adjacent areas. Prepared for U.S. Forest Service, Missoula, MT.

(221) Koch, M. and I.A. Al-Shehbaz. 2004. Taxonomic and phylogenetic evaluation of the American *"Thlaspi"* species: identity and relationship to the Eurasian genus *Noccaea* (Brassicaceae). Systematic Botany 29:375–384.

(222) Koch, M.A., J. Bishop, and T. Mitchel-Olds. 2001. Molecular systematics of the Brassicaceae: Evidence from coding plastids *matK* and nuclear *Chs* sequences. American Journal of Botany 88:534–544.

(223) Krause, D.L. and K.I. Beamish. 1972. Taxonomy of *Saxifraga occidentalis* and *S. marshallii.* Canadian Journal of Botany 50:2131–2141.

(224) Kron, K.A., W.S. Judd, P.F. Stevens, D.M. Crayn, A.A. Anderberg, P.A. Gadek, C.J. Quinn, and J.L. Luteyn. 2002. Phylogenetic classification of Ericaceae: Molecular and morphological evidence. Bot. Rev. 68:335–423.

(225) Kuijt, J. 1982. A flora of Waterton Lakes National Park. University of Alberta Press, Edmonton.

(226) Lackschewitz, K. 1991. Vascular plants of west-central Montana- identification handbook. USDA Forest Service General Technical Report INT-277, Ogden, UT.

(227) Lackschewitz, K., P. Lesica, R. Rosentretter, J. K. Cory and P. F. Stickney. 1982. Noteworthy collections: Montana. Madrono 29:58–60.

(228) Lackschewitz, K, P. Lesica, J. Pierce, J. K. Cory and D. Ramsden. 1984. Noteworthy collections: Montana. Madrono 31:254–257.

(229) Lackschewitz, K., P. Lesica and J.S. Shelly. 1988. Noteworthy collections: Montana. Madrono 35:355–358.

(230) LaFrankie, J.V. 1986. Transfer of the species of *Smilacina* to *Maianthemum* (Liliaceae). Taxon 35:584–589.

(231) LaFrankie, J.V. 2002. *Maianthemum*. Pp. 206–210 *in* FNA Editorial Committee, Flora of North America, Vol. 22, Oxford University Press, New York.

(232) Lamb, S.S. and K.A. Kron. 2003. rbcL phylogeny and character evolution in Polygonaceae. Systematic Botany 28:326–332.

(233) Landolt, E. 2000. Lemnaceae. Pp. 143–153 *in* FNA Editorial Committee, Flora of North America, Vol. 22, Oxford University Press, New York.

(234) Lavin, M. and C. Seibert. 2011. Great Plains Flora? Plant geography of eastern Montana's lower elevation shrub-grass dominated vegetation. In Wambolt, C.L. et al. (compilers), 3–14. Proceedings—Shrublands: wildlands and

wildlife habitats. 15th Wildland Shrub Symposium, 2008 June 17–19, Bozeman, MT. NREI, volume XVI. S.J. and Jessie E. Quinney Natural Resources Research Library, Logan, Utah, USA.

(235) Lellinger, D.B. 1985. A field manual of the ferns and fern allies of the United States and Canada. Smithsonian Institution Press, Washington D.C.

(236) Les, D.H., E.L. Schneider, D.J. Padgett, P.S. Soltis, D.E. Soltis, and M. Zanis. 1999. Classification and floral evolution of water lilies (Nymphaeaceae, Nymphaeales); a synthesis of non-molecular, rbcL, Matk, and 18srDNA data. Systematic Botany 24:28–46.

(237) Lesica, P. 1983. Noteworthy collections: Montana. Madrono 30:196.

(238) Lesica, P. 1986. Vegetation and flora of Pine Butte Fen, Teton County, Montana. Great Basin Naturalist 46:22–32.

(239) Lesica, P. 1989. The vegetation and condition of upland hardwood forests in eastern Montana. Proceedings of the Montana Academy of Sciences 49:45–62.

(240) Lesica, P. 1991. Noteworthy Collections: Montana. Madrono 38:297–298.

(241) Lesica, P. 1993. Using plant community diversity in reserve design for pothole prairie on the Blackfeet Indian Reservation, Montana, USA. Biological Conservation 65:69–75.

(242) Lesica, P. 2002. Flora of Glacier National Park. Oregon State University Press, Corvallis.

(243) Lesica, P. 2009. *Draba calcifuga* (Brassicaceae), a new species from the Rocky Mountains of North America. Novon 19:182–186.

(244) Lesica, P. and R.K. Antibus. 1986. Mycorrhizae of alpine fell-field communities on soils derived from crystalline and calcareous parent materials. Canadian Journal of Botany 64:1691–1697.

(245) Lesica, P. and S.V. Cooper. 1997. Presettlement vegetation of southern Beaverhead County, Montana. Montana Natural Heritage Program, Helena, MT.

(246) Lesica, P. and W. Fertig. 2009. Noteworthy collections: Montana. Madrono 56:67.

(247) Lesica, P. and S.J. Shelly. 1991. Sensitive, threatened and endangered vascular plants of Montana. Montana Natural Heritage Program Occasional Publication No. 1. Helena, Montana.

(248) Lesica, P. and J.S. Shelly. 1995. Effects of reproductive mode on demography and life history in *Arabis fecunda* (Brassicaceae). American Journal of Botany 82:752–762.

(249) Lesica, P. and P.F. Stickney. 1994. Noteworthy collections: Montana. Madrono 41:229–231.

(250) Lesica, P. and P.V. Wells. 1986. Noteworthy collections: Montana. Madrono 33:227–228.

(251) Lesica, P., G. Moore, K.M. Peterson, and J.H. Rumely. 1984. Vascular plants of limited distribution in Montana. Monograph No. 2, Montana Academy of Sciences, Supplement to the Proceedings Vol. 43.

(252) Lesica, P., K. Lackschewitz, J. Pierce, S. Gregory, and M. O'Brien. 1986. Noteworthy collections: Montana. Madrono 33:310–312.

(253) Lesica, P., R.F. Leary, F.W. Allendorf, and D.W. Bilderback. 1988. Lack of genic diversity within and among populations of an endangered plant, *Howellia aquatilis*. Conservation Biology 2:275–282.

(254) Lesica, P., P.F. Stickney, and D. Hanna. 2003. Noteworthy collections: Montana. Madrono 50:214–215.

(255) Lesica, P., S.V. Cooper, and G. Kudray. 2007. Recovery of big sagebrush following fire in southwest Montana. Rangeland Ecology and Management 60:261–269.

(256) Liu, Z., Z. Chen, J. Pan, X. Li, M. Su, L. Wang, H. Li, and G. Liu. 2008. Phylogenetic relationships in *Leymus* (Poaceae: Triticeae) revealed by the nuclear ribosomal internal transcribed spacer and chloroplast trnL-F sequences. Molecular Phylogenetics and Evolution 46:278–289.

(257) Looman, J. 1984. The biological flora of Canada. 5. *Delphinium glaucum* Watson, tall larkspur. Canadian Field-Naturalist 98:345–360.

(258) Löve, A. and D. Löve. 1976. Nomenclatural notes on arctic plants. Botanical Notiser 128:497–523.

(259) Lowry, P.P. 1981. A floristic survey of the Bear Trap Canyon, Madison County, Montana with a discussion of author citations using the connecting words "*in*" or "*ex*." Phytologia 48:81–94.

(260) Luer, C.A. 1975. The native orchids of the United States and Canada excluding Florida. New York Botanical Garden, Bronx.

(261) Lynch, D. 1955. Ecology of the aspen groveland in Glacier County, Montana. Ecological Monographs 25:321–344.

(262) Maguire, B. 1943. A monograph of the genus *Arnica*. Brittonia 4:386–510.

(263) Maguire, B. 1947. Studies in the Caryophyllaceae III. A synopsis of North American species of *Arenaria* sect. *Eremogone* Fenzl. Bulletin of the Torrey Botanical Club 74:38–56.

(264) Mann, C.C. 2005. 1491. New Relevations of the Americas before Columbus. A.A. Knopf, New York.

(265) Mansfield, D.H. 2000. Flora of Steens Mountain. Oregon State University Press, Corvallis.

(266) Martin, F.L. 1950. A revision of *Cercocarpus*. Brittonia 7:91–111.

(267) Mast, A.R. and J.L. Reveal. 2007. Transfer of *Dodecatheon* to *Primula* (Primulaceae). Brittonia 59:79–82.

(268) Mast, A.R., D.M.S. Feller, S. Kelso, and E. Conti. 2004. Buzz-pollinated *Dodecatheon* originated from within the heterostylous *Primula* subgenus Auriculastrum (Primulaceae): A 7-region cpDNA phylogeny and its implications for floral evolution. American Journal of Botany 91:926–942.

(269) Mastrogiuseppe, J., P.E. Rothrock, A.C. Dibble, and A.A. Reznicek. 2002. *Carex* sect. *Ovales*. Pp. 332–378 *in* FNA Editorial Committee, Flora of North America, Vol. 23, Oxford University Press, New York.

(270) McGraw, J.B. and J. Antonovics. 1983. Experimental ecology of *Dryas octopetala* ecotypes. 1. Ecotypic differentiation and life cycle stages of selection. Journal of Ecology 71:879–997.

(271) McGregor, R.L. 1986. Amaranthaceae. Pp. 179–187 *in* Great Plains Flora Association, Flora of the Great Plains. University Press of Kansas, Lawrence.

(272) McLaughlin, S.P. 2007. Tundra to tropics: The floristic plant geography of North America. Botanical Research Institute of Texas, Fort Worth.

(273) McLaughlin, W T. 1935. Notes on the flora of Glacier National Park, Montana. Rhodora 37:362–365.

(274) McNeal, D.W. and T.D. Jacobsen. 2002. *Allium*. Pp. 224–276 *in* FNA Editorial Committee, Flora of North America, Vol. 26, Oxford University Press, New York.

(275) Meerts, P., J.P. Briane, and C. Lefebvre. 1990. A numerical taxonomic study of the *Polygonum aviculare* complex (Polygonaceae) in Belgium. Watsonia 5:177–214.

(276) Merriam, C.H. 1890. Results of a biological survey of the San Francisco Mountain region and desert of the Little Colorado in Arizona. USDA North American Fauna 3:1–136.

(277) Miller, J.M. 1978. Phenotypic variation, distribution and relationships of diploid and tetraploid populations of the *Claytonia perfoliata* complex (Portulacaceae). Systematic Botany 3:322–341.

(278) Miller, J.M. 2003a. *Claytonia*. Pp. 465–475 *in* FNA Editorial Committee, Flora of North America, Vol. 4, Oxford University Press, New York.

(279) Miller, J.M. 2003b. *Montia*. Pp. 485–488 *in* FNA Editorial Committee, Flora of North America, Vol. 4, Oxford University Press, New York.

(280) Miller, J.M. and K.L. Chambers. 2006. Systematics of *Claytonia* (Portulacaceae). Systematic Botany Monographs 78:1–236.

(281) Montagne, C., L.C. Munn, G.A. Nielsen, J.W. Rogers, and H.E. Hunter. 1982. Soils of Montana. USDA-SCS and Montana Agricultural Experiment Station Bulletin 744, Bozeman, MT.

(282) Morgan, D.R. 2006. *Dieteria*. Pp. 395–401 *in* FNA Editorial Committee, Flora of North America, Vol. 20, Oxford University Press, New York.

(283) Morris, M.S., R.G. Kelsey, and D. Griggs. 1976. The geographic and ecological distribution of big sagebrush and other woody *Artemisia* in Montana. Proceedings of the Montana Academy of Sciences 36:56–79.

(284) Morse, C.A. 2006a. *Stenotus*. Pp. 174–177 *in* FNA Editorial Committee, Flora of North America, Vol. 20, Oxford University Press, New York.

(285) Morse, C.A. 2006b. *Tonestus*. Pp. 181–184 *in* FNA Editorial Committee, Flora of North America, Vol. 20, Oxford University Press, New York.

(286) Morton, J.K. 2005a. *Cerastium*. Pp. 74–93 *in* FNA Editorial Committee, Flora of North America, Vol. 4, Oxford University Press, New York.

(287) Morton, J.K. 2005b. *Stellaria*. Pp. 96–114 *in* FNA Editorial Committee, Flora of North America, Vol. 5, Oxford University Press, New York.

(288) Morton, J.K. 2005c. *Silene*. Pp. 166–214 *in* FNA Editorial Committee, Flora of North America, Vol. 5, Oxford University Press, New York.

(289) Moss, E.H. and J.G. Packer. 1983. Flora of Alberta, second edition. University of Toronto Press, Toronto.

(290) Mosyakin, S.L. 2003a. *Corispermum*. Pp. 313–321 *in* FNA Editorial Committee, Flora of North America, Vol. 5, Oxford University Press, New York.

(291) Mosyakin, S.L. 2003b. *Salsola*. Pp. 398–403 *in* FNA Editorial Committee, Flora of North America, Vol. 5, Oxford University Press, New York.

(292) Mosyakin, S.L. 2005. *Rumex*. Pp. 489–533 *in* FNA Editorial Committee, Flora of North America, Vol. 5, Oxford University Press, New York.

(293) Mueggler, W.F. and W.L. Stewart. 1980. Grassland and shrubland habitat types of western Montana. USDA Forest Service General Technical Report INT-66. Ogden, UT.

(294) Muir, P.S. and J.E. Lotan. 1985. Disturbance history and serotiny of Pinus contorta in western Montana. Ecology 66:1658–1668.

(295) Müller, K. and T. Borsch. 2005. Phylogenetics of Amaranthaceae based on matK/trnK sequence data: evidence from parsimony, likelihood, and Bayesian Analyses. Annals of the Missouri Botanical Garden 92:66–102.

(296) Mulligan, G.A. 1972. Cytotaxonomic studies of *Draba* species in Canada and Alaska: *D. oligosperma* and *D. incerta*. Canadian Journal of Botany 50:1763–1766.

(297) Mulligan, G.A. 1976. The genus *Draba* in Canada and Alaska: key and summary. Canadian Journal of Botany 54:1386–1393.

(298) Mulligan, G.A. 1980. The genus *Cicuta* in North America. Canadian Journal of Botany 58:1755–1767.

(299) Mulligan, G.A. 1995. Synopsis of the genus *Arabis* (Brassicaceae) in Canada, Alaska and Greenland. Rhodora 97:109–163.

(300) Mulligan, G.A. and J.N. Findlay. 1970. Sexual reproduction and agamospermy in the genus *Draba*. Canadian Journal of Botany 48:269–271.

(301) Mummenhoff, K., H. Brüggemann, and J.L. Bowman. 2001. Chloroplast DNA phylogeny and biogeography of *Lepidium* (Brassicaceae). American Journal of Botany 88:2051–2063.

(302) Murray, D.F. 2002a. *Carex* sect. *Racemosae*. Pp. 401–414 *in* FNA Editorial Committee, Flora of North America, Vol. 23, Oxford University Press, New York.

(303) Murray, D.F. 2002b. *Carex* sect. *Dornera*. Pp. 528–530 *in* FNA Editorial Committee, Flora of North America, Vol. 23, Oxford University Press, New York.

(304) Myers, A.A. and P.S. Giller (eds). 1988. Analytical biogeography. Chapman and Hall, London.

(305) Nesom, G.L. 2006a. *Ionactis*. Pp. 82–84 *in* FNA Editorial Committee, Flora of North America, Vol. 20, Oxford University Press, New York.

(306) Nesom, G.L. 2006b. *Erigeron*. Pp. 256–348 *in* FNA Editorial Committee, Flora of North America, Vol. 20, Oxford University Press, New York.

(307) Nesom, G.L. 2006c. *Pseudognaphalium*. Pp. 415–425 *in* FNA Editorial Committee, Flora of North America, Vol. 19, Oxford University Press, New York.

(308) Noyes, R.D. and L.H. Riesberg. 1999. ITS sequence data support a single origin for North American Astereae (Asteraceae) and reflect deep geographic divisions in *Aster* s.l. American Journal of Botany 86:398–412.

(309) Noyes, R.D. and D.E. Soltis. 1996. Genotypic variation in agamospermous *Erigeron compositus* (Asteraceae). American Journal of Botany 83:1292–1303.

(310) Oberprieler, C. 2001. Phylogenetic Relationships in *Anthemis* L. (Compositae, Anthemideae) Based on nrDNA ITS Sequence Variation. Taxon 50:745–762.

(311) Ochsmann, J. 2001. On the taxonomy of spotted knapweed (*Centaurea stoebe* L.). Pp. 33–41 in L. Smith (ed.), Proceedings of the First International Knapweed Symposium of the Twenty-first Century, Coeur d'Alene, Idaho. Albany, CA.

(312) O'Kane, S.L. and I.A. Al-Shehbaz. 2003. Phylogenetic position and generic limits of *Arabidopsis* (Brassicaceae) based on sequences of nuclear ribosomal DNA. Annals of the Missouri Botanical Garden 90:603–612.

(313) Owenby, M. 1950. The genus *Allium* in Idaho. Research Studies of the State College of Washington 18:3–39.

(314) Packer, J.G. 2002. *Triantha*. Pp. 61–64 *in* FNA Editorial Committee, Flora of North America, Vol. 26, Oxford University Press, New York.

(315) Padgett, D.J., D.H. Les, and G.E. Crow. 1999. Phylogenetic relationships in *Nuphar* (Nymphaeaceae): evidence from morphology, chloroplast DNA, and nuclear ribosomal DNA. American Journal of Botany 86:1316–1324.

(316) Parfitt, B.D. 1997. *Trolius*. Pp. 189–190 *in* FNA Editorial Committee, Flora of North America, Vol. 3, Oxford University Press, New York.

(317) Park, M.M. and D. Festerling. 1997. *Thalictrum* Pp. 258–271 *in* FNA Editorial Committee, Flora of North America, Vol. 3, Oxford University Press, New York.

(318) Parker, V.T., M.C.Vasey, and J.E. Keeley. 2009. *Arctostaphylos* Pp. 406–445 *in* FNA Editorial Committee, Flora of North America, Vol. 8, Oxford University Press, New York.

(319) Patterson, R.W. 1993. Polemoniaceae. Pp. 824–852 in J. C. Hickman (editor), The Jepson Manual, higher plants of California. University of California Press, Berkeley.

(320) Perry, E.S. 1962. Montana in the geologic past. Bulletin 26, Montana Bureau of Mines and Geology, Butte.

(321) Pfister, R.D., B.L. Kovalchik, S.F. Arno and R.C. Presby. 1977. Forest habitat types of Montana. USDA Forest Service General Technical Report INT-34, Ogden, UT.

(322) Phillips, W. 2003. Plants of the Lewis and Clark Expedition, Mountain Press Publishing Company, Missoula, MT.

(323) Phipps, J.B. 1998. Introduction to the red-fruited hawthorns (*Crataegus*, Rosaceae) of western North America. Canadian Journal of Botany 76:1863–1899.

(324) Phipps, J.B. and R.J. O'Kennon. 1998. Three new species of *Crataegus* (Rosaceae) from western North America: *C. o'kennonii*, *C. okanaganensis* and *C. phippsii*. Sida 18:169–191.

(325) Porsild, A.E. and W.J. Cody. 1980. Vascular plants of continental Northwest Territories, Canada. National Museums of Canada, Ottawa.

(326) Porter, J.M. and L.A. Johnson. 1998. Phylogenetic relationships of Polemoniaceae: inferences from mitochondrial NAD1B intron sequences. Aliso 17:157–188.

(327) Price, R.A. 1993. *Draba*. Pp. 416–420 in J.C. Hickman (ed,), The Jepson Manual, higher plants of California. University of California Press, Berkeley.

(328) Pringle, J.S. 1997. *Clematis*. Pp. 158–176 *in* FNA Editorial Committee, Flora of North America, Vol. 3, Oxford University Press, New York.

(329) Pringle, J.S. 2005. *Gypsophila*. Pp. 153–156 *in* FNA Editorial Committee, Flora of North America, Vol. 5, Oxford University Press, New York.

(330) Pryer, K.M. 1993. *Gymnocarpium*. Pp. 258–262 *in* FNA Editorial Committee, Flora of North America, Vol. 2, Oxford University Press, New York.

(331) Puff, C. 1976. The *Galium trifidum* group (*Galium* sect. *Aparinoides*, Rubiaceae). Canadian Journal of Botany 54:1911–1925.

(332) Rabeler, R.K. and R.L. Hartman. 2005. Caryophyllaceae. Pp. 3–215 *in* FNA Editorial Committee, Flora of North America, Vol. 5, Oxford University Press, New York.

(333) Rabeler, R.K., R.L. Hartman, and F.H. Utech. 2005. *Minuartia*. Pp. 116–136 *in* FNA Editorial Committee, Flora of North America, Vol. 5, Oxford University Press, New York.

(334) Raven, P.H. 1969. A revision of the genus *Camissonia* (Onagraceae) Contributions to the U.S. National Herbarium 37:161–396.

(335) Reese, R.N. 1984. A new variety of *Pedicularis contorta* (Scrophulariaceae) endemic to Idaho and Montana. Brittonia 36:63–66.

(336) Reveal, J.L. 1968. On the names in Fraser's 1813 catalogue. Rhodora 70:25–54.

(337) Reveal, J.L. 2005. *Eriogonum*. Pp. 221–430 *in* FNA Editorial Committee, Flora of North America, Vol. 5, Oxford University Press, New York.

(338) Reveal, J.L. 2009. *Dodecatheon*. Pp. 268–286 *in* FNA Editorial Committee, Flora of North America, Vol. 8, Oxford University Press, New York.

(339) Reveal, J.L. and J.C. Pires. 2002. Phylogony and classification of the Monocotyledons: an update. Pp. 3–36 *in* FNA Editorial Committee, Flora of North America, Vol. 26, Oxford University Press, New York.

(340) Robart, B.W. 2005. Morphological diversification and taxonomy among the varieties of *Pedicularis bracteosa* Benth. (Orobanchaceae). Systematic Botany 30:644–656.

(341) Rogers, C.M. 1968. Yellow-flowered species of *Linum* in Central America and western North America. Brittonia 20:107–135.

(342) Rollins, R.C. 1993. Cruciferae of continental North America. Stanford University Press, Stanford, CA.

(343) Rollins, R.C. and E.A. Shaw. 1973. The genus *Lesquerella* (Cruciferae) in North America. Harvard University Press, Cambridge, MA.

(344) Romaschenko, K., P.M. Peterson, R.J. Soreng, N. Garcia-Jacas, O. Futorna, and A. Susanna. 2008. Molecular phylogenetic analysis of the American Stipeae (Poaceae) resolves *Jarava* sensu lato polyphyletic: evidence for a new genus, *Pappostipa*. Journal of the Botanical Research Institute of Texas 2:165–192.

(345) Romero-Gonzalez, G., A., G.C. Fernandez-Concha, R.L. Dressler, L.K. Magrath, and G.W. Argus. 2002. Orchidaceae. Pp. 490–651 *in* FNA Editorial Committee, Flora of North America, Vol. 26, Oxford University Press, New York.

(346) Rumely, J. and M. Lavin. 1995. The grass flora of Montana. Unpublished manuscript, Montana State University Herbarium.

(347) Rydberg, P.A. 1900. Catalogue of flora of Montana and Yellowstone National Park. Memoirs of the New York Botanical Garden 1:1–492.

(348) Rydberg, P.A. 1917. Flora of the Rocky Mountains and adjacent plains. New Era Printing Co., Lancaster, PA.

(349) Saarela, J.M. and B.A. Ford. 2001. Taxonomy of the *Carex backii* Complex (Section *Phyllostachyae*, Cyperaceae). Systematic Botany 26:704–721.

(350) Saltonstall, K., P.M. Peterson, and R.J. Soreng. 2004. Recognition of *Phragmites australis* subsp. *americanus* (Poaceae: Arundinoideae) in North America: evidence from morphological and genetic analysis. Sida 21(2):683–692.

(351) Schilling, E.E., 2006a. *Helianthus*. Pp. 141–169 *in* FNA Editorial Committee, Flora of North America, Vol. 21, Oxford University Press, New York.

(352) Schilling, E.E., 2006b. *Heliomeris*. Pp. 169–172 *in* FNA Editorial Committee, Flora of North America, Vol. 21, Oxford University Press, New York.

(353) Schwartz, F.C. 2002. *Zigadenus*. Pp. 81–88 *in* FNA Editorial Committee, Flora of North America, Vol. 21, Oxford University Press, New York.

(354) Scott, R.W. 2006. *Brickellia*. Pp. 491–507 *in* FNA Editorial Committee, Flora of North America, Vol. 21, Oxford University Press, New York.

(355) Sears, C.J. 2008. Morphological discrimination of *Platanthera aquilonis*, *P. huronensis*, and *P. dilatata* (Orchidaceae) herbarium specimens. Rhodora 110:389–405.

(356) Selliah, S. and L. Brouillet. 2008. Molecular phylogeny of the North American eurybioid asters (Asteraceae, Astereae) based on the nuclear ribosomal internal and external transcribed spacers. American Journal of Botany 86:901–915.

(357) Semple, J.C. 2006. *Heterotheca*. Pp. 230–256 *in* FNA Editorial Committee, Flora of North America, Vol. 20, Oxford University Press, New York.

(358) Semple, J.C. and J.G. Chmielewski. 1987. Revision of *the Aster lanceolatus* complex, including *A. simplex* and *A. hesperius* (Compositae: Asterae). A multivariate, morphometric study. Canadian Journal of Botany 65:1047–1062.

(359) Semple, J.C. and R.E. Cook. 2006. *Solidago*. Pp. 107–166 *in* FNA Editorial Committee, Flora of North America, Vol. 20, Oxford University Press, New York.

(360) Shechter, Y. and B.L. Johnson. 1966. A new species of *Oryzopsis* (Gramineae) from Wyoming. Brittonia 18:342–347.

(361) Shechter, Y. and B.L. Johnson. 1968. The probable origin of *Oryzopsis contracta*. American Journal of Botany 55:611–618.

(362) Shelly, J.S., P. Lesica, P.G. Wolf, P.S. Soltis, and D.E. Soltis. 1998. Systematic studies and conservation status of *Claytonia lanceolata* var. *flava* (Portulacaceae). Madrono 45:64–74.

(363) Sheviak, C.J. 2002a. *Platanthera*. Pp. 551–571 *in* FNA Editorial Committee, Flora of North America, Vol. 26, Oxford University Press, New York.

(364) Sheviak, C.J. 2002b. *Cypripedium*. Pp. 499–507 *in* FNA Editorial Committee, Flora of North America, Vol. 26, Oxford University Press, New York.

(365) Shinwari, Z.K., R. Terauchi, F.H. Utech, and S. Kawano. 1994. Recognition of the New World *Disporum* Section *Prosartes* as *Prosartes* (Liliaceae) Based on the Sequence Data of the rbcL Gene. Taxon 43:353–366.

(366) Shultz, L.M. 2006. *Artemisia*. Pp. 503–534 *in* FNA Editorial Committee, Flora of North America, Vol. 20, Oxford University Press, New York.

(367) Sinnott, Q.P. 1985. A revision of *Ribes* L. subg. *Grossularia* (Mill.) Pers. sect. *Grossularia* (Mill.) Nutt. (Grossulariaceae) in North America. Rhodora 87:189–286.

(368) Small, E. 1997. Cannabaceae. Pp. 381–387 *in* FNA Editorial Committee, Flora of North America, Vol. 3, Oxford University Press, New York.

(369) Small, E. 1978. A numerical and nomenclatural analysis of morpho-geographic taxa of *Humulus*. Systematic Botany 3:37–76.

(370) Smith, S.G. 2002. *Schoenoplectus*. Pp. 44–60 *in* FNA Editorial Committee, Flora of North America, Vol. 23, Oxford University Press, New York.

(371) Smith, S.G., J.J. Bruhl, M.S. Gonzalez-Elizando, F.J. Menpace. 2002. *Eleocharis*. Pp. 60–120 *in* FNA Editorial Committee, Flora of North America, Vol. 23, Oxford University Press, New York.

(372) Soderstrom, T. et al. 1988. Grass systematics and evolution. Smithsonian Institution Press, Washington, D.C.

(373) Soltis, D.E., P.S. Soltis, M.T. Clegg, and M. Durbin. 1990. rbcL sequence divergence and phylogenetic relationships in Saxifragaceae sensu lato. Proceedings of the National Academy of Sciences 87:4640–4644.

(374) Soltis, D.E., R.K. Kuzoff, E. Conti, R. Gornall, and K. Ferguson. 1996. matK and rbcL Gene sequence data indicate that *Saxifraga* (Saxifragaceae) is polyphyletic. American Journal of Botany 83:371–382.

(375) Soltis, D.E., R.K. Kuzoff, M.E. Mor, M. Zanis, M. Fishbein, L. Hufford, J. Koontz, and M.K. Arroyo. 2001. Elucidating deep-level phylogenetic relationships in Saxifragaceae using sequences for six chloroplastic and nuclear DNA regions. Annals of the Missouri Botanical Garden 88:669–693.

(376) Soltis, P.S. 2006. *Tragopogon*. Pp. 303–306 *in* FNA Editorial Committee, Flora of North America, Vol. 19, Oxford University Press, New York.

(377) Soreng, R. et al. 1990. A phylogenetic analysis of chloroplast DNA restriction site variation in Poaceae subfam. Pooideae. Plant Systematics and Evolution 172:83–97.

(378) Spellenberg R.W. 2003. *Mirabilis*. Pp. 40–57 *in* FNA Editorial Committee, Flora of North America, Vol. 4, Oxford University Press, New York.

(379) Spencer, S.C. and A.E. Spencer. 2003. *Navarettia willamettensis* and *Navarretia saximontana* (Polemoniaceae), new species from ephemeral wetlands of western North America. Madrono 50:196–199.

(380) Spribille, T., B. Heidel, W. Albert, F.J. Tripke, J. Vanderhorst, and G.M. Arvidson. 2002. Noteworthy collections: Montana. Madrono 49:55–58.

(381) Standley, L.A. 1985. Systematics of the Acutae group of *Carex* (Cyperaceae) in the Pacific Northwest. Systematic Botany Monographs 7:1–106.

(382) Standley, P.C. 1921. Flora of Glacier National Park, Montana. Contributions to the U.S. National Herbarium 22:235–438.

(383) Stebbins, G.L. and B. Crampton. 1961. A suggested revision of the grass genera of temperate North America. Recent Advances in Botany 1:133–145.

(384) Stevens, P.F. 2008. Angiosperm Phylogeny Website. Version 9, June 2008. http://www.mobot.org/MOBOT/research/APweb/.

(385) Stewart, R.E. and H.A. Kantrud. 1971. Classification of natural ponds and lakes in the glaciated prairie region. U.S. Fish and Wildlife Service Publication 92.

(386) Strong, M.T. 1994. Taxonomy of *Scirpus*, *Trichophorum*, and *Schoenoplectus* (Cyperaceae) in Virginia Bartonia 58:29–68.

(387) Strother, J.L. and R.R. Weedon. 2006. *Bidens*. Pp. 205–218 *in* FNA Editorial Committee, Flora of North America, Vol. 21, Oxford University Press, New York.

(388) Strother, J.L. and M.A. Wetter. 2006. *Grindelia*. Pp. 424–436 *in* FNA Editorial Committee, Flora of North America, Vol. 20, Oxford University Press, New York.

(389) Stuckey, R.L. 1972. Taxonomy and distribution of the genus *Rorippa* (Cruciferae) in North America. Sida 4: 279–430.

(390) Swab, J.C. 2000. *Luzula*. Pp. 255–267 *in* FNA Editorial Committee, Flora of North America, Vol. 22, Oxford University Press, New York.

(391) Taylor, W.C., N.T. Luebke, D.M. Britton, R.J. Hickey, and D.F. Bruton. 1993. Isoetaceae. Quilwort Family. Pp. 64–75 *in* FNA Editorial Committee, Flora of North America, Vol. 2, Oxford University Press, New York.

(392) Taylor, R.L. 2009. *Lithphragma*. Pp. 77–83 *in* FNA Editorial Committee, Flora of North America, Vol. 8, Oxford University Press, New York.

(393) Thompson, L.S. 1985. The pioneer naturalists, 1805–1864. Montana Magazine, Helena, MT.

(394) Torrecilla, P. and P. Catalán. 2002. Phylogeny of Broad-leaved and Fine-leaved *Festuca* Lineages (Poaceae) based on Nuclear ITS Sequences. Systematic Botany 27:241–251.

(395) Trock, D.K. 2006a. *Achillea*. Pp. 492–494 *in* FNA Editorial Committee, Flora of North America, Vol. 19, Oxford University Press, New York.

(396) Trock, D.K. 2006b. *Packera*. Pp. 570–602 *in* FNA Editorial Committee, Flora of North America, Vol. 20, Oxford University Press, New York.

(397) Trock, D.K. and T.M. Barkley. 1999. *Packera subnuda* comb. nov., a corrected name for *Packera buekii* (Asteraceae: Senecioneae). Sida 18:635.

(398) Tryon, R.M. and A.F. Tryon. 1982. Ferns and allied plants with special reference to tropical America. Springer-Verlag, New York.

(399) Tutin, T.G., V.H. Heywood, N.A. Bunges, D.H. Valentine, S.M. Walters, and D.A. Webb. 1976. Flora Europea. Cambridge University Press, London.

(400) Tweedy, F. 1886. Flora of the Yellowstone National Park. Published by the author. Washington D.C.

(401) Tzvelev, N.N. 1989. The system of grasses (Poaceae) and their evolution. Botanical Review 55:141–203.

(402) Umber, F.E. 1979. The genus *Glandularia* (Verbenaceae) in North America. Systematic Botany 4:72–102.

(403) Urbatsch, L.E., L.C. Anderson, R.P. Roberts, and K.M. Neubig. 2006. *Ericameria*. Pp. 50–77 *in* FNA Editorial Committee, Flora of North America, Vol. 20, Oxford University Press, New York.

(404) USDA Natural Resources Conservation Service (USDA-NRCS). 1999. Montana annual precipitation (map). USDA-NRCS National Cartography & Geospatial Center, Fort Worth, TX.

(405) USDA, NRCS. 2011. The PLANTS Database (http://plants.usda.gov, 30 October 2011). National Plant Data Team, Greensboro, NC 27401-4901 USA.

(406) Utech, F.H. 2002. *Streptopus*. Pp. 145–147 *in* FNA Editorial Committee, Flora of North America, Vol. 26, Oxford University Press, New York.

(407) Valdespino, I.A. 1993. Selaginellaceae. Spike-moss Family. Pp. 38–63 *in* FNA Editorial Committee, Flora of North America, Vol. 2, Oxford University Press, New York.

(408) Vander Kloet, S.P. 1988. The genus *Vaccinium* in North America. Research Branch, Agriculture Canada; Publication 1828.

(409) Vanderpool, S.S. and H.H. Iltis. 2010. *Peritoma*. Pp. 205–208 *in* FNA Editorial Committee, Flora of North America, Vol. 7, Oxford University Press, New York.

(410) Veseth, R. and C. Montagne. 1980. Geologic parent materials of Montana soils. USDA-SCS and Montana Agricultural Experiment Station Bulletin 721, Bozeman, MT.

(411) Vincent, M.A. Molluginaceae. 2003. Pp. 509–512 *in* FNA Editorial Committee, Flora of North America, Vol. 4, Oxford University Press, New York.

(412) Wagner, W.H. and J.M. Beitel. 1993. Lycopodiaceae. Club-moss Family. Pp. 18–37 *in* FNA Editorial Committee, Flora of North America, Vol. 2, Oxford University Press, New York.

(413) Wagner, W.H. and F.S. Wagner. 1986. Three new species of moonworts (*Botrychium* subg. *Botrychium*) endemic in western N. America. American Fern Journal 76:33–47.

(414) Wagner, W.H. and F.S. Wagner. 1993. Ophioglossaceae. Adder's-tongue Family. Pp. 85–106 *in* FNA Editorial Committee, Flora of North America, Vol. 2, Oxford University Press, New York.

(415) Wagner, W.L., R.E. Stockhouse, and W.M. Klein. 1985. The systematics and evolution of the *Oenothera caespitosa* species complex (Onagraceae). Monographs in Systematic Botany of the Missouri Botanical Garden 12:1–103.

(416) Wagner, W.L., P.C. Hoch, and P.H. Raven. 2007. Revised classification of the Onagraceae. Systematic Botany Monographs 83:1–240.

(417) Wahl, H.A. 1954. A preliminary study of the genus *Chenopodium* in North America. Bartonia 27:1–46.

(418) Wambolt, C.L. and M.R. Frisina. 2002. Montana sagebrush guide. Montana Department of Fish, Wildlife, and Parks, Helena.

(419) Warnock, M.J. 1997. *Delphinium*. Pp. 196–240 *in* FNA Editorial Committee, Flora of North America, Vol. 3, Oxford University Press, New York.

(420) Warwick, S.I., I.A. Al-Shehbaz, C.A. Sauder, D.F. Murray, and K. Mummenhoff. 2004. Phylogeny of *Smelowskia* and related genera (Brassicaceae) based on nuclear ITS DNA and chloroplast trnL Intron DNA Sequences. Canadian Journal of Botany 91:99–123.

(421) Watson, L.E. 2006. *Tanacetum*. Pp. 489–491 *in* FNA Editorial Committee, Flora of North America, Vol. 19, Oxford University Press, New York.

(422) Weaver, T.W. 1965. Variation in the spruce complex of the Northern Rocky Mountains *Picea glauca, Picea pungens* and *Picea engelmannii*. M.S. Thesis, University of Montana, Missoula.

(423) Weaver, J.E. 1954. North American prairie. Johnsen Publishing, Lincoln, NE.

(424) Weber, W.A. 1946. A taxonomic and cytological study of the genus *Wyethia*, family Compositae, with notes on the related genus *Balsamorhiza*. American Midland Naturalist 35:400–452.

(425) Weber, W.A. 1987. Colorado flora, western slope. Colorado Associated University Press, Boulder.

(426) Weber, W.A. 2002. *Senecio spribillei* (Asteraceae: Senecioneae), a new species from Montana, U.S.A. Sida 20:511–513.

(427) Weber, W.A. and R.L. Hartman. 1979. *Pseudostellaria jamesiana*, comb. nov., a North American representative of a Eurasian genus. Phytologia 44:313–314.

(428) Wells, E.F. and P.E. Elvander. 2009. Saxifragaceae. Pp. 43–146 *in* FNA Editorial Committee, Flora of North America, Vol. 8, Oxford University Press, New York.

(429) Wells, E.F. and B.G. Shipes. 2009. *Heuchera* Pp. 84–100 *in* FNA Editorial Committee, Flora of North America, Vol. 8, Oxford University Press, New York.

(430) Welsh, S.L. 1974. Anderson's flora of Alaska and adjacent parts of Canada. Brigham Young University, Provo.

(431) Welsh, S.L. 1994. North American *Oxytropis* (L.) DC. (Leguminosae) types at the Natural History Museum and Royal Botanic Garden, England, with nomenclatural comments and a new variety. Great Basin Naturalist 55:271–281.

(432) Welsh, S.L. 2001. Revision of North American species of *Oxytropis* deCandolle (Leguminosae). E.P.S. Inc., Orem, UT.

(433) Welsh, S.L. 2003. *Atriplex.* Pp. 322–381 *in* FNA Editorial Committee, Flora of North America, Vol. 4, Oxford University Press, New York.

(434) Welsh, S.L. and M. Ralphs. 2002. Some tall larkspurs (*Delphinium* —Ranunculaceae) a taxonomic review. Biochemical Systematics and Ecology 30:103–112.

(435) Welsh, S.L., N.D. Atwood, S. Goodrich, and L.C. Higgins. 1993. A Utah flora, second edition. Brigham Young University, Provo, UT.

(436) Western Regional Climate Center (WRCC). Accessed November 2010. Historical climate information http://www.wrcc.dri.edu/CLIMATEDATA.html.

(437) White, W.W. 1956. Native willows found in Montana. Proceedings of the Montana Academy of Sciences 16:21–35.

(438) Whittemore, A.T. 1997a. *Ranunculus.* Pp. 88–135 *in* FNA Editorial Committee, Flora of North America, Vol. 3, Oxford University Press, New York.

(439) Whittemore, A.T. 1997b. *Myosurus.* Pp. 135–138 *in* FNA Editorial Committee, Flora of North America, Vol. 3, Oxford University Press, New York.

(440) Whittemore, A.T. 1997c. *Aquilegia.* Pp. 249–258 *in* FNA Editorial Committee, Flora of North America, Vol. 3, Oxford University Press, New York.

(441) Whittemore, A.T. and A.E. Schuyler. 2002. *Scirpus.* Pp. 8–21 *in* FNA Editorial Committee, Flora of North America, Vol. 3, Oxford University Press, New York.

(442) Wiersema, J.H. and C.B. Hellquist. 1997. Nymphaeaceae. Water-lily family. Pp.66–77 *in* FNA Editorial Committee, Flora of North America, Vol. 3, Oxford University Press, New York.

(443) Wilken, D.H. 1986. Polemoniaceae. Pp. 666–677 *in* Great Plains Flora Association, Flora of the Great Plains, University Press of Kansas, Lawrence.

(444) Wilken, D. and R.L. Hartman. 1991. A revision of the *Ipomopsis spicata* complex (Polemoniaceae). Systematic Botany 16:143–161.

(445) Williams, L.O. 1937. A monograph of the genus *Mertensia* in North America. Annals of the Missouri Botanical Garden 24:17–159.

(446) Wilson, B., R. Brainerd, D. Lytjen, B. Newhouse, and N. Otting. 2008. Field guide to the sedges of the Pacific Northwest. Oregon State University Press, Corvallis.

(447) Windham, M.D. 1993. *Pellaea.* Pp. 175–186 *in* FNA Editorial Committee, Flora of North America, Vol. 2, Oxford University Press, New York.

(448) Windham, M.D. and I.A. Al-Shehbaz. 2006. New and noteworthy species of *Boechera* (Brassicaceae) I: sexual diploids. Harvard Papers in Botany 11:61–88.

(449) Windham, M.D. and I.A. Al-Shehbaz. 2007a. New and noteworthy species of *Boechera* (Brassicaceae) II. apomictic hybrids. Harvard Papers in Botany 11:257–274.

(450) Windham, M.D. and I.A. Al-Shehbaz. 2007b. New and noteworthy species of *Boechera* (Brassicaceae) III. additional sexual diploids and apomictic hybrids. Harvard Papers in Botany 12:235–257.

(451) Wolf, S.J. 2006. *Arnica.* Pp. 366–377 *in* FNA Editorial Committee, Flora of North America, Vol. 4, Oxford University Press, New York.

(452) Western Regional Climate Center (WRCC). 2010. Historical climate information. http://www.wrcc.dri.edu/.

(453) Yuncker, T.G. 1965. *Cuscuta.* North American Flora 4:1–40.

(454) Zika, P.F. 2003. The native subspecies of *Juncus effusus* (Juncaceae) in western North America. Brittonia 55: 150–156.

(455) Zika, P.F. and A.L. Jacobson. 2003. An overlooked hybrid Japanese knotweed (*Polygonum cuspidatum x sachalinense*; Polygonaceae) in North America. Rhodora 105:143–152.

(456) Zimmerman, A.D. and B.D. Parfitt. 2003. *Coryphantha.* Pp. 220–237 *in* FNA Editorial Committee, Flora of North America, Vol. 4, Oxford University Press, New York.

(457) Cialdella, A.M., D.L. Salariato, L. Aagesen, L.M. Giussani, F.O. Zuloaga, and O. Morrone. 2010. Phylogeny of New World Stipeae (Poaceae): an evaluation of the monophyly of *Aciachne* and *Amelichloa*. Cladistics 26:563–578.

(458) Romaschenko, K., P.M. Peterson, R.J. Soreng, O. Futorna, and A. Susanna. 2011. Phylogenetics of *Piptatherum* s.l. (Poaceae: Stipeae): evidence for a new genus, *Piptatheropsis*, and resurrection of *Patis*. Taxon 28:1–14.

(459) Seberg et al. 2010. Diversity, phylogeny, and evolution in the Monocotyledons. Aarhus University Press. Pp. 511–537.

GLOSSARY

>. Greater than, more than.

<. Less than, smaller than.

≤. Less than or equal to, smaller or equal to.

≥. Greater than or equal to, at least, smaller or equal to.

ca. Approximately, about.

Abaxial. The side of the leaf or any other structure facing away from the stem axis during development; it is usually the side of the leaf facing downward at maturity.

Abortive. Imperfectly developed and usually a mere rudiment.

Acaulescent. Without a leaf-bearing stem, or with stem so short that the leaves appear basal.

Achlorophyllous. Without chlorophyll, not green.

Actinorhizal. Forming root nodules with nitrogen fixing actinobacteria.

Acuminate. Tapering gradually to a protracted point.

Adaxial. The side of the leaf or any other structure facing toward the stem axis during development; it is usually the side of the leaf facing upward at maturity.

Adnate. Fused to an organ of a different kind.

Alternate. Borne singly and spaced around and along the axis, applied to leaves or other organs on an axis; see opposite, whorled.

Ament. A dense spike or raceme of apetalous, unisexual flowers; catkin.

Annual. Completing the full cycle of germination to fruiting within a single year and then dying; see biennial, perennial.

Anthesis. The period in the development of the flower during which pollination takes place.

Antrorse. Projected toward the tip, as in the barbs on the hair-like structures in the inflorescence of most species of *Setaria*, which are oriented in the direction of the tip of the hair from which they are borne.

Apetalous. Without petals or a corolla.

Apical. At the top, tip, or end of a structure, terminal.

Apiculate. Having a short, sharply pointed tip.

Apomictic. Reproduction through seeds produced without fertilization.

Appendage. A secondary part attached to a primary structure, an attachment.

Appressed. Lying against an organ or structure. The branches of an inflorescence may be pressed against the main axis, or hairs may be pressed to the surface of a leaf.

Arachnoid. With entangled, cobweb-like hairs.

Areola. Region on a cactus stem bearing the spines or flowers.

Aristate. Having a stiff, bristle-like tip.

Articulated. Jointed. A line of demarcation can usually be seen at which point a separation takes place at maturity.

Ascending. Sloping upward at an angle of about 40-70 degrees.

Attenuate. Gradually tapered.

Auricle. A linear to ear-shaped appendage at the base of a leaf or leaflet.

Auriculate. With auricles.

Auriculate-clasping. Describing an auriculate and sessile leaf base which appears to clasp the stem.

Avalanche slope. Steep slope in a forest landscape where sliding snow prohibits the occurrence of trees.

Awn. A slender, bristle-like projection. Stiff, needle-like pappus element in Asteraceae.

Axil. Angle or space between one part of a plant and another part; between an appendage or branch and a main axis or stem.

Axis. The main stem or structure from which lateral organs arise.

Axillary. Pertaining to or within an axil.

Badlands. Steep, rugged terrain characterized by unvegetated soil.

Banner. Uppermost, often largest, frequently upturned petal of flowers of many members of Fabaceae.

Barbed. Having sharp, normally downward- or backward-pointing projections.

Barbellate. Minutely barbed.

Basifixed. Attached at the base.

Beak. A prominent terminal projection, especially of a carpel or fruit.

Bearded. Furnished with long stiff hairs.

Berry. Fleshy fruit with generally more than 1 seed not encased in a stone.

Bifid. Cleft or two-lobed.

Bilabiate. Two-lipped, with two unequal divisions.

Bilateral. Two-sided, as in spikelets alternating along one of two sides of a rachis, like the inflorescence spikes of Triticeae.

Bilateral symmetry. Divisible into mirror-image halves in only one way.

Bipinnate. Twice pinnately divided.

Bisexual. A flower (floret) having both stamens and pistils, as in a perfect flower.

Biternate. Twice divided into threes.

Blade. Expanded portion of a leaf, petal, or other structure, generally flat.

Bract. Reduced, leaf- or scale-like structure at the base of a flower or inflorescence.

Bracteate. With bracts.

Bracteole. A small bract-like structure borne singly or in a pair on the pedicel or calyx of a flower (*Carex*).

Bristle. A stiff, strong trichome, as in the perianth of some members of the Cyperaceae or Asteraceae.

Bud. An incompletely developed, more or less embryonic shoot (usually covered with bud scales) or flower.

Bulb. Short underground stem and the fleshy leaves or leaf bases attached to and surrounding it.

Bulblet. A small, bulb-like structure that propagates a plant, often in a leaf or bract axil.

Bunched. Several to many upright stems branching from the same root system (i.e., tillers) densely clustered together.

Caducous. Falling or disarticulating readily; falling entirely and immediately after maturation.

Calcareous. Rich in calcium. Soils derived from limestone or dolomite.

Callosity. A thickened, raised area, which is usually hard; a callus.

Callus. Swollen and hardened tissue at the base of the lemma (where callus is derived from rachilla), spikelet (where callus is derived from the pedicel), or inflorescence branch (where callus is derived from the inflorescence rachis). The callus aids in dispersal or burial of the floret, spikelet, or inflorescence branch, by penetrating skin, clothing, hair, or soil.

Calyculi. Bracts simulating a calyx just outside of the sepals.

Calyx. Outermost or lowermost whorl of flower parts, generally green and enclosing inner flower parts in bud.

Campanulate. Bell-shaped; with a flaring tube about as broad as long and a flaring limb.

Canescent. Covered with dense, fine grayish-white trichomes.

Capillary. Very slender and fine.

Capitate. Head-like, densely packed in a head-shaped cluster.

Capsule. A dry fruit formed from two or more united carpels and dehiscing at maturity to release the seeds.

Carnivorous. Consuming animal flesh for nutrition.

Carpel. An organ, generally believed to be leaf-derived, which bears 1 or more ovules, a stigma and sometimes a style.

Caryopsis. A dry, solid minute fruit produced only by members of the grass family, whereby the seed is completely fused to the fruit wall.

Catkin. Spike or spike-like inflorescence of unisexual flowers with inconspicuous perianths, generally wind-pollinated, usually pendent and often with conspicuous bracts. Ament.

Caudex. A short, thick, vertical or branched perennial stem, usually subterranean, or at ground level.

Caulescent. Having an obvious, leaf-bearing, aerial stem.

Cauline. Pertaining to structures, especially leaves, borne along, not confined to the base of, an elongate, above-ground stem.

Chaff. Thin, dry scales.

Channeled. With a deep, longitudinal groove.

Chasmogamous. Describing flowers in which pollination takes place while the flower is open; see cleistogamous.

Cilia. Hairs occurring along margins or edges.

Ciliate. Bearing cilia.

Ciliolate. Minutely ciliate.

Circumscissile. Opening by a transverse line, the top coming off as a lid.

Clasping. Wholly or partly surrounding the stem.

Clavate. Club-shaped, gradually swelling toward the apex.

Claw. The conspicuously narrowed basal part of a flat organ.

Cleistogamous. Flowers which are self-pollinating and set fertile seed without the flower opening.

Collar. The area on the outer or abaxial side of a leaf at the junction of the sheath and blade, opposite the ligule.

Column. Structure at the center of an orchid or mallow flower formed by fusion of stamen(s) and style.

Coma. A tuft of hairs on top of a seed.

Commissure. The (inner) face by which two carpels join one another.

Compound. Composed of two or more anatomically or morphologically equivalent units.

Compressed. Flattened. Laterally compressed is if, for example, a spikelet is squashed or flattened from the sides of the lemmas and glumes. Dorsiventrally compressed is if, for example, a spikelet is squashed or flattened from the backs of the glumes and lemmas.

Concave. Hollowed out or curved inward.

Cone. Reproductive structure composed of an axis (branch) bearing sterile bract-like organs and seed or pollen bearing structures.

Confluent. Blending of one part into another.

Connate. Fused to another organ of the same kind.

Cordate. Heart-shaped; often pertaining to a leaf in which the blade base on both sides of the petiole is rounded and convex.

Corm. An upright underground storage stem.

Cormose. Having the appearance of a corm.

Corymb. Usually a raceme in which the flowers, through unequal pedicels, form a flat-topped or convex inflorescence.

Corymbiform. Having the appearance of a corymb.

Crenate. With obtuse or rounded teeth that either point forward or are perpendicular to the margin.

Cross-corrugate. Wrinkled into alternating furrows and ridges across the axis.

Culm. The stem of a grass, usually hollow, but with solid nodes that represent meristematic tissue. Such meristems are referred to as intercalated.

Cuneate. Wedge-shaped, tapering to the base.

Cyathium. Inflorescence in *Euphorbia* composed of an involucre, one pistil and male flowers with one stamen.

Cyme. An inflorescence in which each flower, in turn, is formed at the tip of a growing axis.

Cymose. Having the appearance of a cyme.

Cypsela. An achene derived from a one-loculed, inferior ovary.

Deciduous. Pertaining to leaves that all fall seasonally or to structures, such as hairs or flower parts, that fall early or readily.

Decumbent. Lying mostly flat on the ground but with tips curving up.

Decurrent. Having the leaf base prolonged down the stem as a winged expansion or rib.

Deltoid. More or less equilaterally triangular, with the corners rounded or not.

Dentate. With sharp, spreading, rather coarse teeth standing out from the margin.

Denticulate. Finely dentate.

Diffuse. Open and much branched; said of an open panicle. When internodes are very much longer than the length of the spikelet.

Digitate. Arranged as the fingers from the palm of a hand; several projections coming from one point.

Dimorphic. Having two different sizes and/or shapes.

Dioecious. Having male and female unisexual flowers on different plants.

Disarticulating. Disjointing or breaking apart at a zone of abscission.

Disciform. In Asteraceae, a head composed of disk flowers and marginal pistillate (or sterile) flowers with minute or missing laminae, superficially similar to a discoid head.

Discoid. In Asteraceae, a head composed entirely of disk flowers.

Disk flower. In Asteraceae, a generally bisexual (occasionally male or sterile), radially symmetrical flower with a 5- (rarely 4-) lobed corolla.

Distal. Away from the point of origin or attachment.

Distichous. Alternating on opposite sides. The glumes and florets of a grass spikelet are distichous in that they are borne alternately (as opposed to oppositely) on one of two opposite sides of the rachilla. The leaves of the grass genus *Distichlis* are often distichous because they are borne alternating up one of two sides of the leafy stem or culm.

Distinct. Separate, not united together.

Distylous. Flowers with short- and long-styled forms.

Dolabriform. Refering to a trichome that is pointed on both ends and attached to the epidermis in between the two ends, often closer to one end than the other.

Dorsal. Pertaining to the surface most facing away from the axis; back of an outer face of organ; lower side of leaf (abaxial).

Ebracteate. Without bracts.

Eglandular. Without glands.

Elaiosome. A fat-bearing seed appendage that attracts insects.

Ellipsoid. A 3-dimensional shape; elliptic in outline and round in cross-section.

Emarginate. Having a broad, shallow notch at the apex.

Entire. Smooth, without indentations or incisions on maigins.

Epidermis. Outermost cell layer or layers of non-woody plant parts.

Epigynous. Sepals, petals, and stamens attached to the floral tube above the ovary with the ovary adnate to the tube or hypanthium.

Equitant. Leaves 2-ranked with overlapping bases, sharply folded along midrib.

Erose. Irregularly, shallowly toothed and/or lobed margins.

Evergreen. With leaves persisting on the plant through at least one winter.

Excurrent. Continuing beyond, as the nerve of a leaf projecting beyond the margin.

Exserted. Projecting beyond surrounding parts.

Exstipulate. Without stipules.

Exudate. Material discharged (exuded) from a plant.

Falcate. Sickle-shaped.

Farinose. Covered with a mealy or powdery substance (farina).

Fascicle. Cluster of secondary leaves borne on a minute, determinate, short shoot in the axil of a primary leaf.

Fellfield. A community of dwarfed, scattered plants in stony soil of an exposed site, often above timberline.

Fen. Plant community characterized by saturated peat hydrated by groundwater.

Filiform. Threadlike, usually flexuous.

Fimbriate. With fringed margins.

Fistulose. Hollow, without pith.

Flag. Scarious appendage on a leaf tip.

Flavonoid. Any of a large group of water-soluble plant pigments that are often yellow.

Flexuous. Bending in a wavy manner.

Foliaceous. Bearing leaves, appearing leaf-like.

Follicle. A dry, dehiscent fruit derived from one carpel that splits along one suture.

Fornices. Small crests or scales in the throat of a corolla.

Fronds. The leaf of a fern or the body of a plant in the Lemnaceae.

Funnelform. Widening from the base more or less gradually through the throat into an ascending, spreading, or recurved limb.

Fusiform. Spindle-shaped; broadest in middle and tapering to each end.

Galea. Hood-like upper lip of some two-lipped corollas.

Gametophyte. A plant bearing sexual organs; in ferns a small discrete plant, subterranean and much smaller than the above-ground plant.

Gemma. A vegetative reproductive bud borne on the stem or frond.

Glabrate. Nearly glabrous.

Glabrous. Smooth; devoid of trichomes or hairs.

Glandular. Bearing small, often spheric bodies on, or embedded in, the epidermis or at the tip of a hair that exudes a generally sticky substance.

Glaucous. Covered with a generally whitish or bluish, waxy or powdery film that is sometimes easily rubbed off.

Globose. Spherical.

Glochids. Barbed trichomes or slender spines, usually in tufts.

Glomerules. An indeterminate dense cluster of sessile or subsessile flowers.

Gynostegium. A structure formed from the union of the anthers and the stigmatic region of the pistil in the Asclepiadaceae.

Hastate. With a pair of basal lobes that flare outwards; usually referring to a leaf.

Haustorium. A specialized organ on the roots of parasitic plants to draw nourishment from host plants.

Helicoid. Coiled like a spiral, as in Boraginaceae; inflorescences that uncoil with maturity.

Hip. An aggregation of achenes surrounded by a berry-like hypanthium as in *Rosa*.

Hirsute. Covered with long, stiff hairs.

Hispid. Covered with long, coarse, stiff trichomes.

Hispidulous. Minutely hispid.

Homostylous. With styles of same sizes or lengths and shapes; see heterostylous.

Hood. A hollow, arched-shaped perianth part, usually an upper petal with a turned down margin.

Humic. Referring to organic, humus-derived, usually acidic soil.

Hyaline. Thin and translucent or transparent.

Hydathode. An opening that exudes water.

Hypanthium. Structure generally in the shape of a tube, cup, or bowl, derived from the fused lower portions of the perianth and stamens and to which the ovary may be fused.

Hypogynous. With the sepals, petals, and stamens joined together below the superior ovary.

Imbricate. Overlapping like roof shingles.

Indurate. Hard or bony in texture.

Indusium. A flap of tissue covering a sorus.

Inferior. Describing an ovary which is below the point at which the sepals, petals and stamens are joined together.

Inflated. Puffed up; somewhat bladdery.

Inflorescence. An aggregation or arrangement of flowers and associated pedicels and bracts and excluding full-sized foliage leaves.

Inrolled. Curled or rolled inward.

Intercostal. Space between two veins or ribs.

Internode. The part of a stem between two successive nodes.

Interrupted. The continuity broken; said of certain dense inflorescences having continuity broken by gaps.

Involucel. Secondary involucre, subtending the secondary umbels in members of Apiaceae.

Involucre. A whorl of bracts subtending an inflorescence.

Irregular. With floral parts within a whorl dissimilar in shape, size, or arrangement; usually bilaterally symmetrical.

Keel. The two united lower petals of a papilionaceous (legume) flower. In grasses, the sharp fold on the back of a compressed sheath, blade, glume, or lemma.

Keeled. Ridged like the keel of a boat.

Lamina. The expanded portion or blade of a leaf or petal.

Lanate. Densely covered with long, intertwined hairs; woolly.

Lanceolate. Lance-shaped, widened at or above the base but below the middle and tapering to the apex.

Lanceoloid. A 3-dimensional shape; lanceolate in outline and round in cross-section.

Lateral. Side, as in the side of a structure.

Laterally. On or at the side.

Layering. Propagating by prostrate stems rooting at the nodes and sending up vertical shoots.

Leaflet. A distinct and separate segment of a leaf, distinguished from a leaf by the absence in its axil of a bud, branch, thorn, or flower.

Legume. A usually dry, dehiscent fruit derived from one carpel that splits along two sutures (Fabaceae).

Lemma. The bract of a spikelet above the glumes; the lowermost bract subtending the grass flower.

Lenticels. A lens-shaped dot or pit on bark, through which gaseous exchange may occur.

Lenticular. Shaped like a biconvex lens, more-or-less two-sided.

Ligulate. Strap-shaped.

Ligule. An outgrowth or projection from the top of the sheath, as in the Poaceae; the strap-shaped portion of a ray or ligulate corolla as in the Asteraceae.

Limb. The expanded, often lobed portion of a corolla distal to the tube and throat; the expanded portion of a petal or sepal distal to the stalk-like base (claw).

Linear. Long and narrow with parallel sides.

Lip. Upper or lower of two parts in a bilaterally symmetrical calyx or corolla; the exceptionally large petal of an orchid flower.

Lobe. A rounded or pointed projection, usually one of two or more, each separated by a fissure or sinus.

Lobed. Bearing lobes for which the intervening sinus is less than half the distance to the midvein.

Loculate. Pertaining to locules.

Locule. Compartment of an ovary or anther.

Loment. A legume that separates transversely between seed sections.

Mealy. Powdery, dry and crumbly.

Medial. Situated in the middle.

Megaspores. A spore that gives rise to a female egg-bearing plant.

Membranous. Thin and translucent like a membrane.

Mericarp. One of the two halves of the fruit in the Apiaceae.

-merous. Parts of a set; e.g., 5-merous corolla has five lobes or petals.

Mesic. Optimal moisture, not dry or wet.

Microsite. A small-scale habitat patch.

Microspores. A spore that gives rise to a sperm-producing plant.

Midvein. The central vein of a leaf or leaflet.

Monocarpic. Flowering and fruiting only once in a lifetime.

Monoecious. With male and female functions in separate flowers but on the same plant.

Monomorphic. All of the same shape and size.

Monophyletic. A group of all organisms that have all evolved from a common ancestor and includes all descendants of that ancestor.

Mucronate. Having an abrupt, short, sharp, terminal tip.

Muricate. Covered with short, hard protuberances.

Mycotrophic. Involved in a symbiotic relationship between a fungus and a root.

Nectary. A specialized nectar-secreting structure or area.

Nerve. A vascular vein of the sheath, blade, glume, lemma, and palea.

Node. Point on the stem where leaves are attached; or the point of branching of the stem.

Nut. A one-seeded, dry, indehiscent fruit with a hard pericarp, usually derived from a one-celled ovary.

Nutlets. Small, dry nut or nut-like fruit, usually several of which are produced by a single flower; e.g., Boraginaceae, Lamiaceae.

Obconic. Cone-shaped, attached at the narrower end.

Obcordate. Inversely heart-shaped, attached at the narrow end.

Oblanceolate. Inversely lance-shaped, attached at the narrow end.

Obovate. Inversely ovate, attached at the narrow end.

Obovoid. A 3-dimensional shape; obovate in outline and round in cross-section.

Obtriangular. Inversely triangular, attached at an apex.

Ochroleucus. Yellowish-white, cream-colored.

Ocrea. A stipular tube surrounding stem above insertion of petiole or blade.

Open. Said of panicles if the branches are long and spreading.

Operculate. With a small lid that becomes detached at maturity.

Opposite. Two leaves or other structures per node, on opposite sides of stem or central axis.

Orbiculate. Circular in outline.

Oribucular. Circular in outline.

Ovary. Ovule-bearing, lower, expanded part of pistil.

Ovate. Egg-shaped in outline with widest axis below middle.

Palate. The raised area in the throat of a sympetalous corolla.

Palea. The inner bract of a floret, usually with two distinct veins or ribs. It is homologous with the prophyll.

Paleae. Chaffy scales.

Palmately. Radiately lobed or divided; with leaflets all attached at the petiole tip.

Panicle. A compound raceme; an indeterminate inflorescence in which the flowers are borne on branches of the main axis or on further branches of these.

Paniculate. Having flowers in a panicle.

Papilionaceous. With large upper petal (banner), two lateral petals (wings), and usually two connate lower petals (keel); as in the Fabaceae.

Papillae. Small protuberances on the surface of an organ.

Papillate. Having papillae.

Papillose. Having minute papillae.

Pappus. Bristly or scaly calyx in the Asteraceae.

Pectinate. Pinnatifid with narrow segments set closely like the teeth of a comb.

Pedicel. Individual flower stalk.

Pedicillate. Having a pedicel.

Peduncle. Main stalk of entire inflorescence.

Pedunculate. Having a peduncle.

Pentagonal. Five-angled.

Perennial. Living more than two years, usually but not always flowering more than once.

Perfect. Describing a flower with both stamens and pistils.

Perfoliate. Having a base completely surrounding the stem.

Perianth. Calyx and corolla collectively, whether or not they are distinguishable.

Pericarp. The ripened walls of the ovary when it becomes a fruit.

Perigynium. Variously shaped, sac-like structure enclosing the ovary and achene in *Carex* and *Kobresia.*

Perigynous. With stamens, petals and sepals borne on a calyx tube (hypanthium) surrounding, but not actually attached to, the superior ovary.

Persistent. Remaining attached. The opposite of deciduous or caducous.

Petal. A corolla member or segment; a unit of the corolla, usually colored.

Petiolate. Having a petiole.

Petiole. Leaf stalk.

Petiolule. Stalk of a leaflet.

Phyllaries. A bract of the involucre that forms the outside of a head, as in the Asteraceae.

Pilose. With soft, shaggy trichomes.

Pinna. A primary division of a fern leaf.

Pinnate. Arranged in one plane along either side of an axis; a leaf is odd-pinnate if there is a terminal leaflet, even-pinnate if there is not.

Pinnule. A secondary division of a fern frond.

Pistil. The female reproductive organ of a flower, with a stigma, style, and ovary.

Pistillate. Having a pistil.

Plantlets. A vegetative bud with rudimentary leaves borne on the stem or frond.

Plicate. Longitudinally folded.

Plumose. Feather-like, with fine hairs on either side of the main bristle.

Pollen. In gymnosperms and angiosperms, minute, grain-like structure containing the sperm, young male gametophyte.

Pollinia. Coherent mass of pollen grains.

Prickles. A sharp-pointed outgrowth from the epidermis of an organ.

Prophyll. One of a pair of small bracts closely subtending the tepals in *Juncus.* In grasses, a scale-like leaf that is produced at the base of an axillary branch and on the side of the axillary branch toward the main axis. It almost always has two distinct ribs or veins. The palea is a prophyll.

Puberulent. Minutely pubescent.

Pubescent. Covered with short, soft hairs.

Punctate. Covered with minute depressions or sunken glands.

Pungent. Terminating in a rigid, sharp point.

Pyramidal. Pyramid-shaped; often applied to a panicle with a conical shape.

Pyriform. Pear-shaped.

Raceme. Unbranched, elongate inflorescence with pedicellate flowers blooming from the bottom up.

Racemose. Having flowers in a raceme, appearing to be a raceme.

Rachilla. The axis of a spikelet, continuous with the pedicel.

Rachis. The main axis of a pinnately compound leaf; major axis of an inflorescence.

Radially symmetrical. Divisible into mirror-image halves across all radii.

Radiate. Describing a inflorescence (head) of Asteraceae with marginal, ligulate ray florets and central, disk florets.

Ranked. A row or column of parts along an axis; e.g., leaves on an erect stem arranged in four vertical rows are 4-ranked.

Ray flower. A generally pistillate or sterile, bilateral flower with a flat, strap- or fan-shaped, often 3-lobed outer portion of the corolla (lamina); appearing in a ring around a central cluster of disk flowers.

Receptacle. The structure to which one or more flowers are attached; in Asteraceae often disk- to cone-shaped.

Regular. Pertaining to flowers with floral parts within a whorl similar in shape and size, radially symmetrical.

Reniform. Kidney-shaped; cordate-based but wider than.

Replum. Partition between two halves of a fruit in the Brassicaceae.

Reticulate. Netted; with veins forming a network.

Reticulation. A network of reticulate veins.

Retrorse. Bent or directed downward or toward the base.

Retuse. With a blunt, slightly notched apex.

Revolute. Margins rolled outward or downward over lower surface.

Rhizome. A horizontal underground stem that can appear to be a root.

Rhizomatous. Having rhizomes.

Rib. A prominent nerve or vein, as in the lemma ribs of *Glyceria.*

Riparian. Pertaining to areas along rivers and streams.

Rootstock. A term applied to miscellaneous types of underground stems or parts, often a rhizome.

Rosulate. With leaves in a rosette.

Rotate. Describing a wheel-shaped corolla with a short tube and wide limb at right angles to the tube.

Rudiment. An imperfectly developed organ or part.

Rugose. Deeply wrinkled.

Runcinate. Pinatifid or cleft with downward- or backward-pointing segments.

Saccate. Pouch-like.

Sagittate. Arrow-shaped with a pair of large, acute, or rounded lobes at the base.

Salverform. Trumpet-shaped; with slender tube and limb nearly at right angles to tube.

Samara. An indehiscent, winged, dry fruit.

Scabrous. Covered with very minute stiff hairs that cause a surface to be rough to the touch.

Scale. Small, non-green leaf or small, scarious, flattened structure within the perianth; applied to small or modified leaves of various kinds, as glumes, lemmas, paleas, prophylls, and reduced leaves on the rhizomes.

Scalloped. Margins with edges having large rounded indentations.

Scape. A flowering stem, naked above the base, with or without a few scale leaves.

Scapiform. Scape-like but with inconspicuous leaves.

Scapose. Scape-like, having a scape.

Scarious. Thin, membranous, and dry, non-green.

Scree. An accumulation of small, broken rock fragments, generally on a slope.

Secund. Flowers or other structures on only one side of an axis.

Seed. A mature, fertilized ovule.

Seleniferous. Having unusually high levels of selenium.

Self-pollinated. Pollination taking place in the bud by pollen from the same flower.

Sepal. Individual member of the calyx, whether fused or not.

Septate. Divided by internal partitions.

Septum. A partition between two chambers.

Sericeous. With long, silky, usually appressed trichomes.

Serotiny. Seed release occurs in response to an environmental trigger, usually fire.

Serrate. With saw-like teeth along the margin.

Serrulate. Minutely serrate.

Sessile. Lacking a stalk, pedicel, petiole or petiolule.

Setae. Bristles.

Setose. Having setae.

Sheath. The lower part of a grass leaf, that part encircling the stem.

Shoot. A young stem or twig and its appendages.

Sigmoid. S-shaped.

Silicle. A dry, dehiscent fruit of the Brassicaceae derived from two carpels that dehisce along two sutures and is no more than twice as long as broad.

Silique. A dry, dehiscent fruit of the Brassicaceae derived from two carpels that dehisce along two sutures and is at least twice as long as broad.

Sinuate. With deep wave-like indentations along the margin.

Sinus. An indentation between adjacent lobes or teeth.

Sorus. A cluster of spore-bearing sacs.

Spadix. Unbranched inflorescence with flowers embedded in the thickened axis.

Spathe. An enlarged bract enclosing an inflorescence. In grasses, a leaf sheath surrounding a portion of or the entire inflorescence is referred to as a spathe.

Spatulate. Oblong or obovate apically with a long-attenuate base, spatula-shaped.

Spiciform. Appearing like a spike.

Spike. Unbranched, elongate inflorescence with sessile flowers.

Spines. A stiff, slender, sharp-pointed structure occurring at a node, modified leaf or stipule.

Spinulose. Bearing small spines.

Sporangiophores. The umbrella-shaped sporangium-bearing unit of the cone-like strobilus.

Sporangium. A spore-bearing sac or case.

Spore. In non-seed plants, a haploid cell generated for dispersal.

Sporocarp. A hard, nut-like structure containing sporangia.

Sporophore. Spore-bearing portion of a leaf in Ophioglossaceae.

Sporophyll. A spore bearing leaf.

Spur. A tubular or pointed projection from the perianth; a short shoot on which flowers and fruits or leaves are borne.

Stamen. Male, pollen-bearing floral organ consisting of an anther and a filament.

Staminate. Bearing one or more stamens.

Staminode. Sterile stamen.

Stellate. Star-like.

Steppe. Low-elevation, treeless, upland vegetation, including grasslands and shrublands.

Sterile. Infertile.

Stigma. Pollen-receptive portion of the pistil.

Stipe. A stalk supporting a structure.

Stipitate. Borne on a stipe or stalk.

Stipulate. Bearing stipules.

Stipule. One of a pair of scales or leaf-like appendages at the base of a petiole.

Stolon. Elongate, above-ground propagative stem, rooting at the tip producing new plants, runner.

Stomate. Opening or pore in leaf epidermis between two guard cells which allows gas exchange.

Strigillose. Minutely strigose.

Strigose. With stiff, straight, sharp, appressed hairs.

Strobilus. A cone-like structure formed from spore-bearing leaves.

Style. Narrow, non-ovule-bearing portion of pistil between stigma and ovary.

Stylopodium. Disk-like enlargement at the base of the style in flowers of the Apiaceae.

Subacaulescent. Nearly stemless.

Submergent. Beneath the surface of the water.

Submersed. Beneath the surface of the water, submerged.

Subpinnae. Divisions of a pinna, secondary or tertiary lobe of a fern frond.

Subplumose. Slightly plumose, with short, fine hairs on either side of the main bristle.

Subshrub. A plant that is woody just above ground level, the upper stems and twigs dying back seasonally.

Subtend. To be below, but associated with.

Subulate. Very narrow and tapering; awl-shaped; linear.

Succulent. Fleshy, juicy, soft in texture.

Sulcus. An elongate depression or furrow, as commonly found on the back of palea.

Superior. Referring to a flower in which the ovary is free and above the point at which sepals, petals, and stamens are united.

Suture. A line of fusion that may break and release pollen or seeds.

Sympetalous. With united petals.

Talus. Relatively stable, sloping accumulation of large rock fragments.

Tawny. Pale brown or brownish- to grayish-yellow.

Teeth. Pointed lobes or divisions.

Tepal. A perianth segment which is not differentiated into distinct sepals and petals.

Terete. Cylindrical and elongate.

Terminal. At the apex or distal end.

Ternate. Lobed or compound into three parts.

Terrestrial. Pertaining to dry land.

Thickets. Dense, shrub-dominated vegetation.

Thorn. A short, sharp-pointed branch.

Throat. Expanded portion of a sympetalous corolla between the tube and the lobed limb.

Tiller. An erect stem. A tiller differs from a stolon in branching off in an ascending or erect manner, and from a rhizome in branching off in an ascending or erect manner and being above ground.

Tomentose. Covered with tomentum.

Tomentum. Dense, soft, interwoven hairs.

Torulose. Cylindrical with slight swellings and constrictions.

Trichome. A hair or hair-like growth on the surface of a plant part.

Trifid. Divided into three parts; two-cleft.

Trifoliolate. Divided into three leaflets.

Tripinnate. Three times pinnately divided or lobed.

Triternate. Three times ternately divided or lobed.

Trophophore. Sterile portion of a leaf in Ophioglossaceae.

Truncate. With an abruptly transverse end as if cut off.

Tube. The cylindric, fused portion at the base of a sympetalous corolla, proximal to the throat and limb.

Tubercle. Small, wart-like projection.

Tuberculate. With tubercles on the surface.

Tufted. In a densely packed cluster.

Turbinate. Top shaped; inversely conic.

Turf. Moist, alpine vegetation dominated by short sedges and forbs.

Turions. An over-wintering bud-like shoot.

Tussock. A tuft or clump of grass or sedges.

Twig. A small, terminal stem segment.

Type locality. The location where the type specimen was collected.

Type specimen. The original specimen from which the description of a new species is made.

Umbel. A flat-topped or convex inflorescence with the pedicels arising at a common point, umbrella-like.

Umbellet. The secondary umbel in a compound umbel.

Uncinate. Hooked at the apex.

Unisexual. Containing only stamens or only pistils, as in the flowers of a monoecious or dioecious plant.

Urceolate. Urn-shaped.

Utricle. A small, bladdery or inflated, one-seeded, dry fruit.

Valve. One of the parts produced by the splitting of a dehiscent fruit when ripe.

Vein. A strand of vascular tissue.

Ventral. Pertaining to the surface nearest the axis; upper surface of a leaf (adaxial).

Verticillasters. A group of cymes arranged around a leaf node.

Verticillate. In 2 or more successive whorls; the whorls separated by distinct internodes.

Vestigial. A much reduced, remnant organ.

Vestiture. Collectively the type of a plant's external hairiness, scaliness, etc.

Villous. Covered with long, soft, crooked but unmatted hairs.

Web. A cluster of long and soft tangled hairs at the base of a floret in species of *Poa*.

Whorl. A circular arrangement of similar parts arising from the same node.

Whorled. Arranged in a whorl.

Wing. Thin, flat extension or appendage of a surface or margin. In Fabaceae the lateral petals.

Winged. Having one or more wings.

Wintergreen. Having green leaves in the winter.

Woodlands. Vegetation dominated by short or scattered trees.

Zygomorphic. Divisible into mirror-image halves in only one way, bilaterally symmetrical.

Abbreviations for states and provinces of the U.S. and Canada.

AB	Alberta	NE	Nebraska
AL	Alabama	NH	New Hampshire
AK	Alaska	NJ	New Jersey
AR	Arkansas	NL	Newfoundland, Labrador
AZ	Arizona	NM	New Mexico
BC	British Columbia	NS	Nova Scotia
CA	California	NT	Northwest Territories
CO	Colorado	NU	Nunavut
CT	Connecticut	NV	Nevada
DE	Deleware	NY	New York
FL	Florida	OH	Ohio
GA	Georgia	OK	Oklahoma
IA	Iowa	ON	Ontario
ID	Idaho	OR	Oregon
IL	Illinois	PA	Pennsylvania
IN	Indiana	PE	Prince Edward Is.
KS	Kansas	QC	Québec
KY	Kentucky	RI	Rhode Island
LA	Louisiana	SC	South Carolina
MA	Massachusettes	SD	South Dakota
MB	Manitoba	SK	Saskatchewan
MD	Maryland	TN	Tennessee
ME	Maine	TX	Texas
MI	Michigan	UT	Utah
MO	Missouri	VA	Virginia
MN	Minnesota	VT	Vermont
MS	Mississippi	WA	Washington
MT	Montana	WI	Wisconsin
NB	New Brunswick	WV	West Virginia
NC	North Carolina	WY	Wyoming
ND	North Dakota	YT	Yukon

Key to Plant Families

1. Plants lacking flowers (conifers, ferns, and fern allies). . . . . Group A
1. Plants reproducing by flowering (occasional plants produce only bulbs). 2

2. Plants truly aquatic with submersed or floating leaves that become limp when withdrawn
   from the water (emergent plants with self-supporting stems are not included) . . . . Group B
2. Plants emergent, not an obligate aquatic. 3

3. Plants herbaceous; leaves unlobed and undivided with parallel veins, often grass-like
   and/or petals and sepals mostly in 3's or 6's or lacking (Monocots) . . . . Group C
3. Plants herbaceous or woody; leaves undivided, divided or lobed, usually with net-veination
   or petals and sepals usually in 2's, 4's, or 5's or both (Dicots). 4

4. Plants trees, shrubs or woody vines, stems woody well above ground level. . . . . Group D
4. Plants herbaceous or woody only at the base 5

5. Flowers without distinct whorls of petals and sepals that are differentiated from each other
   (petals or sepals or both lacking, or petals and sepals identical). . . . . Group E
5. Flowers with petals and sepals different from each other 6

6. Petals separate from each other all the way to the base . . . . Group F
6. Petals united, sometimes only at the base. . . . . .Group G

Group A. Conifers, ferns and fern allies

1. Coniferous trees or shrubs with needle- or scale-like, mostly evergreen leaves .2
1. Herbaceous plants reproducing by spores 5

2. Leaves scale-like, pressed flat to the stem. . . . . .Cupressaceae (p.79)
2. Leaves needle-like 15

3. Fruit a cone of dry, seed-bearing scales. . . . . Pinaceae (p.73)
3. Fruit berry-like, dry or juicy. 16

4. Fruit with a red, fleshy outer covering. . . . . .Taxaceae (p.81)
4. Fruit dry, purplish . . . . .Cupressaceae (p.79)

5. Plants aquatic, growing beneath or on the surface of water for much of the growing season 6
5. Terrestrial plants, sometimes growing in wet soil but not submersed or floating .9

6. Plants floating, branched clusters of overlapping leaves . . . . Salviniaceae (p.73)
6. Plants rooted in the substrate. 7

7. Plants a stemless cluster of linear leaves with white spores at the base where they join
   together . . . . .Isoetaceae (p.55)
7. Plants not as above 8

8. Leaves long-petiolate with 4 leaflets resembling a 4-leaf clover . . . . .Marsileaceae (p.72)
8. Leaves needle-like, crowded on stems conifer-like . . . . Lycopodiaceae (p.52)

9. Stems grooved lengthwise, jointed and easily pulled apart at the nodes; leaves
   reduced to papery scales; plants apparently of only simple or branched stems . . . . . Equisetaceae (p.57)
9. Stems not as above; leaves not papery. 10

10. Leaves small, needle- or scale-like; stems often resembling small conifer branches 11
10. Leaves larger, not-scale-like, often fern-like. 12

11. Plants <2 cm tall; usually on rocks . . . . Selaginellaceae (p.54)
11. Plants mostly >2 cm tall; in soil of forests and meadows . . . . Lycopodiaceae (p.52)

12. Spores clustered on a specialized branch-like portion of the leaf . . . . Ophioglossaceae (p.58)
12. Spores borne on leaves arising directly from the roots; fertile leaves mostly
    similar to the vegetative leaves . . . . .Polypodiaceae (p.64)

Group B. Aquatic plants with submersed or floating leaves

1. Leaves opposite or whorled 2
1. Leaves alternate or all basal or plants stemless. 12

2. Leaves simple, linear with an expanded base and finely dentate margins . . . . Najadaceae (p.590)
2. Leaves not as above 3

3. Leaves whorled 4
3. Leaves opposite 7

4. Leaves deeply divided 5
4. Leaves simple. 6

5. Leaf divisions entire . . . . Haloragaceae (p.328)
5. Leaf divisions with toothed margins . . . . Ceratophyllaceae (p.83)

6. Leaves ≥6 per node . . . . Hippuridaceae (p.427)
6. Leaves 2 to 4 per node. . . . . Hydrocharitaceae (p.584)

7. Leaves filiform, ca. 1 mm wide and 2–10 cm long . . . . Zannichelliaceae (p.591)
7. Leaves wider or longer or both. 8

8. Calyx ≥3 mm long; leaf blades nearly orbicular . . . . Plantaginaceae (p.429)
8. Calyx <3 mm long; leaf blades linear to oblong 9

9. Lowest leaves whorled; fruits on long pedicel-like hypanthium . . . . Hydrocharitaceae (p.584)
9. Leaves all paired; fruits sessile or nearly so. 10

10. Upper portion of stems floating on or near the surface .Callitrichaceae (p.429)
10. Stems prostrate to erect but not freely floating. 11

11. Leaves linear-lanceolate, sheathing the stem . . . . Crassulaceae (p.247)
11. Leaves oblanceolate, short-petiolate . . . . Elatinaceae (p.164)

12. Plants free-floating, not rooted in the substrate 13
12. Plants rooted in the substrate, though sometimes with floating leaves 15

13. Leaves or branches with small bladders; flowers bilabiate. . . . . Lentibulariaceae (p.464)
13. Plants without bladders; flowers inconspicuous or absent. 14

14. Plants with stems and cedar-like overlapping leaves. . . . . Salviniaceae (p.73)
14. Plants stemless; leaves not overlapping cedar-like . . . . Lemnaceae (p.591)

15. Leaves deeply lobed or divided 16
15. Leaves not deeply lobed or divided (sometimes heart-shaped). 18

16. Leaves palmately divided into 4 ovate leaflets; flowers lacking . . . . .Marsileaceae (p.72)
16. Plants not as above 17

17. Petals 4; stamens 6; pistil 1 . . . . .Brassicaceae (p.186)
17. Petals usually >4; stamens >6; pistils >1 . . . . Ranunculaceae (p.83)

18. Leaf-bearing stems lacking; leaves all in a rosette or from a rhizome 19
18. Plants with leafy stems. 24

19. Each leaf with a sac of white spores at the base . . . . .Isoetaceae (p.55)
19. Leaves without a spore sac at the base 20

20. Leaf blades elliptic with a cordate base . . . . .Nymphaeaceae (p.81)
20. Leaves narrower, not cordate. 21

21. Leaves narrow but with a blade distinctly wider than the petiole . . . . Scrophulariaceae (p.462)
21. Leaves grass-like, without a distinct blade and petiole. 22

22. Flowers in an umbel; plants with a heavy, brown rhizome . . . . Butomaceae (p.582)
22. Inflorescence not umbellate; root system different. 23

23. Leaves needle-like, round in cross-section . . . . .Cyperaceae (p.603)
23. Leaves flattened. . . . . vegetative *Sagittaria* or *Sparganium*

24. Some leaf veins branched off the midvein; leaves oblong to ovate 25
24. Main leaf veins parallel to each other; submersed leaves often linear. 26

25. Stem surrounded by a membranous stipule above leaf attachment . . . . Polygonaceae (p.149)
25. Plants without stipules . . . . Ranunculaceae (p.83)

26. Leaves with a pale membranous appendage surrounding the stem (stipule) .27
26. Stipules lacking. 29

27. Flowers and fruits solitary from leaf axils . . . . Pontederiaceae (p.717)
27. Flowers in 2- to many-flowered spikes. 28

28. Flowers 2 per spike; fruits becoming long-stalked . . . . .Ruppiaceae (p.590)
28. Spikes many-flowered; fruits sessile . . . . .Potamogetonaceae (p.585)

29. Flowers sessile, solitary in upper leaf axils; ovary inferior . . . . Campanulaceae (p.465)
29. Flowers in racemes, panicles or dense spikes; ovary superior 30

30. Floating leaves with elliptic or arrow-shaped blades . . . . .Alismataceae (p.583)
30. Floating or emergent leaves linear. 31

31. Leaves V-shaped in cross-section where they meet the stem .Typhaceae (p.601)
31. Leaves flat or broadly concave where they join the stem. 32

32. Plants with a leaf-like bract subtending and longer than the spikelet(s). . . . . .Cyperaceae (p.603)
32. Inflorescence surpassing subtending leaves . . . . Poaceae (p.645)

Group C. Monocots; herbaceous plants with parallel-veined leaves and sepals mostly in 3's or 6's or lacking (some Dicot families will have 3-merous flowers but the leaves will not have parallel veins, others will have parallel veins but not 3-merous flowers)

1. Flowers unisexual, borne in dense clusters 2
1. Flowers bisexual, often not densely clustered 3

2. Stems mostly 3-sided; each flower subtended by 1 (rarely 2), apparent
   scale-like bract (sedges). . . . . Cyperaceae(p.603)
2. Stems round in cross section; flower bracts minute and inconspicuous . . . . .Typhaceae (p.601)

3. Stamens and ovaries enclosed by 1 or 2 bracts, scales or sacs; petals and sepals lacking
   (grasses and sedges). 4
3. Flowers with 3(4) or 6 petals and/or sepals 5

4. Stems mostly solid and 3-sided; each flower subtended by 1 (rarely 2) scale-like bract
   (sedges). . . . . .Cyperaceae (p.603)
4. Stems mostly round in cross-section and swollen at the leaf nodes; each flower
   enclosed by 2 bracts (grasses). . . . . Poaceae (p.645)

5. Ovary inferior 6
5. Ovary superior 9

6. Small leafless plants parasitic on conifers . . . . .Viscaceae (dicot) (p.343)
6. Plants green, leafy and rooted in the soil 7

7. Leaves whorled (≥2 per node) at least below. . . . . Rubiaceae (dicot) (p.468)
7. Leaves 1 or 2 per node. 8

8. Flowers bilaterally symmetrical; i.e., 1 petal different than the other 2. . . . . Orchidaceae (p.734)
8. Flowers radially symmetrical; all petals identical . . . . Iridaceae (p.731)

9. Cauline leaves with main veins branching off of the midvein . . . . Polygonaceae (dicot) (p.149)
9. Leaves with parallel veins. 10

10. Flowers sessile, borne on a spike 11
10. Flowers pedicellate or pedunculate; inflorescence a panicle, raceme, or solitary 13

11. Spike <5 mm in diameter . . . . Juncaginaceae (p.585)
11. Spike ≥8 mm in diameter . . . . 12.

12. Tepals 6; leaves linear, grass-like. . . . . .Acoraceae (p.591)
12. Tepals 4; leaf blades elliptic to obovate . . . . .Araceae (p.591)

13. Fruit a berry 14
13. Fruit a capsule or a circular cluster of follicles 16

14. Vines . . . . Smilacaceae (p.732)
14. Plants erect, not vining. 15

15. Leaves narrow, scale-like; flowers 1 or 2 from leaf axils . . . . .Asparagaceae (p.717)
15. Leaves not scale-like; inflorescence usually terminal. . . . . Liliaceae (p.717)

16. Fruit a circular cluster of follicles 17
16. Fruit a capsule 18

17. Flowers pinkish; leaves linear, tapered to the tip . . . . Butomaceae (p.582)
17. Flowers white; distal portion of emergent leaves at least somewhat wider than
     proximal portion (petiole) . . . . Alismataceae (p.583

18. Fruit a group of 3 1- or 2-seeded capsules. . . . . .Scheuchzeriaceae (p.584)
18. Fruit a single capsule (often 3-chambered) 19

19. Petals and sepals brown or green; leaves grass-like. . . . . .Juncaceae (p.593)
19. Petals and sepals white, yellow or blue 20

20. Inflorescence an umbelliform cyme of ≥5 blue or purplish flowers. . . . . Commelinaceae (p.592)
20. Inflorescence not as above. 21

21. Plants with stiff, thick, spine-tipped, basal leaves; tepals ≥3 cm long . . . . Agavaceae (p.732)
21. Plants not as above . . . . Liliaceae (p.717)

Group D. Trees, shrubs, or woody vines; stems woody well above ground level

1. Trailing or climbing vines 2
1. Trees or shrubs with rigid stems. 11

2. Leaves simple with entire to dentate margins, sometimes cordate-based. 3
2. Leaves lobed or divided into leaflets 6

3. Leaves opposite . . . . Caprifoliaceae (p.471)
3. Leaves alternate. 4

4. Fruit an orange, globose, 3-valved capsule enclosing a red aril . . . . .Celastraceae (p.344)
4. Fruit a berry 5

5. Leaves with parallel veins. . . . . Smilacaceae (p.732)
5. Main leaf veins branch off the midvein . . . . Solanaceae (p.381)

6. Leaves divided into leaflets 7
6. Leaves lobed but not compound 9

7. Fruit feathery-beaked achenes. . . . . Ranunculaceae (p.83)
7. Fruit a berry-like drupe 8

8. Leaves with 3 leaflets . . . . Anacardiaceae (p.351)
8. Leaves with 5–7 leaflets . . . . Vitaceae (p.349)

9. Fruit an achene enclosed in a glandular bract . . . . Cannabaceae (p.107)
9. Fruit a berry 10

10. Leaves with 1–2 lobes only at the base; mature berries red . . . . Solanaceae (p.381)
10. Leaves with lobes maple-like; berries blue. . . . . Vitaceae (p.349)

11. Trees with a single trunk at least 4 m high 12
11. Shrubs usually with multiple stems. 22

12. Leaves opposite 13
12. Leaves alternate. 14

13. Leaves with 5–7 leaflets . . . . .Oleaceae (p.427)
13. Leaves lobed or with 3 leaflets . . . . .Aceraceae (p.351)

14. Fruit a pome or drupe. 15
14. Fruit a dry capsule, samara, nut, pod or nutlets arranged in catkins 17

15. Fruit dry and hard. . . . . Ulmaceae (p.106)
15. Fruit mealy or juicy 16

16. Leaf underside white-mealy with flake-like trichomes . . . . Elaeagnaceae (p.327)
16. Leaves not white-mealy . . . . Rosaceae (p.261)

17. Leaf margins lobed or leaf divided into leaflets 18
17. Leaves with entire to dentate margins 20

18. Leaves pinnately divided into 9–17 leaflets . . . . Fabaceae (p.290)
18. Leaves lobed or with fewer leaflets 19

19. Fruit a nut (acorn). . . . . Fagaceae (p.108)
19. Fruit an oblong samara. . . . . .Aceraceae (p.351)

20. Fruit a capsule with many seeds, each with a tuft of hair. . . . . .Salicaceae (p.175)
20. Fruit a 1-seeded samara or a winged nutlet. 21

21. Fruits winged nutlets arranged in catkins. . . . . Betulaceae (p.108)
21. Fruit an ovate samara. . . . . Ulmaceae (p.106)

22. Leaves apparently absent; stems green, succulent, spiny. . . . . Cactaceae (p.111)
22. Plants not as above 23

23. Leaves opposite or whorled (≥2 per node). 24
23. Leaves alternate (1 per node) or all basal 31

24. Fruit a drupe or berry 25
24. Fruit a capsule or 1-seeded samara. 28

25. Leaves with parallel veins. . . . . .Cornaceae (p.341)
25. Leave or leaflets net-veined 26

26. Leaves mealy with white or brown, flake-like trichomes. . . . . Elaeagnaceae (p.327)
26. Leaves without flake-like trichomes 27

27. Sepals 4; flowers regular . . . . .Rhamnaceae (p.347)
27. Sepals 5; flowers often slightly bilaterally symmetrical .Caprifoliaceae (p.471)

28. Fruit a 1-seeded samara . . . . .Aceraceae (p.351)
28. Fruit a capsule 29

29. Sepals 5; corolla bilabiate. . . . . Plantaginaceae (p.429)
29. Sepals 4; corolla regular with separate petals 30

30. Petals maroon, 1–2 mm long; leaves with serrate margins . . . . .Celastraceae (p.344)
30. Petals white, 1–2 cm long; leaves mostly entire. . . . . .Hydrangeaceae (p.243)

31. Leaves lobed (>1/4-way to midvein) or compound. 32
31. Leaves simple with entire or toothed margins 40

32. Leaves compound (divided into leaflets) 33
32. Leaves lobed but not divided into leaflets. 36

33. Fruit an elongate, many-seeded pod (legume); flower bilaterally symmetrical. . . . . Fabaceae (p.290)
33. Fruit a berry, drupe or aggregate of drupelets; flowers radially symmetrical 34

34. Leaves with stipules at the point of stem attachment. . . . . Rosaceae (p.261)
34. Leaves exstipulate 35

35. Fruit glabrous, blue when mature. . . . . .Berberidaceae (p.101)
35. Fruit glandular-hairy, red when mature. . . . . Anacardiaceae (p.351)

36. Small shrub with leaves deeply divided into linear, mucronate segments . . . . Polemoniaceae (p.387)
36. Plants not as above 37

37. Leaves >10 cm wide with spines on the veins . . . . Araliaceae (p.358)
37. Leaves smaller and unarmed. 38

38. Petals united into a tubular corolla . . . . Asteraceae (p.478)
38. Petals separate (tubular calyx in *Ribes* may appear to be a corolla). 39

39. Petals smaller than the tubular calyx lobes; leaves mostly exstipulate . . . . Grossulariaceae (p.243)
39. Petals often greater than the calyx lobes; leaves with stipules . . . . Rosaceae (p.261)

40. Some or all flowers borne in catkins or dense catkin-like spikes 41
40. Catkins or catkin-like spikes absent. 44

41. Leaves linear, succulent, nearly round in cross-section. 42
41. Leaves flat, not succulent. 43

42. Stems with thorns. . . . . Sarcobataceae (p.127)
42. Stems thornless . . . . Tamaricaceae (p.173)

43. Fruit with numerous seeds, each with a tuft of hair at the tip. . . . . .Salicaceae (p.175)
43. Fruit a 1-seeded, winged or wingless nutlet . . . . Betulaceae (p.108)

44. Fruit a fleshy pome, drupe or berry 45
44. Fruit a dry achene, capsule, follicle or utricle 49

45. Petals united. . . . . Ericaceae (p.225)
45. Petals separate or absent (sepals may appear petal-like) 46

46. Leaves with star- or flake-shaped trichomes . . . . Elaeagnaceae (p.327)
46. Leaves glabrous or with simple hairs 47

47. Flowers with 8 stamens; petals lacking; sepals purple, petal-like . . . . .Thymelaeaceae (p.239)
47. Flowers not as above 48

48. Style 1; with 2 to 4 stigmas. . . . . .Rhamnaceae (p.347)
48. Styles 1 to many; each with only 1 stigma . . . . Rosaceae (p.261)

49. Leaves with scale-like or stellate trichomes . . . . Amaranthaceae (p.113)
49. Leaves glabrous or pubescent with simple hairs 50

50. Stipules (sometimes small) present at the base of leaf petioles. 51
50. Stipules lacking. 54

51. Leaves with entire margins. 52
51. Leaf blades with dentate or serrate margins 53

52. Leaves white-lanate beneath . . . . Rosaceae (p.261)
52. Leaves glabrous . . . . Crossosomataceae (p.289)

53. Inflorescence a densely branched, terminal, many-flowered panicle. . . . . .Rhamnaceae (p.347)
53. Inflorescence a raceme, few-flowered cyme, or flowers solitary . . . . Rosaceae (p.261)

54. Ovary superior; petals arising beneath the ovary. . . . . Ericaceae (p.225)
54. Petals arising on top of the inferior ovary 55

55. Fruit an achene, clustered together in heads . . . . Asteraceae (p.478)
55. Fruit a solitary, sessile capsule in axils of upper leaf-like bracts . . . . Onagraceae (p.329)

Group E. Dicots with petals and/or sepals lacking (minute) or undifferentiated

1. Few to many flowers clustered in heads; each head subtended by a cup- or
   vase-shaped whorl(s) of bracts and appearing like a single flower . . . . Asteraceae (p.478)
1. Flowers not as above 2

2. Plants parasitic on conifers; stems jointed; leaves reduced. . . . . Viscaceae (p.343)
2. Plants not as above 3

3. Plants with milky sap and glabrous foliage. . . . . Euphorbiaceae (p.344)
3. Plants with clear sap or hairy foliage 4

4. Middle and lower leaves opposite or whorled (≥2 per node) 5
4. Middle and lower leaves alternate (1 per node) or all in a basal rosette 14

5. Plants with sharp, stinging hairs on stems and leaves. . . . . .Urticaceae (p.107)
5. Plants without stinging hairs. 6

6. Leaves whorled 7
6. Leaves opposite 8

7. Ovary inferior; styles 2 . . . . Rubiaceae (p.468)
7. Ovary superior; styles 3 . . . . .Molluginaceae (p.132)

8. Middle stem leaves divided or lobed 9
8. Middle stem leaves with entire, toothed or wavy margins 11

9. Plants dioecious; male and female flowers on separate plants . . . . Cannabaceae (p.107)
9. Flowers perfect, containing both male and female function 10

10. Pistils and/or stamens numerous . . . . Ranunculaceae (p.83)
10. Flowers with 1 pistil and 3 to 5 stamens. . . . . Valerianaceae (p.476)

11. Flowers in calyx-like, saucer-shaped involucres. . . . . Nyctaginaceae (p.110)
11. Flowers not in saucer-shaped involucres 12

12. Pistils and/or stamens >10 each . . . . Ranunculaceae (p.83)
12. Pistils and stamens 10 or fewer 13

13. Leaf blades cordate-orbicular, >3 cm wide. . . . . Aristolochiaceae (p.81)
13. Leaf blades smaller and narrower . . . . Caryophyllaceae (p.132)

14. Leaves divided or deeply lobed (>half-way to midvein) 15
14. Leaves with entire, toothed, or wavy margins 21

15. Stamens >10 per flower 16
15. Stamens 1–10 17

16. Plants with milky sap . . . . Papaveraceae (p.102)
16. Plants with clear sap. . . . . Ranunculaceae (p.83)

17. Mature fruit a purple drupe. . . . . Araliaceae (p.358)
17. Fruit not fleshy or purple. 18

18. Flowers green; leaves thick and often covered with white scales . . . . Amaranthaceae (p.113)
18. Flowers with colored sepals or petals; leaves not covered with white scales 19

19. Petals not all the same, united at the base. . . . . Fumariaceae (p.105)
19. Petals all similar, separate to base or nearly so. 20

20. Flowers and fruits in a dense cylindrical spike . . . . Rosaceae (p.261)
20. Inflorescence an umbel . . . . Apiaceae (p.358)

21. Stamens >10 per flower . . . . Ranunculaceae (p.83)
21. Flowers with 4–10 stamens 22

22. Plants without green leaves or stems; leaves scale-like . . . . Ericaceae (p.225)
22. Plants with green leaves and/or stems. 23

23. Flowers in a dense spike 24
23. Inflorescence not in dense spikes 26

24. Inflorescence greenish . . . . Amaranthaceae (p.113)
24. Inflorescence yellowish to blueish at anthesis 25

25. Foliage pubescent; stamens purple or blue . . . . Plantaginaceae (p.429)
25. Foliage glabrous; stamens whitish . . . . Araceae (monocot) (p.591)

26. Fruit nearly orbicular, flattened with an apical notch. .Brassicaceae (p.186)
26. Fruit not as above. 27

27. Flower bracts or sepals with a spine tip . . . . Amaranthaceae (p.113)
27. Flower bracts and sepals without spine tips. 28

28. Membranous appendages (stipules) sheathing the stem above leaf attachment . . . Polygonaceae (p.149)
28. Sheathing stipules lacking 29

29. Rhizomatous plants with fleshy leaves and berry-like fruit. . . . . Santalaceae (p.341)
29. Plants not as above 30

30. Flowers green (reddish); leaves often thick or covered with white scales but
    not hairy . . . . .Amaranthaceae (p.113)
30. Flowers white, yellow or pink, or if green, then foliage hairy 31

31. Flowers pedicellate in umbels 32
31. Flowers not in umbels 33

32. Ovary superior; styles 2 or 3; leaves mostly basal. . . . . Polygonaceae (p.149)
32. Ovary inferior; styles 2; cauline leaves usually present . . . . Apiaceae (p.358)

33. Plants annual; flowers in axillary clusters. . . . . .Urticaceae (p.107)
33. Plants perennial; flowers not clustered in leaf axils . . . . Myrsinaceae (p.240)

Group F. Dicots with both sepals and separate petals

1. Flowers resembling a daisy or dandelion (actually clusters of flowers each with united petals) 2
1. Flowers not as above 3

2. Corolla 4-lobed; stamens 4, separate. . . . . Dipsacaceae (p.478)
2. Corolla usually 5-lobed; stamens 5, united below . . . . Asteraceae (p.478)

3. Flowers bilaterally symmetrical; petals not all identical 4
3. Flowers radially symmetrical: petals all identical 13

4. Flowers with >10 stamens 5
4. Flowers with ≤10 stamens 6

5. Plants annual or biennial; all petals lobed at the tip . . . . Resedaceae (p.225)
5. Plants mostly perennial; not all petals lobed, one hood- or spur-like . . . . Ranunculaceae (p.83)

6. Petals 4, entire or nearly so 7
6. Petals 5 or apparently 3 or 4 with 1 or 2 deeply lobed 8

7. Ovary superior . . . . Fumariaceae (p.105)
7. Ovary inferior . . . . Onagraceae (p.329)

8. Flowers with 5 stamens 9
8. Flowers with 10 stamens 11

9. Petals ≤2 mm long . . . . Saxifragaceae (p.249)
9. Petals >4 mm long 10

10. Glabrous annuals >20 cm high; sepals 3, petaloid. . . . . Balsaminaceae (p.357)
10. Mostly perennial <20 cm high; sepals 5, green . . . . .Violaceae (p.169)

11. Flowers pea-like; petals very different, lowest 2 united to form a canoe-shaped
    keel containing the stamens. . . . . Fabaceae (p.290)
11. Petals saucer- to cup-shaped, petals not very different 12

12. Petals attached at the base of the (superior) ovary . . . . Ericaceae (p.225)
12. Petals attached along the upper surface of the ovary . . . . Saxifragaceae (p.249)

13. Flowers with >10 stamens 14
13. Flowers with 10 or fewer stamens 26

14. Leaves apparently absent; plants with spiny, succulent stems (pads). . . . . Cactaceae (p.111)
14. Plants not as above 15

15. Stem leaves opposite (2 per node). 16
15. Stem leaves alternate (1 per node) or in a basal rosette 17

16. Flowers yellow; sepals glabrous. . . . . Clusiaceae (p.165)
16. Flowers purple; sepals hairy. . . . . .Lythraceae (p.328)

17. Styles several to numerous 18
17. Style 1(2) (sometimes several stigmas) 19

18. Basal portion of sepals united to form a cup fused with at least part of the ovary
    (hypanthium); sepals often with tiny bracts in-between . . . . Rosaceae (p.261)
18. Sepals usually separate to the base; without bracts in-between . . . . Ranunculaceae (p.83)

19. Petals and sepals attached to the top of the inferior ovary. .Loasaceae (p.173)
19. Ovary superior; sepals and petals attached beneath 20

20. Anther filaments united into a tube that surrounds the style. .Malvaceae (p.165)
20. Stamens not fused into a tube 21

21. Leaves simple, not deeply lobed or divided 22
21. Leaves deeply lobed or divided into leaflets. 23

22. Petals >2 cm long; foliage not succulent . . . . .Papaveraceae (p.102)
22. Petals <2 cm long; foliage often succulent. . . . . .Portulacaceae (p.127)

23. Flowers numerous in terminal or axillary racemes. 24
23. Flowers solitary or in few-flowered racemes or umbels 25

24. Leaves 2 or 3 times divided into leaflets. . . . . Ranunculaceae (p.83)
24. Leaves once divided into leaflets . . . . Cleomaceae (p.185)

25. Sepals 2 to 3, often quicky deciduous; leaves lacking stipules . . . . .Papaveraceae (p.102)
25. Sepals 5; leaves stipulate. . . . . .Zygophyllaceae (p.353)

26. Petals and sepals arising from the top of the (inferior) ovary. 27
26. At least the sepals arising from the base of the (superior) ovary 33

27. Leaves opposite (at least below) or whorled 28
27. Leaves alternate or all basal. 31

28. Petals 5; sepals 2. . . . . .Portulacaceae (p.127)
28. Petals and sepals 4- or 6-merous. 29

29. Fruit berry-like; flowers in a hemispheric cluster subtended by 4 petal-like bracts. . . . . .Cornaceae (p.341)
29. Fruit a capsule; inflorescence not as above. 30

30. Sepals alternating with tooth-like appendages at the top of the calyx . . . . .Lythraceae (p.328)
30. Calyx without tooth-like appendages . . . . Onagraceae (p.329)

31. Petals and sepals 4; style 1 . . . . Onagraceae (p.105)
31. Petals and sepals usually ≥5; styles usually ≥2 32

32. Styles 2; leaves simple, sometimes palmately lobed . . . . Saxifragaceae (p.249)
32. Styles usually >2 (rarely 1 or 2); leaves often compound or pinnately lobed . . . . Rosaceae (p.261)

33. Basal portion of sepals united to form a cup fused with the basal part of the ovary (hypanthium) 34
33. Sepals usually separate to the base; hypanthium absent 36

34. Leaves opposite . . . . .Lythraceae (p.328)
34. Leaves alternate or all basal. 35

35. Styles 2; leaves simple, sometimes palmately lobed . . . . Saxifragaceae (p.249)
35. Styles usually >2; leaves often compound or pinnately lobed . . . . Rosaceae (p.261)

36. Styles >5 per flower . . . . Ranunculaceae (p.83)
36. Styles 0–5. 37

37. Sepals 2 to 3. 38
37. Sepals 4 to many 41

38. Leaves pinnately divided . . . . Limnanthaceae (p.357)
38. Leaves entire 39

39. Petals and sepals undifferentiated . . . . Polygonaceae (p.149)
39. Sepals and petals appearing different 40

40. Petals and sepals 2 or 4, the same number for both; flowers sessile . . . . Elatinaceae (p.164)
40. Petals usually more numerous than sepals; flowers rarely sessile . . . . Portulacaceae (p.127)

41. Sepals 4 42
41. Sepals 5 or more 44

42. Leaves palmately divided into 3 or 5 leaflets . . . . Cleomaceae (p.185)
42. Leaves not palmately divided, sometimes palmately lobed or pinnately divided 43

43. Leaves succulent; pistils 4 . . . . Crassulaceae (p.247)
43. Leaves usually not succulent; pistil solitary . . . . .Brassicaceae (p.186)

44. Leaves opposite. 45
44. Leaves alternate or all basal. 48

45. Leaves deeply lobed or divided into leaflets. . . . . .Zygophyllaceae (p.353)
45. Leaves not deeply lobed or divided 46

46. Flowers and fruits sessile or nearly so in leaf axils. . . . . Elatinaceae (p.164)
46. Flowers cymose or solitary, usually pedicellate (nearly sessile in *Silene acaulis*) 47

47. Petals yellow; leaves sometimes minutely dotted . . . . Clusiaceae (p.165)
47. Petals white or pink to purplish; leaves not dotted . . . . Caryophyllaceae (p.132)

48. Leaves divided or lobed at least half-way to the midvein. 49
48. Leaves with entire, toothed or shallowly lobed margins. 52

49. Leaves divided into 4 leaflets evenly spaced around the petiole tip. . . . . Oxalidaceae (p.354)
49. Leaves lobed or pinnately divided into leaflets. 50

50. Styles 5, united. . . . . Geraniaceae (p.354)
50. Styles 1 or 2 51

51. Styles 2. . . . . Saxifragaceae (p.249)
51. Styles 1. . . . . Fabaceae (p.290)

52. Styles and/or pistils 1 or 2 53
52. Styles and/or pistils >2 55

53. Flowers with 2 pistils. . . . . Saxifragaceae (p.249)
53. Flowers with 1 pistil and/or 1 style 54

54. Sepals 5 . . . . Ericaceae (p.225)
54. Sepals 2, 6, 9 or more, but not 5 . . . . Portulacaceae (p.127)

55. Leaves basal, the blades covered with gland-tipped hairs . . . . Droseraceae (p.169)
55. Plants not as above 56

56. Leaves succulent; stamens 4 or 8 to 10 . . . . Crassulaceae (p.247)
56. Leaves not succulent; stamens 5 . . . . Linaceae (p.349)

Group G. Dicots with both sepals and united petals

1. Flowers resembling a daisy or dandelion (actually clusters of flowers each with united petals) 2
1. Flowers not as above 3

2. Corolla 4-lobed; stamens 4, separate . . . . Dipsacaceae (p.478)
2. Corolla usually 5-lobed; stamens 5, united below . . . . Asteraceae (p.478)

3. Flowers bilaterally symmetrical; 1 or more petals different than the others 4
3. Flowers radially symmetrical; petals all the same 19

4. Stamens 10; flowers with apparently 1 petal . . . . Fabaceae (p.290)
4. Stamens <10; petals or corolla lobes >1 5

5. Petals and sepals united on top of the inferior ovary 6
5. Ovary superior 7

6. Stamens 3; cauline leaves opposite . . . . Valerianaceae (p.476)
6. Stamens 5; leaves alternate or all basal . . . . Campanulaceae (p.465)

7. Stamens 8 . . . . .Polygalaceae (p.350)
7. Stamens 2–6 8

8. Leaves opposite 9
8. Leaves alternate or all basal 14

9. Ovary 4-lobed or divided nearly to the base into 4 nutlets 10
9. Ovary not as above 11

10. Corolla barely irregular, deep blue or violet . . . . .Verbenaceae (p.418)
10. Corolla bilabiate (2-lipped) or if barely irregular then corolla pale blue or white . . . . .Lamiaceae (p.418)

11. Fertile stamens 5 . . . . Scrophulariaceae (p.462)
11. Fertile stamens 2 or 4 (sometimes 5 stamens with 1 sterile) 12

12. Calyx strongly 5-angled . . . . Phrymaceae (p.459)
12. Calyx not or weakly 5-angled 13

13. Flowers with 1 sterile stamen; corolla reddish-brown . . . . Scrophulariaceae (p.462)
13. Flowers not as above . . . . Plantaginaceae (p.429)

14. Stamens 6 . . . . Fumariaceae (p.105)
14. Stamens 2–5 15

15. Fertile stamens 5 . . . . Balsaminaceae (p.357)
15. Fertile stamens 2 or 4 (sometimes 5 stamens with 1 sterile) 16

16. Plants without green leaves or leaves all basal and yellowish 17
16. Plants not as above 18

17. Leaves all basal, yellowish, appearing slimy . . . . Lentibulariaceae (p.464)
17. Plants without green leaves . . . . Orobanchaceae (p.447)

18. Corollas with a hood-like upper corolla lip (galea) enclosing the style . . . . .Orobanchaceae (p.447)
18. Corollas 2-lipped (bilabiate), but upper lip not hood-like . . . . Plantaginaceae (p.429)

19. Ovary wholly or partly inferior 20
19. Ovary superior 27

20. Leaves opposite or whorled 21
20. Leaves alternate or all basal 23

21. Inflorescence of paired flowers; stems prostrate, woody . . . . Caprifoliaceae (p.471)
21. Inflorescence of >2 flowers; stems not woody 22

22. Leaves lanceolate to linear-lanceolate with pointed tips, frequently whorled . . . . .Rubiaceae (p.468)
22. Leaves oblong with rounded tips . . . . Valerianaceae (p.476)

23. Stamens 8–10 24
23. Stamens ≤5 25

24. Leaves simple; stems woody . . . . Ericaceae (p.225)
24. Leaves divided and/or lobed; stems not woody . . . . Adoxaceae (p.476)

25. Cauline leaves absent; fens . . . . .Menyanthaceae (p.487)
25. Leafy stems present; drier habitats 26

26. Herbaceous vines; leaves with cordate bases . . . . .Cucurbitaceae (p.173)
26. Plants not as above . . . . Campanulaceae (p.465)

27. Stamens >5 28
27. Stamens 2–5 30

28. Leaves divided into 4 evenly-spaced ovate leaflets . . . . Oxalidaceae (p.354)
28. Leaves simple to lobed but not divided into leaflets 29

29. Leaves lobed . . . . .Malvaceae (p.165)
29. Leaves with entire margins. . . . . Ericaceae (p.225)

30. Leaves opposite 31
30. Leaves alternate or all basal. 37

31. Inflorescence an umbel; stamens united together, adnate to the stigma . . . . .Asclepiadaceae (p.380)
31. Inflorescence not an umbel; stamens not adnate to the stigma. 32

32. Inflorescence spike-like 33
32. Inflorescence cymose or flowers solitary in leaf axils. 34

33. Corolla deep blue or violet . . . . .Verbenaceae (p.418)
33. Corolla pale blue or white. . . . . .Lamiaceae (p.418)

34. Plants with milky sap; fruit a linear follicle containg many seeds with
    long hairs at the tip. . . . . Apocynaceae (p.379)
34. Sap not milky; fruit a capsule 35

35. Style with 3 stigmas . . . . Polemoniaceae (p.387)
35. Style with 1 stigma 36

36. Flowers sessile, if pedicellate, flowers yellow or salmon . . . . Myrsinaceae (p.240)
36. Flowers pedicellate, blue, green or white. . . . . Gentianaceae (p.374)

37. Plants parasitic with climbing stems and lacking green leaves . . . . .Cuscutaceae (p.386)
37. Plants with green leaves. 38

38. Leaves all basal 39
38. Cauline leaves alternate. 40

39. Styles 5; calyx with 5 distinct ribs. . . . . Plumbaginaceae (p.164)
39. Style 1; calyx not ribbed . . . . Primulaceae (p.236)

40. Petals united only at the base 41
40. Corolla with petals united ca. 1/2-way or more. 44

41. Flowers with a short tube at the base. 42
41. Petals almost completely separate. 43

42. Leaf margins entire. . . . . Myrsinaceae (p.240)
42. Margins of at least some leaves lobed, crenate or wavy . . . . Solanaceae (p.381)

43. Styles 5. . . . . Linaceae (p.349)
43. Style 1. . . . . Portulacaceae (p.127)

44. Ovary 4-lobed or divided nearly to the base into 4 nutlets . . . . Boraginaceae (p.403)
44. Fruit a capsule 45

45. Style with 3 stigmas . . . . Polemoniaceae (p.387)
45. Style or stigma simple or 2-lobed. 46

46. Stamens opposite the corolla lobes . . . . Myrsinaceae (p.240)
46. Stamens opposite the calyx lobes 47

47. Ovary partially divided by 2 membranes that do not join in the center. . . . . Hydrophyllaceae (p.398)
47. Ovary completely divided into 2 cells 48

48. Ovules ≤4 . . . . .Convolvulaceae (p.385)
48. Ovules >4 . . . . Solanaceae (p.381)

PTERIDOPHYTES and GYMNOSPERMS

LYCOPODIACEAE: Club-moss Family

Terrestrial evergreen herbs with ascending or (more often) prostrate stems on or just below the ground. **Leaves** with 1 central vein. **Sporangia** near leaf base of sporophylls. **Spores** numerous, all alike. **Gametophyte** subterranean and non-green.

1. Distinct horizontal stems on or just below ground absent; wing-like gemmae often present
   among leaves; sporophylls similar to sterile leaves . . . . *Huperzia*
1. Horizontal stems present; gemmae lacking; sporophylls often different than vegetative leaves 2

2. Sporophylls similar to sterile leaves . . . .*Lycopodiella*
2. Sporophylls distinct from sterile leaves, organized into terminal cone-like structures 3

3. Shoots 4-angled, terete, or flattened; leaves in 4 to 5 ranks . . . . *Diphasiastrum*
3. Shoots terete; leaves in ≥6 ranks. . . . . *Lycopodium*

Diphasiastrum Holub Club-moss, running pine

Low, perennial, evergreen herbs with scattered, upright shoots from spreading, prostrate stems. **Shoots** erect, mostly flattened or 4-angled. **Leaves** narrowly lanceolate, scale-like, appressed, 4–5-ranked, overlapping. **Sporangia** at the base of reduced, lanceolate to fan-shaped sporophylls grouped together to form cone-like strobili.

All three species were formerly placed in *Lycopodium*.

1. Strobili 2 to 3, pedunculate; shoots very flattened . . . . *D. complanatum*
1. Strobili sessile; shoots little flattened 2

2. Ultimate branches (especially below) winged and 4-angled; leaves 4-ranked. . . . . *D. alpinum*
2. Ultimate branches round in cross section; leaves 5-ranked, often spreading . . . .*D. sitchense*

Diphasiastrum alpinum (L.) Holub [*Lycopodium alpinum* L.] **Stems** prostrate, just below ground. **Shoots** clustered, 3–8 cm; branches 4-angled, winged. **Leaves** 2–3 mm long, 4-ranked, overlapping, margins turned under, dorsal and ventral shorter than lateral pair; alternate pairs decurrent. **Strobili** 5–20 mm long, solitary, sessile. **Sporophylls** 2–4 mm long, lanceolate with wavy margins. Moist, humic turf, often with ericaceous subshrubs in areas where snow persists; upper subapine, alpine. AK to WA and MT, Greenland to QC; Europe, Japan.

Diphasiastrum complanatum (L.) Holub [*Lycopodium complanatum* L.] Stems prostrate on or just below ground. **Shoots** 10–25 cm, branches strongly flattened. **Leaves** 4-ranked, appressed, clasping, overlapping; lateral pair linear, dorsal and ventral much shorter. **Strobili** 1 to 3, pedunculate, 1–9 cm long. **Sporophylls** 2–3 mm long, fan-shaped with an abrupt point. Moist forests, especially cedar-hemlock; montane and lower subalpine. AK to Greenland south to OR, MT, WI and NY. (p.56)
Shoots resemble twigs of western redcedar.

Diphasiastrum sitchense (Rupr.) Holub [*Lycopodium sitchense* Rupr.] Similar to *D. alpinum*. **Stems** often on soil surface; branches terete. **Leaves** 5-ranked and somewhat spreading, all nearly similar. Wet humic turf, open forest, often adjacent to water; subalpine. AK to NL south to OR, MT, NY; Japan.
Many of our specimens have erect leaves and can be confused with *D. alpinum* without close inspection.

Huperzia Bernh. Fir-moss

Low, evergreen perennials with clusters of erect or ascending stems, without horizontal stems. **Leaves** ascending or spreading, densely set, narrowly lanceolate; stomates on both sides; small deltoid, 4-leaved gemmae borne among leaves. **Sporangia** on undifferentiated leaves in zones alternating with sterile leaves.

Our three species have been included in the circumboreal *Lycopodium selago* L. [*H. selago* Bernh. ex Schrank & Mart.]. There is evidence that much of the morphological variation in this group is environmentally determined.[187] Specimens without gemmae cannot be confidently identified. Treating these three taxa at the intraspecific level may be more realistic.

1. Gemmae produced in 1 whorl at branch tips . . . . *H. occidentalis*
1. Gemmae produced in >1 whorl 2

2. Mature shoots 10–20 cm long; gemmae in 2 to 3 terminal whorls (absent). . . . . *H. miyoshiana*
2. Mature shoots 3–15 cm long; gemmae throughout mature leaves . . . .*H. haleakalae*

Huperzia haleakalae (Brack.) Holub [*Lycopodium selago* L. misapplied] **Shoots** 3–15 cm. **Leaves** 3–6 mm, entire, ascending or (usually) appressed to the stem, yellowish below, lustrous. **Gemmae** produced throughout. Damp, mossy, humic soil of meadows, turf, fens, often near tree islands where snow lies late, frequently growing with ericaceous subshrubs; subalpine and alpine. AK to WA, CO; Siberia, disjunct in HI.

Huperzia miyoshiana (Makino) Ching [*Lycopodium selago* L. misapplied] **Shoots** to 20 cm. **Leaves** 3–7 mm long, entire, spreading to ascending, light to dark, lustrous. **Gemmae** in 2 to 3 whorls at end of annual growth or absent. Forest (spruce), thickets, streambanks, fens; valleys to lower subalpine. AK to NL south to WA, ID, MT; Japan, Siberia, Korea.

Huperzia occidentalis (Clute) Kartesz & Gandhi [*Lycopodium selago* L. misapplied] **Shoots** 10–30 cm, prominently curved at the base. **Leaves** reflexed to spreading, 4–10 mm, light green, lustrous. **Gemmae** in 1whorl at end of annual growth. Moist to wet montane forest. AK to OR and MT.

Reported for MT,[412] but I have not seen any specimens that strictly conform to the description of this taxon.

Lycopodiella Holub Bog club-moss

Lycopodiella inundata (L.) Holub [*Lycopodium inundatum* L.] Perennial herbs with prostrate stems. **Shoots** upright, to 10 cm, scattered, unbranched. **Leaves** linear-lanceolate, 5–7 mm, densely set, not in distinct ranks, entire, gemmae absent. **Strobili** solitary. **Sporophylls** similar to but longer than vegetative leaves, spreading. Wet, humic soil or shallow water of fens; montane. Flathead and Missoula cos. BC to CA, MT and NL to OH, WV.

Stems creep among bases of sedges; plants overwinter as a bud.

Lycopodium L. Club-moss

Low, perennial, evergreen herbs. **Shoots** scattered, upright, branched or unbranched from spreading prostrate stems or rhizomes. **Leaves** ascending or spreading, narrow, densely set. **Sporangia** on sporophylls spirally-arranged to form terminal cone-like strobili.

1. Strobili pedunculate; sterile leaves minutely hair-tipped. 2
1. Strobili sessile on shoot tips; leaves not hair-tipped. 3

2. Strobili mostly 1 per peduncle; subalpine or alpine . . . . *L. lagopus*
2. Strobili 2 to 4 on a branched peduncle; montane or lower subalpine. *L. clavatum*

3. Upright shoots many-branched and tree-like . . . . *L. dendroideum*
3. Upright shoots not or sparingly branched. . . . . *L. annotinum*

Lycopodium annotinum L. **Shoots** 5–25 cm, solitary, often sparingly branched basally. **Leaves** dark green, spreading, narrowly lanceolate, shallowly dentate, 4–10 mm long. **Strobili** solitary, sessile on stem tips, 1–4 cm long. **Sporophylls** lanceolate, aristate. Mesic to wet forest; valleys to subalpine, occasionally alpine. AK to Greenland south to AZ, NM, OH, TN. (p.56)

A small form of this species can be found growing in *Sphagnum* moss at the margin of fens.

Lycopodium clavatum L. **Shoots** to 20 cm, solitary, often with short, arm-like branches. **Leaves** spreading or ascending, 3–6 mm long, narrow, entire-margined with hair-like tips. **Strobili** 1–4 cm long, 2 to 4 spreading or ascending on a branched peduncle sparsely set with yellowish bracts. **Sporophylls** broadly lanceolate, dentate with a hair-like tip. Moist forest and margins of fens; valleys to montane. BC to NL south to CA, MT, MB, IL, TN.

Similar in stature to the more common *L. annotinum*, but the long leaf tips are distinctive. One collection from Flathead Co. (*C.L. Hitchcock 1868,* MONTU*)* has a solitary strobile but otherwise appears to belong here.

Lycopodium dendroideum Michx. [*L. obscurum* L. var. *dendroideum* (Michx.) D.C. Eaton] **Shoots** 10–20 cm, solitary, many-branched, tree-like. **Leaves** 3–5 mm long, needle-like, spreading to ascending, pale green. **Strobili** 1–5 cm long, solitary, sessile on upright branches. **Sporophylls** yellowish, ca. 3 mm long, ovate with a short point and ragged margins. Moist coniferous forest; valleys to montane. AK to NL south to WA, WY, WV; Asia; known only from Flathead and Glacier cos.

Lycopodium lagopus (Laestadius ex C. Hartman) G. Zinserling ex Kuzeneva-Prochorova [*L. clavatum* L. var. *lagopus* Laestadius ex C. Hartman] Similar to *L. clavatum.* **Shoots** sparingly branched below, 3–10 cm. **Leaves** strongly ascending; **Strobili** solitary and pedunculate or 2 without a peduncle. **Sporophylls** with hair tip. Moist turf near treeline. Circumboreal to BC, MB, MI, NY; disjunct near the center of Glacier National Park.

SELAGINELLACEAE: Spike-moss Family

Selaginella P. Beauv. Spike-moss

Evergreen perennials with leafy, usually prostrate stems on rock or soil. **Shoots** ascending or erect. **Leaves** stiff and spirally-arranged on stems and shoots. **Strobili** cone-like. **Sporophylls** differentiated from vegetative leaves, spirally-arranged, solitary and sessile at the top of fertile shoots. **Megaspores** borne in lower portion of strobili. **Microspores** borne in upper portion.[235]

The varieties of *Selaginella densa* are distinguished by minute characters and are sometimes considered separate species.[407] *Selaginella rupestris* (L.) Spring would key to *S. watsonii* but has leaves in whorls of 6; it is known from the Black Hills and could occur in southeast MT.

1. Leaves thin, soft; strobili cylindric. . . . . *S. selaginoides*
1. Leaves stiff; stobili 4-sided 2

2. Leaves and stem a different color; stems loosely branched. . . . . *S. wallacei*
2. Leaves and stem the same color; stems densely branched. 3

3. Leaves on underside of stem longer than those of upper side. . . . . *S. densa*
3. Leaves the same length around the stem . . . . *S. watsonii*

Selaginella densa Rydb. [*S. rupestris* (L.) Spring var. *densa* (Rydb.) Clute, *S. scopulorum* Maxon, *S. standleyi* Maxon] Plants forming cushions or mats. **Leaves** in whorls of 4 to 6, appressed, linear lanceolate, 2–4 mm long with a median groove and white bristle-tip. **Strobili** 4-sided, 1–3 cm long. **Sporophylls** broadly lanceolate to ovate with a bristle-tip. Soil and rock of grasslands, outcrops, stony slopes, exposed ridges at all elevations. BC to ON south to AZ, NM. (p.56)

1. Sporophylls with cilia on lower half only . . . . var. *scopulorum*
1. Sporophylls ciliate most of their length. 2

2. Bristle-tips of leaves < 1 mm long; alpine. . . . . var. *standleyi*
2. Bristle-tips of many leaves >1 mm long; below alpine . . . . var. *densa*

Selaginella densa var. *densa* occurs plains and valleys to montane and increases in grasslands following drought and overgrazing. *Selaginella densa* var. *scopulorum* (Maxon) R.M.Tryon is found in grasslands, meadows, fellfields, rock outcrops at all elevations; in sheltered habitats the growth form of var. *scopulorum* may approach that of *S. wallacei*. *Selaginella densa* var. *standleyi* (Maxon) R.M.Tryon occurs near or above treeline; similar to var. *scopulorum* but less common, even at high elevations. Paul Standley collected the type specimens for both of the latter two vars. in Glacier National Park. The three vars. are usually well-marked, but intermediates do occur.[235]

Selaginella selaginoides (L.) Link Plants loosely spreading with erect shoots 1–5 cm long. **Leaves** spirally-arranged, lanceolate, 1–4 mm, recurved-dentate, without median groove or bristle-tip. **Strobili** 1–3 cm long, cylindric. **Sporophylls** broadly lanceolate, recurved-dentate, spreading. Moist to wet, mossy soil of seeps and along streams; montane to subalpine. Circumboreal; AK to Greenland south to NV, CO, WI, NY.

Selaginella wallacei Hieron. Plants usually loosely branched. **Leaves** in whorls of 4, 1–3 mm long, stiff, strap-shaped with a whitish tip; the medial groove with distinct ridges; margins ciliate. **Strobili** 1–3 cm long, 4-angled. **Sporophylls** broadly lanceolate, ciliate with a short bristle-tip. Vernally moist, shallow soil, outcrops, talus; at all elevations. BC and AB south to CA and MT.

Selaginella wallacei usually occurs in more protected sites than the other species and generally has a more loosely branched growth form. In exposed sites its growth form can resemble that of *S. densa* var. *scopulorum*. The green leaves contrast sharply with the brown stem where they are joined.

Selaginella watsonii Underw. Plants forming mats. **Leaves** in indistinct whorls of 4, 1–3 mm long, stiff, oblong to linear, with a short, whitish tip; the medial groove with distinct ridges; margins ciliate. **Strobili** 1–3 cm long, 4-angled. **Sporophylls** ovate, ciliate on the lower half or glabrous with a short bristle-tip. Stony, granitic soil of exposed alpine ridges. OR and MT south to CA and UT.

ISOETACEAE: Quillwort Family

Isoetes L. Quillwort

Stemless, aquatic perennials with short, bulbous, 2-lobed rootstocks and fibrous roots. **Leaves** spirally-arranged, grass-like with dilated bases clasping the stock. **Sporangia** borne on the inner face of each leaf base, ≤1/2 covered with a transparent membrane (velum); small, light tan microspores and larger, white megaspores borne on alternating whorls of leaves. **Gametophyte** tiny and non-green.[235,391]

Plants collected early in the growing season have smooth, immature spores that cannot be used for identification. Plants are often found floating after being unearthed by waterfowl.

1. Megaspores densely covered with spines and sharp ridges . . . . *I. echinospora*
1. Megaspores smooth or with bumps and low ridges 2

2. Megaspores ≥0.5 mm across. . . . . *I. occidentalis*
2. Megaspores <0.5 mm across. 3

3. Leaf bases hyaline 1–5 cm above megasporangium; low elevations . . . . *I. howellii*
3. Leaf bases hyaline ≤2 cm above megasporangium; montane or higher . . . . *I. bolanderi*

Isoetes bolanderi Engelm. Plants aquatic. **Leaves** 2–20 cm long, tapering to a long point. **Sporangia** 3–4 mm long. **Megaspores** 0.3–0.5 mm wide. Unconsolidated mud in shallow water of lakes; montane to subalpine. BC and AB south to CA, AZ, NM. (p.56)

Our most commonly collected quillwort species.

Isoetes echinospora Durieu Plants aquatic. **Leaves** 6–15 cm long with a blunt tip. **Sporangia** 3–6 mm long. **Megaspores** 0.4–0.6 mm wide. Unconsolidated mud in shallow water of lakes and sloughs; valleys to lower subalpine. Circumboreal, south to CA, UT, CO, OH, NJ.

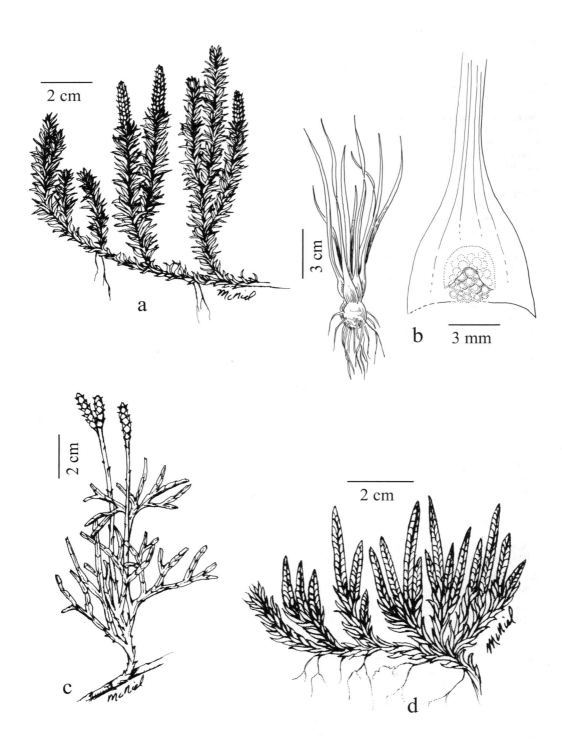

Plate 1. a. *Lycopodium annotinum*, b. *Isoetes bolanderi*, c. *Diphasiastrum complanatum*, d. *Selaginella densa* var. *scopulorum*

Isoetes howellii Engelm. Plants terrestrial or aquatic. **Leaves** to 25 cm long, often black near the base, long-tapering. **Sporangia** 3–7 mm long. **Megaspores** 0.3–0.5 mm wide. Shallow water of ponds, lakes and sloughs; valleys. BC to CA east to ID, MT.

Isoetes occidentalis L.F. Hend. [*I. lacustris* L. var. *paupercula* Engelm.] Plants aquatic. **Leaves** to 20 cm, gradually tapering. **Sporangia** 3–6 mm long. **Megaspores** 0.4–0.8 mm wide. Our one record from shallow water of a montane lake in Missoula Co. AK south to CA, UT, CO.

EQUISETACEAE: Horsetail Family

Equisetum L. Horsetail, scouring rush

Rhizomatous perennials. **Stems** hollow to nearly solid with series of longitudinal ridges terminating at the nodes in papery sheaths with tooth-like projections (sometimes broken off). **Branches** absent or in whorls at the nodes. **Leaves** minute, whorled at the nodes. **Fertile stems** green and similar to sterile ones (monomorphic) or non-green and lacking branches. **Spores** all alike, borne on umbrella-like sporangiophores in successive whorls organized into terminal, cone-like strobili. **Gametophyte** tiny and green.[165]

Hybrids occur sporadically between species; four such hybrids have been reported for Montana: *Equisetum ×ferrissii* Clute (between *E. hymale* and *E. laevigatum*); *E. ×litorale* Kuhlew. ex Rupr. (between *E. arvense* and *E. fluviatile*); *E. ×mackaii* (Newm.) Brichan (between *E. hyemale* and *E. variegatum*); and *E. ×nelsonii* (A.A. Eaton) Schaffn. (between *E. laevigatum* and *E. variegatum*). All have white, misshapen spores and are rare.

1. Stems deep green, leathery, evergreen, all alike, unbranched or branched at the base 2
1. Stems medium or light green, somewhat succulent, annual, branched or unbranched. 4

2. Stems robust, >4 mm in diameter, ridges 14 to 40. . . . .*E. hyemale*
2. Stems <4 mm in diameter, ridges 3 to 12. 3

3. Teeth of stem sheaths 3; stems mostly prostrate and twisted . . . . *E. scirpoides*
3. Teeth of sheaths 4 to 10; stems erect or ascending, not twisted . . . . *E. variegatum*

4. Green stems with whorls of branches at some nodes 5
4. Green stems unbranched or only branched at base 9

5. Branches themselves branched. . . . . *E. sylvaticum*
5. Branches simple. 6

6. Stem central cavity ca. 4/5 diameter of stem . . . . *E. fluviatile*
6. Stem cavity <2/3 diameter of stem. 7

7. Stems monomorphic, fertile stems green and similar to sterile ones; central stem cavity
   <1/3 stem diameter; branches with small central cavity. . . . . *E. palustre*
7. Stems dimorphic; central cavity >1/3 stem diameter; branches solid. 8

8. First branch segment usually longer than the nearest stem sheath. . . . . *E. arvense*
8. First branch segment ca. as long as the nearest stem sheath. . . . . *E. pratense*

9. Teeth of sheaths quickly deciduous, sheaths apparently without teeth;
   moist to dry habitats . . . .*E. laevigatum*
9. Teeth of sheaths apparent; habitat inundated most of growing season . . . . *E. fluviatile*

Equisetum arvense L. COMMON HORSETAIL **Sterile stems** annual, dimorphic, 5–70 cm × 1–5 mm with 10 to 12 ridges, hollow, green. **Sheaths** green with dark teeth. **Branches** usually ascending; first internode longer than subtending stem sheath. **Fertile stems** tan, unbranched, shorter than sterile stems, usually apparent only in spring. **Strobili** 5–40 mm long, blunt. Moist to wet soil of meadows, forests, streambanks, lake shores at all elevations. Cosmopolitan; throughout most of temperate N. America. (p.60)

This extremely variable species can be weedy in moist, disturbed habitats, including gardens. *Equisetum pratense* has distinct spicules on the stem, and first branch internodes are shorter than subtending stem sheath.

Equisetum fluviatile L. **Stems** annual, monomorphic, hollow, branched or not, with 9 to 25 ridges, up to 1 m long, 2–8 mm wide. **Sheaths** green with black teeth. **Branches** with first internode shorter than subtending stem sheath. **Strobili** blunt, 5–30 mm long, maturing in summer. Shallow water of ponds, lakes, sloughs, marshes, fens; plains, valleys to montane. Circumboreal, to WA, IL, PA.

This species often grows in monocultures or with cattail and larger sedges. Unbranched plants resemble *Equisetum laevigatum* but the sheath teeth are not broken off.

Equisetum hyemale L. Scouring rush **Stems** monomorphic, hollow, leathery, dark green, evergreen, unbranched, 18–100 cm × 4–10 mm with 14 to 40 ridges. **Sheaths** green becoming gray with a black lower band and black, deciduous teeth. **Strobili** 5–25 mm long with sharp-pointed tips, maturing in summer. Moist grasslands, meadows, aspen groves, streambanks; plains, valleys to montane. Circumboreal, south to most of temperate N. America.

The large, deep green, leathery stem with ≥14 ridges is diagnostic. This plant can spread rapidly in moist disturbed areas such as roadside ditches, irrigated meadows and river banks.

Equisetum laevigatum A.Braun **Stems** monomorphic, hollow, usually unbranched, annual, 20–100 cm × 2-8 mm with 10 to 32 ridges. **Sheaths** green with a terminal black band, the black teeth quickly deciduous. **Strobili** 10–20 mm, usually blunt. Moist grasslands, woodlands, vernally inundated depressions, roadsides; plains, valleys to montane. BC to QC south to CA, TX, IL.

Stems are lighter green and less leathery than those of *Equisetum hyemale*.

Equisetum palustre L. **Stems** monomorphic, annual, 5–80 cm × 1-4 mm, nearly solid with 4 to 10 ridges. **Branches** whorled on upper half. **Sheaths** green, the dark teeth with hyaline margins. **Strobili** 5–15 mm long with blunt tips, maturing in summer. Wet soil and shallow water, often in forests; valleys to montane, occasionally subalpine. Circumboreal to OR, MT, WI, PA.

Equisetum pratense Ehrh. **Sterile stems** annual, 5–30 cm × 1–2 mm, ascending or erect, nearly solid with 10 to 18 ridges, pale green and branched. **Sheaths** green; the teeth with a dark center and light margins. **Fertile stems** tan, unbranched, occurring in spring, often becoming green and branched later in the season. **Strobili** blunt, 10–20 mm long, uncommon. Shallow water of seeps, swamps, stream margins; valleys to montane. Circumboreal to BC, MT, IL, NY. See *E. arvense*.

Equisetum scirpoides Michx. **Stems** monomorphic, dark green, evergreen, solid, 5–15 cm × ≤1 mm, prostrate or ascending, twisted, mostly unbranched or branched at the base, with 6 ridges. **Sheaths** green below, dark above; the teeth hyaline with dark centers and a bristle-tip. **Strobili** ≤5 mm long, sharp-pointed. Damp, often calcareous soil of spruce forests; valleys to montane. Circumboreal to WA, UT, IL, NY.

Equisetum sylvaticum L. **Sterile stems** annual, 15–70 cm × 2–5 mm with 10 to 18 ridges, green, twice branched, nearly solid. **Sheaths** green below and medium brown above with teeth often united into a few broad lobes below mid-stem. **Fertile stems** tan, unbranched in spring, later becoming green and branched. **Stobili** blunt, 10–30 mm long. Wet meadows, marshes, ponds, along forested streams; valleys to subalpine. Circumboreal to WA, MT, IA, VA.

Equisetum variegatum Schleich. ex F.Weber & D.Mohr **Stems** monomorphic with 3 to 12 ridges, evergreen, 6–30 cm × 1–3 mm, branched at the base, erect or ascending. **Sheaths** green with a terminal black band, teeth white with a black center and hair-like tip. **Strobili** 5–10 mm long with a sharp-pointed tip, maturing in summer. Wet, often calcareous, gravelly soil along seeps, streams, lakes at all elevations. Circumboreal south to OR, UT, IL, NY. Our plants are ssp. *variegatum*. (p.60)

Strobili are rare at high elevations.

OPHIOGLOSSACEAE: Adder's-tongue Family

Perennials. **Stems** unbranched, erect. **Leaf** solitary, divided into a sterile, photosynthetic trophophore and a fertile spore-bearing sporophore (except in *Botrychium paradoxum*). **Trophophore** leaf-like, simple, or pinnately lobed or divided. **Sporophore** simple or pinnately branched. **Spores** all alike. **Gametophyte** fleshy, not green, subterranean.[414]

1. Sterile leaf segment (trophophore) elliptic with entire margins. . . . . *Ophioglossum*
1. Trophophore lobed or divided (or absent) . . . .*Botrychium*

Botrychium Sw. Grapefern, moonwort

Fleshy herbs with fleshy roots. **Trophophore** (sterile blade) 1, spreading or ascending, 1 to 5 times pinnately divided or lobed, attached to common stem at or well above ground level. **Sporophore** (spore-bearing leaf) stalked, usually attached to common stem above ground level, 1 to 3 times pinnate, bearing rows of globose sporangia. **Spores** numerous, yellow.

The Northern Rockies is a center of diversity for the genus. *Botrychium virginianum* and *B. multifidum* (grapeferns) are large and easily seen; all other species (moonworts) are diminutive and inconspicuous. They are difficult to distinguish, and several species often grow together. It is common for some species (e.g., *B. lanceolatum, B. pedunculosum*) to have sporangia on the trophophore. *Botrychium ×watertonense* W.H.Wagner, a hybrid between *B. paradoxum* and *B. hesperium*, is similar to the latter but with spore sacs along the margins of the lobes of the trophophore; it occurs in Glacier National Park. *Botrychium tunax* Stensvold & Farrar and *B. yaaxudakeit* Stensvold & Farrar were recently described from Alaska based on seemingly plastic morphological characters and unpublished molecular evidence and have been reported for MT, but I have seen no specimens.

1. Trophophore (sterile leaf blade) 3 to 4 times pinnate; plants often >15 cm tall 2
1. Trophophore 1 to 2 times pinnate; plants usually <15 cm tall 3

2. Trophophore leathery, joined to the common stem at or below the ground . . . . *B. multifidum*
2. Trophophore thin, joined to common stem well above the ground. . . . . *B. virginianum*

3. Trophophore joined to common stem at or below the ground . . . . *B. simplex*
3. Trophophore joined to common stem above ground level 4

4. Trophophore lacking; sporophore (spore-bearing portion of leaf) double . . . . *B. paradoxum*
4. Trophophore present; sporophore solitary 5

5. Outline of trophophore wider than long . . . .*B. lanceolatum*
5. Trophophore outline longer than wide 6

6. Trophophore usually twice pinnately divided or deeply lobed 7
6. Trophophore usually once pinnate 9

7. Stalk of trophophore >1/4 as long as the rachis. . . . .*B. pedunculosum*
7. Trophophore stalk nearly sessile (<1/3 length of the rachis) 8

8. Basal pair of pinnae noticeably longer than adjacent pair . . . . *B. hesperium*
8. Basal pinnae not much longer than adjacent pair . . . . *B. pinnatum*

9. Lower trophophore pinnae rectangular, not tapered to the union with the rachis. . . . . *B. montanum*
9. Pinnae fan-shaped to linear, at least somewhat narrowed to the base 10

10. Lower trophophore pinnae broadly fan-shaped to semi-circular 11
10. Trophophore lobes narrowly fan-shaped to linear 13

11. Pinnae often overlapping, margins entire to wavy . . . . *B. lunaria*
11. Pinnae well separated, margins dentate to crenulate. 12

12. Pinnae margins finely dentate to minutely crenate. . . . . *B. crenulatum*
12. Margins of pinnae entire to shallowly, few-lobed . . . .*B. minganense*

13. Trophophore pinnae fan-shaped, ascending at a 50–60° angle to the rachis . . . .*B. ascendens*
13. Pinnae not as above. 14

14. Trophophore pinnae linear, 2 to 3 times as long as wide. . . . . *B. lineare*
14. Trophophore narrowly fan-shaped 15

15. Trophophores folded longitudinally, whitish-green 16
15. Trophophores not folded, not whitish-green 17

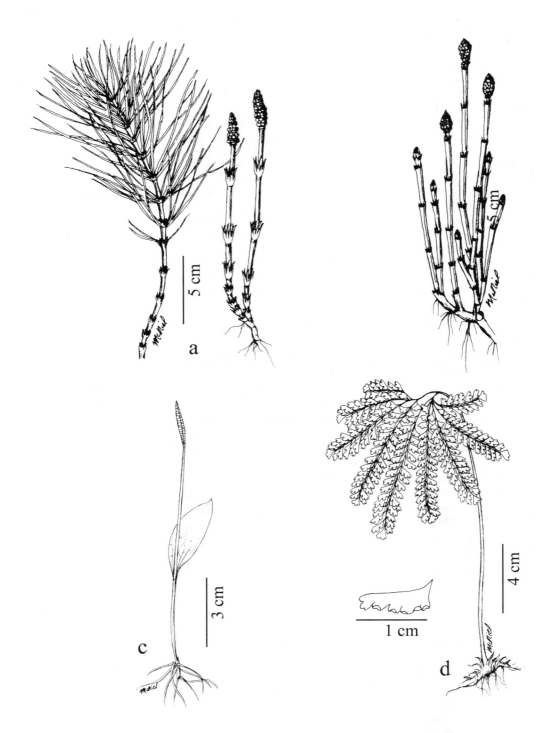

Plate 2. a. *Equisetum arvense*, b. *Equisetum variegatum*, c. *Ophioglossum pusillum*, d. *Adiantum aleuticum*

16. Sporophores <1.5 times as long as the trophophore . . . . *B. campestre*
16. Sporophores 1.5 to 4 times the length of sporophore . . . .*B. pallidum*

17. Basal trophophore pinnae longer than ones above . . . .*B. spathulatum*
17. Basal pinnae ca. same length as those immediately above. . . . .*B. minganense*

Botrychium ascendens W.H.Wagner Plants to 15 cm. **Trophophore** yellow-green; the blade up to 6 cm, oblong with 2 to 5 pairs of strongly ascending, well-separated, fan-shaped, deeply crenulate to dentate pinnae. **Sporophore** 2 times pinnate at the base, once pinnate above, 1 to 2 times length of trophophore. Meadows; montane; known from Lincoln and Lake cos., and reported for Lewis & Clark Co.[413] AK to CA, NV, WY. (p.62)
   Pinnae of *Botrychium minganense* are more nearly perpendicular to the trophophore rachis.

Botrychium campestre W.H.Wagner & Farrar Plants 5-10 cm. **Trophophore** whitish-green, the blade folded longitudinally, oblong, once-pinnate with 2 to 5 or more pairs of well-separated, oblong to fan-shaped pinnae that are entire above, lobed below. **Sporophore** 1 to 2 times pinnate, 1 to 1.5 times length of trophophore. Plants have tiny, spherical gemmae among the roots. Moist meadows; valleys. Known from Flathead Co. AB to MI south to CO, NE, IA, IL.

Botrychium crenulatum W.H.Wagner Plants 5–15 cm tall. **Trophophore** yellow-green, the blade up to 3 cm long, oblong, once-pinnate with 2 to 5 pairs of well-separated, fan-shaped, spreading crenulate-margined pinnae. **Sporophore** 1 to 2 times pinnate, 1 to 3 times length of trophophore. Disturbed meadows, thickets, openings in moist forests; valleys, montane. MT and ID south to CA, AZ. (p.62)
   Botrychium lunaria has more overlapping pinnae; the pinnae of *B. minganense* are not so crenulate; the pinnae of *B. ascendens* are more ascending and more deeply toothed.

Botrychium hesperium (Maxon & R.T.Clausen) W.H.Wagner & Lellinger [*B. matricariifolium* A.Br. ex Doll ssp. *hesperium* Maxon & Clausen] Plants 10–15 cm tall. **Trophophore** gray-green, lanceolate, to 4 cm long with up to 6 pairs of lobed pinnae; the primary lobes overlapping or nearly so; the lowest pair the largest. **Sporophore** 1 to 2 times pinnate, 1 to 2 times length of trophophore. Grasslands or low vegetation in gravelly soil of river terraces and woodlands; montane. BC and AB south to AZ.

Botrychium lanceolatum (S.G.Gmel.) Angstrom Plants 5–15 cm. **Trophophore** dark green to yellow-green, triangular, up to 5 cm long, with 1 to 3 pairs of lobed pinnae; the pinnae ascending and not overlapping. **Sporophore** 1 to 3 times pinnate, 1 to 2.5 times as long as trophophore. Moist meadows, wet forests; montane, subalpine. AK to CA, NM and NL to TN; absent from most of the Great Plains. (p.62)
   This is our only moonwort with the trophophore outline as wide as long. A few sporangia are commonly borne on the trophophore. *Botrychium lanceolatum* ssp. *lanceolatum* and ssp. *angustisegmentum* (Pease & A.H.Moore) R.T.Clausen are separated on the size and color of the trophophore, and both are reported for MT;[414] however, these characters do not seem correlated in our material.

Botrychium lineare W.H.Wagner Plants to 18 cm. **Trophophore** pale green, oblong with 4 to 6 pairs of widely separated, lobed to entire, narrow pinnae. **Sporophore** 1 to 2 times pinnate, 1 to 2 times the trophophore length. Meadows and grasslands; valleys; known from Lincoln Co. ID to QC south to CA, CO.

Botrychium lunaria (L.) Sw. Plants 3–20 cm. **Trophophore** dark green, fleshy, oblong, up to 7 cm long with up to 7 pairs of broadly fan-shaped, closely adjacent, overlapping pinnae with entire or wavy margins. **Sporophore** 1 to 2 times pinnate, 1 to 2 times as long as trophophore. Meadows, forest openings, open slopes at all elevations. AK to Greenland south to CA, NM, WI, PA. (p.62)

Botrychium minganense Vict. Plants 4-20 cm. **Trophophore** dull green, linear, up to 8 cm long, once pinnate with up to 10 pairs of well-separated, fan-shaped, entire to shallowly lobed pinnae. **Sporophore** simple or usually once pinnate, 1.5 to 2.5 times as long as trophophore. Meadows, thickets, forests; montane, subalpine. AK to NL south to CA, CO, WI, NY. (p.62)

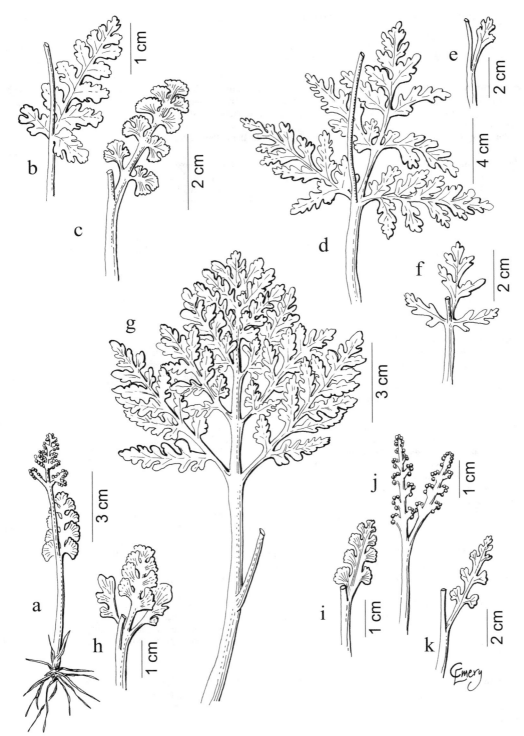

Plate 3. **Botrychium**. **a.** *B. lunaria*, **b.** *B. pinnatum*, **c.** *B. ascendens*, **d.** *B. virginianum*, **e.** *B. montanum*, **f.** *B. lanceolatum*, **g.** *B. multifidum*, **h.** *B. simplex*, **i.** *B. crenulatum*, **j.** *B. paradoxum*, **k.** *B. minganense*

Botrychium montanum W.H. Wagner Plants 4–12 cm tall. **Trophophore** dull gray-green, fleshy, linear, up to 6 cm long, once pinnate with up to 6 pairs of widely separate, fan-shaped to squarish, deeply toothed pinnae; distal pinnae usually broadly attached to the rachis. **Sporophore** once pinnate, 1.5–4.5 times length of trophophore. Sparsely vegetated soil of moist forest, often under *Thuja plicata*; valleys, montane. BC to CA, MT. (p.62)

Botrychium multifidum (S.G.Gmel.) Rupr. Plants 10–45 cm. **Trophophore** shiny green, leathery, evergreen, 4–40 cm long, wider than long with up to 10 pairs of 1 to 2 times pinnate, overlapping pinnae. **Sporophore** 2 to 3 times pinnate, 1 to 2 times longer than trophophore. Moist or wet, organic soils of meadows, forests; valleys, montane. AK to Greenland south to CA, AZ, IL, VA. (p.62)

Botrychium pallidum W.H.Wagner Plants to 12 cm. **Trophophore** folded along the rachis, pale green, glaucous, oblong with up to 5 pairs of non-overlapping, ascending, fan-shaped to rectangular, entire to shallowly toothed pinnae. **Sporophore** 1 to 2 times pinnate, 1.5 to 4 times the length of trophophore. Moist grasslands; valleys, montane; known from Flathead and Lincoln cos. Southern Canada south to MA, MI, CO.

The folded and glaucous trophophore separate this species from the more common *Botrychium minganense*; *B. campestre* has trophophore and sporophore of nearly equal length.

Botrychium paradoxum W.H. Wagner Plants 7–10 cm. **Trophophore** completely converted to a second sporophore. **Sporophore** double, once pinnate, 5–40 mm long, both ca. the same size. Grasslands, meadows, and openings in wet, low-elevation forest; montane, subalpine. BC to SK south to OR, UT. (p.62)

The type locality is in Deer Lodge Co.

Botrychium pedunculosum W.H.Wagner Plants 5–25 cm. **Trophophore** dull green, ovate to oblong with 2 to 5 pairs of deeply lobed pinnae, the lowest nearly overlapping. **Sporophore** 1 to 3 times pinnate, 2 to 4 times the length of the trophophore; the common stem is often reddish-brown. Moist meadows, thickets, wet forest openings; valleys, lower montane. AB, BC south to OR, MT.

Botrychium pinnatum H.St.John [*B. boreale* J. Milde misapplied] Plants 5–20 cm. **Trophophore** green, lanceolate, up to 8 cm long with up to 7 pairs of pinnae; at least the lowest pairs deeply lobed and nearly overlapping the ones above. **Sporophore** 2 times pinnate; 1 to 2 times as long as trophophore. Moist microsites in forests, meadows; montane, subalpine. AK to CA, NV, CO. (p.62)

Botrychium simplex E. Hitchc. Plants 3–15 cm. **Trophophore** light green, lanceolate, to 5 cm long with up to 7 pairs of ascending, fan-shaped pinnae with wavy margins; the lowest often deeply divided and overlapping the ones above. **Sporophore** 1 to 2 times pinnate, 1 to 8 times length of trophophore. Grasslands, meadows; valleys to montane. BC, AB south to CA, CO, and ON, NL to IA, VA. (p.62)

Botrychium spathulatum W.H. Wagner Plants to 10 cm. **Trophophore** yellow-green, leathery, lanceolate with up to 8 pairs of ascending, narrowly spatulate, shallowly lobed to entire, usually well-separated pinnae. **Sporophore** 1 to 2 times pinnate, 1 to 2 times length of trophophore. Moist meadows; valleys; reported for MT.[414] AK to MT and NB south to MI.

Botrychium virginianum (L.) Sw. Virginia grapefern Plants 10–45 cm. **Trophophore** sessile, deltoid, pale green, thin, up to 25 cm long with up to 10 pairs of nearly overlapping pinnae; the basal pairs twice pinnate; the apical merely lobed. **Sporophore** twice pinnate, shorter than to twice the length of trophophore. Dry to moist forests, woodlands, often near water bodies; valleys, montane. AK to NL south to most of N. America and into S. America; Eurasia. (p.62)

Our most common and conspicuous *Botrychium* and the most widespread *Botrychium* species in N. America.

Ophioglossum L. Adder's-tongue

Ophioglossum pusillum Raf. [*O. vulgatum* L. misapplied] Plants 8–30 cm. **Trophophore** elliptic, pale green, thin, to 12 cm long with a short stalk. **Sporophore** undivided, rising 5–15 cm above the trophophore; the conical fertile portion 1–4 cm long, the sterile tip 1–3 mm long. Moist, usually organic soil of meadows, wetlands, roadsides, ditches; valleys to montane. BC to NL south to CA, MT, IA, MD. (p.60)

Plants often grow in dense herbaceous vegetation and may be more common than records indicate.

POLYPODIACEAE: Fern Family

Acaulescent herbaceous perennials from a scaly, horizontal to vertical rhizome or branched caudex. **Leaves** (fronds) petiolate, dimorphic or monomorphic, simple or pinnately compound. **Fertile leaves** bearing groups of sporangia (sori) on the lower surface. **Sori** naked or covered to some degree by a membranous indusium. Gametophytes green, monoecious, usually terrestrial.

Most recent treatments have placed members of the Polypodiaceae *sensu lato* in several different families; however, there does not yet seem to be concensus on the circumscription of these segregate families.[131,178,235,398] Thus I have chosen to retain the traditional inclusive concept of the Polypodiaceae.[100] At the genus and species level I generally followed the concepts presented in the recent Flora of North America treatment.[131]

1. Margins of pinnules inrolled; sori often only along edges of pinnules 2
1. Pinnule margins not inrolled; sori at least partly disposed along veins 8

2. Leaves dimorphic; fertile pinnules narrower than sterile ones 3
2. Fertile and sterile leaves alike 4

3. Fertile pinnules with a mucronate tip . . . . *Aspidotes*
3. Fertile pinnules rounded or acute but not mucronate. . . . . *Cryptogramma*

4. Leaves semi-circular in outline; pinnules discontinuously inrolled . . . . *Adiantum*
4. Leaves narrowly lanceolate to deltoid; pinnules inrolled continuously 5

5. Leaves well-separated along a spreading rhizome . . . . *Pteridium*
5. Leaves clustered on a short rhizome among old petioles 6

6. Pinnules densely hairy beneath . . . . *Cheilanthes*
6. Pinnae and pinnules glabrous or sparsely hairy beneath 7

7. Pinnae or pinnules with entire margins. . . . . *Pellaea*
7. Pinnules with with crennate or lobed margins . . . . *Cheilanthes*

8. Indusium completely lacking or quickly deciduous. 9
8. Indusium present but sometimes inconspicuous 13

9. Leaves once pinnately lobed, evergreen . . . . *Polypodium*
9. Leaves at least partly twice pinnate, deciduous. 10

10. Leaves well-separated on a long rhizome 11
10. Leaves clustered on a short rhizome 12

11. Leaf broader (at the base) than long; primary leaf division ternate . . . . *Gymnocarpium*
11. Leaf longer than wide; primary leaf division pinnate. . . . . *Phegopteris*

12. Petioles ≥3 mm wide at the base; some leaf blades usually >20 cm . . . . *Athyrium*
12. Petiole bases ≤2 mm wide; leaf blades ≤25 cm . . . .*Woodsia*

13. Leaves once pinnate; pinnae dentate to entire but not lobed; sori elongate . . . . *Asplenium*
13. Leaves 2 to 4 times pinnate or once pinnate with pinnae lobed at least basally; sori circular 14

14. Indusium opening into a star shape and exposing all the sori . . . .*Woodsia*
14. Indusium cup- or horseshoe-shaped or circular, partly covering the sori. 15

15. Indusium cup-shaped, attached along the side of the cluster of sori . . . .*Cystopteris*
15. Indusium umbrella- or horseshoe-shaped, attached to the pinnule by a central stalk 16

16. Indusium umbrella-shaped, lacking a lateral slit. . . . . *Polystichum*
16. Indusium horseshoe-shaped with a slit on one side. . . . . *Dryopteris*

Adiantum L. Maidenhair fern

Adiantum aleuticum (Rupr.) C.A.Paris [*A. pedatum* L. var. *aleuticum* Rupr.] Plants with a short, stout, horizontal or ascending rhizome. **Leaves** monomorphic, arching to erect, 10–60 cm long with a glabrous, shiny, black petiole; the blades fan-shaped, appearing palmately divided into several pinnately divided pinnae. **Pinnae** with 15 to 35 light green, fan-shaped to oblong pinnules, 10–25 mm long with jagged apical margins. **Sori** crescent-shaped, borne on edges of pinnules, covered by inrolled margins. Wet rock crevices or shallow soil of cliffs, occasional in damp forest, usually in partial shade; montane, subalpine. AK to NL south to CA, AZ, CO, PA, disjunct in Mexico. (p.60)

Aspidotis (Nutt. ex Hook.) Copel. Lace fern

Aspidotis densa (Brack.) Lellinger [*Cheilanthes siliquosa* Maxon, *Cryptogramma densa* Diels] PODFERN Plants with a short, branched rhizome covered with glossy, brown scales. **Leaves** clustered, dimorphic, 6–20 cm long with long, shiny petioles and lance-shaped to triangular blades. **Sterile leaves** parsley-like, 2 to 3 times pinnate with oblong pinnules. **Fertile leaves** longer, 2 to 3 times pinnate with leathery, linear pinnules. **Fertile pinnules** mucronate with a prominent midrib and pale indusia partly covering the sori from the margins. Dry rock slides and crevices; montane to alpine. BC to CA, UT, WY, disjunct in QC.

More commonly associated with non-calcareous parent material.

Asplenium L. Spleenwort

Plants from short, scaly, vertical to horizontal rhizomes. **Leaves** monomorphic, clustered, linear, once-pinnate; petioles wiry, shorter than the blade. **Pinnae** glabrous or sparsely hairy, ovate with dentate to entire margins. **Sori** elongate, borne along veins near center of pinnae. **Indusium** hyaline, opening along the side. **Gametophyte** green, terrestrial.

1. Rachis of leaves dark, shiny brown to the tip . . . .*A. trichomanes*
1. At least upper portion of rachis greenish . . . . *A. trichomanes-ramosum*

Asplenium trichomanes-ramosum L. [*A. viride* Huds.] **Leaves** 4–15 cm long with petioles green distally. **Pinnae** 6 to 20 pairs with dentate margins. **Sori** 4 to 8 per pinna. Moist to wet, limestone crevices; montane to alpine. Circumboreal south at scattered localities to CA, CO, WI, NY.

Asplenium trichomanes L. **Leaves** 7–35 cm long with a glabrous, shiny, brown rachis. **Pinnae** 12 to 35 pairs, with mainly entire margins. **Sori** 2 to 4 per pinna. Cliff crevices; montane. Circumboreal south to CA, AZ, NM, GA, Mexico. One MT collection made over 100 years ago in Flathead Co.

Athyrium Roth Lady fern

Perennials with short nearly vertical, scaly rhizomes covered with old leaf bases. **Leaves** monomorphic, clustered, 2 to 4 times pinnately divided. **Pinnules** with lobed margins. **Sori** round to narrowly elliptic. **Indusium** persistent, attached at the side or absent.

1. Secondary veins on underside of pinnules obvious; indusium present but soon deciduous;
   plants of forests, wetlands . . . .*A. filix-femina*
1. Secondary pinnule veins obscure; indusium lacking; plants of open, rocky habitats . . . . *A. alpestre*

Athyrium alpestre (Hoppe) Clairv. [*A. distentifolium* Tausch ex Opiz var. *americanum* (Butters) B.Boivin] **Leaves** ascending or erect, densely clustered or widely separated in talus, 15–70 cm long, petiole shorter than the blade. **Leaf blade** narrowly lance-shaped, twice pinnate. **Sori** borne at sinuses of pinnule lobes. **Indusium** absent. Rock slides and stony soil of meadows, often along streams, less common in cliff crevices; subalpine to alpine. Circumboreal south to CA , CO. Our plants are var. *americanum* Butters.

This is our only large fern occurring above treeline; it usually has a distinctive yellowish-green color; more commonly associated with crystalline or weakly calcareous parent material.

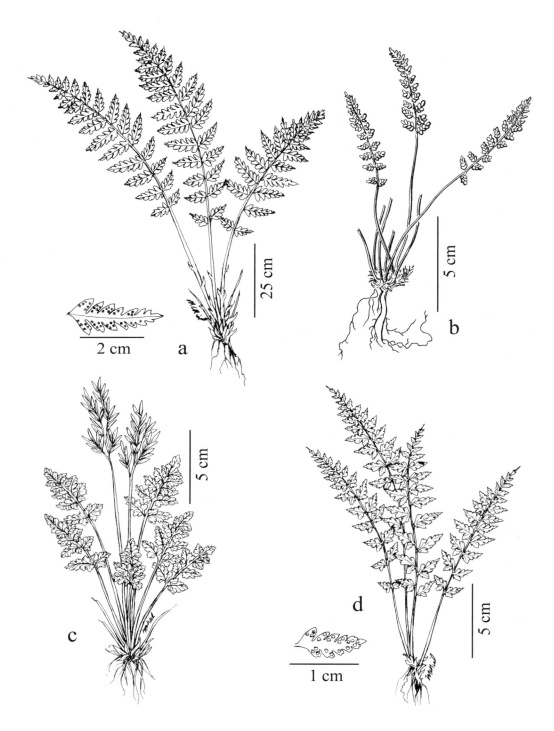

Plate 4. a. *Athyrium filix-femina,* b. *Cheilanthes gracillima,* c. *Cryptogramma acrostichoides,* d. *Cystopteris fragilis*

Athyrium filix-femina (L.) Roth Lady fern **Leaves** ascending or erect, 15–150 cm long, petiole much shorter than the blade. **Leaf blade** narrowly elliptic, twice pinnate. **Sori** borne between midvein and margin of pinnule. **Indusium** often curved with marginal hairs, opening along one edge, deciduous as spores mature. Moist to wet forest, margins of meadows, along streams, avalanche slopes; valleys to subalpine. Circumboreal south to CA, CO. (p.66)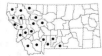

Vegetative specimens of this species can be confused with *Dryopteris filix-mas*, but veinlets of *A. filix-femina* go all the way to the tip of the pinnule lobe but do not in *D. filix-mas*. *Athyrium filix-femina* var. *cyclosorum* Rupr. has yellow spores, while var. *californicum* Butters has brown spores, an impossible distinction for many specimens.

Cheilanthes Sw. Lip fern

Evergreen plants with short, branched rhizomes covered with brown to black scales. **Leaves** clustered, monomorphic; the petiole ca. as long as the blade; blade 2 to 3 times pinnate, narrowly lanceolate. **Pinnules** with reflexed, entire to wavy margins. **Sori** borne under a false indusium continuous with pinnule margins.

1. Leaf rachis and underside of pinnules with hairs and wider hair-like scales . . . .*C. gracillima*
1. Leaf rachis and pinnules with white to tan hairs only . . . . *C. feei*

Cheilanthes feei T. Moore **Leaves** 4–20 cm long. **Leaf blade** subtripinnate at the base with 6 to 12 pairs of pinnae; the rachis hairy but without scales. **Pinnules** oblong to lanceolate, densely white- to brown-villous below. Dry crevices of limestone cliffs; valleys to alpine. AB, BC south to CA, Mexico.

Cheilanthes gracillima D.C.Eaton **Leaves** 5–20 cm long. **Leaf blade** 2 to 3 times pinnate with 9 to 20 pairs of pinnae, the rachis with narrow dark scales. **Pinnules** oblong, nearly glabrous above but covered with hair and narrow brown scales below. Dry to moist rock crevices and steep talus slopes; montane to near treeline. BC, AB south to CA, NV, UT. (p.66)

Plants occur on weakly calcareous sedimentary formations such as the Belt Series as well as granite and other crystalline parent materials. The scales, especially on the rachis are >5 times thicker than the hairs.

Cryptogramma R.Br. Rock brake, parsley fern

Small plants with horizontal rhizomes. **Leaves** glabrous, 2 to 3 times pinnate, dimorphic. **Fertile leaves** with linear pinnules. **Sterile leaves** with broader pinnules, shorter than fertile leaves. **Sori** concealed under inrolled pinnule margins.

1. Leaves scattered; inrolled fertile pinnule margins reflexed. . . . . *C. stelleri*
1. Leaves clustered; inrolled margins of fertile pinnules completely enclosing sori 2

2. Sterile leaf blades thin, translucent when dry, upper surface glabrous . . . . *C. cascadensis*
2. Leaf blades leathery, persistent, upper surface with scattered hairs . . . .*C. acrostichoides*

Cryptogramma acrostichoides R.Br. [*C. crispa* (L.) R.Br. ex Hook. var. *acrostichoides* (R.Br.) C.B.Clarke] **Rhizome** short, covered with old leaf bases. **Sterile leaves** 4–25 cm long, petiole ca. half the length, blades 2 to 3 times pinnate, leathery, parsley-like, ultimate lobes elliptic. **Pinnules** with some secondary veins ending in sunken, white hydathodes. **Fertile leaves** up to 30 cm long, mostly twice pinnate; pinnules linear with yellowish margins rolled under and often meeting in the center before they split at maturity. Shallow soil of rock outcrops, rock slides, less common on wet cliffs; at all elevations. AK to ON south to CA, NM, MN, MI; Asia. (p.66)

Cryptogramma cascadensis Alverson Similar to *C. acrostichoides*. **Sterile leaves** 5–15 cm long, glabrous above, thin and transparent when dry. **Pinnules** without hydathodes. **Fertile leaves** up to 20 cm long. Rock slides and near boulders; subalpine. BC south to CA, ID.

A recently described species segregated from *Cryptogramma acrostichoides* sensu lato.[11]

Cryptogramma stelleri (S.G.Gmel.) Prantl **Rhizome** long and slender, easily broken. **Leaves** well separated. **Sterile leaves** 5–15 cm long; petiole ca. half that length; the blade mostly twice pinnate with lobed or toothed, ovate pinnules. **Fertile leaves** at least half again as long as sterile ones, mostly twice pinnate; pinnules narrowly lanceolate to linear (lobed) with pale margins rolled under but reflexed at the edge. Vernally wet crevices of cliffs and outcrops among moss and other ferns, more common on limestone; montane to subalpine. AK to UT, CO, and ON to NL south to IL, WV.

Cystopteris Bernh. Bladder fern

Plants with scaly rhizomes. **Leaves** monomorphic, 1 to 2 times pinnate. **Pinnules** lobed. **Sori** circular, borne in 1 row on veins of pinnules between midrib and margin. **Indusium** cup-like, attached along midvein side of sorus, deciduous at maturity.

1. Leaf blade nearly as wide as long; rare . . . . *C. montana*
1. Leaf blade 2 to 3 times as long as wide; common . . . . *C. fragilis*

Cystopteris fragilis (L.) Bernh. FRAGILE FERN **Leaves** loosely clustered, glabrous or slightly glandular, 4–35 cm long, petiole shorter to slightly longer than the blade. **Leaf blades** narrowly elliptic to lanceolate, twice pinnate below. **Pinnules** deeply lobed with toothed margins. **Indusium** without glandular hairs. Dry to wet, shallow or stony soil, rock crevices on open slopes, forests, woodlands; all elevations. Cosmopolitan south to CA, NM, NE. (p.66)

At lower elevations this species often grows with the similar *Woodsia* spp., which have more clustered leaves with numerous persistent petioles.

Cystopteris montana (Lam.) Bernh. ex Desv. Plants with long rhizomes. **Leaves** scattered, 10–30 cm long, petiole longer than the blade. **Leaf blades** triangular, twice pinnate below, once above. **Pinnae**: the lowest ca. as large as rest of blade. **Pinnules** deeply lobed with toothed margins; the downward pointing often larger than upward pointing. **Indusium** with gland-tipped hairs on margin. Collected once from wet cliffs near the center of Glacier National Park.[273] Circumboreal to WA, MT, QC, disjunct in CO.

The species superficially resembles *Gymnocarpium* spp. because of the large, lowest pair of pinnae.

Dryopteris Adans. Shield fern

Plants with short, scaly rhizomes, commonly covered with old leaf bases. **Leaves** with scaly petioles; the blades 2 to 3 times pinnate or pinnatifid. **Sori** circular, borne in 1 row on veins midway between midvein and margins of pinnae or pinnules. **Indusium** reniform, attached off-center of the sorus.

The length ratio of downward- to opposing upward-pointing pinnules on the lowest pinnae separates *Dryopteris filix-mas* (ca. 1), *D. carthusiana* (1–2) and *D. expansa* (≥2).

1. Leaves once pinnate, dimorphic, sterile ones arching, fertile ones erect, ... *D cristata*
1. Leaves all alike, twice pinnate to pinnatifid near the base 2

2. Leaves narrowly elliptic or lance-shaped in outline . . . .*D. filix-mas*
2. Leaves broadly ovate to triangular in outline 3

3. First pinnule pointing down on lowest pinnae ≥2 times as long and wide as the
   opposing, upward-pointing one . . . . *D. expansa*
3. First downward-pointing pinnule on lowest pinnae < 2 times as long and wide
   as opposing one. . . . . *D. carthusiana*

Dryopteris carthusiana (Vill.) H.P.Fuchs [*D. austriaca* (Jacq.) Woyn. ex Schinz & Thell. var. *spinulosa* (Muell.) Fiori, *D. spinulosa* (Muell.) Watt] **Leaves** monomorphic, 15–75 cm long, petiole shorter than blade with tan scales toward the base. **Leaf blade** light green, ovate, non-glandular, twice pinnate at the base. **Pinnules** deeply pinnately lobed or toothed; downward-pointing longer than those pointing upward on lower pinnae. **Indusium** lacking glands. Shady, mesic forests, often with *Thuja* or *Tsuga*; montane, subalpine. Circumboreal south to WA, MT, NE, MO, SC. (p.70)

Dryopteris cristata (L.) A.Gray BUCKLER FERN **Leaves** dimorphic, 30–65 cm long; petiole less than half as long as blade. **Leaf blades** narrowly lance-shaped, once pinnate, not glandular. **Sterile leaves** evergreen, arching. **Fertile leaves** erect, longer than the sterile. **Pinnae** deeply pinnately

lobed, fertile ones twisted upward. **Indusium** lacking glands. Wet, open forests, often on margins of fens; montane. Circumboreal south to ID, IL, GA.

This fern is uncommon at the few sites where it occurs. All our other shield ferns have more divided pinnae.

Dryopteris expansa (C.Presl) Fraser-Jenk.& Jermy [*D. dilatata* (Hoffm.) A.Gray ssp. *americana* Hulten] **Leaves** monomorphic, up to 90 cm long; petiole ca. 1/2 as long as blade. **Leaf blade** green, narrowly triangular, twice pinnate, glandular or not. **Pinnules** deeply pinnately lobed, downward-pointing ones on lower pinnae longer than those pointing upward. **Indusium** with or without glands. Mesic, shady forests, often along streams and beneath shrubs; valleys to montane. AK to CA, MT and ON to NL; Greenland, Europe.

Often growing with *Dryopteris filix-mas*, *Athyrium filix-femina* and *Gymnocarpium disjunctum*.

Dryopteris filix-mas (L.) Schott MALE FERN **Leaves** monomorphic, 15–100 cm long, petiole ≤1/3 length of blade. **Leaf blade** narrowly elliptic, green, twice pinnate at the base, not glandular. **Pinnules** downward- and upward-pointing ones ca. equal. **Indusium** lacking glands. Mesic forests and in moist to wet rock crevices, often along streams and beneath tall shrubs; montane to subalpine. Circumboreal to AZ, NM, WI, ON, NS.

Cliff-dwelling plants of exposed sites have more leathery leaves. See *Athyrium filix-femina*.

Gymnocarpium Newman Oak fern

Gymnocarpium dryopteris (L.) Newman [*G. disjunctum* (Rupr.) Ching] Plants with long, slender, scaly rhizomes. **Leaves** to 45 cm long, scattered, erect, petiole longer than blade. **Leaf blade** triangular, twice pinnate below but once above. **Pinnae**: lowest pair similar in size to the rest of blade. **Pinnules** pinnately lobed or toothed, glabrous or slightly glandular. **Sori** circular, borne on veins near margins on underside of fertile pinnules. **Indusium** lacking. Moist forest; montane, subalpine. AK to Greenland south to OR, AZ, NM, IL, WV. (p.70)

Often forming carpets in old-growth forest. Characters used to distinguish *Gymnocarpium disjunctum* from *G. dryopteris* sensu stricto[330] do not consistently covary in northwest Montana, making confident separation difficult.

Pellaea Link Cliff-brake

Plants with short rhizomes, densely covered with reddish-brown scales as well as old leaf bases. **Leaves** monomorphic, 1 to 2 times pinnate with lustrous brown, nearly glabrous petioles. **Pinnules** leathery, usually stalked; the margin rolled under forming a false indusium. **Sori** borne on vein-ends under reflexed pinnule margins.

1. Leaf petioles grooved crosswise near base; old leaf bases many more than green leaves . . . . . . *P. breweri*
1. Leaf petioles without cross grooves; old leaf bases not many more than green leaves . . . . *P. glabella*

Pellaea breweri D.C. Eaton **Leaves** 4–15 cm long with petiole shorter than the blade. **Leaf blade** linear, once pinnate. **Pinnae** ascending, some 2-lobed or mitten-shaped. **Sporangia** with 64 spores. Crevices of limestone cliffs; montane to treeline. WA to MT south to CA, UT.

This species is strongly associated with Paleozoic rock formations.

Pellaea glabella Mett. ex Kuhn **Leaves** 2–40 cm long with petiole shorter than or equal to the blade. **Leaf blade** linear, 1–2 times pinnate below. **Pinnae** ascending, narrowly elliptic, deeply lobed, or pinnate into lanceolate pinnules. **Sporangia** with 32 or 64 spores. Cliff crevices, most commonly on limestone; montane to subalpine. BC to QC south to AZ, TX, TN. (p.70)

More commonly associated with Paleozoic formations. *Pellaea glabella* ssp. *simplex* (Butters) A. Love & D. Love has 32 spores per sporangia and occurs west of the Continental Divide and ssp. *occidentalis* (E.E. Nelson) Windham with 64 spores/sporangium is found east of the Divide,[447] but this character is difficult to assess.

Phegopteris (C.Presl) Fee Beech fern

Phegopteris connectilis (Michx.) Watt [*Thelypteris phegopteris* (L.) Slosson] Plants with long rhizomes. **Leaves** monomorphic, scattered, 15–45 cm long, the petiole longer than blade. **Leaf blade** lance-shaped to deltoid, pinnately divided below. **Pinnae** pinnately lobed, the lowest

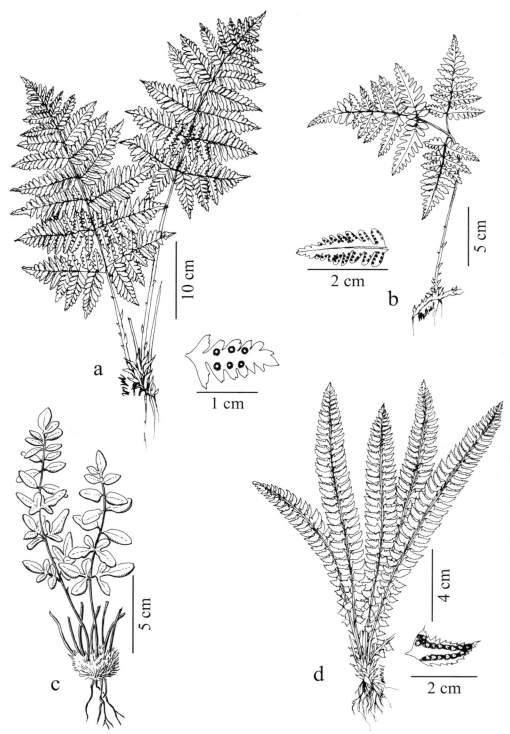

Plate 5. a. *Dryopteris carthusiana*, b. *Gymnocarpium dryopteris*, c. *Pellaea glabella*, d. *Polystichum lonchitis*

pair reflexed, margins and veins beneath with sparse, spreading hairs. **Sori** circular, borne on veins near pinnae margins. **Indusium** absent. Moist soil of mesic, riparian forests or in shaded cliff crevices; montane, subalpine. AK to Greenland south to OR, MT, SK, MO, VA.

Polypodium L. Polypody

Polypodium hesperium Maxon [*P. vulgare* L. misapplied] Small plants with long, creeping, scaly rhizomes. **Leaves** scattered, monomorphic, 3–35 cm long with a petiole shorter than the blade. **Leaf blade** glabrous, somewhat leathery, lanceolate, once pinnately lobed almost to the rachis, with shallowly toothed or wavy margins. **Sori** circular, borne midway between midvein and margins of leaf divisions. **Indusium** absent. Moist to wet, usually moss-covered boulders, cliffs; valleys to subalpine. BC, AB to Mexico.

High elevation plants are dwarfed. In most parts of its range *Polypodium hesperium* is rarely found on limestone,[164] but this does not seem to apply to the weakly calcareous Belt Series in northwest MT. The degree of licorice flavor in the roots is variable. *P. glycyrrhiza* D.C. Eaton, with a pubescent rachis, occurs in adjacent north ID.

Polystichum Roth Sword fern, holly fern

Evergreen perennials with short, stout, scaly, nearly vertical rhizomes. **Leaves** clustered, monomorphic, petioles scaly, much smaller than the blades. **Leaf blades** once pinnate (ours), the pinnae lobed or toothed. **Sori** in 1 to 2 rows on each side of midvein on underside of fertile pinnae. **Indusium** umbrella-like, arising from the center of the sorus.

Polystichum braunii (Spenn.) Fee has been collected in north ID and may occur in northwest MT; it would key to *P. andersonii*, but the latter has bulblets among the pinnae, and the former does not. The sori and indusia of *Dryopteris* spp. are more horseshoe-shaped than those of *Polystichum*.

1. At least lower pinnae deeply lobed 2
1. Pinnae with toothed or spiny margins but not lobed. 4

2. Leaves mostly >25 cm long; larger pinnae ≥3 cm long; bulblets often present among pinnae . *P. andersonii*
2. Leaves mostly ≤30 cm long; larger pinnae ≤3 cm long; bulblets absent 3

3. Pinnae with rounded apices, the teeth spine-tipped. . . . . *P. kruckebergii*
3. Pinnae with acute apices, the teeth mucronate but not spiny . . . .*P. scopulinum*

4. Pinnae >3 times as long as wide, marginal spines parallel to the margin . . . .*P. munitum*
4. Pinnae ≤3 times as long as wide, marginal spines somewhat spreading. . . . . *P. lonchitis*

Polystichum andersonii Hopkins **Leaves** arching, 20–70 cm long, commonly with 1 to several scaly bulblets on central axis among upper pinnae. **Leaf blade** narrowly elliptic or lanceolate. **Pinnae** narrowly lanceolate, the lower deeply divided, 2–10 cm long with sharp-pointed lobes. **Indusium** with sparse marginal hairs. Moist forest, thickets; montane, lower subalpine. AK south to OR, MT.

Polystichum kruckebergii W.H.Wagner **Leaves** erect, 6–20 cm long. **Leaf blade** linear. **Pinnae** overlapping above, ovate, lobed below to toothed above, 5–15 mm long with spreading-spiny margins. **Indusium** with entire margins. Limestone cliffs near treeline. BC to MT south to CA, UT. Known from Deer Lodge and Lake cos.; reported for Carbon and Gallatin cos.[125]

Polystichum lonchitis (L.) Roth HOLLY FERN **Leaves** erect, 5–60 cm long. **Leaf blade** nearly linear. **Pinnae** rarely overlapping, broadly lanceolate to mitten-shaped, 5–30 mm long with spreading-spiny margins. **Indusium** with jagged margins. Wet, rocky slopes, avalanche chutes, cliffs, moist forest; all elevations. AK to Greenland south to CA, AZ, CO, WI, QC. (p.70)

This is our most common sword fern.

Polystichum munitum (Kaulf.) C.Presl **Leaves** arching, 30–120 cm long. **Leaf blade** narrowly lanceolate. **Pinnae** not overlapping, narrowly lanceolate to linear with a shallow lobe at the base, 2–8 cm long with minutely toothed, erect-spiny margins. **Indusium** with fringed margins. Moist forest, often with *Thuja plicata;* valleys to montane. BC to CA, MT.

Polystichum scopulinum (D.C.Eaton) Maxon **Leaves** erect, 15–40 cm long. **Leaf blades** narrowly lanceolate. **Pinnae** overlapping, lanceolate, 1–3 cm long, lobed and toothed with mucronate margins. **Indusium** with entire or fringed margins. Shaded cliffs and rock crevices; montane, subalpine; collected in Ravalli and Sanders cos. BC to MT south to CA, AZ, disjunct in QC and NL.

Pteridium Gled. ex Scop. Bracken

Pteridium aquilinum (L.) Kuhn Perennials with vigorously spreading, scaleless rhizomes. **Leaves** monomorphic, scattered, 30–150 cm long, petiole ca. same length as the blade. **Leaf blade** 3 times pinnate, the first division usually ternate. **Pinnae** 20–50 cm long. **Subpinnae** alternate on the rachis, narrowly lance-shaped. **Pinnules** linear, broadened at the base, upper surface glabrous but hairy below. **Fertile pinnules** with recurved margins partly covering  continuous series of sori. Moist to wet forest, woodlands and meadows, thickets within forested landscapes, often with *Thuja plicata* or along streams; montane to occasionally subalpine. Cosmopolitan. Our plants are var. ***pubescens*** Underw. (p.74)

Woodsia R.Br. Cliff fern

Plants with short, densely scaly rhizomes clothed in old leaf bases. **Leaves** clustered, 1 to 2 times pinnate. **Sori** circular, borne between midvein and margins of pinnules. **Indusium** attached basally and dissected into lobes spreading over the sorus.

The clustered, persistent old leaf bases help distinguish these species from *Cystopteris fragilis*.

1. Leaf rachis and undersides glabrous or with glandular hairs only . . . . *W. oregana*
1. Rachis and leaves with long, non-glandular as well as shorter glandular hairs . . . . *W. scopulina*

Woodsia oregana D.C.Eaton **Leaves** with sparse glandular hairs; the petioles ca. as long as blades and resistant to shattering. **Leaf blades** linear to narrowly elliptic, often twice pinnate below, 5–15 cm long, lower pairs of pinnae widely separated. **Pinnules** lobed or toothed, glabrous or with tack-like glandular hairs. Dry, rocky slopes; valleys to lower subalpine. BC to ON south to CA, OK and MI. Our plants are ssp. ***oregana***.

Woodsia scopulina D.C.Eaton Similar to *W. oregana*. **Leaves** with easily shattering petiole. **Leaf blades** twice pinnate, 4–25 cm. **Pinnules** and rachis with multicellular, non-glandular hairs as well as shorter tack-like glandular hairs. Outcrops, rock slides and thin soil of rocky slopes; montane, subalpine. AK to QC south to CA, NM. (p.74)

Often conspicuous among lichens and mosses in stabilized talus. The density of multicellular hairs varies greatly and can approach that of *Woodsia oregana* in which they are absent. Both ssp. *scopulina* and ssp. *laurentiana* Windham are reported for western MT. Characters separating these subspecies are continuous, and there is no obvious geographic or ecological distinction between them in our area.

MARSILEACEAE: Water-clover Family

Marsilea L. Water clover

Aquatic perennials forming colonies from shallow, branched rhizomes. **Leaves** alternate, the petiole longer than the blade. **Leaf blade** clover-like, palmately divided into 4 equal, broadly ovate pinnae. **Sporocarps** sparsely to densely hairy, globose to ovate with 1 or 2 tooth-like basal projections, borne on unbranched, erect stalks arising near the base of the leaf petioles.[204]

1. Sporocarp with 2 prominent teeth at point of attachment to stalk. . . . . *M. vestita*
1. Sporocarp with 1 prominent tooth, the 2nd blunt or lacking . . . . *M. oligospora*

Marsilea oligospora Goodd. **Leaf petioles** 3–6 cm long. **Pinnae** 6–15 mm long, pilose. **Sporocarps** nodding, 5–6 mm long, long-hairy but glabrous with age; lower tooth ca. 0.5 mm; upper tooth blunt, smaller, or absent. Shallow water; valleys; collected in Lake Co. and the Blackfoot River Valley. [204] WA to MT south to CA, UT.

Marsilea vestita Hook.& Grev. **Leaf petioles** 1–20 cm long (to 40 cm when floating). **Pinnae** 3–19 mm long, sparsely pubescent to glabrous. **Sporocarps** slightly nodding, 3–6 mm long, densely pubescent but glabrous with age; lower tooth ca. 0.5 mm; the upper tooth longer and acute. Fresh to saline, shallow water of ponds, reservoirs, sloughs; plains, valleys. BC to SK south to Mexico, S. America.

SALVINIACEAE: Water-fern Family

Azolla Lam. Mosquito fern.

Azolla filiculoides Lam. Tiny, floating, mat-forming, aquatic annual. **Stems** tightly branched with simple roots. **Leaves** emersed, alternate, overlapping, sessile with 2 ovate lobes <1 mm long; upper lobe green or red, covered with white, flake-like hairs; lower lobe smaller and floating. **Sporocarps** dimorphic, obovoid, apparently in lower leaf axils; megasporocarps with 1 spore < 0.5 mm long; microsporocarps ca. 0.1 mm long with numerous spores clustered in small masses bearing hooked hairs. Shallow water of ponds, sloughs; valleys. BC to CA, AZ, NM, TX, Mexico. Collected in Ravalli Co.

Reports of *A. mexicana* Presl[380] are referable here.

PINACEAE: Pine Family

Monoecious, evergreen or deciduous trees with resinous and aromatic sap. **Leaves** needle-like, spirally-arranged, solitary, or borne in clusters. **Cones** borne on first-year twigs. **Pollen cones** small with spirally-arranged stamens. **Seed cones** large, woody, formed of spirally-arranged scales, maturing in 1 to many seasons. **Fertile scales** bearing 2 seeds on the inner surface and subtended by a papery bract. **Seeds** winged in most species.

The aromatic sap makes these trees highly flammable, especially during late summer of dry years; fire has played a major role in the evolution of many species and their associated communities.

1. Leaves borne in clusters (fascicles) 2
1. Leaves borne singly, not fascicled 3

2. Leaves >5 per cluster, autumn-deciduous . . . .*Larix*
2. Leaves 2-5 per cluster, evergreen . . . . *Pinus*

3. Old needles breaking off above the base causing young branches to be rough where
   needles have fallen. 4
3. Leaves falling from the base, young branches smooth where needles have fallen 5

4. Needles sharp-pointed, not flattened . . . . *Picea*
4. Needles flattened with a rounded tip . . . . *Tsuga*

5. Cones erect and disintegrating on the tree; terminal buds rounded. . . . . *Abies*
5. Cones pendant, falling whole after seeds are shed; terminal buds sharp-pointed. . . . . *Pseudotsuga*

Abies Mill. Fir

Evergreen trees with a conical or spire-like crown. **Bark** thin, smooth. **Branches** whorled; primary limbs branched in one plane to form large, flat sprays. **Leaves** flat, single, spirally-arranged, persisting 5 or more years, leaving a slight, nearly circular depression on twigs after falling. **Buds** globose and resinous. **Cones** borne on 1-year-old twigs. **Pollen cones** clustered. **Seed cones** cylindrical, upright, maturing in 1 season, disintegrating on the tree. **Scales** fan-shaped, pubescent, falling individually. **Seeds** winged.

1. Leaves borne in a single plane; lines of white-resinous stomates on lower surface only;
   montane . . . . *A. grandis*
1. Leaves curving upward, not in 1 plane; stomates on both surfaces; most common in
   subalpine zone . . . . *A. lasiocarpa*

Abies grandis (Douglas ex D.Don) Lindl. GRAND FIR Tree up to ca. 50 m tall with a conical crown and branches that bend down and then out. **Bark** gray with reddish furrows, becoming brown in older trees. **Leaves** 2–4 cm long, blunt-tipped, borne in a single plane opposite each

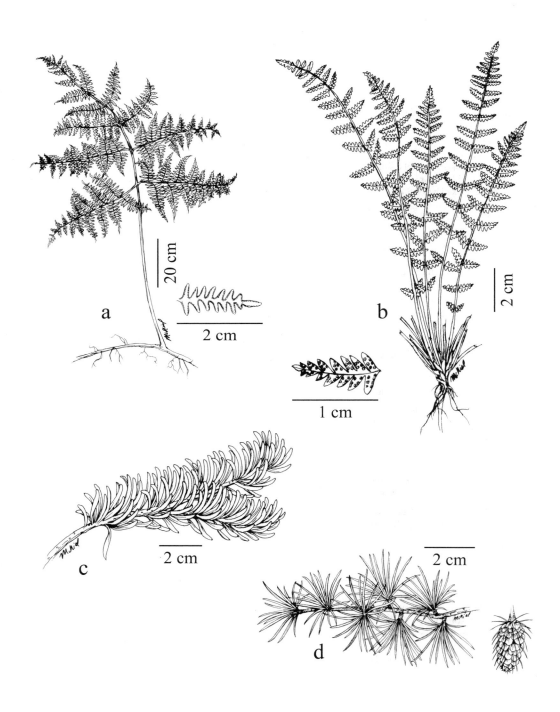

Plate 6. *Pteridium aquilinum*, b. *Woodsia scopulina*, c. *Abies lasiocarpa*, d. *Larix occidentalis*

other on the twigs, erect. **Stomates** on underside only. **Seed cones** usually green, 6–11 cm long. Moist to wet forest; montane. BC to CA, ID, MT.

Grand fir is shade tolerant and grows rapidly but is susceptible to rotting fungi and thus relatively short-lived. Grand fir cones are borne at the tops of the trees and disintegrate in place, so they are rarely seen unless cut by squirrels.

Abies lasiocarpa (Hook.) Nutt. Subalpine fir [*A. bifolia* A.Murray] Generally a small tree up to ca. 30 m tall with a narrow crown. **Bark** gray but splitting to reveal a brownish layer beneath. **Leaves** 1–3 cm long, turned upward, blunt-tipped except on cone-bearing branches. **Stomates** on both surfaces. **Seed cones** deep blue, 3–8 cm long. Moist forests; montane to treeline. YT to CO, AZ, NM. (p.74)

Subalpine fir is shade-tolerant but relatively short-lived and readily killed by fire. It can occur in valleys with cold air drainage, and it commonly forms hedge-like krumholz stands in the alpine zone. Near treeline lower limbs pressed against the ground by snow pack can root and send up new shoots, thus forming clumps of stems. *Abies bifolia* is the Rocky Mountain segregate of *A. lasiocarpa* separable on chemical and minor morphological characters. These two taxa might better be treated as separate subspecies.

Larix Miller Larch

Deciduous trees with open crowns and whorled branches. **Bark** scaly becoming thick with age. **Leaves** pale green, clustered on short, nubbin-like spur shoots. **Pollen cones** solitary on spur shoots. **Seed cones** solitary, pendant, maturing in 1 season. **Scales** thin, ovoid, pubescent. **Bracts** borne outside each scale project beyond it in a long awn. **Seeds** winged.

1. Young branches densely covered with long, tangled hairs; seed cones >35 mm long;
   subalpine and alpine. . . . .*L. lyallii*
1. Young branches glabrous or with short hairs; seed cones ≤40 mm long; valleys and
   montane . . . . *L. occidentalis*

Larix lyallii Parl. Subalpine larch Small trees to 20 m tall with open, often asymmetrical crowns. **Bark** covered with red- to purple-brown flakes. **Leaves** up to 40 per spur, 4-angled, 1–3 cm long. **Seed cones** 30–45 mm long, yellow to purplish-green. Near treeline where snow lies late in areas with stony soil derived from non-calcareous parent material. AB, BC, WA, ID, MT.

Usually the dominant tree where it occurs, forming open woodlands, often with an understory of ericaceous shrubs.

Larix occidentalis Nutt. Western larch Large trees to 60 m; older trees without branches for most of their height. **Bark** of mature trees thick, deeply furrowed, covered with cinnamon-colored plates. **Leaves** 15 to 30 per spur, 2–5 cm long, 3-angled. **Seed cones** brown to reddish, 25–40 mm long. Moist to wet forests, especially on cool slopes; valleys to lower subalpine. BC to OR, ID, MT. (p.74)

Recruitment of western larch occurs only in open ground, usually after fire. It is long-lived, and the thick bark of older trees resists ground fires, so it dominates some old-growth forests.

Picea A. Dietr. Spruce

Evergreen trees with whorled branches and a broadly to narrowly conical crown. **Bark** scaly. **Leaves** 4-angled with stomates on all surfaces, rigid, sharp-pointed, single, spirally-arranged, spreading out in all directions from the twigs. **Pollen cones** clustered. **Seed cones** pendant, ovoid to cylindric, maturing in 1 season, falling entire. **Scales** thin, subtended by a smaller bract. **Seeds** 2 per scale, prominently winged.

Our two species of spruce hybridize extensively. Typical *Picea glauca* is rare in MT, but plants referred to that species are most likely encountered in mountain valleys. Pure *P. engelmannii* is the common spruce of higher elevations. Hybrids with intermediate seed cone characters are common in the montane zones, especially west of the Divide and near the Canadian border. These are referred to *P. engelmannii*.[422]

1. Twigs usually finely pubescent; cone scales with wavy-margined tip. . . . .*P. engelmannii*
1. Twigs glabrous; cones scales with entire margins . . . . *P. glauca*

Picea engelmannii Parry ex Engelm. ENGELMANN SPRUCE Trees up to ca. 50 m tall with narrow,
spire-like crowns. **Bark** gray to reddish brown. **Leaves** blue-green, 1–2 cm long. **Seed cones** 3–6
cm long, yellow- to purple-brown. **Scales** widest above middle with wavy margins along the tip.
Moist to wet forests, often along streams, wetlands; valleys to subalpine. BC and AB to AZ, NM,
Mexico. Our plants are var. **engelmannii**. (p.78)

Occasionally forming krumholz hedges at or above treeline. In poorly developed, calcareous soils, this species will
often dominate timberline forests, while *Abies lasiocarpa* will predominate on non-calcareous soils.

Picea glauca (Moench) Voss WHITE SPRUCE Small tree to 30 m tall with a conical crown. **Bark** gray-brown. **Leaves**
1–2 cm long. **Seed cones** 25–60 mm long. **Scales** fan-shaped, widest near the entire-margined tip. Forests, often along
streams; montane. AK to NL south to MT, CO, SD, WI, NY. One unambiguous specimen from Carbon Co.

Pinus L. Pine

Evergreen trees with conical to flat-topped crowns. **Bark** usually furrowed or plated in older trees.
Leaves needle-like, in groups of 2 to 5 on nubbin-like, spur shoots, these subtended by sheathing scale
leaves at the base. **Pollen cones** densely clustered at the base of current year's growth. **Seed cones** ovoid
to cylindric, maturing the second year, opening at maturity or much later (serotiny). **Scales** numerous,
spirally-arranged, woody, often with a thickened tip. **Seeds** 2 per scale.

Whitebark and limber pines have ascending branches and flat-topped crowns that resemble broad-leaved
trees. These two species as well as western white pine (soft pines) are succumbing to an epidemic of white
pine blister rust, a European disease that has gooseberries and currants (*Ribes* spp.) as intermediate hosts.
All of our species, but especially lodgepole pine, also suffer from periodic outbreaks of mountain pine
beetle.

1. Needles in clusters of 5; cone scales lacking a terminal prickle (soft pines) 2
1. Needles in clusters of 2 to 3; cones scales with a terminal prickle (hard pines). 4

2. At least some needles >7 cm long; cones ca. 3 times as long as wide; lower elevations
   west of the Continental Divide . . . . *P. monticola*
2. Needles <7 cm long; cones ca. 2 times as long as wide; high elevations or east of the Divide. 3

3. Cones 7–14 cm long, falling intact after shedding seed; scales thinner toward the tip . . . .*P. flexilis*
3. Cones 4–8 cm long, seldom falling intact; scales thicker toward the tip. . . . . *P. albicaulis*

4. Needles 4–8 cm long in clusters of 2 . . . . *P. contorta*
4. Needles >8 cm long in clusters of 2 or 3 . . . . *P. ponderosa*

Pinus albicaulis Engelm. WHITEBARK PINE Small tree to 25 m tall with ascending branches and
a rounded or flat-topped crown. **Bark** smooth, light gray. **Leaves** yellow-green, 2–6 cm long, 5
per fascicle. **Seed cones** ovoid, 4–8 cm long, remaining on the tree and closed until opened and/
or dislodged by squirrels or birds. **Scales** thin at the base but thickened toward the tip. **Seeds**
wingless, 7–11 mm long. Usually open forests; subalpine to near or above treeline. BC, AB south
to CA, NV, WY.

Twisted and stunted krumholz forms can often be found in sheltered sites above treeline.

Pinus contorta Douglas ex Loudon LODGEPOLE PINE Small, slender tree to 35 m with whorled
horizontal branches forming a conical crown. **Bark** thin, scaly, brown or gray. **Leaves** yellow-
green, 4–8 cm long, 2 per fascicle. **Seed cones** ovoid but asymmetrical, 2–6 cm long. **Scales**
tongue-shaped with a spine tip. **Seeds** with a conspicuous wing. In mixed or nearly pure forest
stands; valleys to treeline. AK to CA, UT, CO, SD. Our plants are var. **latifolia** Engelm. (p.78)

Lodgepole pine is short-lived and shade intolerant. It is well adapted to fire. Some seed cones open upon maturity and
fall, while others remain on the tree until opened by heat (serotinous). Serotinous cones may remain on the trees for so
long that they become embedded in wood.

Pinus flexilis E.James LIMBER PINE Small tree to 15 m tall, very similar to *P. albicaulis*. **Leaves**
green, 5 per fascicle, 3–7 cm long. **Seed cones** ovoid, 7–14 cm long, falling shortly after opening.
Scales rhombic, thinner towards the tip. **Seeds** 10–15 mm long, wingless. Rocky slopes and
ridges; montane, occasionally subalpine, rarely plains and valleys. BC, AB south to CA, AZ, NM.

Unlike whitebark pine, limber pine cones open, drop their seed, then fall intact. Mature trees usually have a carpet of old cones beneath. Limber pine is most often associated with calcareous parent material, but occurs on soils derived from sandstone and quartzite as well.

Pinus monticola Douglas ex D.Don WESTERN WHITE PINE Large tree to 60 m tall with a conical to rounded crown. **Bark** thin, smooth at first, becoming scaly with age. **Leaves** light blue-green, 4–10 cm long, 5 per fascicle, the tips denticulate. **Seed cones** narrowly elliptic, 15–25 cm long; scales thin, tongue-shaped. **Seeds** with conspicuous wings. Moist forest; valleys to montane. BC to CA, NV, MT.

Western white pine is intolerant of shade and long-lived although prone to blister rust and sensitive to fire.

Pinus ponderosa Douglas ex P.Lawson & C.Lawson PONDEROSA PINE Large trees to 65 m tall with an open, rounded crown and spreading branches. **Bark** of old trees thick, furrowed, covered with scales that resemble pieces of a jigsaw puzzle. **Leaves** yellow-green, 7–25 cm long, 2–3 per fascicle, clustered on branch ends. **Seed cones** broadly ovoid, 7–15 cm long. **Scales** thick with a terminal prickle. **Seeds** with a conspicuous wing. Drier forests as well as rocky exposures (especially sandstone) associated with grasslands; plains, valleys to montane. BC to NE south to Mexico. Our plants are var. *scopulorum* Engelm.

Ponderosa pine is long-lived, fire resistant, shade intolerant, and appears to thrive in drier habitats. Ponderosa pine is virtually absent from the southwest part of the state, possibly because the valleys are too cold during much of the growing season.[14] Trees in the eastern part of the state have mainly 2 needles per fascicle, while those in the west have 3.[14]

Pseudotsuga Carriere Douglas fir

Pseudotsuga menziesii (Mirb.) Franco Large evergreen trees up to 60 m tall with spreading branches and narrow to broadly conical crowns. **Bark** of older trees thick, furrowed and gray with reddish brown between furrows. **Leaves** yellow- to blue-green, single, spirally-arranged and spreading, 15–25 mm long, stomates lacking on upper surface. **Leaf buds** conical. **Seed cones** narrowly ellipsoidal, 3–6 cm long, maturing the first season and shed entire. **Scales** broadly rhombic, stiff, pubescent, subtended by a longer 3-lobed bract. **Seeds** winged. Dry to mesic forests; valleys to subalpine and on the plains along the Missouri River. BC to AB south to CA, TX, Mexico. Our plants are var. *glauca* (Mayr) Franco. (p.78)

Douglas fir is somewhat tolerant of shade, fast-growing, long-lived, and fire resistant when old. The 3-lobed bract with the long, narrow central lobe is diagnostic.

Tsuga (Endl.) Carriere Hemlock

Evergreen trees with conical crowns and often drooping leaders and branches. **Bark** brown, scaly, usually furrowed. **Leaves** single, linear, with a blunt tip and a short narrow petiole mounted on a short peg. **Leaf buds** rounded. **Pollen cones** solitary, globose. **Seed cones** pendent, narrowly ovoid, maturing in 1 year, opening at maturity. **Scales** spirally-arranged, thin, leathery. **Seeds** winged.

1. Leaves arranged in a nearly single plane with whitish lines of stomates on lower surfaces only; seed cones 15–30 mm long. . . . .*T. heterophylla*
1. Leaves surrounding the twigs with stomates on both surfaces; seed cones ≥25 mm long . . .*T. mertensiana*

Tsuga heterophylla (Raf.) Sarg. WESTERN HEMLOCK Medium-size trees to 60 m tall with a narrow crown and drooping leader. **Bark** brown, thin, scaly. **Leaves** flattened, shiny yellow-green above with broad, white stomatal bands below, 8–15 mm long, borne in a single plane or nearly so. **Seed cones** green to brown with age, narrowly ellipsoid, 15–30 mm long. **Scales** oblong-ovoid. Mesic to wet forests; valleys to montane; collected in Flathead and Lake cos. AK to CA, ID, MT. (p.78)

Western hemlock is very shade tolerant but not resistant to fire and is intolerant of drought.

Tsuga mertensiana (Bong.) Carriere. MOUNTAIN HEMLOCK Small to medium-size trees to 45 m tall with a narrow crown and slightly drooping leader. **Bark** gray to reddish brown and deeply fissured. **Leaves** somewhat thickened, whitish on both surfaces, yellow- to blue-green, 12–20 mm long, surrounding the stems or upturned. **Seed cones** purple, 25–60 mm long, nearly linear. **Scales** fan-shaped, as broad as long. Forests in regions of heavy snowfall; subalpine. AK to CA, ID, MT.

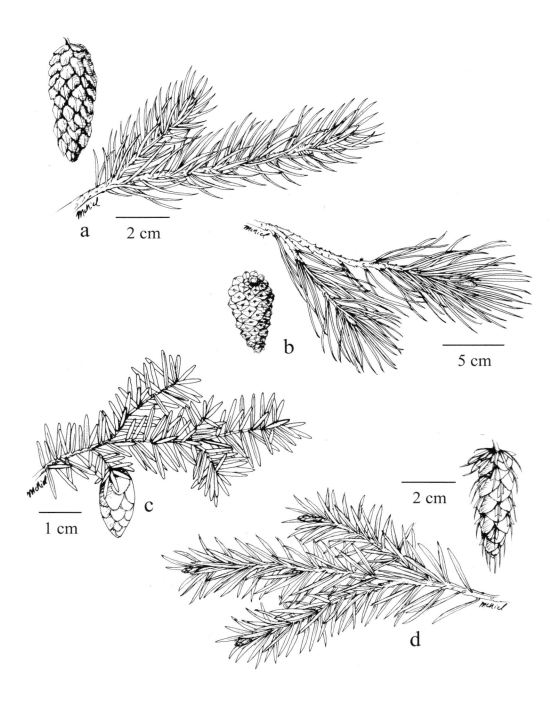

Plate 7. a. *Picea engelmannii*, b. *Pinus contorta*, c. *Tsuga heterophylla*, d. *Pseudotsuga menziesii*

CUPRESSACEAE: Cedar Family

Evergreen shrubs or trees with pollen and seeds borne in separate cones on branch tips. **Leaves** small, simple, thick, scale- or needle-like. **Pollen cones** small, solitary, with umbrella-shaped stamens. **Seed cones** with 4–6 dry scales or 3–8 fleshy, united scales, resembling a berry. **Seeds** usually 1 or 2 borne at the base of each scale.

1. Seed cones fleshy, berry-like; shrubs or small trees . . . . *Juniperus*
1. Seed cones woody, dry, with spreading scales; trees . . . . *Thuja*

Juniperus L. Juniper

Dioecious evergreen shrubs or trees. **Leaves** small, stiff, opposite or in whorls of 2 to 3; those of vigorous, "juvenile" branches longer and needle-like. **Pollen cones** with 12 to 16 sporophylls in 2's or 3's. **Seed cones** globose, fleshy, berry-like with mostly 1 to 5 wingless seeds.

   Although the juniper fruit appears to be a berry, close examination will reveal signs of the joined cone scales. Immature plants of *Juniperus scopulorum* may superficially resemble *J. communis* by having needle-like leaves, but these are not jointed at the base as in *J. communis*. The presence of a nearby adult tree is often the best clue. Leaves of both *J. horizontalis* and *J. scopulorum* have a large, narrowly elliptic, amber-colored gland barely visible on the upper surface. These species are closely related and populations intermediate in stature and leaf shape have been documented in Chouteau Co.

1. Shrubs or small trees, usually with a single trunk and ascending branches 2
1. Shrubs, with spreading branches near the base 3

2. Leaves usually ca. as long as wide with denticulate margins (use 10X) . . . . *J. osteosperma*
2. Leaves usually longer than wide, the margins entire . . . . *J. scopulorum*

3. Leaves <5mm long, some scale-like and appressed, covering the stem . . . .*J. horizontalis*
3. All leaves spreading and needle-like, ≥5 mm long . . . . *J. communis*

Juniperus communis L. COMMON JUNIPER Rounded shrub with prostrate to erect stems up to 1 m. **Bark** brown, fibrous. **Leaves** stiff, needle-like, 5–20 mm long, whitish on lower surface, in whorls of 3, jointed where they meet the stem. **Seed cones** maturing in 2 years, bluish-black, 4–7 mm long, seeds 2 to 3. Drier forests, forest margins, open slopes, outcrops and rock slides at all but the highest elevations. AK to Greenland south to CA, AZ, NM, IL, SC; Eurasia. Our plants are var. *depressa* Pursh. (p.80)

Juniperus horizontalis Moench CREEPING JUNIPER Prostrate shrub with trailing branches. **Bark** brown, stringy. **Leaves** bright to bluish green (often purplish in winter), opposite, scale-like, appressed, ca. 1 mm long, entire-margined with an apiculate tip; juvenile leaves 2–4 mm long, ascending. **Seed cones** maturing in 2 years, blue-black, 3–7 mm long; seeds 1 to 3. Sandy prairie, old gravel bars, limestone outcrops; plains, valleys, montane, rarely up to treeline. AK to NL south to CO, IL, NY.

   Plants in sheltered sites may become semi-erect (see genus description).

Juniperus osteosperma (Torr.) Little UTAH JUNIPER Erect shrub or small tree up to 6 m with stringy, gray-brown bark and spreading or ascending branches. **Leaves** opposite or in whorls of 3, scale-like, 1–2 mm long, yellowish with denticulate margins; juvenile leaves needle-like, 3–5 mm long. **Seed cones** maturing in 1 to 2 years, 1-seeded, blue-brown, 6–8 mm long. Limestone or calcareous sandstone outcrops, slopes and ridges; valleys to montane; Big Horn and Carbon cos. MT south to CA, AZ, NM.

   This is a dominant woodland tree on the south side of the Pryor Mountains where it is at the northern limit of its range.

Juniperus scopulorum Sarg. ROCKY MOUNTAIN JUNIPER Erect shrub or small tree up to ca. 6 m with ascending branches and a conical crown. **Bark** stringy, reddish-brown. **Leaves** green but often with a whitish, glaucous coat, opposite, scale-like, appressed, 1–2 mm long, entire-margined with an acute but not apiculate tip; juvenile leaves needle-like, 2–6 mm long and ascending. **Seed cones** with 1 to 2 seeds, blue with glaucous bloom, 3–6 mm long, maturing in 2 years. Stream

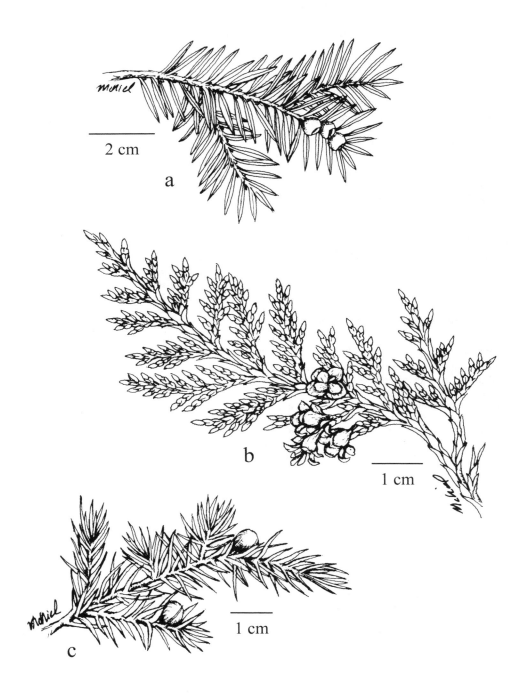

2 cm

a

b

1 cm

c

1 cm

Plate 8. a. *Taxus brevifolia*, b. *Thuja plicata*, c. *Juniperus communis*

terraces, open Douglas fir or ponderosa pine forest, and rocky or barren slopes; plains, valleys to montane. BC, AB, south to AZ, NM.

Rocky Mountain juniper may become the dominant riparian tree along dewatered stream channels.

Thuja L. Arborvitae

Thuja plicata Donn ex D. Don WESTERN REDCEDAR Large evergreen trees to 60 m tall with reddish-brown, fibrous bark and a conical crown, monoecious. **Lower trunk** of older trees broad and often buttressed. **Branchlets** pendant, arrayed in flat sprays. **Leaves** glossy green, scale-like, opposite, overlapping, appressed against the flattened twigs. **Pollen cones** 1–3 mm long, reddish. **Seed cones** ellipsoid, 8–12 mm long, with 4 to 6 pairs of woody, overlapping scales, the top and bottom pairs sterile, maturing the first summer. **Fertile scales** with 2 seeds each. Moist to wet forest, especially along streams; valleys to montane. AK to CA, ID, MT. (p.80)

Western redcedar is long-lived and shade-tolerant and often occurs with cottonwood and spruce along streams and with western hemlock in the wet forests of northwest MT.

TAXACEAE: Yew Family

Taxus L. Yew

Taxus brevifolia Nutt. PACIFIC YEW Dioecious, evergreen shrub or small tree to 10 m tall with spreading or drooping branches forming an open, conical crown. **Bark** thin and brown or purplish. **Leaves** sharp-pointed, yellow-green above, paler beneath, 12–20 mm long, single but borne opposite each other in a single plane. **Pollen cones** solitary, globose, 2–3 mm long. **Seed cones** reduced to a single seed surrounded by a red, fleshy, pea-size, berry-like aril ca. 8 mm across. **Seed** 5–6 mm long. Wet forest; valleys to montane. AK to CA, ID, MT. (p.80)

Yew is very shade-tolerant, grows slowly, and is easily killed by fire; thus, it is usually found only in old forests or sites protected from fire.

DICOTYLEDONS

ARISTOLOCHIACEAE: Dutchman's-pipe Family

Asarum L. Wild ginger

Asarum caudatum Lindl. Rhizomatous, acaulescent, perennial herbs. **Leaves** 2 per node, glabrous or sparsely hairy; the petiole 7–18 cm long; the blade cordate, 4–12 cm long. **Flowers** perfect, solitary, epigynous, pedunculate in leaf axils; petals vestigial; sepals 3, long-attenuate, 2–6 cm, hairy, and brown-purple externally, white and purple internally, spreading, united at the base around the ovary; stamens 12; pistil 6-celled with 6 united styles. **Fruit** a fleshy capsule. Mesic to wet coniferous, often western red-cedar forest, thickets; valleys to montane. BC to CA, ID, MT.

NYMPHAEACEAE: Water-lily Family

Perennial aquatic herbs with thick rhizomes. **Leaves** long-petiolate from the rhizome; blades simple with a cordate base. **Flowers** solitary on long peduncles, perfect, regular with 4 to many separate sepals; 3 to many separate petals; many stamens; and 1 superior ovary with 3 to many carpels. **Fruit** a berry-like capsule with numerous seeds (Wiersema and Hellquist 1997).

Members of this family do not occur in brackish water in MT. The Cabombaceae, including *Brasenia*, is often recognized as a separate family; it does form a distinct clade, but Nymphaeaceae *sensu lato* is monophyletic.[236]

1. Leaf blades without a basal sinus; petioles attached in center . . . . *Brasenia*
1. Leaf blades heart-shaped with petiole attached at a basal sinus. 2

2. Sepals 5–12, yellowish; petals inconspicuous . . . . *Nuphar*
2. Sepals 4, green; petals white, nearly as long as sepals. . . . . *Nymphaea*

Brasenia Schreb. Water shield

Brasenia schreberi J.F. Gmel. **Rhizomes** slender. **Leaves** floating; the blade broadly elliptic, 4–13 cm long; the petiole attached in the center. **Flowers** hypogynous, floating, purple, borne on solitary axillary peduncles; sepals 3, separate, 10–15 mm long; petals 3, separate, slightly longer than sepals; stamens 18 to 36; pistils 9 to 15. **Fruit** a 1-2-seeded capsule 6–10 mm long. Fresh water of lakes, ponds; valleys. BC, AB south to CA, MB and NL south to TX, FL, S. America; Asia, Africa, Australia.

Nuphar Sm. Yellow water lily

Rhizomes large, often branched. **Leaves** submersed to emersed, usually floating; the blades ovate, cordate-based. **Flowers** floating, globose in bud with 5 to 9 yellow or greenish sepals that can turn purplish; numerous inconspicuous, stamen-like petals; and a plate-like stigma.

There appear to be few habitat differences between the two species, although *Nuphar polysepala* does sometimes occur at higher elevations. It has a Cordilleran distribution, while *N. variegatum* is boreal; the two species overlap and intergrade in western Canada and MT. They might be treated as infraspecific taxa but are clearly not closely related to the Old World *N. lutea*.[315]

1. Sepals usually 6, ≤35 mm long. . . . . *N. variegata*
1. Sepals usually 7 to 9, ≥30 mm long . . . . *N. polysepala*

Nuphar polysepala Engelm. [*N. lutea* (L.) Smith ssp. *polysepala* (Engelm.) E.O.Beal] **Leaves** with ovate blades 10–40 cm long. **Sepals** 5 to 12, yellow or greenish, 3–5 cm long. **Fruit** 4–5 cm long. Lakes, ponds, sloughs; valleys, montane, rarely subalpine. AK to CA, NM, CO. (p.84)

Nuphar variegata Engelm. ex Clinton [*N. lutea* (L.) Smith ssp. *variegata* (Durand) E.O. Beal] **Leaves** with broadly ovate blades 6–30 cm long. **Sepals** mostly 6, greenish outside, yellow within, 25–35 mm long. **Fruit** 2–4 cm long. Lakes, ponds, sloughs; valleys, montane. YT to NL south to ID, MT, IA, OH, NY.

Nymphaea L. Water lily

Rhizomes simple or branched. **Leaves** floating; the blades broadly elliptic to orbiculate, cordate-based. **Flowers** floating, ovoid in bud with 4 green sepals; numerous petal-like stamens; and conspicuous white petals borne toward the base of the ovary. **Stigma** with arching, horn-like appendages. **Stamens** yellow.

1a. Sepals ≥3 cm long; petals ≥17 . . . . *N. odorata*
1b. Sepals ≤3 cm long, petals 7 to 15 . . . . *N. leibergii*

Nymphaea odorata Aiton **Leaf blades** nearly orbicular, 8–20 cm long. **Flowers** closing at night; sepals ≥17, 3–7 cm long; petals white, tinged with pink, as long as sepals; stamens 35 to 120. Lakes, ponds; valleys. MB to NL south to TX, FL, Mexico; introduced in BC and AB south to CA, AZ, NM. Our plants are ssp. *odorata*.
Occasionally introduced as an ornamental by people living along lakes.

Nymphaea leibergii Morong [*N. tetragona* Georgi ssp. *leibergii* (Morong) Porsild] **Leaf blades** elliptic to ovate, 5–15 cm long. **Flowers** closing at night; sepals 2–3 cm long; petals 8 to 15, stamen-like, white, as long as sepals; stamens 20 to 40. Lakes, ponds, sloughs; valleys of Flathead and Lake cos. BC to QC south to ID, MT, MI, ME.
Differing from the more northern *Nymphaaea tetragona* by having smaller stigma appendages.[442]

CERATOPHYLLACEAE: Hornwort Family

Ceratophyllum L. Hornwort

Ceratophyllum demersum L. Aquatic, perennial, monoecious herbs without roots. **Stems** flaccid, floating, up to 3 m. **Leaves** 3 to 11 per node, 7–30 mm long, forked into few linear, apiculate-toothed segments. **Flowers** unisexual, solitary in leaf axils; petals and sepals absent; involucre of 8 to 15 simple or forked, linear segments; stamens 3 to 50; pistil solitary. **Fruit** a 1-seeded achene, 4–6 mm long, narrowly elliptic and compressed with 1 terminal and two basal spines, shorter to much longer than the body. Shallow to deep, fresh or brackish water of lakes, ponds, slow streams; plains, valleys. Cosmopolitan, absent from the northernmost parts of N. America.

Plants are usually sterile, reproducing and overwintering by means of vegetative buds (turions).

RANUNCULACEAE: Buttercup Family

Annual or mostly perennial herbs or woody vines. **Leaves** mostly basal and/or alternate. **Flowers** mostly perfect, regular or irregular; sepals mostly 5 to 20, separate, sometimes petal-like; petals mostly (4)5 to 10 or lacking, separate; ovaries 1 to many, 1-celled, superior; stamens 5 to many. **Fruit** an achene or dry carpel (follicle), sometimes a 1- to many- seeded capsule or berry.

Nigella damascena L. (love-in-a-mist), with erect stems and finely dissected leaves, was collected once in Wibaux Co., but it is not clear from the label information whether it was cultivated or naturalized.

1. Fruits dry, 1-chambered, many-seeded, opening at the top and along the edges (follicles),
   in radial clusters of 3 to 20 2
1. Fruit a berry or achene, often with a beak or feather-like style on top 7

2. Leaves 1 to 2 times divided into distinct leaflets. 3
2. Leaves wavy-margined to deeply palmately lobed, without petiolate leaflets 4

3. Petals spurred; sepals >10 mm long . . . . *Aquilegia*
3. Petals without spurs; sepals ≤12 mm long . . . . *Coptis*

4. Flowers regular, dish-shaped 5
4. Flowers bilaterally symmetrical. 6

5. Leaves with wavy margins, not lobed. . . . . *Caltha*
5. Leaves deeply, palmately lobed . . . . *Trollius*

6. Uppermost sepal with a backward-pointing spur . . . .*Delphinium*
6. Spur absent; uppermost sepal helmet-like . . . . *Aconitum*

7. Plants twining vines, woody at least toward the base . . . .*Clematis*
7. Plants herbaceous with erect, creeping or floating stems 8

8. Fruit a red or white berry; petals and sepals inconspicuous; stamens white . . . . *Actaea*
8. Fruit an achene with a persistent style; flowers not as above 9

9. All leaves basal, linear . . . . *Myosurus*
9. Plants with divided or lobed leaves or with stem leaves. 10

10. Stem leafless except for a whorl of 3 divided leaves at the middle . . . . *Anemone*
10. Leaves alternate or all basal. 11

11. Leaves divided into >10 stalked leaflets; male and female flowers sometimes on separate
    plants; sepals and petals inconspicuous or greenish. . . . . *Thalictrum*
11. Leaves entire, lobed, or divided into 3 to 5 stalked leaflets; flowers colored, bisexual. 12

12. Petals lacking; stems ≥50 cm. . . . .*Trautvetteria*
12. Petals present, though sometimes early deciduous; plants often smaller 13

13. Plants terrestrial with leaves 2 to 3 times pinnately dissected into linear segments . . . . *Adonis*
13. Leaves less dissected or aquatic . . . . *Ranunculus*

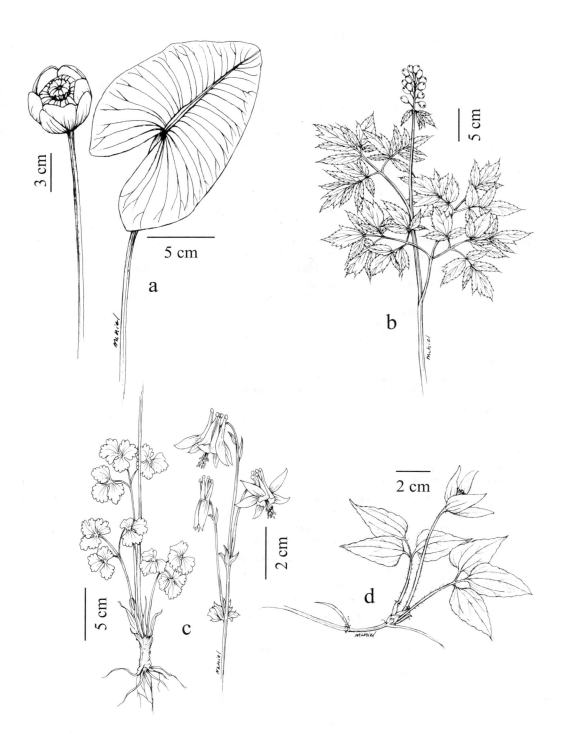

Plate 9. a. *Nuphar polysepala*, b. *Actaea rubra*, c. *Aquilegia flavescens*, d. *Clematis occidentalis*

Aconitum L. Monkshood

Aconitum columbianum Nutt. Perennial with a tuberous crown, fibrous roots and hollow
stems to 180 cm. **Leaves** alternate, petiolate; blades 5–15 cm long, deltoid, deeply divided into
5 to 7 toothed lobes. **Inflorescence** an open leafy-bracteate raceme. **Flowers** irregular; sepals
5, yellow to bluish; uppermost (hood) helmet-like, 15–25 mm high; the lateral pair fan-shaped;
the lower pair much narrower; petals 2, concealed in the hood; stamens many, shorter than the
sepals; pistils 3 to 5; stamens numerous. **Fruit** 3 to 5 follicles, 10–15 mm long, glabrous to
glandular. Moist to wet meadows, open forest, often along streams; montane, subalpine. BC to CA, AZ, TX, Mexico.

Whitish-flowered forms become common south of Missoula; blue-flowered plants are the only form north of
Missoula. The two forms have been considered distinct at the varietal level, but populations with both forms occur
throughout the Bitterroot Range.

Actaea L. Baneberry

Actaea rubra (Aiton) Willd. DOLL'S EYES Fibrous-rooted perennial. **Stems** 30–80 cm. **Leaves**
alternate, 2 to 3 times divided into 3; the ultimate leaflets ovate, 2–12 cm long, with lobed or
toothed margins. **Flowers** ascending-pedunculate in a short raceme that elongates to 10 cm with
age; sepals 3 to 5, white, petal-like, deciduous; petals white, clawed, 2–4 mm long; stamens many,
exceeding the sepals; pistil 1. **Fruit** a several-seeded, red or white berry 4–7 mm long. Moist
forest and thickets, often along streams or with spruce; plains, valleys to lower subalpine. AK to NL south to CA, NM,
SD, OH, NJ. (p.84)

Red- and white-fruited plants may occur in the same population.

Adonis L. Pheasant's eye

Adonis aestivalis L. Glabrous annual. **Stems** 20–60 cm. **Leaves** basal and cauline, ovate in outline, 2 to 3 times
pinnately dissected into linear, ultimate segments. **Flowers** pedicellate, solitary from leaf axils, regular; sepals 5
greenish, 5–10 mm long; petals 6 to 8, orange, 10–17 mm long with a dark, basal spot; stamens and pistils numerous.
Fruit a beaked, wrinkled, glabrous achene, 4–6 mm long borne in a cylindical head 2–3 cm long. Roadsides; valleys. A
garden escape introduced to Granite and Powell cos. and ID south to CA, UT; native to Eurasia.

Anemone L. Windflower, anemone

Herbaceous perennials. **Leaves** basal, petiolate; the blades divided or deeply lobed. **Inforescence** of 1 to
few, long-stalked flowers, subtended by a whorl of leaf-like bracts. **Flowers** lacking petals; sepals 5 to 9,
petal-like; pistils and stamens numerous. **Fruit** an achene with a persistent style (beak) and basal hairs,
clustered in a head.[187]

Anemone quinquefolia L. and *A. virginiana* L. occur in northwestern ND and may occur in adjacent MT;
A. quinquefolia is similar to *A. piperi*; *A virginiana* is similar to *A. cylindrica*, but the achenes are borne in
an ovoid head.

1. Styles of mature achenes >15 mm long; seed heads shaggy; sepals ≥2 cm long. 2
1. Styles <5 mm long; seed heads sometimes cottony but not shaggy; sepals usually <2 cm long 3

2. Sepals blue on the outside; mature styles spreading. . . . . *A. patens*
2. Sepals white (blue-tinged at base); mature styles reflexed . . . . *A. occidentalis*

3. Leaflets of basal leaves ovate to wedge-shaped cut into 0–3 lobes 4
3. Leaflets of basal leaves deeply lobed into several linear segments. 5

4. Leaf-like bracts long-petiolate. . . . . *A. piperi*
4. Bracts sessile on the stem or nearly so . . . . *A. parviflora*

5. Leaf-like bracts at mid-stem sessile or nearly so . . . . *A. canadensis*
5. Bracts at mid-stem short-petiolate 6

6. Cluster of achenes cylindrical, 2–3 cm long. . . . . *A. cylindrica*
6. Head of achenes hemispheric to ovoid, <2 cm long. 7

7. Styles ≤1.5 mm long; ultimate leaf segments with pointed tips . . . .*A. multifida*
7. Styles ≥2.0 mm long; ultimate leaf segments with rounded tips. . . . .*A. drummondii*

Anemone canadensis L. Stems 20–80 cm from extensive rhizomes. **Leaves** appressed-hairy; blades 4–8 cm long with 3 to 5 toothed lobes; ultimate segments 1–4 cm wide. **Inflorescence** of 1 to 3 flowers subtended by 3 sessile, 3-lobed bracts 2–6 cm long. **Flowers** with 4 to 6 white sepals, 10–20 mm long. **Achenes** in an ovoid head, sparsely hairy, 3–6 mm long; beaks straight, ca. 3 mm long. Low areas of moist meadows, around wetlands, roadside ditches; plains. NT to NL south to NM, IA, WV. (p.87)

Anemone cylindrica A.Gray **Stems** hirsute, 20–50 cm from a branched caudex. **Leaves** glaucous, hairy below; blades 2–6 cm long, 3 to 5 times divided to the base into deeply lobed and toothed, oblong segments; ultimate segments 3–10 mm wide. **Inflorescence** of 1 to 4 flowers subtended by ≥3 petiolate bracts similar to the basal leaves. **Flowers** with 5 white (pink tinged) sepals, 8–12 mm long. **Achenes** woolly, 2–3 mm long, in a cylindrical head 2–3 cm long; beaks bent out, ca. 1 mm long. Grasslands, meadows, open forest; plains, valleys, montane. BC to QC south to AZ, KS, IA, PA, NY.

The cylindrical achene head >2 cm long is diagnostic. Sepals are more densely hairy and foliage more glaucous than for *Anemone multifida*.

Anemone drummondii S.Watson [*A. lithophila* Rydb.] **Stems** 5–20 cm from a branched caudex. **Leaves** sparsely long-hairy to glabrous; blades 1–4 cm wide, twice ternately divided and lobed; the ultimate lobes oblong. **Inflorescence** usually 1-flowered, subtended by 2 to 3 petiolate bracts similar to the leaves. **Flowers** with 5 to 7 sepals, white, bluish on the outside, 8–12 mm long. **Achenes** woolly, 2–4 mm long, the beak 2–3 mm long, in a globose head 5–15 mm high. Moist meadows, moraine, stony slopes; near or above treeline. BC, AB south to CA, ID, WY. (p.87)

Rocky Mountain plants have been called var. *lithophila* (Rydb.) C.L.Hitchc. that differs from the more western var. *drummondii* by labile vegetative characters.

Anemone multifida Poir. [*A. globosa* Nutt. ex A.Nelson, *A. tetonensis* Porter ex Britton, *A. multifida* var. *saxicola* B.Boivin] **Stems** 10–50 cm from a branched caudex. **Leaves** sparsely to moderately long-hairy; blades 3–8 cm wide, 2 to 3 times ternately divided and lobed; ultimate segments tapered to the acute tip. **Inflorescence** 1- to several-flowered subtended by 1 to 2 tiers of leaf-like bracts. **Flowers** with 5 to 7 white to rose-purple sepals, 7–15 mm. **Achenes** woolly, 3–4 mm long in a globose head; the beak ca. 1.5 mm long. Grasslands, meadows, open forest; plains, valleys to alpine. AK to NL south to CA, NM, IL, MI, ME. (p.87)

The treatment proposed by Hitchcock, in which two intergradent varieties are distinguished, best reflects our material.[187]

1. Styles straight; ultimate leaf segments ≥3 mm broad; stems to 50 cm . . . . var. *multifida*
1. Styles hooked; ultimate leaf segments ≤3 mm broad; stems ≤20 cm . . . . var. *tetonensis*

Anemone multifida var. *tetonensis* (Porter ex Britton) C.L.Hitchc. occurs at high elevations, while *A. multifida* var. *multifida* occurs up to treeline. Several species and/or varieties have been recognized based on leaf size, style color, number of flowers, density of achene hair and stem height.[53,119] All of these characters seem highly variable within populations and are more likely influenced by environment rather than genetic differences. For example, many specimens have plants with 1 and >1 flowers on the same sheet, and style color may be half yellow and half purple. None of the characters seem strongly correlated in our material.

Anemone occidentalis S.Watson [*Pulsatilla occidentalis* (S.Watson) Freyn] **Stems** from a caudex, long-hairy, 10–60 cm, increasing with maturity. **Leaves** long-hairy; the blades 2–8 cm long, 2 to 3 times ternate to pinnate; ultimate segments narrowly triangular, 1–3 mm wide. **Inflorescence** 1-flowered subtended by 3 petiolate bracts similar to the basal leaves. **Flowers** with 5 to 7, white (tinged with blue) sepals, 2–3 cm long. **Achenes** long-hairy, 3–4 mm long, in an ovoid head 2–8 cm high; beaks plumose, expanding to 2–4 cm at maturity, reflexed down. Meadows, open forest where there is good snow cover; subalpine. BC, AB south to CA, ID, MT. (p.87)

Plants usually bloom close to the ground shortly after snow release, but the mophead-like fruiting heads are conspicuous through much of the short summer.

Anemone parviflora Michx. **Stems** 4–25 cm tall from slender rhizomes. **Leaves** nearly glabrous, the blades, 5–25 mm long, divided into 3 wedge-shaped, shallowly to deeply lobed leaflets. **Inflorescence** 1-flowered, subtended by 2 to 3 deeply lobed, cuneate, sessile bracts 5–30 mm long. **Flowers** with 5 to 6, white or bluish-tinged sepals, 7–25 mm long. **Achenes** woolly, in a globose head 5–12 mm high; style straight, 0.5–2 mm long. AK to NL south to OR, UT, CO. (p.87)

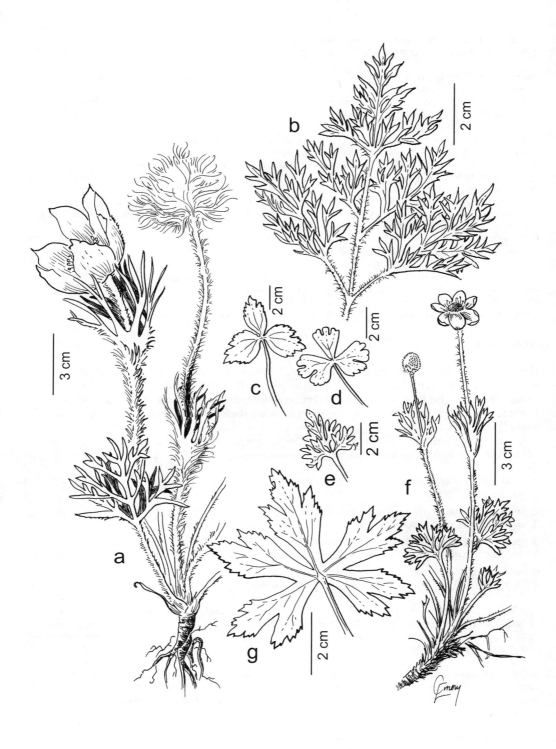

Plate 10. **Anemone**. **a.** *A. patens*, **b.** *A. occidentalis*, **c.** *A. piperi*, **d.** *A. parviflora*, **e.** *A. drummondii*, **f.** *A. multifida* var. *multifida*, **g.** *A. canadensis*

1. Sepals 13–25 mm long; anthers 0.8–1.1 mm long; styles ca. 2 mm long . . . . .var. *grandiflora*
1. Sepals 7–11 mm long; anthers 0.5–0.7 mm long; styles 1–1.5 mm long . . . . .var. *parviflora*

Anemone parviflora var. *grandiflora* Ulbrich occurs on stony, open, calcareous slopes, upper subalpine and alpine; *A. parviflora* var. *parviflora* is found in wet meadows, montane to subalpine. The type specimen of *Anemone parviflora*, collected by Andre Michaux, is in fruit but appears to be the small-flowered form. The large-flowered form occurs mainly north of MT. The two forms intergrade in AK, but are morphologically and ecologically distinct in MT.

Anemone patens L. Pasqueflower, prairie crocus [*A. nuttalliana* DC., *Pulsatilla patens* (L.) Mill.] **Stems** from a branched caudex and woody roots, long-hairy, 5–40 cm, expanding with maturity. **Leaves** long-hairy; blades 3–7 cm wide, twice pinnately divided into linear segments 1–3 mm wide. **Inflorescence** 1-flowered subtended by 3 clasping bracts, 2–4 cm long, deeply divided into linear segments. **Flowers** with 5 to 7 sepals, blue on the outside, blue or whitish on the inside, 2–5 cm long. **Achenes** hairy, 3–4 mm long; beaks plumose, expanding to 2–4 cm at maturity, spreading. Grassland, steppe, open forest; plains, valleys to montane, occasionally to above treeline. Circumboreal south to NM, TX, MO. North American plants are ssp. *multifida* (Pritz.) Hulten. (p.87)

Anemone piperi Britton ex Rydb. **Stems** 10–25 cm from ascending rhizomes. **Leaves** solitary or absent, sparsely hairy to glabrous; the blade with 3 oblanceolate, lobed or toothed leaflets 2–6 cm long. **Inflorescence** 1-flowered, subtended by 3 leaf-like, petiolate, trifoliolate bracts; the oblanceolate leaflets to 7 cm long. **Flowers** with 5 to 7 white sepals 7–15 mm long. **Achenes** in a globose head, short-hairy, 2–4 mm long; the beak ≤1 mm long. Moist coniferous forest; montane to lower subalpine. Endemic to central ID and adjacent BC, OR, MT, disjunct in UT. (p.87)

Aquilegia L. Columbine

Herbaceous perennials from a branched caudex with fibrous roots. **Leaves** long-petiolate, biternate or triternate, the leaflets cuneate and lobed. **Inflorescence** an open, leafy-bracted raceme. **Flowers** regular; sepals 5, petal-like; petals smaller than sepals but expanded behind point of attachment into a nectar-bearing spur; stamens numerous; pistils 5(–10). **Fruit** many-seeded, an aggregate of follicles, opening above.[440]

Flower color is variable within and among species. Length and shape of petals and spurs is a more reliable character to focus on when determining our columbines.

1. Flowers blue to white 2
1. Flowers yellow to red 4

2. Spurs 20–50 mm long; sepals white to pale blue. . . . . *A. coerulea*
2. Spurs ≤15 mm long; sepals blue 3

3. Leaves all basal; petals blue; near or above treeline . . . . *A. jonesii*
3. Leafy stems present; petals whitish; montane . . . .*A. brevistyla*

4. Petal blades ≤5 mm long; sepals usually red. . . . .*A. formosa*
4. Petal blades ≥6 mm long; sepals usually with some yellow . . . . *A. flavescens*

Aquilegia brevistyla Hook. **Stems** 20–60 cm. **Leaves** sparsely hairy; blades twice ternate; ultimate segments 1–4 cm long. **Flowers** few, nodding; sepals yellowish-blue, 10–15 mm long; petals white with a pale blue, hooked spur ca. 5 mm long. **Fruit** 15–25 mm long. Open forest, forest margins; montane. AK to ON south to BC, WY, SD; known only from Judith Basin Co. See *Aquilegia jonesii*.

Aquilegia coerulea E.James **Stems** 30–60 cm. **Leaves** mostly basal, glaucous; blades twice ternate; ultimate segments 1–3 cm long. **Flowers** several, erect; sepals white to pale blue, ca. 3 cm long; petals white with a gently curved spur 3–4 cm long. **Fruit** 2–3 cm long. Moist meadows, open forest; montane, lower subalpine. MT to CA, AZ. Our plants are var. *ochroleuca* Hook.

Aquilegia flavescens S.Watson **Stems** 10–70 cm. **Leaves** glabrous to rarely puberulent with glandular petiolules; blades twice ternate; ultimate segments 15–45 mm long. **Flowers** several, nodding; sepals yellow, 10–20 mm long; petals yellow to deep pink, the straight to incurved spurs 8–12 mm long. **Fruits** glandular, ca. 2 cm long. Moist soil of meadows, open forest, often along streams and on cool, rocky slopes; montane to lower alpine. BC, AB south to OR, UT, CO. (p.84)

Plants with pink petals and spurs are sometimes mistaken for *Aquilegia formosa* which has shorter petals and straight spurs.

Aquilegia formosa Fisch. ex DC. **Stems** 15–50 cm. **Leaves** glabrous; blades twice ternate; ultimate segments 15–40 mm long. **Flowers** several, nodding; sepals red 10–20 mm long; petals yellow, blades ≤6 mm long with straight red spurs 10–15 mm long. **Fruits** 15–25 mm long. Open coniferous forest; montane; known from Beaverhead and Gallatin cos. AK to CA, UT, WY. Our plants are var. **formosa**. See *Aquilegia flavescens*.

Aquilegia jonesii Parry **Stems** 3–10 cm high. **Leaves** glaucous, short-villous, all basal; blades twice ternate; ultimate segments overlapping, 2–10 mm long. **Flowers** solitary, erect, borne near the ground; sepals blue, 10–20 mm long; petals light blue with straight spurs ca. 10 mm long. **Fruits** glabrous, 10–20 mm long with long beaks. Stony, calcareous soil of ridges, outcrops, talus slopes, often in cushion plant communities; alpine, occasionally to the montane zone in exposed sites. Endemic from southern AB to northern WY.

Putative hybrids between *Aquilegia jonesii* and *A. flavescens* occasionally occur in northwest MT and have been called *A. jonesii* ssp. *elatior* Standley; these have been mistaken for *A. brevistyla*.

Caltha L. Marsh marigold

Caltha leptosepala DC. [*C. biflora* DC.] Herbaceous perennials from a simple caudex. **Stems** to 15 cm in flower, expanding in fruit. **Leaves** mainly basal, fleshy, alternate, simple, petiolate; blades 2–8 cm long, narrowly cordate-ovate, crenate. **Flowers** 1 to 2, dish-shaped; sepals 7 to 10, white, often tinged with blue outside, 7–20 mm long; petals absent; stamens ≥10, pistils ≥5. **Fruit** 4 to 15 aggregated, ascending, many-seeded follicles, 1–2 cm long. Wet soil of streambanks, meadows where snow melt accumulates; lower subalpine to alpine. AK south to CA, AZ, NM.

Clematis L. Virgin's bower

Rhizomatous woody vines or herbaceous perennials with prostrate stems. **Leaves** alternate, rarely opposite, cauline, petiolate; blades ternately or pinnately divided. **Inflorescence** axillary or terminal, of 1 to many flowers. **Flowers** perfect or unisexual, radially symmetrical, pedunculate; sepals 4; petals absent; stamens many; pistils 5 to many, 1-seeded. **Fruit** an aggregate of feathery-beaked achenes.[328]

1. Sepals white, 6–10 mm long . . . . *C. ligusticifolia*
1. Sepals blue, >20 mm long 2

2. Leaves pinnate, ultimate segments linear; sepal tips reflexed . . . . *C. hirsutissima*
2. Leaves ternate, larger ultimate leaflets lanceolate; sepals spreading but not reflexed 3

3. Leaves once divided into 3 leaflets . . . . *C. occidentalis*
3. Leaves 2 to 3 times ternate into >3 leaflets . . . . *C. columbiana*

Clematis columbiana (Nutt.) Torr.& A.Gray **Stems** trailing or forming rhizomatous mats. **Leaf blades** 2- to 3-times ternate into lobed, lanceolate leaflets 1–3 cm long. **Flowers** bell-shaped, perfect, solitary, terminal, nodding; sepals deep blue, 25–40 cm long. **Achene beaks** 2–5 cm long. MT, ND south to AZ, NM, TX.

1. Stems trailing; leaf blades twice ternate . . . . var. *columbiana*
1. Stems tufted; leaf blades 3 times ternate . . . . var. *tenuiloba*

Clematis columbiana var. *columbiana* [*C. pseudoalpina* (Kuntze) A.Nelson] occurs in montane open forest. *Clematis columbiana* var. *tenuiloba* (A.Gray) J.S.Pringle [*C. tenuiloba* (A.Gray) C.L.Hitchc.] is found on exposed ridges and in open forests, montane to occasionally alpine. Fruits mature in late summer or autumn; both vars. are strongly associated with calcareous soil.

Clematis hirsutissima Pursh Sᴜɢᴀʀ ʙᴏᴡʟs **Stems** erect, 15–50 cm at maturity. **Leaves** opposite, long-hairy; blades 2 to 3 times pinnately divided into linear segments <5 mm wide. **Flowers** urn-shaped, perfect, nodding, solitary, terminal; sepals violet, 20–45 mm long, tips recurved. **Achene beaks** 2–6 cm long. Grasslands, steppe, open forest; valleys to subalpine. WA to MT south to AZ, NM, OK.

Clematis ligusticifolia Nutt. **Stems** limber, 3–6 m long, sometimes covering shrubs and small trees. **Leaf blades** glabrous, divided into 3 to 7 ovate, lobed leaflets, 2–6 cm long. **Inflorescence** a several-flowered, bracteate panicle from leaf axils. **Flowers** unisexual; sepals spreading, white, 5–10 mm long. **Achene beaks** 2–5 cm long. Open forest, thickets, floodplains; plains, valleys, montane. BC to MB south to CA, NM, NE.

Clematis occidentalis (Hornem.) DC. [*C. columbiana* (Nutt.) Torr.& A.Gray misapplied] **Stems** limber, to 2 m long, often climbing. **Leaf blades** sparsely long-hairy, divided into 3 cordate-lanceolate leaflets, 2–7 cm long with obscurely toothed margins. **Flowers** bell-shaped, perfect, nodding, solitary from leaf nodes; sepals blue, 2–5 cm long. **Achene beaks** 2–5 cm long. Open to closed forest, thickets, often along streams, occasionally on talus; valleys, montane, occasionally higher. BC to NB south to OR, UT, CO, OH. Our plants are var. ***grosseserrata*** (Rydb.) J.S.Pringle. (p.84)

Coptis Salis. Gold thread

Coptis occidentalis (Nutt.) Torr.& A.Gray Rhizomatous, herbaceous perennials with leafless stems 10–25 cm high in fruit. **Leaves** glabrous, petiolate, trifoliolate; leaflets dentate, often 3 to lobed, 2–7 cm long. **Inflorescence** a few-flowered, bractless cyme. **Flowers** perfect or staminate, star-shaped; sepals 5 to 7, white, deciduous, 9–12 mm; petals 5–7, white, clawed with a nectary near the base of the blade, shorter than sepals; stamens 10 to many, pistils 4 to 15, several-seeded. **Fruit** an aggregate of oblong follicles with gently recurved beaks. Moist to wet forests, shrub fields, often with spruce or red cedar; montane, subalpine. Endemic to north ID, adjacent WA, and MT.

Flowers are so early they are rarely seen.

Delphinium L. Larkspur

Herbaceous perennials (one garden-escape annual) from fibrous or tuberous roots. **Leaves** petiolate; the blades orbicular, palmately divided into lobed segments. **Inflorescence** terminal, simple or compound racemes. **Flowers** perfect, bilaterally symmetrical; sepals 5, deciduous, the upper long-spurred; petals 4; the upper pair lanceolate, spurred, enclosed in upper sepal; the lower pair clawed with an ovate to rotund blade; stamens many; pistils 3(–5), several-seeded. **Fruit** an aggregate of beaked follicles.

Characters used to separate species of *Delphinium* vary continuously,[434] and intermediate specimens are not uncommon. *Delphinium bicolor* and *D. nuttallianum* occur in similar habitats and are sometimes difficult to distinguish. Warnock believes that *D. distichum* is the correct name for what has usually been called *D. burkei* because he believes the type of *D. burkei* is a hybrid between *D. depauperatum* and *D. nuttallianum*,[419] but he provides no evidence. I have chosen to conserve the older name[126,187] pending further information. Plants from northern Flathead Co. span the range of variation between *D. burkei* and *D. depauperatum*. Warnock also rejects the name *D. occidentale* because he believes the type is a hybrid between *D. barbeyi* (Huth) Huth and *D. glaucum*; he places plants formerly called *D. occidentale* into *D. glaucum*.[419] However, recent research agrees with older treatments and places our most common tall larkspur in *D. occidentale*.[434] *Delphinium glaucescens* is described as having ample basal leaves divided into narrow ultimate segments;[126,187] however, an isotype at MONTU has less divided leaves with none at the base, seemingly intermediate between *D. occidentale* and "typical" *D. glaucescens*. Such intermediate plants have been collected at higher elevations in the mountains of the Big Hole and Ruby rivers. Dorn merged *D. glaucescens* into *D. glaucum*;[113] however, these two seem distinct in MT. Reports of *Delphinium andersonii* Gray for MT were based on misidentified specimens.

1. Plants annual, garden escapes; ultimate leaf divisions <2 mm wide . . . .*D. ajacis*
1. Plants perennial, native species; ultimate leaf divisions various 2

2. Roots fibrous, often woody. 3
2. Roots relatively short, tuberous 7

3. Stems usually <3 mm wide and solid at mid-length 4
3. Stems obviously hollow and often ≥4 mm wide at mid-length 5

4. Lower pedicels usually much longer than their flowers . . . . *D. bicolor*
4. Pedicels and flowers ca. equal in length . . . . *D. geyeri*

5. Mid-inflorescence flowers shorter than their pedicels. . . . . *D. glaucum*
5. Mid-inflorescence flowers longer their pedicels 6

6. Inflorescence with <30 flowers; basal leaves numerous at flowering. . . . . *D. glaucescens*
6. Inflorescence with >25 flowers; basal leaves withered at flowering . . . . *D. occidentale*

7. Lower petals cleft <1/4 their length; upper petals white with blue lines;
  roots mostly fibrous, extensive . . . . *D. bicolor*
7. Lower petals cleft 1/4–1/2 their length; upper petals not as above; roots tuberous 8

8. Lower flowers shorter than their stalks; inflorescence pyramidal. . . . .*D. nuttallianum*
8. Lower flowers as long or longer than their stalks; inflorescence cylindrical 9

9. Basal leaves present; stem leaves <5; flowers not overlapping. . . . .*D. depauperatum*
9. Basal leaves absent at flowering; stem leaves >5; flowers crowded, overlapping . . . . *D. burkei*

Delphinium ajacis L. [*Consolida ajacis* (L.) Schur] Annual. **Stems** glabrous to sparsely puberulent, 30–80 cm. **Leaves** basal and cauline; the blades circular, 1–5 cm wide; ultimate lobes to 2 mm wide. **Inflorescence** 6- to 30-flowered with leaf-like bracts; pedicels ascending, 1–3 cm long. **Flowers** with spreading, blue (white) sepals 8–18 mm long, ca. equal; spur 12–20 mm long, curved up; petals united into 2, blue to whitish, 2-lobed, 5–8 mm long. **Fruits**, puberulent, 12–25 mm long. Fields, roadsides; valleys. An escaped ornamental introduced in eastern U.S. and sporadically elsewhere; native to Europe.

Delphinium bicolor Nutt. **Stems** glabrous to puberulent and glandular, 10–40 cm from woody, fibrous roots. **Leaves** mostly basal; the blades 2–10 cm wide; ultimate lobes to 6 mm wide. **Inflorescence** 3- to 12-flowered; lower flowers shorter than the puberulent pedicels. **Flowers** with spreading, deep blue sepals 10–25 mm long, the lower pair longer; upper petals white with blue markings; lower petals whitish to deep blue, the blade cleft ≤1/4 the length. **Fruits** puberulent, 12–22 mm long. Grasslands, steppe, open forest, stony ridges. BC to SK south to WA, WY, SD. (p.92)

1. Upper petal tips light blue; veins not obvious. . . . . ssp. *calcicola*
1. Upper petals with deep blue veins . . . . ssp. *bicolor*

Delphinium bicolor ssp. *calcicola* M.J.Warnock & Vanderhorst has larger flowers on average and is endemic to usually shallow, calcareous soils in the montane zone of southwest and south-central Montana. *Delphinium bicolor* ssp. *bicolor* occurs at all elevations and in all types of soil throughout the state.

Delphinium burkei Greene [*D. distichum* Geyer ex Hook.] **Stems** puberulent, 50–80 cm tall from a cluster of tuberous roots. **Leaves** all on stem at flowering time; the blades to 6 cm wide; ultimate segments to 2 mm wide. **Inflorescence** cylindric with 12 to 30 overlapping flowers as long or longer than pedicels. **Flowers** with forward-pointing sepals 7–9 mm long, blue and whitish toward the base; petals blue to white, the lower blades cleft 1/2-way to the base. **Fruits** ca. 15 mm long, puberulent. Moist meadows, steppe; valleys, lower montane. BC to OR east to ID, MT.

Delphinium depauperatum Nutt. **Stems** mostly glabrous, 15–60 cm from a cluster of tuber-like roots. **Leaves** basal and on lower stem; blades, 2–6 cm wide; ultimate lobes to 3 mm wide. **Inflorescence** often branched, 3- to 15-flowered, lower flowers as long to longer than pedicels. **Flowers** with spreading, dark blue, sparsely hairy sepals, 8–12 mm; petals blue to white, the lower blades cleft 1/2-way to base. **Fruits** 11–17 mm, puberulent, often glandular. Moist meadows, often along streams; montane; Beaverhead and Flathead cos. WA to MT south to CA, NV.

Delphinium geyeri Greene **Stems** puberulent, 40–60 cm from woody roots. **Leaves** basal and cauline; the blades 3–10 cm wide; ultimate segments 2–5 mm wide. **Inflorescence** cylindrical with 10–30 overlapping flowers, ca. as long as the pedicels. **Flowers**: sepals 1–2 cm long, blue; lower petals blue, cleft less than 1/4-way to the base; upper petals white with blue tips. **Fruits** 11–15 mm long, puberulent. Grasslands, steppe; valleys in Carbon Co. ID, MT, WY, UT, CO.

Delphinium glaucescens Rydb. **Stems** hollow near the base, glabrous, glaucous, 25–60 cm from woody roots. **Leaves** numerous; blades 4–8 cm wide; ultimate lobes narrow, to 3 mm wide. **Inflorescence** with usually <30 flowers, most longer than their pedicels. **Flowers** with dark blue, nearly glabrous sepals 6–12 mm long, the tips cupped forward; petals deep blue; the upper with white margins; the lower cleft 1–2 mm long. **Fruits** 10–15 mm long, glabrous to puberulent.

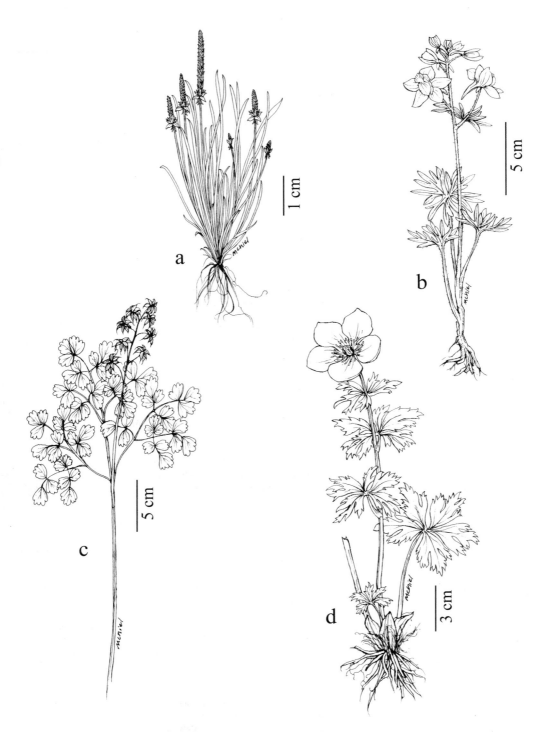

Plate 11. a. *Myosurus minimus*, b. *Delphinium bicolor* var. **bicolor**, c. *Thalictrum occidentale*, d. *Trollius albiflorus*

Meadows, grasslands, steppe; montane to near treeline. Endemic to northwest WY and adjacent ID and MT. See genus discussion.

Delphinium glaucum S.Watson **Stems** hollow, glabrous, glaucous, 35–100 cm from woody roots with sparsely hairy foliage. **Leaves** numerous, mainly cauline; blades to 20 cm wide, deeply lobed into ca. 5-lobed and toothed segments. **Inflorescence** loosely many-flowered with lower pedicels much longer than flowers. **Flowers** with deep purple, glabrous to pubescent sepals 6–8 mm long; upper petals bluish-white; lower petals cleft 1–2 mm. **Fruits** ca. 10 mm long, usually glabrous. Wet tall-herb meadows, thickets, often near streams; upper montane, lower subalpine. AK south to CA, and Mineral Co., MT.

Originally thought to be restricted to the Cascade-Sierra and west, recent studies concluded that it is also found in southeast BC and adjacent MT.[257]

Delphinium nuttallianum Pritz. [*D. nuttallianum* var. *fulvum* C.L.Hitchc., *D. sutherlandii*
M.J.Warnock] **Stems** glabrous to hairy, 15–60 cm from tuber-like roots. **Leaves** few, mostly on lower stem; blades 2–9 cm wide; ultimate lobes to 6 mm wide. **Inflorescence** 4- to 18-flowered; lower flowers shorter than pedicels. **Flowers** with spreading, deep blue sepals 10–15 mm long; the lower ones longer; petals white to sometimes bluish; the lower blades cleft 1/4- to 1/2-way to base. **Fruits** 10–18 mm, hairy. Vernally moist, sometimes shallow soil of meadows, grasslands, rock outcrops, open forest; valleys to alpine. BC, AB south to CA, AZ, CO, NE.

Several varieties as well as *Delphinium sutherlandii* have been recognized based on petal color, vestiture, and inflorescence size.[126,187,419] These characters do not seem correlated in our material and vary within populations.

Delphinium occidentale (S.Watson) S.Watson [*D. glaucum* S.Watson misapplied] **Stems**
glabrous, glaucous, hollow, 60–120 cm from deep fibrous roots. **Leaves** numerous; blades 5–15 cm wide; ultimate lobes to 15 mm wide, mucronate. **Inflorescence** with more than 25 flowers, often branched with lower flowers bracteate, longer than pedicels. **Flowers** with purple to white, puberulent sepals 6–12 mm long, pointing forward; petals blue to whitish, the lower barely cleft. **Fruits** 10–16 mm long, puberulent. Meadows, moist grasslands, steppe; montane, subalpine. OR to MT south to UT, CO.

The white-flowered form has been called var. *cuculatum* Davis; it is common in the southwest part of the state. See discussion under genus.

Myosurus L. Mousetail

Annuals with leafless scapes. **Leaves** all basal, indistinctly petiolate, narrow with entire margins. **Flowers** terminal, solitary, perfect, radially symmetrical; sepals 5, green, narrowly elliptic, not persistent, with a spur about as long as the blade; petals inconspicuous or absent; stamens 5 or 10; pistils 1-seeded, numerous on a conical receptacle. **Fruit** an aggregate of beaked achenes.[439]

Mature fruits are needed for determination of species.

1. Achene beak ≤0.5 mm, inconspicuous. . . . . *M. minimus*
1. Achene beak >0.5 mm, apparent . . . . *M. apetalus*

Myosurus apetalus C.Gay [*M. aristatus* Benth. ex Hook.] **Scapes** 1–5 cm. **Leaf blades** linear
to oblanceolate, 9–30 mm. **Flowers** with scarious-margined sepals ca. 2 mm long. **Achenes** 1–2 mm long with a beak 0.6–1.2 mm long, in a cluster 5–16 mm long. Vernally wet soil around ponds; plains, valleys. BC to MB south to CA, AZ, NM.

1. Head of achenes 4–9 mm; sepals with broad scarious margin . . . . var. *borealis*
1. Achene head 11–26 mm; sepals with narrow scarious margin. . . . . var. *montanus*

Myosurus apetalus var. *borealis* Whittemore occurs in western valleys; *M. apetalus* var. *montanus* (G.R.Campb.) Whittemore has been collected in north-central counties and disjunct in Beaverhead Co.

Myosurus minimus L. **Scapes** 3–10 cm. **Leaf blades** linear, 10–70 mm long. **Flowers** with
sepals 1.5–3 mm long and inconspicuous scarious margins. **Achenes** 1–2 mm long with a beak ≤0.5 mm long, in a cluster 5–25 mm long. Vernally wet soil around ponds; plains, valleys. BC to ON south to CA, Mexico, TN, SC; Europe, Asia, Africa. (p.92)

Ranunculus L. Buttercup, crowfoot

Perennial (occasionally annual) herbs with fibrous, often fleshy (rarely tuberous) roots. **Stems** erect or prostrate (floating) and rooting at the nodes. **Leaves** basal and long-petiolate and/or alternate and reduced up the stem. **Flowers** perfect, radially symmetrical, long-stalked, solitary or in an open, bracteate inflorescence; sepals 5, greenish; petals usually 5, ovate, short-clawed, yellow (rarely white) with a basal nectary; stamens usually numerous; pistils 5-many. **Fruit** a cluster (head) of beaked, ovate achenes.[43,438]

Mature fruits as well as flowers are often necessary for positive identification. *Ranunculus arvensis* L., an introduced annual with spiny achenes, is known from north Idaho and could occur in MT. *Ranunculus flabellaris* Raf., with prostrate stems and more divided leaves than *R. sceleratus*, occurs in adjacent ND. *Ranunculus adoneus* A.Gray, a Southern Rocky Mountain species with finely dissected leaves, is reported for south-central MT,[438] but I have seen no collections to voucher this report. *Ranunculus andersonii* A.Gray with parsley-like leaves and a solitary flower occurs in adjacent east-central ID.

1. Stems floating in water or prostrate and rooting at the nodes or from stolons 2
1. Plants not aquatic; stems ascending or erect 7

2. Leaf blades lance-shaped with entire margins . . . . *R. flammula*
2. Leaf blades toothed, lobed, or divided 3

3. Flowers white; leaf blades divided into thread-like segments; aquatic plants . . . .*R. aquatilis*
3. Flowers yellow; some ultimate leaf segments ≥1 mm wide; aquatic or terrestrial .4

4. Leaf blades merely crenate, not lobed or divided . . . .*R. cymbalaria*
4. Leaf blades lobed or divided 5

5. Leaf blades divided into stalked leaflets; lawn and roadside weed . . . . *R. repens*
5. Some leaf blades lobed but not divided into leaflets; mud or shallow water 6

6. Leaves with 3 to 5 lobes; petals ca. 2 mm long . . . . *R. hyperboreus*
6. Leaves with >5 lobes; petals 3–10 mm long . . . . *R. gmelinii*

7. Basal leaf blades divided into leaflets or lobed ≥1/2-way to midvein 8
7. Margins of some basal leaves entire, toothed, or shallowly lobed 25

8. Achenes hairy 9
8. Achenes glabrous or nearly so 12

9. Plant annual; achene with blade-like beak ca. 3 mm long . . . . *R. testiculatus*
9. Perennial; achene beak <2 mm long 10

10. Achene body ca. 1 mm long; plants <10 cm high . . . .*R. jovis*
10. Achene body 2–3 mm long; most plants >10 cm high 11

11. Basal leaves with 3 obovate lobes; achenes 2–3 mm long . . . .*R. uncinatus*
11. Basal leaves divided into >3 linear lobes; achenes 2 mm long . . . .*R. pedatifidus*

12. Basal leaf blades divided to the midrib into leaflets 13
12. Basal leaf blades entire to lobed, not divided completely to the midrib into leaflets 18

13. Achene with a hooked beak 0.5-1 mm long 14
13. Achene beak straight to slightly curved, 1–5 mm long 16

14. At least some plants with decumbent stems . . . . *R. repens*
14. All stems erect to ascending 15

15. Achene beak ca. 0.5 mm long; sepals spreading . . . . *R. acris*
15. Achene beak ca. 1 mm long; sepals reflexed . . . . *R. acriformis*

16. Petals >8 mm long; achene beak >2 mm long . . . .*R. orthorhynchus*
16. Petals 2–6 mm long; achene beak <2 mm long 17

17. Achene beak <1 mm long; petals shorter than sepals . . . .*R. pensylvanicus*
17. Achene beak ≥1 mm long; petals ca. equal to sepals . . . . *R. macounii*

18. Stems ≤5 cm high; flowers 1 per stem . . . . *R. pygmaeus*
18. Stems >5 cm long 19

19. Beak of achene ca. ≥1 mm long 20
19. Beak of achene ≤0.8 mm long 23

Ranunculus abortivus L. Plants annual or short-lived perennial. **Stems** erect, 10–60 cm with mostly glabrous foliage. **Basal leaf blades** orbiculate, 1–4 cm long with scalloped margins. **Flowers** with petals ca. 2 mm long, shorter than the glabrous, deciduous sepals. **Achenes** glabrous, ca. 1.5 mm long with a curved beak <0.2 mm long; head globose, 3–6 mm high. Moist soil of open, often deciduous woodlands, thickets, frequently in riparian areas; plains, valleys, montane. Most of temperate N. America, excluding the Southwest. (p.98)
   Our plants are sometimes segregated as var. *acrolasius* Fern.

Ranunculus acriformis A.Gray **Stems** 12–60 cm, erect with hirsute foliage. **Basal leaf blades** cordate, 2–6 cm long, ternate, the sessile leaflets twice lobed. **Flowers** with petals 7–13 mm long, ca. twice as long as the hirsute sepals. **Achenes** compressed, 2–3 mm long, glabrous with a hooked beak ca. 1 mm long; heads hemispheric, 5–8 mm high. Moist grassland, meadows, steppe, often near streams; montane, subalpine. ID, MT south to UT, CO. Our plants are var. *montanensis* (Rydb.) L.D.Benson.
   The similar *Ranunculus acris* has a smaller achene beak and more numerous ultimate leaf segments.

Ranunculus acris L. TALL BUTTERCUP. **Stems** 25–80 cm, erect with hirsute foliage. **Basal leaf blades** pentagonal, 3–6 cm long, ternate, the sessile leaflets deeply lobed into oblong, dentate segments. **Flowers** numerous in a diffuse inflorescence; petals 8–10 mm, twice as long as the hirsute sepals. **Achenes** glabrous, 2–3 mm long with a curved beak 0.5 mm long; head globose, 6–9 mm high. Moist meadows, especially irrigated hay fields, and along roads, less abundant in wildland habitats; valleys, occasionally higher. Introduced AK to NL south to CA, AZ, TN, GA; South America, Eurasia, Australia. See *R. acriformis*. (p.98)

Ranunculus alismifolius Geyer ex Benth. **Stems** erect, 7–40 cm with glabrous to hirsute foliage. **Basal leaf blades** 1–10 cm long, lanceolate to narrowly ovate with entire margins. **Flowers** with petals 4–12 mm long, twice as long as the glabrous to hirsute sepals. **Achenes** glabrous to sparsely hairy, ca. 2 mm long with a straight or bent beak 1 mm long; heads hemispheric, 4–6 mm high. Moist to wet meadows, margins of streams, wetlands, rocky places; montane to subalpine. BC south to CA, NV, UT, CO.

1. Stems >3 mm thick, usually >30 cm high; petals ≥8 mm long . . . . var. *alismifolius*
1. Stems ≤3 mm thick, <40 cm high; petals ≤8 mm long 2

2. Foliage glabrous; leaf blades ovate to elliptic. . . . . var. *alismellus*
2. Pedicels and petioles pubescent; leaf blades lanceolate . . . . var. *davisii*

Ranunculus alismifolius var. *alismifolius* is generally found in wet, montane habitats; *R. alismifolius* var. *alismellus* Gray occurs in vernally wet subalpine habitats; *R. alismifolius* var. *davisii* Benson occupies permanently moist to wet subalpine to occasionally montane habitats.

Ranunculus aquatilis L. [*R. longirostris* Godr., *R. subrigidus* W.B.Drew, *R. circinatus* Sibth., *R. aquatilis* var. *capillaceus* (Thuill.) DC.] WATER CROWFOOT Plants aquatic. **Stems** floating with glabrous to sparsely hairy foliage. **Leaf blades** 10–30 mm long, reniform, twice divided into numerous thread-like segments. **Flowers** solitary at leaf nodes; petals white, 4–9 mm long, twice as long as the glabrous, deciduous sepals. **Achenes** glabrous or sparsely hairy, cross-corrugate, 1–2 mm long with a straight beak ca. 0.5 mm long; heads hemispheric, 3–5 mm high. Shallow water of ponds, lakes, streams at all elevations. Circumboreal south to most of N. America. (p.98)
   Our only white-flowered buttercup. *Ranunculus aquatilis* var. *hispidulus* W.B.Drew has been recognized based on the presence of broadly-lobed floating leaves (Whittemore 1997a); however, varietal designation cannot be assigned for completely submersed specimens. Benson recognized *R. longirostris* based on beak length and *R. subrigidus* based on petiole length,[43] characters that vary too continuously in our material to be taxonomically useful.

Ranunculus cardiophyllus Hook. **Stems** hollow, erect, 10–40 cm high with pilose foliage. **Basal leaf blades** 2–5 cm long, cordate with toothed margins. **Flowers** with petals 6–12 mm long, longer than the pilose sepals. **Achenes** short-hairy, ca. 2 mm long with a hooked beak 0.5–1 mm long; heads narrowly ovate, ca. 10 mm long. Moist grasslands; montane. BC to SK south to WA, NM, SD.
   This species is common in the Sweetgrass Hills. Intermediates between *Ranunculus cardiophyllus* and the similar *R. pedatifidus* and *R. inamoenus* are reported for the east side of Glacier National Park.[242]

Ranunculus cymbalaria Pursh Plants stoloniferous. **Stems** erect to decumbent, 2–25 cm with glabrous, entirely basal leaves. **Leaf blades** 1–4 cm long, cordate with scalloped margins. **Flowers** with petals 2–7 mm long, barely longer than the sepals. **Achenes** ribbed, glabrous, 1–2 mm long with a straight beak ca. 0.2 mm long; heads ovoid to cylindric, 5–12 mm high. Mud around ponds, wetlands, streams; plains, valleys, montane. AK to Greenland south to much of N. and S. America; Asia. (p.98)
   Ranunculus cymbalaria var. *saximontanus* Fernald has been segregated by its larger size, but some of our specimens have both vars. on the same sheet.

Ranunculus eschscholtzii Schltdl. **Stems** ascending to erect, 3–30 cm with glabrous foliage. **Basal leaf blades** broadly ovate to reniform, 5–40 mm long, deeply palmately lobed, the segments lobed; stem leaves 0 to 2, divided into linear lobes. **Flowers** few with petals 4–12 mm long, ca. twice as long as the sparsely hairy, often purplish sepals. **Achenes** glabrous, 1–2 mm long with a straight beak 0.5–2 mm long, heads ovoid to cylindric, 5–13 mm high. Moist to wet meadows, stony slopes; subalpine, alpine. AK south to CA, NM; Asia. (p.98)

1. Ultimate basal leaf segments rounded at the tip . . . . var. *eschscholtzii*
1. Ultimate basal leaf segments acute or acuminate 2

2. Basal leaf blades cuneate, tapering to the petiole . . . . var. *eximius*
2. Basal leaf blades cordate, not tapered to the petiole . . . . .var. *suksdorfii*

Ranunculus eschscholtzii var. *eschscholtzii* is usually ≤10 cm high with achene beaks ca. 0.5 mm long and is found near or above treeline; *R. eschscholtzii* var. *suksdorfii* (A.Gray) L.D.Benson [*R. suksdorfii* A.Gray] and *R. eschscholtzii* var. *eximius* (Greene) L.D.Benson [*R. eximius* Greene] are usually >8 cm high with achene beaks 1–2 mm long. Our most common high-elevation buttercup, the 3 vars. are ecologically and geographically similar. *Ranunculus eschscholtzii* var. *trisectus* (Eastw.) L.D.Benson with numerous rounded leaf segments occurs in northwest WY.

Ranunculus flammula L. [*R. reptans* L., *R. flammula* var. *ovalis* (Bigelow) L.D.Benson, *R. flammula* var. *filiformis* (Michx.) Hook., *R. flammula* var. *reptans* (L.) E.Mey.] **Stems** slender with glabrous foliage, prostrate and rooting at the nodes or occasionally erect. **Leaf blades** lanceolate, tapered to the petiole, 5–60 mm long with entire margins. **Flowers** 1 per node; petals 3–5 mm long, slightly longer than the sparsely hairy sepals. **Achenes** glabrous, 1–2 mm long with a beak <0.5 mm long; heads hemispheric, 2–4 mm high. Mud along streams, lakes, ponds, tolerant of alkaline and saline soils; montane to subalpine. Circumboreal south to CA, NM, MN, NJ.

Recognition of varieties based on stature and leaf width seems untenable since these characters are continuous and can vary within populations.

Ranunculus gelidus Kar.& Kir. [*R. verecundus* Robins] **Stems** lax or ascending, 5–20 cm with glabrous foliage. **Basal leaf blades** reniform, 5–20 mm long, deeply lobed into 3 shallowly-lobed segments. **Flowers** with petals 3–6 mm long, barely longer than the sepals. **Achenes** glabrous, 1.5–2 mm long with a hooked or straight beak 0.5 mm long; heads cylindric, 4–12 mm high. Stony, often moist, sparsely vegetated soil of benches, moraine, open slopes; near or above treeline. AK to AB south to OR, ID, MT.

Ranunculus verecundus has been recognized as a species separate from *R. gelidus*.[43,187,430] However, others do not recognize *R. verecundus* as a separate species.[198,325,438] The two species supposedly differ in stem height and achene size: *R. gelidus* is ca. 5 cm high with achenes ca. 2.5 mm; *R. verecundus* is 8–20 cm high with achenes 1–1.5 mm.[43] AK material of *R. gelidus* varies in height with achenes 2–3 mm (fide D. Murray), and MT material is 5–20 cm high with achenes 1.5–2.0 mm. Plants from Glacier National Park have achenes to 2.0 mm long, while those from the Beartooth have achenes ca. 1.5 mm. It appears that this complex forms a gradual and variable cline rather than two distinct entities.[438] See *Ranunculus pedatifidus*.

Ranunculus glaberrimus Hook. SAGEBRUSH BUTTERCUP **Stems** ascending, 5–15 cm with glabrous foliage. **Basal leaf blades** 1–3 cm long, narrowly elliptic to ovate with entire or lobed margins. **Stem leaves** deeply lobed. **Flowers** with petals 6–20 mm long, twice as long as the glabrous sepals. **Achenes** short-hairy, 1–2 mm long with a straight (curved) beak 0.5–1 mm long; heads hemispheric, 5–10 mm high. Grasslands, steppe, open forest, rocky outcrops; plains, valleys, montane, rarely higher. BC to SK south to CA, NM, NE. (p.98)

1. Basal leaf blades elliptic with entire margins; middle lobe of bract the largest. . . . . var. *ellipticus*
1. Basal leaf blade obovate, shallowly lobed; bract lobes ca. equal . . . . var. *glaberrimus*

Ranunculus glaberrimus var. *glaberrimus* is the common form west of the Continental Divide; *R. glaberrimus* var. *ellipticus* Greene is most common east of the Divide.

Ranunculus gmelinii DC. [*R. gmelinii* var. *hookeri* (G.Don) L.D.Benson, *R. gmelinii* var. *limosus* (Nutt.) H.Hara] Plants with glabrous to hirsute foliage, often aquatic. **Stems** hollow, floating or prostrate (rarely erect) and rooting at the nodes. **Leaf blades** reniform, 1–4 cm long, deeply 3-lobed into lobed segments; submerged leaves finely dissected into linear segments. **Flowers** with petals 3–10 mm long, slightly to much longer than the mostly glabrous sepals **Achenes** flattened, 1–2 mm long, glabrous with a straight beak 0.5–1 mm long; heads globose, 4–7 mm high. Mud or shallow water of ponds, lakes, slow streams; plains, valleys, montane, rarely higher. AK to NL south to NM, WY, IL, ME; Asia.

Ranunculus hyperboreus Rottb. [*R. natans* C.A. Mey., *R. natans* var. *intertextus* (Greene) L.D.Benson] **Stems** prostrate, rooting at the nodes, slender with glabrous foliage. **Leaf blades** 3–10 mm long, reniform, 3- to 5-lobed with entire margins. **Flowers** 1 per node; petals ca. 2 mm long, barely longer than the glabrous sepals. **Achenes** glabrous, ca. 1.5 mm long with a curved beak <0.2 mm long; heads hemispheric 3–6 mm high. Wet mud and shallow water of ponds, seeps, springs; montane to alpine. AK to Greenland south in the Rocky Mtns. to CO. Our plants are ssp. *intertextus* (Greene) Kapoor, A.Love & D.Love.

Ranunculus inamoenus Greene **Stems** erect,10–30 cm with hirsute to glabrous foliage. **Basal leaf blades** 1–5 cm long, ovate to fan-shaped, shallowly lobed or scalloped on the upper half. **Flowers** with petals 3–8 mm long, slightly longer than the hirsute sepals. **Achenes** finely hairy, ca. 2 mm long with a hooked beak ≤0.5 mm long; heads cylindric, 5–12 mm high. Moist grasslands, meadows; montane, lower subalpine. BC to SK south to WA, AZ, CO. Our plants are var. *inamoenus*. (p.98)

Ranunculus abortivus has smaller petals; see *R. cardiophyllus*.

Ranunculus jovis A.Nelson **Stems** erect, 2–8 cm with glabrous foliage, arising from tuberous roots. **Basal leaf blades** 15–40 mm long, broadly ovate in outline, deeply divided into 3 to 5 linear-oblong lobes. **Flowers** with petals 7–10 mm long, twice as long as the glabrous sepals.

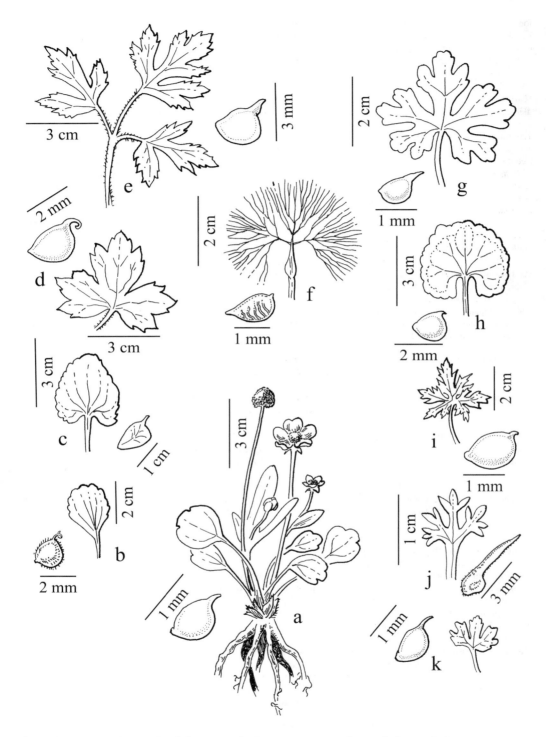

Plate 12. **Ranunculus**. **a.** *R. glaberrimus*, **b.** *R. inamoenus*, **c.** *R. cymbalaria*, **d.** *R. uncinatus*, **e.**
R. macounii, **f.** *R. aquatilis*, **g.** *R. sceleratus*, **h.** *R. abortivus*, **i.** *R. acris*, **j.** *R. testiculatus*,
k. *R. eschscholtzii* var. *eschscholtzii*

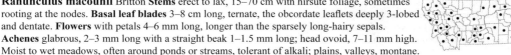

Achenes short-hairy, ca. 1 mm long with a straight beak ca. 0.5 mm long; heads ovate, 3–8 mm high. Sagebrush steppe, open forest; montane, subalpine. ID, MT south to NV, UT, CO.

This plant is rarely seen because it flowers soon after snow release.

Ranunculus macounii Britton **Stems** erect to lax, 15–70 cm with hirsute foliage, sometimes rooting at the nodes. **Basal leaf blades** 3–8 cm long, ternate, the obcordate leaflets deeply 3-lobed and dentate. **Flowers** with petals 4–6 mm long, longer than the sparsely long-hairy sepals. **Achenes** glabrous, 2–3 mm long with a straight beak 1–1.5 mm long; head ovoid, 7–11 mm high. Moist to wet meadows, often around ponds or streams, tolerant of alkali; plains, valleys, montane. AK to NL south to CA, NM, MN. Our plants are var. *macounii*. (p.98)

Standley[382] reports the glabrous form of this plant, var. *oreganus* K.C.Davis for Glacier National Park, but I have not seen specimens.

Ranunculus orthorhynchus Hook. **Stems** 20–75 cm, somewhat hollow, erect or decumbent with hirsute to nearly glabrous foliage. **Basal leaf blades** narrowly ovate, 3–15 cm long, pinnate, the 3 to 5 leaflets deeply 1 to 2 times lobed. **Flowers** with petals 9–14 mm long, almost twice as long as the pilose sepals. **Achenes** flattened, 2–4 mm long, glabrous with straight beak ca. 3 mm long; head ovoid, 8–12 mm high. Wet meadows; montane to lower subalpine. AK south to CA, UT, WY. Our plants are var. *platyphyllus* A.Gray.

Ranunculus pedatifidus Sm. **Stems** erect, 8–25 cm with sparsely hairy foliage. **Basal leaf blades** 1–4 cm long, reniform, 1 to 2 times divided into 7 to 20 narrowly oblong lobes. **Flowers** with petals 8–10 mm long, slightly longer than the pilose, deciduous sepals. **Achenes** ca. 2 mm long, finely hairy with a hooked beak 0.5–1 mm long; heads ovate, ca. 8 mm long. Moist grasslands, turf; montane to alpine. Circumboreal south to NM, AZ. Our plants are var. *affinis* (R.Br.) L.D.Benson.

Ranunculus gelidus has smaller petals, while *R. eschscholtzii* has mostly glabrous, less finely divided leaves; see *R. cardiophyllus*.

Ranunculus pensylvanicus L.f. **Stems**, erect, 20–70 cm with hirsute foliage. **Basal leaf blades** cordate, ternate, 3–10 cm long; the stalked leaflets deeply 3-lobed, dentate. **Flowers** with petals 2–4 mm long, shorter than the pilose sepals. **Achenes** ca. 2 mm long, glabrous with a straight beak ca. 0.7 mm long; head short-cylindrical, 8–15 mm high. Riparian meadows; plains, valleys. AK to NL south to AZ, NM, IA, NY.

Ranunculus populago Greene **Stems** erect, hollow, 8–30 cm with mostly glabrous foliage. **Basal leaf blades** 1–3 cm long, cordate with scalloped margins. **Flowers** with petals 4–8 mm long, twice as long as the glabrous sepals. **Achenes** flattened, glabrous or sparsely hairy, ca. 1.5 mm long with a straight beak <0.5 mm long; heads hemispheric, 2–3 mm high. Wet meadows along streams, wetlands; montane, subalpine. WA to MT south to CA, OR, ID.

Ranunculus pygmaeus Wahl. **Stems** decumbent or erect, 1–4(–6) cm with nearly glabrous foliage. **Basal leaf blades** reniform, to 6 mm long with 3–5 lobes. **Flowers** solitary with petals 2–3 mm long, about as long as the sparsely hairy sepals. **Achenes** glabrous, ca. 1 mm long with a straight beak ca. 0.5 mm long; heads ovoid 3–5 mm high. Moist to wet soil of turf, often on cliffs, cool slopes or along streams or seeps; alpine. Circumpolar south to CO.

Ranunculus repens L. Creeping buttercup **Stems** to 40 cm with hirsute to nearly glabrous foliage, prostrate and rooting at the nodes to occasionally erect. **Basal leaf blades** ovate, 1–8 cm long, ternate; the 3 petiolate leaflets lobed and toothed. **Flowers** with petals 6–15 mm long, twice as long as the hirsute, deciduous sepals. **Achenes** flattened, 2–3 mm long with a hooked beak ca. 1 mm long; heads globose, 5–10 mm high. Lawns, roadsides, pastures, disturbed wetland margins; valleys. Most of western and eastern N. America, S. America, Australia; introduced from Eurasia.

Most of our specimens with erect stems were taken in towns and may be escaped garden cultivars.

Ranunculus rhomboideus Goldie Prairie buttercup **Stems** erect, 5–33 cm with pilose foliage. **Basal leaf blades** 1–4 cm long, ovate, scalloped on the upper half. **Flowers** with petals 6–8 mm long, slightly longer than the pilose sepals. **Achenes** glabrous, ca. 2 mm long with curved beak <0.5 mm long; heads globose, 4–6 mm high. Grasslands, open woodlands; plains. AB to ON south to MT, NE, WI, IA.

I have seen no collections from MT, but it is reported for the state[152,438] and is known from adjacent AB, ND.

Ranunculus sceleratus L. **Stems** erect, hollow, 20–50 cm with mostly glabrous foliage. **Basal leaf blades** 15–50 mm long, reniform, deeply 3-lobed into lobed segments. **Flowers** with petals 3–5 mm long, barely longer than the glabrous sepals. **Achenes** ca. 1 mm long, glabrous with inflated margins and a straight beak ca. 0.1 mm long; heads cylindric, 3–10 mm high. Muddy shores of lakes, streams, wetlands; plains, valleys, montane. Most of N. America north of Mexico; Eurasia. Our plants are var. *multifidus* Nutt. ex Torr.& A.Gray. (p.98)

    Ranunculus sceleratus var. *sceleratus*, with wrinkled achene faces, is native to Eurasia and perhaps eastern N. America and has been found both east and west of MT.

Ranunculus sulphureus Sol. **Stems** erect, 3–20 cm with glabrate foliage. **Basal leaf blades** reniform, 1–3 cm long, 3- to 5-lobed . **Flowers** with petals 8–12 mm long, ca. twice as long as the sepals. **Achenes** glabrous or sparsely brown-hairy, ca. 2 mm long with a curved or straight beak 1.5–2 mm long; heads cylindric, 5–8 mm high. Moist to wet, alpine turf. Circumpolar south to AB, NT, Greenland; disjunct in Carbon Co., MT.

Ranunculus testiculatus Crantz Annual. **Stems** erect, 2–8 cm with villous, entirely basal leaves. **Leaf blades** 5–15 mm long, oblong to spatulate, divided into 3 to 7 linear lobes. **Flowers** solitary; petals quickly falling, 3–6 mm long, as long as the hairy sepals. **Achenes** 1–2 mm long, short-hairy with a blade-like beak ca. 3 mm long; heads cylindric, bristly, to 20 mm long. Disturbed soil of grasslands, along roads; plains, valleys. BC to MB south to CA, NM, KS; introduced from Eurasia. (p.98)

Ranunculus uncinatus D.Don ex G.Don **Stems** erect, 20–70 cm, with hirsute foliage. **Basal leaf blades** 2–10 cm across, divided into 3 ovate, lobed leaflets. **Flowers** with petals 2–4 mm long, as long as sepals. **Achenes** 2–3 mm long, glabrous or hairy with a curved beak 1–1.5 mm long; head globose, 4–6 mm high. Forest, thickets, often near streams or other openings; valleys to rarely lower subalpine. AK to AB south to CA, ID, CO. (p.98)

Thalictrum L. Meadow rue

Rhizomatous herbs. **Leaves** 2 to 4 times ternate or pinnate, ultimate leaflets ovate, stalked, lobed or toothed. **Inflorescence** racemose to paniculate. **Flowers** bisexual or unisexual on separate plants; sepals 4 to 5, green or brown, male longer than female; petals absent; male flowers with numerous stamens; female flowers with 4 to 9 pistils. **Fruit** a cluster of ellipsoid, beaked, hard-coated achenes with prominent parallel veins.

    Thalictrum fendleri Engelm. ex A.Gray with strongly flattened achenes is reported for MT,[317] but I have seen no specimens.

1. Flowers perfect, in a raceme or few-flowered panicles 2
1. Plants dioecious; flowers unisexual in many-flowered panicles 3

2. Plants with 0 to 1 stem leaves; inflorescence a raceme. . . . . *T. alpinum*
2. Plants with leafy stems; inflorescence a leafy, compound panicle . . . . *T. sparsiflorum*

3. Ultimate leaflets with usually 3 acute, entire lobes; achenes hairy. . . . . *T. dasycarpum*
3. Leaflets with rounded lobes that are toothed or lobed; achenes sparsely glandular 4

4. Mature achenes spreading to reflexed; common . . . .*T. occidentale*
4. Mature achenes erect or ascending 5

5. Achenes strongly flattened, 4–6 mm long; stems purplish above. . . . . *T. fendleri*
5. Achenes nearly terete, mostly 3–4 mm long. . . . . *T. venulosum*

Thalictrum alpinum L. Monoecious. **Stems** glaucous, 10–20 cm with glabrous, glaucous foliage. **Leaves** mainly basal, twice ternate to bipinnate; ultimate leaflets leathery, 2–7 mm long, 3- to 5-lobed. **Inflorescence** racemose with 8 to 15 nodding flowers. **Flowers** perfect; sepals purplish, 1–2 mm; anthers ca. 2 mm long, longer than the filaments. **Fruits** 2 to 6 spreading achenes, the body 2–3 mm long, on nodding pedicels. Moist alkaline meadows near streams or seeps, often associated with shrubby cinquefoil and baltic rush on hummocks in wetlands of sagebrush steppe landscapes; montane, lower subalpine in Beaverhead Co. AK to Greenland south at scattered locales to CA, NM, CO; Eurasia.

Thalictrum dasycarpum Fisch.& Ave-Lall. Plants dioecious. **Stems** 60–150 cm with glabrous foliage. **Leaves** sessile, 3–5 times ternate; ultimate leaflets 2–3 cm long, acutely 3-lobed, leathery, distinctly veined, glaucous and sparsely hairy beneath with revolute margins. **Inflorescence** of compound bracteate panicles. **Flowers** with whitish sepals 3–4 mm long; anthers ca. 2 mm long, much shorter than the tapered filaments. **Fruits** ca. 10 pubescent, spreading achenes, the body 2–4 mm long. Wet meadows, thickets, ditch and streambanks; plains, valleys to montane. BC to QC south to AZ, TX, AL, PA.

Thalictrum occidentale A.Gray Plants dioecious. **Stems** 20–80 cm. **Leaves** mostly cauline, 3 to 4 times ternate; ultimate leaflets 10–35 mm long, 3- to 5-lobed, glaucous and glandular beneath. **Inflorescence** a terminal, leafy bracteate panicle. **Flowers** with white to purplish sepals 2–6 mm long; anthers ca. 3 mm, shorter than mature filaments. **Fruits** 6 to 9, spreading to reflexed achenes, the body 5–8 mm long. Coniferous forest, meadows; montane, subalpine. AK to SK south to CA, UT, CO. See *Thalictrum venulosum*. (p.92)

Thalictrum sparsiflorum Turcz. ex Fisch.& C.A.Mey. Monoecious. **Stems** glabrous, 30–90 cm. **Leaves** mostly cauline, bi- to tri-ternate or pinnate; ultimate leaflets glandular-puberulent below, 1–2 cm long, deeply 3-cleft and lobed. **Inflorescence** with few-flowered racemes or panicles from axils of leaf-like bracts. **Flowers** perfect; sepals white to green, ca. 3 mm long; anthers ≤ 1 mm long, much shorter than filaments. **Fruits** 6 to 12 reflexed-stipitate, glandular, flattened achenes, the body 3–5 mm long. Moist to wet meadows, thickets, often along streams, beneath willows; montane, lower subalpine. AK to ON south to CA, UT, CO.

Thalictrum venulosum Trel. Plants dioecious. **Stems** glabrous, 25–70 cm. **Leaves** mostly cauline, 3 to 4 times ternate; ultimate leaflets lobed, glabrous to glandular, 5–20 mm long. **Inflorescence** terminal, paniculate. **Flowers** with greenish-white sepals, 2–4 mm long; anthers 2–4 mm long, shorter than filaments. **Fruits** 5 to 17 erect or ascending achenes, the body nearly terete in cross section, 4–5 mm long. Wetlands, fens, streambanks, thickets, open forest; plains, valleys, montane. YT to QC south to OR, NM, WI.

Many of the ultimate leaf lobes are acute; those of *Thalictrum occidentale* are usually rounded.

Trautvetteria Fisch.& C.A.Mey. False bugbane

Trautvetteria caroliniensis (Walter) Vail Herbaceous perennial from short rhizomes and fibrous roots. **Stems** erect, 50–80 cm. **Leaves** basal and cauline, petiolate; the blade pubescent below, broadly cordate, deeply palmately lobed into 5 to 11 dentate segments; the basal to 30 cm across. **Inflorescence** a terminal corymb with numerous flowers, glandular and covered with dense, small, hooked hairs. **Flowers** perfect, radially symmetrical; sepals 3 to 5, greenish, ca. 3 mm long, deciduous; petals absent; stamens numerous, 4–8 mm long; pistils ca. 15, 1-seeded. **Fruit** a thin-walled and inflated achene (utricle) 2–3 mm long with a hooked beak. Moist coniferous forest, often along streams and with red cedar; valleys, montane. BC to CA, NM, SC, TN, Mexico; Asia.

Trollius L. Globeflower

Trollius albiflorus (A.Gray) Rydb. [*T. laxus* Salisb. var. *albiflorus* A.Gray] Perennial with fibrous roots. **Stems** erect, 5–50 cm high, expanding with maturity. **Leaves** petiolate, the blades broadly cordate, 3–7 cm wide, deeply palmately divided into ca. 5 toothed or lobed, ovate segments. **Flowers** solitary; sepals 5 to 7, petal-like, white, 1–2 cm long; petals inconspicuous, stamen-like; stamens many; pistils 10 to 20. **Fruit** a hemispheric cluster of cylindrical several-seeded follicles 8–15 mm high, opening at the top. Wet meadows, streambanks, often near melting snow; subalpine, lower alpine. BC to AB south to WA, ID, CO. (p.92)

BERBERIDACEAE: Barberry Family

Berberis L. Oregon grape, barberry

Low shrubs with yellowish wood. Stems trailing or erect. Leaves evergreen, glabrous, alternate, simple or pinnately compound. Inflorescence a terminal cluster of bracteate racemes. Flowers prefect, radially symmetrical, each subtended by 3 deciduous bracteoles; petals 6, yellow; sepals 6, yellow, larger than petals; stamens 6; ovary superior. Fruit a blue or red, few-seeded berry.

1. Leaves simple; spines present at leaf nodes . . . . *B. vulgaris*
1. Leaves pinnate; stems not spiny 2

2. Larger leaves with ≥11 leaflets; bud scales 2–4 cm long, tough . . . . *B. nervosa*
2. Leaves with 5 to 9 leaflets; bud scales <1 cm long, membranous 3

3. Leaflets white-papillose (visible at 20X) and glaucous beneath . . . . *B. repens*
3. Leaflets smooth and shiny beneath . . . . *B. aquifolium*

Berberis aquifolium Pursh [*Mahonia aquifolium* (Pursh) Nutt.] Oregon grape Stoloniferous, similar to *B. repens*. **Stems** sometimes erect, 15–100 cm. **Leaves** to 20 cm long, pinnately divided into 5 to 9 ovate leaflets. **Leaflets** more than twice as long as wide, often glossy below, not white-papillose, 3–8 cm long with 12 to 29 prominent spinulose teeth. **Flowers** with sepals 6–8 mm long. **Fruit** blue, 7–14 mm long. Wet coniferous forest; valleys to montane. BC to CA, ID, MT.

Berberis nervosa Pursh [*Mahonia nervosa* (Pursh) Nutt.] Rhizomatous shrubs. **Stems** erect, 10–30 cm. with persistent, leathery bud scales. **Leaves** 20–40 cm long with 9 to 19 leaflets. **Leaflets** up to 7 cm long with spiny-toothed margins, glossy on both surfaces. **Flowers** similar to *B. aquifolium* in racems to 20 cm long. **Fruit** blue, 8–11 mm long. Wet coniferous forest in the valleys. BC to CA, disjunct in ID and Sanders Co., MT.

Berberis repens Lindl. [*B. aquifolium* Pursh var. *repens* (Lindl.) Scoggan, *Mahonia repens* (Lindl.) G.Don] Oregon grape Plants stoloniferous. **Stems** 5–30 cm. **Leaves** to 20 cm long, pinnately divided into 5 to 7 ovate leaflets. **Leaflets** up to twice as long as wide, minutely white-papillose, dull and somewhat glaucous below, 3–9 cm long with 15 to 43 spinulose teeth. **Flowers** with sepals 5–7 mm long. **Fruit** blue, 5–12 mm long. Dry to moist, open or closed forest, brushy slopes; plains, valleys to lower subalpine. BC and AB south to CA, AZ, TX. (p.104)

Berberis vulgaris L. Common barberry **Stems** erect 1–2 m with short axillary shoots and 1 to 3 spines per node. **Leaves** simple, short-petiolate, obovate, 15–50 mm long with spiny margins. **Flowers** with sepals to 8 mm long. **Fruits** red or bluish, ellipsoid, ca. 1cm long. Roadsides, disturbed pastures; plains, valleys. Escaped from cultivation, introduced to much of northern U.S. and adjacent Canada; native to Europe.
  Common barberry has been nearly eradicated because it is the alternate host for black stem rust, a serious disease of wheat.

PAPAVERACEAE: Poppy Family

Herbaceous annuals or perennials. **Leaves** alternate and/or basal, deeply lobed or divided. **Flowers** paniculate, racemose or solitary, perfect, radially symmetrical, hypogynous; sepals 2 to 3, early deciduous; petals colored, 4 to 8, separate; stamens numerous; pistil 1; ovary superior. **Fruit** a many-seeded capsule.
  Eschscholzia californica Cham. (California poppy), with large orange flowers and glabrous, narrowly divided leaves, is sometimes seeded along roads and may persist for a few years.

1. Flowers borne in umbels . . . . *Chelidonium*
1. Flowers solitary or in racemes 2

2. Leaves and stems spiny; flowers white . . . . *Argemone*
2. Foliage glabrous to hispid but not spiny; flowers mostly yellow to red 3

3. Capsule linear, ≥12 cm long at maturity . . . . *Glaucium*
3. Capsule ovate to oblong, <6 cm long . . . . *Papaver*

Argemone L. Prickly poppy

Argemone polyanthemos (Fedde) G.B.Owenby [*A. intermedia* Sweet var. *polyanthemos* Fedde] Taprooted annual or biennial. **Stems** 40–80 cm with prickly foliage. **Leaves** oblanceolate, to 10 cm long, glaucous, deeply pinnately lobed with spiny-toothed margins, the upper clasping. **Inflorescence** bracteate cymose. **Flowers** 5–10 cm across; sepals 3, prickly; petals 6, white;

stamens numerous with yellow filaments. **Capsules** ellipsoid, ca. 2 cm high, yellow-spiny. Roadsides, disturbed grasslands; plains. MT, ND south to TX.

Chelidonium L. Celandine

Chelidonium majus L. Biennial or short-lived perennial with yellow sap, hairy below, glabrous above. **Stems** to 70 cm, branched. **Leaves** petiolate, ovate, to 25 cm long, deeply lobed with toothed margins. **Inflorescence** long-stalked umbels from upper leaf axils. **Flowers** yellow, the pedicels to 4 cm long in fruit; sepals 2, petals 4, 5–10 mm long; stamens at least 12. **Capsules** erect, linear, glabrous, 2–8 cm long. Disturbed riparian corridors; valleys; one thriving population on the north side of Missoula. Introduced to northeast and northwest U.S. and adjacent Canada; native to Eurasia.

Glaucium Miller Horned poppy

Glaucium corniculatum (L.) Rudolph Annual or biennial with white-hairy foliage and yellow sap. **Stems** branched, to 50 cm. **Leaves** oblanceolate, deeply pinnately lobed, to 15 cm long; the upper ovate and clasping. **Flowers** solitary, long-stalked in upper leaf axils; sepals 2, 15–30 mm long; petals 4, orange to red with a black base, 20–35 mm long; stamens many. **Capsules** ascending, linear, slightly curved, 12–20 cm long, pubescent. Collected once in a Stillwater Co. agricultural field. Sporadically introduced in U.S.; native to Europe.

Papaver L. Poppy

Annual or perennial herbs with yellow or milky sap. **Stems** absent or erect. **Leaves** petiolate, cauline and/or basal. **Flowers** paniculate or solitary; sepals 2; petals usually 4; style absent. **Capsules** erect at maturity.[218]

   Papaver dubium L., an introduced garden annual, comes in seed mixes and occasionally escapes.

1. Plants with leafy stems; disturbed sites in valleys 2
1. Plants scapose; subalpine to alpine 4

2. Capsules hispid, oblong . . . .*P. argemone*
2. Capsules glabrous, globose 3

3. Upper leaves petiolate; plants perennial . . . .*P. orientale*
3. Leaves clasping the stem; plants annual . . . . *P. somniferum*

4. Petals >2 cm long . . . .*P. croceum*
4. Petals ≤1 cm long 5

5. Capsule hairs white; petals usually orange to salmon-colored . . . .*P. pygmaeum*
5. Capsule hairs brown; petals yellow . . . .*P. radicatum*

Papaver argemone L. Annual with hispid foliage. **Stems** 15–50 cm., simple or branched. **Leaf blades** twice pinnately divided, to 12 cm long. **Flowers** solitary on branch tips; petals red with a black base, to 25 mm long. **Capsules** oblong, 15–20 mm long with stiff hairs. Escaped from cultivation to roadsides and disturbed fields, though populations do not persist;[226] valleys. Sporadic throughout the U.S.; introduced from Europe; collected in Ravalli Co.

Papaver croceum Ledeb. [*P. nudicaule* L. misapplied] Siberian poppy Taprooted, acaulescent, white-hispid perennial with milky sap. **Leaf blades** lanceolate, 5–20 cm long, deeply lobed. **Flowers** solitary on peduncles 10–40 cm long; petals 4, 1–3 cm long, pale yellowish. **Fruit** strigose with white hairs, many-seeded, globose to ellipsoid capsule, 1–3 cm long. A garden cultivar collected once at an abandoned mine in the Beartooth Mtns. of Stillwater Co.; introduced from Asia.

Papaver pygmaeum Rydb. [*P. radicatum* Rottb. var. *pygmaeum* (Rydb.) S.L.Welsh, *P. alpinum* L. misapplied] Pygmy poppy Tufted, taprooted, acaulescent, white-hispid perennial with milky sap. **Leaf blades** ovate, ≤1 cm long with deep, rounded lobes. **Flowers** solitary on peduncles 5–10 cm long, nodding in bud; sepals 2, falling in flower; petals 4, 5–10 mm long, salmon to orange. **Fruit** a white-bristly, many-seeded, conical capsule, 8–15 mm long. Stony soil of exposed ridges and slopes; alpine. Endemic to the Glacier-Waterton Lakes area of MT and adjacent AB, near or east of the Continental Divide. (p.104)

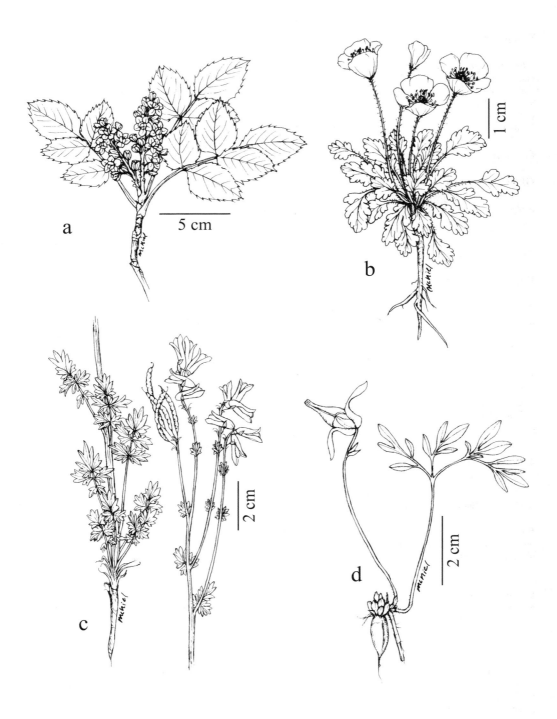

Plate 13. a. *Berberis repens*, b. *Papaver pygmaeum*, c. *Corydalis sempervirens*, d. *Dicentra uniflora*

Stan Welsh believes this plant occurs in Utah and is part of the circumpolar *Papaver radicatum* complex;[435] however, its closest relatives may be in the European *P. alpinum* L. complex.[218] Specimens from a disjunct population in Lewis & Clark Co. have yellow petals that fade to orange. The type locality is in Flathead Co.

Papaver radicatum Rott. [*P. kluanensis* D.Love] Cespitose, taprooted perennial with white-hirsute foliage. **Stems** absent. **Leaves** all basal, the blades lanceolate, 1–3 cm long with 2 to 4 pairs of toothed lobes. **Flowers** solitary on peduncles to 12 cm long; petals yellow, 5–10 mm long. **Fruit** a brown-strigose, oblong capsule, to 12 mm long. Stony slopes, fellfields; alpine. Circumpolar, sporadically south to NM; known from Carbon and Sweet Grass cos.[237] Our plants are ssp. *kluanensis* (D.Love) Murray.

Papaver orientale L. Oriental poppy Perennial with hispid foliage. **Stems** usually simple, 30–70 cm. **Leaves** petiolate, to 30 cm long; blades entire or lobed. **Flowers** long-stalked, solitary from upper leaf axils; petals orange to red, ca. 5 cm long. **Capsules** glabrous, glaucous, globose to ovoid, 15–25 mm long. Roadsides, agricultural fields; valleys. Introduced mainly in the northeast U.S.; native to Eurasia. One collection from Ravalli Co., reports of *P. rhoeas*[113] are referable here.

Papaver somniferum L. Opium poppy Annual with glabrous and glaucous foliage. **Stems** to 1 m, usually unbranched. **Leaves** lanceolate, to 30 cm long. **Flowers** white to purple; petals to 6 cm long with a dark, basal spot. **Capsules** globose, glabrous, 2–5 cm long. Agricultural fields, roadsides; valleys; Flathead Co. Introduced throughout U.S. and southern Canada; native to Eurasia.

FUMARIACEAE: Fumitory Family

Glabrous, annual to perennial herbs. **Leaves** alternate, petiolate, glaucous, pinnately divided into entire-margined segments. **Inflorescence** bracteate. **Flowers** bisexual, bilaterally symmetrical; sepals 2, deciduous; petals 4, separate or united at the base, 1 or 2 sac-like at the base; stamens 6; ovary superior; pistil 1; stigma often 2-lobed. **Fruit** a 1- to many-seeded capsule.

Elaiosomes (oil-bearing appendages) are present on the seeds of some species.

1. Flowers solitary; both outer petals with recurved spurs . . . . *Dicentra*
1. Flowers >1; only 1 outer petal with sac-like spur 2

2. Fruit elongate and many-seeded . . . . *Corydalis*
2. Fruit nearly globose with 1 seed. . . . . *Fumaria*

Corydalis DC. Corydalis

Annual or biennial herbs. **Leaves** petiolate, ovate, 2 to 4 times divided and lobed into narrow, lance-shaped or oblong segments. **Inflorescence** of racemes from upper leaf axils. **Flowers** 2-lipped with 2 lateral wings, the upper petal with cone-shaped spur. **Fruit** a linear, many-seeded capsule.

1. Flowers yellow; capsules nodding, >2 mm wide . . . .*C. aurea*
1. Flowers mainly pink; capsules erect, <2 mm wide. . . . . *C. sempervirens*

Corydalis aurea Willd. Golden smoke **Stems** branched at the base, ascending to prostrate, 5–40 cm long. **Leaves** to 15 cm long. **Flowers** yellow, 10–12 mm long, the spur 2–4 mm long. **Capsules** declined, 15–25 mm long, curved, constricted between seeds. Moist, sparsely vegetated soil of open forest, streambanks, roadsides, eroding slopes; plains, valleys, montane. AK to QC south to CA, TX, LA, OH, VT.

Corydalis sempervirens (L.) Pers. Rock harlequin **Stems** erect, 30–70 cm. **Leaves** to 20 cm long. **Flowers** deep pink with yellow tips, 1–2 cm long with a spur 2–4 mm long. **Capsules** erect, 25–40 mm long. Rocky, disturbed or eroding soil of steep slopes in open forest, often appearing after fire; montane; Flathead and Glacier cos. AK to NL south to BC, MT, MB, GA. (p.104)

Dicentra Bernh.

Dicentra uniflora Kellogg STEER'S HEAD Glabrous, scapose perennial from tuberous roots. **Leaves** long-petiolate, the blade 1–2 cm long, 2 to 3 times pinnately divided into narrowly oblong segments, as long as the stem. **Flowers** solitary on stalks 2–6 cm long, white to pink, 1–2 cm long; both outer petals with recurved spur; inner petals narrowly arrow-shaped, purple-tipped. **Capsules** ovoid, ca. 1 cm long. Vernally moist, often shallow soil of meadows, open forest, rock outcrops; montane, subalpine. WA to MT south to CA, UT, WY. (p.104)

Rarely seen because it flowers very early in habitats that are usually still surrounded by snow and often under a developing canopy of tall herbs.

Fumaria L. Fumitory

Annual herbs. **Stems** erect, branched. **Leaves** petiolate, 3 to 4 times divided and lobed. **Inflorescence** racemes from upper leaf axils. **Flowers** 2-lipped; petals closely appressed, the upper with a sac-like spur. **Fruit** a nearly globose 1-seeded capsule (nut).[54]

1. Corolla 6–10 mm long; the spur ≥2 mm long . . . . *F. officinalis*
1. Corolla 5–6 mm long; the spur <2 mm long . . . . *F. vaillantii*

Fumaria officinalis L. FUMITORY **Stems** 10–40 cm. **Leaves** to 6 cm long. **Flowers** ascending, 6–10 mm long; petals pink, purple near tips. **Capsules** bumpy, 1–2 mm long. Roadsides, disturbed places; valleys. Introduced into most of U.S. and southern Canada; native to Europe.

Fumaria vaillantii Loisel. **Stems** 15–30 cm. **Leaves** to 8 cm long. **Flowers** spreading to ascending, 5–6 mm long; petals pink, purple toward the tip. **Capsules** wrinkled or pitted, ca. 2 mm long. Woodlands, thickets, roadsides; plains, valleys. Sparingly introduced in ND, SD, MT; native to Europe.

ULMACEAE: Elm Family

Shrubs or trees. **Leaves** alternate, simple, petiolate with serrate margins. **Flowers** perfect or unisexual; calyx 4- to 5-lobed; petals lacking; stamens 3 to 9; pistils 1with 2 styles. **Fruit** a 1-seeded drupe or samara.

1. Leaf blade pinnately veined; fruit a samara . . . . *Ulmus*
1. Leaf blade palmately veined; fruit a drupe . . . . *Celtis*

Celtis L. Hackberry

Celtis occidentalis L. Shrubs or stunted trees to 5 m with furrowed and bumpy bark. **Leaf blades** lanceolate, 5–8 cm long with once-serrate margins. **Flowers** unisexual, radially symmetrical. **Male flowers** with 4 to 5 stamens in pendulous clusters. **Female flowers** with 4-lobed calyx. **Fruit** a stalked, blue-black drupe, 7–11 mm across. Riparian forests and other sheltered habitats; plains; collected in Big Horn and Custer cos. MT to QC south to OK and GA.

Our plants are stunted trees at best. Two known locations are in state parks where plants could have originally been cultivated.

Ulmus L. Elm

Trees with spreading crowns and furrowed bark. **Leaves** simple with serrate margins and tufts of hair along veins beneath, alternate, short-petiolate, stipulate. **Inflorescence** fascicles of stalked flowers. **Flowers** on twigs of previous season, perfect, radially symmetrical; calyx with 3 to 9 lobes; stamens 3 to 9; pistils 1with a 2-lobed style; **Fruit** a flattened, membranous-winged samara.

1. Leaves twice serrate, >7 cm long, the upper surface scabrous . . . . *U. americana*
1. Leaves once serrate, <7 cm long, the upper surface smooth . . . . *U. pumila*

Ulmus americana L. Aᴍᴇʀɪᴄᴀɴ ᴇʟᴍ Trees to 15 m with gray to brown, fissured bark. **Twigs**  pubescent with acute buds. **Leaf blades** elliptic with an asymmetric base, 7–12 cm long, scabrous above with twice serrate margins. **Flowers** on drooping pedicels 1–2 cm long; calyx with 7 to 9 lobes; anthers red. **Samaras** ovate, yellowish, ca. 1 cm long. Woodlands in ravines, along rivers and streams; plains. SK to NS south to TX, FL.

Ulmus pumila L. Sɪʙᴇʀɪᴀɴ ᴇʟᴍ Trees to 15 m with gray to brown furrowed bark. **Twigs**  pubescent with ovoid, glabrous buds. **Leaf blades** lanceolate, 2–6 cm long, glabrous above with singly serrate margins. **Flowers** clustered on short pedicels; calyx with 4 to 5 lobes; anthers brownish. **Samaras** orbiculate, yellowish, 10–14 mm across. Roadsides, disturbed fields, streambanks; plains, valleys. Introduced to most of U.S.; native to Asia.

CANNABACEAE: Hemp Family

Dioecious, annual or perennial herbs with pubescent, glandular, aromatic foliage. **Stems** erect or twining, usually branched. **Leaves** petiolate, opposite; the blade deeply lobed to divided with serrate margins. **Flowers** usually unisexual, hypogynous, numerous in axillary and/or terminal inflorescences. **Male flowers** pedicillate; sepals 5, distinct; stamens 5. **Female flowers** nearly sessile; perianth reduced to a membrane surrounding the ovary; petals and sepals absent or obscure; 1 bracteate pistil with 2-lobed stigma. **Fruits** achenes surrounded by membranous perianth.[368]

This family is sometimes included in the Moraceae.

Cannabis L. Hemp, marijuana

Cannabis sativa L. [*C. indica* Lam.] Taprooted annual. **Stems** often branched below, to 1  m. **Leaf blades** palmately divided into 5 to 9 narrowly lanceolate leaflets, up to 9 cm long. **Male flowers** with pedicels to 3 mm long; sepals 2–4 mm long; **Female flowers** enclosed by a glandular bract. **Achenes** greenish, mottled with purple, 2–5 mm long. Fields, roadsides; plains, valleys. Introduced throughout most of U.S.; native to Eurasia.

Humulus L. Hop

Humulus lupulus L. Rhizomatous perennial. **Stems** herbaceous, twining, to 4 m long. **Leaf**  **blades** scabrous, cordate, deeply, palmately 3- to 5-lobed, 4–12 cm long. **Male flowers** white to yellow, paniculate. **Female flowers** in pairs subtended by a bract in cone-like raceme 1–3 cm long. **Fruits** pendulous, long-ovoid, 2–3 cm long, yellow-glandular. Roadsides, streambanks; plains, valleys. Throughout most of U.S. and southern Canada in and east of the Rocky Mtns.

Three vars. are sometimes recognized for N. America.[368] Our material, including some in cultivation, is abundantly glandular on abaxial leaf surfaces but varies in degree of lobing and appears intermediate between var. *neomexicanus* A.Nelson & Cockerell and var. *lupuloides* E.Small, both of which are considered native.

URTICACEAE: Nettle Family

Monoecious (rarely dioecious), annual or perennial herbs. **Leaves** opposite or alternate, simple, petiolate. **Inflorescence** axillary, cymose or paniculate. **Flowers** mostly unisexual; perianth absent or of 3 to 5 distinct tepals. **Male flowers** with 4 stamens. **Female flowers** with 1 superior pistil. **Fruit** an achene.

1. Plants annual, without stinging hairs . . . . *Parietaria*
1. Rhizomatous perennial with stinging hairs on the stem . . . . *Urtica*

Parietaria L. Pellitory

Parietaria pensylvanica Muhl. ex Willd. Annual with sparsely short-hairy foliage. **Stems** simple to branched, 5–30 cm. **Leaves** alternate; the blades 5–40 mm long, narrowly elliptic with entire margins. **Flowers** male and female mixed in axillary clusters subtended by linear bracts 2–5 mm long; female flowers with a tufted stigma. **Achenes** smooth, shiny, ca. 1 mm long. Shaded, often disturbed soil of open forest, streambanks, beneath shrubs or rocks, occasionally epiphytic in humus of deciduous tree crotches; plains, valleys. Southwest Canada through most of U.S. to Mexico.

Urtica L. Nettle

Urtica dioica L. STINGING NETTLE Rhizomatous perennial with nearly glabrous foliage and stinging hairs. **Stems** 4-angled, often unbranched, 50–150 cm. **Leaves** opposite with prominent stipules; the blades lanceolate to ovate, 3–15 cm long with deeply serrate margins. **Flowers** in drooping axillary panicles; female flowers with 2 pairs of tepals; the outer narrow and ca. 1 mm long; the inner broader and longer. **Achenes** flattened-ovoid, ca. 1.5 mm long. Moist, usually disturbed, often organic soil of meadows, open forest, streambanks; plains, valleys to lower subalpine. Most of North America, South America; Eurasia. (p.112)

1. Stems glabrate to sparsely strigose in addition to stinging hairs . . . . ssp. *gracilis*
1. Stems white-pubescent in addition to stinging hairs. . . . . .ssp. *holosericea*

Urtica dioica ssp. *gracilis* (Aiton) Selander is our common form; *U. dioica* ssp. *holosericea* (Nutt.) Thorne is rarely collected in MT.

FAGACEAE: Beech Family

Quercus L. Oak

Quercus macrocarpa Michx. BUR OAK Monoecious trees to 10 m with gray, scaly bark. **Twigs** pubescent, often with corky wings. **Leaves** alternate, simple, petiolate; the blades obovate, 6–15 cm long, pinnately lobed, the lobes rounded with wavy margins. **Flowers** unisexual with a 6-lobed perianth. **Male flowers** borne in a drooping, interrupted, spike-like inflorescence; perianth 6-lobed with 6 to 10 stamens. **Female flowers** in clusters of 1 to 3; the perianth adnate to the ovary, surrounded by a scaly, cup-like involucre; pistil 1 with 3 styles. **Fruit** a hardened, ovoid nut (acorn) 15–50 mm long, surrounded at the base by the hardened, cup-like, scaly and hairy involucre. Ponderosa pine woodlands, sometimes on bentonitic soil; plains of Powder River and Carter cos. SK to NB south to TX, AL, VA.

BETULACEAE: Birch Family

Monoecious, deciduous shrubs or trees. **Leaves** alternate, petiolate; the blades broadly lance-shaped to ovate with toothed margins and hair along veins beneath. **Flowers** unisexual. **Male flowers** of 4 to 6 bracteate stamens in spirally-arranged, drooping, cylindrical catkins; petals and sepals absent. **Female flowers** with 1 pistil, 2 styles, and an inferior ovary subtended by brownish scales (bracts) arranged in ovoid to cylindrical catkins. **Fruit** a winged or wingless nutlet.[56,137]

1. Female catkins ovoid, hardened, persistent, cone-like; terminal bud present . . . . *Alnus*
1. Female catkins cylindric, disintegrating at maturity; terminal bud absent. . . . . *Betula*

Alnus Mill. Alder

Shrubs or rarely trees with smooth gray to brown bark. **Male flowers** 3 per scale with 4 stamens; catkins pendulous. **Female flowers** 2 per scale; scales fleshy, becoming woody; catkins 1–2 cm long, erect, below the male catkins, persistent. **Fruit** a nutlet 2–3 mm long, sometimes with membranous-winged margins.

   Alnus incana is more common at lower elevations, especially west of the Continental Divide, while *A. viridis* is more common at higher elevations. Taxonomy of N. American and Eurasian taxa is not agreed upon. *Alnus rhombifolia* is reported for west-central MT,[56] but I have seen no material to confirm this disjunct occurrence.

1. Winter buds sharply pointed; catkins developing on twigs of the same year . . . .*A. viridis*
1.Winter buds blunt or rounded; catkins developing on twigs of previous summer 2

2. Nutlets without wings; leaf margins not revolute; usually multi-stemmed shrub. . . . . *A. incana*
2. Nutlets wing-margined; leaf margins revolute; trees . . . . *A. rubra*

Alnus incana (L.) Moench THIN-LEAVED ALDER [*A. tenuifolia* Nutt.] Shrubs to 4 m high with grayish-brown bark. **Twigs** finely pubescent with blunt or rounded winter buds. **Leaf blades** 3–8 cm long with doubly serrate margins, paler, pubescent on veins below. **Inflorescence** borne on twigs of previous season. **Female catkins** 1–2 cm at maturity. **Fruit** a nutlet without wing margins.

Moist to wet soil of swamps, thickets, streambanks; valleys, montane. AK to SK south to CA, NM. Our plants are ssp. *tenuifolia* (Nutt.) Breitung.

Alnus rubra Bong. Red ALDER Tree to 20 m with gray bark. **Twigs** reddish with club-shaped buds. **Leaf blades** to 15 cm long, with revolute margins 1- to 2-times serrate, gray, pubescent and glandular below. **Inflorescence** borne on twigs of previous season. **Fruit** with wings to half as wide as nutlet. Moist lower slopes and stream terraces; valleys. AK south to OR, ID, MT. Our plants from Lincoln Co. seem intermediate to *A. incana*.[380]

Alnus viridis (Chaix) DC. Green ALDER, Sitka ALDER [*A. sinuata* (Regel) Rydb. *A. crispa* (Aiton) Pursh ssp. *sinuata* (Regel) Hulten] Shrubs to 3 m high with grayish bark, the stems often sprawling. **Twigs** with sharp-pointed winter buds. **Leaf blades** 3–10 cm long, lower surface slightly paler than above, the margins 1- to 2-times serrate. **Inflorescence** borne on twigs of the season. **Female catkins** 8–15 mm at maturity. **Fruit** with wings at least half as wide as nutlet. Avalanche slopes, cool wet forest, often along streams; montane, subalpine. Boreal Asia and N. America south to CA, WY, WI, NY. Our plants are ssp. *sinuata* (Regel) A.Love & D.Love. (p.112)

Betula L. Birch

Trees or shrubs with peeling bark marked by lenticels. **Twigs** often with elliptical raised glands and whitish lenticels. **Leaves** ovate to nearly orbicular with serrate margins. **Inflorescence**: bracteate catkins borne on twigs of previous season. **Male flowers** 3 per bract with 2 stamens each. **Female flowers** 3 per 3-lobed bract; catkins ascending from below male catkins, disintegrating upon maturity. **Fruit** a wing-margined nutlet ca. 2 mm long.

Betula pendula Roth, the European white birch, is commonly planted as an ornamental, and seedlings may establish in the wild but do not mature.[226] *Betula occidentalis* and *B. papyrifera* hybridize extensively in eastern WA and adjacent ID and OR[187] as well as western MT. *Betula glandulosa* and *B. pumila* L. hybridize throughout most of Idaho (=*B.* ×*sargentii*).[70] The majority of MT plants show some degree of introgression that is not captured in the following key.

1. Shrubs usually <4 m tall; leaf margins crenate; peaty soil of fens, swamps 2
1. Trees (often multi-stemmed) or large shrubs usually > 4m tall; leaf margins dentate; moist,
    often riparian forests. 3

2. Wing of fruit <1/2 as wide as seed; leaves with ≤10 teeth per side, the blades nearly
    orbicular; plants ≤2 m high . . . .*B. glandulosa*
2. Wing of fruit >1/2 as wide as seed; leaves with >10 teeth per side, the blades
    cuneate-based; plants often >2 m high . . . . *B. pumila*

3. Bark of older trees dark; leaves lacking tufts of light brown hair beneath; twigs
    densely glandular . . . . *B. occidentalis*
3. Bark of older trees white; leaves with tufts of brown hair in axils of veins underneath;
    twigs with few glands . . . . *B. papyrifera*

Betula glandulosa Michx. [including *B.* ×*sargentii* Dugle] Bog BIRCH Shrub to 2 m high with brown to black bark. **Twigs** short-pubescent, densely glandular. **Leaf blades** 1–2 cm long, glandular, paler beneath. **Female catkins** 5–20 mm long. **Fruit** with wings 1/4–1/2 as wide as nutlet. Wet, organic soil of swamps, fens; montane to treeline. AK to Greenland south to CA, CO, WI, NL.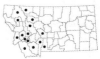

Pure *Betula glandulosa* has small orbicular leaves with less than 10 teeth per side and is known mainly from high elevations or cold montane canyons.[70] The majority of plants south of Glacier National Park and the Flathead Valley show introgression with *B. pumila* and are <3 m tall with cuneate leaves and nutlets with variable-width wings.

Betula occidentalis Hook. Water BIRCH, BLACK BIRCH [*B. fontinalis* Sarg., *B. papyrifera* Marsh var. *occidentalis* (Hook.) Sarg.] Small tree or large shrub to ca. 8 m tall with gray to brown bark, only slightly peeling. **Twigs** glandular, sparsely short-hairy. **Leaf blades** 2–5 cm long, glandular when young, sparsely white-hairy, rounded to pointed at the tip with sharp-dentate margins. **Female catkins** 15–25 mm long. **Fruit** with wings ca. as wide as nutlet. Margins of streams and wetlands; plains, valleys, occasionally montane. AK to MB south to CA, NM, SD.

Plants intermediate between *Betula occidentalis* and *B. papyrifera* occur west of the Continental Divide south to Missoula. These plants have peeling gray bark, nearly glandless twigs and sparse hair on the underside of the leaves.

Plants intermediate between *B. occidentalis* and *B. pumila* occur from Choteau to the Deer Lodge Valley; they are tall shrubs with sharp-pointed leaf margins and cuneate bases.

Betula papyrifera Marsh. [*B. papyrifera* var. *subcordata* (Rydb. ex B.T.Butler) Sarg., *B. papyrifera* var. *commutata* (Regel) Fernald] PAPER BIRCH Small, short-lived tree to 15 m tall with gray or brown bark when young, but white and peeling when older. **Twigs** sparsely glandular and short-hairy. **Leaf blades** 4–7 cm long with sharp-dentate margins and tufts of light brown hair in axils of veins beneath. **Female catkins** 25–35 mm long. **Fruit** with wings as wide or wider than the nutlet. Moist open forest, rocky lower slopes, along streams and lakes; valleys, montane, rarely plains. AK to NL south to WA, CO, MN, PA. (p.112)
On average the leaves are more acuminate than those of *Betula occidentalis*.

Betula pumila L. [*B. glandulosa* Michx. var. *hallii* (Howell) C.L.Hitchc.] Shrubs usually 3–4 m high. **Twigs** short-pubescent, densely glandular. **Leaf blades** cuneate with crenate margins, 1–4 cm long, glandular, paler beneath. **Female catkins** 10–25 mm long. **Fruit** with wings nearly as wide as nutlet. Wet, organic soil of swamps, fens; valleys to lower montane. YT to NS south to CA, CO, IL, NJ. Our plants are var. **glandulifera** Regel. See *Betula glandulosa, B. occidentalis*.

NYCTAGINACEAE: Four-O'Clock Family

Annual or perennial herbs. **Stems** erect or trailing, often branched. **Leaves** opposite, petiolate, simple with entire margins. **Inflorescence** bracteate, the bracts forming a calyx-like involucre. **Flowers** perfect, radially symmetrical, hypogynous; sepals petal-like, united below; petals absent; stamens ca. 5; pistil 1. **Fruits** achenes, sometimes enclosed in the winged calyx base.

1. Lower 1/3 of involucral bracts united; fruit without wings . . . . *Mirabilis*
1. Involucral bracts separate nearly to the base; fruit winged 2

2. Fruits with membranous wings wider than and surrounding the achene . . . . *Tripterocalyx*
2. Fruits with wings ca. as wide as achene above and tapering to a point. . . . . *Abronia*

Abronia Juss. Sand verbena

Abronia fragrans Nutt. ex Hook. Taprooted perennial herbs. **Stems** sparsely pubescent, glandular, ascending, 10–15 cm long. **Leaf blades** glabrous, ovate to lanceolate, 1–3 cm long. **Inflorescence** axillary, long-stemmed clusters subtended by 5 separate, ovate involucral bracts 5–10 mm long. **Flowers:** calyx white, corolla-like, salverform, ca. 12 mm long with 5 bilobed lobes; stamens ca. 5, included; style included; stigma linear. **Fruit** broadly conical, pubescent, ca. 6 mm long with 3- to 5-veined wings above. Sandy or gravelly soil of grasslands, steppe, roadsides; plains, valleys. MT south to NM, TX, Mexico.

1. Wings of mature fruit with pad-like swellings on top. . . . . var. *elliptica*
1. Wings of mature fruit not dilated. . . . . .var. *fragrans*

Abronia fragrans var. *fragrans* is found in the Great Plains; *A. fragrans* var. *elliptica* (A.Nelson) M.E.Jones occurs in intermountain valleys. MT plants are smaller with smaller flowers and fruits than those from the main range of the species.

Tripterocalyx (Torr.) Hook. Sand puffs

Tripterocalyx micranthus (Torr.) Hook. [*Abronia micrantha* Torr.] Annual herbs. **Stems** glandular-pubescent, trailing to ascending, 5–20 cm long. **Leaf blades** ovate, 15–40 mm long, glabrous with ciliate margins. **Inflorescence** axillary, pedunculate clusters subtended by 5 separate, ovate involucral bracts 4–6 mm long. **Flowers** with a white, corolla-like, tubular calyx, ca.10 mm long with 5 flaring, bilobed lobes; stamens 5, included; style included; stigma linear. **Fruit** ellipsoid, scabrous, 1–2 cm long; the achene completely surrounded by the 3 membranous, veiny wings.[138] Sandy soil of grasslands, river terraces; plains, valleys. AB, SK south to CA, NM.

Mirabilis L. Four-O'Clock

Taprooted, perennial herbs. **Stems** from a branched caudex, trailing to ascending. **Inflorescence** terminal or axillary panicles of pedunculate, few-flowered, calyx-like, saucer-shaped involucres, the 5 lobes united below, expanding in fruit. **Flowers** funnelform, 5-lobed; stamens 5, exserted. **Fruit** obovoid, ribbed, hardened.[378]

1. Leaf blades ovate ≤2 times as long as wide; petioles ≥10 mm long. . . . . *M. nyctaginea*
1. Leaf blades linear to narrowly ovate; petioles ≤10 mm long. 2

2. Fruits wrinkled, without tubercles . . . . *M. linearis*
2. Fruits with hair-tipped tubercles . . . . *M. albida*

Mirabilis albida (Walter) Heimeri **Stems** erect or ascending, 20–80 cm, glabrate to puberulent. **Leaves** linear to narrowly ovate, sessile to short-petiolate; the blade 4–10 cm long, glabrate, sometimes glaucous below. **Inflorescence** viscid-pubescent, openly branched; involucres green, ca. 4 mm long, mostly 3-flowered. **Flowers** pink to purple, 8–10 mm long. **Fruits** 5–6 mm long, with hair-tipped tubercles. Sandy soil of grasslands; plains. Throughout most of N. America except the northwest and northeast U.S. One collection from Sheridan Co.

Mirabilis linearis (Pursh) Heimeri [*M. hirsuta* (Pursh) MacMill.] **Stems** ascending, 15–80 cm, glandular-pubescent above, glaucous below. **Leaves** linear to lanceolate, sessile to short-petiolate; the blade 4–9 cm long. **Inflorescence** openly branched; involucres green, ca. 5 mm long, mostly 3-flowered. **Flowers** pink to purple, 6–10 mm long. **Fruits** olive-brown, 4–5 mm long, pubescent, 5-ribbed and cross-ribbed. Sandy or gravelly soil of grasslands, tolerant of saline and alkaline conditions; plains, valleys. MT to MN south to AZ, TX, Mexico. Our plants are var. *linearis*.

Mirabilis nyctaginea (Michx.) MacMill. Wɪʟᴅ ꜰᴏᴜʀ-ᴏ'ᴄʟᴏᴄᴋ **Stems** erect 35–60 cm, mostly glabrous with swollen nodes. **Leaf blades** ovate-lanceolate, glabrous, 4–8 cm long. **Inflorescence** terminal and axillary, openly cymose with umbellate clusters at branch tips; involucres 3-flowered, the bracts 5–10 mm long, light green, veiny. **Flowers** pink, ca. 4 mm long. **Fruits** pubescent, 4–6 mm long, 5-ribbed and warty. Sparsely vegetated gravelly or sandy soil of roadsides; plains, valleys. Most of temperate N. America.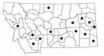

CACTACEAE: Cactus Family

Perennial herbs or subshrubs. **Stems** succulent with spine-bearing areoles, sometimes segmented, often ribbed or conical-tuberculate, spheric to cylindric. **Leaves** vestigial or absent. **Spines** bristle-like to rigid. **Flowers** perfect, radially symmetrical, 1 per areole, epigynous; sepals and petals (tepals) numerous, the outer herbaceous, the inner petal-like, borne on top of hypanthium; stamens numerous; ovary inferior with 1 style and several stigmas. **Fruit** fleshy, many-seeded, berry-like.[44]

1. Stems jointed, cylindric or flattened, not tuberculate . . . . *Opuntia*
1. Stems globose, tuberculate, not jointed 2

2. Flowers arising from base of tubercles; aereoles with white hair at base of spines . . . . *Coryphantha*
2. Flowers arising from near tip of tubercle; aereoles without white hair . . . . *Pediocactus*

Coryphantha (Engelm.) Lem. Pincushion cactus

Stems unsegmented, globose, with spirally-arranged tubercules grooved on the upper surface. **Spines** 5 to 50 per areole, needle-like with white hair at the base. **Flowers** borne at base of tubercles near stem apex; tepals glossy with ragged margins. **Fruits** spineless, ellipsoid.

    Coryphantha may be polyphyletic with respect to *Mammillaria*,[456] and the two are sometimes combined under the latter name.

1. Flowers magenta, remnants persistent on the fruit; some spines brown or red. . . . . *C. vivipara*
1. Flowers yellowish, remnants not persistent; spines white . . . . *C. missouriensis*

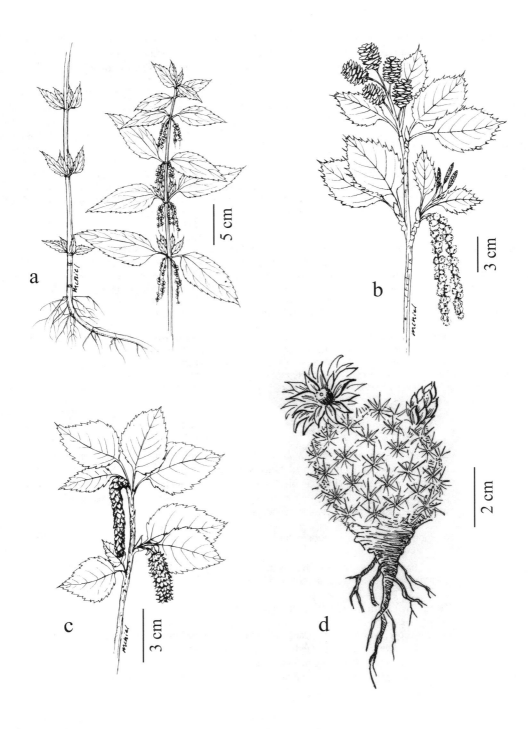

Plate 14. a. *Urtica dioica* ssp. *gracilis*, b. *Alnus viridis*, c. *Betula papyrifera*, d. *Coryphantha vivipara*

Coryphantha missouriensis (Sweet) Britton & Rose [*Mammillaria missouriensis* Sweet] Plants simple to rarely clumped. **Stems** to 5 cm high, deeply buried in soil. **Spines** white, weathering to brown, 6 to 21 per areole, 4–15 mm long; central spines 0 or 1. **Flowers** 1–2 cm long; tepals 13 to 25, yellowish with green to pink midstripe. **Fruit** red, 6–10 mm long. Grassland, steppe; plains. MT, ND south to TX. Our plants are var. *missouriensis*.

Coryphantha vivipara (Nutt.) Britton & Rose Plants simple or clumped, similar to *C. missouriensis*. **Stems** 3–7 cm high, <1/2 buried in soil. **Spines** white, weathering to brown, 11 to 55 per areole, 7–22 mm long; central spines 3 to 6, reddish. **Flowers** 2–3 cm long; tepals 21 to 56, magenta. **Fruit** green to brown, 12–28 mm long. Grassland, steppe; plains, valleys. AB to MB south to CA, NM, TX. Our plants are var. *vivipara*. (p.112)

Opuntia Mill. Prickly pear

Subshrubs with erect to prostrate stems forming clumps or mats. **Stems** jointed, segments glabrous, fleshy and ca. flattened with low tubercules. **Spines**: straight, rigid spines and brownish bristles (glochids) from areoles. **Flowers** yellow, borne near stem tips; anthers yellow; stigma with 5 to 8 short, green lobes. **Fruit** obovoid, short-spiny.

1. Stem segments conspicuously flattened, >5 cm long . . . . *O. polyacantha*
1. Stem segments nearly round in cross-section, mostly <5 cm long . . . . *O. fragilis*

Opuntia fragilis (Nutt.) Haw. JUMP CACTUS **Stems** prostrate, mat-forming; the segments subcylindric, 2–4 cm long, easily detached. **Spines** 3 to 8 per areole, gray, 5–20 mm long. **Flowers** with tepals ca. 15 mm long. **Fruit** tan, 1–3 cm long. Sandy or gravelly soil of grasslands, steppe; plains, valleys, rarely montane. BC to ON south to NM, TX, IL.

Plants rarely flower. Stem segments attach themselves to passing animals and can start new plants after dispersing. Colonies many meters across can occur in very sandy soil.

Opuntia polyacantha Haw. **Stems** prostrate to ascending; segments flattened, obovate, 4–12 cm long. **Spines** yellow-brown, 5 to 11 per areole, 1–4 cm long. **Flowers** with tepals 25–35 mm long. **Fruit** tan to brown, 15–45 mm long. Sandy or fine-textured soil of grasslands, steppe, often in barren habitats; plains, valleys. BC to SK south to NM, TX, NE.

Pediocactus Britton & Rose Hedgehog cactus

Pediocactus simpsonii (Engelm.) Britton & Rose **Stems** unsegmented, solitary or clumped, 7–12 cm high, buried in the soil, globose with spirally-arranged tubercles. **Spines** 20 to 40 per areole; the central 3 to 10, straight, yellow to reddish, 6–12 mm long; marginal spines smaller, whitish. **Flowers** 10–15 mm long, borne near tip of tubercles; tepals yellow-green to magenta, the outer with greenish midstripe. **Fruits** narrow-globose, 6–8 mm long, green-tinged with red. Fine-textured, often stony soil of steppe, grasslands; valleys, lower montane in Beaverhead Co. ID, MT south to NV, AZ, NM. Our plants are var. *simpsonii*.

Similar to *Coryphantha* but can be distinguished by the lack of white hairs at the base of spines.

AMARANTHACEAE: Amaranth Family

Often monoecious or dioecious shrubs, annual or perennial herbs. **Leaves** simple, mostly petiolate below, sometimes sessile above, usually alternate, sometimes farinose. **Flowers** small, greenish, perfect or unisexual; sepals 0-5; petals absent; stamens 1-5; styles 1-5; ovary superior. **Fruit** achenes or utricles, usually enclosed in calyx or pistillate bracts.

Chenopodiaceae and Amaranthaceae have traditionally been considered separate families. However, it appears that some genera in the Chenopodiaceae are more closely related to members of the Amaranthaceae sensu stricto than to other members of the Chenopodiaceae. Expanding the Amaranthaceae to include the Chenopodiaceae results in a monophyletic group.[295] Farinose hairs, typical of many genera are short and inflated but become flake-like with dessication.

1. Stems woody at least at the base. 2
1. Plants herbaceous annuals or perennials 6

Amaranthus L. Pigweed, amaranth

Annual, monoecious or dioecious herbs. **Leaves** alternate, petiolate with usually entire margins.
Inflorescences glomerules arranged in terminal and/or axillary spikes or panicles. **Flowers** hypogynous,
unisexual, inconspicuous, subtended by persistent bracts; petals lacking; sepals 3 to 5, separate; stamens 2
to 5; pistil 1 with 2 to 3 stigmas. **Fruit** a glabrous utricle enclosed by calyx; seeds subglobose to lenticular,
usually smooth and shiny.

These are weeds native to N. America; some may be introduced from farther south on the continent.
Many of the species are reported to hybridize where they occur together.[271] *Amaranthus caudatus* L.
(Love-lies-bleeding), usually with a drooping, red inflorescence, is a common ornamental native to tropical
America; it rarely escapes and does not persist in MT. *Amaranthus spinosus* L., with spiny nodes, was once
collected from potting soil in Flathead Co.

1. Flowers in axillary glomerules; terminal spikes or panicles absent 2
1. Flowers in terminal and axillary spikes or panicles as well as axillary glomerules 4

2. Female flowers with 1 to 3 sepals ≤1 mm long . . . . *A. californicus*
2. Female flowers with 3 to 5 sepals >1 mm long 3

3. Female bracts 2 to 3 times longer than sepals; seeds ≤1 mm long . . . . *A. albus*
3. Female bracts ca. as long as the sepals; seeds ≥1.3 mm long . . . .*A. blitoides*

4. Plants dioecious 5
4. Plants monoecious 6

5. Some female bracts much longer than sepals, greenish . . . . *A. arenicola*
5. Female bracts ca. as long as sepals, reddish . . . .*A. tuberculatus*

6. Sepals rounded or blunt at the tip, often with an excurrent midvein. *A. retroflexus*
6. Sepals acute or tapering to an aristate tip 7

7. Female bracts 5 mm long. . . . . *A. powellii*
7. Female bracts 3–4 mm long. . . . .*A. hybridus*

Amaranthus albus L. Monoecious. **Stems** erect, 10–75 cm with glabrous to pubescent foliage, branched at the base, becoming a tumbleweed. **Leaf blade** 1–3 cm long, longer than the petiole, obovate with a minute spine tip. **Inflorescences** axillary glomerules; bracts lanceolate and long attenuate, spine-tipped, 2–3 mm long. **Flowers**: sepals 3, narrowly lanceolate, acute, 1–2 mm long; stamens 3. **Utricle** 1–2 mm, as long or longer than sepals; seed <1 mm long. Cultivated fields, disturbed pastures, roadsides, streambanks; plains, valleys. Throughout N. America.

Amaranthus arenicola I.M.Johnst. Dioecious. **Stems** erect, 50–100 cm with glabrous foliage. **Leaf blade** oblanceolate, 1–8 cm long, longer than petiole. **Inflorescence** terminal and axillary spikes or panicles; bracts ovate-attenuate, 1–3 mm long. **Flowers**: sepals 5, 2–3 mm long with dark midveins; stamens 5. **Utricle** ca. 1.5 mm long, shorter than sepals. Streambanks, agricultural fields, roadsides; plains, valleys. Most of temperate U.S. Collected in Missoula and Phillips cos.

Amaranthus blitoides S.Watson [*A. graecizans* L. misapplied] Monoecious. **Stems** branched at the base, prostrate to ascending, 5–50 cm with glabrous to pubescent foliage. **Leaf blade** obovate, 10–25 mm long, mostly longer than the petiole. **Inflorescence** axillary glomerules; bracts broadly lanceolate, 2–3 mm long with a spine tip. **Flowers**: sepals 3 to 5, narrowly ovate, acuminate, 2–3 mm long; stamens 3. **Utricle** 2–3 mm, as long as sepals; seed ca. 1.3 mm long. Disturbed soil of fields, roadsides, margins of streams, ponds; plains, valleys. Throughout N. America; native in the west.

Amaranthus californicus (Moq.) S.Watson Monoecious. **Stems** branched at the base, prostrate, mat-forming, 5–40 cm with glabrous, green to reddish foliage. **Leaf blade** obovate to spatulate with wavy margins, 5–15 mm long, longer than the petiole. **Inflorescence** axillary glomerules; bracts linear, spine-tipped, 1–1.5 mm long. **Flowers**: sepals 1 to 3, lanceolate, acuminate, unequal, the longest ca. 1 mm; stamens 3. **Utricle** 2–3 mm long, green-lined, longer than sepals; seed <1 mm long. Vernally wet, often alkaline soil around ponds, roadsides; plains, valleys. AB, SK south to CA, UT, TX.

Amaranthus hybridus L. Monoecious. **Stems** erect, 30–100 cm with glabrous or sparsely hairy foliage. **Leaf blade** lanceolate to ovate, 1–8 cm long, longer than petiole. **Inflorescence** green, dense, cylindric spikes in terminal and sometimes axillary panicles; bracts lanceolate, spine-tipped, 3–4 mm long. **Flowers**: sepals 5, lanceolate, 1.5–2 mm long with excurrent midveins; stamens 4 to 5. **Utricle** 1.5–2 mm long, shorter than sepals. Cultivated fields, roadsides; plains, valleys. Introduced throughout N. America; native to S. America.

Amaranthus powellii S.Watson Monoecious. **Stems** erect, often red-striped, 20–100 cm, branched above with glabrous to sparsely pubescent foliage. **Leaf blade** lanceolate to ovate, 2–8 cm long, mostly as long as the petiole. **Inflorescence** green, terminal and axillary, cylindric panicles; bracts linear-lanceolate, spine-tipped, ca. 5 mm long. **Flowers**: sepals 3 to 5, elliptic, 1.5–4 mm long with excurrent midveins; stamens 3 to 5. **Utricle** 2–3 mm long, shorter than sepals. Cultivated fields, roadsides, streambanks; valleys. Throughout N. America; native to western N. America.

Amaranthus retroflexus L. Monoecious. **Stems** erect, reddish below, 10–100 cm with sparsely pubescent foliage. **Leaf blade** lanceolate to ovate, 2–10 cm long, just longer than petiole. **Inflorescence** often red-tinged, dense, terminal and axillary panicles; bracts lanceolate-attenuate, 3–5 mm long, short-pointed. **Flowers**: sepals 5, ca. 3 mm long with blunt to rounded

tips and excurrent midveins; stamens 4 to 5. **Utricle** ca. 2 mm, nearly as long as sepals. Cultivated fields, roadsides, streambanks; plains, valleys. Throughout N. America. (p.130)

Our most common pigweed; the blunt sepals with excurrent midvein are diagnostic.

Amaranthus tuberculatus (Moq.) J.D.Sauer [*A. tamariscinus* Nutt., *A. rudis* J.Sauer] Dioecious. **Stems** erect, 20–80 cm with glabrous foliage. **Leaf blade** lanceolate to ovate, 2–7 cm long, longer than petiole. **Inflorescences** narrow, open, axillary spikes and panicles; bracts ovate 1–2 mm long, mucronate. **Flowers**: sepals 0 to 5, lanceolate, 2–3 mm long; stamens 5. **Utricle** ca. 1.5 mm, shorter than sepals. Railroad yards; valleys. Most of temperate U.S. Introduced in Europe. Collected in Missoula Co.

Atriplex L. Orach, saltbush

Monoecious or dioecious shrubs, annual, or perennial herbs. **Stems** usually branched. **Leaf blades** often farinose (sessile flake-like vestiture) with entire or dentate margins. **Inflorescence** terminal or axillary, of solitary or clustered flowers in spikes or spike-like panicles. **Male flowers** with a 3- to 5-parted calyx; stamens 3 to 5. **Female flowers** enclosed in 2 small pistillate bracts; calyx absent or minute; stigmas 2. **Fruit** usually with enlarged pistillate bracts; the seeds flattened, mostly vertical.[433]

Mature fruits, usually present in late summer or fall, are needed for confident determinations. *Atriplex prostrata* Boucher ex DC. is reported for MT,[433] but I am not aware of any voucher specimens. Four of our species, *A. argentea*, *A. powellii*, *A. rosea*, and *A. truncata*, are densely gray-farinose. Fine, reticulate veination will be revealed under 10X when the grayish vestiture is scraped away with a sharp blade.

1. Shrubs or subshrubs 2
1. Herbaceous, taprooted annuals 4

2. Stems spiny; leaves nearly as broad as long . . . . *A. confertifolia*
2. Stems without spines; leaves linear to oblong 3

3. Shrubs, often >50 cm high with woody stems . . . . *A. canescens*
3. Subshrubs to 50 cm high, woody only at the base . . . . *A. gardneri*

4. Mature pistillate bracts suborbicular, entire-margined, lacking appendages 5
4. Mature pistillate bracts lanceolate, ovate or deltoid, usually with tooth-like margins or appendages 6

5. Mature pistillate bracts >5 mm long, cordate-based . . . . *A. hortensis*
5. Mature pistillate bracts ≤5 mm long, cuneate-based . . . . *A. heterosperma*

6. Fully expanded leaves green, although sometimes weakly farinose 7
6. Most mature leaves gray, the surface heavily farinose (see genus description) 9

7. Mature pistillate bracts 0.5–2 mm long . . . . *A. suckleyi*
7. Mature pistillate bracts >2 mm long 8

8. Mature pistillate bracts inflated around the seed with a thick, spongy, often
   tuberculate layer of cells . . . . *A. subspicata*
8. Mature pistillate bracts without an inflated spongy layer . . . . *A. patula*

9. Leaves sinuate-dentate, sometimes hastate at the base; pistillate bracts with appendages
   on the face . . . . *A. rosea*
9. Leaves with entire or gently-wavy margins; bracts sometimes with appendages on the
   margin but not on the face 10

10. Mature pistillate bracts elliptic with deeply lobed margins throughout . . . . *A. argentea*
10. Upper portion of mature pistillate bracts with shallowly lobed margins or an apical tooth 11

11. Lower portion of mature pistillate bract tuberculate on the face . . . . *A. powellii*
11. Lower portion of mature pistillate bract smooth . . . . *A. truncata*

Atriplex argentea Nutt. Monoecious annuals with farinose foliage. **Stems** erect, branched, 5–50 cm. **Leaves** often opposite below; the blades lanceolate to triangular, sometimes hastate, 1–4 cm long; margins entire **Inflorescence** axillary glomerules and short terminal spikes, female below male. **Male flowers** 4- to 5-parted. **Female flowers** lacking a calyx; pistillate bracts united to the middle. **Mature pistillate bracts** hardened in the center, 3–6 mm long with lobed margins and often tuberculate faces. Barren, fine-textured to sandy, often saline or alkaline soil of steppe, eroding slopes, badlands; plains, valleys. BC to MB south to CA, TX, Mexico. Our plants are var. *argentea*. (p.118)

The Flathead Co. collection was made over 100 years ago along a railroad grade.

Atriplex canescens (Pursh) Nutt. Four-wing saltbush Dioecious shrubs 40–100 cm tall with gray-scurfy foliage. **Stems** branched with peeling bark below. **Leaves** sessile, linear to narrowly oblong, 2–5 cm long with entire margins. **Inflorescence** glomerules in axillary or terminal spikes. **Male flowers** 5-parted, tawny. **Female flowers** lacking a calyx; pistillate bracts united to the tip. **Mature pistillate bracts** orbiculate, 3–6 mm long with 4 prominent wavy-margined to lobed wings. Sagebrush steppe and saline soil of badlands; plains, valleys to montane. AB to ND south to CA, TX, Mexico. (p.118)

Welsh[433] provides a lengthy discussion of intermediates between *Atriplex canescens* and *A. gardneri* and suggests they might even be thought of as one species. Both are quite variable throughout their large ranges, but the two are usually distinct in MT.

Atriplex confertifolia (Torr.& Frem.) S.Watson Shadscale Dioecious shrubs 40–80 cm tall with scurfy foliage. **Stems** rigid, spiny, intricately branched. **Leaf blades** ovate to orbiculate, 5–20 mm long with entire margins. **Inflorescence** axillary glomerules or dense spike-like panicles. **Male flowers** 4- to 5-parted. **Female flowers** lacking a calyx; pistillate bracts united <1/2-way. **Mature pistillate bracts** ovate to elliptic, 5–13 mm long with entire margins above. Fine-textured, saline soils of stream terraces, badlands, often with sagebrush or greasewood; plains, valleys. OR to ND south to CA, AZ, TX. (p.118)

Atriplex gardneri (Moq.) D.Dietr. [*A. aptera* A.Nelson, *A. falcata* (M.E.Jones) Standl., *A. nuttallii* S. Watson misapplied] Gardner's saltbush Low, mostly dioecious subshrub to 50 cm high. **Stems** prostrate to usually decumbent with gray-scurfy foliage. **Leaves** short-petiolate; the blades linear to spatulate, 5–40 mm long with entire margins, the lowest sometimes opposite. **Inflorescence** glomerules in terminal spikes or dense panicles. **Male flowers** yellow or brown, 5-parted. **Female flowers** lacking a calyx; pistillate bracts united above the middle. **Mature pistillate bracts** lanceolate to orbicular, 2–5 mm long with toothed or winged margins and facial tubercles. Fine-textured, saline soils of stream terraces, badlands, steppe, grasslands; plains, valleys, montane. AB to MB south to CA, AZ, NM. (p.118)

Three vars. of *Atriplex gardneri* are reported for MT:[433] var *gardneri* has globose to flattened pistillate scales with lateral teeth and facial tubercles; var. *aptera* (A.Nelson) S.L.Welsh & Crompton has ellipsoid pistillate scales with 4 lobed lateral wings and is thought to be a hybrid between *A. gardneri* and *A. canescens*; var. *falcata* (M.E.Jones) S.L.Welsh has lanceolate pistillate scales without lateral teeth and low facial tubercles present or not. However, these forms seem to have little ecological distinction in our material, and few of our specimens can be confidently placed.

Atriplex heterosperma Bunge Monoecious annuals, farinose when young but glabrous with age. **Stems** erect, 30–100 cm, branched or simple. **Leaf blades** 2–8 cm long, triangular, hastate with entire to toothed margins. **Inflorescence** a leafy-bracteate, loosely branched, pyramidal panicle. **Male flowers** 5-parted. **Female flowers** lacking a calyx; pistillate bracts distinct. **Mature pistillate bracts** orbiculate with entire margins, of two sizes: 4–6 mm and 2–3 mm. Disturbed soil of roadsides, streambanks; plains, valleys. Introduced BC to ON south to CA, CO, NY. Native to Eurasia.

Atriplex hortensis L. Orach Monoecious annuals with green to red, glabrous foliage. **Stems** erect, 50–150 cm. **Leaf blades** opposite below, the blades lanceolate to broadly ovate, 5–20 cm long with mostly entire margins. **Inflorescence** of spikes in terminal and axillary panicles. **Male flowers** with a 5-parted calyx and 5 stamens. **Female flowers** dimorphic, some bracteate without a calyx, others ebracteate with a 5-parted calyx. **Fruit** either a horizontal seed surrounded by the calyx or a vertical seed surrounded by swollen, wing-like orbicular pistillate bracts 5–10 mm wide. Roadsides, fields; plains, valleys. Introduced to much of northern U.S. and southern Canada. Native to Asia.

Atriplex patula L. [*A. patula* var. *hastata* (L.) A.Gray] Monoecious or dioecious annual, farinose to glabrous. **Stems** erect to ascending, 20–100 cm. **Leaf blades** glabrate, up to 10 cm long, narrowly lanceolate to ovate-hastate with entire to deeply dentate margins. **Inflorescence** axillary and terminal, interrupted spikes of glomerules. **Male flowers** mostly 5-parted. **Female flowers** with a calyx, the pistillate bracts nearly separate. **Mature pistillate bracts** swollen, deltoid, 3–6 mm long with entire to dentate margins and rarely tuberculate faces. Roadsides, fields; plains, valleys, montane. Much of N. America.

Some authors consider this species introduced from Eurasia,[433] while others suggest it is native to N. America.[187] Varieties have been recognized based on leaf shape, but they have little ecological or geographic distinction in MT.

Atriplex powellii S.Watson Dioecious or monoecious annual with farinose foliage. **Stems** erect, branched, 10–40 cm. **Leaf blades** ovate with acute tips and entire margins, 1–3 cm long, 3-nerved from the base. **Inflorescence** axillary glomerules, female below male. **Male flowers** 4- to 5-parted. **Female flowers** lacking a calyx; pistillate bracts united to the tip. **Mature pistillate bracts**

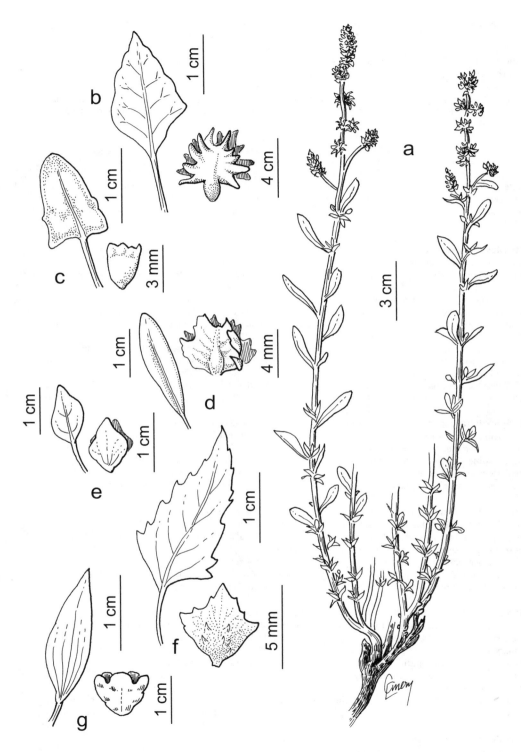

Plate 15. *Atriplex*. **a.** *A. gardneri,* **b.** *A. argentea,* **c.** *A. truncata,* **d.** *A. canescens,* **e.** *A. confertifolia,* **f.** *A. rosea,* **g.** *A. suckleyi*

fiddle-shaped, 3–4 mm long with toothed margins and tuberculate faces below. Barren, fine-textured, often saline or alkaline soil of eroding slopes, badlands, salt flats, drying ponds; plains. AB, SK south to AZ, NM. Our plants are var. *powellii*.

Atriplex prostrata Boucher ex DC. Monoecious annuals with nearly glabrous foliage. **Stems** erect or ascending, 10–100 cm, branched from the base. **Leaf blades** 2–10 cm long, triangular with basal lobes and entire to toothed margins. **Inflorescence** terminal, short spikes of glomerules. **Male flowers** 4- to 5-parted. **Female flowers** lacking a calyx; pistillate bracts distinct. **Mature pistillate bracts** glabrous, triangular, 3–5 mm long. Barren, moist, saline soil along streams, wetlands; plains. Introduced across much of temperate N. America; native to Eurasia. One collection from southern Carbon Co.

Atriplex rosea L. Monoecious annuals with farinose foliage. **Stems** erect, branched, 20–50 cm. **Leaf blades** 1–5 cm long, ovate to triangular, sometimes hastate with sinuate-dentate margins. **Inflorescence** terminal and axillary, of short spikes of glomerules. **Male flowers** 4- to 5-parted. **Female flowers** lacking a calyx; pistillate bracts united below. **Mature pistillate bracts** farinose, triangular, 4–5 mm long with dentate margins. Roadsides, gardens, fields; plains, valleys. Introduced into much of temperate N. America; native to Eurasia. (p.118)

Atriplex subspicata (Nutt.) Rydb. [*A. dioica* Raf. misapplied] Monoecious annuals with sparsely farinose but still green foliage. **Stems** erect, 5–30 cm, branched from the base. **Leaves** sometimes opposite below, the blade 5–30 mm long, ovate to lanceolate, sometimes hastate, with mostly entire margins. **Inflorescence** terminal, short spikes of glomerules. **Male flowers** 4- to 5-parted. **Female flowers** lacking a calyx; pistillate bracts separate to near base. **Mature pistillate bracts** glabrous to sparsely farinose, ovate, 2–8 mm long, with a spongy, often tuberculate layer at the base. Drying mud around saline ponds, wetlands; valleys to montane. AB, SK south to CA, NM, KS.

Welsh [433] combines the eastern *Atriplex dioica* Raf. with *A. subspicata* under the former name. Mature bracts of the similar *A. patula* lack the basal spongy layer.

Atriplex suckleyi (Torr.) Rydb. [*A. dioica* (Nutt.) J.F.Macbr. misapplied] Rillscale Monoecious annuals with mostly glabrous, fleshy, often reddish foliage. **Stems** erect or ascending, branched, 10–30 cm. **Leaves** sessile, lanceolate to narrowly ovate, 1–3 cm long with entire margins. **Male flowers** in short terminal spikes of glomerules, 4- to 5-parted. **Female flowers** solitary in leaf axils below the males; calyx 3- to 4-parted; pistillate bracts ovate, united to the tip. **Mature pistillate bracts** scurfy, ≤2 mm long, united to the tip. Barren, fine-textured, often saline soil of eroding slopes, badlands; plains. AB, SK south to CO, SD. (p.118)

Atriplex truncata (Torr. ex S.Watson) A.Gray Monoecious annuals with farinose foliage. **Stems** erect, simple or branched, 10–50 cm. **Leaf blades** ovate to triangular with entire or wavy margins, 5–20 mm long. **Inflorescence** axillary glomerules. **Male flowers** 3- to 5-parted. **Female flowers** lacking a calyx; pistillate bracts united to the tip. **Mature pistillate bracts** cuneate, 2–3 mm long, rounded at the base, flattened at the summit with a wavy margin. Alkaline soil of steppe, moist meadows; valleys, montane. BC to SK south to CA, NV, UT. (p.118)

Axyris L. Russian pigweed

Axyris amaranthoides L. Monoecious annual herb with sparse, stellate-pubescent foliage. **Stems** erect, branched, 40–100 cm. **Leaves** petiolate, the blades lanceolate, 2–8 cm long with entire margins. **Inflorescence** long-leafy bracteate panicles with spike-like branches. **Flowers** unisexual. **Male flowers** minute, 3-parted, in terminal spikes. **Female flowers** each subtended by 3 leaf-like bracts, below male flowers or in long-pedunculate panicles; calyx 3-parted, ca. 2 mm long, whitish. **Fruit** a flattened utricle, winged above, 2–3 mm long, surrounded by a slightly enlarged calyx. Roadsides, disturbed soil; plains. Introduced to northern U.S. and adjacent Canada; native to Eurasia. One collection from Sheridan Co. over 60 years ago.

Bassia All.

Bassia hyssopifolia (Pall.) Kuntze Annual herbs with long-hairy foliage. **Stems** erect, branched, to 80 cm. **Leaves** sessile, linear-lanceolate, to 2 cm long with entire margins. **Inflorescence** terminal and lateral, bracteate spikes or flowers clustered in leaf axils. **Flowers** mostly perfect; calyx ca. 2 mm across, the 5 segments united below, each with a hooked spine

up to 2 mm long; stamens 5. **Fruit** flattened, enclosed by calyx; seed horizontal. Saline wetlands, roadsides, disturbed fields; plains, valleys. Introduced into most of western, temperate N. America. Native to Eurasia.

Plants are similar to the more common *Kochia scoparia,* which does not have hooked spines on the calyx lobes.

Chenopodium L. Goosefoot

Annual herbs. **Stems** usually branched. **Leaves** petiolate, the blades often farinose; margins entire or dentate. **Inflorescence** terminal spikes or axillary glomerules. **Flowers** mostly perfect, ebracteate; calyx 3- to 5-parted, united below; stamens 1 to 5; pistil with 1 style and 2 stigmas. **Fruit** utricles or achenes, enclosed in calyx; seeds flattened, horizontal or vertical.[91]

Chenopodium pratericola, C. desiccatum, and *C. subglabrum* have sometimes been included under *C. leptophyllum. Chenopodium atrovirens* was included under *C. fremontii.*[187] It would probably key to *C. subglabrum*, though some leaves have one or two dentations. *Chenopodium watsonii* A.Nelson is reported for MT,[91] but I have seen no specimens. *Chenopodium standleyanum* Aellen is reported for MT,[91,151] but I have seen no specimens, and it is a plant of the eastern Great Plains.[152]

1. Leaves below the inflorescence glabrous both above and beneath. 2
1. Leaves farinose at least below (sparse in *C. subglabrum, C. fremontii*) 6

2. Glomerules >4 mm wide, in an ebracteate spike, ebracteate above . . . . *C. capitatum*
2. Glomerules <4 mm wide, often in a bracteate inflorescence 3

3. Leaves linear to narrowly lanceolate with entire margins. . . . . *C. subglabrum*
3. Leaves lanceolate to triangular, some with toothed margins 4

4. Seeds horizontal, flattened on the top . . . . *C. simplex*
4. Seeds vertical in the calyx, flattened on the sides 5

5. Calyx segments united only at the base, becoming reddish in fruit . . . .*C. rubrum*
5. Calyx segments united nearly to their tips, green. . . . .*C. chenopodioides*

6. Stems usually prostrate; leaves farinose beneath but glabrous above . . . . *C. glaucum*
6. Stems erect or ascending; leaves often farinose above. 7

7. Leaves with toothed or lobed margins to near the tip. 8
7. Leaf margins entire around the upper half at least. 10

8. Stems glandular . . . . *C. botrys*
8. Stems not glandular 9

9. Seeds smooth. . . . . *C. album*
9. Seeds honeycomb pitted . . . . *C. berlandieri*

10. Leaf blades with entire margins 11
10. At least some leaf blades toothed or lobed at the base 16

11. Leaf blades ≤4 times as long as wide. 12
11. Leaf blades linear, ≥4 times as long as wide 13

12. Seeds honeycomb pitted . . . . *C. watsonii*
12. Seeds wrinkled . . . . *C. atrovirens*

13. Calyx and leaves glabrous or nearly so . . . . *C. subglabrum*
13. Calyx and often the leaves farinose 14

14. Leaves with 1 vein from the base. . . . . *C. leptophyllum*
14. Leaves with 3 veins from the base. 15

15. Calyx spreading from fruit at maturity. . . . . *C. pratericola*
15. Calyx tightly enclosing fruit at maturity. . . . . *C. desiccatum*

16. Leaves papery-thin when dry, often nearly glabrous; seeds smooth . . . . *C. fremontii*
16. Leaves thicker, farinose below; seeds wrinkled . . . . *C. incanum*

Chenopodium album L. LAMB'S QUARTERS **Stems** erect, simple or branched, farinose, 10–100 cm. **Leaf blades** farinose, lanceolate to ovate, 1–6 cm long, margins at least partly dentate. **Inflorescence** glomerules, 2–4 mm across, in terminal and axillary compound spikes. **Flowers:** calyx segments 5, united at the base, ca 1mm long; stamens 5. **Seed** round, smooth, 0.9–1.6 mm. Roadsides, fields, gardens; plains, valleys. Introduced throughout N. America; native to Europe.

The presumed ubiquity of this species is based in part on the frequent misidentification of *Chenopodium berlandieri.*

Chenopodium atrovirens Rydb. [*C. fremontii* S.Watson var. *atrovirens* (Rydb.) Fosberg] **Stems** erect to ascending, branched, farinose, 10–65 cm. **Leaf blades** farinose beneath, 3-veined, oblong to ovate, 1–3 cm long; margins entire. **Inflorescence** glomerules in terminal and axillary spikes. **Flowers:** calyx segments 5, distinct to the base, ca. 0.8 mm long; stamens 5. **Seed** flattened-globose, wrinkled, ca. 0.9–1.3 mm long. Sparsely vegetated, often disturbed soil; montane, subalpine. AB, SK south to CA, NM.

Chenopodium berlandieri Moq. **Stems** erect to ascending, simple or branched, farinose, 20–100 cm. **Leaf blades** farinose, lanceolate to triangular, 1–8 cm long; margins entire to dentate, sometimes hastate. **Inflorescence** glomerules, more than 4 mm across, in terminal and axillary compound spikes. **Flowers:** calyx segments 5, united at the base, 0.7–1.5 mm long; stamens 5. **Seed** flattened-globose, honeycomb pitted, 1–2 mm long. Roadsides, fields, disturbed grasslands; plains, valleys. A native weed distributed throughout N. America. Our plants are var. *zschackei* (Murr) Murr. (p.130)

Often confused with the introduced *Chenopodium album* which has smooth seed coats.

Chenopodium botrys L. [*Dysphania botrys* (L.) Mosyakin & Clements] JERUSALEM OAK **Stems** erect, branched from the base, 5–50 cm tall with glandular, aromatic foliage. **Leaf blades** oblong to ovate, pinnately lobed with toothed margins, 1–6 cm long. **Inflorescence** of axillary cymes in a long, terminal panicle. **Flowers:** calyx lobes 5, distinct, ca. 1 mm long; stamens 5. **Fruit** a flattened-globose achene; the seed wrinkled, <1 mm across. Disturbed soil along roads, gravelly streambanks; plains, valleys. Introduced throughout temperate N. America; native to Eurasia. (p.122)

Glandular species of *Chenopodium* are sometimes segregated into *Dysphania*;[90] however, there is no evidence that *Chenopodium* sensu lato is not monophyletic; furthermore, *C. simplex* was not placed in *Dysphania* although it has a glandular inflorescence.

Chenopodium capitatum (L.) Ambrosi STRAWBERRY BLIGHT **Stems** erect, 5–50 cm. **Leaf blades** green, fleshy, glabrous, triangular-hastate, to 8 cm long with entire to crenate margins. **Inflorescence** glomerules 3–10 mm in diameter, in upper leaf axils and a terminal spike. **Flowers:** calyx 3- to 5-parted, segments red, nearly distinct, <1 mm long; stamens 3. **Seed** punctate, ca. 1 mm. Roadsides, disturbed fields; plains, valleys. Most of boreal and north-temperate N. America; introduced to Eurasia. (p.122)

1. Glomerules reddish, 6–10 mm across; leaves with strongly dentate margins . . . . var. *capitatum*
1. Glomerules green, ≤5 mm across; leaf margins entire to crenate . . . . var. *parvicapitatum*

Chenopodium capitatum var. **parvicapitatum** Welsh [*C. overi* Aellen] is thought to be native to N. America; it is somewhat morphologically distinct from *C. capitatum* var. **capitatum** which may be introduced.[417] The two vars. intergrade in our area.

Chenopodium chenopodioides (L.) Aellen **Stems** reddish, decumbent to prostrate, 5–25 cm. **Leaf blades** green, fleshy, glabrous, lanceolate to triangular-hastate, 1–6 cm long with dentate margins. **Inflorescence** glomerules 2–3 mm diameter, in upper leaf axils and a terminal spike. **Flowers:** calyx 3- to 5-parted, segments green, almost fully united, <1 mm long; stamen 1. **Seed** ovoid, smooth, 0.6–0.9 mm long. Moist, disturbed soil of roadsides, fields, streambanks; plains, valleys. WA to MT south to CA, CO; S. America.

Chenopodium desiccatum A.Nelson [*C. leptophyllum* (Moq.) Nutt. ex S.Watson var. *oblongifolium* S.Watson] **Stems** erect to ascending, branched from the base, farinose, 5–15 cm. **Leaf blades** densely farinose, somewhat fleshy, narrowly lanceolate, 1–2 cm long, the lower 3-veined with entire margins. **Inflorescence** glomerules in dense terminal and axillary panicles. **Flowers:** calyx of 5 distinct segments, ca. 1 mm long; stamens 5. **Seed** flattened-globose, warty, 0.8–1.1 mm across. Grasslands, steppe; plains. AB south to CA, CO, TX.

The distinction between this plant and *Chenopodium pratericola* is weak.

Chenopodium fremontii S.Watson **Stems** erect to spreading, branched above, farinose, 10–60 cm. **Leaf blades** farinose beneath (sometimes sparse), ovate to triangular, often hastate, 1–4 cm long with entire or wavy margins. **Inflorescence** glomerules 1–2 mm across in interrupted terminal and axillary spikes. **Flowers:** calyx segments 5, distinct to the base, 0.7–1.0 mm long; stamens 5. **Seed** round, smooth, ca. 1 mm. Open soil of outcrops, bare slopes and disturbed habitats; plains, valleys, montane. AB, SK south to CA, NM, TX, Mexico. (p.122)

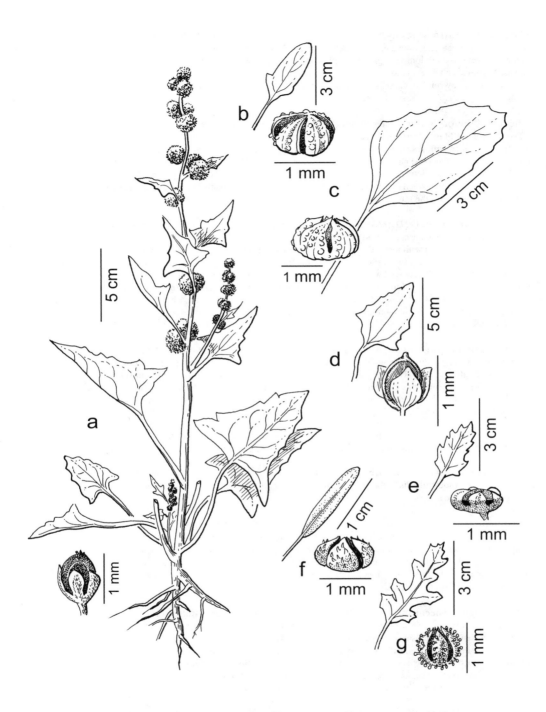

Plate 16. **Chenopodium**. **a.** *C. capitatum*, **b.** *C. fremontii*, **c.** *C. berlandieri*, **d.** *C. rubrum*, **e.** *C. glaucum*, **f.** *C. leptophyllum*, **g.** *C. botrys*

Chenopodium glaucum L. **Stems** prostrate to erect, 10–50 cm. **Leaf blades** fleshy, pale green and glabrous above, white-farinose beneath, lanceolate to ovate, 1–4 cm long with sinuate margins. **Inflorescence** glomerules, 1–2 mm diameter, in bracteate, axillary and terminal spikes. **Flowers:** calyx mostly 5-parted, segments green, nearly distinct, ca. 0.6 mm long; stamen 1. **Seed** flattened-orbicular, punctate, 0.6–1.1 mm, vertical and horizontal. Drying, often saline mud around ponds, lakes, streams, roadsides; plains, valleys, montane. AK to QC south to CA, TX, KY, VA. Our plants are var. *salinum* (Standl.) B.Boivin. (p.122)

    Chenopodium glaucum var. *glaucum*, with ebracteate spikes, is reported for MT,[91] but I have seen no specimens.

Chenopodium incanum (S.Watson) A.Heller [*C. fremontii* S.Watson var. *incanum* S.Watson] **Stems** erect to ascending, simple or branched below, farinose, 10–60 cm. **Leaf blades** farinose beneath, narrowly ovate to triangular, usually hastate, 1–5 cm long with entire or wavy margins. **Inflorescence** glomerules 1–2 mm across in interrupted terminal and axillary, simple or branched spikes. **Flowers:** calyx segments 5, distinct to the base, ca. 1 mm long; stamens 5. **Seed** round, wrinkled, ca. 1 mm. Sparsely vegetated, fine-textured, vernally moist soil along streams, pond margins; plains. AB, SK south to CA, NM, TX, Mexico. Our plants are var. *incanum*.

Chenopodium leptophyllum (Moq.) Nutt. ex S.Watson **Stems** erect, branched, farinose, 5–50 cm. **Leaf blades** farinose beneath, somewhat fleshy, linear, 5–30 mm long, 1-veined with entire margins. **Inflorescence** glomerules in terminal and axillary panicles. **Flowers:** calyx segments 5, distinct to the base, ca. 1 mm long; stamens 5. **Seed** flattened-globose, smooth to lightly wrinkled, ca. 1 mm across. Often sandy soil of grasslands, steppe; plains, valleys, rarely to treeline. AK to SK south to CA, NM, TX. (p.122)

Chenopodium pratericola Rydb. [*C. leptophyllum* (Moq.) Nutt. ex S.Watson var. *pratericola* (Rydb.) F.C.Gates] **Stems** erect, simple or branched, farinose, 15–60 cm. **Leaf blades** farinose, ca. fleshy, linear to lanceolate, 1–4 cm long, mostly 3-veined with entire margins. **Inflorescence** glomerules in terminal and axillary panicles. **Flowers:** calyx 5-parted, segments distinct, ca. 1 mm long, spreading at maturity; stamens 5. **Seed** flattened-globose, wrinkled, 0.9–1.3 mm. Gravelly often disturbed soil in grassland, steppe; plains, valleys, montane. Most of N. America north of Mexico.

    Chenopodium leptophyllum has 1-veined leaves. See *C. dessicatum*.

Chenopodium rubrum L. **Stems** erect to prostrate, 5–50 cm. **Leaf blades** green, fleshy, glabrous, ovate to triangular-hastate, to 6 cm long with entire to dentate margins. **Inflorescence** glomerules, 2–4 mm across, in axillary, simple to compound spikes. **Flowers:** calyx 3- to 4-parted, segments green to red, nearly distinct, ≤1 mm long; stamens 2 to 3. **Seed** smooth, flattened-globose, 0.5–1.0 mm. Moist, sometimes saline or disturbed soil; plains, valleys. AK to NS south to CA, NM, OH, NJ. (p.122)

1. Stems erect to ascending; basal leaf blade margins dentate . . . . var. *rubrum*
1. Stems prostrate; basal leaf blade margins entire on upper half . . . . var. *humile*

Chenopodium rubrum var. *humile* (Hook.) S.Watson is associated with wetlands; *C. rubrum* var. *rubrum* is found on roadsides and in disturbed soil.

Chenopodium simplex (Torr.) Raf. [*C. gigantospermum* Aellen, *C. hybridum* L. var. *simplex* Torr.] **Stems** erect, 15–100 cm. **Leaf blades** glabrous, ovate to triangular, 2–15 cm long, truncate to cordate with sinuate to lobed margins. **Inflorescence** farinose, glandular, open, ebracteate, terminal and lateral panicles of glomerules, 0.5–2 mm diam. **Flowers:** calyx 5-parted, segments united <1/2-way, ca. 1 mm long; stamens 5. **Seed** lenticular, smooth to honeycombed, 1–1.5 mm long. Disturbed soil of woodlands, forests; plains, valleys. Most of temperate and sub-boreal N. America.

Chenopodium subglabrum (S.Watson) A.Nelson [*C. leptophyllum* (Moq.) Nutt. ex S.Watson var. *subglabrum* S.Watson] **Stems** erect, branched, sparsely farinose, 10–50 cm. **Leaf blades** glabrous to sprasely farinose, yellow-green, linear, 1–3 cm long, 1-veined with entire margins. **Inflorescence** widely-spaced glomerules in terminal and axillary panicles. **Flowers:** calyx segments 5, sparsely farinose, ca. 1.5 mm long, united at the base; stamens 5. **Seed** subglobose, smooth, 1.2–1.6 mm long. Sandhills, sandy soil along rivers, sandstone outcrops; plains. AB to MB south to NV, CO, NE, IA.

    Specimens from Cascade Co. approach *Chenopodium leptophyllum*.

Corispermum L. Bugseed

Annual herbs with stellate-hairy to glabrate foliage. **Leaves** sessile, linear with entire margins. **Inflorescence** terminal and axillary, bracteate spikes. **Flowers** perfect, perianth of 0–1 broadly lanceolate, scale-like segment; stamens 1 to 3. **Fruit** flattened, sometimes wing-margined; seed vertical.[290]

North American plants were previously thought to have been introduced from Eurasia, but are now known to be indigenous. Plants intermediate between *Corispermum villosum* and *C. hookeri* Mosyakin occur in MT.[290] Reports of *C. nitidum* Kit. ex Schult. are based on *C. americanum* or hybrids.[290]

1. Fruit wing-margined, as wide as or narrower than the subtending bract . . . . *C. americanum*
1. Fruit without wing margins, wider than subtending bract . . . . *C. villosum*

Corispermum americanum (Nutt.) Nutt. [*C. hyssopifolium* L.misapplied] **Stems** reddish, ascending to erect, 15–60 cm. **Leaves** to 5 × 4 cm. **Inflorescence** with bracts narrowly ovate, 5–9 mm long, as wide or wider than the fruit. **Fruit** ovate, broadly winged, 3–4 mm long. Sand hills and sandy soil of grasslands, steppe; plains, valleys. Throughout N. America. Our plants are var. *americanum*.

Corispermum villosum Rydb. [*C. orientale* Lam. misapplied, *C. hyssopifolium* L.misapplied] **Stems** erect, branched, 10–40 cm. **Leaves** linear, 1–3 cm long. **Inflorescence** with bracts narrowly ovate, 5–15 mm long, as wide or wider than the fruit. **Fruit** ovate, 2–3 mm long, not wing-margined. Sandy soil of grasslands, river banks; plains. AB to QC south to OR, CO, MO. The type is from Gallatin Co.

Cycloloma Moq. Winged pigweed

Cycloloma atriplicifolium (Spreng.) J.M.Coult. Annual herbs with tomentose to glabrous foliage. **Stems** branched, erect, 5–80 cm. **Leaves** short-petiolate; blades 2–7 cm long, lanceolate to narrowly oblong with toothed margins. **Inflorescence** of interrupted spikes in an open, leafy panicle. **Flowers** perfect or female; calyx lobes 5, united to above middle, 2–4 mm across; stamens 5. **Fruit** a subglobose achene, enclosed in the wing-like calyx; the seed flat, horizontal, smooth, 1.5 mm across. Sandy soil of roadsides, streambanks; plains. Most of temperate N. America. One collection from Richland Co.

Grayia Hook.& Arn. Hopsage

Grayia spinosa (Hook.) Moq. Usually dioecious, intricately branched shrub with green and sparsely stellate-hairy foliage. **Stems** erect, to 1 m, with peeling gray bark; lateral branches spine-tipped. **Leaves** petiolate; blades 1–3 cm long, oblanceolate; margins entire, thickened. **Inflorescence** few-flowered clusters in a terminal bracteate spike. **Flowers** unisexual. **Male flowers** of a 4- to 5-parted calyx with 4 to 5 stamens. **Female flowers** enclosed in 2 united bracts; calyx absent. **Fruit** enclosed by swollen orbicular, glabrous, wing-margined bracts 8–15 mm across. Sandy soil of steppe in arid valleys, tolerant of saline conditions; Carbon and Park cos. WA to MT south to CA, AZ, CO.

Halogeton C.A.Mey.

Halogeton glomeratus (M.Bieb.) C.A.Mey. Annual herbs with glabrous, succulent foliage. **Stems** erect, 5–30 cm with spreading lower branches. **Leaves** sessile, terete, bristle-tipped, 6–15 mm long. **Inflorescence** of few-flowered, axillary glomerules. **Flowers** perfect or female, subtended by 2 to 5 fleshy bracts; calyx segments 5, distinct, 2 mm long; stamens 3 to 5. **Fruit:** perfect flowers forming utricles <1 mm long, black; utricles from female flowers 1–2 mm long, brown. Barren usually saline soil of badlands, disturbed steppe in arid valleys. Introduced to WA, MT south to CA, NM; native to Eurasia.

Kochia Roth

Monoecious annual or perennial herbs or subshrubs. **Leaves** sessile, linear. **Inflorescence** few-flowered clusters in axils of leaf-like bracts in a terminal spike or panicle. **Flowers** perfect and female; calyx 5-parted; stamens 5. **Fruit:** a utricle enclosed in the calyx.

1. Perennial subshrub with fleshy foliage . . . . *K. americana*
1. Annuals; foliage not fleshy . . . .*K. scoparia*

Kochia americana S.Watson Subshrub. **Stems** woody at the base, ascending, 8–25 cm. **Leaves** terete, glabrous to tomentose, 8–15 mm long. **Flowers** 2 to 5 in axils of scarcely reduced leaves; calyx segments 1.5 mm long. **Fruit** 1–3 mm, tomentose, scarious-winged, 4–6 mm across at maturity. Often saline soil of stream terraces, moist steppe, often with greasewood; valleys of Beaverhead Co. OR to MT south to CA, NM, TX.

Kochia scoparia (L.) Schrad. Taprooted annual. **Stems** erect, branched, long-hairy, 15–100 cm. **Leaves** linear to narrowly lanceolate, long-ciliate, 5–60 mm long, becoming red with age. **Inflorescence** brown-hairy, diffuse. **Flowers** 2 to 6 in axils of leaf-like bracts; calyx segments ca. 1 mm long. **Fruit** horizontally winged to some degree. Disturbed soil of roadsides, cultivated fields, tolerant of saline soil; plains, valleys. Native to Eurasia and introduced throughout the world.

Krascheninnikovia Gueldenst. Winterfat

Krascheninnikovia lanata (Pursh) A.Meeuse & A.Smit [*Eurotia lanata* (Pursh) Moq., *Ceratoides lanata* (Pursh) J.T.Howell] Monoecious or dioecious subshrub with stellate-pubescent, long-hairy foliage. **Stems** erect, to 40 cm, woody only at the base. **Leaves** short-petiolate, linear, 1–3 cm long with entire margins; smaller leaves in axillary fascicles. **Inflorescence** axillary clusters of small spikes. **Flowers** unisexual, apetalous. **Male flowers** a 4-parted calyx with 4 stamens. **Female flowers** enclosed in 2 partly united bracts; calyx absent. **Fruit** enclosed by densely long-hairy bracts, 4–6 mm long with divergent horn-like tips. Grasslands, steppe; plains, valleys, often with sagebrush. AB to MB south to CA, NM, TX.

Monolepis Schrad. Poverty weed

Monolepis nuttalliana (Schult.) Greene Annual herbs with sparsely farinose, succulent foliage. **Stems** prostrate to ascending, 5–30 cm. **Leaves** petiolate, the blade lanceolate, hastate, 1–3 cm long. **Inflorescence** glomerules in axils of reduced upper leaves. **Flowers** perfect or female; perianth segment 1, 1 mm long; stamen 1. **Fruit:** a utricle 1–1.5 mm long, the pericarp easily separating from the smooth black seed. Moist, often saline or alkaline, sandy to fine-textured soil of wetlands, pond margins, badlands, disturbed pastures, agricultural fields, gardens, roadsides; plains, valleys, montane. AK to MN south to CA, TX, Mexico.

Growth form and habitat are similar to that of *Chenopodium glaucum* which usually has leaves that are white-farinose beneath. *Monolepis pusilla* Torr. ex S.Watson [*Micromonolepis pusilla* (Torr. ex S.Watson) Ulbr.] with erect, highly branched stems, occurs in adjacent ID and WY.

Salicornia L. Glasswort

Salicornia rubra A.Nelson Annual with glabrous, fleshy, reddish foliage. **Stems** erect, branched, 5–15 cm. **Leaves** opposite, minute, scale-like. **Flowers** mostly perfect, sessile in groups of 3 at nodes in a terminal, bracteate spike; calyx 3-parted, pyramidal, 1–2 mm wide, completely closed except for a central slit; stamens 2. **Fruit** flattened, enclosed in spongy calyx; seeds vertical, ca. 0.5 mm across. Moist, saline soil on margins of ponds, wetlands; plains, valleys. BC to MB south to NV, NM, KS.

Salsola L. Russian thistle

Annuals. **Stems** often red-striped. **Leaves** sessile, linear, hispid to glabrous, spine-tipped. **Inflorescence** spike-like with solitary flowers subtended by spine-tipped bracts. **Flowers** perfect, subtended by 2 bracteoles; calyx of 5 distinct, persistent segments; stamens 5. **Fruit:** a utricle enclosed in the calyx that sometimes develops a wing margin.[291]

Fruit does not mature until late September.

1. Mature calyx 4–10 mm across with a broad, 3-parted wing; flower bracts reflexed . . . . *S. tragus*
1. Mature calyx 2–5 mm across without a broad wing; flower bracts ascending . . . . *S. collina*

Salsola collina Pall. **Stems** 10–100 cm, branched. **Leaves** ca. 1 mm wide. **Inflorescence** not interrupted; bracts appressed, overlapping at maturity. **Fruiting calyx** 2–5 mm across without conspicuous wings. Roadsides, cultivated fields; plains, valleys. Introduced to central N. America; native to Eurasia.

Salsola tragus L. [*S. pestifer* A.Nelson, *S. kali* L. misapplied] **Stems** 10–100 cm, branched at the base. **Leaves** ca. 4 cm × 1 mm. **Inflorescence** interrupted at maturity; bracts reflexed at maturity, little overlapping. **Fruiting calyx** 4–10 mm across with 3 conspicuous wings. Roadsides, cultivated fields, dry lake beds; plains, valleys. Introduced throughout temperate N. America; native to Eurasia.
    This is the common tumbleweed of agricultural fields.

Suaeda Forssk. ex J.F.Gmel. Sea blite

Annual or perennial herbs with glabrous foliage. **Leaves** fleshy, sessile or short-petiolate, linear with entire margins. **Inflorescence** terminal, of few-flowered glomerules in bracteate spikes. **Flowers** mostly perfect, radially or slightly bilaterally symmetrical, subtended by 2 scarious bracteoles; calyx fleshy, 5-parted, the segments united below; stamens 5. **Fruit** a utricle enclosed in the calyx; seed usually horizontal.[128]

1. Calyx radially symmetrical; lobes equal, without appendages; plants perennial . . . . *S. nigra*
1. Calyx lobes unequal; at least one segment with a keel-like appendage; plants annual. 2

2. Glomerules 3- to 7-flowered; the bracts lanceolate . . . . *S. calceoliformis*
2. Glomerules 1- to 3-flowered; the bracts more linear . . . . *S. occidentalis*

Suaeda calceoliformis (Hook.) Moq. [*S. americana* (Pers.) Fernald] *[S. depressa* (Pursh) S.Watson misapplied] Annual. **Stems** prostrate to erect, branched above, 5–40 cm. **Leaves** 5–20 mm long. **Glomerules** of 3 to 7 flowers, the bracts broadly lanceolate with scarious basal margins. **Flowers**: calyx segments ca. 1.5 mm, wrinkled at maturity, unequal, at least one with a horn-like appendage. **Seeds** lenticular to flattened, 1–2 mm. Moist saline or alkaline soil along streams, wetlands, roads; plains, valleys. AK to MN south to CA, TX, Mexico; disjunct on northeastern seaboard. See *S. occidentalis*.

Suaeda nigra (Raf.) J.F.Macbride [*S. intermedia* S.Watson., *S. moquinii* (Torr.) Greene] Perennial (annual) herb or subshrub. **Stems** erect to ascending, branched, 8–40 cm. **Leaves** 1–2 cm long, sometimes sparsely hairy. **Glomerules** with 1 to 12 flowers. **Flowers** sometimes unisexual; calyx segments equal, ca. 1.5 mm long. **Seeds** <1 mm, horizontal from perfect flowers, vertical from female flowers. Moist, saline, fine-textured soil of stream terraces, around wetlands; plains, valleys. AB, SK south to CA, TX, Mexico.
    Plants woody at the base with sessile leaves have been segregated as *S. intermedia*.

Suaeda occidentalis (S.Watson) S.Watson Annual. **Stems** erect, branched throughout, 5–30 cm. **Leaves** 5–10 mm long. **Glomerules** of 1 to 3 flowers, the bracts linear to narrowly lanceolate without scarious margins. **Flowers**: calyx segments ca. 1 mm, unequal, wrinkled at maturity, at least 1 with a horn-like appendage. **Seeds** lenticular and black or flat and brown, ca. 1.5 mm. Moist saline or alkaline soil around wetlands; plains; Custer Co. WA to MT south to CA, NV, UT.
    Suaeda calceoliformis is usually a more erect plant, branching more from above middle.

Suckleya A.Gray

Suckleya suckleyana (Torr.) Rydb. Monoecious annual. **Stems** branched, succulent, prostrate to ascending, 10–40 cm. **Leaves** farinose when young, long-petiolate; the blade rhombic to orbicular, 1–3 cm long with crenate margins. **Inflorescence** axillary flower clusters. **Flowers** unisexual. **Male flowers** a 4-parted calyx with 4 stamens. **Female flowers** enclosed in 2 partially united, hirsute bracts; calyx absent. **Fruit** enclosed by swollen, triangular bracts, 4–6 mm long with narrow, crenulate wing margins and divergent tips; seed smooth, brown, ca. 3 mm long. Sparsely vegetated soil of dry, saline lakes and ponds, cultivated fields; plains. AB to NM, TX.

SARCOBATACEAE: Greasewood Family

Sarcobatus Nees Greasewood

Sarcobatus vermiculatus (Hook.) Torr. Monoecious deciduous shrubs to 2 m high with glabrous foliage. **Stems** rigid, with whitish bark and thorns. **Leaves** alternate at least above, mostly on long shoots, sessile, fleshy, linear, 1–3 cm long. **Inflorescence**: spikes bracteate below, 1–3 cm long. **Flowers** unisexual. **Male flowers** above, consisting of 2 to 4 stamens concealed by a peltate scale, congested into a cone-like spike. **Female flowers** perigynous, well separated, solitary in axils of leaf-like bracts, with a cup-like, united calyx; pistil 1. **Fruit** an achene 3–5 mm long surrounded by the calyx, the lobes enlarging into a broad, wavy-margined wing 7–10 mm across. Saline soil of stream terraces, badlands; plains, valleys. AB, SK south to CA, NM, Mexico.

Formerly placed in the Chenopodiaceae or Amaranthaceae; anatomical and molecular studies indicate *Sarcobatus* is a discrete lineage more closely related to the Nyctaginaceae.[42]

PORTULACACEAE: Purslane Family

Annual or perennial herbs, often with succulent foliage. **Leaves** simple with entire margins. **Flowers** perfect, radially symmetrical; sepals 2 to many, separate or united at the base; petals (4)5 to many, separate or united at the base; stamens usually 5; pistil 1; ovary superior (partly inferior in Portulaca). **Fruit** a capsule; seeds 1 to many.

Portulaca grandiflora Hook. (moss rose), a common, introduced ornamental, escapes in adjacent states, but is not reported for MT.

1. Inflorescence capitate; flowers with 4 petals; stems leafless . . . . *Cistanthe*
1. Flowers with >4 petals; not borne in a capitate inflorescence 2

2. Petals yellow; stems prostrate . . . . *Portulaca*
2. Petals white to rose (rarely yellow in cormose plants); stems mostly ascending to erect or absent 3

3. Plants with slender taproot or rhizomes 4
3. Plants with a thickened taproot or nut-like corm. 5

4. Stem leaves 2, opposite. . . . . *Claytonia*
4. Stem leaves usually alternate, usually >2 . . . . *Montia*

5. Capsule breaking along a latitudinal suture near the base; all but one species (*L. triphylla*) with
   swollen taproot . . . .*Lewisia*
5. Capsule breaking along longitudinal sutures; all but one species (*C. megarhiza*) with corms. . . . . *Claytonia*

Cistanthe Spach Pussypaws

Cistanthe umbellata (Torr.) Hershk. [*Spraguea umbellata* Torr., *Calyptridium umbellatum* (Torr.) Greene] Taprooted perennial with a branched caudex, sometimes mat-forming. **Stems** ascending, 1–7 cm, leafless, 1 per rosette. **Leaves** all basal, petiolate; the blades oblanceolate to spatulate, 5–20 mm long. **Inflorescence** hemispheric, umbellate, densely flowered, 1–3 cm across. **Flowers**: petals 4, white, unequal, 3–5 mm long; sepals 2, sometimes pinkish, orbiculate, persistent, scarious, 3–8 mm long, becoming larger in fruit; stamens 3; style exserted. Sandy or gravelly, sparsely vegetated, granitic soil of fellfields, scree slopes; subalpine to alpine, rarely lower. BC to CA, UT, WY.

Morphological and molecular evidence suggests that both *Spraguea* and *Calyptridium* should be included in the S. American genus *Cistanthe*.[75,175]

Claytonia L. Spring beauty, miner's lettuce

Glabrous annuals or perennials. **Basal leaves** petiolate. **Stem leaves** 2, sessile, opposite. **Inflorescence** terminal, racemose. **Flowers**: sepals 2 or 4 to 8, persistent; petals 5, sometimes notched; stamens 5; style divided in 3. **Capsule** 3-chambered; seeds 3 to 6, shiny black with elaiosomes.[78,79,280]

Many of these species have previously been placed in *Montia*. There is evidence for delineating several taxa in the widespread and polymorphic *C. perfoliata* complex, including *Claytonia parviflora* and *C. rubra*.[277]

Claytonia arenicola L.F.Hend. [*Montia arenicola* (L.F.Hend.) Howell] Taprooted annuals. **Stems** ascending to erect, 5–15 cm. **Basal leaf blades** linear to linear-oblong, 15–40 mm long. **Stem leaves** linear, 1–3 cm long. **Inflorescence** bracteate with spreading pedicels to 2 cm long. **Flowers**: sepals 1–2 mm long; petals white, sometimes pink-veined, 4–8 mm long. **Capsule** as long as sepals with 2 to 6 seeds. Vernally moist, stony slopes; montane. Endemic to eastern WA and OR, adjacent ID and northwest MT. One collection from along an old road in Sanders Co.

Claytonia cordifolia S.Watson [*Montia cordifolia* (S.Watson) Pax & K.Hoffm.] Finely rhizomatous perennial. **Stems** erect, 5–30 cm. **Basal leaf blades** cordate to truncate-ovate, 1–7 cm long. **Stem leaves** ovate, 1–5 cm long. **Inflorescence** ebracteate with spreading or recurved pedicels 5–20 mm long. **Flowers**: sepals 2–4 mm long; petals white, 8–12 mm long. **Capsule** as long as sepals with 3 to 6 seeds. Moist spruce or red cedar forests, avalanche slopes, often along streams, seeps; montane, subalpine. BC south to CA, NV, UT.

   Claytonia sibirica is similar but has a bracteate inflorescence.

Claytonia lanceolata Pursh SPRING BEAUTY Perennial from a globose corm, 5–40 mm across. **Stems** erect, 5–20 cm **Basal leaf blades** mostly 1 to 2, lanceolate, to 15 cm long, absent at flowering. **Stem leaves** broadly lanceolate to ovate, 2–7 cm long. **Inflorescence** single-bracteate below with spreading pedicels 1–3 cm long. **Flowers**: sepals 2–6 mm long; petals white with pink lines, 5–14 mm long. **Capsule** as long or longer than sepals with 3 to 6 seeds. Vernally moist soil of grasslands, steppe, meadows, woodlands, thickets, open forest, avalanche slopes; montane to alpine. BC to SK south to CA, NM. (p.130)
   Plants seem to prefer warm slopes where snow accumulates and flower shortly after it melts. See *Claytonia rosea*.

Claytonia megarhiza (A.Gray) Parry ex S.Watson Perennial from a fleshy, branched taproot and caudex. **Stems** lax to ascending, 2–15 cm. **Basal leaf blades** numerous, fleshy, obovate to spatulate, 1–5 cm long. **Stem leaves** linear-oblanceolate, 1–3 cm long. **Inflorescence** bracteate; pedicels 5–20 mm long, ascending **Flowers**: sepals 4–8 mm long; petals white to pink, 5–10 mm long. **Capsule** shorter than sepals with 1 to 6 seeds. Stony, often moist soil of talus slopes, fellfields; alpine. BC, AB south to CA, NV, UT, NM, disjunct in NT.

Claytonia parviflora Douglas ex Hook. [*Montia perfoliata* (Donn ex Willd.) Howell var. *parviflora* (Douglas ex Hook.) Jeps.] Taprooted annuals. **Stems** ascending, 5–20 cm. **Basal leaf blades** linear, 1–10 cm long. **Stem leaves** perfoliate, forming a disk 1–3 cm across. **Inflorescence** 1-bracted with spreading pedicels to 7 mm long. **Flowers**: sepals 1–2 mm long; petals white or pink, 2–4 mm long. **Capsule** as long as sepals with ca. 3 seeds. Vernally moist, sparsely vegetated soil of open forest, grasslands; montane; Flathead and Sanders cos. BC, MT south to CA and AZ. Our plants are var. ***parviflora***.

Claytonia perfoliata Donn ex Willd. [*Montia perfoliata* (Donn ex Willd.) Howell] Taprooted annuals. **Stems** ascending, 3–20 cm. **Basal leaf blades** ovate to rhombic, 1–7 cm long. **Stem leaves** perfoliate, forming a disk.

Inflorescence single-bracteate with spreading pedicels to 5 mm long. **Flowers**: sepals 1–4 mm long; petals white, 2–5 mm long. **Capsule** as long as sepals with ca. 3 seeds. Vernally moist, sparsely vegetated soil along roads. BC to ND south to CA, UT, AZ. Our plants are var. ***perfoliata***. Known from Gallatin and Sanders cos.

Claytonia perfoliata ssp. *intermontana* John M.Mill. & K.L.Chambers, with notched cauline leaves, is reported for MT,[278] but I have seen no specimens. Most plants previously determined to be *C. perfoliata* are now referred to *C. rubra*.

Claytonia rosea Rydb. [*C. multiscapa* Rydb., *C. lanceolata* Pursh var. *flava* (A.Nelson) C.L.Hitchc.] SPRING BEAUTY. Perennial from a globose corm, 1–3 cm across. **Stems** erect, 5–20 cm. **Basal leaf blades** few, linear, to 8 cm long, absent at flowering. **Stem leaves** linear to narrowly lanceolate, 2–8 cm long. **Inflorescence** bracteate below with spreading pedicels 1–3 cm long. **Flowers**: sepals 3–7 mm long; petals white to yellow, 6–12 mm long. **Capsule** as long or longer than sepals with 3 to 6 seeds. Vernally moist soil of grasslands, steppe, meadows, open forest; montane to lower subalpine. BC to SK south to CA, NM.

J. M. Miller separates our plants as *C. multiscapa* based on flower color;[278] however, this is a very labile trait among several congeneric species, and *C. multiscapa* cannot be distinguished on other morphological traits.[362] *Claytonia rosea* is often confused with *C. lanceolata* because they have similar habitats; however, the latter can be distinguished by its broader, 3-veined leaves and emarginate petals.

Claytonia rubra (Howell) Tiedestr. [*C. perfoliata* Donn ex Willd.misapplied, *Montia perfoliata* (Donn ex Willd) Howell misapplied] Annuals. **Stems** ascending, 2–20 cm. **Basal leaf blades** reddish, ovate to deltoid, 5–20 mm long. **Stem leaves** sessile, ovate, perfoliate but not fully united into a disk. **Inflorescence** single-bracted with spreading pedicels to 5 mm long. **Flowers**: sepals 1–2 mm long; petals white to pink, 2–3 mm long. **Capsule** as long as sepals with ca. 3 seeds. Moist, sparsely vegetated soil of open forest, grasslands, often in shade or along streams; valleys. BC to SD south to CA, UT, CO.

1. Leaf blades mostly spatulate; flowers ≤3 mm diam. . . . . ssp. *depressa*
1. Leaf blades more deltoid; flowers 4–6 mm diam. . . . . .ssp. *rubra*

Claytonia rubra ssp. *rubra* is our common form; *C. rubra* ssp. *depressa* (A.Gray) John M.Mill. & K.L.Chambers, is reported for the Mission Mtns.[280]

Claytonia sibirica L. [*Montia sibirica* (L.) Howell] Annual to rhizomatous perennial. **Stems** erect to ascending, 10–30 cm. **Basal leaf blades** elliptic, 1–7 cm long. **Stem leaves** broadly ovate, 1–8 cm long. **Inflorescence** bracteate; pedicels spreading, 1–4 cm long. **Flowers**: sepals 2–4 mm long; petals white, 6–10 mm long. **Capsules** as long as sepals with 3 seeds. Wet forest and along streams, seeps; montane, subalpine. AK south to CA, OR, ID, MT. See *Claytonia cordifolia*.

Lewisia Pursh

Herbaceous perennials with fleshy, glabrous foliage. **Leaves**: basal and/or cauline, the basal petiolate, the cauline sessile. **Flowers** solitary, racemose, cymose or paniculate, each subtended by 2 bracts; sepals 2 or 6 to 9; petals 5 to 10 (or more), quickly withering; stamens 5 to 50; style divided into 3 to 8 stigmas. **Fruits** capsules with 1 to 50 seeds.[176]

Lewisia kelloggii K.Brandegee occurs in east-central ID, close to Beaverhead Co., MT.

1. Petals 15-35 mm long; sepals 4 or more . . . . *L. rediviva*
1. Petals ≤10 mm long; sepals 2 2

2. Stems easily broken off from globose corms . . . . *L. triphylla*
2. Plants with thick, branched roots 3

3. Flowers several-many in a panicle. . . . . *L. columbiana*
3. Flowers solitary on peduncles . . . . *L. pygmaea*

Lewisia columbiana (Howell ex A.Gray) B.L.Rob. **Roots** fleshy and branched with a thick, mostly simple caudex. **Stems** erect, 10–30 cm. **Basal leaf blades** linear-oblanceolate, 1–3 cm long. **Stem leaves** bract-like, small, narrowly lanceolate. **Flowers** paniculate subtended by glandular-toothed bracts; sepals 2, glandular-toothed, ca. 2 mm long; petals 7 to 9, white with pink veins, 6–8 mm long; stamens 5 to 6; style branches 2 to 3. Moist, shady granitic rock outcrops in canyons; montane. BC, MT south to OR, ID. Our plants are var. ***wallowensis*** C.L.Hitchc. which is endemic to central ID, adjacent OR and a single canyon in Ravalli Co.

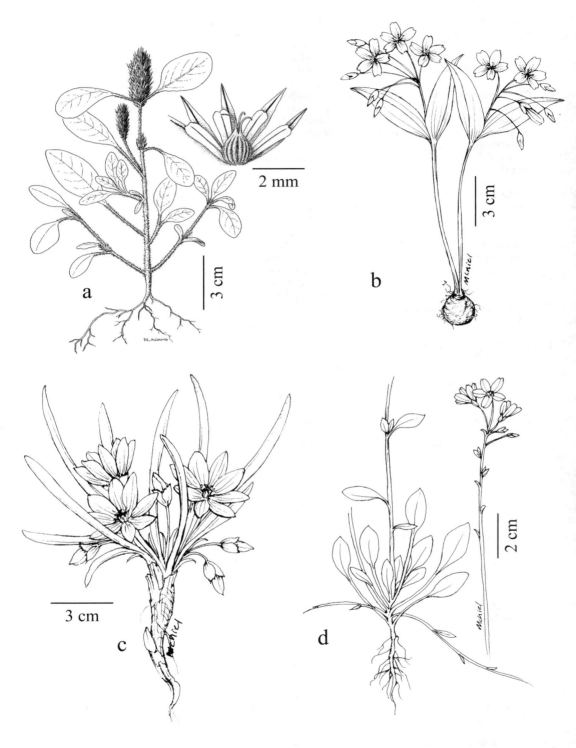

Plate 17. a. *Amaranthus retroflexus*, b. *Claytonia lanceolata*, c. *Lewisia rediviva*, d. *Montia parvifolia*

Lewisia pygmaea (A.Gray) B.L.Rob. [*L. nevadensis* (A.Gray) B.L.Rob.] Pygmy bitterroot **Roots** fleshy and branched with a simple caudex. **Stems** erect or ascending, 1–4 cm with a pair of bracts below the middle. **Basal leaf blades** linear, 1–8 cm long. **Stem leaves** absent. **Flowers** solitary; sepals 2, mostly glandular-toothed, 2–10 mm long, enlarging in fruit; petals 6 to 8, white to magenta, 4–10 mm long; stamens 5 to 8; style branches 3 to 6. Sandy, gravelly or shallow, vernally wet soil of banks, rock ledges, grasslands, meadows, open slopes, often near receding snow banks; montane to alpine. YT to AB south to CA, NM.

Lewisia nevadensis is separated from *L. pygmaea* based on sepal and root characters.[176] Some of our plants have entire sepals but linear leaves, while others have oblanceolate leaves with dentate sepals.

Lewisia rediviva Pursh Bitterroot **Roots** fleshy, branched with a simple or branched caudex. **Stems** leafless, erect or ascending, 1–5 cm with a whorl of membranous bracts below the flower. **Basal leaf blades** tubular, 1–5 cm long, appearing and usually disappearing before flowering. **Flowers** solitary; sepals 6 to 9, unequal, 8–25 mm long with entire margins; petals 12 to 18, white to pink, 15–35 mm long; stamens numerous; style branches 4 to 8. Gravelly or sandy, well-drained, usually sparsely vegetated soil of grasslands; valleys, montane. BC, MT south to CA, AZ, CO. (p.130)

The state flower of MT. Native Americans may have introduced the plant into eastern Flathead Co. and elsewhere.[110] The type locality is south of Missoula.

Lewisia triphylla (S.Watson) R.L.Rob. **Roots** a globose corm 5–15 mm across. **Stems** 1 to many, ascending, 2–6 cm. **Basal leaf blades** linear, 1–6 cm long, absent at flowering. **Stem leaves** opposite or whorled, linear, 1–4 cm long. **Flowers** in bracteate cymes; sepals 2, 2–5 mm long with entire margins; petals 5 to 9, white to pink, 5–7 mm long; stamens 5; style branches 3 to 5. Gravelly, shallow, vernally moist, sparsely vegetated soil of non-calcareous, rock outcrops, open grasslands; montane, subalpine. WA to MT south to CA, CO.

Montia L. Miner's lettuce

Annuals or short-lived perennials. **Leaves** succulent; cauline ≥3, alternate or opposite; sometimes also a basal rosette. **Inflorescence** a terminal or axillary raceme. **Flowers**: sepals 2, persistent, unequal; petals 5; stamens 5; style divided in 3. **Capsule** 3-chambered; seeds 1 to 3, shiny black, minutely tuberculate, sometimes with a minute elaiosome (Miller 2003b).

Montia fontana L., a creeping, annual species with opposite leaves, is reported for MT,[279] but I have seen no specimens.

1. Plants stoloniferous perennials. 2
1. Plants annual, without stolons 3

2. Plants with a basal rosette; stem leaves alternate . . . . *M. parvifolia*
2. Leaves opposite, all cauline . . . . *M. chamissoi*

3. Sepals ca. 2 mm long; petals 1–3 mm long . . . . *M. dichotoma*
3. Sepals 3-4 mm long; petals 4–6 mm long . . . . *M. linearis*

Montia chamissoi (Ledeb. ex Spreng.) Greene Rhizomatous perennial producing sparsely leafy, bulblet-bearing stolons. **Stems** erect, 5–20 cm. **Leaves** all cauline, opposite; the blades oblanceolate, 1–5 cm long. **Flowers**: sepals 2–3 mm long; petals white to pink, 5–7 mm long. Wet meadows, forested streambanks; valleys, montane. AK south to CA, AZ, NM, disjunct in MN, IA, PA.

Montia dichotoma (Nutt.) Howell [*Montiastrum dichotomum* (Nutt.) Rydb.] Annual. **Stems** ascending to erect, 2–7 cm. **Leaves** all cauline, linear, 5–20 mm long. **Flowers**: petals white, 1–3 mm long; sepals ca. 2 mm long, enlarging in fruit. Vernally moist soil of old roads and margins of ponds, seeps in the valleys. BC south to CA, ID, MT.

Montia dichotoma is similar to *M. linearis* but with smaller flowers. Mixed populations are documented, and it's not clear that the two are distinct species.

Montia linearis (Douglas ex Hook.) Greene [*Montiastrum lineare* (Douglas ex Hook.) Rydb.] Annual. **Stems** ascending to erect, 2–15 cm. **Leaves** all cauline, linear, 1–4 cm long. **Flowers**: petals white, 4–6 mm long; sepals 3–4 mm long, enlarging in fruit. Vernally moist, sparsely vegetated soil of grasslands, steppe, temporary wetlands, open forest, rock outcrops, roadsides; plains, valleys. BC to SK south to CA, AZ. See *M. dichotoma*.

Montia parvifolia (Moq. ex DC.) Greene Perennial, often mat-forming from a branched caudex. **Stems** erect to prostrate, 4–25 cm, producing bulblets instead of or in addition to the flowers. **Leaves**: basal leaves petiolate, the blade spatulate to oblanceolate, 5–25 mm long; stem leaves smaller, sessile, alternate, lanceolate. **Flowers**: petals pinkish, 5–12 mm long; sepals 2–3 mm long, unequal. Wet, shallow soil along streams and on rock outcrops, talus slopes; valleys to subalpine. AK to AB south to CA, UT. (p.130)

Portulaca L. Purslane

Portulaca oleracea L. Annual with fleshy, glabrous foliage. **Stems** prostrate to ascending, 5–30 cm. **Leaves** all cauline, alternate, short-petiolate, the blades obovate, 5–20 mm long, rounded at the tip. **Flowers** solitary in leaf axils or in small, terminal clusters; sepals 2, fused to the ovary at the base, 3–4 mm long; petals 5, yellow, ephemeral, as long as sepals; stamens 6 to 12; ovary partly inferior; style branches 4 to 6. Rich disturbed soil of gardens, barn yards, cultivated fields; plains, valleys. Introduced to most of temperate N. America; native of southern Europe.

MOLLUGINACEAE: Carpet-weed Family

Mollugo L. Carpet weed

Mollugo verticillata L. Glabrous annual, mat-forming herb. **Stems** branched at the base, prostrate, to 20 cm long. **Leaves** of basal rosette spatulate, to 25 mm long; cauline leaves 4 to 6 per node, 5–15 mm long, narrowly oblanceolate with entire margins. **Flowers** several in axillary umbels with pedicels 3–8 mm long; sepals 5, separate, white-margined, 2 mm long, oblong, 3-nerved; petals absent; stamens 3; pistil 1 with 3 styles and a superior ovary. **Fruit** a many-seeded, ovoid capsule ca. 2 mm long. Moist sandy soil along rivers, streams; plains, valleys. Most of temperate and tropical N. America, S. America, Eurasia, Africa; collected in Treasure and Yellowstone cos.

This plant may or may not be native to temperate N. America.[411]

CARYOPHYLLACEAE: Pink Family

Annual or perennial, taprooted herbs. **Stems** often swollen at the nodes. **Leaves** stipulate, simple with entire margins, mostly opposite. **Inflorescence** cymose (solitary), terminal and/or axillary. **Flowers** perfect, radially symmetrical; sepals 5, separate or united; petals 5 or absent, separate; stamens 10; ovary superior with 2 to 5 styles. **Fruit** a many-seeded capsule.[332]

Minuartia and *Arenaria* are similar and have sometimes been combined under the latter name; in our area *Arenaria* spp. are usually erect and at least 10 cm high or diminutive annuals, while *Minuartia* spp. are short or lax perennials.

1. At least the lower 1/4 of the sepals united 2
1. Sepals separate to the base. 9

2. Leaves linear, pungent, ≤10 mm long; plants low. 3
2. Leaves wider or shorter, not pungent 4

3. Tube (united portion) ca. 1/4 length of calyx. . . . . *Scleranthus*
3. Tube ca. 1/2 length of the calyx . . . .*Paronychia*

4. Styles 3 to 5 5
4. Styles 2 6

5. Calyx lobes 15–30 mm long, longer than the tube . . . . *Agrostemma*
5. Calyx lobes mostly <15 mm long, shorter than the tube . . . .*Silene*

6. Calyx 1–4 mm long; inflorescence diffuse, cloud-like. . . . .*Gypsophila*
6. Calyx ≥10 mm long. 7

7. Petals white when fresh; sepals acute or rounded. . . . . *Saponaria*
7. Petals usually pink; sepals attenuate 8

8. Petals notched at the tip; calyx not immediately subtended by bracts . . . . *Vaccaria*
8. Petals entire; calyx subtended by 1 to 3 linear to ovate bracts . . . . *Dianthus*

9. Petals absent 10
9. Petals present. 12

10. Sepals ca. 2 mm long; stems glabrous. . . . . *Sagina*
10. Sepals mostly >2 mm long; stems often pubescent or glandular. 11

11. Plants clumped; stems <5 cm long. . . . . *Minuartia*
11. Stems not clumped; often >5 cm long . . . . *Stellaria*

12. Papery bracts (stipules) on the stem at the leaf bases. 13
12. Stipules absent. 14

13. Styles 5; leaves fascicled, appearing whorled . . . . *Spergula*
13. Styles 3; leaves fascicled or not . . . . *Spergularia*

14. Petals 2-lobed at least 1/4 their length 15
14. Petals with rounded to shallowly indented or dentate tips 16

15. Styles mostly 3; fruit capsule ovoid . . . . *Stellaria*
15. Styles mostly 5; capsule cylindric. . . . . *Cerastium*

16. Styles 4 to 5 . . . . *Sagina*
16. Styles 3. 17

17. Flowers borne in an umbel. . . . . *Holosteum*
17. Inflorescence cymose. 18

18. Leaves narrowly oblong, >3 mm wide . . . . *Moehringia*
18. Leaves linear or ovate, <3 mm wide. 19

19. Leaves linear, <10 mm long; capsule splitting into 3 sections . . . . *Minuartia*
19. Leaves ovate or >10 mm long; capsule splitting into 6 sections . . . . *Arenaria*

Agrostemma L. Corn cockle

Agrsotemma githago L. Taprooted annuals with sparsely long-hairy foliage. **Stems** simple or branched, to 100 cm. **Leaves** linear-lanceolate, 3–12 cm long, sessile or united around the stem. **Inflorescence** a few-flowered, terminal, open cyme. **Flowers**: calyx united into a strongly purple-veined tube, 12–15 mm long and linear lobes 15–30 mm long; petals indistinctly clawed, red, smaller than the calyx lobes; stamens 10; styles 5. **Capsule** 15–20 mm long. Cultivated fields, roadsides; plains, valleys. Introduced throughout N. America and worldwide; native to Eurasia.

At one time a common weed of grain fields, it appears to have become less frequent. All our collections were made before 1910.

Arenaria L. Sandwort

Annual or perennial herbs. **Leaves** sessile, often clustered and persistent on short sterile stems, mostly not fascicled. **Inflorescence** bracteate. **Flowers** cup-shaped; sepals separate; petals white with blunt tips; styles 3; stamens 10. **Capsule** ovoid, opening by 6 teeth.[159,263]

Many species often placed in *Arenaria* are here segregated into *Minuartia* and *Moehringia*. All the following species except *A. serpyllifolia* are sometimes placed in the genus *Eremogone*.[159] However, preliminary data supporting this split have not been published, and few other authors have accepted this classification although it was proposed over a century ago.

1. Plants annual; leaves elliptic, ≤5 mm long . . . . *A. serpyllifolia*
1. Plants perennial with a branched caudex; leaves linear, most >5 mm long. 2

2. Leaves mostly <1 cm long . . . . *A. hookeri*
2. Leaves >1 cm long. 3

3. Stems and inflorescence glabrous . . . . *A. congesta*
3. Stem, pedicels, and/or sepals glandular-hairy 4

4. Leaves blue-green glaucous, rigid . . . .*A. aculeata*
4. Leaves green, flexuous 5

5. Sepals ovate, broadly acute . . . . *A. capillaris*
5. Sepals lanceolate, sharply acute or acuminate . . . . *A. kingii*

Arenaria aculeata S.Watson [*Eremogone aculeata* (S.Watson) Ikonn.] Mat-forming perennial from a branched caudex. **Stems** erect, simple, 5–25 cm. **Leaves** glaucous, needle-like, 10–25 mm long, shorter above. **Inflorescence** open, dark-tipped glandular-hairy. **Flowers**: sepals ovate with an acute tip, scarious-margined, 3–5 mm long; petals obovate, 4–10 mm long. **Capsules** glabrous, 5–9 mm long. Sandy or gravelly, granitic soil of exposed ridges, slopes; subalpine and alpine. WA to MT south to CA, NV and UT.

Arenaria capillaris Poir. [*Eremogone capillaris* (Poir.) Fenzl] Perennial from a branched caudex, forming loose mats. **Stems** erect, simple, 5–20 cm. **Leaves** filiform, glabrous, 5–40 mm long, shorter above. **Inflorescence** few-flowered, usually open to somewhat congested, glandular-hairy. **Flowers**: sepals glandular, ovate with a rounded to acute tip, scarious-margined, 3–4 mm long; petals spatulate, 4–8 mm long. **Capsules** glabrous to glandular, 5–8 mm long. Rocky soil of grasslands, meadows, open forest, cliffs; valleys to alpine, mostly middle elevations. AK to AB south to OR, NV, MT. Our plants are var. *americana* (Maguire) Davis. (p.138)
　　Plants with a congested inflorescence could be mistaken for *Arenaria congesta* which has a glabrous inflorescence.

Arenaria congesta Nutt. [*Eremogone congesta* (Nutt.) Ikonn.] Perennial from a branched caudex, forming loose mats. **Stems** erect, simple, 5–30 cm. **Leaves** filiform, flexuous or rigid, glabrous, 2–8 cm long, shorter above. **Inflorescence** many-flowered, congested to sometimes open, glabrous. **Flowers**: sepals glabrous, ovate to lanceolate, scarious-margined, 3–6 mm long; petals oblong, 5–8 mm long. **Capsules** glabrous, 3–6 mm long. Steppe, meadows, open forest. AB, SK south to CA, UT, CO.

1. Inflorescence more open; pedicels 1–6 mm long. . . . . var. *lithophila*
1. Inflorescence congested, capitate; pedicels <0.5 mm long 2

2. Sepal tips rounded . . . . var. *congesta*
2. Sepal tips acute to acuminate, sharp-pointed . . . . var. *cephaloidea*

Arenaria congesta var. *congesta* is known from granite-derived soil at high elevations in the Bitterroot and Absaroka-Beartooth regions; *A. congesta* var. *cepahaloidea* (Rydb.) Maguire is montane, occasionally subalpine, mainly west of the Continental Divide; *A. congesta* var. *lithophila* (Rydb) Maguire is more widespread at all elevations. See *A. capillaris*.

Arenaria hookeri Nutt. [*Eremogone hookeri* (Nutt.) W.A.Weber] Perennial from a branched caudex, forming dense mats. **Stems** erect, simple, 2–6 cm. **Leaves** needle-like with dilated, scarious bases, glabrous, 5–10 mm long; those of sterile stems shorter. **Inflorescence** several-flowered, congested, glandular-hairy. **Flowers**: sepals glabrous, lanceolate, pungent, scarious-margined, 5–8 mm long; petals oblanceolate, 6–8 mm long. **Capsules** glabrous, 3–4 mm long. Stony, sparsely vegetated soil of grassland, steppe, ridge crests, open, eroding slopes; plains, valleys, montane. MT to SD south to NV, MN and OK. Our plants are var. *hookeri*.
　　Arenaria hookeri var. *pinetorum* (A.Nelson) Maguire, loosely matted with longer leaves, occurs in adjacent northeast WY and may occur in southeast MT.

Arenaria kingii (S.Watson) M.E.Jones [*Eremogone kingii* (S.Watson) Ikonn.] Perennial from a loosely branched caudex, forming loose mats. **Stems** erect, simple, 10–30 cm. **Leaves** filiform, sparsely short-hairy, 1–2 cm long, shorter below. **Inflorescence** several-flowered, open, glandular (sparsely). **Flowers**: sepals glabrous to glandular, narrowly ovate with an acute tip, broadly scarious-margined with a deep green center, 3–6 mm long; petals spatulate, sometimes lobed, 5–7 mm long. **Capsules** glabrous, 4–7 mm long. Gravelly, usually calcareous soil of exposed grasslands, steppe, fellfields; montane to alpine in Beaverhead Co., MT, OR, ID, CA, NV, UT. Our plants are var. *glabrescens* (S.Watson) Maguire.

Arenaria serpyllifolia L. Annual with scabrous or short-hairy foliage. **Stems** 5–25 cm, branched at the base. **Leaves** ovate, sharp-pointed, 3–5 mm long, petiolate. **Inflorescence** open, leafy-bracteate, expanding in fruit. **Flowers**: sepals lanceolate ca. 3 mm long; petals ca. 2 mm long. **Capsule** 3–4 mm long, longer than the calyx. Often disturbed soil of grasslands, rock outcrops, along roads, trails; valleys, montane. Throughout temperate N. America. Introduced from Eurasia.

Cerastium L. Mouse-ear chickweed

Annuals or perennials with pubescent foliage. **Leaves** basal and/or cauline, sessile or short-petiolate. **Inflorescence** a terminal, bracteate cyme. **Flowers**: sepals separate, elliptic to ovate; petals white, clawed, the blade 2-lobed; styles mostly 5; capsule cylindric, often curved at maturity, opening by 10 teeth.[286]

    Cerastium tomentosum L. (dusty miller), a common garden plant with white-woolly foliage, occasionally escapes near residences, as in the Rattlesnake Valley of Missoula.

1. Plants perennial, often rhizomatous or mat-forming; petals 1.5 times as long as sepals. 2
1. Taprooted annuals or short-lived perennial with petals ca. equal to sepals 3

2. Inflorescence bracts scarious-margined. . . . . *C. arvense*
2. Bracts of inflorescence not scarious. . . . . *C. beeringianum*

3. Most pedicels longer than sepals 4
3. Most pedicels shorter than sepals 5

4. Pedicel hairs septate but non-glandular; plants biennial or perennial . . . . *C. fontanum*
4. Pedicels with gland-tipped hairs; plants annual . . . . *C. nutans*

5. Sepals 4–5 mm long. . . . . *C. glomeratum*
5. Sepals 6–11 mm long. . . . . *C. dichotomum*

Cerastium arvense L. Field chickweed Perennial forming loose mats. **Stems** trailing or decumbent, branched above, 5–30 cm. **Leaves** lanceolate to oblong, 1–3 cm long with fascicled leaves below. **Inflorescence** glandular, several-flowered. **Flowers**: sepals 4–7 mm long; petals 6–10 mm long. **Capsule** 7–11 mm long. Grasslands, steppe, dry open forest, meadows, turf; all elevations. Circumboreal south to CA, NM, MN, GA. Our plants are ssp. *strictum* Gaudin. (p.138)

Cerastium beeringianum Cham.& Schltdl. [*C. alpinum* L. misapplied] Caespitose perennial similar to *C. arvense* forming small mats with glandular foliage. **Stems** trailing to erect, 4–12 cm. **Leaves** oblong to narrowly elliptic, 5–15 mm long, rarely fascicled. **Inflorescence** glandular, few-flowered. **Flowers**: sepals 4–7 mm long; petals 6–10 mm long. **Capsule** 8–12 mm long. Turf, talus, exposed ridges, moraine, more common on limestone; alpine. AK to NL south to CA, AZ, CO; Eurasia.

Cerastium dichotomum L. Slenderly taprooted annual with viscid-glandular foliage. **Stems** erect, simple or branched at the base, 15–30 cm. **Leaves** narrowly oblanceolate, sessile, 12–30 mm long. **Inflorescence** tightly clustered cymes; pedicels shorter than the sepals. **Flowers**: sepals 6–11 mm long with glandular hairs; petals 8–10 mm long. **Capsule** 10–15 mm long. Agricultural fields; plains. Introduced WA to MT south to CA; native to Europe. One collection from Cascade Co.

Cerastium fontanum Baumg. [*C. vulgatum* L. misapplied] Perennial forming tufts or small mats with pubescent foliage. **Stems** 10–40 cm long, trailing and rooting at the nodes. **Leaves** lanceolate to ovate, 1–3 cm long, rarely fascicled. **Inflorescence** several-flowered with pedicels ≤1 cm. **Flowers**: sepals 5–7 mm long; petals 4–7 mm long. **Capsule** 9–13 mm long. Disturbed soil of lawns, trails, streambanks, roadsides; valleys, montane. Introduced throughout N. America and the world. Our plants are ssp. *vulgare* (Hartm.) Greuter & Burdet.

Cerastium glomeratum Thuill. [*C. viscosum* L. misapplied] Slenderly taprooted annual with hirsute and glandular foliage. **Stems** ascending to erect, branched at the base, 10–30 cm. **Leaves** oblanceolate, petiolate below, 8–25 mm long, rarely fascicled. **Inflorescence** few to many tight clusters or an open cyme; pedicels smaller than the sepals. **Flowers**: sepals 4–5 mm long with non-glandular, multi-cellular hairs; petals 3–5 mm long. **Capsule** 7–10 mm long. Disturbed soil of lawns, fields; valleys, Missoula and Sanders cos. Most of temperate N. America except the Great Plains; introduced from Europe.

Cerastium nutans Raf. [*C. brachypodum* (Engelm. ex A.Gray) B.L. Rob.] Slenderly taprooted annual with glandular-hairy foliage. **Stems** erect, branched at the base, 5–30 cm. **Leaves** oblanceolate, 3–30 mm long, not fascicled. **Inflorescence** several-flowered; the pedicels 3–12 mm long. **Flowers**: sepals 3–5 mm long; petals 3–6 mm long. **Capsule** 8–12 mm long. Vernally moist soil of riparian areas, open forest; valleys, montane. Throughout most of temperate N. America.

1. Pedicels ≤ the calyx . . . . var. *brachypodum*
1. Pedicels > the calyx . . . . var. *nutans*

Cerastium nutans var. ***brachypodium*** Engelm. ex A.Gray is sometimes considered a distinct species;[286] however, it intergrades morphologically with *C. nutans* var. ***nutans***[187] and has a similar range and habitat in our area.

Dianthus L. Pink

Annuals or perennials. **Leaves** sessile or petiolate, united into a sheath below, the blade 1-veined. **Flowers** solitary or in congested to open, bracteate clusters; sepals united below into a tubular calyx; petals usually red or pink, clawed, the blade dentate on the outer margin; styles 2; **Capsule** ovoid to cylindric, opening by 4 teeth.

Our species are European horticultural introductions. *Dianthus barbatus* L. (sweet William) escapes from gardens in Missoula and Ravalli Co. but does not persist.[226]

1. Plants glabrous. . . . . *D. barbatus*
1. Plants pubescent above 2

2. Inflorescence bracts linear-lanceolate; pedicels shorter than the calyx . . . . *D. armeria*
2. Bracts narrowly ovate; some pedicels as long or longer than the calyx . . . . *D. deltoides*

Dianthus armeria L. Deptford pink Annual or biennial. **Stems** 15–50 cm, hairy near nodes. **Leaves** hirsute, linear, 2–8 cm long, those of the stem erect. **Inflorescence** of few, bracteate, several-flowered, tight clusters. **Flowers**: calyx ca. 20-veined, 10–20 mm long, pubescent; petals deep pink with white dots, 15–22 mm long. **Capsule** 10–16 mm long. Grasslands, roadsides; valleys. Throughout temperate N. America and much of the world. Introduced from Europe.
Particularly common in bunchgrass prairie of northwest MT.

Dianthus deltoides L. Maiden pink Perennial with puberulent foliage from a branched caudex, forming loose mats. **Stems** trailing to ascending, 10–40 cm. **Leaves** sessile, linear, erect, 15–25 mm long. **Inflorescence** a few-flowered, linear-bracted, open cyme, rarely 1-flowered. **Flowers**: calyx 25- to 30-veined, 8–17 mm long, glabrous; petals pink, the blade 4–9 mm long. **Capsule** 12–13 mm long. Roadsides, sagebrush steppe; valleys. Much of temperate N. America. Introduced from Europe. One collection from Beaverhead Co.

Gypsophila L. Baby's breath

Perennials from a stout taproot. **Stems** erect, branched above. **Leaves** sessile, cauline, barely united around the stem below. **Inflorescence** a diffusely branched, terminal cyme. **Flowers**: calyx united below with rounded, membranous-margined lobes; petals oblanceolate; stamens 10; styles 2. **Capsule** globose, opening by 4 slits.[329]

Both species are escaped ornamentals introduced throughout north-temperate N. America from Eurasia.

1. Calyx glandular. . . . . *G. scorzonerifolia*
1. Calyx glabrous . . . . *G. paniculata*

Gypsophila paniculata L. Plants glabrous and glaucous. **Stems** 40–90 cm. **Leaves** linear-lanceolate, 2–5 cm long,1-veined. **Flowers**: calyx 1–2 mm long; petals white, 2–3 mm long. **Capsule** smaller than the calyx. Grasslands (especially west of the Continental Divide), fields, roadsides; plains, valleys.

Gypsophila scorzonerifolia Ser. Plants glabrous and glaucous below, glandular above. **Stems** to 70 cm. **Leaves** lanceolate, 3–8 cm long, 1-veined. **Flowers**: calyx 3–4 mm long, glandular; petals 4–6 mm long, pinkish. **Capsule** smaller than the calyx. Collected in calcareous gravels in Teton Co.

Holosteum L. Jagged chickweed

Holosteum umbellatum L. Taprooted, glandular-pubescent annual. **Stems** erect, usually unbranched, 3–20 cm. **Leaves** lanceolate, 1-veined, 2–20 mm long, petiolate in the basal rosette, sessile and few above. **Inflorescence** a terminal, bracteate umbel of few to several flowers; the pedicels 5–20 mm long, reflexed immediately after flowering, but erect in fruit. **Flowers**: sepals separate, lanceolate, nearly glabrous, 2–4 mm long with scarious margins; petals white, clawed, 3–5 mm long with dentate-margined tips; stamens 3 to 5; styles 3. **Capsule** narrowly ovoid, 4–6 mm long, opening by 6 teeth. Vernally moist soil of grasslands, wetland margins; valleys. Throughout temperate N. America. Introduced from Eurasia. Our plants are ssp. *umbellatum*.

This plant is inconspicuous due to its small size and early phenology, and is probably more widespread than records indicate.

Minuartia L. Sandwort

Tufted or mat-forming perennials. **Leaves** linear, often with axillary fascicles. **Inflorescence** terminal, solitary to openly cymose. **Flowers**: sepals separate; petals white with blunt or notched tips; styles 3; stamens 10. **Capsule** ovoid, opening by 3 teeth.[325,333]

These species have often been placed in *Arenaria*. *Minuartia pusilla* (S.Watson) Mattf. occurs in ID and southeast WA and could be expected in dry grasslands of northwest MT.

1. Tip of sepals blunt, hood-like . . . . *M. obtusiloba*
1. Sepals with sharp-pointed tips 2
2. Herbage glabrous. . . . . *M. austromontana*
2. Herbage glandular 3
3. Sepals 1-veined . . . . *M. nuttallii*
3. Sepals 3-veined . . . . *M. rubella*

Minuartia austromontana S.J.Wolf & Packer [*M. rossii* (R.Br. ex Richardson) Graebn. and *Arenaria rossii* R.Br. misapplied] Plants forming small dense mats with glabrous foliage. **Stems** erect to ascending, 1–10 cm. **Leaves** 1-veined, flexuous, bright green, 3–6 mm long with acute tips. **Flowers** solitary; sepals lance-shaped, 2–3 mm long, acute; petals absent, shorter, or as long as the sepals. **Capsules** equal to the calyx. Stony soil of exposed ridges, talus slopes, scree, fellfields; alpine. BC, AB south to OR, UT, WY.

Minuartia nuttallii (Pax) Briq. [*Arenaria nuttallii* Pax] Plants forming loose mats with glandular-hairy foliage. **Stems** ascending or trailing, brittle, 3–25 cm. **Leaves** 1-veined, 5–10 mm long, rigid, mucronate. **Flowers** in few-flowered cymes; sepals 1-veined, 3–5 mm long, lance-shaped, acuminate; petals ≤ sepals. **Capsule** shorter than calyx. Rocky slopes, talus, especially limestone; montane to alpine, more common higher. BC, AB south to CA, UT, WY. Our plants are var. *nuttallii*.

Small plants can be similar to *Minuartia rubella* which has conspicuously 3-veined sepals.

Minuartia obtusiloba (Rydb.) House [*Arenaria obtusiloba* (Rydb.) Fernald, *A. sajanensis* Willd., *A. laricifolia* L.] Plants tufted or forming small mats with glabrous to glandular-hairy foliage. **Stems** erect, 1–10 cm. **Leaves** 1-veined, 4–5 mm long, flexuous with acute tips. **Flowers** solitary; sepals 3-veined, 4–6 mm long, glandular with blunt, hood-like tips; petals 3–8 mm long. **Capsules** as long or longer than calyx. Stony soil of exposed ridges, slopes; upper subalpine, alpine. AK to Greenland south to OR, NM. (p.138)

Glabrous plants occurring in moist, cool, alpine sites in Glacier National Park and the Beartooth have been referred to the arctic *Minuartia biflora* (L.) Schinz & Thell. [=*A. sajanensis*], but are more likely an ecotype of *M. obtusiloba*. [160,425]

Minuartia rubella (Wahlenb.) Hiern. [*Arenaria rubella* (Wahlenb.) Sm., *A. propinqua* Richardson] Plants tufted with glabrous to usually glandular-hairy foliage. **Stems** erect, 2–10 cm tall. **Leaves** 3-veined, 3–10 mm long, rigid, apiculate. **Flowers** few, cymose; sepals 3-veined, lanceolate with papery margins, acuminate, 2–4 mm long; petals elliptic, as long or shorter than the sepals. **Capsules** longer than calyx. Sparsely vegetated, often stony soil of talus slopes, fellfields, gravel bars; valleys to alpine, more common higher. Circumpolar south to CA, AZ, NM.

Moehringia L. Sandwort

Perennials with puberulent foliage from slender rhizomes. **Leaves** mainly cauline, sessile or short-petiolate below, lanceolate to elliptic. **Inflorescence** terminal or axillary, minutely bracteate; flowers solitary or few

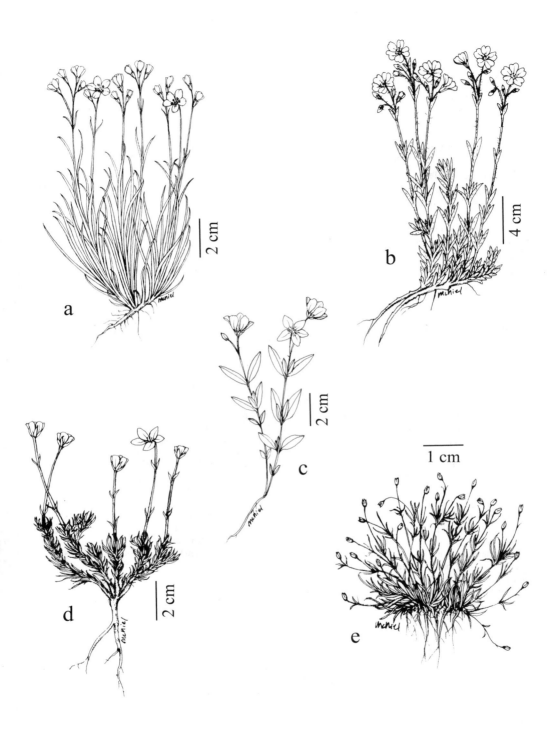

Plate 18. a. *Arenaria capillaris*, b. *Cerastium arvense*, c. *Moehringia lateriflora*, d. *Minuartia obtusiloba*, e. *Sagina saginoides*

in open cymes. **Flowers**: sepals separate; petals white with blunt tips; styles 3; stamens 10; **Capsule** ovoid, opening by 6 teeth.

Our species have often been placed in *Arenaria*.

1. Sepals acute to acuminate; petals ≤1.5 times as long as sepals . . . . *M. macrophylla*
1. Sepals tips rounded; petals 2 to 3 times as long as sepals . . . .*M. lateriflora*

Moehringia lateriflora (L.) Fenzl [*Arenaria lateriflora* L.] **Stems** ascending, mostly simple above, 5–20 cm. **Leaves** thin, 1–3 cm long. **Flowers**: sepals 2–3 mm long, ovate with rounded tips; petals 4–6 mm long. **Capsules** globose, longer than the sepals. Moist soil of meadows, open forests, woodlands, thickets, often along streams; valleys, montane. AK to NL south to CA, NM, NE, PA, VA. (p.138)

Moehringia macrophylla (Hook.) Fenzl [*Arenaria macrophylla* Hook.] **Stems** ascending, mostly simple above, 5–20 cm. **Leaves** lanceolate to narrowly elliptic, 1–5 cm long. **Flowers**: sepals 3–5 mm long, ovate with pointed tips; petals 4–6 mm long. **Capsules** ovoid, as long as the sepals. Thickets, open forest, especially along streams; valleys, montane. NT to NL south to CA, NM, WI, MI.

Paronychia Mill. Nailwort

Paronychia sessiliflora Nutt. Perennial with a branched caudex and scabrous foliage, forming dense mats to 15 cm across. **Stems** ascending, branched, 2–10 cm long. **Leaves** crowded, overlapping, linear-subulate, leathery, 3–7 mm long with white-scarious, attenuate stipules longer than the leaves. **Flowers** terminal, usually solitary, hidden among the leaves; sepals 5, 3–5 mm long, narrowly lanceolate, long-attenuate and awned, united below, yellowish; petals absent or minute; stamens 5; style 1 with 2 stigmas. **Fruit** a utricle enclosed by the calyx. Sparsely vegetated, often stony soil of exposed slopes, ridge crests; plains, valleys to montane. AB, SK south to NV, NM, TX.

The plant vegetatively resembles *Douglasia montana* and occurs in similar habitats.

Sagina L. Pearlwort

Small, tufted, glabrous, short-lived perennial herbs. **Leaves** linear, apiculate, basal and cauline, united around the stem below, axillary fascicles sometimes present. **Inflorescence** flowers solitary on long terminal and axillary peduncles. **Flowers** sepals 4 to 5, separate; petals absent or white, elliptic with entire margins; stamens 4 to 10; styles 4 to 5. **Capsule** opening by 4 to 5 slits.[105]

1. Sepals purple throughout; mature capsule as long as sepals . . . . *S. nivalis*
1. Sepals green to purple-tinged but with white margins; mature capsules longer than sepals 2

2. Sepals 5, stems ascending, not rooting at the nodes. . . . . *S. saginoides*
2. Sepals 4, stems trailing, ascending at the tips, often rooting at the nodes . . . . *S. procumbens*

Sagina nivalis (Lindlblom) Fr. Plants forming small clumps. **Stems** to 4 cm, ascending from axils of basal rosette leaves. **Leaves** to 10 mm long without axillary fascicles. **Flowers**: sepals 4, purplish, 1–2 mm long, erect when mature; petals mostly 4, 1–2 mm long. **Capsule** as long as the sepals. Wet, sparsely vegetated soil of shaded cliffs; alpine. Circumpolar south to Glacier Co., MT.

Sagina procumbens L. Plants often forming mats. **Stems** prostrate, ascending distally, up to 20 cm, often rooting at the nodes. **Leaves** 5–20 mm long, often with axillary fascicles. **Flowers**: sepals 5, ca. 2 mm long, green with white margins spreading after maturation; petals 4, ca. 1 mm long (absent). **Capsule** longer than the sepals. Moist, open soil of streambanks, gardens, roadsides; valleys to montane. Much of temperate N. America; introduced from Eurasia.

Sagina saginoides (L.) H.Karst. Plants forming small tufts. **Stems** 1–10 cm, ascending from a basal rosette. **Leaves** 5–10 mm long without axillary fascicles. **Flowers**: sepals 5, green to purple with white margins, 2–3 mm long, erect when mature; petals mostly 5, 1–2 mm long. **Capsule** longer than the sepals. Moist, sparsely vegetated soil of meadows, cliffs, rock outcrops, often along streams; valleys to alpine. Circumpolar south to CA, NM. (p.138)

Small alpine plants closely resemble *Sagina nivalis*, but the latter have purple-margined sepals.

Saponaria L. Soapwort

Saponaria officinalis L. Bouncing bet, soapwort Rhizomatous perennial with glabrous foliage. **Stems** simple, to 60 cm. **Leaves** lanceolate to obovate, 3–8 cm long, short-petiolate or sessile. **Inflorescence** a terminal, branched, bracteate, many-flowered, open to congested cyme. **Flowers:** calyx 15–20 mm long, united into a veiny tube below and short, triangular lobes above; petals clawed, white to pinkish; the blade 8–15 mm long with erose to entire margins and 2 small adaxial scales; stamens 10; styles 2. **Capsule** ovoid-cylindric,15–20 mm long. Roadsides, gardens, pastures; valleys. Introduced to most of temperate N. America; an escaped garden ornamental native to Europe.

Scleranthus L. Knawel

Scleranthus annuus L. Puberulent annual. **Stems** to 15 cm, ascending to erect, branched below. **Leaves** linear, pungent, 5–10 mm long, the scarious bases united around the stem. **Inflorescence** terminal and axillary, congested cymes. **Flowers:** sepals leathery, 2–4 mm long, united below into a short 10-veined tube with lanceolate lobes; petals absent; stamens 5 to 10; styles 2, exserted from the tube. **Fruit** a utricle enclosed in the hardened calyx. Roadsides, streambanks, disturbed grasslands; valleys. Throughout eastern U.S., Pacific Northwest to CA. Introduced from Europe.

Silene L. Campion, catchfly

Annual or perennial, taprooted, sometimes monoecious or dioecious herbs. **Leaves:** the basal petiolate, the cauline usually sessile. **Inflorescence** flowers solitary or in open to congested, bracteate cymes. **Flowers** perfect or unisexual; sepals united into a tubular, 10- to 30-veined calyx with short lobes; petals clawed, white (pink in *S. acaulis*), 2- or 4-lobed, often with 2 appendages at the base of the blade; stamens 10; styles 3 or 5; ovary mounted on a short, thick stalk. **Capsule** opening by 6 to 10 teeth.[186,288]

Some authorities segregate M*elandrium* and *Lychnis* from *Silene* based on the number of styles, but this character often varies within the same population of a single species. *Silene chalcedonica* (L.) E.H.L.Krause, (Maltese cross), a garden escape with red petals, was collected once in Lincoln Co.; it does not persist.[288]

12. Petal blades 1–2 mm long, barely exserted from the calyx . . . . *S. spaldingii*
12. Petal blades ≥3 mm long, well exserted from the calyx 13

13. Petal blades 2-lobed. . . . . *S. parryi*
13. Petal blades at least 4-lobed . . . . *S. oregana*

14. Petals red to pink 15
14. Petals white to purplish. 16

15. Inflorescence congested, hemispheric; pedicels <1 cm long . . . . *S. chalcedonica*
15. Inflorescence open; some pedicels >1 cm long . . . . *S. dioica*

16. Styles 5. 17
16. Styles 3. 21

17. Plants ≤10 cm high. 18
17. Plants >20 cm high. 19

18. Flowers nodding; calyx ovate, inflated . . . . *S. uralensis*
18. Flowers erect; calyx tubular . . . . *S. hitchguirei*

19. Plants dioecious . . . . *S. latifolia*
19. Plants monoecious or flowers perfect. 20

20. Calyx with short, gland-tipped hairs . . . . *S. drummondii*
20. Calyx short-hairy but not glandular. . . . . *S. douglasii*

21. Calyx not glandular. . . . . *S. douglasii*
21. Calyx glandular. 22

22. Stems decumbent; plants subalpine and alpine. . . . . *S. repens*
22. Stems erect; plants montane . . . . *S. scouleri*

Silene acaulis L. Moss campion Sometimes dioecious, glabrous perennial from a branched caudex, forming mats to 30 cm across. **Stems** erect or ascending, 1–3 cm. **Leaves** crowded below, needle-like, 2–8 mm long. **Flowers** perfect or unisexual, solitary, nearly sessile or on pedicels to 2 cm long; calyx tubular, 4–8 mm long, pinkish, 10-veined; petals rose, the blade 1–6 mm long, 2-lobed; styles 3. **Capsule** smaller than the calyx, opening by 6 slits. Stony soil of tundra and exposed ridges, slopes; alpine, rarely lower. Circumpolar south to OR, AZ, NM. (p.143)

On cool, sheltered slopes plants are more diffuse than the bright green cushions of exposed fellfields. Two vars. are often recognized: the circumpolar var. *exscapa* (All.) DC. & Lam. has a calyx 4–6 mm long, and emarginate petals, whereas the cordilleran var. *subacaulescens* (Williams) Fern.& St.John has a calyx 7–10 mm long and entire petals. However, the varieties have no geographic or ecological integrity in MT, and the distinguishing characters are continuous.

Silene antirrhina L. Puberulent annual. **Stems** erect, often branched above, 10–80 cm with glandular bands below the nodes. **Leaves**: basal oblanceolate, 1–3 cm long; cauline linear to narrowly oblanceolate, 1–8 cm long. **Inflorescence** an open, many-flowered cyme. **Flowers**: calyx 10-veined, ellipsoid, glabrous, 5–10 mm long; petals white to purplish, the blade 2–3 mm long, shallowly bilobed with 2 appendages; styles 3. **Capsule** as long and wide as the calyx, opening by 6 slits. Usually disturbed soil of grasslands, steppe, fields, roadsides; plains, valleys. Throughout temperate N. America, introduced to Europe. (p.143)

Silene conoidea L. Pubescent annual. **Stems** erect, 20–50 cm. **Leaves**: basal oblanceolate, 3–5 cm long, quickly deciduous; cauline lanceolate, 1–5 cm long, united around the stem. **Inflorescence** a glandular, open, 1- to several-flowered cyme. **Flowers**: calyx narrowly ovoid, ca. 25-veined, 10–25 mm long; petals white to pinkish, the blade 5–10 mm long with erose margins and 2 minute appendages; styles 3. **Capsule** 8–20 mm long, opening by slits. Cultivated fields, disturbed grasslands, steppe; valleys. Introduced to AB, SK south to CA, TX, MO; native to Eurasia.

Silene csereii Baumg. Bladder campion Glabrous, short-lived perennial. **Stems** erect, 20–80 cm. **Leaves**: basal few; cauline lanceolate to oblanceolate, 2–7 cm long. **Inflorescence** a narrow cyme. **Flowers**: calyx ca. 20-veined, ellipsoid, 8–11 mm long; petals white, the blade 2-lobed, to 5 mm long with 2 appendages; styles 3, the stigmas exserted. **Capsules** slightly longer and broader than calyx, opening by 6 slits. Roadsides; plains, valleys. Throughout north-temperate N. America; introduced from Europe. See *Silene vulgaris*.

Silene dichotoma Ehrh. Long-hairy annual. **Stems** erect, 40–80 cm. **Leaves**: basal oblanceolate, 3–8 cm long, withering by flowering time; cauline numerous, lanceolate to oblanceolate, reduced upward. **Inflorescence** many-flowered, narrow, apparently racemose. **Flowers** perfect or sometimes unisexual; calyx tubular, setose, 10-veined, 10–15 mm long; petals white, the blades 5–7 mm long, deeply bilobed with 2 appendages; styles 3. **Capsule** equaling the calyx, opening by 6 slits. Roadsides, disturbed grasslands, open forest; valleys. Throughout much of north-temperate N. America; introduced from Europe.

Silene dioica (L.) Clairv. [*Lychnis dioica* L.] Dioecious short-lived perennial with hirsute foliage and a branched caudex. **Stems** erect, 40–80 cm. **Leaves**: basal oblong, withering by flowering time; cauline elliptic, 3–10 cm long. **Inflorescence** an open, glandular, several-flowered cyme. **Flowers** unisexual; calyx 1–2 cm long, purplish, with septate hairs, tubular with lobes 2–4 mm long, becoming ovate in fruit; petals pink to purplish, 2-lobed with 4-lobed appendages, the blade 10–15 mm long; styles 5. **Capsule** ovoid, equaling the calyx, opening by 5 slits. Disturbed soil of moist meadows, open forest; montane. Introduced throughout most of temperate N. America; native to Eurasia. Collected twice in Park Co.

Silene douglasii Hooker Perennial with a finely branched caudex and glabrous to puberulent foliage below the inflorescence. **Stems** ascending, 10–50 cm. **Leaves**: basal oblanceolate, numerous; cauline linear to narrowly oblanceolate, 1–6 cm long. **Inflorescence** a few-flowered cyme. **Flowers**: calyx campanulate, inflated at maturity, sometimes glandular, 12–15 mm long, 10-nerved; petals white, the blade 4–6 mm long, shallowly bilobed with 2 apparent appendages; styles 3 to 5. **Capsule** opening by 3 to 5 slits, equaling the calyx at maturity. Grasslands, stony meadows; valleys to subalpine, more common lower. BC to CA, NV, UT. Our plants are var. ***douglasii***.

The similar *Silene parryi* has glandular foliage below the inflorescence and 4-lobed petals.

Silene drummondii Hook. [*Lychnis drummondii* (Hook.) S.Watson] Puberulent and sometimes glandular perennial from a few-branched caudex. **Stems** erect, unbranched, 20–60 cm. **Leaves**: basal oblanceolate, 3–10 cm long; cauline few, linear. **Inflorescence** a glandular, erect, few-flowered cyme. **Flowers**: calyx tubular, 10–15 mm long, 10-nerved; petals white, the blade narrower than the claw, 1–3 mm long, obscurely bilobed with 2 minute appendages; styles 5. **Capsule** opening by 5 slits, equaling the calyx. Grasslands, steppe, woodlands; plains, valleys, montane. BC to MB south to NM, AZ, MN. (p.143)

Two ssps. are sometimes recognized:[288] ssp. *drummondii*, with petals included in the 4–6 mm wide calyx, and ssp. *striata* (Rydb.) Maguire, with exserted petals and a calyx 6–8 mm wide. All of our plants have a calyx ≤6 mm wide, while the petal length is variable.

Silene flos-cuculi (L.) Clair. Ragged robin Glabrous to scabrous perennial from a branched caudex. **Stems** erect, branched, 30–90 cm. **Leaves** cauline, spatulate to oblanceolate, 2–10 cm long. **Inflorescence** an open, many-flowered cyme. **Flowers**: calyx campanulate, purple, 5–10 mm long, 10-nerved; petals pink, the blade deeply 4-lobed, 6–10 mm long with 2 bilobed appendages; styles 5. **Capsule** opening by 5 slits, equaling the calyx. Moist, disturbed meadows, pastures; valleys; collected in Lincoln Co. Introduced to WA, MT and northeast N. America; a garden escape native to Europe.

Silene hitchguirei Bocquet [*Lychnis apetala* L. var. *montana* (S.Watson) C.L.Hitchc., *S. uralensis* (Rupr.) Bocquet ssp. *montana* (S.Watson) McNeill] Taprooted perennial. **Stems** erect, glandular-hairy, 2–8 cm. **Leaves**: basal numerous, glabrous to short-hairy, oblanceolate 1–4 cm long; cauline few, reduced, linear. **Flowers** solitary, erect; calyx elliptic, glandular, purple-veined, not inflated, 7–10 mm long; petals white to pink, the blade 2–3 mm long, shallowly bilobed; styles 5. **Capsule** opening by 5 slits, equaling the calyx. Stony granitic or volcanic soil of fellfields, turf; alpine. AB to UT, CO. See *Silene uralensis*.

Silene latifolia Poir. White campion [*Lychnis alba* Mill.] Dioecious perennials with hirsute foliage and a branched caudex. **Stems** erect, 40–80 cm. **Leaves**: basal oblong, withering by flowering time; cauline oblanceolate, 3–12 cm long. **Inflorescence** an open, glandular, several-flowered cyme. **Flowers** unisexual; calyx 1–2 cm long, 10-veined in male, 20-veined in female, tubular with lobes 2–4 mm long, becoming ovate in fruit; petals white, 2-lobed with 4-lobed appendages, the blade 10–15 mm long; styles 5. **Capsule** ovoid, equaling the calyx, opening by 5 slits. Disturbed soil of meadows, lawns, streambanks, roadsides; plains, valleys, montane. Throughout most of N. America north of Mexico; introduced from Europe. See *Silene noctiflora*. (p.143)

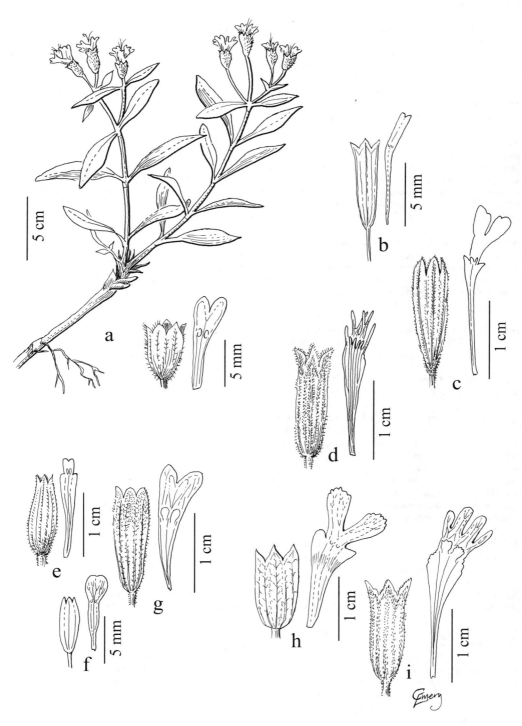

Plate 19. **Silene**. **a.** *S. menziesii*, **b.** *S. acaulis*, **c.** *S. latifolia*, **d.** *S. oregana*, **e.** *S. drummondii*, **f.** *S. antirrhina*, **g.** *S. repens*, **h.** *S. vulgaris*, **i.** *S. parryi*

Silene menziesii Hook. Rhizomatous, mostly dioecious perennials with pubescent to glandular foliage. **Stems** numerous, trailing, branched, 5–40 cm long. **Leaves** mainly cauline, narrowly lanceolate to elliptic, 2–5 cm long. **Inflorescence** few- to many-flowered (solitary) in an open, glandular-hairy cyme. **Flowers** functionally unisexual but with rudiments of the non-functional sex present; calyx obscurely 10-veined, campanulate, 5–8 mm long; petals white, the blade 1–4 mm long, 2-lobed; styles 3. **Capsule** opening by 3 slits, ellipsoid, just longer than the calyx. Forest, woodlands, riparian corridors; valleys, montane. AK to MB south to CA, AZ, NM. (p.143)

Hitchcock recognized two vars.: var. *menziesii* with eglandular lower stems and var. *viscosa* (Greene) C.L.Hitchc.& Maguire with glandular stems,[187] but the degree of glandulosity varies continuously in our material, and there is no ecological or geopgraphic association with this character.

Silene noctiflora L. Taprooted annual, hirsute and glandular above. **Stems** erect, 20–80 cm, mostly simple. **Leaves** mainly cauline, lanceolate to obovate, 3–12 cm long, the lower petiolate. **Inflorescence** an open, glandular, several-flowered cyme. **Flowers**: calyx 12–20 mm long with lobes 5–9 mm long, 10-veined, becoming swollen in fruit; petals white, the blade 2-lobed, 3–10 mm long; styles 5. **Capsules** ovoid, equaling and splitting the calyx, opening by 6 slits. Disturbed fields; valleys. Introduced throughout N. America north of Mexico; native to Europe.

The similar *Silene latifolia* is a dioecious perennial with shorter calyx lobes.

Silene oregana S.Watson Glandular-puberulent perennial from a few-branched caudex. **Stems** simple, 30–60 cm. **Leaves**: basal narrowly oblanceolate, 5–10 cm long; cauline similar, 2–9 cm long. **Inflorescence** a narrow, many-flowered, glandular cyme. **Flowers**: calyx tubular, 10-veined, 10–16 mm long; petals white, the blade 3–5 mm long, deeply 4- to 10-lobed with 4 to 6 appendages; styles 3. **Capsules** ellipsoid just longer than the calyx, opening by 6 slits. Grasslands; valleys to montane. WA to MT south to CA, NV, UT. (p.143)

Silene parryi (S.Watson) C.L.Hitchc.& Maguire Perennial from a branched caudex with short-hairy and glandular herbage. **Stems** 5–50 cm. **Leaves**: basal narrowly oblong, 3–8 cm long; cauline oblanceolate. **Inflorescence** a narrow, 3- to 7-flowered, glandular cyme. **Flowers**: calyx purple, 10-veined, 9–16 mm long, glandular; petals often purple-tinged, the blade 3–6 mm long, 4-lobed; styles 3. **Capsule** as long as the calyx. Grasslands, meadows, steppe, tundra, scree slopes; montane to alpine. BC, AB south to WA, ID, WY. See *Silene douglasii, S. repens*. (p.143)

Silene repens Patrin ex Pers. Perennial from a slenderly-branched caudex; foliage retrose-pubescent. **Stems** trailing but erect distally, 5–25 cm. **Leaves**: basal numerous on short sterile stems, oblanceolate, 2–4 cm long; cauline linear, reduced. **Inflorescence** a compact, densely pubescent cyme. **Flowers** erect; calyx tubular, purplish, 10-veined, 10–15 mm long; petals white, the blade 2-lobed, 3–7 mm long; styles 3, exserted. **Capsules** ovoid, equaling the calyx, opening by 6 slits. Non-calcareous talus, turf; near or above treeline. Circumpolar, south to ID and WY. Our plants are ssp. *australe* C.L.Hitchc.& Maguire. (p.143)

Silene parryi can assume the habit of *S. repens* when growing in rock slides, but the former has 4-lobed petals and is glandular below the inflorescence.

Silene scouleri Hook. Pubescent perennial from a simple or branched caudex. **Stems** simple, 10–40 cm. **Leaves**: basal oblanceolate, 7–15 cm long; cauline lanceolate, 1–12 cm long. **Inflorescence** spike-like with 3 to 6 glandular, congested cymes. **Flowers**: calyx campanulate, 10-veined, 12–18 mm long; petals white, the blade 2–8 mm long, 2–4-lobed with 2 appendages; styles 3. **Capsules** ellipsoid, equaling the calyx, opening by 6 to 8 slits. Grassland, steppe, open forest; montane. BC, AB south to CA, AZ, NM. Our plants are ssp. *hallii* (S.Watson) C.L.Hitchc.& Maguire.

Silene spaldingii S.Watson Viscid, glandular-pubescent perennial from a simple or branched caudex. **Stems** simple, 20–60 cm. **Leaves** all cauline, oblanceolate to lanceolate, 6–7 cm long, short-petiolate below. **Inflorescence** a few- to many-flowered, leafy, compound cyme. **Flowers**: calyx campanulate, 10-veined, 15–20 mm long; petals greenish-white, barely longer than the calyx, the blade ca. 2 mm long, shallowly bilobed, with 4 appendages; styles 3. **Capsules** ellipsoid, equaling the calyx, opening by 6 slits. Deeper soil of fescue grasslands; valleys. BC south to OR, ID, MT.

Listed as threatened under the Federal Endangered Species Act.

Silene uralensis (Rupr.) Bocquet [*Lychnis apetala* L. var. *macrosperma* (A.E.Porsild) B.Boivin] Taprooted perennial with a branched caudex. **Stems** erect, purple-septate hairy, 2–10 cm. **Leaves**: basal numerous, glabrous to short-hairy, oblanceolate, 1–4 cm long; cauline few, reduced, linear. **Flowers** solitary, nodding; calyx ovoid, inflated, glandular, purplish 10–15 mm long; petals purplish, as long as the calyx, the blade 1–3 mm long, shallowly bilobed with 2 appendages; styles 5. **Capsules** equaling the calyx, opening by 10 slits. Stony, calcareous soil of moraine, cliffs, talus; alpine. Circumpolar south to UT, CO.

The similar *Silene hitchguirei* has erect flowers without an inflated calyx.

Silene vulgaris (Moench) Garcke [*S. cucubalus* Wibel] Taprooted perennial with glabrous and glaucous foliage. **Stems** erect, 40–80 cm. **Leaves**: mainly cauline, lanceolate to ovate, 2–6 cm long. **Inflorescence** an open, several- to many-flowered cyme. **Flowers** perfect or sometimes pistillate; calyx green to purplish, 9–14 mm long, inflated in fruit, net-veined; petals white, the blade 4–6 mm long, 2-lobed; styles 3, long-exserted. **Capsules** ovoid, equaling the calyx, opening by 6 slits. Roadsides, streambanks, fields; plains, valleys. Introduced to most of temperate N. America; native to Europe. (p.143)

Silene cserei is similar but has a less inflated calyx with fewer cross veins.

Spergula L. Spurry

Spergula arvensis L. Annual, glabrous below, glandular-hairy above. **Stems** erect to ascending, 10–40 cm. **Leaves** fascicled, appearing whorled, 1-veined; stipules scarious, white, 1 mm long. **Inflorescence** glandular-hairy, terminal, few-flowered cymes, or solitary and axillary. **Flowers**: sepals 5, 2–3 mm long, united at the base, margins scarious; petals 5, white with entire margins, shorter than the calyx; stamens 2 to 10; styles 5. **Capsule** ovoid, 3–5 mm long, opening by 6 slits. Gardens, fields; valleys to montane. Introduced in most of temperate N. America; native to Europe.

Spergularia (Pers.) J.Presl & C.Presl Sand spurry

Small annuals or perennials. **Leaves** cauline, linear, fleshy, apiculate, 1-veined; axillary fascicles sometimes present; stipiules white-scarious. **Inflorescence** terminal, glandular, few-flowered cymes or solitary and axillary. **Flowers**: sepals 5, separate or united at the base, margins scarious; petals 5, as long as the sepals, pink or purplish with entire margins; stamens 2 to 10; styles 3. **Capsule** ovoid, opening by 3 slits.[161]

1. Plants perennial with a woody caudex . . . . *S. media*
1. Plants annual, not woody at the base. 2

2. Leaves fascicled with several smaller leaves in the axils. . . . . *S. rubra*
2. Fascicled leaves generally not present 3

3. Stipules entire; leaves ca.1 mm wide; stamens 2 to 5 . . . . *S. marina*
3. Stipules lacerate; many leaves <1 mm wide; stamens 4 to 7. . . . . *S. diandra*

Spergularia diandra (Guss.) Heldr. Glandular-pubescent annual. **Stems** prostrate, to 15 cm. **Leaves** 10–20 mm long, rarely fascicled; stipules 1–2 mm long. **Flowers** solitary, pedicellate from upper leaf axils or few-flowered, terminal cymes; sepals ca. 3 mm long. **Capsule** ca. as long as calyx; seeds wrinkled. Moist soil of streambanks; plains. Introduced to northwest U.S. and adjacent Canada; native to Europe. One collection from Valley Co. more than 100 years old.

Spergularia marina (L.) Besser [*Spergularia salina* J.Presl & C.Presl] Glandular-hairy annual. **Stems** erect to ascending, 7–20 cm. **Leaves** 5–25 mm long, seldom fascicled; stipules 1–2 mm long. **Flowers** in few-flowered terminal cymes; sepals 2–4 mm long. **Capsule** longer than the calyx; seeds not winged or grooved. Fine-textured saline or alkaline soil around wetlands; plains, valleys. Cosmopolitan, including throughout temperate N. America.

The species may be introduced throughout N. America or native in coastal and inland, naturally saline habitats.

Spergularia media (L.) C.Presl Perennial from a small, woody, branched caudex. **Stems** ascending to prostrate, 10–20 cm. **Leaves** 5–20 mm long, glabrous, few-fascicled; stipules 2–4 mm long. **Flowers** in glandular-hairy, few-flowered, terminal and axillary cymes; sepals 3–5 mm long. **Capsule** longer than the calyx; seeds winged. Moist, alkaline soil of wetland margins; valleys. Introduced worldwide, including northwest and northeast U.S.; native to Europe. One collection from Jefferson Co.

Spergularia rubra (L.) J.Presl & C.Presl Annual. **Stems** prostrate, to 15 cm. **Leaves** glabrous,
3–10 mm long, fascicled; stipules 3–5 mm long. **Flowers** solitary, long-stalked from upper leaf
axils or few-flowered terminal cymes; sepals 3–4 mm long. **Capsule** ca. as long as calyx; seeds
with a sub-marginal groove. Disturbed soil of grasslands, open forest, margins of streams and
wetlands, often along trails and roads; montane. Introduced throughout most of north-temperate N.
America; native to Europe.

Stellaria L. Chickweed

Small annual or perennial herbs with 4-angled stems. **Leaves** mostly sessile. **Inflorescence** axillary and/or terminal;
flowers solitary or in cymes. **Flowers** perfect; sepals separate, usually green with scarious margins; petals white, not
clawed, 2-lobed or absent; styles usually 3; stamens 5 or 10. **Capsule** ovoid, opening by 3 or 6 slits.[287]

1. Plants glandular in the inflorescence 2
1. Plants not glandular 3

2. Leaves linear . . . . *S. jamesiana*
2. Leaves ovate . . . . *S. americana*

3. Stem leaves ciliate-petiolate; stems ciliate in lines. . . . . *S. media*
3. Stem leaves sessile or stems not ciliate. 4

4. Plants taprooted annuals with erect stems. . . . .*S. nitens*
4. Plants finely rhizomatous perennials 5

5. Leaves ovate to narrowly ovate 6
5. Leaves linear to linear-lanceolate. 8

6. Flowers in terminal, leafy-bracteate cymes . . . . *S. calycantha*
6. Flowers solitary from upper leaf axils. 7

7. Sepals ovate with rounded tips. . . . . *S. obtusa*
7. Sepals lanceolate with attenuate tips . . . .*S. crispa*

8. Petals definitely longer than the sepals . . . .*S. longipes*
8. Petals ca. as long or shorter than the sepals 9

9. Leaves and stems scabrous on the margins (use 20X) . . . . *S. longifolia*
9. Leaves and stems smooth 10

10. Flowers solitary from upper leaf axils (bracts of inflorescence green) 11
10. Flowers in terminal, leafy-bracteate cymes (bracts of inflorescence hyaline, scarious). 13

11. At least some leaves >2 cm long . . . . *S. borealis*
11. Leaves ≤2 cm long 12

12. Sepals 2–3 mm long; rare . . . . *S. crassifolia*
12. Sepals 3–5 mm long; common. . . . .*S. longipes*

13. Sepals 2–3 mm long; petals absent or less than 1/2 as long as sepals. . . . .*S. umbellata*
13. Sepals 3–6 mm long; petals about the same length as sepals 14

14. Sepals minutely ciliate . . . . *S. graminea*
14. Sepals glabrous . . . .*S. longipes*

Stellaria americana (Porter ex B.L.Rob.) Standley Glandular-hairy perennial from a slenderly
branched caudex surmounting a taproot. **Stems** branched, trailing, 5–15 cm long. **Leaves** ovate,
1–3 cm long. **Inflorescence** a few-flowered, compact, terminal, leafy cyme. **Flowers**: sepals ovate,
obtuse, 3–5 mm long; petals 4–8 mm long; stamens 5. **Capsule** 5–6 mm long, opening by 3 slits.
Limestone talus; subalpine, alpine. Endemic to MT and adjacent AB. (p.148)
    The long underground stems extend upslope, covered by the shifting talus. The type locality is in Madison Co.

Stellaria borealis Bigelow [*S. calycantha* (Ledeb.) Bong. var. *sitchana* (Steud.) Fernald,
S. calycantha var. *bongardiana* (Fernald) Fernald] Glabrous to sparsely hairy, rhizomatous
perennial. **Stems** branched, prostrate to ascending 12–50 cm. **Leaves** linear to lance-linear, 1–4 cm
long. **Inflorescence** a terminal, few-flowered, leafy cyme or solitary and axillary. **Flowers**: sepals
lanceolate to ovate, 2–4 mm long, obscurely 1–3-veined; petals 1–3 mm long or absent; stamens 5.

Capsule 3–6 mm long, opening by 3 slits. Moist meadows, forests and margins of streams, wetlands; montane to rarely lower subalpine. Circumboreal, south to CA, UT, CO. (p.148)

Both ssp. *sitchana*, (Steud.) Piper & Beattie with 3-veined sepals and lanceolate leaves, and ssp. *borealis* with 1-veined sepals and narrowly elliptic leaves, are reported for MT.[287] However, these two taxa have no ecological or geographic integrity in our area, and the characters separating them are uncorrelated in our material. See *S. calycantha*.

Stellaria calycantha (Ledeb.) Bong. [*S. simcoei* (Howell) C.L.Hitchc.] Rhizomatous perennial. Stems glabrous to pilose, trailing to ascending, to 20 cm. **Leaves** glabrous, narrowly ovate, 8–20 mm long. **Inflorescence** a terminal, few-flowered, leafy cyme. **Flowers**: sepals ovate, 2–3 mm long; petals 1 mm long or absent; stamens 5. **Capsule** 3–5 mm long, opening by 3 slits. Wet turf, cliffs; subalpine, alpine, rarely lower along margins of wetlands. AK south to CA, NM; Asia. (p.148)

The similar *Stellaria borealis* generally occurs at lower elevations; *S. simcoei* is the form with pilose stems.

Stellaria crassifolia Ehrh. Glabrous perennial forming loose mats from slender rhizomes. **Stems** branched, trailing to ascending, 5–15 cm long. **Leaves** lanceolate, 6–16 mm long. **Inflorescence** a delicate, leafy-bracted, few-flowered, terminal cyme. **Flowers**: sepals 2–3 mm long, lanceolate, weakly 3-veined; petals just shorter than sepals; stamens 5 or 10. **Capsules** 4–5 mm long, opening by 6 slits. Wet meadows, thickets, streambanks; montane, subalpine; Beaverhead Co. Circumpolar south to BC, UT, CO.

Stellaria crispa Cham.& Schltdl. Mostly glabrous perennial forming loose mats from slender rhizomes. **Stems** to 30 cm long, trailing to ascending. **Leaves** ovate, 5–20 mm long with wavy margins. **Inflorescence** axillary, of 1 to 2 stalked flowers. **Flowers**: sepals 2–4 mm long, lanceolate with pointed tips and 1 to 3 prominent veins; petals minute or absent; stamens ≤10. **Capsule** 3–6 mm long, opening by 6 slits. Wet rock outcrops, thickets, forest, especially along streams; montane to alpine. AK to CA, ID, MT.

The similar *Stellaria obtusa* has blunt, ovate sepals and rarely occurs very high in the mountains.

Stellaria graminea L. Glabrous, rhizomatous perennial. **Stems** 4-angled, ascending to trailing, 10–100 cm. **Leaves** narrowly lanceolate, sessile, 15–25 mm long, basally ciliate. **Inflorescence** a terminal, bracteate cyme with spreading pedicels. **Flowers**: sepals lanceolate, ciliate, 3-veined, 3–6 mm long; petals 3–7 mm long; stamens 10. **Capsule** 4–6 mm long. Well-irrigated lawns; plains, valleys. Introduced throughout north-temperate N. America; native to Eurasia. (p.148)

Our only collections from the campus of Montana State University over 50 years ago.

Stellaria jamesiana Torr. [*Pseudostellaria jamesiana* (Torr.) W.A.Weber & R.L.Hartm.] Perennial from tuberous-thickened rhizomes, glandular-pubescent above. **Stems** ascending to erect, 15–40 cm. **Leaves** linear-lanceolate, 2–10 cm long. **Inflorescence** open, terminal and axillary, leafy-bracted cymes. **Flowers**: sepals lanceolate, 3–6 mm long, glandular; petals 7–10 mm long, shallowly lobed; stamens 10. **Capsules** 4–5 mm long, opening by 6 slits. Open forest, meadows; subalpine; Beaverhead Co. WA to MT south to CA, AZ, TX.

This species was recently placed in the primarily Asiatic genus *Pseudostellaria* based on the tuberous-thickened rhizomes;[427] however, *P. oxyphylla* (B.L.Rob.) R.L.Hartm. & Rabeller lacks thickened roots. The relationship between our species of *Pseudostellaria* and Asiatic members requires further investigation.[162]

Stellaria longifolia Muhl. ex Willd. Glabrous perennial forming loose mats. **Stems** prostrate to ascending, scabrous on the angles, 10–35 cm. **Leaves** linear, 1–5 cm long, minutely scabrous on the margins. **Inflorescence** a terminal, bracteate, spreading cyme. **Flowers**: sepals ovate, weakly 3-veined, 2–4 mm long; petals equaling the sepals; stamens 5 to 10. **Capsule** 3–6 mm long, opening by 6 slits. Wet soil along streams and wetlands; montane. Circumpolar south to CA, AZ, MO, SC. (p.148)

Stellaria longipes Goldie Glabrous, rhizomatous perennial. **Stems** erect to trailing, 3–35 cm. **Leaves** narrowly lanceolate, 5–35 mm long, sometimes glaucous. **Inflorescence** a terminal, few-flowered cyme or flowers solitary. **Flowers**: sepals narrowly ovate, 3-veined, 3–5 mm long; petals 3–8 mm long; stamens 5 to 10. **Capsule** 4–6 mm long, opening by 6 slits. Moist soil of meadows, talus slopes, along streams; valleys to alpine. Circumpolar, south to CA, NM, SD, NY. (p.148)

1. Flowers in scarious-bracted cymes; sepals distinctly 3-veined . . . . var. *longipes*
1. Flower in leafy-bracted cymes or solitary; sepals obscurely 3-veined . . . . var. *altocaulis*

Stellaria longipes var. *longipes*, >10 cm tall with green foliage, occurs in moist soil of meadows and along streams; valleys to subalpine, more common lower in northwest cos.; *S. longipes* var. *altocaulis* (Hultén) C.L.Hitchc. [*S. monantha* Hultén] is mostly <10 cm tall with glaucous blue-green foliage and occurs in moist turf or talus, subalpine

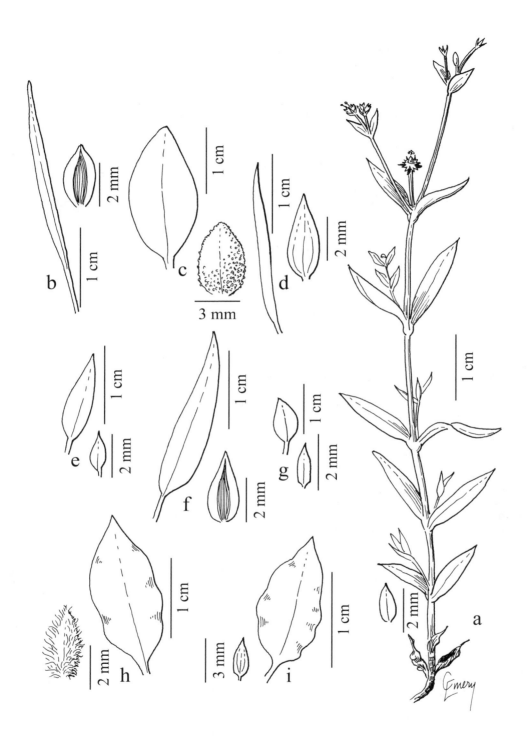

Plate 20. *Stellaria*. **a.** *S. calycantha*, **b.** *S. longifolia*, **c.** *S. americana*, **d.** *S. longipes* var. *longipes*, **e.** *S. umbellata*, **f.** *S. borealis*, **g.** *S. obtusa*, **h.** *S. media*, **i.** *S. crispa*

and alpine, more common higher in southwest cos. Two species from the *Stellaria longipes* complex are reported for our area by some authors,[115,325] whereas others do not even recognize infraspecific taxa because they are all interfertile.[83,287] The two taxa are morphologically and ecologically distinct in MT, so I have chosen to follow Hitchcock[187] in recognizing two vars.

Stellaria media (L.) Vill. COMMON CHICKWEED Annual, sometimes overwintering, with ciliate stems and petioles. **Stems** trailing to ascending, 5–40 cm. **Leaves** ovate, 5–30 mm long, petiolate. **Inflorescence** a terminal, leafy-bracted, long-hairy cyme. **Flowers**: sepals lanceolate, 3–5 mm long, obscurely 1-veined, pubescent; petals short or absent; stamens 3 to 5. **Capsule** 3–5 mm long, opening by 6 slits. Gardens, lawns, disturbed soil along streams, ponds; plains, valleys. Cosmopolitan; introduced throughout N. America; native to Europe. (p.148)

Stellaria nitens Nutt. Slender, nearly glabrous annual. **Stems** erect, simple or few-branched below, 3–20 cm. **Leaves**: basal petiolate with ovate blade 2–5 mm long; cauline linear, 3–15 mm long. **Inflorescence** an erect, terminal, several-flowered cyme. **Flowers**: sepals narrowly lanceolate, 3-veined, glabrous, 3–5 mm long; petals 1–2 mm long or absent; stamens 3 to 5. **Capsules** 2–3 mm long, opening by 3 slits. Grasslands, open forest, pond margins, usually in disturbed soil; valleys. BC south to CA, AZ.

Stellaria obtusa Engelm. Glabrous, mat-forming perennial. **Stems** prostrate, branched, 1–15 cm. **Leaves** 4–12 mm long, ovate, sometimes ciliate below. **Inflorescence**: axillary, solitary flowers. **Flowers**: sepals glabrous, ovate, 1–2 mm long with obscure veins and blunt tips; petals absent; stamens 10. **Capsule** 2–4 mm long, opening by 6 slits. Moist soil along streams, seeps; montane, lower subalpine. BC to AB south to CA, UT, CO. See *Stellaria crispa*. (p.148)

Stellaria umbellata Turcz. Glabrous, rhizomatous perennial. **Stems** ascending to erect, slender, branched, 1–25 cm. **Leaves** lanceolate to narrowly elliptic, petiolate below, 5–15 mm long. **Inflorescence** open, terminal and axillary, scarious-bracted, umbellate cymes. **Flowers**: sepals glabrous, green to purple, 2–3 mm long, lanceolate, obscurely 3-veined; petals absent; stamens 5. **Capsule** 3–5 mm long, opening by 6 slits. Moist, often mossy soil of cliffs and along streams, ponds; subalpine, alpine. AK and NT south to CA, AZ, NM; Asia. (p.148)

Vaccaria Wolf

Vaccaria hispanica (Mill.) Rauschert [*V. segetalis* (Neck.) Garcke ex Asch., *V. pyramidata* Medik., *Saponaria hispanica* Mill.] COW HERB Taprooted, glabrous and glaucous annual. **Stems** erect, 20–80 cm, branched above. **Leaves** mainly cauline, lanceolate to lance-ovate, 2–10 cm long, clasping the stem. **Inflorescence** a terminal, many-flowered, open, flat-topped cyme. **Flowers**: calyx glabrous, 12–15 mm long, becoming ovate and 5-angled; petals pink, clawed, the blade 5–8 mm long, shallowly 2-lobed; styles 2; stamens 10. **Capsule** oblong, shorter than the calyx tube. Fields, roadsides; plains, valleys. Introduced worldwide and throughout temperate N. America; native to Europe.

Most of our collections are >60 years old.

POLYGONACEAE: Buckwheat Family

Annual or perennial, rarely dioecious herbs or rarely subshrubs. **Leaves** basal and/or cauline, mostly alternate with entire margins; membranous stipules (ocrea) present in many genera. **Inflorescence** terminal and/or axillary, cymose, racemose, spicate or capitate; flower clusters sometimes enclosed by a cup- to tube-shaped involucre. **Flowers** usually perfect, dish-shaped to campanulate; sepals and petals undifferentiated; tepals 2 to 6 in 2 whorls, separate or sometimes united at the base; stamens 4 to 9; pistil 1; styles 2 to 3. **Fruit** a glabrous, brown, 1-seeded, 2- to 3-sided achene.

Fagopyrum esculentum Moench (buckwheat of commerce) is reported to be escaped in MT,[181] but I have seen no specimens.

1. At least some cauline leaves with stipules at the base. 2
1. Leaves without stipules 5

2. Basal leaf blades nearly orbicular in outline, as wide as long . . . . *Oxyria*
2. Basal leaf blades linear to elliptic or arrow-shaped, longer than wide 3

3. Tepals 3; diminutive alpine plants <3 cm high . . . .*Koenigia*
3. Tepals 5 to 6; plants often larger 4

4. Tepals 5; achene enclosed by unswollen sepals (achene swollen in *P. phtolaccaefolium*) . . . . . .*Polygonum*
4. Tepals 6; the inner 3 becoming swollen, wing-like and often net-veined with the achene at the base *Rumex*

5. Involucre of 6 bracts separate nearly to the base; annual . . . . *Stenogonum*
5. At least the basal 1/2 of involucre united; annual or perennial . . . . *Eriogonum*

Eriogonum Michx. Wild buckwheat

Annual or perennial herbs or subshrubs. **Leaves** basal, occasionally cauline, petiolate with entire margins, lacking stipules. **Inflorescence** of flower clusters, each subtended by a tubular to campanulate, 5- to 10-lobed involucre, arranged in bracteate racemes or cymes. **Flowers** pedicellate, campanulate to cup-shaped, sometimes with a tubular, stipe-like base; tepals 6, entire, united below; stamens 9; styles 3. **Achene** 3-sided, enclosed by the dried perianth.[337]
    This is the second largest genus endemic to North America.

1. Plants annual or biennial, without a branched caudex 2
1. Plants perennial with a woody, branched caudex. 4

2. Branches of inflorescence glabrous . . . . *E. cernuum*
2. Inflorescence branches villous to tomentose 3

3. Plants scapose or with cauline leaves only near the base . . . .*E.visheri*
3. Flowering stems leafy. . . . . *E. annuum*

4. Tepals hairy or hispid, sometimes sparsely so or only at the base. 5
4. Tepals glabrous or punctate 12

5. Perianth with a tubular base ≥1 mm long that appears to be a continuation of the pedicel 6
5. Perianth with a mostly rounded base 9

6. Plants with 1 involucre per flowering stem . . . . *E. caespitosum*
6. Plants with >1 involucre per flowering stem 7

7. Stems with a whorl of leaves near the middle . . . . *E. heracleoides*
7. Stems leafless except at the base 8

8. Leaves <5 mm wide; flowers pale yellow . . . . *E. androsaceum*
8. Some leaves usually >5 mm wide; flowers bright yellow . . . . *E. flavum*

9. Perianth glandular as well as hairy; inflorescence subtended by 2 linear bracts . . . . *E. pyrolifolium*
9. Perianth not glandular; inflorescence subtended by 0 or >2 bracts 10

10. Inflorescence open. . . . . *E. brevicaule*
10. Inflorescence capitate. 11

11. Perianth white or pink; plants of low elevations . . . . *E. pauciflorum*
11. Perianth light yellow; northwest mountains. . . . . *E. androsaceum*

12. Perianth with a tubular base ≥1 mm long that appears to be a continuation of the pedicel 13
12. Perianth with a mostly rounded base 15

13. Flowering stems <10 cm high; tepals pustulose. . . . . *E. crosbyae*
13. Most flowering stems >10 cm high; tepals smooth. 14

14. Leaf blades >3 times as long as broad; flowering stem with whorl
     of bracts mid-length . . . . *E. heracleoides*
14. Leaf blades ≤3 times as long as broad; flower stems naked . . . . *E. umbellatum*

15. Upper leaf surface sparsely hairy above, margins often turned under. . . . . *E. microthecum*
15. Leaf blades densely hairy on both surfaces. 16

16. Tepals smooth 17
16. Tepals pustulose (bumpy), at least at the base 18

17. Inflorescence open, usually some branches >3 cm long . . . . *E. strictum*
17. Inflorescence capitate; rarely with branches to 3 cm long . . . . *E. ovalifolium*

18. Plants alpine and subalpine; flowers yellow to red. . . . . *E. crosbyae*
18. Plants from valleys to montane; flowers rarely yellow or red 19

19. Plants with 1 involucre per flowering stem . . . . *E. soliceps*
19. Plants with >1 involucre per flowering stem . . . . *E. mancum*

Eriogonum androsaceum Benth. Mat-forming perennial from a spreading, woody caudex.
Flowering stems erect, 2–6 cm, old leaves at the base. **Leaves** basal, crowded, the blade narrowly
oblong, 1–2 cm long, tomentose below, sparsely so above. **Inflorescence** a compact umbel; bracts
linear; involucres tomentose, 3–5 mm long with 5 to 8 erect lobes. **Flowers** 3–6 mm long, light
yellow, fading to orange with a tubular base ca. 1 mm long; tepals sparsely hairy, similar in shape;
stamens exserted. **Achene** 4–6 mm long. Stony, usually calcareous soil of exposed ridges, slopes; upper subalpine,
alpine. Endemic to northwest MT and adjacent BC, AB.

Eriogonum annuum Nutt. Gray-tomentose, taprooted annual or biennial. **Stems** erect, 20–60
cm. **Leaves** basal and cauline, basal withering early; the blades oblanceolate, 1–5 cm long.
Inflorescence an open cyme; bracts triangular; involucres 2–4 mm long with 6 erect lobes.
Flowers 1–3 mm long, white to pink, cup-shaped; tepals glabrous, the outer broader than inner;
stamens included. **Achenes** 1–2 mm long. Sandhills and sandy soil of grasslands, roadsides; plains,
valleys. MT, ND south to NM, TX. (p.152)

Eriogonum brevicaule Nutt. [*E. lagopus* Rydb., *E. pauciflorum* Pursh var. *canum* (S.Stokes)
Reveal] Cespitose, tomentose perennial. **Flowering stems** ascending to erect, 5–20 cm. **Leaves**
mostly basal, the blade linear-oblong, 1–3 cm long. **Inflorescence** an open, umbellate cyme; bracts
scale-like; involucres tomentose, 2–4 mm long with 5 erect lobes. **Flowers** 1–3 mm long, yellow,
cup-shaped; tepals sparsely hairy, similar in shape; stamens exserted. **Achenes** 2–3 mm long. Barren,
calcareous, stony or clayey soil of badlands, exposed ridges, slopes; valleys. MT south to NV, UT, CO. Our plants are
var. *canum* (Stokes) Dorn which is endemic to south-central MT and adjacent WY. (p.152)

Eriogonum caespitosum Nutt. Cushion-forming perennial from a woody, branched
caudex. **Flowering stems** erect, 3–8 cm, covered in old leaf bases. **Leaves** all basal, the blades
oblanceolate, gray-tomentose, 2–10 mm long. **Inflorescence** capitate; bracts absent; involucres
2–4 mm long with 6 to 9 long, reflexed lobes. **Flowers** 3–5 mm long, yellow to red with a tubular
base, functionally unisexual; tepals villous, similar in shape; stamens slightly exserted. **Achenes** 4–5
mm long. Stony, fine-textured, often calcareous soil of sagebrush steppe on exposed slopes, flats; montane. OR, ID, MT
south to CA, AZ.

Eriogonum cernuum Nutt. Taprooted annual. **Stems** simple or branched at the base, erect,
8– 40 cm. **Leaves** all basal or near-basal, the blades ovate to orbicular, 1–3 cm long, tomentose
below, less so above. **Inflorescence** a diffuse, finely-branched cyme with spreading to reflexed,
ultimate branches; bracts scale-like; involucres 1–2 mm long, glabrous with 5 erect lobes. **Flowers**
1–2 mm long, white to pink, campanulate; tepals glabrous, the outer broader than the inner; stamens
exserted. **Achenes** 1–2 mm long. Sandhills and sandy soil of grasslands, streambanks; plains. AB, SK south to AZ, NM.
(p.152)

Eriogonum crosbyae Reveal [*E. capistratum* Reveal var. *muhlickii* Reveal] Cushion-forming
perennial from a woody, branched caudex. **Flowering stems** erect, 1–10 cm, covered in old
leaf bases. **Leaves** all basal, the blades oblanceolate to spatulate, tomentose, 5–10 mm long.
Inflorescence capitate; bracts scale-like; involucre lobes 5 to 7, erect to spreading, tomentose,
2–4 mm long. **Flowers** 2–4 mm long, yellow to red, campanulate; tepals glabrous, basally punctate,
similar in shape; stamens exserted. **Achenes** 2–4 mm long. Cliffs, talus slopes, stony soil of exposed ridges; upper
subalpine, alpine. OR, ID, MT, NV.
    Formerly misidentified as *Eriogonum chrysops* Rydb. See *E. mancum*. This species occurs on both granitic and
calcareous soils.

Eriogonum flavum Nutt. [*E. piperi* Greene] Perennial, often forming cushions from a tightly-
branched caudex. **Flowering stems** erect, white-hairy, 5–30 cm. **Leaves** basal; the blade 1–6
cm long, oblanceolate, tomentose below, sometimes glabrous above. **Inflorescence** capitate to
umbellate; bracts spreading, leaf-like; involucres tomentose, 4–9 mm long with 5 to 8 erect lobes.
Flowers 4–6 mm long, yellow, campanulate with a tubular base; tepals hirsute, similar in shape;
stamens exserted. **Achenes** 3–5 mm long. Stony soil of exposed ridges, moraine, grasslands and meadows. AK to MB
south to OR, CO, NE. (p.152)

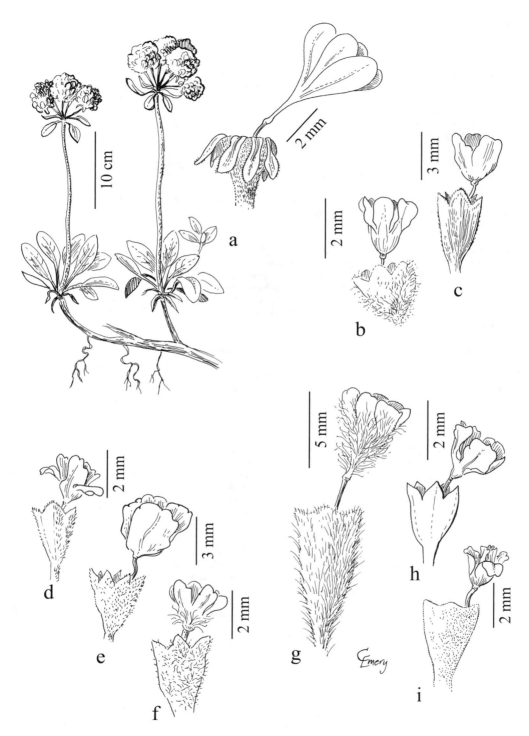

Plate 21. **Eriogonum. a.** *E. umbellatum* var. *majus*, **b.** *E. mancum,* **c.** *E. microthecum,* **d.** *E. brevicaule,* **e.** *E. ovalifolium* var. *ovalifolium* **f.** *E. pauciflorum* **g.** *E. flavum* var. *piperi,* **h.** *E. cernuum,* **i.** *E. annuum*

1. Stipe-like, tubular corolla base ≤1 mm long . . . . var. *flavum*
1. Stipe-like corolla base >1 mm long . . . . var. *piperi*

Eriogonum flavum var. ***piperi*** (Greene) M.E.Jones Occurs montane to alpine in western Montana; a form with ovate leaf blades occurs in grasslands in Lincoln and Flathead cos., and a densely hairy form was collected on the east slopes of the Highland Mountains; *E. flavum* var. *flavum* occurs primarily in plains and valleys east of the Continental Divide. Dwarf, alpine forms [var. *polyphyllum* (Small) M.E.Jones] of both vars. occur and are difficult to distinguish.

Eriogonum heracleoides Nutt. Mat-forming perennial from a woody, branched caudex. **Flowering stems** ascending to erect, 15–40 cm. **Leaves**: basal blades tomentose, less so upward, linear to oblanceolate, 15–50 mm long. **Inflorescence** a compound, umbellate cyme usually subtended by a whorl of leaf-like bracts at mid-stem as well as at the base of the inflorescence; involucres tomentose, 3–5 mm long with 6 to 12 reflexed lobes. **Flowers** 4–8 mm long, white to ochroleucus, campanulate with a long, tubular base; tepals glabrous or sparsely long-hairy, similar in shape; stamens exserted. **Achenes** 3–5 mm long. Grasslands, sagebrush steppe, occasionally open forest; valleys. BC south to CA, UT, CO.

The whorl of bracts at mid-stem, although not always present, is diagnostic. Reveal recognizes var. *leucophaeum* Reveal distinguished by the lack of mid-stem bracts;[336] however, some MT populations contain both forms. Plants from southwest MT have broader leaves than those from the northwest part of the state.

Eriogonum mancum Rydb. Cushion-forming perennial from a woody, branched caudex. **Flowering stems** erect, 1–10 cm, covered in old leaf bases below. **Leaves** all basal, the blades oblanceolate, tomentose, 5–15 mm long. **Inflorescence** capitate with a cluster of several involucres; bracts scale-like; involucres tomentose, 1–4 mm long with 5 erect lobes. **Flowers** 1–3 mm long, cream to yellow, cup-shaped; tepals glabrous, punctate, similar in shape; stamens exserted. **Achenes** 20-3 mm long. Stony, calcareous soil of exposed ridges, warm slopes; valleys to montane. MT, ID, WY, UT. (p.152)

The similar *Eriogonum crosbyae* occurs at higher elevations. The similar *E. soliceps* has only 1 involucre per stem. See *E. pauciflorum*. The type locality is in Beaverhead Co.

Eriogonum microthecum Nutt. [*E. effusum* Nutt] Perennial subshrub with woody stems and a branched caudex. **Flowering stems** ascending, 10–40 cm. **Leaves** all cauline, the blades linear to narrowly obovate, 1–3 cm long, tomentose, densely below, less so above, margins often rolled under. **Inflorescence** an open cyme, often flat-topped; bracts scale-like; involucres glabrate to sparsely tomentose, 5-lobed, 2–3 mm long. **Flowers** 2–4 mm long, white or yellowish, cup-shaped; tepals glabrous, similar in shape; stamens exserted. **Achenes** 2–3 mm long. Grasslands, sagebrush steppe; valleys, montane. WA to MT south to CA, AZ, NM. (p.152)

1. Inflorescence 7–19 cm long . . . . var. *effusum*
1. Inflorescence 1–5 cm long 2

2. Leaf margins revolute; leaf blades nearly tubular. . . . . var. *simpsonii*
2. Leaf margins often curled, but the blades flat. . . . . var. *laxiflorum*

Eriogonum microthecum var. ***laxiflorum*** Hook. is the common form; *E. microthecum* var. *simpsonii* (Benth.) Reveal is known from a single Beaverhead Co. collection; and there is also one collection of *E. microthecum* var. *effusum* (Nutt.) Torr.& A.Gray from Big Horn Co. made more than 50 years ago.

Eriogonum ovalifolium Nutt. Perennial, often forming small cushions from a tightly-branched caudex. **Flowering stems** erect to ascending, tomentose, 2–35 cm, clothed in old leaves at the base. **Leaves** basal, long-petiolate, the blade 2–20 mm long, oblanceolate to orbicular, tomentose. **Inflorescence** capitate to rarely umbellate, bracts scale-like; involucres 1 to many per cluster, tomentose, 2–4 mm long with 5 erect lobes. **Flowers** 2–5 mm long, white, yellow, or rose, cup-shaped; tepals glabrous, the outer much broader than the inner; stamens not exserted. **Achenes** 2–3 mm long. Stony soil of grasslands, steppe, outcrops, exposed ridges and slopes. BC, AB south to CA, AZ, NM. (p.152)

1. Flowering stems mostly ≤5 cm high; leaf blades ≤12 mm long; alpine plants . . . . var. *depressum*
1. Flowering stems ≥5 cm; leaf blades mostly 1–6 cm long; plants of lower elevations. 2

2. Inflorescence umbellate, some branches ≥2 cm long . . . . var. *pansum*
2. Inflorescence capitate, the branches minute 3

3. Flowers yellow . . . . .var. *ovalifolium*
3. Flowers white 4

4. Leaf blades spatulate to oval; flowering stems mostly ≤20 cm. . . . . var. *purpureum*
4. Leaf blades oblanceolate to narrowly elliptic; stems mostly >20 cm . . . . .var. *ochroleucum*

Eriogonum ovalifolium var. *depressum* Blank. is found on exposed ridges, rock slides, barren slopes, subalpine, alpine, more common higher; *E. ovalifolium* var. *pansum* Reveal occurs in sparsely vegetated grasslands, montane, subalpine, more common lower, west of the Divide; *E. ovalifolium* var. *ovalifolium* occurs in montane sagebrush steppe; *E. ovalifolium* var. *purpureum* (Nutt.) Durand is found in steppe, on ridges, outcrops, talus slopes, montane, subalpine; *E. ovalifolium* var. *ochroleucum* (Small ex Rydb.) M.Peck often with brown tomentum and elongate flowering stems, occurs in grasslands, steppe, plains, valleys, montane. See *E. strictum*.

Eriogonum pauciflorum Pursh [*E. multiceps* Nees] Mat-forming, gray-tomentose perennial from a branched caudex. **Flowering stems** ascending, 4–20 cm. **Leaves** all basal, long-petiolate, the blades oblanceolate to spatulate, 1–3 cm long. **Inflorescence** capitate; bracts scale-like; involucres tomentose, 3–5 mm long with 5 erect lobes. **Flowers** 1–3 mm long, white or rose, cup-shaped; tepals pubescent, similar in shape; stamens exserted. **Achenes** 2–3 mm long. Barren, rocky, sandy or fine-textured soil of badlands, steep slopes, ridge crests; plains, valleys. SK, MB south to CO, NE. (p.152)
     Compact plants from calcareous soil in southern Carbon Co. have tepals with sparse hair only on the base of the tepals; they have been mistaken for *Eriogonum mancum*.

Eriogonum pyrolifolium Hook. Perennial with a simple or short-branched caudex and a stout taproot. **Flowering stems** ascending, glabrous to weakly tomentose, 3–15 cm, clothed in old leaf bases below. **Leaves** basal, long-petiolate, the blade 1–4 cm long, elliptic, white- to brown-tomentose below, nearly glabrous above. **Inflorescence** loosely capitate, bracts 2, leaf-like; involucres appearing clustered, sparsely glandular and hairy, 4–5 mm long with 4 to 5 erect lobes. **Flowers** 3–6 mm long, campanulate with a stipe-like base <1 mm long, white to rose; tepals glandular-villous, similar in shape; stamens exserted. **Achenes** 4–5 mm long. Stony, granitic soil of fellfields, outcrops, scree slopes; upper subalpine, alpine in Ravalli and Mineral cos. BC south to CA, ID, MT. Our plants are var. *coryphaeum* Torr.& A.Gray.

Eriogonum soliceps Reveal & Bjork Cushion-forming perennial from a woody, branched caudex. **Flowering stems** erect, 1–6 cm, covered in old leaf bases below. **Leaves** all basal, the blades oblanceolate, tomentose, 5–15 mm long. **Inflorescence** a solitary involucre; bracts absent; involucre tomentose, 2–3 mm long with 5 erect lobes, solitary. **Flowers** 1–3 mm long, white, cup-shaped; tepals glabrous, punctate, similar in shape; stamens slightly exserted. **Achenes** 2–3 mm long. Barren calcareous or non-calcareous soil of ridges, slopes; montane. Endemic to Beaverhead and Deer Lodge cos. and adjacent ID. See *Eriogonum mancum*.

Eriogonum strictum Benth. Mat-forming perennial from a loosely-branched, woody caudex. **Flowering stems** erect to ascending, tomentose, 10–35 cm. **Leaves** basal, long-petiolate, the blade 5–25 mm long, ovate to orbicular, tomentose. **Inflorescence** umbellate to loosely capitate cymes; bracts scale-like; involucres tomentose, 3–4 mm long with 5 erect lobes. **Flowers** 2–3 mm long, white to rose, cup-shaped; tepals glabrous, the outer broader than the inner; stamens not exserted. **Achenes** 3–4 mm long. Sandy or gravelly soil of sagebrush steppe, grasslands, open forest; valleys, montane. WA to MT south to CA, ID. Our plants are var. *proliferum* (Torr.& A.Gray) C.L.Hitchc.
     Vegetatively very similar to *Eriogonum ovalifolium*, but the inflorescences are quite different from all except *E. ovalifolium* var. *pansum*. The Lincoln Co. collection of *E. strictum* may represent an introduction.

Eriogonum umbellatum Torr. Perennial from a branched caudex forming loose mats. **Flowering stems** ascending to erect, 5–40 cm. **Leaves** basal, glabrous to tomentose, the blade oblanceolate to ovate, 3–30 mm long. **Inflorescence** an umbel of capitate cymes subtended by a whorl of leaf-like bracts; involucres tomentose, 4–8 mm long with 6 to 12 reflexed lobes. **Flowers** mostly unisexual, 4–8 mm long, white, yellow, or red, campanulate with a long, tubular base; tepals glabrous, similar in shape; stamens exserted. **Achenes** 2–7 mm long. BC, AB south to CA, AZ, UT. (p.152)

1. Some inflorescence branches with a whorl of bracts at mid-length . . . . .var. *ellipticum*
1. Inflorescence branches without whorled bracts mid-length 2

2. Perianth white, cream or red 3
2. Perianth bright yellow. 5

3. Leaf blades glabrous on both sides . . . . var. *desereticum*
3. Leaf blades densely hairy at least below 4

4. Leaf blades lanate below, appearing white; plants forming compact mats . . . . var. *majus*
4. Leaf blades tomentose below, the green clearly visible; plants forming loose mats . . . var. *dichrocephalum*

5. Leaf blades nearly glabrous below. . . . . var. *aureum*
5. Leaf blades tomentose below. . . . . var. *umbellatum*

Eriogonum umbellatum var. ***aureum*** (Gand.) Reveal is found in valleys to subalpine steppe, meadows, rock slides in southwest cos.; *E. umbellatum* var. *dichrocephalum* Gand. occurs in montane to subalpine steppe, grasslands, rocky slopes; *E. umbellatum* var. *majus* Hook. is found in montane to subalpine grasslands, meadows, steppe, woodlands; *E. umbellatum* var. *ellipticum* (Nutt.) Reveal occurs in montane grasslands in Ravalli Co.; *E. umbellatum* var. ***umbellatum*** and *E.umbellatum* var. *desereticum* Reveal are reported for MT,[336] but I have seen no specimens.

Eriogonum visheri A.Nelson Taprooted annual. **Stems** erect, 3–20 cm. **Leaves** mostly basal, petiolate, the blades elliptic to orbicular, 1–3 cm long, glabrous to sparsely villous. **Inflorescence** a diffuse, finely branched cyme; bracts leaf-like below, scale-like above; involucres 1–2 mm long, glabrous with 5 erect lobes. **Flowers** 1–2 mm long, yellow with dark midveins; tepals sparsely hispid, similar in shape; stamens exserted. **Achenes** ca. 2 mm long. Barren clay soil of badlands, bluffs; plains of Carter Co. MT, SD, ND.

Koenigia L.

Koenigia islandica L. Diminutive, glabrous annual. **Stems** branched or simple, 1–4 cm. **Leaves** cauline, petiolate with a brown stipule (ocrea); the blade spatulate, 1–3 mm long, often reddish. **Inflorescence** of terminal or axillary, few-flowered clusters subtended by a whorl of leaves. **Flowers** campanulate, ca. 1 mm long, of 3 greenish-white, ovate tepals; stamens 3; styles 2. **Achenes** 2-sided, glabrous, 1–2 mm long. Cold, wet soil of granitic seeps; alpine. Circumpolar south to UT and CO. Collected only in Carbon Co.

Oxyria Hill Mountain sorrel

Oxyria digyna (L.) Hill Glabrous, taprooted perennial. **Stems** simple or branched above, 3–35 cm. **Leaves** basal, long-petiolate with sheathing stipules (ocrea); the blade 1–5 cm wide, nearly orbicular, cordate. **Inflorescence** a panicle of clusters of short-pedicellate flowers subtended by stipule-like bracts. **Flowers**: tepals 4, separate, red or green, 1–2 mm long; stamens 6; styles 2. Achenes oval, 3–5 mm wide with broad, winged margins. Stony, moist to wet soil of moraine, rock outcrops, meadows, talus slopes; upper subalpine and alpine. Circumpolar south to CA, AZ, NM. (p.166)

Polygonum L. Knotweed, smartweed

Annual or perennial, usually glabrous herbs. **Stems** prostrate to erect. **Leaves** mainly cauline, petiolate, with sheathing stipules (ocrea) at the petiole base. **Inflorescence** axillary and/or terminal, with solitary or clustered flowers, the clusters organized into racemes. **Flowers** usually perfect; tepals usually 5 (4 to 6), united below, the outer whorl sometimes larger; stamens 8; styles usually 3(2). **Fruit** a 2- to 3-sided achene enclosed in the calyx.[96,179]

Mature achenes are often needed for positive identification. Annuals are predominately self-pollinating. *Polygonum pensylvanicum* L., similar to *P. hydropiper*, was collected once in a Fallon Co. garden. Segregate genera include *Fallopia* Adans., *Persicaria* (L.) Mill. and *Aconogonon* (Meisn.) Rchb., but some of these groups are themselves not monophyletic,[232] so I have maintained the more traditional treatment. *Polygonum spergulariiforme* Meisn. ex Small, similar to *P. polygaloides* but with shiny achenes, is reported for MT,[96] but I have seen no specimens.

1. Leaves nearly all basal, stems with a few reduced leaves. 2
1. Leaves mainly cauline 3

2. Racemes <1 cm wide as pressed; lower flowers replaced by bulblets . . . . *P. viviparum*
2. Racemes >1 cm wide; bulblets not present . . . . *P. bistortoides*

3. Stems twining, lax; leaves cordate. . . . . *P. convolvulus*
3. Stems mostly ascending to erect, not twining; leaves mostly not cordate 4

4. Stipules surrounding the stem to the top, often not torn; many leaves >3 cm long 5
4. Stipules at lower nodes lobed and usually deeply torn; most leaves <3 cm long. 11

5. Leaves >8 cm long; stems usually >1 m high. . . . . *P. cuspidatum*
5. At least some leaves <8 cm long; stems usually <1 m high. 6

6. Achene strongly 3-winged, >4 mm long; plants subalpine and alpine . . . . *P. phytolaccifolium*
6. Achene usually 3-angled but not winged, ≤3 mm long; plants montane or lower. .7

7. Styles 2–4 mm long; tepals rose; plants usually aquatic during part of the year . . . . *P. amphibium*
7. Styles <2 mm long; tepals green, white or pink; plants not truly aquatic 8

8. Inflorescence and/or perianth glandular-punctate 9
8. Inflorescence and perianth lacking glands flush with the epidermis. 10

9. Achenes shiny dark brown or black . . . . *P. punctatum*
9. Achenes dull, light brown . . . . *P. hydropiper*

10. Stipules long-hairy on the sides as well as the top. . . . . *P. persicaria*
10. Stipules sometimes ciliate on the upper margin but glabrous on the sides . . . . *P. lapathifolium*

11. Lower stems with 8 to 16 ribs (section *Polygonum*) 12
11. Stems 4-ribbed to round in cross-section (section *Duravia*). 16

12. Tepals with white to rose margins. 13
12. Teplas with yellowish or greenish margins 14

13. Stems erect; mature achenes included in perianth . . . . *P. ramosissimum*
13. Stems or at least some branches prostrate or ascending; mature achenes exceeding
    the perianth by 1–2 mm . . . . *P. aviculare*

14. Stems sprawling; outer 3 tepals longer than inner 2; leaves oval to obovate . . . . *P. achoreum*
14. Stems somewhat erect; tepals all the same length; leaves linear to oblanceolate 15

15. Stipules disintegrating into brown threads; leaves usually at least 4 times
    as long as wide. . . . . *P. ramosissimum*
15. Stipules silvery-lacerate; leaves mostly ≤3 times as long as wide . . . . *P. erectum*

16. At least pedicels of lower flowers reflexed 17
16. Most pedicels erect or ascending. 20

17. Leaves ovate; leaf margins and stem ribs scabrous (use 20X) . . . . *P. austiniae*
17. Leaves linear to lanceolate or oblanceolate; foliage not scabrous. 18

18. Tepals ca. 2 mm long . . . . *P. engelmannii*
18. Tepals ≥3 mm long 19

19. Many flowers open; anthers ca. 0.5 mm long. . . . . *P. majus*
19. Flowers mostly closed; anthers ca. 0.2 mm long . . . . *P. douglasii*

20. Achenes yellow to brown, shiny or dull. . . . . *P. polygaloides*
20. Achenes shiny black. 21

21. Leaves elliptic to ovate . . . . *P. minimum*
21. Leaves linear to oblanceolate. . . . . *P. sawatchense*

Polygonum achoreum S.F.Blake Annual, often white with mildew. **Stems** prostrate to
ascending, usually branched at the base,10–30 cm. **Leaves** short-petiolate; the blades ovate, 7–25
cm long; stipules 5–8 mm long, lacerate. **Flowers** 3–4 mm long, 1 to 3 in leaf axils; pedicels erect,
1–2 mm long; tepals green with yellowish margins, the outer keeled at the tip. **Achene** 3-angled,
yellow-brown, 2–3 mm long. Disturbed areas of grasslands, roadsides; plains, valleys. AK to QC
south to NV, KS, OH, NY.

Polygonum amphibium L. WATER SMARTWEED [*Persicaria amphibia* (L.) Gray; *Polygonum
coccineum* Muhl.] Rhizomatous, aquatic perennial with glabrous to pubescent foliage. **Stems**
prostrate or floating, rooting at the nodes, 15–100 cm. **Leaves** petiolate; the blades lanceolate with
rounded to pointed tips, 2–20 cm long; stipules 1–2 cm long, glabrous or ciliate. **Flowers** 4–6 mm
long, in a terminal spike-like panicle 1–10 cm long; tepals pink. **Achene** 2-sided, 2–4 mm long, shiny
dark brown. Shallow water of ponds, sloughs; plains, valleys, montane. Circumboreal south to most of U.S.

    Polygonum amphibium var. *emersum* Michx., with hairy leaves, is the terrestrial form that rarely flowers; *P.
amphibium* var. *stipulaceum* N.Coleman has glabrous leaves and is the aquatic form, but these phenotypes are mostly
induced environmentally.[96]

Polygonum austiniae Greene [*P. douglasii* Greene var. *austiniae* (Greene) M.E.Jones]
Annual. **Stems** erect, branched at the base, scabrous, 5–15 cm. **Leaves** short-petiolate; the blades
5–15 mm long, ovate with pointed tips and scabrous margins; stipules 2–4 mm long, lacerate.
Flowers 2–3 mm long, 1–4 in upper leaf axils; pedicels reflexed, 1–3 mm long; tepals green to rose
with white margins. **Achene** 3-sided, 2–3 mm long, shiny black. Stony, often shale-derived soil of
barren banks, slopes; montane. BC, AB south to CA, NV, WY. (p.158)

Polygonum aviculare L. Dooryard knotweed Annual. **Stems** prostrate to ascending, branched
at the base, 5–50 cm. **Leaves** short-petiolate; the blades linear-oblanceolate to elliptic, 3–30 mm
long; stipules 3–15 mm long, lacerate. **Flowers** 2–5 mm long, 1 to 3 in leaf axils; pedicels erect,
2–5 mm long; tepals greenish with white or pink margins. **Achene** 3-sided, brown, 2–5 mm long.
Disturbed soil of grasslands, pond margins, roadsides, yards; plains, valleys to subalpine. Introduced
from Europe, widespread in N. America. (p.158)
　　Six intergrading subspecies have been recognized,[96] four of which are reported for MT; however, the characters
distinguishing them are continuous and not well correlated in our area; thus numerous intermediates exist.[275]

Polygonum bistortoides Pursh [*Bistorta bistortoides* (Pursh) Small, *Persicaria bistortoides*
(Pursh) H.R.Hinds] American bistort Perennial from a corm-like rhizome. **Stems** erect, simple,
15–80 cm. **Leaves**: basal petiolate, the blades 2–15 cm long, narrowly lanceolate; cauline leaves
smaller, linear; stipules 5–40 mm long, glabrous, brown. **Flowers** 3–4 mm long, in a short-cylindric
spike 1–4 cm long; pedicels spreading 2–8 mm long; tepals white to pink; stamens exserted. **Achene**
3-sided, shiny pale brown, 3–4 mm long. Moist grasslands, meadows, turf where wet early but drying later in the season;
montane to alpine. BC, AB south to CA, AZ, NM.

Polygonum convolvulus L. [*Fallopia convolvulus* (L.) A.Löve] Black bindweed Annual with
glabrous to scabrous foliage. **Stems** twining or trailing, 10–100 cm. **Leaves** long-petiolate; the
blades sagittate, 1–4 cm long; stipules glabrous, 2–4 mm long. **Flowers** 3–4 mm long, several
in leaf axils and short, narrow terminal clusters; pedicels spreading 1–3 mm long; tepals greenish,
scabrous, often with thin, white margins. **Achene** 3-sided, 3–4 mm long, black. Disturbed ground of
fields, yards, thickets, roadsides; plains, valleys. Introduced on most continents including most of N. America; native to
Europe.

Polygonum cuspidatum Siebold & Zucc. [*Fallopia japonica* (Hout.) Ronse Decr.] Japanese
knotweed Dioecious, rhizomatous perennial. **Stems** erect, branched, 1–2 m. **Leaves** petiolate;
the blades ovate, 8–20 cm long with pointed tips and a truncate base; stipules brown, glabrous,
4–6 mm long. **Flowers** apparently perfect but functionally unisexual, 5–7 mm long with a tubular
base, numerous in stipulate, axillary panicles; pedicels spreading, 3–5 mm long; tepals white to pink.
Achene shiny-brown, 2–4 mm long, 3-sided, enclosed by an inflated perianth. Gardens, streambanks; valleys. BC to QC
south to most of the U.S.; introduced from Asia as a garden ornamental.
　　Polygonum sachalinense Schmidt [*Fallopia sachalinensis* (F.Schmidt) Ronse Decr.] is similar but has leaves cordate
at the base; it is reported for MT,[135] but I have seen no specimens. *Polygonum ×bohemica* (Chrtek & Chrtkova) Zika &
Jacobson [*Fallopia ×bohemica* (Chrtek & Chrtkova) Bailey] is an intermediate-appearing hybrid between *P. cuspidatum*
and *P. sachalinensis*; many reports of the two parents may have been based on the hybrid.[455] All three species are most
often represented in our area by male plants and are thought to spread by water-borne root fragments.[455]

Polygonum douglasii Greene Annual. Stems erect or ascending, simple or branched, 5–35 cm.
Leaves sessile, 5–40 mm long, linear to elliptic, reduced above; stipules glabrous, 4–10 mm long.
Flowers 3–4 mm long, 1 to 4 in upper leaf axils; pedicels reflexed, 2–5 mm long; tepals green
with white to pink margins. **Achene** 3-sided, 2–3 mm long, shiny black. Sparsely vegetated soil of
streambanks, grasslands, open forest, roadsides; all elevations. AK to CA, AZ, NM, IA, VA. (p.158)

Polygonum engelmannii Greene [*P. douglasii* Greene ssp. *engelmannii* (Greene) Kartesz
& Gandhi] Annual. **Stems** erect or ascending, angled, branched at the base, 2–10 cm. **Leaves**
short-petiolate; the blades 5–15 mm long, linear-oblanceolate, reduced above; stipules glabrous,
2–3 mm long. **Flowers** ca. 2 mm long, 2 to 4 in upper leaf axils; pedicels reflexed, 1–3 mm long;
tepals green with rose margins. **Achene** 3-sided, 1–2 mm long, shiny black. Stony soil of exposed
alpine slopes. BC, AB south to NV, UT, CO. (p.158)

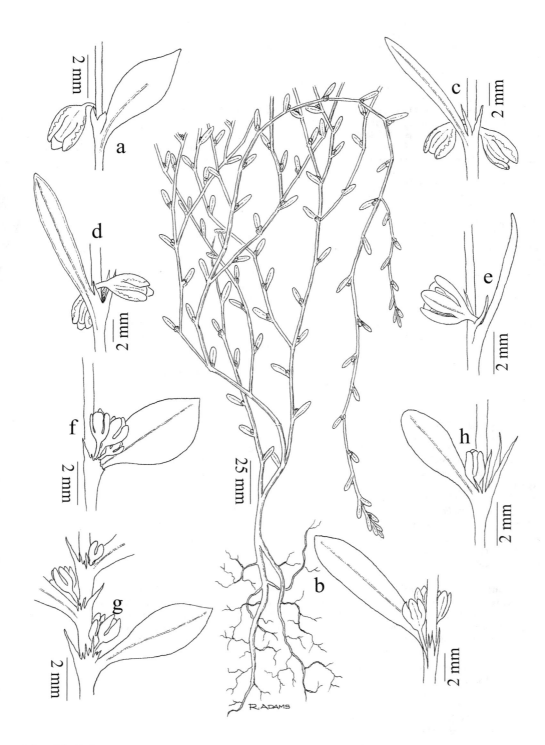

Plate 22. **Polygonum**. **a.** *P. austinae*, **b.** *P. aviculare*, **c.** *P. douglasii*, **d.** *P. engelmannii*, **e.** *P. majus*, **f.** *P. minimum*, **g.** *P. polygaloides* var. *kelloggii*, **h.** *P. ramosissimum*

Polygonum erectum L. Annual. **Stems** erect, branched, 10–50 cm. **Leaves** crowded, lanceolate to elliptic, veiny, short-petiolate, 15–40 mm long; stipules 4–8 mm long, shredding into fibers. **Flowers** 3–4 mm long, 1 to 5 in leaf axils; pedicels erect, 3–7 mm long; tepals green with yellowish margins. **Achene** 3-angled, tan, 2–4 mm long. Disturbed areas of grasslands, fields, roadsides; plains, valleys. AB to QC south to NV, NM, LA, AL.

Polygonum hydropiper L. [*Persicaria hydropiper* (L.) Spach] Glandular-punctate annual. **Stems** erect to prostrate, sometimes rooting at the nodes, branched, 8–80 cm. **Leaves** short-petiolate; the blades lanceolate to elliptic, 2–8 cm long; stipules brown, 5–15 mm long, papery, sparsely ciliate-margined. **Flowers** 2–4 mm long, in narrow terminal racemes 1–7 cm long; pedicels 1–3 mm long, ascending; tepals 4 to 5, densely glandular-punctate, green with white or pink margins, 3-veined. **Achene** 2- to 3-sided, 2–3 mm long, brown with a dull granular surface. Mud along margins of wetlands, streams; plains, valleys. Throughout temperate N. America; introduced from Europe.
  Similar to *Polygonum punctatum* which has shiny black achenes.

Polygonum lapathifolium L. [*Persicaria lapathifolia* (L.) Gray] Sparsely puberulent annual. **Stems** erect to ascending, branched, 10–80 cm. **Leaves** short-petiolate; the blades lanceolate, 2–10 cm long; stipules brown, 4–20 mm long, papery. **Flowers** 2–3 mm long, in sparsely glandular, narrow racemes from upper leaf axils; pedicels 0.5–2 mm long, ascending; tepals usually 4, greenish white to rose with 3 anchor-shaped veins. **Achene** flattened, 2–3 mm long, shiny dark brown. Mud or shallow water along streams, ponds; plains, valleys, montane. Cosmopolitan, throughout N. America.
  Some consider this species native,[179,181] but others believe it is introduced.[187]

Polygonum majus (Meisn.) Piper [*P. douglasii* Greene ssp. *majus* (Meisn.) J.C.Hickman] Annual. **Stems** erect, angled, branched near the base, 8–60 cm. **Leaves** linear-lanceolate, 2–5 cm long; stipules glabrous, 6–10 mm long. **Flowers** 4–5 mm long, 2 to 4 in upper leaf axils; pedicels reflexed, ca. 1 mm long; tepals white to rose. **Achene** 3-sided, 3–5 mm long, black. Grasslands, open forest, along roads; valleys, montane. BC to CA, UT, CO. (p.158)

Polygonum minimum S.Watson Annual. **Stems** branched, ascending, 1–15 cm. **Leaves** ovate, nearly sessile, 3–10 mm long; stipules 2–3 mm long, scabrous. **Flowers** ca. 2 mm long, 1 to 4, concealed in leaf axils; pedicels 2–3 mm long; tepals greenish, pink, or white. **Achene** 3-sided, ca. 2 mm long, shiny black. Shallow, vernally wet soil of outcrops, open slopes, along streams; subalpine, alpine. BC, AB south to CA, CO. (p.158)

Polygonum persicaria L. [*Persicaria maculosa* Gray] LADY'S THUMB Scabrous to sparsely puberulent annual. **Stems** erect to decumbent, rooting at the nodes, branched, 10–80 cm. **Leaves** short-petiolate; the blades lanceolate, 3–12 cm long, usually with a dark spot in the center that becomes obscure with drying; stipules brown below, scarious above, 4–10 mm long, papery, ciliate-margined, strigose to glabrous. **Flowers** 2–3 mm long, in cylindrical, compound racemes 1–3 cm long; pedicels 1–3 mm long, ascending; tepals 4 to 5, green below, white to pink above, 3-veined. **Achene** 2- to 3-sided, 2–3 mm long, shiny, black. Moist soil of fields, gardens, roadsides, wetlands; plains, valleys. Cosmopolitan, introduced throughout N. America; native to Europe.

Polygonum phytolaccifolium Meisn. ex Small [*Aconogonon phytolaccifolium* (Meisn. ex Small) Small] Glabrous perennial from a branched, woody caudex. **Stems** erect, often branched above, 30–90 cm. **Leaves** petiolate; the blades broadly lanceolate, 5–12 cm long; stipules reddish-brown, glabrous, 1–3 cm long. **Flowers** 2–4 mm long, numerous in axillary and terminal, leafy-bracted panicles; pedicels 1–4 mm long; tepals green to white. **Achene** shiny, yellow-brown, 4–7 mm long, 3-winged. Well-drained, non-calcareous soil of meadows, open forest, rock slides, usually on warm slopes; subalpine, alpine. WA to MT south to CA, NV.

Polygonum polygaloides Meisn. Annual. **Stems** erect or ascending, 4-angled, simple or branched at the base, 1–5 cm. **Leaves** sessile, 5–20 mm long, linear; stipules glabrous, 2–5 mm long. **Flowers** 1–3 mm long, axillary and in a dense, terminal, bracteate raceme; bracts linear to ovate; pedicels 0–2 mm long; tepals white to rose with a green midrib. **Achene** 3-sided, 1–3 mm long, shiny or dull brown. BC to SK south to CA, AZ, NM. (p.158)

1. Floral bracts ovate with white margins >0.2 mm wide . . . . .ssp. *confertiflorum*
1. Floral bracts linear with little or no white margin. . . . . ssp. *kelloggii*

Polygonum polygaloides ssp. ***confertiflorum*** (Nutt. ex Piper) J.C.Hickman occurs in vernally wet soil around ponds, wetlands, valleys to subalpine, more common lower; *P. polygaloides* ssp. ***kelloggii*** (Greene) J.C.Hickman is found in moist, open soil of pond margins, meadows, open forest, montane, subalpine, more common higher. These two ssps. are weakly differentiated. *Polygonum polygaloides* ssp. *polygaloides* is reported for MT,[96] but I have seen no specimens.

Polygonum punctatum Elliott [*Persicaria punctata* (Elliott) Small] Glandular-punctate, sometimes rhizomatous annual. **Stems** erect to prostrate, branched, 20–80 cm. **Leaves** short-petiolate; the blades lanceolate to oblanceolate, 2–8 cm long with or without a central dark blotch; stipules brown, 8–10 mm long, papery. **Flowers** 3–4 mm long, in mostly terminal, narrow racemes; pedicels 1–4 mm long, ascending; tepals densely yellow, glandular-punctate, green with white or pink margins. **Achene** 3-sided, 2–3 mm long, shiny dark brown to black. Mud along margins of streams, wetlands; plains, valleys. Throughout temperate and tropical N. America, S. America. See *Polygonum hydropiper*.

Polygonum ramosissimum Michx. Annual. **Stems** erect, branched, 8–70 cm high. **Leaves** short-petiolate, often moldy beneath, linear to narrowly lanceolate, 1–3 cm long,; stipules 5–14 mm long, lacerate, brownish. **Flowers** 2–4 mm long, 1 to 3 in leaf axils; pedicels 1–6 mm long, erect; tepals green with yellow or white to pink margins. **Achene** dark brown, 3-sided. Grasslands, roadsides, ephemeral wetlands; plains, valleys. Throughout N. America. Our plants are ssp. *ramosissimum*. (p.158)
    This plant is native in eastern N. America but possibly introduced west of the Continental Divide.[179] The yellow-margined tepals and erect habit separate this species from *Polygonum aviculare*.

Polygonum sawatchense Small Annual. **Stems** erect, simple or branched at the base, 5–35 cm. **Leaves** short-petiolate; blades 8–20 mm long, linear-oblong, reduced above; stipules lacerate, 2–10 mm long. **Flowers** 3–4 mm long, 1 to 4 in leaf axils; pedicels erect, 1–4 mm long; tepals green with white to pink margins. **Achene** 3-sided, 2–3 mm long, shiny black. Grasslands, steppe; montane to alpine. BC to SK south to CA, AZ, NM. One collection from an alpine, stony, calcareous slope in Beaverhead Co.

Polygonum viviparum L. [*Bistorta vivipara* (L.) Delarbre] ALPINE BISTORT Perennial from a corm-like rhizome. **Stems** erect, simple, 5–30 cm. **Leaves**: basal petiolate, the blades 1–9 cm long, narrowly lanceolate; cauline leaves few, smaller, sessile; stipules 4–22 mm long, glabrous, brown. **Flowers** functionally unisexual, 3–4 mm long, in a narrow spike 2–8 cm long, the lowest replaced by ovoid bulblets; pedicels spreading, 2–5 mm long; tepals pink or green below and white above. **Achene** 3-sided, brown, 2–3 mm long. Moist to wet turf, meadows, especially along streams and on cool slopes; alpine, subalpine, rarely lower. Circumpolar south to OR, NM, MN, NH.
    Stamens are exserted in male flowers but short in female flowers; viable seed is rarely produced.

Rumex L. Dock, sorrel

Annual or perennial, sometimes monoecious or dioecious herbs. **Stems** prostrate or erect. **Leaves** long-petiolate, basal and cauline with sheathing; membranous stipules (ocreae). **Inflorescence** terminal and sometimes axillary panicles or racemes of ocreate clusters. **Flowers** bisexual or unisexual; tepals 6, outer 3 narrow, inner 3 often colored and becoming enlarged and veiny in fruit, sometimes with a bump-like callosity toward the base; stamens 6; styles 3. **Achene** 3-angled, enclosed at base of inner sepals.[292]
    Mature fruit is often needed for positive identification.

1. Flowers unisexual; plants mostly dioecious; basal leaves numerous, cauline leaves reduced 2
1. Flowers perfect; cauline leaves well-developed. 4

2. Leaf blades narrowly elliptic, tapered to the petiole . . . .*R. paucifolius*
2. Basal leaf blades truncate, sagiittate or hastate at the base 3

3. Pedicel jointed in the middle; inner tepals in fruit with a small callosity . . . . *R. acetosa*
3. Pedicel jointed next to the flower; tepals without callosities . . . . *R. acetosella*

4. Plants strongly rhizomatous in sandy soil. . . . .*R. venosus*
4. Plants taprooted, not rhizomatous 5

5. Inner tepals in fruit with marginal teeth or bristles 6
5. Fruits with entire or wavy margins 8

6. Marginal teeth of fruiting inner tepals longer than tepal body width . . . . *R. fueginus*
6. Marginal teeth not as long as tepal body width 7

7. All 3 inner tepals with callosity; leaf blades not cordate . . . . *R. stenophyllus*
7. Usually at least 1 tepal without a callosity; some leaf blades cordate . . . . *R. obtusifolius*

8. Fruiting tepals without callosities 9
8. At least one fruiting tepal with a basal callosity 11

9. Plants with tuberous roots; mature tepals 8–17 mm long. . . . .*R. hymenosepalus*
9. Plants taprooted; mature tepals ≤10 mm long 10

10. Leaves linear to linear-lanceolate, <4 cm wide . . . . *R. salicifolius*
10. Leaves ovate to lanceolate, ≥5 cm wide. . . . .*R. occidentalis*

11. Stems branched at the base or below the inflorescence . . . . *R. salicifolius*
11. Stems unbranched below the inflorescence. 12

12. Leaf blades truncate at the base; callosity <1/2 as long as fruiting tepal . . . . *R. patientia*
12. Leaf blades tapered to the petiole; callosity ≥1/2 as long as fruiting tepal . . . .*R. crispus*

Rumex acetosa L. [*R. lapponicus* (Hiitonen) Czernov, *R. alpestris* Jacq.] Glabrous, dioecious perennial. **Stems** erect, simple, ribbed, 10–100 cm. **Leaf blades** narrowly sagittate to ovate, 2–12 cm long, the basal long-petiolate. **Inflorescence** a narrow panicle 10–30 cm long, of numerous separated clusters of 4 to 8 flowers. **Flowers** unisexual, reddish; outer tepals 2–3 mm long. **Fruit** 2–4 mm long; tepals ovate, cordate, with a small callosity; the pedicel jointed in the middle. **Achene** brown, ca. 2 mm long. Moist meadows, grasslands, steppe; valleys to upper subalpine. Circumboreal south to OR, WY. Our plants are ssp. *alpestris* A.Löve.
Our plants are the arctic-montane segregate of an intergrading circumboreal complex delineated by vegetative characters.[197] The complex is in need of further study,[292] so I have retained the more traditional treatment.

Rumex acetosella L. SHEEP SORREL. Glabrous, dioecious, rhizomatous perennial. **Stems** simple, erect, 8–50 cm. **Leaf blades** lanceolate to commonly hastate, 1–6 cm long. **Inflorescence** a panicle of separate clusters of 5 to 8 flowers. **Flowers** reddish, unisexual; outer tepals 1 mm long. **Fruit** ca. 2 mm long; inner tepals ovate without a callosity; the pedicel jointed next to the flower. **Achene** brown, ca. 1 mm long. Often disturbed soil of grasslands, pastures, open forest, often along trails, roads with introduced weeds; valleys to subalpine. Introduced from Europe into most of N. America. (p.162)

Rumex crispus L. CURLY DOCK Mostly glabrous perennial. **Stem** erect, simple, 30–100 cm. **Leaf blades** narrowly lanceolate, tapered to the petiole, 5–20 cm long with wavy margins. **Inflorescence** a leafy-bracted panicle 15–40 cm long with erect, spike-like branches. **Flowers** perfect; outer tepals 1–2 mm long. **Fruit** 3–5 mm long; inner tepals ovate with a truncate base and a variable-size callosity. **Achene** reddish brown, 2–3 mm long. Moist to wet, often disturbed soil along ditches, roads, streams, wetlands; plains, valleys, montane. Introduced from Europe to most of N. America. (p.162)

Rumex fueginus Phil. [*R. maritimus* L. ssp. *fueginus* (Phil.) Hultén.] Scabrous to puberulent, annual or biennial. **Stems** erect, branched, 5–60 cm. **Leaves** mainly cauline; the blades linear-lanceolate, 2–15 cm long. **Inflorescence** a leafy-bracted panicle of erect, interrupted, spike-like branches. **Flowers** perfect; outer tepals ca. 1 mm long. **Fruit** 1–3 mm long; inner tepals narrowly triangular with linear callosities and marginal, bristle-like lobes. **Achene** light brown, ca. 1 mm long. Drying mud around ponds, wetlands, tolerant of saline and alkaline soil; plains, valleys. Throughout N. America south to S. America. (p.162)

Rumex hymenosepalus Torr. Glabrous perennial with tuberous roots. **Stems** erect, branched, 25–90 cm. **Leaves** mainly cauline; the blades broadly oblanceolate, 5–20 cm long; stipules scarious, conspicuous. **Inflorescence** narrowly paniculate. **Flowers** perfect. **Fruit** 11–16 mm long; inner tepals cordate without a callosity. **Achenes** brown, 4–5 mm long. Sandy soil of grasslands, steppe; valleys. MT south to CA, AZ, TX.
There is one collection from near Helena taken in 1888 (*Anderson 88/66A*, MONTU) ca. 400 km north of the next nearest location.

Rumex obtusifolius L. Puberulent perennial. **Stems** erect, simple, 60–100 cm. **Leaf blades** ovate, cordate, the lower 10–25 cm long. **Inflorescence** a leafy-bracted panicle with ascending, interrupted, spike-like branches. **Flowers** perfect; outer tepals 2–3 mm long. **Fruit** 3–5 cm long; inner tepals triangular with basally toothed margins; a callosity on 1 to 2 tepals. **Achenes** brown,

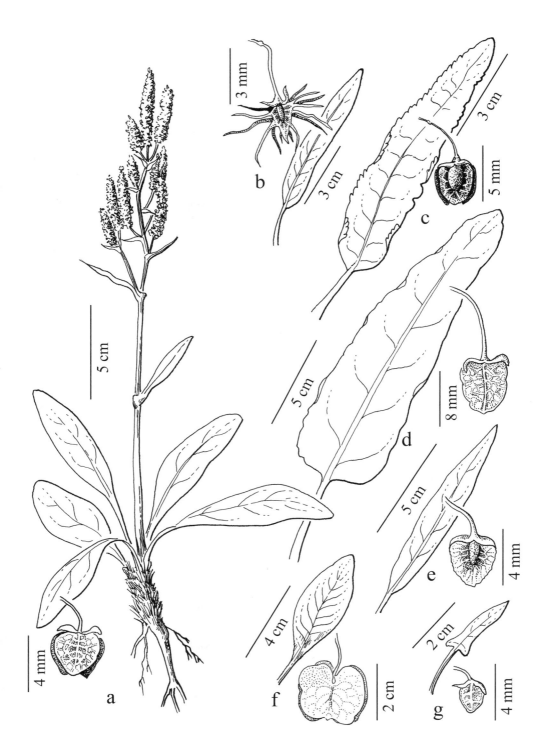

Plate 23. *Rumex*. **a.** *R. paucifolius*, **b.** *R. fueginus*, **c.** *R. crispus*, **d.** *R. occidentalis*, **e.** *R. salicifolius*, **f.** *R. venosus*, **g.** *R. acetosella*

2–3 mm long. Disturbed banks and terraces of rivers, streams; plains, valleys. Introduced into most of N. America; native to Europe.

Rumex occidentalis S.Watson Mostly glabrous perennial. **Stems** erect, simple, 50–120 cm, >6 mm wide. **Leaf blades** 5–25 cm long, lanceolate with a truncate to cordate base and wavy margins. **Inflorescence** 15–40 cm long, a leafy-bracted panicle of ascending branches. **Flowers** perfect; outer tepals 2–3 mm long. **Fruit** 5–10 mm long; inner tepals ovate without a callosity. **Achenes** reddish brown, 3–5 mm long. Fens, wet meadows, thickets; valleys to montane. AK to QC south to CA, NV, NM, SD. (p.162)

Rumex patientia L. Monoecious, glabrous perennial. **Stems** erect, simple below the inflorescence, 15–50 cm. **Leaf blades** elliptic to ovate, 4–20 cm long. **Inflorescence** 8–25 cm long, a narrow, leafy-bracted panicle with whorls of 10 to 20 flowers. **Flowers** unisexual; outer tepals 1–2 mm long. **Fruit** 5–8 mm long; inner tepals orbicular, cordate-based, only 1 with a callosity. **Achenes** brown, 3–4 mm long. Fields, roadsides; plains. Introduced into most of temperate N. America; native to Eurasia.

Rumex paucifolius Nutt. Glabrous, dioecious perennial. **Stems** erect, simple, 10–60 cm from a simple or branched caudex. **Leaves** mostly basal; the blades linear-lanceolate to narrowly elliptic, 2–10 cm long. **Inflorescence** a narrow panicle of erect or ascending, spike-like branches. **Flowers** unisexual, reddish; outer tepals 1–2 mm long. **Fruit** 2–4 mm long; inner tepals orbicular, cordate-based without callosities. **Achenes** brown, 1–2 mm long. Moist grasslands, meadows, steppe, open forest; valleys to alpine. BC, AB south to CA, UT, CO. (p.162)

Rumex salicifolius Weinm. [*R. triangulivalvis* (Danser) Rech.f., *R. utahensis* Rech.f.] Glabrous, perennial. **Stems** ascending to erect, branched, 20–80 cm. **Leaves** mostly cauline; the blades linear-oblanceolate, 3–15 cm long, little reduced above. **Inflorescence** 10–20 cm long, a leafy-bracted panicle of dense, erect, spike-like branches. **Flowers** perfect; outer tepals ca. 2 mm long. **Fruit** ca. 3 mm long; inner tepals ovate with or without a narrowly ovoid callosity. **Achenes** brown, ca. 2 mm long. Moist meadows, thickets, often along streams, ponds; montane, subalpine. AK to QC south to CA, NM, TX, VA; Europe. (p.162)

1. Fruiting tepals with a swollen midvein but without a callosity. . . . . var. *montigenitus*
1. At least 1 fruiting tepal with a distinct callosity . . . . var. *triangulivalvis*

Rumex salicifolius var. *triangulivalvis* (Danser) J.C.Hickman and *R. salicifolius* var. *montigenitus* Jeps. [*R. utahensis*] occur throughout the ecological range of the species in MT. These two vars. are sometimes considered separate species;[292] however the size of the callosity in this species is continuous,[187] and our two taxa show no ecological or geographic integrity.

Rumex stenophyllus Ledeb. Glabrous perennial. **Stems** erect, branched, 25–80 cm. **Leaf blades** lanceolate, 5–25 cm long. **Inflorescence** a narrow, leafy-bracted panicle with erect branches and flowers in whorls of 20 to 25. **Flowers** perfect; outer tepals 1–2 mm long. **Fruit** 3–5 mm long; inner tepals ovate-deltoid with a callosity on each and denticulate margins. **Achene** reddish brown, 2–3 mm long. Along rivers, roadsides; plains. Introduced throughout most of northern U.S. and southern Canada; native to Eurasia.

Rumex venosus Pursh Glabrous, monoecious, rhizomatous perennial with reddish foliage. **Stems** ascending, branched, 15–40 cm. **Leaves** all cauline; the blades thick, lanceolate to ovate, 3–10 cm long; stipules scarious, conspicuous. **Inflorescence** a simple or few-branched panicle with well-separated flower whorls. **Flowers** perfect; outer tepals 3–4 mm long. **Fruit** reddish, 20–25 mm long; inner tepals orbicular, cordate-based without callosities. **Achenes** brown, 5–7 mm long. Sandy soil along rivers, roads, railroads, sandstone outcrops; plains, valleys. AB to MB south to CA, TX, MI. (p.162)

Stenogonum Nutt. Two-whorl buckwheat

Stenogonum salsuginosum Nutt. Slenderly taprooted, glabrous annual. **Stems** ascending, dichotomously branched, 5–20 cm. **Leaves**: basal petiolate, the blade spatulate, 1–2 cm long; cauline sessile, lanceolate, 5–15 mm long. **Inflorescence** solitary, long-stalked involucres from leaf axils; involucral bracts 6, united only at the base, 2–4 mm long.

Flowers yellow, 1–3 mm long; tepals 6, hairy; stamens 9, smaller than the tepals. **Achenes** 2–3 mm long, glabrous, included in the perianth. Saline, clay soil of badlands; valleys. MT south to AZ, NM. Collected in Carbon Co.[249]

PLUMBAGINACEAE: Sea Lavender Family

Limonium Mill. Sea Lavender

Limonium vulgare Mill. Perennial, acaulescent, rhizomatous herbs 10–30 cm. **Leaves** all basal, petiolate; blades 2–10 cm long, leathery, oblanceolate with wavy, entire margins, sparsely farinose. **Inflorescence** a terminal panicle bearing secund 2-flowered spikelets. **Flowers** perfect, regular, hypogynous; calyx 5–7 mm long, purple-ribbed, obconic with white lobes; corolla tubular, ca. 4 mm long; petals beige, united at the base; stamens 5; styles 5; ovary superior, 1-celled. **Fruit** a 1-seeded, inflated achene ca. 5 mm long. Saline soil of grasslands; plains. Introduced to Pondera Co.; native to Eurasia.

The identity of our one collection is not certain.

ELATINACEAE: Waterwort Family

Herbaceous annuals. **Stems** branched at the base, sometimes rooting at the nodes. **Leaves** cauline, opposite, short-petiolate, simple, stipulate. **Flowers** perfect, radially symmetrical, sessile in leaf axils; sepals and petals 2 to 5, separate; pistil 1; ovary superior; stamens 2 to 10. **Fruit** a 2- to 5-chambered capsule with several pitted seeds in longitudinal rows.

1. Stems erect or ascending; plants glandular-pubescent . . . . *Bergia*
1. Plants glabrous, often mat-forming with prostrate to ascending stems . . . . *Elatine*

Bergia L.

Bergia texana (Hook.) Seub. ex Walp. Taprooted and glandular-pubescent. **Stems** erect, often red, 8–20 cm, often with prostrate stems at the base. **Leaf blades** oblong to obovate, serrulate, 5–30 mm long. **Flowers** solitary or few-clustered on short pedicels in leaf axils; sepals 5, acuminate, 3–4 mm long; petals 5, white, hidden by the sepals; stamens 5 or 10. **Capsule** 5-chambered, globose, ca. 3 mm. Drying mud around ponds; plains. WA, MT, SD, IL south to CA, NM, TX, LA. Collected once in Carter Co.

Elatine L. Waterwort

Stems prostrate to ascending, rooting at the nodes. **Leaves** with entire margins. **Flowers** minute, solitary, sessile in leaf axils; sepals and petals 2 to 4; stamens 2 to 8. **Fruit** a 2- to 4-chambered capsule.

1. Seeds strongly curved at least 90 degrees . . . . *E. californica*
1. Seeds straight or slightly curved. 2

2. Seeds with 10 to 15 pits per longitudinal row. . . . . *E. brachysperma*
2. Seeds with 18 to 27 pits per row . . . . *E. rubella*

Elatine brachysperma A.Gray Similar to *E. rubella*. **Leaves** linear to narrowly oblong. **Flowers**: sepals 2 to 3; petals 3; stamens 3. **Capsule** 3-celled. **Seeds** straight or gently curved, with 10 to 15 pits per row. Mud or shallow water of pond margins, tolerant of saline and alkaline soil; plains, valleys. WA, MT to OH, south to CA, TX, GA. Known from Gallatin and Missoula cos.[171]

Elatine californica Gray **Stems** 2–15 cm. **Leaves** linear-oblanceolate, 4–10 mm long. **Flowers** 1–2 mm across; sepals 4; petals 4; stamens 8. **Capsule** 4-celled. **Seeds** curved to a right angle, with 20 to 30 pits in 10 rows. Mud or shallow water of pond margins in the valleys. WA to MT south to CA.

Elatine rubella Rydb. [*E. americana* (Pursh) Arn] [*E. triandra* Schkuhr misapplied] **Stems** 2–6 cm. **Leaves** 2–8 mm long, the blade linear-oblong to narrowly elliptic with blunt or notched tips. **Flowers** 1–2 mm across; sepals 3; petals 3; stamens 3. **Capsule** 3-celled. **Seeds** straight or gently

curved, with 18 to 27 pits in 8 to 10 rows. Mud or shallow water of pond margins, tolerant of saline and alkaline soil; plains, valleys. WA to WI south to CA, TX, Mexico.

CLUSIACEAE: St. John's-wort Family

Hypericum L. St. John's-wort

Perennial herbs. **Stems** glabrous **Leaves** opposite, sessile, translucent- and sometimes black-dotted with entire margins. **Inflorescence** terminal or axillary, bracteate cymes. **Flowers** perfect, regular, hypogynous; sepals 5, sometimes united below; petals 5, separate, entire, yellow; stamens numerous; styles 3(to 5); ovary 3 (to 5)-celled. **Fruit** a many-seeded capsule.

Hypericum was previously placed in the Hypericaceae, a family now merged into the Clusiaceae.

1. Petals not black-dotted on margins, barely longer than sepals 2
1. Petals much greater than sepals, sparsely black-dotted on margins 3

2. Leaves mostly <15 mm long; sepals ovate . . . . *H. anagalloides*
2. Some leaves >15 mm long; sepals attenuate-lanceolate. . . . . *H. majus*

3. Sepals linear-lanceolate, acute . . . .*H. perforatum*
3. Sepals ovate with rounded tips. . . . . *H. formosum*

Hypericum anagalloides Cham.& Schlecht. Mat-forming. **Stems** prostrate below, rooting at the nodes, ascending above and 2–12 cm high. **Leaves** elliptic, 3–10 mm long. **Flowers** few; petals 3–4 mm long; sepals ovate with rounded to acute tips, 2–3 mm long. **Capsule** 5–6 mm long with yellow seeds. Wet soil of meadows along streams, lakes; montane, subalpine. BC, MT south to CA.

Hypericum formosum Kunth [*H. scouleri* Hook.] Rhizomatous and stoloniferous. **Stems** erect, 5–60 cm. **Leaves** broadly elliptic, 5–30 mm long, black-dotted on the margins. **Flowers**: petals 6–12 mm long with black-dotted margins; sepals ovate with rounded tips and amber veins, 3–4 mm long. **Capsule** 5–7 mm long with yellow to brown seeds. Moist to wet, often shallow soil of meadows, streambanks, ledges; valleys to alpine. BC, AB south to CA, WY. (p.166)

1. Stems >20 cm, often branched above; leaves narrowly ovate. . . . . var. *scouleri*
1. Stems <20 cm, mostly simple, leaves broadly ovate . . . . .var. *nortoniae*

Hypericum formosum var. *scouleri* (Hook.) J.M.Coult. occurs in deep soil in the valleys and montane zone; *H. formosum* var. *nortoniae* (Jones) C.L.Hitchc [*H. scouleri* ssp. *nortoniae* (M.E.Jones) J.M.Gillett], the high-elevation ecotype, is often in shallow soil and was named for Gertrude Norton who studied the botany of the Flathead region in the early part of the 20th Century; the type locality is in the Mission Mtns.

Hypericum majus (A.Gray) Britton **Stems** ascending but decumbent, simple or branched, 5–25 cm high. **Leaves** 10–25 mm long, lanceolate with rounded tips. **Flowers**: sepals attenuate-lanceolate, 4–7 mm long; petals as long as sepals. **Capsule** 4–6 mm long with yellow seeds. Shores of rivers, ponds, lakes; valleys, montane. BC to QC south to WA, CO, IL, PA.

Hypericum perforatum L. GOATWEED, KLAMMATH WEED Taprooted and rhizomatous. **Stems** 25–75 cm tall. **Leaves** lanceolate, 1–3 cm long, black-dotted on the margins. **Flowers**: sepals linear-lanceolate, 5–8 mm long; petals 8–14 mm long with black-dotted margins. **Capsule** 5–8 mm long with brown seeds. Grassland, open forest, pastures, roadsides; valleys, montane. Introduced throughout much of temperate N. America; native to Europe.

MALVACEAE: Mallow Family

Annual to perennial, often stellate-pubescent herbs. **Leaves** alternate, lobed, petiolate, stipulate. **Inflorescence** of racemes or flowers axillary. **Flowers** perfect, regular, hypogynous; calyx 5-lobed, sometimes with bracteoles at the base; petals 5, separate, short-clawed, wedge-shaped; stamens numerous, united into a central column; pistil 1; ovary 2- to many-celled; style branched. **Fruit** a hemispheric schizocarp with numerous radial sections.

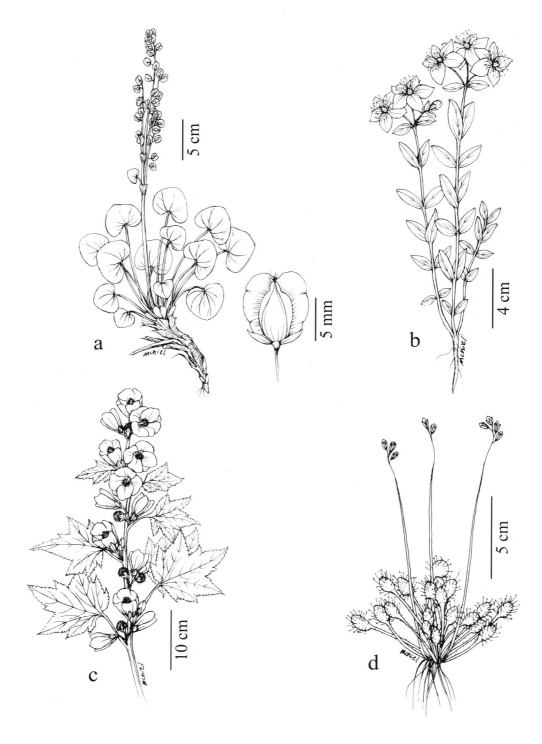

Plate 24. a. *Oxyria digyna*, b. *Hypericum formosum* var. **scouleri**, c. *Iliamna rivularis*, d. *Drosera rotundifolia*

Sida spinosa L., an annual with a spine at the base of the petiole, was collected once 50 years ago in a Missoula railroad yard.

1. Stigmas linear to filiform 2
1. Stigmas swollen at the tip, often capitate or truncate 4

2. Petals >3 cm long . . . . *Alcea*
2. Petals ≤3 cm long 3

3. Small bracts adnate to or closely subtending the calyx . . . . *Malva*
3. Small bracts absent or present at the base of the pedicel . . . . *Sidalcea*

4. Petals white to yellow, sometimes with a purple base 5
4. Petals pink, orange or reddish 6

5. Leaves divided at least 1/2-way to the midvein . . . . *Hibiscus*
5. Leaves entire or toothed . . . . *Abutilon*

6. Petals orange to red; stems 10–70 cm long . . . . *Sphaeralcea*
6. Petals pink to rose; stems 50–200 cm long . . . . *Iliamna*

Abutilon Mill.

Abutilon theophrasti Medik. Velvet-leaf Stellate-pubescent annual. **Stems** erect, simple or branched, 20–100 cm. **Leaf blades** ovate to orbicular, 5–15 cm long. **Flowers** axillary; calyx 8–14 mm high, divided nearly to the base, the lobes acute; petals yellow, 6–15 mm long, shallowly retuse. **Schizocarp** 1–2 cm wide, villous with 10 to 15 sections. Fields, gardens, roadsides; valleys, plains. Introduced to most of temperate N. America; native to Eurasia.

Alcea L. Hollyhock

Alcea rosea L. [*Althaea rosea* (L.) Cav.] Stellate-pubescent biennial or perennial. **Stems** simple, erect, 1–2 m. **Leaf blades** orbicular, 8–30 cm wide, 5- to 7-lobed with a cordate base. **Flowers** in a spike-like raceme; involucel bracts triangular; petals ca. 5 cm long, white to rose or purple. **Schizocarp** with >15 sections. Occasionally escaped ornamental. Introduced from Eurasia. Collected in Lake Co.

Hibiscus L. Rose mallow

Hibiscus trionum L. Flower of an hour Hirsute annual. **Stems** erect, branched, 10–60 cm. **Leaf blades** orbicular, divided nearly to the base; lobes 3 to 7, 1–4 cm long. **Flowers** axillary, subtened by linear bracts; calyx veiny, inflated, 8–14 mm long, with short, acute lobes; petals whitish with a purple base,15–40 mm long. **Schizocarp** 1–3 cm high, hirsute with 5 sections. Fields, gardens, roadsides; plains. SK to NS south to OR, UT, TX, FL; introduced from Europe.

Iliamna Greene Wild hollyhock

Iliamna rivularis (Douglas) Greene [*Sphaeralcea rivularis* (Douglas) Torr.] Sparsely puberulent perennial. **Stems** erect, 50–200 cm. **Leaves** 5–20 cm wide, maple-like with 5 to 7 crenate lobes. **Flowers** pedicillate, 1 to several in upper leaf axils and a terminal panicle; sepals 3–5 mm long, the bracteoles linear; petals 15–25 mm long, pink to rose. **Schizocarp** long-hairy, ca. 1 cm high with ≥10 2- to 4-seeded sections. Along streams and roads, open forest; valleys, montane. AB south to OR, CO. (p.166)

Malva L. Mallow

Annual or perennial herbs with simple or branched pubescence. **Leaves** long-petiolate, shallowly to deeply lobed. **Flowers** in axillary clusters or panicles; calyx shallowly lobed with 3 basal bracteoles; petals cuneate with a short claw and retuse tips; style branches 10 to 15. **Schizocarp** with 10 to 15 1-seeded sections.

All of our species are introduced from Eurasia throughout N. America. *M. rotundifolia* L., with hairy petal claws and longer pedicels, is reported for MT, but these characters are continuous in our area, and our specimens seem intermediate between this and *M. parviflora*, so I have chosen to recognize only the latter.

1. Upper leaves divided nearly to the base . . . . *M. moschata*
1. Leaves with shallowly lobed or crenate margins 2

2. Mature fruit sections smooth although hairy. . . . . *M. neglecta*
2. Outside of mature fruit sections wrinkled or reticulate 3

3. Petals 15–20 mm long; bracteoles ovate to oblong . . . .*M. sylvestris*
3. Petals 3–4 mm long; bracteoles linear . . . .*M. parviflora*

Malva moschata L. Sparsely short-hairy perennial. **Stems** erect, 30–60 cm. **Leaf blades**: the lower shallowly lobed; the upper deeply divided into 3 to 5 deeply toothed lobes. **Flowers** white to pink, 4–5 cm across. **Schizocarp** hairy. Roadsides, gardens; valleys. Known from Granite and Gallatin cos.

Malva neglecta Wallr. Cheeses [*M. rotundifolia* L. misapplied] Short-hairy annual or biennial. **Stems** prostrate to ascending, branched below, 20–50 cm. **Leaf blades** 2–5 cm long, cordate-orbicular, shallowly lobed. **Flowers** white to light blue; calyx 4–7 mm long, inflated in fruit; petals ca. 10 mm long with a pubescent claw. **Schizocarp** 4–7 mm across, the sections puberulent, not reticulate. Disturbed pastures, barnyards, around buildings; plains, valleys.

Malva parviflora L. Puberulent annual or biennial. **Stems** mostly ascending, 10–70 cm. **Leaf blades** 1–5 cm long, cordate-orbicular, shallowly 5- to 7-lobed with crenate margins. **Flowers** whitish; petals 3–4 mm long with a glabrous claw; calyx barely shorter than the petals. **Schizocarp** 4–7 mm across, the sections puberulent and reticulate on the outside margin. Barnyards, fields; plains, valleys.

Malva sylvestris L. Glabrate to hirsute annual or biennial. **Stems** ascending, branched below, 20–50 cm. **Leaf blades** 3–7 cm long, orbicular-cordate, lobed ca. 1/3 the length. **Flowers** purplish; calyx ca. 5 mm long, inflated in fruit; petals 15–30 mm long with a pubescent claw. **Schizocarp** 7–10 mm across, the sections sparsely hairy, wrinkled. Fields, roadsides; plains, valleys. Known from Gallatin and Phillips cos.

Sidalcea A.Gray

Sidalcea oregana (Nutt. ex Torr.& A.Gray) A.Gray Taprooted, stellate-pubescent perennial. **Stems** erect, 50–100 cm. **Leaves** 4–10 cm wide, deeply palmately divided into 5 to 9 narrow, sparsely toothed lobes. **Flowers** dimorphic, the pistillate, with rudementary stamens, smaller than the perfect in a simple terminal raceme; the pedicels 1–5 mm long; calyx 5–7 mm long with linear bracteoles; petals 1–2 cm long, pink to lavender. **Schizocarp** with 5 to 10 smooth-sided, 1-seeded sections, 2–3 mm long. Meadows, grasslands; valleys; Gallatin and Lake cos. WA to MT south to CA, UT, WY. Our plants are var. *procera* C.L.Hitchc.

Sphaeralcea A.St.-Hil. Globemallow

Perennial, stellate-pubescent herbs. **Leaves** petiolate, shallowly to deeply lobed. **Flowers** in bracteate racemes or narrow panicles; calyx deeply lobed with or without bracteoles; petals with a short claw and emarginate tips; style branches 10 to 15. **Schizocarp** with 10 to 15 sections; the upper portion seedless and dehiscent; the lower with reticulate faces and 1 to 2 seeds.

1. Leaves divided to near the base; calyx without bracteoles . . . . *S. coccinea*
1. Leaves shallowly lobed; calyx with bracteoles . . . . *S. munroana*

Sphaeralcea coccinea (Nutt.) Rydb. Scarlet globemallow **Stems** prostrate to ascending, 10–20 cm, numerous from a woody, branched caudex. **Leaf blades** ovate to suborbicular, 1–6 cm long, palmately or pinnately cleft almost to the base into 3 to 5 simple to lobed divisions. **Flowers** in short-pedicellate racemes; calyx 5–8 mm long without bracteoles; petals red-apricot, 10–20 mm long. **Schizocarp** 3–4 mm high, densely pubescent, 1-seeded. Grasslands, steppe, common on roadsides; plains, valleys, montane. BC to MB south to AZ, TX, IA.

Sphaeralcea munroana (Douglas ex Lindl.) Spach ex A.Gray **Stems** erect, 20–70 cm, several from a branched caudex. **Leaf blades** sparsely stellate-pubescent, 2–6 cm long, ovate to suborbicular, truncate at the base, the margins crenate to shallowly lobed. **Flowers** in a narrow panicle; calyx 2–8 mm long with 3 bracteoles; petals pink-apricot, 10–

15 mm long. **Schizocarp** 2–4 mm high, pubescent, the sections mostly 1-seeded. Sparsely vegetated sagebrush steppe, roadsides; valleys, montane. BC, MT south to CA, NV, UT. Known from southern Beaverhead Co.

DROSERACEAE: Sundew Family

Drosera L. Sundew

Insectivorous, perennial herbs. **Leaves** basal, long-petiolate, stipulate; the blade with long, purple, gland-tipped hairs on the upper surface. **Inflorescence** a 1-sided, nodding, raceme-like cyme. **Flowers** regular, perfect; petals white, usually 5, separate; sepals 5; stamens usually 5; ovary superior; styles 3 to 5, 2-lobed to near the base. **Fruit** a many-seeded capsule.

These plants occur only in montane *Sphagnum* fens. Plants overwinter as small rootless turions.

1. Leaf blades orbicular, to 12 mm long . . . . *D. rotundifolia*
1. Leaf blades oblong to linear, 1–3 cm long 2

2. Leaves linear; seeds <1 mm long. . . . .*D. linearis*
2. Leaves narrowly oblong; seeds ≥1 mm long . . . .*D. anglica*

Drosera anglica Huds. **Stems** erect, 8–25 cm. **Leaves** ascending to prostrate; blades narrowly oblong, 10–25 mm long; petiole 2–6 cm long. **Flowers**: petals 4–8 mm long; calyx 4–5 mm long; stamens slightly shorter; styles 4 to 5. **Capsule** 3–6 mm high; seeds 1–1.5 mm long. Circumboreal south to CA, MT, SK.

Drosera linearis Goldie Similar to *D. anglica*. **Leaves** ascending, the blade ca. 2 cm long. **Capsule** 3–5 mm high; seeds <1 mm long. Boreal Canada south to ME, WI, MT. Known from Lewis & Clark and Powell cos.

Drosera rotundifolia L. **Stems** erect, 3–20 cm. **Leaves** spreading to prostrate; blades orbicular, 4–12 mm long; petiole 1–3 cm long. **Flowers**: petals ca. 6 mm long; calyx just shorter; styles usually 3. **Capsule** 4–5 mm high with seeds ca. 1.5 mm long. Circumboreal south to CA, MY, ND, FL. (p.166)

VIOLACEAE: Violet Family

Viola L. Violet

Herbaceous, often stoloniferous perennials or rarely annuals, usually with ascending stems. **Leaves** petiolate, basal and cauline, alternate, simple, stipulate; margins crenate or dentate or rarely dissected. **Flowers** perfect, solitary; long-pedunculate from leaf axils or caudex, usually nodding, bilaterally symmetrical; sepals 5, separate; petals 5, separate, the lowest with a sac-like spur at the base, the upper 2 erect, the lower 3 spreading; stamens 5; style 1, longer than the stamens; ovary superior. **Fruit** a many-seeded, usually glabrous, 3-chambered capsule.

Perennial species produce inconspicuous, cleistogamous flowers at the base. Seeds of many species have an elaisome that attracts ants to disperse them. The cultivated *Viola odorata* L. occurs in lawns but is not truly naturalized. *Viola beckwithii* Torr.& A.Gray occurs near the Centennial Valley in ID, and *V. pedatifida* G.Don is found in western ND; both species have deeply dissected leaves and might be expected in MT.

1. Plants leafy-stemmed annual; stipules large, green, deeply divided . . . . *V. arvensis*
1. Plants perennial, sometimes acaulescent; stipules usually scarious and not leaf-like. 2

2. Petals primarily yellow on the outer surface. 3
2. Petals primarily white or blue 6

3. Leaf blades ovate to lanceolate, longer than wide 4
3. Leaf blades heart-shaped to orbicular, nearly as wide or wider than long 5

4. Leaf blades thick, strongly dentate to lobed, purplish at least on the veins . . . . *V. purpurea*
4. Leaf blades with crenulate to entire margins, not thick or purplish. . . . . *V. nuttallii*

5. Leaf blades orbicular, mostly flat to the ground; stems lacking . . . . *V. orbiculata*
5. Leaf blades with a pointed tip; aerial stems present. . . . . *V. glabella*

Viola adunca Sm. Plants rhizomatous, puberulent to glabrous. **Stems** inconspicuous at first but
elongating to 1–10 cm. **Leaf blades** lanceolate- to ovate-cordate, 5–40 mm wide with crenulate
margins; stipules ragged, narrow, 4–12 mm long. **Flowers** blue, 8–20 mm long; lower petals often
white-based; lateral pair bearded; spur 4–8 mm long; style hairy at the tip. **Capsule** 4–5 mm long.
Moist forest, riparian thickets, meadows, margins of wetlands at all elevations. AK to Greenland south
to CA, CO, SD, MN, NY. (p.172)

Aerial stems of small plants may be inconspicuous. The dwarf alpine form usually has glabrous leaves and has been
designated var. *bellidifolia* (Greene) H.D.Harr.; however, larger glabrous plants occasionally occur at lower elevations,
and some populations consist of both forms. *See Viola nephrophylla.*

Viola arvensis Murray Puberulent annual. **Stems** erect, branched, 10–30 cm. **Leaf blades**
lanceolate to ovate, 1–3 cm long with coarsely toothed margins; stipules deeply divided and
leaf-like. **Flowers** light yellowish, tinged with blue, 5–8 mm long; spur inconspicuous, sepals as
long as petals; style hairy at the tip. **Capsule** 4–7 mm long. Occasional garden escape introduced
throughout most of N. America; native to Europe.

Viola canadensis L. Plants glabrous to puberulent, stoloniferous. **Stems** 8–40 cm. **Leaf blades**
5–8 cm long, broadly cordate to reniform with crenulate margins and pointed tips; stipules
attenuate-lanceolate, 8–16 mm long. **Flowers** white, fading to blue, 10–15 mm long; petals yellow
at the base, the lower with purple lines, lateral pair bearded; spur ca. 2 mm long; style with sparse
long hairs. **Capsule** 5–8 mm long. Moist to wet forest, thickets, often with deciduous trees along
streams, wetlands; plains, valleys, montane. AK to NS south to AZ, NM, TN, SC. (p.172)

Leaves are usually larger than our other white-flowered violets which lack leafy stems. See *Viola glabella.*

Viola glabella Nutt. Plants glabrous to puberulent, rhizomatous. **Stems** 5–20 cm. **Leaf blades**
3–9 cm wide, broadly cordate to reniform, with pointed tips and crenate margins; stipules ovate,
rounded at the tip, 5–10 mm long. **Flowers** yellow, fading to white, 8–14 mm long; the lower
petals with purple lines; the lateral pair bearded; spur 1–2 mm long; style short-hairy at the tip.
Capsule ca. 5 mm long. Moist to wet forest, thickets, often along streams, lakes; valleys to subalpine.
AK to AB south to CA, MT. (p.172)

Leaves of *Viola orbiculata* are rounded at the tip; *V. canadensis* has long-pointed stipules.

Viola macloskeyi F.E.Lloyd Plants mostly glabrous, rhizomatous, stoloniferous. **Stems**
lacking; peduncles 2–9 cm. **Leaf blades** 1–3 cm long, broadly cordate to reniform with crenulate
margins; stipules lanceolate, glandular-toothed. **Flowers** white, 8–12 mm long; the lower petals
with purple lines; the lateral pair mostly glabrous; spur 2–3 mm long; style glabrous. **Capsule** 3–5
mm long. Fens, wet meadows, along streams; montane, subalpine. BC to NL south to most of U.S.
(p.172)

Viola macloskeyi is sometimes separated into two varieties based on leaf size and degree of margin crenulation; these
characters are completely intergradent in our area. Petals of *V. palustris* are larger and bluish.

Viola nephrophylla Greene Plants glabrous, without rhizomes or stolons. **Stems** lacking; peduncles 5–20 cm. **Leaf blades** 2–5 cm wide, broadly cordate with crenate margins; stipules narrow. **Flowers** 12–25 mm long, blue; the lower 3 petals white at the base, bearded; spur 2–3 mm long; style glabrous. **Capsule** 6–8 mm long. Wet, organic soil of meadows, marshes, fens; plains, valleys, montane. BC to NL south to CA, NM, MN, NY. Our plants are var. *cognata* (Greene) C.L.Hitchc. (p.172)

Viola adunca has a longer spur and is usually found in drier habitats; *V. selkirkii* has pubescent leaves with gland-toothed stipules. Plants from green ash woodlands in eastern Montana have more acuminate leaves and may be better referred to *V. pratincola* Greene,[152] now considered a synonym of *V. nephrophylla*.

Viola nuttallii Pursh Glabrous to sparsely puberulent without horizontal rhizomes or stolons. **Stems** mostly subterranean, the aerial portion 2–8 cm. **Leaf blades** 1–10 cm long, narrowly to broadly lanceolate, entire to remotely crenate, the base long-tapered to truncate; stipules adnate to stem. **Flowers** 8–15 mm long, yellow; upper petals brownish on back; the lower 3 with purple lines; spur ca. 1 mm long; style tip hairy. **Capsule** 6–9 mm long. Grasslands meadows, steppe; plains, valleys to lower subalpine. BC to MN south to CA, NM, KS. (p.172)

Leaf shape and degree of pubescence varies in a somewhat clinal manner from east to west. Segregates have been recognized as species based on correlations between these characters and chromosome number.[87,127] Morphological characters intergrade extensively in MT, the different combinations sometimes found in the same population,[152,187] so I have chosen to recognize our taxa at the varietal level.

1. Leaf blades >3 times as long as wide, gradually tapered to the petiole . . . . .var. *nuttallii*
1. Leaf blades mostly <3 times as long as wide, truncate to cuneate at the base 2

2. Some leaf blades deltoid-ovate and truncate at the base . . . . .var. *vallicola*
2. All leaf blades lanceolate and cuneate at the base . . . . .var. *praemorsa*

Viola nuttallii var. *nuttallii* occurs in grasslands, steppe east of the Divide; *V. nuttallii* var. *praemorsa* (Douglas ex Lindl.) S.Watson is found in grasslands, meadows; valleys to subalpine, more common higher; *V. nuttallii* var. *vallicola* (A.Nelson) H.St.John. occurs in grasslands, steppe, meadows west of the Divide and south of the Missouri River; plains to subalpine, more common lower.

Viola orbiculata Geyer ex Holz. Glabrous without rhizomes or stolons. **Stems** lacking; peduncles 2–5 cm long. **Leaf blades** 2–4 cm wide, orbiculate-cordate with crenate margins; stipules lanceolate, 3–9 mm long, sometimes glandular-serrate. **Flowers** 8–12 mm long, yellow; the lower 3 petals with purple lines; the lateral pair bearded; spur ca. 1 mm long; style tip hairy. **Capsule** 5–8 mm long. Moist coniferous forest; valleys to lower subalpine. BC, AB south to OR, ID, MT.

Leaves lay flat on the ground and remain green through the winter.

Viola palustris L. Glabrous, stoloniferous. **Stems** lacking; peduncles 4–15 cm long. **Leaf blades** 1–4 cm long, orbiculate-cordate, margins crenate; stipules lanceolate. **Flowers** white to bluish, 10–15 mm long; lower 3 petals with purple lines; the lateral pair sparsely hairy; spur ca. 2 mm long; style tip glabrous. **Capsule** 4–5 mm long. Wet soil of seeps, streambanks, wet meadows, fens, forests; valleys to rarely subalpine. BC to NL south to CA, UT, CO, NH. See *Viola macloskeyi*. (p.172)

Viola purpurea Kellogg Rhizomatous, puberulent. **Stems** 1–3 cm, mostly subterranean. **Leaf blades** 5–20 mm wide, lanceolate to ovate, thick, purple-tinged, shallowly lobed; stipules lanceolate. **Flowers** yellow, 7–12 mm long; petals with brown lines; the lateral pair bearded; spur 1–2 mm long; style tip hairy. **Capsule** puberulent, 4–5 mm long. Gravelly soil of meadows, grasslands, exposed slopes; montane, subalpine. WA to MT south to CA, AZ, CO. Our plants are var. *venosa* (S.Watson) Brainerd. (p.172)

Viola renifolia A.Gray Sparsely pubescent without rhizomes or stolons. **Stems** lacking; peduncles 5–8 cm long, shorter than the leaves. **Leaf blades** 2–6 cm wide, reniform, broader than long with crenulate margins; stipules lanceolate. **Flowers** white, 10–15 mm long; petals beardless; the lower purple-lined; spur 1–2 mm long; style tip glabrous. **Capsule** 3–5 mm long. Wet alder thickets, spruce forest; valleys, montane. BC to ME south to WA, CO. (p.172)

Both *Viola macloskeyi* and *V. palustris* usually have stolons; their leaves are not wider than long. Many *V. renifolia* plants have only cleistogamous flowers.

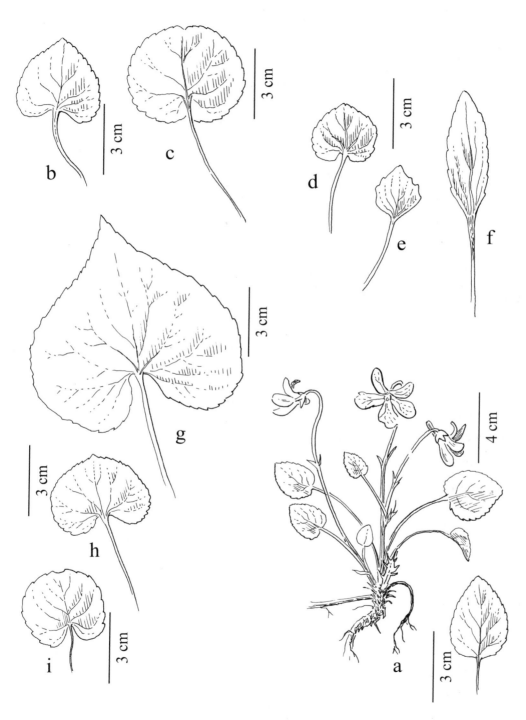

Plate 25. **Viola**. **a.** *V. adunca*, **b.** *V. nephrophylla*, **c.** *V. renifolia*, **d.** *V. macloskeyi*, **e.** *V. purpurea*, **f.** *V. nuttallii* var. *praemorsa*, **g.** *V. canadensis*, **h.** *V. glabella*, **i.** *V. palustris*.

Viola selkirkii Pursh ex Goldie Pubescent without rhizomes or stolons. **Stems** lacking; peduncles 2–5 cm long. **Leaf blades** 1–3 cm wide, broadly cordate with crenulate margins, hairy along veins above; stipules lanceolate, glandular-toothed, adnate to the stem. **Flowers** 8–13 mm long, blue; petals beardless; spur 3–7 mm long; style tip glabrous. **Capsule** 4–6 mm long. Wet montane, riparian forest. Circumboreal south to WA, NM, SD, OH, NH. One collection from Lincoln Co.[380]

Viola septentrionalis Greene Puberulent, without stolons or rhizomes. **Stems** lacking; peduncles 5–10 cm long. **Leaf blades** 2–4 cm wide, broadly cordate with dentate margins; stipules long-attenuate, fringed. **Flowers** blue, 15–25 mm long; lower petals bearded; sepals ciliate; spur 1–2 mm long; style tip glabrous. **Capsule** orbicular. Moist to wet, often riparian forest; valleys; Lake and Missoula cos. BC to NL south to WA, MT, NE, TN, NH.

TAMARICACEAE: Tamarisk Family

Tamarix L. Tamarisk, saltcedar

Tamarix ramosissima Ledeb. Deciduous shrubs. **Stems** slender, numerous from a branched caudex, 1–5 m with thin, reddish-brown bark. **Leaves** alternate, clasping, succulent, broadly lanceolate, scale-like, 1–2 mm long. **Inflorescence** spike-like, bracteate racemes, 15–40 mm long. **Flowers** perfect, regular, hypogenous; sepals 5, separate, ≤1 mm long; petals separate, pink, ovate, ca. 2 mm long; stamens 5; stigmas 3 to 4. **Fruit** a many-seeded capsule 3–4 mm long; seeds with an apical tuft of hairs. Terraces, saline meadows along rivers, streams, ponds; plains. Introduced to most of temperate, western N. American; native to Asia.

Most N. American plants are hybrids between *Tamarix chinensis* Lour. and *T. ramosissima* although they remain distinct in their native range;[139] our plants are more closely related to *T. ramosissima* (fide J. Gaskin).

CUCURBITACEAE: Gourd Family

Monoecious or dioecious, annual or perennial vines with branched tendrils. **Leaves** alternate, simple, petiolate, cordate-based, palmately 5-lobed. **Inflorescence** a raceme, corymb or solitary flowers from leaf axils. **Flowers** unisexual, regular, epigynous; sepals 5, united below; petals 5, united; ovary inferior; stamens 3 to 5, usually 3; style 1 with 1 to 4 stigmas. **Fruit** a many-seeded berry.

1. Fruit bristly or spiny . . . .*Echinocystis*
1. Fruit glabrous . . . . *Bryonia*

Bryonia L. Bryony

Bryonia alba L. Tuberous, dioecious or monoecious perennial. **Leaves** sparsely hairy, 3–10 cm long. **Flowers** greenish-white, 6-lobed; male flowers campanulate, racemose, the petals ca. 8 mm long; female flowers rotate, in corymbs, petals ca. 2 mm long, style 3- to 4-lobed, stamens 3. **Fruit** round, green to black at maturity, 7–10 mm across. Yards, gardens and climbing on riparian shrubs and trees; valleys. Introduced to WA, ID, MT, UT; native to Europe.

Echinocystis Torr.& A.Gray Wild cucumber

Echinocystis lobata (Michx.) Torr.& A.Gray Scabrous, monoecious annual. **Leaves** 5–15 cm long, with toothed margins. **Flowers** greenish-white, 6–10 mm across; male flowers racemose; female flowers solitary; corolla 6-lobed, 3–6 mm long with spreading lobes; stamens 3; style with a lobed stigma. **Fruit** ovoid, 3–5 cm long, inflated, prickly. Riparian forests, thickets; plains, valleys. SK to NL south to TX, FL.

LOASACEAE: Loasa Family

Mentzelia L. Blazing star

Taprooted, annual or perennial herbs. **Stems** whitish. **Leaves** basal and alternate, cauline, without stipules, simple with lobed or toothed margins, pubescent partly with barbed, clinging hairs. **Inflorescence** a bracteate corymb or flowers solitary. **Flowers** perfect, regular, epigynous; sepals 5, united to the top of the inferior ovary; petals 5, separate, sometimes with additional petal-like filaments; stamens numerous; style 1 with 3 stigmas. **Fruit** a many-seeded capsule, opening at the tip.

1. Sepals 1–4 mm long. 2
1. Sepals at least 5 mm long 3

2. Leaves commonly lobed ca. halfway to midvein; inflorescence bracts lanceolate. . . . .*M. albicaulis*
2. Leaves entire or shallowly lobed; inflorescence bracts ovate. . . . . *M. dispersa*

3. Sepals <2 cm long 4
3. Sepals >2 cm long 5

4. Sepals ≥1 cm long; petals ≥20 mm long. . . . . *M. nuda*
4. Sepals ≤1 cm long; petals ≤15 mm long. . . . . *M. pumila*

5. Petals white, apparently 10; floral bracts adherent to the hypanthium. . . . . *M. decapetala*
5. Petals yellow, 5; floral bracts below hypanthium . . . . *M. laevicaulis*

Mentzelia albicaulis (Douglas ex Hook.) Douglas ex Torr.& A.Gray [*Mentzelia montana* (Davidson) Davidson] Annual. **Stems** simple or branched, 8–30 cm, glabrous below. **Leaves** short-petiolate, 2–8 cm long; the blade lanceolate, pinnately lobed, especially below. **Flowers** yellow with linear bracts, diurnal; sepals 2–3 mm long; petals 5, obovate, 3–5 mm long. **Capsule** cylindric, 8–20 mm long; seeds angular, in 2 to 3 columns, papillose under 10X. Sparsely vegetated, often disturbed and sandy soil of grasslands, steppe, open forest; plains, valleys, montane. BC to SD south to CA, NM, TX, Mexico.
   Menzelia albicaulis is similar to the more common *M. dispersa* which has mostly unlobed leaves, broader floral bracts and more finely papillose seeds. *Mentzelia montana* is a large form of *M. albicaulis*.[193]

Mentzelia decapetala (Pursh ex Sims) Urb.& Gilg Usually biennial. **Stems** branched, 40–100 cm. **Leaves** lanceolate, sessile above, 4–15 cm long, deeply serrate. **Flowers** white with pectinate bracts adherent to the hypanthium, nocturnal; sepals 2–4 cm long; petals lanceolate, apparently 10 (5 are staminodes), 3–6 cm long. **Capsule** broadly cylindric, 2–4 cm long; seeds flattened, narrowly winged, horizontal in 3 to 5 columns. Sparsely vegetated, often gravelly soil of steep slopes, roadsides; plains, valleys. AB to MB south to NM and TX. (p.176)
   Fruiting plants can be told from *M. laevicaulis* by the floral bracts on the hypanthium.

Mentzelia dispersa S.Watson Annual. **Stems** simple or branched, brittle, 5–35 cm, puberulent. **Leaves** 1–7 cm long; the blade lanceolate, sometimes shallowly lobed below to entire and sessile above. **Flowers** yellow with ovate bracts, diurnal; sepals ca. 2 mm long; petals 5, obovate, 3–5 mm long. **Capsule** cylindric, ca. 1 cm long; seeds minutely papillose under 20X with grooved angles, pendulous in 1 column. Sparsely vegetated, often sandy or gravelly soil of grasslands, steppe, open forest; valleys. BC to ND south to CA, CO, SD, Mexico. See *Mentzelia albicaulis*. (p.176)

Mentzelia laevicaulis (Douglas ex Hook.) Torr.& A.Gray Usually biennial. **Stems** branched, 20–80 cm. **Leaves** oblanceolate and petiolate below to lanceolate and sessile above, 4–12 cm long, deeply lobed to serrate. **Flowers** light yellow with pinnatifid bracts below the hypanthium, nocturnal; sepals 2–3 cm long; petals 5, lanceolate, 3–6 cm long. **Capsule** broadly cylindric, 2–4 cm long; seeds flattened, broadly winged, horizontal in several columns. Sparsely vegetated soil of steep slopes, roadsides; plains, valleys. BC, MT south to CA, UT, CO. See *Mentzelia decapetala*.

Mentzelia nuda (Pursh) Torr.& A.Gray Biennial or short-lived perennial. **Stems** branched, 15–50 cm. **Leaves** oblanceolate, sessile above, 3–5 cm long, serrate. **Flowers** white with pectinate bracts below the hypanthium, diurnal; sepals 1–2 cm long; petals oblanceolate, apparently ca. 10, 2–4 cm long. **Capsule** cylindric, 15–25 mm long; seeds flattened, winged, horizontal in 3 to 5 columns. Barren, sandy or gravelly soil, often along streams, roadsides; plains, valleys. MT, SD south to CO, TX.

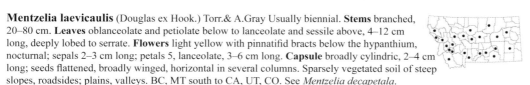

Mentzelia pumila (Nutt.) Torr.& A.Gray Biennial or short-lived perennial. **Stems** branched, 8–30 cm. **Leaves** narrowly lanceolate, short-petiolate, 2–10 cm long, deeply serrate. **Flowers** yellow with lanceolate-pectinate bracts; sepals 5–10 mm long; petals apparently 10, spatulate, 9–15 mm long. **Capsule** broad cylindric, 10–15 mm long; seeds flattened, narrowly winged, horizontal. Barren, sandy soil of steppe, juniper woodlands in desert valleys of Big Horn and Carbon cos. MT, WY, UT.

SALICACEAE: Willow Family

Deciduous, dioecious trees and shrubs. **Leaves** simple, petiolate, alternate, often with deciduous stipules. **Inflorescence** pendulous, cylindric catkins (aments). **Flowers** unisexual, subtended by a scale, without sepals or petals, numerous in unbranched cylindrical spikes (catkins, aments); male flowers of 1 to many stamens; female flowers of 1 ovary with a style and 2 to 4 stigmas. **Fruit** an ovoid capsule; seeds numerous, with a tuft of hair at the tip.

1. Buds with 3 to 10 resinous scales; female catkins penulous; stamens ≥6. *Populus*
1. Bud scales 1; female catkins erect to spreading; stamens 1 to 8. . . . . *Salix*

Populus L. Poplar, cottonwood

Short-lived trees. **Buds** with several scales, resinous. **Leaves** cordate, truncate or cuneate at the base with toothed margins, turning yellow in autumn. **Catkins** appearing before the leaves, pendulous, ebracteate. **Flowers**: male flowers with 7 to 80 stamens; female flowers with 2 to 4 stigmas. **Capsules** ovoid.[56,97]

Seeds of our native species are short-lived and require moist, barren soil to establish. Members of section *Tacamahaca* (*Populus angustifolia, P. balsamifera, P. ×brayshawii*) are more prone to produce clones through root sprouting than is *P. deltoides* (Section *Aigeiros*). It is difficult to determine whether specimens of *P. balsamifera* and *P. detoides* with narrower than average leaves are simply eccentric or have experienced introgression with *P. angustifolia.*

1. Leaf blades white-tomentose beneath; escaped from cultivation. . . . . *P. alba*
1. Leaf blades glabrous beneath; native. 2

2. Most leaf blades ovate to orbicular, ca. as long as wide 3
2. Most leaf blades lanceolate, ≥1.5 times as long as wide 4

3. Leaf tips long-attenuate, marginal teeth >1 mm deep . . . . *P. deltoides*
3. Leaf tips acute, marginal teeth ca. 0.5 mm deep . . . . *P. tremuloides*

4. Some marginal teeth ≥1 mm deep . . . . *P. ×acuminata*
4. Marginal teeth ≤0.5 mm deep. 5

5. Petioles ≤1/3 as long as leaf blade. . . . .*P. angustifolia*
5. Petioles >1/3 as long as blade length. 6

6. Leaf blades somewhat cuneate at the base, mostly >2 times as long as wide . . . .*P. ×brayshawii*
6. Leaf blades broadly rounded or truncate basally, ≤2 times as long as wide . . . . *P. balsamifera*

Populus ×acuminata Rydb. LANCE-LEAF COTTONWOOD Tree 10–30 m with ascending branches. **Bark** tan, becoming furrowed. **Leaf blades** ovate with shallowly cuneate bases, 5–9 cm long, somewhat glaucous below, ≥2.5 times as long as wide, some serrations >0.5 mm deep; petioles ≥1/3 length of blade. **Female catkins** 6–9 cm long; stigmas 2 to 3. **Capsules** 5–7 mm long. Streambanks; plains. AB to ND south to AZ, NM, TX.

Populus acuminata is thought to be a hybrid between *P. deltoides* and *P. angustifolia*; the former has wider leaf blades with truncate to shallowly cordate bases.

Populus alba L. WHITE POPLAR Tree 5–20 m with ascending branches, spreading by horizontal roots. **Bark** white, smooth, becoming gray below. **Leaf blades** 3–5 cm long, ovate with a rounded tip, coarsely sinuate, cordate-based, green above, white-tomentose below. **Female catkins** 2–8 cm long, densely flowered. **Capsules** 3–5 mm long; seeds usually not maturing. Towns, agricultural areas. Introduced throughout much of temperate N. America; native to Eurasia.

Populus angustifolia E.James NARROW-LEAF COTTONWOOD Tree 5–20 m with ascending branches. **Bark** light brown, becoming furrowed. **Leaf blades** lanceolate with cuneate bases, ≥2.5 times as long as wide, 4–11 cm long, glaucous below, with petioles ≤1/3 length of blade. **Female catkins** 3–12 cm long; stigmas 2. **Capsules** 3–6 mm long. Along rivers, streams; valleys, montane. BC to SK south to ID, TX, Mexico. See *Populus acuminata*.

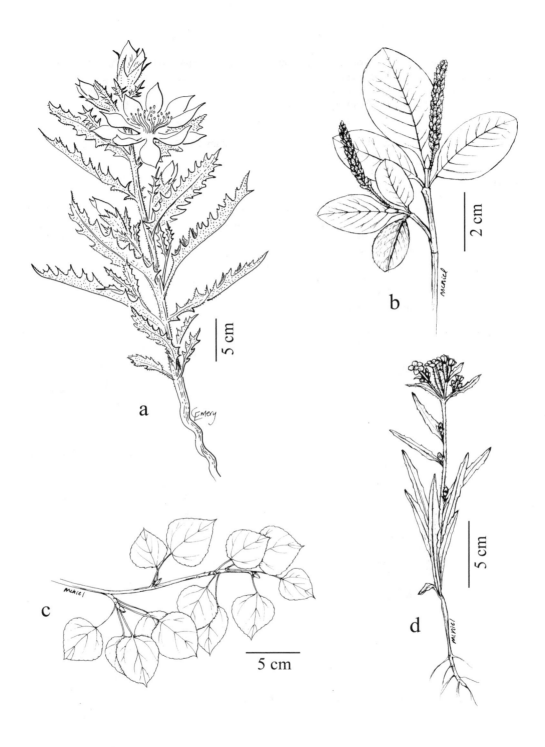

Plate 26. a. *Mentzelia decapetala*, b. *Salix vestita*, c. *Populus tremuloides*, d. *Mentzelia dispersa*

Populus balsamifera L. Black cottonwood, balsam poplar [*P. trichocarpa* Torr.& A.Gray]
Tree to 40 m with ascending branches. **Bark** smooth, white at first, becoming gray and furrowed.
Leaf blades broadly lance-shaped with truncate bases, 5–15 cm long, pale below. **Female catkins**
8–15 cm long; stigmas 3. **Capsules** 5–8 mm long. Forest and gravel bars along rivers, streams,
aspen groves, avalanche slopes, roadsides; valleys to lower subalpine zone. AK to NL south to CA,
WY, WI.

1. Stamens fewer than 20; capsules glabrous, opening by 2 slits . . . . ssp. *balsamifera*
1. Stamens more than 30; capsules puberulent, opening by 3 slits . . . . ssp. *trichocarpa*

Populus balsamifera ssp. **trichocarpa** (Torr.& A.Gray) Brayshaw occurs along rivers and streams, disturbed roadsides
throughout western MT; *P. balsamifera* ssp. *balsamifera* is found mainly in forests, aspen groves, avalanche slopes in
northwest MT. The two taxa are often considered separate species,[13] but the characters separating them are not constant,
and intermediate forms are common.[56] See *Populus ×brayshawii*.

Populus ×brayshawii B.Boivin Tree to 20 m with ascending branches. **Bark** light gray-
brown, shallowly furrowed. **Leaf blades** lanceolate to narrowly ovate with cuneate to rounded
bases, 4–12 cm long, much paler below; serrations ≤0.5 mm deep; petioles >1/3 length of blade.
Female Catkins 2–12 cm long; stigmas 2. **Capsules** 4–6 mm long. Along streams; valleys,
montane. AB to SD south to CO, UT.
    Populus ×brayshawii is completely intergadent with *P. balsamifera* which has wider leaves with truncate to
shallowly cordate bases.

Populus deltoides W.Bartram ex Marshall Plains cottonwood Tree 20–30 m with spreading
branches. **Bark** tan, deeply furrowed. **Leaf blades** 5–10 cm long, deltoid, ca. as long as wide, ca.
the same color above and beneath; the base truncate to cordate; the tip long-attenuate; the margins
with teeth at least 1 mm deep. **Catkins** 5–12 cm long; stigmas 3 to 4. **Capsules** 8–10 mm long.
Along rivers, streams; plains. AB to QC south to CO, TX, FL. Our plants are ssp. *monilifera* (Aiton)
Eckenw. See *P. acuminata*.

Populus tremuloides Michx. Quaking aspen, trembling aspen Tree 2–45 m with spreading
branches. **Bark** smooth, pale, furrowed with age. **Leaf blades** broadly ovate to orbicular, truncate
at the base, 2–8 cm long, pale beneath. **Female catkins** 4–10 cm long; stigmas 2. **Capsules** 4–6
mm long. Along streams, wetlands, topographic depressions, cool slopes where the soil is somewhat
moist; plains, valleys to lower subalpine. AK to NL south to CA, NM, TN, NJ. (p.176)

Salix L. Willow

Shrubs or small trees. **Buds** with a single scale, not resinous. **Leaves** linear to ovate, usually short-
petiolate; margins serrate to entire. **Catkins** sessile or on leafy branchlets, erect or drooping, bracteate,
appearing before or after the leaves. **Flowers**: male flowers with 1 to 8 stamens; female flowers with 2 or 4
stigmas. **Capsules** sessile to stipitate.[69,114,115,116]
    Mature catkins and mature leaves are useful for identification, but both are rarely available at the same
time. Vegetative characters have a good deal of environmentally or developmentally induced variation.
Understanding this variation is required for consistently correct determinations. Characters in the key and
descriptions are for mature leaves and catkins. Reports of *Salix monochroma* C.R.Ball for MT[113] were
based on shade forms of *S. boothii* (fide R. Dorn). *Salix pentandra* L. is reported escaped in MT,[12] but I
have seen no specimens.

1. Plants prostrate shrubs 1–10 cm high with subterranean stems; upper subalpine to alpine 2
1. Plants erect shrubs, mostly >15 cm high; usually plains, valleys to subalpine. 6

2. Leaves ≤10 mm long, not glaucous beneath; capsules glabrous. . . . .*S. rotundifolia*
2. Plants not as above 3

3. Leaf tips rounded, prominently veiny beneath; styles < 0.5 mm long. *S. reticulata*
3. Leaf tips pointed, not prominently veiny; styles 0.3–2 mm long 4

4. Leaf blades 2–6 mm wide, usually not glaucous . . . . *S. cascadensis*
4. Leaf blades mostly >6 mm wide, glaucous beneath. 5

5. Second-year twigs ≥3 mm in diameter; style ≤1 mm long . . . . *S. glauca*
5. Second-year twigs 1–2 mm in diameter; style 1–2 mm long . . . . *S. arctica*

6. Second-year twigs glaucous, sometimes only apparent behind buds 7
6. Second-year twigs without a glaucous bloom 9

7. Catkins tightly flowered, sessile or nearly so; leaves densely silvery-hairy beneath . . . . . *S. drummondiana*
7. Catkins loosely flowered on leafy branchlets; leaves not silvery-hairy. 8

8. Flower bracts yellow to tan, short-hairy . . . . *S. geyeriana*
8. Flower bracts dark brown to black, long-hairy . . . . *S. lemmonii*

9. Leaves linear, >6 times as long as wide, usually <15 mm wide. 10
9. Leaves narrowly lanceolate or broader, usually <6 times as long as wide. .13

10. Leaves lighter below than above; stipes ≥1 mm long. 11
10. Leave ca. equally green on both sides; stipes <1 mm long 12

11. Capsules hairy . . . . *S. petiolaris*
11. Capsules glabrous . . . . *S. eriocephala*

12. Leaves pale green, often hairy; flower bracts lanceolate, acute. . . . . *S. exigua*
12. Leaves bright green above, glabrous; flower bracts blunt-tipped. . . . . *S. melanopsis*

13. Leaves about equally green on both surfaces, not glaucous beneath 14
13. Leaves distinctly lighter beneath, either because of hair or glaucous bloom .20

14. Mature leaves with long-attenuate tips; petioles with glands just below blade. . . . . *S. lasiandra*
14. Leaves not long-attenuate; petioles mostly lacking glands 15

15. Twigs oily; catkins sessile with pubescent capsules. . . . . *S. barrattiana*
15. Twigs not oily; catkins on branchlets or capsules glabrous 16

16 Leaf margins entire; catkins ≤25 mm long. . . . . *S. wolfii*
16. Leaf margins usually denticulate to serrate, rarely entire; usually some catkins >25 mm long 17

17. Leaves essentially glabrous at maturity . . . . *S. boothii*
17. Mature leaves with some hair at least along midvein. 18

18. Petioles and twigs with appressed curly or wavy hairs. . . . . *S. eastwoodiae*
18. Petioles and twigs with spreading straight hairs. 19

19. Mature leaves glabrous to sparsely hairy below; catkins sessile or nearly so . . . . *S. tweedyi*
19. Mature leaves evidently hairy beneath; catkins on leafy branchlets. . . . . *S. commutata*

20. Mature leaf blades densely white- or silvery-hair beneath, green and glabrate to hairy above 21
20. Lower surface of mature leaf blades glaucous but not so obscured by dense hair 23

21. Hairs of lower leaf surface white, long tangled; twigs with felty hair. . . . . *S. candida*
21. Lower leaf hairs silvery, short; twigs short-hairy or glabrate. 22

22. Female catkins sessile; leaf blades lanceolate to elliptic; twigs often glabrous . . . . *S. drummondiana*
22. Female catkins on leafy branchlets; leaf blades oblanceolate to obovate; twigs
    short-hairy (rare "denudata" form of *S. candida* would key here also) . . . . *S. sitchensis*

23. Trees; female catkins with pale flower scales. 24
23. Shrubs; female catkins mostly with dark scales (yellow in *S. petiolaris*, *S. bebbiana*). 27

24. Bud scales with margins overlapping on side facing the branch . . . . *S. amygdaloides*
24. Bud scales cap-like without apparent overlapping margins 25

25. Leaves oblong to lanceolate but without long-attenuate tips . . . . *S. bebbiana*
25. At least some leaves with long-attenuate tips. 26

26. Twigs yellow, brittle at the base; petioles without glands . . . . *S. fragilis*
26. Twigs not brittle; some leaves with glands on petiole just below blade . . . . *S. lasiandra*

27. Buds and twigs oily or sticky; near or above treeline . . . . *S. barrattiana*
27. Buds and twigs not oily; in various habitats 28

28. Twigs of the year purplish with appressed hair; capsules on stipes ≥2 mm long . . . . *S. bebbiana*
28. Plants not as above 29

29. Leaf blades broadest well above middle 30
29. Leaf blades broadest near or below middle 31

30. Plants riparian. . . . . *S. discolor*
30. Plants mainly of upland habitats. . . . . *S. scouleriana*

31. Most leaf blades with entire margins 32
31. Most leaf blades with denticulate to serrate margins 39

32. Fully expanded leaves glabrous or nearly so. 33
32. Fully expanded leaves obviously hairy on at least one side. 36

33. Female catkins sessile. . . . . *S. planifolia*
33. Female catkins on leafy branchlets 34

34. Stipes of capsules 1.5–4 mm long . . . . *S. eriocephala*
34. Stipes <1.5 mm long. 35

35. Capsules glabrous . . . . *S. farriae*
35. Capsules pubescent. . . . . *S. lemmonii*

36. Upper leaf surface dark green with deeply impressed veins . . . . *S. vestita*
36. Veins of upper leaf surface not obviously impressed 37

37. Stipes 1-2 mm long. . . . . *S. geyeriana*
37. Stipes <1 mm long 38

38. Most female catkins ≤2 cm long; petioles 1–3 mm long. . . . . *S. brachycarpa*
38. Most catkins ≥2 cm long; petioles often >3 mm long . . . . *S. glauca*

39. Female catkins sessile 40
39. Female catkins on leafy branchlets 43

40. Mature leaves hairy on upper surface . . . . *S. tweedyi*
40. Upper surface of mature leaves glabrous. 41

41. Capsules glabrous . . . . *S. pseudomonticola*
41. Capsules hairy 42

42. Stipe of capsule ≤1 mm long; upper leaf surface dark green. . . . . *S. planifolia*
42. Stipe of capsule >1 mm long; leaves light green above. . . . . *S. discolor*

43. Female catkins maturing in late summer or early fall; the scales pale and deciduous. . . . .*S. serissima*
43. Female catkins maturing and dispersing seed in late spring or early summer; scales dark
    and persistent. 44

44. Capsules pubescent. 45
44. Capsules glabrous 46

45. Flower bracts dark; styles 0.2–0.9 mm long. . . . . *S. lemmonii*
45. Flower bracts yellowish; styles 0.1–0.4 mm long . . . . *S. petiolaris*

46. Stipe of capsule ≥1.5 mm long; style <1 mm long . . . . *S. eriocephala*
46. Stipe of capsule mostly ≤1.5 mm long; style ≥1 mm long. *S. barclayi*

Salix amygdaloides Andersson Peachleaf willow Small tree or shrub 2–12 m. **Twigs** yellow, glabrous. **Leaf blades** 3–8 cm long, lanceolate, long-acuminate, finely serrate, green above, glaucous beneath. **Female catkins** emerge with the leaves, 3–7 cm long on leafy branchlets 1–3 cm long; scales yellow, long-hairy below, deciduous. **Capsules** glabrous, 2–4 mm long; stipes 1–2 mm long; style ca. 0.3 mm long. Banks of rivers, streams; plains, valleys. BC to QC south to WA, NV, NM, AR, NJ.

Salix arctica Pall. [*S. petrophila* Rydb.] Arctic willow Mat-forming with trailing stems. **Twigs** dark purple, sparsely hairy. **Leaf blades** 7–30 mm long, glaucous beneath, often sparsely hairy, oblanceolate to narrowly elliptic with entire margins. **Female catkins** emerge with the leaves, 1–5 cm long on leafy branchlets 1–3 cm long; scales brown or black, long-hairy. **Capsules** pubescent, 4–7 mm long, sessile; style 1–2 mm long. Moist to wet turf, often along streams or where snow accumulates, sometimes moist scree; upper subalpine, alpine. Circumpolar south to CA, NM. Our plants are var. ***petraea*** (Andersson) Bebb. (p.183)
  Salix reticulata has smaller catkins with fewer capsules and often forms denser mats.

Salix barclayi Andersson Shrub 1–3 m. **Twigs** blackish, mostly glabrous. **Leaf blades** 3–8 cm long, oblanceolate to obovate, finely serrate with glandular teeth, green above, glaucous beneath, glabrous at maturity. **Female catkins** emerge with the leaves, 2–7 cm long on leafy branchlets 7–15 mm long; scales brown to black, long-hairy. **Capsules** glabrous, 4–6 mm long; stipes 0.5–1.5 mm long; style 0.5–1.5 mm long. Along streams, margins of wetlands; montane, subalpine. AK south to WA, OR, ID, WY.

Salix farriae has leaves with entire margins; *S. tweedyi* has sessile female catkins; *S. pseudomonticola* has reddish petioles or midveins.

Salix barrattiana Hook. Shrub 30–100 cm. **Twigs** reddish brown, hairy with oily buds. **Leaf blades** 2–6 cm long, long-hairy on both sides, narrowly elliptic with entire margins. **Female catkins** emerge before or with the leaves, 4–6 cm long, nearly sessile; scales black, long-hairy. **Capsules** hairy, 3–5 mm long; stipes 0.5 mm long; styles ≥1 mm long. Swamps and along small streams; alpine in Carbon and Glacier cos. AK to BC, WY.

The Carbon Co. population straddles the MT-WY border and may be one large male clone.

Salix bebbiana Sarg. BEBB WILLOW Shrubs 2–4 m. **Twigs** purplish, appressed-hairy. **Leaf blades** 2–5 cm long, lanceolate, oblanceolate, or elliptic, green and sparsely hairy above, glaucous beneath; margins usually entire or shallowly toothed. **Female catkins** emerge with the leaves, 1–8 cm long on leafy branchlets 5–10 mm long; scales tan, hairy. **Capsules** 5–8 mm long, hairy, long-beaked; stipes 2–5 mm long; style <0.5 mm long. Moist soil along streams, wetlands, generally in mineral soil; plains, valleys to subalpine. AK to NL south to CA, NM, IN. (p.183)

Salix boothii Dorn [*S. myrtillifolia* Andersson misapplied] Shrubs to 5 m. **Twigs** brown to purple, sparsely hairy. **Leaf blades** 2–7 cm long, lanceolate to narrowly elliptic, nearly glabrous, green above, slightly paler green beneath, margins glandular-serrulate. **Female catkins** emerge with the leaves, 1–5 cm long on leafy branchlets 3–7 mm long; scales dark brown, long-hairy. **Capsules** glabrous, 3–6 mm long; stipes 1–3 mm long; style ca. 0.5 mm long. Along streams, lakes, wetlands; valleys to subalpine. BC to SK south to CA, NM. (p.183)

The leaves with nearly equally green upper and lower surfaces helps distinguish this common species.

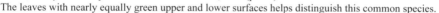

Salix brachycarpa Nutt. Shrubs 30–100 cm. **Twigs** pubescent. **Leaf blades** 1–4 cm long, long-hairy especially beneath, narrowly elliptic with entire margins, glaucous beneath. **Female catkins** emerge with the leaves, 8–30 mm long on leafy branchlets 2–15 mm long; scales yellowish, pubescent. **Capsules** tomentose, 3–5 mm long; stipes <0.3 mm; style ca. 0.5 mm long. Streambanks, fens, rocky slopes, talus (usually limestone); montane to alpine. AK south to OR, UT, NM.

Salix glauca has mostly longer catkins and leaf petioles; leaves of *S. wolfii* are long-hairy but not glaucous beneath. See *S. candida*.

Salix candida Flugge ex Willd. Shrubs 20–100 cm. **Twigs** purple, white-tomentose. **Leaf blades** 3–9 cm long, oblanceolate with entire margins, white-tomentose beneath, sparsely hairy above. **Female catkins** emerge with the leaves, 1–4 cm long, on leafy branchlets 1–5 mm long; scales long-hairy. **Capsules** white-tomentose, 3–8 mm long; stipes ca. 1 mm long; style 1–1.5 mm long, red. Fens, margins of swamps, marshes, usually where calcareous; montane. AK to NL south to WA, ID, CO, SD, IA, NJ. (p.183)

Salix brachycarpa has shorter leaves. The "denudata" form of *S. candida* has nearly glabrous leaves and has been collected in Flathead Co.; it may be a hybrid (fide R. Dorn).

Salix cascadensis Cockerell Dwarf, mat-forming. **Twigs** brown, glabrous. **Leaf blades** 3–20 mm long, narrowly elliptic with entire margins, mostly glabrous. **Female catkins** 6–25 mm long emerge with the leaves on leafy branchlets to 2 cm long; scales dark, long-hairy. **Capsules** villous, 3–5 mm long; stipes ca. 0.5 mm long; styles ca. 0.5 mm long. Moist, stony slopes; alpine. BC to WA, UT, CO.

Specimens from the single known MT population in Deer Lodge Co. seem intermediate between typical *Salix cascadensis* and *S. rotundifolia*.

Salix commutata Bebb Shrubs 20–150 cm. **Twigs** dark, long-hairy, becoming glabrous. **Leaf blades** 2–7 cm long, elliptic with entire to denticulate margins, sparsely long-hairy above and beneath, not glaucous. **Female catkins** emerge with the leaves, 1–4 cm long on leafy branchlets 5–25 mm long; scales light brown, long-hairy. **Capsules** glabrous to rarely pubescent, 4–5 mm long; stipe ≤1 mm long; style 0.5–1 mm long. Along streams, ponds, moist meadows in areas of snow accumulation; subalpine. AK to AB south to OR, ID, MT.

Varieties are sometimes segregated based on the presence of leaf serrations; however, entire-margined leaves can occur with serrate leaves in the same population or even the same plant. *Salix glauca* has narrower leaves that are glaucous beneath.

Salix discolor Muhl. PUSSY WILLOW Shrub to 10 m high. **Twigs** brown and hairy at first, soon black and glabrous. **Leaf blades** 3–8 cm long, lanceolate to obovate, entire to weakly crenate margins, becoming glabrous, glaucous beneath. **Female catkins** emerge before the leaves, 2–6 cm long, sessile; scales brown to black, long-hairy. **Capsules** long-beaked, pubescent, 6–8 mm long; stipe ca. 1 mm long; style ca. 0.5 mm long. Along rivers, streams; plains, valleys. BC to NL south to CO, SD, IN, NC.

Salix scouleriana has mature leaves that are hairy at least below on the midvein and occurs in more upland habitats; *S. farriae* has entire leaves and is found in the mountains.

Salix drummondiana Barratt ex Hook. Shrubs 1–4 m. **Twigs** usually glaucous, especially around buds. **Leaf blades** 2–9 cm long, lanceolate to narrowly elliptic with entire, inrolled margins, green above, moderately to densely silver-hairy below. **Female catkins** emerge before the leaves, 2–7 cm long, sessile; scales black, long-hairy. **Capsules** hairy, 3–6 mm long; stipes ca. 0.5 mm long; styles 1–2 mm long. Along streams, lakes, wetlands, glacial moraine, talus; valleys to upper subalpine. YT to SK south to CA, NM. (p.183)

The silvery felt on the leaf undersides with glaucous twigs is diagnostic; *Salix sitchensis* is similar but the leaves are more oblong, and the catkins occur on leafy branchlets.

Salix eastwoodiae Cockerell ex A.Heller Shrubs 1–2 m. **Twigs** pubescent with curly hair. **Leaf blades** 2–5 cm long, lanceolate to elliptic with glandular-serrate to entire margins, pubescent, not glaucous below. **Female catkins** 1–3 cm long, emerging with the leaves on leafy branchlets 5–15 mm long; scales dark, long-hairy. **Capsules** hairy, 3–5 mm long; stipes to 1.5 mm long; style 0.5–1.0 mm long. Moist meadows and along streams; subalpine, lower alpine. OR to MT south to CA, NV, WY.

Salix boothii has glabrous leaves and capsules; *S. wolfii* has entire leaves and shorter catkins; *S. commutata* has glabrous capsules. Our material seems intermediate between *S. commutata* and *S. wolfii*.

Salix eriocephala Michx. Shrubs 2–6 m. **Twigs** red-brown to yellow, mostly glabrous. **Leaf blades** thin, 2–10 cm long, lanceolate to elliptic with entire or toothed margins, glabrous or nearly so, glaucous beneath. **Female catkins** emerge with the leaves, 2–6 cm long, sessile or on leafy branchlets 2–15 mm long; scales dark brown, glabrous or with an apical fringe of hair. **Capsules** glabrous, 3–6 mm long; stipes 2–4 mm long; style 0.5–1 mm long. Along rivers, streams and in swamps; plains, valleys, montane. Throughout most of Canada and U.S. (p.183)

1. Year-old twigs reddish-brown . . . . .var. *mackenzieana*
1. Year-old twigs green to yellow 2

2. Mature leaves serrate, long-attenuate; catkin axis with straight hairs . . . . var. *famelica*
2. Mature leaves entire to serrulate, acute; catkin axis with tangled hairs .var. *watsonii*

Salix eriocephala var. **mackenzieana** (Hook.) Dorn [*S. rigida* Muhl. var. *mackenzieana* (Hook.) Cronquist, *S. prolixa* Andersson] has lanceolate leaves and is usually found valleys to montane; **S. eriocephala** var. **famelica** (C.R.Ball) Dorn [*S. famelica* (C.R.Ball) Argus] has acuminate-lanceolate leaves and is the common form on the plains; **S. eriocephala** var. **watsonii** (Bebb) Dorn [*S. lutea* Nutt.] has lanceolate to ovate leaves and occurs plains and valleys.

Salix exigua Nutt. SANDBAR WILLOW, COYOTE WILLOW Shrubs to 6 m. **Twigs** brown, sparsely hairy to glabrous. **Leaf blades** 3–12 cm long, linear to narrowly lanceolate with denticulate to nearly entire margins; both surfaces green, glabrous to hairy. **Female catkins** 2–5 cm long, emerging with the leaves on leafy branchlets 8–30 mm long; scales yellow, hairy, deciduous. **Capsules** 3–8 mm long, hairy or glabrous; stipes 0–2 mm long; style <0.2 mm long. Along rivers, ponds, lakes, roads; plains, valleys. AK to NB south to CA, LA, Mexico. (p.183)

1. Capsules 3–5 mm long, mostly sessile; leaves often hairy, entire to denticulate. . . . . ssp. *exigua*
1. Capsules 5–8 mm long, the stipes ≥0.5 mm long; leaves glabrate, remotely spinulose . . . . ssp. *interior*

Salix exigua ssp. **exigua** is most common in the intermountain valleys; **S. exigua** ssp. **interior** (Rowlee) Cronquist [*S. interior* Rowlee, *S. exigua* var. *pedicellata* (Andersson) Cronquist] is more common on the plains. Leaves of the closely related *S. melanopsis* are usually dark green and glabrous.

Salix farriae C.R.Ball Shrubs to 1.5 m. **Twigs** brown, pubescent to glabrous. **Leaf blades** 2–6
cm long, lanceolate to elliptic with entire margins, glaucous beneath. **Female catkins** 1–4 cm
long, emerging with the leaves on leafy branchlets 5–15 mm long; scales dark brown, long-hairy at
least above, sometimes deciduous. **Capsules** glabrous, 3–6 mm long; stipes ca. 0.5 mm long; style
ca. 0.5 mm long. Wet meadows, often near small streams, lakes; subalpine. YT to WA, ID, WY.

Plants are frequently ≤50 cm high. *Salix pseudomonticola* has leaves with toothed margins; S. *planifolia* has shiny
twigs and upper leaf surfaces. Hybrids with *S. planifolia* occur north of Lincoln. See *S. discolor* and *S. barclayi*.

Salix fragilis L. CRACK WILLOW Tree or large shrub to 25 m. **Twigs** tan, pubescent. **Leaf blades**
3–10 cm long, narrowly lanceolate with serrate margins, glabrous to sparsely hairy, glaucous
below. **Female catkins** 2–8 cm long, emerging with the leaves on leafy branchlets 1–3 cm long;
scales pale, long-hairy, deciduous. **Capsules** 3–5 mm long, glabrous; stipes 0.5–1 mm long;
styles ca. 0.5 mm long. Along rivers, streams, around ponds, lakes, wetlands, roads; plains, valleys.
Introduced to much of temperate N. America; native to Eurasia and cultivated as an ornamental or for wildlife habitat.

Our specimens have pubescent twigs and petioles but mostly glabrous leaves and may all be hybrids between *Salix*
fragilis and *S. alba* L.[114] The name *S. rubens* Schrank is available for these plants. *Salix fragilis* is sometimes considered
a hybrid between *S. alba* and *S. euxina* I.V.Belyaeva.[12]

Salix geyeriana Andersson Shrubs to 6 m. **Twigs** pubescent becoming somewhat glaucous.
Leaf blades 2–7 cm long, narrowly lanceolate to elliptic with mostly entire margins, pubescent
on both surfaces, glaucous beneath. **Female catkins** 1–2 cm long, emerging with the leaves on
leafy branchlets up to 1 cm long; scales yellow to tan, short-hairy. **Capsules** hairy, 3–6 mm long;
stipes 1–2 mm long; style <0.5 mm long. Along streams, in wet meadows, fens, swamps; valleys to
subalpine. BC to MT south to CA, NM. (p.183)

The degree of glaucous wax on the twigs is variable. Characters separating *Salix geyeriana* from *S. lemmonii* are
continuous and not always correlated, making determinations difficult.

Salix glauca L. Shrubs 20–100 cm. **Twigs** dark, pubescent. **Leaf blades** 2–5 cm long, elliptic
with entire margins, hairy to glabrous above, sparsely hairy, glaucous below. **Female catkins** 1–6
cm long, emerging with the leaves on leafy branchlets 5–20 mm long; scales brown, long-hairy.
Capsules densely hairy, 4–7 mm long; stipes 0.5–1 mm long; style 0.5–1 mm long. Moist meadows,
limestone talus slopes, often along streams or where snow lies late; subalpine, alpine. Circumpolar
south to UT, NM. Our plants are var. *villosa* Andersson. (p.183)

Leaves of *Salix commutata* are not glaucous beneath; those of *S. farriae* are less hairy; *S. barclayi* has serrate leaf
margins; see *S. brachycarpa*.

Salix lasiandra Benth. [*S. caudata* (Nutt.) A.Heller] WHIPLASH WILLOW Shrub or small tree to
15 m. **Twigs** pubescent, becoming glabrous. **Leaf blades** 3–15 cm long, lanceolate, cuneate-
based, glandular-serrate, acuminate, darker green above than below, glaucous or not. **Female
catkins** 2–8 cm long, emerging with the leaves on leafy branchlets 5–20 mm long; scales yellow,
pubescent below, deciduous. **Capsules** glabrous, 3–5 mm long; stipes ca. 1 mm long; style ≤0.2 mm
long. Along rivers, streams; plains, valleys, montane. AK to CA, NM, SD.

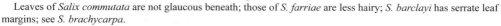

1. Leaves glaucous beneath. . . . . var. *lasiandra*
1. Leaves lighter but not glaucous beneath . . . . var. *caudata*

Salix lasiandra var. *caudata* (Nutt.) Sudw. is the common form in MT; *S. lasiandra* var. *lasiandra* has been collected in
Missoula and Sanders cos. (fide R. Dorn). This species, *S. eastwoodii* and *S. serissima* all have small glands where the
petiole meets the blade.

Salix lemmonii Bebb Shrubs to 3 m. **Twigs** glabrous, sometimes becoming glaucous. **Leaf
blades** 2–6 cm long, lanceolate, mostly with entire or sparsely denticulate margins, glabrous or
sparsely hairy, glaucous beneath. **Female catkins** 10–25 mm long, emerging with the leaves on
leafy branchlets to 1 cm long; scales dark brown to black, long-hairy. **Capsules** hairy, 4–6 mm long;
stipes 0.5–2 mm long; style 0.2–1 mm long. Streambanks, wetlands, usually in landscapes dominated
by mountain big sagebrush; subalpine. WA to MT south to CA and WY.

Salix geyeriana is similar in appearance and habitat, but has tan flower scales, shorter catkins and is much more
common and widespread.

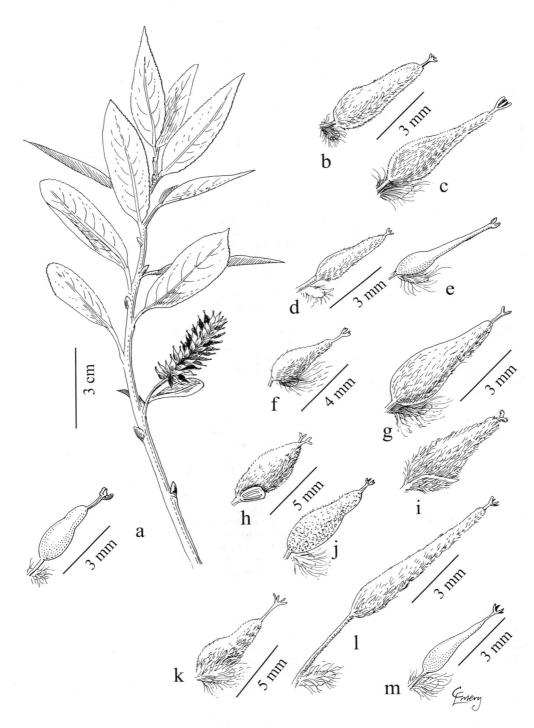

Plate 27. *Salix*. **a.** *S. boothii*, **b.** *S. glauca*, **c.** *S. scouleriana*, **d.** *S. geyeriana*, **e.** *S. wolfii* var. *wolfii*, **f.** *S. drummondiana*, **g.** *S. planifolia*, **h.** *S. reticulata*, **i.** *S. candida*, **j.** *S. exigua* ssp. *exigua*, **k.** *S. arctica*, **l.** *S. bebbiana*, **m.** *S. eriocephala* var. *mackenzieana*

Salix melanopsis Nutt. [*S. exigua* Nutt. ssp. *melanopsis* (Nutt.) Cronquist] Shrubs to 4 m. **Twigs** brown to black, pubescent. **Leaf blades** 2–12 cm long, linear to lanceolate with serrulate to entire margins, dark green on both surfaces, pubescent at first, glabrous at maturity. **Female catkins** 2–4 cm long, emerging with the leaves on leafy branchlets 5–25 mm long; scales yellow, hairy on the margins, deciduous. **Capsules** 4–6 mm long, glabrous; stipes ca. 0.5 mm long; style 0.5 mm long. Along rivers, streams, lakes; plains, valleys to most commonly montane. BC, AB south to CA, CO.

Uncommon forms of *Salix melanopsis* with hairy and/or narrower leaves can be confused with *S. exigua*.

Salix petiolaris Sm. Shrubs to 3 m. **Twigs** brown to black, glabrous. **Leaf blades** 2-15 cm long, linear-lanceolate to narrowly elliptic with serrulate margins, glaucous beneath, glabrate. **Female catkins** 10–35 mm long, emerging with the leaves on leafy branchlets 2–20 mm long; scales yellow, persistent, hairy on the margins. **Capsules** 6–8 mm long, hairy; stipes 1–5 mm long; style <0.5 mm long. Moist to wet meadows; plains. NT to NL south to WA, CO, IL, NJ. One collection from Sheridan Co.

Salix planifolia Pursh [*S. phylicifolia* L. ssp. *planifolia* (Pursh) Hiitonen] Shrubs 0.5–2 m. **Twigs** glabrous, black to reddish, shiny. **Leaf blades** 1–7 cm long, elliptic with entire margins; shiny green above, glaucous below. **Female catkins** 1–4 cm long, emerging before the leaves, sessile; scales black, long-hairy. **Capsules** 3–5 mm long, densely pubescent; stipes up to 1 mm long; style ca. 1 mm long. Along streams, wet meadows, fens; montane to lower alpine. YT to NL south to CA, NM, MN. (p.183)

The shiny twigs and upper leaf surfaces help distinguish this willow. *Salix planifolia* is sometimes divided into var. *planifolia*, a montane form, and var. *monica* (Bebb) C.K.Schneid., a high-elevation form with smaller, more broadly elliptic leaves; however, Dorn does not believe they merit taxonomic recognition.[114]

Salix pseudomonticola C.R.Ball [*S. monticola* Bebb misapplied] Shrubs1–3 m. **Twigs** brown to black, sparsely hairy, eventually glabrous. **Leaf blades** 3–8 cm long, lanceolate to ovate, rounded to truncate basally with serrulate margins, glabrous, glaucous below, often with a red midvein. **Female catkins** 2–6 cm long, sessile, emerging before the leaves; scales dark, long-hairy. **Capsules** glabrous, 4–6 mm long; stipes 0.5–1.5 mm long; style 0.5–1 mm long. Along streams, in swamps; valleys to subalpine. AK to NL south to BC, ID, WY, SD.

Salix reticulata L. S<small>NOW</small> <small>WILLOW</small> [*S. nivalis* Hook.] Mat-forming subshrub. **Twigs** dark, glabrous. **Leaf blades** 5–30 mm long, oblong to obovate with entire margins, green above, glaucous and veiny beneath, glabrous to sparsely hairy. **Female catkins** 5–15 mm long, few-flowered, emerging after the leaves on stem tips; scales yellow-green, nearly glabrous. **Capsules** hairy, 3–6 mm long, sessile; style ≤0.5 mm long. Dry fellfields to moist turf; alpine. Circumpolar south to CA, NM. Our plants are ssp. *nivalis* (Hook.) A.Löve, D.Löve & B.M.Kapoor. (p.183)

Salix arctica has larger catkins; *S. rotundifolia* has smaller leaves and glabrous capsules.

Salix rotundifolia Trautv. [*S. dodgeana* Rydb.] Dwarf, mat-forming. **Twigs** brown, glabrous. **Leaf blades** 5–10 mm long, elliptic to orbicular with entire margins, glabrous, green on both surfaces. **Female catkins** emerge with the leaves, 3–8 mm long, few-flowered, on leafy branchlets 3–8 mm long; scales brown to purple above, glabrous on the back. **Capsules** glabrous, 4–6 mm long; stipe <1 mm long; style ≤0.5 mm long. Stony, calcareous soil of moist turfy slopes; alpine. AK to WY. Our plants are var. *dodgeana* (Rydb.) A.E.Murray.

Salix scouleriana Barratt ex Hook. Shrub or rarely a small tree to 12 m tall. **Twigs** short-hairy. **Leaf blades** 2–12 cm long, obovate to elliptic, widest above the middle with entire to serrulate margins, green above, glaucous and sparsely hairy beneath. **Female catkins** 2–6 cm long, emerging with the leaves on leafless branchlets 3–7 mm long; scales dark brown, long-hairy. **Capsules** 5–8 mm long, densely short-hairy; stipes 1–2 mm long; style ca. 0.5 mm long. Coniferous and riparian forests, avalanche slopes; valleys to subalpine. AK to MB south to CA, NM, SD, Mexico. (p.183)

Our only species that is common in low- and mid-elevation, upland habitats; it often fails to flower in the shade. See *Salix sitchensis*.

Salix serissima (L.H.Bailey) Fernald A<small>UTUMN</small> <small>WILLOW</small> Shrubs 1–2 m. **Twigs** brown, shiny. **Leaf blades** 3–7 cm long, lanceolate to narrowly elliptic with glandular-serrate margins, glabrous, dark green above, glaucous beneath. **Female catkins** 15–20 mm long, emerging with the leaves on leafy branchlets 1–2 cm along; scales yellowish, densely hairy, deciduous. **Capsules**

7–8 mm long, reddish at maturity; stipes 1–2 mm long; style ca. 0.5 mm long. Wet organic soil of fens, swamps; montane. NT to NL south to CO, SD, IN, NJ.

 Our only species with catkins that flowers in spring but matures at the end of the growing season. See *Salix lasiandra*.

Salix sitchensis Sanson ex Bong. Shrubs 1–3 m. **Twigs** short-hairy. **Leaf blades** 2–9 cm long,  oblong-elliptic to obovate with entire margins, green and sparsely hairy above, densely silvery-hairy beneath. **Female catkins** 2–8 cm long, emerging with the leaves on branchlets 5–15 mm long with reduced leaves; scales brown, long-hairy. **Capsules** 3–5 mm long, short-hairy; stipes ca. 1 mm long; style 0.5–1 mm long. Along streams, lakes, wet meadows, swamps; valleys to subalpine. AK south to CA, ID, MT.

 This is our only species with 1 stamen per flower. *Salix drummondiana* has sessile catkins and usually glaucous twigs. A series of collections from near Lolo Pass appears to represent both *S. drummondiana* and *S. sitchensis* species and their hybrids; some plants have various combinations of hairy or glabrous twigs, with or without a glaucous bloom and leafy flowering branchlets. *Salix scouleriana* usually occurs in upland habitats and has leaves more sparsely hairy and glaucous beneath.

Salix tweedyi (Bebb ex Rose) C.R.Ball Shrubs to 4 m. **Twigs** stout, villous. **Leaf blades** 2–8 cm long, broadly lanceolate to ovate with glandular-denticulate margins, glabrous to sparsely hairy, glaucous below. **Female catkins** 3–9 cm long, emerging before the leaves, sessile; scales dark, long-hairy. **Capsules** glabrous, 3–7 cm long; stipes 0.2–1 mm long; style 1–2 mm long. Along streams and in wet meadows around lakes; subalpine. BC, MT south to WA, ID, WY.

 Salix commutata has shorter catkins on leafy branchlets; *S. eastwoodiae* has hairy capsules and catkins on leafy branchlets.

Salix vestita Pursh Shrubs 20–100 cm. **Twigs** brown, first puberulent then glabrous. **Leaf blades** 1–5 cm long, elliptic to ovate or obovate with entire, inrolled margins, rounded at the tip, green with impressed veins above, long, white-hairy and glaucous beneath. **Female catkins** 1–4 cm long, emerging with the leaves on branch tips; scales brown, hairy. **Capsules** 2–5 mm long, long-hairy; stipes ca. 1 mm long; style <0.2 mm long. Stony soil of wet meadows, open slopes, talus, rock ledges, more common on limestone; subalpine, alpine. BC to QC south to WA, MT. (p.176)

Salix wolfii Bebb Shrubs 40–100 cm. **Twigs** pubescent with wavy hairs. **Leaf blades** 1–5 cm long, lanceolate or elliptic with entire margins, densely to sparsely long-hairy on both surfaces, not glaucous. **Female catkins** 10–25 mm long, emerging with the leaves, sessile or on leafy branchlets to 12 mm long; scales brown with long hairs. **Capsules** glabrous or pubescent, 3–5 mm long; stipes 0.5–1 mm long; style 0.5–1 mm long. Streambanks, wet meadows, around ponds; subalpine. OR to MT south to NV, UT and CO. (p.183)

1. Capsules glabrous . . . . var. *wolfii*
1. Capsules pubescent. . . . . .var. *idahoensis*

Plants of **Salix wolfii** var. *wolfii* are less common than *S. wolfii* var. *idahoensis* C.R.Ball. Plants with sparsely hairy capsules may occur with typical var. *idahoensis*. There is little geographic or ecological integrity to the vars. Leaves of *S. boothii*, *S. commutata*, and *S. eastwoodiae* often have denticulate margins; see *S. eastwoodiae*.

CLEOMACEAE: Cleome Family

Malodorous annuals. **Leaves** alternate, petiolate, palmately 3- to 5-foliolate with entire-margined, lanceolate to narrowly elliptic leaflets and bristle-like stipules. **Inflorescence** terminal, bracteate racemes, elongating in flower. **Flowers** regular, perfect; petals 4, separate; sepals 4, separate or united below; ovary superior; pistil 1; stamens 6 to numerous. **Fruit** a stipitate, 1-celled, many-seeded, linear capsule (silique).[193]

 Formerly included in the Capparaceae.

1. Herbage glabrous; stamens 6; capsules pendent . . . . *Cleome*
1. Herbage viscid-hairy; stamens ≥12; capsules erect. . . . . *Polanisia*

Cleome L. Bee-plant

Glabrous annuals. **Leaves** petiolate, palmately 3- to 5-foliolate with bristle-like stipules. **Inflorescence** with mostly linear, setaceous bracts. **Flowers** with sepals united below and 6 stamens. **Fruit** a pendulous, long-stipitate, glabrous capsule.

Recent treatments place these taxa in the genus *Peritoma*.[409]

1. Leaflets 3; flowers pink to purple (white) . . . . *C. serrulata*
1. Leaflets 5; flowers yellow . . . . *C. lutea*

Cleome lutea Hook. [*Peritoma lutea* (Hook.) Raf.] **Stems** simple or branched above, glaucous, 20–50 cm. **Leaves** mostly 5-foliolate; the leaflets 2–5 cm long. **Flowers** yellow; calyx 2–3 mm long; petals 5–10 mm long; stamens ca. twice as long as petals. **Fruit** linear, 15–35 mm long; the stipe 4–15 mm long. Sparsely vegetated, moist, sandy soil of sagebrush or saltbush steppe, roadsides; valleys; Carbon and Big Horn cos. WA to MT south to CA, NM, NE.

Cleome serrulata Pursh [*Peritoma serrulata* (Pursh) DC.] **Stems** simple or branched, 15–80 cm. **Leaves** 3-foliolate; the leaflets 2–5 cm long. **Flowers** pink or purple, rarely white; calyx 2–3 mm long; petals 5–12 mm long; stamens 2 to 3 times as long as petals. **Fruit** broadly linear, 2–6 cm long; the stipe 11–23 mm long. Sparsely vegetated, often saline, sandy to clayey soil of grasslands, steppe, moist meadows, roadsides; plains, valleys. WA to SK south to CA, AZ, NM; introduced in eastern N. America. (p.191)

Polanisia Raf. Clammy weed

Polanisia dodecandra (L.) DC. [*P. trachysperma* Torr.& A.Gray] Glandular-hairy annual. **Stems** simple or branched below, 5–40 cm. **Leaves** petiolate, 3-foliolate; the leaflets narrowly elliptic, 1–5 cm long. **Inflorescence** with simple or 3-foliolate bracts. **Flowers** white to pink, slightly irregular; petals 5–11 mm long, clawed, the blade emarginate; sepals distinct, 2–3 mm long; stamens 12 to 27, ca. twice as long as petals. **Fruit** glandular-pubescent, broadly linear, 1–6 cm long, inflated, erect, short-stipitate. Sparsely vegetated, sandy to clayey soil of sagebrush or saltbush steppe, streambanks, roadsides; plains, valleys. OR to QC south to CA, TX, MD, Mexico. Our plants are the western var. *trachysperma* (Torr.& A.Gray) H.H.Iltis.

BRASSICACEAE: Mustard Family

Annual to perennial herbs. **Leaves** alternate or basal, without stipules. **Vesture** of simple or often branched hairs. **Inflorescence** simple or compound racemes or flowers solitary, terminal and/or axillary, usually ebracteate. **Flowers** perfect, regular, hypogynous; petals 4, separate; sepals 4; stamens usually 6 (4 long, 2 short); ovary superior, 2-celled. **Fruits** many-seeded capsules that are long and narrow (> 3 times as long as broad; siliques) to short and capsule-like (silicles), the 2 chambers separated by a thin septum.[5,193,342]

Flowers are similar across species; mature fruit is essential for identification; sometimes flower color is needed as well. The number of seeds reported includes aborted ovules. *Parrya nudicaulis* (L.) Regel has been found just south of our border in the Beartooth Range and may occur in MT. *Subularia aquatica* L., submersed with subulate leaves, occurs in Yellowstone National Park, WY and may occur in MT.

1. Fruits circular, globose to ovoid, obovoid, <3 times as long as wide (silicles) . . . . Key 1
1. Fruits linear to oblong, ≥3 times as long as wide (siliques) . . . . Key 2

Key 1 Fruits silicles <3 times as long as wide

1. Fruit definitely indented or concave on the top. 2
1. Fruit rounded to pointed on top (obscurely indented). 7

2. Vesture of leaves and stems with at least some branched or stellate hairs 3
2. Leaves and stems with simple hairs only or glabrous 5

3. Cauline leaves auriculate-clasping; base of fruit cuneate . . . . *Capsella*
3. Cauline leaves not auriculate; base of fruit rounded or cordate. 4

4. Plants annual . . . . *Alyssum*
4. Plants perennial (sometimes short-lived) . . . . *Physaria*

5. Mature fruit >10 mm across . . . . *Thlaspi*
5. Mature fruit <6 mm wide. 6

6. Seeds usually 1 per locule . . . . *Lepidium*
6. Seeds at least 2 per locule. . . . . *Noccaea*

7. Silicles not strongly compressed, round or oval in cross-section (sometimes flat when
   pressed) 8
7. Silicles strongly compressed or flattened 15

8. Silicle with a stout beak as long as the body . . . . *Euclidium*
8. Silicle sometimes with a slender style but not stout-beaked 9

9. Vestiture of leaves and stems with at least some branched or stellate hairs .10
9. Leaves and stems with simple hairs only or glabrous 12

10. Plants rosette-forming perennial; flowers yellow . . . .*Physaria*
10. Plants annual; flowers white to pale yellow 11

11. Fruit glabrous; flowers pale yellow . . . . *Camelina*
11. Fruit with branched trichomes; flowers white . . . .*Berteroa*

12. Seeds 1 to 2 per locule. . . . . *Lepidium*
12. Seeds ≥3 per locule 13

13. Fruit ≥4 mm wide; root tuberous. . . . . *Armoracia*
13. Fruit <4 mm wide 14

14. Leaves with entire margins; plants annual . . . . *Hornungia*
14. Leaf margins shallowly toothed to deeply lobed. . . . . *Rorippa*

15. Fruit 1-seeded . . . . *Athysanus*
15. Fruit with at least 1 seed per locule 16

16. Plants pubescent 17
16. Plants glabrous. 20

17. Silicles with >2 seeds per locule. . . . .*Draba*
17. Silicles with 1 to 2 seeds per locule 18

18. Vestiture of simple hairs . . . . *Lepidium*
18. Vestiture of branched, somteimes dolabriform hairs 19

19. Herbage with star-like branched hairs . . . .*Alyssum*
19. Herbage with 2-branched, dolabriform hairs . . . . *Lobularia*

20. Plants acaulescent; peduncles 1-flowered. . . . . *Idahoa*
20. Plants with many-flowered, leafy stems 21

21. Plants with a basal rosette; cauline leaves auriculate . . . . *Noccaea*
21. All leaves cauline, non-auriculate. . . . . *Iberis*

Key 2 Fruits siliques ≥3 times as long as wide

1. Silique 1-seeded, oblong, wing-like, resembling a samara . . . . *Isatis*
1. Silique with more than one seed, not samara-like 2

2. Silique with a large, usually seedless (1-seeded), beak-like, upper portion and a seed-bearing lower
   portion. 3
2. Silique beakless, often with a short style at the tip ≤1 mm long. 11

3. Plants glandular; flowers purple . . . . *Chorispora*
3. Flowers yellow or whitish; plants not glandular 4

4. Siliques strongly constricted between the seeds, with a short sterile portion at
   the base . . . . *Raphanus*
4. Siliques little or not constricted between seeds; basal sterile portion lacking 5

5. Flowers white . . . . *Streptanthella*
5. Flowers yellow 6

6. Beak ≥5 mm long 7
6. Beak ≤5 mm long 9

7. Seeds in 2 rows per locule . . . . .Eruca
7. Seeds in 1 row per locule 8

8. Beak of fruit with 1 seed at the base . . . . .Sinapis
8. Beak of fruit without a seed at the base . . . . .Brassica

9. Basal leaves lobed almost to the midvein . . . . Erucastrum
9. Basal leaves dentate to lobed less than half way to the midvein 10

10. Seeds in 1 row per locule . . . . .Brassica
10. Seeds in 2 rows per locule . . . . .Diplotaxis

11. Seeds in 2 rows in each locule (sometimes rows are imperfect) 12
11. Seeds in 1 row per locule 17

12. Siliques flattened parallel to the septum 13
12. Siliques round or 4-angled in cross-section 14

13. Siliques ≤5 times as long as wide . . . . .Draba
13. Siliques >5 times as long as wide . . . . Boechera

14. Stem leaves entire, auriculate-clasping 15
14. Stem leaves not both entire and auriculate 16

15. Plants ≤15 cm high; petals 2–3 mm long . . . . .Eutrema
15. Plants usually >15 cm high; petals 4–6 mm long . . . . Turritis

16. Vesture with some branched hairs . . . . .Transberingia
16. Vesture of simple hairs or plants glabrous . . . . Rorippa

17. Siliques flattened 18
17. Siliques round or 4-angled in cross-section 21

18. Basal leaves entire to shallowly dentate . . . . Boechera
18. Basal leaves deeply lobed or divided into leaflets 19

19. Vesture of dense simple and branched hairs . . . . Smelowskia
19. Plants glabrous or sparsely hairy with simple hairs . . . . .Cardamine

20. Flowers pale yellow; plants sparsely hairy; style of mature fruit ≥1 mm long . . . . .Diplotaxis
20. Flowers white; plants often glabrous; style ≤1 mm long . . . . .Cardamine

21. Plants glabrous or with sparse hair only at the base of the stem 22
21. Plants sparsely to densely hairy on the lower half at least 27

22. Basal leaves pinnately divided . . . . Barbarea
22. Basal leaves with entire or sinuate margins 23

23. Leaves linear . . . . Schoenocrambe
23. Leaf blades lanceolate to obovate or triangular 24

24. Petals cream to pale yellow; plants annual . . . . Conringia
24. Petals white to purple; plants biennial to perennial 25

25. Cauline leaf blades deltoid to reniform with crenate to dentate margins . . . . Alliaria
25. Cauline leaf blades narrower and entire-margined 26

26. Basal leaf blades lyrate with crenate margins . . . . Arabidopsis
26. Basal leaf blades entire . . . . Thelypodium

27. Flowers yellow 28
27. Flowers white, rose, blue or purple 31

28. Siliques with a stipe ≥1 cm long . . . . .Stanleya
28. Siliques with little or no stipe 29

29. Leaf margins entire to sinuate . . . . .Erysimum
29. Leaves deeply lobed or divided 30

30. Leaves 2–3 times pinnately lobed . . . . Descurainia
30. Leaves once pinnately lobed; the lobes sometimes shallowly dentate . . . . Sisymbrium

31. Leaves lobed nearly to the midvein . . . . *Smelowskia*
31. Leaf margins entire to dentate not more than halfway to the midvein 32

32. Flowers rose to blue or purple 33
32. Flowers white, sometimes fading bluish. 34

33. Vesture of sparse, usually minute, simple hairs . . . . *Hesperis*
33. Vesture of branched hairs. . . . . *Malcolmia*

34. Basal leaves lobed ca. halfway to the midvein. . . . . *Arabidopsis*
34. Margins of basal leaves entire to shallowly lobed 35

35. Petals 2–3 mm long . . . . *Arabidopsis*
35. Petals ≥4 mm long 36

36. Leaves of lower stem entire or slightly crenate . . . . *Arabis*
36. Leaves of lower stem shallowly lobed . . . . *Sandbergia*

Alliaria Heist. ex Fabr.

Alliaria petiolata (M.Bieb.) Cavara & Grande GARLIC MUSTARD Nearly glabrous biennial. **Stems** erect, mostly simple, 15–90 cm. **Leaves** basal and cauline, petiolate; basal blades, reniform, crenate; cauline blades sagittate, dentate, 3–6 cm long. **Inflorescence** an ebracteate raceme, expanding in fruit. **Flowers** white; petals 3-8 mm long. **Fruits** linear siliques, 3-7 cm long, often contracted between seeds; style 1–2 mm long; seeds 3 to 11 per locule; pedicels spreading to ascending, 3–10 mm long. Disturbed soil of woodlands, along roads or near habitations; plains, valleys; collected in Missoula and Wibaux cos. Introduced commonly in temperate eastern N. America, sparingly in west.

Alyssum L. Alyssum

Annuals. **Stems** branched below. **Vesture** of stellate hairs. **Leaves** cauline, oblong with entire margins. **Flowers** yellow, ca. 2 mm long, falling upon opening. **Fruits** circular silicles, flattened parallel to the septum, ca. 3 mm long; seed 1to 2 per locule. Introduced from Europe.

1. Silicles glabrous . . . . *A. desertorum*
1. Silicles pubescent. 2

2. Style ≥1 mm long; silicle with a broad flattened margin; sepals deciduous . . . . *A. simplex*
2. Style ca. 0.5 mm long; margin of silicle narrow; sepals persistent. . . . . *A. alyssoides*

Alyssum alyssoides L. **Stems** to 20 cm. **Leaves** to 2 cm long. **Fruit** pubescent with stellate hairs, narrowly margined; style ca. 0.5 mm long. Grasslands, steppe, open forest, fields, streambanks, roadsides; plains, valleys to montane. Introduced to most of temperate N. America.

Alyssum desertorum Stapf **Stems** to 15 cm. **Leaves** to 2 cm long. **Fruit** glabrous, conspicuously wing-margined; style 0.5–1 mm long. Disturbed soil of grasslands, fields, streambanks, roadsides; plains, valleys. Introduced BC to MB south to CA, UT, MO.

Alyssum simplex Rudolphi [*A. parviflorum* Fisch. ex M.Bieb.] **Stems** 6–20 cm. **Leaves** 1–3 cm long. **Fruit**, 4–6 mm long with a broad winged margin, pubescent with stellate hairs; style 1–1.5 mm long. Disturbed soil of roadsides, open slopes; valleys. Introduced in OR to NE south to CA, NM, KS.

Arabidopsis (DC.) Heynh

Annual or perennial herbs. **Vesture** of simple to stellate hairs. **Leaves** cauline and basal, simple with entire to lobed margins, not auriculate. **Flowers** white to purplish. **Fruit** erect to ascending, glabrous siliques; seeds in 1 row per locule.

1. Basal leaves with entire margins . . . . *A. thaliana*
1. Some basal leaves with lobed margins . . . . *A. lyrata*

Arabidopsis lyrata (L.) O'Kane & Al-Shehbaz [*Arabis lyrata* L., *Arabis kamchatica* (Fisch. ex DC.) Ledeb.] Short-lived perennial. **Stems** to 25 cm from a branched or simple caudex. **Ves,

ture** glabrous or of simple to few-branched hairs. **Basal leaves** oblong with shallowly lobed margins, 1–5 cm long, petiolate. **Stem leaves** oblong, few-lobed or entire-margined. **Flowers** white to purplish; petals 5–8 mm long. **Fruit** 1–4 cm × 1–2 mm. Sparsely vegetated soil of rocky slopes, open forest; montane, lower subalpine. AK to QC, south to WA, MT, MN, GA. Our plants are var. *kamchatica* (Fisch. ex DC.) O'Kane & Al-Shehbaz. One collection from Glacier National Park in Flathead Co.

Arabidopsis thaliana (L.) Heynh. Annual. **Stems** simple or branched, 10–30 cm. **Vesti,

ture**: simple on the stems; simple and branched on the leaves. **Basal leaves** petiolate, oblong to spatulate, 5–25 mm long; margins entire to remotely dentate. **Stem leaves** narrowly oblong to linear. **Flowers** white; petals 2–3 mm long. **Fruit** 8–15 × 0.5–1 mm. Sparsely vegetated soil of grasslands, sagebrush steppe, open forest, roadsides; valleys, montane. Introduced to most of temperate N. America; native to Eurasia.

Arabis L. Rockcress

Taprooted biennial or short-lived perennial. **Leaves** basal and cauline; the basal petiolate. **Vesiture** of simple to stellate hairs. **Inflorescence** an ebracteate raceme. **Flowers** white. **Fruits** erect or ascending siliques, glabrous, nearly terete; seeds winged or not, 1 row in each locule.[193]

Molecular genetic studies have shown that the mostly Eurasian genus *Arabis* (sensu stricto) is distinct from the mainly N. American *Boechera*.[222,312]

1. Cauline leaves auriculate; fruiting pedicels erect; fruits appressed to the stem . . . . *A. hirsuta*
1. Cauline leaves not auriculate; fruiting pedicels spreading; fruits held away from stem . . . . *A. nuttallii*

Arabis hirsuta (L.) Scop. [*A. pycnocarpa* M.Hopkins, *A. eschscholtziana* Andrz.] Biennial or short-lived perennial. **Stems** 15–60 cm, simple or branched at the base. **Basal leaves** oblong with wavy to serrate margins, 2–12 cm long. **Stem leaves** lanceolate, auriculate-clasping. **Vesiture** of simple or branched hairs below, glabrous above. **Petals** 4–9 mm long. **Fruits** 2–5 cm × ≤1 mm, strictly erect. Sparsely vegetated, often disturbed, often stony soil of meadows, open forest; valleys to subalpine. AK to QC south to CA, NM, MN, GA. Our plants are var. *pycnocarpa* (M.Hopkins) Rollins.

Characters used to recognize multiple species in *Arabis hirsuta* sensu lato are continuous, overlapping and not always correlated in our material. One specimen from Sanders Co. has the larger flowers of var. *glabrata* Torr.& A.Gray.

Arabis nuttallii (Kuntz) B.L.Rob. [*A. bridgeri* Jones] Perennial from a simple to branched caudex. **Stems** 4–30 cm, unbranched. **Basal leaves** spatulate, to 3 cm long with entire to crenate margins. **Stem leaves** lanceolate, not auriculate. **Vesiture**: basal leaves glabrous, ciliate or with coarse simple and forked hairs; cauline leaves ciliate. **Petals** 5–8 mm long. **Fruits** 1–4 cm × 1–2 mm, erect on spreading stalks. Vernally moist soil of grasslands, steppe, meadows, open forest; valleys to alpine. BC, AB south to WA, UT, WY. (p.191)

Plants that may have sparse cilia on somewhat succulent basal leaves but are otherwise glabrous occur on stony, calcareous soil near or above treeline. These plants have been called *A. bridgeri*. Rollins believes this is a sporadic, high-elevation form of *Arabis nuttallii* not worthy of taxonomic recognition;[342] however, Gerald Mulligan believes it may be a polyploid derivative of *A. nuttallii*.

Armoracia P.Gaertn.,B.Mey.& Scherb. Horseradish

Armoracia rusticana P.Gaertn.,B.Mey.& Scherb. Glabrous perennial with a thickened taproot. **Stems** erect, branched, 60–100 cm. **Leaves**: the basal petiolate; the blade lanceolate, shallowly lobed and dentate, 12–27 cm long; lower stem leaves pinnatifid into narrowly oblanceolate segments, becoming similar to the basal leaves above. **Inflorescence** an ebracteate raceme. **Flowers** white; petals 4–6 mm long. **Fruit** a subglobose silicle, inflated, 4–6 mm long, glabrous; style ca. 0.5 mm long; seeds in 2 rows per locule; pedicel ascending 8–11 mm long. Fields, roadsides; plains, valleys. Introduced and escaped from cultivation; native to Europe.

Athysanus Greene

Athysanus pusillus (Hook.) Greene Annual. **Stems** slender, usually branched at the base, 5–15 cm. **Leaves** on the lower stem, 4–8 mm long, oblong to obovate with few broad, marginal teeth; short-petiolate below, sessile above. **Vesiture** of simple or 2- to 4-rayed hairs. **Inflorescence** an open raceme of 6 to 20 flowers. **Flowers** white; sepals 1

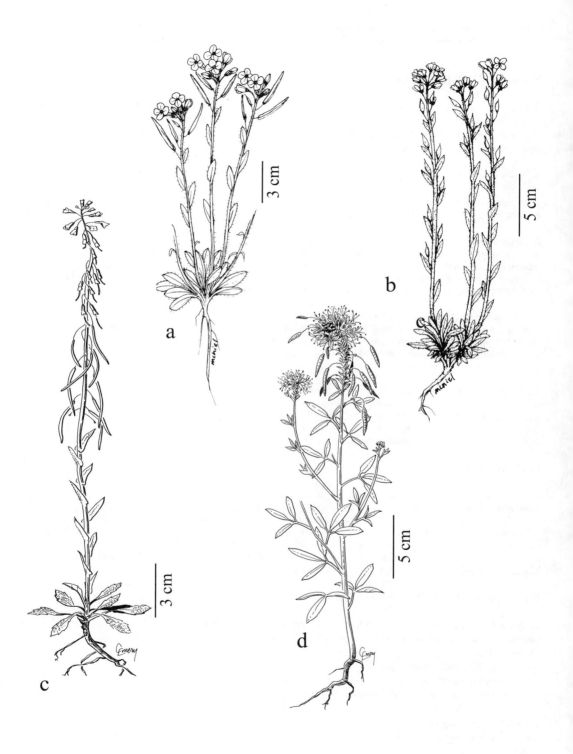

Plate 28. a. *Arabis nuttallii*, b. *Draba aurea*, c. *Boechera retrofracta*, d. *Cleome serrulata*

mm long, deciduous; petals 1–2 mm long or lacking. **Fruit** compressed-ovoid, 1–2 mm wide, hirsute with curved hairs, 1-seeded. Vernally moist, shallow soil of grasslands, cliff ledges; montane. BC, MT south to CA. Known from Ravalli Co.

Barbarea W.T.Aiton Wintercress

Glabrous to sparsely hairy, biennial or short-lived perennial. **Stems** angled, often branched. **Basal leaves** long-petiolate, pinnately divided with a large, ovate terminal lobe; **Stem leaves** auriculate, clasping, lobed to pinnate. **Inflorescence** an ebracteate raceme. **Flowers** yellow. **Fruit** a silique, round or 4-angled in cross-section; seeds wingless, 1 row per locule.

1. Petals 3–6 mm long . . . . *B. orthoceras*
1. Petals 6–8 mm long 2

2. Basal leaves with 1 to 4 pairs of lateral lobes. . . . . *B. vulgaris*
2. Basal leaves with 4 to 10 pairs of lateral lobes. . . . . *B. verna*

Barbarea orthoceras Ledeb. **Stems** erect, 20–50 cm tall. **Basal leaves** 4–12 cm long, oblong with 1 to 6 pairs of lateral lobes with sinuate margins. **Stem leaves** pinnately lobed or divided, reduced upward. **Petals** 3–6 mm long. **Fruits** 15–25 mm long, ≤1 mm wide, 4-angled, erect or spreading, with a beak 2 mm long. Moist soil of thickets, wet meadows, along streams, wetlands, open forest; valleys to lower subalpine. AK to NL south to CA, AZ, MN, NH; Asia.

Barbarea verna (Mill.) Asch. **Stems** erect, 20–80 cm tall. **Basal leaves** 4–12 cm long, oblong with 4–10 pairs of lateral lobes with sinuate margins. **Stem leaves** pinnately lobed or divided, reduced upward. **Petals** 6–8 mm long. **Fruits** 4–8 cm mm long, 1.5–2 mm wide, 4-angled, erect or spreading, with a beak ca. 1.5 mm long. Fields, ditchbanks, roadsides; valleys. Garden herb introduced in eastern and western temperate N. America ; native to Europe. Known from Gallatin Co. (*Whitham s.n.* RM).

Barbarea vulgaris W.T.Aiton Yᴇʟʟᴏᴡ ʀᴏᴄᴋᴇᴛ **Stems** erect, 25–70 cm tall. **Basal leaves** 4–10 cm long, petiolate blades oblanceolate with 1 to 4 pairs of lateral lobes, the margins entire to sparsely toothed. **Stem leaves** becoming lobed upward. **Petals** 6–8 mm long. **Fruits** ascending, 1–4 × 1–1.5 mm, terete with a beak-like style 2–3 mm long. Moist soil of disturbed meadows, roadsides, old fields; valleys. Introduced to most of U.S.; native to Europe.

Berteroa DC.

Berteroa incana (L.) DC. Annual. **Stems** erect, 20–75 cm, usually branched below and above. **Basal leaves** petiolate, oblanceolate with entire margins, soon deciduous. **Stem leaves** similar, sessile, erect, 1–3 cm long, reduced above. **Vestiture** of dense, stellate, appressed trichomes. **Inflorescence** a narrow, simple or compound, many-flowered raceme. **Flowers** white; sepals 2–3 mm long; petals 4–6 mm long. **Fruit** elliptic, inflated, 4–7 mm long, with stellate trichomes; mature style 1–2 mm long; pedicels 4–7 mm long; seeds 3–7 per locule. Fields, roadsides; plains, valleys, montane. Introduced in scattered locations across N. America; native to Europe.

Boechera A.Löve & D.Löve American rockcress

Taprooted biennials or perennials with simple or branched caudex. **Stems** simple. **Leaves** basal and cauline; the basal petiolate. **Vestiture** of simple to stellate hairs. **Inflorescence** an ebracteate raceme. **Flowers** white to purple. **Fruits** siliques, on erect to reflexed pedicels, glabrous, nearly terete with 1 or 2 rows of seeds in each chamber.[193,299,342,448-450]

A mainly N. American genus, all species were formerly placed in *Arabis*.[222] The posture of the siliques, which can change with maturity, is a critical character separating the taxa. There are four recent monographs of this group,[193,299,342,448-450] and there are numerous inconsistencies among them, due, at least in part, to hybridization and apomixis.[112,449,450] For example, *Boechera lyallii* is reported to have auriculate leaves,[448] auriculate or non-auriculate leaves,[193] or all non-auriculate.[342] This confusion makes the following treatment tentative.

Boechera lyallii and *B. microphylla* are distinguished mainly by the width of the siliques. Many high-elevation collections lack mature fruit and have been tentatively assigned to *B. lyallii* because it appears that *B. microphylla* is usually found lower. Recent monographic work segregated four species from what was formerly *Arabis holboellii*: *B. pendulocarpa*, *B. retrofracta*, *B. polyantha* and *B. collinsii*;[448] however, only the first two are now thought to occur in MT (fide M. Windham).

Flowering plants may have one of two forms depending on whether stems are axillary, arising from buds subtended by basal leaves, or terminal, arising from the terminal bud of the rosette.[248]

1. Siliques spreading (horizontal) to reflexed down 2
1. Siliques ascending (above horizontal) to erect. 9

2. Basal leaves often blunt-tipped, felt-like with dense branched hairs . . . . *B. lemmonii*
2. Basal leaves otherwise. 3

3. Outer stems decumbent-based, arising from lateral buds; basal leaves erect. . . . .*B. demissa*
3. Stems arising from terminal bud of the rosette. 4

4. Stem leaves without auricles . . . . *B. pendulocarpa*
4. Stem leaves auriculate 5

5. Stems ≤1 mm wide in the middle; siliques ≤1.5 mm wide . . . . *B. microphylla*
5. Most stems >1 mm wide; siliques often >1.5 mm wide 6

6. Pedicels of mature siliques reflexed down; margins of stem leaves rolled under . . . . *B. retrofracta*
6. Mature siliques on spreading or arching pedicels; stem leaves not rolled .7

7. Pubescence of lower stems sparse and 2- to 3-branched . . . .*B. divaricarpa*
7. Pubescence of lower stems more branched 8

8. Basal leaves often dentate; pedicels with some branched hairs . . . .*B. pauciflora*
8. Basal leaves entire; pedicels glabrate or with only simple hairs. . . . . *B. sparsiflora*

9. Cauline leaves whitish with dense hair. . . . . *B. fecunda*
9. Cauline leaves not whitish-hairy. 10

10. Siliques spreading to ascending. 11
10. Siliques erect 14

11. Caudex usually branched and woody. . . . . *B. microphylla*
11. Caudex usually simple or few-branched, not woody 12

12. Pubescence of lower stems sparse and 2- to 3-branched . . . .*B. divaricarpa*
12. Pubescence of lower stems more branched 13

13. Basal leaves often dentate; pedicels with some branched hairs . . . .*B. pauciflora*
13. Basal leaves entire; pedicels glabrate or with only simple hairs. . . . . *B. sparsiflora*

14. Seeds in 2 rows per locule; flowers sometimes white 15
14. Seeds in 1 row per locule; flowers purple. 17

15. Basal leaves with stalked, 2- to 4-branched hairs . . . . *B. calderi*
15. Basal leaves glabrous or with appressed 2-rayed hairs. 16

16. Caudex usually simple or few-branched, not woody . . . .*B. stricta*
16. Plants with a woody, usually branched caudex . . . . *B. lyallii*

17. Fruits 1–1.5 mm wide; valleys to montane. . . . . *B. microphylla*
17. Fruits 2–3 mm wide; high elevations 18

18. Caudex usually simple or few-branched, not woody . . . . *B. saximontana*
18. Plants with a woody, usually branched caudex . . . . *B. lyallii*

Boechera calderi (G.A.Mulligan) Windham & Al-Shehbaz [*Arabis calderi* G.A.Mulligan]
Biennial or perennial with a simple or rarely few-branched caudex. **Stems** 10–60 cm, terminal.
Vestiture of 3- to 4-branched hairs at the base of the plant. **Basal leaves** oblanceolate to oblong
with entire or dentate margins, 15–40 mm long. **Stem leaves** lanceolate, auriculate-clasping.
Flowers purple, sometimes pale; petals 7–10 mm long. **Fruit** 2–8 cm × 1–2 mm, erect and appressed
to the stem; seeds in 2 rows per locule. Sparsely vegetated, often moist soil of open slopes, cliffs, rock slides, open
forest; valleys to lower alpine. YT to AB south to CA, WY.

Basal leaves of *Boechera stricta* are glabrous or with appressed, 2-branched trichomes.

Boechera demissa (Greene) W.A.Weber [*Arabis demissa* Greene, *Boechera demissa* var. *languida* (Rollins) Dorn] Perennial from a simple or branched caudex. **Stems** 7–20 cm, arising from lateral buds of the rosette. **Basal leaves** narrowly oblong, 5–20 mm long. **Stem leaves** lanceolate, slightly auriculate. **Vestiture** of coarse, branched hairs and marginal cilia, glabrous above. **Flowers** white to pale purple; petals 3–5 mm long. **Fruit** 2–3 cm × 1–2 mm, arcuate, pendulous at maturity; seeds wingless or nearly so, in 1 row per locule. Sparsely vegetated, shallow, calcareous soil of sagebrush steppe; valleys. NV to MT south to AZ, UT, CO.

Our collections are from Carbon Co. and were determined to be *Arabis demissa* var. *languida* Rollins [= *Boechera languida* (Rollins) Windham & Al-Shehbaz] which has auriculate leaves, 3- to 4-rayed trichomes, 10 to 20 flowers, and fruits 30–45 mm long;[450] our plants might just as well be referred to var. *demissa*.

Boechera divaricarpa (A.Nelson) A.Löve & D.Löve [*B. brachycarpa* (Torr.& A.Gray) Dorn, *Arabis divaricarpa* A.Nelson, *A. confinis* S.Watson var. *brachycarpa* (Torr.& A.Gray) S.Watson & J.M.Coult] Biennial or short-lived perennial (monocarpic) with a simple caudex. **Stems** to 80 cm, terminal. **Basal leaves** petiolate, oblanceolate, 1–4 cm long. **Stem leaves** narrowly lanceolate, auriculate-clasping. **Vestiture** appressed simple and 3-branched hairs on lower stems and basal leaves, glabrous above. **Flowers** pink to purple; petals 7–11 mm long. **Fruit** 15–80 mm × 1.5–2 mm, straight to slightly arcuate; pedicels glabrous, spreading at maturity; seeds with an apical wing 0.3 mm wide, in 1 or 2 irregular rows. Often shallow soil of open forest, steppe, grasslands, rock outcrops; montane, lower subalpine. AK to QC south to CA, CO, IA, VT.

This species seems intermediate between *Boechera stricta* and *B. sparsiflora* in both vestiture and posture of siliques; it is thought to be a hybrid that has arisen on numerous independent occasions between *B. stricta* and *B. retrofracta* or *B. sparsiflora*.[112,342] *Boechera grahamii* (Lehm.) Windham & Al-Shehbaz, a hybrid between *B. stricta* and *B. collinsii*, similar to *B. divaricarpa* but with white flowers and narrower siliques, is reported for MT[450] but I have not found any voucher specimens.

Boechera fecunda (Rollins) Dorn [*Arabis fecunda* Rollins] Perennial from a simple to rarely branched caudex. **Stems** 4–25 cm, terminal or axillary. **Basal** leaves spatulate to oblanceolate, 5–20 mm long with entire or few-toothed margins. **Stem leaves** oblong, auriculate on terminal stems but not on axillary stems. **Vestiture** whitish, of dense stalked, few-branched hairs. **Petals** white or tinged with purple, 6–12 mm long. **Fruits** ascending to erect, 3–6 cm × 1–1.5 mm, constricted between seeds; seeds wing-margined, in 1 row per locule. Rocky, calcareous soil of open slopes in grassland, steppe, woodlands, open forest; montane to lower subalpine. Endemic to MT.

The type locality is in Ravalli Co.

Boechera lemmonii (S.Watson) W.A.Weber [*Arabis lemmonii* S.Watson, *A. drepanoloba* Greene, *Boechera drepanoloba* (Greene) Windham & Al-Shehbaz, *B. lemmonii* var. *drepanoloba* (Greene) Dorn] Perennial from a branched caudex. **Stems** 4–25 cm, terminal and axillary. **Basal leaves** spatulate to oblanceolate, 5–25 mm long with mostly entire margins. **Stem leaves** lanceolate, weakly auriculate. **Vestiture** whitish, of dense, branched hairs. **Petals** purple, 5–7 mm long. **Fruits** spreading to pendulous, 2–5 cm × 1–2.5 mm, mostly secund; seeds narrowly winged in 1 row per locule. Stony, sparsely vegetated soil of moraine, talus slopes, ridges, streambanks; montane to alpine. BC, AB south to CA, CO.

Two vars. of *Boechera lemmonii* are usually recognized in MT, var. *lemmonii* and var. *drepanoloba*, with broader siliques and longer stems; however, the characters are continuous and not strongly correlated in our area.[193]

Boechera lyallii (S.Watson) Dorn [*Arabis lyallii* S.Watson, *A. murrayi* G.A.Mulligan] Perennial from a simple or branched caudex. **Stems** 3–20 cm, terminal. **Basal** leaves oblanceolate, 1–3 cm long, fleshy with entire margins. **Stem leaves** lanceolate to ovate, auriculate or not. **Vestiture** of sparse, simple to branched hairs on basal leaves or plants glabrous. **Petals** purple, 6–10 mm long. **Fruits** ascending to erect, 2–5 cm × 2–3 mm; seeds wing-margined, in 1 or 2 rows per locule. Stony soil and cobble of fellfields, moraine; subalpine, alpine, often in areas of late snow release. BC, AB south to CA, UT, WY.

Our material can be divided into two forms: nearly glabrous plants [=*Arabis lyallii* var. *lyallii*] occur in the northwest and south-central regions, while plants with tiny stalked-branched hairs on the basal leaves [=*A. lyallii* var. *nubigena* (J.F.Macbr.& Payson) Rollins] occur in the southwest ranges; both forms occur in the Bitterroot Range. *Boechera stricta* may also have glabrous foliage with erect siliques, but these have 2 rows of seeds per locule, and the flowers are usually white. *Arabis murrayi* was segregated based on the non-auriculate stem leaves and more erect fruits;[299] however, these two characters can vary on the same plant.

Boechera microphylla (Nutt.) Dorn [*B. macounii* (S.Watson) Windham & Al-Shehbaz, *Arabis microphylla* Nutt.] Perennial, sometimes mat-forming from a simple to branched caudex. **Stems** 10–30 cm, slender, terminal. **Basal leaves** linear to narrowly oblanceolate, 7–20 mm long with entire or few-toothed margins. **Stem leaves** lanceolate, weakly auriculate. **Vestiture** of sparse, simple to branched hairs on basal leaves, glabrous above. **Petals** pink to purple, 4–6 mm long. **Fruits** spreading to erect, 2–5 cm × 1–1.5 mm; seeds wing-margined, in 1 row per locule. Stony soil of cliffs, grasslands, open forest. BC, MT south to CA, UT.

1. Siliques erect to ascending. . . . . var. *microphylla*
1. Siliques spreading . . . . var. *macounii*

Boechera microphylla var. **microphylla** is found in rocky soil with partial shade, valleys to montane; ***B. microphylla*** var. *macounii* (S.Watson) Dorn occurs in stony soil and on cliffs, valleys to subalpine.

Boechera pauciflora (Nutt.) Windham & Al-Shehbaz [*B. sparsiflora* (Nutt.) Dorn var. *subvillosa* (S.Watson) Dorn, *Arabis sparsiflora* Nutt. var. *subvillosa* (S.Watson) Rollins] Biennial or short-lived perennial with a simple to few-branched caudex. **Stems** 20–50 cm, terminal. **Basal leaves** oblanceolate, 2–4 cm long with dentate margins. **Stem leaves** linear-lanceolate, auriculate. **Vesture** of dense, small, branched hairs on basal leaves and stem base, becoming glabrous above. **Petals** purple, 5–10 mm long. **Fruits** 25–50 × ≤1 mm, arcuate; pedicels with branched hairs, ascending to weakly descending; seeds narrowly wing-margined, in 1 row per locule. Open forest, cliffs, streambanks; valleys; collected in Missoula Co. BC, MT south to CA, NV, UT, WY.
    Leaves of the similar *Boechera sparsiflora* are usually entire, and its pedicels are glabrous or with simple hairs.

Boechera pendulocarpa (A.Nelson) Windham & Al-Shebaz [*B. exilis* (A.Nelson) Dorn, *Arabis holboellii* Hornem. var *pendulocarpa* (A.Nelson) Rollins] Perennial from a simple or branched caudex. **Stems** 3–25 cm, terminal. **Basal leaves** oblanceolate, 5–17 mm long with entire margins. **Stem leaves** lanceolate, not auriculate. **Vesture** of dense, fine, branched hairs. **Petals** white to purple, 4–8 mm long. **Fruits** pendulous, straight to slightly curved, 2–5 cm × 1.5–2 mm; seeds narrowly wing-margined, in 1 row per locule. Stony soil of grasslands, steppe, rock outcrops; valleys, montane. BC, MT south to CA, UT, CO.
    Similar to *Boechera retrofracta*, but the non-auriculate stem leaves are diagnostic.

Boechera retrofracta (Graham) A.Löve & D.Löve [*B. holboellii* (Hornem.) A.Löve & D.Löve var. *secunda* (Howell) Dorn, *Arabis holboelii* Hornem. var. *secunda* (Howell) Rollins, *A. retrofracta* Graham, *A. holboellii* var. *retrofracta* (Graham) Rydb.] Biennial or short-lived perennial with a simple or sometimes branched caudex. **Stems** to 70 cm, terminal. **Vesture** of dense, fine, branched hairs on basal leaves and stems. **Basal leaves** oblong to oblanceolate with entire to toothed margins, 1–3 cm long. **Stem leaves** narrowly lance-shaped, auriculate-clasping. **Flowers** white to purple; petals 5–9 mm long. **Fruit** 3–8 cm × 1.5–2 mm, straight to slightly curved; pedicels gently to sharply reflexed, glabrous or hairy; seeds narrowly wing-margined, in 1 row per locule. Grasslands, steppe, open forest; valleys to montane. AK to QC south to CA, NV, UT, CO, SD, MI. See *B. pendulocarpa.*(p.191)

Boechera saximontana (Rollins) Windham & Al-Shehbaz [*B. williamsii* (Rollins) Dorn var. *saximontana* (Rollins) Dorn, *Arabis microphylla* Nutt. var. *saximontana* Rollins] Perennial from a usually simple or weakly branched caudex. **Stems** 10–25 cm, terminal. **Basal** leaves oblanceolate, 7–20 mm long with entire margins. **Stem leaves** narrowly lanceolate to narrowly elliptic, auriculate. **Vesture** of sparse to moderately dense, mostly stalked, branched hairs on basal leaves, becoming glabrous above. **Petals** purple, 4–6 mm long. **Fruits** ascending, 2–4 cm long, ca. 2 mm wide; seeds wing-margined, in ca. 1 row per locule. Gravelly soil of alpine slopes; reported for Park Co. ID, MT, WY.
    Boechera lyallii has more erect siliques.

Boechera sparsiflora (Nutt.) Dorn [*Arabis sparsiflora* Nutt.] Biennial or short-lived perennial with a simple to rarely sparingly branched caudex. **Stems** to 75 cm, terminal. **Basal leaves** oblanceolate, 1–4 cm long with entire margins. **Stem leaves** lanceolate, slightly auriculate. **Vesture** of dense, small, branched hairs on basal leaves and stem base, glabrous above. **Petals** white or purple, 5–10 mm long. **Fruits** 3–8 cm × 1.5–2 mm, arcuate; pedicels glabrous or with simple hairs, ascending to weakly descending; seeds narrowly wing-margined, in 1 row per locule. Grasslands and dry, open forests; valleys, montane. BC, MT south to CA, UT, WY.
    Boechera sparsiflora has finer, more branched hairs in the basal leaves and lower stem than *B. divaricarpa*

Boechera stricta (Graham) Al-Shebaz [*Arabis drummondii* A.Gray] Biennial or perennial with a simple or branched caudex. **Stems** 8–60 cm, terminal. **Vesture** lacking or of sparse, sessile, 2-branched hairs at the base of the plant. **Basal leaves** narrowly oblanceolate with entire or weakly dentate margins, 2–6 cm long. **Stem leaves** narrowly lanceolate, auriculate-clasping. **Flowers** white to purplish; petals 6–10 mm long . **Fruit** 2–8 cm × 1–3 mm, erect and appressed to the stem; seeds in

2 rows per locule. Sparsely vegetated, often stony soil of meadows, grasslands, rock slides, open forest, often associated with local disturbance such as ground squirrel digging; valleys to alpine. AK to NL south to CA, CO, OH, DE. (p.198)

Boechera lyallii is also glabrous with erect fruits, but the plants are more likely to have a branched caudex. See *B. divaricarpa*, *B. calderi*.

Brassica L. Mustard

Annual or biennial. **Leaves** basal and cauline; the lower petiolate, pinnately divided or lobed; the upper sessile and less divided. **Vestiture** of simple hairs or plants glabrous and sometimes glaucous. **Inflorescence** an ebracteate raceme. **Flowers** yellow, often with dark veins. **Fruit** a silique, erect or ascending with a distinct, sterile 1-nerved beak at the tip; seeds in 1 row per locule.

Introduced from Europe. Species have been domesticated to various degrees and cultivated for seed or vegetables such as rapeseed, kale, and cauliflower; seeds of *Brassica nigra* and *B. juncea* are ground for "hot" mustard. All species have escaped to become weeds of fields and roadsides throughout temperate N. America.

1. Stem leaves sessile and auriculate . . . .*B. rapa*
1. Stem leaves petiolate or sessile but not auriculate 2

2. Body of silique 2–4 cm long . . . . *B. juncea*
2. Silique body 8–15 mm long . . . . *B. nigra*

Brassica juncea (L.) Czern. [*Raphanus junceus* (L.) Crantz] LEAF MUSTARD Glabrous to sparsely hairy annual. **Stems** erect, branched, 20–80 cm. **Basal leaves** lyrate-pinnate with dentate margins, early deciduous. **Stem leaves** 4–15 cm long, petiolate below, becoming merely lobed and sessile above. **Petals** 7–10 mm long. **Fruits:** the body 2–4 cm long, strongly 2-nerved; the beak 6–10 mm long, conical. Fields, roadsides, disturbed, moist soil; valleys.

Brassica nigra (L.) W.D.J.Koch BLACK MUSTARD Annual. **Stems** erect, branched above, 40–100 cm. **Basal leaves** lyrate-pinnate to lobed with dentate margins, early deciduous. **Stem leaves** reduced above, becoming ovate. **Vestiture** hirsute below to glabrous. **Petals** 6–9 mm long. **Fruits:** strictly erect; the body 8–15 mm long, strongly 2-nerved and angled; the beak 1–3 cm long, conical; pedicel ca. 2 mm long. Fields, roadsides; valleys.

Brassica rapa L. [*B. campestris* L.] FIELD MUSTARD Mostly glabrous and glaucous annual. **Stems** erect, branched above, 30–100 cm. **Basal leaves** lyrate-pinnate with 2–6 lateral lobes and sinuate margins, early deciduous. **Stem leaves** lanceolate with entire margins, 3–10 cm long, becoming sessile, auriculate clasping. **Petals** 7–13 mm long. **Fruits:** the body 3–6 cm long, strongly 2-nerved; the beak 10–15 mm long, conical. Fields, roadsides; valleys.

Braya Sternb.& Hoppe

Braya humilis (C.A.Mey.) R.L.Rob. Short-lived perennial from a usually unbranched caudex. **Stems** 5–10 cm. **Basal leaves** in a rosette, petiolate, 5–15 mm long, oblanceolate with dentate margins. **Stem leaves** similar with entire margins. **Vestiture** of simple or few-branched hairs. **Inflorescence** a compact, ebracteate raceme. **Flowers** white, fading to purple; petals 3–4 mm long. **Fruit** a linear, puberulent silique, 10–15 mm long; style <1 mm long; seeds in 1 row per locule. Sparsely vegetated, vernally moist, calcareous soil of grasslands, steppe, tundra; montane and alpine; known from Beaverhead and Teton cos. AK and NT south to CO, VT; Asia.

A report of *Braya glabella* Richardson for MT[157] is incorrect (fide J. Harris)

Camelina Crantz False flax

Annual. **Leaves** simple, sessile, primarily cauline, gradually reduced upward. **Vestiture** of simple or branched hairs. **Inflorescence** an ebracteate raceme with spreading pedicels. **Flowers** pale yellow. **Fruit** a glabrous, keeled, obovoid silicle with a slender beak; seeds 4–12 per locule.

Both of our species are introduced throughout temperate N. America; native to Europe. *Camelina alyssum* (Mill.) Thell. is reported for MT,[8] but I have seen no specimens and many vouchers have been misidentified.[8]

1. Silicles 5–7 mm long; stems hirsute at the base . . . . *C. microcarpa*
1. Silicles 7–9 mm long; stems glabrous to sparsely hairy at the base . . . .*C. sativa*

Camelina microcarpa Andrz. ex DC. **Stems** 10–90 cm, branched above, hirsute at the base. **Leaves** lanceolate with entire margins, 5–90 mm long. **Petals** 3–4 mm long. **Fruits** 5–7 mm long on pedicels 9–15 mm long; beak 1–2 mm long. Fields, roadsides, grasslands; plains, valleys.

Camelina sativa (L.) Crantz Similar to *C. microcarpa*. **Stems** glabrous or sparsely hairy. **Fruits** 7–9 mm long on pedicels 10–20 mm long. Fields; valleys. Our collections from Flathead and Powell cos. are 100 years old.

Capsella Medik. Shepherd's purse

Capsella bursa-pastoris (L.) Medik. Annual. **Stems** simple or branched, 10–50 cm, glabrous above. **Basal leaves** in a rosette, 1–10 cm long, oblanceolate, pinnately lobed. **Stem leaves** lanceolate, auriculate-clasping with dentate margins. **Vestiture** of simple or stellate hairs. **Inflorescence** an ebracteate raceme with spreading pedicels 7–12 mm long. **Flowers** white; petals 1–3 mm long. **Fruit** an obtriangular, strongly flattened silicle, 5–7 mm long with numerous seeds; style 1 mm long, deciduous. Gardens, fields, roadsides, streambanks; plains, valleys, rarely montane. Introduced throughout N. America, cosmopolitan; native to Eurasia.

Cardamine L. Bittercress

Glabrous to sparsely hairy annual or rhizomatous perennial herbs. **Leaves** basal and cauline, pinnately or rarely palmately lobed, petiolate. **Inflorescence** a raceme with spreading or ascending pedicels. **Flowers** white. **Fruit** a flattened, linear silique, erect or ascending; seeds in 1 row per locule.

Typical specimens of *Cardamine breweri*, *C. oligosperma*, and *C. pensylvanica* are distinct, but immature material appears intermediate and may be difficult to determine.

1. Plants with stems creeping beneath high-elevation limestone talus . . . . *C. rupicola*
1. Plants of vernally moist to wet habitats 2

2. Petals 3.5–6 mm long; plants perennials from rhizomes with fibrous roots . . . . *C. breweri*
2. Petals ≤3.5 mm long; plants fibrous-rooted annuals, not rhizomatous. 3

3. Fruits >1.1 mm wide with 15 to 22 seeds; leaflets all petiolate . . . . *C. oligosperma*
3. Fruits up to 1 mm wide with 24 to 40 seeds; some leaflets without definite
   petioles . . . .*C. pensylvanica*

Cardamine breweri S.Watson Rhizomatous and fibrous-rooted perennial. **Stems** erect to ascending, 6–40 cm, usually simple. **Basal leaves** often undivided, early deciduous. **Stem leaves** usually pinnate with a large, ovate terminal segment 1–5 cm long and sinuate to dentate margins, reduced upward. **Petals** 3–6 mm long. **Fruits** 15–30 mm long, the beak ≤1 mm long; pedicels 8–12 mm long. Wet soil of fens, along streams, lakes, often in water; montane, lower subalpine. AK to MT south to CA, WY.

Both plants with sinuately-margined terminal leaf lobes (var. *breweri*) and with lobed or deeply toothed terminal lobes (var. *leibergii* (Holz.) C.L.Hitchc.) occur in MT; however, these vars. have no geographic or ecological distinctness, and many intermediates occur.

Cardamine oligosperma Nutt. Sparsely hairy annual or biennial. **Stems** erect, 4–20 cm. **Basal leaves** 1–5 cm long, pinnate with ovate, sinuate leaflets. **Stem leaves** pinnate with ovate lateral lobes. **Petals** 1–3 mm long. **Fruits** erect, 15–25 mm long, >1.1 mm wide, the beak <1 mm long; pedicels 3–10 mm long. Wet soil along streams, lakes; valleys to subalpine. BC, MT south to CA, CO; Asia.

1. Raceme umbel-like, the axis 1–2 cm long . . . . .var. *kamtschatica*
1. Raceme more elongate, the axis ≥3 cm long . . . . var. *oligosperma*

Cardamine oligosperma var. *kamtschatica* (Regel) Detling has been collected near treeline in Glacier National Park;[240] *C. oligosperma* var. *oligosperma* has been found in disturbed ground, fens; valleys to lower subalpine; however, our collections seem different than plants from the Pacific slope. The more common *C. pensylvanica* has narrower cauline leaflets.

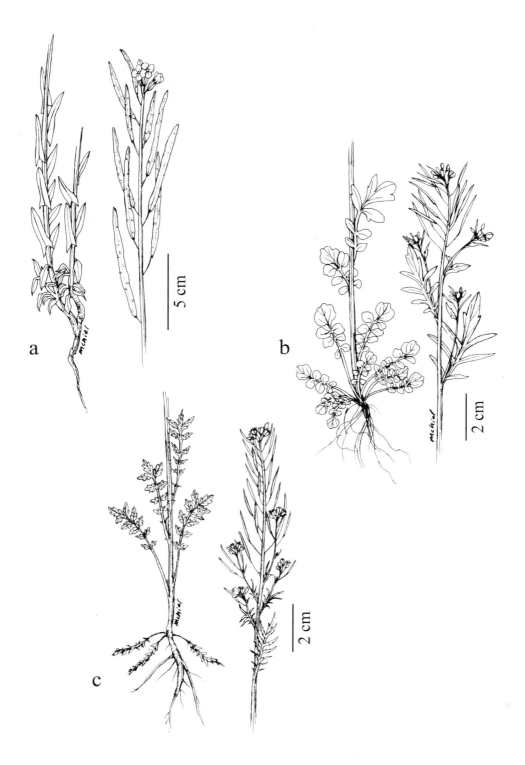

Plate 29. a. *Boechera stricta*, b. *Cardamine pensylvanica*, c. *Descurainia incana* var. *incana*

Cardamine pensylvanica Muhl. ex Willd. Glabrous to sparsely hairy annual or biennial. **Stems** erect to decumbent, 7–40 cm. **Basal leaves** pinnate, 2–7 cm long with oblong to narrowly ovate leaflets, margins dentate. **Stem leaves** with narrower lateral leaflets, not all distinctly petiolate. **Petals** 2–4 mm long. **Fruits** ascending to erect, 10–25 mm long, ca. 1 mm wide, the beak ≤1 mm long; pedicels 5–15 mm long. Wet, sometimes disturbed soil of seeps, wetlands, along streams, lakes; montane. BC to NL south to CA, CO, OK, FL. (p.198)

Cardamine rupicola (O.E.Schulz) C.L.Hitchc. Rhizomatous perennial. **Stems** prostrate to ascending, to 15 cm. **Leaves** thick; those born on rhizome and stem similar, palmately to pinnately divided into 3 to 7 oblanceolate lobes or leaflets 1–4 cm long. **Petals** 7–13 mm long. **Fruits** erect, 15–40 mm × 1–2 mm; the beak 4–6 mm long; pedicels 1–2 cm long. Limestone talus; upper subalpine, alpine. Endemic to MT.

   Subterranean stems may extend downslope under the rocks for a great distance with leaves and ascending stems emerging sporadically. The type locality is in Lake Co.

Chorispora R.Br. ex DC. Blue mustard

Chorispora tenella (Pall.) DC. Annual. **Stems** 7–50 cm, often branched. **Leaves** petiolate; the blade oblong to lanceolate, sinuate-dentate, 3–8 cm long. **Vestiture** of simple, glandular and eglandular hairs. **Inflorescence** a raceme with leaf-like bracts below. **Flowers** blue to lavender; petals 9–12 mm long, clawed with a spreading blade. **Fruit** a curved silique 2–4 cm × ca. 2 mm; beak 10–15 mm long; pedicels stout, spreading, 2–4 mm long. Sparsely vegetated, often saline soil of fields, disturbed grasslands, steppe; plains, valleys. Introduced to much of temperate western N. America; native to Eurasia.

Conringia Heist. ex Fabr. Hare's-ear mustard

Conringia orientalis (L.) C.Presl Glabrous and glaucous, taprooted annual. **Stems** erect, usually unbranched, 20–60 cm. **Basal leaves** petiolate, obovate with sinuate margins, 5–9 cm long, early deciduous. **Stem leaves** lanceolate, auriculate-clasping, 3–12 cm long. **Inflorescence** a few-flowered ebracteate raceme. **Flowers** cream to pale yellow; petals 9–13 mm long. **Fruit** a linear, 4-angled silique 5–12 cm × 1–2 mm, somewhat constricted between seeds, erect; seeds in 1 row per locule; pedicels ascending, 5–15 mm long. Disturbed soil of fields, grasslands, roadsides, eroding soil of badlands; plains, valleys. Introduced throughout much of temperate N. America; native to Europe.

Descurainia Webb & Berthel. Tansy mustard

Taprooted annuals and biennials. **Stems** erect, usually branched. **Basal leaves** pinnately to tri-pinnately lobed or divided, petiolate, forming a rosette, withered at anthesis. **Stem leaves** reduced in size and degree of division upward, becoming sessile. **Vestiture** of branched and simple hairs. **Inflorescence** an ebracteate raceme. **Flowers** small, yellow or white. **Fruit** a linear to fusiform silique; style short.[115]

   Differences between species are subtle; mature fruit is needed for determination.

1. Siliques 14–28 mm long, erect on spreading pedicels . . . . *D. sophia*
1. Siliques mostly 3–15 mm long, spreading to erect 2

2. Siliques clavate, rounded at the tip, 1–2 mm wide . . . . *D. pinnata*
2. Siliques linear, tapered to the tip, acute, ≤1 mm wide 3

3. Siliques strongly torulose; plants <15 cm; subalpine . . . .*D. torulosa*
3. Siliques little or not torulose; well-developed plants >10 cm; lower elevations . . . . *D. incana*

Descurainia incana (Bernh. ex Fisch.& C.A.Mey.) Dorn [*D. incisa* (Engelm. ex A.Gray) Britton, *D. richardsonii* O.E.Schulz] Biennial. **Stems** 10–80 cm. **Stem leaves** 3–8 cm long, ovate to oblanceolate, pinnate, the pinnae lobed to entire. **Vestiture** of minute branched hairs to glabrous or glandular. **Petals** yellow, 1–2 mm long. **Fruit** curved to straight, erect to ascending, 4–15 mm × ≤1 mm, tapering at both ends; style 0.2–0.4 mm long; seeds 4 to 8 per locule in 1 row; pedicels erect to spreading. Grasslands, steppe, open forest, often where disturbed; plains, valleys, montane. AK to QC south to CA, UT, CO, SD, MN. (p.198)

1. Pedicels of mature fruit erect or erect-ascending. 2
1. Pedicels of mature fruit spreading to widely ascending 3

2. Herbage somewhat canescent. . . . . .var. *major*
2. Herbage glabrous to hairy but not canescent. . . . . var. *macrosperma*

3. Glandular hairs present on stem and leaves . . . . var. *viscosa*
3. Herbage not glandular . . . . .var. *incana*

Descurainia incana var. *incana*, with fruits sometimes as little as 6 mm long, occurs valleys to montane; *D. incana* var. *major* (Hook.) Dorn occurs in valleys, Flathead and Glacier cos.; *D. incana* var. *macrosperma* (O.E.Schulz) Dorn occurs in grassland, open forest, montane, subalpine; *D. incana* var. *viscosa* (Rydb.) Dorn [*D. incisa* ssp. *incisa*] is usually found in partial shade, plains, valleys. The vars. have little or no ecological or geographic integrity in MT.

Descurainia pinnata (Walter) Britton Annual. **Stems** 6–70 cm. **Stem leaves** 1–12 cm long, ovate to lanceolate, pinnate to tripinnate. **Vesture** of sparse to dense minute, branched hairs, glabrous above or glandular. **Petals** white to yellow, 1.5–2 mm long. **Fruit** erect to ascending, 5–12 mm × 1–2 mm, clavate, rounded and broader at the tip; style <0.2 mm long; seeds 5 to 20 per locule in 1 or 2 rows; pedicels ascending to spreading, 2–17 mm long. Grasslands, fields; plains, valleys, montane. Throughout temperate N. America.

1. Stems glandular . . . . var. *brachycarpa*
1. Stems not glandular 2

2. Some fruits >12 mm long . . . . .var. *filipes*
2. Fruits ≤12 mm long. 3

3. Pedicels 2–6 mm long . . . . var. *nelsonii*
3. Pedicels 6–13 mm long . . . . .var. *intermedia*

Descurainia pinnata var. *brachycarpa* (Richardson) Fernald is found in grasslands, steppe, plains, valleys; *D. pinnata* var. *intermedia* (Rydb.) C.L.Hitchc. usually occurs in disturbed habitats, plains, valleys; *D. pinnata* var. *filipes* (A.Gray) M.Peck [*D. longepedicellata* (E.Fourn.) O.E.Schulz] occurs from the plains to montane; *D. pinnata* var. *nelsonii* (Rydb.) M.Peck [*D. nelsonii* (Rydb.) Al-Shehbaz & Goodson] is primarily montane. The vars. have little ecological or geographic distinction in MT.

Descurainia sophia (L.) Webb ex Prantl Fɪxᴡᴇᴇᴅ Annual or beinnial. **Stems** 25–75 cm. **Stem leaves** 1–7 cm long, obovate to oblanceolate, bipinnate or tripinnate. **Vesture** of sparse to dense, minute, branched hairs. **Petals** pale yellow, 2–3 mm long. **Fruit** ascending-erect, 14–25 mm × ≤1 mm, linear; style ≤0.1 mm long; seeds 10 to 20 per locule in 1 row; pedicels ascending, 7–16 mm long. Disturbed soil of fields, roadsides; plains, valleys. Introduced throughout temperate N. America; native to Eurasia.

Descurainia torulosa Rollins Biennial. **Stems** 4–15 cm. **Stem leaves** 2–10 cm long, oblanceolate, pinnate, the pinnae lobed to entire. **Vesture** of stellate hairs. **Petals** yellow, 1–2 mm long. **Fruit** curved to straight, torulose, erect, 6–15 × ≤1 mm, tapering at both ends; style 0.1–0.3 mm long; seeds 10 to 18 per locule in 1 row; pedicels erect. Talus slopes; subalpine. Endemic to Park Co., MT (*Rosentretter 1557*, SRP) and western WY.

Diplotaxis DC.

Diplotaxis muralis (L.) DC. Taprooted biennial. **Stems** ascending, usually simple, 10–40 cm. **Leaves** mostly basal, petiolate, oblanceolate with dentate margins, 3–8 cm long. **Vesture** of sparse, stiff retrorse hairs on the lower stem, otherwise glabrous. **Inflorescence** a few-flowered ebracteate raceme. **Flowers** pale yellow; petals 4–6 mm long. **Fruit** a linear silique, 1–4 cm × 1–2 mm, somewhat flattened; seeds in 2 rows per locule; style <2 mm long; pedicels spreading-ascending, 5–20 mm long. Sparsely vegetated, often calcareous soil of roadsides, railroads, streambanks; valleys. Sparingly introduced throughout U.S.; native to Europe.

Draba L. Whitlow-wort

Annuals, biennials or perennials. **Leaves** simple; those of the basal rosette petiolate; those of the stem, if present, similar but sessile. **Vesture** of simple, branched or stellate hairs. **Inflorescence** a short, bracteate or ebracteate raceme. **Flowers** white or yellow. **Fruit** a silicle or short silique, flattened parallel to the septum; seeds in 2 rows per locule.[184,297,342]

Shape of mature fruit and hairs on the leaves are important characters. Many *Draba* spp. are more widespread and abundant in calcareous soil. *Draba borealis* DC. occurs in adjacent WY and could be expected in MT. *Draba fladnizensis* Wulfen, but for the white flowers, is very similar to *D. crassifolia*; a collection from the Anaconda Range may be *D. fladnizensis*, but the specimen is too poor to be certain. Many of the yellow-flowered, cushion-forming species are primarily asexual;[300] as a result, entire populations may possess minor aberrant morphological traits. *Draba pectinipila* Rollins is reported for MT,[9] but I am not aware of any specimens.

1. Plants annual, biennial or short-lived perennial; lacking a branched caudex......................2
1. Plants obviously perennial; some plants with a branched caudex and flower stems from previous years . . .12

2. Fruits pubescent..3
2. Fruits glabrous..7

3. Lowest pedicels subtended by bracts................................4
3. Raceme completely ebracteate...................................5

4. Petals yellow; style of mature fruit ≥0.5 mm long........................*D. aurea*
4. Petals white; style of mature fruit <0.5 mm long........................*D. cana*

5. Upper stem pubescent...*D. praealta*
5. Upper stem glabrous..6

6. Lower pedicels much longer than mature fruit.........................*D. nemorosa*
6. Mature fruit longer than the pedicel................................*D. reptans*

7. Some stem hairs with ≥3 branches.................................8
7. Stems glabrous or with simple or 2-forked hairs........................10

8. Stem leaves entirely lacking....................................*D. verna*
8. Plants with 2 or more stem leaves.................................9

9. Siliques 2–5 mm long; plants annual...............................*D. brachycarpa*
9. Siliques >6 mm long; plants biennial to short-lived perennial................*D. stenoloba*

10. Leaves without branched hairs on the surface, usually glabrous or ciliate..........*D. crassifolia*
10. At least some leaves with branched hairs on the surface...................11

11. Silicles shorter than subtending pedicel; stem leaves lacking................*D. verna*
11. Silicles ca. as long as subtending pedicel; stem usually with ≥1 leaf............*D. albertina*

12. Style nearly obsolete, <0.2 mm long...............................13
12. Style >0.2 mm long..17

13. Stem leaves 0-1...14
13. Stem leaves >1...16

14. Leaves ciliate, otherwise glabrous................................*D. crassifolia*
14. Leaves with hair on the lower surface at least........................15

15. Fruit ovate; lower pedicles 1–4 mm long............................*D. porsildii*
15. Fruit linear; lower pedicels ≥5 mm long.............................*D. albertina*

16. Fruits not twisted; base of stem with some simple hairs...................*D. praealta*
16. Some fruits twisted; base of stem with all branched hairs..................*D. cana*

17. Leaves ciliate, otherwise glabrous................................18
17. Most leaves with some simple to branched hairs on at least lower surface or leaf tips............21

18. Cilia ≥1 mm long, >1/2 as long as width of leaf.......................*D. densifolia*
18. Cilia <1 mm long, <1/2 as long as leaf width.........................19

19. Mature fruit 6-9 mm long......................................*D. crassa*
19. Mature fruit ≤6 mm long.......................................20

20. Lower pedicels 1–3 mm long in fruit; fruits hairy......................*D. globosa*
20. Lower pedicels 3–6 mm long; fruits glabrous; Bitterroot Range...............*D. daviesiae*

21. Leaves with simple or forked cilia >1 mm long on leaf margins as well as simple or branched
    hairs on the surface...22
21. Conspicuous leaf cilia not present................................25

Draba albertina Greene [*D. stenoloba* Ledeb. var. *nana* (O.E.Schulz) C.L.Hitchc.] Taprooted
annual or biennial with a simple caudex. **Stems** 3–25 cm, often branched. **Basal leaves** narrowly
lanceolate, 5–25 mm long with entire to weakly denticulate margins. **Stem leaves** 0 to 3,
lanceolate. **Vestiture** of sparse, simple and forked hairs. **Petals** yellow, 2–3 mm long. **Fruit** linear,
erect to ascending, 5–13 × 1.5–2 mm, glabrous; style <0.2 mm long; lower pedicels 5–12 mm
long. Sparsely vegetated soil of grasslands, open forest, streambanks; montane and subalpine. AK south to CA, NV, AZ.
(p.204)
 Specimens from low elevations are often mistaken for *Draba stenoloba* which has branched rather than simple or
forked hairs on the stem. Plants of *D. albertina* at higher elevations are similar to *D. crassifolia* which lacks branched
hairs on leaf surfaces.

Draba aurea Vahl ex Hornem. Short-lived perennial from a simple caudex. **Stems** 5–40 cm,
rarely branched. **Basal leaves** oblanceolate, 5–30 mm long, usually entire. **Stem leaves** several,
lanceolate, sessile. **Vestiture** of simple, branched and stellate hairs, dense below, less so above.
Petals yellow, 4–6 mm long. **Fruit** erect to ascending, 8–15 × 2–4 mm, pubescent, often twisted;
style 0.5–1 mm long; lower pedicels 3–6 mm long. Sparsely vegetated, rocky soil of grasslands,
open forests; montane to alpine. AK to Greenland south to AZ, NM. (p.191)
 Draba cana has white flowers, and a shorter style.

Draba brachycarpa Nutt. ex Torr.& A.Gray Annual from a simple caudex. **Stems** 5–15 cm, branched at the base.
Basal leaves obovate, 10–15 mm long, entire. **Stem leaves** oblanceolate, sessile. **Vestiture** of dense, sessile, 2- to
4-branched hairs. **Petals** white, 1–3 mm long. **Fruit** spreading-ascending, 2–5 × 1–2 mm, glabrous to finely hairy; style
≤0.1 mm long; lower pedicels 3–6 mm long. Disturbed soil of grasslands, fields; valleys. Native to eastern N. America;
introduced and sporadic in western N. America. Collected once in MT by Francis Kelsey near Helena over 100 years
ago.[184]

Draba calcifuga P.Lesica Mat-forming perennial. **Stems** simple, 1–12 cm. **Basal leaves**
oblanceolate, 3–10 mm long, entire. **Stem leaves** absent. **Vestiture** of pectinate hairs, stalked,
forked or stellate hairs, and simple or forked cilia. **Petals** yellow, 3–6 mm long. **Fruit** ovate,
ascending, 2–6 × 1.5–3 mm, pubescent with simple hairs; style 0.3–1 mm long; lower pedicels
2–12 mm long. Stony, usually non-calcareous soil of fellfields, turf, slopes; alpine. A newly-
described species endemic to southwest MT and adjacent WY, ID.[243]

Draba cana Rydb. [*D. breweri* S.Watson var. *cana* (Rydb.) Rollins, *D. lanceolata* Royle
misapplied] Perennial from a simple or few-branched caudex. **Stems** 5–15 cm, rarely branched.
Basal leaves oblong, 4–15 mm long, entire to weakly dentate. **Stem leaves** several, lanceolate,
sessile, usually dentate. **Vestiture** of dense many-branched hairs. **Petals** white, 2–4 mm long.
Fruit erect, 6–10 × 1–2 mm, pubescent, sometimes twisted; style 0.2–0.4 mm long; lower
pedicels 1–3 mm long. Sparsely vegetated, stony soil of grasslands, meadows, cliffs, open forest, gravel bars; montane
to alpine. AK to QC south to CA, CO, WI, VT. (p.204)
 Rollins believes this plant is better considered a variety of the Sierra Nevada *Draba breweri*.[342] *Draba praealta* does
not have mature fruits that are twisted; see *D. aurea*.

Draba crassa Rydb. Perennial from a simple or branched caudex clothed in old leaf bases. **Stems** simple, 4–12 cm. **Basal leaves** petiolate, oblanceolate, 2–5 cm long, entire. **Stem leaves** several, lanceolate, sessile. **Vestiture** of simple and forked hairs on the stem and sparse cilia on leaf margins. **Petals** yellow, 3–5 mm long. **Fruit** ascending, 5–9 × 3–5 mm, glabrous, often twisted; style 0.3–1 mm long; lower pedicels 2–8 mm long. Talus, cliffs and stony soil of cool slopes, often in the shade of boulders; alpine. MT south to UT and CO.

Draba crassifolia Graham Annual to short-lived perennial from a simple to few-branched caudex. **Stems** 3–8 cm, unbranched. **Basal leaves** narrowly oblanceolate, 4–20 mm long with mostly entire margins. **Stem leaves** 0 to 2. **Vestiture** of simple and forked cilia on leaf margins, otherwise glabrous. **Petals** yellow, 2–3 mm long. **Fruit** ascending, 4–9 × 1.5–2 mm, glabrous; style ≤0.1 mm long; lower pedicels 3–8 mm long. Moist, sparsely vegetated soil of meadows, rocky slopes, often sites of late snowmelt or shade of boulders or cliffs; subalpine to alpine. AK to Greenland south to CA, AZ, CO. See *D. albertina*. (p.204)

Draba daviesiae (C.L.Hitchc.) Rollins [*D. apiculata* Hitchc. var. *daviesiae* C.L.Hitchc.] Mat-forming perennial; the caudex clothed in old leaf bases. **Stems** simple, 1–5 cm. **Basal leaves** oblong, 2–5 mm long, entire. **Stem leaves** lacking. **Vestiture** of simple leaf cilia, otherwise glabrous. **Petals** yellow, 3–5 mm long. **Fruit** ascending, 4–6 mm long, 2–3 mm wide, glabrous; style 0.3–0.5 mm long; lower pedicels 3–6 mm long. Shallow soil of exposed ridges and upper slopes; alpine. Endemic to the Bitterroot Mtns. of Ravalli Co. See *D. densifolia*.

Draba densifolia Nutt. Cushion-forming perennial, the caudex clothed in old leaf bases. **Stems** simple, 3–5 cm. **Basal leaves** oblanceolate, 3–8 mm long, entire, the lower midvein prominent. **Stem leaves** lacking. **Vestiture** of simple leaf cilia ≥1 mm long, otherwise glabrous or leaves with branched hairs below and simple hairs above. **Petals** yellow, 3–5 mm long. **Fruit** ascending, 4–5 × 2–3 mm, pubescent with simple or forked hairs; style 0.5–1 mm long; lower pedicels 2–6 mm long. Gravelly soil of exposed slopes, ridges; montane to alpine. AK to AB south to CA, UT, WY. (p.204)

Plants from sedimentary-derived, usually calcareous soil in the northwest ranges have petals 3–4 mm long and glabrous leaves except for the cilia; plants from granitic soil of southwest MT have petals 4–5 mm long and stalked-branched hairs on the lower leaf surfaces and simple hairs on the upper surfaces in addition to cilia. *Draba daviesiae* has glabrous silicles; *D. globosa* has less conspicuous leaf cilia.

Draba globosa Payson [*D. apiculata* C.L.Hitchc.] Cushion-forming perennial, the caudex clothed in old leaf bases. **Stems** simple, 5–20 mm. **Basal leaves** oblanceolate, curved inward, 2–5 mm long, entire. **Stem leaves** absent. **Vestiture** of simple, short, leaf cilia; otherwise glabrous. **Petals** yellow, fading to white, 3–4 mm long. **Fruit** ascending, 4–6 × 2–3 mm, pubescent with stellate hairs; style 0.5–1 mm long; lower pedicels 1–3 mm long. Moist, sparsely vegetated, stony soil; alpine. ID, MT south to UT, CO. See *D. densifolia*.

Draba incerta Payson Cushion-forming perennial. **Stems** simple, 5–20 cm. **Basal leaves** oblanceolate, 5–15 mm long, entire. **Stem leaves** lacking. **Vestiture** of stalked, pectinate and branched hairs, sometimes ciliate. **Petals** pale yellow, 3–6 mm long. **Fruit** ascending, 5–10 mm long, 2–3 mm wide, pubescent with simple or forked hairs to nearly glabrous; style 0.2–0.5 mm long; lower pedicels 3–12 mm long. Stony soil of fellfields, turf, cliffs, exposed slopes; alpine, occasionally lower on exposed calcareous sites. AK to ID, WY; QC . See *D. oligosperma*. (p.204)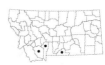

Draba lonchocarpa Rydb. Perennial from a simple or few-branched caudex. **Stems** simple, 3–10 cm. **Basal leaves** oblanceolate, 3–10 mm long, entire. **Stem leaves** usually absent. **Vestiture** of dense, stalked, stellate hairs. **Petals** white, 2–4 mm long. **Fruit** ascending, 5–20 × 1–2 mm, usually glabrous, often twisted; style 0.2–0.5 mm long; lower pedicels 2–10 mm long. Moist soil of cliffs, tundra, cool slopes, talus, shade of boulders; upper subalpine to alpine. AK to OR, ID, CO. Our plants are var. *lonchocarpa*. (p.204)

The type locality is in Lake Co. Reports of *Draba nivalis* Lilj., an arctic species with sessile, stellate hairs, for MT[242] were based on small-fruited specimens of *D. lonchocarpa*.

Draba macounii O.E.Schulz Perennial from a loosely branched caudex clothed in old leaf bases. **Stems** simple, 1–4 cm. **Basal leaves** oblong to obovate, 6–10 mm long, entire. **Stem leaves** absent. **Vestiture** of wrinkled hairs on leaf margins and tips. **Petals** pale yellow, 3–4 mm long. **Fruit** ovate, ascending, 4–8 × 3–5 mm, glabrous, often wrinkled; style 0.3–0.5 mm long; lower pedicels 2–5 mm long. Wet, soil of cool, open slopes, rock outcrops and along streams; alpine. AK to MT, CO. Known from Flathead and Glacier cos.

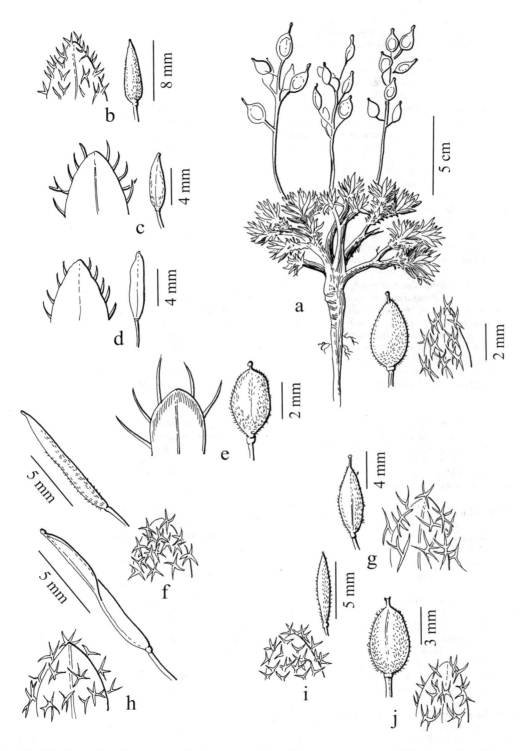

Plate 30. **Draba**. **a.** *D. oligosperma*, **b.** *D. cana*, **c.** *D. albertina*, **d.** *D. crassifolia* **e.** *D. densifolia* **f.** *D. praealta*, **g.** *D. incerta* **h.** *D. lonchocarpa*, **i.** *D. nemorosa*, **j.** *D. paysonii* var. *paysonii*

Draba nemorosa L. Annual from a simple caudex. **Stems** 4–30 cm, unbranched. **Basal leaves** ovate, 6–25 mm long, weakly dentate. **Stem leaves** several, ovate, denticulate. **Vestiture** of stalked-branched and sometimes simple hairs. **Petals** yellow, 1.5–2.5 mm long. **Fruit** spreading-ascending, 4–7 mm × 1.5–2 mm, puberulent; style ≤0.1 mm long; lower pedicels 9–20 mm long. Disturbed soil of grasslands, meadows, steppe, open forest, often associated with pocket gopher or ground squirrel digging; plains, valleys to montane. AK to QC south to CA, CO, MN. (p.204)

Draba oligosperma Hook. Mat-forming perennial. **Stems** simple, 1–12 cm. **Basal leaves** oblanceolate, 3–10 mm long, entire. **Stem leaves** lacking. **Vestiture** of sessile pectinate hairs often aligned parallel to the midvein, rarely with some simple or forked cilia. **Petals** yellow, 3–6 mm long. **Fruit** ovate, ascending, 2–6 mm × 1.5–3 mm, pubescent with simple hairs; style 0.3–1 mm long; lower pedicels 2–12 mm long. Sparsely vegetated, stony soil of grasslands, open forest, rock outcrops, exposed slopes and ridges; montane to alpine. AK south to CA, CO. (p.204)

More common in calcareous soil. The upper surface of some leaves of *Draba oligosperma* may become almost glabrous. Hairs of the similar *D. incerta* are not appressed to the leaf surface. Populations in MT may be either apomictic or sexual.[296]

Draba paysonii J.F.Macbr. [*D. novolympica* Payson & H.St. John] Mat-forming perennial. **Stems** simple, 1–5 cm. **Basal leaves** oblanceolate, 3–8 mm long, entire. **Stem leaves** lacking. **Vestiture** of dense, tangled, stalked branched hairs and long simple and branched hairs and cilia. **Petals** yellow, 3–4 mm long. **Fruit** ovate, ascending, 3–9 mm × 2–5 mm, pubescent with simple or forked hairs; style 0.4–1 mm long; lower pedicels 1–4 mm long. Stony, sparsely vegetated soil of exposed ridges, slopes; montane to alpine. AK south to CA, UT, WY. (p.204)

1. Silicles 5–8 mm long; style ca. 1 mm long . . . . .var. *paysonii*
1. Silicles 3–5 mm long; style <0.8 mm long . . . . var. *treleasii*

Draba paysonii var. *treleasii* (O.E.Schulz) C.L.Hitchc. occurs on soils derived from both sedimentary and crystalline parent material, montane to alpine but more common higher; *D. paysonii* var. *paysonii*, is found only on calcareous soil, subalpine and alpine. The two vars. are usually distinct, but they occur together with intermediates near Red Mountain north of Lincoln. The type locality is in Glacier Co.

Draba porsildii G.A.Mulligan Perennial from a simple or few-branched caudex. **Stems** simple, 2–5 cm. **Basal leaves** oblanceolate, 3–8 mm long, entire. **Stem leaves** absent or 1. **Vestiture** of stalked, mostly 4-branched hairs. **Petals** white, 2–4 mm long. **Fruit** ovate, ascending, 4–6 mm × 1.5–2 mm, glabrous; style 0.1–0.3 mm long; lower pedicels 1–4 mm long. Moist, stony soil of cool upper slopes; upper subalpine. YT south to CO. Our plants are var. *brevicula* (Rollins) Rollins. Known only from the Beartooth Range in Carbon Co.[228]

Draba praealta Greene Biennial or short-lived perennial from a usually simple caudex. **Stems** branched or not, 7–30 cm. **Basal leaves** oblanceolate, 1–3 cm long with dentate margins. **Stem leaves** several, lanceolate. **Vestitiure** of simple to stellate hairs. **Petals** white, 2–3 mm long. **Fruit** erect-ascending, narrowly lanceolate, 8–14 × 1.5–2 mm, puberulent; style ≤0.1 mm long; lower pedicels 4–9 mm long. Sparsely vegetated soil of rocky slopes, cliffs, open forest, streambanks; montane to lower alpine. AK south to CA, WY. See *D. stenoloba, D. cana*. (p.204)

Draba reptans (Lam.) Fernald Annual from a simple caudex. **Stems** 3–12 cm, unbranched, leafless above. **Basal leaves** obovate, 3–20 mm long with entire margins. **Stem leaves** 2 to 5, elliptic. **Vestiture** of dense, simple and branched hairs, becoming glabrous above. **Petals** white, 3–6 mm long or absent. **Fruit** erect, 7–15 mm × 1.5–2.5 mm, sparsely puberulent with simple hairs; style ≤0.1 mm long; lower pedicels 1–6 mm long. Gravelly or sandy soil of grasslands, steppe, open forest; plains, valleys. WA to MA south to CA, TX.

Draba stenoloba Ledeb. Biennial or short-lived perennial from a simple caudex. **Stems** mostly simple, 5–25 cm. **Basal leaves** oblanceolate, 8–15 mm long, weakly dentate. **Stem leaves** few, lanceolate, sessile. **Vestiture** of simple and branched hairs. **Petals** white, 2–3 mm long. **Fruit** spreading, 8–12 mm × 1–1.5 mm, glabrous; style ≤0.1 mm long; lower pedicels 7–10 mm long. Sparsely vegetated soil of grasslands, meadows; montane to subalpine. AK south to WA, MT. Known from Glacier and Ravalli cos.

Draba praealta has pubescent fruit; *D. albertina* has simple or forked hairs rather than branched hairs on the stem.

Draba ventosa A.Gray Mat-forming perennial. **Stems** simple, 1–4 cm. **Leaves** crowded on the lower 1/4 of the stem but not in a rosette, oblanceolate, 5–10 mm long, entire. **Vestiture** of dense, stalked, 4-branched hairs. **Petals** yellow, 4–5 mm long. **Fruit** ovate, ascending, 4–7 × 3–5 mm, pubescent with branched hairs; style 0.6–1.7 mm long; lower pedicels 2–4 mm long. Limestone talus; alpine. YT south to UT and CO; known from Madison Co.

Draba verna L. Annual from a simple caudex. **Stems** 3–15 cm, unbranched. **Basal leaves** oblanceolate, 5–20 mm long, entire to dentate. **Stem leaves** lacking. **Vestiture** of few-branched hairs. **Petals** white, 2-lobed, 2–3 mm long. **Fruit** ascending, 3–8 × 1–2.5 mm, glabrous; style ≤0.2 mm long; lower pedicels 5–20 mm long. Disturbed soil of fields, grasslands, along streams and roads; plains, valleys to montane. Throughout much of eastern and western, temperate N. America.

Some authorities consider this species introduced from Europe,[193,342] but others believe it is circumboreal.[187,327]

Eruca Mill.

Eruca vesicaria (L.) Cav. [*E. sativa* Mill.] Taprooted annual. **Stems** erect, branched, 20–50 cm. **Basal leaves** petiolate, oblanceolate 8–20 cm long, pinnately lobed. **Stem leaves** short-petiolate or sessile, reduced upward. **Vestiture** of stiff simple, often retrorse hairs. **Inflorescence** an ebracteate, sparsely pubescent raceme. **Flowers** pale yellow; petals 3–6 mm long with dark veins. **Fruit** a beaked silique; the body 10–17 × 2–5 mm, reticulate veined; the beak flat, 5–10 mm long; seeds in 2 rows per locule; pedicels erect, 2–4 mm long. Introduced in much of southern U.S. as a salad herb; native to Europe. Our only collection was made in 1898 in Flathead Co. Our plants are ssp. *sativa* (Mill.) Thell.

Erucastrum C.Presl Dog mustard

Erucastrum gallicum (Willd.) O.E.Schulz Taprooted annual or biennial. **Stems** erect, usually branched, 20–50 cm. **Leaves:** basal petiolate, oblong, 3–10 cm long, pinnately to bipinnately lobed; stem leaves more deeply dissected. **Vestiture** of simple, retrorse hairs, becoming glabrous above. **Inflorescence** a raceme, bracteate below; bracts leaf-like. **Flowers** pale yellow; petals 6–9 mm long. **Fruit** a linear silique, 15–40 × 1–2 mm; style 1–3 mm long; seeds in 1 row per locule; pedicels ascending, 5–20 mm long. Disturbed soil of roadsides, agricultural fields; plains, valleys. Introduced into much of northern U.S. and adjacent Canada; native to Europe.

Erysimum L. Wallflower

Taprooted, annual to perennial herbs. **Leaves** basal and cauline similar, simple with entire or dentate margins, gradually reduced upward. **Vestiture** of sessile, forked hairs. **Inflorescence** an ebracteate raceme, often congested in flower but expanding in fruit. **Flowers** yellow. **Fruit** a pubescent, linear silique with a prominent style; seeds in 1 row per locule.

It is helpful to have both flowers and mature fruit for identification.

1. Petals 12–25 mm long 2
1. Petals ≤12 mm long 3

2. Mature siliques spreading; pubescence more sparse on angles, making them appear striped. . *E. asperum*
2. Mature siliques erect-ascending; pubescence uniform . . . .*E. capitatum*

3. Petals 6–10 mm long . . . . *E. inconspicuum*
3. Petals 3–6 mm long 4

4. Fruiting pedicels 2–4 mm long, as thick as the fruit . . . . *E. repandum*
4. Fruiting pedicels ≥6 mm long, narrower than the fruit . . . .*E. cheiranthoides*

Erysimum asperum (Nutt.) DC. Biennial or short-lived perennial; the caudex often clothed in old leaf bases. **Stems** erect, usually simple, 10–50 cm. **Leaves** linear to narrowly oblanceolate, 2–10 cm long with entire to dentate margins. **Petals** 12–25 mm long, sometime orangish. **Fruits** spreading to ascending, 4-angled, 4–10 cm × 1–2 mm; pedicels spreading, 5–12 mm long; style 2–4 mm long. Often sandy soil of grasslands, woodlands, open slopes; plains. MT to MB south to NM and TX.

The similar *Erysimum capitatum* has erect-ascending siliques.

Erysimum capitatum (Douglas ex Hook.) Greene [*E. asperum* (Nutt.) DC. var. *arkansanum* (Nutt.) A.Gray] Biennial or short-lived perennial; the caudex often clothed in old leaf bases. **Stems** erect, usually simple, 5–50 cm. **Leaves** linear to narrowly oblanceolate, 2–8 cm long with entire to dentate margins. **Petals** 13–25 mm long, sometime orangish. **Fruits** ascending, 4-angled, 4–10 cm × 1–2 mm; pedicels ascending, 3–8 mm long; style 1–4 mm long. Stony or sandy soil of grasslands, woodlands, meadows, fellfields, roadsides; valleys to alpine. BC to SK south to CA, TX, Mexico.

    Two vars. are reported for MT:[342] var. *capitatum* and var. *purshii* (Durand) Rollins; however, both forms occur in some populations, and the characters used to separate them are not strongly correlated.

Erysimum cheiranthoides L. Annual. **Stems** simple or branched, 20–75 cm. **Leaves**: the basal withered at flowering; stem leaves narrowly lanceolate, 2–7 cm long with mostly entire margins. **Petals** 3–5 mm long. **Fruit** ascending, 15–25 × ca. 1 mm, weakly 4-angled; pedicels spreading, 5–12 mm long; style 0.5–1 mm long. Disturbed soil of meadows, grasslands, woodlands, avalanche slopes and along streams, wetlands; plains, valleys, montane. AK to NL south to CA, CO, MO, FL.

Erysimum inconspicuum (S.Watson) MacMill. Biennial or short-lived perennial; caudex often clothed in old leaf bases. **Stems** simple or branched, 12–70 cm. **Leaves**: the basal withered by flowering; stem leaves linear to oblanceolate, 1–9 cm long with entire margins. **Petals** 6–12 mm long. **Fruit** erect-ascending, 3–6 cm × 1–2 mm, weakly 4-angled, pubescent; pedicels ascending, 5–10 mm long; style 0.8–1.5 mm long. Grasslands, meadows, open slopes; plains, valleys to subalpine. AK to NL south to OR, CO, MN. (p.212)

Erysimum repandum L. Treacle mustard Annual. **Stems** simple or branched, 20–50 cm. **Leaves** narrowly oblanceolate, 2–8 cm long with entire to sinuate margins. **Petals** 3–6 mm long. **Fruit** ascending to spreading, 4–8 cm × 1–1.5 mm, often curved, 4-angled, constricted between the seeds; pedicels spreading, 2–5 mm long; style 1–3 mm long. Disturbed soil of grasslands, steppe, woodlands; plains, valleys. Throughout most of temperate N. America.

Euclidium W.T.Aiton

Euclidium syriacum (L.) R.Br. Annual. **Stems** erect, branched, 10–40 cm. **Leaves** all cauline, oblanceolate with weakly dentate margins, short-petiolate; blades 2–5 cm long. **Vesture** pubescent with forked hairs. **Inflorescence** axillary, spike-like racemes with stout pedicels. **Flowers** white; petals ca. 1 mm long; stamens 6. **Fruit** an obovoid, hispid, 2-seeded silicle; the body 2–3 mm long, with a stout beak ca. as long. Fields, roadsides; plains. Sparingly introduced into western U.S.; native to Eurasia. One collection from Cascade Co.

Eutrema R.Br.

Eutrema salsugineum (Pall.) Al-Shehbaz & Warwick [*Thellungiella salsuginea* (Pall.) O.E.Schulz, *Arabidopsis salsuginea* (Pall.) N.Busch] Taprooted, glabrous and glaucous annual. **Stems** erect, simple or branched, 5–15 cm. **Leaves** basal and cauline; the basal, petiolate, oblanceolate, 5–10 mm long; the cauline 5–15 mm long, lanceolate with clasping auriculate bases. **Inflorescence** an open, few-flowered, ebracteate raceme. **Flowers** white; petals 2–3 mm long. **Fruit** glabrous, erect siliques, 10–16 × 0.5–1 mm; style <0.2 mm long; seeds in 2 rows per locule. Moist, alkaline or saline soil; valleys. AK south to MT, CO; Asia. Our only collection was made 90 years ago in Beaverhead Co. (*Payson & Payson 1730*, RM).

Hesperis L. Dame's rocket

Hesperis matronalis L. Taprooted perennial. **Stems** 50–100 cm, often branched above. **Leaves** all cauline, lanceolate, 3–10 cm long, sessile above, dentate. **Vesture** of simple and branched hairs. **Inflorescence** an ebracteate, branched raceme. **Flowers** purple to white, fragrant; petals 17–23 mm long. **Fruit** glabrous, linear siliques, constricted between seeds, 5–14 cm × 1–2 mm; style 1–2 mm long; seeds in 1 row per locule; pedicels spreading, 10–25 mm long. Moist, disturbed soil of fields and along roads, streams; plains, valleys. Escaping cultivation throughout much of temperate N. America; introduced from Europe.

Hornungia Rchb.

Hornungia procumbens (L.) Hayek [*Hutchinsia procumbens* (L.) Desv.] Annual. **Stems** 5–15 cm, delicate, branched. **Leaves** all cauline, oblanceolate to spatulate, petiolate below, 5–15 mm long with mostly entire margins. **Vestiture** of sparse, minute hairs or glabrous. **Inflorescence** an ebracteate raceme. **Flowers** white; petals ca. 1 mm long, as long as the sepals. **Fruit** an erect, obovate silicle, flattened perpendicular to the septum, 2–4 mm long; style, 0.2 mm long; seeds 3 to 10 per locule; pedicels spreading, 4–8 mm long. Vernally moist calcareous or saline soil of wet meadows, steppe; valleys, montane. BC, MT south to CA; Eurasia.

Iberis L. Candytuft

Iberis sempervirens L. Glabrous perennial forming loose mats. **Stems** prostrate and erect; erect stems simple, 10–30 cm. **Leaves** all cauline, sessile, narrowly oblanceolate, succulent, 15–35 mm long, entire. **Inflorescence** a compact, ebracteate raceme. **Flowers** white; petals dimorphic 5.5–8.5 mm long or 3.7–5.2 mm long. **Fruit** ovate with a deep, v-shaped sinus, flattened perpendicular to the septum, 6–7 mm long; style 0.1–0.3; seeds 1 per locule; pedicel spreading, 6–8 mm long. Roadsides; valleys. Occasionally introduced from gardens; native to Europe; collected in Ravalli Co.

Idahoa A.Nelson & J.F.Macbr. Scalepod

Idahoa scapigera (Hook.) A.Nelson & J.F.Macbr. Stemless, glabrous annual. **Leaves** basal, long-petiolate; blades 3–5 mm long, ovate, entire or basally lobed. **Flowers** solitary on a scape-like peduncle 2–5 cm long; petals white, 1.5–2 mm long. **Fruit** an orbicular silicle, compressed parallel to the septum, 6–9 mm long, purple-spotted; style ≤0.2 mm long; seeds 3 to 6 per locule. Flowering early in vernally moist, shallow soil of warm slopes in grasslands; valleys, montane; Flathead and Ravalli cos. BC, MT south to CA, NV, ID.

Isatis L. Woad

Isatis tinctoria L. DYER'S WOAD Glaucous biennial with a woody, branched taproot. **Stems** erect, simple, 30–100 cm. **Basal leaves** petiolate, oblanceolate, 5–18 cm long, entire. **Stem leaves** lanceolate, sessile, auriculate. **Vestiture** hirsute with simple hair at the stem base, otherwise glabrous. **Inflorescence** a branched panicle, each raceme ebracteate. **Flowers** yellow; petals 2.5–3.5 mm long. **Fruit** narrowly oblong, blunt-tipped, flattened, 12–18 mm long, lacking a septum; style lacking; seed solitary; pedicel reflexed down, 6–10 mm long. Grasslands, steppe, roadsides; valleys. Introduced and naturalized throughout intermountain U.S.; native to Europe.

Lepidium L. Peppergrass, white top

Annual to perennial herbs. **Leaves** basal and/or cauline, simple, dentate to pinnately lobed, petiolate below, sessile above. **Inflorescence** usually branched, ebracteate racemes. **Flowers** white or yellow; stamens 2, 4, or 6. **Fruit** a silicle, compressed perpendicular to the septum, sometimes notched at the tip; seeds 1 to 2 per locule.

Recent molecular-genetics data indicate that species of *Cardaria* are nested within *Lepidium*.[301] *Lepidium densiflorum*, *L. ramosissimum* and *L. virginicum* are similar and can be difficult to distinguish, partly because there is a great deal of variation within each; the following key will allow determination in most but not all cases.

1. Mature silicles not notched at the tip; plants rhizomatous 2
1. Mature silicles with a notch at least 0.1 mm deep at the tip; not rhizomatous 5

2. Silicles pubescent. 3
2. Silicles glabrous 4

3. Style ≤0.1 mm long. . . . . *L. latifolium*
3. Style >0.4 mm long. . . . . *L. appelianum*

4. Silicles cordate at the base. . . . . *L. draba*
4. Silicles compressed-globose, not basally cordate . . . . *L. chalepense*

5. Stem leaves auriculate to perfoliate 6
5. Stem leaves not auriculate or perfoliate 7

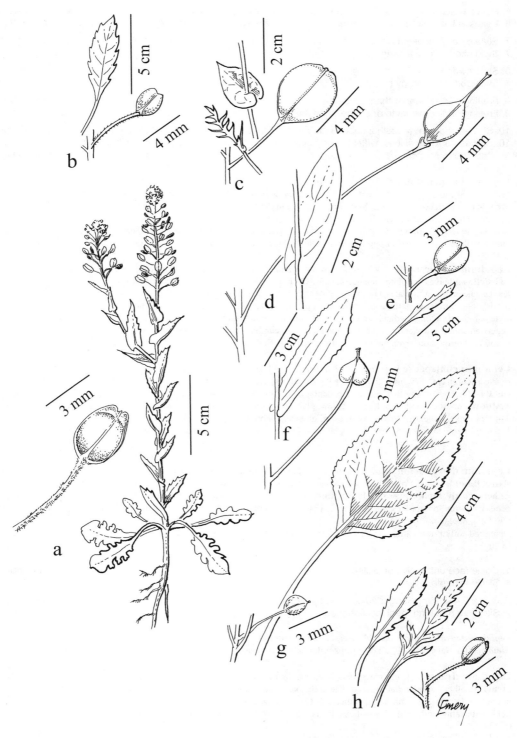

Plate 31. **Lepidium**. **a.** *L. campestre*, **b.** *L. virginicum*, **c.** *L. perfoliatum*, **d.** *L. chalipensis*, **e.** *L. densiflorum*, **f.** *L. draba*, **g.** *L. latifolium*, **h.** *L. ramosissimum* var. *ramosissimum*.

6. Petals white; silicles 4.5–6 mm long; stem leaves never perfoliate . . . . *L. campestre*
6. Petals yellow; silicles 3.5–4.5 mm long; stem leaves sometime perfoliate. . . . .*L. perfoliatum*

7. Style extending beyond the sinus of the silicle, >0.2 mm long; plants perennial . . . . *L. montanum*
7. Style shorter than the sinus; plants annual or biennial. 8

8. Style evident, 0.1–0.5 mm long . . . . *L. sativum*
8. Style absent or nearly so 9

9. Fruiting pedicel longer than the silicle . . . . *L. virginicum*
9. Fruiting pedicel equal to or shorter than the silicle. 10

10. Silicles obovate, widest just above the middle; branches ca. as high as main stem . . . . *L. densiflorum*
10. Silicles narrowly elliptic, widest at or just below the middle; branches often much shorter
   than the main stem. . . . . *L. ramosissimum*

Lepidium appelianum Al-Shebaz [*Cardaria pubescens* (C.A.Mey.) Jarm.] Rhizomatous
perennial. **Stems** ascending to erect, 20–45 cm. **Basal leaves** lanceolate, 3–6 cm long, dentate.
Stem leaves auriculate-clasping. **Vestiture** of sparse to dense, simple hairs. **Flowers** white;
petals 2–4 mm long; stamens 6. **Fruit** obovoid, pubescent, rounded at the base, 2–4 mm long, not
notched; style 0.5–1 mm long; pedicels 4–11 mm long, glabrous. Agricultural fields, roadsides;
plains, valleys. Introduced throughout temperate N. America; native to Asia.

Lepidium campestre (L.) W.T.Aiton Taprooted annual. **Stems** erect, 12–50 cm, branched
above. **Basal leaves** petiolate, oblanceolate, 2–12 cm long with entire to lobed margins. **Stem
leaves** lanceolate, auriculate-clasping, dentate. **Vestiture** of stiff, simple hairs. **Flowers** white;
petals 1.5–3 mm long; stamens 6. **Fruit** ovate, papillose, 4.5–6 mm long, winged and notched;
style 0.4–0.8 mm long; pedicels spreading, pubescent, 4–6 mm long. Disturbed or sparsely
vegetated soil of grasslands, meadows, open forest, streambanks; plains, valleys, montane.
Introduced throughout temperate N. America; native to Eurasia. (p.209)

Lepidium chalepense L. [*Cardaria chalepensis* (L.) Hand.-Mazz., *C. draba* (L.) Desv. var.
repens (Schrenk) O.E.Schulz,] WHITETOP Rhizomatous perennial. **Stems** erect, simple, 20–60
cm. **Basal leaves** oblong to oblanceolate, dentate, 3–12 cm long. **Stem leaves** auriculate clasping.
Vestiture pubescent with simple hairs below, glabrous in the inflorescence. **Flowers** white; petals
3–4 mm long; stamens 6. **Fruits** compressed globose, glabrous, 2–5 mm long, not notched; style
1–2 mm long; pedicels 8–15 mm long, glabrous. Fields, roadsides, streambanks; plains, valleys. Introduced to western
N. America; native to Asia. (p.209)

Lepidium densiflorum Schrad. Annual or biennial. **Stems** erect, 8–50 cm, usually branched
above. **Basal leaves** petiolate, narrowly oblanceolate with sinuate to shallowly lobed margins,
withering early. **Stem leaves** sessile. **Vestiture** of sparse, short, simple hairs. **Flowers** white;
petals ca. 1 mm long or lacking; stamens 2. **Fruit** broadly obovate, glabrous to pubescent, 2.5–3.5
mm long, notched; style lacking; pedicels spreading, glabrous or puberulent, 1.5–3.5 mm long.
Disturbed soil of fields, grasslands, steppe, open forest; plains, valleys, montane. Widespread in temperate N. America.
(p.209)

1. Silicle face uniformly puberulent. . . . . .var. *pubicarpum*
1. Silicle face glabrous 2

2. Silicle 2.4–2.8 mm long; pedicles nearly terete . . . . .var. *densiflorum*
2. Silicle ≥3 mm long; pedicels flattened . . . . .var. *ramosum*

Lepidium densiflorum var. *densiflorum*, *L. densiflorum* var. *pubicarpum* (A.Nelson) Thell**,** and *L. densiflorum* var.
ramosum (A.Nelson) Thell. all occur in similar habitats across the state.

Lepidium draba L. [*Cardaria draba* (L.) Desv.] WHITETOP Perennial similar to *L. chalepense*.
Stems 20–50 cm. **Fruits** glabrous, cordate at the base, broader below than above, 2–4 mm long,
not notched; style 1–1.5 mm long; pedicels 5–12 mm long, glabrous. Most abundant in irrigated
fields with saline soil, roadsides; valleys. Introduced throughout N. America; native to Eurasia.
(p.209)

Lepidium latifolium L. PERENNIAL PEPPERWEED Glabrous, rhizomatous perennial. **Stems** 40–100 cm, erect, branched. **Basal leaves** petiolate; the blade narrowly elliptic, serrate, withered by flowering time. **Stem leaves** becoming sessile. **Flowers** white; petals 1.5–2 mm long; stamens 6. **Fruit** ovate, 2–3 mm long, pubescent, not notched; style ≤0.1 mm long; pedicels 2–5 mm long, ascending, glabrous. Fields, pastures, roadsides; valleys. Introduced throughout temperate N. America; native to Eurasia. (p.209)

Lepidium montanum Nutt. Taprooted biennial or short-lived perennial. **Stems** prostrate to ascending, branched, 10–20 cm. **Basal leaves** 3–10 cm long, petiolate, the blade lanceolate, pinnately lobed with dentate margins. **Stem leaves** similar, reduced upward. **Vestiture** of short, thick hairs. **Flowers** white; petals 2–3 mm long; stamens 6. **Fruit** ovate, glabrous, 2–4 mm long, notched; style 0.2–0.7 mm long; pedicels glabrous, spreading, 3–8 mm long. Moist alkaline meadows; montane in Beaverhead Co. OR to MT south to CA, AZ, NM. Our plants are var. *montanum*.

Lepidium perfoliatum L. Taprooted annual. **Stems** erect, 10–45 cm, simple. **Basal leaves** petiolate, oblanceolate, 1–12 cm long, 2–3-times divided into linear segments. **Stem leaves** 3–30 mm long, broadly ovate, cordate to perfoliate with entire margins, becoming sessile. **Vestiture:** basal leaves with long, simple hair or glabrous; glabrous above. **Flowers** yellow; petals 1–1.5 mm long; stamens 6. **Fruit** ovate, glabrous, 3–4.5 mm long, shallowly notched; style 0.2–0.4; pedicels ascending, glabrous, 3–5 mm long. Sparsely vegetated, vernally moist soil of grasslands, steppe, roadsides and around saline or alkaline wetlands; plains, valleys. Introduced throughout much of temperate N. America; native to Eurasia. (p.209)

Lepidium ramosissimum A.Nelson Annual or biennial. **Stems** erect, 7–50 cm, copiously branched. **Basal leaves** petiolate, oblanceolate, pinnatifid, withering early. **Stem leaves** 12–50 mm long, dentate, becoming linear and entire above. **Vestiture** of small simple hairs. **Flowers** white; petals ca. 1 mm long or lacking; stamens 2. **Fruit** narrowly elliptic, glabrous or sparsely short-hairy, 2.5–3 mm long, narrowly notched; style lacking; pedicels spreading, puberulent, 2–4 mm long. Disturbed soil of grassland, steppe, roadsides; plains, valleys, montane. AK to NL south to CA, UT, CO. (p.209)

1. Silicles glabrous . . . . var. *bourgeauanum*
1. Silicles pubescent. . . . . .var. *ramosissimum*

Lepidium ramosissimum var. *ramosissimum* and *L. ramosissimum* var. *bourgeauanum* (Thell.) Rollins are morphologically distinct but occur in similar habitats.

Lepidium sativum L. PEPPER CRESS Mostly glabrous and glaucous, taprooted annual. **Stems** erect, 20–80 cm, usually branched above. **Basal leaves** withered at anthesis. **Stem leaves** petiolate; blades oblanceolate, bipinnately lobed and toothed, 3–10 cm long, sessile above. **Vestiture** of sparse, simple hairs on upper leaf surfaces. **Flowers** white; petals 1.5–2.5 mm long; stamens 6. **Fruit** broadly eliptic, 4–6 mm long, winged and notched, glabrous; style 0.1–0.5 mm long; pedicels erect-ascending, glabrous, 2–5 mm long. Fields, roadsides; valleys. Introduced across temperate N. America; native to Europe.

Lepidium virginicum L. Annual. **Stems** erect, 8–20 cm, often branched above. **Basal leaves** petiolate, obovate, deeply dentate, early deciduous. **Stem leaves** narrowly oblanceolate, 1–3 cm long, dentate. **Vestiture** of short, simple hairs. **Flowers** white; petals ca. 1 mm long; stamens 2. **Fruit** broadly elliptic, glabrous or pubescent, 2.5–3.5 mm long, notched; style lacking; pedicels spreading, glabrous or pubescent, 3–5 mm long. Disturbed soil of roadsides, fields, grasslands, wetlands; plains, valleys, montane. Throughout temperate N. America south to C. America. (p.209)

1. Pedicels glabrous . . . . .var. *medium*
1. Pedicels puberulent . . . . var. *pubescens*

Lepidium virginicum var. *medium* (Greene) Hitchc. and *L. virginicum* var. *pubescens* (Greene) Thell. are morphologically distinct but occur in similar habitats.

Lobularia Desv.

Lobularia maritima (L.) Desv. Perennial. **Stems** decumbent, branched, 10–35 cm. **Leaves** cauline, linear, 1–6 cm long, entire. **Vestiture** grayish with appressed, dolabriform hairs. **Inflorescence** a simple, narrow, many-flowered raceme. **Flowers** white to pink; sepals 1–2 mm

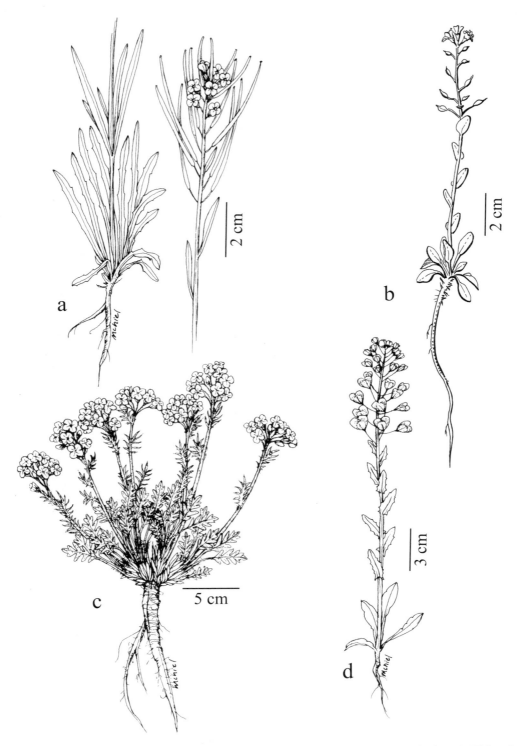

Plate 32. a. *Erysimum inconspicuum*, b. *Noccaea fendleri*, c. *Smelowskia calycina*, d. *Thlaspi arvense*

long; petals 2–3 mm long. **Fruit** elliptic, flattened, 2–4 mm long, sparsely hairy; pedicels 5–11 mm long; style 0.5 mm long; seeds 1 per locule. Roadsides; valleys. Introduced throughout temperate N. America.

Malcolmia W.T.Aiton

Malcolmia africana (L.) W.T.Aiton [*Strigosella africana* (L.) Botsch.] Annual. **Stems** 5–30 cm, often branched. **Leaves** cauline, petiolate; blade lanceolate to oblanceolate, dentate, 4–8 cm long. **Vestiture** of branched hairs. **Inflorescence** a raceme, ebracteate above. **Flowers** pink to rose; petals 6–10 mm long, clawed. **Fruit** a straight or slightly curved silique, 4–6 cm long, erect or spreading on pedicels ca. 1 mm long; seeds in 1 row per locule. Fields, disturbed grasslands; valleys. Introduced to much of western N. America; native to north Africa.

Noccaea Moench

Glabrous perennial herbs. **Leaves** basal and cauline; the basal petiolate, the blade ovate to elliptic, crenate to entire; the cauline broadly lanceolate, auriculate, sessile. **Inflorescence** an ebracteate raceme. **Flowers** white. **Fruits** flattened, obovate to oblanceolate silicles; the apex shallowly 2-lobed; the margins keeled; seeds 1–12 per locule; pedicels ascending to spreading.

Our two species are similar and part of a circumboreal complex. *Noccaea fendleri* may be conspecific with Eurasian *N. montana* (L.) K.K.Mey.[192,193,342] or specifically distinct.[187,221] Molecular evidence suggests that *N. parviflora* may be conspecific with *N. fendleri*.

1. Style ≤0.5 mm long; petals 2–4 mm long . . . . *N. parviflora*
1. Style ≥1 mm long; petals 4–7 mm long . . . . *N. fendleri*

Noccaea fendleri (A.Gray) Holub [*Thlaspi glaucum* (A.Nelson) A.Nelson, *T. fendleri* A.Gray var. *glaucum* (A.Nelson) C.L.Hitchc., *Noccaea montana* (L.) F.K.Mey. and *Thlaspi montanum* L. misapplied] **Stems** 5–35 cm, unbranched from a simple or branched caudex. **Basal leaf blades** 5–20 mm long. **Stem leaves** 5–15 mm long. **Petals** 4.5–7 mm long. **Fruits** 5–10 mm long; styles 1–2 mm long; pedicels 3–10 mm long. Moist soil of grasslands, meadows, open forest, turf; valleys to alpine. WA to MT south to CA, AZ, TX, Mex. Our plants are var. *glauca* (A.Nelson) Al-Shebaz & M.Koch. (p.212)

Noccaea parviflora (A.Nelson) Holub [*Thlaspi parviflorum* A.Nelson] Similar to *N. fendleri*. **Stems** 7–25 cm from a simple caudex. **Basal leaf blades** 5–12 mm long. **Stem leaves** 6–12 mm long. **Petals** 2–4 mm long. **Fruits** 4–6 mm long; styles ≤0.5 mm long; pedicels 2–5 mm long. Moist meadows, turf; montane to alpine. Endemic to southwest MT and adjacent ID, WY.

Physaria (Nutt. ex Torr.& A.Gray) A.Gray Bladderpod

Short-lived, taprooted perennials (annual in *P. arenosa*?) from a simple or branched caudex. **Leaves** basal and cauline; the basal petiolate; stem leaves reduced. **Vestiture** of usually dense, stellate or branched hairs. **Flowers** yellow. **Fruit** a densely pubescent silicle, often short-stipitate, globose or 2-lobed, often inflated; seeds 1–10 per locule.[146,343]

Species of *Lesquerella* are included in *Physaria* based on morphological similarities and unpublished molecular genetics data[7] showing that species of *Physaria* arose from several different lineages of *Lesquerella*; *Physaria* is the older name. *Physaria curvipes*, *P. pachyphylla*, *P. pycnantha*, and *P. spatulata* have all been segregated from what was formerly *Lesquerella alpina*. *Physaria eriocarpa* Grady & O'Kane, described from one population[146] was segregated based on morphological characters that intergrade with those of the more common *P. spathulata* and might best be considered a high-elevation form of the latter species.

1. Mature fruit 2-lobed, at least above; as wide or wider than long 2
1. Mature fruit rounded or pointed at the tip; as long or longer than wide 8

2. Mature fruit ≤7 mm high 3
2. Mature fruit ≥7 mm high 5

3. Mature fruit ≥5 mm high . . . . *P. geyeri*
3. Mature fruit ≤4 mm high 4

4. Leaf trichomes with erect rays; plants montane . . . .*P. klausii*
4. Trichome rays parallel to leaf surface; plants upper subalpine or alpine . . . . *P. humilis*

5. Mature fruit with a deep basal sinus; replum linear to narrowly oblong 6
5. Mature fruit with only a shallow basal sinus; replum elliptic 7

6. Fruit with 2 to 4 ovules per locule (count ovule stalks); some basal leaf blades
   lyrate or dentate . . . . *P. didymocarpa*
6. Fruit with 2 ovules per locule; leaf blades mostly entire . . . .*P. acutifolia*

7. Plants of the plains; replum linear, constricted in the middle . . . .*P. brassicoides*
7. Plants subalpine to alpine; replum narrowly oblong . . . . *P. saximontana*

8. Pedicels recurved 9
8. Pedicels spreading to ascending, sometimes sigmoid 11

9. Style 1–2 mm long; basal leaves spatulate . . . . *P. lesicii*
9. Style ≥2 mm long; basal leaves linear to narrowly oblanceolate 10

10. Raceme secund; outer basal leaf blades oblong to narrowly ovate . . . . *P. arenosa*
10. Raceme not secund; outer basal leaves linear or narrowly oblanceolate . . . .*P. ludoviciana*

11. Mature silicle flattened perpendicular to the septum 12
11. Mature silicle somewhat inflated, not flattened perpendicular to the septum .13

12. Mature silicle orbicular, ca. as wide as high . . . .*P. pulchella*
12. Mature silicle elliptic, longer than wide . . . . *P. carinata*

13. Silicles nearly globose; stems erect . . . .*P. douglasii*
13. Silicles lanceoloid; stems usually decumbent to prostrate 14

14. Basal leaves linear . . . . *P. pycnantha*
14. Basal leaves with distinct blade and petiole 15

15. Style ≤½ length of the silicle . . . . *P. curvipes*
15. Style >½ length of the silicle 16

16. Basal leaf blades nearly 1 mm thick . . . . *P. pachyphylla*
16. Basal leaf blades thinner . . . . *P. spatulata*

Physaria acutifolia Rydb. **Stems** ascending, 5–15 cm from simple caudex clothed in old
leaf bases. **Basal leaves** 2–6 cm long, the blades ovate to obovate, usually entire. **Stem leaves**
spatulate to oblanceolate. **Vestiture** of dense, stellate hairs, the stellae branched. **Petals** 7–12
mm long. **Fruit** of 2 ovate lobes, inflated, 6–10 mm wide; style 4–6 mm long; seeds 2 per locule;
pedicels spreading, straight to sigmoid, 5–10 mm long. Sandy, usually calcareous soil of open slopes
in grasslands, steppe, woodlands; plains, valleys. MT to UT and NM.

Physaria arenosa (Richardson) O'Kane & Al-Shebaz [*Lesquerella arenosa* (Richardson)
Rydb., *L. ludoviciana* Nutt.) S.Watson var *arenosa* (Richardson) S.Watson] **Stems** prostrate to
ascending, 5–20 cm from a simple caudex. **Basal leaves** rosulate, ovate to oblanceolate, 1–5 cm
long, entire. **Stem leaves** linear to narrowly oblanceolate. **Vestiture** of 4- to 6-rayed short-stalked
hairs. **Petals** 6–9 mm long. **Fruit** globose, 3–5 mm high, style 2.5–4 mm long; seeds 4 to 7 per locule;
pedicels curved down, 5–10 mm long. Sparsely vegetated, sandy soil of grasslands, steppe; plains, valleys. AB to MB
south to CO, WY. Our plants are var. **arenosa**. See *P. ludoviciana*.

Physaria brassicoides Rydb. **Stems** ascending, 5–15 cm from simple or branched caudex clothed in old leaf bases.
Basal leaves 2–6 cm long, the blades obovate to rhombic, entire or with a few large teeth. **Stem leaves** spatulate to
oblanceolate. **Vestiture** of dense, stellate hairs, the stellae branched. **Petals** 6–10 cm long. **Fruit** of 2 ovate lobes,
inflated, 7–10 mm wide, the basal sinus small, the apical sinus narrow; style 3–6 mm long; seeds 2 per locule; pedicels
spreading, straight to sigmoid, 5–20 mm long. Sandy soils of badlands, river breaks; plains; Carter and Powder River
cos. MT, ND south to WY, NE.

Physaria carinata (Rollins) O'Kane & Al-Shebaz [*Lesquerella carinata* Rollins var. *languida* Rollins] **Stems**
prostrate to ascending, 3–20 cm from a simple caudex. **Basal leaves** 1–4 cm long, the blade obovate, entire. **Stem
leaves** oblanceolate, becoming sessile above. **Vestiture** of short-stalked, forked hairs. **Petals** 6–11 mm long. **Fruit**
narrowly elliptic, 4–7 mm high, flattened perpendicular to the septum, keeled; style 2–5 mm long; pedicels ascending to
spreading, 3–15 mm long. Sparsely vegetated, stony, calcareous soil of grasslands; valleys to montane; Granite Co. MT,
ID south to WY. (p.216)

Our plants (*Lesquerella carinata* var. *languida*) had been separated from those of WY and ID[342] based on characters that are continuous. *Physaria pulchella* has nearly orbicular fruits and occurs at higher elevations. Reports of *P. paysonii* (Rollins) O'Kane & Al-Shebaz were based on collections of *P. carinata*.

Physaria curvipes (A.Nelson) Grady & O'Kane [*Lesquerella alpina* (Nutt.) S.Watson misapplied] **Stems** ascending, 8–24 cm from a simple caudex. **Basal leaves** spatulate, 25–50 mm long, the blade elliptic to rhombic, entire. **Stem leaves** spatulate. **Vestiture** of dense, appressed, 4- to 5-rayed stellate hairs. **Petals** 4–6 mm long. **Fruit** ovoid, inflated, 5–9 mm high; style 1.5–2.5 mm long; seeds 2 to 4 per locule; pedicels sigmoid-spreading, 4–7 mm long. Stony, calcareous soil of exposed slopes, ridges in grassland, open forest; montane to subalpine. Endemic to Big Horn Co. MT and adjacent WY.

Physaria didymocarpa (Hook.) A.Gray [*P. lanata* (A.Nelson) Rydb., *P. integrifolia* (Rollins) Lichvar] **Stems** ascending, 2–15 cm from simple caudex clothed in old leaf bases. **Basal leaves** 1–6 cm long, the blades obovate, entire to coarsely dentate. **Stem leaves** narrowly oblanceolate, entire or few-toothed. **Vesititure** of dense, stellate hairs, the stellae branched or simple. **Petals** 7–12 cm long. **Fruit** of 2 ovate lobes, inflated, 7–20 mm high, the basal and apical sinuses narrow; style 5–9 mm long; seeds 4 per locule; pedicels spreading, straight to curved, 3–12 mm long. Sparsely vegetated, sandy or clayey soil of eroding slopes, banks, badlands; plains, valleys, montane. AB south to WY. (p.216)

Although *Physaria didymocarpa* usually has 3 to 4 ovules per locule, many fruits, even on the same plant, have only 2. Var. *lanata* A.Nelson has been segregated based on spreading leaf trichomes, a character that seems continuous in our material; var. *integrifolia* Rollins is reported to have large fruits and entire basal leaves, characters that do not correlate in specimens I have seen. See *P. saximontana*.

Physaria douglasii (S.Watson) O'Kane & Al-Shebaz **Stems** erect, 10–30 cm from a simple caudex. **Basal leaves** 2–9 cm long; the blade oblanceolate to obovate with entire to weakly dentate margins. **Stem leaves** numerous, narrowly oblanceolate, entire. **Vesititure** of 4- to 6-rayed sessile hairs. **Petals** 6–9 mm long. **Fruit** globose, inflated, 3–4 mm high; style 3–4 mm long; seeds 2 to 4 per locule; pedicels ascending, straight to sigmoid, 6–20 mm long. Sandy soil of open ponderosa pine forest; valleys. BC, MT south to WA, OR. Known only from Lincoln Co.

Physaria arenosa has down-curved fruiting pedicels; *P. spatulata* has ovoid fruits.

Physaria geyeri (Hook.) A.Gray **Stems** ascending, 3–20 cm from a simple or branched caudex clothed in old leaf bases. **Basal leaves** 1–5 cm long; the blades ovate to orbicular, entire. **Stem leaves** oblanceolate, reduced. **Vesititure** of dense, stellate hairs. **Petals** 7–11 mm long. **Fruit** obcordate, inflated, 5–7 mm high, 2-lobed, with a wide apical sinus; style 3–7 mm long; seeds 2 to 3 per locule; pedicels spreading, sigmoid, 4–12 mm long. Sparsely vegetated, gravelly or sandy soil of grasslands, steppe, open forest; valleys to subalpine. WA, ID and MT. (p.216)

Physaria humilis (Rollins) O'Kane & Al-Shebaz [*Lesquerella humilis* Rollins] **Stems** ascending to prostrate, 2–6 cm from a simple caudex clothed in old leaf bases. **Basal leaves** rosulate, 1–3 cm long; the blade obovate, entire. **Stem leaves** oblanceolate. **Vesititure** of 4- to 6-rayed hairs, the rays branched. **Petals** 5–8 mm long. **Fruit** obcordate, wider than high, 3–4 mm high, flattened perpendicular to the septum; style 1–2 mm long; seeds 2 per locule; pedicels ascending, straight, 3–5 mm long. Fellfields in metamorphic parent material; alpine. Endemic to the Bitterroot Mtns. of Ravalli Co. (p.216)

Physaria klausii (Rollins) O'Kane & Al-Shebaz [*Lesquerella klausii* Rollins] **Stems** ascending, 4–20 cm from a simple caudex. **Basal leaves** rosulate, 1–4 cm long, the blade obovate to orbicular, entire or few-toothed. **Stem leaves** oblanceolate, entire or few-toothed. **Vesititure** of loosely spreading stellate hairs. **Petals** 6–8 mm long. **Fruit** obovate, slightly bilobed above, wider than high, 2–4 mm high, flattened perpendicular to the septum; style 2–3 mm long; seeds 2 per locule; pedicels sigmoid, 3–10 mm long. Sparsely vegetated, gravelly, shale- or limestone-derived soil in grasslands, open forest; montane. Endemic to Lewis & Clark and Meagher cos., MT. (p.216)

Fruits of *Physaria geyeri* are as long as wide. The type locality is in Lewis and Clark Co.

Physaria lesicii (Rollins) O'Kane & Al-Sheebaz [*Lesquerella lesicii* Rollins] **Stems** ascending to prostrate, 5–20 cm from a simple caudex. **Basal leaves** long-petiolate, 1–4 cm long, the blade ovate to elliptic, entire. **Stem leaves** oblanceolate. **Vesititure** of stellate hairs. **Petals** 5.5–7 mm long. **Fruit** globose, 2–4 mm high, inflated; style 1–2 mm long; seeds 3 to 5 per locule; pedicels recurved, 4–11 mm long. Sparsely vegetated, stony, calcareous soil of exposed ridges, upper slopes in grassland, woodlands; montane. Endemic to the Pryor Mtns. of Carbon Co.

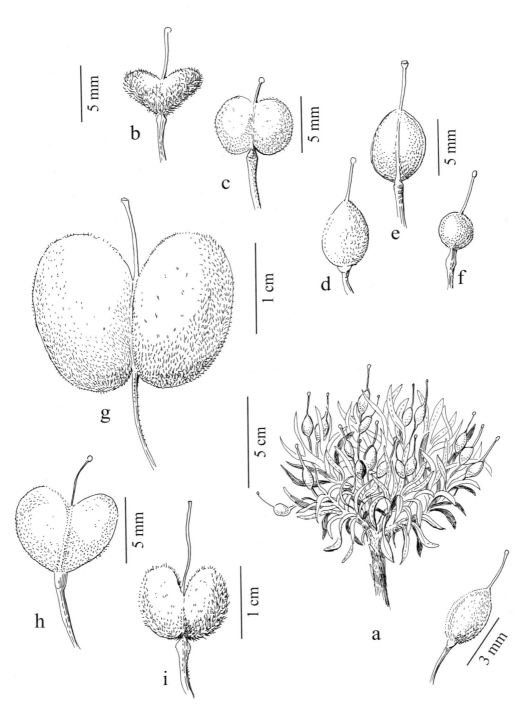

Plate 33. **Physaria**. **a.** *P. pycnantha*, **b.** *P. klausii*, **c.** *P. humilis*, **d.** *P. spathulata*, **e.** *P. carinata*, **f.** *P. ludoviciana*, **g.** *P. didymocarpa*, **h.** *Physaria geyeri*, **i.** *P. saximontana*.

Physaria ludoviciana (Nutt.) O'Kane & Al-Shebaz [*Lesquerella ludoviciana* Nutt.] **Stems** ascending, 15–25 cm from a simple caudex. **Basal leaves** rosulate, erect, linear, 3–7 cm long. **Stem leaves** similar, reduced upward. **Vestiture** of 4- to 7-rayed stellate hairs. **Petals** 6–8 mm long. **Fruit** globose, 3–4 mm high; style 2–5.5 mm long; seeds 2 to 6 per locule; pedicels curved down, 10–16 mm long. Very sandy soil around sandstone outcrops in grasslands; plains. MT to MB south to NV, AZ, CO, OK, IL. (p.216)

The similar *Physaria arenosa* has 1-sided racemes and somewhat shorter stems and broader leaves.

Physaria pachyphylla O'Kane & Grady **Stems** ascending, 2–15 cm from a simple caudex clothed in old leaf bases. **Basal leaves** spatulate to oblanceolate, 15–50 mm long, entire. **Stem leaves** spatulate. **Vestiture** of dense, 5-rayed, stellate hairs. **Petals** 6–8 mm long. **Fruit** narrowly elliptic to ovoid, inflated, 3–6 mm high; style 2.5–4 mm long; seeds 4 per locule; pedicels ascending, curved, 3–10 mm long. Stony soil derived from limestone or calcareous sandstone on exposed slopes, ridges; valleys. Endemic to Carbon Co. MT and probably adjacent WY.

Physaria spatulata has sigmoid rather than up-curved pedicels.

Physaria pulchella (Rollins) O'Kane & Al-Shebaz [*P. carinata* (Rollins) O'Kane & Al-Shehbaz var. *pulchella* (Rollins) O'Kane, *Lesquerella pulchella* Rollins] **Stems** ascending to prostrate, 4–8 cm from a simple caudex. **Basal leaves** rosulate, 1–2 cm long, the blade ovate to elliptic, blunt, entire. **Stem leaves** spatulate. **Vestiture** of dense, stellate hairs. **Petals** 5–9 mm long. **Fruit** broadly elliptic to orbicular, 3–5 mm high, flattened perpendicular to the septum, ridged on each face; style 1.5–3 mm long; pedicels straight to sigmoid, spreading to ascending, 3–6 mm long. Sparsely vegetated, stony, calcareous soil on exposed ridges; subalpine to alpine. Endemic to Beaverhead Co. See *P. carinata*.

Physaria pycnantha Grady & O'Kane [*Lesquerella alpina* (Nutt.) S.Watson var. *condensata* (A.Nelson) C.L.Hitchc misapplied] **Stems** erect to ascending, 1–8 cm from a branched caudex clothed in old leaf bases. **Basal leaves** 4–20 mm long, linear, narrowly oblong, entire. **Stem leaves** similar. **Vestiture** of dense, stellate hairs, the branches raised. **Petals** 3.5–7.5 mm long. **Fruit** ovoid, inflated, 2–3 mm high; style 1.5–3.5 mm long; seeds 2 to 4 per locule; pedicels sigmoid-spreading, 3–5 mm long. Stony, usually calcareous soil of open slopes in grasslands, steppe; valleys to montane. Endemic to MT and adjacent ID. (p.216)

Physaria spatulata has broader basal leaves and more appressed trichomes.

Physaria saximontana Rollins **Stems** ascending to prostrate, 3–10 cm from a simple caudex clothed in old leaf bases. **Basal leaves** 15–40 mm long; the blades orbicular to rhombic, deeply few-toothed or lyrate. **Stem leaves** oblanceolate to spatulate, entire. **Vestiture** of dense, stellate hairs with branched stellae. **Petals** 8–12 mm long. **Fruit** of 2 ovate lobes, inflated, 7–11 mm high, the apical sinus deep, the basal sinus absent; style 5–7 mm long; seeds 2 per locule; pedicels spreading, straight to curved, 5–10 mm long. Sparsely vegetated, stony, usually calcareous soil in grasslands, fellfields on exposed slopes, ridges; montane to alpine. Endemic to MT and WY. Our plants are ssp. *dentata* (Rollins.) O'Kane. (p.216)

Fruits of *Physaria didymocarpa* are lobed at the base with the pedicel entering the basal sinus.

Physaria spatulata (Rydb.) Grady & O'Kane [*Lesquerella alpina* (Nutt.) S.Watson var. *spatulata* (Rydb.) Payson, *P. reediana* O'Kane & Al-Shehbaz ssp. *spatulata* (Rydb.) O'Kane & Al-Shebaz, *Physaria eriocarpa* Grady & O'Kane] **Stems** simple, erect to ascending, 1–15 cm from a simple or branched caudex clothed in old leaf bases. **Basal leaves** 7–30 mm long, oblanceolate to spatulate, entire. **Stem leaves** linear to narrowly oblanceolate. **Vestiture** of dense, appressed, stellate hairs. **Petals** 5–7 mm long. **Fruit** ovoid, inflated, 2–5 mm high; style 2–3.5 mm long; seeds 2 to 4 per locule; pedicels sigmoid-spreading, 4–12 mm long. Sandy or gravelly soil of exposed slopes, ridges in grasslands, steppe, woodlands, fellfields, more common in calcareous soil; plains, valleys to alpine. AB, SK south to WY, SD. See *P. pycnantha*. (p.216)

Raphanus L. Radish

Annual or biennial herbs. **Leaves** petiolate at least below. **Vestiture** of sparse, simple hairs. **Inflorescence** an ebracteate, often branched raceme. **Fruit** a silique with a prominent, conical beak; seeds in 1 row; style absent.

1. Siliques 3–5 mm thick, constricted between seeds; petals white to yellow . . . .*R. raphanistrum*
1. Siliques 5–10 mm thick, little constricted; petals purplish. . . . .*R. sativus*

Raphanus raphanistrum L. **Stems** 30–80 cm. **Leaves** all cauline, petiolate; the blade pinnately lobed with a large terminal lobe or dentate above, 9–23 cm long. **Petals** 13–22 mm long, yellow to white with purple veins. **Fruit** 4–7 cm × 3–5 mm, constricted between seeds; the beak 13–23 mm long; pedicels 10–26 mm long, ascending. Fields; valleys; collected in Pondera and Stillwater cos. Widespread in N. America; introduced from Eurasia.

Raphanus sativus L. **Stems** 40–100 cm from a fleshy taproot (radish). **Leaves** basal and cauline, petiolate below; the blade pinnately lobed with a large terminal lobe, dentate, 10–20 cm long, reduced above. **Petals** 11–18 mm long, purple. **Fruit** 2–7 cm × 5–10 mm, slightly constricted between seeds; the beak 9–20 mm long; pedicels 10–26 mm long, spreading. Fields, roadsides; we have one collection from Madison Co. taken more than 100 years ago. Escaped from cultivation, introduced throughout N. America; native to Eurasia.

Rorippa Scop.

Annual to perennial herbs. **Leaves** basal and/or cauline, dentate to pinnately lobed. **Inflorescence** a raceme, ebracteate or bracteate below, axillary and/or terminal. **Flowers** yellow or white. **Fruit** a silique or silicle, linear to ovoid; seeds usually in 2 rows per locule.[193,389]

The *Rorippa curvipes* complex has been a source of confusion and has several different taxonomic treatments. *Rorippa truncata* is considered a variety of *R. curvipes* by some;[193] however, the fruit shape seems sufficiently distinct for it to be considered a separate species.[389] On the other hand, *R. curvipes* var. *alpina* seems little different from var. *curvipes* except in being longer lived; several high-elevation populations seem to be a mix of the two forms.

1. Fruiting pedicels 1–4 mm long 2
1. Fruiting pedicels ≥4 mm long 7

2. Fruit hairy over the entire surface; plants rhizomatous. . . . .*R. calycina*
2. Fruit glabrous, papillose or hairy on valve margins; non-rhizomatous 3

3. Fruiting pedicels longer than the fruit . . . .*R. palustris*
3. Fruiting pedicel as long as or shorter than the fruit 4

4. Tip of fruit blunt. . . . .*R. truncata*
4. Tip of fruit tapered to the style 5

5. Fruit ≥7 mm long, often hairy on the marginal sutures . . . . *R. curvisiliqua*
5. Fruit ≤7 mm long, papillose or glabrous 6

6. Fruit papillose (use 20X). . . . .*R. tenerrima*
6. Fruit glabrous . . . .*R. curvipes*

7. Pedicels longer than the mature fruit 8
7. Pedicels shorter or as long as the mature fruit. 9

8. Pedicels ≥7 mm long . . . .*R. austriaca*
8. Pedicels ≤6 mm long . . . .*R. palustris*

9. Mature fruit ≥8 mm long 10
9. Fruit mostly <8 mm long 12

10. Fruit >1.5 mm wide. . . . . *R. nasturtium-aquaticum*
10. Fruit ≤1 mm wide 11

11. Petals yellow; plants with erect to ascending stems. . . . .*R. sylvestris*
11. Petals white; stems floating . . . . *R. microphylla*

12. Tip of fruit blunt. . . . .*R. truncata*
12. Tip of fruit tapered to the style 13

13. Stems with short-conical hairs; pedicels 4–9 mm long. . . . . *R. sinuata*
13. Stems glabrous or with sparse fine hair; pedicels 2–5 mm long . . . .*R. curvipes*

Rorippa austriaca (Crantz) Besser Perennial, spreading by horizontal roots or stolons. **Stems** erect to ascending, 40–80 cm. **Leaves** basal and cauline; the basal rosette present only the first year; stem leaves 3–11 cm long, oblanceolate, strongly dentate, auriculate-clasping, sessile. **Vesture** of sparse, simple hairs below, glabrous above. **Petals** yellow, 1–2 mm long. **Fruit** mostly sterile, ovoid silicles 1–2 mm long; the fertile ca. 3 mm long, glabrous; style 0.5–1 mm long; pedicels ascending, 7–10 mm long. Roadsides, fields; valleys. Sporadically introduced throughout temperate N. America; native to Europe; collected in Jefferson and Park cos.

Rorippa calycina (Engelm.) Rydb. Rhizomatous perennial. **Stems** prostrate to ascending, 10–40 cm. **Leaves** all cauline, narrowly lanceolate, 3–7 cm long, pinnately lobed with dentate margins, sessile above. **Vestiture** of simple, straight hairs. **Petals** yellow, ca. 3 mm long. **Fruit** an oblong silicle, 2–4 mm long, pubescent; style 1–1.5 mm long; pedicels ascending, 2–4 mm long. Mud along rivers, ponds; plains. MT, ND, WY.

Rorippa curvipes Greene [*R. alpina* (S.Watson) Rydb., *R. obtusa* (Nutt.) Britton var. *alpina* (S.Watson) Britton] Taprooted annual to perennial. **Stems** prostrate to ascending, 5–20 cm long, branched with racemes terminal and from upper leaf axils. **Leaves** all cauline, petiolate, oblanceolate, 2–8 cm long, crenate-dentate to pinnately lobed, the terminal lobe largest. **Vestiture** of simple hairs. **Petals** yellow, 0.7–2 mm long. **Fruit** an ovoid silicle, 2–6 mm long, 1.3–2 mm wide, glabrous; style 0.4–1 mm long; pedicels spreading, 2–5 mm long. Mud along streams, ponds, lakes. WA and AB south to CA, NM. (p.220)

1. Plants perennial, often with a branched caudex; petals longer than sepals. . . . . var. *alpina*
1. Plants annual or biennial from a simple caudex; petals as long or shorter than sepals. . . . . . . var. *curvipes*

Rorippa curvipes var. *alpina* (S.Watson) Stuckey occurs in the alpine zone; *R. curvipes* var. *curvipes* is found montane to alpine.

Rorippa curvisiliqua (Hook.) Besser ex Britton Taprooted annual to biennial. **Stems** prostrate to ascending, 3–30 cm long, branched with racemes terminal and from upper leaf axils. **Leaves** basal and cauline; the basal withering early; stem leaves oblanceolate, 1–5 cm long, pinnately lobed to deeply dentate, the terminal lobe largest. **Vestiture** of sparse, recurved hair on the stem. **Petals** yellow, 0.6–1.1 mm long, shorter than the sepals. **Fruit** a linear silique, 7–17 mm long, 0.8–1.5 mm wide, usually hispid on the sutures; style 0.1–0.4 mm long; pedicels spreading to ascending, 1–3 mm long. Mud along shores of lakes, ponds, streams; montane, subalpine. AK south to CA, NV, WY. (p.220)
    Infraspecific taxa are difficult to distinguish.[193,342]

Rorippa microphylla (Boenn ex Rchb.) Hyl. ex A.Löve & D.Löve [*Nasturtium microphyllum* Boenn. ex Rchb.] Perennial from fibrous-rooted rhizomes. **Stems** prostrate or floating, hollow, 10–50 cm long. **Leaves** all cauline, petiolate, 2–6 cm long; the blades pinnately compound with lanceolate to elliptic lateral lobes and an elliptic terminal segment. **Petals** white, 4–5 mm long. **Fruit** a linear silique, 8–12 mm long, ca. 0.5 mm wide; style ca. 1 mm long; pedicels spreading to ascending, 8–12 mm long. Lakes, sloughs; valleys. Introduced in much of north-temperate N. America; native to Europe.
    The much more common *Rorippa nasturtium-aquaticum* has wider siliques.

Rorippa nasturtium-aquaticum (L.) Hayek [*Nasturtium officinale* W.T.Aiton, *Cardamine nasturtium-aquaticum* (L.) Borbas] WATER CRESS Nearly glabrous perennial from fibrous-rooted rhizomes. **Stems** prostrate or floating, hollow, 10–70 cm long. **Leaves** all cauline, petiolate, 3–15 cm long; the blades simple and suborbicular or pinnately compound with lanceolate to elliptic lateral lobes and a larger, usually cordate, terminal segment. **Petals** white, 3–5 mm long. **Fruit** a linear silique, 8–18 mm long, 1.8–2.6 mm wide, flattened; style 0.3–1 mm long; pedicels spreading to ascending, 6–15 mm long, the lowest often in leaf axils. Slow water of springs and (often spring-fed) creeks; valleys, montane. Introduced throughout temperate N. America; native to Europe. (p.220)
    In our area water cress often forms solid mats on the surface of water and seems to be confined to unpolluted water that doesn't freeze. Sometimes placed in a separate genus, *Nasturtium*; a recent molecular genetics study,[134] and vegetative morphology suggest that this species is most closely allied to *Cardamine*. See *Rorippa microphyllum*.

Rorippa palustris (L.) Besser [*R. islandica* (Oeder ex Murray) Borbas misapplied] Taprooted annual to short-lived perennial. **Stems** usually erect, 10–100 cm. **Leaves** basal and cauline; the lower leaves 3–15 cm long, petiolate, auriculate-clasping, oblanceolate, pinnately lobed and dentate; upper leaves similar, smaller, less divided and toothed. **Vestiture** of sparse, simple hairs below or plants glabrous. **Petals** yellow, 1.5–3 mm long. **Fruit** an ovoid silicle, 2–5 × 2–3 mm, glabrous; style 0.3–1 mm long; pedicels spreading, 3–10 mm long, longer than the silicle. Wet, often disturbed soil of fields, roadsides and along streams, wetlands; plains, valleys, montane. Throughout most of N. America. (p.220)

1. Stem and lower leaf surfaces with stiff, simple hairs . . . . var. *hispida*
1. Plants glabrous except at the stem base 2

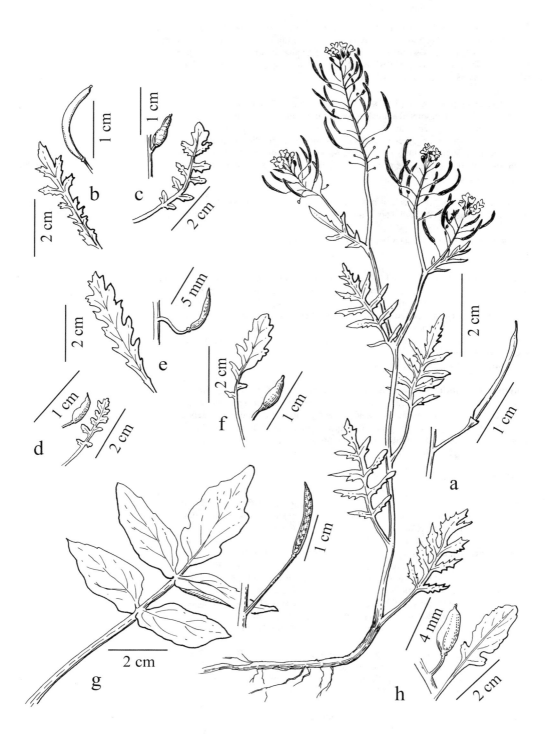

Plate 34. **Rorippa**. **a.** *R. palustris*, **b.** *R. curvisiliqua*, **c.** *R. truncata*, **d.** *R. sinuata*, **e.** *R. tenerrima*, **f.** *R. curvipes*, **g.** *R. nasturtium aquaticum*, **h.** *R. palustris*

2. Stems <40 cm tall, <3 mm in diameter, often purplish . . . . var. *palustris*
2. Stems often >40 cm tall, ≥3 mm in diameter . . . . var. *fernaldiana*

There is little ecological or geographic distinction among ***Rorippa palustris*** var. ***hispida*** (Desv.) Butters & Abbe, ***R. palustris*** var. ***palustris*** and ***R. palustris*** var. ***fernaldiana*** Butters & Abbe.

Rorippa sinuata (Nutt.) Hitchc. Taprooted or rhizomatous perennial. **Stems** ascending, 10–30 cm. **Leaves** all cauline, 2–6 cm long, oblanceolate, pinnately lobed and dentate, the apical lobe similar to the lateral, petiolate to sessile above. **Vestiture** of sparse, globose hairs on stem and lower leaf surfaces. **Petals** yellow, 3–5 mm long. **Fruit** an oblong, slightly curved silicle 4–7 × 1.5–2.5 mm, glabrous; style 0.9–1.5 mm long; pedicels spreading, 4–9 mm long. Moist to wet, sandy or gravelly soil along streams, ponds, lakes; plains, valleys, montane. BC to ON south to CA, NM, AK, LA. (p.220)
  Rorippa sylvestris has narrower siliques.

Rorippa sylvestris (L.) Besser Rhizomatous perennial. **Stems** ascending, branched, 15–50 cm. **Leaves** all cauline, petiolate, 2–15 cm long; the blade oblong, deeply pinnately lobed and dentate, the apical lobe similar to the lateral. **Vestiture** of sparse hairs on the lower stem, otherwise glabrous. **Petals** yellow, 3–5 mm long. **Fruit** a linear silicle, 9–16 × ≤1 mm, glabrous; style 0.8–1.5 mm long; pedicels spreading, 7–11 mm long. Moist soil of fields, roadsides, gravelly riverbanks; plains, valleys. Introduced to BC, ND south to CA, MO; native to Eurasia. (p.220)
  Rorippa austriaca has dentate rather than lobed leaves; see *R. sinuata*.

Rorippa tenerrima Greene Glabrous annual from a thickened taproot. **Stems** prostrate to ascending, 3–20 cm long, branched. **Leaves** basal and cauline; the basal withering early; stem leaves oblanceolate, 1–8 cm long, pinnately lobed with sinuate margins, the terminal lobe largest. **Petals** yellow, 0.5–0.8 mm long, shorter than the sepals. **Fruit** a lanceolate silique, 3–7 × 1–1.5 mm, papillose (use 20X); style 0.2–0.6 mm long; pedicels ascending, 1–3 mm long. Mud along shores of lakes, ponds, streams; plains, valleys, montane. BC to SK south to CA, NM, TX, MO, Mexico. (p.220)

Rorippa truncata (Jeps.) Stuckey [*R. curvipes* Greene var. *truncata* (Jeps.) Rollins] Taprooted annual to short-lived perennial. **Stems** prostrate to ascending, 5–20 cm long, branched with racemes terminal and from upper leaf axils. **Leaves** all cauline, oblanceolate, 2–6 cm long, crenate-dentate to pinnately lobed, the terminal lobe largest. **Vestiture** of sparse simple hairs or glabrous. **Petals** yellow, 0.5–1 mm long, shorter than the sepals. **Fruits** linear-oblong siliques, 4–9 × 1.5–2.4 mm, glabrous; styles 0.2–0.5 mm long; pedicels spreading, 1–5 mm long. Mud around ponds, rivers, lakes; plains, valleys to subalpine. WA to SK south to CA, TX, MO, Mexico. (p.220)
  Similar to *Rorippa curvipes*, but the truncate siliques are distinctive.

Sandbergia Greene

Sandbergia perplexa (L.F.Hend.) Al-Shehbaz [*Halimolobos perplexa* (L.F.Hend.) Rollins] Biennial or short-lived perennial from a simple or branched caudex. **Stems** erect or ascending, simple or few-branched, 12–25 cm. **Leaves** basal and cauline, petiolate, oblanceolate, 1–3 cm long, pinnately lobed. **Vestiture** of stalked 3- to 4-branched hairs. **Inflorescence** an ebracteate raceme. **Flowers** white, fading purplish; petals 5–8 mm long. **Fruit** glabrous to sparsely hairy, erect, linear siliques, 10–16 × 1–1.5 mm, constricted between the seeds; style 0.5–1 mm long; seeds in 1 row per locule; pedicels ascending 5–10 mm long. Gravelly soil of grassland, steppe; montane. Endemic to east-central ID and adjacent MT; one collection from southern Ravalli Co. and reported for Beaverhead Co.[6]

Schoenocrambe Greene

Schoenocrambe linifolia (Nutt.) Greene [*Sisymbrium linifolium* (Nutt.) Nutt.] Glabrous, rhizomatous perennial. **Stems** erect, branched, 20–70 cm from a branched caudex. **Leaves** all cauline, sessile, linear, 25–60 mm long, entire, inrolled. **Inflorescence** an ebracteate raceme. **Flowers** yellow; petals 6–10 mm long. **Fruits** linear siliques, 2–6 cm long; style ≤1 mm long; seeds in 1 row per locule; pedicels ascending, 4–6 mm long. Sagebrush steppe; valleys to montane. BC to AB south to NV, AZ, NM.

Sinapis L. Charlock

Annual. **Stems** erect, branched. **Leaves** cauline, petiolate, pinnately divided or lobed, the upper sometimes merely dentate. **Inflorescence** a simple or compound ebraceate raceme. **Flowers** yellow. **Fruit** a silique, erect to spreading, several-nerved, with a flattened-conical, 1-seeded beak; seeds in 1 row per locule.

Introduced from Eurasia and cultivated for greens and seeds; domesticated *Sinapis alba* is the source of yellow mustard. Similar to *Brassica* which has 2-nerved siliques with non-flattened beaks.

1. Siliques hirsute, 3–5 mm broad . . . . *S. alba*
1. Siliques usually glabrous, <3 mm broad. . . . . *S. arvensis*

Sinapis alba L. [*Brassica hirta* Moench.] Yellow mustard **Stems** 20–60 cm. **Leaves** 8–20 cm long with dentate margins. **Vesture** hirsute with simple hairs to nearly glabrous. **Petals** 10–15 mm long. **Fruit** spreading, 7–15 mm long, hirsute, the beak 1–2 cm long. Fields, roadsides; valleys; Gallatin and Toole cos. Introduced throughout temperate N. America.

Sinapis arvensis L. [*Brassica kaber* (DC.) L.C.Wheeler] **Stems** 10–50 cm. **Leaves** 5–14 cm long, undivided to pinnate with dentate margins. **Vesture** sparse, simple hairs below, glabrous above. **Petals** 8–12 mm long. **Fruit** erect, 2–3 cm long, nearly glabrous, the beak 7–16 mm long. Fields, roadsides; plains, valleys. Introduced throughout temperate N. America.

Sisymbrium L. Tumble mustard

Annual herbs. **Stems** erect, branched. **Leaves** cauline, sometimes basal as well, petiolate, lanceolate to oblanceolate, deeply pinnately lobed. **Vesture** of simple hairs or lacking. **Inflorescence** an ebracteate raceme. **Flowers** yellow to pale yellow. **Fruits** linear siliques; seeds 1 row in each locule; style minute or absent.

All our species introduced from temperate Eurasia.

1. Siliques noticeably thicker than subtending pedicels 2
1. Pedicels ca. as thick as the siliques . . . . *S. altissimum*

2. Fruiting pedicels >5 mm long, ascending. . . . . *S. loeselii*
2. Fruiting pedicels ≤2 mm long, erect . . . . *S. officinale*

Sisymbrium altissimum L. Jim Hill mustard **Stems** 20–80 cm. **Leaves** 1–20 cm long, the lobes linear to oblong, dentate. **Vesture** of long, simple hairs below, glabrous above. **Petals** 6–9 mm long. **Fruits** 4–8 cm × ca. 1 mm; pedicels spreading, 5–10 mm long, as thick as the siliques. Disturbed soil of grasslands, steppe, fields, streambanks, roadsides; plains, valleys, montane. Introduced throughout N. America.

This species invades even ungrazed grasslands by colonizing gopher or ground squirrel diggings. It often branches in a manner resulting in a round plant that can roll for great distances before the wind. It flowers somewhat earlier than *Sisymbrium loeselii*.

Sisymbrium loeselii L. **Stems** 20–90 cm. **Leaves** 1–15 cm long, the lateral lobes linear, curved back, dentate, the terminal lobe larger. **Vesture** of spreading to retrorse, simple hairs. **Petals** 5–7 mm long. **Fruits** 15–40 × ca. 1 mm; pedicels ascending, 7–12 mm long, thinner than the siliques. Disturbed soil of fields, streambanks, roadsides; plains, valleys. Introduced to northern temperate N. America.

Sisymbrium officinale (L.) Scop. Annual or biennial. **Stems** 30–50 cm. **Leaves** 1–20 cm long, the lateral lobes lanceolate to linear, sinuate, the terminal lobe larger, lyrate. **Vesture** of sparse, simple hairs below, glabrous above. **Petals** 2–3 mm long. **Fruits** 10–17 × 1–1.5 mm; pedicels erect, 1–2 mm long, as thick as the siliques. Fields, roadsides; plains, valleys. Introduced throughout temperate N. America.

Brassica nigra is vegetatively similar but has beaked siliques and larger flowers.

Smelowskia C.A. Mey.

Smelowskia calycina (Stephan) C.A. Mey. [*S. americana* Rydb.] Caespitose perennial from a branched caudex clothed in old leaf bases. **Stems** erect, simple, 3–20 cm. **Leaves** basal and cauline, ciliate-petiolate, 1–7 cm long, deeply pinnately lobed into narrowly oblong segments, reduced upward. **Vestiture** of dense simple and branched hairs. **Inflorescence** a raceme, bracteate below. **Flowers** white to pinkish; petals 4–8 mm long. **Fruits** 5–9 mm long, narrowly allipsoid siliques with a prominent midvein; style 0.3–0.5 mm long; seeds in 1 row per locule; pedicels ascending-erect, 4–9 mm long. Stony soil of moraine, cliffs, talus slopes, fellfields; upper subalpine to alpine. AK to OR, CO; Asia. Our plants are var. *americana* (Regel & Herder) Drury & Rollins. (p.212)

Molecular evidence for recognizing *Smelowskia americana* as a separate species[420] seem inconclusive.

Stanleya Nutt. Prince's plume

Biennial to perennial herbs. **Leaves** cauline, sometimes with a basal rosette, petiolate or sessile, simple to pinnately lobed. **Inflorescence** a dense, ebracteate raceme. **Flowers** yellow to white; petals widest at the base and tip. **Fruits** linear siliques with a basal stipe; style minute or absent; seeds in 1 row per locule.

1. Stem leaves sessile, auriculate . . . . *S. viridiflora*
1. Lower stem leaves petiolate. 2

2. Base of petals pubescent; lower stem glabrous to short-hairy. . . . . *S. pinnata*
2. Petals glabrous; lower stem densely long-hairy . . . . *S. tomentosa*

Stanleya pinnata (Pursh) Britton Perennial herb with a woody, branched caudex. **Stems** erect or ascending, 30–80 cm. **Leaves** all cauline, petiolate, 5–15 cm long, lanceolate to elliptic, pinnately lobed to entire. **Vestiture** of sparse, simple hairs. **Petals** yellow, 9–14 mm long, hairy on the inner surface. **Fruit** curved, 4–8 cm long; the stipe short-hairy, 12–21 mm long; pedicels spreading, 6–10 mm long. Clay soil on open slopes of streambanks, badlands; plains, valleys. OR to ND south to CA, AZ, TX.

1. Upper stem leaves broadly elliptic, entire. . . . . var. *integrifolia*
1. Upper stem leaves oblanceolate to lanceolate, entire or lobed . . . . var. *pinnata*

Stanleya pinnata var. *integrifolia* (E.James) Rollins occurs in southern cos.; *S. pinnata* var. *pinnata* is found in the Missouri River drainages. Plants are reported to sequester selenium.[193]

Stanleya tomentosa Parry Perennial (monocarpic?) from a simple or few-branched caudex covered in old leaf bases. **Stems** erect, simple, 40–120 cm. **Leaves** basal and cauline; basal leaves petiolate, 10–30 cm long, oblanceolate, pinnately lobed, the terminal lobe largest; stem leaves similar below, becoming entire above. **Vestiture** of dense, long, simple hairs, at least below. **Petals** pale yellow, 10–15 mm long, glabrous. **Fruit** slightly curved, 3–8 cm long, constricted between the seeds; the stipe 15–25 mm long; pedicels spreading, 10–15 mm long. Stony, calcareous soil of grassland, steppe, juniper woodland; valleys, montane. Endemic to Carbon Co. and north-central WY.

Stanleya viridiflora Nutt. Glabrous and glaucous, biennial or monocarpic perennial from a simple caudex covered in old leaf bases. **Stems** erect, simple or branched above, 50–80 cm. **Leaves** basal and cauline; basal leaves petiolate, the blade 5–13 cm long, oblanceolate, entire or with a few basal teeth or lobes; stem leaves oblanceolate, sessile, auriculate. **Petals** pale yellow, brown at the base, 13–21 mm long, glabrous. **Fruit** downcurved, 3–6 cm long, constricted between seeds; the stipe 10–20 mm long; pedicels spreading, 4–8 mm long. Stony, usually calcareous soil of sparsely vegetated slopes in steppe, woodlands; montane. ID, MT south to CA, NV, UT, CO. Collected in Beaverhead and Madison cos.

Streptanthella Rydb.

Streptanthella longirostris (S.Watson) Rydb. Glabrous, taprooted annual. **Stems** erect, branched, 15–40 cm. **Leaves** all cauline, sessile, narrowly oblanceolate, 3–5 cm long, entire to weakly dentate. **Inflorescence** an ebracteate raceme. **Flowers** white; petals 4–6 mm long. **Fruits** linear siliques, 3–6 cm long with a beak 4–6 mm long; style absent; seeds in 1 row per locule; pedicels reflexed, 1–3 mm long. Sandy soil of sagebrush steppe; valleys; Carbon Co. WA, ID and MT south to CA, AZ, NM, Mexico.

Thelypodium Endl. Thelypody

Often glaucous, taprooted, biennial or perennial herbs. **Leaves** basal and cauline; the basal petiolate; the cauline sessile or short-petiolate. **Vestiture** of simple hairs at the base or glabrous. **Inflorescence** a densely flowered, ebracteate raceme. **Flowers** white to blue. **Fruits** linear siliques with short stipes, constricted between the seeds; seeds in 1 row per locule.[4]

Thelypodium sagittatum and *T. paniculatum* are vegetatively similar and have been considered conspecific. All three species are uncommon in MT.

1. Stem leaves not auriculate at the base . . . . *T. integrifolium*
1. Stem leaves auriculate or sagittate at the base 2
2. Siliques 0.5–1 mm wide; petals 1–3 mm wide . . . . *T. sagittatum*
2. Siliques 1.3–2.3 mm wide; petals 2.5–5 mm wide . . . . *T. paniculatum*

Thelypodium integrifolium (Nutt.) Endl. Glabrous and glaucous, taprooted biennial. **Stems** erect, usually branched, 20–80 cm. **Basal leaf blades** oblong, 1–10 cm long, entire. **Stem leaves** sessile to short-petiolate, reduced, lanceolate. **Petals** white to purple, 4–7 mm long. **Fruits** ascending, 1–3 cm × 0.7–1.1 mm; the stipe 0.3–2 mm long; style 0.4–1.3 mm long; pedicels ascending, 2–8 mm long. Moist to wet, alkaline or saline meadows; plains, valleys, montane. WA to ND south to CA, AZ, NM. Our plants are var. *integrifolium*.

Thelypodium paniculatum A.Nelson [*T. sagittatum* (Nutt. ex Torr.& A.Gray) Endl. var. *crassicarpum* Payson] Biennial or short-lived perennial. **Stems** erect, 25–60 cm. **Basal leaf blades** 2–10 cm long, lanceolate to obovate, entire, withering by flowering. **Stem leaves** lanceolate, auriculate-clasping, 15–30 mm long. **Vestiture** sparse at the base or glabrous. **Petals** lavender, 7–12 × 2.5–4 mm. **Fruits** erect, 25–50 × 1.3–2.3 mm; the stipe 0.5–1 mm long; style 1–3 mm long; pedicels ascending, 4–13 mm long. Moist alkaline meadows; montane. ID, MT, CO, WY. Collected in Gallatin and Madison cos.[4]

Thelypodium sagittatum (Nutt. ex Torr.& A.Gray) Endl. Biennial or short-lived perennial. **Stems** erect, 25–70 cm. **Basal leaf blades** lanceolate to oblong, entire, withering by flowering. **Stem leaves** 2–8 cm long, lanceolate, auriculate-clasping. **Vestiture** sparse at the base or glabrous. **Petals** pale lavender, 6–13 × 1.5–2 mm. **Fruits** ascending, 15–40 × 0.5–1 mm; the stipe 0.2–1 mm long; style 1–3 mm long; pedicels ascending, 6–11 mm long. Moist alkaline meadows; montane; Beaverhead and Madison cos. WA to MT south to NV, UT, CO. Our plants are var. *sagittatum*.

Thlaspi L. Fanweed, pennycress

Thlaspi arvense L. Glabrous, taprooted annual. **Stems** erect, usually branched, 15–60 cm. **Leaves** basal and cauline; the basal withering early; the cauline oblanceolate, 15–55 mm long, petiolate below, auriculate and clasping above, entire to dentate. **Inflorescence** an ebracteate raceme. **Flowers** white; petals 2–4 mm long. **Fruits** flattened, broadly winged, obovate to nearly orbicular silicles 5–13 mm long, deeply notched at the tip; style ≤0.3 mm long; seeds 2 to 8 per locule; pedicels ascending, 5–12 mm long. Disturbed soil of fields, roadsides, grasslands, steppe; plains, valleys, montane. Introduced throughout temperate N. America; native to Europe. (p.212)

Transberingia Al-Shebaz & O'Kane

Transberingia virgata (Nutt.) N.H.Holmgren [*Halimolobos virgata* (Nutt.) O.E.Schulz, *T. bursifolia* (DC.) Al-Shehbaz & O'Kane ssp. *virgata* (Nutt.) Al-Shebaz & O'Kane] Taprooted annual or biennial. **Stems** erect or ascending, simple or branched, 10–50 cm. **Basal leaves** petiolate, oblanceolate, 1–5 cm long, dentate. **Stem leaves** lanceolate, sessile, auriculate, dentate or entire. **Vestiture** simple and branched, short and long hairs. **Inflorescence** a narrow ebracteate raceme. **Flowers** white, petals 2–3 mm long. **Fruit** glabrous, erect, linear siliques, 15–40 mm long, ca. 1 mm wide; style 0.1–0.6 mm long; seeds in 1 to 2 rows per locule; pedicels ascending, 5–17 mm long. Sagebrush steppe, mountain mahogany woodland; valleys, montane in Beaverhead Co. AB to SK south to CA, UT, CO.

Turritis L. Tower mustard

Turritis glabra L. [*Arabis glabra* (L.) Bernh.] TOWER MUSTARD Taprooted annual or biennial (monocarpic) with an unbranched caudex. **Stems** usually unbranched, 40–120 cm. **Vestiture** of simple or branched hairs on basal leaves and stem base, glabrous above. **Basal leaves** petiolate, oblong, shallowly dentate, 2–10 cm long. **Stem leaves** lance-shaped, auriculate, strongly clasping. **Inflorescence** an unbranched raceme. **Flowers** white; petals 4–6 mm long. **Fruits** siliques 5–10 cm × 1–1.5 mm, nearly terete, strictly erect. Disturbed or sparsely vegetated soil of streambanks, meadows, grasslands, open forest; plains, valleys, montane. AK to QC south to CA, NM, MN, GA; Eurasia.

Molecular genetic studies have shown this species is not closely related to *Arabis* sensu stricto or *Boechera*.[222]

RESEDACEAE: Mignonette Family

Reseda L. Mignonette

Glabrous annuals or biennials. **Leaves** alternate, entire or pinnately lobed. **Inflorescence** usually simple, spike-like, bracteate racemes. **Flowers** perfect, irregular; calyx 4- to 6-lobed; petals light yellow, 4 to 6, separate, unequal, lobed at the tip, uppermost the largest; stamens numerous; ovary superior with 3 to 4 carpels; stigmas 3 to 4, sessile. **Fruit** a stipitate, 1-celled, many-seeded capsule.[187]

1. Lower leaves 3-lobed; petals usually 6. . . . . *R. lutea*
1. Leaves entire or crenate but not lobed; petals 4 . . . . *R. luteola*

Reseda lutea L. **Stems** 25–60 cm, branched. **Leaves** mainly cauline, deeply 3-lobed below, pinnately 3- to 7-lobed above into linear-oblanceolate segments. **Flowers**: sepals 6, linear; petals usually 6, ca. 5 mm long, nearly equal, 3-lobed; stamens 15 to 25. **Capsule** oblong, 6–8 mm long. Fields, roadsides; plains, valleys. Introduced across northern U.S. and adjacent Canada; native to Europe.

Reseda luteola L. **Stems** ca. 50 cm, usually simple. **Leaves** basal and cauline, close together, linear-oblanceolate, entire to crenulate. **Flowers**: sepals 4, narrowly ovate, 1–2 mm long; petals 4, the upper ca. 4 mm long, 7- to 11-lobed, the lower smaller; stamens 30 to 50. **Capsule** globose, 3–5 mm long. Fields, pastures; valleys. Introduced sparingly to northeast and west U.S.; native to Europe. Collected once in Sanders Co.

ERICACEAE: Heath Family
Contributed by Peter F. Stickney

Herbaceous perennials, subshrubs, or shrubs. **Leaves** alternate or sometimes opposite, evergreen or deciduous, simple, petiolate or rarely sessile, exstipulate. **Inflorescence**: racemes or flowers from leaf axils or solitary. **Flowers** perfect, regular, hypogynous, rarely epigynous, 5-merous, rarely 4-merous, pedicellate, rarely scapose; sepals partly or wholly distinct or obsolete, persistent in fruit; petals united, separate or lacking; stamens inserted on the floral axis or attached to the base of the corolla; anthers shed pollen through pores, chinks, or slits; stigma capitate or lobed; style usually straight; ovary usually superior with axile placentation. **Fruit** a multi-chambered (usually 5) capsule, or a berry or berry-like; seeds minute, numerous, sometimes winged.

Plants mostly in non-calcareous soils of coniferous forests, mainly in the mountainous portions of the state. Our plants have sometimes been placed in three families,[187,193] and this disposition is also suggested by recent research.[224] The Ericaceae in the broad sense adopted here is a morphologically heterogeneous assemblage of (1) achlorophyllous herbs (Monotropaceae), (2) herbaceous subshrubs with leaves and aerial stems remaining green through at least one dormant season (Pyrolaceae), (3) woody, often mat-forming subshrubs with prostrate principal stems, (4) shrubs with superior ovaries maturing to a capsule or with a berry-like fruit, and (5) shrubs with inferior ovaries maturing to a berry (Vacciniaceae).

1. Plants without green leaves; herbaceous perennials. 2
1. Plants with green leaves; subshrubs, shrubs, or herbaceous perennials. 5

2. Stems red and white striped. . . . . *Allotropa*
2. Stems not red and white striped. 3

3. Plants pale yellow to reddish-orange; infloresence nodding . . . . .Hypopitys
3. Plants not pale yellow to reddish-orange; flower(s) but not inflorescence nodding 4

4. Plants waxy, white; flower solitary . . . . Monotropa
4. Plants reddish-brown, glandular-sticky; flowers many . . . . .Pterospora

5. Plants herbaceous with wintergreen leaves or subshrubs with prostrate and woody
   principal stems 6
5. Shrubs, woody plants with erect or ascending stems. 11

6. Woody subshrubs; principal stems prostrate, trailing or mat-forming. 7
6. Plants with herbaceous erect stems or stems lacking 8

7. Flower/fruit solitary in leaf axils; leaves thin, toothed . . . . Gaultheria
7. Flower/fruit in terminal racemes; leaves thick, entire . . . . Arctostaphylos

8. Aerial leaf-bearing stems absent; leaves borne in rosettes at ground level . . . . Pyrola
8. Aerial (erect) leaf-bearing stems present 9

9. Flower/fruit solitary; leaf-bearing stem short (usually <4 nodes) . . . . Moneses
9. Flower/fruit more than one; stems taller (usually >3 nodes). 10

10. Inflorescence 1-sided; leaves thin, margin finely toothed. . . . . Orthilia
10. Inflorescence not 1-sided; leaves thicker, margin few toothed . . . . .Chimaphila

11. Leaves deciduous. 12
11. Leaves evergreen. 14

12. Ovary inferior, fruit a berry; "true" terminal buds lacking. . . . . Vaccinium
12. Ovary superior, fruit a capsule; "true" terminal buds present 13

13. Corolla ≤5 mm long; leaves dull bluish-green, glandular . . . . Menziesia
13. Corolla >10 mm long; leaves bright to shiny green . . . . .Rhododendron

14. Leaves linear, blade conifer-like. 15
14. Leaf blades broader, not conifer-like 16

15. Leaves opposite, 4-ranked, appressed, (juniper scale-like). . . . . Cassiope
15. Leaves alternate, spreading (fir needle-like) . . . . .Phyllodoce

16. Leaves opposite and sessile . . . . Kalmia
16. Leaves alternate and petiolate 17

17. Lower leaf surface studded with small, yellow, glandular dots . . . . Ledum
17. Lower leaf surface not yellow-glandular . . . . .Arctostaphylos

Allotropa Torr. & A.Gray

Allotropa virgata Torr. & A.Gray CANDYSTICK A chlorophyllous, mycotrophic, firm-fleshy perennial herb to 95 cm. **Stems** erect, simple with red and white, longitudinal stripes (drying brownish). **Leaves** bract-like, alternate, sessile, lance-shaped, white to pink-tinged. **Inflorescence** a densely flowered, elongate, terminal, spike-like raceme; subtending bracts leaf-like, exceeding the flowers. **Flowers** hypogynous, 5-merous, subsessile; sepals separate, ovate, white, stippled with red; petals lacking; stamens 10; ovary superior. **Fruit** a globose, 5-chambered capsule. Dry, coniferous forests; upper montane, lower subalpine; known from Ravalli and Granite cos. BC south to CA, ID, MT.

Arctostaphylos Adans. Manzanita

Evergreen woody plants. **Stems** of the season reddish-brown, minutely glandular-hairy, bark of older stems peeling. **Leaves** alternate, petiolate; blades entire, widest above their mid-length, thick, leathery. **Inflorescence** a nodding, compact, terminal, bracteate raceme. **Flowers** hypogynous, 5-merous, pedicellate; calyx much shorter than the corolla, lobes rounded to orbicular, pinkish; corolla urn-shaped, white, lobes revolute, reddish; stamens 10; ovary superior. **Fruit** a dry, mealy, berry-like drupe with up to 10 stony nutlets.

 Arctous alpina (L.) Nied. [*Arctostaphylos alpina* (L.) Spreng. var. *rubra* (Rehder & E.H. Wilson) Bean] with toothed deciduous leaves and red fruit has been found in an open calcareous fen in the Absaroka Range of WY 11 miles south of MT. It should be looked for in adjacent southern MT.

1. Woody subshrub: principal stems trailing, prostrate; secondary branching, ca. erect, <15 cm high; leaves to 2 cm long . . . . *A. uva-ursi*
1. Shrub: principal stems ascending to erect, >15 cm tall; leaves to 3 cm long . . . . *A. media*

Arctostaphylos media Greene [*A. patula* Greene ssp. *platyphylla* (A.Gray) P.V.Wells] MEDIA MANZANITA Freely branched, open-crown, shrub 15–50 cm, occasionally to 150 cm. **Stems** decumbent to ascending-erect, rigid, stout; bark of older stems smooth, dark reddish-brown. **Leaf blades** green to yellowish-green, dull, widest above or at mid-length. **Corolla** ca. 5 mm long. **Drupe** brownish-red, ca. 9 mm broad. Dry open ridges in ponderosa pine-Douglas-fir forest; montane. BC to OR, MT. Known from Lake and Missoula cos.

Considered by some to be a hybrid between *Arctostaphylos uva-ursi* and *A. columbiana* Piper.[318] These plants were previously reported to be *A. patula*.[250]

Arctostaphylos uva-ursi (L.) Spreng. KINNIKINNICK Stoloniferous, mat-forming, woody subshrub. **Stems**: main stems prostrate, trailing; secondary branching, ascending-erect, to 15 cm, occasionally taller. **Leaf blades** shiny, dark green above, ca. 2 cm long, widest well above their mid-length. **Inflorescence** of ca. 5 flowers. **Corolla** ca. 5 mm long. **Drupe** bright red, shiny, glabrous, ca. 8 mm broad. Dryer coniferous forests, exposed slopes, fellfields, turf; plains, valleys to alpine. Circumboreal south to CA, AZ, NM, SD, IA, IL, VA. (p.228)

Cassiope D.Don Mountain heather

Many-branched, patch-forming, evergreen, dwarf shrubs. **Stems** slender, pliant, ascending in crowded broom-like, fascicled clusters, sometimes layering. **Leaves** opposite, subsessile, lanceolate (juniper scale-like), appressed, 4-ranked, overlapping, completely covering the stem. **Inflorescence** subterminal, 1 to several flowers borne individually from leaf axils. **Flowers** hypogynous, 5-merous, nodding; pedicels erect, slender, stiff, reddish; calyx ovate, reddish; corolla bell-shaped, white; stamens 10, included; ovary superior. **Fruit** an erect, globose, 5-chambered capsule.

1. Dorsal leaf surface with a deep medial groove, leaf margin and groove densely, finely ciliate . . *C. tetragona*
1. Dorsal leaf surface rounded (not grooved), glabrous . . . . *C. mertensiana*

Cassiope mertensiana (Bong.) G.Don WHITE MOUNTAIN HEATHER **Stems** to ca. 20 cm. **Leaves** 2–4 mm long, rounded on the back (dorsal surface), margin entire, glabrous. **Flowers** to 10 mm long, pedicels short, elongating in fruit to ca. 15 mm. **Capsule** ca. 4 mm broad. Forests, turf; subalpine to alpine. AK to AB south to CA, NV, ID, MT. Our plants are var. *gracilis* (Piper) C.L.Hitchc.

Cassiope tetragona (L.) D.Don FOUR-ANGLED MOUNTAIN HEATHER Very similar to *C. mertensiana* in most respects. **Leaves** 3–5 mm long with a prominent medial groove on their dorsal surface, widening toward the base dividing it into 2 small, ear-like lobes; leaf margin and groove copiously, finely ciliate. Stony turf, talus; subalpine, alpine. Circumboreal south to WA, MT, QC. Our plants are ssp. *saximontana* (Small) C.L.Hitchc. (p.228)

Chimaphila Pursh Prince's pine

Rhizomatous, caulescent, evergreen, herbaceous subshrubs. **Stems** herbaceous, unbranched or nearly so, erect-ascending. **Leaves** alternate (crowded internodes appear subopposite or whorled), simple, thick, leathery, margin somewhat serrate. **Inflorescence** pedunculate; flowers in a terminal, corymb-like raceme; pedicels pendent to divergent at anthesis, elongating, erect in fruit. **Flowers** hypogynous, 5-merous, saucer-shaped; calyx lobes united at the base; petals orbicular; stamens 10; ovary superior. **Fruit** a globose, 5-chambered capsule.

1. Leaves widest at their mid-length or below, dull bluish-green; stem reddish; flowers usually ≤3 . *C. menziesii*
1. Leaves widest above their mid-length, glossy green; stem green; flowers usually >3 . . . . *C. umbellata*

Chimaphila menziesii (R.Br.) Spreng. LITTLE PRINCE'S PINE **Stems** to 15 cm, reddish, mostly unbranched. **Leaves** widest at or below their mid-length, bluish-green, apex acute, base rounded. **Flowers** 1 to 3; pedicels with a persistent orbicular bract; sepals triangular; petals white (pinkish

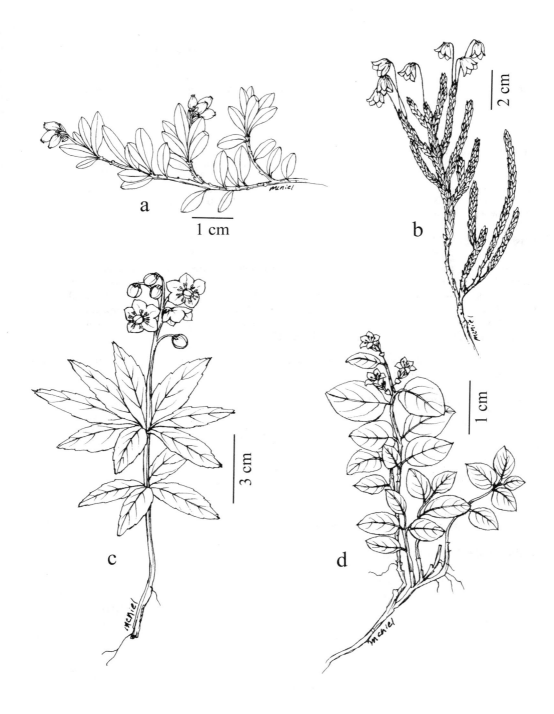

Plate 35. a. *Arctostaphylos uva-ursi*, b. *Cassiope tetragona*, c. *Chimaphila umbellata*,
d. *Gaultheria humifusa*

with age); anthers pink. **Capsule** ca. 5 mm broad. Heavily shaded, sparsely vegetated understories of coniferous forests; montane, subalpine. BC, WA to MT south to CA, NV, UT, CO.

Chimaphila umbellata (L.) W.P.C.Barton PRINCE'S-PINE **Stems** to 30 cm, greenish, occasionally branched. **Leaves** widest above their mid-length, dark green, glossy, apex rounded to weakly pointed. **Flowers** 3 to several, pedicel with a persistent linear bract; sepals oval; petals pinkish; anthers yellow. **Capsule** ca. 6 mm broad. Moist coniferous forest; montane, subalpine. Circumboreal south to CA, NV, AZ, NM, SD, IA, IL, IN, OH, GA. (p.228)

Gaultheria L. Creeping wintergreen

Stoloniferous, mat-forming, evergreen, woody subshrubs. **Stems**: the principal trailing; secondary branching, erect-ascending. **Leaves** alternate, petiolate, leathery, widest below mid-length, margin thickened. **Inflorescence** of solitary flowers in leaf axils of last year's wood. **Flowers** hypogynous, 5-merous; calyx lobes as long as the tube; corolla bell-shaped, white; stamens 10; ovary superior. **Fruit** a thin-walled capsule encased in an enlarged, fleshy calyx forming a red pseudo-berry.

Gaultheria hispidula (L.) Muhl. ex Bigelow with 4-merous flowers and white fruit has been found in north ID and should be looked for in adjacent northwest MT.

1. Calyx glabrous; upper surface leaf veins not indented. . . . . *G. humifusa*
1. Calyx hairy; upper surface leaf veins indented. . . . . *G. ovatifolia*

Gaultheria humifusa (Graham) Rydb. MATTED CREEPING WINTERGREEN **Stems**: secondary occasionally branching, ascending to ca. 5 cm, glabrous. **Leaves** ovate to orbicular, to 15 mm long, entire or nearly so, upper surface smooth; veins not indented. **Flowers** ca. 3 mm long; calyx as long as corolla tube. **Fruit** ca. 6 mm broad. Moist to wet, open forest, turf; subalpine, alpine. BC, AB south to CA, NV, UT, NM. (p.228)

Gaultheria ovatifolia A.Gray SLENDER CREEPING WINTERGREEN **Stems** patch-forming, secondary branching, ascending to ca. 15 cm, with coarse, reddish-brown hairs. **Leaves** ovate, to 25 mm long, dark green, veins of upper surface indented, margin dentate; petioles brown-hairy. **Flowers** ca. 4 mm long; calyx coarse-hairy; corolla white, nearly twice as long as the calyx. **Fruit** ca. 7 mm broad. Moist, upland, coniferous forests; valleys, montane. BC, AB south to CA, ID, MT.

Nearly half of our collections are from old, mechanically disturbed sites such as overgrown road cuts and abandoned logging roads.

Hypopitys Hill Pinesap

Hypopitys monotropa Crantz [*Monotropa hypopitys* L.] FRINGED PINESAP Achlorophyllous, mycotrophic, firm-fleshy perennial herbs. **Stems** erect, to 25 cm, simple, yellowish to reddish-orange, aging dark brown. **Leaves** bract-like, alternate, sessile, ovate-lanceolate, colored as the stem. **Inflorescence** a crowded, few-flowered, terminal raceme nodding at anthesis, elongating and becoming erect in fruit; pedicels elongating in fruit. **Flowers** hypogynous; perianth yellowish; terminal flower 5-merous; those below mostly 4-merous; petals separate, longer than the separate sepals; subtending bracts orange to reddish; stamens 8 or 10; ovary superior. **Fruit** a subglobose, 4- to 5-chambered capsule. Drier coniferous forests; montane, subalpine. Circumboreal south to CA, AZ, NM, TX, AL, FL, Central America. (p.230)

Kalmia L. Laurel

Kalmia polifolia Wangenh. [*K. microphylla* (Hook.) A.Heller, *K. occidentalis* Small] ALPINE LAUREL Rhizomatous, evergreen, low shrub. **Stems** erect-spreading, sparingly branched, to 30 cm, occasionally taller. **Leaves** opposite, subsessile, to 3 cm long, widest at or below their mid-length, leathery, dark green, lustrous above, whitish beneath, margin entire, somewhat revolute. **Inflorescence** clusters of 2 to 10 pedicellate flowers, each borne in axils of reduced leaf-like bracts at the tip of last year's stem; pedicels 15–25 mm long, erect, purple. **Flowers** hypogynous, 5-merous; calyx lobes ovate; corolla saucer-shaped, pink to deep reddish-violet; lobes broadly deltoid with 10 pouches that retain the anthers prior to pollen release; stamens 10; ovary superior. **Fruit** a globose, 5-chambered capsule. Moist meadows, turf, often around lakes; montane to alpine. AK to NF south to CA, CO, PA, NJ. Our plants are var. *microphylla* (Hook.) Rehder. (p.230)

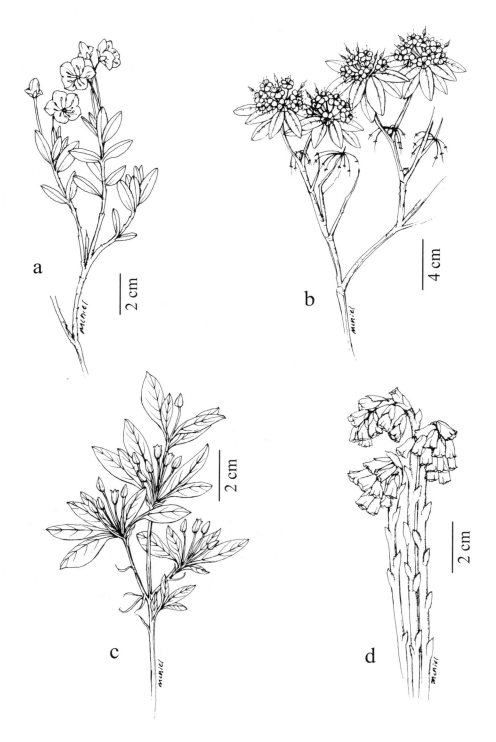

Plate 36. a. *Kalmia polifolia*, b. *Ledum glandulosum*, c. *Menziesia ferruginea*, d. *Hypopitys monotropa*

Ledum L. Labrador tea

Ledum glandulosum Nutt. TRAPPER'S TEA Evergreen, acutely branched shrub. **Stems** erect-ascending, sparingly branched, to 120 cm; floral buds terminal, globose; foliage buds lateral, ovoid, closely crowded beneath the larger floral bud. **Leaves** alternate, petiolate, leathery, to ca. 15 mm long, elliptic to obovate, entire; dark green above, whitish, yellow glandular-dotted below. **Inflorescence** a crowded, umbel-like raceme of many flowers borne from the distal floral bud of last year's stem; pedicels to 2 cm long. **Flowers** hypogynous, 5-merous; sepals, white, glandular, nearly separate; corolla dish-shaped; petals separate, white, oblong; stamens 10; ovary superior. **Fruit** an ovoid, 5-chambered capsule. Moist to wet forests and meadows; valleys to subalpine. BC to AB south to CA, NV, UT, WY. Our plants are var. *glandulosum*. (p.230)

Menziesia Smith Menziesia

Menziesia ferruginea Smith [*M. glabella* A.Gray] FOOL'S HUCKLEBERRY Deciduous, acutely branched shrub. **Stems** erect-ascending, sparingly branched, to 200 cm; floral buds terminal, globose, larger than the lateral, ovoid foliage buds which are closely crowded beneath. **Leaves** alternate, petiolate, thin, to ca. 5 cm long with a musky odor; blades elliptic to obovate, bluish-green, dull, glandular, entire; apex obtuse, midnerve apex a tiny white tooth. **Inflorescence** a crowded, umbel-like raceme of 5 to 10 flowers borne from the distal floral bud of last year's stem; pedicels to 2 cm long. **Flowers** hypogynous, 4-merous; calyx nearly obsolete, shallowly lobed, glandular-ciliate; corolla cylindric-urn-shaped, to 5 mm long, shallowly lobed, coppery-pink; stamens 8; ovary superior. **Fruit** an ovoid, 4-chambered capsule. Cool, moist, upland slopes in coniferous forests, often forming dense thickets on north slopes; valleys to subalpine. BC to AB south to CA, ID, WY. Our plants are var. *glabella* (A.Gray) M.Peck. (p.230)

Moneses Salisb. ex A.Gray One-flowered wintergreen

Moneses uniflora (L.) A.Gray [*Pyrola uniflora* L.] WOODNYMPH Small, caulescent, evergreen herbaceous subshrub from a slender, rhizome-like root. **Stems** herbaceous, unbranched, erect, to ca. 4 cm in flower. **Leaves** in whorls of 3, petiolate, suborbicular, to 2 cm long, dark green. **Inflorescence**: a solitary, nodding flower from a terminal bud on last year's stem, becoming erect in fruit; peduncle elongating in fruit. **Flowers** perfect, regular, hypogynous, 5-merous; corolla saucer-shaped; calyx lobes oval, whitish; petals distinct, ovate, to 10 mm long, white; stamens 10; stigma star-shaped, with 5 erect, marginal lobes; style straight; ovary superior. **Fruit** a globose, 5-chambered capsule. Moist, deeply shaded coniferous forests, often in moss of riparian spruce forest; valleys to subalpine. Circumboreal south to CA, AZ, NM, SD, MN, WS, OH, NY. (p.234)

Monotropa L. Indian pipe

Monotropa uniflora L. Achlorophyllous, mycotrophic, firm-fleshy, perennial herb. **Stems** erect, to 30 cm, unbranched, waxy-white, aging to coal-black; dead stalks hardened, often persisting into the next growing season. **Leaves** bract-like, alternate, sessile, ovate, white. **Inflorescence** a solitary, terminal, nodding flower, becoming erect in fruit. **Flowers** hypogynous, waxy-white, 5-merous; sepals lacking; uppermost stem bracts sepaloid; petals separate, oblong-spatulate; stamens 10; ovary superior. **Fruit** an ovoid, 5-chambered capsule. Moist, deeply-shaded coniferous forests; valleys, montane. NT to NL south to CA, ID, MT, TX, LA, AL, FL, Mexico; Asia.

Orthilia Raf. Sidebells wintergreen

Orthilia secunda (L.) House [*Pyrola secunda* L.] Rhizomatous, caulescent, evergreen, herbaceous subshrub from an elongate rhizome. **Stems** herbaceous, to 25 cm, unbranched. **Leaves** alternate, sometimes appearing subopposite or pseudo-whorled, petiolate; blades to 5 cm long, widest below the mid-length. **Inflorescence** of few to many flowers borne on a terminal, one-sided raceme, arching at anthesis, erect in fruit. **Flowers** hypogynous, 5-merous, nodding; calyx lobes rounded, whitish; corolla cup-shaped; petals distinct, elliptic, to 5 mm long, white to greenish-tinged; stamens 10; stigma shield-shaped, 5-lobed; style straight; ovary superior. **Fruit** a nodding, globose, 5-chambered capsule. Shaded coniferous forests; valleys to subalpine. Circumboreal south to CA, AZ, NE, IA, IL, IN, OH, VA, NY. (p.234)

Phyllodoce Salisb. Mountain heath

Patch-forming, evergreen, dwarf shrubs. **Stems** slender, pliant, ascending-erect, in crowded, many-stemmed, fascicled clusters. **Leaves** alternate, linear, fir-needle-like, crowded, ascending-spreading, tightly revolute, the margin folded under leaving only the midvein of the lower surface exposed; petioles short, persisting after the blades have fallen. **Inflorescence** a corymbiform raceme of few to many flowers terminating last year's stem, each flower axillary to a reduced, leaf-like bract; pedicel glandular-hairy, slender, erect, elevating the flowers above the stem tips. **Flowers** hypogynous, 5-merous, nodding, becoming erect in fruit; calyx united basally; corolla urn-shaped or campanulate, the lobes revolute; stamens 10; ovary superior. **Fruit** a 5-chambered capsule.

When the two following species occur together they frequently produce a sterile hybrid, *Phyllodoce ×intermedia* (Hook.) Rydb., with intermediate floral morphology and pink corollas. *Phyllodoce empetriformis* has longer leaves than *P. glanduliflora.*

1. Corolla bell-shaped, reddish-violet, glabrous; upper leaf surface with a shallow medial furrow . . . .*P. empetriformis*
1. Corolla urn-shaped, pale yellow, glandular; upper leaf surface without a medial furrow . . . . . *P. glanduliflora*

Phyllodoce empetriformis (Sm.) D.Don RED MOUNTAIN HEATH **Stems** to 30 cm. **Leaves** to 12 mm long, dark green with a shallow, medial furrow on the upper surface. **Flowers**: pedicel reddish; calyx lobes ovate, reddish; corolla bell-shaped, lobes short, revolute, light to deep reddish-violet; style exserted. Turf, open forest; subalpine, alpine. AK to AB south to CA, ID, WY.

Phyllodoce glanduliflora (Hook.) Coville YELLOW MOUNTAIN HEATH **Stems** to 15 cm. **Leaves** yellowish-green, to 7 mm long; upper surface slightly convex without a medial furrow. **Flowers**: pedicel, calyx, and corolla densely glandular-hairy; calyx lobes lanceolate, greenish; corolla urn-shaped, the lobes very short, revolute, pale yellow; style included. Forest, turf; subalpine, alpine. AK to AB south to CA, ID, WY. (p.234)

Pterospora Nutt. Pinedrops

Pterospora andromedea Nutt. Achlorophyllous, mycotrophic, perennial herb. **Stems** erect, unbranched, to 170 cm, dark reddish-brown, densely glandular hairy; stalks hardened, persistent, drying dark brown. **Leaves** scale-like, alternate, sessile, lance-shaped, reddish-brown. **Inflorescence** a crowded, elongate, terminal raceme of numerous, nodding flowers. **Flowers** hypogynous, 5-merous; calyx lobes linear-lanceolate, red, glandular-hairy; corolla urn-shaped, lobes revolute, yellowish, glabrous; stamens 10; ovary superior. **Fruit** a globose, 5-chambered capsule. Drier, coniferous forests; valleys, montane. AK, BC to SK south to CA, AZ, NM, Mexico; ON to PE south to MI, NY, NH.

Pyrola L. Wintergreen

Rhizomatous, acaulescent, evergreen, herbaceous subshrubs. **Flowering stems** (scape + inflorescence) annual, herbaceous. **Leaves** alternate, petiolate, closely crowded in ground-level rosettes; blades simple, somewhat leathery, mostly orbicular, margins entire or nearly so. **Inflorescence** a terminal, bracteate raceme, the erect scape arising from the center of the leaf rosette. **Flowers** hypogynous, 5-merous, pedicellate, ca. nodding; calyx united at the base, lobes ovate, often colored; corolla saucer-shaped; petals distinct, obovate-spatulate; stamens 10; style exserted, bent downward (straight in *P. minor*); ovary superior. **Fruit** a depressed-globose, 5-chambered capsule.

1. Leaf blades much longer than wide 2
1. Leaf blades ca. as wide as long 3

2. Blades of principal leaves ca. 3 times longer than wide, cuneate at the base, thick, grayish green . . . .*P. dentata*
2. Blades of principal leaves ca. 1.5 times longer than wide, rounded to an abrupt base, thin, light green. . . . .*P. elliptica*

3. Leaf blade upper surface with whitish veins. . . . .*P. picta*
3. Leaf blade upper surface without whitish veins 4

4. Blades of principal leaves ca. >35 mm long . . . . *P. asarifolia*
4. Blades of principal leaves <35 mm long. 5

5. Style bent down, >2 mm long. . . . . *P. chlorantha*
5. Style straight, ca. 1 mm long . . . . *P. minor*

Pyrola asarifolia Michx. Pink wintergreen **Leaves**: petioles to 15 cm long; blades leatherly, to 9 cm long, orbicular to ovate with a rounded base. **Flowering stems** to 47 cm; raceme 4–13 cm long, 10–25 flowered. **Flowers**: petals pink to red (veins sometimes purple). **Capsules** ca. 8 mm broad. Coniferous, sometimes riparian forests, wetland margins; valleys to subalpine. AK to NF south to CA, UT, NM, SD, IA, WS, IN, PA, MA, ME, NS; Asia. (p.234)

1. Leaf blades orbicular, apex rounded, margin entire; calyx lobes ca.3 mm long; petals white with pink tinge; anthers purple . . . . var. *asarifolia*
1. Leaf blades elliptic-ovate, apex broadly acute, margin minutely dentate; calyx lobes ca. 5 mm long; petals reddish-purple; anthers yellow . . . . var. *purpurea*

Pyrola asarifolia var. *asarifolia* most often found in riparian and perennially moist habitats; *P. asarifolia* var. *purpurea* (Bunge) Fern. occurs most often in upland forest habitats.

Pyrola chlorantha Sw. [*P. virens* Schweigg.] Greenish wintergreen **Leaves** to 6 cm long; blade ovate to orbicular, to ca. 3 cm long, apex rounded to obtuse, base rounded, margin mostly entire; petioles ca. as long as the blade. **Flowering stems** to 28 cm; racemes 2–9 cm long, usually with fewer than 10 flowers. **Flowers**: petals white, tinged with green (greenish in bud). **Capsules** ca. 7 mm broad. Well-drained, upland slopes of coniferous forests; montane, subalpine. Circumboreal south to CA, AZ, NM, NE, MN, MI, OH, VA, RI.

Pyrola dentata Sm. Toothleaf wintergreen **Leaves** to 8 cm long; blades leathery, oblanceolate, to ca. 5 cm long, widest about mid-length, apex rounded to acute, base cuneate, margin entire to occasionally minutely dentate, grayish-green, veins sometimes whitish outlined; petioles ca. as long to shorter than the blades. **Flowering stems** up to 23 cm; racemes to 9 cm long, 7- to 17-flowered. **Flowers**: petals white. **Capsules** ca. 5 mm broad. Drier, coniferous forests; montane, subalpine. BC, MT south to CA, ID, WY.

Pyrola elliptica Nutt. White wintergreen **Leaves** to 7 cm long; blades to ca. 4 cm long, ovate, widest at about mid-length, apex rounded, base rounded to obtuse, margin mostly entire; petioles shorter to ca. as long as the blades. **Flowering stems** to 25 cm; racemes 4–6 cm long, 8- to 14-flowered. **Flowers**: petals white with purple veins. **Capsules** ca. 5 mm broad. Riparian forests, wet to moist, open sites; valleys to lower montane. BC to NF south to AZ, NM, NE, IA, IL, IN, OH, NC, ME, NS.

Pyrola minor L. Lesser wintergreen **Leaves** to 5 cm long; blades suborbicular to ovate, to ca. 3 cm long, apex and base rounded to obtuse, margin mostly entire; petioles shorter than the blades. **Flowering stems** to 19 cm; racemes ca. 3 cm long, 6- to 13-flowered, crowded. **Flowers**: calyx lobes pinkish; petals white; style straight. **Capsules** ca. 5 mm broad. Riparian habitats in coniferous forests; montane, subalpine. Circumboreal south to CA, AZ, NM, MB, WI, MI, NY, ME, NS.

Pyrola picta Sm. White-veined wintergreen **Leaves** to 7 cm long; blades ovate to elliptic, to ca. 5 cm long, apex obtuse to acute, base obtuse to rounded, margins entire, thickened, upper surface dark green, veins white-mottled; petioles red, shorter to ca. as long as the blades. **Flowering stems** to 34 cm; scape red; racemes 10- to 20-flowered. **Flowers**: petals white with yellowish or greenish tinge. **Capsules** ca.7 mm broad. Drier coniferous forests; montane to subalpine. BC, AB south to CA, AZ, NM, SD.

Rhododendron L. Rhododendron

Rhododendron albiflorum Hook. White rhododendron Deciduous, acutely branching shrub. **Stems** erect-ascending, to 150 cm, sparingly branched with coarse, reddish hairs; buds dimorphic; terminal bud foliaceous; floral buds lateral on last year's stem. **Leaves** alternate, short-petiolate, thin; blades obovate, to ca. 8 cm long, bright glossy green when fresh, paler beneath, entire, minutely glandular-ciliate, usually with scattered, coarse, reddish hairs. **Inflorescence** of 1 to 2 flowers from lateral floral buds below the current season's growth; pedicels ca. as long as the flowers. **Flowers** hypogynous, 5-merous; calyx foliaceous; corolla broadly bell-shaped, to 2 cm long, the rounded lobes as long as the

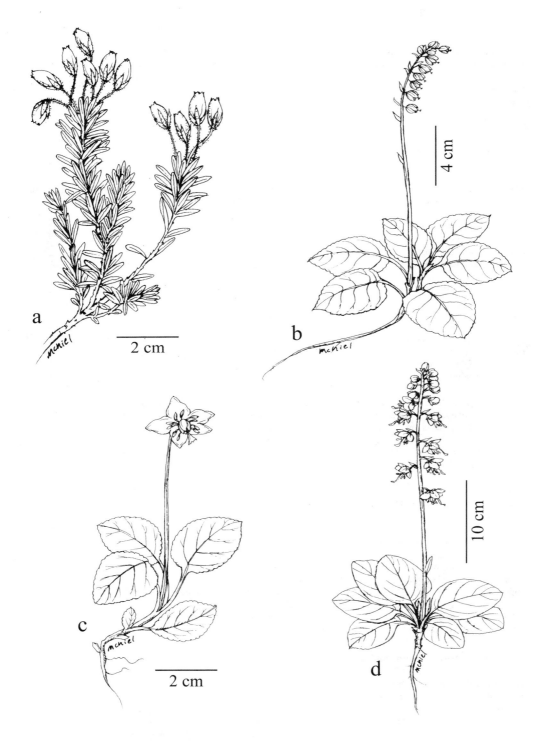

Plate 37. a. **Phyllodoce glanduliflora**, b. **Orthilia secunda**, c. **Moneses uniflora**, d. **Pyrola asarifolia**

tube, white; stamens 10; ovary superior. **Fruit** a 5-chambered capsule. Open, moist, coniferous forests; upper montane, lower subalpine. BC to MT south to OR, ID.

Vaccinium L. Huckleberry, blueberry

Rhizomatous, deciduous, dwarf to medium shrubs. **Stems** branching, determinate, terminal buds lacking. **Leaves** alternate, short-petiolate, thin; blades longer than wide. **Inflorescence** of solitary flowers from the lowest leaf axils of the current season's stem or compact racemes from the uppermost floral bud(s) on last year's stem. **Flowers** epigynous, 5-merous, maturing before the leaves, floral buds absent in all but two species; calyx lobes nearly obsolete in all but two species; petals united forming an urn-shaped corolla; stamens 10; anthers usually with pores on extension tubules and awn-like appendages; ovary inferior. **Fruit** a juicy, edible, many-seeded berry.

Foliage of most of our species turns red or scarlet in autumn. *Vaccinium ovalifolium* Sm., a tall shrub with mostly entire leaves, occurs in both north ID and adjacent southeastern BC. It should be looked for in mesic forests of northwest MT.

1. Flowers several from floral buds at the upper end of last year's stems; leaf blade margin entire 2
1. Flowers solitary from the axils of basal leaves of the current year's stems; leaf blade margins
 finely serrate. 3

2. Leaves and younger stems hirsute; bud scale apex acuminate. . . . . *V. myrtilloides*
2. Leaves and younger stems glabrous; bud scale apex not acuminate . . . . *V. occidentale*

3. Younger stems terete; dwarf shrubs. . . . . *V. caespitosum*
3. Younger stems angled 4

4. 2 to 3 year-old stems green, sharply angled 5
4. 2 to 3 year-old stems not green, not sharply angled 6

5. Stem branching angle narrow, forming broom-like clusters; stems slender, dwarf shrubs;
 subalpine . . . . *V. scoparium*
5. Stem branching wider, not forming broom-like clusters; stems stouter; low shrubs; montane. . . . *V. myrtillus*

6. Principal leaf blades widest at or above mid-length, apex acute to rounded; corolla as wide
 or wider than long. . . . . *V. globulare*
6. Principal leaf blades widest at or below mid-length, apex acuminate; corolla longer
 than wide . . . .*V. membranaceum*

Vaccinium caespitosum Michx. Dwarf huckleberry Dwarf shrub to 30 cm. **Stems** terete. **Leaves** to 3 cm long; blades obovate, apex rounded to broadly acute. **Corolla** cylindric urn-shaped, ca. 5 mm long, pink. **Berry** blue, ca. 8 mm broad. Open forests and forest margins; valleys to subalpine. AK to NF south to CA, ID, NM, SK, MB, MN, WI, MI, NY, Mexico.

Vaccinium globulare Rydb. Globe huckleberry Shrubs 30–120 cm. **Stems** those 2–3 years old yellowish-green to reddish, somewhat angled. **Leaves** to 5 cm long; blades obovate to elliptic, apex rounded to convex-acute. **Flowers** to 6 mm long, as wide or wider than long, pale yellowish-pink. **Berry** waxy-blue, dark purple, or reddish-purplish, to 8 mm broad. Moist, cool, coniferous forests, avalanche slopes; montane, subalpine. BC to AB south to OR, ID, UT.

Montana's most common and widespread huckleberry; it is the species that is picked for fresh fruit and preserves and is harvested commercially by wild crafting. It also is an important staple for bears preparing for hibernation. Greatest berry production is associated with open, older, previously burned forest sites prior to the reestablishment of a coniferous tree overstory. Huckleberry shrubs do not die out with tree overstory reestablishment; rather they tend to grow taller and flower less. The type locality is in Gallatin Co.

Vaccinium membranaceum Douglas ex Torr. Thin-leaf huckleberry Shrubs to 200 cm. **Stems:** those 2–3 years old yellowish-green, reddish, or brown. **Leaves** to 55 mm long; blades ovate-lanceolate to elliptic, acuminate. **Corolla** elliptic urn-shaped, to 6 mm long, longer than wide, pale yellowish-pink. **Berry** dark purple to purplish-black, to 9 mm broad. Moist, coniferous forests; montane. BC to AB south to CA, ID, WY. (p.238)

Superficial resemblance of *Vaccinium globulare* to *V. membranaceum* has led to its reduction by some botanists under the latter species.[408] However, *V. membranaceum* is a taller, coarser shrub of moister, forest habitats with larger, lanceolate, acuminate ("drip tip") leaves and corollas longer than wide compared to *V. globulare*. Due presumably to lack of suitable habitats, the distribution of *V. membranaceum* is sporadic in its MT range.

Vaccinium myrtilloides Michx. VELVET-LEAF BLUEBERRY Clonal, intricately-branched shrubs to 45 cm. **Stems** terete, hirsute with small, white blisters, reddish to brownish; foliage buds small, acuminate with 4 scales; floral buds larger, rotund, of 6 to 8 scales. **Leaves** to 40 mm long; blades elliptic, hirsute, at least on the veins, margins entire, ciliate. **Inflorescence** crowded, few-flowered racemes borne from distal floral buds on last year's stem. **Flowers**: calyx lobes deltoid; corolla cylindrical bell-shaped, to 5 mm long, pale yellow to greenish-white often with narrow, reddish stripes. **Berry** blue, to ca. 9 mm broad; calyx lobes persistent on summit of berry. Coniferous forest; valleys. BC to NF south to MT, IA, IL, IN, OH, WV, PA, CT, ME, NS. Known only from Flathead Co.

Vaccinium myrtillus L. LOW HUCKLEBERRY Low shrubs to ca. 30 cm. **Stems** stout, divergently branched; those 2–3 years old green, sharply angled. **Leaves** to ca. 25 mm long; blades ovate. **Corolla** as wide or wider than long, to ca. 5 mm, greenish-white, tinged with pink. **Berry** dark purple to dark red, to ca. 7 mm broad. Moist, coniferous forest; montane, lower subalpine. BC, AB south to OR, NV, AZ, NM; Europe. (p.238)

Vaccinium occidentale A.Gray [*V. uliginosum* L. ssp. *occidentale* (A.Gray) Hultén] WESTERN BLUEBERRY Shrubs to 60 cm. **Stems** terete; foliage buds small with paired scales; floral buds larger, with 4 to 6 imbricate scales. **Leaves** to 20 mm long; blades elliptic, margins entire, not reticulate-veined beneath. **Inflorescence** of 1 (to 3) flower borne from the uppermost floral buds on last year's stem. **Flowers**: calyx lobes deltoid; corolla elliptic urn-shaped, to ca. 5 mm long, white to pale pink. **Berry** blue, to ca. 5 mm broad; calyx lobes persistent on the summit. Wet meadows to exposed, perennially moist turf; subalpine, alpine. BC, MT south to CA, NV, UT, WY.

Vaccinium scoparium Leiberg ex Coville GROUSE HUCKLEBERRY Dwarf shrubs to ca. 20 cm. **Stems** slender, branching, angle narrow, erect-ascending, broom-like, those 2–3 years old green, sharply angled. **Leaves** to 15 mm long; blades elliptic to narrowly ovate. **Flowers** urn-shaped, to ca. 3 mm long, longer than wide, white-tinged with pink. **Berry** bright, tomato-red, to ca. 4 mm broad. Coniferous forests; subalpine, alpine. BC, AB south to CA, NV, UT, NM, SD.

The principal understory shrub in subalpine lodgepole pine forests.

PRIMULACEAE: Primrose Family

Annual or perennial herbs. **Leaves** simple, entire, unlobed, undivided, basal. **Flowers** bisexual, hypogynous or half-epigynous, pedicellate, radially symmetrical, campanulate or with petals reflexed; petals mostly 5, united at the base or into a tube; sepals 5, united; stamens 5; style 1. **Fruit** a capsule; seeds few to numerous.[187,193]

Anagalis, *Centunculus*, *Glaux*, and *Lysimachia*, formerly placed in Primulace are now placed in the Myrsinaceae.[206]

1. Corolla lobes longer than the tube, reflexed. . . . .*Dodecatheon*
1. Corolla obviously tubular below; lobes not reflexed 2

2. Calyx farinose or glandular. . . . . *Primula*
2. Calyx glabrous or sparsely hairy 3

3. Flowers magenta . . . .*Douglasia*
3. Flowers white, sometimes with a yellow center . . . . *Androsace*

Androsace L. Fairy candelabra

Scapose annual or perennial herbs. **Leaves** all basal, rosulate. **Inflorescence** a terminal, bracteate umbel. **Flowers** long-pedicellate; corolla tubular; calyx hemispheric to campanulate, the lower half united; stamens included; ovary superior. **Capsules** 5-valved.

1. Plants perennial often with a branched caudex . . . . *A. chamaejasme*
1. Plants a tap- or fibrous-rooted annual or biennial 2

2. Plants fibrous rooted; leaves with a definite petiole . . . .*A. filiformis*
2. Plants taprooted; leaves tapered to the base, lacking a petiole. 3

3. Bracts at the base of the flower stalks linear-lanceolate, broadest near the base; leaves
 often dentate. . . . .*A. septentrionalis*
3. Flower bracts narrowly obovate, broadest near the middle; leaves usually entire . . . . *A. occidentalis*

Androsace chamaejasme Wulfen ex Host ROCK JASMINE [*A. lehmanniana* Spreng.] Perennial
forming loose cushions from a branched caudex. **Scapes** solitary, 2–10 cm. **Leaves** oblanceolate,
4–12 mm long, ciliate, sometimes sparsely hairy. **Umbel** few-flowered, congested, hemispheric,
subtended by lanceolate bracts; pedicels 1–4 mm long. **Flowers**: corolla white with a yellow center,
4–8 mm across; calyx subhemispheric, ca. 3 mm long. **Capsules** ca. 2 mm across. Stony, calcareous
soil of rock outcrops, fellfields, exposed slopes; subalpine, alpine, occasionally montane. Circumpolar south to CO. Our
plants are ssp. *lehmanniana* (Spreng.) Hultén.

Androsace filiformis Retz. Mostly glabrous, fibrous-rooted annual. **Scapes** numerous, 1–10
cm. **Leaves** petiolate; the blade lanceolate to ovate, 3–20 mm long, dentate. **Umbel** several-
flowered, diffuse, subtended by minute, lanceolate bracts; pedicels 10–35 mm long, glabrous or
glandular. **Flowers**: corolla white, 2–3 mm across; calyx hemispheric, ca. 2 mm long. **Capsules**
ca. 2 mm across. Vernally moist to wet, sparsely vegetated soil of meadows, along lakes, streams,
ditches; valleys to subalpine. BC, MT south to CA, UT, CO; Eurasia.

Androsace occidentalis Pursh Puberulent, taprooted annual. **Scapes** usually several, 1–6 cm.
Leaves lanceolate to oblanceolate, 4–12 mm long, entire. **Umbel** several-flowered, subtended
by obovate bracts; pedicels 2–35 mm long. **Flowers**: corolla white, 1–2 mm across, included in
the calyx; calyx hemispheric, 3–5 mm long, keeled below each lobe. **Capsules** 2–3 mm across.
Grasslands, steppe, open forest, sometimes in disturbed soil; plains, valleys. BC to ON south to CA,
NM, TX, AR.

Androsace septentrionalis L. Glabrous to glandular-puberulent annual. **Scapes** several, 2–12
cm. **Leaves** lanceolate, 5–25 mm long, entire to dentate. **Umbel** several-flowered; pedicels 3–50
mm long, subtended by linear-lanceolate bracts. **Flowers**: corolla white, 1–3 mm across, exserted
beyond the calyx; calyx 3–4 mm long, turbinate, keeled into the lobes. **Capsules** 2–3 mm across.
Sparsely vegetated soil of grasslands, meadows, steppe, open forest, rocky slopes; plains, valleys to
alpine. Circumboreal south to CA, AZ, NM. (p.238)
 Plants above treeline are greatly dwarfed and may persist for >1 year before flowering.

Dodecatheon L. Shooting star

Scapose perennials, often short-lived. **Leaves** all basal, rosulate, petiolate. **Inflorescence** a terminal,
bracteate umbel or flowers solitary. **Flowers** pink to lavender (sometimes white) long-pedicillate; corolla
short-tubular with long reflexed lobes; calyx short-tubular; stamens exserted, with short, yellow filaments
and anthers sometimes united at the base; ovary superior. **Capsules** cylindric to globose, circumcissile
(operculate) and/or opening by 5 apical slits (Reveal 2009).
 Dodecatheon conjugens and *D. pulchellum* are similar, and both species have glabrous and glandular
forms; *D. conjugens* occurs in generally drier habitats than *D. pulchellum* var. *pulchellum*, but both
species occur together in several areas of the Bitterroot Valley foothills. East of Missoula *D. pulchellum*
var. *cusickii* occurs in a similar habitat to *D. conjugens* var. *viscidum*, and these two taxa along with *D.*
pulchellum var. *pulchellum* occur in close proximity. Molecular evidence suggests that *Dodecatheon* be
submerged in *Primula*,[267,268] but *Dodecatheon* is monophyletic with unique morphology.

1. Stigma ca. twice as wide as the style; petals often sparsely glandular . . . . *D. jeffreyi*
1. Stigma little wider than the style; petals glabrous. 2

2. Connectives uniting base of anther sacs wrinkled horizontally; capsule opening by a
 circular cap (operculum); filaments generally <1 mm long. . . . . *D. conjugens*
2. Connectives smooth or wrinkled longitudinally; capsule opening by 5 slits to the tip;
 filaments generally ≥1.5 mm long. . . . .*D. pulchellum*

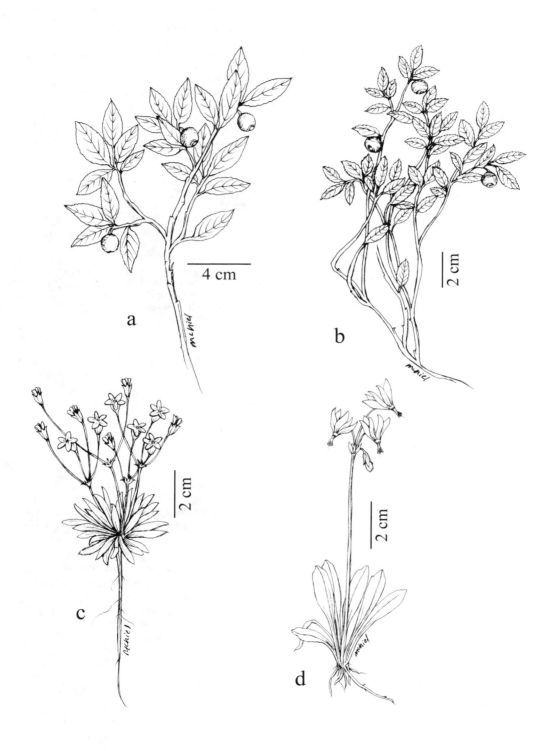

Plate 38. a. *Vaccinium membranaceum*, b. *Vaccinium myrtillus*, c. *Androsace septentrionalis*, d. *Dodecatheon pulchellum*

Dodecatheon conjugens Greene **Scapes** 6–30 cm. **Leaves** glabrous to glandular-hairy; the blade oblanceolate to spatulate, 1–10 cm long, entire. **Umbel** of 1 to 7 flowers; pedicels 1–6 cm long. **Flowers**: corolla lobes 6–18 mm long; connectives purple, horizontally wrinkled; filaments ≤1 mm long; calyx glabrous to glandular. **Capsule** narrowly obovoid to cylindric, 10–15 mm long, operculate. Vernally moist soil of grasslands and dry, open forest; valleys to lower subalpine. BC to SK south to CA, NV, WY.

1. Leaves glabrous or nearly so . . . . var. *conjugens*
1. Leaves distinctly glandular hairy. . . . . var. *viscidum*

Dodecatheon conjugens var. ***viscidum*** (Piper) H.Mason ex H.St.John is more common west of the Continental Divide, valleys to subalpine; *D. conjugens* var. *conjugens* is the more common form east of the Divide, plains, valleys, montane. The type locality is in Lewis and Clark Co.

Dodecatheon jeffreyi Van Houtte **Scapes** 10–60 cm. **Leaves** mostly glabrous; the blade oblong, 3–25 cm long, entire to serrulate. **Umbel** glandular, of 3 to 10 flowers; pedicels 1–4 cm long. **Flowers** sometimes 4-merous; corolla lobes 8–20 mm long, pink to purple (white), darker than the tube, often glandular; base of anther tube purple, horizontally wrinkled; calyx glandular; filaments <1 mm long. **Capsule** ovoid, 6–12 mm long, operculate, valves splitting below the operculum. Meadows, wet, open forest, along streams; upper montane to upper subalpine. AK to CA, ID, MT, WY.

Dodecatheon pulchellum (Raf.) Merr. [*D. cusickii* Greene, *D. pauciflorum* Greene misapplied] **Scapes** 2–40 cm. **Leaves** glabrous to glandular; the blade oblanceolate to obovate, 1–10 cm long, entire. **Umbel** glabrous to glandular, of 1 to 10 flowers; pedicels 5–45 mm long. **Flowers**: corolla lobes 6–16 mm long; connectives purple, smooth or longitudinally wrinkled; calyx glabrous, often purple-spotted. **Capsule** ovoid-cylindrical, 5–12 mm long, opening by 5 valves splitting to the tip. Grasslands to wet meadows; valleys to alpine. AK to MB south to CA, AZ, CO, NE. (p.238)

1. Leaves glandular short-hairy . . . . var. *cusickii*
1. Leaves glabrous (sometimes with sessile, red, crystaline glands). .var. *pulchellum*

Dodecatheon pulchellum var. ***cusickii*** (Greene) Reveal is found in grasslands and dry, open forest, valleys to montane; *D. pulchellum* var. *pulchellum* is found in wet meadows, moist turf, open forest and streambanks, plains, valleys to alpine. Dwarf alpine plants with few flowers have been called var. *watsonii* (Tidestr.) B.Boivin, but this form occurs mixed with typical *D. pulchellum* in many populations.

Douglasia Lindl.

Mat-forming, scapose perennials from a branched caudex. **Leaves** all basal, rosulate, obscurely petiolate. **Inflorescence** a terminal, usually bracteate umbel or flowers solitary. **Flowers** pink to magenta; corolla salverform, 5-lobed; calyx campanulate, keeled; stamens included; ovary superior. **Capsules** globose, opening by 5 valves.

Plants superficially resemble *Phlox* spp. but lack leafy stems. *Douglasia idahoensis* Douglass M. Hend., a recently described species from adjacent ID with several flowers and simple or branched hairs on the scape, could occur in MT along the ID border.

1. Inflorescence of usually a solitary flower subtended by 0–2 bracts . . . . *D. montana*
1. Inflorescence an umbel of 2–10 flowers subtended by >2 bracts . . . . *D. conservatorum*

Douglasia conservatorum Björk Mat-forming. **Scapes** solitary, 12–30 mm long, stellate-pubescent. **Leaves** papillose, linear to linear-elliptic, 7–17 mm long. **Inflorescence** a bracteate umbel with 2 to 10 flowers. **Flowers** on pedicels 3–8 mm long, bractless; calyx pubescent; corolla 10–12 mm long, the 5 lobes 5–6 mm long. **Capsule** 2–3 mm long. Stony soil of exposed ridge tops; subalpine. Currently known from a single population in Sanders Co. and adjacent ID.[45]

Douglasia montana A.Gray Cushion-forming. **Scapes** 1 to 4, erect, 2–15 mm long, pubescent. **Leaves** ciliate to sparsely covered with short hairs, linear-lanceolate, 2–6 mm long. **Flowers** solitary, subtended by 0 to 2 bracts; calyx reddish; corolla 6–8 mm long, the lobes 4–7 mm long. Stony soil of exposed slopes, ridges in grasslands, fellfields; valleys to alpine. AB south to ID, WY.

Primula L. Primrose

Fibrous-rooted, perennial herbs. **Leaves** all basal in a rosette, glabrous or farinose (mealy flavinoid exudate), petiolate. **Inflorescence** a terminal, bracteate umbel. **Flowers** pedicellate, sometimes distylous (with short- and long-styled forms); calyx tubular with erect lobes; corolla salverform with emarginate lobes; ovary superior; stamens included. **Capsule** ovoid, opening by 5 valves.

1. Calyx glandular, not farinose; corolla lobes 5–10 mm long . . . . *P. parryi*
1. Calyx farinose, not glandular; corolla lobes 3–5 mm long 2

2. Corolla white; leaves glabrous or lightly farinose beneath . . . .*P. alcalina*
2. Corolla pinkish; leaves densely farinose beneath . . . . *P. incana*

Primula alcalina Cholewa & Douglass M.Hend. **Scapes** 6–20 cm. **Leaf blades** oblanceolate to obovate, 1–4 cm long, dentate, glabrous above, mostly glabrous below. **Flowers** 3 to 10, distylous; pedicels 1–3 mm long; calyx 3–5 mm long, farinose; corolla white with a yellow center, 4–6 mm long, the lobes 3–5 mm long, bilobed. **Capsule** 5–10 mm long. Hummocks of wet alkaline meadows; montane. Endemic to east-central ID and Beaverhead Co., MT.
　　Similar to the more widespread *Primula incana*. The plant has apparently been extirpated at one of two known MT locations.

Primula incana M.E.Jones **Scapes** 10–35 cm. **Leaf blades** oblanceolate to obovate, 1–5 cm long, dentate to entire, glabrous above, white-farinose below. **Flowers** 3 to 13, homostylous; pedicels 3–10 mm long; calyx 6–9 mm long, farinose; corolla lilac, 6–10 mm long, the lobes 3–5 mm long, bilobed. **Capsule** 5–10 mm long. Moist to wet alkaline meadows, often on hummocks; valleys to montane. AK to QC south to UT, CO, ND. See *P. alcalina*.

Primula parryi A.Gray **Scapes** glandular-puberulent, 10–30 cm from an elongate caudex. **Leaf blades** erect, sparsely glandular, 7–25 cm long, dentate. **Flowers** 3 to 12, homostylous; pedicels 1–6 cm long, glandular; calyx 7–10 mm long, glandular; corolla yellow below, magenta above, 10–13 mm long, the lobes 5–10 mm long, emarginate. **Capsule** 5–8 mm long. Granitic, stony soil or talus slopes; upper subalpine, alpine. MT south to AZ, NM.

MYRSINACEAE: Myrsine Family

Annual or perennial herbs. **Leaves** simple, entire, unlobed, undivided, opposite or alternate above. **Inflorescence**: flowers solitary or in racemes from leaf axils. **Flowers** perfect, hypogynous, pedicellate, radially symmetrical, rotate to campanulate; petals mostly 5, united at the base; sepals 5, united; stamens 5; ovary superior; style 1. **Fruit** a capsule; seeds few to numerous.[84,187]
　　These species were formerly placed in the Primulaceae.[206]

1. Flowers sessile. 2
1. Flowers pedicellate. 3

2. Leaves alternate above . . . .*Centunculus*
2. All leaves opposite . . . . *Glaux*

3. Plants annual; corolla salmon-colored . . . . *Anagallis*
3. Plants perennial; corolla white to yellow. . . . .*Lysimachia*

Anagallis L. Pimpernel

Anagallis arvensis L. SCARLET PIMPERNEL Glabrous annual. **Stems** ascending, prostrate, rooting at basal nodes, 5–40 cm, 4-angled. **Leaves** opposite, sessile, 9–15 mm long, ovate, rounded at the tip. **Inflorescence**: flowers solitary from leaf axils; pedicels ascending at first, then recurved, 1–3 cm long. **Flowers** salmon-colored, rotate; corolla united only at the base, 5–8 mm across, ciliate; sepals linear-lanceolate, 2–4 mm long; filaments hairy. **Capsules** globose, 3–5 mm across, circumcissile. Collected once in Gallatin Co. Introduced throughout most of temperate N. America; native to Eurasia.

Centunculus L. Chaffweed

Centunculus minimus L. [*Anagallis minima* (L.) E.H.L.Krause] Glabrous, fibrous-rooted annuals. **Stems** 2–12 cm, erect to ascending, rooting at lower nodes. **Leaves** all cauline, opposite below, alternate above, obovate, entire, 2–10 mm long. **Flowers** solitary in leaf axils; pedicels ca. 1 mm long; sepals linear-lanceolate, separate to the base, 2–3 mm long; corolla 1–2 mm long, white, the lobes smaller than the tube. **Capsule** globose, circumscissile, ca. 2 mm long. Mud around ponds, lakes, wetlands; plains, valleys. Cosmopolitan.

Glaux L.

Glaux maritima L. [*Lysimachia maritima* (L.) Galasso, Banfi & Soldano] Glabrous, succulent, rhizomatous perennials. **Stems** ascending to prostrate, 4–20 cm, leafless below. **Leaves** opposite, 4–10 mm long, sessile, narrowly elliptic to oblanceolate, entire. **Flowers** solitary, sessile in leaf axils; calyx campanulate, 3–4 mm long; the spreading lobes as long as the tube, petal-like, pink to white; corolla absent; stamens exserted. **Capsules** globose, 2–2.5 mm long, opening by 5 valves. Moist, saline or alkaline meadows, often adjacent to ponds, wetlands; plains, valleys. Circumboreal south to CA, NM, NE.

Lysimachia L. Loosestrife

Rhizomatous perennials. **Leaves** all cauline, opposite, entire. **Flowers** yellow, pedicillate, 5- or 6-parted; calyx divided to near the base; corolla with a short tube and long lobes; stamens 5 to 6, exserted; style exserted. **Capsule** globose, opening by 5 (to 7) valves.

1. Flowers solitary from leaf axils 2
1. Flowers in numerous axillary racemes or spikes 3

2. Stems prostrate; leaves rounded at the tip. . . . . *L. nummularia*
2. Stems erect; leaves pointed at the tip . . . . *L. ciliata*

3. Flowers born in axillary, open, leafy-bracteate panicles; leaves lanceolate . . . . *L. vulgaris*
3. Flowers in dense axillary spike-like racemes; leaves linear-lanceolate . . . . *L. thyrsiflora*

Lysimachia ciliata L. Fʀɪɴɢᴇᴅ ʟᴏᴏsᴇsᴛʀɪꜰᴇ [*Steironema ciliatum* (L.) Baudo] **Stems** erect, 15–80 cm. **Leaves** with ciliate petioles; the blades ovate to lanceolate, 3–12 cm long, glabrous. **Flowers** paired at leaf nodes; pedicels 2–6 cm long; calyx 4–7 mm long; corolla rotate,1–2 cm across, the lobes obovate. **Capsule** 4–6 mm long. Streamside thickets, aspen groves; plains, valleys, montane. Throughout most of temperate N. America.

Lysimachia nummularia L. **Stems** prostrate, rooting at the nodes. **Leaves** sessile below, short-petiolate above; the blades orbicular to elliptic, 1–3 cm long, glandular-punctate. **Flowers** solitary at leaf nodes; pedicels 1–5 cm long; calyx 6–9 mm long; corolla rotate, 2–3 cm across, the lobes obovate, glandular-ciliolate. **Capsule** 4–8 mm long. Shores of lakes, rivers; valleys. Introduced into eastern N. America, sporadic in the west; native to Europe. Collected once in Sanders Co.

Lysimachia thyrsiflora L. **Stems** ascending, rooting at lower nodes, 20–70 cm. **Leaves** sessile, smaller below, narrowly lanceolate, 4–10 cm long, glandular-punctate. **Flowers** short-pedicillate in spike-like, bracteate, ellipsoid racemes, paired at leaf nodes on stalks 2–3 cm long; calyx 2–4 mm long, glandular; corolla campanulate, 4–7 mm long, the lobes linear, purple dotted. **Capsule** 2–3 mm long. Fens, streambanks, marshes; valleys, montane. Circumboreal south to CA, CO, MO, WV. (p.242)

Lysimachia vulgaris L. **Stems** erect, 30–100 cm. **Leaves** short-petiolate in whorls of 2 to 4; the blades lanceolate with entire to wavy margins, 3–12 cm long, puberulent. **Flowers** in axillary racemes; pedicels 5–15 mm long; calyx ca. 3 mm long with red-margined sepals; corolla rotate, 15–20 mm across, orange at the base, the lobes obovate. **Capsule** 3–6 mm long.

Disturbed wet meadows, ditch banks; valleys. A garden ornamental introduced from Europe, escaped in northeast and northwest North America. One collection from Ravalli Co.

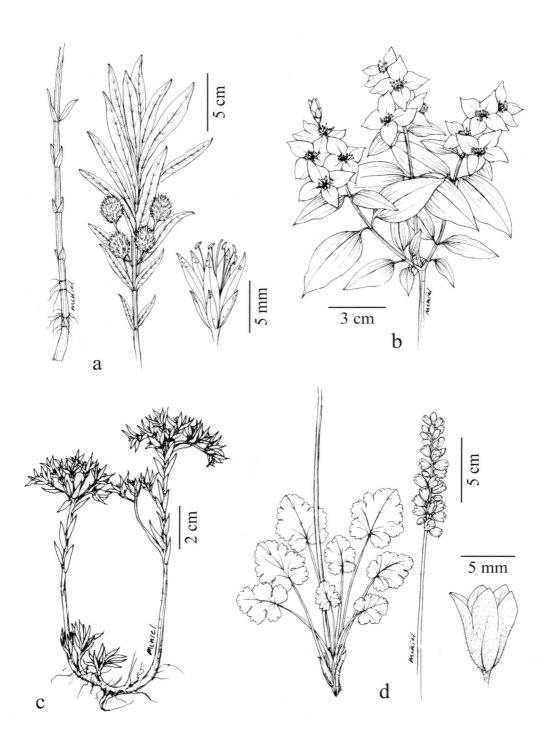

Plate 39. a. *Lysimachia thyrsiflora*, b. *Philadelphus lewisii*, c. *Sedum stenopetalum*, d. *Heuchera cylindrica* var. *cylindrica*

HYDRANGEACEAE: Hydrangea Family

Philadelphus L. Mock orange, syringa

Philadelphus lewisii Pursh Shrub with exfoliating bark, 1–3 m. **Twigs** reddish, glabrous. **Leaves** opposite, short-petiolate, elliptic, the blades 2–6 cm long with mostly entire margins, glabrous to sparsely hairy. **Inflorescence** a several-flowered, terminal raceme on lateral branches. **Flowers** bisexual, regular, fragrant; calyx 4-lobed, forming a hypanthium; petals 4, separate, white, ovate, 1–2 cm long; stamens 20 to 40; stigmas 4; ovary partly inferior. **Fruit** an elliptic, capsule, 6–10 mm long, 4-loculed; seeds numerous. Dry, rocky slopes in grasslands, open forest; valleys, montane. BC, AB south to CA, ID, MT. (p.242)
 The type locality is east of Missoula.

GROSSULARIACEAE: Gooseberry Family

Ribes L. Currant, gooseberry

Shrubs. **Leaves** alternate, lobed, maple-like, often >1 per node, petiolate, usually exstipulate and paler beneath. **Inflorescence** few- to several-flowered, bracteate racemes in leaf axils. **Flowers** bisexual, regular, rotate to tubular; calyx 5-lobed, forming a hypanthium; petals 5, separate, shorter than the calyx lobes; stamens 5; ovary inferior; styles 2, sometimes united below. **Fruit** a many-seeded, globose berry.[367]

 Gooseberry bushes have spiny twigs; currant bushes are unarmed (*Ribes lacustre* has the oxymoronic common name of "spiny currant"). Five of six members of the *R. oxycanthoides* complex occur in MT. Recent authors have followed Sinnott's monograph[367] in considering *R. cognatum*, *R. hendersonii*, *R. irriguum*, and *R. setosum* to be subspecies of *R. oxycanthoides*. *Ribes inerme* was maintained at the species level. Sinnott admits that his treatment is somewhat arbitrary, and all taxa have geographic integrity and are easily distinguishable. Furthermore, a cluster analysis based on 11 morphological charaters indicated that several of the subspecies are as distinct from each other as they are from other taxa accorded specific rank. Consequently I have chosen to recognize all of these taxa as species.

 The hypanthium (calyx tube united to ovary) is often colored and can be mistaken for a corolla tube. *Ribes missouriense* Nutt. is reported for eastern MT,[151,367] but no specimens were cited; however, more recently MT is not listed in the range of this species.[113,152] It would key to *R. inerme*, but would occur way to the east of that species. *Ribes reclinatum* L. [*R. uva-crispa* L.] has been collected in an old mining camp in Flathead Co., but is not known to be naturalized.

1. Stems spineless 2
1. Stems with 1 to few spines at some nodes at least, sometimes bristly between .11

2. Flowers saucer- to shallowly cup-shaped. 3
2. Flowers tubular to campanulate 6

3. Leaves without yellow crystalline glands beneath 4
3. Leaves with sessile, yellow glands beneath. 5

4. Petals purple; petioles with sparse, glandular cilia. . . . . *R. triste*
4. Petals white; petioles without glands . . . .*R. rubrum*

5. Ovary and fruit with sessile, yellow glands. . . . . *R. hudsonianum*
5. Ovary and fruit with stipitate glands . . . .*R. laxiflorum*

6. Leaf blades with sessile or stipitate glands beneath 7
6. Leaf blades not glandular beneath 9

7. Leaf blades with sessile, yellow glands beneath . . . . *R. americanum*
7. Leaf blades with stipitate glands beneath. 8

8. Calyx tube (hypanthium) 2X as long as the lobes; berry red, 4–7 mm long. . . . . *R. cereum*
8. Calyx tube ca. as long as the lobes; berry black, >7 mm long . . . .*R. viscosissimum*

9. Flowers yellow . . . . *R. aureum*
9. Flowers white to pinkish 10

10. Calyx usually glabrous, the lobes 3–4 mm long. . . . .*R. inerme*
10. Calyx glandular-hairy, the lobes 2–3 mm long . . . . *R. cereum*

Ribes americanum Mill. BLACK CURRANT **Stems** erect or ascending, spineless, 1–1.5 m. **Twigs** with sessile, yellow glands, pubescent; the bark gray to black. **Leaf blades** 1–7 cm wide, shallowly cordate to truncate at the base, 3(5)-lobed, shallowly to deeply dentate, the underside sparsely pubescent and glandular. **Inflorescence** 4- to 10-flowered, spreading, ca. as long as the leaves. **Flowers** tubular-campanulate, 6–10 mm long, green to white, pubescent; calyx lobes 2–4 mm long, narrowly oblong; petals white, 2–3 mm long; stamens equaling the petals; styles glabrous, united to the tip. **Berry** glabrous, 5–9 mm long, black, unpalatable. Woodlands, streambanks, riparian forests, swamps; plains, valleys. WA to SK south to CA, NM, SD. (p.246)

Ribes hudsonianum has bowl-shaped, glandular flowers.

Ribes aureum Pursh GOLDEN CURRANT **Stems** erect, spineless, 1–2 m. **Twigs** glabrous to puberulent; the bark brown, becoming gray. **Leaf blades** 1–6 cm wide, cuneate at the base, 3(5)-lobed, the lobes entire to shallowly lobed, mostly glabrous. **Inflorescence** 3–8-flowered, ascending to drooping, as long or longer than the leaves. **Flowers** tubular, 6–8 mm long, yellow; calyx lobes 2–4 mm long, oblong; petals yellow; stamens equaling the petals; styles glabrous, united to the tip. **Berry** glabrous, 4–8 mm long, black, palatable. Woodlands, streambanks, riparian forests; plains, valleys. WA to SK south to CA, AZ, NM, TX. (p.246)

Great Plains plants (var. *villosum* DC.) tend to have a longer hypanthium compared to the sepals than plants from the western U.S. (var. *aureum*), but this distinction is not pronounced in MT. The type locality is near Three Forks.

Ribes cereum Douglas WAX CURRANT **Stems** erect to spreading, spineless, 1–2 m. **Twigs** glandular and/or pubescent, brown becoming gray. **Leaf blades** 1–3 cm wide, shallowly cuneate, shallowly 3-lobed to fan-shaped with crenate margins, glabrous to glandular-pubescent. **Inflorescence** short, 3- to 5-flowered, drooping. **Flowers** tubular, white to pink, 7–13 mm long, glandular-pubescent; calyx lobes 2–3 mm long, ovate; petals dull white, 1–2 mm long; stamens longer than the petals; styles glabrous or pubescent, united to the tip. **Berry** glabrous to glandular, 4–7 mm long, red, insipid. Moist to dry, often stony soil of grasslands, steppe, woodlands, open forest; plains, valleys, montane. BC to ND south to CA, AZ, NM, OK. (p.246)

Ribes hendersonii C.L.Hitchc. [*R. oxyacanthoides* L. ssp. *hendersonii* (C.L.Hitchc.) Q.P.Sinnott] **Stems** sprawling, 30–70 cm with 2 to 3 stout, curved nodal spines. **Twigs** glabrous to puberulent and sparsely bristly, tan becoming gray. **Leaf blades** 1–2 cm wide, cordate, 3- to 5-lobed, crenate, glandular-pubescent. **Inflorescence** of 1 to 2 flowers, shorter than the leaves. **Flowers** campanulate, green to white, 5–8 mm long, nearly glabrous; calyx lobes oblong, 2–3 mm long; petals 1–2 mm long, white; stamens ca. as long as the petals; styles hairy, united ca. half the length. **Berry** reddish, 4–7 mm long, glabrous. Stony soil of open forest, talus slopes; subalpine. MT, ID, NV.

Ribes hudsonianum Richardson STINKING CURRANT **Stems** erect, spineless, 0.5–1.5 m. **Twigs** with sparse, sessile glands, puberulent, tan becoming gray. **Leaf blades** 3–15 cm wide, 3- to 5-lobed, cordate, dentate, crystaline-glandular beneath, pubescent to glabrous, aromatic. **Inflorescence** ascending, 10- to 40-flowered. **Flowers** bowl-shaped, white, 2–5 mm long, pubescent to glabrous, glandular; calyx lobes 2–4 mm long, lanceolate; petals white, 1–2 mm long; stamens as long as petals; styles glabrous, united half way. **Berry** black, 4–10 mm long, mostly glabrous, unpalatable. Moist to wet forest, thickets, often along streams; valleys to lower subalpine. AK to ON south to WA, MT, MN. (p.246)

Varieties have been recognized based on the degree of pubescence and glandulosity, but these characters vary continuously and are not correlated in our material.

Ribes inerme Rydb. **Stems** spreading to erect, 0.5–1.5 cm; nodal spines few or lacking. **Twigs** mostly glabrous, tan becoming gray. **Leaf blades** 2–6 cm wide, cordate to truncate, 3- to 5-lobed, deeply crenate, mostly glabrous, sometimes ciliate. **Inflorescence** spreading, 1- to 4-flowered. **Flowers** campanulate, green to purplish, 6–9 mm long, usually glabrous; calyx lobes lanceolate, 3–4 mm long; petals 1–2 mm long, white to pink; stamens ca. twice as long as the petals; styles hairy, united less than half the length. **Berry** reddish-purple, 5–9 mm long, glabrous, edible. Open forest, thickets, aspen woodlands, rocky slopes, often along streams, lakes, wetlands; valleys, montane. BC, AB south to CA, NM. (p.246)

   Ribes irriguum is similar but has stamens as long as the petals; plants intermediate between these two species are common around Missoula.

Ribes irriguum Douglas [*R. oxyacanthoides* L. ssp. *irriguum* (Dougl.) Q.P.Sinnott] **Stems** erect to ascending, 1–2 m with 1–3 straight nodal spines. **Twigs** puberulent with sparse bristles, brown becoming gray. **Leaf blades** 1–6 cm wide, cordate to truncate, 3- to 5-lobed, deeply crenate, glandular and/or pubescent on both surfaces. **Inflorescence** of 1 to 3 flowers, drooping. **Flowers** campanulate, green to red, 6–11 mm long, glabrous; calyx lobes oblong, 3–5 mm long; petals 2–3 mm long, white to pink; stamens ca. as long as the petals; styles united ca. half their length. **Berry** purple, 6–11 mm long, glabrous. Open forest, woodlands, often along streams; valleys, montane. BC to MT south to OR, ID. (p.246)

   Previous reports of *Ribes cognatum* Greene from Flathead, Lincoln, Missoula and Sanders cos. were based on plants with sparsely hairy flowers; however, the calyx tubes of these flowers are <4 mm long, so they are better referred to *R. irriguum*, although they may represent hybridization between the two species. See *R. oxyacanthoides*, *R. inerme*.

Ribes lacustre (Pers.) Poir. SPINY SWAMP CURRANT **Stems** erect, 0.5–1 m, spiny at the nodes, bristly between. **Twigs** puberulent, red-brown, becoming gray. **Leaf blades** 1–7 cm wide, cordate, 5- to 7-lobed, deeply toothed, nearly glabrous and shiny above, glabrous to sparsely hairy and sometimes glandular on the veins beneath. **Inflorescence** drooping, 5- to 20-flowered. **Flowers** saucer-shaped, 3–4 mm long, greenish to reddish, glandular-hairy; calyx lobes ovate, 2–3 mm long; petals pink, 1–2 mm long; stamens shorter than petals; style glabrous, united more than half way. **Berry** 6–9 mm long, black, bristly-glandular, unpalatable. Moist to wet forest, rock slides, avalanche slopes, cliffs, often along streams; montane, subalpine. AK to NL south to CA, CO, SD, PA.

   Density of spines is quite variable. Our most common species of wet sites in coniferous forest. See *Ribes montigenum*.

Ribes laxiflorum Pursh **Stems** ascending to prostrate, spineless, 0.5–1 m. **Twigs** with sparse sessile glands, puberulent, tan becoming brown. **Leaf blades** 4–8 cm wide, 5-lobed, cordate, serrate, crystalline-glandular and pubescent beneath. **Inflorescence** ascending, 8- to 18-flowered, glandular. **Flowers** bowl-shaped, white to reddish, 3–4 mm long, pubescent, glandular-hairy; calyx lobes 2–3 mm long, oblong; petals reddish, 1–2 mm long; stamens as long as petals; styles glabrous, united half way. **Berry** black, ca. 1 cm long, glandular-hairy. Wet forest, thickets; montane. AK south to CA, ID, MT. Known from Lincoln Co.[380]

   Ribes hudsonianum has white flowers and fruits have sessile rather than stipitate glands.

Ribes montigenum McClatchie **Stems** sprawling to ascending, 30–70 cm, spiny at the nodes, bristly between. **Twigs** puberulent, tan, becoming gray. **Leaf blades** 1–4 cm wide, cordate, 5-lobed, deeply toothed, pubescent and glandular on both sides. **Inflorescence** drooping, 3- to 6-flowered. **Flowers** saucer-shaped, 3–5 mm long, green to pink, glandular-hairy; calyx lobes broadly ovate, 2–3 mm long; petals pink to red, 1–1.5 mm long; stamens longer than petals; style united more than half way, glabrous. **Berry** 5–9 mm long, red, bristly-glandular, palatable. Stony soil of dry forest, rock slides; subalpine. BC to MT south to CA, AZ, NM. (p.246)

   Ribes lacustre occurring in wetter, less exposed habitats is similar but lacks glandular hairs on the upper leaf surface. A report of *R. velutinum* Greene for MT was based on *R. montigenum*.

Ribes oxyacanthoides L. **Stems** ascending to sprawling, 0.5–1.5 m with 1 to 4 straight to curved nodal spines. **Twigs** puberulent, often bristly, yellow becoming gray. **Leaf blades** 1–3 cm wide, cordate, 3- to 5-lobed, deeply crenate, pubescent, often glandular beneath. **Inflorescence** of 1 to 3 flowers, drooping, shorter than the leaves. **Flowers** obconic, green to white, 7–10 mm long, glabrous; calyx lobes oblong, 3–5 mm long; petals 2–3 mm long, white to pink; stamens ca. as long as the petals; styles united ca. half the length. **Berry** purple, 8–12 mm long, glabrous, edible. Woodlands, thickets along streams; plains, valleys, montane. YT to NL south to CO, NE, MN.

   The similar *Ribes irriguum* generally occurs farther west.

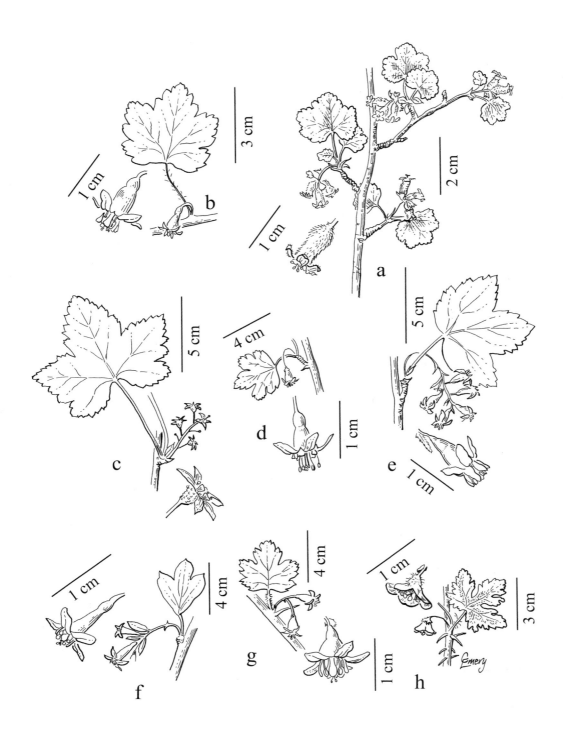

Plate 40. *Ribes*. **a.** *R. cereum*, **b.** *R. setosum*, **c.** *R. hudsonianum*, **d.** *R. inerme*, **e.** *R. americanum*, **f.** *R. aureum*, **g.** *R. irriguum*, **h.** *R. montigenum*

Ribes rubrum L. [*R. sativum* (Rchb.) Syme] RED CURRANT **Stems** erect, spineless, 0.5–1.5 m. **Twigs** glabrous, tan becoming brown, glaucous. **Leaf blades** 2–6 cm wide, 5-lobed, cordate, crenate, pubescent, sparsely so above. **Inflorescence** drooping, 7- to 15-flowered. **Flowers** saucer-shaped, greenish, 2–3 mm long, glabrous; calyx lobes 1–2 mm long, lanceolate; petals white, ca. 1 mm long; stamens as long as the petals. **Berry** red, ca. 7 mm long, glabrous, palatable. Riparian forest, thickets, along streams and ditches, near habitations; valleys. Introduced throughout much of temperate N. America for the edible fruit; native to Eurasia.

Ribes setosum Lindl. [*R. oxyacanthoides* L. ssp. *setosum* (Lindl.) Q.P.Sinnott] **Stems** erect to ascending, 1–1.5 m with 1 to 3 straight to slightly curved nodal spines. **Twigs** puberulent, sometimes bristly, tan becoming gray. **Leaf blades** 1–5 cm wide, cordate, 3- to 5-lobed, deeply crenate, glabrous to pubescent, often glandular beneath. **Inflorescence** of 2 to 3 flowers, drooping, shorter than the leaves. **Flowers** tubular-campanulate, white to pink, 7–12 mm long, glabrous; calyx lobes oblong, 3–5 mm long; petals 2–3 mm long, white to pink; stamens ca. as long as the petals; styles hairy, united ca. half way. **Berry** black, 6–10 mm long, glabrous, edible. Open forest, woodlands, thickets, steppe, often along streams; plains, valleys, montane. ID, MT, WY. (p.246)

This is the most widespread member of the *Ribes oxycanthoides* complex in MT. It can be told by the relatively long, slender hypanthium.

Ribes triste Pall. **Stems** sprawling to ascending, spineless, 0.5–1 m. **Twigs** glabrous to sparsely hairy and glandular, tan becoming brown. **Leaf blades** 2–10 cm wide, 3- to 5-lobed, cordate, dentate, glabrous above, sparsely hairy beneath with sparse, glandular cilia on the petioles. **Inflorescence** drooping, 5- to 10-flowered. **Flowers** bowl-shaped, white to purple, 2–3 mm long, glabrous; calyx lobes 1–2 mm long, fan-shaped; petals purple, <1 mm long; stamens as long as petals; styles glabrous, united half way. **Berry** red, 7–10 mm long, glabrous. Moist forest; montane, subalpine in Granite Co. AK to NL south to OR, SD, VA; Asia.

Ribes viscosissimum Pursh STICKY CURRANT **Stems** erect, spineless, 0.5–1 m. **Twigs** tomentose and sparsely glandular, tan becoming gray. **Leaf blades** 3–12 cm wide, cordate, 3- to 5-lobed, crenate, pubescent and glandular above, more so beneath. **Inflorescence** spreading, 4- to 12-flowered. **Flowers** campanulate, 10–15 mm long, greenish-white, tinged with pink, glandular-hairy; calyx lobes lance-elliptic, 3–5 mm long; petals 3–4 mm long, white; stamens as long as petals; styles united to the tip, glabrous. **Berry** black, 8–15 mm long, glandular-hairy, unpalatable. Moist to wet forests, avalanche slopes, often near streams; montane, lower subalpine. BC to AB south to CA, AZ, CO. Our plants are var. *viscosissimum*.

CRASSULACEAE: Stonecrop Family

Succulent, glabrous, annual or perennial herbs. **Leaves** alternate or opposite, simple, entire. **Inflorescence** a cyme or solitary flowers. **Flowers** perfect, hypogynous, 4- to 5-merous; petals separate, erect or ascending; sepals divided to near the base; stamens 4 or 8 to 10; ovaries 4 to 5, united at the base. **Fruits** follicles narrowed to a terminal style.

1. Diminutive annual; stamens ca. 4. . . . . *Tillaea*
1. Perennials; stamens 8 to 10. . . . . *Sedum*

Sedum L. Stonecrop

Perennials. **Leaves** alternate. **Inflorescence** a terminal or axillary cyme. **Flowers** perfect, hypogynous, usually 5-merous; stamens 10; ovaries 5, united at the base. **Fruits** 5 follicles narrowed to the terminal style; seeds numerous.[88]

Leaves of both *Sedum lanceolatum* and *S. stenopetalum* have often withered by flowering time. *Sedum album* L. (white stonecrop) with white petals, a rare garden escape, was collected once in Missoula Co. (*Pierce 1310* MONTU).

1. Petals and sometimes sepals pink to red; leaves somewhat flattened 2
1. Petals yellow; leaves terete or subterete 3

2. Flowers bisexual; petals 7–10 mm long; sepals 4–6 mm long. . . . . *S. rhodanthum*
2. Flowers mostly unisexual; petals 2–3 mm long . . . . *S. rosea*

3. Basal leaves ovate to suborbicular. 4
3. Basal leaves lanceolate, often withered at flowering; sepals 1–2 mm long 5

4. Basal leaves ovoid, terete; garden escape. . . . . *S. acre*
4. Basal leaves ovate, flattened; cliffs and ridges. . . . . *S. borschii*

5. Leaf surfaces not papillose, with a prominent midvein below; plantlets in upper leaf axils . .*S. stenopetalum*
5. Leaves papillose, without prominent midveins; axillary plantlets absent . . . .*S. lanceolatum*

Sedum acre L. Mat-forming. **Stems** erect, simple or branched, 2–10 cm. **Leaves** ovoid, sessile,
3–5 mm long. **Flowers** 5-merous; petals yellow, lanceolate, 4–5 mm long. **Follicles** spreading,
5–6 mm long; styles straight, ca. 1 mm long. Lawns, roadsides; valleys. Introduced to much of
northern U.S. and adjacent Canada; native to Europe.

Sedum borschii (R.T.Clausen) R.T.Clausen [*S. leibergii* Britton var. *borschii* R.T.Clausen] Short-lived perennial.
Stems ascending from short rootstocks, 4–12 cm. **Leaves** oval to obovate above, 2–6 mm long. **Flowers** 5-merous;
petals yellow, broadly lanceolate, 5–7 mm long. **Follicles** spreading, glandular, 4–6 mm long; styles ca. 1 mm long,
straight. Moist cliffs, stony ridges; montane, subalpine. Endemic to Missoula and Ravalli cos. and adjacent ID.
    Sedum debile S.Watson is reported for MT,[226] but our plants do not have the opposite stem leaves of this species.

Sedum lanceolatum Torr. **Stems** branched, creeping with short, leafy, sterile shoots; fertile
stems erect, 5–20 cm. **Leaves** lanceolate to ovate below, lanceolate above, papillose, 5–20 mm
long, terete. **Flowers** 5-merous; petals yellow, broadly lanceolate, 5–9 mm long. **Follicles** erect,
3–5 mm long; styles 2–3 mm long, divergent. Rocky, often shallow soil of cliffs, grasslands, steppe,
fellfields; valleys to alpine. AK to SK south to CA, NE.
    Sedum stenopetalum is similar but has small, leafy plantlets in upper leaf axils and leaves that are papillose only on
the margins.

Sedum rhodanthum A.Gray [*Rhodiola rhodantha* (A.Gray) H.Jacobsen] Rose crown **Stems**
erect or ascending, 10–35 cm, clustered from a short-branched caudex. **Leaves** all cauline,
oblanceolate, 10–25 mm long, flattened. **Flowers** 5-merous; petals pink, lanceolate, 7–10 mm long.
Follicles erect, 6–8 mm long; styles ca. 1 mm long, barely divergent. Moist turf, fens, often along
streams; upper subalpine, alpine. MT south to UT, CO, AZ.

Sedum rosea (L.) Scop. [*S. integrifolium* (Raf.) A.Nelson, *Rhodiola integrifolia* Raf.] Roseroot,
king's crown, red orpine **Stems** erect or ascending, 2–15 cm, clustered from a short-branched
caudex. **Leaves** all cauline, oblong, 4–20 mm long, flattened, rarely denticulate. **Flowers** usually
5-merous, often imperfect, red; petals ovate, 2–3 mm long. **Follicles** erect, 4–7 mm long; styles <1
mm long, divergent. Moist, stony soil of cliffs, exposed ridges, talus, turf; subalpine, alpine. AK to
Greenland south to CA, CO, NY; Asia. Our plants are var. ***integrifolium*** (Raf.) A.Berger.
    A collection from Fergus Co. with long, yellow petals may be a hybrid with *Sedum lanceolatum*.

Sedum stenopetalum Pursh Similar to *S. lanceolatum*. **Stems** branched, creeping, rooting
with short, leafy, sterile shoots; fertile stems erect, 8–20 cm. **Leaves** papillose on the margins
only, broadly lanceolate, 5–15 mm long, terete with a prominent midvein beneath; upper leaves
with leafy plantlets in the axils. **Flowers** 5-merous; petals yellow, broadly lanceolate, 5–8 mm long.
Follicles spreading, 3–4 mm long; styles 1–2 mm long, straight. Vernally moist, stony, shallow soil of
cliffs, exposed slopes, open forest, woodlands, grasslands; valleys, montane. BC, AB south to CA, MT. (p.242)
    The type locality is south of Missoula.

Tillaea L. Pigmy weed

Tillaea aquatica L. Small glabrous annual. **Stems** prostrate to ascending, branched, 1–4 cm.
Leaves all cauline, opposite, linear, terete, 3–6 mm long, united around the stem. **Flowers** solitary
in leaf axils, 4-merous, calyx ≤1 mm long; petals white, 1–2 mm long. **Follicles** 1–1.5 mm long,
erect, deep red; seeds 1 to few per follicle. Mud or shallow water along ponds, streams; valleys.
Much of temperate N. America; Eurasia.

SAXIFRAGACEAE: Saxifrage Family

Perennial herbs. **Leaves** alternate, simple but often dentate to palmately lobed. **Inflorescence** a raceme, cyme or solitary flowers. **Flowers** perfect, regular, perigynous to epigynous; calyx 5-lobed, the united base partly surrounding or united to the ovary (hypanthium); petals 5, separate; stamens 5 or 10; styles 2; ovary of 2 carpels. **Fruit** a usually 2-lobed capsule or follicle, often with a hollow beak-like style; seeds few to many.[428]

1. Stamens 4; sepals 4; stems prostrate below . . . . *Chrysosplenium*
1. Stamens 5 to 10; sepals 5; stems usually ascending to erect 2

2. Stamens 5 (sometimes also with 5 stamen-like appendages) 3
2. Stamens 10 10

3. Basal leaves glabrous with entire margins . . . . *Parnassia*
3. Basal leaves with lobed or toothed margins 4

4. Stem with at least 1 leaf well below the inflorescence 5
4. Stems leafless 8

5. Upper stem leaves sessile with dilated bases . . . . *Suksdorfia*
5. Most stem leaves petiolate; leaf bases not dilated 6

6. Petals deeply lobed; inflorescence racemose . . . . *Mitella*
6. Petals not lobed; inflorescence paniculate 7

7. Stems brown-hairy and glandular; calyx 3–5 mm long . . . . *Boykinia*
7. Stems glandular but not brown-hairy; calyx 2–3 mm long . . . . *Sullivantia*

8. Lower inflorescence branched or with >1 flower per node . . . . *Heuchera*
8. Lower nodes of inflorescence spike with only 1 flower 9

9. Petals lobed or divided . . . . *Mitella*
9. Petals with entire margins . . . . *Conimitella*

10. Stems leafless (leaf-like bracts may subtend the inflorescence) 11
10. Stem with at least 1 leaf well below the inflorescence 12

11. Petals pinnately divided into several linear lobes; capsule opening widely, appearing like
    a dish of seeds at maturity . . . . *Mitella*
11. Petals not deeply lobed; capsule opening at the tip(s) . . . . *Micranthes*

12. Petals deeply lobed 13
12. Petals not lobed or absent 14

13. Plants <25 cm high; of open habitats . . . . *Lithophragma*
13. Plants >25 cm high; found in moist forest . . . . *Tellima*

14. Leaves leathery with a single clasping leaf at mid-stem . . . . *Leptarrhena*
14. Stem leaves >1, not leathery 15

15. Petals linear; capsule with 2 unequal chambers; forest plants . . . . *Tiarella*
15. Petals oblong to obovate; capsule with equal chambers; plants of open, often stony habitats 16

16. Petals usually purple; basal leaves ≥2 cm wide; lower stems covered in old leaf bases . . . . *Telesonix*
16. Plants not with above combination of characters . . . . *Saxifraga*

Boykinia Nutt.

Boykinia major A.Gray **Stems** brown-hairy and glandular-pubescent, ascending to erect, 20–90 cm from a nearly horizontal rootstock. **Leaves** petiolate, leafy-stipulate; the blade reniform, cordate, deeply 3- to 7-lobed with dentate margins, 2–20 cm wide, sparsely hairy, glandular-hairy on the veins beneath. **Inflorescence** a glandular, many-flowered, hemispheric panicle. **Flowers** campanulate, epigynous; calyx 3–5 mm long with a short tube and lanceolate lobes 1–4 mm long; petals 5–8 mm long, white, clawed; the blade ovate; stamens 5, inserted at the top of hypanthium. **Capsules** 4–5 mm long at maturity. Wet meadows, forests, thickets, often along streams; montane, subalpine. WA to MT south to CA, ID.

Chrysosplenium L.

Chrysosplenium tetrandrum Th.Fr. Glabrous perennial herbs. **Stems** 4–15 mm, branched above, prostrate with erect tips to 10 mm long. **Leaves** alternate, basal or subtending the inflorescence, petiolate; the blades reniform, cordate, crenate, 10–15 mm wide. **Inflorescence** a few-flowered cyme. **Flowers** regular, perfect, perigynous, bowl-shaped, 4-merous; calyx green, ca. 3 mm wide with oval lobes; petals absent; stamens 4, shorter than the calyx. **Fruit** 2 capsules broadly ovoid, ca. 4 mm long, 1/2 inferior; seeds several. Mud or shallow water on margins of small streams; montane; known from Ravalli Co. Circumpolar south to WA, CO.

Conimitella Rydb.

Conimitella williamsii (D.C. Eaton) Rydb. Perennial herbs from simple, ascending, scaly rootstocks. **Stems** simple, leafless, glandular-puberulent, 20–60 cm. **Leaves** long-petiolate in a basal rosette; the blades orbicular, cordate with crenulate margins, 1–5 cm wide, glabrous except for curved, marginal cilia, purplish beneath. **Inflorescence** a 6- to 12-flowered, small-bracted, glandular, spike-like raceme. **Flowers** 5-merous, regular, perfect, perigynous, obconical; calyx 4–7 mm long with oblong lobes ca. 1 mm long; petals white, 4–5 mm long, clawed; stamens 5, equaling the calyx lobes. **Fruit** 2 capsules mostly inferior; seeds numerous. Stony, usually calcareous soil of grasslands, open forest, meadows; montane. Endemic to MT and adjacent AB, ID, WY.

First collected by R.S. Williams over 100 years ago in the Belt Mtns. *Mitella* spp. are usually found in moister habitats, while species of *Heuchera* have more flowers.

Heuchera L. Alumroot

Stems erect to ascending, leafless, glandular above, from a simple or branched caudex covered with old leaf bases. **Leaves** all basal, long-petiolate, stipulate; the blade ovate to reniform, cordate-based, shallowly lobed, crenate to dentate, stiff-ciliate. **Inflorescence** a glandular, raceme-like panicle. **Flowers** campanulate to rotate, sometimes slightly irregular, epigynous; calyx glandular with short lobes (sepals); stamens 5, borne at the edge of the hypanthium with the petals. **Capsules** ovoid with divergent beaks.[429] The species are vegetatively very similar; flowers are necessary for positive identification. See *Conimitella*.

1. Calyx <4 mm long; hypanthium cup-shaped; ovary almost completely inferior . . . . *H. parvifolia*
1. Calyx >4 mm long; hypanthium short-tubular; upper half of capsules visible. 2

2. Stamens longer than the sepals; eastern tier of counties. . . . . *H. richardsonii*
2. Stamens included in calyx; more western 3

3. Petals minute or absent; stems glandular and hairy above . . . .*H. cylindrica*
3. Petals ca. as long as calyx lobes; stems glandular only. . . . .*H. grossulariifolia*

Heuchera cylindrica Douglas **Stems** 10–60 cm. **Leaf blades** broadly ovate, 2–5 cm long, shallowly 5-lobed to twice crenate, glabrous to sparsely hairy above, glabrous to glandular and puberulent beneath. **Flowers** campanulate, regular; calyx 6–10 mm long, green to white; the lobes 2–4 mm long, erect; petals absent or inconspicuous; stamens shorter than the sepals. **Capsules** 5–8 mm long. Stony soil of grasslands, rock outcrops, dry to moist forest, talus slopes; valleys to lower subalpine. BC, AB south to CA, NV, WY. (p.242)

1. Lower stem and petioles densely glandular-hairy . . . . .var. *cylindrica*
1. Lower stem and petioles glabrous or sparsely hairy . . . . var. *glabella*

Heuchera cylindrica var. *glabella* (Torr.& A.Gray) Wheelock. is restricted to, and the most common form in, the southwest mountains; *H. cylindrica* var. *cylindrica* is the only form in the northwest mountains and valleys.

Heuchera grossulariifolia Rydb. **Stems** 6–40 cm. **Leaf blades** ovate to orbicular, 1–4 cm wide, shallowly 5- to 7-lobed, crenate, glabrous to glandular-puberulent. **Flowers** campanulate, slightly irregular; calyx 2–5 mm long, greenish-white; the lobes 2–3 mm long, erect; petals white, 1–2 mm long; stamens shorter than the sepals. **Capsules** 4–7 mm long. Usually granitic soil of crevices, rock slides, fellfields; montane to lower alpine. WA, OR, ID, MT. Our plants are var. *grossulariifolia*.

A dwarf form with glandular leaves occurs above treeline. *Heuchera cylindrica* has less divided leaves and regular flowers. The type locality is in Madison Co.

Heuchera parvifolia Nutt. ex Torr. & A.Gray **Stems** 10–50 cm. **Leaf blades** ovate to reniform, 1–4 cm across, 5- to 7-lobed again shallowly lobed and crenate, glabrous above, glandular beneath. **Flowers** cup-shaped, regular; calyx 2–3 mm long, greenish-white; the lobes 1–1.5 mm long, spreading; petals white, 1–2 mm long; stamens shorter than the sepals. **Capsules** 3–5 mm long. Stony soil of grasslands, steppe, woodlands, rock outcrops, turf, fellfields, open forest; plains, valleys to lower alpine. AB south to NV, AZ, NM.

Characters used to distinguish var. *utahensis* (Rydb.) Garrett from var. *dissecta* M.E.Jones are not strongly correlated in our area; high-elevation plants are smaller, and the leaves are not always strongly dissected.

Heuchera richardsonii R.Br. **Stems** 20–70 cm. **Leaf blades** orbicular to reniforn, 2–9 cm wide, shallowly 5- to 7-lobed, crenate, nearly glabrous above, sparsely hairy beneath. **Flowers** campanulate, irregular; calyx 8–12 mm long, green to pinkish; the lobes 3–4 mm long, erect, yellow to purple; petals 5–7 mm long, purplish; stamens ca. as long as the sepals. **Capsules** 6–8 mm long. Grasslands, woodlands, open forest; plains. AB to MI south to CO, NE, MO.

Leptarrhena R.Br. Leather leaf

Leptarrhena pyrolifolia (D.Don) R.Br. ex Ser. Perennial from horizontal rootstocks. **Stems** simple, 10–30 cm, glabrous to glandular-hairy. **Leaves** glabrous, mostly basal, short-petiolate; the blades obovate, dentate, 2–7 cm long, shiny and leathery; stem leaf solitary, sessile, ovate. **Inflorescence** a terminal, glandular, short-branched panicle. **Flowers** 5-merous, regular, perfect, perigynous, saucer-shaped; calyx barely adnate to the ovary, 2–3 mm across with oblong lobes ca. 1 mm long; petals white, oblanceolate, unequal, 1–2 mm long; ovary slightly inferior; stamens 10, barely longer than the petals. **Fruit** reddish; narrowly ovoid follicles 5–10 mm long including the divergent beaks; seeds numerous. Moist to wet soil of meadows, fens, cliffs, often along streams; subalpine, alpine. AK to WA, ID, MT. (p.252)

Lithophragma (Nutt.) Torr. & A.Gray Prairie star, fringecup

Short-lived perennials, often forming bulblets in the roots and sometimes on the stems. **Stems** glandular. **Leaves** basal and cauline, petiolate, stipulate; the blades deeply lobed, cordate-based. **Inflorescence** a terminal, ebracteate, several-flowered raceme. **Flowers** 5-merous, regular, perfect, perigynous to epigynous; calyx lobes triangular; petals clawed; the blade lobed; stamens 10; styles 3. **Fruit** a 3-beaked capsule; seeds numerous.[392]

1. Petals 3-lobed; calyx obconic; capsule fully enclosed in hypanthium . . . .*L. parviflorum*
1. Petals 5-lobed; calyx cup-shaped; capsule half exposed or flowers replaced by bulblets 2

2. Basal leaves nearly glabrous beneath; bulblets often present. . . . . *L. glabrum*
2. Basal leaves moderately hairy beneath; bulblets absent. . . . .*L. tenellum*

Lithophragma glabrum Nutt. [*L. bulbiferum* Rydb.] **Stems** 5–30 cm, often with purple bulblets in axils of upper leaves or inflorescence branches. **Leaf blades** glabrous below, sparsely hairy above, 5–20 mm wide, reniform, deeply 3-lobed and lobed again. **Flowers** 1 to 6, often replaced by bulblets; calyx cup-shaped, 2–4 mm long; petals white to pink, 5–8 mm long, mostly deeply 5-lobed. **Capsule** half exposed above hypanthium. Meadows, vernally moist grasslands; valleys to subalpine. BC to SK south to CA, UT, CO, SD.

Some populations are composed of plants lacking flowers entirely; populations without bulblets also occur.

Lithophragma parviflorum (Hook.) Nutt. **Stems** 10–50 cm. **Leaf blades** white-hairy and glandular, 1–3 cm wide, suborbicular, twice deeply 3-lobed. **Flowers** 4 to 12; calyx obconic, 4–6 mm long; petals white, 6–9 mm long, 3-lobed. **Capsule** 5–7 mm long, fully enclosed in the hypanthium. Grasslands, meadows, steppe, woodlands, open forest; plains, valleys to subalpine. BC, AB south to CA, CO, SD. (p.252)

Lithophragma tenellum Nutt. **Stems** 5–20 cm. **Leaf blades** glandular-hairy, 1–2 cm wide, ovate to suborbicular, twice deeply 3-lobed. **Flowers** 3 to 9; calyx cup-shaped, 2–4 mm long; petals white to pink, 3–7 mm long, 5-lobed, often with an additional 2 small basal lobes. **Capsule** half exposed above hypanthium. Grasslands, open forest; montane; Gallatin and Sanders cos. BC, AB south to CA, CO, SD.

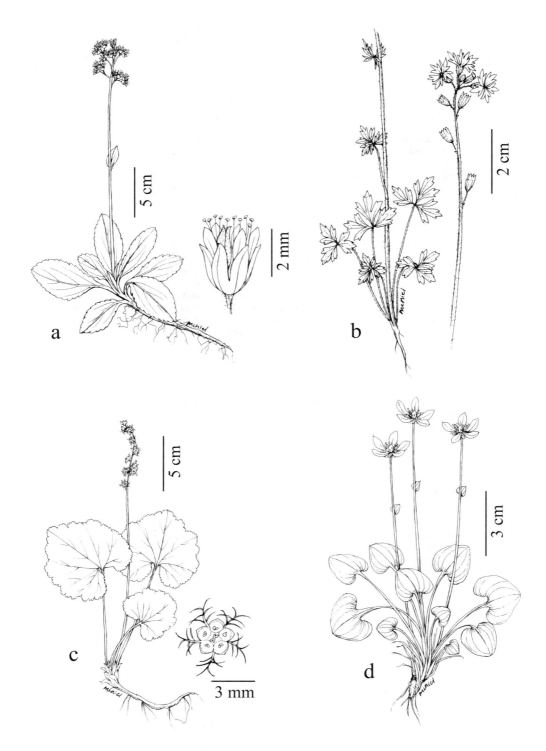

Plate 41. a. *Leptarrhena pyrolifolia*, b. *Lithophragma parviflorum*, c. *Mitella breweri*, d. *Parnassia fimbriata*

Micranthes Haw. Saxifrage

Scapose, perennial (often short-lived) herbs. **Leaves** basal, rosulate, simple. **Inflorescence** a bracteate, often glandular cyme. **Flowers** 5-merous, regular, perfect, hypogynous to epigynous; calyx lobes separate or united basally; petals white to yellow, often clawed; stamens 10; styles 2. **Fruit** a capsule (sometimes appearing like 2 follicles) with 2 spreading beak-like styles; seeds numerous.[64,101,122,289]

These scapose species have been segregated from *Saxifraga* sensu lato based on strong evidence provided by molecular-genetic studies.[374,375] Many of these species resemble each other; plants of *Micranthes nidifica* with glomerate inflorescences can be difficult to distinguish from *M. rhomboidea*, especially near its upper elevational limit. Small plants of *M. hieraciifolia* can resemble *M. apetala*. Large plants of *M. hieraciifolia* resemble *M. oregana*.

1. Basal leaf blades reniform to orbicular, as wide as long. 2
1. Basal leaf blades narrowly oblanceolate to obovate 4

2. Leaf blades tapered to the petiole, not cordate . . . . *M. lyallii*
2. Many leaf blades cordate at the base. 3

3. Petioles glabrous or nearly so . . . .*M. odontoloma*
3. Petioles long-hairy . . . .*Saxifraga mertensiana*

4. Ovary <1/3 inferior at flowering; stamens on a band-like gland surrounding base of ovary. 5
4. Ovary at least ½ inferior at flowering; stamens on a disk covering most of the ovary 8

5. Some flowers replaced by tiny plantlets. . . . . *M. ferruginea*
5. Plantlets absent 6

6. Stems glabrous to sparsely glandular-hairy . . . . *M. lyallii*
6. Stems densely hairy and glandular above 7

7. Many pedicels 2 to 4 times as long as the flowers; filaments clavate, petaloid at anthesis. . . . *M. marshallii*
7. Pedicels 1 to 2 times as long as flowers; filaments subulate . . . .*M. occidentalis*

8. Petals purple or absent (rarely 1or 2 tiny petals present). 9
8. Petals 5, white 11

9. Scapes mostly >10 cm long, inflorescence elongate . . . . *M. hieraciifolia*
9. Scapes mostly <10 cm long 10

10. Petals absent or ≤1 mm long; filaments white . . . . *M. apetala*
10. Petals ≥1 mm long; filaments purplish . . . . *M. nivalis*

11. Leaf blades linear to narrowly oblanceolate 12
11. Leaf blades lanceolate to ovate or obovate 13

12. Plants mostly glabrous, 2–8 cm high . . . . *M. tempestiva*
12. Plants glandular-hairy, 15–70 cm high . . . . *M. oregana*

13. Styles <2 mm long at anthesis; leaf blades without brown hair beneath . . . . *M. nidifica*
13. Styles usually ≥2 mm long at anthesis; many leaves with tiny tufts of brown hair beneath . .*M. rhomboidea*

Micranthes apetala (Piper) Small [*Saxifraga apetala* Piper, *S. integrifolia* Hook. var. *apetala* (Piper) M.E.Jones] Perennial from a simple rootstock. **Stems** solitary, scapose, erect, 3–12 cm, pubescent below, glandular-hairy above. **Leaves** all basal, petiolate, glabrous to sparsely hairy below; the blade ovate to rhombic, 5–25 mm long, entire or obscurely toothed. **Inflorescence** congested-paniculate, glandular. **Flowers:** calyx cup-shaped with lobes 1–2 mm long; hypanthium ≤1 mm long; petals absent or ≤1 mm long; ovary inferior at anthesis but at least half superior at maturity. **Capsule** 2–4 mm long. Moist to wet, gravelly, sparsely vegetated soil of turf or cliffs, often where snow lies late; subalpine, alpine. WA, OR, ID, MT.

Micranthes ferruginea (Graham) Brouillet & Gornall [*Saxifraga ferruginea* Graham] Perennial from a simple or rarely branched rootstock. **Stems** 1 to several, erect, 10–40 cm, pubescent below, glandular above. **Leaves** all basal with broad petioles; the blade 2–10 cm long, spatulate with dentate margins on the upper half. **Inflorescence** an open, many-flowered panicle; some flowers replaced by tiny plantlets. **Flowers:** calyx saucer-shaped with lobes 1–2 mm long; hypanthium absent; petals clawed, white, 3–5 mm long, the upper 3 with 2 yellow, basal spots; ovary superior. **Capsule** 3–6 mm long. Moist rock ledges, fens, wet, gravelly soil along streams, often in moss; valleys to alpine. AK to AB south to CA, ID, MT. (p.255)

One collection from Lincoln Co. lacks plantlets, a trait more common in plants from the northern part of the range.[188]

Micranthes hieraciifolia (Waldst.& Kit. ex Willd.) Haw. [*M. subapetala* (E.E.Nelson) Small, *Saxifraga subapetala* E.E.Nelson, *S. oregana* Howell var. *subapetala* (E.E.Nelson) C.L.Hitchc.] Perennial from a simple or branched rootstock. **Stems** usually solitary, scapose, erect, 10–30 cm, glandular-hairy. **Leaves** all basal, broadly petiolate; the blade mostly glabrous, sparsely ciliate, narrowly lanceolate to ovate, 3–8 cm long with weakly dentate margins. **Inflorescence** glandular, narrow, paniculate; the flowers clustered. **Flowers**: calyx cup-shaped with lobes 1–2 mm long; hypanthium 1–2 mm long; petals usually absent or purplish, shorter than the sepals, falling at anthesis; ovary half inferior at anthesis, becoming more superior at maturity. **Capsule** 2–5 mm long. Moist to wet soil of turf, meadows, grasslands; upper montane to alpine. Circumpolar south to ID, WY.

Many authors have segregated plants from continental U.S. as *Micranthes subapetala*, but the characters separating *M. hieraciifolia* and *M. subapetala* are weak and overlap,[64] and the former was reported for Yellowstone National Park half a century ago.[197]

Micranthes lyallii (Engl.) Small [*Saxifraga lyallii* Engl.] Glabrous, rhizomatous perennial sometimes forming loose mats. **Stems** solitary, scapose, erect, 5–15 cm. **Leaves** all basal, petiolate; the blade obovate to fan-shaped, 1–3 cm long, deeply dentate on the upper half. **Inflorescence** few-flowered, racemose to paniculate. **Flowers**: calyx reddish, saucer-shaped with reflexed lobes 1–3 mm long; hypanthium absent; petals white with 2 green, basal spots, elliptic, 2–4 mm long; ovary superior. **Capsule** 5–8 mm long, follicle-like. Wet soil of gravelly meadows, wet ledges, rock slides, especially along streams; upper subalpine, alpine. AK south to WA, ID, MT. See *M. odontoloma*. (p.255)

Micranthes marshallii (Greene) Small [*M. idahoensis* (Piper) Brouillet & Gornall, *Saxifraga marshallii* Greene, *S. idahoensis* Piper, *S. occidentalis* Wats. var. *idahoensis* (Piper) C.L.Hitchc.] Perennial from a simple rootstock. **Stems** solitary, scapose, 10–25 cm, glandular-hairy. **Leaves** basal (rarely 1 cauline), petiolate; the blade broadly ovate, truncate or rounded to the petiole, 1–4 cm long, prominently dentate, glabrous above, sparsely brown-hairy beneath. **Inflorescence** glandular, spreading-paniculate. **Flowers**: calyx cup-shaped with lobes 1–2 mm long; hypanthium <0.5 mm long; petals obovate, white, 2–3 mm long with 2 yellow spots at the base; ovary superior. **Capsule** 2–4 mm long, follicle-like. Wet rock crevices, talus; montane, subalpine. WA to MT south to CA.

Our plants have been treated as a separate species as well as a variety of *Micranthes occidentalis*, but the petaloid filaments and open inflorescence clearly separate them from *M. occidentalis* and ally them with *M. marshallii*.[223]

Micranthes nidifica (Greene) Small [*Saxifraga nidifica* Greene *S. integrifolia* Hook. var. *leptopetala* (Suksd.) Engl. & Irmsch., *S. integrifolia* var. *columbiana* (Piper) C.L.Hitchc.] Perennial from a simple rootstock, sometimes with bulblets among roots. **Stems** solitary, scapose, erect, 7–30 cm, glandular-hairy. **Leaves** all basal, glabrous, petiolate; the blade ovate, 2–6 cm long, entire to weakly dentate. **Inflorescence** glandular, paniculate, open or congested. **Flowers**: calyx cup-shaped with lobes 1–2 mm long, sometimes purple-tinged; hypanthium 1–2 mm long; petals obovate, white, 1–2 mm long; ovary inferior at anthesis but ca. 1/2 superior at maturity. **Capsule** 2–4 mm long. Vernally moist soil of grasslands; valleys, montane. BC, AB south to NV, ID, UT. (p.255)

Mature ovaries of *Micranthes occidentalis* are >1/2 superior. See *M. rhomboidea*.

Micranthes nivalis (L.) Small [*Saxifraga nivalis* L.] Perennial from a simple rootstock. **Stems** solitary, scapose, erect, 3–8 cm, glandular-hairy. **Leaves** petiolate; blade obovate to ovate, 5–12 mm long, dentate toward the tip. **Inflorescence** few-flowered, glandular, globose-paniculate. **Flowers**: calyx cup-shaped with lobes 1.5–3 mm long, purplish; hypanthium 1.5–3 mm long; petals obovate, purple, 1–2 mm long; ovary at least 1/2 inferior. **Capsule** 3–7 mm long. Rock outcrops, moist fellfields; alpine. Circumpolar south to BC, AB, MT.

This species is thought to be absent from the continental U.S.;[64] however, several of our specimens were annotated as such by J.G. Packer, author of the Flora of Alberta. See *M. occidentalis*.

Micranthes occidentalis (S.Watson) Small [*Saxifraga occidentalis* S.Watson] Perennial from a simple rootstock, often with bulblets among roots. **Stems** usually solitary, scapose, erect, 3–30 cm, glandular-hairy. **Leaves** all basal, petiolate; the blade broadly elliptic to ovate, 1–6 cm long, dentate. **Inflorescence** glandular, paniculate (sometimes congested); the flowers clustered at the branch tips. **Flowers**: calyx cup-shaped with lobes 1–2 mm long; hypanthium ≤0.5 mm long; petals

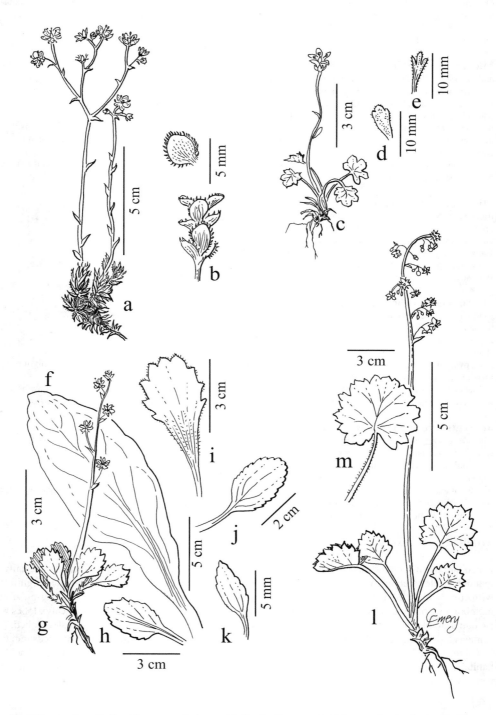

Plate 42. **Micranthes & Saxifraga. a.** *S. bronchialis,* **b.** *S. oppositifolia,* **c.** *S. rivularis,* **d.** *S. adscendens,* **e.** *S. caespitosa,* **f.** *M. oregana,* **g.** *M. lyallii,* **h.** *M. rhomboidea,* **i.** *M. ferruginea,* **j.** *M. occidentalis,* **k.** *M. nidifica,* **l.** *M. odontoloma,* **m.** *M. mertensiana*

obovate, white, 2–3 mm long; ovary at least 3/4 superior. **Capsule** 3–5 mm long. Rock outcrops, vernally moist, usually stony soil of grasslands, meadows, turf; valleys to alpine. BC, AB south to OR, NV, WY. (p.255)

Ovaries of *Micranthes nivalis* and *M. rhomboidea* are ≥1/2 inferior at anthesis; the latter often has red hairs on leaf undersides. See *M. marshallii*.

Micranthes odontoloma (Piper) A.Heller [*Saxifraga odontoloma* Piper, *S. arguta* D.Don and *S. punctata* L. misapplied] Perennial from a horizontal rootstock. **Stems** solitary, scapose, erect, 10–60 cm, glabrous below, glandular-hairy above. **Leaves** all basal, long-petiolate; the blade glabrous, usually ciliate, suborbicular, 2–8 cm wide, dentate. **Inflorescence** glandular, open-paniculate. **Flowers**: calyx saucer-shaped with lobes 1–2 mm long, often purplish; hypanthium <0.5 mm long; petals white with 2 green or yellow, basal spots, broadly elliptic, 2–3 mm long; ovary superior. **Capsule** 4–8 mm long. Along streams, springs, lakes, often on wet, mossy rocks; montane, lower subalpine. BC, AB south to CA, AZ, NM. (p.255)

Reported to hybridize with *Micranthes lyallii* in Alberta;[289] plants with intermediate leaf shape occur in the Bitterroot Range. See *Saxifraga mertensiana*.

Micranthes oregana (Howell) Small [*Saxifraga oregana* Howell] Perennial from a horizontal rootstock. **Stems** usually solitary, scapose, erect, 15–70 cm, glandular-hairy. **Leaves** all basal, broadly short-petiolate; the blade sparsely pubescent, lanceolate to narrowly oblanceolate, 4–20 cm long, minutely dentate. **Inflorescence** glandular, paniculate, open or interruptedly congested. **Flowers**: calyx cup-shaped with lobes 1–2 mm long; hypanthium ≤1 mm long; petals white, ovate, 2–4 mm long; ovary 1/2 inferior at anthesis, becoming more superior at maturity. **Capsule** 2–4 mm long. Wet meadows, fens, streambanks; montane, subalpine. WA to AB south to CA, NV, UT, CO. (p.255)

Micranthes rhomboidea (Greene) Small [*Saxifraga rhomboidea* Greene] Perennial from a simple rootstock. **Stems** usually solitary, scapose, erect, 3–25 cm, glandular-hairy. **Leaves** all basal, petiolate; the blade rhombic to ovate, 1–5 cm long with dentate margins, ciliate, glabrous above, tufted brown-hairy below. **Inflorescence** a glandular, congested cyme. **Flowers**: calyx cup-shaped with lobes 1–2 mm long; hypanthium 1–2 mm long; petals obovate, retuse, white, 2–4 mm long; ovary half inferior at anthesis, becoming more superior at maturity. **Capsule** 3–6 mm long. Moist soil of turf, meadows, grasslands, rock outcrops; subalpine, alpine, rarely montane. MT south to UT, AZ. (p.255)

Plants of *Micranthes nidifica* with a congested inflorescence can be mistaken for *M. rhomboidea*, but they occur at lower elevations and have leaves with more entire margins and lack brown hair beneath. See *M. occidentalis*.

Micranthes tempestiva (Elvander & Denton) Brouillet & Gornall [*Saxifraga tempestiva* Elvander & Denton] Perennial from a simple or branched rootstock. **Stems** solitary, scapose, erect, 2–8 cm, glabrous to sparsely glandular-hairy. **Leaves** all basal, short-petiolate, glabrous; the blade oblanceolate, 5–25 mm long, entire. **Inflorescence** few-flowered, glabrous. **Flowers**: calyx cup-shaped with lobes 1–2 mm long; hypanthium 1–2 mm long; petals white, ca. 1 mm long; ovary inferior at anthesis but at least 1/2 superior at maturity. **Capsule** 3–5 mm long. Moist to wet, gravelly, sparsely vegetated soil of turf, rock slides, ledges, often in areas of late snowmelt; alpine. Endemic to southwest MT.

The type locality is in Deer Lodge Co.

Mitella L. Mitrewort

Rhizomatous and often stoloniferous, perennial herbs. **Stems** slender from a branched caudex, usually glandular at least above. **Leaves** mostly basal, long-petiolate, stipulate, cordate-based, ovate to reniform with crenate to lobed margins, glabrous to sparsely hairy. **Inflorescence** an elongate, glandular-pubescent, membranous-bracted, spike-like raceme. **Flowers** 5-merous, regular, perfect, epigynous; calyx lobes triangular to oblong; petals white to greenish, clawed, the blade deeply palmately to pinnately lobed; stamens 5 or 10; styles 2. **Fruit** a 2-beaked capsule, splitting completely open at maturity; seeds numerous.

There is broad ecological overlap among the species; however, *Mitella stauropetala* occurs in the driest habitats, *M. breweri* is often in areas with deep winter snow cover, and *M. nuda* is usually found in low, wet forest. See *Conimitella*.

1. Stamens 10 . . . . *M. nuda*
1. Stamens 5 2

2. Stamens attached between calyx lobes . . . . *M. pentandra*
2. Stamens in front of calyx lobes. 3

3. Flowers maturing from the top down; 1 to 3 stem leaves present . . . . *M. caulescens*
3. Flowers maturing upward; stems mostly scapose 4

4. Calyx saucer- to cup-shaped; petals with ≥5 lobes . . . . *M. breweri*
4. Calyx obconic or cup-shaped; petals 3-lobed. 5

5. Flowers strongly secund; sepals with sessile, yellow glands . . . . *M. stauropetala*
5. Flowers not secund; sepals glabrous or with few purple glandular hairs . . . . *M. trifida*

Mitella breweri A.Gray **Stems** 12–30 cm. **Leaf blades** 2–8 cm wide, suborbicular to reniform
with crenate-dentate margins; petioles brown-hairy. **Inflorescence** of 8 to 20 flowers, maturing
upward. **Flowers** saucer- to cup-shaped; calyx 2–4 mm wide, the lobes reflexed; petals 1–3 mm
long with 2 to 4 pairs of lobes; stamens 5, opposite the sepals. **Capsule** almost completely inferior.
Cool forests, moist cliffs, meadows, thickets along streams, often where snow lies late; montane to
lower alpine. BC, AB south to CA, ID, MT. (p.252)

Mitella caulescens Nutt. **Stems** 15–35 cm. **Leaf blades** 2–5 cm wide, suborbicular, shallowly
3- to 5-lobed with dentate margins. **Inflorescence** of 5 to 20 flowers maturing from the top down.
Flowers cup-shaped; calyx 3–4 mm wide; the lobes spreading; petals 3–4 mm long with 3 to 5
pairs of lobes; stamens 5, opposite the sepals. **Capsule** half exposed above the hypanthium. Wet
forest, thickets, often along creeks; valleys to montane. BC, MT south to CA, ID.

Mitella nuda L. **Stems** 5–20 cm. **Leaf blades** 1–4 cm wide, suborbicular, with crenate or
shallowly lobed margins. **Inflorescence** of 3 to 10 flowers maturing upward. **Flowers** saucer-
shaped; calyx purplish, 2–4 mm wide, the lobes spreading; petals 2–3 mm long with 3 to 4 pairs of
lobes; stamens 10. **Capsule** half exposed above the hypanthium. Wet riparian forest, often beneath
spruce, growing with *Equisetum* spp.; valleys, montane. AK to NL south to WA, MT, ND.

Mitella pentandra Hook. **Stems** 10–60 cm. **Leaf blades** 1–8 cm wide, suborbicular with
crenate margins. **Inflorescence** of 5 to 25 flowers maturing upward. **Flowers** saucer-shaped;
calyx 2–3 mm wide, the lobes spreading to reflexed; petals 1–3 mm long with 3 to 4 pairs of lobes;
stamens 5, opposite the petals. **Capsule** almost completely inferior. Moist to wet forest, cliffs, stony
wet meadows, often along streams, growing in moss; montane to upper subalpine. AK to AB south to
CA, CO.

Mitella stauropetala Piper **Stems** 15–40 cm. **Leaf blades** 2–9 cm wide, ovate to suborbicular
with crenate margins. **Inflorescence** of 6 to 35 secund flowers, maturing upward. **Flowers**
obconic; calyx 2–3 mm wide; the lobes yellow-glandular, erect with spreading tips; petals white
to purple, 3–4 mm long with 3 linear lobes; stamens 5, opposite the sepals. **Capsule** more than half
inferior. Dry to moist forest; valleys, montane. WA to MT south to UT, CO.

Mitella trifida Graham **Stems** 12–40 cm. **Leaf blades** 2–6 cm wide, broadly ovate to
suborbicular, shallowly 5- to 7-lobed with crenate margins. **Inflorescence** of 6 to 15 flowers
maturing upward. **Flowers** cup-shaped; calyx 2–3 mm wide; the lobes erect, white to purple-tinged
with a few glandular hairs; petals often purplish, 1–3 mm long with 3 oblanceolate lobes; stamens
5, opposite the sepals. **Capsule** half exposed above the hypanthium. Moist forests, thickets, meadows;
montane, subalpine. BC, AB south to CA, MT.

Parnassia L. Grass of Parnassus

Glabrous perennials from short rootstocks. **Stems** scapose with 1(0) sessile, leaf-like bract. **Leaves** basal,
petiolate, simple, entire with parallel veins. **Flowers** solitary, 5-merous, regular, perfect, hypogynous or
nearly so; calyx lobes longer than the tube; petals white; stamens 5, alternating with 5 fringed staminodia
(stamen-like appendages); styles inconspicuous; stigmas 4. **Fruit** a 4-celled, ovoid capsule, opening at the
tip; seeds numerous.

There is strong evidence that this genus should be placed in a family other than the Saxifragaceae (e.g.,
Parnassiaceae).[373]

1. Leaves all basal . . . . *P. kotzebuei*
1. Stems with 1 sessile leaf near the middle 2

2. Petals fringed at the base; basal leaf blades wider than long . . . .*P. fimbriata*
2. Petals not fringed; basal leaves ovate, longer than wide 3

3. Petals 7–12 mm long with ≥7 veins; stem bract cordate-based . . . .*P. palustris*
3. Petals 4–8 mm long with 5 veins; stem bract usually not cordate . . . .*P. parviflora*

Parnassia fimbriata K.D.Koenig **Stems** 1 to several, 8–50 cm with 1 sessile, clasping, ovate, leaf-like bract. **Leaf blades** orbicular to reniform, 1–6 cm broad, deeply cordate. **Flowers:** calyx lobes 4–7 mm long, 5-veined; petals 7–15 mm long, fringed below the middle; staminodia obovate, lobed. **Capsule** 7–11 mm long. Wet meadows, avalanche slopes, along streams, seeps; montane to alpine. AK to AB south to CA, NM. Our plants are var. *fimbriata*. (p.252)

Parnassia kotzebuei Cham. & Schltdl. ex Spreng. **Stems** 1 to several, bractless, 1–12 cm. **Leaf blades** ovate, 6–20 mm long, truncate or cuneate at the base. **Flowers:** calyx lobes 4–8 mm long, 3-veined; petals ovate, barely longer than the setals; staminodia fan-shaped with terminal, capitate lobes. **Capsule** 5–8 mm long. Moist to wet turf, mossy rock ledges; alpine. AK to Greenland south to WA, NV, CO; Asia.

Parnassia palustris L. **Stems** mostly solitary, 10–35 cm with 1, mostly cordate-clasping bract. **Leaf blades** ovate to reniform, truncate to cordate, 8–15 mm long. **Flowers:** calyx lobes 5–9 mm long; petals oblong, 8–12 mm long with 7 to 9 veins; staminodia obovate with 5 to 7 capitate lobes. **Capsule** 7–12 mm long. Fens, wet meadows, thickets, often in moss; plains, valleys to subalpine. Circumboreal south to CA, UT, CO, ND, MN. Our plants are var. *montanensis* (Fernald & Rydb.) C.L.Hitchc., which is intermediate between typical *P. palustris* and *P. parviflora*.[187]

Parnassia parviflora DC. **Stems** mostly solitary, 3–20 cm with a sessile but not clasping, lanceolate to narrowly ovate bract. **Leaf blades** ovate, 1–3 cm long, tapered to the petiole. **Flowers:** calyx lobes purple-tinged, 4–7 mm long, 5-veined, often with a few purple, glandular hairs; petals ovate, 4–7 mm long; staminodia oblong with 5 to 7 capitate lobes. **Capsule** 5–10 mm long. Often calcareous fens, seeps, along streams; montane. BC to QC south to CA, ID, MT, SD, MN. See *P. palustris*.

Saxifraga L. Saxifrage

Perennial (often short-lived) herbs. **Leaves** cauline and sometimes basal as well, alternate, simple. **Inflorescence** a bracteate, often glandular raceme, panicle, cyme, or flowers solitary. **Flowers** 5-merous, regular, perfect, hypogynous to epigynous; calyx lobes separate or united basally; petals white to yellow (purple), often clawed; stamens 10; styles 2. **Fruit** a capsule (sometimes appearing like 2 follicles) with 2 spreading stylar beaks; seeds numerous.[65,104,289]

Scapose species, formerly placed in *Saxifraga* have now been segregated into *Micranthes*.

1. Cauline leaves all near the base of the stem; plants appearing subscapose. . . . . *S. mertensiana*
1. Cauline leaves obvious, on the lower half of the stem at least. 2

2. Lower stems densely leafy; leaves with entire margins (may be ciliate) 3
2. Cauline leaves well separated; at least some basal leaves lobed or deeply toothed. 8

3. Petals white (may be purple spotted). 4
3. Petals yellow or purple 5

4. Leaves subulate, sharp-pointed . . . .*S. bronchialis*
4. Leaves spatulate with rounded tips . . . . *S. tolmiei*

5. Petals purple; leave opposite . . . . *S. oppositifolia*
5. Petals yellow; leaves alternate. 6

6. Plants forming small mats from a branched caudex. . . . .*S. serpyllifolia*
6. Stems solitary from an unbranched caudex. 7

7. Plants with tentacle-like stolons; sepals erect . . . .*S. flagellaris*
7. Plants without stolons; sepals spreading or reflexed . . . . *S. hirculus*

8. All but uppermost flower replaced by tiny red bulblets . . . . *S. cernua*
8. Flowers not replaced by bulblets 9

9. Basal leaf blades wider than long. . . . . *S. rivularis*
9. Basal leaf blades oblong to spatulate, longer than wide 10

10. Leaf blades 3-lobed at least half the length; plants forming small tufts . . . . *S. cespitosa*
10. Leaf blades shallowly lobed or toothed; plants not tufted. . . . . *S. adscendens*

Saxifraga adscendens L. Biennial or short-lived perennial from a simple caudex. **Stems** erect, 1–10 cm, often branched at the base, glandular. **Leaves** short-petiolate, basal and cauline, 3–10 mm long, oblong, often shallowly 3-lobed, glandular-hairy. **Inflorescence** glandular, cymose, of 4 to 12 flowers. **Flowers**: calyx purplish, cup-shaped with lobes 1–2 mm long; hypanthium 2–3 mm long; petals white, 2–4 mm long, obovate with a short claw; ovary inferior. **Capsule** 3–5 mm long. Vernally wet, gravelly, sparsely vegetated microsites in turf, rock outcrops; alpine. AK to OR, UT, CO. (p.255)
     Saxifraga cespitosa has a branched caudex and more deeply lobed leaves; *S. rivularis* and *S. cernua* have palmately lobed leaves with well-differentiated petioles.

Saxifraga bronchialis L. Mat-forming with a branched caudex clothed in old leaves. **Stems** prostrate to ascending, densely leafy; fertile stems 4–15 cm, sparsely glandular-hairy. **Leaves** basal and cauline, needle-like, ciliate, 5–10 mm long, spine-tipped. **Inflorescence** an open, glandular-hairy panicle. **Flowers**: calyx saucer-shaped with lobes 1–2 mm long; hypanthium ≤1 mm long; petals oblong, 4–8 mm long, white, purple-spotted; ovary mostly superior. **Capsule** 3–8 mm long. Talus slopes, rock crevices and in gravelly soil of dry meadows, turf, often along streams; valleys to alpine. BC, AB south to OR, ID, NM. Our plants are ssp. *austromontana* (Wiegand) Piper. (p.255)

Saxifraga cernua L. Short-lived, fibrous-rooted perennial. **Stems** erect, 4–20 cm, glabrous to glandular. **Leaves** basal and cauline, petiolate; the blades reniform, 5–15 mm wide with 3 to 7 shallow lobes, glabrous, with rice-like bulblets in basal axils and small purple bulblets in the cauline axils. **Inflorescence** a glandular raceme or panicle with all but the terminal flower replaced by purple bulblets. **Flower**: calyx cup-shaped with lobes 1–2 mm long; hypanthium ca. 1 mm long; petals white, 5–8 mm long, oblanceolate; ovary inferior. **Capsule** rarely maturing. Moist or wet, gravelly, sparsely vegetated soil of wet cliffs, turf, boulder fields; upper subalpine, alpine. Circumpolar south to WA, NV, NM, SD.

Saxifraga cespitosa L. Forming small cushions from a branched caudex clothed in old leaves. **Stems** ascending, glandular, 1–10 cm. **Leaves** cauline and mostly basal, glabrous to pubescent, narrowly oblong, 3–15 mm long, 3-lobed with an indistinct petiole. **Inflorescence** a few-flowered, glandular cyme. **Flowers**: calyx campanulate with lobes 1–2 mm long; hypanthium ca. 2 mm long; petals white or rarely purple, oblanceolate, 3–5 mm long; ovary inferior. **Capsule** 4–7 mm long. Dry to moist sparsely vegetated soil of rock slides, boulder fields, crevices, turf; alpine. Circumpolar south to CA, NV, NM. See *S. adscendens*. (p.255)

Saxifraga flagellaris Willd. Biennial or short-lived perennial with tentacle-like stolons. **Stems** solitary, erect, 1–10 cm, glandular. **Leaves** basal and cauline, reduced upwards, oblanceolate, ciliate, 5–15 mm long, entire. **Inflorescence** glandular, of 1 to 3 flowers. **Flowers**: calyx campanulate with lobes 2–4 mm long; hypanthium 1–2 mm long; petals yellow with brown veins, 4–7 mm long, oblanceolate; ovary <½ inferior. **Capsule** 6–9 mm long. Moist turf, fellfields, scree, often in sparsely vegetated soil under boulders; alpine. Circumpolar south sporadically to UT, AZ.

Saxifraga hirculus L. Perennial from slender rhizomes. **Stems** solitary, erect, 4–12 cm, long-hairy. **Leaves** basal and cauline, oblanceolate, glabrous, the basal petiolate; the blades 5–20 mm long, entire. **Inflorescence** a solitary flower. **Flowers**: calyx saucer-shaped with lobes 2–5 mm long; hypanthium <0.5 mm long; petals yellow, 7–10 mm long, elliptic; ovary superior. **Capsule** 7–12 mm long. Saturated, organic soil of fens; alpine. Circumpolar south sporadically to UT, CO. Known from one location in Carbon Co.[249]

Saxifraga mertensiana Bong. Perennial with short rhizomes and bulblets among leaf bases. **Stems** solitary, scapose, erect, 10–30 cm, hairy and glandular above. **Leaves** basal and lower-cauline; the petiole long and sparsely long-hairy; the blade sparsely hairy, orbicular, 2–8 cm wide, shallowly lobed, each lobe 3 times crenate. **Inflorescence** glandular, open-paniculate, some flowers replaced by plantlets. **Flowers**: calyx saucer-shaped with reflexed lobes 2–3 mm long; hypanthium absent; petals white, elliptic, 3–4 mm long; ovary superior. **Capsule** 4–7 mm long. Wet cliffs, along small streams; subalpine, alpine. AK south to CA, ID, MT. (p.255)
     Vegetatively similar to species of *Micranthes*; *M. lyallii* has leaf blades longer than wide; *M. odontoloma* has glabrous petioles.

Saxifraga oppositifolia L. Purple mountain saxifrage Glabrous, cushion-forming. **Stems** numerous, prostrate to ascending, densely leafy, 2–4 cm. **Leaves** opposite, 2–4 mm long, ovate, ciliate. **Inflorescence** a solitary flower. **Flowers**: calyx cup-shaped with lobes 2–4 mm long, ciliate; hypanthium absent; petals light purple, oblong, 6–9 mm long; ovary barely inferior. **Capsule** 6–10 mm long including the long beak. Stony slopes, talus, moist cliffs, limestone fellfields; alpine. Circumpolar south to OR, ID, WY. (p.255)

This distinctive plant flowers very early in its exposed habitat.

Saxifraga rivularis L. [*S. debilis* Engelm. ex A.Gray, *S. hyperborea* R.Br.] **Stems** erect to ascending, glabrous to sparsely hairy, 1–15 cm. **Leaves** basal and cauline, long-petiolate, reduced above; the blade glabrous, reniform, 3–20 mm wide, 3- to 5-lobed. **Inflorescence** of 1 to 3 flowers. **Flowers**: calyx with lobes 1–2 mm long; hypanthium 1–2 mm long; petals white, oblanceolate, 2–4 mm long; ovary half inferior. **Capsule** 4–5 mm long. Moist or wet, sparsely vegetated, gravelly soil of rock crevices, scree, turf, often beneath boulders or along streams; alpine. Circumpolar south to CA, AZ, CO. (p.255)

1. Hypanthium U-shaped in longitudinal cross-section; sepals often purple . . . . .ssp. *hyperborea*
1. Hypanthium V-shaped in cross-section; sepals green . . . . .ssp. *debilis*

Saxifraga rivularis ssp. *hyperborea* (R.Br.) Dorn occurs throughout western MT; **S. rivularis** ssp. *debilis* (Engelm. ex A.Gray) Á.Löve, D.Löve & B.M. is found in the southwest cos. *Saxifraga rivularis* forms a circumpolar complex of many forms sometimes treated as separate species.[197] Characters used to delineate taxa are mostly continuous and seem to depend on plant size and phenology, so I prefer to recognize the taxa at a subspecific level. See *S. adscendens*.

Saxifraga serpyllifolia Pursh [*S. chrysantha* A.Gray] Forming small, loose mats from a branched caudex. **Stems** prostrate to ascending, densely leafy at the base; fertile stems 1–4 cm, sparsely glandular. **Leaves** basal and cauline, glabrous, succulent, oblanceolate, entire, 2–7 mm long. **Inflorescence** a solitary flower. **Flowers**: calyx saucer-shaped, the lobes reflexed, 2–4 mm long; hypanthium <1 mm long; petals elliptic, 4–7 mm long, yellow; ovary mostly superior. **Capsule** 5–8 mm long. Moist, gravelly or sandy soil of gentle slopes or among boulders, often where snow lies late; alpine. AK south to UT, NM; Asia. Our plants are var. *chrysantha* (A.Gray) Dorn.

Currently known only from Carbon Co., but a 110 year-old specimen by M. Elrod documents the species for the Bitterroot Mtns where it apparently no longer occurs. Beartooth specimens have glabrous sepals; the Lolo Peak specimen has glandular sepals.

Saxifraga tolmiei Torr. & A.Gray [*Micranthes tolmiei* (Torr. & A.Gray) Brouillet & Gornall] Forming small mats with short, sterile stems from a branched caudex. **Stems** prostrate to ascending, densely leafy at the base; fertile stems 2–6 cm, glabrous. **Leaves** glabrous, succulent, ciliate, spatulate, 3–7 mm long. **Inflorescence** of 1 to 3 flowers. **Flowers**: calyx saucer-shaped with lobes spreading, 2–3 mm long, glabrous; hypanthium <0.5 mm long; petals oblong, 3–5 mm long, white; ovary slightly inferior. **Capsule** 5–9 mm long. Moist rock slides, crevices, stony soil of turf, often where snow lies late; upper subalpine, alpine. BC to MT south to CA, NV, ID; known from Missoula and Ravalli cos.

Molecular-genetic analysis suggests that this species is more closely related to *Micranthes* than other members of *Saxifraga*;[374] however, it is anomalous in either genus and is morphologically more similar to *Saxifraga*.

Suksdorfia Gray

Short-lived perennials with bulblets among the roots. **Stems** simple, glandular. **Leaves** basal and cauline, alternate, the lower petiolate, the cauline stipulate; the blades cordate to reniform, palmately lobed. **Inflorescence** glandular, cymose, somewhat flat-topped. **Flowers** 5-merous, regular, perfect; petals clawed, the blade narrowly elliptic; ovary partly to wholly inferior; stamens 5; styles 2. **Fruit** a capsule with 2 hollow beaks; seeds numerous.

1. Petals white, spreading; stipules membranous . . . .*S. ranunculifolia*
1. Petals usually violet, nearly erect; some stipules green, lobed . . . .*S. violacea*

Suksdorfia ranunculifolia (Hook.) Engel. **Stems** erect, 10–30 cm. **Leaves** fleshy; basal leaves 1–4 cm wide, divided into 3 crenate lobes; upper leaves reduced with an ovate-expanded base. **Flowers**: sepals triangular, 1–2 mm long; petals white, 4–6 mm long, spreading; hypanthium obconic. **Capsule** 2–4 mm long. Shallow soil of moist to wet cliffs, talus, often in moss; montane, lower subalpine. BC, AB south to CA, ID, MT. (p.262)

Suksdorfia violacea A.Gray **Stems** weak, prostrate to ascending, 10–30 cm. **Leaves** 5- to 7-lobed; basal leaves 1–3 cm wide; upper leaves with lobed stipules. **Flowers**: sepals linear-lanceolate, 1–3 mm long; petals violet (white), 5–9 mm long, nearly erect. **Capsule** 4–6 mm long. Moist to wet cliffs, rock slides, often in moss; valleys to lower subalpine. BC, AB south to OR, ID, MT.

Sullivantia Torr. & A.Gray

Sullivantia hapemanii (J.M.Coult. & Fisher) J.M.Coult. Perennial from shallow rhizomes. **Stems** ascending, glandular-hairy, 20–40 cm. **Leaves** basal and cauline, petiolate, reduced upward; the blades glabrous, cordate-orbicular, 2–8 cm wide with 5 to 10 dentate lobes. **Inflorescence** a panicle with spreading branches. **Flowers** 5-merous, regular, perfect; calyx cup-shaped 2–3 mm long with erect lobes; petals white, ca. 3 mm long, obovate; stamens 5; ovary >1/2 inferior. Fruit a capsule ca. 4 mm long; seeds numerous. Wet limestone cliffs; valleys, montane. ID, MT to CO. Our plants are var. *hapemanii* which is endemic to Bighorn and Carbon cos. MT and adjacent WY.

Telesonix Raf.

Telesonix jamesii (Torr.) Raf. [*T. heucheriformis* (Rydb.) Rydb., *Boykinia heucheriformis* (Rydb.) Rosend.] Perennial from a branched caudex covered in old leaf bases. **Stems** simple, erect to ascending, 4–20 cm. **Leaves** petiolate, basal and cauline; the blade cordate-orbicular to reniform with shallowly lobed to crenate margins, 1–5 cm wide, pubescent. **Inflorescence** a leafy-bracteate, glandular panicle. **Flowers** 5-merous, regular, perfect; sepals erect, 3–5 mm long; petals clawed, the blade obovate, purple (white), 3–5 mm long; stamens 10; ovary partly to wholly inferior; hypanthium campanulate, 4–8 mm long; styles 2. **Fruit** a capsule with 2 beaks, 6–10 mm long; seeds numerous. Cool, moist cliffs, rock slides, more frequent on limestone; montane to alpine. AB south to UT, CO, SD. Our plants are var. *heucheriformis* (Rydb.) Bacig.

Tellima R.Br.

Tellima grandiflora (Pursh) Douglas ex Lindl. Perennial from a horizontal rootstock. **Stems** erect, hirsute below, glandular above, 40–70 cm. **Leaves** basal and cauline, sparsely hairy; lower leaves petiolate; the blades cordate-orbicular, 5- to 7-lobed, dentate, 3–12 cm wide; upper leaves smaller, sessile. **Inflorescence** a narrow, glandular raceme. **Flowers** 5-merous, regular, perfect; calyx cup-shaped 6–8 mm long with erect lobes; petals white to reddish, 4–6 mm long with 5 to 7 thread-like lobes at the tip; stamens 10; ovary <1/2 inferior. **Fruit** a capsule 6–8 mm long; seeds numerous. Moist forest along streams; montane. AK south to CA, OR, ID, MT. Known from Flathead and Lincoln cos. (p.262)

Tiarella L. Foamflower

Tiarella trifoliata L. [*T. unifoliata* Hook.] Rhizomatous perennial. **Stems** glandular above, 15–50 cm. **Leaves** mostly basal, petiolate, sparsely hairy; broadly ovate, cordate, 3–8 cm wide, 3- to 5-lobed with crenate margins. **Inflorescence** a narrow, glandular, open panicle. **Flowers** nodding, perfect, 5-merous; calyx irregularly cup-shaped with erect lobes 1–3 mm long; petals white, linear, 2–3 mm long, quickly falling; stamens 10, exserted, unequal; ovary superior. **Fruit** a capsule with 2 unequal, ovoid chambers, 4–6 and 7–10 mm long; seeds 1 per chamber. Moist coniferous forest; montane, lower subalpine. AK south to CA, ID, MT. (p.262)

   Our plants have been called var. *unifoliata* (Hook.) Kurtz. Plants with leaves fully divided into 3 leaflets are occasionally found in Lincoln and Flathead cos. mixed in with the typical form; these could be called var. *trifoliata*, but the distinction is not worth recognizing in our area.

ROSACEAE: Rose Family

Perennial herbs, shrubs, trees. **Leaves** alternate, stipulate, simple or compound. **Inflorescence** a raceme, cyme, or flowers solitary. **Flowers** mostly perfect, regular, 5-merous epigynous to perigynous; hypanthium cup- or saucer-shaped; sepals often with bracteoles between; stamens numerous; ovary partly inferior with ≥10 chambers. **Fruit** an achene, berry, pome or capsule.[104,187]

   Genera are most commonly distinguished on the basis of the fruit, but congeneric species are usually separated by flower or vegetative characters. *Horkelia fusca* Lindl. occurs in north ID[209] and could occur in adjacent MT. *Filipendula rubra* (Hill) B.L.Rob (Queen of the prairie), an eastern N. American species,

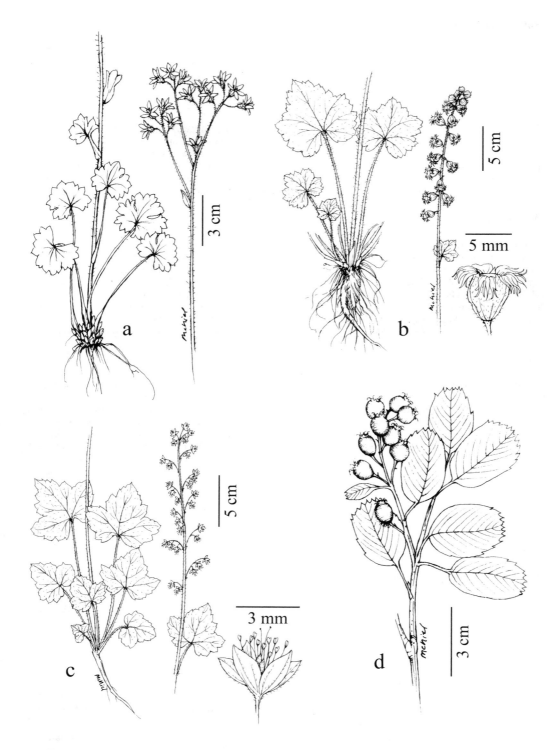

Plate 43. a. *Suksdorfia ranunculifolia*, b. *Tellima grandiflora*, c. *Tiarella trifoliata*, d. *Amelanchier alnifolia*

often cultivated as an ornamental, was collected once in Ravalli Co.; it resembles a herbaceous, pink-flowered spiraea.

1. Plants woody, not dying back to the ground each winter (including cushion plants and plants with trailing woody stems)..................2
1. Plants herbaceous, not woody above ground 23

2. Stems with thorns or prickles..................3
2. Stems unarmed 6

3. Leaves not divided into leaflets; stems with thorns 4
3. Leaves divided into leaflets; stems with prickles 5

4. Leaf margins dentate but not lobed ....*Prunus*
4. Leaf margins lobed and dentate..................*Crataegus*

5. Leaflets 3 (rarely 5); fruit composed of many small segments..................*Rubus*
5. Leaflets 5 to 11; fruit not segmented ....*Rosa*

6. Leaves divided into distinct leaflets 7
6. Leaves with entire or lobed margins, not fully divided into leaflets..................9

7. Leaflets 7 to 15..................*Sorbus*
7. Leaflets 3 to 5, rarely more..................8

8. Margins of leaflets entire ....*Dasiphora*
8. Margins of leaflets toothed ....*Rubus*

9. Plants low subshrubs with trailing stems 10
9. Shrubs with erect stems..................14

10. Leaves deeply lobed..................*Luetkea*
10. Leaves with entire, dentate, or crenate margins..................11

11. Leaves with dentate or crenate margins, at least below..................12
11. Leaves entire 13

12. Flowers solitary; plants of mountains..................*Dryas*
12. Flowers in clusters from the twigs; plants of the Great Plains ....*Prunus*

13. Leaves 2–4 mm long; flowers solitary ....*Kelseya*
13. Leaves 3–12 mm long; flowers in a many-flowered raceme ....*Petrophytum*

14. Leaves with entire, crenate, or serrate margins..................15
14. Leaf margins lobed as well as serrate or dentate..................20

15. Leaves white-lanate beneath..................*Cercocarpus*
15. Leaves not white-woolly beneath..................16

16. Leaves entire-margined ....*Cotoneaster*
16. Leaf margins at least partly serrate 17

17. Basal third of leaf with entire margins..................*Amelanchier*
17. Leaf with finely toothed margins from tip to base..................18

18. Petals ≤2 mm long ....*Spiraea*
18. Petals ≥3 mm long 19

19. Style 1; fruit a 1-seeded drupe..................*Prunus*
19. Styles 3 to 5; fruit a several-seeded pome..................*Pyrus*

20. Leaves deeply 3-lobed like a bird's foot..................*Purshia*
20. Leaves lobed ≤1/2-way to the midvein, like a maple leaf..................21

21. Inflorescence a few-flowered corymb; fruit a raspberry ....*Rubus*
21. Inflorescence a corymb or panicle with numerous (>10) flowers..................22

22. Inflorescence a pyramidal panicle; petals 1–2 mm long..................*Holodiscus*
22. Inflorecence a hemispheric corymb; petals 3–4 mm long ....*Physocarpus*

23. Leaves lobed (sometimes deeply) or crenate but not divided into leaflets..................24
23. Leaves divided into leaflets 28

24. Leaves crenate; prostrate cushion plants..................*Dryas*
24. Leaves lobed; stems erect 25

Agrimonia L. Agrimony

Short-rhizomatous perennials. **Leaves** pinnately compound with unequal, serrate, oblanceolate leaflets; stipules foliaceus, ovate, adnate to the petiole. **Inflorescence** a narrow, bracteate raceme. **Flowers** perigynous; hypanthium 10-grooved below with hooked bristles above; pistils 2, often only 1 developing. **Fruit** an achene enclosed in the hardened hypanthium.

1. Hypanthium hirsute in the grooves; axis of raceme not glandular . . . . A. striata
1. Hypanthium glandular but not hirsute in the grooves; axis of raceme glandular . . . . .A. gryposepala

Agrimonia gryposepala Wallr. **Stems** erect, glandular above, 30–100 cm. **Leaflets** 5 to 9 on larger leaves, 1–4 cm long, glandular beneath; stipules 1–2 cm long, entire or toothed. **Inflorescence** glandular. **Flowers**: hypanthium turbinate to hemispheric, 1–2 mm long, glandular but not hirsute; petals light yellow, 3–6 mm long; stamens ca. 15. **Fruit** a solitary, globose achene ca. 3 mm wide, enclosed in the mature hypanthium 3–5 mm long. Open woodlands, often along streams; plains, valleys of Cascade and Carbon cos. MT to ON south to KS, MO, IN, NC.

Agrimonia striata Michx. **Stems** erect, puberulent above, 40–80 cm. **Leaflets** 5 to 13, 1–6 cm long, glandular beneath; stipules ovate, 5–20 mm long. **Inflorescence** 5–20 cm long. **Flowers**: hypanthium turbinate, 1–2 mm long, equal to the sepals, hirsute in the grooves, often glandular; petals light yellow, 3–5 mm long; stamens 10 to 15. **Fruit** a solitary, globose achene 2–3 mm wide, enclosed in the mature hypanthium 4–6 mm long. Meadows, woodlands, often along streams; plains, valleys, montane. BC to NL south to AZ, NM, IA, NY.

Alchemilla L. Lady's mantle

Alchemilla vulgaris L. Long-hairy, rhizomatous perennial. **Stems** ascending, 15–50 cm. **Leaves** short-petiolate; the blades orbicular, cordate, 3–8 cm wide, with 7 to 9 serrate lobes; stipules fan-shaped, dentate, to 15 mm long. **Inflorescence** a bracteate, paniculate cyme. **Flowers** 4-merous, perigynous; hypanthium obconic, 1–2 mm long, pilose;

sepals ca. 1.5 mm long, spreading, alternating with bracteoles; petals absent; stamens 4; pistil 1. **Fruit** a solitary ovoid achene ca. 2 mm long, enclosed in the hypanthium. Pastures, roadsides; valleys. Introduced to much of eastern U.S., sporadic in the west; native to Eurasia. Collected in Gallatin and Meagher cos.[187]

Amelanchier Medik. Serviceberry, juneberry, saskatoon

Shrubs with smooth, reddish bark, becoming gray. **Leaves** simple, petiolate; the blades ovate to broadly elliptic, rounded to blunt at the tip, dentate on the upper half; stipules linear. **Inflorescence** a short, bracteate raceme terminating lateral shoots. **Flowers** epigynous; hypanthium campanulate; sepals 5; petals 5, spatulate, white; stamens 15 to 20; pistil 1. **Fruit** a 4- to 10-seeded pome.

Amelanchier humilis Wiegand, an eastern species with more acute leaf tips, is reported for Carter and McCone cos.[151] We have specimens from Fallon, Sheridan, Carbon, and Valley cos. that might fit this description, but the distinction seems dubious in our area.[152]

1. Leaf blades sparsely hairy; pome purple at maturity; petals 8–18 mm long. . . . . *A. alnifolia*
1. Leaf blades lanate beneath; pome white to orange; petals 5–10 mm long . . . . *A. utahensis*

Amelanchier alnifolia (Nutt.) Nutt. ex M.Roem. [*A. pumila* Nutt. ex Torr. & A.Gray] **Stems** 1–4 m high. **Leaves** 2–5 cm long, sparsely puberulent, glaucous beneath. **Raceme** of 5 to 15 flowers. **Flowers**: hypanthium 2–3 mm long; sepals reflexed, 2–5 mm long; petals 8–18 mm long; styles 4 to 5, united below. **Pome** 5–10 mm long, ca. 10-seeded, edible. Moist to dry forest, grasslands, meadows, woodlands, avalanche slopes; plains, valleys to lower subalpine. AK to QC south to CA, AZ, NM, MN. (p.262)

The three vars. reported for MT are distinguished by petal length and pubescence of the hypanthium, characters that vary continuously in MT, often in the same population. There is a tendency for plants from east of the Continental Divide to have smaller petals.

Amelanchier utahensis Koehne **Stems** 1–2 m high, intricately branched. **Leaves** 1–3 cm long, tomentose beneath. **Raceme** of 3 to 6 flowers. **Flowers**: hypanthium 1–2 mm long; sepals reflexed, 1–4 mm long; petals 5–10 mm long; styles 2 or 3, separate. **Pome** 5–10 mm long, 3- to 6-seeded. Woodlands, open forest; valleys. WA to MT south to CA, AZ, NM. Our plants are var. *utahensis*. Known from Gallatin and Stillwater cos.

Cercocarpus Kunth Mountain mahogany

Shrubs with hard wood and smooth bark. **Leaves** short-petiolate, stipulate, simple, leathery, clustered on short lateral shoots. **Inflorescence** of 1 to 3 short-pedicellate flowers on lateral shoots. **Flowers** 5-merous, perigynous; hypanthium tubular below, becoming cup-shaped above; sepals deciduous, without bracteoles; petals absent; stamens 15 to 45; pistil 1. **Fruit** a leathery achene enclosed in the hypanthium with an exserted, coiled, plumose style.[266]

1. Leaf blades dentate . . . . *C. montanus*
1. Leaf blades entire with inrolled margins . . . . *C. ledifolius*

Cercocarpus ledifolius Nutt. CURL-LEAF MOUNTAIN MAHOGANY Evergreen shrubs 1.5–5 m. **Stems** reddish, tomentose when young, becoming gray. **Leaf blades** linear, 1–3 cm long, nearly glabrous above, white-lanate beneath with entire, inrolled margins. **Flowers**: hypanthium 4–6 mm long, lanate, the limb reddish; sepals 1–2 mm long; stamens 15 to 25. **Achene** 5–10 mm long; the style 2–4 cm long. Stony slopes, cliffs, rock outcrops, most common on limestone or sandstone; plains, valleys, montane. WA to MT south to CA, AZ, UT, CO. Our plants are var. *intercedens* C.K.Schneid.

Browsing by deer often causes plants to become intricately branched and impenetrable.

Cercocarpus montanus Raf. BIRCH-LEAF MOUNTAIN MAHOGANY Deciduous shrubs 2–4 m high. **Stems** red-brown becoming grayish. **Leaf blades** obovate, 12–25 mm long, finely serrate, hirsute on both surfaces, white-lanate beneath, prominently veiny. **Flowers**: hypanthium 10–15 mm long, finely hairy; sepals 1–2 mm long; stamens 22 to 44. **Achene** 9–12 mm long; the style 5–9 cm long. Stony slopes, open pine forest; plains. OR to SD south to CA, AZ, TX, Mexico. One extensive population in Treasure Co. Our plants are var. *montanus*.

Chamaerhodos Bunge

Chamaerhodos erecta (L.) Bunge Biennial or short-lived, monocarpic, taprooted perennial. **Stems** usually solitary, branched, 4–30 cm. **Herbage** villous, glandular. **Leaves** basal and cauline, 1–2 cm long, petiolate; blades obovate, twice deeply divided into linear lobes, reduced above. **Inflorescence** an open, often flat-topped, leafy-bracteate cyme. **Flowers** perfect, perigynous; hypanthium campanulate, 2–3 mm long; sepals ca. 2 mm long, setose, glandular; petals white, as long as sepals; stamens 5; pistils 5 to 10. **Fruit** an ovoid achene ca. 1.5 mm long. Stony or shallow soil of grassland, streambanks, roadsides; plains, valleys, montane. AK to MB south to UT, CO, SD.

Comarum L.

Comarum palustre L. [*Potentilla palustris* (L.) Scop.] Rhizomatous perennial. **Stems** ascending, prostrate, or floating, reddish, glabrous to sparsely glandular, 20–50 cm. **Leaves** sheathing-petiolate, pinnately divided into 3 to 7 leaflets; leaflets oblong, sparsely hairy, glaucous beneath, dentate, 2–7 cm long. **Inflorescence** a leafy, few- to several-flowered, glandular-pubescent cyme. **Flowers** perfect, perigynous; hypanthium saucer-shaped; sepals ovate, purple, 7–15 mm long, alternate with narrower and shorter bracteoles; petals dark red, 3–7 mm long; stamens usually ≥20; styles numerous. **Fruit** a cluster of ovoid achenes, each ca. 1.5 mm long enveloped by the calyx; styles deciduous at maturity, attached laterally. Wet or inundated, organic soil of fens, marshes; montane. Circumboreal south to CA, WY, OH.

　　Formerly considered a species of *Potentilla*, but now thought to be more closely related to *Sibbaldia*.[123]

Cotoneaster Medik.

Cotoneaster lucidus Schltdl. Shrubs to 2 m tall. **Stems** erect to arching, pubescent when young. **Leaves** simple, petiolate, stipulate; the blades 2–5 cm long, ovate, entire, glabrous, shiny above. **Inflorescence** terminal, few-flowered cymes. **Flowers** small, perfect, perigynous, 5-merous; hypanthium saucer-shaped; sepals ca. 2 mm long; petals pinkish, as long as sepals; stamens 10 to 20; styles 2 to 5. **Fruit** a black, globose, 3- to 4-seeded pome 7–9 mm long. Moist riparian forest; valleys. Common ornamental rarely escaped; native to Eurasia. Collected in Missoula Co.

Crataegus L. Hawthorn

Shrubs or small trees, armed with thorns. **Leaves** petiolate, simple with serrate margins, deciduous. **Inflorescence** a corymb or cyme on branch tips. **Flowers** perfect, epigynous; hypanthium cup-shaped; sepals 5; petals 5, white; stamens 5 to 20; pistil 1 with 1 to 5 carpels. **Fruit** a globose pome enclosing 1 to 5 nutlets.[115, 323,324]

　　Many hawthorn plants produce seed agamospermously, resulting in a myriad of slightly different forms. Many of the taxa proposed by Phipps and O'Kennon may be good species, but I am unable to distinguish them in our collections. My treatment is conservative pending more investigations into the group. Mature fruit may be required for determination.

1. Some leaves lobed >1/2-way to the midvein . . . . *C. monogyna*
1. Leaves unlobed or shallowly lobed. 2

2. Thorns <3 cm long; mature fruit dark purple or black. 3
2. Many thorns ≥3 cm long; mature fruit red. 5

3. Inflorescence long-hairy; fruit with scattered hairs . . . . *C. castlegarensis*
3. Inflorescence and fruit glabrous 4

4. Stamens 10; fruit pulp yellow to orange; thorns usually >15 mm long . . . . *C. douglasii*
4. Stamens 20; fruit pulp green; thorns usually ≤15 mm long . . . .*C. suksdorfii*

5. Leaves tapered gradually to the petiole with purple, gland-tipped serrations,
　 often extending down the petiole . . . . *C. chrysocarpa*
5. Leaves rounded at the base, serrations without conspicuous glands . . . . *C. macracantha*

Crataegus castlegarensis J.B.Phipps & O'Kennon Shrubs 2–8 m. **Stems** shiny-brown becoming gray; thorns 13–25 mm long, straight to curved. **Leaf blades** ovate to obovate, 3–6 cm long, glandular-serrate; sometimes shallowly lobed, appressed-hairy above, sparsely hairy below. **Inflorescence** 5- to 25-flowered, sparsely long-hairy. **Flowers**: hypanthium sparsely hairy to glabrous externally; sepals 2–3 mm long, glandular-serrate; petals 5–8 mm long; stamens 10; styles 3 or 4. **Pome** 8–12 mm wide, black with scattered hairs at maturity. Fields, riparian thickets, along ditches; valleys. BC to SK south to OR, UT, WY.

Crataegus chrysocarpa Ashe [*C. williamsii* Eggl., *C. orbicularis* J.B.Phipps & O'Kennon, *C. columbiana* Howell misapplied] Shrub 2–4 m high. **Stems** light brown, becoming gray; thorns mostly 3–6 cm long, straight to curved. **Leaf blades** ovate, 2–6 cm long, glandular-serrate, shallowly lobed, the lobes apiculate, sparsely appressed-hairy. **Inflorescence** 5- to 12-flowered, long-hairy. **Flowers:** hypanthium villous; sepals with glandular margins, 2–3 mm long; petals 7–10 mm long; stamens 10; styles 3 or 4. **Pome** 8–10 mm wide, red to burgundy, sparsely pubescent. Riparian forest, woodlands, thickets, often along ditches; plains, valleys. BC to NL south to OR, UT, CO, SD.

Plants with densely hairy inflorescences and hairy fruit have been called *Crataegus chrysocarpa* var. *piperi* (Britton) Kruschke; however, it and var. *chrysocarpa* are often not discernable,[323] and both forms occur in similar habitats throughout the state.

Crataegus douglasii Lindl. [*C. rivularis* Nutt., *C. okennonii* J.B.Phipps] Shrubs 4–6 m high. **Stems** tan when young becoming gray; thorns 15–25 mm long, straight to curved. **Leaf blades** obovate, 3–7 cm long, twice serrate and shallowly lobed above, appressed-hairy, glabrous and glaucous beneath. **Inflorescence** 5- to 25-flowered, glabrous. **Flowers:** hypanthium glabrous externally; sepals 2–4 mm long; petals 5–8 mm long; stamens 10; styles 3 or 4. **Pome** 6–12 mm wide, purple to black, glabrous at maturity. Fields, riparian forest, thickets; valleys. AK to SK south to CA, ID, WY, SD. (p.268)

Crataegus okennonii has been segregated on the basis of larger flowers and smaller thorns; however, these characters are continuous and not strongly correlated in our material.

Crataegus macracantha Lodd. ex Louden [*C. succulenta* Schrad. ex Link. var. *occidentalis* (Britton) E.J.Palmer, *C. phippsii* O'Kennon] Shrub 2–6 m. **Stems** purplish-brown, becoming gray; thorns 3–5 cm long, curved. **Leaf blades** ovate, 2–5 cm long, serrate, mostly unlobed, sparsely hairy above, glabrous beneath. **Inflorescence** 10- to 15-flowered, villous. **Flowers:** hypanthium villous; sepals 3–4 mm long, glandular-dentate; petals 6–8 mm long; stamens 10; styles 2 to 4. **Pome** 7–10 mm wide, red, pubescent. Riparian thickets, woodlands; plains, valleys. BC to QC south to OR, AZ, NM. Our plants are var. *occidentalis* (Britton) Eggl.

Crataegus monogyna Jacq. Shrub or small tree 5–8 m high. **Stems** reddish at first, becoming gray; thorns 1–2 cm long, straight. **Leaf blades** ovate, 2–4 cm long, serrate, deeply 3- to 7-lobed, glabrous to sparsely hairy. **Inflorescence** 10- to 20-flowered, pubescent. **Flowers:** hypanthium pubescent; sepals 1–2 mm long; petals 5–7 mm long; stamens 20; style 1. **Pome** 8–10 mm wide, red, glabrous. Fields, along ditches; valleys. Escaped from cultivation throughout most of temperate N. America; collected in Missoula Co.; native to Europe.

Crataegus suksdorfii (Sarg.) Kruschke Shrubs 3–5 m. **Stems** tan becoming gray; thorns 15–25 mm long, straight to curved. **Leaf blades** glandular, elliptic to obovate, 3–6 cm long, glandular-serrate; sometimes shallowly lobed, appressed-hairy above, glabrous beneath. **Inflorescence** 7- to 15-flowered, glabrous, punctate. **Flowers:** hypanthium glabrous externally; sepals ca. 4 mm long; petals 6–8 mm long; stamens 20; styles 4 or 5. **Pome** 1 cm wide, black and glabrous at maturity. Riparian thickets; valleys. AK south to CA, ID, MT.

In the Rocky Mountains, the diploid *Crataegus suksdorfii* generally has smaller thorns and fruit and twice as many stamens as the polyploid *C. douglasii*.[71,111]

Dasiphora Raf. Shrubby cinquefoil

Dasiphora fruticosa (L.) Rydb. [*Potentilla fruticosa* L., *Pentaphylloides fruticosa* (L.) O.Schwarz, *P. floribunda* A.Löve,] Highly branched shrub 10–100 cm high. **Stems** reddish-brown, villous, becoming glabrous. **Leaves** petiolate; the blade sparsely hairy, glaucous beneath, with 5 narrowly lanceolate pinnate leaflets 7–20 mm long with entire, inrolled margins; stipules membranous, narrowly ovate. **Inflorescence** 1- to 5-flowered, terminal cymes or more commonly solitary in leaf axils. **Flowers** perfect, perigynous; hypanthium cup-shaped, villous; sepals 5,ovate, 4–7 mm long with narrow bracteoles as long as the sepals; petals yellow, 7–12 mm long; stamens 20 to 25; styles numerous. **Fruit** a long-hairy achene, 1–2 mm long. Moist grasslands, meadows, fens, open forest, exposed slopes, ridges; valleys to alpine. Circumboreal south to CA, NM, MN, NJ. (p.268)

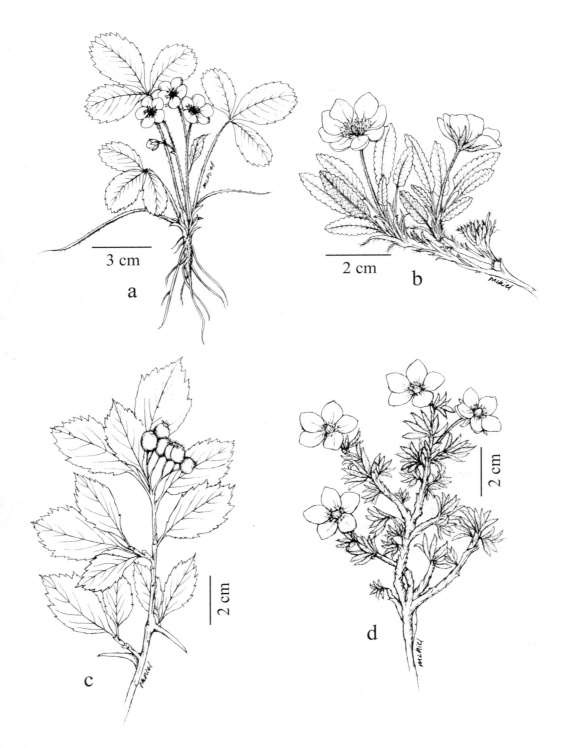

Plate 44. a. *Fragaria virginiana*, b. *Dryas octopetala*, c. *Crataegus douglasii*, d. *Dasiphora fruticosa*

Dryas L. Mountain avens

Mat-forming subshrubs. **Stems** woody, trailing, branched. **Leaves** petiolate, simple, leathery, veiny above, white-woolly beneath, margins inrolled. **Inflorescence** a solitary flower borne on a long scape. **Flowers** perfect, perigynous; hypanthium saucer-shaped; sepals 8 to 10 without bracteoles, villous and with dark, nail-like hairs; petals 8 to 10; stamens and pistils numerous. **Fruit** an achene with a plumose style.

The feathery styles curl when moist and straighten when dry, helping to work the achene into the soil. The yellow-flowered species is rarely found at high elevations while the white-flowered species is rare below subalpine.

1. Flowers with yellow petals and small bracts on the stalks . . . .*D. drummondii*
1. Flowers white, without bracts on the stalks 2

2. Leaf margins crenate the entire length. . . . .*D. octopetala*
2. Leaf margins entire at least on upper half of blade . . . .*D. integrifolia*

Dryas drummondii Richardson ex Hook. **Leaf blades** elliptic, 15–35 mm long with crenate margins, sparsely tomentose above, white-lanate beneath. **Scapes** 5–20 cm. **Flowers** nodding; hypanthium villous within; calyx lobes ovate, 5–6 mm long; petals yellow, elliptic, 8–13 mm long, nearly erect. **Achenes** 3–4 mm long; style 15–25 mm long. Riparian gravel bars, moraine; montane, less commonly to alpine. AK to QC south to OR, MT. More common on calcareous parent material.

Dryas integrifolia Vahl Similar to the more common *D. octopetala*. **Leaf blades** lanceolate, 1–2 cm long, entire or crenate only at the base. **Scapes** 1–5 cm high. **Flowers** erect; calyx lobes 4–6 mm long; petals white, elliptic, ca. 1 cm long, spreading. **Achenes** ca. 3 mm long; styles ca. 2 cm long. Stony, calcareous fellfields; alpine. AK to Greenland south to MT, NH.

Our collections from Fergus Co. have leaves crenate below, perhaps suggesting introgression with *D. octopetala* (fide W. A. Weber).

Dryas octopetala L. MOUNTAIN AVENS, ALPINE DRYAD **Leaf blades** lanceolate, 8–25 mm, crenate the entire length. **Scapes** 1–6 cm high. **Flowers** erect; calyx lobes 5–8 mm long; petals white, elliptic, 10–15 mm long, spreading. **Achenes** ca. 3 mm long; styles 15–25 mm long. Stony soil of exposed slopes, fellfields, moist turf; alpine, less common subalpine. Circumpolar south to OR, ID, CO. Our plants are ssp. *hookeriana* (Juz.) Hultén. (p.268)

Genetically distinct ecotypes occur in dry fellfields and moist tundra.[270] The species occurs on acidic as well as calcareous soils but is more abundant on the latter.

Drymocallis Fourr. ex Rydb.

Herbaceous perennials. **Leaves** petiolate, stipulate, basal and cauline; the blades pinnately compound with shallowly lobed or dentate margins. **Inflorescence** a leafy-bracted cyme. **Flowers** perfect, perigynous; hypanthium saucer-shaped; sepals alternate with bracteoles; stamens usually ≥20; styles numerous. **Fruit** a cluster of ovoid achenes enveloped by the calyx; styles tapered at both ends, attached on the bottom half of the achene, deciduous at maturity.

Formerly considered part of *Potentilla*; our species are separated mainly on continuous qualitative characters, making consistent determinations difficult. *Drymocallis fissa* (Nutt.) Rydb. [*Potentilla fissa* Nutt.] is reported for MT,[52,113] but I have seen no specimens.

1. Inflorescence open; lateral branches ascending to spreading . . . .*D. glandulosa*
1. Inflorescence congested; lateral branches erect or strongly ascending. . . . . *D. arguta*

Drymocallis arguta (Pursh) Rydb. [*D. convallaria* (Rydb.) Rydb., *Potentilla arguta* Pursh] Perennial from a simple or branched caudex. **Stems** erect, brown-hirsute and glandular, 25–80 cm. **Leaf blades** oblong, 6–12 cm long, pinnately divided into 5 to 9 obovate, shallowly lobed, sparsely hairy leaflets. **Inflorescence** a narrow, stiffly erect, flat-topped, glandular-hairy cyme. **Flowers**: sepals ovate, 6–7 mm long; bracteoles narrower and shorter; petals white to pale yellow, 7–9 mm long. **Achenes** smooth, ca. 1 mm long. Grasslands, steppe, dry meadows, thickets, woodlands; plains, valleys, montane. BC to QC south to NV, AZ, NM, MO, NJ.

Plants of the western Cordillera have been called *Drymocallis convallaria*, while those of the Great Plains *D. arguta*; however, the distinction is not apparent in our area.[152] *Drymocallis glandulosa* is often smaller with a more spreading inflorescence and usually occurs in stonier soil.

Drymocallis glandulosa (Lindl.) Rydb. [*D. pseudorupestris* (Rydb.) Rydb., *D. glabrata* Rydb. *Potentilla glandulosa* Lindl.] Perennial from a mostly branched caudex. **Stems** erect to ascending, hirsute and glandular-pubescent, 10–70 cm. **Leaf blades** oblanceolate, pinnately divided into 5 to 7 obovate to suborbicular, glandular, dentate leaflets, 6–60 mm long. **Inflorescence** a glandular cyme with spreading to ascending branches. **Flowers**: sepals lanceolate, 4–7 mm long; bracteoles shorter than the sepals; petals 5–10 mm long. **Achenes** smooth, <1 mm long. Rocky soil of open forest, grasslands, meadows, outcrops; valleys to alpine. BC, MT south to CA, AZ, UT, Mexico.

1. Petals no longer than the sepals; uncommon . . . . ssp. *glandulosa*
1. Petals longer than the sepals; common 2

2. Petals white to ochroleucus; lateral leaflets ca. as wide as long . . . . ssp. *pseudorupestris*
2. Petals yellow; lateral leaflets longer than wide. . . . . ssp. *glabrata*

Drymocallis glandulosa* ssp. *pseudorupestris (Rydb.) Soják is found in rocky soil of grasslands, open forest, fellfields, outcrops, valleys to alpine; ***D. glandulosa* ssp. *glabrata*** (Rydb.) Soják [=*P. glandulosa* var. *intermedia* (Rydb.) Hitchc.] is found in moist grasslands, meadows, fellfields, valleys to subalpine; ***D. glandulosa* ssp. *glandulosa*** occurs in stony soil of grasslands, open forest, valleys, montane. See *D. arguta*.

Fragaria L. Strawberry

Herbaceous, stoloniferous, scapose perennials. **Leaves** long-petiolate, stipulate, with 3 leaflets, ovate to obovate, dentate, glaucous beneath. **Inflorescence** a few-flowered, bracteate cyme. **Flowers** perfect, perigynous; hypanthium saucer-shaped; sepals 5 with 5 bracteoles; petals 5, white, longer than the sepals; stamens ca. 20; pistils numerous. **Fruit** hemispheric, red, juicy, berry-like receptacle with scattered seeds imbedded.

1. Terminal tooth of central leaflet extending beyond the adjacent ones; upper surface prominently
   veiny. . . . . *F. vesca*
1. Terminal tooth of middle leaflet shorter than adjacent ones; upper surface not prominently
   veiny. . . . . *F. virginiana*

Fragaria vesca L. Woodland strawberry **Leaflets** 1–8 cm long, green, sparsely hairy above, sericeous beneath, prominently veiny. **Inflorescence** usually longer than the leaves. **Flowers**: sepals lanceolate 4–7 mm long; petals 5–10 mm long. **Fruit** ca. 1 cm across. Thickets, coniferous and riparian forest; valleys, montane. Temperate Europe and N. America. Our plants are var. *bracteata* (A.Heller) R.J.Davis.

Less common than *Fragaria virginiana* and more confined to forested habitats.

Fragaria virginiana Mill. Wild strawberry Similar to *F. vesca*. **Leaflets** 2–6 cm long, glabrous, yellow-green, inconspicuously veiny above. **Inflorescence** 5–20 cm long, as long or longer than the leaves. **Flowers**: sepals lanceolate, 4–6 mm long; petals 7–12 mm long. **Fruit** ca. 1 cm across. Meadows, grasslands, forest; plains, valleys to subalpine. Native to most of temperate N. America. (p.268)

1. Petiole and scape sparsely appressed-hairy . . . . var. *glauca*
1. Petiole and scape spreading-pilose . . . . var. *platypetala*

Fragaria virginiana* var. *platypetala (Rydb.) H.M.Hall is mainly confined to the northwest counties; *F. virginiana* var. *glauca* S.Watson occurs throughout the state with little difference in habitat between the two.

Geum L. Avens

Herbaceous perennials from an often-branched rootstock clothed in old leaf bases. **Leaves** basal and cauline, petiolate, leafy-stipulate, deeply pinnately divided with small segments among the main ones; stem leaves reduced. **Inflorescence** bracteate cymes, corymbs, or flowers solitary. **Flowers** perfect, perigynous; sepals 5 with 5 bracteoles; petals 5, obovate; stamens and pistils numerous. **Fruit** an achene with a persistent, elongate style, numerous on a hemispheric receptacle.

Geum triflorum fruit heads are cottony; those of *G. aleppicum, G. macrophyllum* and *G. rivale* appear spiny with achene beaks that serve to distinguish among them. *Potentilla* spp. have non-persistent styles.

1. Stems with several well-developed leaves. 2
1. Stems scapose with 1 or 2 reduced leaves (not including leaf-like bracts) 5

2. Sepals purple; petals pinkish . . . . *G. rivale*
2. Sepals green; petals white or yellow 3

3. Lower portion of style sparsely, minutely glandular . . . . *G. macrophyllum*
3. Lower portion of style glabrous or sparsely hairy but not glandular 4

4. Petals white; receptacle hirsute with hairs 2/3 as long as achenes . . . . *G. canadense*
4. Petals yellow; receptacle with short pubescence . . . . *G. aleppicum*

5. Flowers nodding; petals erect; stem leaves opposite; mature styles >1 cm long. . . . .*G. triflorum*
5. Flowers erect; petals spreading; stem leaves (if present) alternate; styles shorter . . . . *G. rossii*

Geum aleppicum Jacq. **Stems** erect or ascending, 25–90 cm, hirsute. **Leaf blades** obovate, sparsely hairy, 7–15 cm long, divided into 5 to 9 ovate, dentate main leaflets; the terminal lobe or leaflet little wider than the adjacent ones. **Inflorescence** an open, erect, leafy-bracteate cyme. **Flowers** erect; hypanthium saucer-shaped; sepals green, 3–7 mm long, reflexed, tomentose on the margins; petals yellow, 5–8 mm long. **Achenes** 3–4 mm long, long-hairy; mature persistent style 3–5 mm long, glabrous; the deciduous, spreading upper portion hirsute. Wet meadows, fens, marshes, thickets, moist to wet forests, often along streams; valleys, montane. Circumboreal south to CA, NM, MN, PA.

Geum canadense Jacq. **Stems** erect or ascending, 30–100 cm, sparsely hairy to glabrous. **Leaf blades** obovate, 8–15 cm long, divided into 2 to 4 ovate, dentate main lateral leaflets; the terminal leaflet wider than the laterals. **Inflorescence** an open, erect, pubescent, leafy-bracteate cyme. **Flowers** erect; hypanthium saucer-shaped; sepals green, 4–8 mm long, reflexed, tomentose on the margins; petals white, 5–9 mm long. **Achenes** 2–4 mm long, long-hairy; persistent style 4–7 mm long, glabrous; the deciduous, hooked upper portion sparsely hairy. Riparian woodlands, thickets; plains in Rosebud and Powder River cos. MT to NS south to TX, AL, GA.

Geum macrophyllum Willd. **Stems** erect or ascending, 20–100 cm, hirsute, glandular above. **Leaf blades** oblong, sparsely hirsute, 8–25 cm long, divided into 4 to 8 ovate, dentate main lateral leaflets; the terminal leaflet cordate-orbicular, 3- to 5-lobed, much larger than the lateral ones. **Inflorescence** a glandular, open, leafy-bracteate cyme. **Flowers** erect; hypanthium saucer-shaped; sepals green, 2–5 mm long, reflexed, tomentose on the margins; petals yellow, 4–7 mm long. **Achenes** 2–3 mm long, villous; persistent style 3–5 mm long, glandular; the deciduous, spreading upper portion hirsute. Wet meadows, thickets, woodlands, moist forest openings, often along streams, wetlands; valleys to lower subalpine. AK to NL south to CA, NM, MN, NY; Asia. Our plants are var. *perincisum* (Rydb.) Raup. (p.273)

Some plants from the northwest counties have unlobed terminal leaflets and would be referred to var. *macrophyllum*, but they have glandular inflorescences.

Geum rivale L. **Stems** erect or ascending, 25–80 cm, sparsely hirsute below. **Leaf blades** oblong, hirsute, 8–20 cm long, divided into 2 to 6 ovate, dentate main lateral leaflets; the terminal leaflet cuneate-ovate, lobed. **Inflorescence** an open, leafy-bracteate cyme. **Flowers** ascending to nodding; hypanthium cup-shaped; sepals purple, 7–11 mm long, erect; petals pinkish with purple veins, 6–10 mm long. **Achenes** 3–4 mm long, long-hairy; persistent style 8–12 mm long, long-hairy and glandular below; the deciduous, spreading upper portion hirsute. Wet soil of meadows, thickets, fens, marshes, spruce forest; valleys, montane. Circumboreal south to WA, NM, IN, NJ.

Geum rossii (R.Br.) Ser. Plants with branched rootstocks, often forming mats. **Stems** erect or ascending, subscapose, 2–20 cm, pubescent. **Leaf blades** 2–10 cm long, pubescent, sometimes glandular, narrowly lanceolate, with numerous pinnate, lobed divisions. **Inflorescence** a 1- to 4-flowered cyme. **Flowers** erect; hypanthium cup-shaped; sepals erect, green to purple, 3–6 mm long; petals yellow, 5–9 mm long. **Achenes** 2–4 mm long, pubescent; style 3–4 mm long, glabrous. Turf, fellfields; alpine. AK south to AZ, NM, CO; Asia.

1. Calyx green; leaflets ovate, not lobed >1/2-way to midvein. . . . . var. *rossii*
1. Calyx purplish; leaflets lanceolate or deeply divided into linear lobes . . . . .var. *turbinatum*

Geum rossii var. *turbinatum* (Rydb.) C.L.Hitchc. is abundant in the southwest and south-central ranges, but is absent in the northwest ranges dominated by rocks of the Belt Series; *G. rossii* var. *rossii* occurs sparingly in Lake and Missoula cos.; plants that appear intermediate occur rarely in the Bitterroot Range.

Geum triflorum Pursh PRAIRIE SMOKE, OLD MAN'S WHISKERS **Rootstock** branched, spreading below ground. **Stems** erect, pilose, 10–50 cm, with 1 pair of reduced, opposite leaves. **Leaf blades** 4–15 cm long, long-hairy, oblanceolate with numerous pinnate, lobed divisions. **Inflorescence** a mostly 3-flowered cyme. **Flowers** nodding; hypanthium cup-shaped; sepals erect, dark pink, 7–11 mm long with slightly longer bracteoles; petals cream, ca. as long as the sepals. **Achenes** ca. 2 mm long, short-hairy; style 2–5 cm long, plumose. Mesic grasslands, meadows, woodlands, turf; plains, valleys to alpine. BC to NL south to CA, NM, IL, NY. (p.273)

Both var. *ciliatum* (Pursh) Fassett, with linear-lobed leaflets and longer styles, and var. *triflorum*, with shorter styles and less divided leaflets, are reported for MT. However, these characters are continuous and without geographic or ecological distinctiveness in our area, and some populations contain both forms.

Holodiscus (K.Koch) Maxim. Ocean spray

Holodiscus discolor (Pursh) Maxim. MOUNTAIN SPRAY Shrubs to 3 m tall. **Stems** erect; bark reddish-tan, becoming gray. **Leaves** simple, petiolate, exstipulate; the blades 2–9 cm long, ovate, lobed, dentate, pubescent, sparsely so above. **Inflorescence** a many-flowered, highly branched pyramidal panicle at branch tips. **Flowers** small, perfect, perigynous, 5-merous; hypanthium saucer-shaped; sepals 1.5– 2 mm long; petals white, as long as sepals; stamens 20; styles 5. **Fruit** a hirsute, 1-seeded achene, ca. 1.5 mm long; style ca. 1 mm long. Rocky soil of dry, open forest, brushy slopes; valleys to lower subalpine. BC to CA, ID, MT. (p.273)

Ivesia Torr. & A.Gray

Herbaceous, taprooted perennials from woody, branched rootstocks clothed in old leaf bases. **Stems** glandular-hairy above, erect, simple. **Leaves** mostly basal, short-petiolate; the blades linear, pinnately divided into numerous, overlapping, obovate, deeply lobed leaflets. **Inflorescence** a congested, glandular-hirsute cyme. **Flowers** perfect, perigynous, 5-merous; sepals erect with narrow bracteoles; petals yellow; stamens 5; styles 2 to 4. **Fruit** a glabrous, ovoid achene with a straight style.

1. Petals longer than calyx; hypanthium cup-shaped . . . . *I. tweedyi*
1. Petals just shorter than the calyx; hypanthium obconic . . . . *I. gordonii*

Ivesia gordonii (Hook.) Torr. & A.Gray **Stems** 5–15 cm. **Leaves** mostly basal; the blades glandular-puberulent, the blade 2–10 cm long; leaflets obovate, deeply lobed, 2–8 mm long. **Flowers:** hypanthium obconic, 3–4 mm long; sepals 2–4 mm long; petals barely shorter than the sepals. **Fruit** ca. 2 mm long. Stony, sparsely vegetated soil of rock outcrops, stream beds, grasslands, fellfields; montane to alpine. WA to MT south to CA, UT, WY.

Ivesia tweedyi Rydb. **Stems** 5–20 cm. **Leaves** basal and few cauline; the blade glabrous to glandular-puberulent, 3–8 cm long; leaflets obovate, deeply lobed, 4–8 mm long. **Flowers:** hypanthium cup-shaped, 2–3 mm long; sepals 2–4 mm long; petals barely shorter than the sepals. **Fruit** ca. 2 mm long. Stony, sparsely vegetated soil of rock outcrops, grasslands; subalpine. Endemic to eastern WA, ID, MT. Known from Mineral Co.

Kelseya Rydb.

Kelseya uniflora (S.Watson) Rydb. Shrubs forming large, solid cushions to 80 cm across. **Stems** intricately branched, crowded, prostrate, clothed in old leaves; erect shoots 1–2 cm. **Leaves** cauline, crowded, overlapping, exstipulate, sessile, obovate, sericeous, 2–4 mm long, entire. **Inflorescence** a solitary, terminal flower. **Flowers** perfect, perigynous, 5-merous; hypanthium campanulate, deep pink; sepals ca.2 mm long; petals deep pink, 2–3 mm long; stamens ca. 10; styles 3 to 5. **Fruit** 1-seeded, ellipsoid follicles that open completely, ca. 3 mm long. Crevices of vertical limestone cliffs; montane, subalpine. Endemic to MT and adjacent WY, ID.

Petrophytum caespitosum is vegetatively similar but has numerous flowers in spike-like clusters. The type locality is in Lewis and Clark Co.

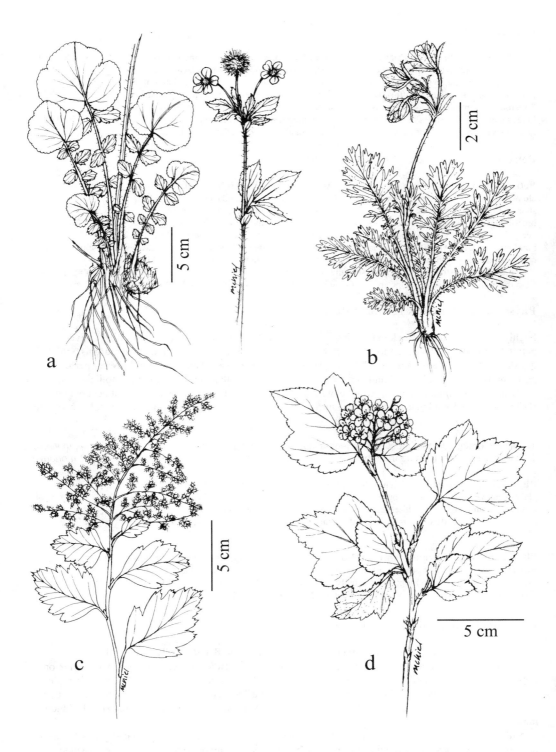

Plate 45. a. *Geum macrophyllum*, b. *Geum triflorum*, c. *Holodiscus discolor*, c. *Physocarpus malvaceus*

Luetkea Bong.

Luetkea pectinata (Pursh) Kuntze PARTRIDGE FOOT Evergreen subshrub forming loose mats. **Stems** mainly prostrate with erect shoots 6–16 cm, pubescent above. **Leaves** basal and cauline, petiolate, exstipulate; the blades 5–10 mm long, fan-shaped, twice ternately divided into oblong lobes, often black-punctate. **Inflorescence** a leafy-bracted raceme. **Flowers** perfect, perigynous, 5-merous; hypanthium obconic; sepals triangular, ca. 2 mm long; petals white, spatulate, 2–4 mm long; stamens ca. 20; styles 5. **Fruit** a follicle 2–3 mm long; seeds several; style <1 mm long. Often shallow soil of moist to wet, open forest, often where snow lies late; subalpine. AK south to CA, ID, MT.

Petrophytum (Nutt. ex Torr. & A.Gray) Rydb.

Petrophytum caespitosum (Nutt.) Rydb. Mat-forming shrub. **Stems** branched, prostrate, clothed in old leaf bases; erect stems scapose, 3–10 cm. **Leaves** basal and cauline, exstipulate, short-petiolate, oblanceolate, glabrate to sericeous, entire, 3–12 mm long; cauline reduced, bract-like. **Inflorescence** a dense, cylindrical raceme, 1–4 cm long. **Flowers** perfect, 5-merous, perigynous; hypanthium cup-shaped, sericeous; sepals 1–2 mm long, deltoid; petals white, 1.5–2.5 mm long; stamens 20 to 40; styles usually 5. **Fruit** a 1- to 2-seeded follicle, ca. 3 mm long. Crevices of cliffs and rock outcrops, usually on limestone; montane to alpine. OR to SD south to CA, AZ, NM, TX, Mexico.

Plants from limestone throughout most of the state are sericeous with sepals and petals ca. 1.5–2 mm long; plants from the granitic Bitterroot Range are glabrate with sepals ca. 1 mm long and petals ≥2 mm long. See *Kelseya*.

Physocarpus (Cambess.) Raf. Ninebark

Highly branched shrubs. **Bark** exfoliating forming reddish and brown longitudinal stripes. **Leaves** petiolate, simple; the blade glaucous beneath, ovate, cordate, 3 to 5-lobed with dentate margins; stipules deciduous. **Inflorescence** a terminal, hemispheric, bracteate, densely hairy corymb. **Flowers** perfect, perigynous, 5-merous; hypanthium cup-shaped; stamens 20 to 40; styles 1 to 3. **Fruit** a follicle.

Physocarpus capitatus (Pursh) Kuntze, with 3 to 5 styles, is similar to *P. malvaceus*; it occurs in north ID and could be found in northwest MT. Our two species are very similar and can only be told apart with confidence in fruit.

1. Follicles erect, adjacent most of their length; some leaf blades >3 cm long . . . . *P. malvaceus*
1. Follicles divergent from near the base; most leaves ≤2 cm long . . . . *P. monogynus*

Physocarpus malvaceus (Greene) Kuntze **Stems** erect, 50–150 cm. **Leaf blades** 2–7 cm long, glabrate above, sparsely stellate-pubescent beneath. **Flowers**: hypanthium 1–2 mm long; sepals ovate, 2–4 mm long; petals white, ca. 4 mm long; styles usually 2. **Follicle** hairy, 3–4 mm long; seeds 1 to 2. Dry coniferous forest; valleys, montane. BC to AB south to OR, UT, WY. (p.273)

Physocarpus monogynus (Torr.) J.M.Coult. **Stems** erect, ca. 1 m. **Leaf blades** 1–2 cm long, sparsely stellate-pubescent beneath. **Flowers**: hypanthium 1–2 mm long; sepals ovate, 2–3 mm long; petals white, 3–4 mm long; styles usually 2. **Follicle** hairy, 3–4 mm long; seeds 1 to 3. Rocky limestone-derived soil of dry, open forest; montane. MT, SD south to NM, TX.

Potentilla L. Cinquefoil

Herbaceous annuals to perennials. **Leaves** petiolate, stipulate, basal and usually cauline; the blades palmately or pinnately compound with lobed or dentate leaflets. **Inflorescence** a leafy-bracted cyme or flowers solitary in leaf axils. **Flowers** perfect, perigynous; hypanthium saucer-shaped; sepals alternate with bracteoles; stamens usually ≥20; styles numerous. **Fruit** a cluster of ovoid achenes enveloped by the calyx; styles tapered from the base, attached apically or subapically (laterally in *P. anserina*), deciduous at maturity.[124]

A difficult genus due to frequent asexual reproduction. Members of the closely related genus *Geum* have persistent styles that make the clusters of achenes appear spiny or cottony. *Comarum, Dasiphora* and *Drymocallis* have been segregated out of *Potentilla* sensu lato based on recent molecular data.[123] There is a great deal of variation in several of the taxa outlined here (e.g., *P. glaucophylla, P. gracilis*,

P. pensylvanica, P. arenosa), most likely the result of hybridization and agamospermy. I have not seen a classification that is entirely satisfactory for plants in our area.

1. Basal leaves pinnately divided; at least 1 pair of leaflets clearly below the tip of the rachis 2
1. Basal leaves ternate or palmately divided; all primary leaflets attached to the tip of the rachis. 14

2. Leaves covered with felt-like tomentum beneath 3
2. Leaves glabrous to sericeous beneath, but not tomentose 7

3. Plants stemless, spreading by stolons; flowers solitary . . . .*P. anserina*
3. Plants with leafy stems, without stolons; flowers >1. 4

4. Stipules deeply lobed; bracteoles equal or longer than sepals; styles ≤1.2 mm long . . . . . . *P. pensylvanica*
4. Stipules entire or shallowly lobed; bracteoles shorter than sepals; styles ≥1.5 mm long. 5

5. Petals ca. as long as the sepals; inflorescence with >5 flowers. . . . .*P. hippiana*
5. Petals longer than sepals; inflorescence usually of 1 to 5 flowers 6

6. Style ≥2 mm long . . . . *P. macounii*
6. Styles ca. 1 mm long . . . . *P. rubricaulis*

7. Leaves glandular beneath 8
7. Leaves not glandular 9

8. Leaves glandular but not long-hairy beneath . . . . *P. brevifolia*
8. Leaves long-hairy beneath. . . . .*P. jepsonii*

9. At least some leaflets ≥2 cm long. . . . . *P. drummondii*
9. Leaflets <2 cm long 10

10. Plants annual to short-lived perennial; leaflets lobed <1/2-way to midvein . . . . *P. paradoxa*
10. Plants with a branched rootstock; leaflets deeply lobed. 11

11. Leaves subpinnate; leaflets attached very close together 12
11. Leaves clearly pinnate; leaflets shorter than length of leaflet-bearing rachis 13

12. Leaflets often glaucous, the ultimate lobes often >2 mm wide. . . . .*P. glaucophylla*
12. Leaflets not glaucous; ultimate segments <2 mm wide . . . .*P. multisecta*

13. Hairs of lower stem spreading and wrinkled; plants of dry, exposed habitats . . . . *P. ovina*
13. Hairs of lower stem straight and appressed; plants of moist meadows . . . . *P. plattensis*

14. At least some leaves palmately divided into ≥5 leaflets 15
14. Leaves all with only 3 (rarely 5) leaflets 25

15. Leaves covered with felt-like tomentum beneath 16
15. Leaves glabrous to sericeous beneath, but not tomentose 20

16. Leaves with inrolled margins and deep green upper surfaces. . . . .*P. argentea*
16. Leaf margins flat; upper surfaces green but not dark green. 17

17. Some basal leaves with only 3 leaflets. . . . . *P. arenosa*
17. Basal leaves with 5 to 7 leaflets 18

18. Stems >10 cm; leaflets >2 cm long . . . . *P. gracilis*
18. Stems <15 cm; leaflets ≤2 cm long 19

19. Styles ≤1.2 mm long. . . . . *P. rubricaulis*
19. Styles ≥1.5 mm long. . . . . *P. concinna*

20. Taprooted annual or biennial; petals shorter than the sepals. . . . . *P. rivalis*
20. Perennial, often with a branched, woody caudex; petals longer than sepals. 21

21. Anthers ≥0.8 mm long; leaflets of many leaves >2 cm long 22
21. Anthers <0.8 mm long; leaflets often <2 cm long 23

22. Stems and petioles hispid with straight, stiff, perpendicular hairs . . . .*P. recta*
22. Stems and petioles with curly or appressed hairs . . . . *P. gracilis*

23. Leaflets often glaucous, the ultimate lobes often >2 mm wide. . . . .*P. glaucophylla*
23. Leaflets not glaucous or glaucous bloom obscured by hairs; ultimate segments <2 mm wide 24

24. Stems somewhat prostrate; pedicels ca. recurved in fruit . . . .*P. multisecta*
24. Stems ascending to erect; pedicels remaining straight in fruit. . . . .*P. glaucophylla*

25. Leaves glabrous or nearly so; inflorescence glandular . . . . *P. flabellifolia*
25. Leaves variously hairy, at least beneath; inflorescence often not glandular. 26

26. Leaves tomentose beneath 27
26. Leaves variously hairy but not tomentose 29

27. Lower stem and petioles hirsute and puberulent, lacking felt-like tomentum . . . .*P. hookeriana*
27. Lower stems and petioles tomentose. 28

28. Petioles and lower stem with felt-like tomentum; other longer hairs lacking . . . . *P. nivea*
28. Petioles and lower stem with mostly straight, long hairs in addition to tomentum . . . .*P. uniflora*

29. Plants alpine; stems <10 cm high; petals longer than sepals. . . . . *P. hyparctica*
29. Plants of valleys to montane elevations; stems usually >10 cm; petals smaller than sepals. 30

30. Stem and leaves glandular. . . . . *P. biennis*
30. Stem and leaves villous or hirsute but not glandular 31

31. Lower stem stiffly hirsute . . . . *P. norvegica*
31. Lower stem villous . . . .*P. rivalis*

Potentilla anserina L. [*Argentina anserina* (L.) Rydb.] Silverweed Stoloniferous, stemless with a growth form similar to strawberry. **Leaves** basal, arising singly or in tufts from stolon nodes. **Leaf blades** oblong, 5–20 cm long, divided into 15 to 25 oblanceolate, dentate leaflets, glabrous to sparsely hairy above, white-tomentose beneath. **Inflorescence** of solitary flowers on long, sericeous peduncles. **Flowers**: sepals spreading, 2–5 mm long; bracteoles as long or longer; petals yellow, 5–9 mm long. **Achenes** lightly wrinkled, ca. 2 mm long; styles attached laterally. Wet or vernally wet meadows, margins of streams, ponds, wetlands; plains, valleys, montane. Circumboreal south to CA, AZ, NM, NE, IN, NY. Our plants are ssp. *anserina*. (p.278)

Potentilla arenosa (Turcz.) Juz. [*P. hookerana* Lehm. misapplied] Perennial from a branched caudex clothed in old leaf bases. **Stems** scapose or with 1 or 2 leaf-like bracts, erect to ascending, sericeous below. **Leaf blades** ovate with 3 to 5 ovate, coarsely dentate to pinnately lobed leaflets, 5–15 mm long, white-tomentose beneath, green and sparsely sericeous above. **Inflorescence** an open, 2- to 5-flowered cyme, sericeous, often also with short, curled hairs. **Flowers**: sepals lanceolate, 3–5 mm long; bracteoles shorter and narrower; petals yellow, 3–5 mm long, barely longer than the sepals. **Achenes** smooth, ca. 1.5 mm long. Stony soil of limestone fellfields; alpine. Circumpolar south to UT, CO.
    This species is part of a complex with *Potentilla nivea* and *P. uniflora*.

Potentilla argentea L. Perennial from a woody, branched caudex. **Stems** ascending, sericeous and hirsute, 10–30 cm. **Leaf blades** suborbicular, palmately divided into 5 to 9 oblanceolate, lobed leaflets, 1–2 cm long, white-tomentose beneath, deep green and sparsely hairy above with inrolled margins. **Inflorescence** a hemispheric cyme. **Flowers**: sepals ovate, 2–3 mm long; bracteoles ca. as long; petals yellow, 2–4 mm long. **Achenes** wrinkled, <1 mm long. Often compacted soil of roadsides, trails, pastures, around lakes, habitations; valleys. Introduced to much of north-temperate N. America; native to Europe.

Potentilla biennis Greene Taprooted annual or biennial. **Stems** erect, simple or branched, villous, glandular, 10–40 cm. **Leaf blades** ovate with 3 obovate, sharply lobed, glandular-hairy leaflets, 1–2 cm long. **Inflorescence** a leafy, glandular-hairy cyme with ascending branches. **Flowers**: sepals deltoid, 2–4 mm long; the bracteoles nearly as large; petals yellow, 1–2 mm long, smaller than the sepals. **Achenes** smooth, <1 mm long. Moist, often disturbed soil on the margins of streams, lakes, wetlands; plains, valleys. BC to SK south to CA, AZ, CO.

Potentilla brevifolia Nutt. Perennial from a long-branched rootstock, clothed in old leaves. **Stems** ascending, glandular, 5–15 cm. **Leaf blades** cordate-ovate, 10–15 mm long, pinnately divided into 3 to 7 deeply lobed, fan-shaped, glandular-puberulent leaflets. **Inflorescence** a several-flowered, glandular cyme with ascending branches. **Flowers**: sepals ovate, 3–4 mm long with shorter, lanceolate bracteoles; petals yellow, 5–6 mm long, retuse. **Achenes** smooth, 1–1.5 mm long. Granitic talus, cool rocky slopes; alpine in Madison Co. OR to MT south to NV, WY. (p.278)

Potentilla concinna Richardson Perennial from a branched caudex clothed in old leaf bases. **Stems** prostrate to ascending, sericeous, 2–10 cm. **Leaf blades** cordate-orbicular to ovate, palmately divided into usually 5-lobed, pinnately toothed leaflets 5–20 mm long, sparsely to densely sericeous above, white-tomentose beneath. **Inflorescence** a few-flowered, sericeous, cyme with spreading branches. **Flowers**: sepals broadly lanceolate, 5–7 mm long with narrower bracteoles

3–4 mm long; petals yellow, 5–8 mm long, truncate-tipped. **Achenes** smooth, ca. 2 mm long. Sparsely vegetated soil of exposed grassland ridges, slopes. AB to MB south to CA, AZ, CO, SD. (p.278)

1. Leaflets divided at least halfway to the midvein . . . . .var. *divisa*
1. Leaflets divided less than halfway to the midvein. . . . . var. *concinna*

Potentilla concinna var. *concinna* occurs plains, valleys, montane; *P. concinna* var. *divisa* Rydb. occurs plains, valleys. *P. ovina* is similar and occurs in similar habitats, but the leaves are not tomentose beneath.

Potentilla drummondii Lehm. Perennial from a branched caudex. **Stems** prostrate to ascending, glabrous to appressed-hairy, 8–25 cm. **Leaf blades** oblanceolate, 2–10 cm long, pinnately divided into 5–9 obovate, palmately lobed, nearly glabrous leaflets. **Inflorescence** a few-flowered, sericeous cyme with ascending branches. **Flowers**: sepals lanceolate, 5–9 mm long; bracteoles shorter; petals yellow, 5–7 mm long. **Achenes** 1.5–2 mm long, smooth. Rocky soil of meadows, open forest; subapine. AK south to CA, ID, MT. Known from Lincoln Co.

Potentilla flabellifolia Hook. ex Torr. & A.Gray Perennial from a branched, spreading caudex. **Stems** erect, glabrous to puberulent, 10–30 cm. **Leaf blades** reniform with 3 obovate, shallowly lobed, nearly glabrous leaflets 1–3 cm long. **Inflorescence** a few-flowered, glandular-pubescent cyme. **Flowers**: sepals lanceolate 3–6 mm long; bracteoles shallowly lobed, shorter than the sepals; petals yellow, 7–11 mm long. **Achenes** ca. 1.5 mm long, smooth. Moist to wet non-calcareous soil of meadows, open forest, often around lakes, wetlands; subalpine. BC, AB south to CA, OR, ID, MT.

Potentilla glaucophylla Lehm. [*P. diversifolia* Lehm.] Perennial from a branched caudex. **Stems** ascending to nearly prostrate, sparsely hairy (rarely sericeous), 10–40 cm. **Leaf blades** cordate-orbicular to ovate, palmately (rarely pinnately) divided into 5 to 7 oblong, toothed to lobed, glabrous to sparsely hairy, usually glaucous leaflets, 1–3 cm long. **Inflorescence** an open, sericeous cyme with spreading branches. **Flowers**: sepals narrowly deltoid, 3–6 mm long; bracteoles narrower and shorter; petals yellow, 6–9 mm long. **Achenes** ca. 1.5 mm long, weakly wrinkled. Grasslands, turf, fellfields; subalpine, alpine. YT to SK south to CA, UT, NM. (p.282)

1. Leaflets dentate less than halfway to the midvein . . . . var. *glaucophylla*
1. Leaflets deeply divided into linear lobes. . . . . var. *perdissecta*

Potentilla glaucophylla var. *glaucophylla* is ubiquitous at high elevations; *P. glaucophylla* var. *perdissecta* (Rydb.) Soják occurs mostly on calcareous soils in southwest cos. Extreme forms of the two vars. are distinct, but intermediate specimens are common.

Potentilla gracilis Douglas ex Hook. [*P. pulchella* R.Br., *P. pectinisecta* Rydb.] Perennial from a mostly branched caudex. **Stems** erect or ascending, glabrate to pubescent, subscapose, 10–60 cm. **Leaf blades** suborbicular, palmately divided into 5 to 9 oblanceolate, dentate to deeply lobed, glabrous to tomentose leaflets 2–8 cm long. **Inflorescence** a many-flowered, sericeous, sometimes glandular, open cyme with ascending branches. **Flowers**: sepals lanceolate, 3–8 mm long; bracteoles narrower and shorter; petals yellow, 5–10 mm long. **Achenes** smooth, 1–1.5 mm long. Meadows, grasslands, steppe; plains, valleys to subalpine. AK to MB south to CA, AZ, NM.

1. Leaves glabrate to sericeous beneath but not woolly-tomentose 2
1. Leaves white to gray, woolly-tomentose beneath. 3

2. Calyx densely glandular, sparsely hairy . . . . var. *brunnescens*
2. Calyx sparsely glandular or eglandular, often sericeous . . . . var. *fastigiata*

3. Leaflets dissected less than halfway to the midvein. . . . . .var. *pulcherrima*
3. Leaflets dissected more than halfway to the midvein. 4

4. Leaflets pinnately cut nearly to the midvein into linear teeth . . . . var. *elmeri*
4. Leaflets cut ca. 3/4-way to the midvein into narrowly lanceolate teeth. . . . . var. *flabelliformis*

Potentilla gracilis var. *fastigiata* (Nutt.) S.Watson [*P. gracilis* var. *permollis* (Rydb.) C.L.Hitchc., *P. gracilis* var. *glabrata* (Lehm.) C.L.Hitchc.] occurs in moist grasslands, meadows, open forest, plains, valleys, montane; *P. gracilis* var. *brunnescens* (Rydb.) C.L.Hitchc. is found in grasslands, valleys, montane. *Potentilla gracilis* var. *flabelliformis* (Lehm.) Nutt. ex Torr. & A.Gray is found in grasslands, steppe, moist meadows, valleys, montane; it is transitional to var. *pulcherrima*; *P. gracilis* var. *pulcherrima* (Lehm.) Fernald is found in moist to wet meadows, rarely rocky, dry sites, valleys to subalpine, mainly in southwest cos.; it is transitional to var. *elmeri*; *P. gracilis* var. *elmeri* occurs in grasslands

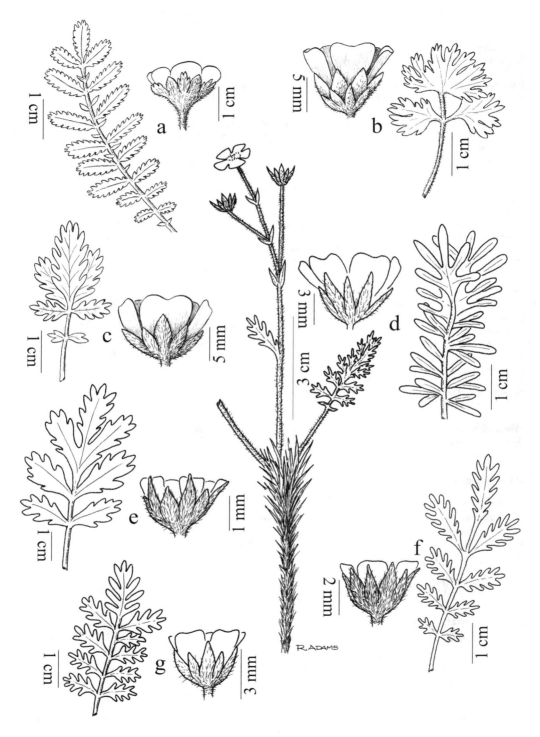

Plate 46. **Potentilla**. **a.** *P. anserina*, **b.** *P. breviflora*, **c.** *P. concinna* var. *concinna*, **d.** *P. ovina*, **e.** *P. paradoxa*, **f.** *P. pensylvanica*, **g.** *P. plattensis*

to wetlands, valleys, montane. The size of flowers is variable within varieties. The extremes are distinctive, the vars. are connected by intergrading forms. Hairs on the stem and petioles of the similar *P. recta* are stiffly spreading.

Potentilla hippiana Lehm. [*P. effusa* Douglas ex Lehm.] Perennial from a thick, branched caudex. **Stems** erect to ascending, tomentose, 15–40 cm. **Leaf blades** oblong, pinnately divided into 5 to 13 oblong, lobed leaflets, 1–3 cm long, sericeous and/or tomentose above, tomentose beneath. **Inflorescence** many-flowered, sericeous and tomentose, open with ascending branches. **Flowers**: sepals lanceolate, 3–5 mm long; bracteoles narrower and shorter; petals yellow, 4–7 mm long, barely longer than the sepals. **Achenes** smooth, ca. 1.5 mm long. Stony or sandy soil of grasslands, steppe; plains, valleys, montane, rarely higher. BC to SK south to AZ, NM, NE.

Both var. *hippiana*, with a sericeous calyx, and var. *effusa* (Dougl. ex Lehm.) Dorn with a tomentose calyx are reported for MT, but the majority of our specimens are intermediate.

Potentilla hookeriana Lehm. Perennial from a branched caudex. **Stems** several, erect to ascending, 5-30 cm. **Leaves**: petioles with long, spreading hairs and sparsely puberulent or glandular; blades of 3(5) ovate, coarsely dentate to pinnately lobed leaflets, 5-20 mm long, white-tomentose beneath, with stiff long hairs above and beneath. **Inflorescence** an open, 2- to 7-flowered cyme. **Flowers**: sepals lanceolate, 3–8 mm long, stiff-hairy, glandular; bracteoles shorter and narrower; petals yellow, 4–5 mm long, barely longer than the sepals. **Achenes** smooth, ca. 1.5 mm long. Limestone cliffs, bluffs, fellfields; valleys, alpine. MT south to CO.

This species is generally considered an arctic-alpine species, but many of our specimens come from the Clark Fork River canyon in Granite Co.

Potentilla hyparctica Malte [*P. nana* Willd. ex Schltdl. misapplied] Perennial with a long-branched caudex clothed in old stipules. **Stems** erect, scapose, strigose, 1–2 cm. **Leaf blades** ovate with 3 obovate, dentate, sparsely hairy leaflets, 5–10 mm long. **Inflorescence** glandular, of 1 or 2 flowers. **Flowers**: sepals ovate, 3–4 mm long; bracteoles shorter; petals yellow, 4–6 mm long. **Achenes** smooth, ca. 1.5 mm long. Stony soil of moist turf; alpine. Circumpolar south to BC, WY. Our plants are ssp. *elatior* (Abrom.) Elven & D.F.Murray. Collected in Carbon Co.

Our plants are "morphologically deviant" from more northern populations.[124]

Potentilla macounii Rydb. [*P. concinna* Richardson var. *macounii* (Rydb.) C.L.Hitchc.] Perennial with a branched caudex clothed in old leaf bases. **Stems** prostrate to ascending, sericeous, 3–10 cm. **Leaf blades** cordate-ovate, pinnately divided into 5 to 7 pinnately toothed leaflets, 5–15 mm long, sparsely to densely sericeous above, white-tomentose beneath. **Inflorescence** a few-flowered, sericeous cyme with spreading branches. **Flowers**: sepals broadly lanceolate, 4–7 mm long with narrower bracteoles 3–5 mm long; petals yellow, 5–8 mm long, truncate. **Achenes** smooth, ca. 2 mm long. Sandy to gravelly, shallow, often calcareous soil of grasslands, woodlands, outcrops; valleys, montane. Endemic to MT and adjacent AB and WY.

Leaflets of *Potentilla concinna* are all attached at the tip of the petiole; some leaflets of *P. macounii* are affixed well below the tip.

Potentilla multisecta (S.Watson) Rydb. [*P. diversifolia* Lehm. var *multisecta* S.Watson] Perennial from a branched caudex. **Stems** nearly prostrate to ascending, sparsely hairy, 4–10 cm. **Leaf blades** ovate,1–3 cm long, palmately to subpinnately divided into 5 to 7 oblong, lobed, glabrate to sparsely sericeous leaflets. **Inflorescence** a few-flowered, sericeous cyme. **Flowers**: sepals lanceolate, 3–5 mm long; bracteoles narrower and shorter; petals yellow, 5–7 mm long. **Achenes** ca. 2.5 mm long, smooth. Stony soil of exposed slopes, fellfields; upper subalpine, alpine. ID, MT south to NV, UT.

Potentilla nivea L. Mat-forming perennial from a thick, branched caudex clothed in old leaf bases. **Stems** erect to ascending, tomentose and appressed-hairy, 1–10 cm. **Leaf blades** ovate with 3 elliptic, coarsely dentate to pinnately lobed leaflets, 5–12 mm long, white-tomentose beneath, sparsely long-hairy above. **Inflorescence** an open, 1- to 3-flowered cyme. **Flowers**: sepals lanceolate, 4–5 mm long; bracteoles as long; petals yellow, 4–6 mm long, equal to much longer than the sepals. **Achenes** smooth, ca. 1.5 mm long. Stony soil of fellfields, turf, rockslides; upper subalpine, alpine. Circumpolar south to BC, NV, UT, CO. (p.282)

Petals often have an orange spot that fades with drying. *Potentilla uniflora* (sensu lato) lacks tomentum on the petioles and lower stem. Plants with very sparse tomentum in addition to being sericeous are referred to *P. arenosa* (Turcz.) Juz. and are considered more closely related to *P. nivea*.[124]

Potentilla norvegica L. Taprooted biennial or perennial from a simple or branched caudex. **Stems** erect, hirsute, 15–60 cm. **Leaf blades** reniform with 3 obovate, coarsely dentate, sparsely hairy leaflets, 1–5 cm long. **Inflorescence** a leafy, narrow, several-flowered cyme. **Flowers**: sepals narrowly lanceolate, 3–5 mm long; bracteoles similar in size and greener; petals light yellow, 2–4 mm long, shorter than the sepals. **Achenes** net-veined, ca. 1 mm long. Margins of wetlands, roadsides, gardens; valleys, montane. Circumboreal south to much of temperate N. America; possibly introduced in the Pacific Northwest.[187]

Potentilla ovina J.M.Macoun Perennial from a branched caudex clothed in old stipules, sometimes forming loose mats. **Stems** ascending to lax, with sparse, wrinkled, spreading hair 3–15 cm. **Leaf blades** sericeous, linear, 1–5 cm long, pinnately divided into 9 to 21 deeply lobed leaflets, the ultimate segments linear. **Inflorescence** few-flowered, appressed-hairy, open with spreading branches. **Flowers**: sepals lanceolate, 3–5 mm long; bracteoles narrower and shorter; petals yellow, 4–6 mm long, longer than the sepals. **Achenes** smooth, ca. 1.5 mm long. Sparsely vegetated, exposed ridges, fellfields; plains, valleys to alpine. BC to SK south to CA, UT, MN. (p.278)

   This plant is more common in calcareous soil and replaces *Geum rossii* var. *turbinatum* as a dominant fellfield plant in the limestone ranges of southwest MT.[94] *Potentilla macounii* has fewer, less divided leaflets; *P. plattensis* is found in wet habitats.

Potentilla paradoxa Nutt. Taprooted, annual, biennial or short-lived perennial. **Stems** ascending to prostrate, sparsely hirsute, 10–30 cm. **Leaf blades** oblong, glabrate to hirsute, pinnately divided into 5 to 9 narrowly obovate, dentate to pinnately lobed leaflets, 1–2 cm long. **Inflorescence** a leafy, compact, villous cyme. **Flowers**: sepals deltoid, 3–4 mm long; bracteoles nearly as long and wide; petals yellow, 1–3 mm long, retuse. **Achenes** shallowly wrinkled, <1 mm long with a lateral wedge-shaped wing. Wet or vernally wet soil along margins of lakes, ponds, rivers; plains, valleys, montane. Throughout much of temperate N. America. (p.278)

Potentilla pensylvanica L. [*P. bipinnatifida* Douglas, *P. litoralis* Rydb.] Perennial from a woody, branched caudex clothed in old stipules. **Stems** ascending to erect, pubescent to tomentose, 10–30 cm. **Leaf blades** oblong, 3–8 cm long, pinnately divided into 5 to 11 lanceolate, pinnately lobed leaflets, sparsely hairy above, tomentose to sericeous beneath. **Inflorescence** a hirsute, tomentose, glandular, compact cyme. **Flowers**: sepals lanceolate, 3–6 mm long; bracteoles equal or longer; petals yellow, 4–7 mm long, ca. as long as the sepals. **Achenes** smooth, granular on top, ca. 1.5 mm long. Stony soil of grasslands, steppe, meadows; plains, valleys to rarely lower subalpine. AK to QC south to CA, NM, NE, MN. (p.278)

   Forms with fewer, less dissected leaflets and no tomentum occur in the same populations as more typical forms. *Potentilla bipinnatifida* and *P. litoralis* have been segregated from *P. pensylvanica* sensu lato based on characters of vesiture, leaf shape and calyx bracteoles,[124] but the characters do not seem to be correlated in our area.[152] One collection from above treeline in Teton Co. (*Lesica 4052*, MONTU) approaches *P. jepsonii* Ertter but is similar in most regards to some low-elevation *P. pensylvanica* that recognition of the former species in our area seems unwarranted at this time. See *P. hippiana*.

Potentilla plattensis Nutt. Perennial from a branched caudex. **Stems** ascending to lax, sparsely appressed-hairy, 7–20 cm. **Leaf blades** glabrate to sparsely hairy, linear, 2–5 cm long, pinnately divided into 11 to 23 deeply lobed leaflets, the ultimate segments linear. **Inflorescence** several-flowered, appressed-hairy, open with spreading to recurved branches. **Flowers**: sepals lanceolate, 3–5 mm long; bracteoles narrower and shorter; petals yellow, 4–7 mm long, longer than the sepals. **Achenes** smooth, ca. 1.5 mm long. Moist, alkaline meadows, often on hummocks with *Dasiphora fruticosa*; plains, montane, lower subalpine. AB to MB south to AZ, NM, CO, SD. (p.278)

   The similar *Potentilla ovina* has stems with loose, wrinkled hairs and occurs in dry, exposed habitats.

Potentilla recta L. SULPHUR CINQUEFOIL Perennial from a simple to branched caudex. **Stems** erect, hirsute to hispid, 20–50 cm. **Leaf blades** suborbicular, palmately divided into 5 to 7 lanceolate, dentate leaflets, 2–7 cm long, spreading-hairy. **Inflorescence** open, flat-topped, hispid. **Flowers**: sepals lanceolate, strongly veined, 4–7 mm long; bracteoles similar; petals light yellow, 6–10 mm long, longer than the sepals. **Achenes** ridged with thickened margins, ca. 1 mm long. Grasslands, meadows, pastures, disturbed forests; plains, valleys, montane. Introduced to much of north-temperate N. America; native to Eurasia.

   Potentilla gracilis has pubescence that is more appressed or wrinkled, not projecting at right angles from the stems and petioles.

Potentilla rivalis Nutt. Taprooted annual or biennial. **Stems** erect, branched, villous, 10–30 cm. **Leaf blades** broadly obovate, pinnately divided into 3 to 5 oblong, coarsely dentate, villous leaflets, 1–3 cm long. **Inflorescence** a leafy, villous, many-flowered cyme. **Flowers**: sepals deltoid, 2–3 mm long; bracteoles narrower, ca. as long; petals light yellow, 1–2 mm long, shorter than the sepals. **Achenes** smooth, <1 mm long. Pond margins, disturbed moist meadows; plains, valleys. BC to SK south to CA, AZ, NM, TX, MO.

Potentilla norvegica has hirsute stems; *P. biennis* has glandular stems.

Potentilla rubricaulis Lehm. [*P. concinna* Richardson var. *rubripes* (Rydb.) C.L.Hitchc., *P. modesta* Rydb.] Perennial from a branched caudex clothed in old leaf bases. **Stems** prostrate to ascending, sericeous with wrinkled hairs, 5–15 cm. **Leaf blades** cordate-orbicular to ovate, subpinnately divided into 5 to 7 deeply lobed leaflets 5–12 mm long, glabrate to sericeous above, white-tomentose beneath. **Inflorescence** a few-flowered, sericeous, cyme with spreading branches. **Flowers**: sepals broadly lanceolate, 3–5 mm long with shorter, narrower bracteoles; petals yellow, 5–7 mm long, truncate. **Achenes** smooth, ca. 2 mm long. Sparsely vegetated, stony soil of exposed ridges, slopes, usually on limestone; subalpine, alpine. AK to Greenland south to NV, UT, CO.

Potentilla rubricaulis (sensu lato) is thought to be of hybrid origin and encompasses numerous forms that may reproduce asexually and have been divided into many segregate taxa.[124] Our material has not been analyzed in relation to Eurasian plants, so I have chosen to maintain the broad concept until further analyses are completed.

Potentilla uniflora Ledeb. [*P. ledebouriana* A.E.Porsild, *P. subgorodkovii* Jurtzev] Mat-forming perennial from a thick, branched caudex clothed in old leaf bases. **Stems** scapose, erect to ascending, closely tomentose and with longer straight to twisted hairs. **Leaf blades** reniform with 3 ovate, coarsely dentate to pinnately lobed leaflets, 5–15 mm long, white-tomentose beneath, green and long-hairy above. **Inflorescence** an open, tomentose, 1- to 3-flowered cyme. **Flowers**: sepals lanceolate, 3–5 mm long; bracteoles nearly as long, but narrower; petals yellow, 4–8 mm long, longer than the sepals. **Achenes** smooth, ca. 1.5 mm long. Stony, sparsely vegetated soil of exposed slopes, ridges; upper subalpine, alpine. AK to QC south to OR, CO; Siberia.

This is part of a difficult circumpolar complex, probably due in large part to hybridization and apomixis. European researchers have segregated out numerous taxa; our plants would probably be referred to *Potentilla subgorodkovii*.[124] At this point I prefer to use the more inclusive and recognizable name. See *P. nivea*, *P. arenosa*.

Prunus L. Cherry, plum

Shrubs or small trees. **Bark** often with white spot-like lenticels. **Leaves** stipulate, simple with dentate margins, often with small glands at the top of the petiole, glaucous beneath. **Inflorescence** racemose, cymose, or flowers solitary. **Flowers** perfect, perigynous; sepals 5; petals 5, white, short-clawed; stamens 20 to 30; style 1; fruit a 1-seeded drupe.

Prunus mahaleb L. (sweet cherry) and *P. cerasus* L. (sour cherry), occasionally escape cultivation, but I am not aware that they ever reproduce in the wild.

1. Inflorescence an elongate raceme with >12 flowers . . . . *P. virginiana*
1. Inflorescence corymbose or umbellate with <12 flowers 2

2. Flowers on pedicels attached directly to the stem 3
2. Inflorescence a pedunculate corymb 5

3. Plants prostrate; leaves 1-3 cm long . . . . *P. pumila*
3. Plants erect shrubs or trees; some leaves >3 cm long. 4

4. Some branches thorn-tipped; calyx pubescent; dried fruit >1 cm long. . . . . *P. americana*
4. Stems not thorny; calyx glabrous; fruit ≤1 cm long. . . . . *P. cerasus*

5. Corymb with a leaf-like bract near the base . . . .*P. mahaleb*
5. Corymb ebracteate. 6

6. Leaf tips acuminate . . . . *P. pensylvanica*
6. Leaves rounded or acute but not long-tapered to the tip . . . .*P. emarginata*

Prunus americana Marshall Wild plum Shrub to 5 m, spreading by rhizomes and forming thickets. **Twigs** glabrous to puberulent, reddish, becoming gray, forming thorns. **Leaf blades** ovate to lanceolate, 5–10 cm long, acuminate, glabrous beneath. **Inflorescence** 2 to 5 pedicellate flowers on short branch tips before leaves appear. **Flowers**: hypanthium obconic, glabrous, ca. 3

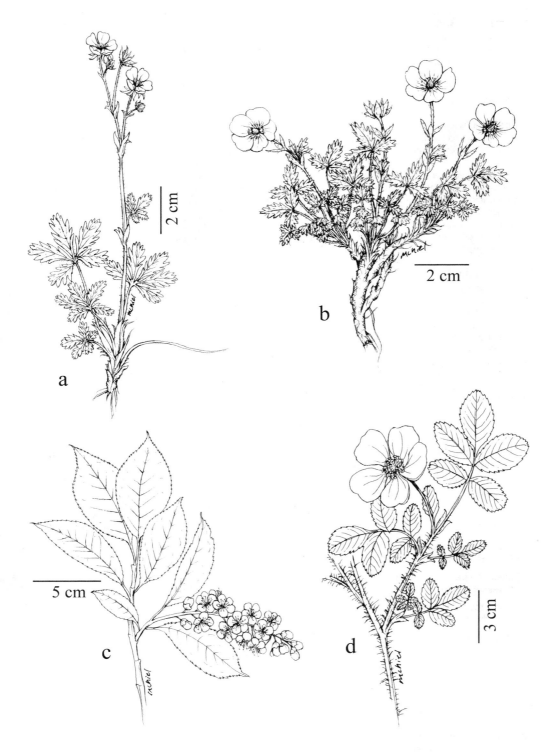

Plate 47. a. *Potentilla glaucophylla* var. *glaucophylla*, b. *Potentilla nivea*, c. *Prunus virginiana*
d. *Rosa acicularis*

mm long; sepals lanceolate, puberulent, glandular, 3–4 mm long, reflexed; petals 5–7 mm long. **Drupe** a plum, 2–3 cm long, yellowish to purple, glaucous. Riparian thickets, woodlands, cool slopes; plains, valleys. SK to QC south to AZ, NM, LA, FL.

Populations west of the Continental Divide in MT may have been introduced by Native Americans.

Prunus emarginata (Douglas) Eaton Shrub or small tree 2–5 m. **Twigs** purple, glabrous. **Leaf blades** oblong to oblanceolate, rounded to the tip, 2–8 cm long, glabrous or puberulent beneath. **Inflorescence** a hemispheric, 5- to 8-flowered corymb, appearing with the leaves. **Flowers**: hypanthium 2–3 mm long, obconic, glabrous or hairy; sepals ovate, reflexed, 2–3 mm long; petals 4–7 mm long. **Drupe** a red to black cherry, 6–8 mm long when dry. Open forest, thickets, along streams; valleys, montane. BC, MT south to CA, NV, UT.

Most of our plants would be considered var. *emarginata*; however, plants from Lincoln and Sanders cos. have pubescent calyces and leaf undersides and would be assigned to var. *mollis* (Douglas) A.E.Murray.

Prunus pensylvanica L.f. PIN CHERRY Shrub or small tree 3–5 m, similar to *P. emarginata*. **Twigs** reddish, glabrous. **Leaf blades** lanceolate to elliptic, short-acuminate, 2–6 cm long, glabrous. **Inflorescence** a hemispheric, 4- to 6-flowered corymb appearing with the leaves. **Flowers**: hypanthium obconic, 2–3 mm long, glabrous; sepals ovate, reflexed, 2–3 mm long; petals 5–7 mm long, sometimes hairy at the base. **Drupe** a red cherry, 4–7 mm long. Open forest, woodlands, thickets; plains, valleys to lower subalpine. BC to NL south to MT, CO, SD, LA, VA.

Missoula Co. specimens were collected on roadsides. See *P. emarginata*.

Prunus pumila L. SAND CHERRY Prostrate shrubs, spreading by subsurface stems. **Twigs** glabrous, red, becoming gray. **Leaf blades** elliptic, 1–3 cm long, acute, glabrous with entire margins below, glaucous beneath. **Inflorescence** 2 to 4 pedicellate flowers from terminal or subterminal buds, appearing with the leaves. **Flowers**: hypanthium cup-shaped, 2–4 mm long, glabrous; sepals oblong, 1–2 mm long, glandular-dentate; petals 4–6 mm long. **Drupe** a dark purple cherry 8–9 mm long dried. Stony soil of exposed ridges, upper slopes; plains. MT to NB south to CO, NE, KS, NC. Known from Fallon and McCone cos. Our plants are var. ***besseyi*** (Bailey) Waugh. which is restricted to the Great Plains.

Prunus virginiana L. CHOKECHERRY Shrub or rarely a tree to 6 m. **Twigs** reddish-brown, glabrous. **Leaf blades** elliptic, short-acuminate, 4–12 cm long, glabrous. **Inflorescence** a cylindric, many-flowered raceme 5–12 cm long. **Flowers**: hypanthium cup-shaped, 1–2 mm long, glabrous; sepals deltoid, glandular-dentate, ≤1 mm long; petals 3–6 mm long. **Drupe** a red cherry becoming black, 5–8 mm long dry. Riparian thickets, forests, pine and ash woodlands, rocky slopes, stony soil of grasslands; plains, valleys, montane. BC to NL south to CA, NM, TX, TN, NC. Our plants are var. *melanocarpa* (A.Nelson) Sarg. (p.282)

Purshia DC. ex Poir.

Purshia tridentata (Pursh) DC. BITTERBRUSH Intricately branched shrub to 2 m. **Stems** rigid, erect to ascending; bark brown, tomentose, becoming gray. **Leaves** fascicled on tips of short shoots, simple, sessile, fan-shaped, 7–18 mm long, shallowly 3-lobed at the tip, tomentose, sparsely so above, dense beneath, margins inrolled. **Inflorescence** of solitary, sessile flowers from tips of short shoots. **Flowers** perfect, perigynous; hypanthium obconic, tomentose, glandular, 4–6 mm long; sepals 5, ovate, 2–3 mm long; petals 5, clawed, yellow, 4–7 mm long; stamens ca. 25; pistil 1. **Fruit** a pubescent, ellipsoid achene, 9–14 mm long, partly enclosed in the hypanthium; style ca. 1 mm long. Stony or sandy soil of grasslands, steppe, open ponderosa pine forest; valleys, montane. BC, MT south to CA, AZ, NM, SD.

The type locality is in Powell Co.

Pyrus L. Pear, apple

Pyrus malus L. [*Malus pumila* Miller] APPLE Small tree to 10 m. **Stems**: twigs purplish; bark grayish. **Leaves** simple, petiolate, stipulate; the blades 1–10 cm long, elliptic to ovate, serrate, tomentose beneath, sparsely so above. **Inflorescence** umbellate. **Flowers** perfect, perigynous, 5-merous; hypanthium saucer-shaped; sepals 5–10 mm long; petals white, 12–25 mm long; stamens numerous; styles ca. 5. **Fruit** a glabrous pome usually 2–4 cm long with ca. 10 seeds. Roadsides, fields; valleys, plains; collected in Missoula and Stillwater cos. The apple of commerce introduced throughout temperate N. America; native to Eurasia.

Rosa L. Rose

Shrubs. **Stems** spiny, branched above. **Leaves** cauline, petiolate, stipulate, pinnately divided into 7 to 11 dentate leaflets; stipules green, leaf-like, adherent to the petiole base. **Inflorescence** a small cyme or flowers solitary. **Flowers** perfect, perigynous; hypanthium almost completely enclosing the ovary; sepals 5; petals pink to rose, shallowly lobed at the tip; stamens numerous; pistils >10. **Fruit** a swollen, pulpy hypanthium (hip) enclosing the hairy achenes.

Characters separating the species are not as unambiguous as the following key might suggest. Good representation of inflorescences and second-year stems is needed for confident determination.

1. Stems with stout, hooked prickles just below leaf attachments
   (fine straight prickles may also be present)..2
1. Stems bristly with fine straight prickles, but those below leaf nodes not noticeably different............3

2. All, or nearly all, flowers and fruits solitary ....*R. nutkana*
2. Many inflorescences with ≥2 flowers ....*R. woodsii*

3. Sepals not persistent on fruit; petals mostly <15 mm long.........................*R. gymnocarpa*
3. Sepals persistent on fruit; petals usually >15 mm long 4

4. Stems usually <40 cm high; plants of grasslands ....*R. arkansana*
4. Stems mostly >40 cm; plants of forests ....*R. acicularis*

Rosa acicularis Lindl. Prickly rose [*R. sayi* Schwein.] **Stems** lax, 50–150 cm, densely to sparsely covered with fine, straight prickles; those at nodes similar. **Leaves** with 5 to 7, narrowly elliptic leaflets, pubescent on veins beneath, 1.5–5 cm long; the rachis glabrate to glandular-pubescent; **Flowers** mostly solitary; hypanthium glabrous, 3–6 mm long; sepals 15–20 mm long, spreading, dilated at the tip, puberulent, sometimes stipitate-glandular, persistent in fruit; petals 15–30 mm long. **Hip** globose to pyriform, 1–2 cm long, reddish-purple. Moist to wet coniferous forest; valleys to lower subalpine. Circumboreal south to BC, CO, NE, MN, VT. Our plants are ssp. *sayi* (Schwein.) W.H.Lewis. (p.282)

Usually flowers earlier than *Rosa woodsii*; *R. nutkana* usually has larger, hooked, nodal prickles.

Rosa arkansana Porter Prairie rose Subshrub. **Stems** to 40 cm from rhizomes or horizontal roots, often partially dying back in winter, sparsely to densely covered with fine, straight to slightly curved prickles; those at nodes similar. **Leaves** with 7 to 11 narrowly elliptic leaflets, 1–4 cm long, glabrate above, pubescent beneath and on rachis. **Flowers** 1 to 4 in a corymb; hypanthium glabrous, 4–5 mm wide; sepals 12–15 mm long, usually stipitate-glandular, spreading, persistent in fruit; petals 15–25 mm long. **Hip** globose, 8–15 mm long, purplish-red. Grasslands; plains, valleys, montane. BC to MB south to NM, TX, MO.

The more common *Rosa woodsii* may occur in similar habitats but is usually taller and more shrub-like.

Rosa gymnocarpa Nutt. Baldhip rose **Stems** lax, 30–100 cm, usually sparsely and finely prickly; nodal prickles similar. **Leaves** with 5 to 9 elliptic leaflets, glabrous, glandular-dentate, 5–30 mm long; the rachis sparsely stipitate-glandular. **Flowers** usually solitary on glandular pedicles; hypanthium glabrous, 3–4 mm long; sepals 5–10 mm long, deciduous; petals 7–12 mm long. **Hip** elliptic, 8–12 mm long, red. Dry to wet coniferous forest; montane, lower subalpine. BC, MT south to CA, OR, ID.

Rosa nutkana C.Presl **Stems** 50–150 cm, with paired, large, hooked, nodal prickles, otherwise sparsely prickly. **Leaves** with 5 to 7 elliptic leaflets, 2–5 cm long; the rachis glandular, often puberulent. **Flowers** usually solitary at tips of new twigs; hypanthium glabrous to minutely glandular, 5–9 mm wide; sepals 15–35 mm long, glabrous to pubescent, usually stipitate-glandular, persistent, dilated at the tip; petals 20–35 mm long. **Hip** globose, red to purple, 12–20 mm long. Open forest, thickets, often along streams, ponds, lakes; montane. BC, MT south to CA, OR, UT, CO. Our plants are var. *hispida* Fernald. See *R. acicularis*.

Rosa woodsii Lindl. Woods' rose **Stems** 50–150 cm, sparsely to densely covered with fine prickles as well as larger, hooked to straight nodal prickles. **Leaves** with 5 to 9, ovate, sometimes glandular-dentate leaflets 1–4 cm long. **Flowers** ≥2 in small corymbs; hypanthium glabrous, 4–5 mm long; sepals 8–20 mm long, persistent, pubescent, especially on the margins, sometimes stipitate-glandular; petals 15–25 mm long. **Hip** globose, 6–15 mm long, red. Open forest, woodlands, riparian thickets, snow-catchment areas of grasslands, steppe; plains, valleys, montane. YT to QC south to CA, NM, TX.

Our most common species of moist, open sites and disturbed areas. Two vars. are sometimes recognized: var. *ultarmontana* (S.Watson) Jeps., with larger leaflets is the cordilleran ecotype, and var. *woodsii*, with smaller leaflets 1–2 cm long, is the Great Plains ecotype. This trend can be seen in our specimens, but it seems too weak to merit formal recognition.

Rubus L. Raspberry, blackberry, bramble

Shrubs or subshrubs. **Stems** erect or trailing, prickly in some species, biennial; first-year canes bear leaves; second-year canes bear leaves and flowers. **Leaves** cauline, petiolate, stipulate, palmately lobed or divided into 3(to 5) leaflets with toothed margins. **Inflorescence** a few-flowered cyme or solitary. **Flowers** perfect (unisexual in *R. ursinus*), perigynous; sepals 5, becoming reflexed; petals 5; stamens and pistils numerous. **Fruit** an aggregate of juicy, 1-seeded drupelets.

Berries in which the druplets cling to the fibrous receptacle are called blackberries or dewberries, while those that are thimble-like and come away from the receptacle when mature are raspberries.

1. Stems unarmed 2
1. Stems prickly 5

2. Stems erect, >20 cm; leaves lobed but not divided . . . . *R. parviflorus*
2. Subshrubs often with trailing stems, <20 cm high; leaves divided into 3 to 5 leaflets 3

3. Petals pink to rose; leaflets rounded at the tip . . . .*R. acaulis*
3. Petals white; leaflets acute. 4

4. Most leaves with 3 leaflets; plants stoloniferous. . . . .*R. pubescens*
4. Most leaves with 5 or seemingly 5 leaflets (lateral pair deeply divided); plants without stolons . . *R. pedatus*

5. Some leaflets lobed almost to the midvein. . . . .*R. laciniatus*
5. Leaflets dentate or shallowly lobed 6

6. Stems trailing; flowers unisexual (female with aborted anthers) . . . . *R. ursinus*
6. Stems erect to arching 7

7. Prickles straight and nearly terete . . . . *R. idaeus*
7. Main prickles flattened, curved or hooked 8

8. Sepals ovate, 4–6 mm long; fruit a blackberry . . . . *R. discolor*
8. Sepals lanceolate, long-attenuate, 6–10 mm long; fruit a raspberry . . . .*R. leucodermis*

Rubus acaulis Michx. Rhizomatous subshrub. **Stems** herbaceous, unarmed, glabrous to sparsely hairy, 5–20 cm. **Leaves** with 3 ovate leaflets, glabrous above, sparsely hairy beneath, 15–50 mm long; stipules narrowly ovate, entire. **Flowers** solitary on stem tips; hypanthium glabrous, 3–4 mm long; sepals narrowly lanceolate, 5–8 mm long, puberulent; petals erect, pink to red, 8–13 mm long. **Fruit** a red, globose raspberry ca. 1 cm long. Hummocks in *Sphagnum* fens, spruce swamps; montane. AK to NL south to BC, CO, MN.

Rubus discolor Weihe & Nees [*R. procerus* Boulay] HIMALAYAN BLACKBERRY Shrub. **Stems** arching or trailing, 1–2 m, pubescent, with stout, flattened, hooked prickles. **Leaves** with 3 to 5 lanceolate, coarsely dentate, ovate leaflets, 4–9 cm long, glabrous above, tomentose beneath. **Flowers** several in a prickly, open cyme at branch tips; hypanthium ca. 2 mm long, tomentose; sepals ovate, apiculate, 4–5 mm long; petals white, 6–8 mm long, longer than the sepals. **Fruit** a globose blackberry ca. 15 mm wide. Roadsides; valleys, montane. Introduced throughout much of temperate N. America; native to Eurasia. One location in Missoula Co.

Rubus idaeus L. RED RASPBERRY Shrubs. **Stems** erect or lax, glabrous to pubescent, 20–150 cm. **Leaves** with 3 to 5 lanceolate, coarsely dentate, ovate leaflets, 2–9 cm long, usually tomentose beneath. **Flowers** solitary or in few-flowered cymes from leaf axils; hypanthium stipitate-glandular, tomentose, 1–2 mm long; sepals acuminate-lanceolate, 6–11 mm long; petals white, spreading, 4–6 mm long. **Fruit** a hemispheric, red raspberry, 8–12 mm across. Open forest, woodlands, along streams, on brushy, often burned-over slopes, especially rock slides; plains, valleys to subalpine. Circumboreal south to OR, NM, SD, MO. (Pl.48)

1. Leaflets white-tomentose beneath . . . . var. *strigosus*
1. Leaflets green and glabrate beneath . . . . var. *peramoenus*

Rubus idaeus var. *strigosus* (Michx.) Maxim. is our common form, plains to subalpine; *R. idaeus* var. *peramoenus* (Greene) Fernald is uncommon in forests, valleys to montane, mainly in northwest cos.

Rubus laciniatus Willd. EVERGREEN BLACKBERRY Shrub. **Stems** ascending to lax, 1–3 m, with stout, flattened, recurved prickles. **Leaves** persistent, with (3)5 coarsely toothed to sharply lobed, ovate leaflets 2-6 cm long. **Flowers** in a hemispheric, several-flowered cyme; calyx lanate, prickly; sepals lanceolate, 8–15 mm long; petals pink, 9–14 mm long, 3-lobed. **Fruit** a globose to ovoid, black blackberry, 10–15 mm across. Escaped from cultivation in north-temperate N. America; introduced from Europe. Collected in Deer Lodge and Glacier cos.

Rubus leucodermis Douglas ex Torr. & A.Gray BLACK RASPBERRY Shrub. **Stems** erect to arching, 1–3 m, armed with stout, curved, flattened prickles. **Leaves** with 3 to 5 narrowly ovate leaflets, 15–70 mm long, dentate to shallowly lobed, densely white-tomentose beneath, glabrous above. **Flowers** several in a prickly, hemispheric cyme; hypanthium tomentose, often glandular; sepals lanceolate, acuminate, 6–10 mm long; petals white, 4–8 mm long, shorter than the sepals. **Fruit** a purple, puberulent, hemispheric raspberry, 8–12 mm wide. Open coniferous forest; valleys, montane. BC, MT south to CA, NV, UT.

Rubus parviflorus Nutt. THIMBLEBERRY Shrubs. **Stems** unarmed, 50–150 cm, puberulent and stipitate-glandular at first, older bark thin and shredding. **Leaves** undivided; the blade cordate-suborbicular, 5-lobed, 6–15 cm, glabrate, stipitate-glandular on the veins beneath. **Flowers** few in an open corymb; hypanthium stipitate-glandular; sepals ovate with acuminate tips, 1–2 cm long, glandular, pubescent on the margins; petals white, 15–25 mm long. **Fruit** a red, hemispheric, hairy raspberry, 1–2 cm wide. Moist to wet, open forest, thickets, avalanche slopes; valleys to subalpine. AK to ON south to CA, NM, SD, Mexico.

Rubus pedatus Sm. Trailing subshrubs. **Stems** prostrate, unarmed, to 1 m long; erect branches herbaceous, 1–2 cm long with 1 to 3 leaves. **Leaves** long-petiolate with 3 leaflets 15–25 mm long, glabrate; the lateral ones divided almost to the base, giving 5 narrowly obovate segments. **Flowers** solitary, long-stalked from stem tips; hypanthium 1–2 mm long, glabrate; sepals narrowly lanceolate, ciliate, 7–12 mm long; petals white, 6–8 mm long, spreading. **Fruit** a red dewberry composed of few druplets. Wet coniferous forest, *Sphagnum* fens; montane. AK to AB south to OR, ID, MT.

Rubus pubescens Raf. Trailing subshrub. **Stems** trailing, unarmed; erect shoots, 5–30 cm, bearing 2 to 4 leaves. **Leaves** with 3 narrowly ovate, sparsely pilose leaflets, 2–8 cm long. **Flowers** 1 to 4 at shoot tips or directly from the main stem; hypanthium pilose, sparsely glandular; sepals lanceolate, 3–5 mm long; petals white, 5–8 mm long, erect. **Fruit** a globose, red dewberry, 5–12 mm wide. Wet soil of spruce forests, shady margins of wetlands, seeps; valleys, montane. BC to NL south to WA, CO, IN. (Pl.48)

Rubus ursinus Cham. & Schltdl. PACIFIC BLACKBERRY Shrub. **Stems** trailing, often rooting at the tips, armed with retrorse prickles; floral branches erect, 5–15 cm, leafy. **Leaves** glabrate with 3 lanceolate to ovate leaflets 2–8 cm long; the terminal lobed and larger than the lateral pair. **Flowers** unisexual, in few-flowered cymes; calyx villous, purple stipitate-glandular; sepals narrowly lanceolate, 5–8 mm long; petals white, 10–16 mm long on staminate flowers; pistillate flowers with abortive anthers and oval petals 6–10 mm long. **Fruit** a cylindrical, red blackberry 2–3 cm long. Wet forest, often along roads, streams; valleys. BC, MT south to CA, ID.

Sanguisorba L. Burnet

Annual or perennial herbs. **Leaves** petiolate, lobed-stipulate, pinnately divided into toothed or lobed leaflets. **Inflorescence** a dense, cylindrical spike. **Flowers** perfect or unisexual, perigynous, each subtended by a scarious, ovate, ciliate bract; hypanthium urn-shaped, 4-angled; sepals 4, petal-like; bracteoles lacking; petals lacking; stamens 2 to 12, exserted; pistils 1 or 2. **Fruit** an achene enclosed in a thickened hypanthium.

1. Leaflets lobed almost to the midvein; stamens 2 . . . . *S. annua*
1. Leaflets deeply dentate, <1/2-way to the midvein; stamens ca. 12 . . . . *S. minor*

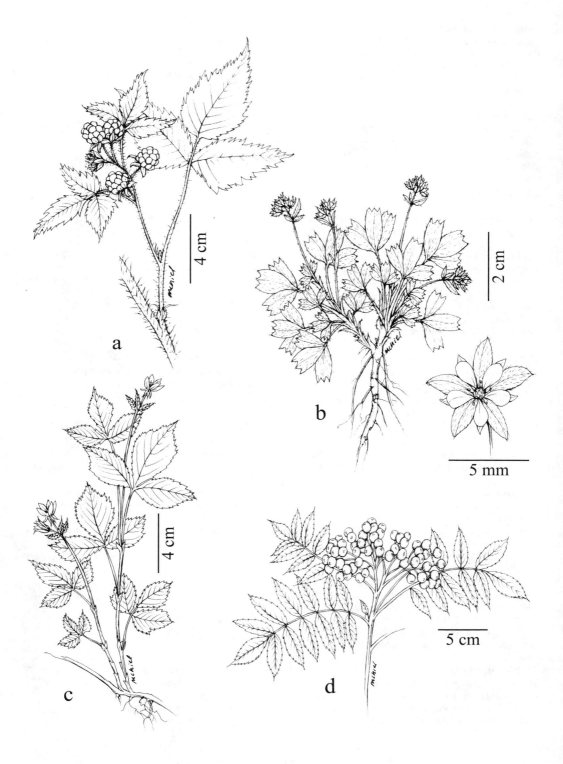

Plate 48. a. ***Rubus idaeus***, b. ***Sibbaldia procumbens***, c. ***Rubus pubescens***, d. ***Sorbus scopulina***

Sanguisorba annua (Nutt.) Nutt. [*S. occidentalis* Nutt. ex Torr. & A.Gray] Glabrous annual or biennial. **Stems** erect, often branched, 15–40 cm. **Leaflets** 7 to 15, oblanceolate, 6–14 mm long, deeply pinnately divided into linear lobes. **Flowers** perfect; sepals deltoid, 1–3 mm long, green with white or pink margins; stamens 2; pistil 1. **Fruit** ovoid, 2–3 mm long with unequal wings on the angles. Vernally moist, often compacted soil of grasslands, roadsides, wetlands; valleys. BC, MT south to CA, ID.

Sanguisorba minor Scop. Perennial from a simple or branched caudex. **Stems** erect or ascending, glabrate, simple or branched above, 25–40 cm. **Leaflets** 11 to 21, elliptic, glabrate, dentate, 5–12 mm long. **Flowers** unisexual or perfect; sepals ovate, 4–5 mm long, green with reddish margins; stamens ca. 12; pistils 2. **Fruit** pear-shaped, 3–5 mm long, wrinkled between ridges. Roadsides; valleys. Introduced to much of U.S.; native to Eurasia.

Sibbaldia L.

Sibbaldia procumbens L. Low, mat-forming, perennial herb from a branched caudex. **Stems** trailing to ascending, 2–12 cm. **Leaves** petiolate, stipulate, sparsely strigose; leaflets 3, oblong, 5–20 mm long with blunt, 3-toothed tips. **Inflorescence** a compact, bracteate, strigose cyme of 2–15 flowers. **Flowers** perfect, perigynous; hypanthium saucer-shaped; sepals 5, ovate, 2–5 mm long; bracteoles narrower, shorter; petals 5, light yellow, 1–2 mm long, shorter than the sepals; stamens 5; pistils 5 to 15. **Fruit** an achene ca. 1.5 mm long; style lateral. Sparsely vegetated, sometimes disturbed soil of meadows, turf, often under krummholz or where snow accumulates; upper subalpine, alpine. Circumpolar south to CA, UT, CO, NH. (Pl.48)

Sorbus L. Mountain ash

Shrubs or small trees. **Leaves** petiolate, exstipulate, pinnately divided into 11 to 17 dentate leaflets. **Inflorescence** a many-flowered, densely branched, flat-topped to hemispheric cyme. **Flowers** perfect, epigynous; hypanthium urn-shaped; sepals 5, deltoid, persistent; petals 5, white, short-clawed; stamens 15 to 20; pistils 2 to 5. **Fruit** a small, globose, few-seeded pome.

1. Introduced trees; some leaves with ≥13 leaflets. . . . . *S. aucuparia*
1. Native shrubs; leaves with <13 leaflets 2

2. Twigs and inflorescence sparsely red-hairy; leaflets rounded at the tip . . . . *S. sitchensis*
2. Twigs and inflorescence white-hairy; leaflets acuminate . . . .*S. scopulina*

Sorbus aucuparia L. Rowan tree, European mountain ash Tree to 6 m tall. **Stems** purple, grayish-pilose; buds gray-lanate. **Leaves** glabrous to strigose with 11 to 15 lanceolate leaflets 2–6 cm long. **Inflorescence** flat-topped to hemispheric. **Flowers**: calyx pilose; sepals <1 mm long; petals 4–5 mm long. **Pomes** red, ca. 10 mm long. Moist open forest near human habitations; valleys; Flathead and Missoula cos. Introduced as an ornamental throughout temperate N. America; native to Europe.

Birds, especially waxwings and robins, eat the fruit from planted trees and spread the seed to adjacent forests.

Sorbus scopulina Greene Shrubs. **Stems** 1–3 m, glabrate to gray-pilose, becoming brown. **Leaves** mostly glabrous, green above, paler beneath with 9 to 13 narrowly lanceolate, acuminate leaflets 3–7 cm long. **Inflorescence** flat-topped to hemispheric, white-pilose. **Flowers**: calyx white-pilose; sepals 1–1.5 mm long; petals 4–6 mm long. **Pome** usually orange (red), 5–10 mm long. Moist forest, avalanche slopes, margins of lakes, streams; montane, subalpine. AK to AB south to CA, NM, SD. Our plants are var. *scopulina*. (Pl.48)

Sorbus sitchensis M.Roem. Shrub. **Stems** 1–3 m, purplish; buds reddish-hairy. **Leaves** glabrate to sparsely red-hairy below with 7 to 11, narrowly oblong leaflets 2–6 cm long. **Inflorescence** hemispheric to pyramidal, sparsely reddish-hairy. **Flowers**: calyx glabrous; sepals 1–2 mm long; petals 4–5 mm long. **Pome** red, glaucous, 8–10 mm long. Open forest, wet meadows, often near streams, lakes; subalpine. AK south to CA, OR, MT. Our plants are ssp. *sitchensis*.

Spiraea L. Spiraea, Meadowsweet

Shrubs. **Leaves** petiolate, exstipulate, oblong to ovate, dentate. **Inflorescence** a densely branched, bracteate cyme at stem tips. **Flowers** perfect, perigynous; hypanthium cup-shaped; sepals 5; bracteoles absent; petals 5; stamens ≥15; pistils 5. **Fruit** a glabrous, fusiform follicle splitting along the inner suture; seeds 2 to 4 per follicle.

1. Inflorescence conical, at least as long as broad. 2
1. Inflorescence flat-topped to hemispheric, broader than long 3

2. Flowers pink to rose; inflorescence ≥3 times as long as wide . . . . *S. douglasii*
2. Flowers pink-tinged; inflorescence 1 to 2 times as long as wide . . . . *S. pyramidata*

3. Flowers white; leaves glabrous . . . . *S. betulifolia*
3. Flowers rose; leaves hairy on the margins. . . . . *S. splendens*

Spiraea betulifolia Pall. [*S. lucida* Douglas ex Greene] Rhizomatous. **Stems** 20–80 cm, glabrous, reddish-brown, shredding with age. **Leaf blades** 2–9 cm long, glabrous, serrate above, basally entire, green above, paler beneath. **Inflorescence** flat-topped, glabrous, 3–15 cm across. **Flowers**: sepals <1 mm long, glabrous; petals white, ca. 2 mm long. **Follicles** 3 mm long. Moist to dry forest, forest margins, avalanche slopes; valleys to subalpine. BC to SK south to OR, WY, SD. Our plants are var. *lucida* (Douglas ex Greene) C.L.Hitchc.

Spiraea douglasii Hook. **Stems** 1–2 m, pilose, becoming brown, glabrous. **Leaf blades** 2–8 cm long, glabrous, paler beneath, serrate on the upper half. **Inflorescence** a conical, pilose panicle 4–20 cm long. **Flowers**: sepals hairy to glabrous, 0.5 mm long; petals pink or rose, ca. 1.5 mm long. **Follicles** 2–3 mm long. Thickets, wet meadows, usually along streams, wetlands; valleys. AK south to MT, ID, CA. Our plants are var. *menziesii* (Hook.) C.Presl.

Spiraea pyramidata Greene Rhizomatous. **Stems** 40–100 cm, glabrous to sparsely hairy, reddish-brown. **Leaf blades** 3–7 cm long, glabrous, serrate on the upper half. **Inflorescence** pyramidal, glabrous, 3–8 cm across. **Flowers**: sepals <1 mm long, sparsely white-hairy; petals white, ca. 2 mm long. **Follicles** ca. 3 mm long. Open forest, stream banks; valleys to lower subalpine. BC, MT south to OR, ID.

Generally considered to be a hybrid between *Spiraea betulifolia* and *S. dougalsii*, Hitchcock believes it is a self-perpetuating entity rather than a sporadic result of the two parental species being sympatric.[187]

Spiraea splendens Baumann ex K.Koch [*S. densiflora* Nutt. ex Torr. & A.Gray, *S. betulifolia* Pall. var. *rosea* A.Gray] Rhizomatous. **Stems** 50–100 cm, glabrous to pubescent, light brown. **Leaf blades** 2–6 cm long, glabrous, glandular-serrate above, entire near the base, green above, paler beneath. **Inflorescence** hemispheric, 15–50 mm across. **Flowers**: sepals ca. 1 mm long; petals rose, ca. 2 mm long. **Follicles** 2–3 mm long. Stony soil of open forest, rock outcrops, moist meadows, avalanche slopes, often along streams, lakes; subalpine, rarely montane. BC, AB south to OR, ID, WY. (p.304)

Waldsteinia Willd.

Waldsteinia idahoensis Piper Rhizomatous, scapose perennial herb. **Leaves** basal, with long, hirsute petioles, sheathing-stipulate; the blades cordate-orbicular, shallowly 3- to 5-lobed, crenate, 2–4 cm long. **Inflorescence** a few-flowered, glandular-hairy cyme born on ascending to erect, bracteate peduncles 6–12 cm long. **Flowers** perfect, perigynous; hypanthium obconic, 2–3 mm long; sepals lanceolate, 2–4 mm long; bracteoles absent; petals pale yellow, 5–7 mm long, suborbicular; stamens numerous; pistils 2 to 4, puberulent. **Fruit** a cylindrical achene. Openings of coniferous forest; montane. Endemic to central ID and adjacent Missoula Co., MT.

CROSSOSOMATACEAE

Glossopetalon A.Gray Grease bush

Glossopetalon spinescens A.Gray [*G. nevadense* A.Gray] Intricately branched, glabrous shrubs. **Stems** to 2 m high, thorny, ribbed, bearing old petiolar bases, green becoming tan. **Leaves** alternate, short-petiolate, narrowly oblanceolate, entire, 3–10 mm long; stipules lobed, adnate to the petiole. **Flowers** solitary at nodes of leaf-like bracts, perfect, perigynous; sepals 5, ovate, 1–2 mm long; petals 5, linear, ca. 3 mm long; stamens 6 to 10; pistil 1; ovary superior. **Fruit**

an asymmetrical-ovoid, ribbed, leathery follicle 3–6 mm long with 1 seed. Crevices of granitic cliffs; montane. WA to MT south to CA, AZ, NM, TX, Mexico. One location in Ravalli Co. Our plants are var. *aridum* M.E.Jones.

This plant was formerly placed in the Celastraceae.[188]

FABACEAE (Leguminosae): Pea Family

Annual to perennial herbs, shrubs or trees. **Leaves** stipulate, alternate, mostly pinnately or palmately divided into distinct leaflets. **Inflorescence** a terminal or axillary spike or raceme. **Flowers** perfect, perigynous mostly zygomorphic (bilaterally symetrical); hypanthium usually cup-shaped; corolla usually papilionaceous; i.e., of 5 separate dissimilar petals: upper petal (banner) usually largest and bent up in the middle; lower petals (keel) united along the bottom to form a canoe-like envelope enclosing the style and stamens; lateral petals (wings) are similar, usually overlapping the keel; sepals 5; stamens 10, separate or united into a sheath; pistil 1; ovary superior. **Fruit** a 1- to many-seeded pod (legume) or a linear series of individually enclosed seeds (loment).[34,201]

Identification may be difficult without both flowers and mature fruits. Members of this family form symbiotic relationships with root bacteria, allowing them to use atmospheric nitrogen and thrive in infertile soils of banks, gravel bars and moraine. Measurements were taken across the arc of the reflexed banner.

1. Plants woody-stemmed shrubs or trees 2
1. Plants herbaceous, sometimes with woody caudexs 6

2. Leaflets 1 to 5 . . . .*Cytisus*
2. Leaflets ≥5 3

3. Herbage canescent; corolla reduced to just the banner petal . . . . *Amorpha*
3. Herbage glabrous to sparsely pubescent; corolla papilionaceous 4

4. Trees, often spiny-twigged; inflorescence densely-flowered, pendent . . . . *Robinia*
4. Unarmed shrubs; inflorescence not as above 5

5. Herbage glandular-punctate; corollas white . . . . *Dalea*
5. Herbage not glandular-punctate; corollas yellow . . . .*Caragana*

6. Terminal leaflet reduced to a bristle or curling tendril 7
6. Terminal leaflet similar to lateral ones 8

7. Leaflets 2 to 8, not including stipules at base of petiole; if 8 then flowers yellowish . . . .*Lathyrus*
7. Leaflets 8 to 14; flowers blue-purple . . . . *Vicia*

8. Leaves simple . . . . *Astragalus*
8. Leaves compound 9

9. Leaves palmate with 3 to 12 leaflets (not including leaf-like stipules at petiole base) 10
9. Leaves pinnate with ≥5 leaflets, at least 1 pair attached well below rachis tip 20

10. Pod ovoid, sometimes with only 1 or 2 seeds 11
10. Pod elongate with >2 seeds, sometimes curved or coiled 16

11. Herbage glandular-punctate 12
11. Herbage glabrous to pubescent but not glandular-punctate 13

12. Sepals 1–2 mm long; pod glandular-dotted . . . . *Psoralidium*
12. Sepals (at least in fruit) >2 mm long; pod not glandular . . . .*Pediomelum*

13. Inflorescence >4 cm long, loosely flowered . . . . *Melilotus*
13. Inflorescence <3 cm long 14

14. Inflorescence of 1–2 flowers among the leaves . . . . *Astragalus*
14. Inflorescence of >2 flowers, terminating a stem 15

15. Pod curved or coiled, longer than calyx . . . . *Medicago*
15. Pod ovoid, mostly hidden by the calyx . . . . *Trifolium*

16. Inflorescence an umbel or flowers solitary . . . . *Lotus*
16. Inflorescence a few- to many-flowered raceme 17

17. Some leaves with >3 leaflets . . . . *Lupinus*
17. Leaflets 3 18

18. Pod curved or coiled, ≤1 cm long . . . . *Medicago*
18. Pod straight or curved, ≥3 cm long 19

19. Inflorescence of 1–2 flowers among the leaves . . . . *Astragalus*
19. Inflorescence terminating the stem, of >2 flowers . . . . *Thermopsis*

20. Leaves glandular-punctate, at least beneath 21
20. Leaves not punctate 22

21. Pod 3–4 mm long with 1 or 2 seeds . . . . *Dalea*
21. Pod 1–2 cm long, many-seeded, covered with hooked bristles . . . .*Glycyrrhiza*

22. Inflorescence an umbel . . . . *Coronilla*
22. Inflorescence a raceme 23

23. Fruit 1-seeded, flattened, 5–7 mm long; calyx lobes linear, longer than the tube . . . . *Onobrychis*
23. Pods with ≥2 seeds, often not flattened 24

24. Pod papery, inflated, ovoid; corolla brick-red, drying purple . . . . *Sphaerophysa*
24. Pod not papery-inflated and/or corollas not brick red 25

25. Pod constricted between seeds, flattened; keel petal squared off in front . . . . *Hedysarum*
25. Pod not constricted between seeds; keel rounded or pointed 26

26. Keel petal with beak-like point at the tip; pod with sharp-pointed tip at least 1 mm long . . . . *Oxytropis*
26. Keel petal rounded at the tip; pod often without sharp-pointed tip . . . . *Astragalus*

Amorpha L.

Amorpha canescens Pursh LEAD PLANT Rhizomatous shrub. **Stems** 20–80 cm, tomentose, becoming glabrate. **Leaves** pinnate with 13 to 21 oblong, canescent, entire leaflets; stipules inconspicuous. **Inflorescence** narrow, axillary, bracteate, pubescent racemes, 7–15 mm long. **Flowers**: calyx obconic with sessile glands; sepals 3–5 mm long; corolla reduced to the violet banner, 4–6 mm long, enclosing the basaly-united stamens. **Legume** 1-seeded, 3–4 mm long, canescent with sessile grands. Grasslands, woodlands, often in sandy soil; plains. MT to MB south to NM, TX, IL. Collected once in Carter Co. ca. 70 years ago.

Astragalus L. Milkvetch, locoweed

Perennial (annual) herbs. **Leaves** mostly odd-pinnate or trifoliolate; stipules paired, attached to the stem, distinct or united. **Inflorescence** axillary, obscurely bracteate racemes. **Flowers** papilionaceous with banner usually as long or longer than wings or keel; stamens 9 united, 1 separate. **Fruit** a linear to globose legume usually with an acute tip and 2 to many seeds.

The largest genus in the Fabaceae and second largest in MT, *Astragalus* has many narrowly endemic species in western N. America. Having flowers and mature fruit is often essential for positive identification. *Astragalus amnis-amisi* Barneby, *A. amblytropis* Barneby, *A. aquilonius* (Barneby) Barneby, *A. beckwithii* Torr. & A.Gray var. *sulcatus* Barneby, *A. calycosus* Torr. ex S.Watson var. *scaposus* (A.Gray) M.E.Jones, and *A. paysonii* (Rydb.) Barneby occur in east-central ID and could occur in adjacent MT.

1. Plants low; leaflets 3–9 mm long with a hard, sharp-pointed tip . . . . *A. kentrophyta*
1. Plants not as above 2

2. Leaflets linear, the terminal one confluent with the rachis 3
2. Leaves simple or leaflets usually wider, the terminal one always jointed to the rachis 6

3. Ovary and legume strigose . . . . *A. convallarius*
3. Ovary and legume glabrous; the mature legume ≥5 mm wide 4

4. Inflorescence with <10 flowers, <10 mm long; legume inflated, mottled . . . . *A. ceramicus*
4. Inflorescence with >10 flowers, ≥12 mm long; legume hard, leathery 5

5. Leaflets ca. 1 mm wide; calyx black-strigose . . . . *A. pectinatus*
5. Leaflets 2–3 mm wide; calyx white-strigose . . . . *A. grayi*

6. Leaflets 1 to 3 7
6. Leaflets of most leaves 5 or more 11

7. Some or all leaves simple . . . . *A. spatulatus*
7. Leaves all trifoliolate 8

8. Calyx tube 1–5 mm long. 9
8. Calyx tube >5 mm long. 10

9. Banner ca. 6 mm long; plants of undeveloped limestone soils. . . . .*A. aretioides*
9. Banner ≥10 mm long; plants of clay soils. . . . . *A. barrii*

10. Banner 15–30 mm long; leaflets often >10 mm long . . . . *A. gilviflorus*
10. Banner 12–15 mm long; leaflets ≤10 mm long. . . . .*A. hyalinus*

11. Hairs on leaves and stems all or in part of dolabriform hairs (use 20X). . . . . Group A
11. Hairs of leaves and stems all basifixed. 12

12. Ovary and legume glabrous. . . . . Group B
12. Ovary and pod sparsely to densely hairy. 13

13. Banner 5–15 mm long . . . . .Group C
13. Banner ≥15 mm long . . . . .Group D

Group A: some or all hairs dolabriform (sessile, 2-forked, like a 2-bladed propeller)

1. Plants subacaulescent; stems (excluding peduncles) <4 cm long. 2
1. Most plants with stems >4 cm long 4

2. Calyx tube <7 mm long; banner 9–11 mm long . . . . *A. lotiflorus*
2. Calyx tube ≥7 mm long; banner ≥15 mm long 3

3. Racemes shorter than subtending leaves; legume maroon-mottled . . . .*A. chamaeleuce*
3. Racemes longer than subtending leaves; legume not mottled. . . . . *A. missouriensis*

4. Leaflets almost as wide as long . . . . *A. oreganus*
4. Leaflets ≥twice as long as wide 5

5. Mature legumes pendent 6
5. Mature legumes erect. 7

6. Legume strongly curved; flowers ochroleucus; stems >40 cm. . . . . *A. falcatus*
6. Legume mostly straight; flowers white and blue; stems ≤30 cm. . . . . *A. miser*

7. Flowers purple . . . . *A. adsurgens*
7. Flowers white to ochroleucus. 8

8. Stems arising singly or few together from a rhizome; sepals 2–4 mm long . . . . *A. canadensis*
8. Stems clustered from a tightly branched caudex; sepals 1–2 mm long . . . .*A. terminalis*

Group B: ovary and legume glabrous

1. Banner 15–25 mm long (measured across the arc). 2
1. Banner 5–15 mm long 6

2. Mature legumes erect. 3
2. Mature legumes spreading or pendent. 4

3. Mature legumes and stipes strictly erect; flowers white . . . . *A. atropubescens*
3. Legumes erect but stipes spreading; flowers ochroleucus. . . . . *A. scaphoides*

4. Stems sprawling to ascending; mature legumes orbicular. . . . . *A. crassicarpus*
4. Stems erect; legumes linear. 5

5. Legume grooved beneath with a stipe 6–10 mm long . . . . *A. drummondii*
5. Legume triangular in cross section with a stipe 3–5 mm long . . . .*A. racemosus*

6. Flowers pink to purple 7
6. Flowers white to yellow. 9

7. Racemes with 5 to 10 flowers. . . . . *A. miser*
7. Racemes with >10 flowers 8

8. Keel petal 9–13 mm long; leaflet tips rounded; legume 2-grooved below . . . . *A. bisulcatus*
8. Keel petal 5–8 mm long; leaflet tips blunt to emarginate; legume terete . . . .*A. flexuosus*

9. Mature legumes strictly erect . . . .*A. atropubescens*
9. Legumes spreading to pendent 10

10. Mature legume inflated, terete or flattened perpendicular to the suture 11
10. Mature legume flattened parallel to the suture 12

11. Stems erect; mature legumes lacking a significant flattened beak . . . . *A. americanus*
11. Stems sprawling to ascending; mature legume with a flattened beak 3–5 mm long . . . . *A. lentiginosus*

12. Calyx lobes ca. 1 mm long 13
12. Calyx lobes 2–6 mm long 14

13. Mature legume 7–15 mm long, stipitate . . . . *A. tenellus*
13. Mature legume 15–30 mm long, without a stipe arising from the base of the calyx . . . . *A. miser*

14. Flowers 7–10 mm long . . . .*A. australis*
14. Flowers >12 mm long . . . .*A. racemosus*

Group C: legume pubescent, banner 5–15 mm long

1. Taprooted annuals . . . . *A. geyeri*
1. Perennial with a branched caudex 2

2. Mature legume globose to ovoid, ≤2X as long as wide, often inflated 3
2. Mature legume narrowly elliptic to linear, >2X as long as wide 8

3. Mature legume >15 mm long; maroon-mottled . . . . *A. platytropis*
3. Mature legume <15 mm long 4

4. Flowers ≥12 mm long . . . .*A. cicer*
4. Flowers <10 mm long 5

5. Mature legume leathery, not inflated 6
5. Mature legume papery and inflated 7

6. Legume compressed perpendicular to the suture; leaflet tips blunt to emarginate . . . . *A. gracilis*
6. Legume compressed parallel to the suture; leaflet tips acute . . . . *A. vexilliflexus*

7. Stems prostrate at the base; calyx tube <2.5 mm long . . . . *A. microcystis*
7. Stems ascending to erect; calyx tube ≥2.5 mm long . . . . *A. eucosmus*

8. Legumes erect . . . . *A. agrestis*
8. Legumes spreading to pendent 9

9. Mature legume with a stipe ≥1 mm long 10
9. Mature legume not stipitate 12

10. Stems 5–30 cm, often decumbent or sprawling; petals purple with white wings . . . .*A. alpinus*
10. Stems erect, 20–70 cm; wing petals purplish or (rarely) all petals white 11

11. Mature legume not grooved beneath, densely strigose . . . .*A. robbinsii*
11. Mature legume 2-grooved beneath, sparsely strigose . . . . *A. bisulcatus*

12. Mature legume nearly terete or triangular in cross section 13
12. Mature legume strongly flattened parallel to the sutures 14

13. Stems 2–10 cm; legume 9–11 mm long, triangular in cross section; alpine . . . . *A. lackschewitzii*
13. Stems >20 cm; legume 10–20 mm long, nearly terete; plains, valleys . . . .*A. flexuosus*

14. Legume black-strigillose . . . . *A. bourgovii*
14. Legume white-strigillose 15

15. Keel 3–6 mm long; legume 6–13 mm long; lower stems often prostrate . . . . *A. vexilliflexus*
15. Keel 6–10 mm long; legume 15–30 mm long; plants erect . . . . *A. miser*

Group D: legume pubescent; banner 15–25 mm long

1. Herbage green, hairs sparse 2
1. Herbage woolly or silvery-hairy 5

2. Legume subglobose 3
2. Legume elliptic in outline 4

3. Calyx white-strigose . . . .*A. plattensis*
3. Calyx black strigose . . . .*A. cicer*

4. Sepals ca. 1 mm long; stipules connate around the stem; legume spreading
   to pendent. . . . . *A. cibarius*
4. Sepals 2–4 mm long; stipules connate; legume erect . . . . *A. agrestis*

5. Plants with stems ≥10 cm . . . . *A. inflexus*
5. Plants subacaulescent; stems ≤6 cm. 6

6. Herbage villous with spreading, tangled hairs . . . . *A. purshii*
6. Vesture of dense, appressed, mostly straight hairs 7

7. Legume 25–45 mm long; leaflets often >4 mm wide; dry habitats . . . . *A. shortianus*
7. Legumes 15–24 mm long; leaflets usually <4 mm wide; vernally moist sites . . . . *A. argophyllus*

Astragalus adsurgens Pall. [*A. striatus* Nutt. ex Torr. & A.Gray] Caespitose perennial.
Herbage sparsely to densely strigose with dolabriform hairs. **Stems** 15–40 cm from a branched
caudex. **Leaves** with 9 to 21 narrowly elliptic leaflets, 1–2 cm long; stipules lanceolate, united
around the stem at the base. **Inflorescence** spike-like with 15 to 80 erect to ascending flowers.
Flowers purple; sepals 1–4 mm long; calyx white and black dolabriform-hairy; banner barely
reflexed, 11–16 mm long; keel 8–13 mm long. **Legume** sessile, erect, 6–12 mm long, linear-ovoid
with appressed, basifixed, white hairs, grooved on 1 side. Often stony soil of grasslands, steppe, open forest; plains,
valleys, montane. Circumboreal south to WA, NM, NE. Our plants are var. ***robustior*** Hook. (p.297)

Astragalus agrestis Douglas ex G.Don [*A. dasyglottis* Fisch. ex DC.] Perennial from rhizome-like caudex branches.
Herbage sparsely strigose. **Stems** single or loosely tufted, 5–30 cm, lax to erect. **Leaves** with 11 to 23 narrowly elliptic
leaflets, 5–20 mm long; stipules ovate to lanceolate, lower ones united around the stem. **Inflorescence** ovoid, with 5 to
15 erect flowers. **Flowers** purple (rarely ochroleucus) with pale wing tips and banner base; sepals 2–4 mm long; calyx
pilose with black or white hairs; banner barely reflexed, 13–22 mm long; keel 11–12 mm long. **Legume** erect, narrowly
ellipsoid, 4–10 mm long, white-hairy, grooved on 1 side. Moist soil of grasslands, steppe, meadows, open forests,
thickets, often near streams, wetlands; plains, valleys, montane. AK to SK south to NM, KN, IA; Asia. (p.297)

Astragalus alpinus L. Perennial with a widely branched caudex. **Herbage** sparsely strigose.
Stems 5–30 cm, ascending to erect. **Leaves** with 13 to 25 elliptic leaflets with rounded tips, 8–20
mm long; stipules ovate, distinct. **Inflorescence** loose, with 8 to 25 spreading flowers. **Flowers**
purple with white wings and petal bases; sepals 1–2 mm long; calyx black-strigose; banner reflexed,
7–12 mm long; keel ca. as long. **Legume** narrowly elliptic, black-strigose, 8–14 mm long, shallowly
grooved, pendent with a stipe 1–2 mm long. Stony, soil of tundra, moist meadows, gravel bars, forest openings; montane
to alpine. Circumboreal south to NV, NM, MB. Our plants are var. ***alpinus***. (p.304)
   Gravel bar plants may form large mats as much as 60 cm across.

Astragalus americanus (Hook.) M.E.Jones Perennial with a woody caudex. **Herbage**
glabrate to sparsely strigose. **Stems** erect, hollow, 50–100 cm. **Leaves** with 11 to 15 narrowly
elliptic leaflets, 2–5 cm long; stipules oblong, reflexed, 1–2 cm long, distinct. **Inflorescence**
loose with 15 to 40 nodding flowers. **Flowers** ochroleucus; sepals ciliate, <1 mm long; banner,
11–15 mm long; keel nearly as long. **Legume** 15–25 mm long, glabrous, narrowly ellipsoid,
terete, pendent with a stipe 4–7 mm long. Open forest, woodlands, thickets, moist meadows, often
along streams; valleys to lower subalpine. AK to ON south to BC, CO, MB.

Astragalus aretioides (M.E. Jones) Barneby [*Orophaca aretioides* (M.E. Jones) Rydb.] Cushion-forming from a
much-branched caudex. **Herbage** sericeous with dolabriform hairs. **Stems** compressed, <2 cm long. **Leaves** trifoliolate,
the leaflets narrowly elliptic, 3–8 mm long; stipules hyaline, ciliate, connate, clothing the stem. **Inflorescence** 2-flowered
on an ascending peduncle 5–10 mm long. **Flowers** purple; sepals 1–2 mm long; calyx pilose; banner slightly reflexed,
ca. 6 mm long; keel 4–5 mm long. **Legume** ascending, ovoid, 4–5 mm long, sericeous. Stony, limestone-derived soil of
exposed ridges; montane in Big Horn and Carbon cos. MT south to CO.

Astragalus argophyllus Nutt. Nearly acaulescent, often mat-forming perennial from a
branched, woody caudex. **Herbage** densely silvery-strigose. **Leaves** with 7 to 21 narrowly
oblong to elliptic leaflets, 5–15 mm long. **Inflorescence** of 2–8 ascending flowers on a peduncle
1–5 cm long. **Flowers** purple; sepals 2–4 mm long; calyx black- and white-hairy; banner slightly
reflexed, 15–24 mm long; keel 12–20 mm long. **Legume** ascending, narrowly ovoid with a
flattened beak, silvery pilose, 15–20 mm long. Sparsely vegetated soil of moist, alkaline meadows,
streambanks; montane, lower subalpine. ID, MT south to NV, AZ, UT, WY. Our plants are var. ***argophyllus***.

Astragalus argophyllus var. *martini* M.E. Jones, with ochroleucus flowers, is reported for MT,[188] but I have seen no specimens, and it is not reported for MT by other authorities.[33,34,201]

Astragalus atropubescens J.M.Coult. & Fisch. [*A. stenophyllus* Torr. & A.Gray misapplied] Perennial from a tightly-branched caudex. **Herbage** strigose. **Stems** erect, 15–50 cm. **Leaves** with 19 to 29 linear-oblanceolate, obtuse leaflets, 5–20 mm long with emarginate tips; stipules lanceolate, 2–4 mm long. **Inflorescence** loose with 10 to 25 nodding to spreading flowers. **Flowers** white to cream; calyx black-hairy; sepals 1–2 mm long; banner partly erect, 12–18 mm long; keel 10–12 mm long. **Legume** linear-ellipsoid, glabrous, 13–20 mm long, tightly erect with stipes 2–6 mm long. Grasslands, sagebrush steppe; montane. Endemic to southwest MT and east-central ID.

The type locality is in Powell Co.

Astragalus australis (L.) Lam. [*A. aboriginum* Richardson ex Sprengel] Herbaceous perennial. **Herbage** densely to sparsely villous. **Stems** 6–40 cm, spreading or ascending from a branched caudex. **Leaves** short-petiolate with 7 to 15 narrowly lanceolate to elliptic leaflets 5–25 mm long; lower stipules ovate and united opposite the petiole, becoming lance-attenuate and distinct above. **Inflorescence** with 5 to 25 spreading to declined flowers. **Flowers** white; calyx black-strigillose; sepals 2–3 mm long; banner upturned, 7–10 mm long; keel 5–8 mm long, purple-tipped. **Legume** crescent-shaped, laterally compressed, 1–2 cm long, glabrous, nodding with a stipe 2–8 mm long. Sparsely vegetated, stony soil of grasslands, fellfields, exposed slopes; plains, valleys to alpine. Circumboreal south to OR, CO, ND. N. American plants are var. *glabriusculus* (Hook.) Isley. (p.297)

Plants from the Great Plains tend to have longer, broader, and more nearly glabrous leaflets; plants from the western part of the state are usually associated with limestone.

Astragalus barrii Barneby [*Orophaca barrii* (Barneby) Isely] Cushion-forming from a much-branched caudex. **Herbage** densely sericeous with dolabriform hairs. **Stems** compressed, <2 cm long. **Leaves** trifoliolate; leaflets linear-lanceolate, 5–10 mm long; stipules hyaline, ciliate, connate. **Inflorescence** of 2 to 5 ascending flowers on an ascending peduncle 5–20 mm long. **Flowers** purple; sepals 1–3 mm long; calyx pilose; banner slightly reflexed, 10–16 mm long; keel 7–10 mm long. **Legume** ascending, ovoid, 4–8 mm long, sericeous. Sparsely vegetated, clay soil of grasslands, steppe, pine-juniper woodlands on hills, bluffs, badlands; plains. Endemic to southeast MT, adjacent SD, WY, NE.

Astragalus bisulcatus (Hook.) A.Gray Shrub-like perennial from a usually branched caudex. **Herbage** sparsely strigose. **Stems** erect, 20–70 cm, hollow at the base. **Leaves** with 15 to 23 narrowly elliptic leaflets, 1–2 cm long with rounded tips; stipules lanceolate, 3–6 mm long. **Inflorescence** dense with 20 to 80 nodding flowers. **Flowers** purple (white); sepals 1–4 mm long; calyx black- and/or white-hairy; banner becoming reflexed 10–15 mm long; keel 9–13 mm long. **Legume** 10–15 mm long, glabrous or strigose, linear with 2 grooves beneath, pendulous with a stipe 2–4 mm long. Grasslands, steppe, usually on clay soils; plains, valleys. AB south to NM, ID, WY, NE, KS. Our plants are var. *bisulcatus*. (p.297)

This species is reported to occur only in seleniferous soil (in MT derived from marine shales) and to have the rank odor of selenium when crushed; however, several of our populations occur on limestone- or granite-derived soil.

Astragalus bourgovii A.Gray Perennial with a branched caudex. **Herbage** strigose. **Stems** 5–30 cm, erect to ascending. **Leaves** with 13 to 23 lanceolate leaflets 5–12 mm long; stipules deltoid, 1–3 mm long, basally connate. **Inflorescence** longer than the leaves, open with 5 to 10 ascending flowers. **Flowers** purple; calyx black-strigose; sepals 1–3 mm long; banner 8–11 mm long, reflexed; keel 7–9 mm long. **Legume** ascending to pendent, black-strigillose, flattened, keeled on the sides, 10–18 mm long, narrowly elliptic. Stony soil of meadows, moraine, grasslands, fellfields, open forest; subalpine, alpine. Endemic to southern BC, AB, adjacent ID, MT. (p.304)

This relatively nondescript, common species is often mistaken for several other species with small purple flowers, including *Astragalus alpinus*, *A. australis*, *A. miser*, and *A. agrestis*.

Astragalus canadensis L. Rhizomatous perennial. **Herbage** sparsely to densely strigose with dolabriform hairs. **Stems** erect, 15–60 cm. **Leaves** with 11 to 23, narrowly elliptic leaflets, 1–4 cm long with blunt to rounded tips; stipules lanceolate-attenuate, 5–10 mm long. **Inflorescence** longer than the leaves, crowded with 30 to 100 spreading to declined flowers. **Flowers** ochroleucus; calyx black- and/or white-strigose; sepals 2–4 mm long; banner12–15 mm long, becoming moderately reflexed; keel 8–12 mm long, sometimes purple-tipped. **Legume** erect, elliptic to cylindric, 9–15 mm long, crowded. Grasslands, steppe, open forest, thickets; plains, valleys, montane. BC to QC south to CA, CO, TX, AR, VA. (p.297)

1. Legume glabrous, not grooved. . . . . var. *canadensis*
1. Legume pubescent, grooved beneath longitudinally 2

2. Leaves silvery-hairy . . . . var. *brevidens*
2. Leaves sparsely strigose . . . . .var. *mortonii*

Astragalus canadensis var. *canadensis* is found in moist meadows, grasslands on the plains; *A. canadensis* var. *mortonii* (Nutt.) S.Watson occurs in open forests to moist meadows, valleys to montane; *A. canadensis* var. *brevidens* (Gand.) Barneby is usually found in moist meadows along streams, valleys to montane.

Astragalus ceramicus E.Sheld. PAINTED MILKVETCH Perennial from rhizome-like rootstock branches. **Herbage** densely strigose. **Stems** solitary, lax to ascending, 10–25 cm. **Leaves** with 1 to 7 linear leaflets, 1–12 cm long; terminal leaflet longest, confluent with the rachis; stipules lanceolate, 2–6 mm long, often united basally. **Inflorescence** among the leaves, open, with few spreading to ascending flowers. **Flowers** white to light purple; calyx white- and/or black-strigose; sepals 2–3 mm long; banner reflexed, 8–11 mm long; keel 6–8 mm long. **Legume** ovoid, inflated, 15–40 mm long, maroon-mottled, glabrous, papery, pendent. Sand or very sandy soil of sandhills, below sandstone outcrops; plains, valleys, montane. ID to ND south to AZ, NM, OK.

1. Pods with a stipe 1–3 mm long. . . . . var. *filifolius*
1. Pods sessile . . . . var. *apus*

Astragalus ceramicus var. *filifolius* (A.Gray) F.J.Herm. and *A. ceramicus* var. *apus* Barneby occur in the same types of habitat, but the former is found in eastern cos., and the latter is endemic to Beaverhead Co. and adjacent ID.

Astragalus chamaeleuce A.Gray Nearly stemless perennial with a short-branched caudex. **Herbage** with dense, appressed, dolabriform and sometimes basifixed hairs. **Stems** <2 cm, clothed in stipules. **Leaves** with 7 to 11 obovate leaflets, 4–13 mm long with rounded to truncate tips; stipules deltoid, 2–7 mm long, distinct. **Inflorescence** short with 4 to 9 spreading to ascending flowers, shorter than the leaves, prostrate in fruit. **Flowers** lavender; calyx black- and/or white-strigose; sepals 2–4 mm long; banner 16–26 mm long, barely reflexed; keel 15–20 mm long. **Legume** ovoid with a subulate tip, purplish-mottled, strigillose, 2–4 cm long, inflated. Sparsely vegetated soil of sagebrush or shadscale steppe; valleys. MT south to UT, CO. Our plants are var. *chamaeleuce*. Known from southern Carbon Co.
The similar and more common *Astragalus missouriensis* does not have mottled fruits.

Astragalus cibarius E.Sheld. Perennial from a short-branched caudex. **Herbage** glabrate to strigose. **Stems** ascending, 5–30 cm. **Leaves** with 11 to 19 elliptic leaflets, 5–20 mm long with rounded to truncate tips; stipules ovate, 2–5 mm long, membranous, mostly distinct. **Inflorescence** ca. as long as the leaves, dense with 4 to 15 ascending flowers. **Flowers** magenta with yellowish wings; calyx black-strigose; sepals ca. 1 mm long; banner 15–19 mm long, reflexed; keel 9–13 mm long. **Legume** erect, narrowly elliptic, curved, compressed perpendicular to the septum, white-strigillose, 2–3 cm long. Grasslands, steppe, open forest; plains, valleys, montane. ID, MT south to NV, UT, CO. (p.297)

Astragalus cicer L. Perennial from rhizome-like caudex branches. **Herbage** strigose. **Stems** ascending to erect, 30–70 cm. **Leaves** with 17 to 29 linear-elliptic leaflets 5–35 mm long with acute tips; stipules 2–8 mm long, the lower sheathing, the upper lanceolate and distinct. **Inflorescence** dense with 10 to 30 ascending flowers. **Flowers** ochroleucus; calyx black-strigose; sepals 2–4 mm long; banner 12–16 mm long, little reflexed; keel 9–11 mm long. **Legume** subglobose, pilose, beaked, inflated, 10–15 mm long with long white and short black hairs. Streambanks, roadsides; plains, valleys. Introduced as a forage crop sporadically in the western U.S.; native to Eurasia.

Astragalus convallarius Greene [*Astragalus diversifolius* A.Gray misapplied] Perennial from branched rootstocks. **Herbage** densely strigose. **Stems** erect, sparsely leafy, 20–60 cm. **Leaves** with 1 to 9 (mostly 1) filiform leaflets 2–6 cm long, terminal leaflet confluent with the rachis; stipules deltoid, 2–7 mm long, the lower connate. **Inflorescence** higher than the leaves, open with 2 to 8 spreading flowers. **Flowers** ochroleucus; calyx black- and/or white-strigose; sepals ca. 1 mm long; banner erect, 9–11 mm long with purple lines; keel 7–10 mm long. **Legume** pendent, linear, compressed, strigose, 2–3 cm long. Grasslands, sagebrush steppe, pine woodlands; valleys, montane. ID, MT south to NV, UT, CO. Our plants are var. *convallarius*.

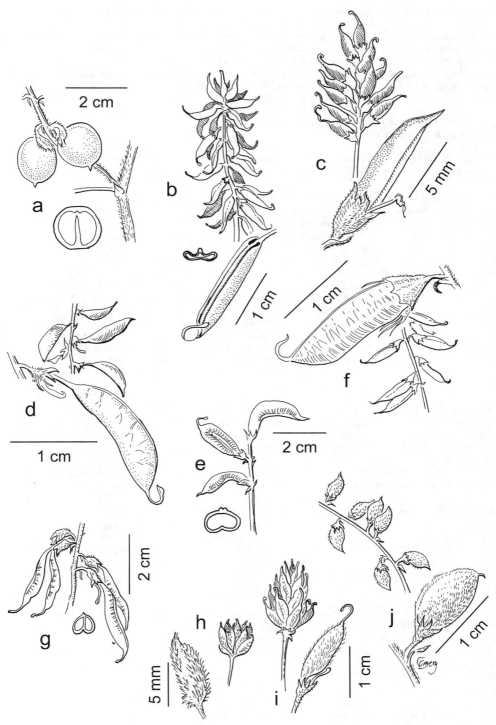

Plate 49. *Astragalus* I. **a.** *A. crassicarpus*, **b.** *A. bisulcatus*, **c.** *A. canadensis* var. *mortonii*, **d.** *A. australis*, **e.** *A. cibarius*, **f.** *A. flexuosus*, **g.** *A. drummondii*, **h.** *A. agrestis*, **i.** *A. adsurgens*, **j.** *A. eucosmus*

Astragalus crassicarpus Nutt. GROUND PLUM Perennial from a branched caudex. **Herbage** sparsely to densely strigose. **Stems** prostrate to ascending, 5–40 cm. **Leaves** with 13 to 21 narrowly oblong leaflets, 5–20 mm long with rounded tips; stipules lanceolate, 3–8 mm long, distinct. **Inflorescence** shorter than the leaves, crowded, of 5 to 20 spreading to ascending flowers. **Flowers** purple or white and purple-tipped; calyx black- and/or white-strigose; sepals 2–4 mm long; banner 22–30 mm long, moderately reflexed; keel 14–20 mm long. **Legume** spreading, globose, glabrous, 13–20 mm long, fleshy, becoming hard, leathery. Grasslands, sagebrush steppe; plains, valleys. AB, SK south to AZ, NM, TX, AR. (p.297)

1. Flowers predominately white with purple-tipped petals . . . . .var. *paysonii*
1. Flowers predominately purple . . . . var. *crassicarpus*

Astragalus crassicarpus var. *paysonii* (E.H.Kelso) Barneby occurs primarily in the western half of the state; *A. crassicarpus* var. *crassicarpus* is found on the Great Plains.

Astragalus drummondii Douglas ex Hook. Perennial from a tightly branched caudex. **Herbage** villous. **Stems** erect, 30–60 cm. **Leaves** with 15 to 33, linear-oblanceolate leaflets, 20–35 mm long with rounded tips. **Inflorescence** exceeding the leaves, dense with 20 to 50 slightly declined flowers. **Flowers** white; calyx black-villous; sepals 2–5 mm long; banner 15–22 mm long, moderately reflexed; keel 11–15 mm long. **Legume** pendent, linear, glabrous, grooved beneath, 15–30 mm long with a stipe 6–10 mm long. Grasslands, sagebrush steppe, woodlands; plains, valleys. AB, SK south to ID, UT, NM, KS. (p.297)

Astragalus eucosmus B.L.Rob. Perennial from a tightly branched caudex. **Herbage** sparsely strigose. **Stems** erect, 25–60 cm long. **Leaves** with 9 to 15 linear-elliptic leaflets, 10–25 cm long with rounded tips; stipules ovate to lanceolate, 4–9 mm long, distinct. **Inflorescence** open with 10 to 30 spreading flowers. **Flowers** purple; sepals ca. 1 mm long; calyx gray-pilose; banner 6–9 mm long, slightly reflexed; keel 4–8 mm long. **Legume** pendent, ovoid, black- and white-strigose, 8–12 mm long, triangular in cross section. Moist meadows along streams; montane. BC, AB south to UT, CO. (p.297)

The similar *Astragalus gracilis* has leaflets with blunt tips; *A. robbinsii* and *A. flexuosus* have larger flowers and fruits.

Astragalus falcatus Lam. Perennial from a branched caudex. **Herbage** of sparse, dolabriform hairs. **Stems** ascending to erect, 40–70 cm, branched, tangled together above. **Leaves** with 19 to 37 linear-elliptic leaflets, 10–25 mm long; stipules lanceolate, 7–12 mm long, distinct. **Inflorescence** dense with 20 to 50 declined flowers. **Flowers** ochroleucus; sepals ≤1 mm long; calyx sparsely strigose; banner nearly straight, 9–11 mm long; keel 8–10 mm long. **Legume** pendent, leathery, sharp-pointed, strigillose, deeply grooved beneath, ca. 2 cm long, narrowly lanceolate, strongly curved. Grasslands; valleys. Introduced to western U.S. as a forage crop and sporadically naturalized; native to Eurasia; known from Gallatin and Lincoln cos.

Astragalus flexuosus Douglas ex G.Don Perennial from a branched caudex. **Herbage** moderately to densely strigillose. **Stems** erect or ascending, 25–50 cm. **Leaves** with 11 to 21, linear to oblong leaflets, 3–19 mm long with blunt or emarginate tips. **Inflorescence** ca. as long as the leaves, narrow, open, with 10–30 spreading flowers. **Flowers** pink to purple; calyx gray-villous; sepals <1 mm long; banner 8–11 mm long, reflexed; keel 5–8 mm long. **Legume** pendent, linear to linear-elliptic, nearly terete, 1–2 cm long, glabrate to white-strigose. Often sandy or gravelly soil of grasslands; plains, valleys. AB to MB south to AZ, NM, NE. Our plants are var. *flexuosus*. (p.297)

Astragalus geyeri A.Gray Taprooted annual. **Herbage** glabrate to strigillose. **Stems** erect, 5–15 cm, branched. **Leaves** with 5 to 13 narrowly oblong leaflets, 5–20 mm long with rounded tips. **Inflorescence** much shorter than the leaves, with few spreading flowers. **Flowers** pale purple; calyx white- and black-strigose; sepals 1–2 mm long; banner 5–8 mm long, slightly reflexed; keel 3–5 mm long. **Legume** ovoid, inflated, curved, 15–24 mm long, sparsely strigose, papery at maturity. Sandy soil of grasslands, sagebrush steppe; plains, valleys; known from Carbon, Dawson, and Garfield cos. WA to MT south to CA, AZ, UT. Our plants are var. *geyeri*.

Astragalus gilviflorus E.Sheld. [*Orophaca triphylla* (Pursh) Britton] Cushion-forming from a branched caudex. **Herbage** silvery-sericeous with basifixed and dolabriform hairs. **Stems** <3 cm long, clothed in stipules. **Leaves** trifoliolate; leaflets oblanceolate, 5–20 mm long, acute; stipules hyaline, ciliate, 6–13 mm long, united behind the petiole. **Inflorescence** of 1 or 2 flowers, shorter than leaf petioles. **Flowers** ochroleucus; sepals 2–4 mm long; calyx strigose; banner slightly reflexed, 15–30 mm long; keel 10–22 mm long, often purple-tipped. **Legume** ascending, ovoid, 6–10 mm long, sericeous,

enclosed in the calyx. Sandy or gravelly soil of grasslands, sagebrush steppe, barren slopes; plains, valleys. AB to MB south to UT, CO, NE, OK. See *A. hayalinus*.

Astragalus gracilis Nutt. Perennial from a widely branched rootstock. **Herbage** strigose. **Stems** branched, lax to ascending, 20–50 cm. **Leaves** with 9 to 13 narrowly oblong leaflets, 5–15 mm long with blunt to emarginate tips, glabrous above; stipules deltoid, 1–4 mm long, distinct. **Inflorescence** narrow with 10 to 20 ascending flowers, 1 to 2 times as long as the leaves. **Flowers** lavender; calyx white-strigose; sepals <1 mm long; banner 5–8 mm long, reflexed; keel 4–6 mm long. **Legume** pendent, ovoid, slightly curved, white-strigose, 6–10 mm long, compressed perpendicular to the septum. Sandy soil of grasslands, sagebrush steppe; plains, valleys. MT, ND south to NM, TX, OK. (p.301)

Fruits of *Astragalus flexuosus* are longer and compressed parallel to the septum.

Astragalus grayi Parry ex S.Watson Broom-like, possibly seleniferous (Isley 1998) perennial from a tightly branched caudex. **Herbage** strigose. **Stems** erect, 20–40 cm, basally leafless. **Leaves** with 5–11 linear leaflets, 15–50 mm long; terminal leaflet confluent with the rachis; stipules deltoid, 4–7 mm long, the lower ones connate. **Inflorescence** of 10–30 spreading to ascending flowers, ca. as long as the leaves. **Flowers** ochroleucus; calyx densely white-strigose; sepals 1–3 mm long; banner 12–20 mm long, moderately reflexed; keel 8–13 mm long. **Legume** erect, ellipsoid, 9–15 mm long, glabrous. Silty to clayey soil of open slopes in sagebrush or shadscale steppe, often along streams; valleys. Endemic to Carbon and Big Horn cos. and adjacent WY.

Astragalus hyalinus M.E. Jones [*Orophaca hyalina* (M.E. Jones) Isely] Cushion-forming from a branched caudex. **Herbage** silky-strigose with dolabriform hairs. **Stems** <3 cm long, clothed in stipules. **Leaves** trifoliolate; leaflets oblanceolate, 3–10 mm long, acute; stipules hyaline, 2–7 mm long, united behind the petiole. **Inflorescence** of 1 or 2 flowers, shorter than leaf petioles. **Flowers** ochroleucus; sepals 3–4 mm long; calyx white-sericeous; banner slightly reflexed, 12–15 mm long; keel 8–12 mm long, purple-tipped. **Legume** ascending, ovoid, 4–8 mm long, sericeous, hidden among leaves. Sandy or stony soil of sparsely vegetated sagebrush steppe, pine woodlands; plains, valleys. MT to NE south to CO.

The similar, more common *Astragalus gilviflorus* has a larger, less hairy calyx.

Astragalus inflexus Douglas ex Hook. Perennial from a tightly branched caudex. **Herbage** densely villous. **Stems** ascending, 8–30 cm. **Leaves** with 15 to 29 narrowly elliptic leaflets with rounded to acute tips; stipules lanceolate, 1–3 cm long, distinct. **Inflorescence** just longer than the leaves, dense, of 6 to 19 spreading to ascending flowers. **Flowers** magenta; calyx villous; sepals 4–5 mm long; banner 18–23 mm long, reflexed; keel 12–16 mm long. **Legume** narrowly ovoid, compressed perpendicular to the septum, curved, sharp-beaked, 10–25 mm long, densely silky. Grasslands, sagebrush steppe, open pine woodlands; valleys, montane. WA, MT, ID, OR. (p.301)

Astragalus purshii has fewer, often yellow flowers and shorter stems.

Astragalus kentrophyta A.Gray Mat-forming perennial from an intricately branched caudex. **Herbage** strigillose to villous. **Stems** basally prostrate; ascending portion 1–3 cm. **Leaves** with 3 to 7 narrowly lanceolate leaflets 3–9 mm long with mucronate tips; terminal leaflet confluent with the rachis; stipules 2–5 mm long, basally connate. **Inflorescence** of 1 to 3 flowers, shorter than the leaves. **Flowers** purple to white; calyx white- and/or black-strigillose; sepals 1–3 mm long; banner reflexed, 4–8 mm long; keel 3–7 mm long. **Legume** ovoid, 3–5 mm long, compressed, strigose. Stony or gravelly, sparsely vegetated soil of grasslands, steppe, fellfields; plains, valleys to alpine. AB, SK south to CA, AZ, NM, NE.

1. Hairs all basifixed; flowers predominately purple, ≥5 mm long. . . . . var. *tegetarius*
1. Some dolabriform hairs mixed with basifixed; petals predominately white, 4–5 mm long . . . var. *kentrophyta*

Astragalus kentrophyta var. *kentrophyta* is found in sandy or gravelly soil, often around ponds on the plains; *A. kentrophyta* var. *tegetarius* (Wats.) Dorn occurs in fellfields, morraine, scree near or above treeline. See *A. vexilliflexus*.

Astragalus lackschewitzii Lavin & H.Marriott Perennial from a branched caudex. **Herbage** glabrate to strigose. **Stems** prostrate to ascending, 2–10 cm. **Leaves** with 15 to 23 lanceolate leaflets, 2–10 mm long with acute tips, glabrous above, ciliate; stipules acuminate, 2–5 mm long, basally connate. **Inflorescence** just above the leaves, of 1 to 5 spreading to ascending flowers. **Flowers** purple; calyx black- or white-strigillose; sepals 1–2 mm long; banner 9–12 mm long, reflexed; keel 8–10 mm long. **Legume** spreading, curved-ellipsoid, strigillose, 9–11 mm long, triangular in cross section. Stony calcareous soil of fellfields, meadows; alpine; Endemic to Pondera and Teton cos.

Astragalus lentiginosus Douglas ex Hook. Perennial from a branched caudex. **Herbage** glabrate to sparsely strigose. **Stems** prostrate to ascending, 10–40 cm. **Leaves** with 11 to 19 obovate leaflets, 5–15 mm long with blunt to emarginate tips; stipules lanceolate, 2–3 mm long, distinct. **Inflorescence** shorter than the leaves, dense with 10–25 spreading to ascending flowers. **Flowers** ochroleucus, rarely purple-tipped; calyx white-strigose; sepals 1–2 mm long; banner 8–13 mm long, slightly reflexed; keel 6–8 mm long. **Legume** spreading, glabrous, ovoid, 15–20 mm long with a flattened beak 3–5 mm long. Often silty soil of sagebrush steppe; valleys. WA to MT south to CA, AZ, NM, TX, Mexico; known from Beaverhead Co. Our plants are var. *salinus* (Howell) Barneby; var. *platyphyllidum* (Rydb.) M.Peck is reported for MT,[201] but I have seen no specimens.

Astragalus leptaleus A.Gray Perennial from rhizome-like caudex branches. **Herbage** glabrate to sparsely strigose. **Stems** erect to ascending, 5–25 cm. **Leaves** with 13 to 19 linear-lanceolate leaflets, 3–15 mm long with acute tips; stipules 1–4 mm long, connate. **Inflorescence** sparse with 1 to 5 spreading flowers, shorter than the leaves. **Flowers** white; calyx black strigose; sepals 1–3 mm long; banner 8–12 mm long, moderately reflexed; keel 6–8 mm long with a purple tip. **Legume** pendent, narrowly elliptic, flattened, sparsely black- and white-strigose, 8–14 mm long. Moist, often alkaline meadows along streams, wetlands; valleys, montane. AB south to ID, CO.
   Astragalus miser is similar but usually caespitose.

Astragalus lotiflorus Hook. Annual to usually perennial from a simple or branched caudex. **Herbage** strigose with dolabriform hairs. **Stems** erect to ascending, 1–3 cm, shorter than the mostly basal leaves. **Leaves** with 5–15 linear-elliptic leaflets with acute tips; stipules lanceolate, 2–5 mm long, distinct. **Inflorescence** nearly sessile with 2–8 spreading flowers, shorter or longer than the leaves. **Flowers** ochroleucus; calyx white-strigose; sepals 2–4 mm long; banner 9–11 mm long, nearly straight; keel 6–8 mm long. **Legume** narrowly ellipsoid, beaked, inflated, compressed perpendicular to the septum, 15–25 mm long, densely silky. Grasslands, sagebrush steppe; plains, valleys. AB to MN south to NM, TX, IA. (p.301)
   Flowers borne at the base of the plant are cleistogamous; these seem to be more fertile than the chasmogamous flowers in our material. Plants may produce a few cleistogamous fruits their first year.

Astragalus microcystis A.Gray Perennial from an intricately branched caudex. **Herbage** strigose to villous. **Stems** prostrate, ascending at the tip, 10–50 cm. **Leaves** with 9 to 15 linear-elliptic leaflets, 5–15 mm long with rounded to acute tips; stipules 3–5 mm long, basally connate, turning black with drying. **Inflorescence** open with 5 to 10 spreading flowers. **Flowers** pink to purple, often with white wings; calyx white- or black-strigillose; sepals 1–2 mm long; banner 6–9 mm long, reflexed; keel 4–5 mm long. **Legume** pendent, obovoid, inflated, papery, white-strigillose, 5–10 mm long. Sandy or gravelly soil of grasslands, fellfields or on streambanks; valleys to alpine. WY, ID, MT. (p.301)
   Astragalus vexilliflexus is similar in flower,[33] but differs by having stipules that do not blacken upon drying; however, fruits are needed for positive determination.

Astragalus miser Douglas ex.Hook. Perennial with a branched caudex. **Herbage** glabrate to strigose or villous with dolabriform hairs. **Stems** 5–30 cm, prostrate to ascending. **Leaves** with 9 to 17 linear to ovate leaflets, 5–25 mm long with rounded to acute tips; stipules lanceolate, 2–4 mm long, basally connate. **Inflorescence** barely longer than the leaves, open with 5 to 10 spreading flowers. **Flowers** white to purple; calyx black- and/or white-strigose; sepals ca. 1 mm long; banner 8–12 mm long, reflexed; keel 6–10 mm long, often purple-tipped. **Legume** pendent, glabrous to white-strigillose, 15–30 mm long, flattened, linear. Grasslands, sagebrush steppe, open forest; valleys to subalpine. BC, AB south to AZ, CO, SD. (p.301)

1. Vestiture of basifixed and dolabriform hairs 2
1. Pubescence entirely of basifixed hairs 4

2. Herbage and legume villous with loose, twisted hairs . . . . var. *crispatus*
2. Pubescence of straight, stiff hairs. 3

3. Leaflets linear to linear lanceolate; flowers white with a purple keel tip . . . . var. *praeteritus*
3. Leaflets elliptic to oblanceolate; petals mainly purple. . . . . var. *decumbens*

4. Leaflets equally strigose above and beneath . . . . var. *miser*
4. Leaflets glabrous to glabrate on the upper surface 5

5. Stems <15 cm long; keel 8–10 mm long . . . . var. *hylophilus*
5. Stems ≥15 cm long; keel 6–8 mm long . . . . var. *serotinus*

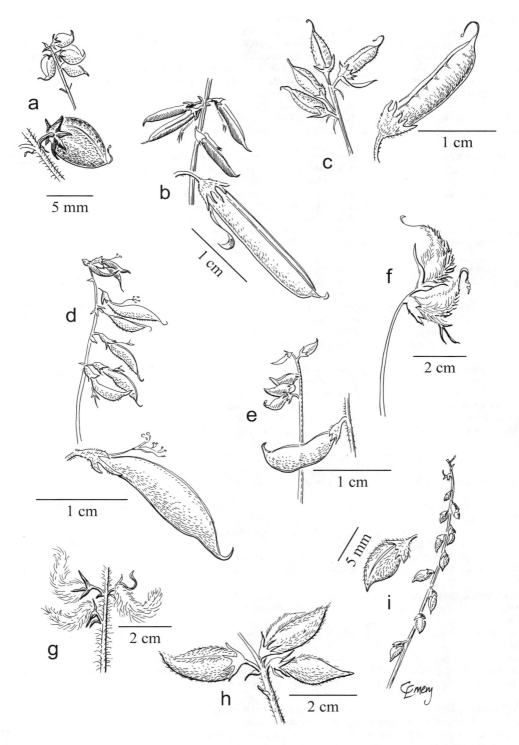

Plate 50. *Astragalus* **II**. **a.** *A. microcystis*, **b.** *A. miser* var. *hylophilus*, **c.** *A. missouriensis*, **d.** *A. robbinsii*, **e.** *A. vexilliflexus*, **f.** *A. pushii* var. *purshii*, **g.** *A. inflexus*, **h.** *A. lotiflorus*, **i.** *A. gracilis*

Astragalus miser var. *crispatus* (M.E. Jones) Cronquist is 4–10 cm high with leaflets 5 to 9 times longer than wide, occurring in granitic soil of montane grasslands, pine forest; endemic to the southern Bitterroot Range of MT, ID; *A. miser* var. *praeteritus* Barneby is 4–15 cm high with leaflets 6 to 10 times longer than wide and is found in stony, calcareous soil of montane grasslands, steppe, woodlands; endemic to east-central ID and adjacent MT, WY; *A. miser* var. *decumbens* (Nutt ex Torr. & A.Gray) Cronquist is 3–8 cm high with leaflets 4 to 6 times longer than wide, occurring in calcareous soil of montane coniferous woodlands; endemic to south-central MT and adjacent WY; *A. miser* var. *miser* is 10–30 cm high with leaflets 6 to 12 times longer than wide, found in valley to montane grasslands, pine woodlands; widespread; *A. miser* var. *hylophilus* (Rydb.) Barneby is 5–20 cm high with leaflets 3 to 10 times longer than wide and is found in open forest or less commonly grasslands, valleys to subalpine; widespread; *A. miser* var. *serotinus* (A.Gray ex Cooper) Barneby is 20–35 cm high with leaflets 5 to 6 times longer than wide; widespread in valley grasslands.

Astragalus missouriensis Nutt. Often nearly stemless perennial with a short-branched caudex. **Herbage** with dense, appressed, dolabriform hairs. **Stems** 1–8 cm, ascending to erect. **Leaves** with 9 to 17 elliptic leaflets, 5–15 mm long with rounded tips; stipules lanceolate, 2–6 mm long, distinct. **Inflorescence** longer than the leaves, short, dense with 3 to 12 spreading to ascending flowers, often prostrate in fruit. **Flowers** lavender; calyx black- and/or white-strigose; sepals 3–5 mm long; banner 15–22 mm long, moderately reflexed; keel 14–20 mm long. **Legume** linear-oblong with a subulate tip, 15–27 mm long, strigillose, turgid, becoming leathery. Grasslands, sagebrush steppe, pine woodlands; plains, valleys. AB to MN south to UT, NM, TX. Our plants are var. *missouriensis*. See *A. chamaeleuce*. (p.301)

Astragalus oreganus Nutt. Rhizomatous perennial. **Herbage** strigose with dolabriform hairs. **Stems** ascending, zig-zag, 10–20 cm. **Leaves** with 5 to 15 obovate leaflets, 5–15 mm long with blunt to rounded tips; lower stipules connate 5–10 mm long. **Inflorescence** barely longer than the leaves, crowded with 15 to 40 spreading flowers. **Flowers** ochroleucus; calyx black- and/or white-strigose; sepals 1–3 mm long; banner 10–15 mm long, becoming moderately reflexed; keel 10–13 mm long. **Legume** ascending, linear-oblong, weakly villous, leathery, 10–16 mm long. Sandy, calcareous soil of sagebrush steppe; valleys. Endemic to central WY and adjacent Carbon Co., MT.

The similar *Astragalus canadensis* is taller without zig-zag stems.

Astragalus pectinatus (Hook.) Douglas ex G.Don Perennial from a loosely branched caudex. **Herbage** strigose. **Stems** ascending, 15–50 cm, basally leafless. **Leaves** with 9 to 15 linear leaflets, 2–5 cm long; the terminal confluent with the rachis; stipules deltoid, 4 to 10 mm long, the lower ones sheathing. **Inflorescence** of 10 to 30 spreading to ascending flowers, ca. as long as the leaves. **Flowers** white; calyx densely black-strigose; sepals 1–3 mm long; banner 15–23 mm long, moderately reflexed; keel 11–14 mm long. **Legume** pendent, linear-ovoid, 1–2 cm long, glabrate, leathery. Grasslands, pine woodlands; plains, valleys. AB, SK south to CO, KS.

Astragalus grayi is similar but has erect pods and a white-hairy calyx.

Astragalus plattensis Nutt. Perennial from widely branched caudex. **Herbage** villous. **Stems** prostrate to ascending, 10–20 cm. **Leaves** with 13 to 25 oblong to narrowly elliptic leaflets, 4–15 mm long with rounded tips; stipules lanceolate, 3–8 mm long, the lower connate. **Inflorescence** shorter than the leaves, crowded, of 6 to 12 spreading to ascending flowers. **Flowers** purple; calyx white-strigose; sepals 2–4 mm long; banner 15–20 mm long, moderately reflexed; keel 10–15 mm long. **Legume** spreading, ovoid, villous, fleshy, becoming hard, leathery, 15–20 mm long including the filiform beak 3–6 mm long. Grasslands, sagebrush steppe; plains. MT, ND south to TX.

Astragalus crassicarpus has larger flowers, glabrous fruits and a more tightly branched caudex.

Astragalus platytropis A.Gray Nearly acaulescent perennial with a woody, simple or branched caudex. **Herbage** densely strigose, canescent. **Stems** <1 cm, covered in stipules. **Leaves** with 9 to 15 linear-elliptic leaflets, 3–8 mm long with rounded tips; stipules lanceolate, ca. 1 mm long, connate. **Inflorescence** mostly shorter than the leaves, dense with 3 to 7 spreading flowers. **Flowers** ochroleucus; calyx white-strigose; sepals 1–2 mm long; banner 7–9 mm long, moderately reflexed; keel 7–9 mm long, purple-tipped. **Legume** pendent, white-strigillose, 15–30 mm long, inflated, papery, ovoid, maroon- and yellow-mottled. Silty, stony, often calcareous soil of sagebrush steppe; valleys, montane. CA, NV; disjunct in southwest MT and adjacent ID.

Astragalus purshii Douglas ex Hook. Nearly acaulescent perennial from a branched caudex. **Herbage** woolly-villous. **Stems** prostrate, 5–10 cm, clothed in stipules. **Leaves** with 7 to 17 obovate to lanceolate leaflets with rounded to acute tips; stipules lanceolate, 3–10 mm long, distinct. **Inflorescence** sessile, up to as long as the leaves, of 3 to 10 spreading to ascending flowers. **Flowers** dull white to purple; calyx black- and/or white-villous; sepals 2–5 mm long; banner 18–25 mm long, moderately reflexed; keel 15–20 mm long. **Legume** narrowly ovoid, compressed perpendicular to the septum, curved to nearly straight, sharp-beaked, 12–25 mm long, densely silky. Grasslands, sagebrush steppe; plains, valleys, montane. BC, AB south to CA, NM, UT, CO, SD. (p.301)

1. Flowers ochroleucus to pale lavender, 18–25 mm long; pods nearly straight . . . . var. *purshii*
1. Flowers magenta, 18–20 mm long; pods curved . . . . var. *concinnus*

Astragalus purshii var. *purshii* occurs in grasslands, sagebrush steppe of plains, valleys; *A. purshii* var. *concinnus* Barneby is found in montane sagebrush steppe in southwest cos.

Astragalus racemosus Pursh Seleniferous perennial from a branched caudex. **Herbage** strigose. **Stems** ascending to erect, 10–50 cm. **Leaves** with 11 to 29 oblong leaflets, 1–3 cm long with acute tips; stipules lanceolate, 3–12 mm long, basally connate. **Inflorescence** longer than the leaves, dense, of 20 to 50 spreading to reflexed flowers. **Flowers** ochroleucus; calyx white-strigose; sepals 2–6 mm long; banner 13–20 mm long, moderately reflexed; keel 10–15 mm long, purple-tipped. **Legume** pendent, narrowly oblong, 15–30 mm long with a stipe 3–5 mm long, glabrate, triangular in cross section. Clay soil derived from Pierre shale in grasslands, sagebrush steppe, often in badlands; plains of Carter and Fallon cos. MT, ND south to UT, NM, TX, OK. Our plants are var. *racemosus*.

Astragalus robbinsii (Oakes) A.Gray Perennial from a loosely branched caudex. **Herbage** glabrate. **Stems** ascending to erect, 20–50 cm. **Leaves** with 7 to 17 lanceolate to oblanceolate leaflets, 10–25 mm long with rounded tips. **Inflorescence** longer than the leaves, secund with 7 to 12 spreading to declined flowers. **Flowers** pale purple; calyx black-strigose; sepals 1–2 mm long; banner reflexed, 9–10 mm long; keel 7–9 mm long. **Legume** pendent, linear-elliptic, triangular in cross section, 13–20 mm long, black- and white-strigose with a stipe 2–5 mm long. Open forest, thickets, meadows, gravel bars, especially along rivers, streams; valleys, montane. AK to NL south to WA, ID, CO, VT. Our plants are var. *minor* (Hook.) Barneby. (p.301)

Astragalus eucosmus has more ovate pods, and pods of *A. australis* are glabrous.

Astragalus scaphoides (M.E. Jones) Rydb. Perennial from a tightly-branched caudex. **Herbage** strigose. **Stems** erect, 15–50 cm. **Leaves** with 15 to 21 oblanceolate leaflets, 5–20 mm long with rounded tips; stipules lanceolate, 1–7 mm long, distinct. **Inflorescence** exceeding the leaves, dense at first with 15–30 nodding to spreading flowers. **Flowers** ochroleucus; calyx black-strigose; sepals 2–3 mm long; banner erect, 16–24 mm long; keel 12–16 mm long. **Legume** narrowly ovoid, glabrous, 15–24 mm long, erect with a stipe 4–5 mm long that arches out and then up. Sagebrush steppe; montane. Endemic to Beaverhead Co. and adjacent ID. The type locality is in Beaverhead Co.

The somewhat more widespread *Astragalus atropubescens* has strictly erect fruit with whiter flowers.

Astragalus shortianus Nutt. Acaulescent perennial from a simple or branched caudex. **Herbage** densely silvery-strigose. **Leaves** with 7 to 17 obovate leaflets, 5–20 mm long. **Inflorescence** of 7 to 16 ascending flowers on a peduncle 2–15 cm long. **Flowers** purple; sepals 3–6 mm long; calyx villous; banner slightly reflexed, 19–22 mm long; keel 15–17 mm long. **Legume** ascending, ovoid to ellipsoid with a flattened beak, silvery-pilose, 25–45 mm long. Stony soil of grasslands, steppe; valleys. MT to NE, CO. One roadside collection from Carbon Co.

Astragalus spatulatus E.Sheld. Acaulescent perennial from a tightly branched caudex, sometimes forming small cushions. **Herbage** strigose to canescent with dolabriform hairs. **Stems** lacking. **Leaves** simple (rarely trifoliolate), petiolate; the blade linear-oblanceolate, 1–3 cm long; stipules connate. **Inflorescence** exceeding the leaves, of 3 to 12 ascending flowers. **Flowers** purple; calyx white- or black-strigose; sepals 1–2 mm long; banner 6–9 mm long, moderately erect; keel 5–7 mm long. **Legume** erect, linear, flattened, slightly curved, white-strigose, 7–15 mm long. Sparsely vegetated, stony soil of grasslands, sagebrush steppe; plains, valleys. AB to SK south to UT, CO, NE.

Astragalus tenellus Pursh Perennial from a spreading, branched caudex. **Herbage** glabrate to strigillose. **Stems** ascending, 15–40 cm. **Leaves** with 11 to 21 linear to oblanceolate leaflets, 6–15 mm long with rounded tips; stipules connate, 3–6 mm long. **Inflorescence** ca. as long as the leaves, open, of 7 to 20 spreading flowers. **Flowers** ochroleucus; calyx sparsely black- and/or white-strigose; sepals ca. 1 mm long; banner 6–10 mm long, reflexed; keel 5–7 mm long, sometimes purple-tipped. **Legume** pendent on a stipe 2–3 mm long, narrowly elliptic, 7–15 mm long, flattened, glabrous, purple-mottled. Often sandy or gravelly soil of grasslands, sagebrush steppe, badlands, river banks; plains, valleys. YT south to NV, NM, NE, MN. Our plants are var. *tenellus*.

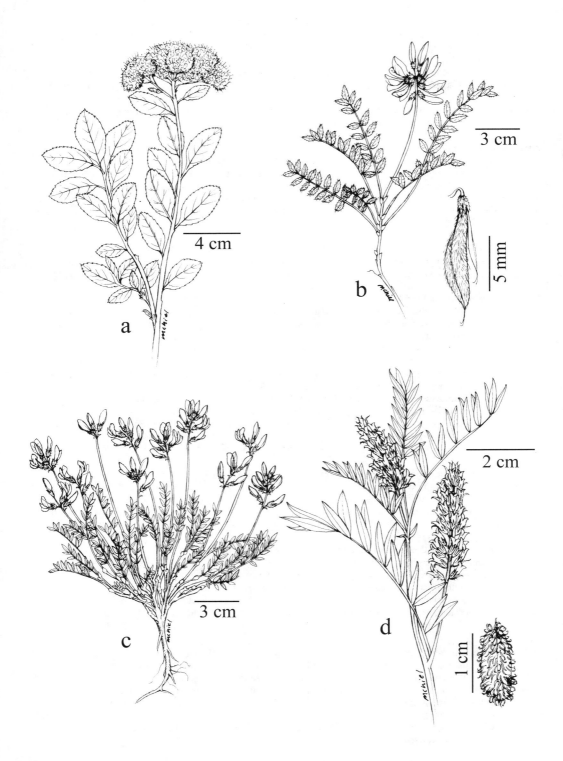

Plate 51. a. *Spiraea splendens*, b. *Astragalus alpinus*, c. *Astragalus bourgovii*, d. *Glycyrrhiza lepidota*

Astragalus terminalis S.Watson Perennial from a tightly branched caudex. **Herbage** canescent with dolabriform hairs. **Stems** erect, 5–30 cm. **Leaves** with 13 to 21 oblanceolate to obovate leaflets, 6–15 mm long with rounded to emarginate tips; stipules lanceolate, 2–5 mm long, distinct. **Inflorescence** exceeding the leaves, dense at first with 15 to 30 spreading flowers. **Flowers** white to pale blue; calyx white- and/or black-strigose; sepals 1–2 mm long; banner reflexed, 10–16 mm long; keel 8–11 mm long, purple-tipped. **Legume** erect, linear-ovoid, glabrous, 12–18 mm long, becoming woody. Stony, usually calcareous soil of sagebrush steppe, grasslands; valleys to alpine. Endemic to Beaverhead and Madison cos. and adjacent ID, WY. See *Astragalus atropubescens*.

The type locality is in Beaverhead Co.

Astragalus vexilliflexus E.Sheld. Perennial often forming loose mats from a branched caudex. **Herbage** glabrate to densely strigillose. **Stems** prostrate to ascending, 5–30 cm. **Leaves** with 7 to 11 linear-lanceolate to narrowly elliptic leaflets, glabrate above, 5–12 mm long with acute tips; stipules lanceolate, 2–4 mm long, basally connate. **Inflorescence** ca. as long as the leaves, with 5 to 10 spreading flowers. **Flowers** white to purple; calyx black- and/or white-strigose; sepals 1–2 mm long; banner 6–8 mm long, erect; keel 3–6 mm long, purple-tipped. **Legume** spreading, narrowly ovate, white-strigillose, 6–13 mm long, flattened but swollen around the seeds. Stony, barren ground in grasslands, steppe, gravel bars, fellfields; montane to alpine. BC to SK south to ID, WY, SD. (p.301)

Riparian plants often have longer leaves with broader leaflets; plants from the Bighorn Basin have small flowers with large fruits. Leaves of *Astragalus kentrophyta* have leaflets confluent with the rachis.

Caragana Fabr.

Caragana arborescens Lam. Siberian pea tree Many-stemmed shrub to 4 m. **Stems** branched, pubescent. **Leaves** often fascicled, becoming glabrate, even-pinnate with 4 to 10 oblong-elliptic leaflets 1–3 cm long with rounded or truncate, mucronate tips; stipules often with nodal spines. **Inflorescence** of 1 to 4 flowers, fascicled on spur branches. **Flowers** papilionaceous, yellow; calyx pubescent, 6–7 mm long; sepals ca. 1 mm long; banner ca. 2 cm long, reflexed, barely longer than the wings and keel. **Fruit** a glabrous, pendent, linear-oblong legume, 3–5 cm long, narrowed to a slender beak; seeds numerous. Rarely escaping into fields, pastures, roadsides; plains, valleys; collected in Missoula Co. Introduced as an ornamental and windbreak; native to Asia.

Coronilla L.

Coronilla varia L. Crown vetch Glabrous, extensively branched perennial. **Stems** lax to ascending, often clambering, 15–80 cm. **Leaves** odd-pinnate with 15 to 23 narrowly oblong leaflets 1–2 cm long with rounded, mucronate tips; stipules linear to oblong, 1–5 mm long, distinct. **Inflorescence** a hemispheric, ebracteate umbel with 6 to 25 spreading flowers on an erect peduncle exceeding the leaves. **Flowers** white to light purple, papilionaceous; calyx tube ca. 2 mm long; sepals ca. 1 mm long; banner 9–13 mm long, moderately reflexed; keel as long; stamens 9 united, 1 separate. **Fruit** ascending to erect, 2–6 cm long, linear, constricted between the 3 to 10 seeds. Roadsides, rarely escaping into fields; valleys. Introduced across N. America for fodder and to stabilize road cuts; native to Europe.

Cytisus L. Broom

Cytisus scoparius (L.) Link Scotch broom Deciduous shrubs 1–3 m. **Stems** strongly angled, green. **Herbage** glabrous to pubescent. **Leaves** nearly sessile, trifoliolate below, simple above; the elliptic leaflets 6–15 mm long; stipules absent. **Inflorescence**: 1 to 3 axillary, pedicellate flowers. **Flowers** papilionaceous, yellow; calyx tubular, 5–6 mm long, glabrous; sepals unequal; corolla 15–22 mm long; styles curved beyond the keel; stamens 10, 4 longer than the others. **Legume** linear-oblong, flattened, 25–40 mm long, ciliate on the sutures; seeds several. Moist roadsides; valleys. Introduced as an ornamental and escaped on the east and west coasts of the U.S.; native to Europe. Known from a single Sanders Co. population just east of the ID border.

Dalea L.

Perennial from a simple to branched caudex. **Herbage** glandular-punctate. **Leaves** odd-pinnate; stipules linear, distinct. **Inflorescence** terminal and axillary, bracteate spikes. **Flowers** nearly regular; calyx 10-ribbed with 5 unequal sepals; corolla of 5 somewhat unequal, separate clawed petals; stamens 5 to 10; **Fruit** a punctate-glandular, obovate legume with 1 or 2 seeds.

1. Spike loosely flowered, the axis visible between flowers . . . .*D. enneandra*
1. Spike densely flowered. 2

2. Calyx glabrous or glabrate . . . .*D. candida*
2. Calyx strigose to sericeous. 2

3. Leaflets 3 to 7. . . . .*D. purpurea*
3. Leaflets >7 . . . .*D. villosa*

Dalea candida Michx. ex Willd. [*Petalostemon candidum* Michx.] WHITE PRAIRIE CLOVER **Root** mostly with a simple crown. **Stems** erect, 20–50 cm, branched above. **Herbage** glabrous. **Leaves** with 5 to 9 oblong leaflets 5–20 mm long with rounded to emarginate tips. **Spikes** dense, tubular, 1–5 cm long. **Flowers** white; calyx tube ca. 2 mm long, glabrous; sepals ca. 1 mm long, ciliate; petals 3–7 mm long; stamens 5. **Legume** glabrate, 3–4 mm long, exserted from the calyx. Grasslands, open slopes; plains. AB to MB south to AZ, TX, KS, TN, AL. Our plants are var. *oligophylla* (Torr.) Shinners. See *D. purpurea*.

Dalea enneandra Nutt. ex Fraser **Roots** yellow. **Stems** erect, 50–90 cm, unbranched below. **Herbage** glabrous, glaucous. **Leaves** with 7 to 11 narrowly oblanceolate leaflets, 4–12 mm long with rounded tips. **Spikes** loose, 2–10 cm long with 5 to 35 flowers. **Flowers** white, enfolded in a pale, ovate bract; calyx tube 2–3 mm long, sericeous; sepals 3–4 mm long; petals 2–6 mm long; stamens 9. **Legume** glabrous, 3–4 mm long, included in the calyx. Sandy or gravelly soil of grasslands, juniper woodlands; plains, valleys. MT, ND south to NM, TX, KS, OK.

Reported to be more common in calcareous soils.[152]

Dalea purpurea Vent. [*Petalostemon purpureum* (Vent.) Rydb.] PURPLE PRAIRIE CLOVER **Root** sometimes with a branched crown. **Stems** erect, 15–50 cm. **Herbage** glabrous to villous. **Leaves** with 3 to 7 linear to narrowly oblanceolate leaflets, 10–15 mm long with acute tips. **Spikes** ovoid-cylindric, 1–4 cm long, villous. **Flowers** bright purple; calyx tube 1–2 mm long, sericeous with straight hairs; sepals ca. 1 mm long; petals 2–6 mm long; stamens 5. **Legume** glabrous, ca. 2 mm long, included in the calyx. Grasslands; plains, valleys. AB to MB south to NM, TX, LA, AL.

Dalea candida has rounded to emarginate leaf tips.

Dalea villosa (Nutt.) Spreng. [*Petalostemon villosus* Nutt.] SILKY PRAIRIE CLOVER **Root** orange with a branched crown. **Stems** ascending, 15–35 cm. **Herbage** densely sericeous. **Leaves** with 11–21 narrowly elliptic leaflets, 5–8 mm long with acute tips. **Spikes** dense, oblong-cylindric, 1–5 cm long, villous. **Flowers** white to light purple; calyx tube ca. 2 mm long, spreading-pilose; sepals ca. 1 mm long; petals 2–5 mm long; stamens 5. **Legume** densely villous, 2–3 mm long, barely exserted. Sandy soil of grasslands, sandhills or below sandstone outcrops; plains. SK, MB south to NM, TX, KS.

Glycyrrhiza L. Licorice

Glycyrrhiza lepidota Nutt. ex Pursh Rhizomatous, perennial herb. **Roots** woody, aromatic. **Stems** erect, 30–120 cm, little branched. **Herbage** pubescent, punctate-glandular, sometimes with stalked glands; stipules small, deciduous. **Leaves** odd-pinnate with 7 to 15 lanceolate leaflets, 1–4 cm long with acute tips. **Inflorescence** a dense, spike-like, bracteate, pedunculate raceme, 1–6 cm long, as high as the subtending leaves. **Flowers** papilionaceous, dull white; calyx tube 3–4 mm long; sepals 2–4 mm long; banner moderately erect, 9–12 mm long; keel 7–11 mm long, as long as the wings; stamens 9 united, 1 separate. **Fruit** an ellipsoid legume, 1–2 cm long, covered with hooked bristles 2–3 mm long; seeds numerous. Gravel bars, moist meadows, thickets along streams, rivers; plains, valleys. BC to ON south to CA, TX, AK, Mexico. (p.304)

Both var. *glutinosa* S.Watson and var. *lepidota*, with and without stalked glands respectively, are reported for MT; however, the trait is somewhat continuous, and both forms occur together in some populations.

Hedysarum L. Sweet vetch

Perennial herbs from woody, branched caudexs. **Herbage** minutely glandular-punctate. **Leaves** odd-pinnate; stipules brown, long, connate. **Inflorescence** a narrow, axillary, pedunculate, bracteate raceme with spreading or declined flowers. **Flowers** papilionaceous, the keel prow-like, longer than wings and (usually) the reflexed banner; calyx with 2 tiny bracteoles; stamens 9 united, 1 separate. **Fruit** a flattened, linear-oblanceolate loment, transversely constricted into ovate, wing-margined, 1-seeded segments.

The constricted pods and blunt keel separate *Hedysarum* from *Astragalus* and *Oxytropis*.

1. Flowers yellow or whitish . . . .*H. sulphurescens*
1. Flowers pink to magenta 2

2. Calyx lobes linear, nearly equal; loment mostly veined across the face. *H. boreale*
2. Some calyx lobes triangular, shorter; loment net-veined 3

3. Loment segments 6–10 mm wide; keel 15–20 mm long . . . . *H. occidentale*
3. Loment segments 3–6 mm wide; keel 10–16 mm long . . . . *H. alpinum*

Hedysarum alpinum L. **Stems** erect, 25–60 cm. **Herbage** glabrate to sparsely strigose. **Leaves** with 15 to 21 lanceolate leaflets, 1–3 cm long with rounded tips; stipules 5–30 mm long. **Flowers** pink; calyx tube strigose, 2–3 mm long; sepals deltoid, ca. 1 mm long, subequal; keel 10–15 mm long. **Loment** sparsely strigose with 1 to 5 reticulate-veined segments, 3–6 mm wide. Gravels bars, moist meadows of aspen parkland; plains, valleys, montane. AK to ON south to BC, WY, SD, VT. Our plants are var. *philoscia* (A.Nelson) Rollins.

Hedysarum boreale Nutt. **Stems** ascending to erect, 15–40 cm. **Herbage** sparsely strigillose to canescent. **Leaves** with 7 to 15 lanceolate leaflets 1–2 cm long with rounded tips; stipules 2–10 mm long. **Flowers** pink to purple; calyx tube 1–3 mm long; sepals linear, 3–5 mm long, subequal; keel 10–18 mm long. **Loment** puberulent with 2 to 6 segments, 4–7 mm wide; the veins mostly perpendicular to the axis. Sparsely vegetated soil of badlands, river banks, grasslands, steppe; plains, valleys, montane. AK to NL south to OR, NM, TX. Our plants are var. *boreale*.

Hedysarum occidentale Greene **Stems** erect, 20–80 cm. **Herbage** sparsely strigillose. **Leaves** with 9 to 21 lanceolate to ovate leaflets 1–3 cm long with rounded tips; stipules 5–25 mm long. **Flowers** deep pink to purple; calyx tube 2–3 mm long, strigose; sepals deltoid to linear, unequal; keel 15–20 mm long. **Loment** sparsely strigose with 1 to 4 reticulate segments, 6–10 mm wide. Moist, open forest, meadows; montane, subalpine. WA to MT south to ID, WY, CO.

Hedysarum sulphurescens Rydb. **Stems** ascending to erect, 15–50 cm. **Herbage** glabrate to sparsely strigose. **Leaves** with 9 to 17 lanceolate to obovate leaflets 1–4 cm long with rounded tips; stipules 10–15 mm long. **Flowers** light yellow to dull white; calyx tube 2–3 mm long, glabrous; sepals 1–2 mm long, slightly shorter above than below; keel 10–18 mm long. **Loment** with 1 to 4 reticulate segments 5–8 mm wide. Grasslands, gravel bars, calcareous soil of moraine, fellfield, turf; montane to alpine. BC, AB south to WA, WY. (p.308)

Plants with light pink flowers have been collected at higher elevations near the Continental Divide south of Glacier National Park and may be hybrids with *Hedysarum occidentale*.

Lathyrus L. Sweet pea

Glabrous perennial herbs. **Stems** ascending or weak, twining. **Leaves** even-pinnate, the rachis prolonged into a tendril or bristle; stipules usually prominent, often with a basal lobe. **Inflorescence** few- to several-flowered, ebracteate, axillary racemes. **Flowers** papilionaceous; calyx 10-veined; banner reflexed, shorter or equal to the wings; stamens 9 united, 1 separate; style with a line of hairs on the inner surface. **Fruit** a linear, somewhat flattened, glabrous legume.

Lathyrus pauciflorus Fernald with several pairs of linear leaflets occurs in adjacent north ID. *Lathyrus ochroleucus* is our only common native species. See *Vicia*.

1. Leaf rachis prolonged into a bristle. . . . . *L. bijugatus*
1. Leaf rachis prolonged into a curling tendril. 2

2. Leaflets 6 to 8; flowers ochroleucus . . . . *L. ochroleucus*
2. Leaflets 2; flowers pink to reddish 3

3. Stems terete, without wings; leaflets <4 cm long with rounded tips . . . .*L. tuberosus*
3. Stems winged; some leaflets >4 cm long, acute 4

4. Stipules 3–5 cm long; legume 6–10 cm long . . . . *L. latifolius*
4. Stipules 1–3 cm long; legume 4–6 cm long . . . . *L. sylvestris*

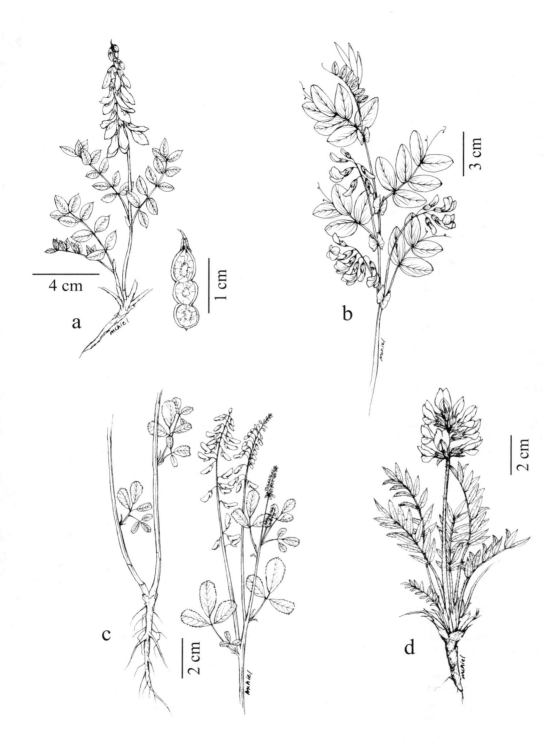

Plate 52. a. *Hedysarum sulphurescens*, b. *Lathyrus latifolius*, c. *Melilotus officinalis*, d. *Oxytropis campestris* var. *cusickii*

Lathyrus bijugatus T.G.White [*L. lanszwertii* Kellogg var. *sandbergii* (T.G.White) Broich] Rhizomatous. **Stems** ascending, 15–35 cm. **Leaves** with 2 to 4 linear to narrowly elliptic leaflets, 2–8 cm long; rachis prolonged into a bristle; stipules linear, 5–10 mm long. **Flowers** blue; calyx tube 2–4 mm long, glabrous; sepals 1–2 mm long, unequal; banner 10–12 mm long. **Legume** 3–4 cm long, slightly curved. Open, park-like, pine forest; valleys. Endemic to ID, adjacent WA and MT. Known from Flathead and Lincoln cos.

A recent treatment places these plants in the more widespread *Lathyrus lanszwertii*[58] but provides no convincing evidence for the change.

Lathyrus latifolius L. Rhizomatous. **Stems** winged, climbing, 60–100 cm. **Leaves** with 2 lance-elliptic leaflets 1–14 cm long; rachis prolonged into a tendril; stipules ovate, 3–5 cm long. **Flowers** pinkish; calyx tube 2–4 mm long; sepals unequal, longer than the tube; banner 18–25 mm. **Legume** 6–10 cm long. Roadsides, fields; valleys; Lake and Sanders cos. Introduced as an ornamental, escaped throughout U.S.; native to Europe.

Lathyrus ochroleucus Hook. Rhizomatous. **Stems** ascending or climbing, 30–80 cm. **Leaves**  with 4 to 8 elliptic leaflets 2–6 cm long; rachis prolonged into a tendril; stipules ovate, entire, 2–3 cm long. **Flowers** ochroleucus; calyx tube 3–5 mm long, glabrous; sepals ciliate, 1–5 mm long, the upper shorter, broader; banner 10–17 mm long. **Legume** 3–5 cm long. Thickets, meadows, moist, open forest especially aspen; valleys, montane. AK to QC south to WA, WY, NE, OH, PA. (p.308)

Lathyrus sylvestris L. Rhizomatous. **Stems** winged, climbing, 6–20 cm. **Leaves** with 2 lanceolate leaflets, 5–12 cm long; rachis prolonged into a tendril; stipules lanceolate, 1–3 cm long. **Flowers** red (drying blue); calyx tube 4–5 mm long; sepals unequal, shorter than the tube; banner ca. 15 mm long. **Legume** 4–6 cm long. Roadsides, fields; valleys; Ravalli and Sanders cos. Introduced as an ornamental, escaped throughout U.S.; native to Europe.

Lathyrus tuberosus L. Tuberous and rhizomatous. **Stems** ascending, 20–60 cm. **Leaves** with 2 oblanceolate leaflets, 2–4 cm long with a mucronate tip; rachis prolonged into a tendril; stipules linear, ca. 1 cm long. **Flowers** red; calyx 5–7 mm long; sepals subequal; banner 12–16 mm. **Legume** 2–4 cm long. Roadsides; valleys. Introduced sporadically in N. America; native to Europe.

Lotus L. Bird's-foot trefoil

Annual or perennial. **Leaves** odd-pinnate with 3 to 5 leaflets; stipules minute. **Inflorescence** axillary pedunculate umbels or flowers solitary. **Flowers** papilionaceous; sepals subequal; banner shorter than the other 4 subequal petals; stamens 9 united, 1 separate. **Fruit** a linear, glabrous, terete or slightly flattened legume.

Lotus nevadensis (S.Watson) Greene var. *douglasii* (Greene) Ottley is similar to *L. corniculatus* but has short-petiolate leaves; it occurs in north ID and could be found in northwest MT.

1. Plants annual; stems erect; flowers solitary . . . . *L. purshianus*
1. Plants mat-forming perennial with lax stems; flowers in umbels . . . . *L. corniculatus*

Lotus corniculatus L. [*L. tenuis* Waldst. & Kit. ex Willd.] Taprooted perennial with a branched crown. **Stems** simple or branched, ascending to prostrate, 15–60 cm. **Herbage** glabrous to sparsely strigose. **Leaves** sessile, with 5 oblanceolate leaflets, 5–15 mm long. **Inflorescence** a long-pedunculate, few-flowered umbel subtended by a trifoliolate bract. **Flowers** yellow, sometimes red-tinged; calyx tube 2–3 mm long with linear sepals 1–3 mm long; banner 6–10 mm long; keel 8–12 mm long. **Legume** 2–4 cm long. Roadsides, fields, pastures; plains, valleys. Introduced for forage and to stabilize road banks, throughout much of N. America; native to Eurasia.

Lotus purshianus (Benth.) F.Clements & E.Clements [*L. unifoliolatus* (Hook.) Benth.] Taprooted annual. **Stems** often simple, erect, 5–30 cm. **Herbage** villous. **Leaves** nearly sessile, with 3 (1 or 5) narrowly elliptic leaflets, 5–20 mm long. **Inflorescence** of solitary pedicellate flowers subtended by a leaflet-like bract. **Flowers** pale yellow; calyx tube ca. 1 mm long; sepals linear, 2–5 mm long; keel 4–8 mm long. **Legume** 15–35 mm long. Sparsely vegetated soil of grasslands, woodlands, roadsides; plains, valleys. BC to MN south to CA, TX, AL, Mexico. Our plants are var. ***helleri*** (Britton) Isely.

Lupinus L. Lupine, lupin

Annual and perennial herbs. **Leaves** palmately 5- to 12-foliate; leaflets oblanceolate; stipules attached to the petiole. **Inflorescence** of narrow, terminal racemes. **Flowers** papilionaceous; calyx 2-lipped, the upper lip 2-lobed; banner reflexed, mostly white to yellow in the center; keel crescent-shaped; wings usually longer than the keel and sometimes joined at the tips; stamens 9 united, 1 separate. **Fruit** a flattened, white-becoming amber-pubescent, linear-elliptic legume with 2 chambers, often slightly constricted between seeds.[34,187,201]

Albino plants or even populations of typically blue-flowered species occur sporadically. Banner is measured across its curviture. Vesture often turns from white to amber in herbarium specimens. *Lupinus plattensis* Wats., with rhizome-like caudex branches, is reported for MT,[113] but I have seen no specimens.

1. Plant taprooted annual of sandy soil . . . .*L. pusillus*
1. Perennial from a tightly to loosely branched caudex 2

2. Plants subacaulescent; racemes overtopped by the leaves; flower bracts persistent . . . .*L. lepidus*
2. Racemes higher than the leaves; flower bracts deciduous at anthesis 3

3. Banner conspicuously pubescent on much of the upper surface. 4
3. Banner glabrous or nearly so 5

4. Stems velvety with spreading, long and short hairs; banner reflexed <45 degrees; racemes
   dense . . . .*L. leucophyllus*
4. Stems sericeous with appressed hairs; banner reflexed >60 degrees. . . . . *L. sericeus*

5. Calyx asymmetrical behind, prolonged into sac-like spur. 6
5. Base of calyx symmetrical, not spurred 7

6. Spur of lower flowers ≥1 mm long; tip of leaflets rounded, mucronate; wing petals often
   with hairs near the tip . . . .*L. arbustus*
6. Spur <1 mm long; leaflets acuminate; wing petals not hairy at the tip . . . .*L. argenteus*

7. Banner reflexed from the top of the wings <45 degrees. . . . .*L. argenteus*
7. Flowers widely gaping, the banner reflexed from top of wings >60 degrees 8

8. Banner darker in the center at anthesis; stems arising from rhizome-like caudex
   branches. . . . . *L. plattensis*
8. Banner lighter in the center at anthesis; plants with a short-branched caudex . . . .*L. polyphyllus*

Lupinus arbustus Douglas ex Lindl. [*L. laxiflorus* Douglas ex Lindl.] Perennial from
a shallow, branched caudex. **Stems** erect to ascending, 30–70 cm, glabrate to strigillose, little
branched. **Leaves** mainly cauline; basal petioles 8–15 cm long; leaflets 7 to 11, 3–10 cm long
with acute to mucronate tips, glabrate to strigose. **Racemes** loose to moderately dense, higher than
the leaves, 6–20 cm long. **Flowers** blue and yellow; calyx tube 1–2 mm long, distended behind the
pedicel into a spur 1–2 mm long; upper lip 2–3 mm long; lower lip slightly longer; banner 8–11 mm long, sparsely
pubescent around the upper calyx lobe, reflexed <45; wings 7–12 mm long, sometimes hairy at the tip. **Legume** 15–35
mm long. Grasslands, dry, open forest; valleys, montane. WA to MT south to CA, NV, UT. (p.312)

White-flowered plants have been called var. *calcaratus* (Kellogg) Isely, but they occur sporadically in this and other species and probably do not merit taxonomic recognition. *Lupinus arbustus* seems to merge seamlessly into *L. argenteus* var. *rubricaulis*[34] in our area and might just as well be called *L. argenteus* var. *laxiflorus* (Lindl.) Dorn.[115]

Lupinus argenteus Pursh Perennial from a shallow, branched caudex. **Stems** erect to
ascending, 10–90 cm, simple or branched above, glabrate to sericeous. **Leaves** mostly cauline;
lower petioles 2–15 cm long; leaflets 5 to 11, strigose to sericeous, sometimes glabrate above,
2–7 cm long with acute tips. **Racemes** moderately dense, greater than the leaves, 2–15 cm long.
Flowers blue; calyx tube 1–3 mm long, flush with the pedicel or distended into a spur <1 mm
long; upper lip 1–4 mm long; lower lip slightly longer; banner 4–11 mm long, glabrous or nearly
so, reflexed <45; wings 5–13 mm long. **Legume** 8–25 mm long. Grasslands, sagebrush steppe, open forest, woodlands,
fellfields; plains, valleys to alpine. BC to SK south to CA, AZ, NM, OK. (p.312)

1. Plants of dry habitats or poorly developed soils; leaflets usually folded; herbage usually sericeous. 2
1. Plants of meadows, open forest, moist steppe; leaflets usually flat and glabrate on the upper surface 3

2. Plants mostly >25 cm high; plants of dry low-elevation habitats . . . . var. *argenteus*
2. Plants <25 cm high; plants subalpine, alpine . . . . var *depressus*

3. Flowers (wings) 5–7 mm long. . . . . var. *parviflorus*
3. Flowers 7–10 mm long 4

4. Stems mostly unbranched with 1 raceme. . . . . var. *rubricaulis*
4. Stems branched above with >1 raceme . . . . var. *argentatus*

Lupinus argenteus var. ***argentatus*** (Rydb.) Barneby has stems 30–90 cm; lower petioles 3–6 cm long; leaflets 2–4 cm long, glabrate to strigose above; flowers 8–10 mm long; sagebrush steppe, grasslands, open forest; plains, valleys, montane; *L. argenteus* var. *argenteus* has stems 15–70 cm; lower petioles 2–8 cm; leaflets 1–4 cm long, strigose to sericeous above; flowers 8–11 mm long, grasslands; steppe; plains, valleys, montane; *L. argenteus* var. *depressus* (Rydb.) C.L.Hitchc. has stems 10–25 cm; lower petioles 2–6 cm long; leaflets 1–4 cm long, sericeous above; flowers 8–11 mm long; meadows, grasslands, fellfields; subalpine, alpine; *L. argenteus* var. *parviflorus* (Nutt. ex Hook. & Arn.) C.L.Hitchc. has stems 40–60 cm; lower petioles 2–8 cm long; leaflets 1–5 cm long, glabrous to strigose above; flowers 5–7 mm long; moist meadows, sagebrush steppe; montane, lower subalpine; *L. argenteus* var. *rubricaulis* (Greene) S.L.Welsh has stems 20–50 cm; lower petioles 2–15 cm long; leaflets 15–70 mm long, glabrate above; flowers 6–13 mm long; open forest; montane, subalpine; *L. argenteus* var. *argophyllus* (A.Gray) S.Watson [*L. caudatus* Kellogg var. *argophyllus* (A.Gray) S.L.Welsh] is reported for MT,[187,201] but I have seen no specimens. Several specimens of var. *depressus* from the east half of Glacier National Park have flowers with well-reflexed banners. The taxonomy of this group is confusing to say the least. Our five vars. are somewhat ecologically divergent, and the extremes of each are distinctive, but they are joined by many intermediate specimens. I have chosen to closely follow Barneby's treatment;[34] other treatments[115] may be just as realistic. The type locality is northeast of Lincoln.

Lupinus lepidus Douglas ex Lindl. Low perennial from a usually simple or branched caudex, sometimes forming small mats. **Stems** simple, ascending, ≤1 cm, clothed in stipules. **Leaves** mostly basal; lower petioles 2–8 cm long; leaflets 5 to 9, sericeous to villous, 1–2 cm long with acute tips. **Racemes** dense, 2–5 cm long, hidden by the leaves. **Flowers** blue; calyx tube 1–2 mm long, sericeous, not spurred; sepals ca. 1 mm long; upper lip bilobed, 3–5 mm long; lower lip nearly entire; banner 5–8 mm long, glabrous, reflexed <45°; wings 7–10 mm long. **Legume** ca. 1 cm long. Moist, stony soil of meadows, streambanks, grasslands; montane to alpine. BC, MT south to CA, NV, UT, CO. Our plants are var. *utahensis* (S.Watson) C.L.Hitchc. (p.312)

Variety *lobbii* (A.Gray ex S.Watson) C.L.Hitchc., with racemes not concealed by the leaves, is reported for southwest MT,[201] but I have not seen any specimens. *Lupinus argenteus* var. *depressus* is often mistaken for *L. lepidus*; the latter retains the flower bracts, but the former does not.[34]

Lupinus leucophyllus Douglas ex Lindl. Perennial from a branched caudex. **Stems** erect, 30–70 cm, simple or branched above, tomentose and spreading-hairy. **Leaves** mostly cauline; lower petioles 3–10 cm long; leaflets 7 to 10, velvety-sericeous, 3–8 cm long with acute tips. **Racemes** dense, greater than the leaves, 5–25 cm long. **Flowers** pale blue; calyx tube ca. 2 mm long, densely villous, not spurred; upper lip ca. 3 mm long; lower lip slightly longer; banner 8–10 mm long, pubescent on the back, reflexed <45°; wings 8–11 mm long. **Legume** 12–26 mm long. Sagebrush steppe, grasslands; valleys, montane. BC, MT south to CA, NV, UT. Our plants are var. *leucophyllus*. (p.312)

The velvety vestiture with densely aggregated pale blue flowers makes this a distinctive lupine in our area. Some plants from the Bozeman area have the dense inflorescence typical of *Lupinus leucophyllus* but vestiture more typical of *L. polyphyllus* var. *burkei*, *L. arbustus* or *L. argenteus* var. *rubricaulis*.[187]

Lupinus polyphyllus Lindl. [*L. wyethii* S.Watson] Perennial from a tightly to loosely branched caudex. **Stems** erect, 1–80 cm, glabrate to sericeous. **Leaves** basal and cauline; lower petioles 6–30 cm long; leaflets 7 to 11, strigose to sericeous beneath, glabrous to sericeous above, 3–10 cm long with acute tips. **Racemes** solitary, dense, higher than the leaves, 2–25 cm long. **Flowers** dark blue; calyx tube strigose to sericeous, 1–2 mm long, not spurred; upper lip 3–6 mm long, equal to the lower; banner glabrous, 8–12 mm long, reflexed >50°, usually with a yellow center; wings 9–15 mm long. **Legume** 10–25 mm long. BC, AB south to CA, NV, UT, NM. (p.312)

1. Leaflets glabrous to glabrate above; lower half of stems hollow; moist habitats . . . . var. *burkei*
1. Leaflets sericeous on the upper surface; stems barely hollow; dry sites . . . . var. *humicola*

Lupinus polyphyllus var. ***burkei*** (S.Watson) C.L.Hitchc. occurs in moist meadows, streambanks; valleys to subalpine; *L. polyphyllus* var. *humicola* (A.Nelson) Barneby [*L. wyethii*] is found in usually gravelly, well-drained soils of grasslands, steppe, open forest; plains, valleys to subalpine. The two vars. are mostly distinct in our area; the few intermediate specimens with sparsely strigose leaflets could be referred to var. *prunophilus* (M.E.Jones) L.Phillips.[201] A large form with an upper calyx lobe 7–8 mm long and a banner ca. 15 mm long occurs in upper montane sites of the

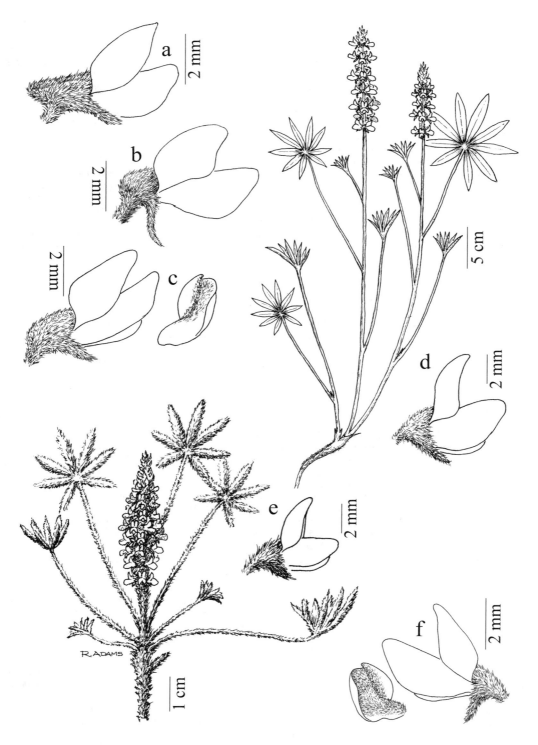

Plate 53. *Lupinus*. **a**. *L. arbustus*, **b**. *L. argenteus* var. *argenteus*, **c**. *L. leucophyllus*,
d. *L. polyphyllus* var. *burkei*, **e**. *L. lepidus*, **f**. *L. sericeus*

Pryor Mtns. *Lupinus arbustus* has the glabrate leaflets of var. *burkei* but has a more open raceme and a spurred calyx and occurs in dry, open forests.

Lupinus pusillus Pursh Taprooted annual. **Stems** erect, 5–20 cm, spreading-villous, often branched at the base. **Leaves** cauline; petioles 1–5 cm long; leaflets 5 to 7, narrowly oblong, glabrous above, sparsely villous beneath, 1–3 cm long with rounded tips. **Racemes** somewhat open, ca. as high as the leaves, 2–6 cm long. **Flowers** blue; calyx tube ca. 2 mm long, villous; upper lip ca. 1 mm long; lower lip much longer; banner 5–9 mm long, glabrous; reflexed ca. 45°; wings 7–10 mm long. **Legume** 10–15 mm long with 2 seeds and a flattened beak 1–3 mm long. Sandy soil of grasslands, sagebrush steppe; plains, valleys. WA, AB, ND south to CA, AZ, NM, NE. Our plants are var. ***pusillus***.

Lupinus sericeus Pursh SILKY LUPINE Perennial from a branched caudex. **Stems** erect to ascending, 20–60 cm, silky-strigose, branched above. **Leaves** basal and cauline; lower petioles 3–10 cm long; leaflets 7 to 9, oblanceolate, sericeous on both surfaces, 2–5 cm long with acute tips. **Racemes** somewhat open, higher than the leaves, 5–15 cm long. **Flowers** blue; calyx tube 1–2 mm long, villous; upper lip 3–5 mm long, ca. as long as the lower; banner 8–12 mm long, pubescent on much of the back, reflexed ≥60°; wings 9–12 mm long. **Legume** 2–3 cm long. Grasslands, sagebrush steppe, woodlands, dry open forest; plains, valleys to subalpine. BC, AB south to AZ, NM, SD. (p.312)

Lupinus polyphyllus var. *humicola* is similar but has glabrous banners, often with more contrast between the blue background and yellow or white center.

Medicago L. Medick

Annual or perennial herbs. **Leaves** petiolate, trifoliolate; leaflets serrate; stipules adnate to petiole. **Inflorescence** pedunculate, axillary, bracteate, capitate racemes. **Flowers** papilionaceous; calyx nearly as long as corolla, sepals ca. equal; banner erect, longer than wings and keel; stamens 9 united, 1 separate. **Fruit** a coiled or curved legume with 2 to many seeds.

Our species are European, introduced widely in N. America for forage and hay.

1. Deep-rooted perennials; flowers ≥6 mm long; leaflets 2–4 cm long 2
1. Shallow-rooted annual or short-lived perennial; flowers 2–5 mm long; leaflets ≤2 cm 3

2. Flowers purple or white; legume coiled . . . . *M. sativa*
2. Flowers yellow; legume curved but not coiled . . . .*M. falcata*

3. Inflorescence of 10 to 35 flowers; legume puberulent . . . . *M. lupulina*
3. Inflorescence of 2 to 5 flowers; legume spiny. . . . .*M. polymorpha*

Medicago falcata L. Deep-rooted perennial similar to *M. sativa*. **Stems** ascending, 40–100 cm. **Herbage** sparsely strigose. **Leaflets** oblanceolate 2–4 cm long. **Inflorescence** 1–2 cm long with 10 to 50 flowers. **Flowers** yellow, 6–8 mm long; calyx 3–5 mm long. **Legume** linear-elliptic, curved, not coiled, 6–10 mm wide, glabrous. Meadows, fields, roadsides; plains, valleys.

Medicago lupulina L. BLACK MEDICK Annual to perennial. **Stems** prostrate to ascending, 10–45 cm. **Herbage** sparsely villous. **Leaflets** obovate, 5–15 mm long. **Inflorescence** 5–10 mm long with 10 to 35 nodding flowers. **Flowers** yellow, 2–3 mm long; calyx ca. 1.5 mm long. **Legume** 1-seeded, reniform, becoming black, sparsely glandular-hairy, veiny, 2–3 mm long. Roadsides, fields, lawns, grasslands, woodlands, meadows; plains, valleys.

Medicago polymorpha L. [*M. hispida* Gaertn.] BUR CLOVER Taprooted annual. **Stems** prostrate to ascending 10–40 cm. **Herbage** glabrate to sparsely strigose. **Leaflets** broadly oblong, 1–2 cm long. **Inflorescence** of 2 to 5 flowers. **Flowers** yellow, 4–5 mm long; calyx ca. 3 mm long. **Legume** circular in 2 to 7 coils, 4–8 mm long, covered with straight or hooked spines. Grasslands, fields, roadsides in Judith Basin and Yellowstone cos.

Medicago sativa L. ALFALFA, LUCERNE Deep-rooted perennial. **Stems** ascending, 30–80 cm. **Herbage** glabrate to sparsely villous. **Leaflets** oblanceolate, 1–3 cm long. **Inflorescence** 1–5 cm long with 10 to 40 flowers. **Flowers** purple or white, 7–12 mm long; calyx 4–7 mm long; sepals longer than the tube. **Legume** circular in 2 to 3 spirals, 3–5 mm wide, veiny, sparsely villous. Fields, roadsides; plains, valleys.

Melilotus Mill. Sweet clover

Sweet-scented, taprooted biennials. **Herbage** nearly glabrous. **Leaves** petiolate, stipulate; leaflets 3, oblong, serrate. **Inflorescence** narrow, minutely bracteate, axillary racemes, 5–16 cm long, of 20 to 80 nodding flowers. **Flowers** papilionaceous; calyx lobes subequal; banner moderately reflexed; stamens 9 united, 1 separate. **Fruit** an ovoid, mostly 1-seeded legume.

Our two species are European, introduced widely in N. America and throughout MT for forage, hay, roadside stabilization. There is a tendency for *Melilotus alba* to be more common along streams and in somewhat moister habitats than *M. officinalis*. Flowers of *M. officinalis* will fade with age and appear whitish. *Melilotus indicus* (L.) All., an annual sweetclover, was collected once in MT from an unknown location nearly 100 years ago.

1. Flowers white . . . .*M. alba*
1. Flowers yellow . . . .*M. officinalis*

Melilotus alba Medik. WHITE SWEET CLOVER **Stems** erect, 50–100 cm. **Leaflets** 1–3 cm long. **Flowers** white, 3–4 mm long; calyx ca. 2 mm long with linear lobes. **Legume** 3–4 mm long. Roadsides, fields, streambanks; plains, valleys.

Melilotus officinalis (L.) Pall. YELLOW SWEET CLOVER Similar to *M. alba*; **Flowers** yellow, 4–6 mm long; calyx ca. 2 mm long. **Legume** ca. 3 mm long. Fields, roadsides, grasslands, open slopes in badlands; plains, valleys. (p.308)

Onobrychis Mill. Sainfoin

Onobrychis viciifolia Scop. Deep-rooted perennial with a branched caudex. **Stems** erect, 30–70 cm. **Herbage** glabrous to sparsely villous, minutely punctate. **Leaves** petiolate, odd-pinnate with 15 to 21 narrowly elliptic to oblong leaflets 1–2 cm long with rounded, mucronate tips; stipules brown-hyaline, 5–11 mm long, connate below with deltoid lobes. **Inflorescence** dense, axillary, bracteate racemes 3–12 cm long with spreading flowers. **Flowers** papilionaceous, magenta; calyx tube ca. 2 mm long, the narrowly lanceolate lobes 3–5 mm long; banner reflexed, 10–12 mm long, ca. equal to the prominent, prow-shaped keel; wings 3–4 mm long; stamens of lower flowers 9 united, 1 separate, those of upper flowers all united. **Fruit** 1-seeded, flattened, ovate, veiny, 5–7 mm long. Fields, roadsides; plains, valleys. Widely introduced in N. America for forage and hay; native to Europe.

Hedysarum spp. have similar flowers but loments have 2 to 7 seeds.

Oxytropis DC. Crazyweed, locoweed

Perennial, often acaulescent herbs with persistent leaf bases. **Leaves** petiolate, odd-pinnate, stipulate; leaflets asymmetrical at the base. **Inflorescence** terminal or sometimes axillary, narrow, bracteate racemes on naked peduncles. **Flowers** papilionaceous; banner erect; keel with a subulate beak hidden by the wings; stamens 9 united, 1 separate. **Fruit** a hairy, narrowly ovate legume with a ventral groove on 1 side and a sharp-pointed beak, usually without a stipe.[34,431,432]

Oxytropis campestris and *O. sericea* can be difficult to distinguish; number of leaves and size of flowers are the characters most often used to separate the two, but they overlap; mature pods or old pods from the previous season are required. The varieties of these two species also intergrade.

1. Pods pendulous; plants often caulescent. 2
1. Pods erect to ascending; plants acaulescent 3

2. Leaf petioles villous; legumes pilose . . . . *O. deflexa*
2. Leaf petioles and legumes sparsely strigose . . . . *O. riparia*

3. Herbage with short glandular hairs among the longer hairs. . . . . *O. borealis*
3. Herbage not glandular 4

4. Racemes with mostly 1 to 3 flowers. 5
4. Most racemes with >3 flowers 7

5. Legume stipitate. . . . .*O. podocarpa*
5. Legume sessile. 6

6. Flowers purple; banner 7–10 mm long. . . . .*O. parryi*
6. Flowers white to yellow; banner ≥10 mm long . . . . alpine forms of *O. campestris* or *O. sericea*

7. Petals white to yellow. 8
7. Petals blue to purple. 9

8. Banner 10–20 mm long; dried pod papery. . . . . *O. campestris*
8. Banner 16–25 mm long; dried pod firm . . . . *O. sericea*

9. Leaflets whorled on the rachis . . . . *O. splendens*
9. Leaflets opposite on the rachis. 10

10. Herbage strigose with dolabriform hairs. . . . . *O. lambertii*
10. Herbage sericeous to villous with basifixed hairs. 11

11. Calyx with long, spreading white hairs, not concealing the surface . . . . *O. besseyi*
11. Calyx densely sericeous with smaller black hairs among the white, especially on the lobes. . . . *O. lagopus*

Oxytropis besseyi (Rydb.) Blank. Acaulescent. **Herbage** silvery-strigose with long and short
hairs. **Leaflets** 7 to 21, lanceolate, 5–15 mm long. **Stipules** membranous, adnate to the petiole,
pilose. **Inflorescence** of 5 to 30 flowers, 8–20 cm high including the peduncle. **Flowers** magenta,
fading to blue; calyx hirsute with white hairs; sepals 2–6 mm long, narrowly lanceolate; banner 15–
23 mm long; wings 2-lobed. **Legumes** ascending, 15–20 mm long, white-villous, inflated, rupturing
the persistent calyx. Sparsely vegetated, stony soil of grasslands, steppe, dry woodlands; plains, valleys to subalpine.
MT south to UT, CO. (p.316)

1. Racemes 2–8 cm long, loosely 8- to 20-flowered; leaflets 9 to 21 . . . . var. *besseyi*
1. Racemes 5–20 mm long, congested, 3- to 10-flowered; leaflets 5 to 9 . . . . var. *argophylla*

Oxytropis besseyi var. *besseyi* occurs plains, valleys to subalpine; *O. besseyi* var. *argophylla* (Rydb.) Barneby is found
in calcareous soil, valleys, montane and is endemic to MT and adjacent WY. *Oxytropis besseyi* flowers several weeks
after *O. lagopus* at the same elevation. Plants of var. *besseyi* from the Sapphire Range east of Hamilton have short black
hairs on the calyx similar to *O. lagopus* but are taller and bloom later than *O. lagopus*. Plants of var. *besseyi* from high
elevations in the Beartooth and Crazy mtns. have small flowers. The type locality is in Gallatin Co.

Oxytropis borealis DC. [*O. viscida* Nutt.] Acaulescent. **Herbage** villous, glandular-pubescent.
Leaflets 15 to 35, lanceolate to ovate, 8–15 mm long. **Stipules** white-papery, deltoid, 3–16 mm
long. **Inflorescence** of 3 to 20 flowers, 5–25 cm long including the peduncle. **Flowers** red-purple;
calyx glandlar-hairy; sepals narrowly lanceolate, 2–4 mm long; banner 11–15 mm long. **Legume**
erect, 10–15 mm long, viscid, black- and/or white-strigose. Sparsely vegetated, gravelly soil of
grasslands, exposed ridges at all elevations. AK to CA, CO. Our plants are var. *viscida* (Nutt.) S.L.Welsh. (p.316)

Oxytropis campestris (L.) DC. [*O. columbiana* H.St.John] Acaulescent. **Herbage** appressed-
villous. **Leaflets** 7 to 29, lanceolate, 3–23 mm long. **Stipules** membranous, adnate to the petiole,
villous. **Inflorescence** of 3 to 40 flowers, 4–40 cm long including the peduncle. **Flowers** white to
ochroleucous; calyx white-villous to black-strigose; sepals lanceolate, 1–3 mm long; banner 10–20
mm long. **Legume** erect, sessile, 1–2 cm long, white- and/or black-pilose; walls ≤0.5 mm thick,
papery when dry. Stony soil of grasslands, gravel bars, fellfields; valleys to alpine. Circumboreal south to WA, CO, ND.
(p.308)

1. Corolla white; keel strongly purple-spotted; leaflets mostly ≤17. . . . . var. *columbiana*
1. Corolla ochroleucous; keel immaculate; leaflets sometimes >17. 2

2. Plants usually >15 cm high; leaflets mostly ≥17; calyx white-sericeous. . . . . var. *spicata*
2. Plants mostly ≤15 cm high; leaflets mostly ≤17; calyx black- and white-strigose. . . . . var. *cusickii*

Oxytropis campestris var. *spicata* Hook. [*O. campestris* var. *gracilis* (A.Nelson) Barneby] occurs in grasslands, gravel
bars; plains, valleys to montane; *O. campestris* var. *cusickii* (Greenm.) Barneby occupies stony soil of exposed ridges,
slopes; subalpine, alpine; *O. campestris* var. *columbiana* (H.St.John) Barneby is similar to *O. sericea* var. *sericea,* found
on gravelly shores around Flathead Lake; also reported for the North Fork Flathead River[121] where plants usually have a
purple-spotted keel, but are better referred to var. *spicata* (fide S. L. Welsh).

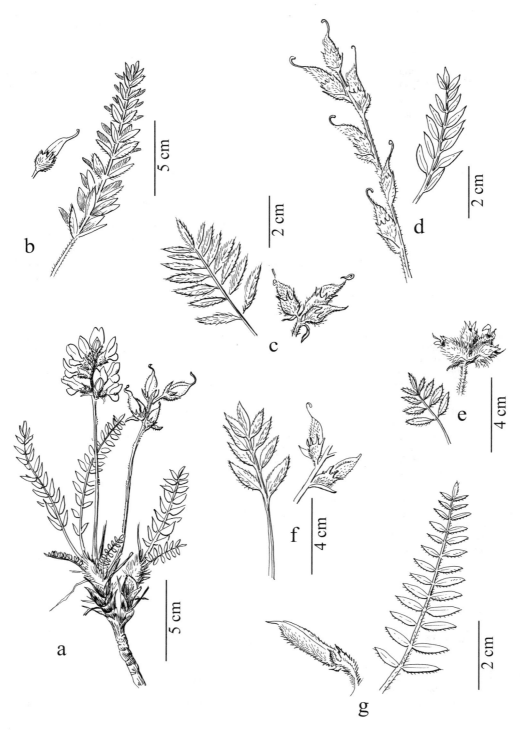

Plate 54. **Oxytropis**. **a.** *O. borealis*, **b.** *O. splendens*, **c.** *O. besseyi*, **d.** *O. lambertii*, **e.** *O. lagopus* var. *conjugens*, **f.** *O. sericea* var. *speciosa*, **g.** *O. deflexa* var. *sericea*

Oxytropis deflexa (Pall.) DC. **Stems** absent or erect, 0–15 cm. **Herbage** villous. **Leaflets** 15 to 30, ovate to lanceolate. **Stipules** herbaceous, lanceolate, distinct, 6–12 mm long, basally adnate to the petiole. **Inflorescence** of 2 to 40 spreading flowers, 5–30 cm long including the peduncle. **Flowers** white to bluish; calyx white-sericeous and black-strigose; sepals linear, 1–5 mm long; banner 6–10 mm long. **Legumes** drooping, 1–2 cm long with a short stipe, black- and white-strigose, compressed perpendicular to the grooved upper suture. Stony soil of meadows, woodlands, thickets, fellfields; montane to alpine. Circumboreal south to CA, NV, UT. (p.316)

1. Plants acaulescent; inflorescence with 2 to 5 flowers . . . . var. *foliolosa*
1. Plants with stems 3–20 cm; inflorescence with 7 to 40 flowers . . . . var. *sericea*

Oxytropis deflexa var. *sericea* Torr. & A.Gray is found in meadows, sagebrush steppe, open forest, often along streams, montane; *O. deflexa* var. *foliolosa* (Hook.) Barneby occurs in calcareous soil of meadows, fellfields, upper montane to alpine in southwest cos. The former var. resembles *Astragalus robbinsii* and *A. eucosmus*, but the latter two are not as shaggy and lack the grooved upper suture on their pods.

Oxytropis lagopus Nutt. Acaulescent. **Herbage** silky-sericeous with mostly long, white hairs. **Leaflets** 7 to 15, lanceolate, 5–15 mm long. **Stipules** membranous, adnate to the petiole, pilose, basally connate. **Inflorescence** of 5 to 20 flowers, 5–12 cm high including the peduncle. **Flowers** magenta, often fading purple; calyx densely white-villous, with shorter appressed black hairs; sepals 3–5 mm long, narrowly lanceolate; banner 13–18 mm long. **Legumes** spreading, 1–2 cm long, white-pilose, inflated, enclosed in or exserted from the calyx. Sparsely vegetated, stony soil of grasslands, steppe, dry woodlands, often on limestone; valleys, montane. AB to ID and WY. (p.316)

1. Leaflets 5 to 9; rachis as long as longest leaflet. . . . . var. *conjugans*
1. Leaflets 9 to 17; rachis longer than leaflets 2

2. Calyx inflated, deciduous with the enclosed pod . . . . var. *lagopus*
2. Calyx little inflated, ruptured by developing pod. . . . . var. *atropurpurea*

Oxytropis lagopus var. *conjugans* Barneby occurs on limestone, valleys to montane, endemic to central MT and adjacent AB; *O. lagopus* var. *lagopus* is found from the valleys to subalpine; *O. lagopus* var. *atropurpurea* (Rydb.) Barneby is found from the valleys to alpine. See *O. besseyi.*

Oxytropis lambertii Pursh Acaulescent. **Herbage** silvery-strigose partly with dolabriform hairs. **Leaflets** 9 to 19, linear-lanceolate, 1–3 cm long. **Stipules** membranous, villous, 7–24 mm long, the free portion deltoid. **Inflorescence** of 6 to 18 flowers, 10–30 cm long including the peduncle. **Flowers** magenta, drying purple; calyx densely white-strigose; sepals 2–3 mm long, linear; banner 15–22 mm long. **Legume** erect, sessile, 1–2 cm long, including the long beak, leathery or woody, white-strigose. Grasslands, steppe, woodlands; plains. BC to MB south to AZ, NM, TX. Our plants are var. *lambertii.* (p.316)

Occasional hybrids with *Oxytropis sericea* are multi-colored.

Oxytropis parryi A.Gray Acaulescent. **Herbage** densely sericeous. **Leaflets** 9 to 19, lanceolate, 3–6 mm long. **Stipules** membranous, adnate to the petiole, villous, 2–5 mm long, basally connate. **Inflorescence** of 1 to 3 flowers, 2–4 cm high including the peduncle. **Flowers** purplish, blooming early; calyx densely black- and white-sericeous; sepals 1–2 mm long, lanceolate; banner 7–10 mm long. **Legumes** ascending, 1–2 cm long, densely black- and white-strigose, thin, leathery. Calcareous fellfields; alpine in Beaverhead and Madison cos. ID, MT south to CA, NV, UT.

Oxytropis podocarpa A.Gray Acaulescent. **Herbage** white-villous. **Leaflets** 9 to 15, linear, 5–8 mm long. **Stipules** membranous, 4–10 mm long, adnate to the petiole, connate below. **Inflorescence** of 1 or 2 flowers, 1–3 cm high including the peduncle. **Flowers** purple; calyx black- and/or white-hirsute; sepals 2–3 mm long; banner 11–14 mm long. **Legumes** stipitate, inflated, papery, 2–3 cm long including the beak, black- and/or white-strigose. Stony, barren, calcareous soil of exposed ridges, slopes; alpine in Glacier and Teton cos. AB to CO, Baffin Island to NL.

Oxytropis riparia Litv. **Stems** ascending to sprawling, 20–60 cm. **Herbage** glabrate to sparsely strigose. **Leaflets** 11 to 15, narrowly ovate, 1–4 cm long. **Stipules** herbaceous, lanceolate, 4–10 mm long, adnate to the petiole. **Inflorescence** axillary racemes 1–25 cm long with 3 to 30 flowers, longer than the leaves. **Flowers** purple; calyx white- and/or black-strigose; sepals 1–2 mm long; banner 6–7 mm long. **Legumes** pendent, 1–2 cm long with a short stipe, white-strigose, somewhat compressed perpendicular to the grooved upper suture. Moist alkaline, riparian meadows; valleys. Introduced for forage sporadically to MT, WY, ID; native to Eurasia. *Oxytropis deflexa* var. *sericea,* is also caulescent but strongly villous.

Oxytropis sericea Nutt. Acaulescent. **Herbage** appressed white-villous. **Leaflets** 11 to 21, lanceolate to narrowly ovate, 5–32 mm long. **Stipules** white-membranous, villous, adnate to the petiole, 8–22 mm long. **Inflorescence** of 5 to 20 flowers, 5–30 cm high including the peduncle. **Flowers** white to ochroleucous; calyx pilose with white and/or black hairs; sepals linear to lanceolate, 2–4 mm long; banner 16–25 mm long. **Legumes** erect, sessile, 15–25 mm long, black-and/or white-strigose, thick and rigid, the walls 0.5–1 mm thick. Grasslands and exposed ridges, slopes at all elevations. YT south to NV, NM, TX. (p.316)

1. Flowers white; keel tip purple-spotted; raceme elongate in fruit. . . . . var. *sericea*
1. Flowers white to ochroleucous; keel immaculate; raceme remaining compact . . . . var. *speciosa*

Oxytropis sericea var. *sericea* occurs in grasslands, sagebrush steppe, plains, valleys, montane; *O. sericea* var. *speciosa* (Torr. & A.Gray) S.L.Welsh [*O. sericea* var. *spicata* (Hook.) Baneby] is found in grasslands, sagebrush steppe, meadows, fellfields at all elevations. Alpine plants usually occur on limestone. It may not always be possible to distinguish *O. campestris* var. *cusickii* from high-elevation forms of *O. sericea* var. *speciosa* (fide R. Barneby).[432] Plants with light blue flowers from eastern MT are assumed to be hybrids with *O. lambertii*.

Oxytropis splendens Douglas ex Hook. Acaulescent. **Herbage** densely villous. **Leaflets** lanceolate, 7–25 mm long in 7 to 15 whorls of 3 to 4. **Stipules** membranous, 10–15 mm long, sericeous, adnate to the petiole, connate below. **Inflorescence** of 9 to 35 flowers, 10–35 cm high including the peduncle. **Flowers** magenta; calyx densely white-villous; sepals linear-lanceolate, 1–6 mm long; banner 10–15 mm long. **Legumes** erect, 1–2 cm long, sessile, thin-walled, villous. Grasslands, meadows, woodlands; montane. AK to ON south to NM, ND, MN. (p.316)

Pediomelum Rydb.

Perennial herbs. **Stems** naked below except for stipules. **Leaves** petiolate, palmately compound; stipules connate below the petiole. **Inflorescence** axillary and/or terminal, spike-like racemes. **Flowers** papilionaceous; calyx basally asymmetrical; banner moderately reflexed, ca. as long as wings; stamens 9 united, 1 separate. **Fruit** a 1-seeded, ovoid, beaked legume, enclosed in the calyx.

   Species were formerly placed in *Psoralea. Pediomelum cuspidatum* (Pursh) Rydb., with glandular herbage, is reported for MT,[152] but I not aware of any voucher specimens.

1. Flowers <10 mm long; stems >15 cm. . . . . *P. argophyllum*
1. Flowers ≥10 mm long; stems mostly 0–20 cm 2

2. Herbage densely hirsute; plants caulescent. . . . . *P. esculentum*
2. Herbage appressed-strigose; plants acaulescent above ground . . . . *P. hypogaeum*

Pediomelum argophyllum (Pursh) J.W.Grimes [*Psoralea argophylla* Pursh] **Stems** arising from rhizomes, 15–60 cm. **Herbage** densely appressed-sericeous. **Leaflets** 3 to 5, obovate, apiculate, 1–4 cm long. **Stipules** linear, 3–10 mm long. **Racemes** axillary, 2–9 cm long including the peduncle, as long as the leaves. **Flowers** blue, fading brown, 4–6 mm long; calyx densely sericeous, sepals lanceolate, 2–3 mm long; the lowest much longer. **Legume** tomentose, ca. 8 mm long; the beak 3–4 mm long. Grasslands, sagebrush steppe; plains, valleys. AB to MB south to NM, TX, MO.

Pediomelum esculentum (Pursh) Rydb. [*Psoralea esculenta* Pursh] INDIAN BREADROOT **Stems** 4–20 cm, from a tuberous, swollen root. **Herbage** hirsute; upper leaflet surfaces glabrate; stems with densely spreading hairs. **Leaflets** 5, narrowly elliptic, 2–4 cm long. **Stipules** lanceolate, 3–8 mm long. **Racemes** 5–12 cm long including the peduncle, longer than the subtending leaves. **Flowers** blue fading to dull white, 14–18 mm long; calyx densely hirsute; sepals 5–7 mm long. **Legume** glabrate, 5–7 mm long; the beak 1–2 cm long. Often sandy soil of grasslands, pine woodlands; plains. AB to MB south to CO, OK, MO. (p.320)
   Pediomelum hypogaeum has appressed rather than spreading hairs on the lower stem.

Pediomelum hypogaeum (Nutt.) Rydb. [*Psoralea hypogaea* Nutt.] Plants nearly acaulescent, arising from long, vertical, rhizome-like branches of the swollen taproot. **Herbage** white-strigose. **Leaflets** 5 to 7, narrowly oblanceolate or lanceolate, 2–6 cm long. **Stipules** lanceolate, 10–25 mm long, densely strigose. **Raceme** 1–5 cm long including the peduncle, shorter than the leaf petioles. **Flowers** lavender to purple, drying tan, 10–13 mm long; calyx appressed hirsute; sepals 5–7 mm long, linear-lanceolate, the lowest longer and wider. **Legume** hirsute, 5–6 mm long; the beak 8–13 mm long. Sandy soil of grasslands, river bluffs; plains. MT south to NM, TX. Our plants are var. ***hypogaeum***. See *P. esculentum*.

Psoralidium Rydb. Scurf-pea

Perennial from deep, branched roots. **Herbage** brown to black, glandular-punctate. **Leaves** petiolate, palmately compound with 3 to 5 short-petiolate leaflets. **Stipules** herbaceous, adnate to the petiole, distinct. **Inflorescence** axillary, bracteate, loosely flowered racemes. **Flowers** papilionaceous; banner moderately reflexed, ca. as long as wings; stamens 9 united, 1 separate. **Fruit** a 1-seeded, indehiscent, short-beaked, glandular-punctate legume.

Species were formerly placed in *Psoralea*. *Psoralidium tenuiflorum* generally has wider-shaped leaflets than *P. lanceolatum*.

1. Flowers purple; legume longer than wide; at least lowest sepal >1 mm long . . . .*P. tenuiflorum*
1. Flowers white; legume orbicular; sepals ≤1 mm long. . . . .*P. lanceolatum*

Psoralidium lanceolatum (Pursh) Rydb. [*Psoralea lanceolata* Pursh] **Stems** ascending to erect, 10–50 cm. **Herbage** glabrate to sparsely strigose. **Leaflets** linear to oblanceolate, sometimes apiculate, 1–4 cm long. **Raceme** 1–4 cm long including the peduncle. **Flowers** white to ochroleucus with a purple-tipped keel, 5–7 mm long; calyx strigose, glandular; sepals deltoid, ciliate, ≤1 mm long. **Legume** orbicular, pubescent, glandular, 4–6 mm long, not enclosed in the calyx. Sandhills, sandy soil of streambanks, around sandstone outcrops; plains, valleys, montane. WA to SK south to CA, TX.

Psoralidium tenuiflorum (Pursh) Rydb. [*Psoralea tenuiflora* Pursh] **Stems** sprawling to erect, 30–70 cm. **Herbage** strigose. **Leaflets** oblanceolate, 1–4 cm long. **Raceme** 1–8 cm long including the peduncle. **Flowers** lavender, drying tan, 3–6 mm long; calyx strigose, glandular; sepals ciliate, 1–2 mm long, the lowest longer. **Legume** ellipsoid, glabrous, glandular, 7–8 mm long. Sandy soil of grasslands, pine woodlands; plains. MT to MN south to NM, TX, Mexico.

The Mineral Co. record was collected along a roadside and may have been introduced.

Robinia L. Locust

Robinia pseudoacacia L. Black locust Trees to 20 m with furrowed bark. **Herbage** glabrate to puberulent. **Stems** sometimes with nodal spines. **Leaves** petiolate, odd-pinnate with 9 to 17 ovate to narrowly elliptic leaflets, 2–6 cm long with pubescent petiolules; stipules linear, early-deciduous. **Inflorescence** axillary, pendent, densely flowered raceme 5–20 cm long. **Flowers** papilionaceous, white; calyx pubescent; sepals deltoid, 1–3 mm long, lower 2 larger; banner 14–19 mm long with a yellow center, ca. as long as the wings and keel; stamens 9 united, 1 separate. **Fruit** a linear, glabrous legume flattened parallel to the sutures, 4–10 cm long, slightly indented between the seeds. River banks, fields; valleys; collected in Lake and Missoula cos. Introduced from eastern N. America to much of the U.S. and Europe.

Sphaerophysa DC.

Sphaerophysa salsula (Pall.) DC. [*Swainsona salsula* (Pall.) Taubert] Rhizomatous herb. **Stems** single or clustered, erect, 40–90 cm. **Herbage** short-strigose. **Leaves** petiolate, stipulate, odd-pinnate with 15 to 25 elliptic leaflets 5–15 mm long. **Inflorescence** loosely flowered, axillary racemes 5–18 cm long. **Flowers** papilionaceous; brick red, fading to brown; calyx white-strigose; sepals deltoid, ca. 1 mm long; banner 10–14 mm long, erect, ca. as long as the wings and keel; stamens 9 united, 1 separate. **Fruit** a papery, inflated, ovoid legume, 15–25 mm long. Roadsides, streambanks; valleys. Introduced sparingly in western U.S.; native to Asia.

This plant has the growth habit of *Glycyrrhiza*, but the fruit and leaves resemble those of *Astragalus*.

Thermopsis R.Br. Golden pea

Rhizomatous perennials. **Leaves** cauline, petiolate, trifoliolate; stipules dimorphic, lower ones at leafless nodes papery and connate, upper ones adnate to the petiole and resembling leaflets. **Inflorescence** terminal and axillary, narrow, bracteate racemes. **Flowers** papilionaceous, yellow; banner erect; stamens 10, separate. **Fruit** a linear legume, compressed parallel to the suture, slightly indented between the seeds. [187]

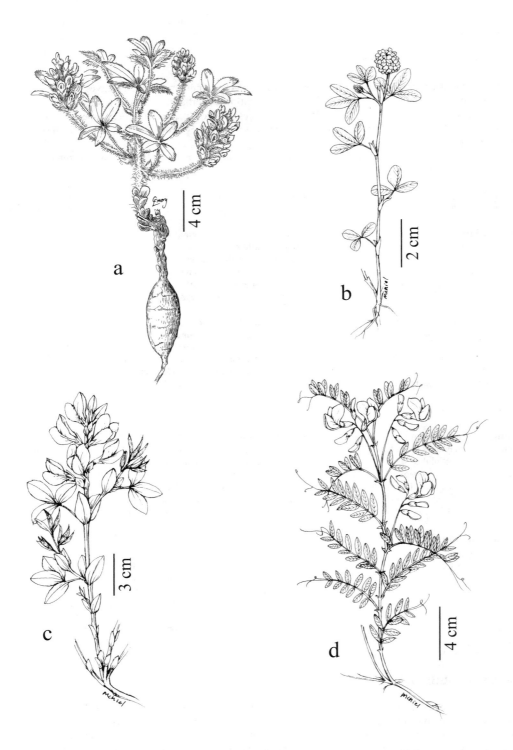

Plate 55. a. **Pediomelum esculentum**, b. **Trifolium aureum**, c. **Thermopsis rhombifolia**, d. **Vicia americana**

The two species are distinct in MT, but less so in other portions of their range; they are sometimes treated as conspecific.[34,201]

1. Legume curved, ascending to spreading on the inflorescence rachis . . . . *T. rhombifolia*
1. Legume straight, erect or nearly so . . . .*T. montana*

Thermopsis montana Nutt. [*T. rhombifolia* var. *montana* (Nutt.) Isely] **Stems** erect, often clustered, hollow below, 40–90 cm. **Herbage** glabrate to strigose. **Leaflets** oblanceolate to obovate, 3–7 cm long; stipules broader but shorter. **Raceme** dense, 10–20 cm long, longer than the leaves. **Flowers** 18–30 mm long; calyx pubescent; sepals 3–4 mm long. **Legume** erect, pubescent, straight, 4–6 cm long with 5 to 15 seeds. Moist meadows, thickets, sagebrush steppe, forest, often along streams; valleys, montane. BC, MT south to CA, UT, CO.

1. Leaflets 4 to 6 times as long as broad . . . . var. *montana*
1. Leaflets <4 times as long as broad. . . . . .var. *ovata*

Thermopsis montana var. **ovata** (B.L.Rob. ex Piper) H.St. John is found in valley riparian forests in Sanders Co.; *T. montana* var. **montana** occurs in sagebrush steppe to riparian forest and is widespread; valleys, montane.

Thermopsis rhombifolia (Nutt. ex Pursh) Richardson **Stems** ascending to erect, often single, 15–40 cm. **Herbage** glabrate to strigose. **Leaflets** ovate, 15–50 mm long; stipules shorter, wider. **Racemes** moderately dense, 3–10 cm long, longer than the leaves. **Flowers** 15–25 mm long; calyx glabrous to sparsely strigose; sepals 2–4 mm long. **Legume** ascending to spreading, pubescent, curved, 3–8 cm long with 3 to 10 seeds. Grasslands, sagebrush steppe, badlands, river banks; plains, valleys. AB to ND south to CO. (p.320)

Trifolium L. Clover

Annual or perennial herbs. **Leaves** petiolate, trifoliolate; leaflets with toothed margins; stipules adnate to the petiole, sometimes connate below. **Inflorescence** axillary and/or terminal, pedunculate, capitate racemes. **Flowers** papilionaceous; banner moderately reflexed, longer than wings and keel; stamens 9 united, 1 separate. **Fruit** a globose to reniform legume enclosed in the dried flower; seeds 1 to 10.

Although native clovers are a conspicuous part of the native montane flora in southern MT and farther south in the Rocky Mountains, they are absent from northern MT through the Canadian Rockies.

1. Plants annual from a slender taproot 2
1. Plants perennial, rhizomatous or from a heavy taproot and often branched crown 8

2. Raceme closely subtended by a saucer-shaped involucre 3
2. Racemes not subtended by an involucre 5

3. Involucre villous . . . . *T. microcephalum*
3. Involucre glabrous 4

4. Sepals with many-branched spine tips. . . . . *T. cyathiferum*
4. Sepals tipped with a simple spine . . . .*T. variegatum*

5. Calyx villous; sepals plumose-hairy . . . . *T. arvense*
5. Calyx glabrous or nearly so 6

6. Racemes <10 mm long . . . .*T. dubium*
6. Racemes ≥10 mm long. 7

7. Petiolule (leaflet stalk) of terminal leaflet twice as long as for lateral leaflets. . . . . *T. campestre*
7. Petiolule of leaflets all ca. the same length . . . .*T. aureum*

8. Plants subacaulescent; stems (not including peduncles) 0–3 cm 9
8. Well-developed plants with leafy stems >3 cm long. 13

9. Calyx villous 10
9. Calyx glabrous 11

10. Leaflets sharply toothed . . . . *T. gymnocarpon*
10. Leaflets entire. . . . .*T. dasyphyllum*

Trifolium arvense L. Annual. **Stems** erect, 10–40 cm. **Herbage** strigose to villous. **Leaflets** linear-oblanceolate, 8–20 mm long with entire margins; stipules 5–10 mm long, free portion linear. **Racemes** axillary and terminal, thimble-shaped, non-involucrate, 5–20 mm long, just longer than the leaves. **Flowers** white to pink, 5–6 mm long; calyx villous; sepals linear, plumose, 3–5 mm long, longer than the corolla. **Legume** 1–2 mm long, 1-seeded. Roadsides, gravel pits; valleys. Introduced to eastern U.S. and Pacific states; native of Eurasia.

Trifolium aureum Pollich [*T. agrarium* L.] HOP CLOVER, YELLOW CLOVER Annual. **Stems** erect or ascending, 15–50 cm. **Herbage** glabrate. **Leaflets** oblanceolate to narrowly elliptic, 5–20 mm long, distal half serrulate; stipules 5–15 mm long, free portion linear-lanceolate, often longer than the petioles. **Racemes** ovoid, axillary, non-involucrate, 1–2 cm long, longer than the leaves. **Flowers** yellow, drying brown, 5–7 mm long; calyx glabrate; sepals 1–2 mm long, upper longer than lower. **Legume** 2–3 mm long, 1-seeded. Along roads, in disturbed meadows, grasslands; valleys, lower montane. Introduced to northern U.S. and adjacent Canada; native to Europe. See *T. campestre, T. dubium.* (p.320)

Trifolium beckwithii Brewer ex S.Watson Perennial with a branched caudex. **Stems** erect, 4–15 cm. **Herbage** glabrous. **Leaflets** narrowly lanceolate to elliptic, serrulate, 2–5 cm long; stipules lanceolate, 1–3 cm long. **Racemes** globose, terminal, 2–4 cm long, long-peduncled, not involucrate. **Flowers** declined, pink to purple, 12–17 mm long; calyx glabrous; sepals linear-lanceolate, 2–4 mm long. **Legume** short-stipitate, ca. 2 mm long with 1 to 4 seeds. Moist riparian meadows, open forests; valleys. OR to MT south to CA, NV, UT. (p.324)

Trifolium campestre Schreb. [*T. procumbens* L.] Annual. **Stems** prostrate to ascending, 10–30 cm. **Herbage** glabrate to villous. **Leaflets** oblanceolate with retuse tips, 8–10 mm long, serrulate; stipules 4–9 mm long, ovate, shorter than the petioles. **Racemes** ovoid, axillary, not involucrate, 10–15 mm long, ca. as long as the leaves. **Flowers** yellow, 4–6 mm long, the banner striate; calyx glabrate; sepals ca. 1 mm long, upper shorter than lower. **Legume** short-stipitate, glabrous, 2–3 mm long, 1-seeded. Roadsides, gardens; valleys. Introduced throughout much of temperate N. America; native to Eurasia.
    Trifolium campestre can be distinguished from *T. dubium* and *T. aureum* by the striate corolla. Our only collection was made over 100 years ago in Gallatin Co.

Trifolium cyathiferum Lindl. Annual. **Stems** ascending to erect, 1–6 cm. **Herbage** glabrous. **Leaflets** obovate, 5–15 mm long, spiny-serrulate; stipules lanceolate, 4–10 mm long. **Racemes** axillary, 8–15 mm long, barely longer than the leaves, subtended by a spiny-margined saucer-shaped involucre. **Flowers** white to pink, 6–10 mm long; calyx glabrous; sepals spiny-lobed, 3–5 mm long, ca. as long as the corolla. **Legume** short-stipitate, ≤1 mm long with 1 or 2 seeds. Wet meadows, sandy streambanks, roadsides; valleys, montane; Missoula and Ravalli cos. BC, MT south to CA, NV, ID.

Trifolium dasyphyllum Torr. & A.Gray Subacaulescent perennial from a heavy taproot and branched crown. **Stems** ascending, 1–3 cm. **Herbage** strigose to sericeous. **Leaflets** narrowly lanceolate to oblanceolate, entire, 7–20 mm long; stipules awn-tipped, ca. 10 mm long. **Racemes** globose, terminal, 10–25 mm long, long-pedunculate, not involucrate. **Flowers** yellow and purple, 10–15 mm long; calyx villous; sepals linear, 3–9 mm long, the lower the longest. **Legume** pubescent, 4–6 mm long with 1 to 3 seeds. Stony soil of turf, fellfields; alpine. MT to UT, CO.

Trifolium dubium Sibth. Annual. **Stems** prostrate to erect, 5–40 cm. **Herbage** glabrous, villous on the stem. **Leaflets** obovate, 5–10 mm long, upper half denticulate, retuse-tipped; stipules 2–5 mm long, ovate. **Racemes** axillary, ebracteate, umbel-like, 6–7 mm long, longer than the leaves. **Flowers** yellow, 3–5 mm long; calyx glabrous; sepals linear, ≤1 mm long. **Legume** 1–2 mm long, 1-seeded. Roadsides, lawns, fields; valleys. Introduced to much of temperate N. America; native of Europe.

Trifolium campestre has larger heads and a striate corolla. *Medicago lupulina* is superficially similar but has rounded leaflet tips and larger stipules.

Trifolium eriocephalum Nutt. [*T. eriocephalum* var. *piperi* J.S.Martin] Perennial with a branched caudex and stout taproot. **Stems** erect, 3–15 cm. **Herbage** glabrate to villous. **Leaflets** narrowly ovate to elliptic, serrulate, 1–2 cm long; stipules 2–5 cm long, free portion lanceolate. **Racemes** globose, mostly terminal, 1–3 cm long, long-peduncled, not involucrate. **Flowers** white to purple, 8–15 mm long, corolla bent down; calyx villous; sepals linear, plumose, 2–4 mm long. **Legume** pubescent, 2–3 mm long with 1 to 4 seeds. Moist riparian meadows; montane. WA to MT south to CA, NV, UT. Our plants are var. **arcuatum** McDermott, known from Ravalli Co.

Trifolium fragiferum L. Stoloniferous perennial. **Stems** prostrate, rooting at the nodes, 5–40 cm. **Herbage** glabrate. **Leaflets** obovate, 1–2 cm long, serrulate with retuse tips; stipules 10–15 mm long, membranous, acuminate. **Racemes** axillary, globose, ca. 1 cm wide in flower, 20 mm wide in fruit, involucrate, long-pedunculate. **Flowers** pink, 4–6 mm long; calyx villous, becoming inflated and veiny; sepals linear, 2–3 mm long. **Legume** ca. 2 mm long, 1-seeded. Lawns, moist meadows, streambanks; valleys. Introduced to western U.S.; native to Eurasia.

Trifolium gymnocarpon Nutt. Subacaulescent perennial with a branched caudex and stout taproot. **Stems** erect, 1–3 cm with persistent stipules. **Herbage** appressed-villous. **Leaflets** narrowly obovate, 5–15 mm long, sharply serrate with pointed tips; stipules scarious, 5–10 mm long. **Racemes** umbellate, terminal, 1–2 cm wide, long-peduncled, non-involucrate, shorter than the leaves. **Flowers** ochroleucus, 8–10 mm long; calyx villous; sepals lanceolate, 2–3 mm long. **Legume** short-stipitate, glabrous, 2–3 mm long, mostly 1-seeded. Sagebrush steppe, open ponderosa pine forest; montane; Granite and Ravalli cos. OR to MT south to CA, AZ, NM.

Trifolium haydenii Porter Subacaulescent perennial, often forming tight to loose mats from a branched caudex and thick taproot. **Stems** prostrate to ascending, 1–2 cm, covered in old stipules. **Herbage** glabrous. **Leaflets** oblong to obovate, denticulate, 4–20 mm long; stipules scarious, lanceolate, 3–8 cm long. **Racemes** umbellate, terminal, 1–3 cm wide, non-involucrate, longer than the leaves. **Flowers** declined, whitish to light purple, 10–15 mm long; calyx glabrous; sepals narrowly lanceolate, 2–4 mm long. **Legume** short-stipitate, glabrous, ca. 3 mm long with 1 or 2 seeds. Sparsely vegetated, stony soil of meadows, turf, fellfields; subalpine, alpine. Endemic to southwest MT and adjacent WY. (p.324)

The type locality is in Madison Co. A specimen from the Little Belt Mtns. at the north edge of the range has exceptionally small flowers.

Trifolium hybridum L. Short-lived perennial. **Stems** prostrate to ascending, 15–50 cm. **Herbage** glabrate. **Leaflets** ovate, 1–3 cm long, serrulate with rounded to retuse tips; stipules 5–20 mm long, acuminate, membranous below. **Racemes** axillary, globose, 15–25 mm wide, non-involucrate, shorter or longer than the leaves. **Flowers** white to pink, 5–10 mm long; calyx glabrous but sparsely ciliate in the sinuses; sepals linear, 1–2 mm long. **Legume** short-stipitate. 3–4 mm long with 2 to 4 seeds. Roadsides, fields, lawns; plains, valleys. Introduced for forage to most of temperate N. America; native of Europe. (p.324)

Trifolium repens may also have pink flowers but lacks the ciliate calyx sinuses.

Trifolium latifolium (Hook.) Greene Rhizomatous perennial. **Stems** clustered, erect, 5–20 cm. **Herbage** glabrate to sparsely pubescent. **Leaflets** narrowly ovate to obovate, serrulate, 1–5 cm long; stipules lanceolate, 1–2 cm long. **Racemes** axillary and terminal, hemispheric, 2–3 cm wide, non-involucrate. **Flowers** pink, 12–16 mm long, ascending, becoming declined; calyx villous; sepals linear-lanceolate, 3–6 mm long. **Legume** strigose, 3–4 mm long with 1 to 2

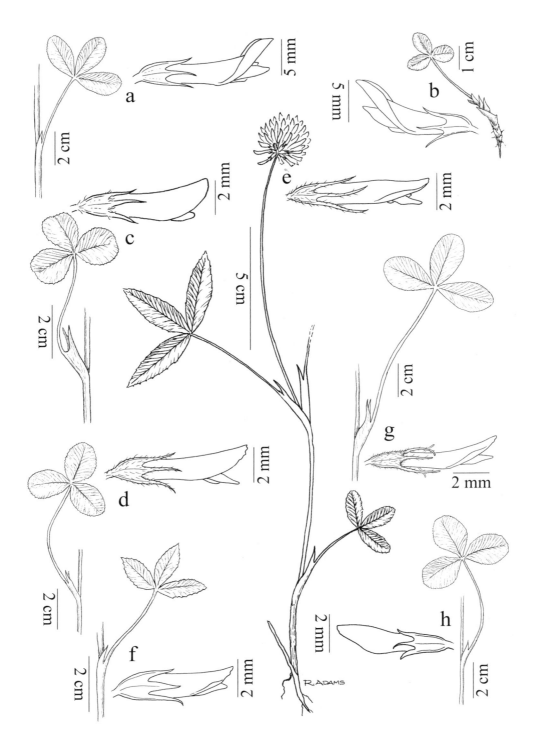

Plate 56. *Trifolium*. **a.** *T. beckwithii,* **b.** *T. haydenii,* **c.** *T. hybridum,* **d.** *T. latifolium,*
e. *T. longipes,* **f.** *T. parryi,* **g.** *T. pratense,* **h.** *T. repens*

seeds. Moist forest openings, meadows; valleys. Endemic to Lincoln and Sanders cos., eastern WA, adjacent ID, OR. (p.324)

Trifolium longipes Nutt. Rhizomatous perennial from a branched caudex. **Stems** ascending, 5–30 cm. **Herbage** pubescent. **Leaflets** narrowly lanceolate to elliptic, denticulate, 2–6 cm long; stipules lanceolate, 1–3 cm long. **Racemes** globose, axillary and terminal, 1–3 cm long, long-peduncled, non-involucrate. **Flowers** becoming reflexed, white to purple, 11–15 mm long; calyx villous; sepals linear, 4–6 mm long. **Legume** short-stipitate, with 1 to 4 seeds. Moist meadows, woodlands, often along streams, wetlands; valleys to subalpine. BC, MT south to CA, UT, CO. Our plants are var. *reflexum* A.Nelson. (p.324)

Variety *pedunculatum* (Rydb.) C.L.Hitchc., with obconic heads, occurs in central ID and could occur in adjacent MT. One collection from Beaverhead Co. has linear leaflets, short sepals.

Trifolium microcephalum Pursh Annual. **Stems** ascending to erect, 6–15 cm. **Herbage** villous to hirsute. **Leaflets** obovate, 6–15 mm long, serrate distally; stipules lanceolate, 5–10 mm long. **Racemes** axillary, 5–10 mm long, longer than the leaves, subtended by a villous, saucer-shaped involucre with spine-tipped lobes. **Flowers** white to pink, 4–5 mm long; calyx villous basally; sepals deltoid, spine-tipped, 3–5 mm long, ca as long as the corolla. **Legume** larger than the calyx, 2–3 mm long with 1 or 2 seeds. River banks, disturbed soil of fields; valleys; Missoula and Ravalli cos. BC, MT south to CA, AZ, Mexico.

The type locality is south of Missoula.

Trifolium nanum Torr. Mat-forming, subacaulescent perennial with a branched caudex from a heavy taproot. **Stems** ascending, 1–2 cm. **Herbage** glabrous. **Leaflets** oblanceolate, serrulate above, 5–8 mm long; stipules lanceolate, sheathing the stem. **Racemes** terminal, of 1 to 4 flowers, barely longer than the leaves subtended by 4 small bracts united below. **Flowers** magenta, 13–18 mm long; calyx glabrous; sepals deltoid, 1–3 mm long, subequal. **Legume** glabrous, 5–15 mm long with 1 to 4 seeds. Non-calcareous soil of fellfields, turf; alpine. MT to UT, NM.

Trifolium parryi A.Gray Subacaulescent perennial from a branched caudex and thick taproot. **Stems** ascending, 1–5 cm. **Herbage** glabrate to brown-pubescent. **Leaflets** obovate, weakly serrulate, 1–3 cm long; stipules scarious, 5–15 mm long. **Racemes** above the leaves, globose, terminal, 15–25 mm long subtended by 6 to 12 scarious, ovate bracts. **Flowers** erect, purple, 11–18 mm long; calyx glabrous; sepals linear-lanceolate, 2–5 mm long. **Legume** short-stipitate, glabrous, 5–6 mm long with 1 to 4 seeds. Non-calcareous soil of meadows, turf; alpine. ID, MT south to UT, NM. Our plants are var. *montanense* (Rydb.) S.L.Welsh. (p.324)

Trifolium pratense L. Short-lived perennial. **Stems** ascending to erect, 30–70 cm. **Herbage** glabrate to sparsely villous. **Leaflets** ovate to elliptic, 2–5 cm long, serrulate with rounded tips; stipules 1–3 cm long, lanceolate, green-veined. **Racemes** axillary, short-pedunculate, globose, 2–4 cm long, non-involucrate, ca. as long as the leaves. **Flowers** ascending, reddish-purple, 10–15 mm long; calyx hirsute; sepals linear-lanceolate, 2–5 mm long. **Legume** 2–3 mm long with 1 or 2 seeds. Meadows, fields, lawns, roadsides, river banks; plains, valleys, montane. Introduced for forage and hay throughout most of temperate N. America; native to Europe. (p.324)

Trifolium repens L. Stoloniferous perennial. **Stems** prostrate to ascending, 5–60 cm. **Herbage** glabrate. **Leaflets** obovate, 5–30 mm long, serrulate with rounded to retuse tips; stipules 3–10 mm long, united most of their length. **Racemes** axillary, globose, 15–25 mm wide, not involucrate, long-pedunculate, longer than the leaves. **Flowers** white to pink, 5–10 mm long; calyx with purple sinuses, sparsely hairy at the base; sepals linear, 1–3 mm long, the upper longer. **Legume** 3–5 mm long with 3 or 4 seeds. Meadows, moist grasslands, roadsides, fields; plains, valleys, rarely subalpine. Introduced for a cover crop or forage throughout temperate N. America; native to Eurasia. See *T. hybridum*. (p.324)

Trifolium variegatum Nutt. Annual. **Stems** prostrate to erect, 5–30 cm. **Herbage** glabrous. **Leaflets** obovate, 5–15 mm long, serrate; stipules ovate, lacerate. **Racemes** axillary, umbellate, 1–2 cm wide, longer than the leaves, subtended by a glabrous, saucer-shaped involucre with spine-tipped lobes. **Flowers** purple, 6–10 mm long; calyx glabrous; sepals narrowly deltoid, spine-tipped, 3–5 mm long, shorter than the corolla. **Legume** often rupturing the calyx, 2–3 mm long with 1 or 2 seeds. Fields, gardens; possibly introduced in valleys of Missoula Co. BC, MT south to CA, AZ, UT, WY.

Vicia L. Vetch

Annual or perennial twining herbs. **Leaves** short-petiolate, stipulate, odd-pinnate, the terminal leaflet modified into a tendril. **Inflorescence** an axillary raceme, sometimes reduced to 1 to 3 pedicillate flowers. **Flowers** papilionaceous; banner reflexed, longer than the keel; stamens 9 united, 1 separate; style with a ring of hairs below the stigma. **Fruit** a several-seeded, linear to narrowly oblong legume, flattened parallel to the sutures.

Lathyrus is our other Fabaceous genus with twining stems.

1. Flowers 12 to 70 in a crowded, secund raceme. 2
1. Flowers 1 to 8 in loosely flowered racemes 3

2. Some calyx teeth >2 mm long; herbage with long spreading hairs . . . .*V. villosa*
2. Calyx teeth ≤2 mm long; herbage glabrate to strigose. . . . . *V. cracca*

3. Flowers 5–6 mm long . . . . *V. tetrasperma*
3. Flowers 10–22 mm long 4

4. Sepals 1–3 mm long; racemes with 2 to 8 flowers . . . . *V. americana*
4. Sepals 3–4 mm long; inflorescences with 1 or 2 flowers . . . . *V. sativa*

Vicia americana Muhl. ex Willd. AMERICAN VETCH Rhizomatous perennial. **Stems** prostrate to ascending or climbing, 10–100 cm. **Herbage** glabrate to sparsely strigose. **Leaves** with a simple or branched tendril; the 6 to 14 leaflets ovate to linear, 1–4 cm long, mucronate with rounded to retuse tips; stipules dentate, 2–8 mm long. **Racemes** secund, loose with 2 to 8 flowers. **Flowers** blue to purple, 14–22 mm long; calyx glabrous to sparsely villous; the deltoid sepals 1–3 mm long, the upper shorter. **Legume** stipitate, narrowly elliptic, 2–3 cm long. Grasslands, meadows, woodlands, open forest; plains, valleys, montane. AK to ON south to CA, NM, TX, MO, OH, Mexico. (p.320)

1. Flowers 5 to 10; tendrils usually branched; leaflets mostly >7 mm wide . . . . var. *americana*
1. Flowers 2 to 5; tendrils simple; most leaflets <7 mm wide . . . . var. *minor*

Vicia americana var. *americana* [*V. americana* var. *truncata* (Nutt.) W.H.Brewer ex S.Watson] is found in forest, wet meadows in northwest cos.; *V. americana* var. *minor* Hook. is found in grasslands, woodlands throughout MT.

Vicia cracca L. Rhizomatous perennial. **Stems** clambering or climbing, 50–100 cm. **Herbage** glabrate to strigose. **Leaves** with a branched tendril; the 12 to 18 leaflets linear, 15–20 mm long; stipules toothed to entire, 7–15 mm long. **Racemes** secund, dense with 20 to 70 flowers. **Flowers** violet, 10–15 mm long; calyx strigose; sepals deltoid to lanceolate, 0.5–2 mm long, the upper shorter. **Legume** stipitate, oblong, 15–20 mm long. Roadsides, fields; valleys. Introduced for forage and roadside stabilization; native to Europe. Known from one Ravalli Co. collection.

Vicia sativa L. Annual. **Stems** prostrate or climbing, 20–90 cm. **Herbage** glabrate to sparsely strigose. **Leaves** with a branched tendril; the 6 to 16 leaflets oblanceolate with retuse-mucronate tips, 14–30 mm long; stipules deeply dentate, 2–8 mm long. **Racemes** of 1 or 2 flowers. **Flowers** blue to magenta, 10–18 mm long; calyx sparsely strigose; sepals linear-lanceolate, 3–4 mm long, subequal. **Legume** linear, 2–7 cm long. Roadsides, fields; valleys. Introduced throughout temperate N. America as a cover crop and forage; native to Europe.

Vicia tetrasperma (L.) Schreb. Annual. **Stems** prostrate or climbing, 30–70 cm. **Herbage** glabrate. **Leaves** with a simple tendril; leaflets 8–10, linear, 15–20 mm long; stipules linear, 2–4 mm long. **Racemes** of 2 or 3 flowers. **Flowers** blue, 5–6 mm long, the banner barely reflexed; calyx glabrate; sepals linear-lanceolate, ca. 1 mm long. **Legume** stipitate, narrowly elliptic, 10–15 mm long. River banks; valleys. Introduced throughout much of temperate U.S.; native to Europe. Collected once in Ravalli Co.

Vicia villosa Roth Annual. **Stems** prostrate or climbing, 20–100 cm. **Herbage** villous. **Leaves** with a branched tendril; the 10 to 16 leaflets linear-lanceolate, 1–2 cm long; stipules entire to dentate, 6–13 mm long. **Racemes** secund, dense with 12 to 30 flowers. **Flowers** blue and white, 12–18 mm long; calyx sparsely villous; the linear lower sepals 2–6 mm long, the upper much shorter. **Legume** stipitate, elliptic, 15–30 mm long. Roadsides, fields, grasslands; valleys. Introduced throughout temperate N. America for forage; native to Europe.

ELAEAGNACEAE: Oleaster Family

Actinorhizal shrubs or small trees. **Herbage** with star- or flake-shaped trichomes. **Leaves** short-petiolate, entire-margined. **Inflorescence** of 1 to 4 short-pedicellate flowers in leaf axils or on short first-year twigs. **Flowers** bisexual or unisexual, perigynous, 4-merous; hypanthium 4-lobed; petals absent; style 1; stamens 4 or 8. **Fruit** a solitary achene surrounded by the swollen hypanthium base which becomes mealy or fleshy

1. Leaves opposite; berry orange to red, juicy . . . . *Shepherdia*
1. Leaves alternate; berry silvery, dry. . . . . *Elaeagnus*

Elaeagnus L.

Shrubs or small trees. **Stems** thorny or not. **Leaves** alternate. **Flowers** perfect, borne in leaf axils; stamens 4. **Fruit** drupe-like, dry, whitish with scale-like trichomes.

1. Stems thorny; twigs silvery; leaves narrowly lanceolate. . . . . *E. angustifolia*
1. Stems unarmed; twigs brownish; leaves ovate to elliptic . . . . *E. commutata*

Elaeagnus angustifolia L. RUSSIAN OLIVE Large shrub or small tree to 8 m. **Stems** thorny; twigs silvery-mealy becoming orange-brown. **Leaf blades** narrowly lanceolate, 3–10 cm long, white-mealy, silvery beneath, less so above. **Flowers:** hypanthium tubular, 5–6 mm long, silvery; sepals deltoid, 2–4 mm long, yellow within. **Fruit** ovoid, 8–15 mm long, becoming green. Woodlands, thickets, riparian forests, moist meadows around wetlands, somewhat tolerant of saline soil; plains, valleys. Introduced to north-temperate N. America as a windbreak or ornamental; native to Eurasia.

Elaeagnus commutata Bernh. ex Rydb. SILVERBERRY, WOLF WILLOW Shrub 1–3 m high, spreading by roots. **Stems** unarmed; twigs mealy, becoming brown, then gray. **Leaf blades** narrowly ovate to elliptic, 2–7 cm long, white-mealy, silvery beneath. **Flowers** sweetly scented; hypanthium campanulate, 6–10 mm long, silvery outside; sepals 3–5 mm long, yellowish inside. **Fruit** ellipsoid, 9–12 mm long. Forming thickets on banks of rivers, streams; plains, valleys. AK to QC south to ID, UT, ND, MN. (p.330)

Shepherdia Nutt.

Dioecious shrubs. Stems thorny or not. **Leaves** opposite. **Flowers** unisexual, borne on short, first-year twigs; stamens 8. **Fruit** berry-like, juicy, red to orange.

1. Stems thorny; leaves oblanceolate, white-mealy beneath . . . . *S. argentea*
1. Stems unarmed; leaves narrowly ovate, brown-mealy beneath. . . . . *S. canadensis*

Shepherdia argentea (Pursh) Nutt. BUFFALOBERRY Erect shrub 2–4 m. **Stems** with thorn-tipped side shoots; twigs white-mealy, becoming gray. **Leaf blades** oblanceolate, 2–4 cm long, white-mealy on both sides. **Flowers:** hypanthium mealy, 1–3 mm long; sepals golden on the inside; male sepals ca. 0.5 mm long; female sepals ca. 1 mm long. **Fruit** ovoid, red, juicy, 5–7 mm long, edible. Cool slopes in grasslands, riparian thickets, forests; plains. BC to MB south to CA, AZ, NM, NE.

Shepherdia canadensis (L.) Nutt. SOAPBERRY, CANADA BUFFALOBERRY Shrub 1–2 m. **Stems** unarmed; twigs brown-mealy, becoming gray. **Leaf blades** narrowly ovate, 2–10 cm long, green above, white- and brown-mealy beneath. **Flowers:** hypanthium brown-mealy, 2–4 mm long; sepals yellow-green on the inside, male ca. 1 mm long, female 1–2 mm long, stellate-pubescent at the base. **Fruit** red, ovoid, 5–7 mm long, insipid. Forests, woodlands; plains, valleys, montane. AK to NL south to CA, AZ, NM, SD, NY. (p.330)

Shepherdia canadensis increases with fire.

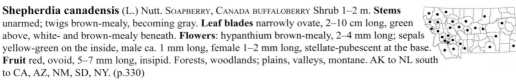

HALORAGACEAE: Water milfoil Family

Myriophyllum L. Water milfoil

Submersed, glabrous, aquatic herbs. **Stems** flaccid. **Leaves** 4-whorled, deeply pinnately divided into linear segments. **Inflorescence** a terminal bracteate spike, with whorled, sessile flowers. **Flowers** regular, perigynous, small, unisexual, male above female, 4-merous; petals sometimes absent; stamens 4 or 8; ovary inferior; stigmas 2 to 4, feathery. **Fruit** 4 nut-like mericarps.[76]

Native species overwinter as tight balls of unexpanded leaves (turions).

1. Floral bracts pinnately divided . . . .*M. verticillatum*
1. At least upper floral bracts entire to dentate. 2

2. Flower bracts >5 mm long . . . . *M. quitense*
2. Flower bracts <4 mm long 3

3. Leaves with 4 to 14 pairs of segments; segments mostly spreading . . . . *M. sibiricum*
3. Leaves with 14 to 24 pairs of segments; segments ascending . . . . *M. spicatum*

Myriophyllum quitense Kunth **Stems** simple or branched. **Leaves** 4- to 5-whorled, to 25 mm long, divided into 13 to 21 filiform segments. **Bracts** 7–10 mm long, larger than fruits. **Flowers** male and female similar; sepals white; petals 2–3 mm long, quickly deciduous; bracts pectinate, 1–2 mm long; stamens 8. **Mericarps** ca. 2 mm long.

Rivers, streams; valleys, montane. BC south to CA, AZ, WY, Mexico, S. America. Known from Gallatin Co.

Myriophyllum sibiricum Kom. [*M. exalbescens* Fernald, *M. spicatum* L. var. *exalbescens* (Fernald) Hultén] **Stems** whitish when dry, branched. **Leaves** 4-whorled, 1–4 cm long, divided into 10 to 24 filiform segments. **Bracts** 1–3 mm long ca. as long as the fruits. **Female flowers** with a vestigial calyx and corolla; bracts dentate, smaller than the flowers. **Male flowers** with small, pink petals; bracts narrowly ovate, entire, smaller than the flowers; stamens 8. **Mericarps** 2–3 mm long.
Shallow to rather deep water of ponds, lakes; plains, valleys, montane. AK to NL south to CA, AZ, TX, IL, WV; Europe (p.330).

Myriophyllum spicatum L. Eurasian water milfoil **Stems** branched, tawny when dry. **Leaves** 4- to 5-whorled, to 25 mm long, divided into 24 to 50 filiform segments. **Bracts** 1–3 mm long, smaller than fruits. **Female flowers** with a vestigial calyx and corolla; bracts entire to pectinate. **Male flowers** with small, deciduous, pink petals; bracts narrowly ovate, entire; stamens 8. **Mericarps** 2–3 mm long. Open water of reservoirs; valleys. Introduced throughout N. America; native to Eurasia; reported for several cos.

Myriophyllum verticillatum L. **Stems** whitish, branched. **Leaves** 4-whorled, to 45 mm long, divided into 18 to 34 filiform segments. **Bracts** 3–8 mm long; longer than the fruits, pectinate, 1 to 2 times as long as flowers. **Female flowers** with sepals <1 mm long; petals absent; bracts minute. **Male flowers** with small, yellowish petals; bracts dentate; stamens 8. **Mericarps** 2–3 mm long. Shallow to rather deep water of ponds, lakes; plains, valleys to montane. AK to NL south to CA, AZ, TX, IL, WV; Europe. Collected in Lincoln, Flathead, and Sweet Grass cos.; reported for Sanders and several other cos.

LYTHRACEAE: Loosestrife Family

Annual or perennial herbs. **Leaves** opposite (at least below), simple with entire margins, sessile or nearly so; stipules vestigial. **Inflorescence** of clusters of sessile flowers in leaf axils or a leafy-bracteate spike. **Flowers** perfect, regular, perigynous, 4- or 6-merous; sepals alternating with tooth-like appendages; petals distinct; stamens 4 to 12; ovary superior; style 1. **Fruit** a capsule with numerous seeds.

1. Perennials; hypanthium cylindric; petals 5 to 7 . . . . Lythrum
1. Annuals; hypanthium campanulate; petals 0 to 4. 2

2. Leaves clasping . . . . Ammannia
2. Leaves sessile to short-petiolate but not clasping . . . . Rotalla

Ammannia L.

Ammannia robusta Heer & Regel [*A. coccinea* Rottb. ssp. *robusta* (Heer & Regel) Koehne]
Fibrous-rooted, glabrous annual. **Stems** erect, 3–20 cm. **Leaves** opposite, linear-oblanceolate,
1–4 cm long, sessile, clasping the stem. **Flowers** 4-merous, 1 to 3 per leaf axil; hypanthium short-
campanulate, 8-ribbed, 2–3 mm long, enlarging in fruit; petals purple, ca. 1 mm long; stamens 4.
Capsule membranous, globose, ca. 4 mm long. Moist, often saline or alkaline soil of wetland margins,
fields; plains. Throughout much of central and western N. America.

Lythrum L. Loosestrife

Perennials. **Stems** erect, branched, clustered. **Leaves** opposite, sometimes alternate above, sessile, entire.
Flowers: hypanthium cylindric, striate, purplish, puberulent, 4–5 mm long; sepals with linear appendages
in the sinuses; petals purple; stamens 6 to 12. **Capsule** ovoid, included in the hypanthium.

1. Flowers in terminal spike-like panicles . . . . *L. salicaria*
1. Flowers solitary in axils of reduced upper leaves . . . . *L. alatum*

Lythrum alatum Pursh **Stems** 4-angled, 40–120 cm. **Herbage** glabrate. **Leaves** lanceolate to ovate, 15–60 mm long.
Inflorescence of solitary, short-pedicellate flowers in axils of reduced upper leaves. **Flowers**: hypanthium puberulent,
5–7 mm long; petals 6, 3–6 mm long; stamens 6. Wet meadows, streambanks; plains. MT to QC south to TX, MS, FL.
Our plants are var. ***alatum***. One unconfirmed collection from Carter Co.[147]

Lythrum salicaria L. Purple loosestrife **Stems** 50–150 cm. **Herbage** glabrate to pubescent.
Leaves cordate-based, narrowly lanceolate, 3–10 cm long. **Inflorescence** a terminal, leafy-
bracteate interrupted spike; flowers several in bract axils. **Flowers**: hypanthium puberulent, 4–5
mm long; sepals ca. 0.5 mm long with longer appendages; petals 7–11 mm long; stamens ca.
12, exserted. Marshes, pond margins, streams; valleys. Introduced as an ornamental and escaped
throughout northern U.S. and adjacent Canada; native to Eurasia.

Rotala L. Toothcup

Rotala ramosior (L.) Koehne Glabrous, fibrous-rooted annual. **Stems** ascending to erect, 3–10
cm, angled. **Leaves** opposite, short-petiolate; the blades narrowly oblanceolate, 5–15 mm long.
Inflorescence of 1 to 3 flowers in leaf axils. **Flowers**: 4-merous, hypanthium campanulate, 1–3
mm long, enlarging in fruit; sepals <0.5 mm long; petals white, ca. 1 mm long; stamens 4, included.
Capsule globose, ca. 3 mm long, included in the hypanthium. Shores of ponds, lakes, wetlands;
plains, valleys. Throughout U.S. south to S. America.

THYMELAEACEAE: Mezereum Family

Daphne L.

Daphne mezereum L. Mezereum Deciduous shrub. **Stems** erect, gray-brown, 50–100 cm. **Herbage** glabrous. **Leaves**
alternate, simple, entire, oblanceolate, 4–8 cm long. **Inflorescence** of sessile, few-flowered clusters just below new stem
growth. **Flowers** perfect, regular, perigynous; hypanthium tubular, 7–9 mm long, hirsute; sepals 4, purple, 5–7 mm
long; petals absent; stamens 8; style 1. **Fruit** a 1-seeded, red, ovoid drupe, 7–8 mm long. Riparian woodlands; valleys.
Introduced as an ornamental and escaped mainly in eastern N. America; native to Eurasia. Collected in Missoula Co.

ONAGRACEAE: Evening-primrose Family

Annual, perennial, rarely suffruticose herbs. **Leaves** usually simple, opposite or alternate; stipules absent
or vestigial. **Inflorescence** bracteate racemes or spikes. **Flowers** perfect, epigynous, usually regular,
4-merous; hypanthium elongate; petals separate; stamens mostly 4 or 8; ovary inferior, usually 4-celled;
style 1. **Fruit** a many-seeded capsule.

The inferior ovary combined with the 4-merous flowers distinguish this family. Recent molecular-
genetics studies have prompted the division of *Camissonia* into several segregate genera and the placement

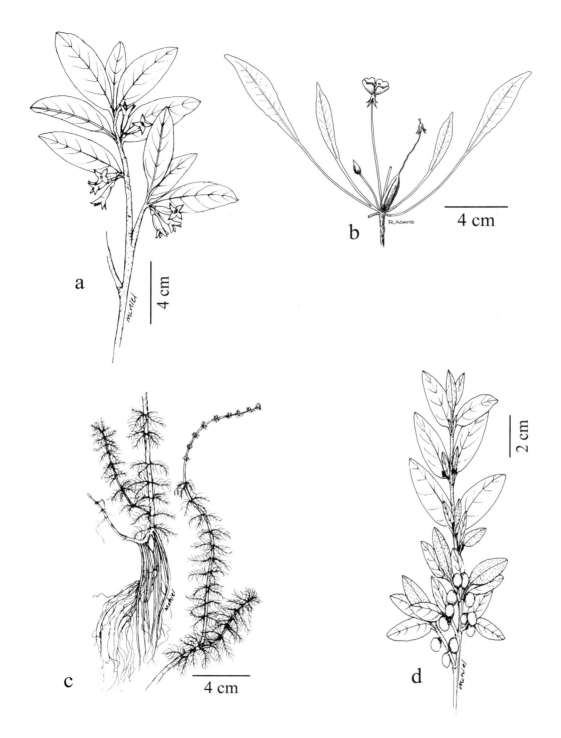

Plate 57. a. *Elaeagnus commutata*, b. *Camissonia subacaulis*, c. *Myriophyllum sibiricum*, d. *Shepherdia canadensis*

of *Gaura* and *Calylophus* into *Oenothera*;[416] however, these taxonomic changes seem premature at this time.

1. Petals and sepals 2; fruit with hooked hairs . . . . *Circaea*
1. Petals and sepals 4; fruit without hooked hairs 2

2. Capsule hardened, 5–9 mm long, 1- to 4-seeded . . . . *Gaura*
2. Capsule many-seeded, >9 mm long or not hardened 3

3. Stem branches fine and hair-like; ovary with 2 chambers . . . . *Gayophytum*
3. Branches not so slender; ovary 4-chambered 4

4. Seeds with a tuft of hair at the tip 5
4. Seeds without tuft of hair 6

5. Petals >1 cm long, completely separate, unlobed . . . . *Chamerion*
5. Petals <1 cm long, united at the very base, lobed at the tip . . . . *Epilobium*

6. Stigma with 4 linear lobes ≥3 mm long 7
6. Stigma round or discoid, sometimes slightly lobed; lobes ≤2 mm long 8

7. Petals pink to purplish, often deeply lobed (unlobed in *C. rhomboidea*) . . . . *Clarkia*
7. Petals yellow to white (fading to pink), notched but not deeply lobed . . . . *Oenothera*

8. Petals white, pink, purple; seeds with a deciduous tuft of hair . . . . *Epilobium*
8. Petals yellow (rarely white); seeds without hair 9

9. Plants a subshrub, woody at the base; stigma short-lobed; seeds in 8 rows . . . . *Calyophus*
9. Plants herbaceous; stigma usually unlobed; seeds in 4 rows . . . . *Camissonia*

Calylophus Spach

Calylophus serrulata (Nutt.) P.H.Raven [*Oenothera serrulata* Nutt.] Perennial subshrub with a woody, branched caudex. **Stems** ascending, 10–40 cm. **Herbage** strigose. **Leaves** all cauline, alternate, short-petiolate; blades simple, linear to oblanceolate, serrulate, 1–3 cm long with axillary fascicles of smaller leaves. **Inflorescence**: solitary, sessile flowers in axils of upper leaf-like bracts. **Flowers** 4-merous; hypanthium trumpet-shaped, 5–7 mm long; sepals 3–5 mm long; petals yellow, obovate, 7–10 mm long; stamens 8; stigma discoid, shallowly lobed. **Fruit** a sessile, linear, 4-loculate capsule 15–25 mm long; seeds in 2 rows per locule. Often sandy or gravelly soil of grasslands; plains. AB to MB south to AZ, NM, TX, MO, Mexico.

Camissonia Link

Annual or perennial herbs. **Leaves** alternate or all basal. **Inflorescence** a usually bracteate spike or raceme. **Flowers** 4-merous; petals yellow (white in *C. minor*), obovate, short-clawed; stamens 8; stigma discoid to globose. **Fruit** a straight to curved, elongate, 4-loculate capsule; seeds lacking a coma of hair.[334]

Species with globose stigmas formerly in *Oenothera* have been placed in *Camissonia*.[334]

1. Plants acaulescent perennials 2
1. Plants slender-stemmed annuals 4

2. Leaf blades entire at least in the distal half . . . . *C. subacaulis*
2. Leaf blades pinnately lobed most of the length 3

3. Hypanthium 6–15 mm long; sepals 3–5 mm long . . . . *C. breviflora*
3. Hypanthium 2–5 cm long; sepals 6–13 mm long . . . . *C. tanacetifolia*

4. Leaf blades ovate, <3 times as long as wide; fruits pedicellate . . . . *C. scapoidea*
4. Leaf blades linear to oblanceolate, >3 times as long as wide; fruits sessile 5

5. Petals white . . . . *C. minor*
5. Petals yellow 6

6. Leaves 1–3 mm wide, clustered beneath the inflorescence; stems naked below . . . . *C. andina*
6. Leaves ca. 1 mm wide, ca. evenly distributed along the stem . . . . *C. parvula*

Camissonia andina (Nutt.) P.H.Raven (*Oenothera andina* Nutt., *Neoholmgrenia andina* (Nutt.) W.L. Wagner & Hoch) Taprooted annual. **Stems** slender, erect, 3–10 cm, sometimes branched from the base. **Herbage** glabrate to finely strigose. **Leaves** obscurely petiolate, clustered below the inflorescence, linear, 5–15 mm long. **Inflorescence** a short, leafy-bracteate spike. **Flowers**: hypanthium ca. 0.5 mm long; sepals ca. 1 mm long, reflexed; petals ca. 1 mm long; stigma globose. **Capsule** sessile, linear-ovoid, 4-angled, strigose, 4–11 mm long. Sandy or gravelly soil of grasslands, woodlands, open slopes; valleys. BC, MT south to CA, NV, UT, WY. Known from Carbon and Missoula cos.

Camissonia breviflora (Torr. & A.Gray) P.H.Raven [*Oenothera breviflora* Torr. & A.Gray, *Taraxia breviflora* (Torr. & A.Gray) Nutt.ex Small] Taprooted, short-lived, acaulescent perennial, sometimes spreading by roots. **Herbage** glabrate to strigose. **Leaves** petiolate, all basal; the blade narrowly lanceolate, deeply pinnately lobed, 3–10 cm long. **Flowers** sessile in axils of leaves; hypanthium 6–15 mm long, the upper expanded portion 1–2 mm long; sepals 3–5 mm long, reflexed; petals 3–8 mm long; stigma globose, slightly lobed. **Capsule** sessile, narrowly ovoid, 10–15 mm long, slightly curved with a beak 5–10 mm long. Moist, often calcareous soil of ephemeral wetlands, stream and pond margins; montane, subalpine. BC to SK south to NV, UT, CO.

Camissonia minor (A.Nelson) P.H.Raven [*Oenothera minor* A.Nelson, *Eremothera minor* (A.Nelson) W.L. Wagner & Hoch] Taprooted annual. **Stems** slender, erect, 5–20 cm, branched from the base. **Herbage** glabrate, glandular above. **Leaves** petiolate, mostly basal; the blade oblanceolate, 1–5 cm long. **Inflorescence** a leafy-bracteate spike. **Flowers**: hypanthium 1–3 mm long; sepals 1–2 mm long, reflexed; petals 1–2 mm long, white; stigma globose. **Capsule** sessile, linear, straight to contorted, 4-angled, glandular-puberulent, 8–25 mm long. Sandy soil of open slopes, drainageways; valleys. WA to MT south to CA, NV, UT, CO. Known from Carbon Co.

Camissonia parvula (Nutt. ex Torr. & A.Gray) P.H.Raven [*Oenothera parvula* Nutt. ex Torr. & A.Gray] Slender, taprooted annual. **Stems** erect, 3–15 cm, branched from the base. **Herbage** glabrous. **Leaves** short-petiolate, linear, 1–3 cm long. **Inflorescence** a contorted, leafy-bracteate, few-flowered spike. **Flowers**: hypanthium 1–2 mm long; sepals 1–2 mm long, reflexed; petals 2–3 mm long; stigma globose. **Capsule** sessile, clavate, obscurely glandular-hairy, 20–25 mm long. Sandy soil of open slopes; valleys. WA to MT south to CA, AZ, UT, CO. Known from Carbon Co.

Camissonia scapoidea (Torr. & A.Gray) P.H.Raven [*Oenothera scapoidea* Torr. & A.Gray, *Chylismia scapoidea* (Torr. & A.Gray) Small] Taprooted annual. **Stems** erect, branched at the base, 5–20 cm. **Herbage** glabrate. **Leaves** long-petiolate, mostly basal; the blade ovate, 1–5 cm long with sinuate margins, sometimes with a few small lobes isolated below. **Inflorescence** an elongating raceme. **Flowers**: hypanthium 2–4 mm long; sepals 2–4 mm long; petals 3–5 mm long, minutely red-spotted basally; stigma discoid. **Capsule** pedicellate, spreading-erect, clavate, 1–2 cm long, straight to slightly curved, glabrous. Sandy or clay soil of open slopes, sagebrush steppe; plains, valleys. WA to MT south to AZ, NM, CO. Known from Carbon and Chouteau cos.

Camissonia subacaulis (Pursh) P.H.Raven [*Oenothera subacaulis* (Pursh) Garrett, *Taraxia subacaulis* (Pursh) Rydb.] Taprooted, short-lived, acaulescent perennial. **Herbage** glabrous. **Leaves** petiolate, all basal; the blade lanceolate to oblanceolate, sometimes pinnately lobed below, 4–15 cm long. **Flowers** sessile in axils of leaves; hypanthium 2–5 cm long, the upper expanded portion 1–3 mm long; sepals 6–15 mm long, reflexed; petals 7–15 mm long; stigma globose, slightly lobed. **Capsule** sessile, ovoid, 13–25 mm long, 4-sided. Moist meadows, grasslands; montane, subalpine. WA to MT south to CA, NV, UT, CO. (p.330)

Camissonia tanacetifolia (Torr. & A.Gray) P.H.Raven [*Oenothera tanacetifolia* Torr. & A.Gray, *Taraxia tanacetifolia* (Torr. & A.Gray) Piper] Taprooted, acaulescent perennial. **Herbage** glabrate. **Leaves** petiolate, all basal; the blade narrowly lanceolate, deeply pinnately lobed, 5–20 cm long. **Flowers** sessile in axils of leaves; hypanthium 2–5 cm mm long, the upper expanded portion 3–7 mm long; sepals 6–13 mm long, reflexed; petals 1–2 cm long; stigma globose, slightly lobed. **Capsule** sessile, narrowly ovoid, 4-angled, 10–25 mm long, glabrate, slightly curved with a beak 5–15 mm long. Meadows, sagebrush steppe; montane. WA to MT south to CA, NV, ID. Collected in Gallatin and Madison cos.

Chamerion Raf. ex Holub, Fireweed

Perennial herbs. **Leaves** mostly alternate, sessile or short-petiolate; margins entire. **Inflorescence** a terminal bracteate raceme. **Flowers** 4-merous slightly irregular; hypanthium absent; petals obovate, the lower slightly narrower than upper; stigma prominently 4-lobed; stamens 8. **Fruit** a linear, puberulent capsule with 4 locules; seeds numerous with a tuft of fine, white hair (coma) at the tip.[101]

1. Inflorescence leafy-bracteate with mostly 2 to 12 flowers . . . . *C. latifolium*
1. Bracts of inflorescence not leafy; flowers numerous . . . .*C. angustifolium*

Chamerion angustifolium (L.) Holub Fireweed [*Epilobium angustifolium* L.] Rhizomatous. **Stems** unbranched, 20–120 cm. **Herbage** glabrous below, puberulent in the inflorescence. **Leaves** narrowly lanceolate, 5–15 cm long. **Racemes** many-flowered with minute bracts, nodding in bud. **Flowers** magenta; sepals spreading, 7–15 mm long; petals 10–16 mm long. **Capsule** 1–8 cm long; the coma 10–17 mm long. Open slopes, streambanks, avalanche chutes, open forest; valleys to subalpine. Circumboreal south to CA, NM, SD, OH, NC. (p.334)

Both diploid (ssp. *angustifolium*) and polyploid (ssp. *circumvagum* (Mosquin) Hoch) plants occur in MT, but cannot be reliably distinguished on morphological characters. The windborne seeds quickly colonize burned over forest and plants persist until the new tree canopy closes over, a process that may take centuries in some subalpine areas.

Chamerion latifolium (L.) Holub Alpine fireweed [*Epilobium latifolium* L.] Perennial from a branched caudex. **Stems** simple, 5–60 cm. **Herbage** glabrous to puberulent. **Leaves** lanceolate to linear-lanceolate, 2–5 cm long, opposite below. **Racemes** short with 2 to 12 flowers and leaf-like bracts. **Flowers** rose; sepals 1–2 cm long; petals 12–25 mm long. **Capsule** 2–10 cm long; the coma 9–15 mm long. Often calcareous gravel bars or less commonly on stony slopes; montane to alpine. AK to Greenland south to CA, CO; Asia. Our plants are ssp. *latifolium*.

One of the first plants, along with *Dryas drummondii*, to colonize and stabilize fresh river gravels. Low-elevation plants often have narrower leaves than those near treeline.

Circaea L. Enchanter's nightshade

Circaea alpina L. Rhizomatous perennial. **Stems** erect, 5–30 cm. **Herbage** glabrous to strigose, glandular. **Leaves** all cauline, opposite, petiolate; the blades simple, 1–8 cm long, ovate, entire to serrate. **Inflorescence** terminal and axillary, minutely bracteate racemes. **Flowers** white, 2-merous; hypanthium ca. 0.5 mm long; sepals 1–2 mm long, reflexed; petals 2-lobed, 1–2 mm long; stamens 2. **Fruit** an obovoid 1-celled capsule, 1–2 mm long, with hooked hairs; seed solitary. Moist forest, thickets, often in moderately disturbed soil; valleys to subalpine. Circumboreal south to CA, CO, GA. (p.334)

1. Leaf blades cordate-based, serrate . . . . ssp. *alpina*
1. Leaf blades subentire with a more rounded base. . . . . .ssp. *pacifica*

Circaea alpina ssp. *alpina* is found in deep shade of mesic montane forest west of the Divide; *C. alpina* ssp. *pacifica* (Asch & Magnus) Raven is more widespread and generally occurs higher or in less shaded sites.

Clarkia Pursh

Slender annuals. **Leaves** cauline, simple, opposite or alternate, often with axillary fascicles of smaller leaves. **Flowers** solitary in upper leaf axils, 4-merous; petals (ours) clawed at the base; fertile stamens 4 or 8; stigma 4-lobed. **Fruit** a 4-loculate, slender capsule; seeds lacking a coma.

1. Petals 1–3 cm long, deeply lobed; leaves linear . . . .*C. pulchella*
1. Petals 6-8 mm long, unlobed; leaves lanceolate. . . . . *C. rhomboidea*

Clarkia pulchella Pursh Elkhorns, ragged robin **Stems** erect, 10–50 cm. **Herbage** puberulent. **Leaves** alternate, linear, obscurely petiolate, 1–6 cm long. **Flowers** slightly irregular, pedicellate, magenta (white); hyapnthium 1–5 mm long; sepals 6–15 mm long, often joined at the tips; petals 1–3 cm long, deeply 3-lobed with a pair of small lateral lobes on the claw; stamens 8, only 4 fertile; stigma lobes 1–3 mm long. **Capsule** linear, vaguely 4-angled, 1–2 cm long. Grasslands, pine woodlands; valleys. BC, MT south to OR, ID, WY, SD. (p.334)

Clarkia rhomboidea Douglas ex Hook. **Stems** erect, 10–40 cm, nodding at the tip. **Herbage** glabrate to sparsely strigose. **Leaves** subopposite, short-petiolate; the blade lanceolate, 1–4 cm long. **Flowers** slightly irregular, pedicellate, pink-purple; hyapnthium 1–2 mm long; sepals 5–7 mm long, reflexed; petals 6–8 mm long, the blade elliptic; stamens 8; stigma lobes ca. 1 mm long. **Capsule** linear, terete, 15–25 mm long. Forest openings; valleys; collected in Lincoln and Sanders cos. WA to MT south to CA, AZ, UT.

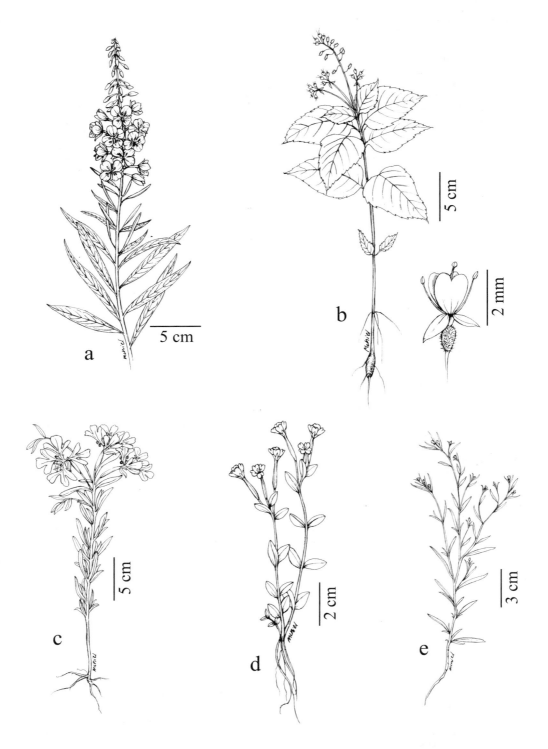

Plate 58 a. *Chamerion angustifolium*, b. *Circaea alpina* ssp. *pacifica*, c. *Clarkia pulchella*, d. *Epilobium anagallidifolium*, e. *Gayophytum diffusum*

Epilobium L. Willow herb

Annual or perennial herbs or 1 subshrub, sometimes forming bulb-like turions at the base. **Leaves** basal and cauline, sessile or short-petiolate, usually opposite below but often alternate above. **Inflorescence** a usually bracteate spike or raceme, often nodding in bud. **Flowers** 4-merous; petals 2-lobed; stamens 8; stigma usually capitate to 4-lobed. **Fruit** an elongate, 4-loculate capsule; seeds mostly with a deciduous tuft of fine, white hair (coma) at the tip.

Large-flowered species formerly placed in *Epilobium* are now placed in *Chamerion*. Species formerly placed in *Boisduvalia* are included here.[191] The treatment herein follows that of Peter Hoch.[104,289] Hybridization occurs among many of the species.[189] Species can be confidently distinguished using microscopic seed coat characters; however, many species appear similar in the field, and most of the characters used to separate species (e.g., presence of turions, vesture, flower size) are subject to environmentally-induced variation.[189] *Epilobium anagallidifolium, E. clavatum, E. hornemani,* and *E. lactiflorum* were previously considered conspecific.[187] There is one specimen annotated to *E. foliosum* (Torr.& A.Gray) Suksd. by P. Hoch (*Lackschewitz 2698* MONTU), but it does not seem to possess the characters that are reported to distinguish it from *E. minutum.*[190] *Epilobium leptocarpum* Hausskn., similar to *E. ciliatum* but with tawny coma, occurs to the north and south of MT.

1. Stigma with 4 long lobes; petals pale yellow to white. . . . .*E. suffruticosum*
1. Stigma round or slightly lobed; petals white to purple 2

2. Plants annual; taprooted or if fibrous-rooted then leaves alternate below 3
2. Plants perennial; leaves mostly opposite below. 6

3. Capsules sessile; seeds without a coma 4
3. Capsules pedicellate; seeds with a coma. 5

4. Leaves sparsely villous to glabrate; sepals ca. 1 mm long; capsules with 2 rows of seeds per
   locule . . . . *E. pygmaeum*
4. Leaves sericeous; sepals 2–6 mm long; seeds in 1 row per locule . . . . *E. densiflorum*

5. Stems glabrous on at least the lower half. . . . . *E. brachycarpum*
5. Stems puberulent below. . . . . *E. minutum*

6. Inflorescence, especially fruits, canescent with minute hairs . . . . *E. palustre*
6. Inflorescence glandular-hairy or glabrous, not canescent 7

7. Stems glabrous below the inflorescence 8
7. Stems pubescent below, often in faint lines from leaf bases 10

8. Upper leaves serrate; turions often present at stem base . . . . *E. leptocarpum*
8. Upper leaves entire to weakly serrulate; turions lacking 9

9. Herbage glaucous; plants without leafy stolons . . . . *E. glaberrimum*
9. Herbage not glaucous; thin, leafy stolons often present. . . . . *E. oregonense*

10. Stems erect at the base 11
10. Stems curved or angled upward at the base 14

11. Pedicel in fruit 1–5 mm long . . . .*E. saximontanum*
11. Pedicel in fruit 7–40 mm long 12

12. Leaves sometimes entire, mostly sessile; inflorescence unbranched . . . . *E. halleanum*
12. Leaves serrate, short-petiolate; inflorescence sometimes branched. 13

13. Coma white or off-white, readily deciduous . . . . *E. ciliatum*
13. Coma tawny, persistent . . . . *E. leptocarpum*

14. Petals white to pale pink, 2–5 mm long; seeds smooth . . . . *E. lactiflorum*
14. Petals pink to purple or >5 mm long; seeds sometimes papillose 15

15. Stems mostly 15–30 cm; leaves broadly lanceolate, dentate. . . . . *E. hornemannii*
15. Stems mostly <15 cm; leaves narrowly lanceolate, entire to weakly dentate. 16

16. Mature fruit 17–36 mm long, linear. . . . . *E. anagallidifolium*
16. Fruit 10–20 mm long, slightly clavate . . . . *E. clavatum*

Epilobium anagallidifolium Lam. [*E. alpinum* L. var. *alpinum*] Perennial often forming loose mats from fine, white roots. **Stems** often clustered, ascending, 2–12 cm. **Herbage** glabrate. **Leaves** opposite, short-petiolate; the blades 6–25 mm long, oblanceolate to narrowly elliptic, entire. **Inflorescence** a leafy-bracted, few-flowered raceme. **Flowers** pink to purple; hypanthium ca. 1 mm long, glabrate; sepals erect, ca. 2 mm long; petals 3–4 mm long. **Capsule** pedicellate, linear, 17–36 mm long, glabrate; seeds minutely papillose with a deciduous coma. Wet, gravelly, often mossy soil of seeps, meadows, along streams, often near melting snow; alpine, subalpine. Circumboreal south to CA, CO, NH. (p.334)

Epilobium brachycarpum C.Presl [*E. paniculatum* Nutt. ex Torr. & A.Gray] Taprooted annual. **Stems** erect, mostly 10–60 cm, branched above. **Herbage** glabrous below, glandular-puberulent above. **Leaves** alternate, short-petiolate, often with axillary fascicles, early deciduous below; blades narrowly lanceolate, entire to weakly serrulate, to 15 cm long with an apical gland. **Inflorescence** of racemes terminating the branches. **Flowers** white to purple; hypanthium 0.5–2 mm long, glabrous; sepals 1–3 mm long; petals 3–5 mm long. **Capsule** short-pedicellate, beaked, 12–32 mm long, glabrous or glandular-puberulent; seeds minutely papillose with a deciduous coma. Disturbed soil of meadows, grasslands, steppe, woodlands, streambanks, roadsides; valleys, montane. BC to QC south to CA, NM, SD, MN.

Epilobium minutum has stems puberulent to the base.

Epilobium ciliatum Raf. [*E. adenocaulon* Hausskn., *E. glandulosum* Lehm. misapplied, *E. watsonii* Barbey misapplied] Perennial sometimes with basal turions and rhizomes. **Stems** erect, 10–80 cm, glabrous to puberulent, sometimes glandular above. **Herbage** glabrate. **Leaves** opposite, short-petiolate; blades lanceolate, 2–8 cm long, serrulate, alternate in the inflorescence. **Inflorescence** leafy-bracted racemes. **Flowers** white to purple; hypanthium 1–2 mm long; sepals 1–3 mm long, often pink; petals 2–8 mm long. **Capsule** pedicellate, 2–6 cm long, puberulent; seeds papillose in ridges with a deciduous coma. Moist soil along streams, lakes; montane, subalpine. North America, S. America; Asia.

1. Inflorescence mostly simple; stems glabrous below with basal turions; petals 5–10 mm long. . . . . .ssp. *glandulosum*
1. Inflorescence branched; stems puberulent below; turions absent; petals 2–5 mm long . . . . . . .ssp. *ciliatum*

Epilobium ciliatum ssp. *ciliatum* occurs in wet meadows, streambanks; plains, valleys, montane; *E. ciliatum* ssp. *glandulosum* (Lehm.) Hoch & P.H.Raven is found in moist open forest; montane, subalpine.

Epilobium clavatum Trel. [*E. alpinum* L. var. *clavatum* (Trel.) C.L.Hitchc.] Perennial from dark, wiry roots without turions. **Stems** ascending, 3–12 cm. **Herbage** sparsely strigillose, glandular above. **Leaves** opposite, sessile to short-petiolate; the blades ovate to elliptic, entire to weakly dentate, 5–20 mm long. **Inflorescence** a few-flowered, leafy-bracted raceme. **Flowers** pink to purple; hypanthium ca. 1 mm long, sparsely glandular; sepals, 2–3 mm long; petals 4–7 mm long. **Capsule** pedicellate, narrowly clavate, 2–4 cm long, glandular-puberulent; seeds minutely papillose with a deciduous coma. Moist to wet, stony soil of moraine, talus slopes; subalpine, alpine. AK to CA, CO.

Epilobium densiflorum (Lindl.) Hoch & P.H.Raven [*Boisduvalia densiflora* (Lindl.) S.Watson] Annual. **Stems** erect, 10–60 cm. **Herbage** sericeous, sometimes glandular above. **Leaves** alternate, sessile, lanceolate, serrulate, 15–50 mm long. **Inflorescence** a crowded leafy-bracteate spike, sometimes flowers in axils of basal leaves. **Flowers** rose to pink; hypanthium 1–4 mm long; sepals erect, 2–6 mm long; petals 3–8 mm long. **Capsule** sessile, fusiform, 5–10 mm long, pilose, glandular with a beak <1 mm long; seeds reticulate, in 4 rows, without a coma. Streambanks, moist meadows; valleys. BC, MT south to CA, UT, Mexico; one collection from Sanders Co.

Epilobium glaberrimum Barbey Perennial with dark, wiry roots. **Stems** ascending, 5–40 cm. **Herbage** glabrous, glaucous. **Leaves** opposite, sessile or subsessile, clasping, lanceolate to ovate,1–3 cm long, entire to weakly serrate. **Inflorescence** leafy-bracted racemes, erect in bud. **Flowers** pink to purple; hypanthium glabrous, ca. 1 mm long; sepals 2–3 mm long; petals 4–6 mm long. **Capsule** long-pedicellate, 15–60 mm long, glabrous; seeds papillose with a deciduous coma. Moist, stony soil of cliffs and along streams, trails; montane, subalpine. BC, AB south to CA, UT, WY. Our plants are ssp. *fastigiatum* (Nutt.) Hoch & P.H.Raven. See *E. halleanum*, *E. oregonense*.

Epilobium halleanum Hausskn. [*E. watsonii* Barbey misapplied] Perennial with turions on fine horizontal roots. **Stems** erect, 5–50 cm. **Herbage** sparsely puberulent, glandular above. **Leaves** opposite, mostly sessile, narrowly elliptic, entire to serrulate, 2–8 cm long. **Inflorescence** a few-flowered raceme, nodding in bud. **Flowers** white to pink; hypanthium 0.5–1 mm long; sepals glabrous to glandular-puberulent, 1–2 mm long; petals 2–4 mm long. **Capsule** pedicellate, 15–45 mm long, glabrous or glandular-hairy; seeds reticulate or papillose with a deciduous coma. Wet meadows, along streams, wetlands, moist grasslands, rocky slopes; montane, subalpine. BC to SK south to CA, AZ, CO.

Leaves of *Epilobium ciliatum* are usually more serrate; *E. glaberrimum* has glaucous herbage.

Epilobium hornemannii Rchb. [*E. alpinum* L. var. *nutans* Hornem.] Fibrous-rooted perennial. **Stems** ascending, 10–25 cm, puberulent in lines. **Herbage** glabrate, glandular above. **Leaves** opposite, petiolate below; the blades 1–3 cm long, lanceolate to elliptic, usually entire. **Inflorescence** a leafy-bracted, few-flowered raceme. **Flowers** pink to purple; hypanthium ca.1.5 mm long, glabrous; sepals 2–4 mm long; petals 3–8 mm long. **Capsule** pedicellate, linear, 25–50 mm long, glabrous; seeds minutely papillose with a coma. Wet, mossy, stony soil of meadows, forests, often along streams; montane, subalpine. Circumboreal south to CA, NM, NH. Our plants are ssp. *hornemannii*.

Epilobium lactiflorum Hausskn. [*E. alpinum* L. var. *lactiflorum* (Hausskn.) C.L.Hitchc., *E. hornemannii* Rchb. var. *lactiflorum* (Hausskn.) D.Löve] Perennial with fine, white roots. **Stems** ascending, 5–30 cm with fine lines of hair from the leaf bases. **Herbage** glabrate. **Leaves** opposite, petiolate to sessile; the blades lanceolate to ovate, 1–4 cm long, entire. **Inflorescence** a few-flowered raceme, nodding in bud. **Flowers** white to light pink; hypanthium 1–2 mm long; sepals sparsely glandular-pubescent, 2–3 mm long; petals 3–8 mm long. **Capsule** long-pedicellate, 2–6 cm long, glandular-hairy; seeds minutely reticulate with a deciduous coma. Wet meadows, cliffs, streambanks, roadsides; upper montane, subalpine. AK to NL south to CA, CO, NH.

Epilobium minutum Lindl. ex Lehm. Taprooted annual. **Stems** erect, 3–25 cm, branched above. **Herbage** puberulent. **Leaves** opposite below, short-petiolate; blades lanceolate, entire, 5–20 mm long. **Inflorescence** of few-flowered racemes terminating branches. **Flowers** white to pink; hypanthium 0.5–1 mm long, glabrous; sepals 1–2 mm long; petals 2–3 mm long. **Capsule** pedicellate, beaked, 1–2 cm long, glabrous; seeds finely papillose with a deciduous coma. Cliffs, talus slopes; valleys, montane. BC, MT south to CA, NV.

Epilobium oregonense Hausskn. [*E. alpinum* L. var. *gracillimum* (Trel.) C.L.Hitchc.] Perennial with fine roots and often thin, leafy stolons. **Stems** slender, simple, erect, 5–30 cm. **Herbage** glabrous. **Leaves** nearly erect, sessile, linear-lanceolate, entire, 5–25 mm long. **Inflorescence** a few-flowered, glabrate raceme. **Flowers** white to purple; hypanthium ca. 1 mm long, puberulent; sepals 2–3 mm long; petals 5–6 mm long. **Capsule** pedicellate, 1–7 cm long, glabrate; seeds smooth with a deciduous coma. Fens, wet meadows; montane, subalpine. BC, MT south to CA, NV, UT, CO.

Epilobium glaberrimum and *E. leptocarpum* generally have wider leaves.

Epilobium palustre L. [*E. leptophyllum* Raf.] Perennial with fine, white roots. **Stems** erect, 7–40 cm. **Herbage** strigilose to glabrate. **Leaves** opposite, linear to lanceolate, sessile, 1–7 cm long, entire to remotely serrulate. **Inflorescence** a few- to many-flowered, canescent-strigose raceme. **Flowers** white to pink; hypanthium ca. 1 mm long, puberulent; sepals 2–4 mm long; petals 3–6 mm long. **Capsule** pedicellate, 3–9 cm long, canescent; seeds papillose with a persistent coma. Wet meadows, fens, wetlands; valleys to subalpine. Circumboreal south to CA, CO, SD, MI, PA.

1. Leaves linear to lanceolate, often serrulate, glabrate on the upper surface . . . . var. *palustre*
1. Leaves linear, entire, puberulent on the upper surface . . . . var. *gracile*

Epilobium palustre var. *palustre* is found around sloughs, marshes, lakes; valleys to montane; *E. palustre* var. *gracile* (Farw.) Dorn occurs in fens; montane, subalpine.

Epilobium pygmaeum (Speg.) Hoch & P.H.Raven [*Boisduvalia glabella* (Nutt.) Walp.] Fibrous-rooted annual. **Stems** ascending to erect, puberulent, 5–25 cm, branched at the base. **Herbage** glabrate. **Leaves** alternate, sessile, lanceolate to oblanceolate, weakly dentate, 5–20 mm long. **Inflorescence** a crowded, leafy-bracteate spike. **Flowers** pink; hypanthium ca. 1 mm long; sepals erect, ca. 1 mm long; petals 2–3 mm long. **Capsule** sessile, clavate, 6–8 mm long, puberulent with a beak ≥1 mm long; seeds reticulate, in >4 rows, without a coma. Mossy shores of shallow ponds, ephemeral wetlands; plains, valleys. BC to SK south to CA, UT, WY, SD, Mexico.

Epilobium saximontanum Hausskn. [*E. watsonii* Barbey misapplied] Perennial with basal turions. **Stems** erect, 20–40 cm. **Herbage** glabrate. **Leaves** opposite, lanceolate, sessile or short-petiolate, sometimes clasping, 2–5 cm long, serrulate. **Inflorescence** a leafy-bracted raceme, erect or nodding in bud. **Flowers** white to purple; hypanthium 1–2 mm long, puberulent; sepals 1–4 mm long; petals 2–5 mm long. **Capsule** pedicellate, 20–55 mm long, glandular-pubescent; seeds papillose with a deciduous coma. Meadows, streambanks; plains, valleys. AB to NL south to CA, NM.

Epilobium suffruticosum Nutt. Perennial from a woody branched caudex. **Stems** prostrate to ascending, 10–25 cm, branched above. **Herbage** strigillose. **Leaves** mostly opposite, sessile, 1–2 cm long, lanceolate, entire. **Inflorescence** a leafy-bracted raceme. **Flowers** yellow or white; hypanthium 1–3 mm long; sepals 3–5 mm long; petals 7–9 mm long, 1 slightly larger; stigma 4-lobed. **Capsule** slightly curved, puberulent, 10–25 mm long; seeds reticulate with a deciduous coma. Riparian gravel bars; valleys. ID, MT, WY.

Gaura L. Butterfly weed

Annual to perennial herbs. **Leaves** basal and/or cauline, simple, alternate. **Inflorescence** a terminal, small-bracteate spike. **Flowers** 4-merous, irregular; hypanthium woolly within; sepals reflexed; petals clawed, mostly held above horizontal, white, aging pink; stamens 8; style exserted. **Fruit** a hardened, few-seeded, narrowly ovoid capsule.

1. Perennial with clustered stems; leaves mostly 1–4 cm long; sepals 5–9 mm long . . . . *G. coccinea*
1. Single-stemmed annual; lower leaves >4 cm long; sepals 1–4 mm long. . . . . *G. parviflora*

Gaura coccinea Pursh [*Oenothera suffrutescens* (Ser.) W.L.Wagner & Hoch] SCARLET GAURA, SCARLET BUTTERFLY WEED. Perennial from a branched caudex. **Stems** ascending, clustered, 10–40 cm. **Herbage** strigose. **Leaves** linear to lanceolate, 1–4 cm long, entire to dentate. **Spikes** 4–15 cm long, often nodding. **Flowers** irregular; hypanthium canescent, 4–11 mm long; sepals 5–10 mm long; petals 3–10 mm long. **Capsule** canescent, 4-angled, 5–9 mm long. Often sandy or gravelly soil of grasslands, sagebrush steppe; plains, valleys. BC to MB south to CA, AZ, NM, TX, Mexico.
    A collection from calcareous hills near Bozeman has glabrous, narrowly elliptic, entire leaves.

Gaura parviflora Dougl. ex Lehm. Taprooted, single-stemmed annual. **Stems** erect, 30–80 cm, branched above. **Herbage** glandular, villous. **Leaves** lanceolate, 3–11 cm long, weakly dentate. **Spikes** 10–30 cm long. **Flowers** slightly irregular; hypanthium 2–4 mm long; sepals 1–3 mm long; petals 1–3 mm long. **Capsule** mostly glabrous, 5–8 mm long, 4-angled, rounded between the angles. Fields, roadsides, streambanks; plains, valleys. West and central U.S., Mexico.

Gayophytum A.Juss. Ground smoke

Taprooted annuals. **Leaves** cauline, alternate, simple, entire, sessile or short-petiolate, linear to narrowly oblanceolate. **Inflorescence** a leafy-bracted, often branched spike or raceme. **Flowers** 4-merous; hypanthium minute; sepals reflexed, ca. as long as petals; petals white with 1–2 green to yellow basal spots, aging pink; stamens 8; stigma globose. **Fruit** a 2-loculed but 4-valved, linear to clavate capsule, often somewhat constricted between seeds; seeds in 1 to 2 rows per locule.
    Determination is impossible without mature fruit.

1. Flowers and fruits sessile or subsessile; pedicels ≤2 mm long. 2
1. Flowers and fruits with pedicels >2 mm long. 4

2. Plants branched throughout . . . . *G. decipiens*
2. Plants branched only in the lower half. 3

3. Seeds at an oblique angle in the locule; capsule glabrous . . . . *G. humile*
3. Seeds vertical, end-to-end; capsule glabrous to pubescent . . . . *G. racemosum*

4. Capsule short, mostly 3–6 mm long, spreading to reflexed; pedicels at least as long as capsules . . . . *G. ramosissimum*
4. Capsules mostly longer on ascending to erect pedicels shorter than the capsule 5

5. Seeds vertical and staggered between 2 locules; capsule constricted between seeds . . . . *G. diffusum*
5. Seeds vertical and adjacent in the 2 locules; capsule little constricted . . . . *G. decipiens*

Gayophytum decipiens F.H.Lewis & Szweyk. **Stems** erect, 10–40 cm, spreading-branched at the base. **Herbage** glabrous to pubescent. **Leaves** erect, 5–20 mm long. **Petals** 1–2 mm long. **Capsule** short-pedicellate, erect, 6–15 mm long; seeds vertical and side-by-side in the 2 locules. Meadows, often along streams; montane, subalpine. WA to MT south to CA, AZ, CO. One collection from Beaverhead Co.

Gayophytum diffusum Torr. & A.Gray [*G. diffusum* var. *strictipes* (Hook.) Dorn] **Stems** erect, 10–40 cm, with ascending branches. **Herbage** glabrate. **Leaves** spreading, 1–4 cm long. **Petals** 1–3 mm long. **Capsule** ascending, 5–20 mm long, glabrous to pubescent; pedicels 2–6 mm long; seeds 3 to 18, glabrous to puberulent, vertical, staggered between locules. Open slopes in grasslands, sagebrush steppe, forest, roadsides; montane, subalpine. BC, MT south to CA, AZ, NM, CO, SD. Our plants are ssp. *parviflorum* F.H.Lewis & Szweyk. (p.334)

Gayophytum humile A.Juss. **Stems** erect, 5–10 cm, with erect branches. **Herbage** glabrous. **Leaves** erect, 1–2 cm long. **Petals** ca. 1 mm long. **Capsule** subsessile, erect, 6–10 mm long, flattened; seeds 24 to 50, glabrous, oblique and side-by-side between locules. Stony soil of forest openings; montane, lower subalpine. WA to MT south to CA, NV, UT, WY, Chile.

Gayophytum racemosum Torr. & A.Gray **Stems** erect, 2–15 cm, with ascending branches. **Herbage** glabrous. **Leaves** erect, 5–15 mm long. **Petals** ca. 1 mm long. **Capsule** erect, 6–12 mm long, barely flattened; pedicels 1–2 mm long; seeds 14 to 34, glabrous, vertical and end-to-end in each locule. Meadows, sagebrush steppe; upper montane, subalpine. WA to AB south to CA, NV, UT, CO, SD.

Gayophytum ramosissimum Torr. & A.Gray **Stems** erect, 10–40 cm, with ascending branches. **Herbage** glabrous. **Leaves** spreading, 1–4 cm long. **Flowers** long-pedicellate; petals 0.5–1.5 mm long. **Capsule** spreading to reflexed, 3–6 mm long, glabrous; pedicels 3–6 mm long; seeds 6 to 16, glabrous, vertical in 1 or 2 staggered rows in each locule. Sagebrush steppe, woodlands; valleys, montane. WA to MT south to CA, AZ, NM.

Oenothera L. Evening primrose

Biennial to perennial herbs. **Leaves** usually basal and sometimes cauline, alternate, simple, petiolate. **Inflorescence** a leafy-bracteate spike or sessile in axils of basal rosette leaves. **Flowers** 4-merous, yellow or white and aging pink; hypanthium tubular, elongate; sepals narrow, reflexed; petals obovate, not clawed; stamens 8; stigma distinctly 4-lobed. **Fruit** a sessile 4-loculed capsule.[415]

Oenothera perennis L., with erect or ascending stems and winged, clavate capsules, is native to eastern N. America and was once collected from a roadside south of Hamilton (*Cory 1913* MONTU) but does not seem to persist.

1. Plants acaulescent; all leaves basal. 2
1. Plants with leafy stems. 3

2. Flowers white; sepals 1–3 cm long; capsule mostly bumpy. . . . .*O. cespitosa*
2. Petals yellow; sepals 6–12 mm long; capsule not bumpy . . . . *O. flava*

3. Petals yellow. 4
3. Petals white 5

4. Petals 2–4 cm long. . . . . *O. elata*
4. Petals <2 cm long. . . . . *O. villosa*

5. Inflorescence glandular . . . .*O. nuttallii*
5. Inflorescence strigose to hispid but not glandular 6

6. Taprooted annual; capsule erect; seeds in 2 rows in each locule . . . . *O. albicaulis*
6. Mostly perennials; capsule ascending; seeds in 1 row per locule . . . . *O. pallida*

Oenothera albicaulis Pursh Taprooted annual. **Stems** erect, 5–30 cm, simple or branched at the base. **Herbage** strigose, sparsely hispid. **Leaves** basal and cauline; basal blades oblanceolate, entire to lobed below, 1–6 cm long; cauline leaves more deeply lobed, reduced upward. **Flowers** white; hypanthium 15–30 mm long, sepals 15–25 mm long; petals 15–40 mm long; stigma lobes 5–7 mm long. **Capsule** spreading to erect, linear, obscurely 4-angled, 15–30 mm long, sparsely

hispid. Sandy soil of grasslands, sagebrush steppe, ponderosa pine woodlands; plains, valleys. MT, ND south to AZ, NM, TX, Mexico.

Oenothera cespitosa Nutt. GUMBO LILY, SAND LILY, ROCK ROSE Acaulescent perennial with a simple or branched caudex surmounting a taproot. **Herbage** strigose. **Leaves** oblanceolate, 2–15 cm long with entire to serrate margins. **Flowers** white, opening at night, turning pink the following morning; hypanthium 2–8 cm long; sepals 1–3 cm long; petals 15–35 mm long; stigma lobes 3–6 mm long. **Capsule** erect, narrowly ovoid to ellipsoid, 12–40 mm long, woody, vaguely 4-angled, usually bumpy, sometimes pedicellate, often partly below ground. Sandy or clay, usually sparsely vegetated soil of open slopes, badlands, sandhills, roadsides; plains, valleys to subalpine. WA to SK south to CA, NM, TX, Mexico. Our plants are ssp. *caespitosa*.

High-elevation plants have obovate leaves with mostly entire margins.

Oenothera elata Kunth [*O. hookeri* Torr. & A.Gray ssp. *hirsutissima* (A.Gray ex S.Watson) Munz, *O. biennis* L. var *hirsutissima* A.Gray ex S.Watson] Taprooted biennial or short-lived perennial. **Stems** erect, often branched at the base, 30–100 cm. **Herbage** strigose, hirsute, glandular. **Leaves** basal and cauline, lanceolate to oblanceolate; the basal 5–15 cm, entire to weakly dentate; cauline leaves reduced upward. **Flowers** light yellow; hypanthium 2–4 cm long, pubescent; sepals 2–4 cm long; petals 2–4 cm long, aging orange; stigma lobes 4–5 mm long. **Capsule** ascending, linear, 2–4 cm long, 4-angled, strigose. Streambanks, roadsides; valleys; collected once in Gallatin Co. WA to MT south to CA, AZ, NM, TX.

The more common *Oenothera villosa* has smaller flowers.

Oenothera flava (A.Nelson) Garrett Acaulescent perennial from a simple caudex. **Herbage** glabrate or sparsely strigose. **Leaves** linear-oblanceolate, pinnately lobed below, 5–25 cm long. **Flowers** light yellow, drying pinkish; hypanthium 3–10 cm long; sepals 6–12 mm long, free or remaining united at tips; petals 1–3 cm long; stigma lobes ca. 3 mm long. **Capsule** erect, ovoid, 4-ribbed, 1–3 cm long, glabrate to strigose. Moist meadows, gravelly soil of roadsides; plains, valleys, montane. WA to SK south to CA, NM, TX, Mexico. Our plants are ssp. *flava*.

Flowers open at dusk and close by morning. Plants are primarily self-pollinated in spite of the showy flowers.[104]

Oenothera nuttallii Sweet Perennial with rhizome-like caudex branches. **Stems** erect, horizontally branched, 40–60 cm with white, peeling epidermis, glandular above. **Herbage** strigose. **Leaves** cauline, linear-lanceolate or oblanceolate, entire to weakly serrate, 2–10 cm long with fascicles in some axils. **Flowers** white; hypanthium 15–30 mm long, glandular; sepals 1–2 cm long, united to one side; petals 15–20 mm long; stigma lobes 4–6 mm long. **Capsule** erect, linear, ca. 2 cm long, glandular. Sandy soil of grasslands, sagebrush steppe; plains. BC to ON south to CO, KS, MN.

Oenothera pallida Lindl. Annual or rhizomatous perennial. **Stems** erect, simple or branched below, 15–30 cm, with white, peeling epidermis. **Herbage** strigose, sparsely hispid. **Leaves** cauline, linear to linear-oblanceolate, pinnately lobed, 2–6 cm long. **Flowers** white; hypanthium strigose, sparsely hispid, 15–30 mm long; sepals ca. 2 cm long, united to one side; petals 15–25 mm long; stigma lobes 4–6 mm long. **Capsule** ascending, linear, slightly curved, 15–35 mm long, strigose, hispid. Sandy soil, sandhills; plains, valleys, montane. BC south to AZ, NM.

1. Calyx glabrous; plants rhizomatous . . . . .ssp. *pallida*
1. Calyx long-hairy; plants sometimes taprooted annuals . . . . ssp. *trichocalyx*

Oenothera pallida ssp. *pallida* [*O. latifolia* (Rydb.) Munz, *O. pallida* var. *idahoensis* Munz] occurs in Beaverhead Co.; *O. pallida* ssp. *trichocalyx* (Nutt.) Munz & W.M.Klein is found in the eastern half of MT. Plants with gray-strigilose foliage were once considered var. *idahoensis,* endemic to southwest MT and adjacent ID, but the distinction is no longer considered significant (fide W. L. Wagner).

Oenothera villosa Thunb. [*O. strigosa* Willd. ex Spreng., *O. biennis* L. misapplied] Taprooted biennial or short-lived perennial. **Stems** erect, simple or branched at the base, 30–100 cm. **Herbage** strigose, sparsely hirsute, glandular above. **Leaves** basal and cauline, lanceolate to oblanceolate; the basal 7–20 cm, deciduous, entire to weakly dentate; cauline leaves reduced upward. **Flowers** light yellow; hypanthium 2–4 cm long, glandular-puberulent; sepals 1–2 cm long; petals 8–15 mm long; stigma lobes 4–7 mm long. **Capsule** ascending, linear, 2–5 cm long, cylindric, strigose, hirsute. Usually disturbed soil of roadsides, fields, streambanks; plains, valleys. BC to MB south to CA, AZ, OK. (p.342)

Nearly all of our plants would be assigned to ssp. *strigosa* (Rydb.) W.Dietr. & P.H.Raven by virtue of their glandular inflorescences; plants with few glandular trichomes in the inflorescence and referable to ssp. *villosa* occur throughout

MT. The degree of glandularity and the abundance of red-based hairs appears to have little association with geographic or ecological gradients in our area. Flower size is highly variable. Flowers are often borne near the base of the plant.

CORNACEAE: Dogwood Family

Cornus L. Dogwood

Shrubs or subshrubs. **Leaves** opposite or whorled, exstipulate, elliptic with entire margins and prominent parallel veins. **Inflorescence** umbellate to hemispheric, terminal cymes. **Flowers** small, regular, epigynous, perfect; sepals 4, inconspicuous; petals 4, separate, white; stamens 4; ovary inferior; style 1; ovary 2-carpellate. **Fruit** a globose, 1- to 2-seeded drupe.

1. Creeping subshrub to 20 cm tall; leaves whorled. . . . . *C. canadensis*
1. Shrub >1 m tall; leaves opposite . . . . *C. sericea*

Cornus canadensis L. Bunchberry Rhizomatous subshrub. **Stems** erect, 5–20 cm with paired, narrowly elliptic, membranous bracts below. **Herbage** strigose. **Leaves** subsessile, paler beneath, 1–7 cm long, 4 to 7, whorled at stem tips, usually smaller pair below. **Inflorescence** a dense, capitate, hemispheric cyme subtended by 4 white, ovate, petal-like bracts, 6–25 mm long. **Flowers**: petals 1–2 mm long. **Drupe** red, 4–8 mm long. Deep, moist, coniferous forest, rarely on hummocks in fens; valleys to lower subalpine. AK to Greenland south to CA, NM, OH, NJ; Asia. (p.342)

Cornus sericea L. Red-osier dogwood [*C. stolonifera* Michx.] Branching shrubs to 3 m. **Twigs** reddish, strigose at first becoming glaucous and then glabrous. **Herbage** sparsely strigose. **Leaves** petiolate; the blades pale beneath, 3–15 cm long with acuminate tips. **Inflorescence** a flat-topped compound cyme 3–10 cm across. **Flowers**: hypanthium 1–2 mm long; petals 2–4 mm long. **Drupe** white to bluish, 1-seeded, 5–7 mm long. Moist forest, thickets, especially along streams, avalanche slopes; plains, valleys to montane. AK to NL south to Mexico. Our plants are *Cornus sericea* ssp. *sericea*. (p.342)
   Cornus sericea ssp. *occidentalis* (Torr. & A.Gray) Fosberg, with rough-hairy foliage, is reported for MT, but I have seen no specimens.

SANTALACEAE: Sandalwood Family

Glabrous, rhizomatous, partially parasitic, perennial herbs. **Leaves** alternate, short-petiolate, simple with entire margins, exstipulate. **Inflorescence** a cyme or flowers axillary. **Flowers** perfect or unisexual, regular, epigynous; hypanthium campanulate, the top forming a disk; sepals 5, petal-like; petals absent; stamens 5; style 1; ovary with 1 locule. **Fruit** a berry-like drupe or nut with persistent sepals; seeds 1 to 4.

1. Fruit nut-like; leaves linear . . . . *Thesium*
1. Fruit a drupe; leaves lanceolate to elliptic 2

2. Flowers several in a terminal inflorescence; fruits green to blue . . . . *Comandra*
2. Flowers usually 3, axillary near the middle of the stem; fruits red . . . . *Geocaulon*

Comandra Nutt. Bastard toad-flax

Comandra umbellata (L.) Nutt. [*C. pallida* A.DC.] **Stems** erect, simple, 10–25 cm, usually clustered. **Leaves** sessile, thick, glaucous, lanceolate, 1–3 cm long. **Inflorescence** a hemispheric cyme, terminal and from uppermost leaf axils. **Flowers** perfect; hypanthium green, 2–3 mm long; sepals lanceolate, 2–3 mm long, puberulent on the inner surface. **Fruit** a drupe, green becoming blue to brown, 4–8 mm long. Grasslands, sagebrush steppe; plains, valleys, montane. BC to NL south to CA, AZ, TX, GA. Our plants are var. *pallida* (A.DC.) M.E.Jones. (p.342)

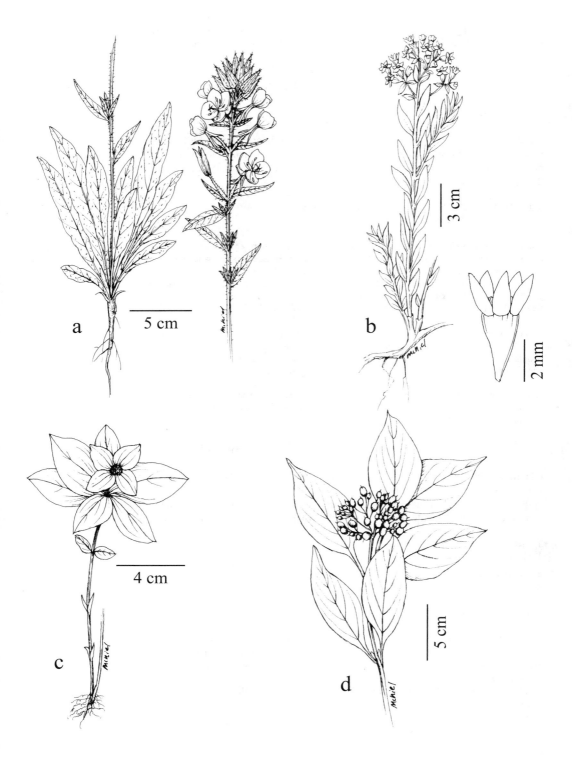

Plate 59. a. *Oenothera villosa*, b. *Comandra umbellata*, c. *Cornus canadensis*, d. *Cornus sericea*

Geocaulon Fernald

Geocaulon lividum (Richardson) Fernald [*Comandra livida* Richardson] **Stems** erect, simple,10–30 cm, usually solitary. **Leaves** short-petiolate, narrowly elliptic, 1–5 cm long. **Inflorescence**: 3 pedicellate flowers per axil of middle leaves. **Flowers** all apparently perfect, but some functionally male; hypanthium green to purple, ca. 1 mm long; sepals deltoid, 1–1.5 mm long. **Fruit** a red drupe, 5–10 mm long, solitary in leaf axils. Moist coniferous, often spruce, forest; valleys, montane. AK to NL south to WA, MT, MN, NY.

Thesium L.

Thesium arvense Horv. Herbaceous perennial with a branched crown from a deep-seated root. **Stems** ascending to erect, 10–30 cm, highly branched. **Leaves** sessile, linear, 1–5 cm long. **Inflorescence** an open-branched panicle with minutely glandular-serrate, leaf-like bracts. **Flowers** perfect, yellow-green; hypanthium 1–2 mm long; sepals 0.5–1 mm long. **Fruit** an ellipsoid nut, 2–3 mm long, ribbed longitudinally. Moist grasslands, wet meadows; valleys, montane. Introduced to western MT; native to Eurasia.

Related species have become pests in the Old World by parasitizing crop plants.

VISCACEAE: Mistletoe Family

Arceuthobium M.Bieb. Dwarf mistletoe

Weakly green, parasitic perennials. **Stems** angular, densely branched. **Leaves** opposite, minute scales. **Inflorescence** minutely bracteate, compound, axillary spikes (male) or racemes (female). **Flowers** unisexual, epigynous, regular, petals absent; male flowers with 3 sepals ca. 1 mm long and 3 stamens; female flowers with 2 sepals; ovary inferior. **Fruit** a berry, brown distally and green or blue proximally; seed solitary, sticky.[166]

1. Plants parasitic on Douglas fir (*Pseudotsuga*) . . . . *A. douglasii*
1. Plants parasitic on pines (*Pinus*) or larch (*Larix*) 2

2. Plants parasitic on larch . . . .*A. laricis*
2. Plants parasitic on pine 3

3. Plants parasitic on 5-needle pines (*P. albicaulis, P. flexilis*) . . . .*A. cyanocarpum*
3. Plants parasitic on 2- to 3-needle pines (*P. contorta, P. ponderosa*) . . . . *A. americanum*

Arceuthobium americanum Nutt. ex Engelm. **Stems** yellowish, 1–8 cm, clustered; branches apparently whorled. **Berry** bluish, 3–4 mm long. Parasitizing *Pinus contorta*, rarely on *P. ponderosa*, causing a diffuse witch's broom; montane, subalpine. BC to ON south to CA, CO. (p.346)

Arceuthobium cyanocarpum (A.Nelson ex Rydb.) A.Nelson [*A. campylopodum* Engelm. f. *cyanocarpum* (A.Nelson ex Rydb.) L.S.Gill] **Stems** yellowish, 2–4 cm, densely clustered. **Berry** blue, 3–4 mm long. Parasitizing *Pinus flexilis* and *P. albicaulis*, causing a compact witch's broom; valley to subalpine; collected in Gallatin and Big Horn cos. OR to MT south to CA, NV, UT, CO.

Arceuthobium douglasii Engelm. **Stems** olive-green, 3–20 mm long, often forming lines on the twigs. **Berry** green, 3–5 mm long. Parasitizing *Pseudotsuga menziesii*, causing a diffuse witch's broom; valleys, montane. BC to AB south to CA, AZ, NM, CO, Mexico.

Arceuthobium laricis (Piper) H.St.John [*A. campylopodum* Engelm f. *laricis* (Piper) L.S.Gill] **Stems** purplish-yellow, 1–4 cm, densely clustered or not. **Berry** purple, 3–4 mm long. Parasitizing *Larix occidentalis* and rarely *L. lyallii*, causing a diffuse witch's broom; valleys, montane. BC, MT, WA, OR, ID.

CELASTRACEAE: Staff Tree Family

Woody vines or shrubs. **Leaves** petiolate, alternate or opposite, simple. **Inflorescence** an axillary panicle or cyme. **Flowers** perfect or unisexual, regular, rotate, 4- or 5-merous; petals distinct; stamens inserted on perigynous disk; style 1; ovary imbedded in the disk. **Fruit** a capsule enclosing an aril.

1. Vines with alternate leaves; plains in eastern MT . . . . Celastrus
1. Low shrubs with opposite leaves; western MT. . . . . Paxistima

Celastrus L. Bittersweet

Celastrus scandens L. Woody vines spreading by creeping roots. **Stems** climbing or sprawling, to 18 m. **Leaves** alternate; blades elliptic to obovate, acuminate, serrulate, 3–10 cm long; stipules linear, early deciduous. **Inflorescence** an axillary panicle. **Flowers** unisexual, green, 5-merous; sepals 1–2 mm long; petals 3–4 mm long; ovary 3-celled; stigma 3-lobed. **Fruit** an orange, globose, 3-valved capsule, 8–12 mm wide, enclosing a red aril; seeds 3 to 6. Riparian woodlands, forest; plains; collected once in Dawson Co. SK to QC south to WY, TX, TN, NC.

Paxistima Raf.

Paxistima myrsinites (Pursh) Raf. MOUNTAIN LOVER, MOUNTAIN BOX Glabrous, evergreen shrub. **Stems** prostrate to ascending, 20–60 cm. **Leaves** short-petiolate, opposite; blades shiny, deep green, narrowly elliptic to obovate, serrulate, 1–4 cm long; stipules minute. **Inflorescence** few-flowered axillary cymes. **Flowers** perfect, 4-merous; sepals ca. 1 mm long; petals maroon, elliptic, 1–2 mm long; stamens 4. **Fruit** a dry, ovoid capsule, 3–6 mm long; seeds 1 or 2 enclosed in a white aril. Moist coniferous forest; valleys to subalpine. BC, AB south to CA, AZ, Mexico. (p.346)

EUPHORBIACEAE: Spurge Family

Euphorbia L. Spurge

Annual or perennial herbs with latex-like sap. **Leaves** sessile or short-petiolate, simple, cauline, alternate or opposite, usually stipulate. **Inflorescence** a cup-shaped, lobed involucre (cyathium) with 4 or 5, usually colored glands between the lobes, containing 1 female and several male flowers, borne in leaf axils, sometimes organized into terminal cymes with opposite, leaf-like bracts. **Flowers** unisexual, lacking a perianth; male flowers of 1 stamen; female flowers a 3-celled, 3-ovulate ovary with 3 2-lobed styles. **Fruit** a 3-celled, ellipsoid to globose capsule; 1 seed per cell.[152]

    The genus is composed of two distinct types in MT: plants with alternate leaves and leafy-bracteate umbellate inflorescences, and plants with opposite, asymmetrical leaves and 1 to few cyathia in leaf axils. The latter is sometimes segregated as *Chamaesyce*, but there is no evidence that *Euphorbia* sensu lato is not monophyletic. *Euphorbia hexagona* Nutt. ex Spreng. and *E. commutata* Engelm. ex A.Gray are reported for MT,[113] but I have seen no specimens.

1. Leaves opposite; stems prostrate or ascending. 2
1. At least the lower leaves alternate; stems erect to ascending 7

2. Leaves linear, ≥4 times as long as wide, entire . . . . E. missurica
2. Leaves ovate to oblong, often serrulate above. 3

3. Stems pubescent; leaves often with a dark spot . . . . E. supina
3. Plants glabrous. 4

4. Seeds smooth under 10X. 5
4. Seeds ridged or wrinkled under 10X 6

5. Cyathia 1–1.5 mm long. . . . . E. geyeri
5. Cyathia 0.5–1 mm long. . . . . E. serpens

6. Seeds with parallel transverse ridges; leaves obscurely serrulate above . . . . .E. glyptosperma
6. Seeds reticulate wrinkled; leaf margins distinctly serrate toward the tip . . . . E. serpyllifolia

7. Upper leaves and inflorescence bracts white-margined. . . . . E. marginata
7. Leaves and bracts without conspicuous white margins 8

Euphorbia agraria M.Bieb. Glabrous perennial with spreading roots. **Stems** erect, 30–70 cm. **Leaves** alternate, oblong to ovate, entire, cordate, 25–45 mm long. **Inflorescence** a terminal umbel with yellow-green, reniform bracts and solitary, bracteate cyathia in upper leaf axils. **Cyathia** 2–3 mm long; the 4 yellowish glands with a lunate appendage. **Capsule** 2–3 mm long, rugose. Grasslands, pastures, roadsides; plains, valleys. Introduced sporadically to northern U.S.; native to Europe; known from Sweet Grass and Wheatland cos.

Euphorbia cyparissias L. Cypress spurge Glabrous perennial with spreading roots. **Stems** erect, 15–30 cm, branched above. **Leaves** alternate, linear, entire, numerous, 1–3 cm long, whorled at the base of the umbel. **Inflorescence** a compact umbel with ovate-cordate bracts. **Cyathia** ca. 3 mm long; glands with an orangish, lunate appendage. **Capsule** ca. 3 mm long, smooth. Roadsides, yards, gardens; plains, valleys. Introduced ornamental escaped throughout U.S.; native of Eurasia.

Euphorbia esula L. Leafy spurge Glabrous perennial with spreading roots. **Stems** erect, 30–80 cm. **Leaves** alternate, linear-oblanceolate, entire, 2–6 cm long. **Inflorescence** a terminal umbel with yellow-green, ovate bracts 8–16 mm long; solitary, bracteate cyathia in upper leaf axils. **Cyathia** 2–3 mm long; the 4 yellowish glands with a lunate appendage. **Capsule** ca. 4 mm long, nearly smooth. Grasslands, meadows, woodlands, riparian forest; plains, valleys. Introduced ornamental escaped over most of north-temperate N. America; native to Eurasia. (p.346)

Euphorbia geyeri Engelm. [*Chamaesyce geyeri* (Engelm.) Small] Glabrous annual. **Stems** prostrate, 5–45 cm. **Leaves** opposite, oblong-ovate, obtuse to emarginate at the tip, rounded at the base, entire, 4–12 mm long. **Inflorescence**: solitary cyathia in leaf axils. **Cyathia** 1–1.5 mm long; glands with white to reddish margins. **Capsule** 1–2 mm long; seeds smooth. Sandy soil of grasslands, sagebrush steppe; plains. MT to WI south to NM, TX. Reported for Custer Co.[151]

Euphorbia glyptosperma Engelm. [*Chamaesyce glyptosperma* (Engelm.) Small] Glabrous annual. **Stems** prostrate, 5–20 cm, highly branched. **Leaves** opposite, oblong to linear-oblong, obtuse to emarginate at the tip, asymmetrical at the base, weakly serrulate above, 3–10 mm long. **Inflorescence**: solitary cyathia in branch axils. **Cyathia** 0.5–1 mm long; the 4 pink glands with thin white margins. **Capsule** 1–2 mm long; seeds parallel-ridged. Grasslands, fields, roadsides, open slopes, streambanks; plains, valleys. Throughout temperate N. America. See. *E. serpyllifolia*.

Euphorbia helioscopia L. Annual. **Stems** erect, 20–50 cm, simple or branched below but not above. **Herbage** glabrous to pilose. **Leaves** alternate, oblong to spatulate, serrulate, 15–30 mm long. **Inflorescence** a terminal compound umbel with leaf-like bracts, subtended by a whorl of leaves. **Cyathia** 2–3 mm long, villous; the 4 oval, yellow lobes without appendages. **Capsule** 2–3 mm long, smooth. Grasslands, fields, pastures; plains, valleys. Introduced in eastern and northwest U.S.; native to Eurasia. Known from Flathead and Daniels cos.

Euphorbia marginata Pursh Snow on the mountain Annual. **Stems** erect, 20–70 cm, unbranched below inflorescence. **Herbage** glabrous to sparsely villous. **Leaves** alternate, sessile, ovate to elliptic, 3–8 cm long, entire, acute, mucronate, whorled at the base of the umbel. **Inflorescence** a terminal, compound umbel with white-margined, leaf-like bracts. **Cyathia** 3–4 mm long, villous; with 4 white, reniform glands. **Capsule** 4–6 mm long, pubescent. Fields, badlands, roadsides; plains. MT to MN south to NM, TX.

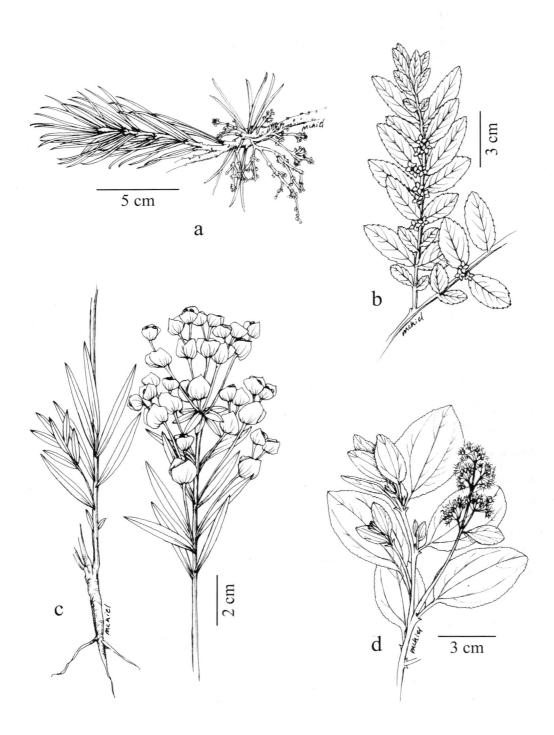

Plate 60. a. *Arceuthobium americanum*, b. *Paxistima myrsinites*, c. *Euphorbia esula*, d. *Ceanothus velutinus*

Euphorbia missurica Raf. [*Chamaesyce nuttallii* (Engelm.) Small, *C. missurica* (Raf.) Shinners] Glabrous annual. **Stems** ascending to erect, 20–80 cm, highly branched. **Leaves** linear to linear-oblong, rounded to emarginate at the apex, oblique at the base, entire, 5–20 mm long. **Inflorescence**: a solitary cyathia in leaf axils. **Cyathia** 1–2 mm long; glands with white margins. **Capsule** 2–3 mm long; seeds smooth to obscurely wrinkled. Streambanks; plains, valleys. MT to MN south to NM, TX.

Euphorbia peplus L. Glabrous annual. **Stems** erect, 10–30 cm, branched in the inflorescence. **Leaves** alternate, petiolate; blades obovate with rounded tips, entire, 5–25 mm long. **Inflorescence** a dichotomously branched umbel with leaf-like bracts. **Cyathia** ca. 1 mm long, glands with attenuate-lunate appendages. **Capsule** ca. 2 mm long, smooth. Gardens, roadsides; valleys. Introduced throughout much of N. America; native to Eurasia.

Euphorbia robusta (Engelm.) Small [*E. brachycera* Engelm. var. *robusta* (Engelm.) Dorn] Small Perennial from a branched caudex. **Stems** erect, 8–20 cm, simple. **Herbage** glabrous to puberulent (Carbon Co.). **Leaves** alternate, ovate with acute tips, entire, 6–15 mm long. **Inflorescence** a dichotomously branched umbel with leaf-like bracts. **Cyathia** 2–3 mm long, glands with yellow to red, lunate appendages. **Capsule** 3–5 mm long, papillose. Usually sandy or gravelly soil of grasslands, sagebrush steppe, woodlands; plains, valleys. MT to MN south to AZ, NM.

Euphorbia serpens Kunth [*Chamaesyce serpens* (Kunth) Small] Glabrous annual. **Stems** prostrate, 5–20 cm, highly branched. **Leaves** opposite, ovate to orbicular, rounded at the tip, asymmetrical at the base, entire, 2–6 mm long. **Inflorescence**: solitary cyathia in branch axils. **Cyathia** 0.5–1 mm long; glands green to pink with white margins. **Capsule** 1–1.5 mm long; seeds smooth. Fields, streambanks; plains, valleys. MT to ON south to AZ, TN, FL, Mexico.

Euphorbia serpyllifolia Pers. [*Chamaesyce serpyllifolia* (Pers.) Small] Annual. **Stems** prostrate to ascending, 1–15 cm, highly branched. **Herbage** glabrous to sparsely villous. **Leaves** opposite, oblong to obovate, sometimes curved, serrate in the upper third, asymmetrical at the base, 3–12 mm long. **Inflorescence**: 1 to few cyathia in leaf axils. **Cyathia** ≤1 mm long; glands pink with white margins. **Capsule** 1.5–2 mm long; seeds lightly reticulate-wrinkled. Grassland, sagebrush steppe, pine woodlands, roadsides; plains, valleys. BC to MI south to CA, TX, Mexico.
  Euphorbia glyptosperma has less distinctly serrate leaves and distinctly parallel-ridged seeds.

Euphorbia spathulata Lam. Glabrous annual. **Stems** erect, 5–30 cm, branched in the inflorescence. **Leaves** alternate, oblong with rounded tips, serrulate, 1–3 cm long. **Inflorescence** a dichotomously branched umbel with leaf-like bracts. **Cyathia** ≤1 mm long, membranous; glands small, yellowish, without appendages. **Capsule** 2–3 mm long, distinctly papillose. Sparsely vegetated soil of grasslands, sagebrush steppe, pine woodlands; plains. WA to MN south to TX, Mexico.

Euphorbia supina Raf. [*Chamaesyce supina* (Raf.) Moldenke, *Euphorbia maculata* L.] Annual. **Stems** prostrate, 5–20 cm. **Herbage** villous. **Leaves** opposite, oblong, acute, asymmetrical at the base, serrulate toward the tip, 5–15 mm long, often with a dark spot on the upper surface. **Inflorescence**: solitary cyathia in leaf axils. **Cyathia** ca. 1 mm long; glands with white or pink margins. **Capsule** 1–2 mm long, hairy. Roadsides, lawns; valleys in Sanders Co. Introduced sporadically to western U.S.; native to Eastern N. America.

RHAMNACEAE: Buckthorn Family

Shrubs or small trees, sometimes monoecious or dioecious. **Leaves** alternate or rarely opposite, simple with serrate margins, petiolate; stipules small. **Inflorescence** a panicle or few-flowered cymes, terminal and/or axillary. **Flowers** perfect, 4- to 5-merous, regular, perigynous to epigynous; sepals deltoid; petals separate; ovary 1, superior (sometimes appearing inferior). **Fruit** a few-seeded capsule or firm-fleshed, berry-like drupe.

1. Inflorescence axillary; fruit a berry-like drupe . . . .*Rhamnus*
1. Inflorescence terminal; fruit a capsule . . . .*Ceanothus*

Ceanothus L.

Shrubs. **Leaves** alternate, deciduous or evergreen, pubescent beneath, glandular-serrate; petioles 1–2 cm long. **Inflorescence** densely branched, terminal panicles. **Flowers** perfect, white, 5-merous; petals long-clawed, the blade cupped; ovary embedded in a lobed disk; stigmas 3. **Fruit** a 3-chambered capsule.

The presence of older, partially brown leaves on *Ceanothus velutinus* helps separate it from *C. sanguineus*.

1. Leaves lanceolate, ≤2 cm wide . . . . *C. herbaceus*
1. Leaves ovate, some >2 cm wide 2

2. Leaves shiny and sticky on the upper surface, persistent for >1 year . . . . *C. velutinus*
2. Leaves deciduous, not sticky or shiny above . . . . *C. sanguineus*

Ceanothus herbaceus Raf. Stems ascending to erect, 50–100 cm, highly branched. **Leaves** deciduous; blades lanceolate to narrowly ovate, 2–6 cm long, sparsely villous beneath. **Flowers** ca. 3 mm across. **Capsule** 3-lobed, 2–4 mm wide. Ponderosa pine woodland; plains. Eastern and central N. America, west to MT, WY; collected in Powder River Co. 60 years ago. Our plants are var. **pubescens** (Torr. & A.Gray ex S.Watson) Shinners.

Ceanothus sanguineus Pursh Red-stem ceanothus, wild lilac **Stems** erect 1–2 m; twigs purplish. **Leaves** deciduous; blades 2–8 cm long; stipules quickly deciduous. **Flowers** 3–4 mm across. **Capsule** globose, 2–3 mm wide. Dry, open forest, especially in recently burned areas; montane. BC, MT south to CA.

Ceanothus velutinus Douglas ex Hook. Buckbrush, tobacco brush **Stems** ascending to erect, 50–150 cm; twigs green, pubescent at first. **Leaves** fragrant, evergreen, 3–7 cm long, shiny above, pale beneath; stipules ca. 1 mm long. **Flowers** 4–5 mm across. **Capsule** 3–5 mm wide, 3-lobed with a low ridge on the back of each lobe. Open forest, especially in recently burned areas; montane, lower subalpine. BC, AB south to CA, CO, SD. (p.346)

Rhamnus L. Buckthorn

Shrubs or small trees, often monoecious or dioecious. **Leaves** alternate or opposite, deciduous, ovate to oblong. **Inflorescence** axillary, 1 to few pedicellate flowers or several-flowered umbels. **Flowers** green, perfect or unisexual, 4- to 5-merous; petals shorter than the sepals or lacking; ovary partly embedded in a perigynous disk; stigmas 2 to 4. **Fruit** a 2- to 4-seeded berry-like, black drupe.

1. Many leaves of young twigs opposite or subopposite; flowers 4-merous. . . . . *R. cathartica*
1. Leaves alternate; flowers 5-merous 2

2. Axillary inflorescences of 4 to 40 flowers; lateral leaf veins >8 per side. . . . . *R. purshiana*
2. Axillary inflorescences of 2 to 5 flowers; lateral leaf veins ≤8 per side. . . . . *R. alnifolia*

Rhamnus alnifolia L'Her. Dioecious shrub. **Stems** erect or ascending, 50–200 cm; twigs gray. **Leaves** alternate, 3–12 cm long, glandular-serrate, puberulent on the veins below; petioles 5–15 mm long. **Inflorescence** few axillary, pedicellate flowers. **Flowers** functionally unisexual, 5-merous; sepals 1–2 mm long; petals lacking; stigmas 3. **Drupe** 3-seeded, 5–9 mm wide, poisonous. Wet forest openings or forming thickets along the margins of marshes, swamps, lakes, fens; valleys, montane. BC to NL south to CA, WY, NE, OH, NJ. (p.352)

Rhamnus cathartica L. Common buckthorn Dioecious shrub or small tree. **Stems** erect, 2–6 m; twigs sometimes ending in thorns. **Leaves** opposite, 2–6 cm long, glandular-serrate, glabrous; petioles 1–2 cm long. **Inflorescence** several pedicellate flowers in leaf axils. **Flowers** functionally unisexual, 4-merous; sepals 1–3 mm long; petals erect; stigmas 4. **Drupe** 4-seeded, 5–7 mm across. Fields, vacant lots, riparian corridors; plains, valleys. Introduced ornamental escaped across much of N. America; native to Eurasia.

Rhamnus purshiana DC. Shrub or small tree, sometimes monoecious. **Stems** erect, 2–4 m; twigs, petioles and veins brown-puberulent. **Leaves** alternate, 4–12 cm long, serrulate; petioles 5–20 mm long. **Inflorescence** several-flowered, axillary umbels. **Flowers** perfect or unisexual, 5-merous; sepals 1–2 mm long, puberulent; petals minute; stigmas 2–3. **Drupe** 2- to 3-seeded, 6–9 mm across. Wetland thickets, wet forest openings; valleys. BC, MT south to CA, ID.

VITACEAE: Grape Family

Woody vines. **Leaves** petiolate, alternate; blade palmately lobed or divided; stipules deciduous; tendrils branched, opposite upper leaf axils. **Inflorescence** axillary racemes or panicles. **Flowers** perfect or unisexual, regular, hypogynous, 5-merous; calyx reduced, obscurely lobed; petals separate; stamens 5; ovary superior; style 1. **Fruit** a globose, 2-celled berry with 2 seeds per cell.

1. Leaves lobed up to ca. halfway to the midvein. . . . . *Vitis*
1. Leaves palmately compound with distinct leaflets . . . . *Parthenocissus*

Parthenocissus Planch. Virginia creeper

Leaves palmately divided; leaflets serrate, attenuate-tipped, turning red in autumn. **Inflorescence** a panicle from upper leaf nodes. **Flowers** perfect. **Berry** dark blue with 1 to 4 seeds.

1. Panicle with a central axis; tendrils with adhesive disks. . . . . *P. quinquefolia*
1. Panicle dichotomously branched; tendrils without adhesive disks . . . . *P. vitacea*

Parthenocissus quinquefolia (L.) Planch. **Leaflets** 5 to 7, lanceolate to obovate, 3–13 cm long, puberulent below. **Tendrils** 3- to 8-branched with adhesive disks. **Inflorescence** a hemispheric panicle of cymes with a distinct central axis. **Flowers**: petals 2–3 mm long. **Berry** 5–8 mm wide. Riparian forest; valleys. Introduced in western U.S. where it occasionally escapes cultivation; native to eastern N. America. Known from Missoula Co.

Parthenocissus vitacea (Knerr) Hitchc. [*P. inserta* (J.Kern.) Fritsch] **Leaflets** 5, narrowly obovate, 3–10 cm long, glabrate. **Tendrils** 3- to 5-branched without adhesive disks. **Inflorescence** a dichotomously branched, hemispheric panicle. **Flowers**: petals 2–3 mm long. **Berry** 6–10 mm wide. Riparian forest; plains. MT to QC south to CA, AZ, UT, TX.

Vitis L. Grape

Vitis riparia Michx. Plants dioecious. **Leaves** long-petiolate; blades cordate-ovate, acuminate, shallowly lobed, serrate, 5–12 cm long, glabrous above, puberulent on veins below. **Tendrils** branched, without adhesive disks. **Inflorescence** a branched panicle from upper leaf nodes. **Flowers** functionally unisexual; petals early deciduous. **Berry** purple, glaucous, 7–11 mm across, edible. Riparian forest; plains. MT to QC south to NM, TX, TN.

LINACEAE: Flax Family

Linum L. Flax

Glabrous, annual or perennial herbs. **Leaves** cauline, alternate, simple, entire; stipules inconspicuous or absent. **Inflorescence** cymose or racemose, leafy-bracteate, terminal. **Flowers** perfect, regular, 5-merous; sepals overlapping, awned; petals fan-shaped, separate or united basally; stamens 5; ovary superior; styles 5, distinct above. **Fruit** a 5-loculate, ovoid capsule; seeds 1 or 2 per cell.

1. Plants perennial from a branched rootstock. . . . . *L. lewisii*
1. Plants annual 2

2. Flowers blue; capsules 6–8 mm long . . . .*L. usitatissimum*
2. Flowers yellow; capsules 3–5 mm long 3

3. Style 3–6 mm long; petals 7–16 mm long . . . .*L. rigidum*
3. Style 1–3 mm long; petals 5–9 mm long . . . . *L. australe*

Linum australe A.Heller Annual. **Stems** erect, 10–40 cm, branched above, puberulent below. **Leaves** linear, 5–25 mm long with small, purple stipular glands. **Inflorescence** a panicle with erect branches; upper leaves and bracts awned and glandular-serrulate. **Flowers** yellow; sepals lanceolate, 4–7 mm long, glandular-serrulate; petals 5–9 mm long, pubescent at the base; styles 2–3 mm long, united to above middle. **Capsule** 3–5 mm long, splitting into 5 2-seeded segments. Grasslands, roadsides, pine woodlands; plains. AB south to AZ, UT, TX, Mexico.

Linum lewisii Pursh [*L. perenne* L. var. *lewisii* (Pursh) Eaton & Wright] WILD BLUE FLAX Perennial from a branched caudex. **Stems** ascending, 5–60 cm, simple. **Leaves** numerous, linear, 5–20 mm long. **Inflorescence** a leafy raceme. **Flowers** blue; sepals ovate, 4–5 mm long; petals 9–16 mm long, quickly falling. **Capsule** 5–8 mm long, splitting into 10 1-seeded sections. Grasslands, sagebrush steppe, badlands, woodlands, meadows, fellfields, roadsides at all elevations. AK to QC south to CA, TX, WI. (p.352)

The alpine ecotype, with shorter leaves and smaller flowers, has been called var. *alpicola* Jeps.

Linum rigidum Pursh Annual. **Stems** erect, 5–50 cm, branched above. **Leaves** linear, 1–3 cm long; bracts awned, glandular-serrate, sometimes with small, purple stipular glands. **Inflorescence** a leafy-bracteate, often flat-topped panicle. **Flowers** yellow; sepals lanceolate, 5–9 mm long, glandular-serrate; petals 7–16 mm long, pubescent, often red at the base; styles 3–6 mm long, united to above middle. **Capsule** 4–5 mm long, splitting into 5 2-seeded segments. Grasslands, sagebrush steppe; plains, valleys. AB to MB south to NM, TX.

1. Stipular glands present; styles 4–6 mm long; plants ≥20 cm . . . . var. *rigidum*
1. Stipular glands absent; styles 3–4 mm long; plants ≤20 cm. . . . . .var. *compactum*

Linum rigidum var. *compactum* (A.Nelson) C.M.Rogers and *L. rigidum* var. *rigidum* intergrade[341] and are similar ecologically and geographically in MT.

Linum usitatissimum L. COMMON FLAX Annual. **Stems** 20–80 cm, mostly unbranched. **Leaves** linear-lanceolate, 1–3 cm long. **Flowers** blue (white); sepals ovate, 6–8 mm long, the inner ciliate; petals 10–15 mm long; styles distinct. **Capsule** 6–8 mm long, splitting into 10 1-seeded segments. Fields, roadsides; plains, valleys. Introduced from Europe, cultivated for fiber and oil; sporadic in western N. America.

POLYGALACEAE: Milkwort Family

Polygala L. Milkwort

Glabrous annual or perennial herbs. **Leaves** simple, entire, linear to linear-oblanceolate, alternate or whorled, exstipulate. **Inflorescence** terminal or axillary, spike-like racemes. **Flowers** perfect, irregular, hypogynous, resembling pea flowers; sepals 5, 2 large, petal-like (wings) and 3 small; petals 3, united below, the lower one keel-shaped; stamens 8; style 2-lobed. **Fruit** an ellipsoid, biloculate capsule; seeds enclosed in an aril, 1 per locule.

Flowers superficially resemble papilionaceous flowers of the Fabaceae.

1. Plants perennial from a branched caudex; lower leaves alternate. . . . . *P. alba*
1. Plants taprooted annuals; leaves whorled below . . . .*P. verticillata*

Polygala alba Nutt. Perennial from a branched caudex. **Stems** ascending to erect, simple, 10–30 cm. **Leaves** alternate, 5–25 mm long. **Inflorescence** terminal. **Flowers** white with a green center; wings 2–4 mm long; keel 2–3 mm long with 4 lobes on each side. **Capsule** 2–3 mm long. Grasslands, ponderosa pine woodlands; plains. WA to MN south to AZ, TX, Mexico.

Polygala verticillata L. WHORLED MILKWORT Annual. **Stems** erect, 5–20 cm, branched above. **Leaves** whorled or alternate above, 1–2 cm long. **Inflorescence** axillary. **Flowers** white to pink; wings 1–2 mm long; keel 1–1.5 mm long with 1 to 2 lobes on each side. **Capsule** ca. 2 mm long. Grasslands; plains. MB to MA south to UT, TX, FL.

ACERACEAE: Maple Family

Acer L. Maple

Monoecious or dioecious, deciduous shrubs or trees. **Leaves** opposite, petiolate, palmately lobed or divided; stipules absent. **Inflorescence** racemes or panicles, axillary or terminal on short shoots. **Flowers** regular, unisexual, perigynous, 5-merous; petals distinct or absent; stamens 4 to 10, usually 8; ovary superior, 2-lobed and 2-celled; stigmas 2. **Fruit** a double samara, each oblong-winged mericarp 1-seeded.

Bigtooth maple (*Acer grandidentatum* Nutt.) is reported for south-central MT,[113] but I have seen no specimens.

1. Leaves divided into 3 or 5 leaflets; petals absent. . . . . *A. negundo*
1. All or some leaves simple, lobed; petals present 2

2. Shrubs or small trees; petioles with clear sap; seed ca. 5 mm long. . . . . *A. glabrum*
2. Trees; sap of petioles milky; seed 10–15 mm long. . . . . *A. platanoides*

Acer glabrum Torr. ROCKY MOUNTAIN MAPLE Large monoecious or dioecious shrubs or small trees up to 15 m. **Twigs** glabrous, red to purple. **Leaves** ovate-cordate, glabrous, 3–12 cm across, 3- to 5-lobed, doubly serrate; petioles red. **Inflorescence** a few-flowered panicle on short leafy side shoots. **Flowers** functionally unisexual; sepals and petals 3–4 mm long, yellowish, distinctly veined; stamens 8 to 10. **Samara** 2–4 cm long; the seed ca. 5 mm long. Moist, open, coniferous forests, riparian forests, avalanche slopes; valleys to lower subalpine. AK to AB south to CA, NM, NE. (p.352)

Our plants have been separated into two vars., var. *douglasii* (Hook.) Dippel, with deeply lobed to divided leaves and red twigs, and var. *glabrum*, with shallowly lobed leaves and gray twigs. However, these characters show no ecological or geographic integrity and intergrade in southwest and south-central parts of MT.

Acer negundo L. BOX ELDER Small dioecious trees to 15 m. **Twigs** glabrate to pilose. **Leaves** 3- or (rarely) 5-foliolate; leaflets pubescent on veins beneath, lanceolate to ovate, coarsely few-toothed, 3–8 cm long; lateral pair asymmetric. **Inflorescence**: female flowers in pendent racemes from axillary buds; male flowers fascicled. **Flowers** unisexual, green; sepals ca. 1 mm long; petals absent. **Samara** 2–3 cm long; the seed ca. 1 cm long. Riparian forest, woodlands, roadsides, fields; plains, valleys. Throughout temperate N. America.

1. Twigs glabrous . . . . var. *negundo*
1. Twigs pilose . . . . var. *interius*

Acer negundo var. *interius* (Britton) Sarg. is native to MT; *A. negundo* var. *negundo* is indigenous to eastern N. America and introduced as a street tree and escapes near urban areas.

Acer platanoides L. NORWAY MAPLE Monoecious trees to 30 m. **Twigs** purplish, glabrate. **Leaves** glandular at first, cordate, 5- or 7-lobed ≤1/2-way to the midvein, 6–20 cm wide, deeply toothed, deep green; petioles with white sap. **Inflorescence** erect panicles from short, side shoots. **Flowers** unisexual, yellow; sepals 4–5 mm long; petals 3–5 mm long. **Samara** 30–55 mm long; the seed 10–15 mm long. Riparian forest, lawns, gardens; valleys of Missoula Co. An introduced ornamental escaping in eastern N. America, sporadically in the west; native to Europe.

ANACARDIACEAE: Cashew Family

Dioecious or polygamo-dioecious shrubs or vines. **Leaves** petiolate, alternate, trifoliate or pinnately compound, turning red in autumn; stipules absent. **Inflorescence** terminal or axillary racemes or panicles. **Flowers** small, regular, hypogynous, mostly unisexual, 5-merous; sepals united below; petals distinct; stamens 5; ovary superior; stigmas 3. **Fruit** a 1-seeded drupe.

1. Drupes red; shrubs. . . . . *Rhus*
1. Drupes yellowish-white; vines or subshrubs. . . . . *Toxicodendron*

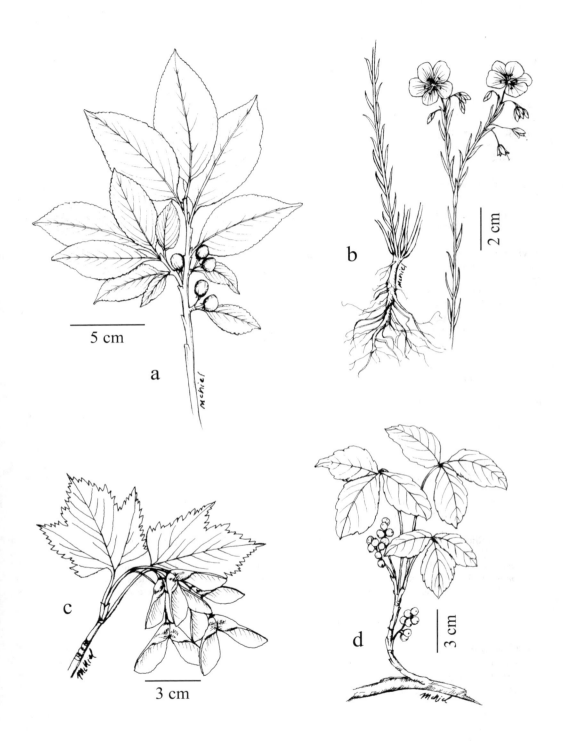

Plate 61. a. *Rhamnus alnifolia*, b. *Linum lewisii*, c. *Acer glabrum*, d. *Toxicodendron rydbergii*

Rhus L. Sumac

Polygamo-dioecious shrubs. **Leaves** 3-foliolate or pinnately compound. **Inflorescence** terminal or axillary panicles. **Flowers** subtended by small bracts; female flowers with 5 to 10 vestigial stamens. **Fruit** a red, glandular-hairy drupe.

1. Leaves trifoliate . . . . *R. aromatica*
1. Leaves odd-pinnate . . . . *R. glabra*

Rhus aromatica Aiton [*R. trilobata* Nutt.] SKUNKBUSH SUMAC Shrub 1–2 m. **Herbage** glabrate to puberulent, fragrant. **Leaves** trifoliate; leaflets ovate with rounded lobes; central leaflet the largest, 2–4 cm long. **Inflorescence** terminal and axillary, globular, hairy-bracteate racemes. **Flowers**: sepals reddish, ca. 1 mm long; petals yellow, ca. 2 mm long, pubescent below. **Drupes** subglobose, 5–7 mm wide, crowded on stem tips. Grasslands, woodlands, riparian forest; plains, valleys. Throughout most of temperate N. America. Our plants are var. *trilobata* (Nutt.) A.Gray.

Rhus glabra L. SMOOTH SUMAC Shrub 1–3 m, spreading by roots. **Herbage** glabrate. **Leaves** odd-pinnate; leaflets 9 to 17, lanceolate, serrate, 4–8 cm long, paler below. **Inflorescence** a dense, terminal, pyramidal, branched panicle, 5–10 cm long, hirsute. **Flowers** green; sepals 1–2 mm long; petals ca. 2 mm long, pubescent on the inner face. **Drupe** compressed-globose, red, 3–4 mm long, crowded. Grasslands, ponderosa pine woodlands; plains, valleys. BC to QC south to OR, NM, TX, AR.

Toxicodendron Mill. Poison ivy

Toxicodendron rydbergii (Small ex Rydb.) Greene [*Rhus radicans* L.] Rhizomatous subshrub, shrub, or (rarely) vine. **Stems** erect, 10–50 cm, pubescent, branched and bearing leaves above. **Herbage** glabrous, sparsely pubescent on veins, allergenic. **Leaves** trifoliate, long-petiolate; leaflets ovate, remotely serrate to entire, 2–12 cm long, shiny above; lateral leaflets asymmetrical. **Inflorescence** an axillary, few-branched panicle. **Flowers** white; petals 2–3 mm long, dark-veined. **Drupes** crowded, globose, greenish-yellow, 4–6 mm long. Streambanks, woodlands, open forest; plains, valleys. Much of temperate N. America. (p.352)

In western MT *Toxicodendron rydbergii* usually occurs in stony soil of steep slopes, streambanks or lake shores; it is most common in riparian woodlands in eastern MT.

ZYGOPHYLLACEAE: Caltrop Family

Annual or perennial herbs. **Leaves** petiolate, alternate or opposite, stipulate, pinnately lobed or compound. **Inflorescence**: solitary, axillary, pedunculate flowers. **Flowers** perfect, regular, hypogynous, 5-merous; petals and sepals separate; stamens 10 to 15; ovary superior, 3- to 5-loculate; style 1. **Fruit** a capsule or schizocarp.

1. Leaves alternate, deeply lobed into linear segments . . . . *Peganum*
1. Leaves opposite, divided into distinct, oblong to ovate leaflets 2

2. Fruit a spiny schizocarp; leaves with 6 to 16 leaflets . . . . *Tribulus*
2. Fruit an oblong, unarmed capsule; leaves with 2 leaflets. . . . . *Zygophyllum*

Peganum L.

Peganum harmala L. Glabrous annual. **Stems** erect, branched, 20–50 cm. **Leaves** alternate; blades 1 or 2 times pinnately divided into linear segments 1–3 cm long. **Flowers**: sepals linear, simple or lobed, 1–2 cm long; petals white, oblanceolate, 14–18 mm long; ovary with 3 carpels; stamens 15. **Fruit** a membranous, globose, many-seeded capsule 10–15 mm long. Grasslands, pastures; plains. Introduced to WA and MT south to CA, AZ, NM, TX; native to Europe. Known from Phillips and Valley cos.

Tribulus L. Caltrop

Tribulus terrestris L. Puncture vine Taprooted, mat-forming annual. **Stems** prostrate 10–40 cm. **Herbage** strigose. **Leaves** opposite, even-pinnate; leaflets 6 to 16, ovate, ca. 5 mm long. **Flowers**: petals yellow, 3–5 mm long; stamens 10; ovary 5-lobed. **Fruit** a schizocarp, each 2-seeded mericarp with two large spines 2–6 mm long and numerous fine bristles. Sandy or gravelly soil of streambanks, roadsides; plains, valleys. Introduced throughout the U.S.; native to southern Europe.

Zygophyllum L. Beancaper

Zygophyllum fabago L. Glabrous annual. **Stems** ascending, branched, 20–50 cm. **Leaves** opposite, compound; leaflets 2, oblong, 1–2 cm long. **Flowers**: sepals oblanceolate, ca. 6 mm long; petals yellow, separate, barely exceeding the sepals; ovary superior with 5 carpels; stamens 10; filaments basally winged. **Fruit** a pendent, oblong, many-seeded capsule 2–3 cm long. Fields, roadsides; valleys. Introduced in most of western U.S.; native to Eurasia. Known from Beaverhead and Lewis & Clark cos.

OXALIDACEAE: Wood Sorrel Family

Oxalis L. Wood sorrel

Short-lived perennials. **Leaves** alternate, long-petiolate, stipulate, trifoliate; the leaflets obcordate, folding up at night. **Inflorescence** axillary, pedunculate, few-flowered, bracteate umbels with spreading, upturned pedicels. **Flowers** perfect, regular, hypogynous, 5-merous; sepals and petals distinct or united basally; stamens 10; ovary 5-loculate; styles 5; stigmas capitate. **Fruit** an erect, linear-oblong, 5-celled capsule; seeds with transverse ridges.

1. Stems creeping, rooting at the nodes . . . . *O. corniculata*
1. Stems ascending, not rooting at the nodes or only at stem bases 2

2. Inflorescence with septate hairs; capsules sparsely hirsute . . . . *O. stricta*
2. Plants without septate hairs; capsules strigose. . . . . *O. dillenii*

Oxalis corniculata L. **Stems** prostrate to ascending at the tips, rooting at the nodes, 1–10 cm. **Herbage** purple-tinged, glabrate to strigose. **Leaflets** 1-3 cm long; stipules 1-4 mm long. **Flowers**: sepals 2-4 mm long; petals yellow, 4-8 mm long. **Capsule** pubescent 15-25 mm long; seed ridges not white. Lawns, gardens; plains, valleys. Introduced worldwide; native to Europe.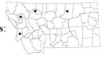

Oxalis dillenii Jacq. **Stems** ascending, 3–20 cm. **Herbage** strigose to villous with non-septate, long-pointed hairs. **Leaflets** 5–12 mm long; stipules 0.5–3 mm long. **Flowers**: sepals 3–5 mm long; petals yellow, 6–8 mm long. **Capsule** 1–2 cm long, pubescent with retrorse hairs; seed ridges white. Lawns, gardens, fields, streambanks; plains, valleys. Most of temperate N. America.

Oxalis stricta L. Rhizomatous. **Stems** ascending to erect, 10–50 cm, villous with contorted, purple-septate hairs. **Leaflets** glabrate, 1–2 cm long; stipules absent. **Flowers**: sepals 3–5 mm long; petals 4–9 mm long. **Capsule** 5–15 mm long, villous with septate hairs; seed ridges obscurely white. Disturbed soil of moist forest; valleys. Most of temperate N. America; collected in Missoula Co.

GERANIACEAE: Geranium Family

Annual or perennial herbs. **Leaves** petiolate, deeply lobed or divided. **Inflorescence** axillary, long-pedunculate umbels or cymes. **Flowers** perfect, regular, hypogynous, rotate, 5-merous; sepals and petals separate; stamens 5 or 10; ovary 5-loculate and lobed; styles 5, united; stigmas 5. **Fruit** a 5-celled, 5-seeded capsule with an elongate style column.

1. Leaf blade oblong, pinnately divided into >5 leaflets . . . . *Erodium*
1. Leaf blade palmately lobed or divided . . . . *Geranium*

Erodium L'Her. ex Aiton Crane's bill

Erodium cicutarium (L.) L'Her. ex Aiton FILAREE Taprooted annual. **Stems** ascending to prostrate, 5–60 cm. **Herbage** villous to hirsute, sometimes glandular above. **Leaves** basal and opposite on the stem; blades odd-pinnate, 1–10 cm long; leaflets lanceolate to ovate, deeply serrate. **Inflorescence** an axillary, minutely-bracteate umbel with 2 to 5 flowers. **Flowers** long-pedicellate; sepals ovate, strigose, awn-tipped, 3–6 mm long; petals pink to rose, 3–7 mm long; stamens 5. **Capsule** 4–6 mm long with an elongate style column 2–3 cm long at maturity, separating into 5 carpels; styles becoming twisted. Disturbed soil of grasslands, roadsides, lawns, fields, streambanks; valleys. Introduced throughout much of temperate N. America; native to Eurasia.

Geranium L. Geranium

Annual to perennial herbs with swollen nodes. **Leaves** long-petiolate, alternate below but opposite in the inflorescence, palmately lobed or divided. **Inflorescence** axillary 2- or 3-flowered umbels. **Flowers** pedicellate; sepals imbricate, lanceolate to ovate, bristle-tipped; petals obovate; stamens 10, 5 sometimes sterile. **Capsules** splitting from the base into 5 sections, each curving upward to expel the solitary seed.

    Geranium oreganum Howell [*G. incisum* (Nutt.) Torr. & A.Gray], similar to *G. viscosissimum* but with ciliate petals, is reported for MT[155] but is otherwise not thought to occur east of the Cascades.[187] Occasional hybrids between *Geranium richardsonii* and *G. viscosissimum* can be found in the aspen parkland along the east edge of Glacier National Park.

1. Plants annual 2
1. Plants perennial 6

2. Leaves divided into 3 or 5 leaflets; petals >9 mm long. . . . .*G. robertianum*
2. Leaves lobed but not fully divided into leaflets; petals ≤9 mm long 3

3. Sepals lacking a bristle tip 4
3. Sepals tipped with a bristle 1–2 mm long. 5

4. Fertile stamens 10; ovary glabrous . . . . *G. molle*
4. Fertile stamens 5; ovary pubescent . . . .*G. pusillum*

5. Fruiting pedicel much longer than calyx; beak of style column 4–5 mm long. . . . . *G. bicknellii*
5. Fruiting pedicel ca. equal to the calyx; stylar beak ≤3 mm long . . . . *G. carolinianum*

6. Flowers white; capsule 20–25 mm long, including the style column . . . . *G. richardsonii*
6. Flowers rose; capsule 25–30 mm long, including the style column . . . .*G. viscosissimum*

Geranium bicknellii Britton Annual or biennial. **Stems** erect, branched, 10–40 cm. **Herbage** sparsely hirsute, glandular in the inflorescence. **Leaf blades** 2–6 cm wide, 5- or 7-lobed; each lobe shallowly lobed. **Flowers:** sepals 4–7 mm long, lanceolate, bristle-tipped; petals 5–9 mm long, pink (white); fertile stamens 10. **Capsule** 15–25 mm long including the style column. Coniferous forest and (uncommonly) disturbed soil of streambanks, roadsides, often abundant for 2 to 4 years following fire; valleys, montane. AK to NL south to CA, UT, CO, SD, IN, MA.

Geranium carolinianum L. Annual. **Stems** erect, branched, 15–40 cm. **Herbage** glabrate to hirsute, glandular in the inflorescence. **Leaf blades** cordate, orbicular, 2–7 cm wide, cleft nearly to the midvein into 3 to 5 lobed segments. **Flowers:** sepals ovate, 4–8 mm long, bristle-tipped; petals as long as the sepals, pink to rose; fertile stamens 10. **Capsule** 1–2 cm long including the style column. Moist, disturbed soil of fields, thickets, roadsides; valleys; Fergus and Missoula cos. Throughout temperate N. America.

Geranium molle L. Annual. **Stems** ascending, 10–40 cm. **Herbage** hirsute, glandular. **Leaf blades** cordate, orbicular, 6–14 cm wide, cleft halfway to the midvein into 5 to 7 lobed segments. **Flowers:** sepals 2–4 mm long; petals 3–5 mm long, pink; fertile stamens 10. **Capsule** 7–10 mm long including the style column. Lawns, gardens, disturbed riparian areas; valleys. Introduced to much of temperate N. America; native to Europe. One collection from Fergus Co.

Geranium pusillum L. Annual. **Stems** prostrate to erect, 7–50 cm. **Herbage** glabrate to hirsute, glandular. **Leaf blades** cordate, orbicular, 1–6 cm wide, cleft at least halfway to the midvein into 5 to 9 lobed segments. **Flowers:** sepals 2–4 mm long; petals ca. as long, purple; fertile stamens 5. **Capsule** 7–11 mm long including the style column. Fields, disturbed grasslands, riparian areas; valleys, montane. Introduced BC to MA south to OR, KS, NC; native to Europe.

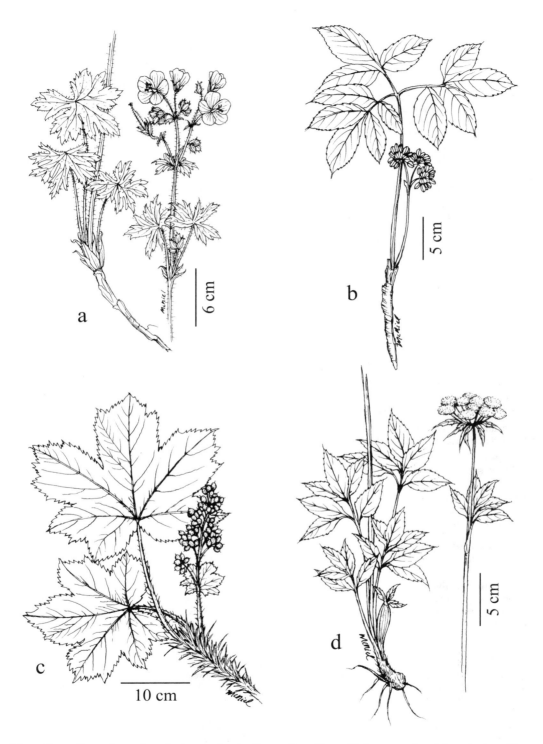

Plate 62. a. *Geranium viscosissimum*, b. *Aralia nudicaulis*, c. *Oplopanax horridus*, d. *Angelica dawsonii*

Geranium richardsonii Fisch. & Trautv. WHITE GERANIUM Perennial. **Stems** erect, 20–90 cm. **Herbage** sparsely strigose, glandular in the inflorescence. **Leaf blades** cordate, orbicular, 3–14 cm wide, cleft nearly to the midvein into 3 to 7 lobed segments. **Flowers**: sepals 5–10 mm long, bristle-tipped; petals 9–18 mm long, white, purple-veined; fertile stamens 10. **Capsule** 20–25 mm long including the style column. Meadows, thickets, moist open forest, woodlands; valleys, montane. YT south to CA, NM, SD.

Geranium robertianum L. Annual. **Stems** prostrate to ascending, 10–30 cm. **Herbage** villous, glandular in the inflorescence, scented. **Leaf blades** deltoid in outline, 4–10 cm wide, palmately divided into 3 or 5 deeply twice-lobed leaflets. **Flowers**: sepals 5–6 mm long; petals ca. 12 mm long, rose; fertile stamens 5. **Capsule** 2–3 cm long including the style column. Moist, disturbed soil of riparian forests; valleys of Missoula Co. Introduced to north-temperate N. America; native to Eurasia.

Geranium viscosissimum Fisch & C.A.Mey. STICKY GERANIUM Perennial. **Stems** erect, 15–90 cm. **Herbage** strigose to villous, glandular, especially in the inflorescence. **Leaf blades** cordate, orbicular, 5–14 cm wide, deeply cleft into 5 or 7 lobed segments. **Flowers**: sepals 8–12 mm long, bristle-tipped; petals 12–20 mm long, rose to purple; fertile stamens 10. **Capsule** 25–40 mm long including the style column. Grasslands, meadows, woodlands, open forest; valleys to lower subalpine. BC to SK south to CA, CO, SD. (p.356)

Both *Geranium viscosissimum* var. *viscosissimum* with short, glandular hairs on the petioles, and var. *incisum* (Torr. & A.Gray) N.H.Holmgren, lacking glandular petioles, occur in MT, but they lack geographic or ecological distinction in our area.

LIMNANTHACEAE: Meadow-foam Family

Floerkea Willd.

Floerkea proserpinacoides Willd. FALSE MERMAID Fleshy, glabrous, fibrous-rooted annual. **Stems** prostrate to ascending, 5–25 cm. **Leaves** alternate, petiolate; blades 1–6 cm long, odd-pinnate; leaflets narrowly elliptic, 5–20 mm long. **Inflorescence** solitary, long-pedicellate flowers from leaf axils. **Flowers** perfect, regular, hypogynous, 3-merous; sepals united at the base, 2–3 mm long; petals separate, white, spatulate, 1–2 mm long; stamens 3 to 6; ovary with 2 or 3 carpels; style 1. **Fruit** 2 or 3 bumpy, globose nuts, 1–3 mm long, united at the base, partially enclosed by the calyx. Wet, shady soil of open forest, meadows, often along streams; valleys, montane. Throughout most of temperate N. America.

BALSAMINACEAE: Touch-me-not Family

Impatiens L. Touch me not, jewelweed

Glabrous annuals. **Stems** hollow, succulent with swollen nodes. **Leaves** alternate, simple, exstipulate, petiolate; blades ovate, crenate; the serrations mucronate. **Inflorescence** few-flowered, pedunculate racemes from upper leaf axils. **Flowers** perfect (or cleistogamous), irregular, hypogynous; sepals 3, petaloid, one larger and saccate; petals apparently 3, the upper broader than long, the lateral ones 2-lobed (each lobe considered a petal; total petals 5); stamens 5; style 1; ovary 5-celled. **Fruit** a clavate, pod-like capsule that, when ripe, catapults few to several seeds upon touch.

1. Flowers with a recurved spur 5 mm long . . . .*I. aurella*
1. Flowers without a spur . . . .*I. ecalcarata*

Impatiens aurella Rydb. [*I. noli-tangere* L. misapplied] **Stems** erect, branched, 20–70 cm. **Leaf blades** 3–12 cm long; petioles 2–4 cm long. **Flowers** yellow, often with orange spots, 11–15 mm long; saccate sepal with a recurved spur 6–10 mm long. **Capsule** 1–2 cm long. Wet, usually organic soil of marshes, ditches; valleys. AK to OR, ID, MT.

Impatiens ecalcarata Blank. **Stems** erect, branched, 30–80 cm. **Leaf blades** 2–8 cm long; petioles 1–4 cm long. **Flowers** orange, 10–18 mm long; saccate sepal not spurred. **Capsule** 8–15 mm long. Wet, usually organic soil of thickets, swamp forest, marshes, streambanks, ditches; valleys. BC, MT south to OR.

The type locality is in Sanders Co.

ARALIACEAE: Ginseng Family

Perennial herbs or shrubs. **Leaves** alternate, petiolate, deeply lobed or divided. **Inflorescence** corymbs or racemes of dense umbellate clusters. **Flowers** small, perfect, regular, epigynous, 5-merous; sepals 5 or absent; petals separate; ovary 5-loculate, covered by a disk; styles 1 to 5. **Fruit** a drupe with usually 5 seeds.

1. Spiny shrubs ≥1 m tall; leaves lobed . . . . *Oplopanax*
1. Low herbs; leaves twice divided into leaflets . . . . *Aralia*

Aralia L.

Aralia nudicaulis L. Wıʟᴅ sᴀʀsᴀᴘᴀʀıʟʟᴀ Rhizomatous perennial, appearing acaulescent. **Stems** erect, short, barely above ground level. **Herbage** glabrous to puberulent. **Leaves** solitary; petiole 15–25 cm long; blade 15–25 cm long, twice divided, first into 3 then 3 to 5 parts; leaflets obovate to lanceolate or ovate, acuminate, 3–15 cm long, serrate, sometimes with asymmetrical bases. **Inflorescence** of usually 3 umbels; rays 2–7 cm long; the peduncle shorter than the leaf. **Flowers** greenish-white; sepals obsolete; petals 1–2 mm long, reflexed; stamens longer than petals. **Drupes** becoming dark purple, 4–8 mm wide. Moist forest; valleys, montane. BC to NL south to WA, CO, MO, GA. (p.356)

Oplopanax (Torr. & A.Gray) Miq.

Oplopanax horridus (Sm.) Miq. Dᴇᴠıʟ's ᴄʟᴜʙ Deciduous shrubs. **Stems** densely spiny, 1–2 m, usually unbranched. **Herbage** villous with spines along the leaf veins and petioles. **Leaf blades** orbicular, cordate, 10–30 cm wide with 5 to 9 acuminate, dentate lobes. **Inflorescence** dense umbels in a raceme 8–20 cm long. **Flowers** greenish; sepals minute; petals 2–3 mm long; stamens 5. **Drupes** red, 4–6 mm wide. Low areas of moist to wet forest, uncommon on avalanche slopes; montane, lower subalpine. AK to AB south to OR, ID, MT; MI and ON. (p.356)

APIACEAE (Umbelliferae): Carrot or Parsley Family

Annual to perennial herbs. **Leaves** alternate or basal, petiolate, usually divided or deeply lobed. **Inflorescence** a simple or compound umbel; each umbel or umbellet often subtended by involucel bracts. **Flowers** mostly perfect, regular, epigynous, 5-merous; sepals small or obsolete; petals distinct; stamens 5; ovary inferior, 2-loculate; style 1 per locule, swollen at the base forming a stylopodium. **Fruit** a shizocarp of 2 1-seeded mericarps joined together on their inner faces (commissure); each mericarp flattened or hemispheric in cross section with 5 ribs or wings on the outer face and oil tubes in the intervals between.[104]

Flower morphology throughout the family is relatively undifferentiated; mature fruit and often also flowers are required for genus and species identification.

1. Basal leaves undivided. 2
1. Lowest leaves divided 4

2. Leaves with spine-tipped margins . . . . *Eryngium*
2. Leaf margins not spine-tipped 3

3. Basal leaves heart-shaped, toothed; some stem leaves divided . . . .*Zizia*
3. Leaves entire-margined . . . . *Bupleurum*

4. Fruit with straight or hooked prickles 5
4. Fruit lacking prickles. 6

5. Fruit with hooked prickles; herbage glabrous. . . . .*Sanicula*
5. Fruit with straight prickles; stems hirsute below. . . . . *Daucus*

6. At least some ultimate leaf segments >10 mm wide 7
6. Ultimate leaf segments (lobes) <10 mm wide; leaves often fern-like. 16

7. Leaves with 3 leaflets. . . . . *Heracleum*
7. Some leaves with >3 leaflets 8

8. Fruit linear, at least 3 times as long as wide. . . . . *Osmorhiza*
8. Fruit ovate or elliptic, <2 times as long as wide 9

9. Leaves once pinnately divided; leaflets sometimes lobed 10
9. At least the lowest leaves twice or more pinnate 13

10. Flowers yellow; fruit obovate to elliptic. . . . . *Pastinaca*
10. Flowers white; fruit ovoid to nearly circular in outline. 11

11. Stems ≥5 mm wide, hollow. . . . . *Angelica*
11. Stems <5 mm wide. 12

12. Involucral bracts absent; plants terrestrial . . . . *Pimpinella*
12. Involucral bracts linear; plants growing in water. . . . . *Berula*

13. Base of stem thickened and hollow with horizontal chambers. . . . . *Cicuta*
13. Stem not greatly swollen or hollow at the base 14

14. Ultimate leaflets all deeply lobed . . . . *Ligusticum*
14. Ultimate leaflets dentate or only a few lobed 15

15. Fruits winged on margins and dorsal faces . . . . *Angelica*
15. Fruits not winged on the dorsal faces. . . . . *Lomatium*

16. Fruit linear to club-shaped; leaflets ≥1 cm long . . . . *Osmorhiza*
16. Fruit narrowly elliptic or wider or leaflets ≤1 cm long 17

17. Ultimate leaflets linear, >5 times as long as wide. 18
17. Leaflets linear-oblong or wider, <5 times as long as wide 25

18. Plants with small bulbs in the upper leaf axils . . . . *Cicuta*
18. Plants lacking bulbils 19

19. Mericarps flattened parallel to the commissure . . . .*Lomatium*
19. Mericarps rounded on the back 20

20. Margin of leaflets serrate . . . . *Sium*
20. Margin of leaflets entire 21

21. Flowers white. 22
21. Flowers yellow 24

22. Plants scapose; involucel bracts ≤0.5 mm long . . . . *Orogenia*
22. Plants with cauline leaves below; involucel bracts 1–3 mm long 23

23. Leaves pinnate, ultimate segments >2 cm long; mericarps hemispheric. . . . . *Perideridia*
23. Leaves 2 to 3 times pinnate, ultimate divisions <2 cm; mericarps ellipsoid . . . . *Musineon*

24. Herbage glaucous; plants usually >30 cm; garden escape . . . . *Anethum*
24. Herbage not glaucous; plants ≤30 cm . . . . *Musineon*

25. Mericarps lance-linear; garden escape . . . . *Anthriscus*
25. Mericarps hemispheric, elliptic, or ovoid. 26

26. Leaves once pinnate 27
26. At least some leaves 2 or more times pinnate 29

27. Plants of stony habitats; 2–8 cm tall; forming mats . . . . *Shoshonea*
27. Plants of wet habitats; stems ≥20 cm; not mat-forming 28

28. Leaflets >2 cm long . . . . *Sium*
28. Many leaflets <2 cm long . . . . *Berula*

29. Base of stem swollen, hollow with horizontal chambers. . . . . *Cicuta*
29. Base of stem little if at all swollen. 30

30. Stylopodium conic to hemispheric 31
30. Stylopodium absent to obscure 34

31. Stems hollow, purple-spotted at least below . . . . *Conium*
31. Stems not purple-spotted; usually not hollow. 32

32. Caudex covered with stringy fibers. . . . . *Ligusticum*
32. Caudex not string-fibrous 33

33. Mericarps with winged margins . . . .*Conioselinum*
33. Mericarps ribbed but not wing-margined . . . . *Carum*

34. Mericarps rounded on the back (dorsal surface) . . . . *Musineon*
34. Mericarps flattened on the dorsal surface 35

35. Meicarps winged on the dorsal surface . . . . *Cymopterus*
35. Mericarps winged on the margin but not on the dorsal surface . . . .*Lomatium*

Anethum L. Dill

Anethum graveolens L. Taprooted annual. **Stems** erect, branched, 20–150 cm. **Herbage** glabrous, glaucous, scented. **Leaves** petiolate, pinnately dissected into numerous linear segments 2–20 mm long. **Umbels** terminal and axillary, compound, to 15 cm in diameter; involucre and involucel absent. **Flowers** yellow; stylopodium conic; petals 1–1.5 mm long. **Mericarps** ovoid, beaked, ca. 6 mm long, glabrous; ribs nearly obsolete. Roadsides; valleys. Garden escape introduced throughout N. America. One collection from Gallatin Co.

Angelica L.

Taprooted perennials. **Stems** glabrous. **Leaves** mostly bipinnately divided; petioles sheathing the stem; leaflets serrate, sometimes reduced to sheaths in the inflorescence. **Inflorescence** compound umbels; involucel bracts linear or absent. **Flowers** white, pinkish or yellow; stylopodium low-conical; sepals minute. **Mericarps** flattened, elliptic to ovate; the ribs winged or simply thickened; oil tubes solitary in the intervals.

1. Flowers yellow; involucral bracts leafy . . . . *A. dawsonii*
1. Flowers white to pink; involucre absent or nearly so 2

2. Ovaries, fruits and pedicels scabrous. . . . . *A. roseana*
2. Ovaries, fruits and pedicels smooth 3

3. Leaves bipinnate, ovate in outline . . . . *A. arguta*
3. Leaves lanceolate in outline; mainly once-pinnate; lowest pair sometimes bipinnate . . . . *A. pinnata*

Angelica arguta Nutt. **Stems** erect, 50–150 cm. **Herbage** anise-scented, glabrate or sparsely pubescent on veins beneath. **Leaves** bipinnate; leaflets ovate to narrowly elliptic, 3–10 cm long, serrate, sometimes the basal few-lobed. **Umbels**: rays unequal, 2–8 cm long; involucre and involucel absent. **Flowers** white; petals 1–2 mm long. **Mericarps** glabrous, obovate, 4–7 mm long; lateral ribs broadly winged; dorsal ribs narrowly winged. Meadows, wet forests, woodlands, avalanche slopes, especially in aspen groves, along streams; montane, subalpine. BC, AB to CA, UT, WY.

Angelica dawsonii S.Watson YELLOW ANGELICA **Stems** erect, 15–70 cm. **Herbage** glabrous. **Leaves** bipinnate; leaflets lanceolate, finely serrate, 2–8 cm long. **Umbel** solitary; rays 1–3 cm long, subequal; involucral bracts leafy, deeply toothed; involucel bracts smaller, few-toothed. **Flowers** yellow; petals ca. 1 mm long. **Mericarps** glabrous, broadly elliptic, 5–7 mm long; ribs winged, the lateral slightly wider than the dorsal. Meadows, open forest, often with aspen or beargrass; montane, subalpine. Endemic to northwest MT and adjacent BC, AB, ID. (p.356)

Angelica pinnata S.Watson **Stems** erect, 30–80 cm. **Herbage** glabrous. **Leaves** once-pinnate to subbipinnate; leaflets lanceolate, 2–8 cm long, serrate; only the lowest sometimes lobed. **Umbels**: rays unequal, 1–6 cm long; involucre and involucel absent. **Flowers** white to pink; petals ca. 1 mm long. **Mericarps** glabrous, elliptic, 3–5 mm long; lateral ribs broadly winged; dorsal ribs barely winged. Thickets, meadows, often along streams; montane, lower subalpine. MT south to UT, NM.

Angelica roseana L.F.Hend. **Stems** ascending to erect, 15–80 cm. **Herbage** scabrous. **Leaves** bipinnate; leaflets lanceolate to ovate, 1–4 cm long, sharply serrate, veiny. **Umbels**: rays unequal, 1–8 cm long; involucre absent; involucel bracts linear, 2–9 mm long or absent. **Flowers** white to pink; petals barely 1 mm long. **Mericarps** scabrous, elliptic, 4–6 mm long; ribs thick, winged, the lateral barely wider than the dorsal. Non-calcareous, talus slopes and rocky soil of meadows, open forest; subalpine. MT to ID, UT.

One specimen from the Mission Mtns. has glabrous fruits but scabrous pedicels and could represent hybridization between *Angelica arguta* and *A. roseana*.

Anthriscus Pers.

Anthriscus cerefolium (L.) Hoffm. CHERVIL Taprooted annual or biennial. **Stems** erect, branched, 20–50 cm. **Herbage** glabrous to sparsely hirsute. **Leaves** petiolate, tripinnate; blades 2–10 cm long, parsley-like; the leaflets with rounded lobes. **Umbels** axillary, compound; rays 1–2 cm long; involucre absent; involucel bracts linear, 1–3 mm long. **Flowers** white; stylopodium conic; petals 1–1.5 mm long. **Mericarps** linear, beaked, ca. 6 mm long, glabrous; ribs nearly obsolete. Disturbed, moist soil of open riparian forest; valleys. Introduced for culinary purposes and escaped in the more temperate and humid areas of N. America; native to Europe. Collected in Missoula Co.

Berula W.D.J.Koch Water parsnip

Berula erecta (Huds.) Coville Glabrous, fibrous-rooted, sometimes stoloniferous, aquatic perennial. **Stems** erect to prostrate, flaccid, 10–50 cm. **Leaves** once-pinnate with 7 to 21 lanceolate to ovate, serrate to lobed leaflets 1–5 cm long; leaflets of upper leaves narrower. **Umbels** compound, axillary and terminal; rays 1–2 cm long; involucral and involucel bracts lance-linear, serrate, 2–9 mm long. **Flowers** white; stylopodium low-conic; petals ca. 0.5 mm long. **Mericarps** hemispheric, 1–2 mm long, glabrous; ribs inconspicuous; oil tubes many in the intervals. Slow-moving water of springs, streams, marshes; valleys. Throughout temperate N. America, Europe. Our plants are var. *incisa* (Torr.) Cronquist.

Bupleurum L.

Bupleurum americanum J.M.Coult. & Rose Glabrous and glaucous perennial with a woody, branched caudex. **Stems** erect to ascending, 5–50 cm. **Leaves** simple, linear to lance-linear, entire, 2–12 cm long. **Umbels** compound, terminal and axillary; rays 1–5 cm long; involucral bracts lanceolate, 5–15 mm long; involucel of ovate bracts 2–5 mm long, united below; umbellets compact. **Flowers** yellow; stylopodium hemispheric; petals <1 mm long. **Mericarps** narrowly elliptic, 2–3 mm long, glabrous, glaucous; ribs raised but not winged; oil tubes many in the intervals. Stony soil of grasslands, meadows, turf, fellfields; montane to alpine. AK to ID, WY; Asia. (p.364)
  Our only member of the Apiaceae with entire leaves. The stylopodia of high-elevation plants are often purple.

Carum L. Caraway

Carum carvi L. Glabrous, taprooted biennial. **Stems** erect, 20–80 cm. **Leaves** basal and cauline with long, sheathing petioles; blades lanceolate in outline, 8–15 cm long, 3 to 4 times pinnately divided into short, linear segments. **Umbels** compound, terminal and axillary; rays unequal, 1–4 cm long; involucral and involucel bracts minute or absent. **Flowers** white; stylopodium hemispheric; petals ca. 1 mm long. **Mericarps** glabrous, elliptic, slightly compressed perpendicular to the commissure, 3–4 mm long; ribs raised; oil tubes 1 per interval. Moist to wet meadows, fields; valleys, montane. Introduced to north-temperate N. America; native to Europe.

Cicuta L. Water hemlock

Glabrous perennials with thickened roots and inflated stem bases. **Leaves** bi- or tri-pinnate. **Umbels** compound; involucral and involucel bracts few and linear or absent. **Flowers** white; sepals evident; stylopodium low-conic. **Mericarps** ovoid, compressed parallel to the commissure, glabrous; ribs thick and raised, wider than the intervals; oil tubes 1 per interval.[298]
  The roots of *Cicuta douglasii* and *C. maculata* are highly poisonous; the primary lateral veins end in the sinuses between serrations, and this distinguishes them from *Sium*. These two species are morphologically very similar and are sometimes combined under the latter name. There is a suite of characters distinguishing them, but most are microscopic.[298] *Cicuta douglasii*, a polyploid, is most common west of the Continental Divide in wet habitats, while *C. maculata* occurs mainly east of the Divide, often in somewhat more mesic habitats.

1. Small bulbs present in axils of upper leaves . . . . *C. bulbifera*
1. Bulbils lacking. 2

2. Tertiary venation enclosing sections of leaflets ca. twice as long as wide . . . . *C. douglasii*
2. Tertiary venation sections ca. as long as wide . . . . *C. maculata*

Cicuta bulbifera L. **Stems** usually solitary, erect, slender, 30–80 cm. **Leaves** cauline; leaflets linear, nearly entire, 1–3 cm long; uppermost leaves simple or once-pinnate with axillary bulblets. **Umbel** solitary and terminal or absent, usually sterile; rays 1–2 cm long. **Mericarps** 1–2 mm long. Fens; montane. AK to NL south to OR, MT, NE, LA, FL.

Cicuta douglasii (DC.) J.M.Coult. & Rose **Stems** erect, 30–100 cm. **Leaves** basal and cauline; leaflets lanceolate to ovate, serrate, 15–60 mm long; tertiary venation enclosing sections of leaf ca. twice as long as wide, reticulations often not meeting each other. **Umbels** terminal and axillary; rays 1–5 cm long. **Mericarps** 2–4 mm long; dorsal ribs wider than the darkened intervals. Wet meadows, marshes, along streams, wetlands; valleys, montane. AK to CA, NV, ID, MT. (p.364)

Cicuta maculata L. Similar to *C. dougalsii*. **Stems** erect, 30–100 cm. **Leaves** basal and cauline; leaflets lanceolate, serrate, 2–7 cm long; tertiary venation enclosing sections of leaf ca. as long as wide, reticulations usually completely closed. **Umbels** terminal and axillary; rays 1–5 cm long. **Mericarps** 2–4 mm long; dorsal ribs as wide or wider than the darkened intervals. Moist meadows, woodlands, streambanks; plains, valleys. Throughout N. America. Our plants are var. *angustifolia* Hook.

Conioselinum Fisch. ex Hoffm. Hemlock parsley

Conioselinum scopulorum (A.Gray) J.M.Coult. & Rose Glabrate perennial with tuberous and fibrous roots. **Stems** hollow, erect, mostly simple, 30–100 cm. **Leaves** basal and cauline; petioles expanded basally; blade deltoid in outline, 5–20 cm long, 1 to 2 times pinnately dissected into lobed, ovate to oblanceolate segments 1–4 cm long. **Umbels** usually terminal, compound; peduncles 5–15 cm long; involucral bracts absent or inconspicuous; involucel bracts linear, scarious, 5–15 mm long. **Flowers** white; sepals absent; stylopodium conic; styles reflexed. **Mericarps** elliptic, glabrous, 3–6 mm long, compressed parallel to the commissure; marginal ribs low-winged; oil tubes 2 to 4 in the intervals. Meadows, often along streams; valleys, collected once in Big Horn Co., MT south to AZ, NM.

Conium L. Poison hemlock

Conium maculatum L. Glabrous, highly poisonous biennials. **Stems** stout, hollow, erect, branched, 0.5–3 m, purple-spotted. **Leaves** basal and cauline, petiolate; blades ovate in outline, 15–30 cm long, tripinnate; leaflets lobed; ultimate segments <1 mm long. **Umbels** terminal and axillary, compound; involucral and involucel bracts lanceolate to ovate. **Flowers** white; stylopodium low-conic; styles reflexed. **Mericarps** ovoid, slightly compressed parallel to the commissure, glabrous, 2–3 mm long; ribs barely winged; oil tubes many in the intervals. Moist, disturbed soil along streams, ditches, cool open slopes; plains, valleys. Introduced throughout most of temperate N. America; native to Eurasia.

Cymopterus Raf.

Taprooted perennials with a thick taproot. **Leaves** basal or clustered on a pseudoscape and appearing basal, petiolate, pinnate or bipinnate. **Umbels** terminal, long-pedunculate, compound; involucre often absent; involucel present. **Flowers** yellow or white to purple; sepals small or obsolete; stylopodium inconspicuous. **Mericarps** slighlty flattened parallel to the commissure; ribs winged; oil tubes 1 to several per interval.

   Cymopterus douglassii R.L.Hartm. & Constance occurs in east-central ID and could occur in the mountains of southern Beaverhead Co.; *C. evertii* R.L.Hartm. & Kirkp. is endemic to northwest WY and could occur in adjacent MT; *C. williamsii* R.L.Hartm. & Kirkp. is found in the central Big Horn Range in northern Wyoming and could occur in the Big Horn or Pryor Mtns. of MT.

1. Leaves whorled at the top of a naked pseudoscape (sometimes buried) 2
1. Leaves and peduncles arising from a woody caudex covered in old leaf bases 3

2. Foliage glaucous; petals yellow . . . . *C. glaucus*
2. Foliage green; petals white. . . . .*C. acaulis*

3. Petals white; herbage scabrous . . . . *C. nivalis*
3. Petals yellow; herbage glabrous. 4

4. Leaves ovate in outline; herbage aromatic; 1 stem leaf usually present . . . . *C. terebinthinus*
4. Plants acaulescent; leaves lanceolate in outline. 5

5. Fruits scabrous; sepals ≤0.5 mm long; involucel bracts 3–5 mm long . . . . *C. hendersonii*
5. Fruits glabrous; sepals >0.5 mm long; involucel bracts ≥5 mm long . . . . *C. longilobus*

Cymopterus acaulis (Pursh) Raf. **Stems** solitary or few pseudoscapes, 0.5–8 cm, barely reaching above ground. **Herbage** glabrous. **Leaves** with long petioles, dilated basally, whorled at the top of the pseudoscape; blades ovate in outline, bipinnate, 1–10 cm long; ultimate divisions linear, uncrowded. **Umbels** usually several, dense; peduncles 2–12 cm long; rays 2–7 mm long; involucre absent; involucel bracts lanceolate, 2–8 mm long, united basally on one side of the umbellet. **Flowers** white, flowering early; sepals ca. 0.5 mm long. **Mericarps** 7–11 mm long, elliptic, glabrous; wings 1–2 mm wide, wavy; oil tubes several in the intervals. Sandy or clay soil of grasslands, sagebrush steppe; plains. SK to MN south to OR, AZ, TX, Mexico.

Cymopterus glaucus Nutt. **Stems** solitary or few pseudoscapes 2–15 cm. **Herbage** glabrous and glaucous. **Leaves** whorled at the top of the pseudoscape; petioles dilated basally; blades 2–9 cm long, lanceolate to ovate in outline, bipinnate with crowded, lobed segments. **Umbels** solitary or few; peduncles 2–15 cm long; rays unequal, 5–30 mm long; involucral and involucel bracts linear, scarious, 1–3 mm long. **Flowers** yellow; sepals 0.5 mm long. **Mericarps** 5–7 mm long, elliptic, glabrous; wings ca. 0.5 mm wide, purplish; 3 or 4 oil tubes per interval. Gravelly soil of grasslands, open forest; valleys to subalpine. Endemic to central ID and adjacent MT.

Cymopterus hendersonii (J.M.Coult. & Rose) Cronquist Acaulescent with a woody, branched  caudex covered in stiff, old petioles. **Herbage** glabrous. **Leaves** with basally dilated petioles; blades 1–4 cm long, linear-lanceolate in outline, bipinnate; ultimate segments lanceolate, crowded, 2–3 mm long. **Umbels** solitary; peduncles 2–10 cm long, prostrate to ascending; rays crowded, 1–3 mm long; involucre absent or a solitary bract; involucel bracts linear, 3–5 mm long. **Flowers** yellow; sepals ≤0.5 mm long. **Mericarps** scabrous, elliptic, 3–5 mm long; wavy wings 0.5–1.5 mm wide; oil tubes 1 to 5 per interval. Talus and stony soil of fellfields, limber pine woodlands; subalpine, alpine. Endemic to southwest MT and adjacent ID. See *C. longilobus.*

Cymopterus longilobus (Rydb.) W.A.Weber Acaulescent with a woody, branched caudex covered in stiff, old petioles. **Herbage** glabrous. **Leaves** with long petioles, swollen basally; blades 2–10 cm long, linear-lanceolate in outline, bipinnate; ultimate segments linear, well-separated, 2–5 mm long. **Umbels** solitary; peduncles 2–30 cm long, erect; rays unequal, 5–30 mm long; involucral and involucel bracts linear, 5–10 mm long. **Flowers** yellow; sepals scarious-margined, ca. 1 mm long. **Mericarps** glabrous, elliptic, 3–6 mm long; ribs nearly plane, 1–1.5 mm wide; oil tubes 1 to 5 per interval. Talus and stony soil of open slopes; subalpine to alpine; Beaverhead and Carbon cos. MT south to NV, UT, NM.

Some authors consider *Cymopterus longilobus* conspecific with *C. hendersonii*;[104] however, the latter has scabrous fruits, compact umbels and lacks the several long involucral bracts.

Cymopterus nivalis S.Watson [*C. bipinnatus* S.Watson] Acaulescent with a woody, branched  caudex covered in old petioles. **Herbage** scabrous. **Leaves** long-petiolate; blades 1–5 cm long, lanceolate in outline, bipinnate; ultimate segments crowded, 1–2 mm long. **Umbels** solitary, compact; peduncles 3–17 cm long, prostrate to ascending; rays 2–4 mm long; involucre absent; involucel bracts acuminate-ovate, 2–4 mm long, scarious-margined, united at the base. **Flowers** white; sepals ca. 0.5 mm long. **Mericarps** narrowly elliptic, 3–7 mm long; marginal wings wavy, 1–1.5 mm wide, wider than the dorsal ribs; oil tubes 5 to 6 per interval. Gravelly soil of grasslands, sagebrush steppe, woodlands, open forest, fellfields; montane to alpine. OR to MT south to NV, UT, WY. (p.364)

The name *Cymopterus bipinnatus* has been applied to our plants, but these cannot be confidently distinguished from *C. nivalis*.[104,158]

Cymopterus terebinthinus (Hook.) Torr. & A.Gray Subacaulescent with a woody, branched caudex. **Herbage** glabrous, aromatic. **Leaves** long-petiolate; blades 2–15 cm long, ovate in outline, tripinnate; ultimate segments crowded, linear to lance-ovate. **Umbels** solitary; peduncles 7–25 cm long, erect to ascending; rays unequal, 3–30 mm long; involucre absent; involucel bracts linear, scarious, 1–3 mm long, united at the base. **Flowers** yellow; sepals 0.5–1 mm long. **Mericarps** glabrous, ovate, 5–8 mm long; ribs with plane to wavy wings 0.5–1 mm wide; oil tubes 3 to 12 per interval. Grassland, sagebrush steppe, woodlands; plains, valleys, montane. WA to MT south to CA, NV, UT, CO.

1. Ultimate segments of leaves linear, >3 times as long as wide; mericarp wings plane . . . . var. *foeniculaceus*
1. Ultimate leaf segments lanceolate, ≤3 times as long as wide; mericarp wings often wavy . . . . var. *albiflorus*

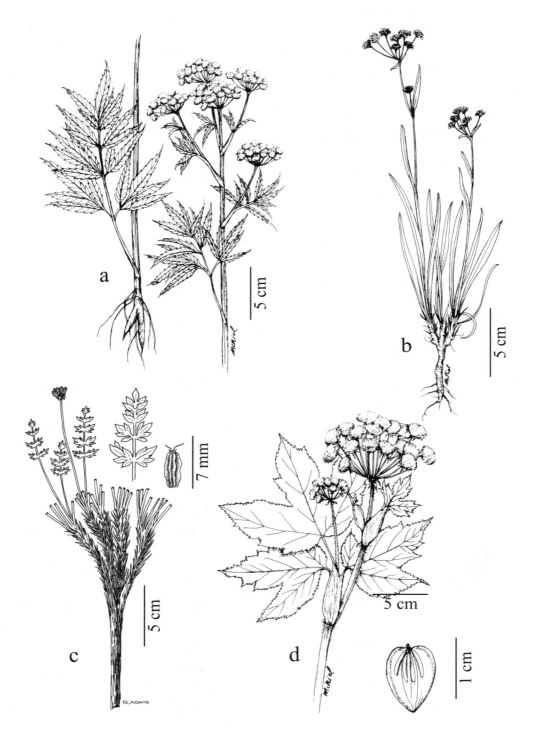

Plate 63. a. *Cicuta douglasii*, b. *Bupleurum americanum*, c. *Cymopterus nivalis*, d. *Heracleum lanatum*

Cymopterus terebinthinus var. *albiflorus* (Torr. & A.Gray.) M.E.Jones [*C. terebinthinus* var. *calcareous* (M.E.Jones) Cronquist] occurs in calcareous soil, valleys, montane; *C. terebinthinus* var. *foeniculaceus* (T & G.) Cronq. is found in forests and grasslands; plains, valleys, montane.

Daucus L. Carrot

Daucus carota L. WILD CARROT, QUEEN ANNE'S LACE Biennial with a stout taproot. **Stems** erect, simple or branched, 20–100 cm. **Herbage** glabrate to hirsute. **Leaves** basal and cauline, petiolate; blades lanceolate to ovate in outline, 5–20 cm long, bipinnate; leaflets deeply lobed, crowded. **Umbels** compound; peduncles 10–30 cm long; rays unequal, 2–4 cm long; involucral bracts 1–4 cm long, pinnate with widely-spaced, linear lobes; involucel bracts linear or 3-lobed, 4–15 mm long, ciliate. **Flowers** white; central flower of central umbel often purple; sepals minute; petals ca. 0.5 mm long; stylopodium conical. **Mericarps** elliptic, slightly compressed parallel to the commissure, 3–4 mm long; ribs with barbed prickles; oil tubes solitary in the intervals. Fields, roadsides; valleys. Introduced throughout N. America; native to Eurasia.

Eryngium L.

Eryngium planum L. Glabrous, taprooted biennial or perennial. **Stems** erect, branched above, 30–100 cm. **Leaves** basal and cauline; lower leaves petiolate, simple, ovate-cordate; upper leaves smaller with spine-tipped lobes. **Umbels** dense, hemispheric; involucral bracts similar to upper leaves. **Flowers** white to light blue. **Mericarps** scabrous, ovoid, moderately flattened lengthwise; ribs absent. Fields, roadsides; plains, valleys. Introduced to southern Canada and adjacent U.S.; native to Eurasia. One collection from Glacier Co.

Heracleum L.

Heracleum lanatum Michx. [*H. sphondylium* L. var. *lanatum* (Michx.) Dorn] COW PARSNIP Robust perennial with a stout taproot. **Stems** erect, hollow, 1–2 m. **Herbage** pubescent on veins and stems. **Leaves** basal and cauline, divided into 3 cordate leaflets; each 10–30 cm long, serrate and 3- to 7-lobed; petioles dilated basally. **Umbels** compound, glandular-villous, 10–20 cm across; rays 2–10 cm long; involucral and involucel bracts linear, 3–10 mm long. **Flowers** white; sepals minute; petals ca. 2 mm long; outer petals of outer flowers in each umbellet larger, lobed; stylopodium conical. **Mericarps** glabrous, obcordate to obovate, 7–12 mm long, strongly compressed parallel to the commissure; marginal wings 0.5–1 mm long; dorsal ribs raised; oil tubes solitary in the intervals, visible as dark lines reaching halfway to the base. Moist soil of avalanche slopes, thickets, open forest, woodlands, often along streams; valleys to subalpine. AK to NL south to CA, AZ, OH, and MO. (p.364)

Heracleum maximum W.Bartram is the oldest name applied to American plants but it was, until recently, considered invalid; however, our plants have also been considered conspecific with European *H. sphondylium*.[68] I have chosen to retain the name with the longest usage in our area until a global study of the species has been completed.[104]

Ligusticum L. Lovage

Perennials with a stout taproot and a simple crown clothed in old petioles. **Herbage** aromatic. **Leaves** basal and cauline, petiolate; blades ovate in outline, 2 to 4 times pinnately divided. **Umbels** compound; involucre and involucel absent or nearly so. **Flowers** white; stylopodium hemispheric. **Mericarps** elliptic, little compressed, glabrous; ribs evident; oil tubes 1 to 6 per interval.

Ligusticum filicinum and *L. tenuifolium* appear to intergrade in areas of contact, such as the Bitterroot Range.

1. Ultimate leaf segments ≥3 times as long as wide. 2
1. Ultimate leaf segments ≤3 times as long as wide. 3

2. Fruits ≥4 mm long; stem with at least 1 well-developed leaf . . . . *L. filicinum*
2. Fruits <4 mm long; stem leaf solitary and reduced or absent. . . . . *L. tenuifolium*

3. Ultimate leaflets 2–5 cm wide, mainly dentate . . . . *L. verticillatum*
3. Ultimate leaflets mostly 5–20 mm wide, more lobed 4

4. Plants ≥50 cm high; rays of terminal umbel 15 to 40 . . . . *L. canbyi*
4. Plants 20–60 cm; rays of terminal umbel 7 to 14 . . . .*L. grayi*

Ligusticum canbyi J.M.Coult. & Rose **Stems** erect, 50–120 cm, branched. **Herbage** glabrous. **Leaf blades** subtripinnate, 6–25 cm long; leaflets lanceolate, 2–7 cm long, pinnately lobed into lanceolate ultimate segments ≤3 times as long as wide. **Umbels**: rays 15 to 40, 1–5 cm long, scabrous. **Mericarps** 2–5 mm long; marginal and dorsal wings ≤0.5 mm wide, slightly wavy. Moist forests, wet meadows, often along streams; montane. BC, MT south to OR, ID.

The type locality is in Missoula Co. See *Ligusticum verticillatum*.

Ligusticum filicinum S.Watson **Stems** erect, branched, 40–100 cm. **Herbage** glabrous. **Leaf blades** tripinnate, 12–25 mm long; leaflets linear to deeply lobed, 1–4 cm long; ultimate segments mostly >3 times as long as wide. **Umbels**: rays 12 to 20, 2–7 cm long, scabrous. **Mericarps** 4–7 mm long; marginal and dorsal wings <0.5 mm wide, plane. Open coniferous forest, moist meadows, aspen woodlands; montane, subalpine. ID, MT, UT, WY.

Ligusticum grayi J.M.Coult. & Rose **Stems** erect, nearly scapose, 20–60 cm. **Herbage** glabrous. **Leaf blades** bipinnate; leaflets ovate, deeply lobed, 1–3 cm long; ultimate segments mostly <3 times as long as wide. **Umbels**: rays 7 to 14, 20–35 mm long. **Mericarps** 4–6 mm long, narrowly winged. Moist meadows, streambanks; subalpine. WA to MT south to CA, NV, UT. One collection from Ravalli Co.

Ligusticum tenuifolium S.Watson **Stems** subscapose, ascending to erect, 10–70 cm, simple or few-branched. **Herbage** glabrous. **Leaf blades** bipinnate, 2–15 cm long; leaflets linear to deeply lobed, 2–7 mm long; ultimate segments mostly >3 times as long as wide. **Umbels**: rays 5 to 13, 5–30 mm long. **Mericarps** 2–3 mm long; marginal and dorsal wings ≤0.2 mm wide, plane. Moist to wet meadows, open forest; subalpine. OR, ID, UT, MT, WY. (p.372)

Ligusticum verticillatum (Hook.) J.M.Coult. & Rose **Stems** erect, 100–200 cm, branched. **Herbage** glabrous. **Leaf blades** subtripinnate; leaflets ovate, 3–8 cm long, pinnately lobed into toothed, ovate, ultimate segments. **Umbels**: rays 15 to 30, 4–8 cm long, scabrous. **Mericarps** 4–6 mm long; marginal and dorsal wings narrow. Moist forest, streambanks; montane. Endemic to northern ID, adjacent BC and MT. One collection from Ravalli Co.

Easily confused with large specimens of *Ligusticum canbyi*.

Lomatium Raf. Biscuitroot, desert parsley

Perennials with tuberous or woody roots. **Leaves** cauline and/or basal; petioles with dilated bases; blades 2 to 4 times dissected into small segments. **Umbels** compound; involucre inconspicuous or absent; involucel evident to absent. **Flowers** yellow or white (purple); sepals minute or absent; stylopodium inconspicuous. **Mericarps** flattened parallel to the commissure; marginal ribs winged; dorsal ribs raised or thickened but not winged; oil tubes 1 to several in the intervals.

Reports of *Lomatium farinosum* were based on a mislabeled specimen. *Lomatium nudicaule* (Pursh) J.M.Coult. & Rose, with ovate leaflets, occurs in southeast BC and adjacent AB[187] and could occur in MT.

1. At least some ultimate leaf segments >1 cm long, >5 times as long as wide. 2
1. Ultimate leaf segments <1 cm long, often <5 times as long as wide 6

2. Flowers white or ochroleucus. 3
2. Flowers yellow 4

3. Leaflets all linear; southwest MT . . . .*L. cusickii*
3. Some leaflets lobed; northwest MT . . . .*L. geyeri*

4. Involucel absent . . . . *L. ambiguum*
4. Involucel bracts present although sometimes inconspicuous. 5

5. Stems puberulent; ultimate leaf segments 1–12 cm long. . . . . *L. triternatum*
5. Stems glabrous; ultimate leaf segments 1–3 cm long . . . . *L. nuttallii*

6. Herbage, at least the peduncles, pubescent 7
6. Herbage glabrous or scabrous 9

7. Flowers yellow; fruits pubescent. . . . . *L. foeniculaceum*
7. Flowers white; fruits glabrous. 8

8. Involucel bracts ≤4 mm long; fruits 6–9 mm long . . . . *L. orientale*
8. Some involucel bracts >5 mm long; fruits 10–15 mm long. . . . .*L. macrocarpum*

9. Stems ≥4 mm thick near the base; plants usually >30 cm high . . . . *L. dissectum*
9. Stems <4 mm thick; plants mostly <40 cm high 10

10. Fruit ≥3 times as long as wide . . . . *L. bicolor*
10. Fruit ca. twice as long as wide 11

11. Ultimate leaf segments linear or linear-oblong . . . . *L. sandbergii*
11. Ultimate leaf segments oblong to ovate 12

12. Involucel bracts ovate or obovate, united below; peduncles glabrous . . . .*L. cous*
12. Involucel bracts lance-linear, distinct; peduncles scaberulous . . . .*L. attenuatum*

Lomatium ambiguum (Nutt.) J.M.Coult. & Rose Plants from an unbranched, tuberous or bulbous taproot. **Herbage** glabrous. **Leaf blades** obovate in outline, biternate or triternate; ultimate segments well-separated, linear, 1–5 cm long, 2–5 mm wide, >5 times as long as wide; dilated petioles heavily veined. **Umbels:** peduncles ascending, 5–20 cm; rays unequal, 1–10 cm long; involucel absent. **Flowers** yellow. **Mericarps** lanceolate to narrowly elliptic, 5–8 mm long, glabrous; marginal wings ≤0.5 mm wide; oil tubes solitary in the intervals. Stony soil of grasslands, steppe, rock outcrops; valleys, montane. BC, MT south to OR, UT, WY. (p.368)
The similar *Lomatium triternatum* has small, linear involucel bracts.

Lomatium attenuatum Evert Plants subacaulescent from an elongate thickened root and simple or branched crown. **Herbage** minutely scabrous. **Leaf blades** ovate in outline; bipinnate with lobed leaflets; ultimate segments ovate to oblong, crowded, 2–5 mm long. **Umbels:** peduncles ascending, 10–30 cm; rays unequal, 1–4 cm long; umbellets compact; involucel bracts lance-linear, 1–3 mm long, scarious-margined. **Flowers** yellow. **Mericarps** elliptic, glabrous, 6–8 mm long; marginal wings 0.5–1 mm wide; oil tubes prominent, 1 per interval. Stony soil of grasslands, sagebrush steppe, talus slopes, woodlands, often on limestone, often with mountain mahogany; montane. Endemic to Beaverhead and Madison cos. and adjacent WY. See *L. cous*. (p.368)

Lomatium bicolor (S.Watson) J.M.Coult. & Rose [*L. leptocarpum* (Torr. & A.Gray) J.M.Coult. & Rose] Plants caulescent from a linear-thickened taproot. **Stems** ascending to erect, 2–15 cm. **Herbage** minutely scabrous. **Leaf blades** ovate in outline, ca. triternate or tripinnate; ultimate segments crowded, linear 1–3 × 0.2-0.5 mm. **Umbels:** peduncle 7–20 cm; rays unequal, 2–8 cm long; umbellets congested; involucel bracts absent or solitary, linear, 1–2 mm long. **Flowers** yellow. **Mericarps** glabrous, lance-elliptic, 7–12 mm long; marginal wings ca. 0.5 mm wide; oil tubes inconspicuous. Shallow soil of rock outcrops in meadows, open forest; montane, lower subalpine; Ravalli and Sanders cos. ID, MT south to UT, CO. (p.368)
Our plants appear to be *Lomatium bicolor* var. *bicolor* with finely dissected leaflets and glabrous mericarps; however, the lack of an involucel and the disjunct distribution from other populations suggests they are not typical of that variety in its main range in southern ID and UT.

Lomatium cous (S.Watson) J.M.Coult. & Rose Plants subacaulescent from a tuberous-thickened root and simple to branched crown. **Herbage** glabrous. **Leaf blades** ovate in outline; ca. tripinnate; ultimate segments narrowly elliptic, crowded, 2–5 mm long. **Umbels:** peduncles prostrate to ascending, 2–20 cm; rays becoming unequal, 1–6 cm long; involucel bracts 1–3 mm long, obovate to elliptic, scarious-margined, united below and on 1 side of the compact umbellet. **Flowers** yellow. **Mericarps** elliptic, glabrous, 5–9 mm long; marginal wings 0.5–1 mm wide; oil tubes inconspicuous, 1 to 4 per interval. Stony, often shallow soil of grasslands, fellfields at all elevations, most common at the lowest and highest elevations. AB, SK to OR, ID, UT, WY. (p.368)
Lomatium attenuatum has linear-lanceolate involucel bracts; several specimens of *L. cous* from southwest MT approach *L. attenuatum* by having scabrous peduncles and rays.

Lomatium cusickii (S.Watson) J.M.Coult. & Rose Plants subacaulescent from a branched caudex. **Herbage** glabrous to slightly scaberulous. **Leaf blades** obovate in outline, biternate or triternate; ultimate leaflets linear, widely spaced, 1–6 cm × 1–3 mm. **Umbels:** peduncles ascending to erect, 3–25 cm; rays becoming unequal, 1–3 cm long; umbellets compact; involucel bracts lanceolate, 0.5–2 mm long, united below. **Flowers** white or ochroleucus. **Mericarps** oblong, glabrous, 6–12 mm long; marginal wings 0.5–1.5 mm long; oil tubes 1 or 2 per interval. (p.368)
Sandy or gravelly, usually granitic soil of meadows, fellfields, open forest; subalpine, alpine. OR, ID, MT.

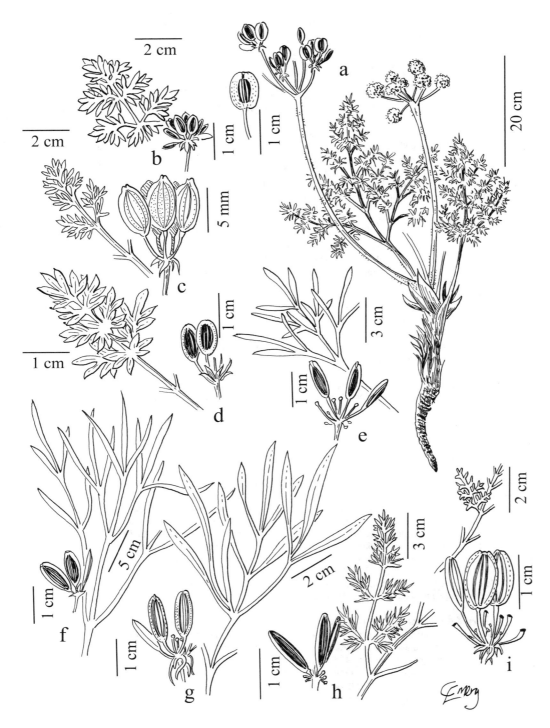

Plate 64. **Lomatium**. **a.** *L. foeniculaceum*, **b.** *L. cous*, **c.** *L. sandbergii*, **d.** *L. attenuatum*, **e.** *L. ambiguum*, **f.** *L. cusickii*, **g.** *L. triternatum var. triternatum*, **h.** *L. bicolor*, **i.** *L. dissectum*

Lomatium dissectum (Nutt.) Mathias & Constance FERN-LEAVED DESERT PARSLEY Plants
caulescent. **Stems** ascending, 15–100 cm from a mostly simple caudex. **Herbage** glabrous to
puberulent, mildly aromatic. **Leaf blades** 8–25 cm long, broadly ovate, 3 to 4 times pinnately
divided; ultimate segments crowded, linear, 2–8 × ca. 1 mm. **Umbels:** peduncles 8–50 cm;
rays ca. equal 2–6 cm long; involucel bracts linear, 5–15 mm long. **Flowers** yellow. **Mericarps**
glabrous, elliptic, 7–13 mm long, marginal wings ca. 0.5 mm long; oil tubes inconspicuous. Stony
soil of open slopes, grasslands, thickets, open forest; valleys to lower subalpine. BC to SK south to CA, AZ, CO. Our
plants are var. *multifidum* (Nutt.) Mathias & Constance. (p.368)

Lomatium foeniculaceum (Nutt.) J.M.Coult. & Rose Plants acaulescent from a thick taproot.
Herbage gray-puberulent. **Leaf blades** ovate in outline, 3 to 4 times pinnately dissected; ultimate
segments crowded, linear 1–3 mm long; petiole sheaths purplish. **Umbel:** peduncles ascending,
2–25 cm; rays unequal, 1–4 cm long; umbellets compact; involucel bracts lanceolate, 4–6 mm long,
separate to united. **Flowers** yellow. **Mericarps** puberulent, elliptic, 7–10 mm long; marginal wings
1–1.5 mm wide; oil tubes 1 to several per interval. Sandy or clay soil of grasslands, sagebrush steppe, woodlands,
badlands; plains, valleys, montane. BC to MB south to OR, AZ, TX, MO. (p.368)
    Both *Lomatium foeniculaceum* var. *foeniculaceum*, with involucel bracts separate nearly to the base and short dense
puberulence, as well as *L. foeniculaceum* var. *macdougalii* (J.M.Coult. & Rose) Cronquist, with united involucel bracts
and longer, less dense vestiture, are reported for MT; however, these characters do not seem strongly correlated in our
material nor do the vars. have geographic integrity.

Lomatium geyeri (Wats.) Coult. & Rose Plants acaulescent from a bulbous-swollen taproot. **Herbage** glabrous or
puberulent on the peduncle. **Leaf blades** ovate in outline, bipinnate; ultimate segments uncrowded, linear, 5–15 mm
long. **Umbel:** peduncles erect, 7–25 cm; rays unequal, 1–4 cm long; umbellets compact; involucel bracts linear, 1–3
mm long. **Flowers** white. **Mericarps** glabrous, elliptic, 7–12 mm long; marginal wings 1–2 mm wide; oil tubes 1 to 8
per interval. Vernally moist, often shallow soil of rock outcrops in open forest; valleys. Endemic to eastern WA, ID, and
Lincoln Co., MT.

Lomatium macrocarpum (Nutt. ex Torr. & A.Gray) J.M.Coult. & Rose Plants subacaulescent
from a thickened taproot. **Herbage** pubescent. **Leaf blades** ovate in outline, bipinnate, divided
into lanceolate, deeply pinnately lobed leaflets; ultimate segments 2–5 × ≤2 mm. **Umbel:**
peduncles ascending, 5–25 cm; rays ca. equal, 1–5 cm long; involucel bracts linear, 3–13 mm long,
united on 1 side of the compact umbellet. **Flowers** white (pink). **Mericarps** lance-elliptic, 10–15 mm
long, glabrous; marginal wings 1–2 mm wide; oil tubes solitary in each interval. Grasslands or shallow soil of outcrops,
open slopes; plains, valleys, montane. BC to MB south to CA, CO, SD.

Lomatium nuttallii (A.Gray) J.F.Macbr. Plants acaulescent from a woody, branched caudex. **Herbage** glabrous. **Leaf
blades** lanceolate in outline, 1 or 2 times pinnate; ultimate segments uncrowded, linear, 5–30 × 1–2 mm, mucronate.
Umbel: peduncles erect, 10–40 cm; rays unequal, 1–3 cm long, spreading at maturity; involucel bracts linear, 1–4 mm
long, scarious. **Flowers** yellow. **Mericarps** glabrous, oblong, 5–8 mm long; marginal wings 0.5–1 mm wide; oil tubes 3
to 5 per interval. Rocky soil of pine woodlands; plains. MT to NE, WY, UT. Known from Bighorn and Rosebud cos.

Lomatium orientale J.M.Coult. & Rose Plants subacaulescent to caulescent from a slender
taproot. **Herbage** puberulent. **Leaf blades** lanceolate in outline, bi- to tri-pinnate, 3–8 cm long;
ultimate segments crowded, linear, 2–5 mm long. **Umbel:** peduncles 5–20 cm long; rays becoming
unequal, 5–30 mm long; involucel bracts linear, 2–4 mm long, scarious-margined. **Flowers** white.
Mericarps glabrous, elliptic to obovate, 6–9 mm long; marginal wings 0.5–1 mm wide; oil tubes 1 to
4 per interval. Grasslands, sagebrush steppe, ponderosa pine woodlands; plains, valleys. MT, MB, MN south to AZ, NM,
TX, KS, IA.

Lomatium sandbergii (J.M.Coult. & Rose) J.M.Coult. & Rose Plants subacaulescent to
caulescent from a slender taproot. **Stems** ascending, 1–15 cm in fruit. **Herbage** glabrous to
scabrous. **Leaf blades** lanceolate to ovate in outline, 1–7 cm long, 2 or 3 times pinnate; ultimate
segments linear, 1–4 mm long; petiole sheaths scarious. **Umbels:** peduncles 2–15 cm long;
rays unequal, 1–7 cm long; involucel bracts filiform 2–3 mm long. **Flowers** yellow. **Mericarps**
scaberulous, narrowly elliptic, 5–8 mm long; marginal wings ca. 0.5 mm wide; oil tubes 4 to 5 per interval. Shallow soil
of outcrops, open slopes, ridges in grasslands, fellfields, forest openings; upper montane to alpine. Endemic to northwest
MT, adjacent ID, BC, AB. (p.368)

Lomatium triternatum (Pursh) J.M.Coult. & Rose [*L. simplex* (Nutt. ex S.Watson) J.F.Macbr.] Plants subacaulescent from a slender taproot and usually simple crown. **Herbage**: puberulent. **Leaf blades** obovate in outline, 4–20 cm long, triternate or tripinnate; ultimate segments well-separated, linear (rarely oblong), 1–12 cm long. **Umbels**: peduncles ascending, 2–40 cm; rays unequal, 1–7 cm long; involucel bracts few, linear, 2–4 mm long. **Flowers** yellow. **Mericarps** lanceolate to narrowly elliptic, 7–15 mm long, glabrous; marginal wings 1–1.5 mm wide; oil tubes solitary in the intervals. Grasslands, sagebrush steppe, meadows, open forest, pine woodlands; valleys, montane. BC, AB south to CA, CO. (p.368)

1. Ultimate leaf divisions narrowly oblong, some >1 cm wide . . . . .var. *anomalum*
1. Ultimate leaf divisions linear, mostly ≤5 mm wide 2

2. Fruit broadly elliptic; marginal wings as wide or wider than the body. . . . . var. *platycarpum*
2. Fruit narrowly elliptic; marginal wings as wide or narrower than the body . . . . .var. *triternatum*

Lomatium triternatum var. *platycarpum* (Torr.) B.Boivin occurs mainly in southwest cos.; *L. triternatum* var. *anomalum* (M.E.Jones ex J.M.Coult. & Rose) Mathias is known from Sanders Co.; *L. triternatum* var. *triternatum* is more common in northwest cos. Some plants from the southwest cos. have a branched caudex. The fruit wing-width character may reflect maturity rather than a genetic disposition.

Musineon Raf.

Perennials with a thickened taproot. **Leaves** petiolate, pinnately or ternately dissected; ultimate segments linear. **Umbels** compound; involucre absent; involucel bracts lance-linear, distinct. **Flowers** yellow or white, often fading white; sepals inconspicuous; stylopodium absent. **Mericarps** ovate, little compressed, ribbed but not winged; oil tubes 1 to 5 per interval.

1. Peduncles and most leaves arising from a pseudoscape; ultimate leaflets ovate, sharply
    lobed. . . . .*M. divaricatum*
1. Leaves alternate; ultimate leaflets linear or lobed into linear segments. . . . .*M. vaginatum*

Musineon divaricatum (Pursh) Raf. Subacalescent, often with a short pseudoscape. **Herbage** glabrate. **Leaves** subopposite; blades ovate in outline, 2–12 cm long, pinnate with sharply lobed leaflets; ultimate segments crowded, 3–10 mm long. **Umbels**: peduncles ascending, 1–10 cm long, ca. as long as the leaves; rays 1–2 cm long; involucel bracts, 1–7 mm long, scarious-margined. **Flowers** yellow. **Mericarps** 2–4 mm long, scabrous; oil tubes 3 to 4 per interval. Often stony soil of grasslands, sagebrush steppe, woodlands; plains, valleys. AB, SK south to NV, WY, NE.

Musineon vaginatum Rydb. Caulescent. **Stems** prostrate to ascending, 15–30 cm. **Herbage** glabrous. **Leaves** alternate; blades ovate in outline, subtripinnate; ultimate segments linear to oblong, 3–25 mm long. **Umbels**: peduncles 5–25 cm long, longer than the leaves; rays 5–40 mm long; involucel bracts 2–3 mm long. **Flowers** yellow or white. **Mericarps** 3–5 mm long, glabrous or scaberulous; oil tubes 1 to 4 per interval. Stony, usually calcareous soil or talus in open, often Douglas fir forest, woodlands; valleys to subalpine. Endemic to MT, north-central WY.
    The type locality is in Gallatin Co.

Orogenia S.Watson Turkey peas

Orogenia linearifolia S.Watson Perennial with a globose or tuberous root and a naked, subsurface stem. **Leaves** basal; petiole with a sheathing base; blades ovate in outline, 2–6 cm long, 1 to 2 times divided into few, remote, linear segments 1–4 cm × 1–5 mm. **Umbel** compound; peduncle 2–8 cm long; rays few, 2–5 mm long; umbellets compact; involucre absent; involucel bracts absent or few, narrowly linear. **Flowers** white; sepals absent; stylopodium absent. **Mericarps** glabrous, hemispheric, 3–4 mm long; dorsal ribs thickened; marginal ribs thickened; oil tubes several per interval. Grasslands, meadows, forest openings; montane, subalpine. WA to MT south to ID, UT, WY.

Osmorhiza Raf. Sweet cicely

Perennials with fragrant roots. **Leaves** petiolate, bipinnate or biternate; leaflets ovate, serrate to sharply lobed. **Umbels** compound, axillary and/or terminal; rays few; involucre and involucel present or absent.

Flowers: sepals absent; stylopodium low-conic to discoid. **Mericarps** linear to clavate, barely compressed; dorsal and marginal ribs narrow; oil tubes obscure.

All of our species except *Osmorhiza occidentalis* are vegetatively similar; mature fruits are required for confident determination.

1. Petals yellow; fruits glabrous . . . . *O. occidentalis*
1. Petals white; fruit stiff-hairy. 2

2. Involucel bracts present; style 1.5–3 mm long . . . . *O. longistylis*
2. Involucel absent; style ≤0.5 mm long 3

3. Fruit club-shaped, the tip rounded; rays spreading . . . . *O. depauperata*
3. Fruit linear, concavely narrowed to the beak-like tip; rays ascending 4

4. Stylopodium as broad as high; mature fruit 11–15 mm long . . . . *O. chilensis*
4. Stylopodium broader than high; fruit 8–12 mm long. . . . . *O. purpurea*

Osmorhiza chilensis Hook. & Arn. **Stems** erect, often branched above, 25–80 cm. **Herbage** glabrous to sparsely hirsute, non-aromatic. **Leaves** biternate; leaflets coarsely toothed to incised, 2–8 cm long. **Umbels**: peduncles becoming 10–25 cm long; rays ascending, 3–10 cm long; involucre and involucel absent. **Flowers** white; stylopodium conic, beak-like. **Mericarps** erect-hispid, linear-oblong, 11–15 mm long, concavely narrowed to the tip. Dry to moist forests, thickets, woodlands; valleys to lower subalpine. AK to NL south to CA, AZ, MI, NH; S. America.

Osmorhiza depauperata Phil. **Stems** erect, often branched, 10–60 cm. **Herbage** glabrous to sparsely hirsute, non-aromatic. **Leaves** biternate; leaflets coarsely toothed to incised, 1–6 cm long. **Umbels**: peduncles becoming 6–12 cm long; rays spreading 2–12 cm long; involucre and involucel absent. **Flowers** white; stylopodium discoid. **Mericarps** hispid, 8–12 mm long, narrowly clavate, rounded at the tip. Dry to moist forests, woodlands, often along streams; valleys, montane. AK to NL south to CA, NM, MI, VT; S. America.

Osmorhiza longistylis (Torr.) DC. **Stems** ascending to erect, 20–60 cm. **Herbage** sparsely hirsute, aromatic. **Leaves** biternate; leaflets 2–8 cm long, coarsely serrate to lobed. **Umbels**: peduncles becoming 3–10 cm long; rays ascending, 1–5 cm long; involucral bracts 1 to several, lanceolate, serrate, 5–15 mm long; involucel bracts 3–10 mm long. **Flowers** white; stylopodium conic. **Mericarps** sparsely hispid, linear-oblong, 10–15 mm long, concavely narrowed to the tip; style 2–3 mm long, reflexed. Riparian forest, woodlands; plains, valleys. SK to QC south to CO, TX, GA.

Osmorhiza occidentalis (Nutt. ex Torr. & A.Gray) Torr. **Stems** erect, often clustered, 30–120 cm, villous at the nodes. **Herbage** puberulent, aromatic. **Leaves** biternate to bipinnate; leaflets toothed to lobed, 2–10 cm long. **Umbels**: peduncles 4–20 cm long; rays becoming 2–8 cm long; involucre and involucel absent. **Flowers** yellow; stylopodium low-conic. **Mericarps** linear, glabrous, 12–20 mm long. Moist meadows, woodlands, thickets, avalanche slopes, open forest, often along streams; valleys to subalpine. BC, AB to CA, CO. (p.372)

Osmorhiza purpurea (J.M.Coult. & Rose) Suksd. **Stems** erect, often branched above, 15–30 cm. **Herbage** glabrous to sparsely hirsute, non-aromatic. **Leaves** biternate; leaflets coarsely toothed to incised, 1–6 cm long. **Umbels**: peduncles becoming 2–8 cm long; rays ascending, 2–5 cm long; involucre and involucel absent. **Flowers** white; stylopodium discoid. **Mericarps** hispid, linear-oblong, 8–10 mm long, concavely narrowed to the tip. Moist coniferous forest, thickets, often along streams; valleys, montane. AK to CA, ID, MT.

Pastinaca L. Parsnip

Pastinaca sativa L. WILD PARSNIP Biennial or short-lived perennial from a stout taproot. **Stems** erect, 30–80 cm. **Herbage** glabrous to puberulent. **Leaves** basal and cauline, petiolate, 10–30 cm long; blades lanceolate, pinnate; leaflets narrowly ovate, toothed, shallowly few-lobed, 2–8 cm long. **Umbel** compound; peduncle 3–15 cm long; rays unequal, 2–10 cm long; involucre and involucel absent. **Flowers** yellow; sepals absent; stylopodium low-conic. **Mericarps** glabrous, elliptic to suborbicular, 5–8 mm long, flattened parallel to the commissure; marginal wings ca. 0.5 mm long; dorsal ribs inconspicuous; oil tubes solitary in the intervals, visible as darkened lines. Wet, often riparian meadows; valleys. Introduced to much of N. America; native to Europe.

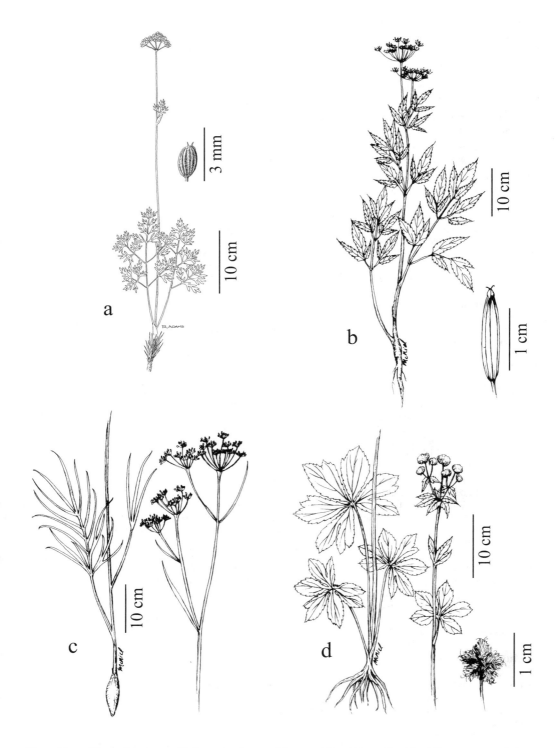

Plate 65. a. *Ligusticum tenuifolium*, b. *Osmorhiza occidentalis*, c. *Perideridia montana*,
d. *Sanicula marilandica*

Perideridia Rchb. Yampah

Perideridia montana (Blank.) Dorn [*P. gairdneri* (Hook. & Arn.) Mathias misapplied]
Perennial with tuberous roots. **Stems** erect, 30–80 cm. **Herbage** glabrous. **Leaves** cauline, long-
petiolate; blades pinnate; leaflets linear to filiform, 4–12 cm long, often withered at flowering.
Umbel compound; peduncle 3–10 cm long; rays 1–5 cm long; involucral bracts linear, 1–4
mm long or absent; involucel bracts ca. 1 mm long. **Flowers** white; sepals present; stylopodium
conic. **Mericarps** glabrous, hemispheric, 2–3 mm long; ribs raised but not winged; oil tubes solitary in the intervals.
Grasslands, meadows, woodlands; valleys to lower subalpine. BC to SK south to CA, SD. (p.372)
   Type locality is in the Bridger Range north of Bozeman.

Pimpinella L.

Pimpinella saxifraga L. Taprooted perennial. **Stems** erect, 30–80 cm. **Herbage** puberulent. **Leaves** pinnate; leaflets
ovate, lobed, serrate, 1–4 cm long, more deeply incised on upper leaves. **Umbel**: rays 2–4 cm long; involucre absent;
involucel bracts linear, 2–5 mm long. **Flowers** white; sepals minute; stylopodium low-conic. **Mericarps** ovoid, little
compressed, 1–2 mm long, glabrous; ribs inconspicuous; oil tubes 3 per interval. Fields; valleys. Sporadically introduced
in N. America; native to Eurasia; one location in Ravalli Co.

Sanicula L. Sanicle, snakeroot

Glabrous perennials. **Leaves** pinnately or palmately divided; leaflets lobed or toothed. **Umbels** compound;
umbellets congested; rays few; involucral bracts opposite, leaf-like; involucel bracts united at the base.
Flowers perfect or staminate; male flowers pedicellate; perfect flowers sessile; sepals present; stylopodium
inconspicuous. **Mericarps** ovoid, little compressed and covered with hooked prickles; ribs inconspicuous.

1. Lowest leaves with 3 deeply divided leaflets; flowers yellowish. . . . . *S. graveolens*
1. Lower leaves palmately divided into 5 to 7 toothed leaflets; flowers white. . . . . *S. marilandica*

Sanicula graveolens Poepp. ex DC. Taprooted. **Stems** ascending to erect, solitary, 5–30 cm,
branched at the base. **Leaves** petiolate; blades ovate in outline, ternate, 2–4 cm long; leaflets
deeply 3-lobed and toothed. **Umbel** compact; peduncle 2–30 cm long; rays few, 5–20 mm long;
involucel bracts 6 to 10, lance-linear, 1–4 mm long. **Flowers** light yellow. **Mericarps** 2–4 mm long.
Vernally moist, rocky soil of grasslands, meadows, open forest; valleys, montane. BC, AB to CA, WY.

Sanicula marilandica L. Black snake-root Fibrous-rooted. **Stems** solitary, erect, 30–80 cm.
Leaves petiolate below, sessile above; blades reniform in outline, 6–20 cm wide, palmate; leaflets
5- to 7-lobed and/or coarsely toothed, 3–10 cm long. **Umbel**: peduncle 4–15 cm long; rays 1–5
cm long; involucel bracts oblanceolate, 1–2 mm long. **Flowers** greenish-white. **Mericarps** 4–7 mm
long. Meadows, willow thickets, moist woodlands, open forest; plains, valleys, lower montane. BC to
NL south to WA, NM, NE, FL. (p.372)

Shoshonea Evert & Constance

Shoshonea pulvinata Evert & Constance Mat-forming, acaulescent perennial 2–8 cm with a densely branched
caudex. **Herbage** glabrous to scaberulous. **Leaves**: petioles basally dilated and white-membranous; blades pinnate, 1–3
cm long; ultimate segments linear, 3–10 mm long. **Umbels** compact; rays few, 1–5 mm long; involucre absent; involucel
bracts linear-lanceolate, 2–5 mm long. **Flowers** yellow, perfect or staminate; perfect flowers sessile; staminate flowers
pedicellate; sepals ca. 1 mm long, persistent; stylopodium absent. **Mericarps** ellipsoid, scabrous, 2–3 mm long; ribs
raised and thickened, not winged; oil tubes 2 to 6 in each interval. Stony, calcareous soil of Douglas fir woodlands,
fellfields; montane, lower subalpine. Endemic to Carbon Co. and adjacent WY.

Sium L.

Sium suave Walter Water parsnip Glabrous, fibrous-rooted perennial. **Stems** ascending to
erect, 40–100 cm. **Leaves** pinnate; leaflets lance-linear, serrate, 2–9 cm long; submersed leaves
narrower or bi- to tri-pinnate. **Umbels**: peduncles 5–15 cm long; rays 1–5 cm long; involucral
bracts lanceolate, reflexed, 4–12 mm long; involucel bracts linear, 1–6 mm long, scarious-margined.
Flowers white; sepals minute; stylopodium low-conical. **Mericarps** ellipsoid, glabrous, little

compressed, 2–4 mm long; ribs raised, corky, not winged; oil tubes 1 to 3 per interval. Shallow, fresh water and banks of ponds, swamps, marshes, slow streams; plains, valleys, montane. BC to NL south to CA, TX, SC. (p.378)

Lateral leaf veins do not always terminate between marginal teeth as in the poisonous *Cicuta* spp.

Zizia W.D.J.Koch

Zizia aptera (A.Gray) Fernald HEART-LEAVED ALEXANDERS Glabrous, fibrous-rooted perennial. **Stems** erect, 10–50 cm. **Leaves** serrate, dimorphic; basal leaves petiolate, simple; blade cordate, 1-8 cm long; upper leaves sessile, ternate with ovate leaflets. **Umbels** compact; rays 1–3 cm long, sometimes scabrous; involucral bracts absent or 1; involucel bracts linear, 2–5 mm long. **Flowers** yellow; sepals present; stylopodium absent. **Mericarps** glabrous, 2–4 mm long, ellipsoid, moderately compressed parallel to the commissure; ribs raised, not winged; oil tubes solitary in the intervals. Moist, often alkaline meadows, woodlands; plains, valleys, montane. BC to QC south to OR, CO, GA.

The difference between basal and stem leaves is diagnostic. Reports of *Zizia aurea* for MT were apparently based on misidentified *Z. aptera*.

GENTIANACEAE: Gentian Family

Annual to perennial herbs. **Leaves** simple, entire, mostly opposite, exstipulate. **Inflorescence** cymose. **Flowers** perfect, regular, 4- to 5-merous; calyx deeply to shallowly lobed; corolla of united petals; stamens 4 or 5; style 1; ovary superior, 2-carpellate. **Fruit** a capsule opening by 2 valves.

Gentianella and *Gentianopsis* were previously placed in *Gentiana*.

1. Corolla 4-merous; each petal with a backward-pointing spur . . . . *Halenia*
1. Corolla 4- or 5-merous; petals not spurred 2

2. Corolla rotate, dish- or bowl-like; petals separate >1/2-way to the base 3
2. Corolla tubular or campanulate, united >1/2-way from base 5

3. Plants annual, lacking a cluster of basal leaves . . . .*Lomatogonium*
3. Plants biennial to perennial 4

4. Corolla blue to purple . . . .*Swertia*
4. Corolla white to greenish-white . . . . *Frasera*

5. Corolla 35–50 mm long . . . . *Eustoma*
5. Corolla smaller 6

6. Corolla salverform, lobes spreading at right angles to the tube; anthers coiled . . . . *Centaurium*
6. Corolla campanulate, the lobes erect to ascending 7

7. Corolla with conspicuous folds (plaits) between the lobes . . . . *Gentiana*
7. Corolla lacking plaits 2

8. Corolla >2 cm long; corolla lobes with fringed margins . . . . *Gentianopsis*
8. Corolla <2 cm long; corolla lobes entire, although sometimes fringed within . . . . *Gentianella*

Centaurium Hill

Glabrous, fibrous-rooted annuals. **Stems** erect. **Leaves** basal and/or cauline, opposite, sessile. **Inflorescence** a panicle. **Flowers** 5(4)-merous, salverform; calyx united below; sepals linear; stamens exserted; style and stigma 1. **Capsule** cylindric, 7–15 mm long.

1. Basal leaves forming a small rosette . . . .*C. erythraea*
1. Leaves all cauline . . . . *C. exaltatum*

Centaurium erythraea Rafn [*C. umbellatum* Gilib. misapplied] **Stems** simple or branched at the base, 10–50 cm. **Leaves** basal and cauline, oblong, veiny, 1–4 cm long. **Inflorescence** cymose, subtended by 2 bracts. **Flowers** 5-merous; calyx 4–6 mm long; corolla yellowish; the tube 8–12 mm long; the lobes 4–6 mm long; anthers twisted. **Capsule** cylindric, 7–15 mm long. Disturbed meadows, roadsides; valleys. Introduced from Europe to eastern and western N. America. Collected once in Sanders Co.[380]

Centaurium exaltatum (Griseb.) W.Wight ex Piper **Stems** simple to few-branched, 8–40 cm. **Leaves** all cauline, lance-linear, 1–3 cm long. **Inflorescence** few-flowered. **Flowers** 5(4)-merous; calyx 6–10 mm long; sepals 5–8 mm long; corolla tube 6–13 mm long; the lobes 2–4 mm long, pink to white. Moist alkaline soil of shallow wetlands; plains. BC, MT south to CA, NV, UT, CO.

Our only collection from Yellowstone Co. was made over 100 years ago and the population may have been introduced.[152]

Eustoma Salisb. Prairie gentian

Eustoma grandiflorum (Raf.) Shinners [*E. exaltatum* (L.) Salisb. ex G.Don ssp. *russellianum* (Hook.) Kartesz] Glabrous and glaucous, taprooted annual. **Stems** erect, simple or branched, 25–60 cm. **Leaves** opposite, short-petiolate; blades lance-ovate to elliptic, 2–7 cm long. **Inflorescence** a panicle. **Flowers** 5-merous, campanulate; calyx lobes linear, 12–20 mm long; corolla 35–50 mm long, deeply lobed. **Capsule** ellipsoid, 1–2 cm long. Meadows and moist grasslands; plains. SD to TX, Mexico.

Our one collection from an agricultural field in McCone Co. suggests it may have been introduced.

Frasera Walter

Taprooted perennial herbs. **Leaves** thick, white-margined, basal and opposite on the stem, entire. **Inflorescence** a leafy-bracted, simple or compound raceme. **Flowers** rotate, 4-merous; calyx deeply divided; petals each with hair-like scales and 1 to 2 depressed, fringed and hooded glands at the base; stigma 2-lobed. **Capsule** with few to many seeds.

1. Flowers white; sepals 4–6 mm long . . . . *F. albicaulis*
1. Flowers pale green, blue-spotted; sepals 8–15 mm long . . . .*F. speciosa*

Frasera albicaulis Griseb. **Stems** erect, 10–40 cm from a branched caudex. **Herbage** glabrous to puberulent. **Leaves**: basal erect, linear-oblanceolate, 4–15 cm long; cauline shorter, linear. **Inflorescence** a compound raceme of several whorls, each subtended by opposite, linear bracts. **Flowers** white to light purple; sepals 4–6 mm long; corolla lobes ovate, 6–8 mm long, each with 1 gland at the base. **Capsule** ellipsoid, 10–15 mm long. Meadows, sagebrush steppe; valleys to subalpine. BC, MT south to CA, NV, ID. Our plants are var. *albicaulis*.

Frasera speciosa Douglas ex Griseb. GREEN GENTIAN, CENTURY PLANT **Stems** simple, erect, 20–150 cm. **Herbage** glabrous to puberulent. **Leaves**: basal oblanceolate, 7–25 cm long; cauline whorled, much smaller and more linear. **Inflorescence** a raceme composed of numerous whorls of flowers, each subtended by gradually reduced, foliaceous bracts. **Flowers** pale green with blue spots; sepals 8–15 mm long; corolla lobes broadly ovate, 1–2 cm long, each with 2 adjacent glands at the base. **Capsule** oblong, 1–2 cm long. Meadows, sagebrush steppe, moist grasslands, more common in calcareous soil; montane, lower alpine. WA to MT, SD south to CA, AZ, NM, TX, Mexico.

Plants grow for many years as rosettes before flowering once and then dying.

Gentiana L. Gentian

Annual to perennial herbs. **Leaves** opposite, sessile, entire. **Inflorescence** a terminal cyme or solitary flower. **Flowers** 4- to 5-merous; calyx lobes short compared to the tube; corolla with folds (plaits) between the lobes; ovary stipitate; style short or absent. **Capsule** ellipsoid, short-stalked.[142]

1. Largest cauline leaves >15 mm long 2
1. Cauline leaves <15 mm long 4

2. Flowers white with purple lines. . . . . *G. algida*
2. Flowers deep blue 3

3. Flowers solitary and terminal . . . . *G. calycosa*
3. Flowers several in a leafy-bracted panicle . . . . *G. affinis*

4. Stems erect; flowers 5-merous, >1 per stem . . . . *G. glauca*
4. Stems prostrate to ascending; flowers usually 4-merous, solitary 5

5. Flowers white; leaves with white margins ca. 0.5 mm wide; plants montane. . . . . *G. aquatica*
5. Flowers blue; leaf margins hardly white; plants alpine. . . . . *G. prostrata*

Gentiana affinis Griseb. PRAIRIE GENTIAN Perennial from a branched caudex. **Stems** ascending to erect, simple, 10–30 cm. **Herbage** glabrous to puberulent. **Leaves** cauline, lanceolate to ovate, 1–4 cm long, the lowest smaller. **Inflorescence** a leafy-bracteate panicle. **Flowers** 5-merous, funnelform; calyx 5–17 mm long, the lobes unequal, acute; corolla deep blue, 23–40 mm long with fringed plaits between spreading lobes. **Capsule** 2–4 cm long. Moist soil of meadows, grasslands; plains, valleys, montane. BC to MB south to CA, AZ, CO, SD.

Gentiana algida Pall. WHITE GENTIAN Glabrous perennial from a tightly branched caudex. **Stems** ascending, simple, 5–15 cm. **Leaves** basal and cauline, linear-oblanceolate, 2–5 cm long. **Inflorescence** 1 to 3 sessile terminal flowers. **Flowers** 5-merous, funnelform; calyx 19–28 mm long, the lobes linear, 7–8 mm long; corolla white with purple longitudinal lines, 3–5 cm long, strongly plicate. **Capsule** 3–4 cm long. Moist tundra, fellfields, usually in non-calcareous soil; alpine. AK to UT, NM; Asia.

Gentiana aquatica L. [*G. fremontii* Torr.] Glabrous annual or biennial similar to *G. prostrata*. **Stems** prostrate to ascending, simple or branched at the base, 2–10 cm. **Leaves** basal and cauline; the basal suborbicular, 5–10 mm long; the cauline oblanceolate, 4–7 mm long. **Inflorescence** a solitary, terminal flower. **Flowers** 4- to 5-merous, funnelform; calyx 4–12 mm long; corolla white, 7–15 mm long with spreading lobes 1–2 mm long. **Capsule** 4–7 mm long, opening by 2 valves on an elongating stipe. Moist alkaline meadows; montane. AK south to CA, AZ, NM.

This species has been included in *Gentiana prostrata* by other authors.[188] Our plants seem to fit the description provided by Weber[425] rather than Gillet[142] with mature capsules exserted from the dried corolla in *G. aquatica* but not *G. prostrata*.

Gentiana calycosa Griseb. EXPLORER'S GENTIAN Glabrous perennial from a branched caudex. **Stems** ascending, simple, 10–30 cm. **Leaves** cauline, ovate, 1–3 cm long; the lowest smaller. **Inflorescence** a solitary, terminal flower. **Flowers** 5-merous, funnelform to campanulate; calyx 12–22 mm long, the lobes rounded; corolla deep blue, 32–44 mm long, plicate with fringed plaits between erect to spreading lobes. **Capsule** 15–25 mm long. Moist to wet soil of meadows, cliffs, open forest, often where snow lies late; subalpine, alpine. BC, AB south to CA, NV, AZ, WY. Our plants have been called var. *obtusiloba* C.L.Hitchc., the type locality of which is in Glacier National Park. (p.378)

Gentiana glauca Pall. Glabrous, mat-forming perennial. **Stems** simple, erect, 3–10 cm. **Leaves** basal and cauline, ovate, fleshy, 5–15 mm long. **Inflorescence** a terminal, bracteate cluster of 3 to 4 short-pedicellate flowers. **Flowers** tubular, 5-merous; calyx bluish, 5–10 mm long, the lobes unequal, rounded; corolla blue, 15–20 mm long, the lobes acuminate. **Capsule** ca. 1 cm long. Wet soil of rock ledges; alpine; one site in Flathead Co. AK to WA, MT; Asia.

The only known population in the U.S. Rocky Mountains is in Glacier National Park. Marcus Jones first discovered our small population in 1910, and it was still extant in 2010.

Gentiana prostrata Haenke MOSS GENTIAN Light green, glabrous biennial or short-lived perennial with fleshy, fibrous roots. **Stems** prostrate to ascending, simple or branched at the base, 2–12 cm. **Leaves** basal and cauline; basal suborbicular, 5–12 mm long; cauline oblanceolate, 4–7 mm long. **Inflorescence** a solitary, terminal flower. **Flowers** 4-merous, funnelform; calyx 6–10 mm long; corolla blue, 1–2 cm long with spreading lobes 1–2 mm long. **Capsule** 4–7 mm long, opening by 2 valves on an elongating stipe. Moist, low turf; alpine. AK to CA, UT, CO.

Flowers open only on sunny afternoons.

Gentianella Moench Gentian

Glabrous annual or biennial herbs. **Leaves** opposite, basal and cauline, sessile or subsessile. **Inflorescence** a terminal cyme or flowers axillary. **Flowers** 4(5)-merous; calyx lobed nearly to the base; corolla funnelform; the lobes lacking folds between. **Capsule** globose.[141]

1. Corolla lobes not fringed; alpine. . . . . *G. propinqua*
1. Corolla lobes fringed within; alpine or lower. 2

2. Pedicels shorter than the flowers . . . . *G. amarella*
2. Pedicels longer than the flowers. . . . . *G. tenella*

Gentianella amarella (L.) Börner [*Gentiana amarella* L.] **Stems** erect, 10–40 cm. **Leaves** lanceolate, 1–4 cm long. **Inflorescence** a leafy-bracteate panicle. **Flowers**: calyx 4–11 mm long, the lobes subequal, linear-lanceolate; corolla pink(white) to pale violet, 1–2 cm long, the lobes fringed on the inside. **Capsule** ca. as long as corolla. Grasslands, wet meadows, open forest, thickets, often where disturbed; montane to alpine. Circumboreal south to CA, NM, SD, MN, VT. Our plants are ssp. *acuta* (Michx.) J.M.Gillett. (p.378)

Gentianella propinqua (Richardson) J.M.Gillett [*Gentiana propinqua* Richardson] **Stems** ascending to erect, angled, 2–10 cm. **Leaves** 5–20 mm long, oblong below, ovate above. **Inflorescence** pedicellate flowers terminal and paired in leaf axils. **Flowers**: calyx 5–10 mm long with unequal, ovate lobes; corolla purple 1–2 cm long, lower ones smaller, rarely fully open, the lobes not fringed. **Capsule** 1–2 cm long. Inconspicuous in moist turf; alpine. AK to NL south to MT, WY.

Gentianella tenella (Rottb.) Börner [*Gentiana tenella* Rottb.] **Stems** ascending to erect, 4–13 cm. **Leaves** 5–20 mm long, oblanceolate. **Inflorescence** long-pedicellate flowers solitary and paired in leaf axils. **Flowers**: calyx 5–12 mm long with subequal, lanceolate lobes; corolla white to blue, 5–9 mm long, the lobes fringed within. **Capsule** 5–12 mm long. Moist turf; alpine. Circumpolar south to CA, AZ, NM.

Gentianopsis Ma Fringed gentian

Glabrous annuals or perennials. **Leaves** basal and cauline, opposite, entire. **Inflorescence** a solitary, pedicellate flower terminating branch tips. **Flowers** 4-merous; calyx deeply lobed, the lobes lanceolate, slightly unequal; corolla funnelform to campanulate, deep blue (white), twisted in bud, the lobes erect to spreading; stigmas 2. **Capsule** narrowly ovoid, stipitate.[200]

The three species are very similar; the difference between *Gentianopsis detonsa* and *G. macounii* seems rather trivial. Populations of these two species will always have some simple-stemmed plants; however, populations of *G. simplex* have only simple-stemmed plants.

1. Stems simple; anthers 1–1.5 mm long . . . . *G. simplex*
1. Stems of well-developed plants branched; anthers >2 mm long 2

2. Base of calyx papillose on the ribs . . . . *G. macounii*
2. Base of calyx glabrous . . . . *G. detonsa*

Gentianopsis detonsa (Rottb.) Ma [*G. thermalis* (Kuntze) H.H.Iltis, *Gentiana detonsa* Rottb., *Gentianella detonsa* (Rottb.) G.Don] Annual **Stems** erect, usually branched, 5–25 cm. **Leaves**: the lower oblong, 1–3 cm long, short-petiolate; the upper leaves lanceolate to oblanceolate, 1–4 cm long. **Flowers**: pedicels 3–13 cm long; calyx 15–30 mm long, keeled with a membranous flap in the sinuses within; corolla 25–45 mm long, the spreading obovate lobes with fringed margins. **Capsule** 10–35 mm long. Calcareous wet meadows; montane. Cicumboreal south to CA, NV, UT, NM, Mexico. Our plants are var. *elegans* (A.Nelson) N.H.Holmgren.

Gentianopsis macounii (T.Holm) H.H.Iltis [*Gentianella crinita* Bercht. & J.Presl ssp. *macounii* (T.Holm) J.M.Gillett, *Gentiana detonsa* Rottb. misapplied] Annual. **Stems** erect, simple or branched, 5–20 cm. **Leaves**: the basal oblong, short-petiolate, 5–15 mm long; stem leaves linear-lanceolate, 1–3 cm long, sessile. **Flowers**: pedicels 1–12 cm long; calyx 10–25 mm long, keeled with a membranous flap in the sinuses within; corolla 2–4 cm long, the spreading obovate lobes with fringed margins. **Capsule** 15–30 cm long. Calcareous fens, wet meadows; valleys, montane. AB to QC south to MT, SD.

Gentianopsis simplex (A.Gray) H.H.Iltis [*Gentiana simplex* A.Gray, *Gentianella simplex* (A.Gray) J.M.Gillett] Rhizomatous perennial (Groff 1989). **Stems** erect, simple, 5–15 cm. **Leaves** cauline, sessile, lanceolate, 1–2 cm long. **Flowers**: pedicels 2–8 cm long; calyx 12–25 mm long without a membranous flap in the sinuses; corolla funnelform, 25–40 mm long, the lobes entire to minutely denticulate. **Capsule** 1–2 cm long. Non-calcareous, wet meadows, *Sphagnum* fens; montane, lower subalpine. OR to MT south to CA, NV.

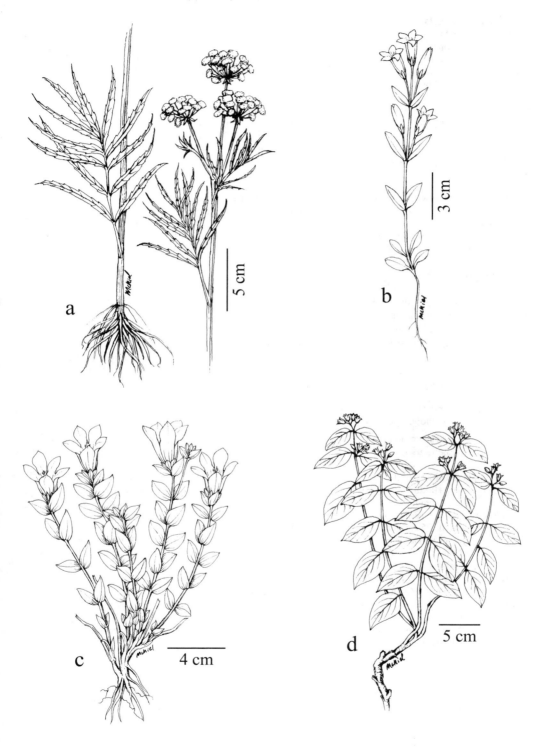

Plate 66. a. *Sium suave*, b. *Gentianella amarella*, c. *Gentiana calycosa*, d. *Apocynum androsaemifolium*

Halenia Borkh. Spurred gentian

Halenia deflexa (Sm.) Griseb. Glabrous, taprooted annual. **Stems** erect, 4-angled, mostly simple, 10–40 cm. **Leaves**: the basal oblanceolate, petiolate, 15–40 mm long; cauline ovate or obovate, sessile, 2–5 cm long. **Inflorescence** a terminal, few-flowered cyme with paired flowers in upper leaf axils. **Flowers** 4-merous; pedicels 5–25 mm long; calyx lobed nearly to the base, 3–6 mm long, lobes oblong; corolla green or purple, divided nearly to the base, 8–12 mm long including the linear, backward-pointing spurs on each petal; stigma 2-lobed. **Capsules** oblanceolate 8–12 mm long. Moist, often disturbed meadows, thickets, forest openings; valleys, montane; Flathead and Lincoln cos. BC to NS south to MT, SD, MI, NY.

Lomatogonium A.Braun

Lomatogonium rotatum (L.) Fr. ex Fernald Glabrous annual or biennial. **Stems** erect, simple or branched, 10–30 cm. **Leaves** cauline, opposite, entire, lanceolate, 1–2 cm long. **Inflorescence** of few, pedicellate, axillary and terminal flowers. **Flowers** rotate, 4-merous; pedicels 5–25 mm long; calyx deeply divided, the linear sepals ca. as long as the petals; corolla blue, divided nearly to the base, each petal 6–15 mm long, narrowly ovate with 2 fimbriate glands at the base; style lacking. **Capsule** ovoid, 5–12 mm long. Moist, calcareous meadows; montane in Beaverhead Co. AK to Greenland south to NM; Asia.

Swertia L.

Swertia perennis L. Glabrous perennial with short rhizomes and fibrous roots. **Stems** ascending to erect, simple, 5–40 cm. **Leaves** oblanceolate; basal petiolate, 2–25 cm long; cauline sessile, 1–15 mm long. **Inflorescence** of few-flowered axillary cymes. **Flowers** 5-merous, rotate; pedicels 1–5 cm long; calyx 4–8 mm long, divided to near the base, the lobes lanceolate; corolla blue to violet, divided to near the base; petals 8–12 mm long, each with 2 dark, fimbriate glands at the base; style short; stigma 2-lobed. **Capsule** ovoid, 10–13 mm long. Wet, usually non-calcareous soil of meadows, fens, thickets, often along streams, around ponds; montane to alpine. Circumboreal south to CA, NV, UT, NM.

APOCYNACEAE: Dogbane Family

Apocynum L. Dogbane, Indian hemp

Rhizomatous, perennial herbs with milky sap. **Leaves** cauline, opposite, entire, short-petiolate, lighter below than above. **Inflorescence** a highly branched cyme. **Flowers** regular, perfect, 5-merous; calyx deeply divided; corolla tubular to campanulate, the lobes ca. as long as the tube; ovary superior; pistil 1; stamens 5. **Fruits** paired, linear, terete follicles; seeds numerous with a long coma.

In spite of the following key, there do not appear to be more than two distinct entities in our area: *Apocynum androsaemifolium* with pinkish flowers and drooping leaves with longer petioles, and *A. cannabinum* with greenish-white flowers and ascending leaves with very short petioles. Plants referred to *A. sibiricum* in our area appear to be intermediate to *A. cannabinum*. Hybrids, with characters of both species, between *A. androsaemifolium* and either *A. cannabinum* or *A. sibiricum* are common and referred to *A. floribundum* Greene [*A. medium* Greene]. *Vinca minor* L., common periwinkle, was collected once in a forest opening in Glacier National Park.[242]

1. Corolla pinkish, ≥5 mm long; leaves drooping; petioles of main leaves
    often >5 mm long . . . . *A. androsaemifolium*
1. Corolla green-white, ≤5 mm long; leaves spreading to ascending; petioles 0–5 mm long. 2

2. Leaves of lower main stem cordate-based; follicles 4–11 cm long. . . . . *A. sibiricum*
2. Leaves rounded to the petiole, not cordate; follicles 12–16 cm long . . . .*A. cannabinum*

Apocynum androsaemifolium L. **Stems** ascending to erect, 25–100 cm, usually branched. **Herbage** glabrous to pubescent. **Leaves** drooping; blades ovate to lance-ovate, mucronate, 2–10 cm long, lighter beneath. **Inflorescence** axillary and terminal cymes. **Flowers** tubular to campanulate; calyx 2–3 mm long; corolla white with pink lines, tube 3–5 mm long, lobes 2–3 mm long. **Follicles** pendulous or ascending, 5–13 cm long. Forest openings, meadows, roadsides; plains, valleys. Throughout most of N. America. (p.378)

1. Corolla campanulate, the mouth much wider than the base; follicles becoming
   pendulous; coma of seed 15–17 mm long . . . . var. *androsaemifolium*
1. Corolla tubular; follicles ascending to erect; coma 12–15 mm long . . . . var. *pumilum*

Apocynum androsaemifolium var. *androsaemifolium* and *A. androsaemifolium* var. *pumilum* A.Gray lack ecological integrity in MT. Very few specimens have both flowers and fruits so it is difficult to determine whether the characters distinguishing the varieties covary in our material.

Apocynum cannabinum L. **Stems** erect, 30–120 cm, usually branched. **Herbage** glabrate. **Leaves** ascending; blades lanceolate, apiculate, 4–8 cm long. **Inflorescence** bracteate panicles at tips of upper side branches. **Flowers** tubular; calyx 2–3 mm long; corolla greenish-white, tube 2–3 mm long, lobes 1–2 mm long. **Follicles** pendulous, 12–16 cm long. Rocky soil, often along streams; plains, valleys. Throughout most of N. America.

Apocynum sibiricum Jacq. **Stems** ascending to erect, 50–100 cm, usually branched. **Herbage** glabrous. **Leaves** ascending; blades lanceolate, the lower cordate-based, mucronate, 4–10 cm long. **Inflorescence** bracteate panicles terminal and at tips of upper side branches. **Flowers** tubular; calyx 2–3 mm long; corolla greenish-white to pale yellow, tube 2–3 mm long, lobes 1–2 mm long. **Follicles** pendulous, 4–11 cm long. Moist soil of grasslands, meadows; plains, valleys. Throughout most of N. America.

ASCLEPIADACEAE: Milkweed Family

Asclepias L. Milkweed

Perennial herbs with milky sap. **Leaves** cauline, alternate, opposite or whorled, entire, short-petiolate. **Inflorescence** simple, axillary and/or terminal umbels. **Flowers** regular, perfect, 5-merous; calyx deeply divided; sepals usually reflexed; petals united at the base, usually reflexed; stamens 5, attached to the base of corolla, united together and adnate to the stigma (gynostegium), each anther with a hood-like appendage and usually an inserted or exserted horn; pollen in 5 pollinia formed of adjacent pairs of anthers; ovaries 2, superior; stigma 5-lobed. **Fruit** a pair of fusiform to ovoid follicles.

1. Leaf blades linear to filiform 2
1. Leaf blades narrowly lanceolate to ovate 4

2. Umbels nearly sessile; peduncles ≤5 mm long . . . . *A. stenophylla*
2. Peduncles >5 mm long 3

3. Peduncles 5–10 mm long; follicles 4–8 cm long . . . . *A. pumila*
3. Peduncles 1–4 cm long; follicles 8–10 cm long . . . . *A. verticillata*

4. Hoods 8–15 mm long; petals 7–12 mm long . . . . *A. speciosa*
4. Hoods 2–5 mm long; petals 4–7 mm long 5

5. Umbels sessile; leaves ovate . . . . *A. ovalifolia*
5. Umbels pedunculate 6

6. Horns absent; follicles 7–10 cm long . . . . *A. viridiflora*
6. Horns present, arching over the gynostegium; follicles 5–8 cm long . . . . *A. incarnata*

Asclepias incarnata L. SWAMP MILKWEED Fibrous-rooted. **Stems** simple to branched, 70–100 cm. **Herbage** glabrous to puberulent on the stem. **Leaves** opposite, spreading; blades lanceolate, 4–15 cm long. **Umbels** of 10 to 40 flowers; peduncles 1–7 cm long. **Flowers** 9–11 mm high; sepals villous, 1–2 mm long; petals purplish, glabrous, 5–6 mm long; gynostegium pink, glabrous, 1–2 mm high; hoods oblong, 2–3 mm long; horns exserted, arching. **Follicles** erect, fusiform, 5–8 cm long, smooth, mostly glabrous. Meadows, streambanks; plains. SK to NS south to NM, FL. Our plants are var. *incarnata*. Known from Carbon and Wibaux cos.

Asclepias pumila (A.Gray) Vail PLAINS MILKWEED Rhizomatous. **Stems** simple or branched at the base, 5–20 cm. **Herbage** glabrous to puberulent. **Leaves** crowded, alternate, erect or spreading, filiform, 1–5 cm long. **Umbels** of 5 to 20 flowers; peduncles 5–10 mm long. **Flowers** 5–8 mm high; sepals villous, ca. 2 mm long; petals white to pink, glabrous, 3–4 mm long; gynostegium greenish-white, glabrous, 1–1.5 mm high; hoods oblong, 1–2 mm long;

horns exserted, arching. **Follicles** erect to ascending, fusiform, smooth, puberulent, 4–8 cm long. Grasslands; plains. MT, ND south to NM, TX, IA.

Asclepias verticillata has longer peduncles.

Asclepias ovalifolia Decne. Rhizomatous. **Stems** simple, 20–60 cm. **Herbage** puberulent. **Leaves** opposite, ascending; blades ovate, 4–8 cm long. **Umbels** of 8 to 20 flowers; sessile. **Flowers** 4–10 mm high; sepals puberulent, ca. 2 mm long; petals green-white with a purple center, 4–6 mm long, pubescent; gynostegium white, glabrous, ca. 2 mm high; hoods oblong, 4–5 mm long with a shallow marginal lobe at midlength; horns exserted, arched. **Follicles** erect, fusiform, smooth, tomentose, 6–8 cm long. Grasslands, ponderosa pine woodlands; plains. SK, MB south to WY, SD, IL. Collected in Carter and Sheridan cos.

Asclepias speciosa Torr. SHOWY MILKWEED Rhizomatous. **Stems** simple, 50–100 cm. **Herbage** tomentose. **Leaves** opposite, ascending to spreading; blades lanceolate to ovate, sometimes cordate-based, 5–20 cm. **Umbels** of 10 to 40 flowers; peduncles 2–8 cm long. **Flowers** 15–25 mm high; sepals tomentose, 3–6 mm long; petals rose, 7–12 mm long, pubescent; gynostegium pinkish, glabrous, 2–4 mm high; hoods lanceolate, spreading, 8–15 mm long; horns exserted, strongly curved. **Follicles** erect to ascending, ovoid, tuberculate, tomentose, 5–10 cm long. Grasslands, meadows, fields, roadsides, marshes; plains, valleys. BC to MB south to CA, NM, TX, IL. (p.384)

Asclepias stenophylla A.Gray Taprooted with a simple crown. **Stems** slender, simple or branched, 15–30 cm. **Herbage** puberulent. **Leaves** spreading to erect, alternate to subopposite; blades linear, 4–15 cm long. **Umbels** of 10 to 20 flowers; peduncles 0–4 mm long. **Flowers** 4–5 mm high; sepals 2–3 mm long, puberulent; petals, greenish-white, glabrous, 3–5 mm long; gynostegium greenish-white, glabrous, ca. 1.5 mm high; hoods oblong, lobed, 3–4 mm long; horns adnate to the hood and included within it. **Follicles** erect, fusiform, smooth, glabrate, 9–12 cm long. Very sandy soil of grasslands; plains. MT to IL south to CO, TX, LA. Known from Carter Co.

Asclepias verticillata L. WHORLED MILKWEED Fibrous-rooted with a branched crown. **Stems** usually simple, 30–60 cm. **Herbage** glabrate to puberulent. **Leaves** spreading, whorled or closely alternate; blades linear, 2–6 cm long. **Umbels** of 6 to 20 flowers; peduncles 1–4 cm long. **Flowers** 5–8 mm high; sepals 1–3 mm long, pubescent; petals greenish-white, purple-tinged, 2–5 mm long, glabrous; gynostegium greenish-white or purplish, glabrous, ca. 1.5 mm high; hoods oblong, 1–2 mm long; horns exserted, arching. **Follicles** ascending to erect, fusiform, smooth, puberulent, 8–10 cm long. Sandy or stony soil of grasslands, badlands, woodlands; plains. SK to VT south to AZ, TX, FL.

Asclepias viridiflora Raf. GREEN MILKWEED Taprooted with a usually simple crown. **Stems** usually simple, 10–50 cm. **Herbage** puberulent to tomentose. **Leaves** spreading to erect, opposite to subopposite; blades lance-linear to ovate, 4–15 cm long. **Umbels** of 20 to 80 flowers; peduncles 2–15 mm long. **Flowers** 9–12 mm high; sepals 2–3 mm long, glabrate; petals green, puberulent, 4–7 mm long; gynostegium light green, glabrous, 2–3 mm high; hoods oblong, 4–5 mm long with a pair of shallow, marginal lobes at midlength and a dorsal, petal-like appendage; horns absent. **Follicles** erect, ovoid-fusiform, smooth, glabrate to puberulent, 7–10 cm long. Sandy soil of grasslands; plains. MT to MB south to AZ, TX, GA, Mexico.

SOLANACEAE: Potato Family

Annual to perennial herbs or shrubs. **Leaves** alternate, petiolate, exstipulate. **Inflorescence** a bracteate raceme or axillary cymes of 1 to few flowers. **Flowers** perfect, regular to slightly irregular, 5-merous; calyx of sepals united below; petals united; corolla rotate, campanulate or tubular; stamens 5; style 1; stigma 2-lobed; ovary superior. **Fruit** a berry or capsule.

1. Corolla ≥2 cm long 2
1. Corolla <2 cm long 4

2. Corolla 5–8 cm long; capsule often spiny. . . . . *Datura*
2. Corolla 2–5 cm long; capsule not spiny 3

3. Corolla white, narrowly funnelform; petioles not winged . . . . *Nicotiana*
3. Corolla campanulate, yellow and purple; lower petioles winged . . . . *Hyoscyamus*

4. Shrubs; some leaves fascicled. . . . . *Lycium*
4. Herbaceous annuals or perennials or vines; leaves not fascicled 5

5. Corolla nearly without lobes . . . . *Physalis*
5. Corolla lobes as long or longer than the tube. . . . . *Solanum*

Datura L. Jimson weed

Datura stramonium L. Annual. **Stems** erect, branched, 30–150 cm. **Herbage** glabrate to puberulent, especially on the veins, poisonous. **Leaf blades** ovate, 6–18 cm long, sharply lobed. **Inflorescence** of solitary, axillary flowers. **Flowers** showy; calyx tubular, 3–5 cm long with short lanceolate lobes; corolla white to purple, funnelform, the tube 5–8 cm long, the limb 3–5 cm across with shallow, mucronate lobes. **Fruit** an erect, spiny to unarmed, ovoid capsule 2–4 cm long, subtended by a collar formed from the calyx base. Roadsides, railroad rights-of-way, yards, gardens; valleys. Introduced to much of the U.S., more common in the south; native to tropical America.

Hyoscyamus L. Henbane

Hyoscyamus niger L. Annual or biennial. **Stems** erect, usually branched, 25–100 cm. **Herbage** viscid-villous, especially on the stems and veins, malodorous, poisonous. **Leaves** winged-petiolate below to sessile above; blades ovate, 3–30 cm long, pinnately lobed to dentate, overlapping. **Inflorescence** a branched, leafy-bracted raceme. **Flowers** slightly irregular; calyx campanulate, 10–12 mm long, greatly expanded in fruit; corolla funnelform, yellowish with purple veins, the tube 15–30 mm long, the limb 20–25 mm wide with 5 unequal lobes. **Fruit** a globose capsule 1–2 cm long, enclosed by the veiny, glandular calyx. Roadsides, fields, pastures; plains, valleys. Introduced throughout much of temperate N. America; native to Europe.

Lycium L. Desert thorn

Lycium barbarum L. [*L. halimifolium* Mill.] MATRIMONY VINE Shrub. **Stems** ascending, arching or climbing, 1–3 m. **Herbage** glabrous. **Leaves** short-petiolate, solitary or fascicled; blades oblanceolate, entire, 2–4 cm long. **Inflorescence** axillary, few-flowered cymes; pedicels 5–15 mm long. **Flowers** funnelform; calyx 3–4 mm long with 3 ovate lobes as long as the tube; corolla pink to purple, drying brown, the tube 6–10 mm long, the lobes just shorter than the tube. **Fruit** a red ellipsoid berry, ca. 1 cm long, glabrous. Fields, roadsides, vacant lots; valleys. Introduced as an ornamental throughout temperate N. America; native to Eurasia.

Nicotiana L. Tobacco

Nicotiana attenuata Torr. ex S.Watson COYOTE TOBACCO Taprooted annual. **Stems** simple or branched, 30–80 cm. **Herbage** glabrate but stems and inflorescences glandular-pubescent. **Leaves** basal and cauline, petiolate; blades lanceolate, 5–8 cm long, becoming short-petiolate, narrower up the stem. **Inflorescence** axillary, few-flowered cymes and a terminal panicle. **Flowers** narrowly funnel-form; calyx 5–7 mm long, teeth unequal and shorter than the tube; corolla white, 2–3 cm long, the limb 10–14 mm wide. **Fruit** a 4-valved, ovoid capsule, 8–12 mm long, mostly enclosed in the pustulate calyx. Sandy or gravelly soil of streambanks, railroad tracks; valleys. BC, MT south to CA, AZ, NM, Mexico. Known from Lewis & Clark Co. and 2 collections from unknown southwest cos.

Physalis L. Ground cherry, tomatillo

Perennial herbs, sometimes with a woody caudex. **Leaves** cauline, petiolate, usually entire to dentate. **Inflorescence** of 1 to 3 axillary flowers on reflexed pedicels. **Flowers** 5-merous, broadly campanulate; calyx lobes shorter than the tube; corolla yellowish with purple markings in the throat. **Fruit** a many-seeded, yellowish berry enclosed in the swollen calyx.

　　Physalis pumila Nutt. is reported for Fergus Co.,[113] but I have seen no specimens.

1. Stems with glandular hairs 2
1. Stems without glandular hairs 3

2. Plants pubescent with hairs ≤1 mm long . . . . *P. hederaefolia*
2. Foliage with some hairs >1 mm long . . . . *P. heterophylla*

3. Flowering calyx villous . . . . *P. pumila*
3. Flowering calyx strigose . . . .*P. longifolia*

Physalis hederaefolia A.Gray **Stems** ascending to erect, usually branched, 20–40 cm. **Herbage** pubescent with short glandular hairs and few or no long, multicellular hairs. **Leaves**: blades ovate, often cordate, crenate to dentate, 2–4 cm long; petioles 3–25 mm long. **Flowers**: pedicels 3–8 mm long, calyx tube 3–4 mm long, ca. as long as the lobes; corolla 10–15 mm long. **Berry** 8–10 mm long, enclosed in a 10-ribbed calyx 2–3 cm long. Grasslands; plains. MT south to CA, AZ, NM, TX. Our plants are var. *hederaefolia.* Reported for Big Horn Co.[151]

Physalis heterophylla Nees Clammy ground cherry **Stems** erect, simple or branched, 20–40 cm. **Herbage** pubescent; the stem with long, multicellular hairs and short, glandular hairs. **Leaves**: blades ovate, basally asymmetrical, entire to crenate, 2–6 cm long; petioles 1–4 cm long. **Flowers**: pedicels 8–15 mm long, calyx tube 7–12 mm long, longer than the lobes; corolla 12–18 mm long, limb 12–18 mm wide. **Berry** 1–2 cm long, the swollen calyx 3–4 cm long. Grasslands, fields, roadsides; plains, valleys. MT to NS south to UT, TX, FL.

Populations of this plant west of the Continental Divide may be adventive.

Physalis longifolia Nutt. **Stems** ascending to erect, usually branched, 8–40 cm. **Herbage** glabrate to strigose. **Leaves**: blades lanceolate, entire to sinuate, 2–7 cm long; petioles 5–30 mm long. **Flowers**: pedicels 5–15 mm long; calyx tube 6–12 mm long, twice as long as the lobes, short-hairy on the veins and margins; corolla 10–15 mm long, limb 10–15 mm wide. **Berry** ca. 15 mm long, the swollen calyx 2–3 cm long. Grasslands, fields, roadsides; plains. MT to ON south to AZ, UT, LA, VA.

Solanum L. Nightshade

Annual or perennial herbs, sometimes woody at the base and vine-like. **Leaves** cauline, petiolate, crenate to deeply lobed. **Inflorescence** racemose or of few axillary flowers. **Flowers** 5-merous; calyx lobes reflexed; corolla funnelform, the spreading to reflexed limb longer than the tube; anthers united around the style, exserted. **Fruit** a many-seeded berry with a persistent calyx.

Most or perhaps all of our species are poisonous to some extent.

1. Plants vining; inflorescence with >7 flowers. . . . . *S. dulcamara*
1. Stems prostrate to erect but not twining; axillary cymes with usually <7 flowers 2

2. Leaf margins sinuate to shallowly lobed. 3
2. Leaf margins lobed ≥1/2-way to the midvein 4

3. Stems densely glandular-villous. . . . . *S. sarrachoides*
3. Stems glabrous to sparsely strigose. . . . . *S. nigrum*

4. Stems and leaves spiny; corolla lobes 5–10 mm long . . . . *S. rostratum*
4. Herbage not spiny; corolla lobes 2–3 mm long . . . . *S. triflorum*

Solanum dulcamara L. Bittersweet nightshade Rhizomatous perennial. **Stems** often woody at the base, prostrate, ascending or clambering, up to 3 m long. **Herbage** glabrate. **Leaf blades** ovate, truncate to cordate-based, often with 1 to 2 deep basal lobes, 3–10 cm long. **Inflorescence** axillary, branched cymes, 7- to 14-flowered; peduncle 1–5 cm long. **Flowers**: calyx lobes ca. 1 mm long; corolla purple, lobes reflexed, lanceolate, 5–9 mm long with a pair of yellow spots at the base of each; anthers 4–6 mm long. **Berry** becoming red, 8–12 mm long; persistent calyx not swollen. Riparian forests, woodlands, thickets, often along streams, ditches, around buildings; plains, valleys. Introduced throughout much of temperate N. America; native to Eurasia. (p.384)

Solanum nigrum L. [*S. interius* Rydb.] Black nightshade Annual. **Stems** ascending to erect, branched, 15–60 cm. **Herbage** glabrous to strigose. **Leaf blades** ovate, entire to sinuate, 2–8 cm long; petioles wing-margined. **Inflorescence** axillary, few-flowered corymbs; peduncle ca. 1 cm long. **Flowers**: calyx lobes 0.5–1.5 mm long; corolla white, lobes spreading, 1–3 mm long; anthers 1–2 mm long. **Berry** becoming black, 7–10 mm long, persistent calyx not swollen. Fields, gardens, streambanks; plains, valleys. Cosmopolitan.

Most of our collections labeled *Solanum nigrum* were *S. sarrachoides.* Many segregates of *S. nigrum* sensu lato have been recognized; *S. interius*, a native western N. American form is told from the European *S. nigrum* sensu stricto by having 2 globose concretions smaller than the seeds in each berry. I have chosen not to recognize the distinction because the character is not apparent in our few specimens.

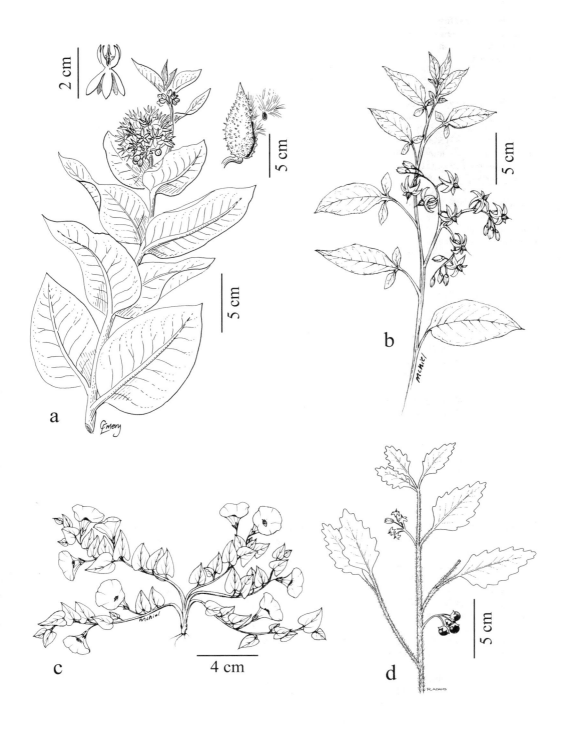

Plate 67. a. *Asclepias speciosa*, b. *Solanum dulcamara*, c. *Convolvulus arvensis* d. *Solanum sarachoides*

Solanum rostratum Dunal BUFFALO BUR Annual. **Stems** ascending to erect, branched, 10–40 cm. **Herbage** with stellate hairs, beset with spines on the stem and veins. **Leaf blades** oblong, 2–8 cm long, deeply pinnately lobed, lobes rounded, sinuate to lobed. **Inflorescence** several-flowered racemes, pedicels erect in fruit; peduncle 1–4 cm long, borne between the leaves. **Flowers**: calyx lobes 3–6 mm long, densely spiny; corolla yellow; lobes spreading, 5–10 mm long; anthers 7–9 mm long, 1 anther longer and purple. **Berry** 7–10 mm long, completely enclosed in the swollen calyx. Grasslands, streambanks, pastures, roadsides; plains, valleys. Sporadic throughout much of U.S. and Mexico; native to the Great Plains states, perhaps introduced west of the Divide.

Solanum sarrachoides Sendtn. [*S. physalifolium* Rusby] Annual. **Stems** ascending to erect, branched, 5–40 cm. **Herbage** glandular-villous. **Leaf blades** lanceolate to ovate, sinuate to dentate, 2–6 cm long. **Inflorescence** axillary, few-flowered corymbs; peduncle 5–10 mm long. **Flowers**: calyx lobes ca. 1 mm long; corolla white, lobes spreading, ca. 2 mm long; anthers 1–2 mm long. **Berry** becoming yellowish, 6–7 mm long, half enclosed in the swollen calyx. Fields, roadsides, streambanks; plains, valleys. Introduced throughout much of the U.S.; native to S. America. (p.384)

Solanum triflorum Nutt. Mat-forming annual. **Stems** prostrate to ascending, branched at the base, 5–40 cm. **Herbage** sparsely strigose. **Leaf blades** lanceolate to ovate, pinnately lobed, 2–5 cm long. **Inflorescence** axillary, few-flowered umbels; peduncle 5–15 mm long. **Flowers**: calyx lobes 2–4 mm long, linear-lanceolate; corolla white, lobes puberulent, spreading, 2–3 mm long; anthers 2–3 mm long. **Berry** greenish, 7–12 mm long, persistent calyx not swollen. Disturbed soil of grasslands, prairie-dog towns, streambanks, roadsides; plains. BC to MN south to CA, NM, TX, Mexico.

CONVOLVULACEAE: Morning glory Family

Annual or perennial herbs, some vining. **Leaves** alternate, exstipulate. **Inflorescence** of axillary, pedicellate flowers or cymose. **Flowers** perfect, regular, 5-merous; sepals distinct or basally united; corolla funnelform, twisted in bud; stamens 5; ovary superior; styles 1 or 2; stigmas 2. **Fruit** a 1- or few-seeded, ovoid capsule.

1. Corolla <1 cm long; peduncle <5 mm long . . . . *Evolvulus*
1. Corolla >1 cm long; peduncle >8 mm long 2

2. Stems ascending to erect; corolla lavender . . . .*Ipomaea*
2. Stems twining, prostrate or trailing 3

3. Calyx enclosed by 2 bracts; corolla 3–7 cm long . . . . *Calystegia*
3. Bracts linear, below the calyx, not enclosing it; corolla 1–3 cm long . . . . *Convolvulus*

Calystegia R.Br. Bindweed

Calystegia sepium (L.) R.Br. [*Convolvulus sepium* L.] HEDGE BINDWEED Rhizomatous perennial. **Stems** prostrate or twining, 1–3 m long. **Herbage** glabrous to sparsely pubescent. **Leaves** sagittate, acuminate, cordate-based, 3–8 cm long; petiole 1–7 cm long. **Flowers** solitary; peduncle 1–10 cm long; calyx 8–17 mm long, the segments lanceolate, completely enveloped by 2 subtending, leaf-like bracts; corolla white, 3–7 cm long; stigmas 2-lobed; stamens included. **Capsule** ca. 1 cm long, 1-locular. Fields, gardens, thickets; plains, valleys. Introduced throughout temperate N. America; native to Eurasia.

Convolvulus L. Bindweed

Convolvulus arvensis L. FIELD BINDWEED Rhizomatous perennial. **Stems** pubescent, prostrate or twining, branched at the base, 20–100 cm long. **Leaves** glabrous, sagittate, cordate-based, mostly rounded at the tip, 1–5 cm long; petiole 5–25 mm long. **Flowers** solitary; peduncle 1–5 cm long with a pair of small bracts just below the flower; calyx 3–5 mm long, the segments obovate, overlapping; corolla white or pinkish, 15–25 mm long; stamens included. **Capsule** 4–7 mm long, 2-locular. Fields, vacant lots, roadsides; plains, valleys. Introduced throughout temperate N. America; native to Europe. (p.384)

Evolvulus L.

Evolvulus nuttallianus Roem. & Schult. Perennial with a simple or branched caudex. **Stems** ascending to erect, branched and often woody below, 7–20 cm long. **Herbage** brownish- or grayish-hirsute. **Leaves** short-petiolate; blades lanceolate to oblanceolate, 8–15 mm long. **Flowers** solitary; peduncle 1–4 mm long with 2 small basal bracts; calyx 3–5 mm long, the segments linear-lanceolate; corolla light lavender, 4–7 mm long; stigmas 2-lobed; stamens included. **Capsule** 3–5 mm long, membranous. Often sandy soil of grasslands, sagebrush steppe; plains, valleys. MT, SD south to AZ, TX, TN.

Ipomoea L. Morning glory

Ipomoea leptophylla Torr. Bush morning glory Glabrous perennial with an enlarged root. **Stems** ascending to erect, 30–100 cm. **Leaves** short-petiolate; blades linear-lanceolate, entire, 3–12 cm long. **Inflorescence** few-flowered, axillary cymes; peduncles ca. 1 cm long. **Flowers**: calyx 5–10 mm long, lobes ovate, the inner longer and wider; corolla lavender, 5–7 cm long; stamens unequal, included. **Capsule** 10–15 mm long, beaked. Barren slopes in grasslands, steppe; plains, valleys. MT, SD south to NM, OK.

CUSCUTACEAE: Dodder Family

Cuscuta L. Dodder

Glabrous, achlorophyllous, parasitic annuals with an ephemeral root system. **Stems** white to orange or pink, twining, attaching to host plants by penetrating branches (haustoria). **Leaves** alternate, scale-like, inconspicuous. **Inflorescence** compact few- to many-flowered cymes, sometimes appearing globular. **Flowers** regular, perfect, 4- to 5-merous, sessile or short-pedicellate; sepals distinct or basally united; corolla cup-shaped to campanulate, short-lobed, the tube with fimbriate appendages below the staminal insertion; stamens 4 or 5, barely exserted between corolla lobes; styles 2. **Fruit** a globose to ovoid capsule at least partly surrounded by the withered corolla.[102,453]

1. Stigma linear, appearing to be a continuation of the style 2
1. Stigma capitate, differentiated from the style 3

2. Calyx membranous; corolla tube ca. 2 mm long . . . . *C. epithymum*
2. Calyx fleshy; corolla tube ca. 1.5 mm long. . . . . *C. approximata*

3. Capsule ovoid to elliptic, longer than wide, somewhat pointed at the tip . . . . *C. megalocarpa*
3. Capsule globose, not longer than wide, not pointed. 4

4. Capsule surrounded by the withered corolla only at the base . . . . *C. pentagona*
4. All but the tip of the capsule enclosed in the withered corolla 5

5. Flowers mostly 5-merous, 2–4 mm long. . . . . *C. indecora*
5. Flowers mostly 4-merous, 1.5–2 mm long . . . .*C. corylii*

Cuscuta approximata Bab. **Stems** yellow to orange. **Inflorescence** of few- to several-flowered glomerules. **Flowers** sessile, 2–3 mm long, 5-merous; calyx turgid, yellow, the lobes broader than long, rounded at the tip; corolla tube as long as calyx, appendages almost reaching the top of the tube; stamens barely surpassed by the corolla lobes; stigmas elongate. **Capsule** circumscissile near the base. On legumes and other forbs in fields, grasslands; plains, valleys. Introduced into western N. America; native to Europe and Africa.

Cuscuta coryli Engelm. **Stems** yellow. **Inflorescence** of few- to several-flowered, loose glomerules. **Flowers** 2–3 mm long, usually 4-merous, on pedicels 1–3 mm long; calyx lobes deltoid, acute; corolla tube shorter than the calyx, lobes acuminate, inflexed at the tip, appendages small, dentate wings; stamens ca. as long as the corolla lobes; stigmas globose. **Capsule** not circumscissile; styles in a thickened recession. On riparian shrubs, forbs, especially *Glycyrrhiza*; plains, valleys. Throughout much of central and eastern U.S.

Cuscuta epithymum Murray **Stems** yellow to purplish. **Inflorescence** of compact, few- to many-flowered glomerules. **Flowers** sessile, 2–3 mm long, 5-merous; calyx membranous, purplish, the lobes ovate; corolla tube longer than the calyx, the lobes acute, appendages reaching the middle of the tube; stamens barely surpassed by the corolla lobes; stigmas linear. **Capsule** circumscissile. Hosts are legumes and other often introduced forbs in grasslands, forest openings; valleys. Introduced into western N. America; native to Europe.

Cuscuta indecora Choisy **Stems** yellow. **Inflorescence** of loose few-flowered glomerules. **Flowers** short-pedicellate, 3–4 mm long, 5-merous; calyx not fleshy, yellowish, the lobes broadly ovate; corolla papillose, the tube longer than the calyx, appendages reaching the top of the tube; stamens surpassed by the corolla lobes; stigmas globose. **Capsule** not circumscissile; styles in a thickened recession. On alfalfa; valleys. Western N. America south to Mexico, S. America. Known from Carbon Co.

Cuscuta megalocarpa Rydb. [*C. umbrosa* Beyr. ex Hook.] **Stems** yellow. **Inflorescence** of compact, many-flowered glomerules. **Flowers** sessile or subsessile, 1–3 mm long, 5-merous; calyx not fleshy, yellowish, the lobes ovate; corolla tube as long as the calyx, appendages nearly reaching the top of the tube; stamens barely surpassed by the corolla lobes; stigmas globose, slightly beaked. **Capsule** not circumscissile; styles in a terminal recession. On plants of the Asteraceae and other forbs; valleys; Meagher and Park cos. N. America, S. America.

Cuscuta pentagona Engelm. **Stems** yellow. **Inflorescence** of compact, many-flowered glomerules. **Flowers** sessile or subsessile, 1–3 mm long, 5-merous; calyx not fleshy, yellowish, the lobes ovate; corolla tube as long as the calyx; appendages nearly reaching the top of the tube; stamens barely surpassed by the corolla lobes; stigmas globose. **Capsule** not circumscissile; styles in a terminal recession that is not thickened. On introduced clovers and other forbs in fields, gardens; plains, valleys. N. and S. America.

MENYANTHACEAE: Buckbean Family

Menyanthes L. Buckbean, bogbean

Menyanthes trifoliata L. Glabrous, acaulescent perennial with a thick rhizome. **Scapes** ascending, 10–30 cm, just surpassing the leaves. **Leaves**: petioles 5–30 cm, basally swollen and sheathing; blades trifoliolate; leaflets elliptic to oblong, entire, 3–10 cm long. **Inflorescence** a bracteate raceme, 3–10 cm long. **Flowers** perfect, regular, 5-merous; sepals 4–7 mm long, united basally; corolla white, salverform, the tube 4–7 mm long, lobes lanceolate, 6–12 mm long, fimbriate ventrally; anthers purple, exserted; ovary ca. 1/3 inferior; style 1; stigma 2-lobed. **Fruit** an ovoid capsule 6–10 mm long with numerous seeds. Fens and wet, peaty soil of marshes, swamps, often with sedges and *Sphagnum* moss; valleys to lower subalpine. Circumboreal south to CA, CO, MO, PA. (p.390)

POLEMONIACEAE: Phlox Family

Annual to perennial herbs or subshrubs. **Leaves** cauline and often basal, alternate or opposite, entire to variously dentate to lobed or divided. **Inflorescence** cymose or flowers solitary. **Flowers** perfect, regular, usually 5-merous; sepals united below into a tube; corolla campanulate to salverform; stamens alternate with corolla lobes; ovary superior; style 1; stigmas 3. **Fruit** a globose to ovoid, 3-loculate capsule with 1 to many seeds per locule.

There has been a good deal of research on the phylogeny within this family in the past 20 years. Most researchers agree that genera must be realigned, but how revised genera should be circumscribed awaits further study. For example, it seems clear that *Gilia* sensu lato is probably not a monophyletic clade;[205,326] however, recent proposals to recognize several segregate genera (*Aliciella, Microgilia, Latrocasis*) are not strongly supported by molecular data or by more traditional morphological data.[150] Similarly, it seems clear that *Linanthus* sensu lato is probably not a natural grouping with some species more closely related to *Leptodactylon* than other species of *Linanthus*. However, I believe there is currently not enough information to accurately recircumscribe these genera. Consequently I have chosen to follow a more traditional taxonomy for these groups,[319,443] knowing that the inevitable realignment awaits more thorough studies.

1. Leaves all opposite (sometimes appearing whorled) throughout 2
1. Leaves alternate above, sometimes opposite below 6

2. All true leaves appearing whorled immediately beneath the inflorescence . . . . *Gymnosteris*
2. Leaves borne along the stem 3

3. Leaves entire, usually narrow . . . . *Phlox*
3. Leaves deeply divided (sometimes appearing whorled) 4

4. Plants thin-stemmed annuals . . . . *Linanthus*
4. Plants perennial, woody at the base 5

5. Leaves 10-15 mm long . . . . *Linanthastrum*
5. Leaves 3-7 mm long . . . . *Leptodactylon*

6. Leaves lobed into needle-like or spine-tipped segments 7
6. Leaf segments sometimes linear, but not needle-like or spine-tipped 8

7. Plants annual . . . . *Navarretia*
7. Plants perennial, woody at the base . . . . *Leptodactylon*

8. Flowers red; corolla tube 2–3 cm long . . . . *Ipomopsis*
8. Flowers white, bluish, yellowish, or pink, not red; corolla tube mostly <2 cm long 9

9. Calyx mostly of a nearly uniform texture, not membranous between sepal midveins 10
9. Calyx white and membranous between green sepal midveins 11

10. Leaves pinnately compound with many leaflets . . . . *Polemonium*
10. Leaves entire or 3-lobed . . . . *Collomia*

11. Leaves opposite below but alternate in the inflorescence . . . . *Microsteris*
11. Leaves all alternate 12

12. Inflorescence congested, capitate . . . . *Ipomopsis*
12. Inflorescence of 1–few pedicellate flowers in axils of upper leaves 13

13. Sepals and often upper portion of plant glandular . . . . *Gilia*
13. Sepals tomentose but not glandular . . . . *Ipomopsis*

Collomia Nutt.

Taprooted, annual or perennial herbs. **Leaves** cauline, alternate, sometimes opposite below, simple, entire to few-lobed. **Inflorescence** of dense, few- to many-flowered clusters at tips or in forks of branches. **Flowers** funnelform; calyx glandular. **Capsule** with 1 seed per locule.

1. Perennials with trailing stems . . . . *C. debilis*
1. Annuals with erect stems 2

2. Corolla tube 2–3 cm long . . . . *C. grandiflora*
2. Corolla tube 6–15 mm long 3

3. Inflorescence congested, head-like; stamens unequally inserted in the corolla . . . . *C. linearis*
3. Inflorescence diffuse; stamens equally inserted . . . . *C. tinctoria*

Collomia debilis (S.Watson) Greene Perennial with a deeply buried taproot. **Stems** mostly prostrate, naked, buried beneath stones with only the ascending leafy tips in evidence, 2–10 cm long. **Herbage** puberulent to glandular-pubescent. **Leaves** short-petiolate; blade linear-lanceolate to oblong, 7–25 mm long. **Inflorescence** clusters of several sessile flowers at stem tips. **Flowers**: calyx 5–9 mm long; corolla white to blue, the tube 8–13 mm long, lobes 3–10 mm long; stamens exserted, all attached at the same level. Scree, talus slopes. WA to MT south to CA, NV, UT, WY.

1. Leaf blades linear-oblanceolate, entire, gradually tapered to the petiole . . . . .var. *camporum*
1. Leaf blades oblong to oblanceolate, some often lobed; petiole more distinct . . . . var. *debilis*

Collomia debilis var. *camporum* Payson is found in the montane zone; *C. debilis* var. *debilis* occurs montane to alpine. The species varies greatly in lobing of the leaves as well as flower size and color, but these characters do not seem to be correlated with each other or with geographic or ecological factors.

Collomia grandiflora Douglas ex Lindl. Annual, similar to the more common *C. linearis*. **Stems** simple, erect, 15–60 cm. **Herbage** glabrate to puberulent, glandular above. **Leaves** sessile, alternate, lanceolate, entire, 2–7 cm long. **Inflorescence** a dense, leafy-bracted, terminal cluster of sessile flowers. **Flowers**: calyx 5–7 mm long, the tube white; corolla yellow to light pink, tube 2–3 cm long, lobes 6–10 mm long; stamens unequally attached, mostly included in the corolla. Rocky soil of forest openings; montane. BC, MT south to CA, AZ, UT.

Collomia linearis Nutt. Annual. **Stems** simple, erect, 7–50 cm. **Herbage** puberulent, glandular-pubescent above. **Leaves** sessile, alternate, linear-lanceolate, entire, 1–10 cm long. **Inflorescence** a dense, terminal cluster of sessile flowers. **Flowers**: calyx 3–5 mm long; corolla pink, tube 6–9 mm long, lobes 1–3 mm long; stamens unequally attached, mostly included in the corolla. Often disturbed soil of grasslands, sagebrush steppe, meadows, pastures, open forest, thickets, rock outcrops, along streams; valleys, montane. BC to ON south to CA, NM, MO, IL. (p.390)

The superficially similar *Microsteris gracilis* has a calyx with hyaline intercostal membranes and opposite lower leaves.

Collomia tinctoria Kellogg Annual. **Stems** profusely branched, erect, 5–15 cm. **Herbage** glandular-puberulent. **Leaves** alternate above, often opposite below, linear, 5–15 mm long. **Inflorescence** few-flowered clusters at tips and in axils between branches. **Flowers**: calyx 2–3 mm long, setaceous; corolla pink, tube 8–14 mm long, lobes 1–3 mm long; stamens equally attached, barely exserted from the corolla. WA to MT south to CA, NV, ID. Collections from Flathead and Cascade cos. were made over 100 years ago.

Gilia Ruiz & Pav.

Annuals. **Leaves** basal and/or alternate on the stem, simple, entire to pinnately lobed, short-petiolate below, sessile above. **Inflorescence** of 1 to few pedicellate flowers in axils of upper leaves. **Flowers** perfect, regular, 5-merous; sepals glandular, united, membranous between the ribs (intercostal membrane); corolla campanulate to salverform; stamens equally inserted in the corolla, filaments ca. as long as the anthers. **Capsule** with 1 to several seeds per locule.

1. Leaf margins entire. 2
1. Leaf margins dentate to lobed 3

2. Corolla tube 2–3 mm long; seeds several per locule . . . . *G. minutiflora*
2. Corolla tube 1–2 mm long; seeds 1 or 2 per locule . . . .*G. tenerrima*

3. Base of stem with cobweb-like pubescence; corolla tube 4–5 mm long . . . . *G. tweedyi*
3. Base of stem not cobwebby; corolla tube ca. 2 mm long. . . . .*G. leptomeria*

Gilia leptomeria A.Gray [*Aliciella leptomeria* (A.Gray) J.M.Porter] **Stems** erect, 5–20 cm, branched above. **Herbage** puberulent, stipitate-glandular on stems. **Leaves**: basal oblanceolate, 1–5 cm long, coarsely dentate with spinulose teeth; cauline few, reduced. **Inflorescence** of solitary flowers on ascending to erect pedicels in an open-branched, bracteate cyme. **Flowers**: calyx ca. 2 mm long; corolla salverform, pink, with a yellow throat, tube ca. 2 mm long, lobes ca. 1 mm long. **Capsule** ca. 3 mm long with several seeds per locule. Sandy, sparsely vegetated soil of sagebrush steppe, streambanks; montane; Carbon Co. WA to MT south to CA, NM, CO. Our plants are var. *leptomeria*.

Gilia minutiflora Benth. [*Ipomopsis minutiflora* (Benth.) V.E.Grant, *Microgilia minutiflora* (Benth.) J.M.Porter & L.A.Johnson] **Stems** erect, 10–40 cm, often branched above. **Herbage** glabrate to glandular, puberulent. **Leaves** all cauline, 1–4 cm long, entire, linear or the lowest ternate into linear segments. **Inflorescence** of solitary pedicellate flowers in an open bracteate, branched raceme. **Flowers**: calyx ca. 2 mm long; corolla campanulate, white to pink, tube 2–3 mm long, lobes 1–2 mm long. **Capsule** 3–4 mm long. Grasslands; valleys of Ravalli Co. BC, MT south to OR, ID.

Larger and less diffusely branched than the more common *Gilia tenerrima*.

Gilia tenerrima A.Gray [*Lathrocasis tenerrima* (A.Gray) L.A.Johnson] **Stems** ascending to erect, 5–20 cm, branched, often intricately so. **Herbage** with stalked, purple glands on stems. **Leaves**: basal oblanceolate, entire, 1–3 cm long; cauline narrower. **Inflorescence** of solitary flowers on ascending to drooping pedicels in a bracteate cyme. **Flowers**: calyx 1–2 mm long; corolla campanulate, white to pink, tube 1–2 mm long, lobes ca. 1 mm long. **Capsule** 1–2 mm long. Steep, eroding slopes of grasslands, sagebrush steppe, woodlands; valleys. OR to MT south to NV, UT, WY. See *G. minutiflora*.

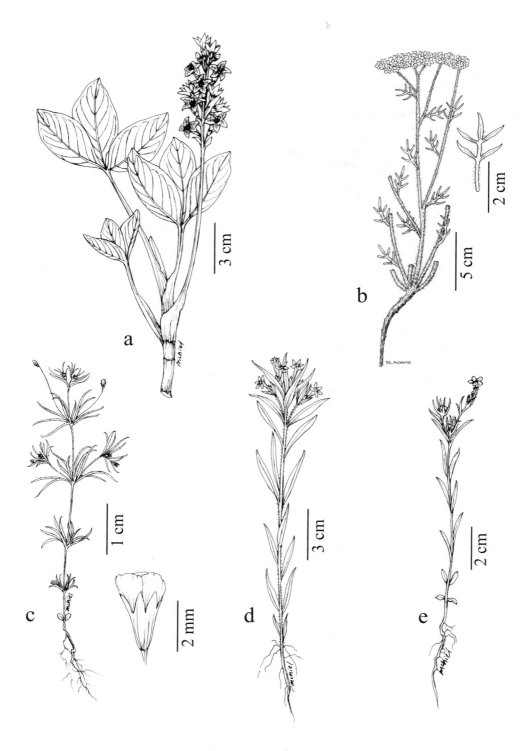

Plate 68. a. *Menyanthes trifoliata*, b. *Ipomopsis congesta* var. *congesta*, c. *Linanthus septentrionalis*, d. *Collomia linearis*, e. *Microsteris gracilis*

Gilia tweedyi Rydb. [*G. sinuata* Douglas ex Benth. var. *tweedyi* (Rydb.) Cronquist, *G. inconspicua* (Sm.) Sweet var. *tweedyi* (Rydb.) Cronq.] **Stems** erect, 5–25 cm, usually branched above. **Herbage** loosely tomentose, stipitate-glandular especially above. **Leaves**: basal oblanceolate, 2–5 cm long, deeply pinnately lobed, lobes lance-linear, mucronate; cauline few. **Inflorescence** terminal, leafy-bracteate cymes with ascending pedicels. **Flowers**: calyx 3–5 mm long, purple spotted, swollen in fruit; corolla funnelform, blue, with a yellow throat, tube 4–5 mm long, lobes ca. 1 mm long. **Capsule** 4–5 mm long with several seeds per locule. Sandy, sparsely vegetated soil of sagebrush steppe, grasslands; valleys, lower montane. OR to MT south to NV, ID, WY.

Gymnosteris Greene

Gymnosteris parvula A.Heller Diminutive, glabrous annuals. **Stems** simple, erect, 1–4 cm. **Leaves** linear-lanceolate, sessile, 3–13 mm long, all subtending the inflorescence. **Inflorescence** a terminal cluster of 1 to several, sessile flowers. **Flowers** funnelform; calyx 2–4 mm long, with a hyaline tube and green lobes; corolla pinkish with a yellow throat, tube 2–5 mm long, lobes ca. 1 mm long; stamens subsessile, included. **Capsule** with several seeds per locule. Sparsely vegetated soil of sagebrush steppe; montane. OR to MT south to CA, NV, UT, CO. Known from Beaverhead and Gallatin cos.[171]

Ipomopsis Michx. Gilia

Annual to perennial herbs. **Leaves** basal and/or alternate on the stem, simple, entire to pinnately lobed, short-petiolate. **Inflorescence** densely branched, capitate, bracteate, terminal and axillary cymes. **Flowers** perfect, regular, 5-merous; sepals united, membranous between the ribs; corolla campanulate to salverform; stamens equally inserted in the corolla. **Capsule** with 1 to several seeds per locule.[149,444]

These species have sometimes been placed in *Gilia*.[102,187,435]

1. Corolla red; the tube 2–3 cm long . . . . *I. aggregata*
1. Corolla white to light blue; the tube 3–9 mm long. 2

2. Plants annual; corollas bluish. . . . .*I. pumila*
2. Plants perennial (biennial); corollas white 3

3. Filaments shorter than anthers; style well-included in the corolla . . . . *I. spicata*
3. Filaments longer than anthers; style ca. as long as corolla tube . . . .*I. congesta*

Ipomopsis aggregata (Pursh) V.E.Grant [*Gilia aggregata* (Pursh) Spreng.] SCARLET GILIA Biennial or monocarpic, short-lived perennial. **Stems** erect, simple, 20–80 cm. **Herbage** malodorous, sparsely villous, glandular especially above. **Leaves** ovate in outline, 2–10 cm long; the blade pinnate with linear lobes, becoming simple in the inflorescence. **Inflorescence** a simple or branched raceme, 5–30 cm long. **Flowers**: calyx 4–8 mm long; corolla scarlet with white spots, salverform, tube 2–3 cm long, lobes 6–10 mm long; stamens barely exserted. **Capsule** ca. 5 mm long with 1 to 2 seeds per locule. Often sandy soil of grasslands, sagebrush steppe, ponderosa pine woodlands, streambanks, roadsides; valleys, montane. BC to MT south to CA, AZ, NM, TX. Our plants are var. *aggregata*.

Ipomopsis congesta (Hook.) V.E.Grant [*Gilia congesta* Hook., *Ipomopsis crebrifolia* (Nutt.) Dorn] Perennial with a woody, branched (rarely simple) caudex. **Stems** simple, prostrate to erect, 4–20 cm. **Herbage** tomentose to glabrous. **Leaves** 5–60 mm long, simple or pinnately lobed into linear segments. **Inflorescence** terminal and axillary, capitate, minutely-bracteate cymes. **Flowers**: calyx 3–5 mm long; corolla white, salverform, tube 3–5 mm long barely greater than the calyx, lobes 1–3 mm long; stamens exserted, filaments longer than the anthers. **Capsule** 2–4 mm long with 1 or 2 seeds per locule. Sandy or clay soil of grasslands, sagebrush steppe, badlands, pine woodlands; plains, valleys, montane. OR to ND south to CA, AZ, NM, CO. (p.390)

1. Leaves all pinnate. . . . . ssp. *congesta*
1. At least basal leaves linear, entire 2

2. Leaves green and glabrous . . . . ssp. *crebrifolia*
2. Leaves sparsely villous. . . . . ssp. *pseudotypica*

Ipomopsis congesta ssp. *crebrifolia* (Nutt.) A.G.Day has prostrate to ascending stems and occurs in sandy-clay soil of eroding, montane slopes in Beaverhead Co.; *I. congesta* ssp. *pseudotypica* (Constance & Rollins) A.G.Day has more erect stems and is found in sandy, sparsely vegetated soil of grasslands in Powder River Co.; *I. congesta* ssp. *congesta* has erect stems and is widespread.

Ipomopsis pumila (Nutt.) V.E.Grant Annual. **Stems** erect, 3–20 cm, usually branched below, branches often as high as the main stem. **Herbage** villous on leaves; stems tomentose. **Leaves**: all cauline, 1–4 cm long; blades oblong in outline ternately to pinnately lobed; lobes linear, mucronate. **Inflorescence** few- to many-flowered, congested, axillary cymes. **Flowers**: calyx 5–6 mm long; corolla funnelform, light blue, tube 6–9 mm long, lobes 1–3 mm long. **Capsule** enclosed in the calyx, 3–5 mm long with several seeds per locule. Sparsely vegetated, sandy soil of sagebrush steppe; valleys. ID, MT south to AZ, NM, TX. Known from Carbon Co.

Ipomopsis spicata (Nutt.) V.E.Grant [*Gilia spicata* Nutt.] Perennial with a usually simple or branched caudex. **Stems** simple, erect, 3–35 cm. **Herbage** glandular-puberulent to tomentose to glabrous. **Leaves** 1–5 cm long, deeply ternately to pinnately lobed into linear segments, the basal leaves sometimes linear, withering at flowering. **Inflorescence** a terminal, congested, spicate-capitate, bracteate cyme. **Flowers**: calyx 3–6 mm long; corolla white, salverform, tube 6–9 mm long, surpassing the calyx, lobes 3–5 mm long; stamens included with filaments shorter than the anthers. **Capsule** 3–5 mm long with several seeds per locule. Sandy or gravelly, often calcareous soil of eroding slopes in grasslands, steppe, woodlands; valleys to alpine. ID to SD south to UT, NM, NE, KS.

1. Hairs of inflorescence ≤0.7 mm long . . . . ssp. *spicata*
1. Hairs of inflorescence >1 mm long . . . . ssp. *orchidacea*

Ipomopsis spicata ssp. *orchidacea* (Brand.) Wilken & R.L.Hartman occurs from valleys to alpine; *I. spicata* ssp. *spicata* is found valleys, montane, known from Carbon Co.

Leptodactylon Hook. & Arn. Prickly phlox

Taprooted subshrubs. **Leaves** opposite or alternate, deeply palmately divided into rigid, linear, mucronate segments, often with axillary fascicles. **Inflorescence** solitary, sessile, axillary or terminal flowers. **Flowers** 4- to 5-merous, opening at night; calyx with hyaline intercostal membranes, lobes spine-tipped; corolla funnelform to salverform, white; stamens equally attached in the corolla, barely exserted to included. **Capsule** with several seeds per locule.

1. Plants forming cushions; flowers 4-merous; leaves opposite . . . . *L. caespitosum*
1. Subshrubs with ascending to erect stems; flowers 5-merous; leaves alternate above . . . .*L. pungens*

Leptodactylon caespitosum Nutt. [*Linanthus caespitosus* (Nutt.) J.M.Porter & L.A.Johnson] Densely caespitose from a many-branched caudex, forming mats 10–60 cm across. **Stems** ascending to erect, simple, 2–6 cm. **Herbage** glabrate to sparsely glandular and ciliate. **Leaves** opposite above, closely overlapping, mostly ternately divided to the base, 3–7 mm long. **Flowers** terminal, 4-merous; calyx 4–7 mm long, lobes sparsely villous, unequal; corolla tube 10–15 mm long, lobes 3–5 mm long. Barren, eroding slopes of Chugwater sandstone in sparse steppe, juniper woodlands; valleys of Carbon Co. MT south to NV, UT, WY, NE.

Leptodactylon pungens (Torr.) Nutt. [*Linanthus pungens* (Torr.) J.M.Porter & L.A.Johnson] Openly branched subshrub. **Stems** ascending to erect, 10–30 cm. **Herbage**: stems glandular-puberulent, the leaves less so. **Leaves** with axillary fascicles, alternate below, 4–10 mm long, divided to the base into 3 to 9 linear segments. **Flowers** 5-merous, solitary in upper leaf axils; calyx 6–9 mm long; corolla tube 10–15 mm long, lobes 5–9 mm long. **Capsule** 3–5 mm long with 5 to 10 seeds per locule. Grasslands, sagebrush steppe; montane. BC, MT south to CA, AZ, NM, NE, Mexico. Our plants are var. *pungens*.

Linanthastrum Ewan

Linanthastrum nuttallii (A.Gray) Ewan [*Linanthus nuttallii* (A.Gray) Greene, *Leptosiphon nuttallii* (A.Gray) J.M.Porter & L.A.Johnson] Taprooted perennial from a woody, branched crown. **Stems** ascending to erect, usually simple, 5–20 cm. **Herbage** glabrate, puberulent on the stems. **Leaves** sessile, opposite, 10–15 mm long, deeply palmately lobed into 5 to 9 linear, mucronate segments, often with axillary fascicles. **Inflorescence** terminal, compact, leafy-bracted cymes. **Flowers** 5-merous; calyx 6–7 mm long, puberulent, tube green throughout but with a hyaline membrane between the lobes; corolla white, salverform, tube 6–7 mm long, puberulent, lobes 3–5 mm long. **Capsules**

4–6 mm long, usually with 1 seed per locule. Talus, cliffs, forest openings; montane, subalpine in Ravalli Co. WA to MT south to CA, AZ, NM, CO, Mexico. Our plants are var. ***nuttallii***.

Linanthus Benth.

Linanthus septentrionalis H.Mason [*Leptosiphon septentrionalis* (H.Mason) J.M.Porter & L.A.Johnson] Taprooted annual. **Stems** thin, 2–25 cm, sometimes branched above. **Herbage** glabrous, puberulent on stems. **Leaves** sessile, 3–15 mm long, palmately divided to the base into 3 to 7 linear lobes. **Inflorescence** of flowers paired in upper leaf axils; pedicels 4–20 mm  long. **Flowers** 5-merous; calyx 2–3 mm long with hyaline intercostal membranes; corolla white, campanulate, 1.5–2 times as long as the calyx, tube ca. 2 mm long, lobes ca. 2 mm long; stamens equally inserted at the top of the tube among a line of short hairs, well exserted. **Capsule** with 2 to 8 seeds per locule, 2–3 mm long. Vernally moist, often shallow soil of grasslands, sagebrush steppe, forest openings, talus, rock outcrops, wetland margins; plains, valleys to subalpine. BC, AB south to CA, NV, UT, CO. (p.390)

Reports of the similar but more southern *Linanthus harknessii* (Curran) Greene in MT[382] are referable here.[242]

Microsteris Greene

Microsteris gracilis (Hook.) Greene [*Phlox gracilis* (Hook.) Greene] Taprooted annual. **Stems** erect, simple or branched, 2–30 cm. **Herbage** glabrate, glandular-puberulent on the stems. **Leaves** sessile, entire, opposite below, alternate above, lanceolate to oblanceolate, 5–35 mm long. **Inflorescence** loose to glomerate, of single or paired flowers at branch tips. **Flowers** 5-merous; calyx 5–9 mm long with hyaline intercostal membranes; corolla salverform, the tube white to yellow, 7–10 mm long, the lobes white to deep pink and 1–3 mm long; stamens included in the tube, unequally inserted with filaments shorter than the anthers. **Capsule** 3–5 mm long with 1 seed per locule. Often disturbed soil of grasslands, steppe, woodlands, rock outcrops, forest openings; plains, valleys, montane. BC, AB south to CA, AZ, NM, S. America. (p.390)

Navarretia Ruiz & Pav.

Taprooted annuals. **Leaves** alternate, short-petiolate; blades ovate to oblong in outline, 5–20 mm long, deeply bipinnately divided into spine-like segments. **Inflorescence** terminal, congested, leafy-bracteate heads. **Flowers** sessile, 5-merous; calyx with hyaline intercostal membranes, the lobes spine-tipped, unequal and sometimes trifid; corolla salverform; stamens equally inserted on short filaments at the mouth of the tube. **Capsule** with several seeds per locule, opening by breaks in the lower wall.[379]

Our three species were previously all considered conspecific with *Navarretia intertexta*.

1. Each corolla lobe with 3 veins . . . .*N. divaricata*
1. Corolla lobes with 1 vein each 2

2. Style 5–8 mm long, exserted from corolla . . . .*N. intertexta*
2. Style ca. 3 mm long, included. . . . . *N. saximontana*

Navarretia divaricata (Torr. ex A.Gray) Greene **Stems** erect, 2–5 cm; ascending lateral branches longer than main stem. **Herbage** viscid-villous above. **Leaves** 8–25 mm long with linear lobes, the central one much the longest. **Flowers**: calyx 5–10 mm long, asymmetrical; corolla white to lavender, shorter than the calyx, tube 3–5 mm long, lobes ca. 1 mm long; style included; stamens not or little exserted. **Capsule** 2–4 mm long. Fields, meadows; valleys. BC to CA, NV, MT. One collection from Sanders Co.

Reports of *Navarretia squarrosa* (Eschsch.) Hook. & Arn.[113] are referable here.

Navarretia intertexta (Benth.) Hook. **Stems** erect, 2–12 cm, simple or branched, the branches prostrate to ascending. **Herbage** glabrate, stems puberulent. **Leaves** 5–20 mm long with 6 to 13 primary lobes on the upper leaf. **Flowers**: calyx 5–8 mm long, villous within; corolla pale blue, ca. as long as the calyx, tube 5–7 mm long, lobes ca. 2 mm long; style 5–8 mm long; stamens exserted well beyond the top of the tube. **Capsule** 2–3 mm long. Vernally wet soil of meadows, margins of ponds, streams; valleys; Lincoln and Sanders cos. WA to MT south to CA, AZ, NV, ID. Our plants are ssp. *propinqua* (Suksd.) A.G.Day.

Navarretia saximontana S.C.Spencer **Stems** erect, 2–15 cm, simple or branched, the branches prostrate to ascending. **Herbage** glabrate, stems puberulent. **Leaves** 5–10 mm long with 5 to 7  primary lobes on the upper leaf. **Flowers**: calyx 6–9 mm long, the lobes villous; corolla pale blue, shorter than the calyx, tube ca. 4 mm long, lobes ca. 2 mm long; style ca. 3 mm long; stamens ca.

as long as the tube. **Capsule** 2–3 mm long. Margins of shallow ponds, lakes; plains, valleys. ID to SK south to AZ, UT, CO; mainly east of the Rocky Mtns.

Phlox L. Phlox

Perennial herbs, most species forming dense to loose mats. **Leaves** opposite, sessile, simple, entire, often with axillary fascicles of smaller leaves. **Inflorescence**: flowers solitary or in few-flowered cymes on branch tips. **Flowers** 5-merous; calyx with hyaline intercostal membranes; corolla salverform with conspicuous lobes; stamens unequally inserted in the tube; anthers near the mouth of the tube. **Capsule** with usually 1 seed per locule.[102]

Many of the species are variable and appear to intergrade with each other. For example, large-leaved forms of *Phlox pulvinata* begin to approach small-leaved forms of *P. kelseyi* var. *missoulensis*; small, compact plants of *P. hoodii* are similar to *P. muscoides*. Habitat often seems to be a good character to distinguish species.

1. Inflorescence with ≥3 pedicellate flowers; some leaves >25 mm long; stems erect 2
1. Flowers mostly sessile and <3 per stem; leaves usually <25 mm long 3

2. Styles 1–2 mm long; corolla lobes lobed at the tip. . . . .*P. speciosa*
2. Styles 6–15 mm long; corolla lobes with entire margins. . . . .*P. longifolia*

3. Hyaline area between green calyx ribs (intercostal membrane) keeled. . . . .*P. austromontana*
3. Intercostal membrane flat, not keeled 4

4. Most leaves 2–5 mm long, triangular, tightly appressed, forming 4-sided shoots . . . . *P. muscoides*
4. Most leaves ≥5 mm long (except *P. albomarginata* with white-margined leaves), ovate
   to linear, not forming 4-sided shoots. 5

5. Leaves lanceolate to oblanceolate, usually noticeably white-margined and basally ciliate 6
5. Leaves linear, obscurely white-margined or not, ciliate or not 7

6. Leaves 2–6 mm long . . . . *P. albomarginata*
6. Some leaves 9–20 mm long. . . . . *P. alyssifolia*

7. Most leaves 5–10 mm long; tangled-villous at the base. . . . .*P. hoodii*
7. Largest leaves >10 mm long; ciliate and/or glabrate to villous but not tangled-villous. 8

8. Rhizomatous plants of sandy soil on the plains . . . . *P. andicola*
8. Taprooted plants of the mountains and intervening valleys 9

9. Larger leaves >1 mm wide at midlength. 10
9. Leaves ≤1 mm wide at midlength. 11

10. Leaves minutely scabrous; flowers sometimes >1 per stem . . . . *P. multiflora*
10. Some leaves ciliate, hirsute, or glandular; flowers mostly 1 per stem . . . . *P. kelseyi*

11. Calyx glabrate to villous with septate hairs but not glandular. . . . . *P. diffusa*
11. Calyx sparsely to densely glandular. 12

12. Stems loosely erect, 4–10 cm; plants of grasslands or montane forest. . . . . *P. caespitosa*
12. Plants forming shorter, more compact cushions near or above treeline. . . . . *P. pulvinata*

Phlox albomarginata Jones Taprooted, loosely mat-forming. **Stems** prostrate to ascending, 2–5 cm, sparsely hirsute, glandular above. **Leaves** stiff, ovate to oblanceolate, 2–6 mm long, mostly puberulent, basally ciliate, white-margined. **Flowers** solitary; pedicels 0–10 mm long; calyx glandular-hirsute, 6–10 mm long with flat intercostal membranes; corolla white to light blue, the tube 9–12 mm long, lobes 6–9 mm long; style 5–8 mm long. Stony, calcareous soil of grasslands, rock outcrops; valleys to (rarely) subalpine. Endemic to southwest MT and adjacent ID. (p.396)

The type locality is near Helena. The more widespread *Phlox alyssifolia* has larger, glabrate leaves.

Phlox alyssifolia Greene [*P. variabilis* Brand] Taprooted, mat-forming. **Stems** prostrate to ascending, 4–15 cm, sparsely hirsute, glandular above. **Leaves** firm, linear-oblanceolate to elliptic, 9–20 mm long, mostly glabrous, basally ciliate, usually white-margined. **Flowers** 1 to 3; pedicels 0–15 mm long; calyx glandular-hirsute, 9–11 mm long with flat intercostal membranes; corolla white, the tube 10–17 mm long, lobes 8–12 mm long; style 6–12 mm long. Stony, sparsely vegetated soil of grassland, pine woodlands; plains, valleys to (rarely) lower alpine. BC to SK south to CO, SD. See *P. albomarginata*. (p.396)

Phlox andicola E.E.Nelson Rhizomatous, forming loose to dense mats. **Stems** prostrate to ascending, 4–15 cm, puberulent to loosely tomentose. **Leaves** firm, linear, 8–25 mm long, glabrate to villous. **Flowers** 1 to 3; pedicels 0–10 mm long; calyx loosely tomentose, 6–10 mm long with flat intercostal membranes; corolla white, the tube 10–13 mm long, lobes 5–8 mm long; style 5–9 mm long. Open, sandy soil of grasslands, pine woodlands, often near sandstone outcrops; plains. MT, ND south to CO, KS.

Phlox austromontana Coville Taprooted, forming loose to dense mats. **Stems** ascending to erect, 4–10 cm, glabrous to pubescent. **Leaves** firm, linear, 7–15 mm long, glabrate to villous. **Flowers** solitary; calyx sparsely villous, 6–9 mm long with keeled intercostal membranes; corolla white, the tube 8–12 mm long, lobes 4–8 mm long; style 2–5 mm long. Sagebrush steppe; valleys to montane. OR to MT south to CA, AZ, UT, CO, Mexico. One collection from Beaverhead Co. (p.396)

Phlox caespitosa Nutt. Taprooted, mat-forming. **Stems** ascending to erect, 4–12 cm, glandular-puberulent. **Leaves** stiff, linear, 5–10 mm long, spine-tipped, sparsely glandular-puberulent, often with axillary fascicles. **Flowers** solitary; calyx sparsely villous, glandular, 5–10 mm long with flat intercostal membranes; corolla white to blue, the tube 7–12 mm long, lobes 4–10 mm long; style 3–7 mm long. Grasslands, open forest, sagebrush steppe, rock outcrops; valleys, montane. BC, MT south to OR, ID. (p.396)

The type specimen is from the lower Flathead Valley. See *Phlox diffusa*.

Phlox diffusa Benth. Taprooted, loosely mat-forming. **Stems** prostrate to ascending, 2–10 cm, glabrous to minutely glandular above. **Leaves** firm, linear, 5–15 mm long, glabrous, ciliate. **Flowers** solitary; calyx villous, 5–9 mm long with flat intercostal membranes; corolla white to blue, the tube 9–12 mm long, lobes 5–9 mm long; style 4–8 mm long. Stony, non-calcareous fellfields, open slopes, rock outcrops, talus, forest openings; upper montane to alpine. BC, MT south to CA, NV, ID; known from Missoula and Ravalli cos. (p.396)

Phlox pulvinata and *P. caespitosa* usually have a glandular calyx; the former occurs at high elevations while the latter does not.

Phlox hoodii Richarson Taprooted, forming loose to dense mats. **Stems** prostrate to ascending, 1–6 cm, glabrous or glandular above. **Leaves** stiff, linear, thinly white-margined, 4–12 mm long, glabrous to glandular, tangled-villous at the base and along the margins. **Flowers** solitary; calyx tangled-villous, sometimes glandular, 4–8 mm long with flat intercostal membranes; corolla white to bluish, the tube 6–11 mm long, lobes 3–7 mm long; style 2–5 mm long. Grasslands, sagebrush steppe, woodlands, most common on exposed sites or where overgrazed; plains, valleys to subalpine. AK to SK south to CA, UT, CO, NE. (p.396)

The degree of villosity and glandularity is variable; glandular plants approach *Phlox kelseyi* var. *missoulensis* but generally have shorter leaves. See *P. muscoides*.

Phlox kelseyi Britton [*P. missoulensis* Wherry] Taprooted, forming loose to dense mats. **Stems** prostrate to ascending, 3–12 cm, glabrous or sparsely hirsute and glandular. **Leaves** firm to succulent, linear, 8–25 mm long, glabrous to glandular-puberulent, often ciliate. **Flowers** solitary; calyx glabrate to glandular-villous, 6–12 mm long with flat intercostal membranes; corolla white to blue, the tube 9–16 mm long, lobes 5–9 mm long; style 4–7 mm long. ID, MT south to NV, CO. (p.396)

1. Leaves succulent, relatively soft; plants of moist to wet meadows. . . . . .var. *kelseyi*
1. Leaves firm to rigid, not succulent; plants of exposed slopes and ridges. . . . . var. *missoulensis*

Phlox kelseyi var. *missoulensis* (Wherry) Cronquist occurs in valleys to (rarely) the subalpine zone; endemic to west-central and central MT; *P. kelseyi* var. *kelseyi* is found in valley to montane wetlands. The two vars. have sharply distinct habitats, but intergrade morphologically and have a similar range in MT. See *P. hoodii* and *P. pulvinata*. The type locality for the species is near Helena.

Phlox longifolia Nutt. Taprooted with creeping subterranean branches. **Stems** loosely tufted, erect, simple, 8–35 cm, glabrous below to glandular-pubescent above. **Leaves** linear to lance-linear, 2–10 cm long, glabrate. **Flowers** ≥3 in upper in a leafy-bracted cymes; pedicels 7–30 mm long; calyx 8–12 mm long with keeled intercostal membranes; corolla white to pink, the tube 12–15 mm long, lobes 6–12 mm long; style 6–15 mm long. Sagebrush steppe, grasslands, forest openings, roadsides; valleys, montane. BC, MT south to CA, AZ, NM.

The Flathead Co. record is from a roadside and may have been introduced. See *P. speciosa*.

Plate 69. **Phlox**. **a.** *P. albomarginata*, **b.** *P. alyssifolia*, **c.** *P. austromontana*, **d.** *P. caespitosa*, **e.** *P. diffusa*, **f.** *P. hoodii*, **g.** *P. kelseyi* var. *kelseyi*, **h.** *P. muscoides*, **i.** *P. multiflora*, **j.** *P. pulvinata*

Phlox multiflora A.Nelson Taprooted, loosely mat-forming. **Stems** prostrate to ascending, 3–12 cm, glabrous below, villous, glandular above. **Leaves** pliable, linear, 12–30 mm long, glabrate, minutely scabrous, sometimes sparsely villous. **Flowers** 1 to 3, very fragrant; calyx glabrous to villous, 9–12 mm long with flat intercostal membranes; corolla white, the tube 8–15 mm long, lobes 6–11 mm long; style 5–8 mm long. Sagebrush steppe, grasslands, open forests; montane to lower alpine. ID, MT south to NV, UT, CO. (p.396)

Phlox muscoides Nutt. [*P. bryoides* Nutt., *P. hoodii* Richardson ssp. *muscoides* (Nutt.) Wherry] Taprooted, cushion-forming. **Stems** ascending to erect, 1–3 cm, tangled-villous. **Leaves** closely overlapping, stiff, lanceolate, 2–4 mm long, tangled-villous, white-margined. **Flowers** solitary; calyx woolly-villous, 3–6 mm long with flat intercostal membranes; corolla white to bluish, the tube 4–10 mm long, lobes 2–4 mm long; style 2–5 mm long. Often gravelly, calcareous soil of rock outcrops, exposed ridges and slopes in sagebrush steppe, grasslands, open forest; valleys, montane. OR to MT south to NV, UT, CO, NE. (p.396)
    Wherry considered *Phlox muscoides* to be a depressed form of *P. hoodii*. Plants resemble mat-forming *Selaginella* spp., and the stems have a whitish aspect.

Phlox pulvinata (Wherry) Cronquist Taprooted, mat-forming. **Stems** ascending, 1–6 cm, glandular-pubescent. **Leaves** stiff, linear, 5–12 mm long, ciliate, glandular-pubescent. **Flowers** solitary, sessile; calyx glandular-villous, 5–8 mm long with flat intercostal membranes; corolla white to blue, the tube 9–13 mm long, lobes 4–7 mm long; style 2–5 mm long. Fellfields, turf, rock outcrops; subalpine, alpine. OR to MT south to NV, UT, CO. (p.396)
    Our most common phlox above treeline but replaced by *Phlox diffusa* in the Bitterroot and nearby ranges. Large plants of *P. pulvinata* resemble small plants of *P. kelseyi* var. *missoulensis*.

Phlox speciosa Pursh Taprooted with woody, spreading branches. **Stems** ascending to erect, branched at the base, 15–35 cm, glabrous below, puberulent above. **Leaves** linear-lanceolate, 2–8 cm long, glabrous to puberulent. **Flowers** ≥5 in a terminal, leafy-bracted cyme; pedicels 5–35 mm long; calyx 8–14 mm long with nearly flat intercostal membranes; corolla white to blue, the tube 10–15 mm long, lobes notched, 9–12 mm long; style 1–2 mm long. Grasslands, open forest; montane. BC, MT south to CA, NV, ID.
    The more common *Phlox longifolia* has entire corolla lobes and carinate intercostal calyx membranes.

Polemonium L.

Glandular annual to perennial herbs. **Leaves** alternate, petiolate, basal and cauline, pinnately divided into entire or lobed leaflets. **Inflorescence**: flowers solitary and axillary or in terminal, leafy-bracted cymes. **Flowers** 5-merous; calyx herbaceous; corolla blue, campanulate to funnelform; stamens equally inserted, exserted beyond the tube. **Capsule** with 1 to 10 seeds per locule.
    Plants, especially of *Polemonium viscosum*, often have a skunk-like odor, but the intensity is variable among populations.

1. Plants annual; corolla no longer than the calyx . . . . *P. micranthum*
1. Plants perennial; corolla surpassing the calyx 2

2. Rhizomatous plants; basal leaves scattered; riparian and wetland areas . . . .*P. occidentale*
2. Plants tufted with a well-developed clusters of basal leaves; upland plants 3

3. Corolla >15 mm long; leaflets deeply divided, appearing whorled. . . . . *P. viscosum*
3. Corolla <12 mm long; leaflets with entire margins . . . .*P. pulcherrimum*

Polemonium micranthum Benth. Taprooted annual. **Stems** erect, simple, 5–20 cm. **Herbage** glandular-puberulent. **Leaves** 1–5 cm long, cauline; blades with 7–15 lanceolate, entire leaflets. **Inflorescence** of solitary flowers in upper leaf axils. **Flowers**: calyx 4–5 mm long, glandular-pubereulent; corolla campanulate, tube and lobes 1–2 mm long; stamens and style shorter than the corolla. **Capsule** 4–5 mm long, enclosed in the swollen calyx. Disturbed soil of grasslands, sagebrush steppe, pine woodlands, lawns, gardens, often beneath shrubs; valleys. BC, MT south to CA, NV, ID.

Polemonium occidentale Greene Rhizomatous perennial. **Stems** erect, simple, not tufted, 20–80 cm. **Herbage** glabrate below, sparsely glandular-pubescent above. **Leaves** long-petiolate, 5–20 cm long; blades with 11 to 27 lanceolate, entire leaflets, 1–5 cm long. **Inflorescence** a terminal, leafy-bracted cyme. **Flowers**: calyx 5–8 mm long; corolla campanulate, tube 4–7 mm long with a ring of hairs at the level of filament insertion, lobes 5–9 mm long; stamens ca. as long as corolla; style exserted. **Capsule** 3–5 mm enclosed in the calyx. Wet meadows, thickets, open forest, often along streams; valleys to subalpine. AK to CA, NV, UT, MN.

Polemonium pulcherrimum Hook. Jacob's ladder Tufted perennial from a simple or branched caudex. **Stems** erect, simple or basally branched, 5–30 cm. **Herbage** glabrous, glandular-puberulent. **Leaves** mostly basal, 5–15 cm long; blades with 11–23 lanceolate, entire leaflets 5–10 mm long. **Inflorescence** a terminal, leafy-bracted cyme. **Flowers**: calyx 4–7 mm long; corolla broadly funnel-form, tube yellow at the apex, 4–5 mm long, lobes 4–8 mm long; stamens and style shorter than the corolla. **Capsule** 3–5 mm long, enclosed in the calyx. Stony soil of outcrops, grasslands, meadows, talus slopes; valleys to subalpine, more common lower. AK to CA, AZ, NM. Our plants are var. **pulcherrimum**.

Polemonium viscosum Nutt. Sky pilot Tufted perennial from a simple or usually branched caudex. **Stems** erect, simple, 5–25 cm. **Herbage** densely glandular-villous. **Leaves** all or mostly basal; blades with 10 to 25 leaflets; leaflets deeply palmately divided giving the appearance of whorled leaflets; the 2 to 5 ovate lobes 1–6 mm long. **Inflorescence** a terminal, leafy-bracted cyme. **Flowers**: calyx 7–13 mm long; corolla funnel-form, tube 10–16 mm long, lobes 5–10 mm long; stamens and style ca. as long as the corolla. Fellfields, rocky slopes, talus; alpine. BC, AB south to OR, NV, NM. (p.401)

HYDROPHYLLACEAE: Waterleaf Family

Annual or perennial herbs. **Leaves** petiolate, mostly alternate, usually pinnately lobed, exstipulate. **Inflorescence** narrow, helicoid cymes that straighten and elongate with maturity or flowers solitary. **Flowers** perfect, regular, usually 5-merous; calyx deeply lobed; corolla campanulate to funnelform; stamens 5; ovary superior, 1-loculate; style 2-lobed. **Fruit** an ovoid to globose capsule with 1 to many seeds.

1. Leaf blades fan-shaped, palmately lobed. . . . . *Romanzoffia*
1. Leaf blades ovate to lance-shaped in outline, entire to pinnately lobed. 2

2. Leaf blades entire. 3
2. Leaf blades lobed. 5

3. Flowers in few- to many-flowered cymes . . . .*Phacelia*
3. Flowers solitary on naked peduncles or in leaf axils 4

4. Plants acaulescent perennials; flowers borne on naked peduncles. *Hesepochiron*
4. Plants small, caulescent annuals; flowers solitary in leaf axils. . . . . *Nama*

5. Flowers in few- to many-flowered cymes 6
5. Flowers solitary in leaf axils 7

6. Flower clusters shorter than the leaves . . . .*Hydrophyllum*
6. Flower clusters among and above the leaves . . . .*Phacelia*

7. Calyx with reflexed appendages between calyx lobes. . . . . *Nemophila*
7. Calyx lacking calyx appendages . . . .*Ellisia*

Ellisia L.

Ellisia nyctelea (L.) L. Taprooted annuals. **Stems** ascending to erect, usually branched, 3–20 cm. **Herbage** strigose, petioles ciliate. **Leaves** petiolate; blades 1–6 cm long, deeply pinnately lobed; lobes entire to dentate. **Inflorescence**: solitary, axillary, pedicellate flowers. **Flowers** funnelform; pedicels 5–15 mm in flower; calyx hispid, divided to the base, 4–6 mm long; corolla blue, tube 2–4 mm long, lobes 2–3 mm long; stamens included. **Capsule** 5–6 mm long, hispid, partly enclosed or subtended by the swollen, veiny calyx, 4-seeded. Often disturbed and partly shaded soil of grasslands, woodlands, streambanks, talus, roadsides; plains, valleys, montane. AB to MI south to NM, TX, LA.

    Nemophila breviflora has reflexed appendages between corolla lobes.

Hesperochiron S.Watson

Taprooted acaulescent perennials. **Leaves** all basal, petiolate, simple, entire. **Inflorescence** of pedunculate flowers arising with the leaves from the caudex. **Flowers**: calyx divided to the base; corolla rotate to funnelform, pubescent within the tube, lobes ca. as long as the tube; stamens ca. as long as the tube. **Capsule** many-seeded.

1. Leaf blades glabrous on the lower surface; sepals ciliate . . . . *H. pumilus*
1. Leaf blades hirsute on both surfaces; sepals villous . . . . *H. californicus*

Hesperochiron californicus (Benth.) S.Watson **Leaf blades** hirsute on both surfaces, 1–5 cm long, elliptic to lanceolate, ca. as long as the petioles. **Flowers**: peduncles 5–45 mm long; calyx 7–8 mm long; corolla campanulate to funnelform, white to purplish, tube 5–6 mm long, lobes 4–10 mm long. Moist to wet, sometimes alkaline meadows; montane. WA to MT south to CA, NV, UT, WY.

Hesperochiron pumilus (Griseb.) Porter Sometimes short-rhizomatous. **Leaf blades** hirsute on margins and upper surface, 1–4 cm long, lanceolate to oblanceolate, ca. as long as the petioles. **Flowers**: peduncles 1–5 cm long; calyx 5–9 mm long; corolla rotate, white with purple lines, 7–12 mm long, tube much shorter than the lobes. Moist to wet, non-calcareous meadows; montane, subalpine. WA to MT south to CA, NV, UT.

Hydrophyllum L. Waterleaf

Hydrophyllum capitatum Douglas ex Benth. BALLHEAD WATERLEAF Fibrous-rooted perennials. **Stems** ascending to erect, 5–25 cm. **Herbage** sparsely hirsute or strigose. **Leaves** long-petiolate; blades 3–10 cm long, pinnately divided into 7 to 11 sessile, ovate, simple or lobed leaflets. **Inflorescence** a densely-branched, capitate cyme, shorter than the subtending leaf. **Flowers** campanulate; calyx hirsute, lobed to near the base, 4–6 mm long; corolla light blue, 7–10 mm long, tube shorter than the lobes, ciliate at points of filament attachment; stamens and style exserted. **Capsule** with 1 to 3 seeds. Vernally moist, sometimes disturbed soil or talus in open forest, woodlands, grasslands, meadows, thickets; valleys to lower subalpine. BC, AB south to CA, CO. Our plants are var. *capitatum*. (p.401)

Nama L.

Nama densum Lemmon Dichotomously branched, taprooted annual. **Stems** prostrate to ascending, spreading-branched at the base, forming small mats 3–10 cm across. **Herbage** hirsute. **Leaves** alternate, entire, ca. sessile, 1–3 cm long, oblanceolate. **Inflorescence**: solitary, axillary and terminal, sessile flowers. **Flowers** funnelform; calyx 4–6 mm long, divided into long, linear lobes; corolla white to lavender, 5–7 mm long, the tube much longer than the lobes; filaments unequally inserted; stamens and style included. **Capsule** 2–3 mm long with several seeds. Sparsely vegetated, sandy soil of sagebrush steppe; valleys. WA to MT south to CA, NV, UT, CO. Known from Carbon Co. Our plants are var. *parviflorum* (Greenm.) C.L.Hitchc.

Nemophila Nutt.

Nemophila breviflora A.Gray Taprooted annuals. **Stems** weak, ascending to erect, simple or branched, angled, 10–30 cm. **Herbage** strigose; petioles ciliate. **Leaves** alternate, petiolate; blades ovate in outline, 1–3 cm long, deeply pinnately divided into 5 lanceolate lobes. **Inflorescence**: solitary, axillary, reflexed-pedicellate flowers. **Flowers** campanulate; calyx stiff-ciliate, 2–4 mm long, divided to the base with small, deltoid, reflexed appendages between the lobes; corolla lavender, 1–3 mm long, shorter than the calyx, the tube longer than the lobes; stamens and style included. **Capsule** 1-seeded, partly enclosed by the swollen calyx. Vernally moist, often disturbed soil in open forest, meadows, thickets; valleys, montane. BC, MT south to CA, NV, UT, CO. See *Ellisia*.

Phacelia Juss. Phacelia

Annual or perennial herbs. **Leaves** alternate, petiolate, entire to pinnately divided. **Inflorescence** axillary and terminal, helicoid, ebracteate cymes, capitate at first, elongating with maturity. **Flowers**: calyx divided to near the base; corolla rotate to campanulate; stamens equal with paired scales where they join the corolla tube; style 2–lobed. **Capsule** with 4–many seeds.[102]

1. Annual with flowers >7 mm across . . . . *P. linearis*
1. Plants either perennial or flowers smaller. 2

2. Plants annual 3
2. Plants perennial 6

3. Leaves entire to dentate. 4
3. Leaves deeply lobed. 5

4. Flowers yellow; leaves crenate to dentate . . . . *P. lutea*
4. Flowers white; leaves entire . . . . *P. incana*

5. Calyx lobes linear; leaves with >3 pairs of lateral lobes. . . . . *P. ivesiana*
5. Calyx lobes ovate to lanceolate; leaves with ≤3 pairs of lateral lobes . . . . *P. thermalis*

6. Leaves with 0–2 pairs of lateral lobes 7
6. Leaves with >2 pairs of lateral lobes 8

7. Plants monocarpic, usually with a single stem from a simple caudex; lower leaves often
    with 1 or 2 pairs of basal, lateral lobes . . . . *P. heterophylla*
7. Strong perennials; well-developed plants with a branched caudex; leaves rarely
    with 1 pair of lateral, basal lobes . . . . *P. hastata*

8. Plants perennial; larger plants with a branched caudex. 9
8. Plants biennial (rarely annual) from a simple caudex. 11

9. Seeds ≤4; corolla white to light blue; stems lax; plants of sagebrush habitats. . . . . *P. ramosissima*
9. Seeds >4; corolla blue to purple; stems erect; plants mostly upper montane to alpine 10

10. Leaves sericeous, divided to the midvein into mostly linear ultimate segments. . . . . *P. sericea*
10. Leaves green, divided ca. halfway to the midvein into deltoid segments. . . . . *P. lyallii*

11. Style cleft ≤1/2-way; ovules numerous; usually moist habitats. . . . . *P. franklinii*
11. Style cleft >1/2-way; ovules 4; dry habitats . . . . *P. glandulosa*

Phacelia franklinii (R.Br.) A.Gray Annual or usually biennial. **Stems** erect, usually simple, 10–70 cm. **Herbage** hirsute, sparsely glandular, more glandular above. **Leaves** basal and cauline, the blade 1–9 cm long, lanceolate, deeply pinnately divided into lanceolate to oblong, mostly entire lobes. **Inflorescence** of terminal and axillary, compact cymes. **Flowers**: calyx hispid, 3–7 mm long; corolla purple, campanulate, 4–9 mm long, pubescent, lobes 3–6 mm long; stamens pubescent, exserted; style exserted, divided <1/2-way. **Seeds** numerous. Usually disturbed soil or talus of sagebrush steppe, meadows, woodlands, open forest, roadsides, often along streams; valleys, montane. YT to ON south to ID, UT, WY, MI.

Phacelia glandulosa Nutt. Annual. **Stems** erect, often branched, 10–30 cm. **Herbage** malodorous, short-hirsute, brown-stipitate-glandular. **Leaves** basal and cauline, the blade linear-lanceolate, 2–10 cm long, deeply pinnately divided into ovate, crenate to lobed lobes. **Inflorescence** terminal and axillary cymes. **Flowers**: calyx glandular, 3–4 mm long; corolla campanulate, purple, 5–8 mm long, lobes 3–5 mm long; stamens glabrous, exserted; style exserted, divided >1/2-way. **Seeds** 2 to 4, grooved, pitted. Eroding soil of slopes in sagebrush steppe; montane. ID, MT south to UT, CO.

Phacelia hastata Douglas ex Lehm. [*P. leucophylla* Torr., *P. leptosepala* Rydb., *P. hastata* var. *alpina* (Rydb.) Cronquist, *P. hastata* var. *leptosepala* (Rydb.) Cronquist] Perennial, usually with a branched caudex. **Stems** ascending to erect, usually simple, 5–40 cm. **Herbage** silvery-hirsute, hispid with thick hairs. **Leaves** basal and cauline; the blade 2–10 cm long, oblanceolate to elliptic, entire or rarely with a pair of basal lobes. **Inflorescence** of terminal and axillary cymes. **Flowers**: calyx hispid and hirsute, 3–4 mm long; corolla campanulate, white to purplish, 4–7 mm long, lobes 2–4 mm long; stamens pubescent, exserted; style exserted, deeply divided. **Seeds** ca. 2. Stony, usually sparsely vegetated soil or talus

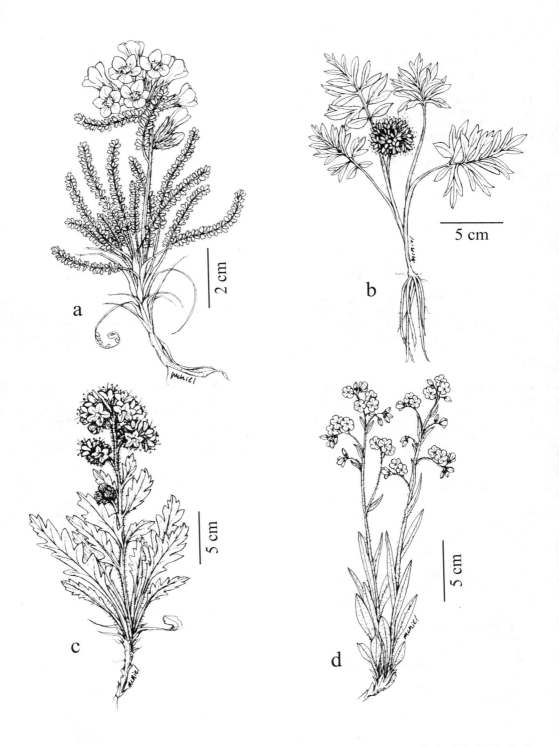

Plate 70. a. *Polemonium viscosum*, b. *Hydrophyllum capitatum*, c. *Phacelia lyallii*, d. *Hackelia micrantha*

in grasslands, sagebrush steppe, woodlands, meadows, fellfields, open forest; valleys to alpine. BC, AB south to CA, NV, UT, CO, NE.

Montana plants have been classified into three taxa based on vesture, size and flower color. Extreme forms are distinctive, but most populations are intermediate; the first two characters vary continuously; and none are correlated with ecology or distribution in our area.

Phacelia heterophylla Pursh Taprooted biennial or short-lived, monocarpic perennial. **Stems** erect, usually simple, 20–80 cm. **Herbage** hispid, hirsute. **Leaves** basal and cauline, prominently veiny; the blade 2–10 cm long, lanceolate to oblanceolate, entire; some lower leaves with 1 to 2 pairs of basal lobes. **Inflorescence** long, narrow, of compact, terminal and axillary cymes. **Flowers**: calyx hispid, glandular, 3–4 mm long; corolla campanulate, white to blue, 4–7 mm long, lobes 2–3 mm long; stamens pubescent, exserted; style exserted, deeply divided. **Seeds** ca. 2. Sparsely vegetated soil of grasslands, sagebrush steppe, pine woodlands, roadsides; plains, valleys, montane. BC, MT south to CA, AZ, NM, Mexico.

Phacelia heterophylla hybridizes with the similar P. hastata which can usually be distinguished by its branched caudex.[102] It seems possible that P. heterophylla is a monocarpic form of P. hastata.

Phacelia incana Brand. Annual. **Stems** erect, simple or branched, 3–10 cm. **Herbage** glandular-puberulent. **Leaves** basal and cauline; blades ovate to elliptic, 5–15 mm long, entire. **Inflorescence** few-flowered, terminal cymes. **Flowers**: calyx hirsute, glandular, 3–5 mm long, enlarged in fruit; corolla funnelform, white, 4–6 mm long, lobes ca. 1 mm long; stamens included; style short-branched, included. **Seeds** numerous, reticulate. Stony, calcareous soil of sagebrush steppe, mountain mahogany woodlands; valleys, montane; Beaverhead Co. ID, MT south to NV, UT, WY.

Phacelia ivesiana Torr. Annual. **Stems** ascending to erect, branched, 2–10 cm; outer branches often longer than the central axis. **Herbage** hirsute, glandular above. **Leaves** cauline; blades lanceolate to oblanceolate, 1–4 cm long, deeply pinnately divided into entire, lanceolate lobes. **Inflorescence** terminal cymes. **Flowers**: calyx hispid, 2–5 mm long, enlarged in fruit; corolla white, funnelform, 2–4 mm long, lobes <1 mm long; stamens and style included. **Seeds** several, cross-corrugated. Sandy soil of sagebrush steppe, juniper woodlands; valleys of Carbon Co. ID, MT south to CA, NV, UT, CO.

Phacelia linearis (Pursh) Holz. Annual. **Stems** erect, simple or branched, 5–40 cm. **Herbage** strigose, hirsute; stems puberulent. **Leaves** cauline, short-petiolate; blades lanceolate to linear, 1–5 cm long, often with 1 or 2 pairs of basal lobes. **Inflorescence** a crowded cluster of few-flowered, axillary and terminal cymes. **Flowers**: calyx hispid, 5–7 mm long, enlarged in fruit; corolla rotate-campanulate, lavender, 8–12 mm long, lobes 4–6 mm long; stamens exserted; style short-branched, exserted. **Seeds** numerous, reticulate. Sparsely vegetated soil of grasslands, sagebrush steppe, dry open forest; plains, valleys, montane. BC to AB south to CA, UT, WY.

Phacelia lutea (Hook. & Arn.) J.T.Howell [*P. scopulina* (A.Nelson) J.T.Howell] Annual. **Stems** prostrate to ascending, branched at the base, forming small mats 2–10 cm across. **Herbage** short-hirsute. **Leaves** basal and cauline; blades obovate, 1–4 cm long, crenate to dentate. **Inflorescence** few-flowered, terminal cymes. **Flowers**: calyx puberulent, 2–4 mm long; corolla campanulate, yellow, ca. as long as calyx, lobes ca. 0.5 mm long; stamens included; style short-branched, pubescent, included. **Seeds** numerous, reticulate, cross-corrugated. Eroding, calcareous soil in sagebrush steppe; valleys, montane. OR to MT south to CA, AZ, UT, CO. Collected in Silver Bow Co. over 100 years ago.[102]

Phacelia lyallii (A.Gray) Rydb. Perennial, often with a branched caudex. **Stems** ascending to erect, usually simple, 5–25 cm. **Herbage** sparsely strigose, hirsute above. **Leaves** mostly basal; blade 2–8 cm long, oblanceolate, pinnately lobed ca. halfway into deltoid segments. **Inflorescence** a densely flowered, compound, terminal cyme. **Flowers**: calyx hirsute, 4–6 mm long; corolla campanulate, blue, pubescent, 5–9 mm long, lobes 3–5 mm long; stamens exserted, ca. 2 times as long as corolla; style exserted, divided <1/2-way. **Seeds** several, reticulate. Meadows, along streams, rocky slopes; subalpine, alpine. Endemic to BC, AB, MT. (p.401)

The similar *Phacelia sericea* has more narrowly and deeply divided leaf blades and more strongly exserted stamens.

Phacelia ramosissima Douglas ex Lehm. Perennial. **Stems** brittle, erect to ascending, often branched, 50–100 cm. **Herbage** hirsute, glandular, malodorous. **Leaves** mainly cauline; the blade 5–20 cm long, lanceolate, deeply pinnately divided into oblong, lobed and dentate lobes. **Inflorescence** of terminal and axillary cymes. **Flowers**: calyx hispid, 4–6 mm long; corolla pale lavender, narrowly campanulate, 6–9 mm long, pubescent, lobes 2–5 mm long; stamens pubescent, ca. twice as long as corolla; style exserted, divided >1/2-way. **Seeds** numerous, pitted. Sandy soil of sagebrush steppe; montane; one collection from Silver Bow Co. WA to MT south to CA, NV, AZ.

Phacelia sericea (Graham) A.Gray Perennial, often with a branched caudex. **Stems** ascending to erect, usually simple, 10–45 cm. **Herbage** strigose to sericeous. **Leaves** mostly basal; the blade 2–8 cm long, lanceolate to ovate, deeply pinnately lobed into linear to oblanceolate, entire to lobed segments. **Inflorescence** a densely flowered, compound, terminal cyme. **Flowers**: calyx hispid, 3–5 mm long; corolla campanulate, purple, pubescent, 5–8 mm long, lobes 2–4 mm long; stamens >2 times as long as corolla; style exserted, divided <1/2-way. **Seeds** several; reticulate. Talus slopes, fellfields, rock outcrops, grasslands, open forest, stream banks, roadsides; montane to alpine. BC, AB south to CA, CO. Our plants are var. *sericea*.

Many low-elevation plants have strigose foliage but lack the ciliate petioles of var. *ciliosa* Rydb. *Phacelia sericea* flowers earlier than *P. lyallii* when they occur together.[382]

Phacelia thermalis Greene Annual. **Stems** erect to ascending, branched at the base, 5–20 cm. **Herbage** glandular-pubescent. **Leaves** cauline; blades obovate, 1–5 cm long, crenate to deeply lobed. **Inflorescence** a terminal cyme. **Flowers**: calyx glandular, 3–4 mm long, swollen in fruit; corolla funnelform, lavender, 3–5 mm long, lobes ca. 1 mm long; stamens included; style cleft to below the middle, pubescent, included. **Seeds** numerous; reticulate. Barren clay slopes of river bluffs; valleys. OR, ID south to CA, disjunct in Garfield and Phillips cos., MT.

Romanzoffia Cham. Mist maiden

Romanzoffia sitchensis Bong. Perennial. **Stems** weak, prostrate to ascending or hanging, 5–20 cm. **Herbage** glabrous with sparsely villous petioles and stems. **Leaves** basal, long-petiolate with bulblets in the swollen bases; blades reniform with shallowly lobed margins, 1–3 cm wide. **Inflorescence** a sparsely flowered, racemose cyme; pedicels 6–20 mm long. **Flowers** campanulate; calyx divided to near the base, 2–4 mm long; corolla white, 6–10 mm long, lobes 2–4 mm long, yellow at the base; stamens included; style 2–5 mm long, undivided. **Capsule** with numerous seeds. Wet, shallow soil of cliffs, streambanks; subalpine, alpine. AK to CA, ID, MT.

BORAGINACEAE: Borage Family

Annual or perennial herbs. **Leaves** alternate, simple, mostly entire, exstipulate. **Herbage** often hispid or coarsely hirsute. **Inflorescence** helicoid cymes or racemes, expanding with maturity. **Flowers** perfect, regular, usually 5-merous; calyx deeply to shallowly lobed; corolla campanulate to salverform, usually with appendages (fornices) at the inside top of the tube alternate with the lobes; stamens 5; ovary superior, 2-carpellate; style simple to 4-lobed. **Fruit** 4 nutlets, joined at the base or along the ventral surface, producing a scar in the area of union.

The ornamentation of mature fruits is often important for identification. A small escaped population of the garden flower lungwort, *Pulmonaria officinalis* L., was once collected in Ravalli Co. *Borago officinalis* L., a medicinal herb, was collected once in a Ravalli Co. yard.

1. Flowers white, yellow or orange 2
1. Flowers blue to reddish-purple 11

2. Flowers white. 3
2. Flowers yellow or orange. 9

3. Herbage glabrous . . . . *Heliotropum*
3. Herbage strigose to hirsute or hispid. 4

4. Nutlets with marginal prickles . . . . *Hackelia*
4. Nutlets lacking prickles 5

5. Leaves and stems strigose, villous, or (rarely) sparsely hirsute, not hispid. 6
5. Leaves and stems hirsute and also usually hispid 7

6. Calyx uncinate-hispid . . . . *Myosotis*
6. Calyx hispid with straight not hooked hairs . . . . *Plagiobothrys*

7. Scar of nutlet a longitudinal groove . . . . *Cryptantha*
7. Nutlet scar basal 8

8. Plant perennial; stems stout, 30–70 cm tall; corolla tube 12–14 mm long . . . . *Onosmodium*
8. Annuals; stems ≤30 cm; corolla tube 5–8 mm long. . . . . *Lithospermum*

9. Inflorescence completely ebracteate; leaves lanceolate to ovate . . . . *Symphytum*
9. Inflorescence bracteate, at least toward the base; leaves narrower 10

10. Annuals; nutlets 2–3 mm long . . . . *Amsinckia*
10. Perennials; nutlets 4–6 mm long . . . . *Lithospermum*

11. Nutlets with prickles 12
11. Nutlets smooth, wrinkled or bumpy but not prickly 14

12. Flowers magenta; nutlets 4–7 mm long. . . . . *Cynoglossum*
12. Flowers blue; nutlets 2–5 mm long 13

13. Plants annual, mostly <15 cm tall; pedicels nearly erect in fruit . . . .*Lappula*
13. Plants biennial or perennial, usually >20 cm tall; pedicels reflexed in fruit . . . . *Hackelia*

14. Mature calyx compressed, 1–2 cm wide . . . . *Asperugo*
14. Mature calyx not compressed, <1 cm across 15

15. Nutlets smooth. 16
15. Nutlets muricate (bumpy) or rugose (ridged). 18

16. Corolla funnelform, the limb ascending . . . . *Mertensia*
16. Corolla salverform, the limb perpendicular to the tube 17

17. Cushion-forming perennials ≤10 cm high . . . . *Eritrichium*
17. Annual or perennial with stems >10 cm; not cushion-forming. . . . . *Myosotis*

18. Nutlets muricate. . . . . *Borago*
18. Nutlets rugose 19

19. Corolla irregular, campanulate. . . . . *Echium*
19. Corolla regular, salverform . . . . *Anchusa*

Amsinckia Lehm. Tarweed

Taprooted annuals. **Herbage** hirsute, hispid with broad-based trichomes. **Leaves** basal and cauline, linear to linear-oblanceolate, obscurely petiolate to sessile. **Inflorescence** helicoid, basally-bracteate, short-pedicellate racemes. **Flowers**: calyx divided to the base; corolla salverform; stamens included in the corolla tube; style included, unlobed. **Nutlets** bumpy with a ventral keel; scar basal or along lower keel.
  Amsinckia menziesii has more lax stems and is less hispid than *A. lycopsoides*.

1. Fornices hairy; stamens inserted at or below middle of the tube . . . . *A. lycopsoides*
1. Fornices absent; stamens inserted near the top of the tube . . . . *A. menziesii*

Amsinckia lycopsoides Lehm. **Stems** erect, usually simple, 10–60 cm. **Leaves** 2–10 cm long. **Flowers**: calyx 6–10 mm long; sepals unequal; corolla yellow to orange, tube 7–9 mm long, lobes 1–2 mm long, fornices hairy; stamens inserted at or below middle of the tube. **Nutlets** ovoid, 2–3 mm long, green to brown. Lawns, roadsides; valleys. BC, MT south to CA. Collected once in Gallatin Co.

Amsinckia menziesii (Lehm.) A.Nelson & J.F.Macbr. **Stems** erect, usually branched, 10–80 cm. **Leaves** mainly cauline, 2–12 cm long. **Flowers**: calyx 2–5 mm long; sepals equal; corolla yellow, tube 3–5 mm long, lobes ca.1 mm long, fornices absent; stamens inserted at the top of the tube. **Nutlets** ovoid, 2–3 mm long, green to black. Roadsides, pastures; valleys. AK south to CA, NV, UT, WY; perhaps introduced in MT.[187]

Anchusa L. Alkanet

Annual or perennial herbs. **Herbage** spreading-hispid. **Leaves** basal and cauline. **Inflorescence** helicoid, bracteate, short-pedicellate cymes. **Flowers**: calyx deeply divided; corolla blue, salverform; stamens inserted near the mouth of the tube; style barely exserted, unlobed; fornices prominent, erect, ciliate. **Nutlets** rugose, rectangular, ridged, attached at the base; scar surrounded by a rim.
  All of our species are native to Europe and introduced across temperate N. America.

1. Corolla limb 3–5 mm across; plants annual . . . .*A. arvensis*
1. Corolla limb 6–20 mm across; plants perennial 2

2. Corolla limb 12–20 mm across; nutlets 5–9 mm long. . . . .*A. azurea*
2. Corolla limb 6–11 mm across; nutlets 3–4 mm long. . . . . *A. officinalis*

Anchusa arvensis (L.) Bieb. **Stems** erect, simple or few-branched, 30–80 cm. **Leaves** basal lanceolate to ovate, petiolate, becoming sessile above, 2–15 cm long. **Flowers**: bracts leaf-like; calyx ca. 5 mm long with acute lobes; corolla tube curved, 4–7 mm long, lobes unequal, 2–3 mm long. **Nutlets** erect, 2–4 mm long. Roadsides, agricultural fields; plains, valleys.

Anchusa azurea Mill. Perennial. **Stems** erect, simple, 15–70 cm. **Leaves**: the basal oblanceolate, petiolate, 5–30 cm long, becoming lance-ovate and sessile above. **Flowers**: bracts leaf-like; calyx ca. 10 mm long, expanded in fruit, divided to below the middle; corolla tube 9–11 mm long, lobes 8–10 mm long. **Nutlets** erect, 5–9 mm long. Fields, gardens; valleys; collected in Lake Co.

Anchusa officinalis L. Perennial. **Stems** erect, simple or branched, 30–100 cm. **Leaves**: the basal oblanceolate, short-petiolate, becoming sessile above, 3–20 cm long. **Flowers**: bracts leaf-like; calyx 4–6 mm long, divided to mid-length; corolla tube 6–8 mm long, lobes 3–6 mm long. **Nutlets** 3–4 mm long. Roadsides, fields; valleys; collected in Flathead Co.

Asperugo L. Madwort

Asperugo procumbens L. Weak-stemmed annual. **Stems** ascending, lax, 25–100 cm. **Herbage** hispid; stems retrorse-prickly. **Leaves**: the basal petiolate, oblanceolate to spatulate, 2–8 cm long, becoming smaller and sessile on the stem. **Inflorescence** of solitary or paired, axillary flowers. **Flowers** reflexed-pedicellate; calyx divided ca. halfway, becoming compressed and 1–2 cm across in fruit with 2 small teeth between each pair of lobes, veiny and prickly; corolla funnelform, blue, ca. 2 mm long, the lobes ca. 1 mm long, fornices apparent; stamens included; style short, unlobed. **Nutlets** ovate, minutely muricate, compressed, 2–3 mm long, hidden by the calyx; scar above midlength. Lawns, gardens, riparian thickets, vacant lots; valleys. Introduced throughout much of northern U.S. and adjacent Canada; native to Eurasia.

Cryptantha Lehm. ex G.Don Cryptantha, miner's candle

Annual to perennial herbs with mostly erect stems. **Herbage** hirsute and often hispid with stiff, long trichomes from swollen bases. **Leaves** cauline; biennials and perennials also with basal rosettes. **Inflorescence** axillary and terminal, helicoid cymes of sessile or short-pedicellate flowers. **Flowers**: calyx divided to the base, enlarging in fruit; corolla salverform, white or rarely yellow, fornices yellow; stamens included. **Nutlets** smooth or bumpy (muricate) but not spiny; scar indicating line of attachment on the ventral surface.[102,180]

   Cryptantha celosioides and *C. spiculifera* are closely related and difficult to distinguish; *C. celosioides* usually has an unbranched caudex and dies after flowering once, while *C. spiculifera* is a true perennial, often with a branched caudex. Annual species are similar vegetatively but differ in nutlet characters. *Cryptantha cinerea* (Greene) Cronquist [*C. jamesii* (Torr.) Payson] was reported for MT,[151] but I have seen no specimens.

1. Plants biennial or perennial; stems ≥3 mm thick; caudex often branched 2
1. Plants annual without a caudex; stems 1–2 mm thick 8

2. Corolla tube >6 mm long, noticeably longer than the calyx . . . .*C. flavoculata*
2. Corolla tube ≤6 mm long, ca. as long as the calyx. 3

3. Ventral surface (side facing axis) of the nutlets smooth. 4
3. Ventral surface of the nutlets ridged, wrinkled or muricate. 5

4. Nutlets laterally separated from each other; plains . . . . *C. cinerea*
4. Nutlets adjacent laterally; montane to alpine . . . . *C. sobolifera*

5. Nutlets usually solitary; herbage not hispid . . . . *C. cana*
5. Nutlets 2 to 4 in most flowers; herbage hispid 6

6. Plants monocarpic biennial or short-lived perennial; caudex usually simple although stems
     sometimes branched at the base . . . . *C. celosioides*
6. Plants long-lived perennial with a branched caudex 7

7. Nutlet scar open most of its length; style ca. as long as the nutlets. . . . .*C. humilis*
7. Nutlet scar closed except at the very base; style longer than the nutlets. . . . . *C. spiculifera*

8. Some or all of the nutlets with muricate dorsal surfaces 9
8. Nutlets with smooth surfaces 12

9. Dorsal surface of 1 nutlet smooth, the other 3 nutlets muricate. 10
9. All four nutlets muricate 11

10. Cymes bracteate nearly to the tip. . . . . *C. minima*
10. Cymes ebracteate or bracteate just at the base. . . . . *C. kelseyana*

11. Nutlets ovate, >0.7 mm wide . . . . *C. ambigua*
11. Nutlets linear-lanceolate, ≤0.7 mm wide. . . . . *C. scoparia*

12. Scar of nutlets on the edge of the ventral surface . . . .*C. affinis*
12. Scar of the nutlet in the middle of the ventral surface 13

13. Nutlets with narrowly winged, knife-edged margins . . . .*C. watsonii*
13. Margins of nutlets rounded. 14

14. Nutlets lanceolate, ≤0.7 mm wide; plants of sandhill habitats . . . . *C. fendleri*
14. Nutlets ovate, >0.7 mm wide; plants of sandy or gravelly soil but not restricted to pure sand . .*C. torreyana*

Cryptantha affinis (A.Gray) Greene Annual. **Stems** slender, simple or branched, 5–30 cm. **Herbage** hirsute and sometimes hispid. **Leaves** linear-oblanceolate to oblong, 1–3 cm long. **Inflorescence** ebracteate or bracteate below. **Flowers**: calyx hispid and hirsute, 3–5 mm long in fruit; corolla 1–2 mm long. **Nutlets** 4, ovate, ca. 2 mm long, smooth, shiny; scar closed, on edge of ventral surface. Often gravelly soil of open forest, grasslands; valleys, montane. BC south to CA, NV, UT, WY. (p.407)

Cryptantha ambigua (A.Gray) Greene Annual. **Stems** usually branched, 5–25 cm. **Herbage** hirsute and hispid. **Leaves** linear, 5–35 mm long. **Inflorescence** ebracteate. **Flowers**: calyx hispid, hirsute, 5–6 mm long in fruit; corolla 2–4 mm long. **Nutlets** 4, ovate, 1–2 mm long, muricate; scar closed, widened at the base. Sandy or gravelly soil or talus of grasslands, sagebrush steppe, woodlands, outcrops; valleys, montane, rarely subalpine. BC, MT south to CA, UT, CO. (p.407)

Cryptantha cana (A.Nelson) Payson Caespitose perennial from a branched caudex covered in old leaf bases. **Stems** erect to ascending, 4–15 cm. **Herbage** sericeous. **Leaves** mostly basal, petiolate, linear to oblanceolate, acute, 7–30 mm long, becoming sessile above. **Inflorescence** bracteate below. **Flowers**: calyx hispid, hirsute, 5–8 mm long in fruit; corolla 3–4 mm long. **Nutlets** usually solitary, asymmetric, muricate, ca. 3 mm long, sharp-margined; scar widened at the base. Sandy, calcareous soil of open slopes in juniper woodlands; valleys; Big Horn and Carbon cos. MT to SD south to CO, NE.

Cryptantha celosioides (Eastw.) Payson Biennial or short-lived, monocarpic perennial, often branched at the base but with a simple caudex. **Stems** erect, 6–35 cm. **Herbage** tomentose, hispid. **Leaves**: basal petiolate, spatulate to oblanceolate, rounded to obtuse, 1–6 cm long; cauline narrower, acute. **Inflorescence** bracteate below. **Flowers**: calyx hispid, hirsute, 7–10 mm long in fruit; corolla 3–6 mm long, 6–11 mm across the limb, fornices ochroleucus. **Nutlets** 4, ovate, rugose dorsally, 3–4 mm long; scar closed at the base. Stony or sandy, often sparsely vegetated soil of grasslands, sagebrush steppe; plains, valleys, montane. BC to ND south to OR, CO, NE. (p.407)

Cryptantha fendleri (A.Gray) Greene Annual. **Stems** slender, usually branched, 5–25 cm. **Herbage** hirsute, hispid. **Leaves** linear, 1–3 cm long. **Inflorescence** ebracteate. **Flowers**: calyx hispid, hirsute, 4–6 mm long in fruit; corolla 2–3 mm long. **Nutlets** 4, lanceolate, ca. 2 mm long, smooth, often brown-mottled at maturity; scar long, closed, widened at the base. Sandhills; plains, montane; Beaverhead and Sheridan cos. Sporadic from AB, SK south to AZ, UT, CO. (p.407)

Cryptantha flavoculata (A.Nelson) Payson Perennial with a branched caudex. **Stems** erect, 7–20 cm. **Herbage** appressed-hirsute. **Leaves** basal and cauline, petiolate, oblanceolate, acute to rounded, 2–5 cm long, becoming sessile above. **Inflorescence** of compact cymes, leafy-bracteate below. **Flowers**: calyx hispid, tomentose, 6–10 mm long in fruit; corolla 7–10 mm long with prominent yellow fornices. **Nutlets** 3–4, ovate, 3–4 mm long, ridged and muricate; scar open its entire length. Sparsely vegetated, sandy soil of juniper woodlands; valleys, montane; known from Carbon Co. ID, MT south to CA, NV, UT, CO.
  Plants have one of two kinds of flowers: those with included anthers and a slightly exserted style or slightly exserted anthers with a shorter style.

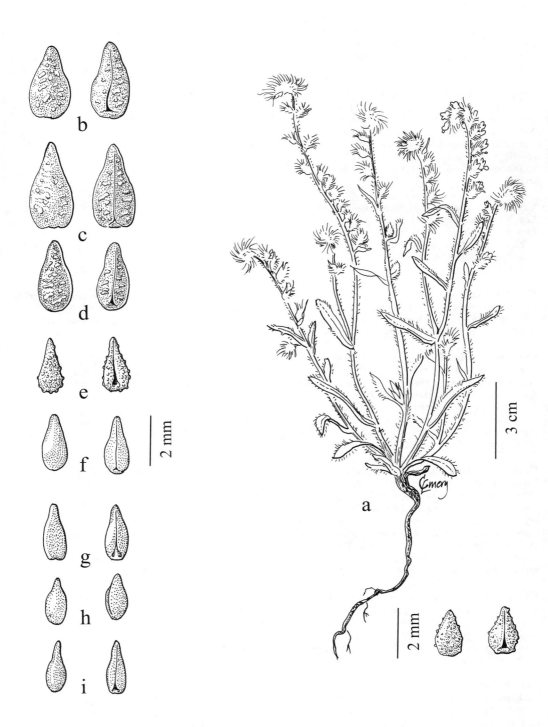

Plate 71. **Cryptantha**. **a.** *C. ambigua*, **b.** *C. spiculifera*, **c.** *C. sobolifera*, **d.** *C. celosioides*, **e.** *C. kelseyana*, **f.** *C. watsonii*, **g.** *C. torreyana*, **h.** *C. affinis*, **i.** *C. fendleri*

Cryptantha humilis (Greene) Payson Perennial with a branched caudex. **Stems** erect, 5–15 cm. **Herbage** hispid, tomentose. **Leaves** basal and cauline, petiolate, oblanceolate, acute, 1–3 cm long. **Inflorescence** densely cylindrical, of compact cymes, leafy-bracteate below. **Flowers**: calyx hispid, tomentose, 5–7 mm long in fruit; corolla 3–5 mm long, fornices sometimes yellow. **Nutlets** 1 to 4, ovate, densely muricate-tuberculate, 2–4 mm long; scar open at least below. Sagebrush steppe, juniper woodland; valleys, montane. OR to MT south to CA, NV, UT, CO. Known from Beaverhead and Jefferson cos.

Cryptantha kelseyana Greene Annual. **Stems** slender, usually branched at the base, 5–20 cm. **Herbage** appressed-hirsute, hispid. **Leaves** linear, 1–2 cm long. **Inflorescence** ebracteate or bracteate at the base. **Flowers**: calyx hispid, hirsute, 5–7 mm long in fruit; corolla ca. 2 mm long. **Nutlets** 4, lance-ovate, ca. 2 mm long, 1 smooth and slightly larger, the other 3 sparsely muricate; scar closed, widened at the base. Sagebrush steppe; valleys, montane. MT, SK south to ID, UT, CO. (p.407)

The type locality is in Powell Co. The similar *Cryptantha minima* has an inflorescence that is bracteate to near the apex.

Cryptantha minima Rydb. **Stems** branched, 5–15 cm. **Herbage** hirsute, strongly hispid. **Leaves** linear to oblanceolate, 5–30 mm long. **Inflorescence** linear, bracteate to near the apex. **Flowers**: calyx hispid, hirsute, 5–7 mm long in fruit; corolla 2–3 mm long. **Nutlets** 4, lance-ovate, 2–2.5 mm long, 1 smooth and slightly larger, the other 3 sparsely muricate; scar closed, widened at the base. Sandy soil of grasslands, sagebrush steppe; plains, valleys. SK south to NM, TX. See *C. kelseyana*.

Cryptantha scoparia A.Nelson **Stems** mostly simple, 8–25 cm. **Herbage** hispid and strigose. **Leaves** linear, 1–2 cm long. **Inflorescence** ebracteate. **Flowers**: calyx hispid, hirsute, 5–6 mm long in fruit; corolla ca. 2 mm long. **Nutlets** 4, narrowly lanceolate, densely muricate, ca. 2 mm long; scar closed except at the base. Sparsely vegetated, sandy soil in sagebrush steppe; valleys. WA to MT south to NV, UT, WY. Known from Carbon Co.

Cryptantha sobolifera Payson [*C. nubigena* (Greene) Payson misapplied] Perennial with a branched caudex. **Stems**: flowering stems erect, 5–15 cm; short sterile stems 1–3 cm. **Herbage** villous, hispid. **Leaves** petiolate, 1–3 cm long; basal and sterile stem leaves spatulate; cauline leaves oblanceolate. **Inflorescence** short, leafy-bracteate below. **Flowers**: calyx hispid, puberulent, 5–6 mm long in fruit; corolla 3–5 mm long, fornices yellow. **Nutlets** 3 or 4, lance-ovate, obscurely ridged dorsally, smooth ventrally, 3–4 mm long; scar long, barely open at the base. Sparsely vegetated, stony soil of exposed slopes, ridges; subalpine, alpine, rarely lower. OR to MT south to CA. (p.407)

The type locality is in Glacier National Park.

Cryptantha spiculifera (Piper) Payson [*C. macounii* (Eastw.) Payson, *C. interrupta* (Greene) Payson misapplied] Perennial with a branched caudex. **Stems** erect, 3–20 cm. **Herbage** strigose, hispid. **Leaves** basal and cauline, petiolate, linear-oblanceolate, acute, 1–6 cm long, becoming sessile above. **Inflorescence** leafy-bracteate below. **Flowers**: calyx hirsute, hispid, 5–8 mm long in fruit; corolla 5–9 mm long, 6–8 mm across the limb, fornices ochroleucus. **Nutlets** 1 to 4, ovate to lance-ovate, ridged-rugose, 2–3 mm long; scar closed. Sandy or stony soil of ridges, open slopes in grasslands, sagebrush steppe; plains, valleys, montane. WA to MT south to CA, NV, WY. (p.407)

Cryptantha macounii has been separated out based on the length of the nutlets, but this character is continuous and varies within populations. If these two taxa are separated, our plants would be *C. macounii*.

Cryptantha torreyana (A.Gray) Greene Annual. **Stems** slender, lax, simple or branched, 8–35 cm. **Herbage** hispid, strigose. **Leaves** linear to oblanceolate, 1–5 cm long. **Inflorescence** ebracteate. **Flowers**: calyx hispid, hirsute, 3–5 mm long in fruit; corolla ca. 2 mm long. **Nutlets** 4, ovate, 1.5–2 mm long, smooth; scar closed, widened at the base. Sparsely vegetated soil, talus of woodlands, open forest, grasslands, sagebrush steppe, often in partial shade around the base of trees; plains, valleys, montane. BC, MT south to CA, NV, UT, WY. (p.407)

Cryptantha watsonii (A.Gray) Greene Annual. **Stems** simple or branched, 10–25 cm. **Herbage** hispid. **Leaves** linear, 1–4 cm long. **Inflorescence** ebracteate. **Flowers**: calyx hispid, hirsute, 3–5 mm long in fruit; corolla ca. 2 mm long. **Nutlets** 4, lanceolate, smooth, ca. 2 mm long with sharp, often winged margins, brown-mottled; scar long, closed, widened at the base. Sandy, gravelly, or clayey soil of sagebrush steppe, grasslands, open forest, woodlands; valleys, montane. WA to MT south to CA, NV, UT, CO. (p.407)

Cynoglossum L. Hound's tongue

Cynoglossum officinale L. Taprooted biennial or short-lived monocarpic perennial. **Stems** erect, 30–100 cm. **Herbage** villous. **Leaves** basal and cauline, petiolate, 7–25 cm long, becoming sessile above; blades oblanceolate to lanceolate. **Inflorescence** of axillary and terminal racemes; pedicels short, spreading to reflexed at maturity. **Flowers**: calyx deeply divided, sepals oblong, 5–7 mm long in fruit; corolla salverform, reddish-purple, 4–5 mm long, 6–9 mm across; stamens alternate with the fornices; style short, entire. **Nutlets** flattened-ovoid, 4–7 mm long, with barbed prickles, spreading at maturity; scar ovate. Disturbed ground of pastures, fields, roadsides, grasslands, meadows, woodlands, riparian thickets; plains, valleys. Introduced throughout temperate N. America; native to Europe.

Echium L. Viper's bugloss

Echium vulgare L. Blueweed Taprooted biennial. **Stems** erect, 20–80 cm. **Herbage** strigose, hispid with broad-based trichomes. **Leaves** basal and cauline, petiolate, 6–20 cm long, becoming sessile above; blades linear to linear-oblanceolate. **Inflorescence** an elongate, bracteate raceme of helicoid cymes. **Flowers** showy; calyx hispid, deeply divided, 5–8 mm long; corolla campanulate, blue, sometimes reddish-tinged, 15–20 mm long with 5 unequal lobes; stamens unequal, 4 exserted; style deeply 2-lobed, hairy. **Nutlets** 2–3 mm long, bullet-shaped, rugose; scar basal, surrounded by a rim. Roadsides, fields, vacant lots; valleys. Introduced to north-temperate N. America; native to Europe.

Eritrichium Schrad. ex Gaudin Dwarf forget-me-not

Small, cushion-forming perennials from a highly branched caudex. **Leaves** crowded, basal and cauline, linear to linear-oblanceolate. **Inflorescence** compact, terminal, flat-topped, ebracteate to bracteate cymes. **Flowers**: calyx divided to the base; corolla blue, salverform, fornices yellow; stamens alternate with fornices and below them; style short, included, undivided. **Nutlets** 1–4, ovate to obovate with a flattened, smooth or fringed dorsal face; scar basal.

Our two species are similar in appearance and occupy similar habitats but at different elevations.

1. Calyx 3–4 mm long; herbage silvery-strigose hiding leaf surface; valleys to montane . . . .*E. howardii*
1. Calyx 2–2.5 mm long; herbage long-sericeous, green of leaf surface apparent; alpine . . . . *E. nanum*

Eritrichium howardii (A.Gray) Rydb. **Stems** erect to ascending, 1–9 cm. **Herbage** densely silvery-strigose. **Leaves** 5–20 mm long, acute. **Flowers**: calyx 3–4 mm long; corolla 4–5 mm long, the limb 5–9 mm wide. **Nutlets** ca. 2 mm long, dorsally smooth or minutely muricate. Gravelly, calcareous soil of exposed ridges and flats in grasslands, woodlands; valleys, montane. Endemic to MT, adjacent WY.

Eritrichium nanum (Vill.) Schrad. ex Gaudin **Stems** erect to ascending, 1–10 cm. **Herbage** sericeous. **Leaves** 4–10 mm long, rounded at the tip; hairs denser toward the tip. **Flowers**: calyx 2–2.5 mm long; corolla 2–4 mm long, the limb 4–8 mm wide. **Nutlets** ca. 2 mm long, smooth. Fellfields, rock outcrops; alpine. Circumpolar south to OR, UT, NM. Our plants are var. *elongatum* (Rydb.) Cronquist.

Hackelia Opiz Stickseed

Annual to perennial herbs. **Leaves**: the basal petiolate; cauline narrower and becoming sessile. **Inflorescence** ebracteate or basally bracteate, axillary or terminal racemes that expand with age; pedicels reflexed in fruit. **Flowers**: calyx hispid, divided to the base; corolla salverform, white or blue, fornices apparent; stamens included in the tube. **Nutlets** with a central, mostly ovate scar on the ventral surface; margins and sometimes dorsal ridge with hooked prickles.

1. Plants annual, biennial or short-lived monocarpic perennials, without a branched caudex or
   persistent basal rosette of leaves. 2
1. Plants long-lived perennials with a branched caudex and rosettes of green basal leaves 3

2. Basal leaves oblong; corolla limb 3–4 mm across; nutlet body 2–3 mm long . . . . *H. deflexa*
2. Basal leaves narrowly oblanceolate; corolla limb 4–7 mm; nutlet body 3–4 mm long . . . . *H. floribunda*

3. Flowers white; nutlets cup-shaped, the marginal prickles united below and forming a
curved-in margin. . . . . *H. cinerea*
3. Flowers blue or blue-tinged; marginal prickles distinct not forming a curved-in border 4

4. Corolla blue with a whitish eye; nutlet body 3–5 mm long . . . . *H. micrantha*
4. Corolla white, tinged with blue; nutlet body 2–3 mm long . . . . *H. patens*

Hackelia cinerea (Piper) I.M.Johnst. Perennial from a simple or branched caudex. **Stems**
ascending to erect, 15–70 cm. **Herbage**: stems strigose; leaves hirsute and hispid-ciliate. **Leaves**:
basal oblong, 5–20 cm long; cauline lanceolate to elliptic. **Flowers**: calyx 2–4 mm long; corolla
white, 3–4 mm long, 6–11 mm across the limb with yellow, pubescent fornices. **Nutlet** body 3–5
mm long; prickles united at the base and forming a curled-in margin; dorsal face with smaller prickles.
Vernally moist cliffs, talus slopes in grasslands, open forest; valleys, montane. Eastern WA to western MT. See *H. patens*.

Hackelia deflexa (Wahlenb.) Opiz Taprooted annual. **Stems** simple, erect, 15–75 cm. **Herbage**
strigose, hirsute on the stems. **Leaves** mostly cauline, 2–15 cm long, oblong below, lanceolate to
elliptic above. **Flowers**: calyx 1–2 mm long; corolla blue, ca. 2 mm long, 2–3 mm across the limb.
Nutlet body 2–3 mm long; marginal prickles distinct; dorsal face mostly unarmed. Thickets, riparian
woodlands, open forest; plains, valleys. Circumboreal south to WA, ID, CO, IA, VT. Our plants are
var. ***americana*** (A.Gray) Fernald & I.M.Johnst.
    Hackelia floribunda has narrower leaves with a distinct basal rosette.

Hackelia floribunda (Lehm.) I.M.Johnst. Biennial or monocarpic, short-lived perennial from
a simple caudex. **Stems** erect, usually simple, 30–120 cm. **Herbage** strigose to hirsute. **Leaves**
narrowly oblanceolate, 4–20 cm long, the basal withered at flowering. **Flowers**: calyx 1–2 mm
long; corolla blue, 1.5–2 mm long, 4–7 mm across the limb with yellow, papilose fornices. **Nutlet**
body 3–4 mm long; marginal prickles distinct; dorsal face mostly unarmed, papilose to hirsute.
Meadows, thickets, open forest, often along streams; valleys, montane. BC to SK south to CA, AZ, NM, SD.
    Hackelia micrantha has barbed nutlet faces and persistent basal rosettes; see *H. deflexa*.

Hackelia micrantha (Eastw.) J.L.Gentry [*H. jessicae* (E.A.McGregor) Brand] Perennial,
usually from a branched caudex, otherwise similar to *H. floribunda*. **Stems** erect, usually simple,
30–80 cm. **Herbage** hirsute, strigose. **Leaves** 4–25 cm long. **Flowers**: calyx 1–1.5 mm long;
corolla ca. 2 mm long, 6–9 mm across the limb, blue with yellow or white fornices. **Nutlet** body
3–5 mm long; marginal prickles distinct; dorsal face with 4 to 8 smaller prickles. Meadows, moist
grasslands, sagebrush steppe; montane, subalpine. BC, AB south to CA, UT, CO. (p.401)

Hackelia patens (Nutt.) I.M.Johnst. Perennial from a simple or branched caudex. **Stems** erect,
15–50 cm. **Herbage** strigose. **Leaves**: basal, linear-oblanceolate to oblanceolate, 2–15 cm long.
Flowers: calyx 2–4 mm long; corolla bluish-white, 3–4 mm long, 5–9 mm across the limb with
yellow, pubescent fornices. **Nutlet** body 2–3 mm long; prickles distinct; dorsal face hirsute with few
minute prickles. Sandy or gravelly soil of slopes in sagebrush steppe, open forest; montane, subalpine.
ID, MT south to NV, UT, WY. Our plants are var. ***patens***.

Heliotropium L. Heliotrope

Heliotropium curassavicum L. Glabrous and glaucous, rhizomatous perennial. **Stems** 10–80
cm, prostrate to ascending. **Leaves** cauline, oblanceolate to spatulate, short-petiolate, 1–6 cm
long, the lowest reduced to scales. **Inflorescence** a branched, terminal, helicoid spike. **Flowers**:
calyx deeply divided, 3–4 mm long; corolla white, salverform, 5–6 mm long, 5–9 mm across the
limb; stamens included; stigma disc-like. **Nutlets** broadly ovate, 1–3 mm long, smooth; scar ovate in
the center of the ventral face. Margins of saline ponds, lakes; plains, valleys. Temperate N. America to S. America.

Lappula Moench Stickseed

Taprooted annuals. **Herbage** strigose to short-hirsute. **Leaves**: basal petiolate, oblanceolate; cauline more
linear, becoming sessile above. **Inflorescence** bracteate, terminal racemes that expand with age. **Flowers**:
calyx divided to the base; corolla funnelform, fornices apparent; stamens included. **Nutlets** with marginal
barbed prickles, muricate to rugose on the dorsal face; scar ovate at the base of a ventral keel.

1. Marginal prickles of nutlets in 1 row, well separated . . . . *L. redowskii*
1. Marginal prickles in 2 to 3 rows, close together and often touching. . . . .*L. squarrosa*

Lappula redowskii (Hornem.) Greene **Stems** simple or branched, 5–40 cm. **Leaves** 1–6 cm long. **Flowers**: calyx 1–2 mm long; corolla blue or white, 3–4 mm long, 2–3 mm across the limb. **Nutlet** body ca. 2 mm long; marginal prickles in 1 row, well-separated. Sparsely vegetated, often disturbed soil of grasslands, sagebrush steppe, woodlands, open forest, fields, roadsides; plains, valleys, montane. AK to MB south to CA, TX, MO.

1. Marginal prickles distinct to the base or nearly so . . . . .var. *redowskii*
1. Marginal prickles united basally to form a cup-like rim around the nutlet. . . . . var. *cupulata*

Lappula redowskii var. *redowskii* and *L. redowskii* var. *cupulata* (A.Gray) M.E.Jones have little ecological or geographical integtrity in MT; supposedly both native and introduced forms occur in N. America.[102]

Lappula squarrosa (Retz.) Dumort. [*L. echinata* Gilib.] Similar to *L. redowskii*. **Stems** 5–40 cm. **Leaves** 1–7 cm long. **Flowers**: calyx 2–3 mm long; corolla blue, 3–4 mm long, 1–3 mm across the limb. **Nutlet** body 2–3 mm long; marginal prickles in 2 to 3 rows of different lengths, distinct at the base. Sparsely vegetated, often disturbed soil of grasslands, sagebrush steppe, woodlands, streambanks, roadsides; plains, valleys. Introduced in much of temperate N. America; native to Eurasia, possibly native in the Rocky Mountain region.[102]

Lithospermum L. Stoneseed, puccoon, gromwell

Annual or perennial herbs. **Herbage** strigose to hirsute. **Leaves** cauline, linear-lanceolate, mostly sessile. **Inflorescence** axillary and terminal, bracteate racemes; pedicels short or absent. **Flowers**: calyx deeply divided; corolla funnelform to salverform, white to yellow, fornices present or absent; anthers included; style unlobed, included to barely exserted. **Nutlets** gray, ovate, smooth to rugose; scar basal, often rimmed; ventral surface keeled.

1. Flowers white; plants annual; corolla limb 2–4 mm across . . . . *L. arvense*
1. Flowers yellow; plants perennial; corolla limb ≥7 mm across. 2

2. Corolla ≥15 mm long, fornices present; herbage strigose . . . . *L. incisum*
2. Corolla <8 mm long, fornices absent; herbage hirsute. . . . . *L. ruderale*

Lithospermum arvense L. Taprooted annual. **Stems** erect, 10–70 cm, usually branched. **Herbage** strigose. **Leaves** linear-oblong, 2–5 cm long, the lowest withered at flowering. **Inflorescence** becoming open; pedicels short. **Flowers**: calyx 4–5 mm long; corolla white, sometimes bluish, 5–8 mm long with longitudinal lines of hair within the tube, 3–4 mm across the limb, fornices absent; style included. **Nutlets** 2.5–4 mm long, rugose; scar unrimmed. Disturbed soil of grasslands, fields, roadsides; valleys. Introduced throughout most of temperate N. America; native to Eurasia.

Lithospermum incisum Lehm. Perennial from a branched caudex. **Stems** prostrate to erect, 5–25 cm, branched. **Herbage** strigose. **Leaves** linear to linear-lanceolate, 1–5 cm long. **Inflorescence** leafy-bracteate, few-flowered racemes. **Flowers**: calyx 6–9 mm long; corolla yellow, 2–4 cm long, 7–12 mm across the limb, lobes erose, fornices present; style slightly exserted. **Nutlets** 3–4 mm long, shiny; basal scar rimmed. Often sandy soil of grasslands, sagebrush steppe; plains, valleys. BC to ON south to UT, TX, MO, IN, Mexico.

Flowers produced late in the season have corollas smaller than the calyx, short styles, and are presumably cleistogamous. One Lincoln Co. population is completely cleistogamous.

Lithospermum ruderale Douglas ex Lehm. Perennial from a branched caudex. **Stems** ascending to erect, 12–50 cm, branched. **Herbage** strigose; stems hirsute. **Leaves** linear to linear-lanceolate, 2–10 cm long, the lowest reduced. **Inflorescence** leafy-bracteate, few-flowered racemes. **Flowers**: calyx 3–5 mm long; corolla light yellow, 4–9 mm long, short-glandular at the summit of the tube, 5–12 mm across the limb, fornices absent; style included. **Nutlets** 4–6 mm long, shiny, smooth; basal scar slightly rimmed. Grasslands, open forest; valleys, montane. BC to SK south to CA, CO. (p.416)

Lowest, earliest flowers are often larger than those that bloom later in the same raceme.

Mertensia Roth Bluebells

Perennial herbs. **Stems** usually simple. **Herbage** glabrous or strigose. **Leaves** cauline, with or without a basal rosette, petiolate below, sessile above. **Inflorescence** ebracteate, terminal or sometimes axillary cymes. **Flowers**: calyx lobed ≥1/2-way to the base; corolla blue, usually funnelform, much longer than calyx, the limb campanulate, fornices usually evident; anthers exserted from the tube but not the limb; style unlobed, included or exserted. **Nutlets** ovate, rugose.[445]

   Bluebells are divided into "tall" and "short" species; groups that are usually, but not always, distinct. *Mertensia oblongifolia*, *M. lanceolata* and *M. viridis*, three of the short species, are very similar and intergrade throughout their range.[187] *Mertensia viridis* has a line of hair inside the corolla tube, while *M. oblongifolia* does not; otherwise these two species seem morphologically and ecologically indistinguishable in our area.

1. Well-developed plants ≥40 cm tall; lower cauline leaves with evident lateral veins .2
1. Plants <40 cm tall; lower cauline leaves without evident lateral veins 3

2. Sepals ciliate, rounded at the tip; basal leaves tapering to the petiole. . . . . *M. ciliata*
2. Sepals strigose, acute; basal leaves rounded or truncate at the base. . . . . *M. paniculata*

3. Plants from a corm or tuberous root; without basal leaves. 4
3. Plants with woody, fibrous roots; basal leaves usually present 5

4. Flowers campanulate, without a distinct tube and limb . . . . *M. bella*
4. Flowers funnel-form divided into a distinct tube and limb. . . . . *M. longiflora*

5. Filaments shorter than the anthers; plants subalpine or alpine 6
5. Filaments equal to or longer than the anthers; plants usually of valleys to lower subalpine 7

6. Anther tips just reaching the fornices; corolla tube glabrous within . . . . *M. alpina*
6. Anthers elevated above the fornices; corolla hairy within. . . . . *M. perplexa*

7. Corolla tube hairy toward the base within. . . . . *M. viridis*
7. Corolla glabrous within 8

8. Corolla tube >1.2 times as long as the limb . . . . *M. oblongifolia*
8. Corolla tube shorter or equal to the limb . . . . *M. lanceolata*

Mertensia alpina (Torr.) G.Don Plants with a branched caudex. **Stems** prostrate to ascending, 5–20 cm. **Herbage** glabrous, strigose on upper leaf surfaces. **Leaves** basal and cauline; blades lanceolate to elliptic, 1–5 cm long; lateral veins obscure. **Flowers**: calyx ciliate, 2–4 mm long, divided to the base; corolla 6–9 mm long, limb ca. as long as the tube; filaments short; anthers just reaching the fornices; style no higher than the anthers. **Nutlets** ca. 2 mm long. Moist meadows, turf; alpine. ID, MT south to NM, CO.
   Hybrids with *Mertensia ciliata* are reported for Carbon Co.[187]

Mertensia bella Piper Plants with a cormous root. **Stems** solitary, weak, ascending to erect, 10–20 cm. **Herbage** sparsely strigose. **Leaves** all cauline; blade ovate, 1–5 cm long; lateral veins apparent. **Flowers**: calyx strigose, 2–4 mm long, divided to the base; corolla 6–8 mm long, campanulate, limb and tube not differentiated, fornices absent; style higher than the anthers. **Nutlets** 1–2 mm long. Wet meadows, often on slopes; lower subalpine. Endemic to eastern OR, north ID, adjacent MT, disjunct in western OR. There is one location in Missoula Co. (p.413)

Mertensia ciliata (James ex Torr.) G.Don Plants with a branched caudex. **Stems** ascending to erect, 20–90 cm. **Herbage** mostly glabrous. **Leaves** basal and cauline; blades lanceolate to ovate, 3–12 cm long; lateral veins obvious. **Flowers**: calyx 1–3 mm long, divided to the base, lobes rounded at the tip, ciliate; corolla 10–15 mm long, tube with or without hairs inside, 1–1.2 times as long as the limb; filaments inserted at the fornices, ca. as long as the anthers; style higher than the anthers. **Nutlets** 2–3 mm long. Wet meadows, thickets, moist open forest, talus, often along streams; montane, lower alpine. OR to MT south to CA, NM, CO. Our plants are var. *ciliata*. (p.413)
   Putative hybrids with *Mertensia viridis* occur sporadically. See. *M. paniculata*.

Mertensia lanceolata (Pursh) DC. Plants with a branched caudex, similar to *M. oblongifolia*. **Stems** lax to ascending, 10–30 cm. **Herbage** glabrous to minutely strigose on leaf margins. **Leaves** basal and cauline; blades oblanceolate, lanceolate, or elliptic, 2–7 cm long; lateral veins of cauline leaves obscure. **Flowers**: calyx strigose, ciliate, 3–6 mm long, divided 2/3-way to the base;

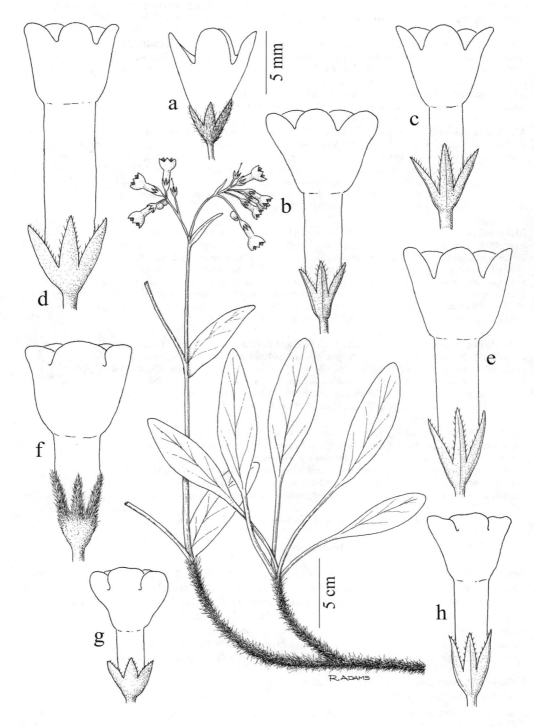

Plate 72. *Mertensia*. **a.** *M. bella*, **b.** *M. ciliata*, **c.** *M. lanceolata*, **d.** *M. longifolia*,
e. *M. oblongifolia*, **f.** *M. paniculata*, **g.** *M. perplexa*, **h.** *M. viridis*

corolla 9–14 mm long, tube glabrous within, equal to or shorter than the limb; filaments inserted at the fornices; style barely longer than the anthers. **Nutlets** ca. 3 mm long. Meadows, woodlands; plains. SK south to NM, NE. (p.413)

Mertensia longiflora Greene Plants with a tuberous root. **Stems** solitary, weak, ascending to erect, 5–15 cm. Herbage glabrous to spiculate. **Leaves** all cauline; blade oblong to obovate, 1–4 cm long without lateral veins. **Flowers**: calyx 4–6 mm long, divided to the base, glabrous to ciliate; corolla 14–20 mm long, tube glabrous within, 2 to 3 times as long as the limb; filaments attached at the fornices; style higher than the anthers. **Nutlets** ca. 2 mm long. Vernally, often shallow soil of grasslands, cliffs, open forests; valleys, montane. BC, MT south to CA, OR, ID. (p.413)

Mertensia oblongifolia (Nutt.) G.Don Plants with a branched caudex. **Stems** ascending to erect, 5–30 cm. **Herbage** glabrous to strigose, upper leaves often ciliate. **Leaves** basal and cauline; blades lanceolate to oblanceolate, 1–8 cm long; lateral veins obscure. **Flowers**: calyx 4–7 mm long, divided nearly to the base; corolla 13–17 mm long, tube glabrous within, 1.2 to 2 times as long as the limb; filaments ca. as long as the anthers, inserted at the fornices; style barely longer than the anthers. **Nutlets** 2–3 mm long. Grasslands, meadows, outcrops, woodlands, open forest; valleys to alpine. WA to MT south to NV, UT, WY. (p.413)

Mertensia paniculata (Aiton) G.Don Plants with a branched caudex. **Stems** ascending to erect, 25–100 cm. **Herbage** sparsely strigose to minutely puberulent. **Leaves** basal and cauline; blades lanceolate to ovate, acute to acuminate, 3–14 cm long; the basal subtruncate; lateral veins obvious. **Flowers**: calyx strigose, 2–6 mm long, divided to below the middle with acute tips; corolla 9–15 mm long, limb 1.2 to 1.6 times as long as the tube, tube with or without hairs within; filaments inserted at the fornices, ca. as long as the anthers; style higher than the anthers, sometimes slightly exserted. **Nutlets** 3–4 mm long. Wet forests, thickets, usually along streams; valleys to montane. AK to QC south to OR, ID, MT. (p.413)

The more common *Mertensia ciliata* has more rounded, less hairy sepals.

Mertensia perplexa Rydb. Plants with a branched caudex, similar to *M. alpina*. **Stems** lax to ascending, 5–20 cm. **Herbage** glabrous to strigose. **Leaves** basal and cauline; blades lanceolate to linear-oblanceolate, 15–40 mm long; lateral veins obscure. **Flowers**: calyx 2–3 mm long, divided to the base; corolla 7–11 mm long, tube with a basal ring of hairs, limb ca. as long as the tube; filaments shorter than the anthers, inserted at the fornices; style no higher than the anthers. **Nutlets** 1–2 mm long. Meadows, turf, talus slopes; alpine. MT to CO. (p.413)

Mertensia viridis (A.Nelson) A.Nelson Plants with a branched caudex. **Stems** lax to ascending, 5–30 cm. **Herbage** glabrous to strigose. **Leaves** basal and cauline; blades oblanceolate to narrowly ovate, 2–7 cm long; lateral veins obscure. **Flowers**: calyx 4–6 mm long, sparsely ciliate, divided nearly to the base; corolla 8–12 mm long, tube pubescent within, 1 to 1.2 times as long as the limb; filaments ca. as long as the anthers, inserted at the fornices; style barely longer than the anthers. **Nutlets** ca. 3 mm long. Grasslands, sagebrush steppe, meadows, talus; valleys to subalpine. OR to MT south to CA, NV, NM, CO. (p.413)

Northern populations are glabrous; those farther south are strigose. See *Mertensia ciliata*.

Myosotis L. Forget-me-not, scorpion grass

Fibrous-rooted, annual to perennial herbs. **Herbage** glabrous to hirsute. **Leaves** cauline and sometimes basal, oblanceolate, petiolate below, becoming sessile and more lanceolate above. **Inflorescence** terminal and axillary, helicoid racemes, becoming elongate at maturity. **Flowers**: calyx divided ≤1/2-way to the base; corolla salverform, blue (white), often with a yellow or white center, tube ca. as long as the calyx, fornices apparent; stamens included; style unlobed. **Nutlets** ovoid, smooth and shiny; scar basal.

The *Myosotis sylvatica* complex is thought to contain one to several species. I have taken an intermediate position by recognizing a low-elevation and an arctic-alpine species.

1. Calyx with forward-pointing, appressed hair only. 2
1. Calyx with some spreading, hooked hairs 3

2. Corolla 2–5 mm across spreading lobes . . . . *M. laxa*
2. Corolla 5–10 mm across. . . . . *M. scorpioides*

3. Plants perennial; corolla 4–8 mm across 4
3. Plants annual; corolla 2–5 mm across 5

4. Plants short-lived perennials; stems often branched below; low-elevations. . . . . *M. sylvatica*
4. Perennial sometimes with a branched caudex but not branched below;
   high-elevations . . . . *M. alpestris*

5. Corolla white; 2 sepals longer than the others . . . . *M. verna*
5. Corolla blue; calyx symmetrical 6

6. Fruiting pedicels longer than the calyx . . . . *M. arvensis*
6. Fruiting pedicels shorter than the calyx . . . . *M. micrantha*

Myosotis alpestris F.W.Schmidt [*M. sylvatica* Ehrh. ex Hoffm. var. *alpestris* (F.W.Schmidt) Koch, *M. asiatica* (Vesterg.) Schischkin & Sergievskaja] ALPINE FORGET-ME-NOT Perennial, sometimes with a branched caudex. **Stems** erect, simple, 5–25 cm. **Herbage** sparsely hirsute. **Leaves** basal and cauline, 1–7 cm long. **Flowers**: calyx 3–5 mm long in fruit, uncinate-hirsute; corolla blue, limb 3–9 mm across. **Nutlets** black, 1.5–2.5 mm long. Meadows, grasslands, turf, fellfields; montane to alpine. AK to BC, ID, WY, SD; Asia. (p.416)
    Small plants are sometimes mistaken for *Eritrichium*.

Myosotis arvensis (L.) Hill Annual. **Stems** simple or branched, 10–40 cm. **Herbage** strigose, hirsute. **Leaves** mainly cauline, 1–5 cm long. **Flowers**: calyx 3–5 mm long in fruit, uncinate-hispid; corolla blue, limb 2–4 mm across. **Nutlets** brown, 1–2 mm long. Disturbed soil along streams, ditches, gardens; valleys. Introduced to northern U.S. and adjacent Canada; native to Europe.

Myosotis laxa Lehm. Annual to short-lived perennial. **Stems** lax, ascending, 7–60 cm, rarely rooting at the nodes. **Herbage** strigose, sometimes sparsely so. **Leaves** mainly cauline, 1–8 cm long. **Flowers**: calyx 3–5 mm long in fruit, strigose; corolla blue, limb 2–5 mm across. **Nutlets** black, ca. 2 mm long. Mud or shallow water along rivers, streams, lakes; valleys. Circumboreal south through much of north-temperate U.S.

Myosotis micrantha Pall. ex Lehm. BLUE SCORPION GRASS Annual. **Stems** often branched at the base, 3–20 cm. **Herbage** hirsute to strigose. **Leaves** mainly cauline, 5–30 mm long. **Flowers**: calyx ca. 4 mm long in fruit, uncinate-hispid; corolla blue, limb 1–2 mm across. **Nutlets** brown, ca. 1 mm long. Shallow or disturbed soil of grasslands, rock outcrops, roadsides and margins of wetlands, streams; plains, valleys. Introduced to much of north-temperate N. America; native to Eurasia.

Myosotis scorpioides L. SCORPION GRASS Annual to short-lived perennial. **Stems** lax, prostrate to ascending, 10–40 cm, often rooting at the nodes. **Herbage** sparsely strigose. **Leaves** mainly cauline, 1–8 cm long. **Flowers**: calyx 3–4 mm long in fruit, strigose; corolla blue, limb 5–10 mm across. **Nutlets** black, ca. 2 mm long. Margins of slow streams, sloughs, ponds or disturbed soil of wet forests; valleys. Introduced to much of temperate N. America; native to Europe.

Myosotis sylvatica Hoffm. GARDEN FORGET-ME-NOT Short-lived perennial sometimes with a branched caudex. **Stems** erect or ascending, simple, 5–40 cm. **Herbage** sparsely hirsute. **Leaves** basal and cauline; blades 2–10 cm long. **Flowers**: calyx 3–6 mm long in fruit, uncinate-hirsute; corolla blue, limb 4–8 mm across. **Nutlets** black, 1–2 mm long. Moist disturbed forest; valleys. A garden escape, introduced throughout much of northern N. America; native to Europe. Collected once in Flathead Co.

Myosotis verna Nutt. Annual. **Stems** often branched above, 5–15 cm. **Herbage** hirsute. **Leaves** mainly cauline, 1–3 cm long. **Flowers**: calyx 4–6 mm long in fruit, asymmetrical, uncinate-hispid; corolla white, limb 1–2 mm wide. **Nutlets** brown, white-speckled, 2–3 mm long. Moist to wet soil of meadows, streambanks; valleys. Indigenous to eastern U.S. and adjacent Canada, but probably introduced to the Pacific Northwest. Collected in Lake and Flathead cos.

Onosmodium Michx.

Onosmodium molle Michx. Perennial from a woody, branched caudex. **Stems** ascending to erect, simple, 30–70 cm. **Herbage** densely hirsute, hispid. **Leaves** cauline, sessile, lanceolate, 3–8 cm long. **Inflorescence** branched, bracteate, terminal, helicoid cymes. **Flowers**: calyx hispid, deeply divided into unequal segments, 6–12 mm long; corolla villous, white with green lobes, tubular-funnelform, 12–14 mm long, lobes 2–3 mm long, fornices absent; anthers at the top of the tube; style exserted; anthers and style maturing before the corolla opens. **Nutlets** 1–4, ovate, smooth, 3–5 mm long;

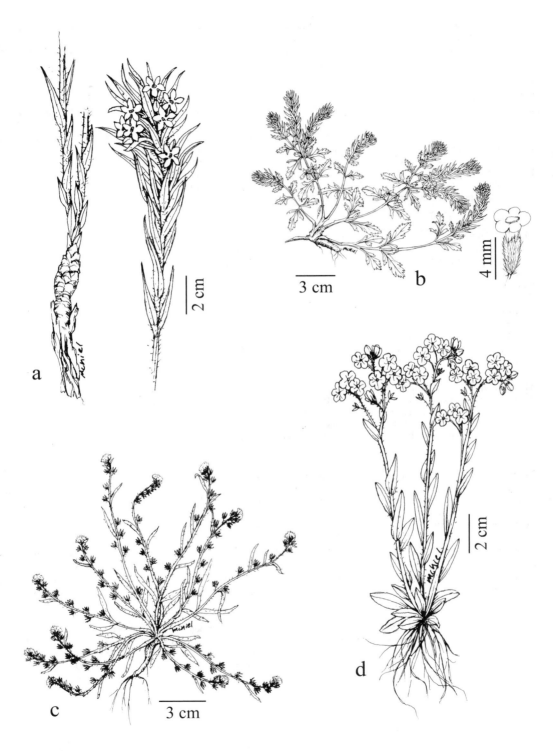

Plate 73 a. *Lithospermum ruderale*, b. *Verbena bracteata*, c. *Plagiobothrys scouleri*, d. *Myosotis alpestris*

scar basal without a rim. Grasslands, woodlands; plains, valleys to montane. AB to MB, NY south to TX, MO, VA. Our plants are var. *occidentale* (Mack.) I.M.Johnst.

Plagiobothrys Fisch. & C.A.Mey. Popcorn flower

Annual or perennial herbs. **Stems** usually branched at the base. **Leaves** mainly cauline, sessile, linear to linear-oblanceolate, opposite below, alternate above. **Inflorescence** terminal, bracteate or ebracteate, helicoid racemes. **Flowers**: calyx hispid, divided to below the middle; corolla white, funnelform, limb shorter than the tube, the tube shorter than the calyx; stamens and style included. **Nutlets** 1 to 4, lanceolate to narrowly ovate, rugose and tuberculate; scar at the base of the ventral keel.

All but one of our species have each been collected only once. In most cases they were associated with roads or man-made reservoirs, suggesting these disjunct populations may have been introduced by human activity; *P. scouleri* is our only common species.

1. Corolla limb 5–10 mm across; plants perennial; stoloniferous; herbage villous . . . . *P. mollis*
1. Corolla limb 2–4 mm wide; annuals with strigose to hirsute herbage. 2

2. Herbage hirsute with spreading hairs. . . . .*P. salsus*
2. Lower leaves and stems strigose with appressed hairs 3

3. Fruiting calyx lobes not elongate, 2–4 mm long . . . . *P. scouleri*
3. Fruiting calyx lobes becoming elongate, 4–7 mm long. 4

4. Stems prostrate; calyx lobes becoming curved to one side . . . .*P. leptocladus*
4. Stems ascending to erect; calyx lobes symmetrical . . . .*P. stipitatus*

Plagiobothrys leptocladus (Greene) I.M.Johnst. Annual. **Stems** prostrate, 5–30 cm. **Herbage** sparsely strigose. **Leaves** 1–3 cm long. **Inflorescence** bracteate. **Flowers** calyx 4–7 mm long in fruit, the lobes curved to one side; corolla 2–3 mm long, limb 1–3 mm across. **Nutlets** ovate, 2–3 mm long with bristle-tipped hairs. Drying mud around wetlands; plains. OR to SK south to CA, NV, SD, Mexico. Known from Phillips Co.

Plagiobothrys mollis (A.Gray) I.M.Johnst. Stoloniferous perennial. **Stems** prostrate to ascending, 4–20 cm. **Herbage** villous. **Leaves** 2–6 cm long. **Inflorescence** ebracteate. **Flowers**: calyx 3–5 mm long, symmetrical; corolla ca. 3 mm long, limb 5–10 mm across. **Nutlets** ovate, ca. 2 mm long, glabrous. Drying mud or margins of lakes, ponds; valleys. OR to MT south to CA, NV. Known from Sanders Co.

Plagiobothrys salsus (Brandegee) I.M.Johnst. Annual. **Stems** prostrate, 5–30 cm. **Herbage** sparsely hirsute. **Leaves** 1–2 cm long. **Inflorescence** leafy-bracteate. **Flowers**: calyx 4–5 mm long in fruit, symmetrical; corolla ca. 2 mm long, limb 2–3 mm across. **Nutlets** lanceolate, ca. 1.5 mm long, glabrous. Drying mud of shallow wetlands; montane. NV and adjacent OR, disjunct in MT. Known from Beaverhead Co.

Plagiobothrys scouleri (Hook. & Arn.) I.M.Johnst. Annual. **Stems** prostrate to ascending, 3–25 cm. **Herbage** strigose below; hispid in the inflorescence. **Leaves** 1–6 cm long. **Inflorescence** bracteate below. **Flowers**: calyx 2.5–4 mm long, symmetrical; corolla 1.5–2.5 mm long, limb 2–3 mm across. **Nutlets** lanceolate, ca. 1.5 mm long, mostly glabrous. Drying mud around wetlands, streambanks, roadsides; plains, valleys, montane. AK to MB south to CA, NM, MN. (p.416)

Plagiobothrys stipitatus (Greene) I.M.Johnst. Annual. **Stems** ascending to erect, 5–25 cm. **Herbage** strigose below, becoming hispid in the inflorescence. **Leaves** 1–5 cm long. **Inflorescence** minutely bracteate. **Flowers**: calyx ca. 5 mm long in fruit, symmetrical; corolla ca. 2 mm long, limb 2–4 mm across. **Nutlets** lanceolate, ca. 2 mm long, glabrous. Roadsides and margins of lakes, ponds; valleys. OR to MT south to CA, NV. Our plants are var. *micranthus* (Piper) I.M.Johnst. Known from Lake Co.

Symphytum L. Comfrey

Symphytum officinale L. Taprooted perennial. **Stems** erect, 30–100 cm. **Herbage** hirsute, hispid. **Leaves** basal and cauline, petiolate; blades narrowly ovate to lanceolate, 5–25 cm long, becoming smaller and sessile above. **Inflorescence** small, axillary and terminal, ebracteate cymes. **Flowers**: calyx ca. 5 mm long, divided >1/2-way; corolla yellowish, funnelform, 1–2 cm long, tube longer than limb; stamens inserted at the fornices; style exserted. **Nutlets** ovoid, ca. 4 mm long, smooth, black; scar below the ventral keel. Fields, lawns, disturbed riparian areas; valleys. Introduced for medicinal purposes to much of N. America; native to Europe.

VERBENACEAE: Vervain Family

Verbena L. Vervain

Annual or perennial herbs. **Leaves** all cauline, opposite, exstipulate, short-petiolate, simple, toothed to pinnately dissected. **Inflorescence** terminal spikes, each flower subtended by a lance-linear bract. **Flowers** perfect, slightly irregular; calyx divided into 5 unequal lobes; corolla funnelform, the limb unequally 5-lobed; stamens 4, inserted in the tube; ovary superior, 4-chambered, sometimes 4-lobed; style 1, 2-lobed but only 1 lobe stigmatic. **Fruit** 4 nutlets enclosed in the calyx.

 Verbena bipinnatifida Nutt. [*Glandularia bipinnatifida* (Nutt.) Nutt.] is reported for northeast MT,[113] but I have seen no specimens, and it is not reported for MT by other sources.[402]

1. Calyx 6–10 mm long. . . . . *V. bipinnatifida*
1. Calyx 2–6 mm long. 2

2. Plants branched at the base, the branches prostrate to ascending . . . . *V. bracteata*
2. Stems erect, not branched at the base. 3

3. Calyx 2–4 mm long; herbage strigose . . . .*V. hastata*
3. Calyx ≥4 mm long; herbage hirsute . . . .*V. stricta*

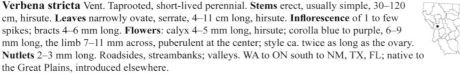

Verbena bracteata Lag. & Rodr. Taprooted annual or mostly short-lived perennial. **Stems** prostrate to ascending, branched at the base, 6–50 cm, sparsely hirsute. **Leaves** ovate in outline, deeply lobed, sharply dentate, strigose to hirsute, 1–4 cm long with winged petioles. **Inflorescence** spike-like with spreading bracts 5–10 mm long. **Flowers**: calyx 3–5 mm long, hispid; corolla blue, slightly 2-lipped, 4–6 mm long, the limb 2–4 mm across; style ≤1 mm long. **Nutlets** ca. 2 mm long. Disturbed soil of roadsides, margins of shallow wetlands; plains, valleys. Throughout temperate N. America. (p.416)

Verbena hastata L. Fibrous-rooted perennial. **Stems** erect, usually simple, 20–100 cm, hirsute. **Leaves** lanceolate, serrate, hirsute, strigose, 3–12 cm long. **Inflorescence** a flat-topped panicle of narrow spikes; bracts 2–3 mm long. **Flowers**: calyx 2–4 mm long, strigose; corolla blue to violet, slightly irregular, 3–5 mm long, the limb 5–7 mm wide; style ca. 1 mm long. **Nutlets** ca. 2 mm long. Riparian meadows, thickets; valleys. Throughout most of temperate N. America.

Verbena stricta Vent. Taprooted, short-lived perennial. **Stems** erect, usually simple, 30–120 cm, hirsute. **Leaves** narrowly ovate, serrate, 4–11 cm long, hirsute. **Inflorescence** of 1 to few spikes; bracts 4–6 mm long. **Flowers**: calyx 4–5 mm long, hirsute; corolla blue to purple, 6–9 mm long, the limb 7–11 mm across, puberulent at the center; style ca. twice as long as the ovary. **Nutlets** 2–3 mm long. Roadsides, streambanks; valleys. WA to ON south to NM, TX, FL; native to the Great Plains, introduced elsewhere.

LAMIACEAE (Labiatae): Mint Family

Annual or perennial herbs. **Stems** usually 4-angled. **Leaves** opposite, simple, mostly petiolate, exstipulate. **Herbage** often punctate, aromatic. **Inflorescence** spike-like, of several small cymes forming verticillasters at nodes of upper leaves or bracts. **Flowers** perfect; calyx campanulate to tubular, 5-lobed; corolla usually irregular, 5-lobed, often bilabiate and 3-lobed below, occasionally 4-lobed and nearly regular; stamens 2 or 4; ovary superior, 2-carpellate basally, 4-lobed apically; style usually bilobed. **Fruit** 4, 1-seeded nutlets, joined at the base.

 Origanum vulgare L. (oregano), with small, bilabiate, equal-lobed corollas, occasionally escapes gardens into lawns.

1. Corollas cup- or funnel-shaped, nearly regular 2
1. Corollas distinctly irregular with a distinct upper and/or lower lip. 4

2. Fertile stamens 2 . . . . *Lycopus*
2. Fertile stamens 4 3

Agastache Clayton ex Gronov. Giant hyssop

Perennial herbs from a branched, woody caudex. **Leaves** mostly cauline; blades serrate. **Inflorescence** terminal, spike-like, of crowded verticillasters. **Flowers**: calyx with ≥15 veins; corolla tubular, bilabiate, upper lobes short; stamens 4, exserted, upper pair longer than the lower; style shortly 2-lobed. **Nutlets** rounded.

Agastache urticifolia is our only common species.

2. Corollas blue; leaves canescent beneath. . . . .*A. foeniculum*
2. Corollas white or pinkish; leaves puberulent but not canescent beneath. . . . . *A. urticifolia*

Agastache cusickii (Greenm.) A.Heller **Stems** erect to lax, 5–20 cm, usually simple. **Herbage** puberulent, pleasantly aromatic. **Leaf blades** ovate, 10–15 mm long, crenate. **Inflorescence** 1–4 cm long. **Flowers:** calyx puberulent, glandular, 6–8 mm long, teeth narrowly lanceolate; corolla white, 8–10 mm long. Limestone talus slopes, often with mountain mahogany; montane, lower subalpine. NV, OR, ID, MT. Known from Beaverhead Co.
    Caudex branches occur beneath the shifting talus.

Agastache foeniculum Kuntze LAVENDER HYSSOP **Stems** erect, 50–120 cm, simple or few-branched, puberulent at the nodes. **Leaf blades** cordate-ovate, glabrous above, puberulent below, 4–9 cm long, acute, serrate. **Inflorescence** 3–8 cm long with purplish bracts. **Flowers:** calyx conical, puberulent, 4–7 mm long, teeth deltoid, purplish; corolla blue, 15–20 mm long. Ash or pine woodlands; plains. AB to ON south to MT, TX, KS, MO. Known from Carter and Richland cos.

Agastache urticifolia (Benth.) Kuntze **Stems** erect, 40–120 cm, simple or branched above. **Herbage** glabrous to puberulent, punctate, aromatic. **Leaf blades** cordate, ovate to deltoid, 3–10 cm long, crenate-serrate. **Inflorescence** 3–15 cm long with purplish bracts. **Flowers:** calyx puberulent, glandular, 9–12 mm long, teeth narrowly lanceolate, purplish, veins obvious; corolla white to pink, 10–16 mm long. Grasslands, shrub steppe, meadows, thickets, often near streams; valleys to lower subalpine. BC, MT south to CA, CO.

Ajuga L. Bugle

Ajuga reptans L. Fibrous-rooted perennial, forming loose mats of stolons. **Stems** erect, 10–30 cm, mostly simple, pubescent. **Leaves** petiolate, becoming sessile above; blades obovate, crenate, 2–10 cm long, glabrate. **Inflorescence** crowded; verticillasters in axils of upper leaves and bracts. **Flowers:** calyx villous, 4–5 mm long, the equal lobes just shorter than the tube; corolla blue, 9–15 mm long, the upper lip inconspicuous, middle lobe of lower lip large and bifid; stamens 4, exserted; style branches short, subequal. Roadsides; valleys in Flathead Co. An introduced ornamental, escaped sporadically throughout N. America; native to Europe.

Dracocephalum L. Dragon's head

Annuals or short-lived perennials. **Leaves** cauline, petiolate; blades serrate. **Inflorescence** terminal; verticillasters in axils of leaf-like bracts. **Flowers:** calyx with ≥15 veins, upper lobe wider than the others; corolla tubular, bilabiate, upper lobes united into a hood; stamens 4, included in the hood; style shortly 2-lobed. **Nutlets** smooth.

1. Verticillasters crowded; calyx hirsute . . . . *D. parviflorum*
1. Verticillasters forming an interrupted spike; calyx puberulent. . . . . *D. thymiflorum*

Dracocephalum parviflorum Nutt. [*Moldavica parviflora* (Nutt.) Britton] Usually biennial or short-lived, monocarpic perennial. **Stems** erect, 15–80 cm, simple or branched. **Herbage** puberulent. **Leaf blades** lanceolate to narrowly ovate, strongly serrate, spinescent, 25–60 mm long. **Inflorescence** crowded, 2–10 cm long; flowers partly hidden by ciliate, subulate-lobed bracts. **Flowers:** calyx sparsely hirsute, reticulate-veined, 10–14 mm long, teeth and tube ca. equal; corolla villous, purple, 2–3 mm longer than the calyx. Forests following fire and along streams, roads, trails; valleys, montane. AK to NL south to OR, NM, NE, WI, NY. (p.423)

Dracocephalum thymiflorum L. [*Moldavica thymiflora* (L.) Rydb.] Annual. **Stems** erect, 20–50 cm, simple or few-branched below, puberulent. **Leaf blades** lanceolate, serrate, 1–2 cm long, glabrous above, pubescent below. **Inflorescence** interrupted, 5–20 cm long. **Flowers:** calyx purple, pubescent, 5–8 mm long, teeth 1–2 mm long, awn-tipped, the upper wider; corolla blue, 6–9 mm long. Roadsides, streambanks; plains, valleys. Sparingly introduced from Eurasia to much of northern U.S.

Galeopsis L. Hemp nettle

Galeopsis tetrahit L. Annual. **Stems** erect, 15–70 cm, hispid. **Leaf blades** ovate, serrate, strigose, hirsute, glandular, 3–10 cm long. **Inflorescence** of verticillasters at upper leaf nodes. **Flowers**: calyx 8–12 mm long, lobes ca. equal, the tube, spine-tipped; corolla purple, spotted, 15–23 mm long, bilabiate, upper lip hood-like, lower lip with a pair of vertical projections; stamens 4. Disturbed soil of open, wet forest, woodlands, meadows, often along streams, trails; valleys. Introduced from Eurasia to much of northern U.S. and adjacent Canada.

Populations are often small. *Galeopsis bifida* Boenn [*G. tetrahit* var. *bifida* (Boenn.) Lej. & Courtois], with an emarginate lower corolla lobe, is also reported for MT, but I have not seen specimens that I can positively assign to this taxon.

Glechoma L. Ground ivy

Glechoma hederacea L. Gill over the ground Perennial herbs with stolons or shallow rhizomes. **Stems** creeping to ascending, 5–30 cm, glabrous to scabrous. **Leaves** cauline; blades ovate to reniform, cordate, hirsute above, crenate, 1–3 cm long. **Inflorescence** of several pedicellate flowers at leaf nodes. **Flowers** perfect or sometimes female; calyx with 15 veins, 5–7 mm long, scabrous, teeth subulate, the upper longer; corolla ca. 1 cm long, blue with purple spots, bilabiate, middle lobe of lower lip the largest; stamens 4, upper pair longer; style 2-lobed. Moist soil of lawns, streambanks; valleys. Introduced to most of temperate N. America; native to Eurasia.

Hedeoma Pers. Mock pennyroyal

Taprooted, annual or perennial, aromatic herbs. **Leaves** sessile, entire, linear to oblanceolate. **Inflorescence** of verticillasters in upper leaf axils. **Flowers**: calyx ca. 13-veined, slightly bilabiate, widened at the base; corolla tubular, bilabiate, upper lip shorter; stamens 2, barely exserted; style unequally 2-lobed. **Nutlets** glaucous.

1. Plants annual; calyx teeth spreading in fruit. . . . .*H. hispida*
1. Plants perennial; calyx teeth convergent in fruit. . . . .*H. drummondii*

Hedeoma drummondii Benth. Perennial with a woody base. **Stems** usually several, branched, ascending to erect, 4–25 cm, puberulent. **Leaves** 5–15 mm long, puberulent. **Verticillasters** with 2 to 5 flowers. **Flowers**: calyx hispid, glandular, 5–6 mm long, the teeth closing in fruit; corolla lavender, 7–11 mm long. Stony soil of outcrops in grasslands, sagebrush steppe, woodlands; plains, valleys, montane. MT to MN south to NM, TX, OK, Mexico.

Hedeoma hispida Pursh Annual. **Stems** erect, branched below, 4–20 cm, strigose. **Leaves** 10–15 mm long, ca. glabrous, sometimes cilliate. **Verticillasters** of 4 to 10 flowers. **Flowers**: calyx hispid, 4–6 mm long; corolla lavender, 4–5 mm long. Grasslands, sagebrush steppe; plains. AB to VT south to MT, TX, MS.

Plants described above are reported to be cleistogamous.[152] Chasmogamous flowers have corollas 6–7 mm long, but I have seen no such plants in our material.

Hyssopus L. Hyssop

Hyssopus officinalis L. Rhizomatous perennial. **Stems** erect, 30–60 cm, puberulent, woody at the base. **Leaves** subsessile, narrowly lanceolate, 1–3 cm long, entire. **Inflorescence** of congested verticillasters in axils of leaf-like bracts. **Flowers**: calyx 6–8 mm long, 15-nerved, the lobes ca. half as long as tube; corolla deep blue, ca. 1 cm long, bilabiate, lower lip longer; stamens 4. Roadsides, fields; valleys. Introduced from Eurasia for medicinal purposes and sporadically escaped in temperate N. America. Known from Lake and Flathead cos.

Lamium L. Dead nettle

Taprooted annuals. **Leaves** serrate to shallowly lobed, cordate-based. **Inflorescence** few-flowered verticillasters in axils of upper leaves and leaf-like bracts. **Flowers**: calyx obscurely veined, with equal hirsute to hispid lobes; corolla tubular, pubescent, bilabiate, the upper lip forming a hood, lower lip with a cleft central lobe and obscure marginal lobes; stamens 4, lower pair longer; style subequally cleft.

1. Upper leaves sessile and clasping. . . . . *L. amplexicaule*
1. Upper leaves petiolate . . . .*L. purpureum*

Lamium amplexicaule L. **Stems** ascending, branched at the base. **Herbage** strigose. **Leaves**
sessile above; blades orbicular, crenate, 5–25 mm long. **Inflorescence** of few verticillasters
subtended by leaf-like bracts. **Flowers**: calyx 5–8 mm long; corolla purplish, 12–18 mm long,
upper lip 2–5 mm long. Lawns, fields; plains, valleys. Introduced to most of N. America; native to
Eurasia.
Plants with small cleistogamous flowers occur sporadically.

Lamium purpureum L. **Stems** prostrate to ascending, branched at the base, 8–40 cm. **Herbage** pubescent. **Leaves**
petiolate; blades deltoid to ovate, 7–30 mm long. **Inflorescence** congested, verticillasters subtended by ovate, leaf-
like bracts. **Flowers**: calyx 5–6 mm long; corolla light purple, 10–15 mm long, upper lip 2–4 mm long. Lawns, fields;
valleys. Introduced to much of N. America; native to Eurasia; known from Gallatin and Sanders cos.

Leonurus L. Motherwort

Leonurus cardiaca L. Short-rhizomatous perennial. **Stems** simple, erect, 40–100 cm,
puberulent on the angles. **Leaves** petiolate; blades palmately lobed halfway to the midvein and
serrate, 3–10 cm long, puberulent, gradually reduced upwards. **Inflorescence** of many-flowered
verticillasters in axils of upper leaves. **Flowers**: calyx glandular, 5-veined, 5–7 mm long, the
aristate lobes as long as the tube, the lower 2 reflexed down and somewhat longer; corolla pink,
8–12 mm long, bilabiate, the upper lip hood-like, villous, the lower lobes short; stamens 4. **Nutlets**
3-sided, pubescent on top. Gardens, fields, streambanks; plains, valleys. Introduced for medicinal purposes from Asia to
much of temperate N. America.

Lycopus L. Water horehound

Rhizomatous perennials. **Herbage** glandular-punctate, mildly aromatic. **Leaves** cauline, petiolate
or sessile, serrate to lobed. **Inflorescence** of dense verticillasters in upper leaf axils. **Flowers**: calyx
campanulate with 4 or 5 equal lobes ca. equal to the tube, veins 4 to 5; corolla campanulate, nearly regular,
pubescent within, upper 2 lobes united into 1; stamens 2, just included in the corolla. **Nutlets** ridged on the
angles.
See *Mentha arvensis*.

1. Calyx lobes rounded to acute, ca. as long as the mature nutlets. . . . .*L. uniflorus*
1. Calyx lobes subulate, surpassing the mature nutlets 2

2. Leaves sessile or nearly so; mature nutlets >1.5 mm long . . . . *L. asper*
2. Leaves short-petiolate; mature nutlets ≤1.5 mm long. . . . . *L. americanus*

Lycopus americanus Muhl. ex W.P.C.Barton **Rhizomes** elongate. **Stems** erect, 15–50 cm,
usually simple. **Herbage** pubescent at the nodes and on veins. **Leaves** petiolate, blades lanceolate,
2–7 cm long, lobed ca. halfway to the midvein. **Flowers**: calyx 1–3 mm long with 5 subulate lobes
surpassing the nutlets; corolla white, 2–3 mm long. **Nutlets** 1–1.5 mm long. Marshes, meadows,
thickets, along streams, wetlands; plains, valleys. Native to much of N. America.

Lycopus asper Greene **Rhizomes** tuberous-thickened. **Stems** erect, 20–80 cm, usually simple,
sometimes stoloniferous. **Herbage** sparsely strigose; stems hirsute. **Leaves** sessile, lanceolate,
3–8 cm long, serrate. **Flowers**: calyx 2–5 mm long with 5 subulate lobes surpassing the nutlets;
corolla white, 3–5 mm long. **Nutlets** 1.5–2 mm long. Marshes, wet meadows, shores of lakes,
streams; plains, valleys. BC to ON south to CA, CO.

Lycopus uniflorus Michx. **Rhizomes** slender. **Stems** with a tuberous base, erect, 10–50
cm. **Herbage** glabrous; stems sparsely puberulent. **Leaves** short-petiolate; blades lanceolate
to narrowly ovate, serrate, 2–8 cm long. **Flowers**: calyx 1–2 mm long with 5 rounded lobes not
exceeding the mature nutlets; corolla white to pink, 2–3 mm long. **Nutlets** 1–2 mm long. Wet,
organic soil of fens, marshes, swamps; valleys, montane. AK to NL south to CA, WY, NE, AR, NC.
(p.423)

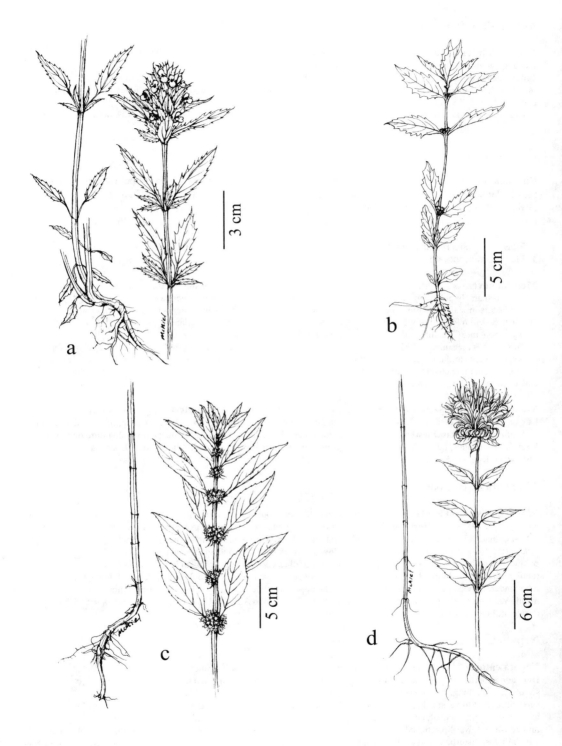

Plate 74. a. *Dracocephalum parviflorum*, b. *Lycopus uniflorus*, c. *Mentha arvensis*, d. *Monarda fistulosa*

Marrubium L. Horehound

Marrubium vulgare L. Taprooted perennial. **Stems** ascending to erect, simple, 30–100 cm. **Herbage** white-woolly. **Leaves** cauline, short-petiolate; blades ovate, crenate, 1–3 cm long. **Inflorescence** of dense, head-like verticillasters in leaf axils. **Flowers**: calyx 4–6 mm long, woolly with long hairs within the throat, the 10 hooked-spinulose teeth shorter than the campanulate tube; corolla white, 5–7 mm long, bilabiate, the 2 lips ca. equal in length; stamens 4. **Nutlets** minutely reticulate. Roadsides, fields, streambanks; plains, valleys. Sporadically introduced throughout N. America; native to Europe.

Mentha L. Mint

Rhizomatous perennials. **Herbage** pleasantly aromatic, glandular-punctate. **Leaves** petiolate to nearly sessile, lanceolate to narrowly ovate, serrate. **Inflorescence** of dense verticillasters in axils of upper leaves or reduced bracts. **Flowers**: calyx 10-nerved with 5 nearly equal lobes; corolla nearly regular with a short tube and 4 lobes, 1 lobe emarginate and slightly larger; stamens 4, exserted. **Nutlets** ellipsoid.

1. Flowers in axils of unreduced leaves . . . . *M. arvensis*
1. Flowers in bracteate terminal spikes . . . . *M. spicata*

Mentha arvensis L. Field mint **Stems** ascending or erect, 10–50 cm, mostly simple. **Herbage** densely to sparsely pubescent. **Leaves** short-petiolate; blade 2–7 cm long. **Inflorescence** of verticillasters in leaf axils. **Flowers**: calyx pubescent, 2–4 mm long, lobes acuminate; corolla purplish, 3–7 mm long, rarely 5-lobed. Wet meadows, fens, shores of streams, wetlands; plains, valleys, montane. Circumboreal south to CA, NM, TX, MO, WV. (p.423)

This plant is the source of the familiar mint odor when walking along shores of ponds and streams. *Lycopus* spp. are vegetatively similar but have angled nutlets. Varieties of *Mentha arvensis* have been described based on degree of pubescence, but this character varies continuously and has no geographic relationship; N. American plants have been called var. *canadensis* (L.) Kuntze.

Mentha spicata L. Spearmint **Stems** erect, 30–100 cm, often branched above. **Herbage** glabrous except hirsute on leaf veins. **Leaves** sessile or subsessile, 2–7 cm long. **Inflorescence** terminal spikes 3–12 cm long, of verticillasters in axils of reduced bracts. **Flowers**: calyx 1–3 mm long, the lobes ciliate, the tube glabrous; corolla 2–4 mm long, pale lavender. Roadsides, fields, ditches; valleys. Introduced for cultivation and escaped throughout N. America; native to Europe; collected in Missoula Co.

Monarda L. Bee balm

Monarda fistulosa L. Wild bergamot, horsemint Rhizomatous perennial. **Stems** numerous, erect, simple, 30–80 cm. **Herbage** glabrous to puberulent, punctate, pleasantly aromatic. **Leaves** short-petiolate; blades lanceolate to narrowly ovate, serrate, 2–8 cm long. **Inflorescence** a terminal, bracteate, head-like cluster of sessile flowers. **Flowers**: calyx ca. 15-veined, glandular-puberulent below, hispid above, hairy within, 5–11 mm long, the 5 lobes very short, equal, glandular; corolla tubular, hirsute, 15–30 mm long, bilabiate, lavender, upper lip arched, hood-like, lower lip reflexed down, the middle lobe longer, turned up; stamens 2, exserted from the upper lip; style unequally 2-lobed. **Nutlets** oblong. Grasslands, meadows, open forest; plains, valleys, montane. BC to QC south to AZ, TX, GA. Our plants are var. *menthifolia* (Graham) Fernald. (p.423)

Nepeta L. Catnip

Nepeta cataria L. Taprooted perennial. **Stems** erect, 30–100 cm, usually branched above. **Herbage** puberulent to canescent, aromatic. **Leaves** petiolate; blades cordate, ovate to deltoid, serrate, 2–7 cm long. **Inflorescence** terminal, spike-like clusters of inconspicuously bracteate verticillasters. **Flowers**: calyx 15-veined, glandular-hirsute, 5–7 mm long, sometimes purplish, lobes narrow and unequal; corolla white, purple-spotted, hirsute, 8–12 mm long, bilabiate, the upper lip hood-like, lower lip reflexed down, the middle lobe largest; stamens 4, enclosed in the upper lip; style subequally lobed. Fields, disturbed grasslands, around buildings, along roads and streams; plains, valleys. Introduced to most of N. America; native to Eurasia.

Physostegia Benth. Obedient plant

Physostegia parviflora Nutt. ex A.Gray [*Dracocephalum nuttallii* Britton] Rhizomatous perennial herbs. **Stems** erect, 20–80 cm, usually simple. **Leaves** 3–10 cm long, sessile, lanceolate to oblanceolate, serrate, the lowest deciduous. **Herbage** glabrous below the inflorescence, non-aromatic. **Inflorescence** of bracteate, pedicellate flowers in a terminal, simple or branched, glandular-puberulent raceme. **Flowers**: calyx 4–6 mm long, glandular,10-veined, the subequal 5 lobes shorter than the tube; corolla 10–15 mm long, lavender, bilabiate, upper lip hood-like, 3–5 mm long, lower lip reflexed down with 3 nearly equal lobes; stamens 4, included in the upper lip; style equally lobed. Moist, often stony soil on banks of streams, wetlands; valleys. BC to MN south to OR, ID, WY, NE.

A garden escape of the larger-flowered *Phystostegia virginiana* (L.) Benth is reported for central MT.[74]

Prunella L. Self heal

Prunella vulgaris L. Heal all Herbaceous perennial from a short rhizome. **Stems** prostrate to erect, 4–30 cm, simple or branched below. **Herbage** glabrous to sparsely villous. **Leaves** petiolate; blades lanceolate to ovate, entire to shallowly crenate, 2–7 cm long. **Inflorescence** of verticillasters in a dense, terminal spike 2–5 cm long; bracts purplish. **Flowers**: calyx 10-veined, sparsely villous, bilabiate, 6–10 mm long, upper lip with 3 shallow lobes, lower lip with 2 apiculate lobes; corolla bilabiate, purple, 10–15 mm long, upper lobe hood-like, central lower lobe large and ciliate; stamens 4, enclosed in the hood; style equally 2-lobed. Circumboreal south to most of U.S. (p.428)

1. Leaf blades ≥2 times as long as wide; stems ascending to erect, mostly ≥4 cm . . . . var. *lanceolata*
1. Leaf blades ≤2 times as long as wide; stems prostrate to ascending, usually <4 cm . . . . var. *vulgaris*

The native ***Prunella vulgaris*** var. ***lanceolata*** (W.P.C.Barton) Fernald occurs in moist forest openings, meadows, along streams; valleys, montane. The introduced *P. vulgaris* var. ***vulgaris*** is found in disturbed soil along roads and in lawns.

Salvia L. Sage

Annual to perennial herbs. **Herbage** glandular-punctate, usually aromatic. **Leaves** petiolate below, becoming sessile above. **Inflorescence** of few-flowered verticillasters in an interrupted, bracteate spike. **Flowers**: calyx 10- to 15-veined, bilabiate; corolla bilabiate, upper lobe hood-like; stamens 2; style unequally 2-lobed. **Nutlets** rounded.

1. Corolla >15 mm long; inflorescence bracts aristate . . . .*S. sclarea*
1. Corolla ≤15 mm long; inflorescence bracts acute to apiculate but not aristate 2

2. Calyx glabrous with equal lobes. . . . . *S. verticillata*
2. Calyx glandular to puberulent. 3

3. Taprooted annual; leaf blades cuneate at the base . . . . *S. reflexa*
3. Rhizomatous perennial; leaf blades truncate-based . . . . *S. nemorosa*

Salvia nemorosa L. Rhizomatous perennial. **Stems** erect, 50–80 cm, simple or branched above. **Herbage** puberulent to hirsute, aromatic. **Leaves** petiolate, sessile upward; blades lanceolate to ovate, truncate, 4–10 cm long, serrate. **Inflorescence** of congested, purple-bracted verticillasters. **Flowers**: calyx glandular-hirsute, 5–8 mm long, bilabiate, lower lip with 3 apiculate lobes; corolla blue to purple, 9–12 mm long, glandular, puberulent, lower lip small, the middle lobe largest; stamens included in upper lip. Fields, roadsides; plains, valleys. Introduced horticulturally and occasionally escaped; native to Eurasia.

Salvia pratensis L., with green inflorescence bracts, commonly forms hybrids with *S. nemorosa* in Europe,[177] and some of our plants may be such hybrids.

Salvia reflexa Hornem. Taprooted annual. **Stems** erect, 10–40 cm, usually branched. **Herbage** strigose, unpleasantly aromatic. **Leaves** short-petiolate; blades lanceolate to oblong, 1–5 cm long, obscurely crenate. **Inflorescence** of solitary or paired, short-pedicellate flowers at nodes of an interrupted raceme. **Flowers**: calyx puberulent, 5–9 mm long, bilabiate with 3 apiculate lobes; corolla blue, 8–12 mm long, upper lip small, lateral lobes of lower lip small; stamens included in upper lip. Roadsides, fields; plains, valleys. MT to OH south to AZ, TX, AR, Mexico.

Salvia sclarea L. Clary Taprooted biennial. **Stems** erect, 50–100 cm, usually branched. **Herbage** glandular-hirsute. **Leaves** long-petiolate; blades ovate, truncate, 7–20 cm long, reduced above. **Inflorescence** of well-separated verticillasters subtended by sessile, entire, aristate bracts. **Flowers**: calyx glandular, ca. 1 cm long, lobes aristate; corolla 20–25 mm long, blue and yellow, upper lip erect, middle lobe of lower lip much broader than the lateral lobes; stamens exserted from the upper lip. Roadsides, fields; valleys; Ravalli Co. Introduced as an ornamental and occasionally escaped in eastern and western U.S.

Salvia verticillata L. Perennial. **Stems** erect, 40–80 cm, branched above. **Herbage** puberulent, minutely glandular. **Leaves** long-petiolate; blades cordate, lanceolate, 3–10 cm long, reduced above. **Inflorescence** of clustered verticillasters subtended by purplish, ovate bracts. **Flowers**: calyx hirsute, glandular, ca. 6 mm long, lobes unequal; corolla 9–12 mm long, blue, glandular, puberulent; stamens included in the upper lip. Grasslands, fields; valleys. Introduced to northeast U.S., adjacent Canada and MT; native to Eurasia. Known from one location in Lake Co.

Satureja L. Savory

Annual or perennial herbs. **Leaves** cauline, petiolate. **Flowers**: calyx 10- to 15-nerved; corolla bilabiate, lips subequal, upper lobe flat, lower lip reflexed down; stamens 4, included; style equally 2-lobed.

1. Plants annual; leaf blades <12 mm long . . . . *S. acinos*
1. Rhizomatous perennials; some leaf blades >12 mm long 2

2. Stems prostrate; flowers 7–10 mm long, 1 to 2 in axils of upper leaves . . . . *S. douglasii*
2. Stems erect; flowers 10–12 mm long, clustered in verticillasters. . . . . *S. vulgaris*

Satureja acinos (L.) Scheele [*Acinos arvensis* (Lam.) Dandy] Fibrous-rooted annual. **Stems** erect, often branched at the base, 5–20 cm. **Herbage** puberulent. **Leaves** short-petiolate; blades ovate, 5–12 mm long, entire or nearly so. **Inflorescence** of 1 to 3 pedicellate flowers in upper leaf axils. **Flowers**: calyx hispid, 4–6 mm long, lobes subequal; corolla pale purple, 7–10 mm long. Roadsides in open forest; valleys, montane. Introduced sparingly to northern U.S.; native to Europe. Known from Powell Co.

Satureja douglasii (Benth.) Briq. [*Clinopodium douglasii* (Benth.) Kuntze] Rhizomatous perennial. **Stems** 10–30 cm, prostrate with ascending to erect, simple branches. **Herbage** sparsely puberulent. **Leaves** short-petiolate; blades ovate, remotely crenate, 1–3 cm long. **Inflorescence** of 1 or 2 axillary, pedicellate flowers. **Flowers**: calyx 3–5 mm long, 12- to 15-veined, lobes subequal, puberulent; corolla white, 5–9 mm long, puberulent, upper lip purplish. Moist coniferous forest; valleys, montane. BC, MT south to CA. Known from Sanders and Ravalli cos.

Satureja vulgaris (L.) Fritsch [*Clinopodium vulgare* L.] Wild basil Rhizomatous perennial. **Stems** erect, 20–40 cm, mostly simple. **Herbage** sparsely hirsute. **Leaves** petiolate; blades lanceolate to ovate, 1–3 cm long, crenate. **Inflorescence** of dense bracteate verticillasters, terminal or in axils of uppermost leaves. **Flowers**: calyx tubular, 8–10 mm long, villous, lobes unequal, setaceous, upper longer than lower; corolla purple, 10–12 mm long, campanulate, indistinctly bilabiate. Roadsides; valleys. Native to Europe, eastern N. America, introduced elsewhere. Known from Flathead Co.

Scutellaria L. Skullcap

Scutellaria galericulata L. Rhizomatous perennial. **Stems** erect, 15–70 cm, simple or branched. **Herbage** glabrous to puberulent. **Leaves** sessile to short-petiolate, lanceolate, truncate, shallowly serrate, 2–6 cm long. **Inflorescence** of 1 or 2, pedicellate, axillary flowers. **Flowers**: calyx 3–4 mm long, bilabiate, lips entire with a prominent ridge-like appendage on the upper lip; corolla conical, 12–20 mm long, bilabiate, upper lip hood-like, lower lip with reduced lateral lobes; stamens 4, included; style unlobed. **Nutlets** ovoid. Fens, marshes, thickets, wet meadows; plains, valleys, montane. Circumboreal south to CA, AZ, MO, DE. (p.428)

Stachys L. Hedge nettle

Stachys palustris L. Woundwort Rhizomatous perennial. **Stems** erect, 20–80 cm, simple or branched. **Herbage** hirsute, aromatic. **Leaves** short-petiolate to sessile; blades 3–9 cm long, lanceolate, truncate, serrate. **Inflorescence** of verticillasters in a terminal, interrupted, bracteate spike. **Flowers**: calyx glandular-villous, 5–9 mm long, lobes equal, purplish; corolla purple, spotted, 10–15 mm long, bilabiate, the upper lip hood-like, the lower 3-lobed; stamens 4, included; style subequally 2-lobed. **Nutlets** rounded above. Moist or wet soil of marshes, wet meadows, often around streams, ponds; plains, valleys, montane. Circumboreal south to OR, AZ, OH, NY. Our plants are ssp. *pilosa* (Nutt.) Epling. (p.428)

Teucrium L. Germander

Teucrium canadense L. Rhizomatous perennial. **Stems** erect, 20–70 cm, mostly simple. **Herbage** strigose to hirsute. **Leaves** petiolate; blades lanceolate to ovate, serrate, 3–8 cm long, deciduous below. **Inflorescence** a terminal, spike-like, bracteate raceme. **Flowers**: calyx glandular, hirsute to villous, purplish, 5–7 mm long, the lower 2 lobes longer; corolla light purple, ca. 1 cm long, seemingly 1-lipped with 2 pairs of lateral lobes and a larger ovate middle lobe; stamens 4, exserted. **Nutlets** rounded on top, almost completely united. Moist meadows around streams, lakes; plains, valleys. Throughout temperate N. America. Our plants are var. *boreale* (E.P.Bicknell) Shinners.

Thymus L. Thyme

Thymus pulegioides L. Perennial. **Stems** woody at the base, creeping to ascending, puberulent in lines. **Leaves** all cauline, sessile to short-petiolate, entire, elliptic to ovate, 3–10 mm long. **Inflorescence** short-spiciform, of several crowded, bracteate whorls. **Flowers** perfect; calyx with 10 to 13 veins, ca. 3 mm long, bilabiate, the upper lobes longer; corolla 4–5 mm long, purple, bilabiate, lower lip 3-lobed, middle lobe the largest; stamens 4, exserted; style 2-lobed. Fields, lawns, roadsides; valleys. Introduced ground cover escaped in northeast U.S. and adjacent Canada as well as the Pacific Northwest.

Confused taxonomy makes determination of our plants problematic.

OLEACEAE: Olive Family

Fraxinus L. Ash

Fraxinus pennsylvanica Marshall Green ash Dioecious tree to 12 m with an open crown and spreading branches. **Bark** dark gray, shallowly furrowed; twigs green to brown. **Leaves** opposite, petiolate, ovate in outline, odd-pinnate with 5 to 7 leaflets; leaflets short-petiolate, lanceolate, serrate, 5–12 cm long, mostly glabrous. **Inflorescence** a panicle, appearing before the leaves. **Flowers** unisexual; calyx 1–1.5 mm long, cup-like with 4 shallow lobes; corolla absent; male flowers with 2 stamens, 2–4 mm long; female flowers with 2 aborted stamens and 1 pistil; ovary superior; stigma 2-lobed. **Fruit** a 1-seeded, oblong, flattened, distally-winged samara, 2–4 cm long. Along rivers, streams, ephemeral drainages, cool north-facing slopes; plains. MT to QC south to TX, LA, FL.

Plants are susceptible to heart rot but resprout from stumps to produce multi-stemmed trees. *Fraxinus pennsylvanica* is native east of the Continental Divide in MT; it has been introduced west of the Divide and occasionally becomes established along streams.

HIPPURIDACEAE: Mare's-tail Family

Hippuris L. Mare's tail

Hippuris vulgaris L. Glabrous, rhizomatous, at least partly submerged perennial. **Stems** erect, 10–70 cm, usually simple. **Leaves** linear, sessile, 1–5 cm long, 6- to 12-whorled; nodes closer together on emerged portion of stem. **Flowers** sessile, solitary in leaf axils, perfect; calyx green, tubular; petals absent; style 1; stamen 1; ovary elliptic, ca. 1 mm long. **Fruit** 1-seeded achene, 1.5–2 mm long, enclosed in the calyx. Shallow water of sloughs, ponds, lakes; plains, valleys. Circumboreal south to CA, NM, NE, MN, NY. (p.428)

Submerged leaves are flaccid, while those above water are more rigid. Many systematists believe that the Hippuridaceae should be placed in the Plantaginaceae[384] based on molecular evidence. The morphological differences between the two groups make it difficult for me to adopt this classification.

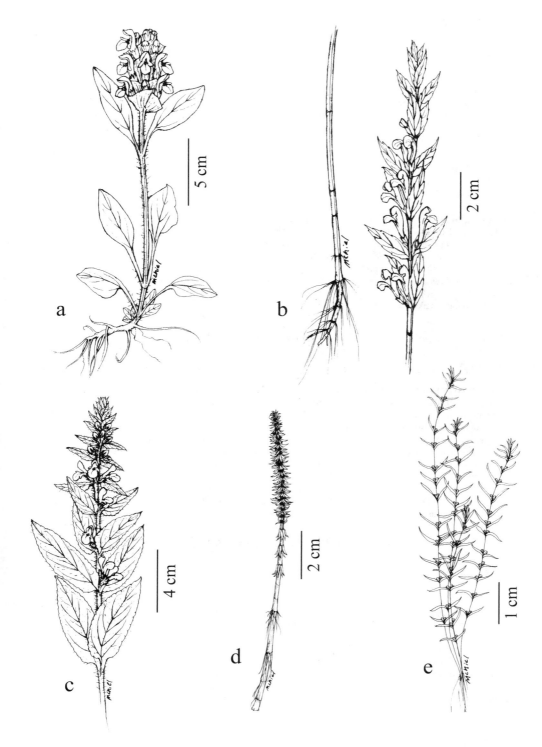

Plate 75. a. *Prunella vulgaris*, b. *Scutellaria galericulata*, c. *Stachys palustris*, d. *Hippuris vulgaris*, e. *Callitriche hermaphroditica*

CALLITRICHACEAE: Water-starwort Family

Callitriche L. Water-starwort

Delicate, glabrous, annual, aquatic plants, fibrous-rooted at the lower nodes. **Leaves** opposite, entire, emarginate at the tip. **Flowers** 1 to 3, sessile in leaf axils, small, green, unisexual; sepals and petals absent; male flower of 1 stamen; female flower a single pistil; ovary with 2 carpels; styles 2. **Fruit** flattened with a vertical groove on each face, 4-lobed separating into 4 1-seeded achenes.

Leaves on floating portion of the stem are closer together than those that are submerged. Plants sometimes occur in damp mud. Mature fruit is required for positive determination of some species. Many systematists believe that the Callitrichaceae should be placed in the Plantaginaceae[384] based on molecular evidence. The morphological differences between the two groups make it difficult for me to follow suit.

1. Fruits with a conspicuous, white wing-margin around the entire circumference . . . . *C. stagnalis*
1. Fruits narrowly wing-margined only above or not at all 2

2. Plants completely submerged; leaves all linear, not basally joined around the stem;
   fruit nearly circular . . . . *C. hermaphroditica*
2. Shoot tips usually floating; leaf pairs joined together around the stem; floating
   leaves oblong; fruit elliptic to oblong. 3

3. Fruit ca. as long as wide, somewhat oblong, not winged above . . . . *C. heterophylla*
3. Fruit longer than wide, elliptical, narrowly winged above . . . .*C. palustris*

Callitriche hermaphroditica L. [*C. autumnalis* L.] Plants usually wholly submerged. **Stems** 5–20 cm. **Leaves** linear, 5–10 mm long, not joined together around the stem. **Flowers** solitary in leaf axils; bracteoles absent. **Fruit** 1–2 mm long, round in outline, sometimes with narrowly winged margins at the apex. Shallow water of ponds, lakes, slow streams; plains, valleys, montane. Circumboreal south to CA, NM, NE, MN, VT. (p.428)

Callitriche heterophylla Pursh Plants submerged below, floating above. **Stems** 2–25 cm. **Leaves** joined together around the stem; submerged leaves linear, 3–15 mm long; floating leaves oblong, 2–5 mm wide. **Flowers** usually 1 male and 1 female per bracteate axil. **Fruit** oblong, ca. 1 mm wide and long, not winged. Shallow water of ponds, lakes at all elevations. Across most of northern U.S. and adjacent Canada.
   Callitriche palustris, with fruits longer than wide, is vegetatively similar and more common.

Callitriche palustris L. [*C. verna* L.] Plants mainly submerged. **Stems** 3–30 cm. **Leaves** joined together around the stem; submerged leaves linear, 5–25 mm long; floating leaves oblong, to 4 mm wide. **Flowers** usually 1 male and 1 female per bracteate axil. **Fruit** elliptic, ca. 1 mm wide, >1 mm long; narrowly winged above. Shallow water of ponds, lakes; valleys, montane. Circumboreal through much of N. America. See *C. heterophylla*.

Callitriche stagnalis Scop. Plants submerged to terrestrial. **Stems** 3–15 cm. **Leaves** joined together around the stem, linear to oblong, 5–10 mm long. **Flowers** usually 1 male and 1 female per bracteate axil. **Fruit** nearly round in outline, ca. 1.5 mm long with a conspicuous, whitish, wing-margin around the entire circumference. Shallow water of sloughs, borrow pits; valleys. Eastern and northwest N. America, Europe. This species may be introduced in Missoula and Ravalli cos.

PLANTAGINACEAE: Plantain Family

Annual or perennial herbs. **Leaves** basal and/or cauline, alternate, opposite or whorled, mostly simple, rarely pinnately lobed, exstipulate. **Flowers** perfect, zygomorphic, 4- to 5-merous; corolla sympetalous, often tubular and bilabiate; stamens 2 to 5; ovary superior, 2-loculed; style 1, simple or 2-lobed. **Fruit** a rotund to ovoid, few- to many-seeded capsule.

   Bacopa, Besseya, Chaenorhinum, Chionophila, Collinsia, Digitalis, Gratiola, Linaria, Penstemon, Synthyris, and *Veronica* were formerly placed in the Scrophulariaceae. *Kickxia spuria* (L.) Dumort., with broadly ovate leaves and snapdragon-like flowers, was collected once in a Choteau garden; it is introduced from Europe.

1. Leaves all basal (well-differentiated bracts in *Synthyris*) 2
1. Plants with leafy flowering stems 3

2. Stamens 2; flowers blue to purplish; peduncles bracteate . . . . *Synthyris*
2. Stamens 4 or 5; flowers white; peduncles naked . . . . *Plantago*

3. Cauline leaves alternate 4
3. Cauline leaves opposite 8

4. Corolla absent; filaments (anther stalks) red to purple . . . .*Besseya*
4. Flowers with a corolla 5

5. Corolla bell-shaped, 4–6 cm long, purplish . . . . *Digitalis*
5. Corolla tubular at the base and 2-lipped or 4-lobed at the mouth 6

6. Corolla 4-lobed at the mouth; flowering stalk with reduced bracts only . . . . *Synthyris*
6. Corolla 2-lipped; cauline leaves present 7

7. Flowers solitary from upper leaf axils . . . . *Chaenorhinum*
7. Flowers in a terminal, bracteate raceme . . . . *Linaria*

8. Anther-bearing stamens 2 9
8. Anther-bearing stamens 4 10

9. Corolla saucer-shaped with 4 distinct lobes . . . .*Veronica*
9. Corolla tubular with 4–5 small lobes at the mouth . . . . *Gratiola*

10. Stamens 5, 4 with anthers and 1 sterile . . . . *Penstemon*
10. Fertile stamens 4, sterile stamens absent 11

11. Herbage glandular-pubescent, at least in the inflorescence 12
11. Herbage glabrous to puberulent, but not glandular 13

12. Inflorescence a terminal, secund raceme; corolla 9–13 mm long . . . . *Chionophila*
12. Flowers solitary in leaf axils; corolla 5–6 mm long . . . .*Chaenorhinum*

13. Corolla campanulate, barely irregular, whitish, 5-lobed, ... *Bacopa*
13. Corolla bilabiate, blue and white, seemingly 4-lobed . . . . *Collinsia*

Bacopa Aubl. Water hyssop

Bacopa rotundifolia (Michx.) Wettst. Aquatic to emergent, glabrous, fibrous-rooted perennials. **Stems** prostrate, glabrous to puberulent, 5–25 cm, rooting at the lower nodes. **Leaves** opposite, sessile, obovate to orbicular, glabrous, 1–3 cm long. **Inflorescence** 1 to 4 pedicellate flowers at leaf nodes. **Flowers**: calyx 3–6 mm long, divided nearly to the base into 5 unequal lobes; corolla campanulate, white with a yellow throat, 5-lobed, barely irregular, 5–7 mm long; stamens 4; style 2-lobed. **Capsule** orbicular, 3–6 mm long, opening by 4 valves. Shallow water or mud on margins of ponds; plains. MT to MN south to CA, TX, LA, MS.

Besseya Rydb. Kittentails

Fibrous-rooted perennials. **Leaves**: basal leaves petiolate, blades ovate to elliptic, sometimes cordate, serrate; cauline leaves reduced, alternate, sessile. **Inflorescence** a dense cylindrical, bracteate spike, expanding at maturity. **Flowers**: calyx cup-shaped, deeply 2- to 4-lobed; corolla absent or vestigial; stamens 2, exserted, conspicuous; style capitate. **Capsule** tomentose, shallowly notched apically.

1. Calyx 3- to 4-lobed, fully enclosing the ovary; stamens red, 4–6 mm long . . . . *B. rubra*
1. Calyx 2- to 3-lobed, not enclosing the ovary; stamens purple, 5–10 mm long . . . . *B. wyomingensis*

Besseya rubra (Douglas ex Hook.) Rydb. **Stems** erect, 15–40 cm, simple. **Herbage** puberulent to sparsely villous. **Basal leaf blades** sometimes cordate, 4–12 cm long. **Inflorescence** 5–25 cm long; bracts oblanceolate. **Flowers**: calyx 4-lobed, 2–4 mm long, septate-villous mainly on the margins, fully enclosing the ovary; stamens dark red, 4–6 mm long. **Capsule** 4–7 mm long, ovoid, longer than wide. Grasslands, open ponderosa pine forest; valleys, montane. WA to MT south to OR.

Besseya wyomingensis (A.Nelson) Rydb. **Stems** erect, 5–30 cm, simple. **Herbage** puberulent. **Basal leaf blades** 2–7 cm long. **Inflorescence** 2–10 cm long; bracts lanceolate. **Flowers**: calyx 2-lobed, 2–4 mm long, densely septate-puberulent, not fully enclosing the ovary; stamens purple, 5–10 mm long. **Capsule** 3–5 mm long, globose, wider than long. Grasslands, open forest, fellfields, turf at all elevations. BC to SK south to UT, CO, NE. (p.432)

Chaenorhinum (DC.) Rchb.

Chaenorhinum minus (L.) Lange Annual. **Stems** ascending to erect, branched, 5–30 cm. **Herbage** glandular-pubescent. **Leaves** cauline, short-petiolate, opposite below, alternate above; blades oblanceolate, 1–3 cm long. **Inflorescence** of pedicellate flowers solitary in leaf axils. **Flowers**: calyx 3–5 mm long, divided to the base into 5 oblanceolate lobes; corolla 7–8 mm long including the ca. 2 mm backward-pointing spur, bilabiate, upper lip blue, straight, lower lip yellow with 3 spreading lobes and a raised base (palate); stamens 4; style capitate. **Capsule** globose, 4–6 mm long. Roadsides, streambanks; valleys. Introduced to much of northern U.S. and adjacent Canada; native to Europe.

Chionophila Benth.

Chionophila tweedyi (Canby & Rose) L.F.Hend. Perennial from a simple or few-branched caudex. **Stems** erect, simple, 3–30 cm. **Herbage** glabrous below, glandular-pubescent above. **Leaves**: the basal petiolate, the blades oblanceolate, 1–4 cm long; cauline leaves opposite, 1 or 2 pairs, lanceolate, sessile, 5–12 mm long. **Inflorescence** a terminal, secund raceme of 4 to 10 solitary flowers in axils of paired bracts. **Flowers**: calyx 3–6 mm long with 5 lobes ca. as long as the tube; corolla white and lavender, 9–13 mm long, tubular, bilabiate, the 5 lobes nearly equal; stamens 4; stigma capitate. **Capsule** ovate, 6–9 mm long. Granitic soil of open forest, meadows, rock outcrops, most common in whitebark pine or subalpine larch forests; subalpine, alpine. Endemic to southwest MT, adjacent ID. The type locality is in Beaverhead Co.

Collinsia Nutt. Blue-eyed Mary

Collinsia parviflora Lindl. Annual. **Stems** erect, 2–30 cm, often branched. **Herbage** glabrate, stems puberulent. **Leaves** all cauline, opposite, narrowly oblong, 7–40 mm long. **Inflorescence** cymose with 1 to several pedicellate flowers in axils of upper leaves and leaf-like bracts. **Flowers**: calyx 3–6 mm long with 5 lanceolate lobes; corolla 4–9 mm long, blue and white, bilabiate with a broad spur at the base, upper lip erect, lower lip 3-lobed, saccate, the middle lobe tiny; stamens 4; style capitate. **Capsule** ovate, opening by 4 valves, 2–4 mm long with 2 to 4 seeds. Sparsely vegetated soil of forest openings, grasslands, meadows, rock outcrops; valleys to subalpine. AK to ON south to CA, CO.

Digitalis L. Foxglove

Digitalis purpurea L. Taprooted biennial. **Stems** erect, simple, 50–100 cm. **Herbage** pubescent, glandular above. **Leaves** basal and alternate, petiolate; the blade 2–15 cm long, lanceolate, serrate, reduced upward. **Inflorescence** a secund, bracteate raceme; bracts entire. **Flowers** short-pedicellate; calyx of 5 separate, ovate sepals, 10–18 mm long; corolla narrowly campanulate, pale purple-mottled, 3–5 cm long; stamens 4; stigmas 2. **Capsule** opening by 4 valves. Roadsides, often adjacent to moist forest; valleys. Flathead and Lake cos. Introduced BC, MT south to CA; native to Europe.

Gratiola L. Hedge hyssop

Fibrous-rooted annuals. **Stems** usually branched. **Leaves** cauline, opposite, entire, sessile. **Inflorescence** of 1 to 2 pedicellate flowers in axils of upper leaves. **Flowers**: calyx divided to the base into 5 lanceolate lobes (appearing to be 7-lobed in *G. neglecta*), expanding in fruit; corolla tubular, nearly regular, appearing to be shallowly 4-lobed, the upper lobe shallowly emarginate; stamens 2, included; style 2-lobed. **Capsule** globose, opening by 4 valves.

1. Calyx glandular, subtended by 2 sepal-like bracts, making the calyx appear 7-lobed . . . . *G. neglecta*
1. Calyx glabrate, without sepal-like bracts . . . . *G. ebracteata*

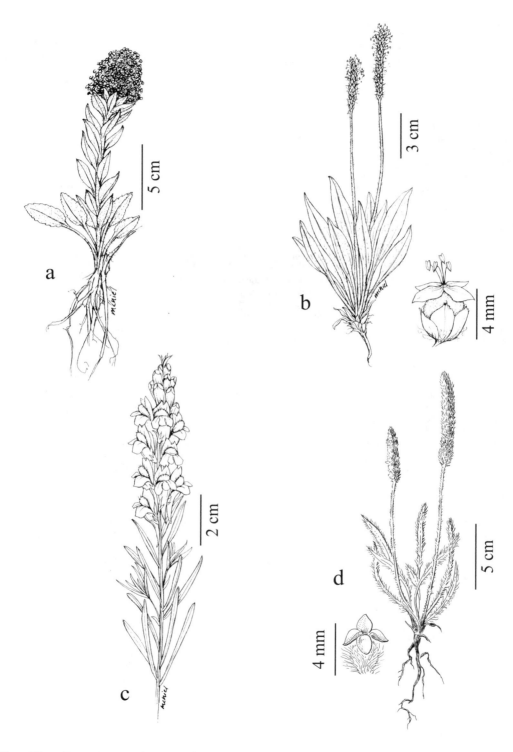

Plate 76. a. ***Besseya wyomingensis***, b. ***Plantago canescens***, c. ***Linaria vulgaris***, d. ***Plantago patagonica***

Gratiola ebracteata Benth. **Stems** prostrate to ascending, 3–15 cm. **Herbage** glabrous. **Leaves** lanceolate, 1–3 cm long. **Flowers**: calyx 4–8 mm long, glabrous, not subtended by bracts; corolla 6–10 mm long, whitish with a yellow tube. **Capsule** 4–6 mm long. Drying mud of pond and lake margins; valleys, montane. BC, MT south to CA.

Gratiola neglecta Torr. **Stems** ascending to erect, 4–25 cm, fistulose below. **Herbage** glabrate to glandular-puberulent. **Leaves** lanceolate to oblanceolate, 1–3 cm long. **Flowers**: calyx 2–6 mm long, glandular, subtended by 2 sepal-like bracts; corolla 6–13 mm long, white to pale blue with a yellow tube. **Capsule** 4–6 mm long. Drying mud along ponds, streams, rivers; plains, valleys. BC to QC south to CA, AZ, TX, MS, GA.

Linaria Mill. Toadflax

Taprooted annuals or perennials. **Leaves** alternate, entire. **Herbage** glabrous. **Inflorescence** a terminal, bracteate raceme. **Flowers**: calyx divided to the base with equal lobes; corolla tubular with a long, backward-pointing spur, bilabiate, the lower lip with 3 reflexed lobes and a raised base (palate); stamens 4; stigma capitate. **Capsule** ovoid.

1. Corolla blue; slender annuals. . . . . *L. canadensis*
1. Corolla yellow; plants rhizomatous perennials 2

2. Leaves ovate, <8 times as long as wide, clasping the stem. . . . .*L. dalmatica*
2. Leaves linear, >8 times as long as wide. . . . . *L. vulgaris*

Linaria canadensis (L.) Dum.Cours. [*Nuttallanthus texanus* (Scheele) D.A.Sutton] Annual. **Stems** erect, mostly simple, 15–30 cm, sometimes with prostrate, basal, usually sterile stems. **Leaves** dimorphic; leaves of prostrate stems opposite or 3-whorled, oblanceolate, 5–10 mm long; cauline leaves becoming alternate above, linear, 1–3 cm long. **Flowers**: calyx 2–4 mm long, lobes linear; corolla blue, 6–13 mm long, the curved spur an additional 2–11 mm long, lower lip 7–11 mm long, straight, palate white. **Capsule** 2–3 mm long. Sparsely vegetated, sandy soil of grasslands, woodlands; plains. Dawson and Carter cos. Throughout most of U.S. and adjacent Canada.

Our plants have the smaller flowers of *Linaria canadensis* var. *canadensis* but the tuberculate seeds of var. *texana* (Scheele) Pennell.

Linaria dalmatica (L.) Mill. Dalmation toadflax Rhizomatous perennial, producing prostrate, rooting stems from the caudex. **Stems** erect, branched, 40–80 cm. **Leaves** ovate, acute, waxy, 2–5 cm long, sessile, clasping. **Flowers**: calyx 6–12 mm long, lobes lanceolate; corolla yellow, 12–24 mm long, the straight spur an additional 11–17 mm long, lower lip 5–11 mm long, reflexed down, palate white- or orange-tomentose. **Capsule** 7–8 mm long. Fields, roadsides, grasslands, often in stony soil; plains, valleys. A garden escape, introduced in temperate N. America, sporadic in the east but becoming common in the west; native to Europe.

Linaria vulgaris Hill Butter and eggs Taprooted perennial. **Stems** clustered, erect, mostly simple, 15–60 cm. **Leaves** numerous, linear-oblanceolate, 1–5 cm long. **Flowers**: calyx 3–5 mm long, lobes lanceolate; corolla yellow, 10–18 mm long, the straight spur an additional 8–14 mm long, lower lip 6–9 mm long, reflexed down with a high orange palate. **Capsule** 6–8 mm long. Roadsides, meadows, grasslands; plains, valleys, montane. A garden escape, introduced to much of temperate N. America; native to Eurasia. (p.432)

Penstemon Schmidel Beardtongue

Perennial herbs or subshrubs. **Stems** mostly simple, clustered on a branched, woody caudex. **Leaves** simple; the basal petiolate; the cauline opposite, often sessile. **Inflorescence** of several cymes forming verticillasters at nodes of a bracteate panicle or raceme. **Flowers**: calyx deeply lobed into 5 equal sepals, expanding in fruit; corolla tubular, bilabiate, the lips spreading, upper lip 2-lobed, lower lip 3-lobed, lobes ca. equal; stamens 5, 4 fertile, the other (staminode) without an anther but usually yellow-hairy; stigma unlobed. **Capsules** ovoid, beaked.[102]

The largest genus endemic to N. America. Flowers are required for determination. Calyx measurements are from flowers at anthesis. *Penstemon venustus* Douglas ex Lindl., was once collected from a roadside seeding in Flathead Co. A report of *Penstemon cyananthus* Hook.[113] was based on a cultivated specimen.

The report of *P. payettensis* A.Nelson & J.F.Macbr. (Lackschewitz 1991)[226] was based on a misidentified specimen (fide C. Freeman).

1. Anthers villous with long, contorted, often tangled hairs. 2
1. Anthers glabrous to puberulent, but not long-hairy. 6

2. Foliage below the inflorescence glandular . . . .*P. montanus*
2. Foliage below the inflorescence glabrous to puberulent 3

3. Stem leaves narrowly lanceolate, ≥5 times as long as wide 4
3. Stem leaves elliptic to obovate, 2 to 4 times as long as wide. 5

4. Leaves all cauline, obscurely serrate . . . .*P. lyallii*
4. Basal leaves present, entire. . . . .*P. caryi*

5. Stems erect to ascending from the base, woody much of their length. . . . .*P. fruticosus*
5. Stems prostrate and woody only at the base with ascending, herbaceous stems . . . .*P. ellipticus*

6. Inflorescence sparsely to densely glandular-pubescent. 7
6. Inflorescence glabrous to puberulent but not glandular 20

7. Corolla primarily white 8
7. Corolla rose, blue, or purple. 9

8. Basal leaves present; blades mostly entire; plains. . . . .*P. albidus*
8. Leaves all cauline; blades distinctly serrate; mountains. . . . . *P. deustus*

9. Ovaries and capsules glandular on top 10
9. Ovaries and capsules glabrous 11

10. Staminode long-hairy for much of its length; leaves often serrate . . . .*P. eriantherus*
10. Staminode beared only at the tip; leaves entire . . . .*P. whippleanus*

11. Stem leaves linear 12
11. Stem leaves narrowly lanceolate to ovate 13

12. Corolla pink to rose; basal leaves filiform. . . . .*P. laricifolius*
12. Corolla blue; basal leaves linear to linear-oblanceolate. . . . .*P. aridus*

13. All the basal or lowest leaves entire 14
13. At least some basal or lowest leaves denticulate to serrate. 17

14. Leaves puberulent 15
14. Leaves below the inflorescence glabrous. 16

15. Basal leaves present; corolla ≤16 mm long . . . . *P. humilis*
15. Leaves all cauline; corolla ≥16 mm long. . . . . *P. radicosus*

16. Verticillasters open, sparsely glandular-pubescent; stem leaves often rounded at the tip . . . . . *P. albertinus*
16. Verticillasters more compact or densely glandular-pubescent; stem leaves tapered to the
    acute tip . . . . *P. attenuatus*

17. Leaves all cauline; anther sacs remaining U-shaped. . . . .*P. diphyllus*
17. Basal leaves present; anther sacs widely spreading after dehiscing 18

18. Corolla pale lilac to pale violet . . . . *P. gracilis*
18. Corolla deep blue. 19

19. Stem leaves oblanceolate to lanceolate, not clasping . . . .*P. albertinus*
19. Stem leaves broadly lanceolate to ovate, the upper often clasping . . . . *P. wilcoxii*

20. Anther sacs hairy on the back 21
20. Anther sacs glabrous except sometimes along the openings. 25

21. Sepals ovate to obovate with rounded, erose tips 22
21. Sepals with long, acuminate tips 23

22. Calyx 2–3 mm long; plants of the Great Plains . . . .*P. glaber*
22. Calyx 4–6 mm long; plants of the mountains and valleys. . . . . *P. cyaneus*

23. Leaves puberulent, especially beneath . . . . *P. lemhiensis*
23. Leaves glabrous 24

24. Stem leaves ovate; anther sacs short-hairy. . . . . *P. cyananthus*
24. Stem leaves linear-lanceolate to lanceolate, anther sacs long-hairy . . . .*P. caryi*

Penstemon albertinus Greene [*P. caelestinus* Pennell] **Stems** ascending to erect, 10–40 cm. **Herbage** glabrous. **Basal leaf blades** lanceolate to narrowly ovate, entire to weakly serrate, 1–4 cm long. **Stem leaves** narrowly lanceolate to oblanceolate, entire to weakly serrate, 15–45 mm long. **Inflorescence** with 3 to 8 open verticillasters, glabrous to weakly glandular. **Flowers:** calyx 3–5 mm long; sepals ovate, acute to acuminate, barely scarious-margined; corolla blue, 12–18 mm long, glabrous to glandular; anthers glabrous, opening across their full length. **Capsule** 3–6 mm long. Grasslands, rock outcrops, dry, open forest; valleys to lower subalpine. BC, AB, ID, MT. (p.437)

Typical specimens of *Penstemon albertinus* and *P. wilcoxii* are easily distinguished. However, their ranges overlap in western Montana, and intermediates between the two species are common in this region. These intermediate forms have serrate, oblanceolate upper stem leaves. They seem to be most common near Missoula and have been called *P. caelestinus*. Taxonomic recognition of intermediates between these closely related species in their zone of contact seems unwarranted. See *P. attenuatus*.

Penstemon albidus Nutt. **Stems** ascending to erect, 15–30 cm. **Herbage** puberulent. **Basal leaf blades** oblanceolate to obovate, entire to denticulate, 3–5 cm long. **Stem leaves** lanceolate, entire to denticulate, 2–7 cm long. **Inflorescence** with 3 to 6 verticillasters, densely glandular-pubescent. **Flowers:** calyx 4–7 mm long; sepals lanceolate with thin, scarious margins; corolla white, 13–20 mm long, glandular; anthers glabrous, opening across their full length. **Capsule** 4–9 mm long. Often sandy soil of grasslands, sagebrush steppe; plains, valleys. AB to MB south to NM, TX.

Penstemon angustifolius Nutt. ex Pursh **Stems** erect, 15–30 cm. **Herbage** glaucous, glabrous. **Leaf blades** mostly cauline, linear, entire, 2–9 cm long. **Inflorescence** crowded with 3 to 6 glabrous verticillasters. **Flowers:** calyx 4–8 mm long; sepals lanceolate, acuminate, slightly scarious-margined; corolla blue to lavender, 13–18 mm long; anthers glabrous, opening across their length. **Capsule** 6–10 mm long. Sandy soils of grasslands, pine woodlands; plains. Carter and Dawson cos. MT, ND south to AZ, NM, CO, OK. Our plants are var. **angustifolius**.

Penstemon arenicola A.Nelson **Stems** erect, 8–20 cm. **Herbage** glabrous, glaucous. **Basal leaf blades** thick, linear-oblanceolate, entire, 3–6 cm long. **Stem leaves** lanceolate, 2–4 cm long. **Inflorescence** of 2 to 6 nearly contiguous, dense, glabrous verticillasters; bracts lanceolate. **Flowers:** calyx 5–7 mm long; sepals lanceolate, acuminate, slightly scarious-margined; corolla blue to lavender, 11–15 mm long; anthers glabrous, opening across their length. **Capsule** 7–10 mm long. Rocky or sandy soil of sagebrush steppe, grassland; valleys. MT, ID, WY, UT, CO.

Several specimens appear intermediate to *Penstemon nitidus*.

Penstemon aridus Rydb. **Stems** 5–25 cm. **Herbage** glabrous to puberulent on the stems.
Basal leaf blades densely clustered, linear or linear-oblanceolate, entire, 1–5 cm long. **Stem**
leaves linear, entire, 1–5 cm long. **Inflorescence** of 1 to 5, open, few-flowered, sparsely glandular
verticillasters. **Flowers:** calyx often bluish, 3–6 mm long; sepals lanceolate to ovate, sometimes
erose- and scarious-margined; corolla blue, 10–18 mm long, glandular; anthers glabrous, opening
across their full length. **Capsule** 4–7 mm long. Rocky soil of grasslands, sagebrush steppe, woodlands, exposed ridges,
slopes; montane, subalpine. Endemic to southwest MT, adjacent ID, WY. See *P. humilis*. (p.437)
    The type locality is south of Bozeman.

Penstemon attenuatus Douglas ex Lindl. **Stems** ascending to erect, 5–70 cm. **Herbage**
glabrous to puberulent on the stems. **Basal leaf blades** oblanceolate to elliptic, entire, 2–10 cm
long. **Stem leaves** narrowly lanceolate to oblanceolate, entire to rarely denticulate, 2–10 cm long.
Inflorescence with 2 to 4 verticillasters, glandular. **Flowers:** calyx 3–6 mm long; sepals lanceolate
to narrowly ovate, acuminate, scarious-margined; corolla blue, 1–2 cm long, glandular; anthers
glabrous. **Capsule** 3–6 mm long. Stony soil; montane to alpine. WA to MT south to OR, ID, WY.

1. Sepals densely glandular, with a narrow scarious margin; plants ≥30 cm high . . . . .var. *attenuatus*
1. Sepals sparsely glandular, conspicuously scarious basally; plants <40 cm high . . . . . . var. *pseudoprocerus*

Our common form is ***Penstemon attenuatus*** var. ***pseudoprocerus*** (Rydb.) Cronquist which occurs in grasslands,
meadows, open forests, fellfields; montane to rarely alpine; *P. attenuatus* var. ***attenuatus*** is found in meadows, open
forest; montane to subalpine in northwest cos. Reports of *P. attenuatus* var. *militaris* (Greene) Cronquist were based on
a misidentified specimen (fide C. Freeman). Entire-leaved forms of *P. albertinus* are similar to var. *pseudoprocerus* but
they usually have rounded to shortly acute cauline leaves and more open verticillasters. *P. attenuatus* is thought to be
derived from *P. albertinus*.[102]

Penstemon caryi Pennell **Stems** erect, 15–35 cm. **Herbage** glabrous, often glaucous. **Basal leaf blades** linear-
oblanceolate, entire, 2–7 cm long. **Stem leaves** lanceolate to linear-lanceolate, 4–8 cm long, clasping the stem.
Inflorescence crowded, of 3 to 7 few-flowered verticillasters. **Flowers:** calyx 7–10 mm long; sepals ovate, acuminate,
erose- and scarious-margined; corolla blue to lavender, 22–32 mm long; anthers villous, not opening near the filament;
staminode sometimes nearly glabrous. **Capsule** 5–12 mm long. Stony calcareous soil of Douglas fir forest, juniper
woodlands, sagebrush steppe; montane, lower subalpine. Endemic to Carbon Co., MT and adjacent WY.
    Penstemon cyaneus is usually taller with more blunt-tipped sepals.

Penstemon confertus Douglas ex Lindl. **Stems** erect, 10–50 cm. **Herbage** glabrate. **Basal**
leaf blades lanceolate to oblanceolate, entire, 2–10 cm long. **Stem leaves** narrowly lanceolate to
ovate, 1–8 cm long. **Inflorescence** of 2 to 8 dense verticillasters, glabrous, well-separated below;
bracts scarious, erose, acuminate. **Flowers:** calyx 2–5 mm long; sepals obovate, erose- and scarious-
margined with an acuminate tip; corolla ochroleucous to whitish, 8–11 mm long; anthers glabrous,
opening across their full length. **Capsule** 3–5 mm long. Usually stony soil of grasslands, woodlands, open forest; valleys
to subalpine. BC, AB, OR, ID, MT. (p.437)
    Pink-flowered populations occur sporadically throughout the range and may indicate introgression with the closely
related *Penstemon procerus*.

Penstemon cyaneus Pennell **Stems** erect, 20–60 cm. **Herbage** glabrous, often glaucous. **Basal**
leaf blades lanceolate to oblanceolate, entire, 3–12 cm long. **Stem leaves** linear-lanceolate to
narrowly ovate, 3–10 cm long, clasping the stem. **Inflorescence** crowded, of 3 to 10 open, glabrous
verticillasters. **Flowers:** calyx 4–6 mm long; sepals ovate with rounded but erose tips, scarious-
margined; corolla blue, 24–30 mm long; anthers hispid, twisted, not opening near the filament.
Capsule 10–13 mm long. Sagebrush steppe, grasslands, often on roadcuts; valleys, montane. Endemic to southern ID,
adjacent MT, WY. (p.437)
    Penstemon lemhiensis has acuminate sepals and occurs immediately to the north of *P. cyaneus*. See *P. caryi*.

Penstemon deustus Douglas ex Lindl. Hot-rock penstemon **Stems** clustered, sprawling to
erect, 10–30 cm, some sterile. **Herbage** glabrous. **Leaves** all cauline, oblanceolate, lanceolate,
or ovate, sharply serrate, 2–6 cm long, petiolate below, sessile above. **Inflorescence** of 2 to 5,
glabrous to obscurely glandular verticillasters. **Flowers:** calyx 4–6 mm long; sepals lanceolate,
distally scarious-margined; corolla whitish, 9–14 mm long, obscurely glandular-puberulent; anthers
glabrous, opening across their full length. **Capsule** 3–6 mm long. Rock outcrops, talus slopes, stony soil of sagebrush
steppe, woodlands; montane. WA to MT south to CA, NV, UT, WY. Our plants are var. **deustus**. (p.437)

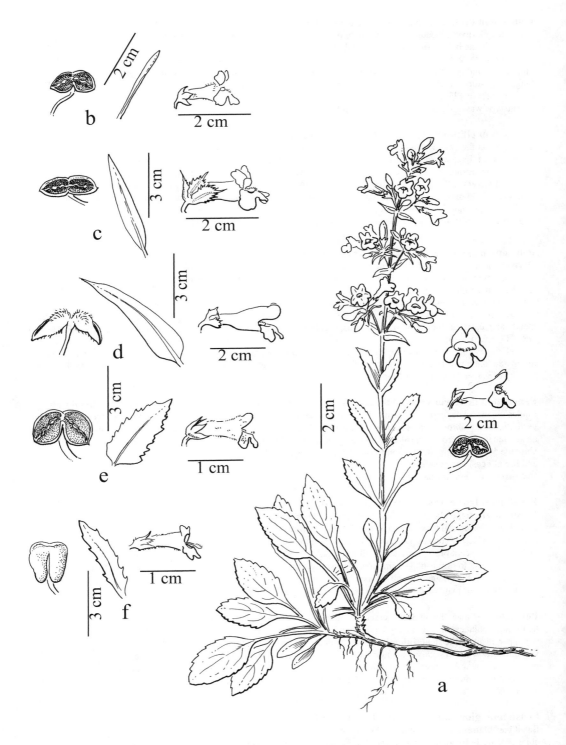

Plate 77. **Penstemon** I. **a.** *P. albertinus*, **b.** *P. aridus*, **c.** *P. confertus*, **d.** *P. cyaneus*, **e.** *P. deustus*, **f.** *P. diphyllus*

Penstemon diphyllus Rydb. **Stems** numerous, erect, 20–70 cm. **Herbage** glabrous to puberulent. **Leaves** all cauline, lanceolate to oblong, dentate to shallowly lobed, 1–6 cm long, sessile, petiolate below. **Inflorescence** glandular-pubescent, open-paniculate; bracts becoming entire. **Flowers**: calyx 3–5 mm long; sepals lanceolate, not scarious; corolla blue to violet, 12–16 mm long, glandular; anthers U-shaped, glabrous, opening only near the filament. **Capsule** 5–7 mm long. Talus slopes, rock outcrops, stony soil of open forest, streambanks, roadcuts; montane. Endemic to central ID, adjacent MT. (p.437)

The type was collected west of Thompson Falls.

Penstemon ellipticus J.M.Coult. & E.Fisch. **Stems** sterile and fertile, mat-forming, prostrate and woody at the base, ascending distally, 5-15 cm. **Herbage** glabrous, puberulent on the stems. **Leaves** all cauline, serrate; those of sterile stem petiolate, blades elliptic,1–3 cm long; those of fertile stems sessile, elliptic to obovate, serrate. **Inflorescence** a few-flowered, glandular-pubescent raceme. **Flowers**: calyx 8–15 mm long; sepals linear to lanceolate, 7–9 mm long; corolla light lavender, 25–35 mm long, glabrous; anthers woolly-villous, opening across their full length. **Capsule** 8–11 mm long. Rock outcrops, moraine, exposed slopes; subalpine, alpine. BC, AB, ID, MT. (p.454)

Sometimes mistaken for *Penstemon fruticosus* which has narrower leaves and occurs further southeast. See *P. lyallii*.

Penstemon eriantherus Pursh FUZZY-TONGUE PENSTEMON **Stems** ascending to erect, 5–30 cm. **Herbage** puberulent. **Basal leaf blades** oblanceolate to linear-oblanceolate, usually serrate, 2–6 cm long. **Stem leaves** lanceolate to linear-lanceolate, 2–6 cm long. **Inflorescence** crowded, of 2 to 6 glandular-villous verticillasters. **Flowers**: calyx 6–12 mm long; sepals lanceolate, not scarious; corolla lavender to pinkish with purple lines, yellow-bearded within, 15–30 mm long, upper portion of tube expanded; anthers glabrous, opening across their full length; staminode long-hairy. **Capsule** 6–12 mm long, glandular-pubescent. Stony soil of grasslands, sagebrush steppe, woodlands; plains, valleys, montane. BC, AB south to OR, CO, NE. (p.440)

Both var. *eriantherus* and var. *redactus* Pennell & D.D.Keck are reported for MT. However, the characters purported to separate them overlap, are not correlated, and the vars.occur together in the southwestern part of the state. The type locality of var. *redactus* is near Monida.

Penstemon flavescens Pennell **Stems** erect, 8–30 cm. **Herbage** glabrous to puberulent on the stem. **Basal leaf blades** oblanceolate, entire, 1–6 cm long. **Stem leaves** lanceolate to oblong, 2–5 cm long. **Inflorescence** of 1 to 3 dense verticillasters, glabrous; bracts scarious. **Flowers**: calyx 4–8 mm long; sepals ovate, erose- and scarious-margined with an acuminate tip; corolla ochroleucus to light yellow, 12–15 mm long; anthers glabrous, opening across their full length. **Capsule** 5–7 mm long. Stony, granitic soil of open forest, meadows; subalpine. Endemic to Bitterroot Mtns. of Mineral and Ravalli cos. and adjacent ID.

Penstemon confertus has smaller flowers.

Penstemon fruticosus (Pursh) Greene **Stems** sterile and fertile, woody, erect to ascending, 15–30 cm. **Herbage** glabrous to puberulent on the stems. **Leaves** all cauline, clustered near the base; lower leaves and those of sterile stems petiolate with elliptic, mostly entire blades 1–6 cm long; those of fertile stems sessile, elliptic, mostly entire. **Inflorescence** a few- to several-flowered, glandular-pubescent bracteate raceme. **Flowers**: calyx 6–10 mm long; sepals lanceolate, acuminate; corolla lavender to purplish, 3–5 cm long, glabrous; anthers woolly-villous, opening across their full length. **Capsule** 5–12 mm long. Stony, granitic soil of open forest, outcrops, talus slopes; montane, lower subalpine. BC, MT south to OR, ID, WY. Our plants are var. *fruticosus*. See *P. ellipticus*. (p.440)

Penstemon glaber Pursh **Stems** erect, 10–50 cm. **Herbage** glabrous to puberulent. **Basal leaf blades** oblanceolate to oblong, entire, 2–8 cm long. **Stem leaves** lanceolate, 3–12 cm long, the upper clasping the stem. **Inflorescence** crowded above, of 2 to 10 few-flowered, glabrous verticillasters. **Flowers**: calyx 2–3 mm long; sepals obovate, scarious-margined with a short acute tip; corolla blue to lavender, 22–35 mm long; anthers hispid, not opening near the filament. **Capsule** 8–12 mm long. Stony, shale-derived soil of grasslands, pine woodlands; plains. MT, ND south to NM, NE. Our plants are var. *glaber*.

Penstemon globosus (Piper) Pennell & D.D.Keck **Stems** erect, 10–40 cm. **Herbage** glabrous. **Basal leaf blades** elliptic to oblanceolate, entire, 2–7 cm long. **Stem leaves** lanceolate to oblanceolate, 1–10 cm long, the upper clasping the stem. **Inflorescence** of 1 to 3 glabrous, well-separated verticillasters. **Flowers**: calyx 4–7 mm long; sepals ovate, erose and widely scarious-margined below with an acuminate tip; corolla blue to violet, 14–16 mm long, glabrous; anthers

glabrous, opening narrowly with pouch-like tips, not confluent across the connective. **Capsule** 6–7 mm long. Moist to wet meadows; subalpine. Endemic to central ID, adjacent OR, MT. (p.440)

Dehisced anther sacs of *Penstemon rydbergii* open widely and fully across their length.

Penstemon gracilis Nutt. **Stems** erect, 15–50 cm. **Herbage** glabrous to puberulent on the stem. **Basal leaf blades** oblanceolate, remotely denticulate, 2–6 cm long. **Stem leaves** narrowly lanceolate, 2–8 cm long, clasping, denticulate. **Inflorescence** of 3 to 5 verticillasters, glandular-pubescent. **Flowers**: calyx 5–6 mm long; sepals lanceolate, barely scarious; corolla blue to pale violet, 14–20 mm long, glandular; anthers glabrous, opening across their full length. **Capsule** 4–8 mm long. Grasslands, woodlands; plains, valleys. BC to ON south to NM, CO, NE.

A Great Plains species, but reported for Sanders Co. (Hitchcock et al 1959).

Penstemon grandiflorus Nutt. **Stems** erect, 40–90 cm. **Herbage** glabrous, glaucous. **Basal leaf blades** oblong, entire, 2–12 cm long. **Stem leaves** spatulate to oribucular, 1–9 cm long, the upper clasping the stem. **Inflorescence** interrupted, of 3 to 7 few-flowered, glabrous verticillasters. **Flowers**: calyx 7–11 mm long; sepals lanceolate, acuminate, not scarious; corolla blue, 35–48 mm long; anthers glabrous, opening across their full length. **Capsule** 16–20 mm long. Grasslands, open cottonwood forests; plains. MT to IN south to TX. One collection from Custer Co.; a collection from Phillips Co. represents a reclamation planting.

Penstemon humilis Nutt. ex A.Gray **Stems** 5–25 cm. **Herbage** densely puberulent. **Basal leaf blades** oblanceolate, entire, 1–5 cm long. **Stem leaves** narrowly oblanceolate, entire, 1–5 cm long. **Inflorescence** of 2 to 7 glandular verticillasters. **Flowers**: calyx 3–5 mm long; sepals lanceolate to ovate, acuminate, thinly scarious-margined; corolla blue, 10–16 mm long, glandular; anthers glabrous, opening across their full length. **Capsule** 3–6 mm long. Sagebrush steppe; montane. WA to MT south to CA, NV, UT, CO. One collection from Beaverhead Co.; reported for Madison Co.[125]

Penstemon aridus has glabrous, narrower leaves. Some specimens of *P. humilis* from southern Beaverhead and Madison cos. have exceptionally puberulent and linear leaves.

Penstemon laricifolius Hook. & Arn. **Stems** 5–20 cm. **Herbage** glabrous, often with puberulent stems. **Basal leaf blades** densely clustered, filiform, entire, 1–4 cm long. **Stem leaves** similar. **Inflorescence** an open, glabrous raceme or panicle. **Flowers**: calyx 3–6 mm long; sepals lanceolate, acuminate, basally scarious-margined; corolla rose to burgundy, 11–20 mm long, glabrous; anthers glabrous, opening across their full length. **Capsule** ca. 5 mm long. Stony, calcareous soil of grasslands, juniper woodlands; montane. Carbon and Big Horn cos., MT to CO.

Penstemon lemhiensis (D.D.Keck) D.D.Keck & Cronquist **Stems** erect, 20–70 cm. **Herbage** puberulent. **Basal leaf blades** lanceolate to linear-lanceolate, entire, 3–12 cm long. **Stem leaves** linear-lanceolate to narrowly ovate, 3–10 cm long, clasping the stem. **Inflorescence** crowded, of 3 to 8 open, glabrous verticillasters. **Flowers**: calyx 6–11 mm long; sepals lanceolate, acuminate with scarious margins; corolla blue, 22–35 mm long; anthers hispid, twisted, not opening near the filament; staminode glabrous. **Capsule** 6–12 mm long. Usually sparsely vegetated soil of grasslands, sagebrush steppe, open forest, often along roads; montane. Endemic to southwest MT, adjacent ID. See *P. cyaneus*. (p.440)

Penstemon lyallii A.Gray **Stems** ascending to erect, often woody at the base, 10–60 cm. **Herbage** puberulent at least on the stems. **Leaves** all cauline, narrowly lanceolate, obscurely serrate, 2–12 cm long, mostly sessile, smaller and petiolate below. **Inflorescence** glandular-pubescent, open-paniculate. **Flowers**: calyx 9–18 mm long; sepals narrowly lanceolate, not scarious; corolla pale lavender, 25–40 mm long, glabrous, hairy within; anthers woolly-villous, opening across their full length; staminode glabrous. **Capsule** 1–2 cm long. Stony, sparsely vegetated soil of talus, rock outcrops, open slopes, often along roads; montane, subalpine. BC, AB, WA, ID, MT. (p.440)

Specimens with lanceolate leaves 3 to 4 times as long as wide have been collected in the area of the Bob Marshall Wilderness; these may represent hybrids with *Penstemon ellipticus*.[187]

Penstemon montanus Greene **Stems** sterile and fertile, sometimes mat-forming, prostrate to ascending, 5–30 cm. **Herbage** glandular-pubescent. **Leaves** all cauline, greatly reduced below, ovate, serrate, short-petiolate below to clasping above, 1–3 cm long. **Inflorescence** a few-flowered, glandular-pubescent bracteate raceme. **Flowers**: calyx 6–13 mm long; sepals narrowly lanceolate, not scarious; corolla pale lavender, 25–33 mm long, glabrous; anthers woolly-villous, opening across their full length. **Capsule** 7–13 mm long. Talus and open, often eroding slopes; montane, subalpine. ID, MT, WY, UT.

Penstemon fruticosus and *P. ellipticus* do not have glandular foliage.

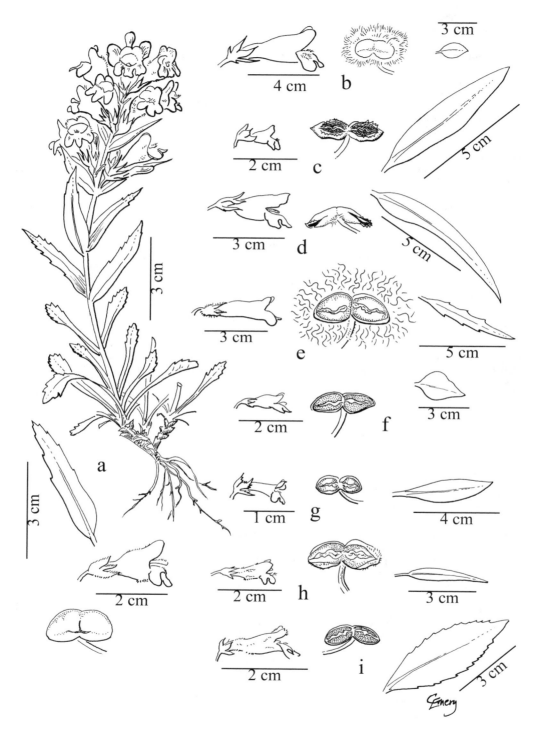

Plate 78. **Penstemon II. a.** *P. eriantherus*, **b.** *P. fruticosus*, **c.** P. globosus, **d.** *P. lemhiensis*, **e.** *P. lyallii*, **f.** *P. nitidus*, **g.** *P. procerus*, **h.** *P. radicosus*, **i.** *P. wilcoxii*

Penstemon nitidus Douglas ex Benth. **Stems** erect, 10–35 cm. **Herbage** glabrous, glaucous. **Basal leaf blades** thick, oblanceolate to oblong, entire, 2–10 cm long. **Stem leaves** lanceolate to obovate, 1–6 cm long. **Inflorescence** of 3 to 10 nearly contiguous, dense, glabrous verticillasters; bracts lanceolate to ovate. **Flowers**: calyx 3–6 mm long; sepals lanceolate, acuminate, slightly scarious-margined; corolla blue to lavender, 12–16 mm long; anthers glabrous, opening across their length. **Capsule** 6–15 mm long. BC to MB south to WA, CO, ND. (p.440)

1. Inflorescence bracts lanceolate to narrowly ovate, mostly not clasping . . . . var. *polyphyllus*
1. Bracts ovate and clasping . . . . var. *nitidus*

Penstemon nitidus var. *nitidus* is found in barren sites in grasslands, sagebrush steppe, coniferous woodlands; plains, valleys to rarely subalpine throughout the area; *P. nitidus* var. *polyphyllus* (Pennell) Cronquist occurs in similar habitats, valleys to montane in the southern cos. The type locality for var. *polyphyllus* is near Missoula. High-elevation plants from Teton Co. approach var. *polyphyllus*. See *P. arenicola*.

Penstemon procerus Douglas ex Graham [*P. confertus* Douglas ex Lindl. var. *procerus* (Douglas ex Graham) Coville] **Stems** erect, 5–50 cm. **Herbage** glabrous or puberulent on the stems. **Basal leaf blades** oblanceolate, entire, 2–6 cm long. **Stem leaves** lanceolate to linear-lanceolate, 1–8 cm long. **Inflorescence** of 2 to 4 well-separated, dense, glabrous verticillasters; bracts scarious-margined, acuminate. **Flowers**: calyx 3–4 mm long; sepals lanceolate, erose- and scarious-margined, acuminate; corolla blue, sometimes with a white tube, 6–10 mm long, glabrous; anthers glabrous, opening across their full length. **Capsule** 3–5 mm long. Moist meadows, grasslands, open forest; valleys to alpine. AK to MB south to CA, CO. Our plants are var. *procerus*. (p.440)
    Ochroleucus- to white-flowered plants of *Penstemon procerus* occur sporadically in the northwest portion of MT; these may be the result of introgression with *P. confertus*.

Penstemon radicosus A.Nelson **Stems** ascending to erect, 20–40 cm, woody at the base. **Herbage** puberulent. **Leaves** all cauline, narrowly lanceolate, entire, 2–6 cm long, sessile. **Inflorescence** of 2 to 5 glandular-pubescent, open verticillasters. **Flowers**: calyx 4–7 mm long; sepals lanceolate, acuminate, scarious-margined; corolla blue, 16–21 mm long, glandular; anthers glabrous, opening across their full length, spreading, V-shaped. **Capsule** 5–8 mm long. Sagebrush steppe, open forest; montane. ID, MT south to NV, UT, CO. (p.440)

Penstemon rydbergii A.Nelson **Stems** erect, 20–40 cm. **Herbage** glabrous to puberulent on the stem. **Basal leaf blades** oblanceolate to obovate, entire, 2–7 cm long. **Stem leaves** lanceolate to oblanceolate, 2–7 cm long, the lower petiolate. **Inflorescence** of 2 to 7 glabrous, dense verticillasters. **Flowers**: calyx 3–7 mm long; sepals obovate, erose and widely scarious-margined with an acuminate tip; corolla blue to violet, 11–16 mm long, glabrous; anthers glabrous, opening across their full length. **Capsule** 5–10 mm long. Sagebrush steppe, open forest; montane. WA to MT south to CA, AZ, NM.
    Penstemon procerus hybridizes with *P. globosus* in northern Beaverhead Co.[86] I believe these hybrid plants should be referred to *P. procerus*. See *P. globosus*.

Penstemon whippleanus A.Gray **Stems** ascending to erect, 20–60 cm. **Herbage** glabrous. **Basal leaf blades** lanceolate to narrowly ovate, entire to weakly serrate, 3–8 cm long. **Stem leaves** lanceolate to oblanceolate, entire, 2–6 cm long, clasping. **Inflorescence** with 2 to 5 few-flowered, glandular-pubescent verticillasters. **Flowers**: calyx 7–10 mm long; sepals narrowly lanceolate, acuminate, basally scarious-margined; corolla whitish to dull maroon, 20–27 mm long, glandular; anthers glabrous, opening across their full length. **Capsule** 6–9 mm long, glandular. Meadows, grasslands, open forest; subalpine. Beaverhead and Gallatin cos., MT, adjacent ID south to AZ, NM.

Penstemon wilcoxii Rydb. **Stems** ascending to erect, 30–80 cm. **Herbage** glabrous to sparsely puberulent. **Basal leaf blades** lanceolate to oblanceolate, serrate, 3–9 cm long. **Stem leaves** lanceolate to ovate, serrate to denticulate, 3–7 cm long, often clasping the stem, larger than the basal blades. **Inflorescence** with 3 to 8, open verticillasters, glabrous to sparsely glandular. **Flowers**: calyx 2–5 mm long; sepals lanceolate to ovate, acute, scarious-margined; corolla blue, 12–21 mm long, glandular; anthers glabrous, opening across their full length. **Capsule** 4–7 mm long. Rocky soil of woodlands, open forest, shrub fields; valleys, montane. Endemic to eastern OR, WA to northwest MT. (p.440)
    The type locality is near Kalispell. See *Penstemon albertinus*.

Plantago L. Plantain

Taprooted annual or fibrous-rooted perennial herbs. **Leaves** all basal, simple, entire, petiolate, prominently veined. **Inflorescence** a terminal, bracteate spike; bracts hyaline-margined. **Flowers** slightly irregular, perfect, 4-merous; sepals hyaline-margined, separate to near the base; corolla sympetalous, hyaline-white, the tube enclosed by sepals, the lobes spreading; stigma 2-lobed; stamens 4, exserted. **Capsule** few-seeded.

1. Plants taprooted annuals 2
1. Fibrous-rooted perennials with a stout caudex. 3

2. Bracts and sepals glabrous . . . .*P. elongata*
2. Bracts and sepals villous . . . .*P. patagonica*

3. Scape (peduncle) ≥5 times as long as the spike . . . . *P. lanceolata*
3. Scape usually ≤4 times as long as the spike 4

4. Leaf blades ovate to elliptic, <2.5 times as long as wide . . . . *P. major*
4. Leaf blades lanceolate to oblanceolate, >2.5 times as long as wide 5

5. Corolla lobes 2–3 mm long, erect in fruit . . . . *P. hirtella*
5. Corolla lobes 1–2 mm long, remaining reflexed in fruit 6

6. Summit of caudex brown woolly-villous . . . .*P. eriopoda*
6. Caudex not brown-woolly. 7

7. Leaves villous; corolla lobes ca. 2 mm long. . . . . *P. canescens*
7. Leaves glabrous; corolla lobes ca. 1 mm long . . . .*P. tweedyi*

Plantago canescens Adams Perennial. **Herbage** villous. **Leaf blades** lanceolate, 2–8 cm long. **Inflorescence**: scape 5–15 cm; spike 2–6 cm long, cylindrical; bracts ca. 2 mm long, glabrous to ciliate. **Flowers**: sepals ovate, ca. 2 mm long; corolla lobes 1.5–2 mm long. **Capsules** 2–3 mm long. Fescue grasslands, limber pine savanna; valleys to lower subalpine. AK to BC, AB, MT; Asia. (p.432)

Plantago elongata Pursh Annual. **Herbage** puberulent. **Leaf blades** linear, thread-like, 1–5 cm long. **Inflorescence**: scape 2–8 cm; spike glabrous, linear, 5–65 mm long; bracts 1.5–2 mm long. **Flowers** partly unisexual with non-functional stamens and pistils; sepals glabrous; corolla lobes 0.5–1 mm long. **Capsules** 2–3.5 mm long. Vernally moist, often saline soil of meadows, grasslands, margins of wetlands; plains, valleys. BC to MB south to CA, TX, Mexico. Our plants are var. *elongata*.

Plantago eriopoda Torr. Perennial. **Herbage** succulent, glabrous to sparsely villous; caudex brown-woolly. **Leaf blades** oblanceolate to elliptic, 3–20 cm long. **Inflorescence**: scape 4–25 cm; spikes 3–20 cm long, narrowly cylindric; bracts glabrous, 2–3 mm long. **Flowers**: sepals glabrous; corolla lobes 1–2 mm long. **Capsule** 3–4 mm long. Fine-textured, moist, saline or alkaline soil, often around wetlands, streams; plains, valleys, montane. YT to MN south to CA, NV, UT, WY, NE.

Plantago hirtella Kunth Perennial. **Herbage** succulent, glabrous to villous; sparingly brown-woolly on the caudex. **Leaf blades** oblanceolate to narrowly elliptic, 5–20 cm long. **Inflorescence**: scape 5–40 cm; spikes congested to somewhat sparse, 5–25 cm long, narrowly cylindric; bracts ca. 3 mm long. **Flowers**: sepals glabrous; corolla lobes 2–3 mm long. **Capsule** 2–5 mm long, covered by the persistent, erect corolla lobes. Moist saline soil around springs, wetlands; valleys. Coastal WA south to S. America; collected once in Ravalli Co. where it was probably introduced.

Plantago lanceolata L. Eɴɢʟɪsʜ ᴘʟᴀɴᴛᴀɪɴ Perennial. **Herbage** glabrous to villous; caudex sparingly brown-woolly. **Leaf blades** narrowly lanceolate, 2–15 cm long. **Inflorescence**: scape 10–60 cm; spikes congested, 5–50 mm long, conical, becoming cylindric; bracts 1.5–2.5 mm long, glabrous, acute to acuminate. **Flowers**: sepals apically ciliate, the outer 2 united; corolla lobes 1.5–2.5 mm long. **Capsule** 2–3 mm long. Roadsides, trails, streambanks, disturbed grasslands, around buildings, lawns; plains, valleys. Cosmopolitan; introduced from Eurasia.

Plantago major L. Cᴏᴍᴍᴏɴ ᴘʟᴀɴᴛᴀɪɴ Perennial. **Herbage** glabrous to puberulent. **Leaf blades** ovate, 3–15 cm long. **Inflorescence**: scape 5–25 cm; spikes congested to somewhat sparse, 3–25 cm long, narrowly cylindric; bracts 1.5–3 mm long. **Flowers**: sepals glabrous; corolla lobes 0.5–1

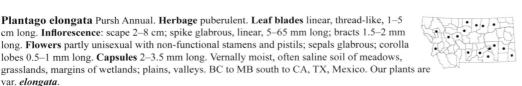

mm long. **Capsule** 2–4 mm long. Roadsides, trails, streambanks, around buildings, lawns; plains, valleys. Cosmopolitan; introduced from Eurasia.

Plantago patagonica Jacq. Annual. **Herbage** villous. **Leaf blades** linear to narrowly oblanceolate, 1–10 cm long. **Inflorescence**: scape 3–15 cm; spike villous, cylindrical, 5–120 mm long; bracts narrowly lanceolate, 2–12 mm long. **Flowers**: sepals villous, 2–4 mm long; corolla lobes 1.5–2 mm long. **Capsules** 2–4 mm long. Grasslands, sagebrush steppe, fields, roadsides; plains, valleys. BC to SK south to CA, AZ, NM, TX, Mexico. (p.432)

Plants with flower bracts greatly exceeding the flowers occur sporadically throughout the range of the species and have been called var. *spinulosa* (Dcne.) A.Gray. They are often confused with *P. aristata* Michx., which occurs east of MT and has glabrous upper leaf surfaces.

Plantago tweedyi A.Gray Perennial. **Herbage** succulent, glabrous, sparsely pubescent on the scapes, barely woolly on the caudex. **Leaf blades** lanceolate to oblanceolate, sometimes remotely crenate, 3–13 cm long. **Inflorescence**: scape 5–20 cm; spikes 2–10 cm long, cylindric; bracts deltoid, glabrous, 1–2 mm long. **Flowers**: sepals glabrous, 2–3 mm long; corolla lobes 0.5–1 mm long. **Capsule** 3–4 mm long. Grasslands, sagebrush steppe, moist meadows; montane, lower subalpine. ID, MT south to AZ, UT, CO.

Synthyris Benth. Kittentails

Fibrous-rooted, scapose perennials. **Leaves** basal, petiolate, pinnately or palmately lobed. **Inflorescence** a spike-like, bracteate raceme surmounting an alternate-bracteate scape, lengthening in fruit. **Flowers**: calyx divided into 4 sepals; corolla blue, subrotate, 4-lobed, upper lobes larger; stamens 2; stigma capitate. **Capsule** orbicular, compressed, shallowly notched at the style.

1. Leaves pinnately divided to the midvein. . . . . *S. pinnatifida*
1. Leaves lobed but not divided to the midvein 2

2. Leaf margins more crenate than lobed; scape longer than leaves. . . . . *S. missurica*
2. Leaves lobed >1/2-way to midvein; scape sometimes shorter than the leaves . . . . *S. canbyi*

Synthyris pinnatifida S.Watson [*S. dissecta* Rydb.] **Herbage** glabrate to villous. **Leaf blades** ovate to lanceolate, 1–6 cm long, pinnately divided; leaflets deeply lobed; ultimate segments deltoid. **Inflorescence**: scape 3–12 cm long, as long or longer than the leaves; bracts ciliate. **Flowers**: calyx 3–4 mm long; corolla 4–8 mm long, lobes ca. as long as the tube. **Capsule** 4–6 mm long, puberulent. Stony, usually non-calcareous soil of fellfields; alpine. WA to MT south to ID, UT, WY. Our plants are var. **canescens** Cronquist.

Recent molecular studies suggest that our plants may be distinct from more southern plants and should be called *Synthyris dissecta*.[195]

Synthyris canbyi Pennell **Herbage** glabrous to sparsely villous. **Leaf blades** ovate-cordate to orbicular, 1–4 cm long, 1 to 2 times palmately lobed >1/2-way to the midvein, ultimate segments dentate. **Inflorescence** villous; scape 4–10 cm long, shorter to longer than the leaves; bracts serrate. **Flowers**: calyx 3–6 mm long; corolla 5–10 mm long, lobes ca. as long as the tube. **Capsule** 4–6 mm long, pubescent. Stony, calcareous soil of fellfields, talus slopes; alpine. Endemic to Lake and Missoula cos.

Specimens from the southern end of the Mission Range appear intermediate between *Synthyris canbyi* and *S. pinnatifida*.

Synthyris missurica (Raf.) Penell **Herbage** glabrous. **Leaf blades** orbicular-cordate, 2–6 cm wide, shallowly lobed, crenate. **Inflorescence**: scape 6–20 cm long, longer than the leaves. **Flowers**: calyx 1–3 mm long; corolla 3–7 mm long, lobes longer than the tube. **Capsule** 3–6 mm long, pubescent. Moist meadows in forest openings; subalpine. WA to MT south to CA. Known from Ravalli Co.

Veronica L. Speedwell, betony

Herbaceous annuals or perennials. **Leaves** cauline, simple, opposite. **Inflorescence** axillary or terminal racemes with alternate, usually minute bracts. **Flowers** rotate; calyx deeply 4-lobed; corolla irregular with a short tube, 4-lobed, the upper lobe largest, the lowest smaller; stamens 2, exserted; stigma capitate.

Capsule compressed, obovate to orbicular, shallowly to deeply notched at the style, opening by 2 or 4 valves.

The shape of the mature capsules is often helpful in distinguishing species.

1. Flowers in racemes (often paired) arising from axils of upper leaves 2
1. Flowers solitary and axillary or in spikes or racemes terminating the stem 7

2. Lower leaves and stem hairy; plants of lawns and disturbed uplands 3
2. Plants glabrous; occurring around streams and ponds 4

3. Leaves sessile with 10 to 22 marginal teeth; corolla 9–12 mm across. . . . . *V. chamaedrys*
3. Leaves short-petiolate with ≥24 marginal teeth; corolla 4-7 mm across. . . . . *V. officinalis*

4. Leaves petiolate at mid-stem . . . . *V. americana*
4. Leaves sessile 5

5. Capsules wider than high, deeply notched. . . . . *V. scutellata*
5. Capsules as high or higher than wide, obscurely notched 6

6. Leaves 2 to 3 times as long as wide; mature pedicels upturned . . . . *V. anagallis-aquatica*
6. Leaves 3 to 5 times as long as wide; mature pedicels not upturned . . . . *V. catenata*

7. Plants annual, often with a slender taproot; flowers often just 2–3 mm across 8
7. Plants perennial, with slender rhizomes or stolons; flowers 3–10 mm across 12

8. Upper stem leaves with 1 to many lobes . . . . *V. verna*
8. Stem leaves sometimes toothed but not lobed. 9

9. Fruiting pedicels ca. 1 mm long 10
9. Fruiting pedicels >2 mm long 11

10. Leaves strap-shaped; flowers white. . . . . *V. peregrina*
10. Leaves ovate; flowers blue. . . . . *V. arvensis*

11. Fruiting sepals ≤5 mm long; fruiting pedicels <10 mm long . . . . *V. biloba*
11. Fruiting sepals .5 mm long; fruiting pedicels >12 mm long. . . . . *V. persica*

12. Leaves sharply serrate; stems ≥40 cm. . . . . *V. longifolia*
12. Leaves entire to shallowly crenate; stems mostly <40 cm 13

13. Styles 4–10 mm long . . . . *V. cusickii*
13. Styles <4 mm long 14

14. Stems trailing at the base, short-hairy . . . . *V. serpyllifolia*
14. Stems with long hairs, erect or curved at the base, not trailing . . . . *V. wormskjoldii*

Veronica americana Schwein. ex Benth. AMERICAN BROOKLIME Rhizomatous perennial. **Stems** often branched, basally prostrate, rooting at the nodes, terminally ascending, 5–40 cm. **Herbage** glabrous. **Leaves** short-petiolate; blades 1–6 cm long, lanceolate to ovate, serrate. **Inflorescence** of axillary racemes, glabrous; pedicels 5–12 mm long. **Flowers**: calyx 2–4 mm long; corolla blue, 5–9 mm across; style 2–3 mm long. **Capsule** broadly ovate, 2–4 mm long, glabrous, barely notched. Slow, shallow water or wet banks of streams, ditches, wetlands, wet meadows; plains, valleys, montane. AK to NL south to CA, AZ, NM, TX, NC, Mexico. (p.445)

Veronica anagallis-aquatica L. Rhizomatous perennial. **Stems** ascending, simple or branched, 10–50 cm. **Herbage** glabrous. **Leaves** sessile, 2–6 cm long, lanceolate, crenate, the upper clasping. **Inflorescence** glabrous to sparsely glandular-pubescent, axillary racemes; pedicels 3–7 mm long. **Flowers**: calyx 2–3 mm long; corolla blue, 3–5 mm across; style 2–3 mm long. **Capsule** obovoid, 2–4 mm long, glabrous, barely notched. Mud or shallow water along streams, rivers, lakes; plains, valleys. Introduced to most of N. America; native to Eurasia. (p.445)

Fully open flowers are uncommon. *Veronica catenata* often has whitish flowers.

Veronica arvensis L. Annual. **Stems** prostrate to erect, simple or branched, 5–25 cm. **Herbage** puberulent to glandular-pubescent. **Leaves** short-petiolate; blades 5–12 mm long, ovate, crenate to deeply serrate. **Inflorescence** a terminal, glandular-pubescent to hirsute raceme; pedicels ca. 1 mm long. **Flowers**: calyx 2–3 mm long; sepals unequal; corolla deep blue, 2–4 mm across; style ca. 0.5 mm long. **Capsule** obcordate, deeply notched, 2–4 mm high, sparsely glandular-villous. Disturbed areas in grasslands, roadsides; valleys. Introduced to temperate N. America; native to Eurasia.

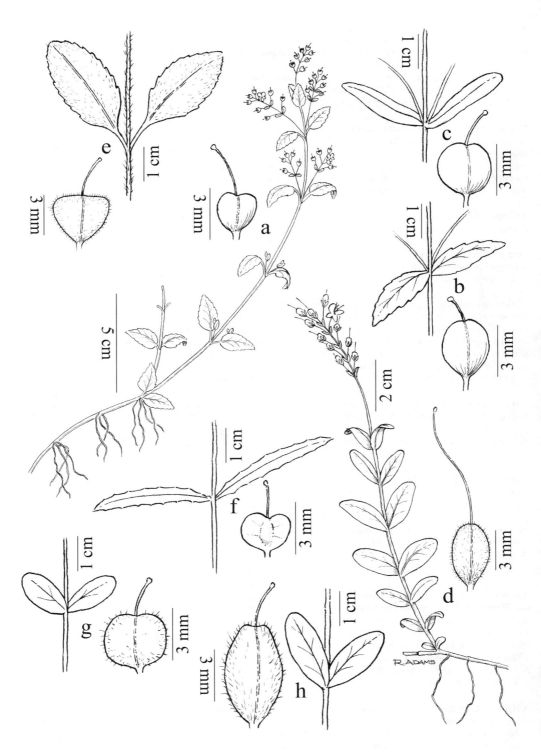

Plate 79. *Veronica*. **a.** *V. americana*, **b.** *V. anagalis-aquatica*, **c.** *V. catenata*, **d.** *V. cusickii*,
e. *V. officinalis*, **f.** *V. scutellata*, **g.** *V. serpylifolia* var. *humifusa*, **h.** *V. wormskjoldii*

Veronica biloba L. Annual. **Stems** erect, simple or branched, 5–20 cm. **Herbage** glandular-pubescent. **Leaves** few, short-petiolate, deciduous; blades 5–20 mm long, elliptic, serrate. **Inflorescence** a terminal, viscid raceme; bracts lobed, leaf-like; pedicels 4–15 mm long. **Flowers**: calyx 2–4 mm long, swelling in fruit; corolla blue, 2–4 mm across; style 0.5–1 mm long. **Capsule** reniform to obcordate, deeply lobed, 3–4 mm high, glandular-hispid. Disturbed soil of grasslands, roadsides; valleys. Introduced in the western U.S.; native to Eurasia.

Flowers open for a very short time or not at all.

Veronica catenata Pennell Rhizomatous perennial. **Stems** ascending, simple or branched, 10–45 cm. **Herbage** glabrate. **Leaves** 2–6 cm long, sessile, clasping, lanceolate, entire to obscurely serrate. **Inflorescence** of glabrous to sparsely glandular, axillary racemes; pedicels 3–7 mm long. **Flowers**: calyx 2–3 mm long; corolla white to blue, 3–5 mm across; style 1–2 mm long. **Capsule** globose, 2–4 mm long, obscurely notched. Slow water along streams; plains, valleys, montane. Much of temperate N. America, Europe. See *V. anagallis-aquatica*. (p.445)

Veronica chamaedrys L. Rhizomatous perennial. **Stems** often branched, basally prostrate, rooting at the nodes, terminally ascending 10–30 cm. **Herbage** sparsely pubescent. **Leaves** opposite, 1–3 cm long, sessile or short-petiolate, ovate, deeply serrate. **Inflorescence** of puberulent axillary racemes; pedicels 5–7 mm long. **Flowers**: calyx 3–4 mm long; corolla blue, 9–12 mm across; style 3–5 mm long. **Capsules** not maturing in our material. Lawns, valleys; Missoula Co. Introduced to eastern and northwest N. America; native to Europe.

The more common *Veronica officinalis* has more finely toothed leaves and smaller flowers.

Veronica cusickii A.Gray Rhizomatous perennial. **Stems** ascending to erect, simple, 5–25 cm. **Herbage** sparsely villous on the stem. **Leaves** opposite, ovate to obovate, entire, 8–25 mm long, sessile. **Inflorescence** a terminal, glandular-villous raceme; pedicels 2–6 mm long. **Flowers**: calyx 2–5 mm long, the sepals unequal; corolla dark blue, 7–12 mm across; style 6–10 mm long. **Capsule** obovate, 4–6 mm long, glandular-pubescent, obviously notched. Moist to wet meadows, turf, often along streams; subalpine, alpine. WA to MT south to CA, OR, ID.

Veronica wormskjoldii has pubescent foliage and a shorter style. (p.445)

Veronica longifolia L. Stoloniferous perennial. **Stems** 40–100 cm, erect, creeping at the base. **Herbage** puberulent. **Leaves** opposite or whorled, lanceolate, serrate, petiolate, 4–12 cm long. **Inflorescence** a terminal, densely flowered, puberulent raceme; pedicels 1–2 mm long. **Flowers**: calyx 2–4 mm long; corolla blue to violet, 6–9 mm across; style 6–10 mm long. **Capsule** ovoid, 2–4 mm long, glabrous, shallowly notched. Open forest, roadsides, fields. Introduced sporadically across northern U.S. and adjacent Canada. One collection from Sanders Co.

Veronica officinalis L. Rhizomatous perennial. **Stems** often branched, basally prostrate and rooting at the nodes, terminally ascending 5–25 cm. **Herbage** pubescent to hirsute. **Leaves** short-petiolate; blades 15–40 mm long, ovate to obovate, serrulate. **Inflorescence** of glandular-pubescent, axillary racemes; pedicels ca. 1 mm long. **Flowers**: calyx 1–3 mm long; corolla light blue, 4–7 mm across; style 3–4 mm long. **Capsule** obtriangular, not notched, 3–5 mm long, glandular. Lawns, roadsides and open to closed, usually disturbed forest; valleys, montane. Introduced to eastern and northwest N. America; native to Europe. See *V. chamaedrys*. (p.445)

One of the few exotics that invades a closed forest canopy.

Veronica peregrina L. Annual. **Stems** ascending to erect, simple or branched below, 2–30 cm. **Herbage** glandular-pubescent. **Leaves** few, sessile, 5–20 mm long, oblanceolate, obscurely crenate to entire. **Inflorescence** a terminal, glandular raceme; pedicels ≤1 mm long. **Flowers**: calyx 2–3 mm long; corolla white to lavender, 2–3 mm across; style <0.5 mm long. **Capsule** obcordate, shallowly notched, 2–4 mm high, glandular-puberulent. Disturbed, sparsely vegetated soil of grasslands, roadsides and drying mud of wetlands, streambanks; plains, valleys, montane. Throughout much of N. and S. America; Asia. Our plants are var. **xalapensis** (Kunth) Pennell.

Corollas rarely open fully.

Veronica persica Poir. Annual. **Stems** prostrate to ascending, simple or branched, sometimes rooting at lower nodes, 10–30 cm. **Herbage** pubescent to hirsute. **Leaves** short-petiolate; blades 7–25 mm long, ovate, serrate to shallowly lobed. **Inflorescence** of solitary flowers in upper leaf axils; pedicels 10–25 mm long. **Flowers**: calyx 5–7 mm long; sepals ciliate; corolla blue, 8–12 mm across; style 2–3 mm long. **Capsule** reniform, widely and shallowly notched, 3–5 mm wide, pubescent. Lawns; valleys. Introduced sporadically across N. America; native to Asia.

Veronica scutellata L. Rhizomatous perennial. **Stems** often branched, basally prostrate, rooting at the nodes, terminally ascending 5–40 cm. **Herbage** glabrous. **Leaves** 2–5 cm long, sessile, linear-lanceolate to lanceolate, entire to obscurely serrate. **Inflorescence** of glabrous, axillary racemes; pedicels 5–12 mm long. **Flowers**: calyx 1–2 mm long; corolla blue, 3–5 mm across; style 1–2 mm long. **Capsule** reniform, flattened, 3–5 mm wide, deeply notched, glabrous. Mud along rivers, lakes, ponds; valleys. Circumboreal south to CA, CO, ND, VA. (p.445)

Veronica serpyllifolia L. Rhizomatous perennial. **Stems** simple, basally prostrate, terminally ascending 5–25 cm. **Herbage** puberulent on the stems. **Leaves** opposite, 5–15 mm long, short-petiolate below, sessile above, ovate, entire to obscurely crenate. **Inflorescence** a puberulent to glandular-pubescent, terminal raceme; pedicels 3–5 mm long. **Flowers**: calyx 2–4 mm long; corolla white to blue, 4–7 mm across; style 2–4 mm long. **Capsule** reniform, notched, 3–4 mm long, glandular-pubescent.

1. Flowers blue; pedicels with septate, glandular hairs . . . . var. *humifusa*
1. Flowers white to pale blue; pedicels puberulent but not glandular. . . . . var. *serpyllifolia*

Veronica serpyllifolia var. *humifusa* (Dicks.) Vahl occurs in moist, sparsely vegetated soil especially along streams, lakes, trails; valleys to subalpine; circumboreal south to much of N. America; *V. serpyllifolia* var. *serpyllifolia* occurs in lawns, along roads; valleys, montane; introduced to eastern and northwest N. America; native to Europe. (p.445)

Veronica verna L. **Stems** erect, simple or branched, 3–15 cm. **Herbage** strigose with curled hairs. **Leaves** short-petiolate; blades 2–10 mm long, lanceolate to ovate, some pinnately lobed. **Inflorescence** a terminal, glandular-pubescent raceme; bracts leaf-like; pedicels ca. 1 mm long. **Flowers**: calyx 2–3 mm long; corolla blue, 2–3 mm across; style ca. 0.5 mm long. **Capsule** obovate, deeply notched, 2–3 mm high, glandular-villous. Disturbed soil of grasslands, roadsides; valleys. Introduced to northeast and northwest U.S., adjacent Canada; native to Eurasia.

Veronica wormskjoldii Roem. & Schult. Rhizomatous perennial. **Stems** erect, simple, 4–30 cm. **Herbage** villous on the stem. **Leaves** opposite, lanceolate to ovate, obscurely crenate, 1–3 cm long, nearly sessile. **Inflorescence** a terminal, glandular-villous raceme; pedicels 2–6 mm long. **Flowers**: calyx 2–4 mm long; corolla deep blue, 4–8 mm across; style ca. 1 mm long. **Capsule** obovate, 5–7 mm long, glandular-puberulent, barely notched. Moist meadows, turf, often near streams; subalpine, alpine. AK to Greenland south to CA, NM. See *V. cusickii*. (p.445)

OROBANCHACEAE: Broomrape Family

Completely or partially parasitic, annual or perennial herbs. **Leaves** alternate, exstipulate. **Flowers** perfect, zygomorphic; calyx 4- to 5-lobed; corolla tubular, 5-lobed or bilabiate; stamens 4; ovary superior, 2-loculed; style 1; stigma simple, or 2- or 4-lobed. **Fruit** a many-seeded, 2-valved capsule.

Castilleja, Cordylanthus, Euphrasia, Melampyrum, Orthocarpus, Pedicularis, and *Rhinanthus* were formerly placed in the Rhinantheae tribe of the Scrophulariaceae recognized by the hood-like upper corolla lip (galea). Members of these genera are non-obligate hemiparasites; their roots attach to those of neighboring plants in order to gain water and nutrients. Plants in the genus *Orobanche* are complete and obligate parasites on the roots of other plants.

1. Plants non-green; leaves scale-like; corolla 5-lobed, not bilabiate. . . . . *Orobanche*
1. Plants with green leaves; corolla strongly bilabiate, the upper lip forming a hood (galea). 2

2. Leaves opposite 3
2. Leaves alternate 5

3. Corolla yellow; calyx inflated, veiny, nearly circular in fruit. . . . . *Rhinanthus*
3. Corolla white; calyx little inflated. 4

4. Leaves entire or nearly so, linear-lanceolate . . . . *Melampyrum*
4. Leaves ovate, serrate. . . . . *Euphrasia*

5. Leaves simple or pinnately lobed; margins crenate to serrate . . . . *Pedicularis*
5. Leaves simple to deeply lobed; margins of leaves or lobes entire 6

6. Plants perennial and/or the galea noticeably longer than the lower lip . . . . *Castilleja*
6. Plants annual; galea ca. as long as the lower lip 7

7. Inflorescence with clusters of 2–5 flowers . . . . *Cordylanthus*
7. Inflorescence with mostly more than 5 flowers. . . . .*Orthocarpus*

Castilleja Mutis ex L.f. Paintbrush

Perennials from a branched caudex, rarely annuals. **Leaves** alternate, sessile, entire to pinnate. **Inflorescence** dense, terminal, bracteate spikes; bracts becoming colored upward. **Flowers**: calyx usually colored as the bracts, tubular, with 4 equal lobes or unequally cleft; corolla green or greenish-yellow, tubular, bilabiate, the upper lip (galea) arched and beak-like, lower lip 3-lobed; stamens 4, enclosed in the galea; style usually capitate. **Capsule** obovoid to globose.[102]

Several species are endemic to the region centered in southwest MT. Closely related species of *Castilleja* are often difficult to distinguish due to interspecific gene flow via hybridization. For example, *C. miniata* intergrades with *C. rhexifolia*, which intergrades with *C. occidentalis*, which intergrades with *C. sulphurea*, etc.[102,289] High-elevation specimens from Lewis & Clark Co. appear intermediate between *C. pulchella* and *C. crista-galli*. Many species are polyploid, and *C. miniata*, *C. rhexifolia*, and *C. sulphurea* have multiple levels of ployploidy, and some species may be of hybrid origin stabilized by polyploidy.[170] As a result the following key is approximate rather than absolute. Only *C. sessiliflora* and *C. exilis* are unmistakable.

1. Galea 2–9 mm long (measure several) 2
1. Galea 8–20 mm long 13

2. Green leaves on at least upper third of stem lobed 3
2. Green leaves (except sometimes the uppermost) entire 9

3. Clefts between calyx lobes all ca. the same length 4
3. Calyx cleft more deeply in front and back compared to on the sides. 5

4. Bracts of inflorescence woolly-villous; plants alpine. . . . . *C. nivea*
4. Bracts of inflorescence puberulent to sparsely villous; montane to subalpine . . . . *C. longispica*

5. Ultimate calyx lobes rounded 6
5. Ultimate calyx lobes acute to acuminate 7

6. Calyx 20–28 mm long; corolla enclosed in calyx; montane to subalpine . . . . . : . . . . *C. cusickii*
6. Calyx 13–23 mm long; corolla exserted; subalpine to alpine . . . . *C. pulchella*

7. Lower corolla lip ≥1/2 as long as the galea . . . .*C. pallescens*
7. Lower corolla lip ≤1/3 as long as galea or less. 8

8. Herbage glabrous to sparsely puberulent. . . . . *C. cervina*
8. Herbage gray-puberulent, often with retrorse hairs . . . . *C. flava*

9. Plants annuals of wet meadows; bracts red. . . . . *C. exilis*
9. Plants perennial from a branched caudex; bracts primarily white or yellow. .10

10. Uppermost leaves or central lobe of uppermost leaves linear . . . . *C. pallescens*
10. Uppermost leaves or central lobe lanceolate or broader 11

11. Plants mostly 10–20 cm high; alpine or upper subalpine . . . . *C. occidentalis*
11. Plants mostly 20–50 cm; valleys to lower subalpine. 12

12. Galea 4–8 mm long, <1/2 the length of the corolla tube; bracts lobed. . . . .*C. lutescens*
12. Galea 6–10 mm long, usually >1/2 as long as corolla tube; lower bracts often entire . . . . *C. sulphurea*

13. Corolla exserted ≥20 mm beyond the bracts; corolla strongly arched, galea much
    shorter than the tube . . . . *C. sessiliflora*
13. Plants not as above 14

14. Inflorescence yellow 15
14. Inflorescence primarily orangish, reddish, or purplish 16

15. Stems mostly solitary, prostrate at the base, often rooting from the nodes . . . .*C. gracillima*
15. Stems clumped, erect or ascending at the base . . . . *C. sulphurea*

16. Calyx conspicuously more deeply cleft on the side adjacent to the corolla lip than in back. . . . *C. linariifolia*
16. Primary clefts of calyx ca. equal. 17

17. Green leaves on at least upper third of stem lobed 18
17. Green leaves (except sometimes the uppermost) entire 21

Castilleja angustifolia (Nutt.) G.Don Perennial. **Stems** erect, 8–35 cm, simple. **Herbage** puberulent to hispid. **Leaves** 1–5 cm long, linear to linear-lanceolate, the upper with 1 to 2 pairs of linear lobes. **Inflorescence** villous, sometimes glandular; bracts pink to red, lanceolate with 1 to 2 pairs of lateral lobes. **Flowers**: calyx 14–20 mm long, primary clefts slightly deeper adjacent to the galea than in front, lobes 1–4 mm long, rounded to obtuse; corolla 17–29 mm long, galea 6–12 mm long, lower lip minute. Sagebrush steppe; valleys. ID, MT south to NV, UT, WY. Our plants are var. *angustifolia*.

The closely related *Castilleja chromosa* A.Nelson occurs south of our area; reports for MT are based on *C. angustifolia*. Plants from Carbon Co. generally have red inflorescences, less hispid foliage, a shorter corolla and eglandular bracts, whereas plants from Beaverhead Co. have pink-purple and more glandular flowers with a longer corolla.

Castilleja covilleana L.F.Hend. Perennial. **Stems** erect or ascending, 10–20 cm, simple. **Herbage** sparsely villous. **Leaves** 1–4 cm long, linear with 1 or 2 pairs of linear lobes. **Inflorescence** puberulent, sparsely villous; bracts red, linear with 2 to 3 pairs of long, linear lobes. **Flowers**: calyx 15–25 mm long, primary clefts slightly deeper adjacent to the galea than in front, lobes acute to acuminate, 1–7 mm long; corolla 25–35 mm long, galea 8–13 mm long, lower lip green, minute. Barren, granitic soil, talus; subalpine. Endemic to central ID and adjacent Ravalli Co., MT.

Castilleja crista-galli Rydb. Perennial. **Stems** erect, simple, 20–50 cm. **Herbage** puberulent. **Leaves** 2–7 cm long, linear to linear-lanceolate, the upper with 1 pair of linear lobes. **Inflorescence** glandular-puberulent, villous; bracts bright red, lanceolate with 1 or 2 pairs of lateral lobes. **Flowers**: calyx 20–28 mm long, primary clefts deeper in front than in back, lobes acute to acuminate, 3–8 mm long; corolla 26–36 mm long, galea 10–13 mm long, lower lip green, minute. Grasslands, meadows, turf; montane to alpine. Endemic to southwest MT and adjacent ID, WY.

Castilleja linariifolia has the calyx much more deeply cleft adjacent to the corolla lip, and flowers become more widely separated at maturity. It is speculated that *C. crista-galli* is a hybrid between *C. miniata* and *C. linariifolia*.[170] The type locality is in Gallatin Co.

Castilleja cusickii Greenm. Perennial. **Stems** ascending to erect, simple, 10–35 cm. **Herbage** viscid-villous. **Leaves** suberect, 2–5 cm long, narrowly lanceolate, the lower entire, the upper with 1 or 2 pairs of lateral lobes. **Inflorescence** glandular-villous; bracts yellow or yellow-tipped, lanceolate to narrowly ovate with 1 to 3 pairs of pointed lobes. **Flowers**: calyx 20–28 mm long, longer than the flowers, primary clefts slightly deeper adjacent to the corolla lip than the back, lobes rounded, 1–4 mm long; corolla yellow, ca. as long as the calyx, galea 3–5 mm long, lower lip 3–4 mm long. Moist meadows, grasslands, sagebrush steppe; montane, subalpine. BC, AB south to NV, ID, WY. (p.450)

Castilleja flava and *C. pallescens* have smaller lower corolla lips.

Castilleja exilis A.Nelson Annual. **Stems** erect, unbranched, 15–50 cm. **Herbage** glandular-puberulent. **Leaves** linear to linear-lanceolate, 2–4 cm long, entire. **Inflorescence** elongate, glandular-pilose; bracts linear-lanceolate, entire, red-tipped, longer than the leaves and flowers. **Flowers**: calyx 1–2 cm long, primary clefts equal, lobes acute, 1–2 mm long; corolla 15–20 mm long, galea 5–7 mm long, lower lip minute. Moist to wet alkaline meadows; valleys. WA to MT south to CA, AZ, NM.

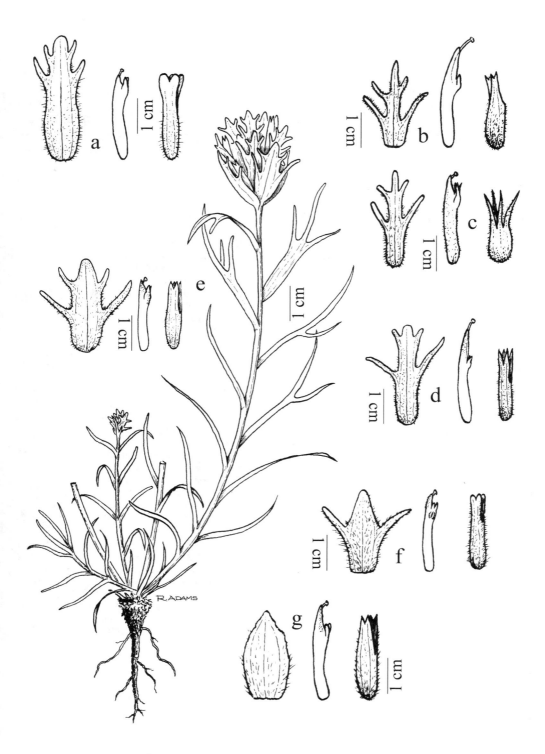

Plate 80. *Castilleja*. **a.** *C. cusickii*, **b.** *C. flava*, **c.** *C. longispica*, **d.** *C. lutescens*, **e.** *C. pallescens*, **f.** *C. pulchella*, **g.** *C. sulphurea*

Castilleja flava S.Watson [*C. rustica* Piper] Perennial. **Stems** erect, simple or branched, 10–50 cm. **Herbage** puberulent. **Leaves** 2–6 cm long, linear, at least the uppermost with 1 or 2 pairs of linear lobes. **Inflorescence** villous; bracts usually yellow (red), lanceolate with usually 1 pair of narrow lobes. **Flowers**: calyx 14–23 mm long, primary clefts subequal to slightly deeper adjacent to the corolla lip than the back, lobes acute, 2–5 mm long; corolla green to yellow, 16–25 mm long, galea 5–9 mm long, lower lip much smaller than the galea. Grasslands, sagebrush steppe, woodlands; montane, subalpine. WA to MT south to NV, UT, CO. (p.450)

1. Primary clefts of calyx much deeper adjacent to the corolla lip than in the back; lower corolla lip ca. 1–2 mm long . . . . var. *flava*
1. Primary clefts of calyx subequal; corolla lower lip ca. 2–4 mm long. . . . . var. *rustica*

Castilleja flava var. *rustica* (Piper) N.H.Holmgren is frequently associated with sagebrush in southwest cos.; *C. flava* var. *flava* is more widespread. *Castilleja cervina* Greenm. is similar and occurs in adjacent ID and BC. One specimen from Flathead Co. seems intermediate between the two species.

Castilleja gracillima Rydb. Perennial. **Stems** appearing solitary, simple, basally prostrate, often rooting at the nodes, ascending, 20–50 cm. **Herbage** glabrous to villous. **Leaves** 2–7 cm long, linear-lanceolate to lanceolate, entire. **Inflorescence** villous; bracts yellow to orange, oblong, entire or with 1 pair of small lobes. **Flowers**: calyx 15–22 mm long, primary clefts subequal, lobes acute, 2–4 mm long; corolla yellow, 20–30 mm long, galea 8–12 mm long, lower lip greener, much smaller than the galea. Wet meadows; valleys, montane. BC south to ID, WY.
 Castilleja sulphurea generally occurs in drier habitats and has stems clumped on a woody caudex.

Castilleja hispida Benth. Perennial. **Stems** erect or ascending, usually simple, 15–50 cm. **Herbage** hispid to villous. **Leaves** oblanceolate, 2–7 cm long, at least the upper with 1 or 2 pairs of narrow lateral lobes. **Inflorescence** glandular, villous to hispid; bracts red-tipped, ovate with 1 or 2 pairs of narrow lateral lobes. **Flowers**: calyx 19–24 mm long, primary clefts subequal, ultimate lobes rounded to acute, 2–7 mm long; corolla green to red, 20–35 mm long, galea 9–15 mm long, lower lip 1–3 mm long. Grassland, dry open forest; valleys, montane. BC, AB south to OR, ID, MT.
 Both *Castilleja hispida* var. *hispida* and var. *acuta* Owenby occur in MT; however, they have little geographic integrity and characters distinguishing them are only weakly correlated here. Terminal leaf lobes of *Castilleja angustifolia* are ca. as wide as the lateral ones; *C. miniata* has entire leaves and occurs in somewhat moister habitats.

Castilleja linariifolia Benth. Perennial. **Stems** erect, 15–50 cm, sometimes branched above. **Herbage** glabrate. **Leaves** 2–6 cm long, linear to linear-lanceolate, entire or with a pair of linear lobes. **Inflorescence** villous to tomentose; bracts red to orange, narrowly lanceolate with a pair of linear lobes. **Flowers**: calyx 19–26 mm long, curved back, primary clefts much deeper adjacent to the corolla lip, ultimate lobes 1–3 mm long, acute; corolla 23–35 mm long, galea 10–20 mm long, lower lip 1–2 mm long. Grasslands, sagebrush steppe, talus slopes, coniferous woodlands; montane. OR to MT south to CA, AZ, NM.
 The Ravalli Co. collection is a hybrid with *Castilleja miniata* (fide N. Holmgren). See *C. crista-galli*.

Castilleja longispica A.Nelson Perennial. **Stems** ascending to erect, simple or branched, 10–30 cm. **Herbage** puberulent to hirsute. **Leaves** linear to lanceolate, 2–4 cm long, the lower entire, upper 3-lobed. **Inflorescence** puberulent to sparsely villous; bracts green to yellow or purplish, lanceolate with 1 or 2 pairs of lateral lobes. **Flowers**: calyx lanceolate, 12–20 mm long, subequally cleft into 4 linear-lanceolate lobes, 4–8 mm long; corolla 14–23 mm long, galea 4–8 mm long, lower lip 2–4 mm long. Sagebrush steppe, grasslands, woodlands; montane. OR to MT south to CA, WY. Known from Carbon and Madison cos. (p.450)
 Castilleja nivea also has 4 equal calyx lobes but is confined to high-elevation habitats.

Castilleja lutescens (Greenm.) Rydb. Perennial. **Stems** erect, simple or branched, 12–50 cm. **Herbage** puberulent. **Leaves** linear-lanceolate, 3–6 cm long, mostly entire. **Inflorescence** glandular-puberulent to villous; bracts yellow-tipped, lanceolate with 1 to 3 pairs of narrow lateral lobes, shorter than the galea. **Flowers**: calyx 14–22 mm long, primary clefts ca. equal, ultimate lobes acute to acuminate, 1–3 mm long; corolla 16–25 mm long, galea 4–8 mm long, lower lip 3–5 mm long. Grasslands, meadows, open forest; valleys, montane. BC, AB south to OR, ID, MT. (p.450)
 Castilleja cusickii has lobed leaves; *C. sulphurea* has a larger galea and sometimes unlobed bracts.

Castilleja miniata Douglas ex Hook. Perennial. **Stems** erect, simple or branched, 15–80 cm. **Herbage** glabrous to puberulent. **Leaves** linear to lanceolate, 3–7 cm long, entire. **Inflorescence** villous, glandular-puberulent; bracts orange to red, ovate, entire or with a pair of small lateral lobes. **Flowers**: calyx 16–28 mm long, primary clefts slightly greater adjacent to the corolla lip, ultimate lobes narrowly acute, 2–8 mm long; corolla green to red, 27–38 mm long, galea 12–15 mm long, lower lip 1–3 mm long. Moist grasslands, meadows, thickets, forest openings, often near streams; valleys to subalpine. AK to MB south to CA, NM.

This species occurs in more habitats than any of our other paintbrushes. A white-flowered form occurs sporadically along the Canadian border in the western half of the state. *Castilleja hispida* has lobed leaves; *C. rhexifolia* is very similar but has purplish flowers and mostly occurs at higher elevations.

Castilleja nivea Pennel & Owenby Perennial. **Stems** erect, simple, 5–12 cm. **Herbage** puberulent to villous. **Leaves** linear, 2–4 cm long, the lower entire, upper 3-lobed. **Inflorescence** villous-tomentose; bracts yellow- to purple-tipped, lanceolate with a pair of linear lobes. **Flowers**: calyx yellow, linear-lanceolate, 15–20 mm long, subequally cleft into 4 linear-lanceolate lobes, 4–5 mm long; corolla 10–22 mm long, galea 4–6 mm long, lower lip 2–3 mm long. Turf, fellfields; alpine. Endemic to southwest MT, adjacent WY. The type locality is in Carbon Co.

Castilleja occidentalis Torr. Perennial. **Stems** erect, simple, 5–20 cm. **Herbage** glabrate to puberulent. **Leaves** narrowly lanceolate, 1–3 cm long, usually entire, the lowest reduced. **Inflorescence** viscid-villous; bracts yellowish-white to green, sometimes purplish-tipped, narrowly ovate, entire or with a pair of small lateral lobes. **Flowers**: calyx 15–23 mm long, primary clefts subequal, ultimate lobes rounded, 1–3 mm long; corolla whitish, 18–25 mm long, galea 5–9 mm long, lower lip 2–4 mm long. Moist meadows, moraine; upper subalpine, alpine. BC, AB south to UT, NM.

Hybridization with *Castilleja rhexifolia* results in populations with flowers of every shade between white and purple. See *C. sulphurea*.

Castilleja pallescens (A.Gray) Greenm. Perennial. **Stems** erect, simple, 5–20 cm. **Herbage** puberulent. **Leaves** linear-lanceolate, 1–4 cm long, entire, upper often with linear lobes. **Inflorescence** puberulent to pilose; bracts green, yellow or purple-tipped, lanceolate with 1 or 2 pairs of narrow, lateral lobes. **Flowers**: calyx yellow, 12–22 mm long, primary clefts slightly deeper adjacent to the galea, ultimate lobes acute, 0.5–3 mm long; corolla 16–21 mm long, galea 2–4 mm long, lower lip 2–3 mm long. Grasslands, sagebrush steppe, fellfields, open forests, woodlands; valleys to alpine. OR to MT south to NV, WY. Our plants are var. *pallescens*. (p.450)

One specimen from Beaverhead Co. approaches var. *inverta* (A.Nelson & J.F.Macbr.) Edwin. *Castilleja pulchella* has rounded calyx lobes and a galea ca. twice as long as the lower lip of the corolla; *C. nivea* and *C. longispica* have a calyx with four equal clefts.

Castilleja pulchella Rydb. Perennial. **Stems** ascending, simple, 4–15 cm. **Herbage** glandular-pubescent. **Leaves** linear to lanceolate, 1–3 cm long, the lower entire, upper 3-lobed. **Inflorescence** puberulent, viscid-villous; bracts yellow to purplish, narrowly ovate, entire or with a pair of lateral lobes. **Flowers**: calyx 15–21 mm long, primary clefts subequal, ultimate lobes rounded, 1–3 mm long; corolla 18–25 mm long, galea 3–5 mm long, lower lip 1–3 mm long. Turf, fellfields; subalpine, alpine. Endemic to southwest MT, adjacent ID, WY, disjunct in northern UT. See *C. pallescens*. (p.450) The type locality is in Madison Co.

Castilleja rhexifolia Rydb. Perennial. **Stems** erect or ascending, usually simple, 10–30 cm. **Herbage** glabrate. **Leaves** lanceolate, 3–6 cm long, mostly entire. **Inflorescence** viscid-villous; bracts reddish-purple, ovate, entire or with a pair of small lateral lobes. **Flowers**: calyx 16–25 mm long, primary clefts slightly greater adjacent to the corolla lip, ultimate lobes rounded to acute, 2–4 mm long; corolla green to red, 24–32 mm long, galea 8–12 mm long, lower lip 1–4 mm long. Moist meadows; subalpine, alpine. BC, AB south to OR, UT, CO.

Very similar to the yellow-flowered *Castilleja sulphurea*; see *C. occidentalis, C. miniata*. The type locality is in Madison Co.

Castilleja sessiliflora Pursh Perennial. **Stems** erect or ascending, simple, 10–30 cm. **Herbage** villous-tomentose. **Leaves** linear, 2–7 cm long, entire below, those above with 1 or 2 pairs of linear lobes. **Inflorescence** viscid-villous; bracts green or yellow- to purple-tipped, narrowly lanceolate with 1 or 2 pairs of linear lobes. **Flowers**: calyx 26–38 mm long, primary clefts subequal, ultimate lobes purplish, linear-lanceolate, 6–15 mm long; corolla yellow to purplish, 35–50 mm long, galea 8–12 mm long, lower lip 3–5 mm long. Grasslands, sagebrush steppe; plains, valleys, montane. AB to SK south to AZ, TX, IL.

Castilleja sulphurea Rydb. [*C. septentrionalis* Lindl. misapplied] Perennial. **Stems** ascending to erect, simple or branched, 15–35 cm. **Herbage** glabrous to puberulent. **Leaves** 2–6 cm long, narrowly lanceolate, entire. **Inflorescence** glandular-villous; bracts yellow, ovate, entire, ca. as long as the flowers or with a pair of small lateral lobes. **Flowers**: calyx 15–23 mm long, primary clefts slightly deeper adjacent to the corolla lip than the back, lobes rounded to acute, 1–3 mm long; corolla yellow, 20–24 mm long, galea 6–10 mm long, lower lip 1–3 mm long. Grasslands, dry meadows, hummocks in fens; valleys to subalpine. BC, AB south to UT, NM, SD. (p.450)

    Castilleja sulphurea completely intergrades with *C. occidentalis* with increasing elevation in Alberta;[289] *C. sulphurea* approaches *C. gracillima* in habit when growing in fens. See *C. lutescens*. The type locality is in Park Co., MT or WY.

Cordylanthus Nutt. ex Benth. Bird's beak

Cordylanthus ramosus Nutt. ex Benth. Annual. Stems erect, 8–20 cm, usually branched. **Herbage** puberulent. **Leaves** all cauline, alternate, 1–5 cm long, linear, entire or with a pair of linear lobes. **Inflorescence** of 2 to 5 flowers, terminal, head-like, bracteate spikes; flower bracts of 2 kinds: outer leaf-like, linear; inner bracts green, as long as the calyx. **Flowers**: calyx 10–18 mm long, green, bract-like with short apical lobes, the calyx and opposing bract appearing to form a deeply divided bilabiate calyx; corolla 11–20 mm long, tubular, bilabiate, 2 lips ca. equal, nearly unlobed, galea whitish, lower lip purplish; stamens 4; anthers 2-celled. **Capsule** asymmetrical, 7–10 mm long. Sagebrush steppe; valleys, montane; Beaverhead and Madison cos. OR to MT south to CA, NV, UT, CO.

    Cordylanthus capitatus Nutt. ex Benth. is reported for southwest MT,[187] but I am not aware of any specimens.

Euphrasia L. Eyebright

Euphrasia subarctica Raup [*E. arctica* Lange ex Rostrup var. *disjuncta* Cronquist] Diminutive annual. **Stems** erect, simple or branched, 2–6 cm. **Herbage** puberulent. **Leaves** cauline, opposite, ovate, serrate, 3–7 mm long. **Inflorescence** a short, terminal, leafy-bracted spike. **Flowers**: calyx with 4 lanceolate lobes 2–3 mm long, longer in fruit, purple-veined; corolla bilabiate, 2–3.5 mm long, white and purple-spotted, lower lip longer than the galea, 3-lobed; stamens 4, stigma capitate. **Capsule** 4–7 mm long, sparsely pubescent. Inconspicuous in moist turf; alpine; Glacier Co. AK to Greenland south to BC, MT, MI, ME.

Melampyrum L.

Melampyrum lineare Desr. Cow wheat Fibrous-rooted annual. **Stems** erect, 7–30 cm, usually branched above. **Herbage** puberulent. **Leaves** opposite, short-petiolate; blades 2–5 cm long, linear-lanceolate, entire or few-toothed. **Inflorescence** terminal, leafy-bracted spike. **Flowers**: calyx 3–5 mm long, glandular-puberulent with 4 unequal, lanceolate lobes; corolla bilabiate, 5–10 mm long, white, lower lip 3-lobed, yellow near the base, as long as the galea; stamens 4; stigma capitate. **Capsule** ovate, flattened, beaked, 6–10 mm long. Moist, open, coniferous forest; valleys, montane. BC to NL south to WA, MT, MN, GA. (p.454)

Orobanche L. Broomrape

Achlorophyllous perennial (annual?) herbs. **Stems** erect, mainly subterranean. **Leaves** alternate, scale-like, exstipulate. **Flowers** perfect; calyx deeply 5-lobed; corolla tubular, 5-lobed, slightly bilabiate; stamens 4; anthers glabrous to woolly; ovary superior; stigma capitate to short-lobed. **Fruit** a many-seeded, 2-valved, obovoid capsule.

    Plants have swollen stem bases that envelope the roots of neighboring green plants and obtain all their water and nutrition parasitically. It is possible that *Orobanche uniflora* is sometimes an annual, but little is known about the life history of these plants.

1. Flowers borne on pedicels ≥2 cm long 2
1. Pedicels <2 cm long 3

2. Flowers 1 to 3, bluish-purple; calyx lobes longer than the tube . . . . *O. uniflora*
2. Flowers >3, brownish; calyx lobes shorter or equal to the tube . . . . *O. fasciculata*

3. Corolla lobes 2–5 mm long . . . . *O. ludoviciana*
3. Corolla lobes >6 mm long . . . . *O. corymbosa*

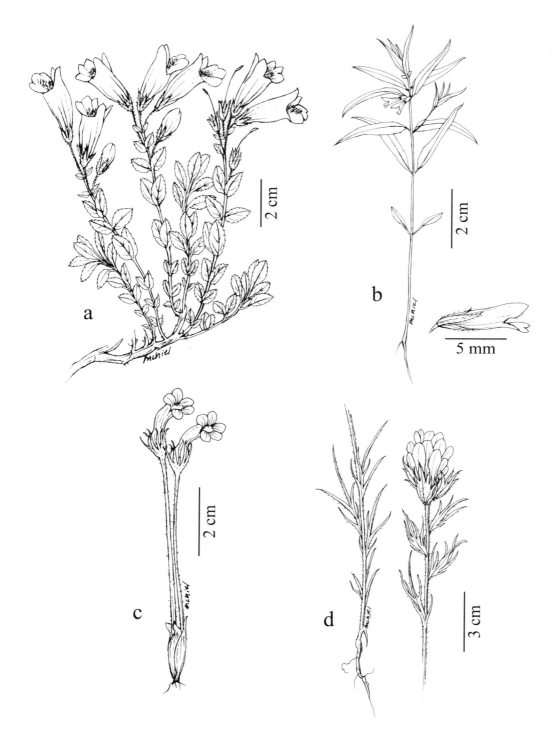

Plate 81. a. ***Penstemon ellipticus***, b. ***Melampyrum lineare***, c. ***Orobanche uniflora***,
d. ***Orthocarpus tenuifolius***

Orobanche corymbosa (Rydb.) R.S.Ferris **Stems** 1–4 cm above ground. **Herbage** purplish, puberulent. **Inflorescence** a glandular, crowded, bracteate corymb; pedicels ascending to erect, 3–9 mm long. **Flowers**: calyx 11–17 mm long; sepals linear, 7–14 mm long; corolla 20–25 mm long, light purple, yellowish within the mouth, the lobes 6–8 mm long. **Capsule** 8–14 mm long. Sagebrush steppe; valleys, montane. WA to MT south to CA, NV, UT.

Parasitic primarily on *Artemisia*. The lower corolla lip of many of our collections is longer than typically described.

Orobanche fasciculata Nutt. **Stems** 3–8 cm above ground. **Herbage** yellow to purplish, glandular-puberulent. **Inflorescence** a short, bracteate, glandular raceme of 3 to 10 flowers; pedicels erect, 2–8 cm long. **Flowers**: calyx 7–10 mm long; sepals deltoid, 3–6 mm long; corolla 15–30 mm long, brownish-red, yellow within the mouth, the lobes 2–6 mm long. **Capsule** 10–13 mm long. Grasslands, sagebrush steppe; plains, valleys. YT south to Mexico, OK, IN.

Usually parasitic on *Artemisia* or other members of Asteraceae.

Orobanche ludoviciana Nutt. **Stems** 2–5 cm above ground. **Herbage** purplish, glabrous to puberulent. **Inflorescence** a glandular, crowded, bracteate spike, raceme or panicle; pedicels ascending, 0–6 mm long. **Flowers**: calyx 8–14 mm long; sepals linear-lanceolate, 5–10 mm long; corolla 11–21 mm long, purple, yellow within the mouth, the lobes 2–5 mm long. **Capsule** 7–10 mm long. Grasslands, sagebrush steppe; plains, valleys, montane. WA to MB south to Mexico, MO, IN.

Plants from the sandhills of southern Beaverhead Co. have unusually small flowers.

Orobanche uniflora L. **Stems** ca. 1 cm above ground. **Herbage** whitish, glandular-pubescent. **Inflorescence** glandular-pubescent, of 1–2 (3–5) pedicellate flowers; pedicels erect, 2–8 cm long. **Flowers**: calyx 5–10 mm long; sepals acuminate-lanceolate, 3–7 mm long; corolla 13–20 mm long, purple, the lobes 2–4 mm long. **Capsule** 8–10 mm long. Stony, often vernally moist, often shallow soil of grasslands, meadows, open forest; valleys to subalpine. Much of temperate N. America. Our plants are var. *minuta* (Suksd.) Beck. (p.454)

Parasitic on many hosts belonging to the Saxifragaceae, Asteraceae, Crassulaceae, Ranunculaceae.[273]

Orthocarpus Nutt. Owl clover

Taprooted annual herbs. **Leaves** all cauline, alternate, sessile. **Inflorescence** dense, terminal, bracteate spikes, becoming expanded in fruit; bracts often wider than the leaves. **Flowers**: calyx divided above into 4 acute lobes, the median clefts deeper than the lateral; corolla tubular, bilabiate, the upper lip (galea) beak-like, lower lip obscurely 3-lobed, pouch-like, nearly as long as the galea; stamens 4, enclosed in the galea; style usually capitate. **Capsule** obovoid to globose.[85]

1. Bracts of upper flowers pink; leaves of mid-stem divided. . . . . *O. tenuifolius*
1. Flower bracts green; mid-stem leaves undivided 2

2. Galea projecting a tiny beak beyond lower lip; stems usually branched above . . . . *O. tolmiei*
2. Galea as long as lower lip; stems usually simple . . . . *O. luteus*

Orthocarpus luteus Nutt. **Stems** erect, usually simple, 7–25 cm. **Herbage** hirsute and glandular-pubescent. **Leaves** numerous, linear, entire, 1–3 cm long. **Inflorescence bracts** dark green, lanceolate to narrowly ovate with a pair of linear lobes. **Flowers**: calyx 5–8 mm long, ultimate lobes 1–3 mm long, those adjacent to the galea longer; corolla 9–14 mm long, yellow, galea 2–4 mm long, lower lip equal to the upper lip. **Capsule** 4–7 mm long. Grasslands, sagebrush steppe; plains, valleys. BC to ON south to CA, NM, NE, MN.

Orthocarpus tolmiei is usually branched and has more yellowish herbage.

Orthocarpus tenuifolius (Pursh) Benth. **Stems** erect, mostly simple, 10–25 cm. **Herbage** puberulent. **Leaves** linear with 1 or 2 pairs of linear lobes, 1–3 cm long. **Inflorescence bracts** ovate, entire or with a pair of linear lobes, ciliate, pink-tipped. **Flowers**: calyx 5–9 mm long, ultimate lobes linear, 2–6 mm long; corolla 15–18 mm long, yellow, galea 4–7 mm long, upper lip projecting beyond the lower lip. **Capsule** 5–8 mm long. Grasslands, sagebrush steppe; valleys, montane. BC, MT south to OR, ID. The type locality is south of Missoula. (p.454)

Orthocarpus tolmiei Hook. & Arn. **Stems** erect, usually simple, 10–30 cm. **Herbage** glandular-puberulent. **Leaves** linear to linear-oblanceolate, entire, 2–4 cm long. **Inflorescence bracts** light green, lanceolate with a pair of linear lobes, glandular. **Flowers**: calyx 5–8 mm long, lobes 2–4 mm long, those adjacent to the galea longer; corolla 10–15 mm long, yellow, galea 3–5 mm long, upper lip hooked and projecting beyond the lower lip. **Capsule** 3–5 mm long. Grasslands, meadows; montane. ID, MT, UT, WY. Known from Gallatin Co. See *O. luteus*.

Pedicularis L. Lousewort

Perennial, fibrous-rooted herbs. **Stems** erect, simple. **Herbage** mostly glabrous below the inflorescence. **Leaves** basal and/or alternate, serrate to pinnately lobed, at least the lower petiolate, becoming reduced and sessile above. **Inflorescence** a terminal, green-bracteate spike-like raceme, expanding in fruit. **Flowers** short-pedicellate; calyx tubular, mostly 5-lobed, usually more deeply cleft in front; corolla tubular, bilabiate, upper lip (galea) hood-like sometimes tapered to a beak, lower lip 3-lobed, reflexed down; stamens 4, included in the galea; stigma capitate. **Capsule** flattened.[187,335,340]

1. Leaves serrate but not lobed 2
1. Leaves deeply pinnately lobed 3

2. Flowers red to burgundy; plants of riparian meadows . . . . *P. crenulata*
2. Flowers white; plants of montane to subalpine forests . . . .*P. racemosa*

3. Galea with a beak ≥4 mm long curved up or down 4
3. Galea with a beak 0–3 mm long 5

4. Galea beak 5–13 mm long, curled upward . . . . *P. groenlandica*
4. Galea beak 5–8 mm long, curved down in a semicircle . . . . *P. contorta*

5. Leaves all or mostly cauline; plants 8–75 cm high . . . .*P. bracteosa*
5. Leaves primarily basal; plants 5–40 cm high 6

6. Corolla yellow . . . .*P. oederi*
6. Corolla reddish to purple 7

7. Galea beak 1–2 mm long . . . . *P. parryi*
7. Galea beak absent 8

8. Bracts similar in shape to the leaves; plants ≤10 cm high . . . .*P. pulchella*
8. Bracts linear-lobed; plants usually >10 cm high . . . . *P. cystopteridifolia*

Pedicularis bracteosa Benth. **Stems** 8–75 cm, clustered on the caudex. **Leaves** mainly cauline, 2–15 cm long; blades deeply pinnately lobed into narrowly oblong, serrate segments, 5–60 mm long. **Inflorescence** densely flowered, 3–20 cm long; bracts lanceolate, ciliate. **Flowers**: calyx 6–13 mm long, 5-lobed, the lobes 4–8 mm long, the closest to the galea shorter; corolla yellow or purplish, 12–26 mm long, galea 4–12 mm long, arched, beak tip small or lacking. **Capsule** 4–13 mm long. BC, AB south to CA, CO. (p.458)

1. Sepals acuminate, the tips filiform; corolla yellow or burgundy . . . . .var. *bracteosa*
1. Sepals lanceolate; corolla yellow 2

2. Galea beakless . . . . var. *paysoniana*
2. Galea with a small beak ≥1 mm long . . . . var. *canbyi*

Pedicularis bracteosa var. *bracteosa* occurs in montane to alpine moist forests, meadows, avalanche slopes, turf, more common north; *P. bracteosa* var. *paysoniana* (Pennell) Cronquist averages taller with larger flowers than the other varieties and occurs in moist subalpine forest and meadows, more common south; *P. bracteosa* var. *canbyi* (A.Gray) Cronquist is found in subalpine and alpine moist forest and meadows. The vars. intergrade; var. *siifolia* (Rydb.) Cronquist has been separated from var. *canbyi* based on the glabrous inflorescence; however, the degree of long-hairiness of bracts varies continuously and cannot be confidently used to separate the two taxa,[340] so I have combined them under the older name.

Pedicularis contorta Benth. Parrot's beak **Stems** 8–30 cm, clustered on the caudex. **Leaves** mainly basal, 3–14 cm long; blades deeply pinnately lobed into narrowly oblong, serrate segments 2–10 cm long. **Inflorescence** sparsely flowered, 2–20 cm long; bracts lanceolate with 1 or 2 pairs of linear lobes. **Flowers**: calyx 6–10 mm long, 5-lobed, the lobes 1–3 mm long, the one closest to the galea shorter; corolla white to reddish-purple, 10–16 mm long, galea 5–8 mm long with a long beak curved into a semicircle, lower lip wide, the 2 outer lobes much wider than the central one. **Capsule** 6–10 mm long. Common in grasslands, meadows, open forest; montane, subalpine. BC, AB south to CA, ID, WY. (p.458)

1. Corolla white; calyx light green. . . . . .var. *contorta*
1. Corolla pink to purple; calyx green to reddish 2

2. Flower bracts villous-ciliate; calyx green . . . . var. *ctenophora*
2. Flower bracts glabrous; calyx red- or purple-blotched . . . . var. *rubicunda*

Pedicularis contorta var. *contorta* is found in montane to alpine open forest, dry meadows, sagebrush steppe; *P. contorta* var. *ctenophora* (Rydb.) A.Nelson & J.F.Macbr. occurs in montane to subalpine fescue grasslands; it is endemic to the Big Horn Mtns. of WY and southern Beaverhead Co., MT where the type was collected; *P. contorta* var. *rubicunda* Reese is found in subalpine open forests, moist meadows; it is endemic to Ravalli Co. and adjacent ID.

Pedicularis crenulata Benth. Stems 15–40 cm, tomentose in lines, often clustered on the caudex. **Leaves** basal and cauline, sessile, 3–12 cm long; blades narrowly lanceolate, crenate. **Inflorescence** densely flowered, 2–10 cm long, leafy-bracteate. **Flowers**: calyx 8–12 mm long, villous; sepals 1–2 mm long fused into 2 lobes; corolla red to burgundy, 20–26 mm long, galea 9–15 mm long with an inconspicuous beak, lower lip inconspicuously lobed. **Capsule** 1–2 cm long. Moist riparian meadows; valleys, montane. MT, WY south to CA, NV, UT, CO. Known from Beaverhead Co. (p.458)

Pedicularis cystopteridifolia Rydb. Stems 10–40 cm, often solitary. **Leaves** basal and cauline, 5–12 cm long; blades deeply pinnately lobed into lobed and serrate, lanceolate to ovate segments 2–10 mm long. **Inflorescence** glandular-villous, densely flowered, 2–12 cm long; bracts obovate, long-acuminate, sometimes with a pair of short lobes. **Flowers**: calyx 7–11 mm long, 5-lobed, the lobes 3–5 mm long, the closest to the galea shorter; corolla purple, 16–25 mm long, galea 8–12 mm long with an inconspicuous beak. **Capsule** 8–13 mm long. Moist grasslands, meadows, sagebrush steppe, turf; montane to alpine. Endemic to southwest MT, adjacent WY. (p.458)

The type was collected in Madison Co.

Pedicularis groenlandica Retz. ELEPHANT'S HEAD **Stems** 10–70 cm, clustered on the caudex. **Leaves** basal and cauline, 3–15 cm long; blades deeply pinnately lobed into lanceolate, serrate segments 2–10 cm long. **Inflorescence** densely flowered, 3–15 cm long; bracts small, linear-lobed, glabrous. **Flowers**: calyx 4–6 mm long, 5-lobed, the lobes subequal, deltoid, 0.5–2 mm long, prominently veined; corolla burgundy or magenta, 9–15 mm long, galea 4–13 mm long, tubular, curved upward. **Capsule** 6–9 mm long. Wet meadows, fens, often along streams; montane to lower alpine. BC to NL south to CA, NM. (p.458)

Pedicularis oederi Vahl **Stems** 4–15 cm, mostly solitary. **Leaves** mainly basal, 3–10 cm long; blades deeply pinnately lobed into serrate, ovate segments 2–10 mm long. **Inflorescence** villous, densely flowered, 2–7 cm long; bracts leaf-like. **Flowers**: calyx 9–12 mm long, 5-lobed, the lobes 1–3 mm long; corolla yellow, 15–25 mm long, galea 7–10 mm long without a beak. Alpine turf. Circumpolar south to central BC, disjunct in Carbon and Stillwater cos. and adjacent WY.

Pedicularis parryi A.Gray **Stems** 5–25 cm, often clustered on the caudex. **Leaves** basal and cauline, 2–10 cm long; blades deeply pinnately lobed into serrate, lanceolate segments 2–5 mm long. **Inflorescence** villous, densely flowered, 1–10 cm long; bracts linear with 1 to 2 pairs of linear lobes. **Flowers**: calyx 7–11 mm long, 5-lobed, the lobes equal, 2–3 mm long; corolla purple, 13–22 mm long, galea 6–10 mm long with a short beak 1–2 mm long. **Capsule** 8–12 mm long. Grasslands, meadows, open forest, turf; upper montane to alpine. ID, MT south to AZ, NM. Our plants are var. **purpurea** Parry. (p.458)

Pedicularis parryi var. *parryi* has yellow flowers and glabrous stems; it is reported as far north as the Big Horn Mtns., WY; a specimen from southern Beaverhead County appears intermediate between the two vars.

Pedicularis pulchella Pennell **Stems** 3–15 cm, often clustered on the caudex. **Leaves** mainly basal, 2–5 cm long; blades deeply pinnately lobed into lobed and serrate, ovate segments 1–7 mm long. **Inflorescence** villous, densely flowered, 2–5 cm long; bracts leaf-like. **Flowers**: calyx purple-veined, 8–10 mm long, 5-lobed, the lobes equal, 2–5 mm long; corolla purple, 18–24 mm long, galea 8–13 mm long, obscurely beaked. **Capsule** 4-10 mm long. Turf, fellfields, meadows; upper subalpine, alpine. Endemic to southwest MT, adjacent WY. (p.458)

The type was collected in the Anaconda Range.

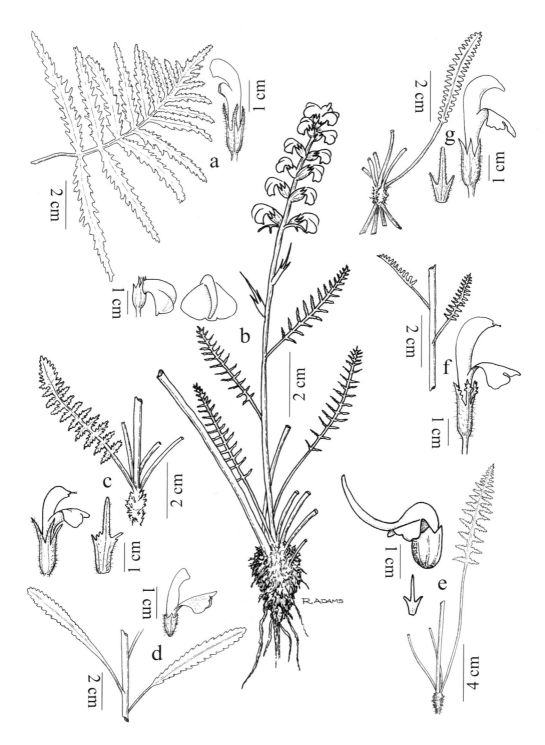

Plate 82. **Pedicularis**. **a.** *P. bracteosa* var. *bracteosa*, **b.** *P. contorta* var. *contorta*,
c. *P. cystopteridifolia*, **d.** *P. crenulata*, **e.** *P. groenlandica*, **f.** *P. pulchella*, **g.** *P. parryi*

Pedicularis racemosa Dougl. ex Benth. **Stems** 15–60 cm, clustered on the caudex. **Leaves** mainly cauline, 2–8 cm long; blades narrowly lanceolate, serrate, not lobed. **Inflorescence** sparsely flowered, 3–7 cm long; bracts leaf-like. **Flowers**: calyx 4–8 mm long, 2-lobed, more deeply cleft adjacent to the lower lip, the lobes 1–3 mm long; corolla white, 12–15 mm long, galea 5–9 mm long with a long beak curved into a semicircle, lower lip wide, the 2 outer lobes much wider than the central one. **Capsule** 8–14 mm long. Coniferous forest; montane, subalpine. BC, AB south to CA, AZ, NM. Our plants are var. ***alba*** (Pennell.) Cronquist.

Rhinanthus L.

Rhinanthus crista-galli L. [*R. minor* L.] YELLOW RATTLE Annual. **Stems** erect, mostly unbranched, 10–50 cm. **Herbage** puberulent. **Leaves** cauline, opposite, sessile, lanceolate, serrate, 2–5 cm long. **Inflorescence** a spike with paired flowers subtended by upper leaves or leaf-like bracts. **Flowers**: calyx compressed, broadly ovate with 4 short lobes, inflated and 1–2 cm long in fruit; corolla barely exserted, yellow, bilabiate, galea 2–4 mm long, lower lip shorter; stamens 4; stigma entire. **Capsule** orbicular, flattened, 3–9 mm long, enclosed in calyx. Grasslands, meadows, dry open forest, streambanks, roadsides, often where disturbed; valleys, montane. Circumboreal south to OR, CO, NY. (p.463)

Our plants are part of a circumpolar complex and are sometimes considered distinct from European material, but the definitive study has not been done.

PHRYMACEAE: Lopseed Family

Mimulus L. Monkey flower

Annual or perennial herbs. **Leaves** opposite, simple, exstipulate. **Inflorescence** of pairs of pedicellate flowers in upper leaf axils. **Flowers** perfect, zygomorphic, 5-merous; calyx tubular to campanulate, expanding in fruit, 5-angled with 5 short lobes (teeth); corolla tubular, raised inside on the bottom (palate), at least somewhat bilabiate, upper lip 2-lobed, often erect, lower lip 3-lobed, spreading; stamens 4; ovary superior, 2-loculed; style 1; stigmas 2. **Fruit** a many-seeded, ellipsoidal capsule.

Mimulus was formerly placed in the Scrophulariaceae. *Mimulus glabratus* Kunth was reported for MT[148] based on a misidentified specimen of *M. guttatus*.

1. Plants fibrous-rooted annuals; stems usually ≤15 cm 2
1. Plants rhizomatous and/or stoloniferous perennials; stems often >15 cm. 10

2. Flowers purplish. 3
2. Flowers yellow 5

3. Corolla ≤10 mm long . . . . *M. breweri*
3. Corolla 10–25 mm long 4

4. Pedicels 1–3 mm long; capsule ovoid . . . . *M. nanus*
4. Pedicels ≥3 mm long; capsule tubular . . . . *M. clivicola*

5. Upper calyx lobe obviously larger than the others . . . . *M. guttatus*
5. Calyx lobes ca. equal 6

6. Stems and calyx viscid-villous with multi-cellular hairs. . . . . *M. floribundus*
6. Stems and calyx viscid-puberulent with unicellular hairs 7

7. Leaves sessile; calyx ca. as long as the pedicel . . . . *M. suksdorfii*
7. Leaves petiolate; calyx mostly shorter than the pedicel. 8

8. Corolla 5–8 mm long . . . . *M. breviflorus*
8. Corolla usually ≥8 mm long 9

9. Pedicel 2 to 3 times as long as the calyx; calyx teeth ca. 0.5 mm long . . . . *M. ampliatus*
9. Pedicel 3 to 4 times as long as the calyx; calyx teeth ca. 1 mm long. . . . . *M. hymenophyllus*

10. Flowers red to purple 11
10. Flowers yellow 12

11. Calyx 10–16 mm long; corolla <30 mm long; plains. . . . . *M. ringens*
11. Calyx >16 mm long; corolla ≥30 mm long; valleys to subalpine. . . . . *M. lewisii*

12. Flowers 1 per stem on a long pedicel. . . . . *M. primuloides*
12. Stems with >1 flower 13

13. Herbage viscid-villous; calyx teeth ca. equal . . . . *M. moschatus*
13. Herbage glabrous to glandular-puberulent; upper calyx tooth longer 14

14. Stems ≤20 cm high with 1 to 5 flowers; high subalpine to alpine. . . . .*M. tilingii*
14. Stems often >20 cm high, often with >5 flowers; valleys to lower subalpine . . . .*M. guttatus*

Mimulus ampliatus A.L.Grant Annual. **Stems** erect, simple or branched, 2–12 cm. **Herbage** glabrate to glandular-puberulent. **Leaves** petiolate; blade ovate, dentate, 2–10 mm long. **Flowers**: calyx purplish, 4–8 mm long, ca. 0.5 mm long; pedicel 2 to 3 times as long as calyx; corolla yellow, 10–15 mm long, bilabiate, lower lip longer, little deflexed from the upper lip. **Capsule** 3–6 mm long. Vernally moist, often mossy soil of cliffs, slopes, streambanks; valleys to subalpine. Endemic to central ID and western MT. (p.461)

    Mimulus ampliatus has been mistaken for the closely related *M. patulus* Pennel, *M. washingtonensis* Gand. and *M. floribundus*.

Mimulus breviflorus Piper Annual. **Stems** erect, simple or branched, 3–8 cm. **Herbage** glandular-puberulent. **Leaves** petiolate, the blade narrowly elliptic, entire, 5–10 mm long. **Flowers**: calyx 3–5 mm long, teeth equal, ≤1 mm long; pedicel 1 to 3 times as long as clayx; corolla yellow, 5–8 mm long, slightly bilabiate, lobes subequal. **Capsule** 5–6 mm long. Vernally moist, shallow soil of rock outcrops; montane. Flathead and Glacier cos. WA to MT south to CA, NV. (p.461)

    Sometimes occurs with *M. ampliatus* which has ovate leaf blades.

Mimulus breweri (Greene) Coville Annual. **Stems** erect, usually simple, 3–15 cm. **Herbage** stipitate-glandular. **Leaves** petiolate; blades narrowly lanceolate, entire, 5–20 mm long. **Flowers**: calyx purplish, 3–7 mm long, teeth equal, ca. 1 mm long; pedicel 1 to 2 times as long as clayx; corolla purple to red, 5–10 mm long, moderately bilabiate; stigma tips unequal. **Capsule** 5–7 mm long. Vernally moist, sparsely vegetated soil of rock outcrops, streambanks, avalanche slopes; valleys to subalpine. Ravalli and Stillwater cos. BC, MT south to CA, NV, UT. (p.461)

Mimulus clivicola Greenm. Annual. **Stems** ascending to erect, mostly simple, 3–10 cm. **Herbage** glandular-pubescent. **Leaves** sessile; blades oblong, entire, ca. 1 cm long. **Flowers** few; calyx 6–8 mm long, longer than the pedicel, teeth subequal, 1–2 mm long; corolla purple, 10–15 mm long, bilabiate, lobes unequal with yellow marks on the palate. **Capsule** 5–8 mm long, tubular. Vernally moist soil of partially wooded slopes; montane. Eastern WA, OR, adjacent ID and Sanders Co., MT.

Mimulus floribundus Lindl. Annual. **Stems** prostrate to erect, simple or branched, 4–12 cm. **Herbage** glandular-villous with septate hairs. **Leaves** petiolate; the blade ovate, dentate, 4–12 mm long. **Flowers**: calyx 4–7 mm long, teeth equal, ≤1 mm long; pedicel 2 to 3 times as long as calyx; corolla yellow, 7–11 mm long, moderately bilabiate, lower lobes larger, palate with red spots. **Capsule** 3–6 mm long. Vernally moist cliffs, streambanks; valleys. BC, AB south to CA, NM. (p.461)

    Mimulus ampliatus does not have septate hairs. See *M. breviflorus*.

Mimulus guttatus DC. Annual to perennial, sometimes stoloniferous. **Stems** prostrate to erect, simple, sometimes rooting at the nodes, 4–70 cm. **Herbage** glabrous to sparsely glandular-pubescent above. **Leaves** short-petiolate below, sessile above; blades ovate, serrate, 0.5–9 cm long. **Flowers**: calyx 6–14 mm long, teeth acute, 0.5–3 mm long, upper tooth longer; pedicel 2 to 3 times as long as calyx; corolla 15–40 mm long, yellow, red-spotted, strongly bilabiate, lips spreading, palate closing the throat. **Capsule** 5–12 mm long. Shallow water of streams or vernally moist soil of wet cliffs, seeps, springs; valleys to lower subalpine. Throughout most of N. America. (p.461)

    Plants of permanently moist sites are large and perennial, while those of vernally moist sites may be small and annual. Diminutive annuals could be mistaken for *Mimulus floribundus* but for the asymmetrical calyx. *Mimulus nasutus* Greene, with the upper calyx lobe 3 times as long as the other lobes, is generally subsumed under *M. guttatus* and is reported to occur in western MT (fide L. Fishman). See *M. tilingii*.

Mimulus hymenophyllus Meinke **Stems** ascending, usually simple or branched at the base, 2–10 cm. **Herbage** sparsely glandular-pubescent. **Leaves** long-petiolate; the blade ovate, obscurely dentate, 5–12 mm long. **Flowers**: calyx 3–5 mm long, teeth equal, ca. 1 mm long; pedicel 3 to 4 times as long as calyx; corolla yellow, 7–20 mm long, nearly regular. **Capsule** 3–6 mm long. Cool, moist cliffs; montane. Northeast OR and western MT. (p.461)

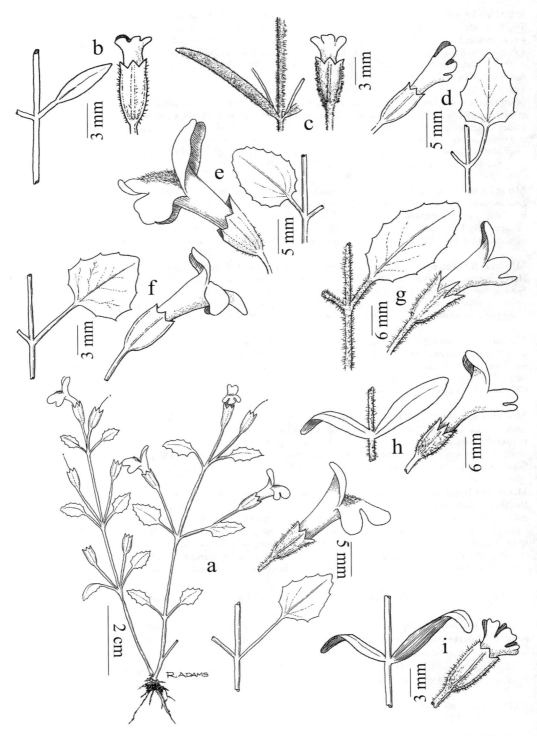

Plate 83. *Mimulus*. **a.** *M. ampliatus*, **b.** *M. breviflorus*, **c.** *M. breweri*, **d.** *M. floribundus*, **e.** *M. hymenophyllus*, **f.** *M. guttatus*, **g.** *M. moschatus*, **h.** *M. nanus*, **i.** *M. suksdorfii*

Mimulus lewisii Pursh LEWIS' MONKEY FLOWER Rhizomatous perennial. **Stems** erect, simple, 15–70 cm. **Herbage** glabrate to glandular pubescent. **Leaves** sessile to subsessile, lanceolate to ovate, serrate, 1–7 cm long. **Flowers**: calyx 15–25 mm long, ≤½ as long as the pedicel, teeth acute, subequal; corolla 3–5 cm long, bilabiate, the limb rose to magenta, tube yellow and purple-spotted, palate bearded, 2-ridged, yellow. **Capsule** 12–16 mm long. Wet meadows, along streams; montane, subalpine. AK south to CA, UT, WY. (p.463)

Mimulus moschatus Lindl. Rhizomatous perennial. **Stems** prostrate and ascending, rooting at the nodes, often branched at the base, 3–40 cm. **Herbage** viscid-villous. **Leaves** petiolate, blades ovate, serrate, 1–6 cm long. **Flowers**: calyx 7–11 mm long, teeth acute, subequal; pedicel 1 to 2 times as long as calyx; corolla 12–24 mm long, light to dark yellow, bilabiate, palate bearded, purple-lined. **Capsule** 3–7 mm long. Wet forest openings, muddy soil of meadows or along streams, ponds; valleys, montane. BC, MT south to CA, UT, CO. (p.461)

Mimulus nanus Hook. & Arn. Annual. **Stems** ascending to erect, branched, 2–10 cm. **Herbage** glandular-puberulent. **Leaves** petiolate below, sessile above; blades oblanceolate, entire, 5–15 mm long. **Flowers** few, clustered; calyx 5–8 mm long, at least twice as long as the pedicel, teeth equal, 1–2 mm long; corolla magenta, 1–2 cm long, bilabiate, lobes subequal with hair and yellow spots on the palate. **Capsule** 7–11 mm long. Vernally moist, sandy soil of open slopes; montane to lower subalpine in Gallatin and Ravalli cos. WA to MT south to CA, NV, ID, WY. (p.461)

Mimulus primuloides Benth. Rhizomatous, stoloniferous, mat-forming perennial. **Stems** erect, 5–15 mm, simple. **Herbage** glabrate. **Leaves** crowded, subsessile, oblanceolate to elliptic, serrate, 7–20 mm long. **Flowers** solitary; calyx 5–8 mm long, teeth acute, ciliate, equal, ≤1 mm long; peduncle 3 to 10 times as long as clayx; corolla 10–17 mm long, yellow, red-spotted, obscurely bilabiate, palate bearded. **Capsule** 5–7 mm long. Wet, mossy, organic, granitic soil of fens, spring-meadows; montane, subalpine in Ravalli and Beaverhead cos. WA to MT south to CA, AZ.

Mimulus ringens L. Mat-forming, rhizomatous perennial. **Stems** ascending to erect, simple or branched, 4-angled, 20–100 cm. **Herbage** glabrous. **Leaves** sessile, lanceolate to oblanceolate, serrate, 2–8 cm long. **Flowers**: calyx 10–16 mm long, teeth acute, subequal; pedicel ca. twice as long as calyx; corolla 20–27 mm long, purple, bilabiate, palate bearded, closing the throat. **Capsule** 10–12 mm long. Streambanks; plains. SK to NS south to ID, CO, TX. Known from Chouteau Co. along the Missouri River.

Mimulus suksdorfii A.Gray Annual. **Stems** erect, usually branched, 2–10 cm. **Herbage** glabrate to glandular-puberulent. **Leaves** sessile, linear-lanceolate, serrate, 5–15 mm long. **Flowers**: calyx purplish, 3–5 mm long, ca. as long as the pedicel, teeth equal, ≤0.5 mm long; corolla yellow, 4–6 mm long, obscurely bilabiate, lobes subequal. **Capsule** 4–5 mm long. Sagebrush steppe, grasslands, woodlands; montane. WA to ID south to CA, AZ, UT, CO. (p.461)

Mimulus tilingii Regel Rhizomatous perennial. **Stems** ascending to erect, simple, 3–15 cm. **Herbage** glabrous below, glandular-pubescent above. **Leaves** petiolate; blades ovate, serrate, 1–2 cm long. **Flowers**: calyx 12–15 mm long, often purplish, teeth acute, upper tooth longer; pedicel 1 to 2 times as long as calyx; corolla 2–3 cm long, yellow, red-spotted, strongly bilabiate, lips spreading, palate bearded, closing the throat. **Capsule** 7–10 mm long. Wet gravel, shallow water of streams, wet cliffs; upper subalpine to alpine. AK south to CA, NM. Our plants are var. *tilingii*.
  Similar to *Mimulus guttatus*, but the flowers are noticeably larger compared to the leaves.

SCROPHULARIACEAE: Figwort Family

Annual or perennial herbs. **Leaves** alternate or opposite, simple, exstipulate. **Inflorescence** a raceme or panicle. **Flowers** perfect, zygomorphic, 5-merous; corolla sympetalous, bilabiate to rotate; stamens (4)5; ovary superior, 2-loculed; style 1; stigma capitate. **Fruit** a many-seeded, 2-celled capsule.

  Most genera formerly placed in this family have been transferred to the Orobanchaceae or Plantaginaceae based on strong molecular-genetic evidence.

1. Corolla bilabiate; leaves opposite; inflorescence open-paniculate. . . . . *Scrophularia*
1. Corolla rotate; leaves alternate or all basal; inflorescence spike-like or flowers solitary 2

2. Leaves mainly basal; flowers solitary on elongate peduncles . . . . *Limosella*
2. Leaves alternate; inflorescence racemose. . . . . *Verbascum*

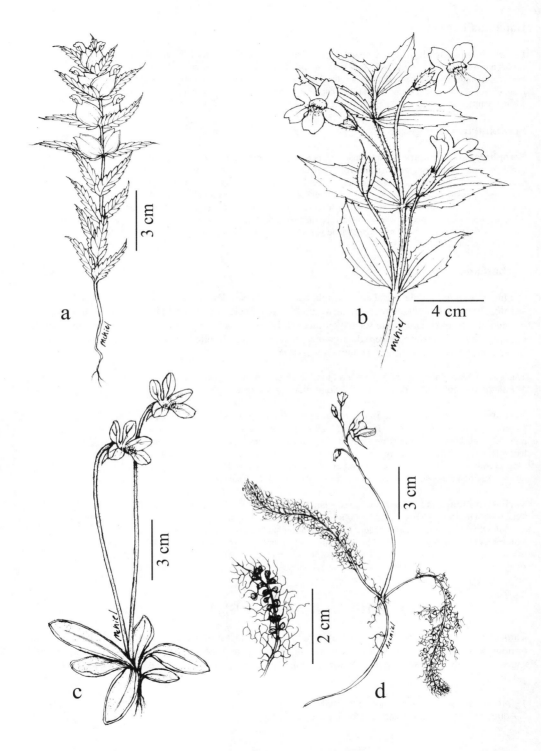

Plate 84. a. *Rhinanthus crista-galli*, b. *Mimulus lewisii*, c. *Pinguicula vulgaris*, d. *Utricularia vulgaris*

Limosella L. Mudwort

Limosella aquatica L. Glabrous annual. **Leaves** basal (alternate); petioles 1–25 cm long; blades narrowly oblong, entire, 4–25 mm long. **Scapes** 5–20 mm long. **Flowers**: calyx 1–3 mm long, campanulate with 5 deltoid lobes shorter than the tube; corolla campanulate, white, 2–3 mm long, ca. regular, 5-lobed; stamens 4; stigma capitate. **Capsule** ovoid, 2–3 mm long. Drying mud around ponds, streams, wetlands; plains, valleys. Circumboreal south to CA, NM, NE, MN.

Scrophularia L. Figwort

Scrophularia lanceolata Pursh Herbaceous perennial. **Stems** erect, 4-angled, 40–150 cm, usually simple. **Herbage** glandular-puberulent, glaucous. **Leaves** cauline, opposite, petiolate; blades lanceolate to deltoid, 5–13 cm long, serrate. **Inflorescence** a minute-bracteate panicle of cymes, opposite below, alternate above. **Flowers**: calyx cup-like, 2–3 mm long, deeply and unequally 5-lobed into rounded, erose segments; corolla reddish-brown, ovoid, 7–10 mm long, bilabiate, upper lip 2-lobed, lower lip shorter; stamens 5, 4 fertile; staminode flattened, without a filament; stigma capitate. **Capsule** ovoid, 5–10 mm long. Meadows, thickets, often along streams; plains, valleys to subalpine. BC to NS south to CA, NM, CO, OK, VA.

Verbascum L. Mullein

Biennials. **Leaves** simple, basal and cauline; basal leaves petiolate, forming a rosette; cauline leaves sessile, alternate. **Inflorescence** a bracteate, spike-like panicle or raceme. **Flowers** rotate; calyx deeply divided into 5 equal, lanceolate sepals; corolla nearly regular with a short tube, 5-lobed, the lower lobes slightly longer; stamens 5, exserted; stigma capitate. **Capsule** 2-celled, broadly ellipsoid. Introduced throughout temperate N. America; native to Eurasia.

1. Lower leaves glabrous; inflorescence glandular-hairy; flowers pedicellate . . . . *V. blattaria*
1. Foliage densely hairy; plants not glandular; flowers nearly sessile . . . . *V. thapsus*

Verbascum blattaria L. Moth Mullein **Stems** erect, simple or branched, 30–80 cm. **Herbage** glabrous below the inflorescence. **Basal leaf blades** oblanceolate, 4–15 cm long, shallowly pinnately lobed, crenate, usually pressed flat to the ground. **Stem leaves** reduced upward, lanceolate, serrate. **Inflorescence** a narrow, glandular-puberulent raceme. **Flowers**: calyx 5–7 mm long; corolla white, 15–30 mm across; anthers purple, villous. **Capsule** 5–10 mm long, glandular. Disturbed, often stony soil of grasslands, streambanks, roadsides; plains, valleys.

Verbascum thapsus L. Common Mullein **Stems** erect, simple, 40–150 cm. **Herbage** yellow-tomentose with stellate hairs. **Basal leaf blades** oblanceolate to obovate, 5–40 cm long, entire to shallowly serrate. **Stem leaves** reduced upward, lanceolate, basally decurrent on the stem. **Inflorescence** a spike-like panicle. **Flowers**: calyx 4–10 mm long; corolla yellow, 15–20 mm across; filaments of upper anthers yellow-villous. **Capsule** 7–10 mm long, stellate-hairy. Disturbed, often stony soil of grasslands, meadows, open forest, talus slopes, roadsides; plains, valleys, montane.

LENTIBULARIACEAE: Bladderwort Family

Carnivorous perennial herbs. **Leaves** basal or alternate. **Inflorescence** a few-flowered, ebracteate raceme or flowers solitary. **Flowers** perfect, zygomorphic; calyx 2- or 5-lobed; corolla tubular, bilabiate, basally spurred; stamens 2; ovary superior. **Fruit** an orbicular to ovoid, many-seeded capsule opening by 2 or 4 valves.

1. Terrestrial plants; leaves simple, all basal . . . . *Pinguicula*
1. Aquatic plants with alternate, divided leaves . . . . *Utricularia*

Pinguicula L. Butterwort

Pinguicula vulgaris L. [*P. macroceras* Link] Plants fibrous-rooted, scapose. **Leaves** succulent, short-petiolate; blades1–4 cm long, oblong with inrolled, entire margins; upper surface viscid. **Inflorescence** a solitary flower terminating an ebracteate peduncle 3–12 cm long. **Flowers**: calyx 3–5 mm long, the 5 lobes longer than the tube; corolla purple, funnelform, somewhat bilabiate, 15–25 mm long, the spur 7–11 mm long. **Capsule** globose, ca. 4 mm high. Wet soil of fens, ledges, along streams; subalpine, rarely alpine in Flathead and Glacier cos. Circumboreal south to CA, MT, MI, NY. Our plants are var. *macroceras* (Link.) Herder. (p.463)

Utricularia L. Bladderwort

Aquatic, submersed, rootless herbs. **Stems** flaccid, prostrate; small, bladder-like traps borne on branches or leaves. **Leaves** alternate, 1 to 3 times palmately or pinnately dissected into linear segments. **Inflorescence** an erect or ascending, few-flowered, bracteate raceme. **Flowers**: calyx 2-lobed; corolla yellow, strongly bilabiate, spurred, upper lip barely 2-lobed, lower lip entire or obscurely lobed. **Capsule** globose.

    Plants flower only sporadically and overwinter as large, bristly vegetative buds (turions). Our species do not occur in saline water.

1. Inflorescence with ≥5 flowers; leaves pinnately divided . . . . *U. vulgaris*
1. Inflorescence with <6 flowers; leaves palmately divided 2

2. Bladders borne on leafless branches; leaf segments often serrate . . . . *U. intermedia*
2. Bladders on borne on leafy branches; leaf segments mostly entire. . . . . *U. minor*

Utricularia intermedia Hayne **Stems** slender. **Turions** ovoid, 6–15 mm long. **Leaves** 5–20 mm long, the segments flattened, entire to obscurely serrate. **Bladders** 3–5 mm long, on separate, leafless branches. **Inflorescence** of 1 to 4 flowers; scape 8–25 cm. **Flowers**: calyx 3–5 mm long; corolla 10–15 mm long; spur 5–9 mm long. *Sphagnum* fens; montane. Circumboreal south to CA, NV, ID, MT, OH, DL. Known from Flathead Co.

    Plants are inconspicuous, creeping on peat beneath shallow water between emergent plants. Our specimens are vegetative.

Utricularia minor L. **Stems** slender. **Turions** ovoid, 2–9 mm long. **Leaves** 2–8 mm long, palmately divided, the segments flat, entire. **Bladders** 1–3 mm long; occurring on the leafy stems. **Inflorescence** of 2 to 5 flowers; scape 5–10 cm. **Flowers**: calyx 1–2 mm long; corolla 5–10 mm long; spur ca. 2 mm long. **Capsule** ≤1 mm long. Shallow water of fens, fen lakes, ponds; valleys, montane. Circumboreal south to CA, CO, IA, NJ.

Utricularia vulgaris L. **Stems** coarse. **Turions** ovoid, 15–25 mm long. **Leaves** crowded, 1–5 cm long, pinnately divided, the segments entire. **Bladders** numerous, 2–5 mm long, attached to leaf segments only. **Inflorescence** of 5 to 12 flowers; scape 10–40 cm. **Flowers**: calyx 4–5 mm long; corolla 10–18 mm long; spur 5–8 mm long. **Capsule** 2–5 mm long. Lakes, ponds, sloughs; plains, valleys, montane. Circumboreal south to most of U.S. (p.463)

CAMPANULACEAE: Bellflower Family

Annual to perennial herbs. **Leaves** basal and/or alternate, simple, exstipulate. **Flowers** perfect, epigynous; calyx united to the ovary (hypanthium); sepals 5; corolla 5-lobed, sympetalous, regular or bilabiate; stamens 5; ovary inferior; style 1 with 2 to 5 stigmas. **Fruit** a many-seeded capsule opening by valves or pores.

    Lobelia and related genera are sometimes placed in a separate family, the Lobeliaceae.

1. Flowers regular, campanulate; stamens separate 2
1. Flowers irregular, at least somewhat bilabiate; stamens united. 6

2. Plants perennial; flowers distinctly pedicellate (except in *C. glomerata*) . . . . *Campanula*
2. Plants annual; flowers sessile or nearly so. 3

3. Hypanthium linear-obconic. 4
3. Hypanthium campanulate. 5

4. Corolla smaller than sepals; stems <10 cm . . . . *Githopsis*
4. Corolla longer than sepals; stems often >10 cm . . . . *Triodanis*

5. Leaves oblanceolate, clasping . . . . *Githopsis*
5. Leaves ovate, sessile to short-petiolate . . . . *Heterocodon*

6. Capsules <8 mm long. . . . .*Lobelia*
6. Capsules ≥10 mm long when mature. 7

7. Plants aquatic, submersed; stems flaccid . . . .*Howellia*
7. Plants of vernally wet habitats; stems ascending to erect, fistulose. . . . . *Downingia*

Campanula L. Bellflower

Perennial herbs with milky sap. **Leaves** basal and cauline. **Flowers** regular, blue; corolla campanulate or funnelform; stigmas 3 to 5. **Capsule** ovoid to hemispheric, usually ribbed, opening by 3 or 5 pores.

1. Corollas ≥2 cm long; plants naturalized ornamentals. 2
1. Corollas <2 cm long; plants rarely naturalized ornamentals. 4

2. Flowers in hemispheric clusters . . . . *C. glomerata*
2. Flowers distinctly pedicellate 3

3. Corolla 2–3 cm long . . . . *C. rapunculoides*
3. Corolla 3–5 cm long . . . . *C. medium*

4. At least some stems with >1 flower 5
4. Flowers solitary. 7

5. Stem leaves lanceolate, serrate; stems often >2 mm wide . . . . *C. rapunculoides*
5. Stem leaves linear, entire; stems <2 mm wide. 6

6. Basal leaves cordate (often deciduous before flowering); stems 6–60 cm . . . . *C. rotundifolia*
6. Basal leaves linear-oblong; stems 2–10 cm. . . . . *C. scabrella*

7. Stems and leaf margins erect- to retrorse-puberulent . . . . *C. scabrella*
7. Stems and leaf surfaces glabrous 8

8. Hypanthium glabrous to short-puberulent; leaf margins scabrous. . . . .*C. parryi*
8. Hypanthium and lower sepals sparsely villous; leaf margins smooth. *C. uniflora*

Campanula glomerata L. Clustered bellflower Taprooted. **Stems** erect, branched, 30–55 cm. **Herbage** sparsely hispid. **Basal leaves** petiolate; blades lanceolate, serrate, 5–10 cm long, deciduous at flowering. **Stem leaves** becoming sessile and clasping. **Inflorescence** of hemispheric clusters, terminal and in upper leaf axils. **Flowers**: sepals linear, ciliate, 7–11 mm long, erect; corolla conical to campanulate, 2–3 cm long; style shorter than corolla; stigmas 3. Roadsides; valleys. Introduced sporadically throughout temperate N. America; a garden escape native to Eurasia; collected in Flathead and Ravalli cos.

Campanula medium L. Taprooted biennial. **Stems** erect, simple, 30–60 cm. **Herbage** hispid. **Basal leaves** petiolate; blades oblong, serrate, 3–6 cm long. **Stem leaves** becoming sessile and lanceolate. **Inflorescence** a bracteate raceme. **Flowers**: sepals lanceolate, ca. 1 cm long, erect; corolla campanulate, sometimes white, 3–5 cm long; style shorter than corolla; stigmas 3. Roadsides; valleys. Introduced as an ornamental and escaped in northeast and northwest N. America; native to Eurasia; collected in Lake Co.

Campanula parryi A.Gray Fibrous-rooted with a slender-branched crown. **Stems** slender, ascending to erect, simple, 5–30 cm. **Herbage** glabrous; leaves with ciliate petiole bases, margins scabrous. **Basal leaves** petiolate; blades oblanceolate, entire, 1–2 cm long. **Stem leaves** narrower, longer, becoming sessile. **Inflorescence** a solitary, terminal flower. **Flowers**: hypanthium glabrous to puberulent; sepals linear-lanceolate, 3–8 mm long, erect; corolla campanulate, 9–15 mm long; style shorter than corolla; stigmas 3(to 5). **Capsule** erect, 7–11 mm long. Non-calcareous soil of fellfields, talus, meadows, open forest; upper subalpine, alpine. WA to MT south to AZ, NM, CO. Our plants are var. *idahoensis* McVaugh.
Our plants all appear to be intermediate to *Campanula scabrella* (fide N. Morin).

Campanula rapunculoides L. Creeping bellflower Tuberous-rhizomatous. **Stems** erect, simple, 40–80 cm. **Herbage** glabrous to puberulent. **Basal leaves** short-petiolate; blades lanceolate to ovate, serrate, 5–8 cm long. **Stem leaves** similar, becoming sessile above. **Inflorescence** a raceme; bracts small, linear. **Flowers** nodding; sepals lanceolate, 5–9 mm long, reflexed; corolla campanulate, 12–30 mm long; style ca. as long as the corolla; stigmas 3. **Capsule** ca. 4 mm long. Roadsides, lawns; valleys. Introduced throughout temperate N. America; native to Europe.

Campanula rotundifolia L. Harebell Plants from slender caudex branches. **Stems** ascending to erect, simple, 6–60 cm. **Herbage** glabrous. **Basal leaves** long-petiolate; blades ovate to orbicular, weakly dentate, 5–25 mm long, usually withered. **Stem leaves** becoming linear and sessile above, 1–7 cm long. **Inflorescence** a terminal, few-flowered raceme. **Flowers** nodding; sepals narrowly lanceolate, 3–12 mm long, erect; corolla campanulate, 10–25 mm long; style ca. as long as the corolla; stigmas 3 (2 to 4). **Capsule** nodding, 3–6 mm long. Grasslands, meadows, open forest, cliffs, exposed slopes, ridges; all elevations. Circumboreal south to CA, TX, Mexico, IA, PA. (p.470)

Campanula scabrella Engelm. Taprooted with a simple or slender-branched caudex. **Stems** ascending to erect, simple, 2–10 cm. **Herbage** spreading-puberulent on stems and leaf undersides. **Basal leaves** petiolate; blades oblanceolate, entire, 5–30 mm long. **Stem leaves** narrower, becoming sessile. **Inflorescence** a solitary, terminal flower. **Flowers** erect; hypanthium puberulent; sepals linear-lanceolate, 3–4 mm long, erect; corolla campanulate, 7–10 mm long; style ca. as long as the corolla; stigmas 3. **Capsule** erect, 4–6 mm long. Non-calcareous soil of stony turf, fellfields, talus; alpine. WA to MT south to CA.

Campanula uniflora L. Arctic bellflower Taprooted with a branched caudex. **Stems** ascending, simple, 2–15 cm. **Herbage** glabrous. **Basal leaves** petiolate; blades narrowly oblong, entire, 1–2 cm long. **Stem leaves** becoming linear-oblanceolate. **Inflorescence** a solitary, terminal flower. **Flowers** erect; hypanthium sparsely villous; sepals lanceolate, 2–7 mm long, erect; corolla campanulate, 5–11 mm long; style included; stigmas 3. **Capsule** erect, ca. 15 mm long. Non-calcareous soils of moist meadows, turf, cool slopes; alpine. Circumpolar south to CO.

Downingia Torr.

Downingia laeta (Greene) Greene Glabrous, fibrous-rooted annual. **Stems** ascending, mostly simple, fistulose, 5–40 mm. **Leaves** cauline, sessile, linear-lanceolate, entire, 2–15 mm long. **Inflorescence** a terminal, 1- to few-flowered, leafy-bracteate spike. **Flowers**: hypanthium tubular; calyx with 5 linear-lanceolate sepals 2–4 mm long; corolla blue, bilabiate, 2–5 mm long, the lower 3 lobes whitish at the base; ovary inferior; anthers and style united into an exserted tube; stigmas 2. **Capsule** cylindric, 1–2 cm long. Vernally wet mud on shores of ponds, pools; valleys. OR to MT south to CA, NV, UT, CO.

Githopsis Nutt.

Githopsis specularioides Nutt. Taprooted annual. **Stems** erect, mostly branched, 5–10 cm. **Herbage** pubescent. **Leaves** cauline, sessile, oblanceolate, 5–15 mm long, dentate. **Inflorescence** of solitary, sessile flowers on branch tips. **Flowers**: hypanthium narrowly obconic; calyx 5-10 mm long with 5 linear sepals; corolla blue, regular, campanulate, ca. 5 mm long; ovary inferior; style included; stigma 3-lobed. **Capsule** obconic, ribbed, 5–10 mm long. Vernally moist, shallow soil of rock outcrops; valleys. WA to MT south to CA. Known from Sanders Co.[254]

Heterocodon Nutt.

Heterocodon rariflorum Nutt. Fibrous-rooted annual. **Stems** ascending to erect, mostly branched, 5–20 cm. **Herbage** glabrate to sparsely ciliate. **Leaves** cauline, sessile, broadly ovate, clasping, dentate, 3–9 mm long, reduced below. **Inflorescence** a terminal, leafy-bracteate spike. **Flowers**: hypanthium shortly obconic, sparsely villous; calyx with 5 ovate, dentate sepals 1–3 mm long; corolla blue, regular, campanulate, 3–4 mm long; ovary inferior; hypanthium; stigma 3-lobed. **Capsule** 2–3 cm long. Vernally moist soil of rock ledges, streambanks in grasslands, open forest; valleys. BC, MT south to CA, NV, WY.

Howellia A.Gray

Howellia aquatilis A.Gray Glabrous, fibrous-rooted, aquatic annual. **Stems** flaccid, floating or submersed, branched, 10–60 cm. **Leaves** cauline, linear, entire, 1–4 cm long. **Inflorescence** of solitary, sessile, axillary flowers. **Flowers**: hypanthium tubular; calyx with 5 lanceolate sepals 1-3 mm long, expanding in fruit; corolla white, bilabiate, 3–4 mm long; ovary inferior; anthers and style united into a tube; stigma 2-lobed. **Capsule** cylindric, 1–2 cm long, with 2 to 5 large seeds. Shallow, fresh water of ponds, sloughs; valleys of Lake and Missoula cos. WA to MT south to CA.

Submersed flowers do not open and are self-fertilized.[253] Listed as threatened under the Endangered Species Act.

Lobelia L.

Fibrous-rooted perennials, often with milky sap. **Leaves** basal and cauline, simple. **Inflorescence** a bracteate raceme. **Flowers** upside down on twisted pedicels; calyx with 5 sepals; corolla bilabiate, with the 2-lobed lip above and split halfway down the tube; stamens united around the style; ovary inferior; stigma 2-lobed. **Capsule** obconic to globose.

1. Stem leaves ≤3 mm wide; inflorescence few-flowered. . . . . *L. kalmii*
1. Some stem leaves >4 mm wide; inflorescence many-flowered . . . . *L. spicata*

Lobelia kalmii L. Stems erect, usually simple, 10–40 cm. **Herbage** glabrate. **Basal leaves** petiolate, spatulate, entire to obscurely dentate, 1–3 cm long. **Stem leaves** becoming linear-lanceolate, 1–4 cm long. **Inflorescence** of <10 flowers; pedicels bracteate. **Flowers**: sepals lanceolate, 2–3 mm long; corolla 5–10 mm long, blue, whitish at top of the tube. **Capsule** obconic, 4–7 mm long. Organic soil of wet meadows, fens; valleys, montane. YT to NL south to WA, MT, MN, PA.

Lobelia spicata Lam. Stems erect, usually simple, glabrate, 15–60 cm. **Basal leaves** petiolate, obovate, serrate, ciliate, 15–50 mm long. **Stem leaves** oblanceolate, 1–4 cm long, serrate. **Inflorescence** many-flowered. **Flowers**: sepals linear, 2–4 mm long; corolla 4–6 mm long, light blue, lower lip pubescent. **Capsule** globose, 2–4 mm long, sparsely hispid. Wet meadows, prairie; plains. SK to QC south to TX, LA, MS, GA.

Triodanis Raf. Venus' looking glass

Taprooted annuals. **Stems** 5-angled. **Leaves** cauline, simple. **Inflorescence** a leafy-bracteate, spike-like raceme. **Flowers** regular, epigynous, dimorphic; chasmogamous flowers with 5 sepals and a funnelform corolla, lobes longer than the tube; stamens distinct; ovary inferior; stigma 3- to 5-lobed; cleistogamous flowers with 3 to 5 sepals and rudimentary corollas. **Capsule** surmounted by aristate sepals, opening by 1 to 3 pores or slits.

1. Flower bracts linear to lanceolate, >6 times longer than wide . . . . *T. leptocarpa*
1. Flower bracts ovate, <3 times as long as wide. . . . . *T. perfoliata*

Triodanis leptocarpa Nieuwl. Stems erect, simple or branched, 5–40 cm. **Herbage** sparsely short-hispid. **Leaves** lanceolate or oblanceolate, entire to obscurely crenate, sessile, 7–25 mm long. **Inflorescence** with clusters of 1 to 3 sessile flowers in axils of linear bracts; lower flowers cleistogamous. **Chasmogamous flowers**: sepals linear, 6–12 mm long; corolla deep blue, 5–8 mm long. **Capsule** linear, 14–20 mm long for chasmogamous flowers, 8–15 mm long for cleistogamous flowers. Grasslands; plains. MT to IN south to TX, AR.

Triodanis perfoliata (L.) Nieuwl. Stems erect, usually branched, 10–40 cm. **Herbage** sparsely hispid. **Leaves** ovate to orbicular, crenate, sessile, cordate-clasping above, 5–20 mm long. **Inflorescence** with clusters of 1 to 3 sessile flowers in axils of leaf-like bracts; lower flowers cleistogamous. **Chasmogamous flowers**: sepals lanceolate, 3–7 mm long; corolla purple, 6–10 mm long. **Capsule** ellipsoid, 5–8 mm long for chasmogamous flowers, 3–4 mm long for cleistogamous flowers. Disturbed soil of grasslands, roadsides; plains, valleys, montane. Temperate N. America south to S. America.

Small plants from Lincoln Co. appear to be entirely cleistogamous.

RUBIACEAE: Madder Family

Annual to perennial herbs. **Leaves** simple, entire, stipulate, opposite or whorled. **Inflorescence** terminal and/or axillary, cymose. **Flowers** mostly perfect, epigynous, mostly 4(3 to 5)-merous; calyx lobed or obsolete; corolla regular, sympetalous, rotate to funnelform; stamens as many as corolla lobes; ovary

inferior with 2 carpels; styles 2; stigmas capitate or short-lobed. **Fruit** a schizocarp splitting into 2 mericarps. Vestiture of the mature fruits is often necessary for identification.

1. Leaves opposite; sepals small but present. . . . . *Kelloggia*
1. Leaves mostly in whorls of 4 to 8 (often opposite in *G. bifolium*); sepals absent . . . . *Galium*

Galium L. Bedstraw

Annual or perennial. **Stems** 4-angled. **Leaves** linear to narrowly lanceolate, sessile, rarely opposite or in whorls of 4 to 8. **Flowers** bisexual, rotate; sepals absent; corolla mostly 4-lobed, lobes longer than the tube, usually white; stamens 4; stigma capitate. **Mericarp** glabrous to hairy.

    A report of *Galium palustre* L.,[113] which has 4-lobed corollas, was probably based on *G. trifidum*.

1. Leaf tips cuspidate with an attenuate tip ca.1 mm long 2
1. Leaf tips rounded to acute without a long tip 6

2. Flowers yellow; leaves revolute, linear. . . . . *G. verum*
2. Flowers white; leaves not revolute 3

3. Fruits glabrous; stems ascending to erect . . . . *G. mollugo*
3. Fruits with hooked or straight hairs; stems lax, prostrate or climbing. 4

4. Mature fruits >2 mm high; taprooted annuals. . . . . *G. aparine*
4. Fruits 1–2 mm high; perennials 5

5. Hooked hairs of fruit <0.5 mm long; inflorescence irregularly branched . . . . *G. mexicanum*
5. Hooked hairs of fruit ≥5 mm long; inflorescence branched in 3's . . . .*G. triflorum*

6. Taprooted annuals with erect stems. 7
6. Rhizomatous perennials with erect to lax stems 8

7. Flowers yellow; pedicels <2 mm long, hidden beneath the leaves. . . . .*G. pedemontanum*
7. Flowers white; pedicels ≥5 mm long in fruit . . . . *G. bifolium*

8. Stems erect; inflorescence dense; upland habitats . . . . *G. boreale*
8. Stems lax; inflorescence diffuse; wet habitats . . . . *G. trifidum*

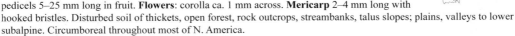

Galium aparine L. Cᴌᴇᴀᴠᴇʀs, ɢᴏᴏsᴇ ɢʀᴀss Annual. **Stems** lax, prostrate or climbing, simple or branched, 10–80 cm. **Herbage** hooked-scabrous on stem angles and leaf margins. **Leaves** linear to narrowly oblanceolate, cuspidate, 5–20 mm long, in whorls of 6(8). **Inflorescence** a few-flowered umbel terminating axillary peduncles and subtended by a whorl of leaf-like bracts; pedicels 5–25 mm long in fruit. **Flowers**: corolla ca. 1 mm across. **Mericarp** 2–4 mm long with hooked bristles. Disturbed soil of thickets, open forest, rock outcrops, streambanks, talus slopes; plains, valleys to lower subalpine. Circumboreal throughout most of N. America.

Galium bifolium S.Watson Annual. **Stems** erect, simple or branched, 3–25 cm. **Herbage** glabrate. **Leaves** narrowly lanceolate to oblanceolate, 5–25 mm long, opposite or in whorls of 4. **Inflorescence**: solitary flowers from leaf axils; pedicels 5–30 mm long in fruit. **Flowers**: corolla 3-lobed, ≤1 mm across. **Mericarp** 2–4 mm long with hooked bristles. Moist meadows, grasslands; montane, subalpine. BC, MT south to CA, AZ, UT, CO.

Galium boreale L. Nᴏʀᴛʜᴇʀɴ ʙᴇᴅsᴛʀᴀw Rhizomatous perennial. **Stems** erect, branched, 8–60 cm, often sterile. **Herbage** glabrate to scabrous. **Leaves** narrowly lanceolate, 1–4 cm long, in whorls of 4. **Inflorescence** dense, terminal, many-flowered, bracteate panicles; pedicels 1–2 mm long in fruit. **Flowers**: corolla 3–5 mm across. **Mericarp** ca. 1 mm long, hirsute. Grasslands, meadows, open forest; plains, valleys to subalpine. Circumboreal south to CA, TX, MO, OH. (p.470)

Galium mexicanum Kunth [*G. asperrimum* A.Gray] Rhizomatous perennial. **Stems** lax, prostrate or climbing, mostly branched, 15–60 cm. **Herbage** sparsely hooked-scabrous on stem angles and leaf margins. **Leaves** narrowly lanceolate to elliptic, cuspidate, 1–3 cm long, in whorls of 5 or 6. **Inflorescence** a terminal, open, many-flowered, bracteate cyme; pedicels 2–5 mm long in fruit. **Flowers**: corolla 2–3 mm across. **Mericarp** ca. 1 mm long with short, hooked bristles. Forests, thickets; valleys to subalpine. WA to MT south to CA, NM, Mexico. Our plants are ssp. ***asperulum*** (A.Gray) Dempster.

    Galium aparine has smaller inflorescences and larger fruits.

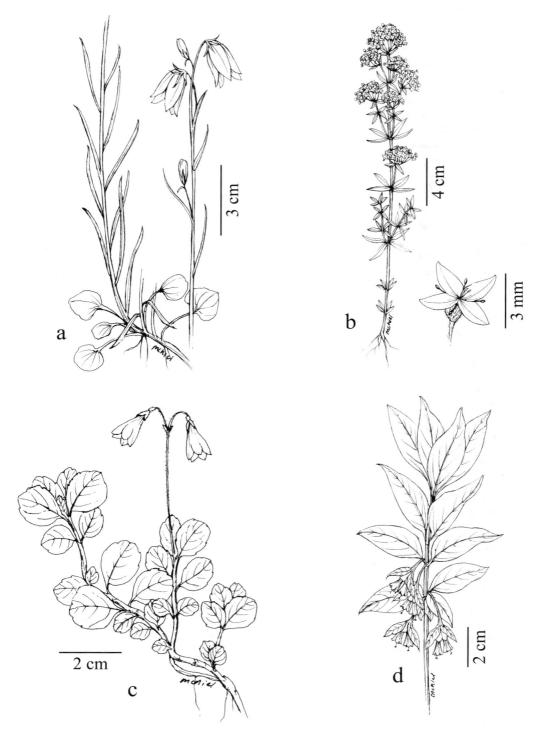

Plate 85. a. ***Campanula rotundifolia***, b. ***Galium boreale***, c. ***Linnaea borealis***, d. ***Lonicera involucrata***

Galium mollugo L. Rhizomatous perennial. **Stems** ascending to erect, branched, 30–100 cm. **Herbage** sparsely scabrous on stems and leaf margins. **Leaves** oblanceolate, cuspidate, 1–3 cm long, in whorls of 6 to 8. **Inflorescence** an oppositely-branched, open, cymose panicle; pedicels 2–3 mm long. **Flowers**: corolla ca. 2 mm across. **Mericarp** ca. 1 mm long, glabrous. Roadsides; valleys, montane; Broadwater and Missoula cos. Commonly introduced in eastern N. America, sporadic in the west; native to Eurasia.

Galium pedemontanum (Bellardi) All. [*Cruciata pedemontana* (Bellardi) Ehrend.] Annual. **Stems** erect, simple or branched, 10–35 cm. **Herbage** hirsute with simple or stellate hairs. **Leaves** ovate to elliptic, 5–15 mm long, in whorls of 4. **Inflorescence** 2- to 3-flowered cymes, curled under the leaves; pedicels <1 mm long in fruit. **Flowers**: corolla yellow, 4-lobed, ≤1 mm across. **Mericarp** <1 mm long, glabrous. Pastures, grasslands; valleys. Introduced to eastern and northwest N. America; native to Europe. Known from Missoula and Lake cos.

Galium trifidum L. Rhizomatous perennial. **Stems** lax, prostrate or climbing, mostly branched, 8–35 cm. **Herbage** glabrous to hooked-scabrous on stem angles and leaf margins. **Leaves** linear to narrowly oblanceolate, 5–15 mm long, in whorls of 4 (to 6). **Inflorescence** axillary 1- to 3-flowered umbels; pedicels 2–15 mm long in fruit. **Flowers**: corolla ca. 1 mm across, 3- to 4-lobed. **Mericarp** 1–2 mm long, sparsely spiculose, the mericarps becoming separate. Wet meadows, marshes, fens; montane to lower subalpine. Circumboreal south to CA, TX, GA, Mexico.

1. Fruiting pedicels ≤7 mm long . . . . ssp. *subbiflorum*
1. Fruiting pedicels ≥7 mm long 2

2. Corolla ≥2 mm in diameter; some leaves ≥3 mm wide . . . . ssp. *columbianum*
2. Corolla <2 mm wide; leaves 1–3 mm wide . . . . ssp. *trifidum*

Galium trifidum ssp. *subbiflorum* (Wiegand) Puff is widespread and somewhat less scabrous; *G. trifidum* ssp. *trifidum* is more densely spiculose and confined to the northwest part of MT; *G. trifidum* ssp. *columbianum* (Rydb.) Hultén [*G. cymosum* Wiegand] is difficult to discern from the latter and has a similar distribution.[331] The three subspecies appear to be without ecological distinction in MT.

Galium triflorum Michx. Sweet-scented bedstraw Rhizomatous perennial. **Stems** lax, prostrate or climbing, mostly simple, 20–70 cm. **Herbage** glabrous except sparsely scabrous on stem angles, leaf margins. **Leaves** narrowly elliptic, cuspidate, 2–5 cm long, in whorls of 6. **Inflorescence** axillary, 3-flowered umbels or 3-parted compound umbels; pedicels 6–15 mm long in fruit. **Flowers**: corolla 1–2 mm across. **Mericarp** 1–2 mm long not including the hooked bristles. Moist to wet, coniferous forest, thickets; valleys to lower subalpine. Circumboreal south to CA, NM, OK, FL.

   Plants from exposed sites have broader leaves. Foliage has a faint vanilla smell upon drying. *Galium mexicanum* often has narrower leaves and more than 3 flowers per inflorescence.

Galium verum L. Rhizomatous perennial. **Stems** erect, branched, 40–80 cm. **Herbage** glabrate. **Leaves** linear, cuspidate, 1–3 cm long, in whorls of 6 to 10, margins revolute. **Inflorescence** a dense, elongate cymose panicle; pedicels 1–2 mm long. **Flowers** yellow; corolla 2–4 mm across. **Mericarp** ca. 1 mm long, glabrous. Fields, grasslands, open forest; valleys. A garden escape, introduced to north-temperate N. America; native to Europe. Also reported for Glacier Co.[242]

Kelloggia Torr. ex Benth.

Kelloggia galioides Torr. Rhizomatous perennial. **Stems** ascending to erect, simple, 10–60 cm. **Herbage** glabrous. **Leaves** sessile, opposite, entire, lanceolate, 15–50 cm long. **Inflorescence** a terminal, open, few-flowered, cyme; pedicels 6–30 mm long. **Flowers** funnelform, 4(3 to 5)-merous; calyx lobes tiny; corolla funnelform, 4–8 mm long, the lobes as long as the tube; style 2-lobed; stamens and style barely exserted. **Mericarp** 3–4 mm long with hooked bristles. Forest openings; valleys, montane. WA to MT south to CA, AZ, UT, WY, Mexico. Known from Mineral Co.

CAPRIFOLIACEAE: Honeysuckle Family

Erect or trailing shrubs or vines. **Leaves** opposite, petiolate; stipules small or absent. **Inflorescence** cymes or 1–2 flowers in leaf axils. **Flowers** perfect, epigynous, mostly 5-merous; corolla gamopetalous, regular or bilabiate; stamens attached to the tube; ovary inferior; style 1. **Fruit** usually a berry or drupe (achene).

   Viburnum and *Sambucus* are sometimes placed in the Adoxaceae.[21] Stipules of *Sambucus* and *Viburnum* act as extrafloral nectaries.[102]

1. Plants shrubs with erect stems. 2
1. Plants trailing subshrubs or climbing vines. 6

2. Leaves divided or obviously lobed 3
2. Leaves of older stems with mostly entire margins (leaves of new shoots sometimes lobed). 4

3. Leaves divided into 5 to 7 leaflets . . . . *Sambucus*
3. Leaves 3-lobed, maple-like. . . . . *Viburnum*

4. Inflorescence an umbel-like cyme . . . . *Viburnum*
4. Inflorescence a few-flowered raceme. 5

5. Corollas to 7 mm long; berries white . . . . *Symphoricarpos*
5. Flowers ≥9 mm long; berries red to black. . . . . *Lonicera*

6. Trailing subshrub; leaves shallowly toothed on the tip . . . . *Linnaea*
6. Vine; leaves with entire margins. . . . . *Lonicera*

Linnaea L. Twinflower

Linnaea borealis L. Evergreen subshrub. **Stems** slender, trailing; short erect shoots 1–6 cm. **Herbage** glabrate to pubescent, glandular above. **Leaf blades** ovate to suborbicular, shallowly dentate above, 7–20 mm long. **Inflorescence** terminal, of 2 pedicellate, nodding flowers opposite each other on a scape 3–8 cm long. **Flowers** regular, 5-merous; sepals linear-lanceolate, 2–4 mm long; corolla funnelform, light pink, hairy within, 8–14 mm long; stamens 4. **Fruit** a round, glandular-hairy, ovate achene 1–2 mm long. Moist, cool, coniferous forest; valleys to lower subalpine. Circumboreal south to CA, NM, IN, WV. Our plants are var. *longiflora* Torr. (p.470)

Lonicera L. Honeysuckle

Vines or erect shrubs. **Leaves** usually short-petiolate, simple, mostly entire. **Inflorescence** axillary, 2-flowered peduncles or a terminal umbel; base of flowers bracteate. **Flowers**: calyx small or obscure, shallowly lobed; corolla barely bilabiate, tubular to funnelform, 5-lobed, tube slightly spurred at the base, hairy within; stamens slightly exserted; stigma globose. **Fruit** a several-seeded berry.

1. Vining plants; flowers orange; inflorescence umbellate . . . . *L. ciliosa*
1. Erect shrubs; flowers yellow or white; flowers paired in leaf axils 2

2. Flowers with purplish bracts; berry black; leaves long-pointed . . . . *L. involucrata*
2. Flower bracts green; berry red; leaves not acuminate 3

3. Twigs puberulent and/or sparsely villous 4
3. Twigs glabrous 5

4. Leaves oblanceolate to obovate, widest in distal half; wet habitats . . . . *L. caerulea*
4. Leaves narrowly ovate, widest near the base; upland sites . . . . *L. morrowii*

5. Corolla white, lobes shorter than the tube; common native . . . . *L. utahensis*
5. Corolla pink, lobes longer than the tube; garden escape . . . . *L. tatarica*

Lonicera caerulea L. [*L. cauriana* Fernald] Shrubs. **Stems** erect, branched 20–80 cm; twigs sparsely villous. **Leaf blades** oblanceolate to obovate, 2–7 cm long, rounded at the tip, sparsely villous beneath. **Inflorescence**: peduncles 2–8 mm long; lower 2 bracts linear, green; upper bracts brown, enclosing the ovaries. **Flowers**: corolla yellow, 9–13 mm long, glabrate, lobes ca. as long as the tube. **Berry** red (glaucous-blue), ca. 1 cm long, borne in the cup formed by the bracts. Organic soil of wet meadows, fens, thickets; montane, subalpine. Circumboreal south to CA, NV, WY, MN, PA. Our plants are var. *cauriana* (Fernald) B.Boivin.

Lonicera ciliosa (Pursh) Poir. ex DC. Vines. **Stems** lax or twining, hollow, branched; twigs glaucous. **Leaves** sessile above; blades glaucous beneath, ciliate, elliptic, 4–10 cm long. **Inflorescence** a terminal, sessile, several-flowered umbel subtended by a perfoliate, leaf-like bract 4–10 cm across. **Flowers**: corolla orange, 23–32 mm long, the flared lobes 6–10 mm long. **Berry** red, ca. 8 mm across. Openings in dry to moist forest; valleys, montane. BC, MT south to CA.

Lonicera involucrata (Richarson) Banks ex Spreng. Bearberry, black twinberry Shrubs. Stems erect, 50–150 cm; twigs 4-angled, glabrate. **Leaf blades** 4–15 cm long, oblong to ovate with acuminate tips, glabrate to sparsely hirsute beneath. **Inflorescence**: peduncles 1–4 cm long, surmounted by 4 purple, glandular bracts reflexed in fruit. **Flowers**: corolla yellow, 10–15 mm long, glandular-pubescent, lobes shorter than the tube. **Berry** black, 6–10 mm wide. Moist forest, meadows, avalanche slopes, especially along streams; montane, subalpine. BC to QC south to CA, NM, WI. Our plants are var. *involucrata*. (p.470)

Lonicera morrowii A.Gray Shrubs. **Stems** erect, branched, 2–4 m; twigs puberulent. **Leaf blades** lanceolate with acute tips, 2–6 cm long, pubescent beneath. **Inflorescence**: peduncles 10–15 mm long; bracts linear, 4–6 mm long. **Flowers**: sepals ciliate; corolla white, ca. 15 mm long, pubescent, lobes longer than the tube. **Berry** red. Urban streambanks, fields; valleys. Introduced mainly to northeastern U.S.; native to Japan.

Lonicera tatarica L. **Stems** erect, branched, 2–4 m; twigs glabrous. **Leaf blades** ovate to lanceolate with rounded tips, 2–6 cm long, glabrate. **Inflorescence**: peduncles 8–16 mm long; bracts linear, 2–4 mm long. **Flowers**: corolla pink, 10–16 mm long, glabrous, lobes longer than the tube. **Berry** orange to red, 5–6 mm long. Roadsides, streambanks; valleys. Introduced across northern U.S. and adjacent Canada; native to Eurasia.

Lonicera utahensis S.Watson Shrubs. **Stems** spreading to erect, branched, 50–150 cm; twigs glabrous. **Leaf blades** ovate to elliptic with rounded tips, 2–6 cm long, sparsely puberulent beneath. **Inflorescence**: peduncles 5–20 mm long; bracts linear, ca. 2 mm long. **Flowers**: corolla ochroleucus, 12–20 mm long, glabrous, lobes shorter than the tube. **Berry** red, 8–10 mm long. Moist forest; montane, subalpine. BC, AB south to CA, UT, WY.

Often flowering before the leaves are fully expanded.

Sambucus L. Elderberry

Shrubs. **Stems** with soft pith and glaucous twigs. **Leaves** petiolate, pinnately compound; leaflets lanceolate, serrate, asymmetrical at the base, glabrate, glaucous beneath. **Inflorescence** terminal, densely flowered, cymose. **Flowers** 5-merous; calyx inconspicuous or absent; corolla white, regular, rotate, lobes longer than the tube; style obscure; stigma 3- to 5-lobed. **Fruit** a few-seeded berry.

1. Berries black; inflorescence pyramidal. . . . . *S. racemosa*
1. Berries waxy blue; inflorescence flat-topped to broadly hemispheric. . . . . *S. caerulea*

Sambucus caerulea Raf. [*S. mexicana* C.Presl ex DC. ssp. *caerulea* (Raf.) E.Murray, *S. nigra* L. ssp. *cerulea* (Raf.) Bolli] Blue elderberry **Stems** 2–4 m. **Leaflets** 5 to 9, 5–15 cm long. **Inflorescence** a flat-topped, compound umbel-like cyme 8–20 cm across. **Flowers**: corolla 4–6 mm across. **Berry** blue, glaucous, 4–6 mm wide. Open forest, thickets; valleys, montane. BC, MT south to CA, AZ, NM.

The fruits are edible. Plants traditionally called *Sambucus caerulea* are part of a complex that includes the European *S. nigra*, the eastern *S. canadensis* L. and the more southern *S. mexicana*.

Sambucus racemosa L. [*S. melanocarpa* A.Gray] Black elderberry **Stems** 50–200 cm. **Leaflets** 5 to 7, 3–12 cm long. **Inflorescence** a pyramidal to hemispheric panicle, 3–10 cm long. **Flowers**: corolla 4–5 mm across, the lobes commonly reflexed. **Berry** black, 3–5 mm wide. Moist open forest, thickets, avalanche slopes, often along streams; valleys to subalpine. Circumboreal south to CA, NM, IL, GA. Our plants are var. *melanocarpa* (A.Gray) McMinn.

One or two specimens from west-central MT have reddish-black fruit and may be intermediate to var. *pubens* (Michx.) S.Watson, with red or yellow berries. (p.475)

Symphoricarpos Duhamel Snowberry

Shrubs with spreading rootstocks. **Leaves** petiolate, simple, entire or occasionally shallowly lobed. **Inflorescence** terminal or axillary, 2- to several-flowered racemes. **Flowers** 5-merous; calyx shallowly lobed; corolla campanulate, white or pinkish, hairy within; style with a capitate or barely 2-lobed stigma. **Fruit** a white, 2-seeded berry.

Juvenile stems commonly have shallowly lobed leaves.

1. Corolla longer than wide, lobes <1/2 as long as the tube; pith solid . . . . *S. oreophilus*
1. Corolla ca. as long as wide, lobes >1/2 as long as the tube; pith hollow 2

2. Style ≥3 mm long, exserted from corolla tube; twigs usually puberulent . . . . *S. occidentalis*
2. Style ≤3 mm long, included in the corolla tube; twigs mostly glabrous. . . . . *S. albus*

Symphoricarpos albus (L.) S.F.Blake **Stems** spreading to erect, highly branched, 50–150 cm; twigs brown, glabrous. **Leaves** ovate, 1–5 cm long, sparsely villous beneath. **Inflorescence** mostly with ≤6 flowers. **Flowers:** corolla campanulate, 4–7 mm long, lobes shorter than the tube; style glabrous, 2–3 mm long; style and stamens included. **Berry** 8–15 mm long. Moist to dry forest; plains, valleys to lower subalpine. AK to NS south to CA, CO, NE, VT. Our plants are var. *laevigatus* (Fernald) S.F.Blake. (p.475)

Symphoricarpos occidentalis Hook. **Stems** spreading, simple or branched, 30–100 cm; twigs brown, finely puberulent. **Leaves** ovate, 1–5 cm long, pubescent beneath. **Flowers:** corolla bowl-shaped, 4–10 mm long, lobes as long as the tube or longer; style glabrous or hairy at mid-length, 3–8 mm long; style and stamens exserted. **Berry** 6–9 mm long. Grasslands, meadows, thickets, or dry, open forest; plains, valleys. BC to ON south to WA, NM, OK, MI.

This plant spreads more widely and vigorously than our other two species; it is sometimes called buckbrush in the eastern part of the state.

Symphoricarpos oreophilus A.Gray **Stems** spreading to erect, highly branched, 30–150 cm; twigs glabrous to puberulent. **Leaves** narrowly elliptic to ovate, 1–3 cm long, pubescent beneath. **Flowers:** corolla tubular-campanulate, 5–10 mm long, lobes <1/2 as long as the tube; style glabrous, 2–4 mm long; stamens barely included. **Berry** 7–10 mm long. Dry, open forest, grasslands, sagebrush steppe, talus; montane, subalpine. BC, MT south to CA, NM, Mexico. Our plants are var. *utahensis* (Rydb.) A.Nelson.

Viburnum L.

Shrubs. Leaves petiolate, simple, serrate to lobed; stipules absent or small. **Inflorescence** terminal, umbel-like, cymose, expanding in fruit. **Flowers** perfect, regular or marginal ones sterile and irregular, 5-merous; calyx shallowly lobed; corolla white, campanulate to rotate; stamens mostly exserted. **Fruit** a 1-seeded drupe.

Leaves of all but *Viburnum lantana* turn red or purple in autumn.

1. Many leaves 3-lobed, maple-like 2
1. Leaves unlobed 3

2. Leaves minutely and sparsely glandular beneath; flowers all similar. . . . . *V. edule*
2. Leaves not glandular; outer flowers of cyme much larger than the inner. *V. opulus*

3. Leaves and twigs glabrous. . . . .*V. lentago*
3. Leaves and twigs puberulent with stellate hairs . . . .*V. lantana*

Viburnum edule (Michx.) Raf. MOOSEBERRY LOW-BUSH CRANBERRY. **Stems** spreading to erect, branched, 50–200 cm; twigs brown, glabrous. **Leaves** exstipulate; blades broadly ovate, mostly 3-lobed above, serrate, 3–10 cm long, sparsely hirsute, sunken-glandular beneath; lobes acute. **Inflorescence** 1–3 cm long. **Flowers:** corolla campanulate to rotate, deeply lobed, 2–3 mm long. **Drupe** red or orange, 7–10 mm long. Moist forest, often near streams; montane, lower subalpine. AK to NL south to OR, CO, MN, PA. (p.475)

Viburnum lantana L. WAYFARING TREE **Stems** erect, branched, 2–3 m; twigs gray-puberulent with stellate hairs. **Leaves:** blades ovate to obovate, serrate, unlobed, 5–10 cm long, puberulent. **Flowers:** corolla subrotate, ca. 4 mm across. **Drupe** red, 8–10 mm long. Riparian forest, thickets; valleys. Ornamental, occasionally introduced in northern U.S. and adjacent Canada. Collected in Missoula Co.

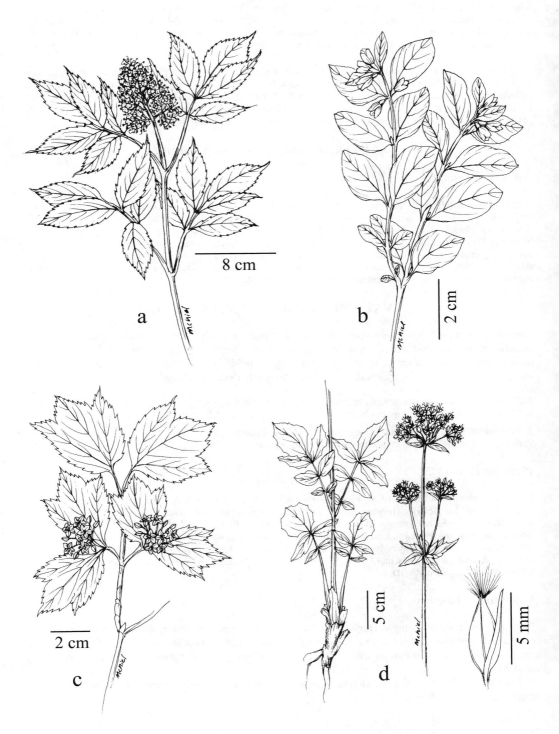

Plate 86. a. *Sambucus racemosa*, b. *Symphoricarpos albus*, c. *Viburnum edule*, d. *Valeriana sitchensis*

Viburnum lentago L. Nannyberry **Stems** erect, branched, 50–150 cm; twigs gray-brown, glabrous. **Leaves** exstipulate; blades ovate to lanceolate, acuminate, serrate, unlobed, 3–9 cm long, glabrous. **Inflorescence** 4–10 cm across. **Flowers**: corolla subrotate, 2–4 mm long. **Drupe** blue, glaucous, 5–10 mm long. Riparian forest, thickets; plains; Bighorn and Richland cos. MB to QC south to CO, NE, MO, GA.

Viburnum opulus L. High-bush cranberry **Stems** spreading to erect, branched, 1–3 m; twigs green-brown, glabrous. **Leaves** stipulate; blades ovate, 3-lobed above, dentate or lobed again, 4–10 cm long, hirsute beneath; lobes acuminate. **Inflorescence** 6–10 cm across. **Flowers**: corolla subrotate, 2–3 mm long; outermost ring of flowers rotate, sterile, 1–2 cm across. **Drupe** red, 10–15 mm long. Riparian forest; valleys. BC to NL south to WA, MT, SD, IL, PA; Europe. Our plants are var. *americanum* Aiton. and are probably introduced in Missoula and Ravalli cos.

ADOXACEAE: Moschatel Family

Adoxa L. Moschatel

Adoxa moschatellina L. Glabrous, rhizomatous perennials. **Stems** erect, 5–20 cm. **Leaves** long-petiolate; basal leaves petiolate, ternate, primary segments petiolate, 2- to 3-times lobed, 2-4 cm long, ultimate segments rounded; cauline leaves opposite, similar to primary segments of the basal leaves. **Inflorescence** a terminal, few-flowered, capitate cyme. **Flowers** partially epigynous; sepals 2 or 3; corolla rotate, yellowish, 5–8 mm across; the central flower 4-lobed, lateral flowers 5-lobed; stamens 8 to 10; ovary half-inferior; styles 3 to 5; stigma capitate. **Fruit** a 3- to 5-seeded berry. Open coniferous forest; montane. Circumboreal south to UT, CO, SD, IA, NY.

VALERIANACEAE: Valerian Family

Annual or perennial herbs. **Leaves** petiolate, basal and cauline, exstipulate; stem leaves opposite. **Inflorescence** many-flowered, bracteate panicles. **Flowers** epigynous, perfect or unisexual; calyx absent, or of numerous bristles expanding in fruit; corolla regular or bilabiate, tubular, 5-lobed at the mouth with a short basal spur; stamens 3, exserted; ovary inferior; style 1. **Fruit** an achene.

1. Plants perennial; calyx segments expanded and feathery in fruit . . . . *Valeriana*
1. Plants annual; calyx segments absent 2
2. Achene glandular-puberulent; inflorescence interrupted spike-like . . . . *Plectritis*
2. Achene glabrous; flowers in glomerules at paired branch tips . . . . *Valerianella*

Plectritis (Lindl.) DC.

Plectritis macrocera Torr. & A.Gray Fibrous-rooted annual. **Stems** erect, 6–30 cm, simple or branches paired in the axils. **Herbage** glabrous. **Basal leaves** petiolate, oblong, entire; blades 1–2 cm long, withering at flowering. **Stem leaves** sessile, oblong, 7–40 mm long. **Inflorescence** a dense, glandular, interrupted spike. **Flowers** perfect; calyx absent; corolla white, bilabiate, short-spurred, 2–4 mm long, sparsely glandular, lobes shorter than the tube; stigma 2-lobed. **Achene** compressed-orbicular, beaked, winged, 2–4 mm long, glandular-puberulent. Grasslands, open ponderosa pine forest; valleys. BC, MT south to CA, ID, UT.

Valeriana L. Valerian

Perennials. **Stems** erect, simple. **Herbage** glabrous. **Leaves** entire to pinnately divided. **Inflorescence** paniculate, expanding in fruit. **Flowers**: calyx of numerous plumose bristles expanding in fruit; corolla regular; stigma 3-lobed. **Fruit** an achene with longitudinal nerves and surmounted by the persistent, feathery calyx.

    Valeriana occidentalis and *V. dioica* intergrade in our area. The distinction between the two is not recognized by all authorities.

1. Corolla ≥4 mm long 2
1. Corolla <4 mm long 3

2. Basal leaves reduced or absent on flowering stems; lateral lobes of middle stem
   leaves ovate, often dentate . . . .*V. sitchensis*
2. Basal leaves prominent; lateral lobes of stem leaves oblanceolate, entire . . . . *V. acutiloba*

3. Terminal lobe of stem leaves as narrow as the lateral lobes; plants with a taproot . . . . *V. edulis*
3. Terminal lobe of stem leaves broader than the lateral lobes; plants with short,
   fibrous-rooted rhizomes 4

4. Lateral lobes of stem leaves linear-lanceolate . . . .*V. dioica*
4. Lateral lobes of stem leaves oblanceolate to ovate . . . . *V. occidentalis*

Valeriana acutiloba Rydb. Short-rhizomatous. **Stems** 10–50 cm, puberulent. **Basal leaves**
long-petiolate; blades 2–8 cm long, ovate to obovate, entire, on separate shoots. **Stem leaves**
smaller than the basal leaves; blades entire or with 1 or 2 pairs of oblanceolate lateral lobes <1 cm
wide. **Inflorescence** hemispheric, 1–3 cm across, puberulent. **Flowers** perfect; corolla white, 4–7
mm long, lobes ca. half as long as the tube. **Achenes** lanceolate, 3–6 mm long, puberulent or glabrous.
Moist meadows; subalpine. OR to MT south to CA, AZ, NM. Our plants are var. *pubicarpa* (Rydb.) Cronquist.
   Valeriana sitchensis usually lacks the clusters of basal leaves.

Valeriana dioica L. Short-rhizomatous. **Stems** 10–70 cm. **Basal leaves** long-petiolate; blades
2–10 cm long, oblanceolate to elliptic, entire. **Stem leaves** short-petiolate; blades 2–8 cm long
with 1 to 3 pairs of linear-lanceolate, lateral lobes and a larger terminal lobe. **Inflorescence** a
compound, glabrous to puberulent panicle, hemispheric in flower, 1–3 cm across. **Flowers** perfect
and female; corolla white, 2–3 mm long, lobes ca. as long as the tube. **Achenes** lanceolate, 4–5 mm
long, glabrous, sometimes purple-spotted. Dry to moist forest, grasslands, meadows; valleys to subalpine. YT to NL
south to WA, WY. Our plants are var. *sylvatica* S.Watson.
   Valeriana occidentalis generally occurs in moister sites and has leaves with wider lateral lobes. High-elevation plants
from southern Beaverhead Co. have puberulent achenes.

Valeriana edulis Nutt. ex Torr. & A.Gray Taprooted with a simple or branched caudex. **Stems**
10–50 cm. **Basal leaf blades** linear-oblanceolate, 7–15 cm long, entire or sometimes with 1 to 3
pairs of linear lobes. **Stem leaves** 1 pair or absent, nearly sessile; blades linear, 1–5 cm long with 2
to 5 pairs of linear lateral lobes. **Inflorescence** a glandular panicle. **Flowers** perfect and unisexual;
corolla white, 1–3 mm long, lobes shorter than the tube. **Achenes** ovate, 3–4 mm long, hirsute. Moist,
often calcareous soil of grasslands, sagebrush steppe, meadows, stony limestone fellfields; montane to alpine. BC, MT
south to Mexico, also MN, IA, ON.
   A few populations from wet, non-calcareous meadows have more spatulate basal leaves.

Valeriana occidentalis A.Heller Short-rhizomatous. **Stems** 30–70 cm. **Basal leaves** long-
petiolate; blades 2–6 cm long, ovate to elliptic, entire or with a pair of basal lobes. **Stem leaves**
short-petiolate; blades 3–10 cm long with 1 to 3 pairs of oblanceolate to ovate lateral lobes and
a larger terminal lobe. **Inflorescence** hemispheric, 1–3 cm across, sparsely glandular-puberulent.
Flowers perfect and female; corolla white, 3–4 mm long, lobes ca. as long as the tube. **Achenes**
lanceolate, 4–6 mm long, mostly glabrous. Moist forests, thickets, wet meadows; montane, subalpine. OR to MT south
to CA, AZ, CO, SD. See *V. dioica*.

Valeriana sitchensis Bong. Plants from a short, often branched rhizome. **Stems** 25–75 cm,
glabrous except at the nodes. **Basal leaves** absent or similar to but smaller than the cauline. **Stem
leaves** petiolate; blades 6–15 cm long, with 1 to 4 pairs of lanceolate to ovate, crenate lateral
lobes and a somewhat larger terminal lobe. **Inflorescence** hemispheric, 2–6 cm across, puberulent.
Flowers perfect; corolla white to pink, sparsely hairy, 5–8 mm long, lobes <half as long as the tube.
Achenes lance-ovate, 4–6 mm long, glabrous. Meadows, open forest, avalanche slopes, often along streams or habitats
with abundant snow cover; montane, subalpine. AK to QC south to CA, ID, MT. See *V. acutiloba*. (p.475)

Valerianella Mill.

Valerianella locusta (L.) Betcke Taprooted annual. **Stems** erect, 10–40 cm, dichotomously branched, sparsely
scabrous. **Basal leaves** petiolate, oblong, entire; blades 1–3 cm long, withering at flowering. **Stem leaves** glabrous,
oblong to lanceolate, weakly dentate, 1–5 cm long. **Inflorescence** hemispheric clusters, 7–18 mm wide, terminating the
branches. **Flowers** perfect; calyx absent; corolla white, campanulate, short-spurred, 1–2 mm long, glabrous, lobes ca. as
long as the tube; stigma 3-lobed. **Achene** compressed-ovoid, winged, 2–3 mm long, glabrous. Meadows, fields; valleys;
known from Lincoln Co. Introduced to eastern and northwest U.S.; native to Europe.

DIPSACACEAE: Teasel Family

Biennial or perennial herbs. **Leaves** exstipulate, basal and opposite on the stem. **Inflorescence** dense terminal clusters (heads); the flowers sessile on a receptacle subtended by bracts (involucre). **Flowers** perfect, epigynous; calyx with 4 to 12 teeth; corolla slightly irregular, narrowly funnelform; stamens 4; ovary inferior; style 1. **Fruit** an achene enclosed in the persistent calyx.

1. Stem prickly; involucral bracts awl-like, longer than the inflorescence. . . . . *Dipsacus*
1. Stem not prickly; involucral bracts lanceolate, shorter than the inflorescence . . . . *Knautia*

Dipsacus L. Teasel

Dipsacus fullonum L. [*D. sylvestris* Huds.] Taprooted biennial. **Stems** erect, branched, angled, 50–200 cm. **Herbage** glabrous except for hooked prickles on stem and midribs. **Basal leaves** oblanceolate, crenate, withering with flowering. **Stem leaves** lanceolate, 5–30 cm long, dentate below, becoming connate above. **Heads** ovoid, 2–8 cm high; involucral bracts linear, prickly, unequal, 1–15 cm long; bracts between flowers lanceolate, awned. **Flowers**: calyx silky, 4-angled; corolla 4-lobed, light blue above, white below, pubescent, 10–15 mm long; lobes ≤1 mm long; stigma entire; stamens exserted. **Achene** 4-angled, cylindrical, 4–6 mm long. Moist meadows, margins of wetlands, thickets; valleys. Introduced throughout temperate N. America; native to Europe.

Knautia L.

Knautia arvensis (L.) Coult. Taprooted perennial. **Stems** erect, branched above, 30–80 cm. **Herbage** hirsute. **Basal leaves** petiolate; blades oblanceolate, crenate, 3–15 cm long. **Stem leaves** oblanceolate, 4–20 cm long, deeply pinnately lobed; the lobes linear-lanceolate. **Heads** hemispheric, 1–3 cm across; involucral bracts lanceolate, 5–12 mm long; receptacle bracts absent. **Flowers**: calyx hirsute with 8 to 12 aristate sepals 2–4 mm long; corolla 4-lobed, light purple, glabrous to sparsely hairy, 6–8 mm long, lobes 1–3 mm long; stigma shallowly lobed; stamens barely exserted. **Achene** conical, 5–6 mm long, hirsute. Fields, pastures, roadsides; valleys, montane. Introduced to northern U.S., adjacent Canada; native to Europe.

ASTERACEAE (Compositae): Sunflower Family

Herbs and shrubs. **Inflorescence** of sessile flowers clustered in heads on a common receptacle, tightly surrounded by a cup- or tube-like involucre of bracts (phyllaries) in 1 to many series of different heights; receptacle often with small bracts subtending each flower (paleae, scales). **Flowers** epigynous, perfect or unisexual, usually 5-merous; calyx modified into pappus of bristles, scales or lacking; corolla regular, tubular and apically lobed (disk flower) or irregular with apical portion (ligule) prolonged, petal-like, shallowly 3-lobed to entire; anthers united into a tube around the style; ovary inferior; style 1, 2-lobed. **Fruit** an achene (cypsela).

This is the largest family in MT, containing some of the showiest wildflowers as well as some of the worst weeds. The flower heads are often mistaken for flowers. Flower heads are of 3 types: **ligulate**–all flowers (ray flowers) have a modified limb, a strap-shaped, often shallowly 3-lobed lamina; **discoid**–all flowers have tubular corollas; and **radiate**–central flowers are tubular, and the outer are ray flowers. The phyllaries and pappus are important characters for delineating genera.

Molecular genetic studies have prompted proposed realignments of many genera within this family. Some of these proposals are well-supported by data; others are more tentative. I have adopted a conservative philosophy and accepted only changes that seem strongly supported by published data. A particularly difficult situation is presented by the New World asters. It seems clear that New World asters are not closely related to European asters where the genus name *Aster* has priority.[308] While it is clear that New World asters are not in the genus *Aster*, their disposition in segregate genera is not confidently known. Our New World asters have been segregated into 7 genera (*Almuaster, Canadanthus, Eucephalus, Eurybia, Ionactis, Oreostemma, Symphyotrichum*), the latter five of which were erected over a century ago.

Key to the groups of genera

1. Shrubs or subshrubs; woody above-ground stems present . . . . Group A
1. Plants herbaceous; roots or caudex may be woody 2

2. White-woolly herbs with discoid heads; upper portion of involucral bracts scarious . . . . Group B
2. Plants not as above 3

3. Heads ligulate; disk flowers absent 4
3. Heads with disk flowers 5

4. Flowers yellow or orange (sometimes drying pink) . . . . Group C
4. Flowers white, pink, blue or purple . . . . Group D

5. Heads discoid; ray flowers absent 6
5. Heads radiate; both ray flowers (sometimes inconspicuous) and disk flowers present 7

6. Flowers yellow or orangish . . . . Group E
6. Flowers green, white, pink, purple (rarely yellow at the base) . . . . Group F

7. Rays yellow or orangish 8
7. Rays white, pink, blue, purple 9

8. Receptacle naked, without scales, awns or bristles mounted on it (pappus is mounted
   on achenes) . . . . Group G
8. Receptacle with scales, awns, or bristles between or enwrapping achenes . . . . Group H

9. Pappus a low crown, short teeth, or absent . . . . Group I
9. Pappus of bristles or bristle-like scales . . . . Group J

Group A. Shrubs and subshrubs

1. Pappus of disk flowers of erose scales or absent 2
1. Pappus of capillary bristles 4

2. Pappus of erose scales; ray flowers present . . . . *Gutierrezia*
2. Pappus absent 3

3. Phyllaries in 2 to 3 subequal series; inflorescence capitate to densely corymbiform . . . . *Sphaeromeria*
3. Phyllaries in 4 to 7 unequal series; inflorescence usually a raceme or panicle . . . . *Artemisia*

4. Heads with a few ray flowers . . . . *Ericameria*
4. Ray flowers absent 5

5. Twigs white-tomentose 6
5. Twigs glabrous, resinous, or puberulent but not tomentose 7

6. Upper leaves fascicled, sometimes with spines . . . . *Tetradymia*
6. Upper leaves mostly not fascicled, spines absent . . . . *Ericameria*

7. Leaves resin-dotted, many twisted; stems glabrate to puberulent . . . . *Chrysothamnus*
7. Plants not as above 8

8. Leaves linear; twigs greenish . . . . *Ericameria*
8. Leaves linear-lanceolate; epidermis of twigs whitish . . . . *Lorandersonia*

Group B. White-woolly herbs with discoid heads; upper portion of involucral bracts papery

1. Plants annual 2
1. Plants perennial 5

2. Leaves opposite . . . . *Psilocarphus*
2. Leaves alternate 3

3. Receptacle naked; basal leaves usually present at flowering . . . . *Gnaphalium*
3. Receptacle with chaff between or surrounding the flowers (sometimes appearing to be
   an involucre); basal leaves no persistent into flowering 4

4. Stems often branched from the base; heads clustered on branch tips . . . . *Evax*
4. Stems mostly simple; heads in leaf axils . . . . *Filago*

5. Involucral bracts almost entirely papery, shiny-white, glabrous . . . . *Anaphalis*
5. At least part of involucral bracts colored or hairy 6

6. Heads with flowers of only one sex (male heads sometimes absent) . . . . *Antennaria*
6. Female flowers on outside; bisexual flowers on inside of same head . . . . *Gnaphalium*

Group C. Heads ligulate; flowers yellow or orange

1. Plants scapose; stems leafless (rarely 1 leaf) or absent 2
1. Stems with usually ≥2 leaves (sometimes near the base) 7

2. Heads solitary. 3
2. Heads ≥2 per stem 5

3. Outer series of involucral bracts much shorter than the inner . . . . *Taraxacum*
3. Outer series of involucral bracts nearly as long the inner. 4

4. Some pappus bristles tapered to a broad base (narrow scales) . . . . *Nothocalais*
4. All pappus bristles capillary, not broadened at the base. . . . . *Agoseris*

5. Pappus partly of plumose capillary bristles . . . . *Hypochaeris*
5. Pappus of barbed or smooth capillary bristles 6

6. Plants with a mostly vertical taproot. . . . . *Crepis*
6. Plants with a short, fibrous-rooted, nearly horizontal rhizome . . . . *Hieracium*

7. Pappus plumose bristles or scales with plumose bristle tips 8
7. Pappus of capillary bristles. 10

8. Pappus with a broadened, scale-like base. . . . . *Microseris*
8. Pappus bristles not broadened at the base 9

9. Leaves entire; involucral bracts in 1 equal series. . . . . *Tragopogon*
9. Some leaves with linear lobes . . . . *Scorzonera*

10. Achenes flattened; many species with prickles on leaves or stems. 11
10. Achenes ca. terete; plants without prickles. 12

11. Involucres cylindrical, ca. 2 times as long as wide . . . . *Lactuca*
11. Involucres campanulate, ca. as long as wide . . . . *Sonchus*

12. Annual with pinnately lobed leaves. . . . . *Malacothrix*
12. Plants not as above 13

13. Plants fibrous-rooted, sometimes with a horizontal caudex or stolons. . . . . *Hieracium*
13. Plants taprooted. . . . . *Crepis*

Group D. Ligulate heads; flowers white, pink, blue or purple

1. Pappus bristles plumose 2
1. Pappus bristles smooth or with short barbs or absent 3

2. Involucre 20–30 mm high; phyllaries in 1 series. . . . . *Tragopogon*
2. Involucre 7–14 mm high; phyllaries in 2 series, the outer reduced . . . . *Stephanomeria*

3. Plants scapose; heads 1 per stem . . . . *Agoseris*
3. Plants with leafy stems; heads ≥2 per stem 4

4. Flowers white 5
4. Flowers pink to blue 7

5. Plants glabrous. . . . . *Prenanthes*
5. Plants villous or hirsute at least toward the base 6

6. Stem stipitate-glandular; leaves wing-petiolate; lawn weed. . . . . *Lapsana*
6. Herbage not glandular; petioles absent or not winged; forests . . . . *Hieracium*

7. Achenes flattened. . . . . *Lactuca*
7. Achenes nearly terete, not compressed. 8

8. Pappus of tiny scales . . . . *Cichorium*
8. Pappus of capillary bristles. 9

9. Achenes with a beak 5–6 mm long. . . . . *Chondrilla*
9. Achenes sometimes fusiform but not beaked. 10

10. Stems spine-tipped. . . . . *Pleiacanthus*
10. Stems not spiny . . . . *Lygodesmia*

Group E. Heads discoid; flowers yellow

1. Leaves opposite 2
1. Leaves alternate 5

2. Pappus of capillary bristles or scales divided into bristles 3
2. Pappus of awns or absent 4

3. Involucral bracts with tear-like, sessile, amber glands . . . . *Dyssodia*
3. Involucral bracts without sessile glands . . . . *Arnica*

4. Outer phyllaries leaf-like, longer than inner . . . . *Bidens*
4. Outer phyllaries often reflexed, shorter than the inner . . . . *Thelesperma*

5. Pappus absent or an inconspicuous crown 6
5. Pappus of scales or bristles 11

6. Inner phyllaries spine-tipped; leaves spiny-dentate . . . . *Carthamus*
6. Phyllaries and leaves not spiny 7

7. Heads solitary at branch tips . . . . *Matricaria*
7. Heads in a definite inflorescence 8

8. Heads in an open, flat-topped to broadly hemispheric inflorescence. . . . . *Tanacetum*
8. Heads in a spike-like to capitate inflorescence. 9

9. Inflorescence elongate . . . . *Artemisia*
9. Inflorescence capitate. 10

10. Leaves mostly 3-lobed at the tip. . . . . *Sphaeromeria*
10. Leaves entire or pinnate but not 3-lobed at the tip . . . . *Artemisia*

11. Pappus of scales 12
11. Pappus of bristles 13

12. Leaf margins spiny-dentate . . . . *Carthamus*
12. Leaves pinnately divided into linear lobes . . . . *Hymenopappus*

13. Pappus bristles clear white. 14
13. Pappus bristles tawny to dirty white 15

14. Phyllaries weakly imbricate in 2 to 4 series . . . . *Conyza*
14. Phyllaries equal or subequal in 1(2) series. . . . . *Senecio*

15. Involucre puberulent, not glandular . . . . *Pyrrocoma*
15. Involucre glandular. 16

16. Receptacle naked; involucre 8–11 mm high. . . . . *Triniteurybia*
16. Receptacle bristly; involucre 6–9 mm high. . . . . *Xanthisma*

Group F. Heads discoid; flowers green, white, pink, purple (rarely yellow at the base)

1. Mature involucres forming spiny or hispid burs 2
1. Involucres not bur-forming 4

2. Burs <8 mm long . . . . *Ambrosia*
2. Burs >10 mm long 3

3. Burs ellipsoid in leaf axils . . . . *Xanthium*
3. Burs globose in panicles on branch tips. . . . . *Arctium*

4. Leaves with spines >2 mm long on the margins 5
4 Leaves not spiny or spines <1 mm long 7

5. Receptacle naked. . . . . *Onopordum*
5. Receptacle bristly. 6

6. Pappus bristles plumose . . . . *Cirsium*
6. Pappus bristles barbed but not plumose . . . . *Carduus*

7. Receptacle densely bristly; involucral bracts fringed or spiny . . . . *Centaurea*
7. Receptacle naked or with fine scales or pubescent 8

8. Flowers white 9
8. Flowers green, pink, blue, purple 16

9. Pappus absent 10
9. Pappus of scales or bristles 11

10. Leaves opposite at least below . . . . *Iva*
10. Leaves alternate . . . . *Adenocaulon*

11. Pappus of scales . . . .*Chaenactis*
11. Pappus of simple to plumose bristles 12

12. Involucre glandular 13
12. Involucre glabrate to pubescent but not glandular 14

13. Leaves broadly lanceolate to deltoid . . . . *Ageratina*
13. Leaves lanceolate to oblanceolate . . . . *Brickellia*

14. Leaves linear (inconspicuous ray flowers usually present) . . . . *Symphyotrichum*
14. Leaves lanceolate to deltoid 15

15. Leaves glandular-punctate beneath . . . . *Brickellia*
15. Leaves white-tomentose beneath . . . . *Petasites*

16. Pappus an inconspicuous crown or absent 17
16. Pappus of smooth to plumose bristles 18

17. Receptacle conical, 8–45 mm high; flowers purple above . . . . *Rudbeckia*
17. Receptacle much smaller, not conical; flowers greenish . . . . *Artemisia*

18. Leaves in whorls of 3 to 4 . . . . *Eupatorium*
18. Leaves alternate 19

19. Receptacle with fine, flat scales among the flowers . . . .*Acroptilon*
19. Receptacle naked or pubescent 20

20. Leaves glandular or punctate beneath 21
20. Leaves glabrate to tomentose but not glandular or punctate 22

21. Phyllaries with prominent veins; corollas not glandular . . . . *Brickellia*
21. Phyllaries not prominently veined; corollas glandular . . . . *Liatris*

22. Leaves lanceolate to narrowly ovate, cuneate at the base . . . . *Saussurea*
22. Basal leaves sagittate or deltoid, cordate-based . . . . *Petasites*

Group G. Heads radiate; rays yellow; receptacle naked, without scales, awns or bristles

1. Leaves opposite 2
1. Leaves alternate or basal 4

2. Leaves without sessile glands . . . . *Arnica*
2. Leaves with scattered, sessile, amber glands 3

3. Leaves divided into 3(5) lobes . . . . *Picradeniopsis*
3. Leaves 1 to 2 times pinnately divided into >3 lobes . . . . *Dyssodia*

4. Pappus of bristles (rarely also scales) 5
4. Pappus of scales or awns, capillary bristles absent 14

5. Involucral bracts in 1 series (tiny bracts often at base of involucre) . . . . *Senecio*
5. Involucral bracts in ≥2 series 6

6. Pappus bristles in 2 series, the outer short and inconspicuous 7
6. Pappus bristles in 1 series 8

7. Leaves linear . . . . *Erigeron*
7. Leaves oblanceolate to obovate . . . . *Heterotheca*

8. Heads >5 per stem 9
8. Heads 1 to 5 on most stems 10

9. Leaves punctate; heads clustered in small glomerules in the inflorescence . . . . *Euthamia*
9. Leaves not punctate; heads not glomerate . . . .*Solidago*

10. Involucre 4–6 mm high. . . . . *Erigeron*
10. Involucre ≥7 mm high. 11

11. Stem leaves greatly reduced compared to the basal . . . . *Stenotus*
11. Stems leafy, cauline leaves gradually reduced upward 12

12. Basal leaves distinctly petiolate . . . .*Pyrrocoma*
12. Petioles of basal leaves short, tapered to the blade. 13

13. Plants of plains. . . . . *Stenotus*
13. Plants of high elevations in the mountains . . . . *Tonestus*

14. Receptacle hemispheric, ca. as high as wide; ray lamina reflexed . . . . *Helenium*
14. Receptacle flat to convex, wider than high; rays usually not reflexed 15

15. Leaves deeply lobed. 16
15. Leaves entire to shallowly lobed ≤1/2-way to the midvein 17

16. Ray lamina deeply 3-lobed. . . . . *Hymenoxys*
16. Rays entire or very shallowly lobed . . . . *Eriophyllum*

17. Stems scapose; cauline leaves absent . . . .*Tetraneuris*
17. Stem leaves present. 18

18. Involucral bracts glabrous, resinous, with recurved tips. . . . . *Grindelia*
18. Involucral bracts not as above 19

19. Leaves shallowly lobed . . . . *Hulsea*
19. Leaves entire 20

20. Involucre <8 mm high. . . . . *Eriophyllum*
20. Involucre >8 mm high. 21

21. Involucre glandular. . . . . *Platyschkuhria*
21. Involucre tomentose but not glandular . . . . *Hymenoxys*

Group H. Heads radiate; rays yellowish; receptacle with scales or awns or bristles between achenes (sometimes enwrapping achenes)

1. Pappus of conspicuous scales, awns, or bristles. 2
1. Pappus absent or a low crown or inconspicuous teeth 7

2. Pappus of tawny bristles. . . . . *Xanthisma*
2. Pappus of awns or scales. 3

3. Pappus of stiff, sharp-pointed awns 4
3. Pappus of soft, more membranous scales. 5

4. Involucral bracts in 1 or 2 subequal series. . . . .*Verbesina*
4. Involucral bracts in 2 unequal series; the outer longer, leaflike; the inner membranous . . . . *Bidens*

5. Disk flowers purple-tipped; ray lamina deeply 3-lobed. . . . . *Gaillardia*
5. Disk flowers all yellow; rays not deeply lobed 6

6. Achenes strongly compressed, thin-edged. . . . . *Helianthella*
6. Achenes barely or moderately compressed, not thin-edged . . . . *Helianthus*

7. Cauline leaves opposite at least below 8
7. Leaves alternate or nearly all basal 11

8. Leaves deeply divided into linear segments. . . . .*Coreopsis*
8. Leaves simple, not lobed 9

9. Herbage glandular, aromatic . . . .*Madia*
9. Herbage strigose to hirsute, not glandular 10

10. Plants annual; ligules 2–6 mm long . . . .*Lagophylla*
10. Plants perennial; ligules 7-20 mm long. . . . . *Viguiera*

11. Stems nearly scapose; cauline leaves greatly reduced compared to basal. . . . .*Balsamorhiza*
11. Cauline leaves well represented. 12

12. Leaves with entire margins or nearly so. 13
12. Leaves lobed 15

13. Ray lamina 2–6 mm long . . . . *Lagophylla*
13. Ray lamina >5 mm long 14

14. Disk flowers purple. . . . . *Rudbeckia*
14. Disk flowers yellow. . . . . *Wyethia*

15. Herbage villous; pappus absent; ray lamina 1–12 mm long. 16
15. Herbage glabrous to strigose; pappus of short teeth or a low crown; rays 1–4 cm long 17

16. Ray lamina <2 mm long . . . . *Achillea*
16. Ray lamina >5 mm long . . . . *Anthemis*

17. Each ray flower with small, hirsute clasping scale . . . . *Ratibida*
17. Ray flowers not enfolded by a receptacle scale. . . . . *Rudbeckia*

Group I. Heads radiate; rays white, pink, blue or purple; pappus low crown, short teeth or absent

1. Receptacle naked (may be pitted) 2
1. Receptacle with awns, scales, or bristles between or clasping the achenes. 5

2. Involucral bracts equal in length in 1 series . . . . *Bellis*
2. Involucral bracts in 2 to 5 series, often unequal. 3

3. Leaves shallowly toothed or lobed ≤1/2-way to the midvein . . . . *Leucanthemum*
3. Leaves deeply lobed or divided 4

4. Leaves divided into lanceolate, serrate or sharply lobed leaflets. . . . . *Tanacetum*
4. Leaves divided into long, linear segments . . . . *Matricaria*

5. Receptacle with hair-like bristles between the achenes. . . . . *Centaurea*
5. Receptacle with awns or scales between or enfolding the achenes 6

6. Leaves simple with entire to serrate margins. 7
6. Leaves deeply divided or lobed 9

7. Leaves opposite. . . . . *Galinsoga*
7. Leaves alternate. 8

8. Rays pink to purplish; receptacle with stiff awns . . . . *Echinacea*
8. Rays white; receptacle with scales clasping the achenes . . . . *Wyethia*

9. Ray lamina 1–5 mm long . . . . *Achillea*
9. Ray lamina 5–15 mm long . . . . *Anthemis*

Group J. Heads radiate; rays white, pink, blue or purple; pappus of bristles or bristle-like scales

1. Leaf blades cordate or sagittate. . . . . *Petasites*
1. Leaf blades linear to ovate but not cordate-based. 2

2. Receptacle bristly; involucral bracts erose, fimbriate or spiny . . . . *Centaurea*
2. Receptacle naked or with scales; involucral bracts not appendaged. 3

3. Ray lamina ≤1 mm long, ca. as long as pappus bristles 4
3. Ray lamina >1 mm long 5

4. Rays absent; involucre 5–10 mm high . . . . *Symphyotrichum*
4. Rays present; lamina ≤1 mm long; involucre 2–5 mm high . . . . *Conyza*

5. Involucral bracts narrow, ca. equal in length, not overlapping like shingles (imbricate). 6
5. Involucral bracts in two or more unequal series, at least somewhat imbricate. 8

6. Leaves clasping above; involucre glandular. . . . . *Canadanthus*
6. Plants not as above 7

7. Stems short or absent; involucre mostly ≥8 mm high. . . . . *Townsendia*
7. Plants not as above . . . . *Erigeron*

8. Leaf margins deeply lobed . . . . *Machaeranthera*
8. Leaf margins not spiny or lobed. 9

9. Involucre 7–14 mm high, the bracts long-pointed. . . . . *Xylorhiza*
9. Involucre smaller or with acute to rounded tips 10

10. Pappus with an outer series of short bristles in addition to the longer ones; leaves ≤4 mm wide . . *Ionactis*
10. Pappus lacking an outer series of short bristles; leaves often wider 11

11. Plants taprooted 12
11. Plants fibrous-rooted or rhizomatous 14

12. Involucre glandular. . . . . *Machaeranthera*
12. Involucre glabrate, villous or strigose but not glandular 13

13. Plants perennial; stems 2–10 cm long, decumbent . . . . *Oreostemma*
13. Plants not as above . . . . *Townsendia*

14. Involucre and peduncles glandular (often minutely so) 15
14. Involucre and peduncles not glandular. 18

15. Rays white (rarely pale blue), drying pinkish . . . . *Almutaster*
15. Rays blue or violet 16

16. Achenes 7- to 12-veined. . . . . *Eurybia*
16. Achenes 2- to 5-veined. 17

17. Involucral bracts purple-tipped . . . . *Eucephalus*
17. Involucral bracts not purple-tipped . . . . *Symphyotrichum*

18. Involucral bracts strongly imbricate, the tips often purple-tinged . . . . *Eurybia*
18. Involucral bracts somewhat imbricate, the tips usually not purplish. . . . . *Symphyotrichum*

Achillea L. Yarrow

Aromatic, rhizomatous perennials. **Leaves** basal and alternate, mostly pinnately dissected. **Inflorescence** corymbiform; heads several to many. **Heads** radiate; phyllaries scarious in 2 to 4 series; receptacle hemispheric with small scales (paleae) between the flowers. **Ray flowers** white or yellow, female, fertile. **Disk flowers** perfect, white or yellow (pink); style branches flattened. **Pappus** absent. **Achenes** oblong to obovate, compressed.

Achillea alpina L. [A. sibirica Ledeb.] is reported for Gallatin Co.,[113] but specimens I have seen have been of cultivated plants.

1. Disk flowers yellow. . . . . *A. filipendulina*
1. Disk flowers white. 2

2. Leaf blades 2 to 3 times pinnately divided 3
2. Leaf blades entire to serrate. 4

3. Involucre 3–5 mm high; ray lamina 2–3 mm long. . . . . *A. millefolium*
3. Involucre 2–3 mm high; ray lamina 1–2 mm long. . . . . *A. nobilis*

4. Ray lamina 4–5 mm long; leaves entire to serrulate . . . . *A. ptarmica*
4. Ray lamina 1–3 mm long; leaves 1 to 2 times serrate . . . .*A. alpina*

Achillea filipendulina Lam. Fernleaf yarrow **Stems** erect, simple, 40–100 cm. **Herbage** glandular, pubescent. **Leaves** basal and cauline, sessile, narrowly lanceolate, 4–30 cm long, twice pinnately dissected; ultimate lobes ca. 1 mm long and wide. **Inflorescence** flat-topped to hemispheric, up to 10 cm across with numerous heads. **Involucre** sparsely glandular, 3–4 mm high, ca. 2 mm wide; bracts with scarious margins, villous. **Rays** yellow, few; ligules ca. 1 mm long. **Disk flowers** yellow, 30 to 40. **Achenes** 1–2 mm long. Gravelly soil along roads, fields; valleys. Introduced as an ornamental; escaped in Missoula and Beaverhead cos.; native to Europe.

Achillea millefolium L.[A. lanulosa Nutt.] Yarrow **Stems** ascending to erect, simple, 6–60 cm. **Herbage** sparsely- to tangled-woolly. **Leaves** basal and cauline, petiolate below; blades narrowly oblanceolate, 1–15 cm long, finely 3 times pinnately dissected; ultimate lobes 1–2 × 1 mm. **Inflorescence** hemispheric, 2–15 cm across with numerous heads. **Involucre** 3–5 × 2–4 mm; bracts in 3 series; outer bracts green with brown, fimbriate margins, sparsely glandular. **Rays** 5 to 12, white (rarely pink); ligules 2–3 mm long, not quite as wide. **Disk flowers** white,10 to 30; corolla 2–5 mm long. **Achenes** glabrous, 1–2 mm long, obscurely winged. Grasslands, meadows, sagebrush steppe, open forest, talus, fellfields; all elevations. Circumboreal and throughout temperate N. America. (p.488)

Our plants have been called *Achillea millefolium* ssp. *lanulosa* (Nutt.) Piper, and higher-elevation plants with darker phyllaries have been called var. *alpicola* (Rydb.) Garrett. These differences may be due to phenotypic plasticity.[395]

Achillea nobilis L. **Stems** ascending to erect, simple, 20–60 cm. **Herbage** villous below, tomentose above. **Leaves** mainly cauline, sessile; blades ovate, 1–6 cm long, twice pinnately dissected; ultimate lobes ca. 1 mm long and wide. **Inflorescence** hemispheric, 2–10 cm across with numerous heads. **Involucre** sparsely glandular, 2–3 mm high, ca. 2 mm wide; bracts in 2 series; outer bracts light green, scarious, villous. **Rays** white, 8 to 10; ligules 1–2 mm long, wider than long. **Disk flowers** white,10 to 30; corolla 1–2 mm long. **Achenes** ca. 1 mm long. Gravelly soil along roads, streambanks; valleys. Introduced to MT more than 100 years ago; native to Europe. Collected in Flathead and Lake cos.
Leaves of *Achillea millefolium* are more feathery.

Achillea ptarmica L. **Stems** erect, branched above, angled, 30–60 cm. **Herbage** glabrous below, pubescent above. **Leaves** cauline, sessile; blades linear-lanceolate, 3–10 cm long, serrulate. **Inflorescence** leafy-bracteate, of few heads. **Involucre** 4–6 × 4–8 mm; bracts in 3 series; outer bracts villous, green with brown margins. **Rays** white, 8 to 10; ligules 4–5 mm long and wide. **Disk flowers** white, 10–30; corolla ca. 3 mm long. **Achenes** 1–2 mm long. Roadsides, fields, gardens; valleys. Introduced across northern U.S. and adjacent Canada; native to Eurasia. Collected in Flathead Co.

Acroptilon Cass.

Acroptilon repens (L.) DC. [*Centaurea repens* L.] RUSSIAN KNAPWEED Rhizomatous perennial.
Stems ascending to erect, branched, 30–70 cm. **Herbage**: stems tomentose; leaves glabrate, punctate beneath. **Leaves** cauline, alternate, petiolate; blades oblanceolate, entire to pinnately lobed, 3–10 cm long, becoming smaller and sessile above. **Heads** discoid; involucres broadly ovoid, 9–15 mm high; phyllaries broadly ovate, green or tan below, scarious above, sharp-pointed, well imbricate in several series; receptacle flat with fine scales. **Disk flowers** perfect, light purple (white); corolla 11–14 mm long, slender with an expanded limb; style branches united almost to the tips. **Pappus** deciduous flattened bristles with plumose tips. **Achenes** ellipsoid, glabrous, 2–4 mm long. Fields, grasslands, meadows, roadsides, most often on stream terraces; plains, valleys. Introduced to most of temperate N. America.
Sometimes included in *Centaurea*, *Acroptilon repens* differs by having monomorphic flowers and a plumose pappus.

Adenocaulon Hook. Trail plant

Adenocaulon bicolor Hook. PATHFINDER Fibrous-rooted perennial. **Stems** erect, simple, 30–70 cm, tomentose. **Leaves** alternate, petiolate; blades 3–15 cm long, deltoid, cordate, dentate to shallowly lobed, glabrous above, tomentose beneath. **Inflorescence** an open panicle, minutely bracteate, stipitate-glandular. **Heads** discoid; involucre 1–2 mm long; phyllaries ca. 5, ovate, green, ciliate; receptacle hemispheric, naked. **Disk flowers** white, the outer female, the inner functionally male; corolla ca. 1 mm long; style branches short or absent. **Pappus** absent. **Achenes** linear-oblong, 5–8 mm long, stipitate-glandular above. Moist to wet forest, especially in lightly disturbed areas such as along trails; valleys, montane. BC to ON south to CA, MT and SD. (p.488)

Ageratina Spach

Ageratina occidentalis (Hook.) R.M.King & H.Rob. [*Eupatorium occidentale* Hook.] Rhizomatous perennial with a woody, branched caudex. **Stems** erect, 15–60 cm, simple, puberulent. **Leaves** cauline, alternate, short-petiolate, simple; blades minutely punctate-glandular, deltoid to lanceolate, serrate, 2–5 cm long. **Inflorescence** paniculate. **Heads** discoid; involucre tubular, 2–6 mm high; phyllaries in 2 to 5 series, subequal, linear-lanceolate, glandular-puberulent; receptacle convex, naked. **Disk flowers** perfect, tubular, 9 to 21, white to purple; corolla 4–8 mm long; style branches clavate. **Achenes** obconic, 5-angled, glandular, 2–4 mm long; pappus 1 series of capillary bristles. Stony soil of talus slopes, moist forest openings; montane, lower subalpine. WA to MT south to CA, NV, UT. Collected in Mineral and Ravalli cos.
Ageratina differs from *Eupatorium* sensu stricto by having a longer corolla tube, a glabrous style base, and a different base chromosome number.[219]

Agoseris Raf. Mountain dandelion

Annual or mostly perennial, taprooted, scapose herbs with milky sap. **Leaves** basal, simple, tapering to the petiole. **Heads** solitary, long-pedunculate, ligulate; phyllaries lanceolate, green, in 2 to 5 series, the inner longer and narrower than the outer; receptacle flat, naked. **Ray flowers** perfect, yellow, orange or purple, 5-lobed at the tip. **Pappus** 50 to 125 capillary, obscurely barbellate bristles. **Achenes** fusiform to linear-conical, 10-ribbed, mostly beaked.[23]

Achene beaks elongate as fruits develop; completely mature achenes may be needed for positive identification.

1. Plants annual; involucres villous with dark-septate hairs . . . .*A. heterophylla*
1. Perennials; involucres glabrous to villous with transparent, septate hairs 2

2. Corollas orange to pink, usually drying light purple . . . . *A. aurantiaca*
2. Corollas yellow 3

3. Leaf blades entire or with a few shallow lobes . . . . *A. glauca*
3. Leaf blades lobed >1/2-way to the midvein 4

4. Mature achene beak <6 mm long; involucral bracts glabrous or tomentose . . . . *A. parviflora*
4. Mature achene beak >8 mm long; involucral bracts villous on the margins . . . . *A. grandiflora*

Agoseris aurantiaca (Hook.) Greene [*A. lackschewitzii* Douglass M.Hend & R.K.Moseley, *A. carnea* Rydb.] Perennial 6–60 cm. **Leaves** oblanceolate, 5–35 cm long, entire, dentate, or with pinnate, linear lobes. **Herbage** glabrous to sparsely villous. **Involucre** 1–3 cm high; phyllaries lanceolate, in 2 to 4 indistinct series, green or purple-spotted, glabrate to villous. **Rays** orange to pink; ligules 6–12 mm long. **Pappus** 6–15 mm long. **Achene** body 5–9 mm long; beak 3–7 mm long. Meadows, rocky slopes. BC to QC south to CA, AZ, NM.

1. Fresh flowers orange; body of achene abruptly joined to the beak .var. *aurantiaca*
1. Fresh flowers dark pink; achene gradually tapered to the beak. . . . . .var. *carnea*

Agoseris aurantiaca var. *aurantiaca* occurs in dry meadows, grasslands, open slopes; montane to alpine; *A. aurantiaca* var. *carnea* (Rydb.) P. Lesica (ined.) is found in moist to wet meadows; montane to subalpine.

Agoseris glauca (Pursh) Raf. Perennial 8–50 cm. **Leaves** linear to oblanceolate, 3–35 cm long, entire or with 1 to 3 pairs of shallow lobes. **Herbage** glabrous to sparsely villous. **Involucre** 1–2 cm long; phyllaries linear-lanceolate to lanceolate, in 2 to 3 indistinct series, green, glabrous to villous, sometimes glandular, often darkened along the midvein. **Rays** yellow; ligules 6–22 mm long. **Pappus** 8–18 mm long. **Achene** body 4–9 mm long; beak 0.5–3 mm long. Grasslands, meadows, sagebrush steppe, turf; all elevations. BC to ON south to CA, AZ, NM, SD, MI. (p.488)

1. Peduncles and involucre mostly glabrous . . . . var. *glauca*
1. Peduncles and involucres puberulent to villous . . . . var. *dasycephala*

Agoseris glauca var. *glauca*, with more linear leaves and achene beaks <1 mm long, occurs in grasslands, meadows; plains, valleys, montane: *A. glauca* var. *dasycephala* (Torr.& A.Gray) Jeps.,with more oblanceolate leaves and achene beaks 2–3 mm long, is found in grasslands, meadows, sagebrush steppe, turf at all elevations.

Agoseris grandiflora (Nutt.) Greene Perennial 25–60 cm. **Leaves** linear-oblanceolate, 10–25 cm long, deeply pinnatifid. **Herbage** glabrate to villous. **Involucre** 2–4 cm high; phyllaries in 4 to 5 series, green, ciliate, purplish medially. **Rays** yellow; ligules 3–7 mm long. **Pappus** 7–15 mm long. **Achene** body 3–7 mm long; beak 9–16 mm long. Meadowy forest openings; valleys, montane. BC, MT south to CA, NV, UT.

Agoseris heterophylla (Nutt.) Greene Annual 4–25 cm. **Leaves** oblanceolate, 2–10 cm long, entire to remotely dentate. **Herbage** puberulent to villous. **Involucre** 1–2 cm high; phyllaries in 2 or 3 series, the outer often spotted, lanceolate, villous with dark-septate hairs. **Rays** yellow; ligules 2–4 mm long. **Pappus** 4–9 mm long. **Achene** body 2–5 mm long; beak 7–10 mm long. Grasslands, shallow soil of rock ledges; valleys. BC, MT south to CA, AZ, NM.

Agoseris parviflora (Nutt.) D.Dietr. [*A. glauca* (Pursh) Raf. var. *laciniata* (D.C.Eaton) Smiley] Perennial 5–20 cm. **Leaves** linear-lanceolate, 3–15 cm long, deeply pinnately divided into 5–8 pairs of linear lobes. **Herbage** sparsely villous to tomentose. **Involucres** 2–3 cm long; phyllaries narrowly lanceolate, in ca. 2 series, green, glabrous to tomentose with a purple midvein. **Rays** yellow; ligules 8–15 mm long. **Pappus** 1–2 cm long. **Achene** body 5–7 mm long; beak 3–5 mm long. Grasslands, sagebrush steppe; montane. OR to MT south to CA, AZ, NM, SD.

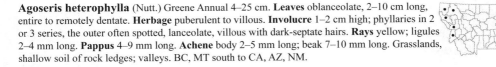

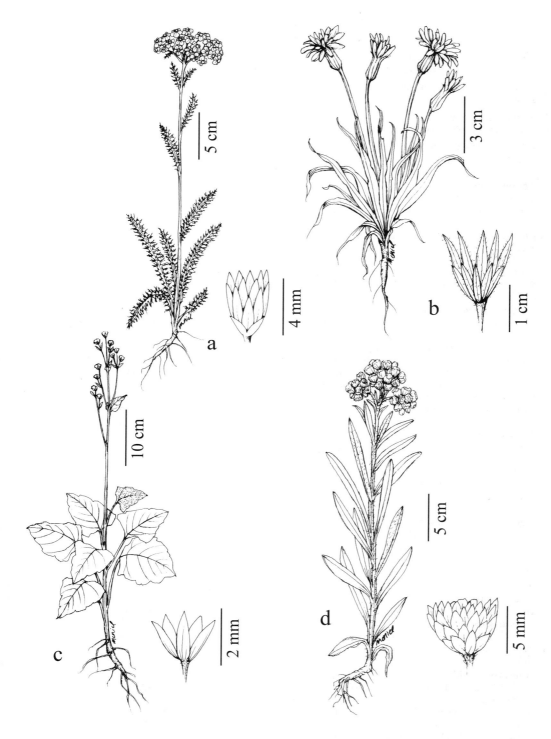

Plate 87. a. *Achillea millefolium*, b. *Agoseris glauca*, c. *Adenocaulon bicolor*, d. *Anaphalis margaritacea*

Almutaster A. Löve & D. Löve

Almutaster pauciflorus (Nutt.) A. Löve & D. Löve [*Aster pauciflorus* Nutt.] Rhizomatous perennial. **Stems** ascending, 12–40 cm. **Herbage** glabrate; stems glandular-pubescent above. **Leaves** basal and cauline, petiolate below, clasping above; blades linear, 3–12 cm long, entire. **Inflorescence** paniculate with few heads; peduncles stipitate-glandular. **Heads** radiate; involucre narrowly campanulate, 5–8 mm high; phyllaries linear to lanceolate, glandular, green, sometimes purplish; receptacle flat, glabrous. **Ray flowers** female, 15 to 30, white or pale blue, ligules 5–10 mm long. **Disk flowers** perfect, 40 to 50, pale yellow; corolla ca. 5 mm long. **Pappus** of numerous capillary bristles. **Achenes** obconic, ca. 1 mm long, sparsely hairy with 7 to 10 veins. Wet, usually calcareous soil of fens, wet meadows; plains; collected in Sheridan and Wheatland cos. NT south to CA, AZ, NM, TX, ND.

Ambrosia L. Ragweed, bursage

Annual or perennial herbs. **Leaves** petiolate, mostly 1 to 2 times divided or lobed. **Inflorescence** narrowly racemose; male heads above; female heads clustered. **Heads** unisexual, discoid. **Female heads** 1- or 2-flowered; phyllaries 12 to 30 in 5 to 8 series, becoming a winged or spiny bur. **Male heads**: phyllaries 5 to 16, connate; receptacle ca. flat with scarious scales (paleae). **Male florets** 5 to 60; corollas funnelform, white or purple. **Pappus** absent. **Achenes** ovoid, enclosed in a bur.

Though native, all of our species are weedy. *Ambrosia artemisiifolia* and *A. trifida* are native to eastern N. America and the Great Plains but probably introduced west of the Continental Divide; the former is a problematic weed in Europe.

1. Leaves unlobed or with 1 or 2 pairs of palmate lobes >1 cm wide . . . . *A. trifida*
1. Leaves pinnatifid; ultimate lobes mostly <5 mm wide 2

2. Leaves tomentose beneath . . . . *A. tomentosa*
2. Leaves glabrous or strigose to hispid beneath 3

3. Leaves and stems hispid with short, stiff, conical hairs . . . . *A. acanthicarpa*
3. Leaves and stems strigose 4

4. Taprooted annual; lower leaves 2 to 3 times pinnately divided . . . . *A. artemisiifolia*
4. Rhizomatous perennial; lower leaves mostly once pinnate . . . . *A. psilostachya*

Ambrosia acanthicarpa Hook. Taprooted annual. **Stems** erect, simple or branched, 10–50 cm. **Herbage** hispid, strigose. **Leaves** mostly alternate; blade broadly ovate, 2–8 cm long, twice pinnatifid into ovate ultimate segments. **Female heads** 1-flowered. **Male heads** cup-shaped, black-veined, 3–5 mm wide with 6 to 12 florets, spreading on peduncles 1–2 mm long. **Bur** 3–5 mm long, oblong, villous with stout spines. River banks; plains, valleys. WA and AB. to MN south to CA, AZ, TX.

Ambrosia artemisiifolia L. COMMON RAGWEED Taprooted annual. **Stems** erect, branched, 10–60 cm. **Herbage** strigose, glandular. **Leaves** opposite below; blade ovate, 1–8 cm long, ca. twice pinnatifid. **Female heads** 1-flowered. **Male heads** cup-shaped, villous, glandular, 2–3 mm wide with 12 to 20 florets, spreading on peduncles ca. 0.5 mm long. **Bur** 2–3 mm long, ovoid, beaked, glandular, villous with few short spines around the middle. Roadsides, fields; plains, valleys. Throughout N. America.

Ambrosia psilostachya DC. WESTERN RAGWEED Rhizomatous perennial. **Stems** erect, simple or branched, 10–50 cm. **Herbage** strigose, sometimes glandular. **Leaves** opposite below; blade ovate, 2–8 cm long, once pinnatifid into dentate lobes. **Female heads** 1-flowered. **Male heads** cup-shaped, 2–3 mm wide, hispid, glandular with 5 to 15 florets, spreading on peduncles ≤1 mm long. **Bur** 2–3 mm long, ovoid, glandular, hispid. Fields, roadsides, river terraces; plains. Throughout N. America.

Ambrosia tomentosa Nutt. Rhizomatous perennial. **Stems** ascending to erect, simple or branched, 10–40 cm, puberulent. **Leaves** alternate; blades oblanceolate, 3–15 cm long, white-tomentose beneath, thinly strigose above, ca. twice pinnately divided and lobed; ultimate segments deltoid. **Female heads** 2-flowered. **Male heads** saucer-shaped with black veins, 3–6 mm wide with 25 to 40 florets, spreading on peduncles ca. 1 mm long. **Bur** 4–6 mm long, 2-beaked with erect spines. River banks; plains. MT to MI south to AZ, NM, TX.

Ambrosia trifida L. Giant ragweed Taprooted annual. **Stems** erect, simple or branched, 30–150 cm, strigose to hispid. **Leaves** opposite; blade lanceolate to broadly ovate, 4–15 cm long, serrate, often 3-lobed, hispid, glandular. **Female heads** 1-flowered. **Male heads** saucer-shaped, black-veined, 2–4 mm wide, villous, hispid, glandular with 3 to 25 florets, nodding on peduncles 1–3 mm long. **Bur** 3–5 mm long, deltoid, ribbed, glabrous with ca. 5 erect spines. Roadsides, fields, streambanks; plains, valleys. Throughout N. America.

Anaphalis DC. Everlasting

Anaphalis margaritacea (L.) Benth. & Hook. f. Pearly everlasting Rhizomatous perennial. **Stems** erect, simple, 20–70 cm. **Herbage** white-woolly. **Leaves** all cauline, alternate, sessile, entire, narrowly lanceolate, 3–10 cm long. **Inflorescence** paniculate, hemispheric, 2–15 cm across. **Heads** discoid; involucre 4–7 mm high; phyllaries in 8 to 12 series, white, papery, villous, brown at the base, nearly obscuring the flowers; receptacle naked. **Flowers** numerous, yellow, unisexual; corolla 3–5 mm long; female flowers peripheral; male flowers central. **Pappus** 10 to 20 deciduous bristles. **Achenes** oblong, ca. 1 mm long, scabrous. Often lightly disturbed meadows, open forests, roadsides; valleys to subalpine. AK to NL south to CA, NM, KS, NC. (p.488)

Antennaria Gaertn. Pussytoes

Fibrous-rooted, perennial herbs. **Stems** simple. **Herbage** villous to white-tomentose. **Leaves** basal and alternate, simple, entire; basal leaves petiolate; stem leaves sessile above, sometimes with a darkened, twisted, scarious appendage (flag). **Inflorescence** racemose, corymbiform or paniculate, rarely solitary. **Heads** discoid; phyllaries imbricate, at least partly scarious, white, green, rose or black, in 3 to 6 series, usually villous at the base; receptacle naked; male involucres usually shorter than female. **Flowers** unisexual, male and female on separate plants; corollas tubular, male wider than female. **Pappus** of bristles, those of male with dilated tips. **Achenes** ellipsoid to ovoid.

   In some species male plants are rare, and seed is produced asexually; this and the occurrence of polyploidy makes species delineation difficult.[39]

1. Many or all stems with 1 or 2 heads. 2
1. Most or all stems with >2 heads. 3

2. Plants mat-forming; phyllaries brown; montane or lower . . . . *A. dimorpha*
2. Plants not forming mats; phyllaries blackish-green; alpine. . . . .*A. monocephala*

3. Plants with leafy stolons; often forming loose or dense mats. 4
3. Plants without stolons; not mat-forming or forming loose mats from a branched caudex 14

4. Heads long-pedunculate in an open raceme or panicle. . . . . *A. racemosa*
4. Heads in a dense, hemispheric, corymbiform cluster. 5

5. Outer involucral bracts (phyllaries) brown to black in the lower half. 6
5. Outer phyllaries white or pink to light brown below 11

6. Upper portion of outer phyllaries greenish-black, pointed 7
6. Upper portion of phyllaries tan to brown, rounded 8

7. Middle and upper stem leaves with dark, scarious flags on the tips. . . . .*A. alpina*
7. Middle stem leaves without flags . . . . *A. media*

8. Stems and leaves with tiny, stalked amber glands amidst tomentum . . . . *A. aromatica*
8. Herbage without glands 9

9. Basal leaves linear-oblanceolate, some >15 mm long. . . . .*A. corymbosa*
9. Basal leaves oblanceolate to narrowly spatulate, ≤15 mm long. 10

10. Basal leaves spatulate, <1 cm long . . . .*A. densifolia*
10. Basal leaves oblanceolate, some ≥1 cm long . . . . *A. umbrinella*

11. Oldest basal leaves becoming glabrous or at least distinctly less hairy on the upper
      surface . . . .*A. neglecta*
11. Basal leaves ca. equally hairy on upper and lower surfaces 12

12. Female involucres ≥7 mm high; basal leaves broadly spatulate . . . . *A. parvifolia*
12. Female involucres ≤7 mm high; basal leaves oblanceolate to narrowly spatulate. 13

13. Achenes mostly glabrous; phyllaries often pink; male plants absent, drier habitats. . . . . *A. rosea*
13. Achenes somewhat pappilose; phyllaries white; male plants present; moist sites. . . . . *A. microphylla*

14. Plants with a branched caudex, forming small or loose mats. 15
14. Plants rhizomatous, not mat-forming 17

15. Leaves ≤12 mm long; plants of high elevations . . . . *A. alpina*
15. Some basal leaves >12 mm long; plants often of low elevations. 16

16. Phyllaries light brown with white tips . . . . *A. luzuloides*
16. Phyllaries white with a black spot at the base . . . .*A. corymbosa*

17. Upper half of outer phyllaries pure white . . . .*A. anaphaloides*
17. Most or all of upper half of phyllaries black to dirty white. 18

18. Stems ≤20 cm . . . . *A. lanata*
18. Stems >20 cm . . . . *A. pulcherrima*

Antennaria alpina (L.) Gaertn. Sparsely stoloniferous. **Stems** erect, 4–12 cm. **Basal leaves** spatulate to oblanceolate, 5–12 mm long, 1-nerved, usually less hairy above. **Stem leaves** linear, 5–20 mm long, flagged. **Inflorescence** corymbiform with 3 to 6 heads; male plants absent. **Involucre**: female 4–7 mm high. **Phyllaries** dark brown to black. **Corolla**: female 3–5 mm long. **Pappus** 3–6 mm long. **Achenes** 1–2 mm long, papillose. Moist turf, meadows on cool slopes, often where snow lies late; upper subalpine, alpine. Circumpolar south to WY.
 Antennaria monocephala, which lacks stolons and has only 1 or 2 heads per stem, sometimes occurs with *A. alpina*. *Antennaria media* has basal leaves tomentose on both surfaces and lacks flags on the mid-stem leaves.

Antennaria anaphaloides Rydb. Rhizomatous, non-stoloniferous. **Stems** erect, 6–50 cm. **Basal leaf blades** oblanceolate to narrowly elliptic, 3–12 cm long, 3- to 5-nerved. **Stem leaves** linear to narrowly lanceolate, 1–8 cm long, usually flagged. **Inflorescence** corymbiform with numerous heads. **Involucre**: female 4–8 mm high; male 4–6 mm high. **Phyllaries** brown at the base contrasting with a white upper portion. **Corolla**: female 2–5 mm long; male 1.5–3 mm long. **Pappus** 3–5 mm long. **Achenes** 1–2 mm long, glabrous. Grasslands, meadows, open forest; valleys to alpine. BC to SK south to OR, UT, CO.
 Antennaria pulcherrima, with outer phyllaries that are distally brown to gray rather than pure white, generally occurs in wetland habitats. The type locality is in Gallatin Co.

Antennaria aromatica Evert Stoloniferous, mat-forming from a branched caudex. **Stems** ascending to erect, 2–10 cm. **Herbage** glandular, aromatic. **Basal leaves** spatulate, 5–12 mm long, 1-nerved. **Stem leaves** linear, 3–15 mm long. **Inflorescence** corymbiform with 2 to 5 heads. **Involucre**: female 5–8 mm high; male 4–5 mm high. **Phyllaries** brown. **Corolla**: female 3–4 mm long; male 2–3 mm long. **Pappus** 3–6 mm long. **Achenes** 1–2 mm long, pappilose. Limestone talus, exposed stony soil of fellfields; upper montane to alpine. AB to ID, WY.
 The type locality is in Carbon Co.

Antennaria corymbosa E.E.Nelson Stoloniferous, loosely mat-forming. **Stems** erect, 6–25 cm. **Basal leaves** oblanceolate, 1–3 cm long, 1-nerved. **Stem leaves** linear, 1–3 cm long. **Inflorescence** corymbiform with 3 to 12 heads. **Involucre**: female and male 4–6 mm high. **Phyllaries** white with a dark basal spot. **Corolla**: female 2–4 mm long; male 2–3 mm long. **Pappus** 2–5 mm long. **Achenes** ≤1 mm long, nearly glabrous. Moist to wet meadows; montane to alpine. WA to MT south to CA, NV, UT, NM.

Antennaria densifolia A.E.Porsild Stoloniferous, mat-forming. **Stems** ascending to erect, 2–5 cm. **Basal leaves** short-spatulate, 3–7 mm long, 1-nerved, 2 times as long as wide. **Stem leaves** linear, 2–7 mm long; the upper flagged. **Inflorescence** corymbiform with 2 to 6 heads. **Involucre**: female 4–6 mm high; male 3–5 mm high. **Phyllaries** brown. **Corolla**: female 2–4 mm long; male 2–3 mm long. **Pappus** 2–4 mm long. **Achenes** ca. 1 mm long, glabrous. Limestone talus, open, calcareous soil of steep slopes; alpine. AK to MT. A single widely disjunct population is known from Deer Lodge Co.
 Antennaria alpina, *A. media,* and *A. umbrellata* have basal leaves 2 to 6 times as long as wide. See *A. aromatica*.

Antennaria dimorpha (Nutt.) Torr. & A.Gray Cushion pussytoes Mat-forming from a much-branched caudex. **Stems** inconspicuous, erect, 5–25 mm. **Basal leaves** linear-spatulate, 1–3 cm long, 1-nerved. **Stem leaves** linear, 5–15 mm long. **Inflorescence** a solitary head. **Involucre**: female 10–13 mm high; male 6–8 mm high. **Phyllaries** brown. **Corolla**: female 6–9 mm long;

male 3–6 mm long. **Pappus** 5–12 mm long. **Achenes** 2–4 mm long, pubescent. Well-drained soil of grasslands; plains, valleys, montane. BC to SK south to CA, UT, NM, NE.

Antennaria lanata (Hook.) Greene Rhizomatous, without stolons. **Stems** erect, 3–20 cm. **Basal leaves** oblanceolate, 1–9 cm long, 3-nerved. **Stem leaves** linear-oblanceolate, 1–5 cm long, flagged. **Inflorescence** densely corymbiform with 3 to 10 heads. **Involucre**: female 5–8 mm high; male 3–6 mm high. **Phyllaries** dark brown to blackish-green. **Corolla**: female 2–4 mm long; male 3–5 mm long. **Pappus** 3–5 mm long. **Achenes** 1–2 mm long, pappilose. Meadows, turf, often where snow lies late; subalpine, alpine. BC, AB south to CA, UT, WY.

Antennaria luzuloides Torr. & A.Gray Loosely mat-forming from a branched caudex. **Stems** ascending to erect, 10–40 cm. **Basal leaves** erect, linear-oblanceolate, 3–9 cm long, 3-nerved. **Stem leaves** linear to lanceolate, 1–7 cm long. **Inflorescence** open- to dense-corymbiform with numerous heads. **Involucre**: female 5–6 mm high; male 4–5 mm high. **Phyllaries** light brown with white tips. **Corolla**: both sexes 2–3 mm long. **Pappus** 3–5 mm long. **Achenes** 1–2 mm long, nearly glabrous. Shallow soil of grasslands, outcrops; valleys to lower subalpine. BC, AB to CA, NV, UT, CO, NE.

Some populations have plants bearing small plantlets among the upper leaves.

Antennaria media Greene [*A. alpina* (L.) Gaertn. var. *media* (Greene) Jeps.] Stoloniferous, mat-forming. **Stems** erect, 3–12 cm. **Basal leaves** oblong to oblanceolate, 5–20 mm long, 1-nerved. **Stem leaves** linear, 1–2 cm long, uppermost sometimes flagged. **Inflorescence** corymbiform with 3 to 7 heads; male plants rare. **Involucre**: female 4–7 mm high. **Phyllaries** greenish-black. **Corolla** 2–4 mm long. **Pappus** 3–6 mm long. **Achenes** 0.5–1.5 mm long, glabrous to pappilose. Meadows, turf; subalpine, alpine. AK to CA, AZ, NM.

These plants are sometimes considered conspecific with *Antennaria alpina*; however, *A. media* as described above is a polyploid, hybrid complex.[39] See *A. alpina*.

Antennaria microphylla Rydb. Stoloniferous, loosely mat-forming. **Stems** ascending to erect, 8–30 cm, glandular above beneath the tomentum. **Basal leaves** spatulate to oblanceolate, 5–15 mm long, 1-nerved. **Stem leaves** linear, 8–25 mm long. **Inflorescence** dense- to open-corymbiform with 6 to 13 heads. **Involucre**: female 5–7 mm high; male 4–5 mm high. **Phyllaries** white. **Corolla**: female 3–4 mm long; male 2–3 mm long. **Pappus** 3–5 mm long. **Achenes** ca. 1 mm long, glabrous. Moist grasslands, moist to wet meadows; plains, valleys, montane. AK to QC south to CA, AZ, NM, NE, MN.

Antennaria rosea has narrower basal leaves and usually occurs in drier habitats. The type locality is in Gallatin Co.

Antennaria monocephala DC. Forming small clumps from a few-branched caudex; stolons absent. **Stems** erect, 1–10 cm. **Basal leaves** oblanceolate, 3–12 mm long, 1-nerved, less hairy above. **Stem leaves** linear, 4–11 mm long, flagged. **Inflorescence** a solitary (2 to 3) head; male plants absent. **Involucre**: female 5–8 mm high. **Phyllaries** blackish-green. **Corolla**: female 2–4 mm long. **Pappus** none. **Achenes** ca. 1 mm long, glabrous. Moist turf of cool slopes; alpine. AK to Greenland south to BC, WY; Asia.

Most plants of *Antennaria alpina* have stolons and more than two heads.

Antennaria neglecta Greene [*A. howellii* Greene] Stoloniferous, loosely mat-forming. **Stems** ascending to erect, 6–35 cm. **Basal leaves** oblong to spatulate, 1–4 cm long, 1-nerved, gray-villous to glabrous above. **Stem leaves** linear-oblanceolate, 1–3 cm long, the upper flagged. **Inflorescence** corymbiform with 4 to 12 heads; male plants rare. **Involucre** 6–9 mm high. **Phyllaries** white. **Corolla** 3–6 mm long. **Pappus** 4–9 mm long. **Achenes** 1–2 mm long, papillose. Grasslands, woodlands, open forest; plains, valleys, montane. YT to QC south to CA, UT, CO, OK, KY, SC.

Antennaria neglecta, as described above, consists of the sexual *A. neglecta* (sensu stricto) and apomictic populations derived from *A. neglecta* and several other sexual species.[36,39] These apomictic populations have been segregated into several subspecies of *A. howellii*.[37] I recognize the value of maintaining the sexual progenitors of this apomictic complex as separate species. However, the only firm morphological character separating *A. neglecta* (sensu stricto) from *A. howellii* is the presence of male plants in the population, a character rarely documented on herbarium specimens. Thus I have chosen to follow older treatments and combine them into an expanded concept of the older name.[103] Our plants can be divided into two groups. Plants from northwest MT with basal leaves completely glabrous on the upper surface correspond to *A. howellii* ssp. *howellii*. Plants with basal leaves becoming nearly glabrous on the upper surface with age would be *A. howellii* ssp. *petaloidea* (Fernald) R.J.Bayer or *A. howellii*. ssp. *neodioica* (Greene) R.J.Bayer. *Antennaria neglecta* (sensu stricto) is reported for MT,[39] but I have seen no specimens to verify this report.

Antennaria parvifolia Nutt. Stoloniferous, mat-forming. **Stems** ascending to erect, 3–15 cm. **Basal leaves** spatulate, 5–15 mm long with rounded-mucronate tips, 1-nerved. **Stem leaves** linear, 1–2 cm long. **Inflorescence** corymbiform with 3 to 7 heads; male plants absent. **Involucre** 7–11 mm high. **Phyllaries** white above, green to pink below. **Corolla** 5–7 mm long. **Pappus** 4–9 mm long. **Achenes** 1–2 mm long, glabrate. Grasslands, sagebrush steppe, open forest, stony calcareous soil of exposed slopes; plains, valleys to montane. BC to QC south to WA, AZ, TX, NE, MI.

Antennaria rosea has narrower basal leaves.

Antennaria pulcherrima (Hook.) Greene Rhizomatous, non-stoloniferous. **Stems** erect, 20–50 cm. **Basal leaf blades** oblanceolate, 4–15 cm long, 3- to 5-nerved. **Stem leaves** linear to linear-oblanceolate, 1–12 cm long, often flagged. **Inflorescence** corymbiform with 4 to 20 heads. **Involucre**: female 7–11 mm high; male 6–8 mm high. **Phyllaries** brown to black at the base, paler but not pure white above. **Corolla**: female 5–6 mm long; male 4–5 mm long. **Pappus** 4–10 mm long. **Achenes** 1–2 mm long, glabrous. Wet soil of meadows, fens, turf; montane to lower alpine. AK to NL south to WA, CO, MB. Our plants are ssp. *pulcherrima*. See *A. anaphaloides*.

Antennaria racemosa Hook. Stoloniferous, mat-forming. **Stems** ascending to erect, 5–50 cm, glandular above. **Basal leaf blades** elliptic, 1–6 cm long, 3-nerved, glabrous above. **Stem leaves** lanceolate to oblanceolate, 1–4 cm long. **Inflorescence** racemose to paniculate with 3 to 20 heads on glandular peduncles 5–25 mm long. **Involucre**: female 6–9 mm high; male 4–7 mm high. **Phyllaries** narrow, dirty white to light brown. **Corolla**: female 4–6 mm high; male 3–4 mm long. **Pappus** 3–7 mm long. **Achenes** 1–2 mm long, glabrous. Moist to dry forests, occasionally meadows; valleys to subalpine. BC, AB to OR, WY. (p.494)

Antennaria rosea Greene Rosy pussytoes [*A. microphylla* Rydb. misapplied] Stoloniferous, mat-forming. **Stems** erect, 4–35 cm. **Basal leaves** oblong to linear-oblanceolate, 5–25 mm long, 1-nerved. **Stem leaves** linear, 3–30 mm long. **Inflorescence** corymbiform with 3 to 20 heads; male plants rare. **Involucre** 4–7 mm high. **Phyllaries** white, pink, green, or brown. **Corolla** 3–5 mm long. **Pappus** 3–6 mm long. **Achenes** 1–2 mm long, glabrous or papillose. Grasslands, meadows, sagebrush steppe, open forest, fellfields, outcrops at all elevations. AK to NL south to CA, NM, MB. (p.494)

Antennaria rosea is a complex of asexual, polyploid hybrids derived from as many as eight other species.[38] Bayer[39] recognizes four subspecies, all of which occur in MT. Relatively tall, low-elevation plants with larger involucres are ssp. *rosea* or ssp. *arida* (A.Nelson) R.J.Bayer; the latter occurring in moister sites. Smaller plants in exposed sites at all elevations are ssp. *confinis* (Greene) R.J.Bayer or ssp. *pulvinata* (Greene) R.J.Bayer; the former with smaller involucres. The extremes of these four taxa are distinct, but the characters separating them are continuous, and many specimens are difficult to determine with confidence. See *A. microphylla*.

Antennaria umbrinella Rydb. Stoloniferous, mat-forming from a branched caudex. **Stems** erect, 4–20 cm. **Basal leaves** oblanceolate to linear-spatulate, 5–15 mm long, 1-nerved. **Stem leaves** linear, 7–15 mm long. **Inflorescence** corymbiform with 3 to 10 heads. **Involucre**: female 4–7 mm high; male 3–5 mm high. **Phyllaries** rounded, dirty white above dark brown below. **Corolla**: female 3–4 mm long; male 2–3 mm long. **Pappus** 3–5 mm long. **Achenes** ca. 1 mm long, glabrous. Often stony or shallow soil in grasslands, fellfields, sagebrush steppe; valleys to alpine. BC to MB south to CA, AZ, CO. See *A. aromatica*.

The type locality is in the Little Belt Mtns.

Anthemis L. Chamomile

Annual or perennial herbs. **Leaves** alternate, short-petiolate, 2 to 3 times pinnately dissected. **Inflorescence**: heads on long peduncles from upper leaf axils. **Heads** radiate. **Involucre** hemispheric; phyllaries imbricate in 3 to 5 series, margins scarious; receptacle with scarious scales (paleae). **Rays** white or yellow, female or sterile. **Disk flowers** yellow; style branches flattened. **Pappus** absent. **Achenes** obconic, 4-angled.

1. Perennials; rays yellow, fertile . . . . *A. tinctoria*
1. Plants annual; rays white 2

2. Rays sterile; achenes with tuberculate ridges . . . .*A. cotula*
2. Rays fertile; achenes glandular but not tuberculate . . . .*A. arvensis*

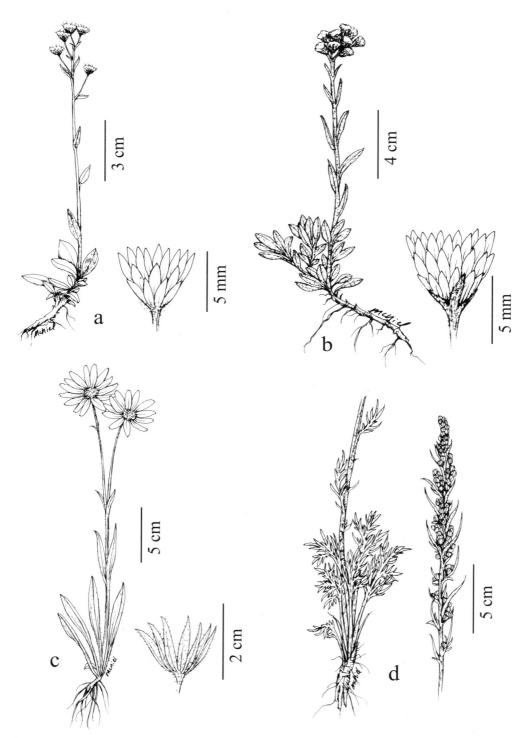

Plate 88. a. *Antennaria racemosa*, b. *Antennaria rosea*, c. *Arnica sororia*, d. *Artemisia campestris*

Anthemis arvensis L. Taprooted annual. **Stems** ascending to erect, usually branched, 10–40 cm. **Herbage** glabrate to villous, non-aromatic. **Leaves** oblong, 15–35 mm long; ultimate segments linear to narrowly deltoid. **Peduncles** 4–15 cm long. **Involucre** 6–13 mm wide, villous. **Rays** 5 to 20, white, fertile; ligules 5–15 mm long. **Disk corolla** 2–3 mm long. **Achenes** ca. 2 mm long, glandular, angles tuberculate. Fields, roadsides; valleys. Introduced throughout much of temperate N. America; native to Eurasia. Collected in Lake Co. more than 100 years ago.

Anthemis cotula L. Dogfennel Taprooted annual. **Stems** erect, simple or branched, 10–60 cm. **Herbage** glabrate to sparsely villous, aromatic. **Leaves** oblong, 15–50 mm long; ultimate segments linear. **Peduncles** 2–7 cm long. **Involucre** 7–11 mm wide, sparsely villous. **Rays** 10 to 15, white, sterile; ligules 5–10 mm long. **Disk corolla** 2–3 mm long, glandular. **Achenes** 1–2 mm long, sparsely glandular, angles glabrous. Roadsides, fields; valleys. Introduced throughout N. America; native to Europe.

Matricaria maritima is similar but has odorless herbage and fertile ray flowers.

Anthemis tinctoria L. [*Cota tinctoria* (L.) J.Gay ex Guss.] Yellow chamomile Short-lived perennial. **Stems** erect, simple or branched, 30–60 cm. **Herbage** villous, non-aromatic. **Leaves** oblong, 2–4 cm long; ultimate segments deltoid, spine-tipped. **Peduncles** 4–12 cm long. **Involucre** 1–2 cm wide, sparsely villous. **Rays** 20 to 30; ligules yellow, fertile, 6–12 mm long. **Disk corolla** 3–4 mm long. **Achenes** ca. 2 mm long. Roadsides; valleys. Introduced across temperate N. America; native to Europe.

Arctium L. Burdock

Monocarpic biennial or perennial herbs. **Leaves** basal and alternate, similar, petiolate; blades cordate-deltoid to ovate, serrate. **Inflorescence** leafy-bracteate, corymbiform to paniculate. **Heads** discoid. **Involucre** globose; phyllaries in 9 to 17 series, narrow, tipped with spreading, hooked spines above; receptacle with subulate scales. **Disk flowers** elongate-tubular; style hairy below the stigmas. **Pappus** of deciduous bristles. **Achenes** compressed-obovoid, glabrous.

Both species occur in disturbed, fertile soil of fields, pastures, thickets and woodlands, often along streams; plains, valleys. Introduced throughout U.S. and adjacent Canada; native to Eurasia.

1. Inflorescence corymbiform; involucres 3–4 cm wide . . . . *A. lappa*
1. Inflorescence racemose; involucres 15–30 mm wide. . . . .*A. minus*

Arctium lappa L. **Stems** erect, branched, 50–150 cm. **Leaves** with solid petioles; blades 10–50 cm long, tomentose beneath. **Inflorescence** corymbiform, heads short-pedunculate. **Involucre** 3–4 cm wide, glabrate. **Disk flowers** ca. 40; corolla purple, 9–14 mm long. **Pappus** 2–5 mm long. **Achenes** 6–7 mm long.

Arctium minus (Hill) Bernh. **Stems** erect, branched, 30–120 cm. **Leaves** with hollow petioles; blades 5–20 cm long, glandular, sparsely tomentose beneath, becoming glabrate. **Inflorescence** racemose, heads on peduncles 5–20 mm long. **Involucre** 15–30 mm wide, glabrate to loosely tomentose. **Disk flowers** ca. 30; corolla pink to purple, 7–12 mm long. **Pappus** 1–4 mm long. **Achenes** 4–6 mm long.

Arnica L.

Perennial herbs with rhizomes or elongate rootstocks. **Stems** simple below the inflorescence. **Leaves** simple, opposite; basal leaves often on separate shoots. **Inflorescence** of solitary heads, terminal or on axillary peduncles. **Heads** radiate or discoid, often hemispheric; phyllaries green, in 2 equal series; receptacle convex, naked. **Ray flowers** female, yellow or orange, conspicuous or absent, usually minutely lobed at the tip. **Disk flowers** yellow, perfect; style branches flattened. **Pappus** of white to brown bristles, barbed to subplumose (i.e., barbs longer than thickness of the bristle). **Achenes** cylindrical, 5- to 10-ribbed.[187451]

Vestiture of the achenes is a useful character; the shape of the heads can be helpful in the field. *Arnica diversifolia* is a hybrid complex and includes many different-appearing plants that are, to some extent, intermediate between *A. mollis, A. latifolia* and *A. amplexicaulis*. Confident identification of all specimens may not be possible. *Arnica lonchophylla* Greene, similar to *A. angustifolia* but taller and leafier, occurs to the north and south of MT.

1. Rays lacking; lowest flowers usually nodding. . . . . *A. parryi*
1. Rays present on most plants 2

2. Stems with 4 or fewer pairs of leaves. 3
2. Stem leaves ≥4 pairs on well-developed plants 10

3. Pappus light brown, subplumose (barbs much longer than bristle width) 4
3. Pappus white, the bristles short-barbed. 5

4. Heads hemispheric, often with >90 disk flowers; leaves entire to denticulate . . . .*A. mollis*
4. Heads turbinate with <90 disk flowers; leaves usually dentate . . . .*A. diversifolia*

5. Basal leaf blades broadly lanceolate to broadly cordate-ovate, dentate 6
5. Basal leaf blades lanceolate, entire 7

6. Achene hairy or stipitate-glandular to the base; involucre villous, glandular . . . . *A. cordifolia*
6. Achene glabrous at the base; involucre glandular-puberulent (sparsely villous) . . . .*A. latifolia*

7. Plants densely long-woolly; high elevations . . . . *A. angustifolia*
7. Plants glabrous to short-hairy. 8

8. Most heads with 7 to 10 rays; leaves glabrate to glandular-pubescent; plants usually at
   high elevations . . . . *A. rydbergii*
8. Heads with >10 rays; leaves densely glandular-pubescent; plants usually lower 9

9. Tufts of brown hair among old leaf bases at ground level . . . .*A. fulgens*
9. Hair among leaf bases white or lacking . . . .*A. sororia*

10. Involucral bracts acute with a small tuft of white hair at the tip. . . . . *A. chamissonis*
10. Tip of involucral bracts attenuate, no hairier than body 11

11. Leaf margins obviously dentate; stems mostly solitary. . . . .*A. amplexicaulis*
11. Leaves entire; stems usually clustered. . . . . *A. longifolia*

Arnica amplexicaulis Nutt. [*A. lanceolata* Nutt. ssp. *prima* (Maguire) Strother & S.J.Wolf] **Stems** solitary or clustered, erect, 20–80 cm. **Herbage** puberulent to villous. **Leaves** mostly cauline, sessile, 4 to 7 pairs, dentate, lanceolate to elliptic, 5–15 cm long. **Heads** 3 to 10; involucre campanulate, 8–15 mm high, glandular, septate-villous. **Rays** 8 to 14; ligules 1–2 cm long. **Pappus** subplumose, brownish. **Achenes** glabrous, 4–8 mm long. Moist forest, woodlands, often along streams; montane. AK south to CA, NV, UT, WY.

   This species is often confused with its hybrid derivative *Arnica diversifolia,* which has petiolate leaves less than three times as long as wide and is mainly subalpine. *Arnica amplexicaulis* is sometimes considered conspecific with *A. lanceolata*;[451] the two are similar but also widely disjunct. Until new data are presented I choose to follow the more familiar treatment.

Arnica angustifolia Vahl [*A. alpina* Salisb., *A. tomentosa* J.M.Macoun] **Stems** solitary, erect, 5–25 cm. **Herbage** glandular, villous. **Leaves:** basal and cauline similar; blades entire, narrowly lanceolate, 2–7 cm long; basal leaves petiolate; stem leaves 1 to 3 pairs, becoming sessile. **Heads** solitary; involucre campanulate, 10–15 mm high, densely villous. **Rays** 9 to 14; ligules 1–2 cm long. **Pappus** short-barbed, white. **Achenes** hirsute, 3–8 mm long. Stony, calcareous soil of exposed ridges, talus slopes; upper subalpine to alpine. Circumpolar south to MT. Our plants are ssp. *tomentosa* (J.M.Macoun) G.W.Douglas & Ruyle-Douglas. (p.498)

Arnica chamissonis Less. **Stems** solitary, erect, 20–80 cm. **Herbage** glabrate to glandular-puberulent. **Leaves** mostly cauline, 5 to 10 pairs, entire to denticulate, lanceolate to narrowly ovate, 4–15 cm long, short-petiolate below. **Heads** 3 to 10; involucre hemispheric, 5–12 mm high, septate-villous, glandular. **Rays** 10 to 15; ligules 8–15 mm long. **Pappus** short-barbed to subplumose, tan. **Achenes** glabrate to glandular, 3–8 mm long. Moist meadows, open forest, woodlands, often along streams; valleys, montane. AK to QC south to CA, AZ, NM.

   The species has been divided into several subspecies and varieties based on pappus color and branching, leaf margins and size.[262] These characters do not seem strongly correlated in our area, so I have chosen not to recognize infraspecific taxa.[451]

Arnica cordifolia Hook. Heart-leaf arnica **Stems** solitary or clustered, ascending to erect, 10–50 cm. **Herbage** villous, glandular-puberulent. **Leaves**: basal long-petiolate on separate shoots; blades ovate, cordate to truncate, dentate, 4–12 cm long; cauline 2 to 4 pairs, petiolate, smaller and less cordate upward. **Heads** 1 to 3; involucre villous, glandular, campanulate, 13–20 mm high. **Rays** 10 to 15; ligules 1–3 cm long. **Pappus** short-barbed, white. **Achenes** hirsute, 5–10 mm long, often glandular. Dry to moist forest; valleys to subalpine. AK south to CA, NM, MI. (p.498)

The dwarf, high-elevation form has been called *Arnica cordifolia* var. *pumila* (Rydb.) Maguire. One specimen collected from standing water has nearly glabrous phyllaries and solitary heads. *Arnica latifolia* is similar but the involucre is barely villous.

Arnica diversifolia Greene [*A. ovata* Greene misapplied] **Stems** solitary or clustered, erect, 15–50 cm. **Herbage** glabrate to glandular-puberulent. **Leaves**: basal petiolate on separate shoots; blades ovate, cordate to truncate, dentate, 2–6 cm long; cauline 2 to 4 pairs, short-petiolate, the middle the largest. **Heads** 1 to 5; involucre turbinate, 8–15 mm high. **Rays** 8 to13; ligules 10–15 mm long. **Pappus** subplumose, tawny. **Achenes** sparsely hirsute, glandular, 5–7 mm long. Moist, stony soil of meadows, open slopes, thickets, often along streams; montane, subalpine, rarely lower or higher. AK to CA, UT, CO. (p.498)

Arnica diversifolia is thought to be a hybrid complex involving *A. cordifolia*, *A. latifolia*, *A. mollis* and *A. amplexicaulis*; some plants resemble *A. cordifolia*, while others are more like *A. mollis*. It is not surprising that Cronquist[103] believes the type of *A. ovata* is best referred to *A. mollis*, while Wolf believes *A. ovata* is the same as *A. diversifolia* and takes precedence because it was published first.[451]

Arnica fulgens Pursh **Stems** solitary, 20–40 cm with tufts of long brown hair at the base and amongst the roots. **Herbage** glandular-pubescent. **Leaves**: basal oblanceolate, short-petiolate; blades entire, 3–10 cm long; cauline 2 to 4 pairs, sessile above. **Heads** solitary; involucre hemispheric, 12–15 mm high. **Rays** 10 to 23; ligules 10–25 mm long. **Pappus** short-barbed, white. **Achenes** hirsute, glandular, 3–7 mm long. Grasslands, woodlands, meadows, sagebrush steppe; plains, valleys, montane. BC to MB south to CA, NV, UT, CO. (p.498)

Arnica sororia occurs in the same habitat, and is nearly identical but has narrower phyllaries and lacks the brown hair that surmounts the caudex.

Arnica latifolia Bong. [*A. gracilis* Rydb.] **Stems** mostly solitary, ascending to erect, 8–60 cm. **Herbage** glabrate to weakly tomentose, sometimes glandular-pubescent. **Leaves**: basal long-petiolate on separate shoots; blades lanceolate to sagittate, dentate, 2–12 cm long; cauline 2 to 4 pairs, lanceolate to ovate, denticulate, sessile to short-petiolate. **Heads** 1 to 9; involucre turbinate, 9–18 mm high. **Rays** 5 to 15; ligules 15–30 mm long. **Pappus** short-barbed, white. **Achenes** 5–10 mm long, hirsute or sometimes glandular above, glabrous below. Moist forest, meadows, rocky slopes. AK south to CA, UT, CO. (p.498)

1. Middle stem leaves petiolate; heads 1 to 9, relatively narrow; involucre <13 mm high . . . . .var. *gracilis*
1. Middle stem leaves sessile; heads 1 to 3; involucre 10–18 mm high. . . . . .var. *latifolia*

Arnica latifolia var. *latifolia* occurs from valleys to subalpine; *A. latifolia* var. *gracilis* (Rydb.) Cronquist occurs most often in subalpine habitats. The two vars. are often considered separate species; however, characters separating them are continuous and overlapping, and the range of the latter is contained within the range of the former. A third form found at high elevations has short stems with small, mostly solitary heads, and short-petiolate to sessile cauline leaves; it is referred to var. *latifolia* on the basis of the few heads and sessile leaves. See *A. cordifolia*.

Arnica longifolia D.C.Eaton **Stems** solitary or clustered, erect, 20–70 cm. **Herbage** glandular-puberulent. **Leaves** cauline, sessile, 5 to 7 pairs, entire, narrowly lanceolate, 5–15 cm long. **Heads** 3 to 12; narrowly campanulate, 7–10 mm high, glandular-puberulent. **Rays** 8 to 13; ligules 8–17 mm long. **Pappus** short-barbed, tan to brown. **Achenes** 3–7 mm long, glandular-puberulent. Very stony soil of meadows, talus slopes, often along streams; subalpine, alpine, rarely lower. BC, AB south to CA, NV, UT, CO.

Arnica mollis Hook. [*A. ovata* Greene] **Stems** solitary or clustered, erect, 15–60 cm. **Herbage** glandular, puberulent to villous. **Leaves**: basal petiolate, sometimes on separate shoots; blades oblong to elliptic, denticulate, 3–12 cm long; cauline 3 to 4 pairs, sessile above. **Heads** 1 to 3; involucre hemispheric, 1–2 cm high, glandular-pubescent. **Rays** 12 to 18; ligules 1–2 cm long.

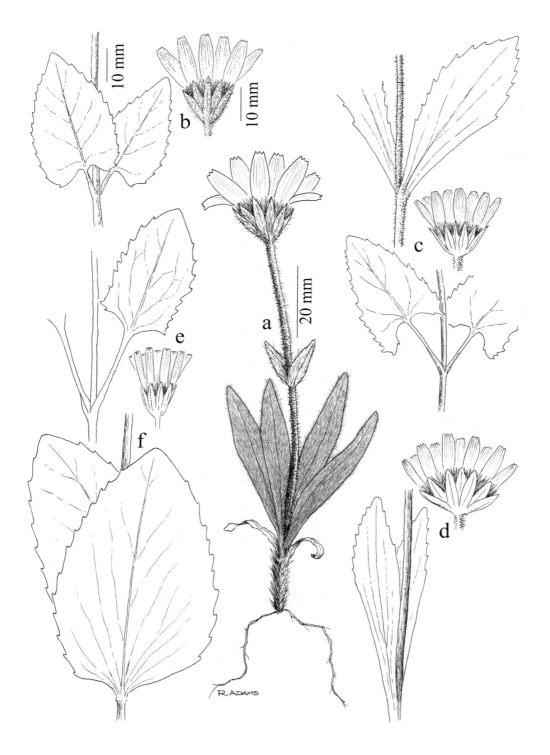

Plate 89. **Arnica**. **a.** *A. angustifolia*, **b.** *A. cordifolia*, **c.** *A. diversifolia*, **d.** *A. fulgens*, **e.** *A. latifolia* var. *gracilis*, **f.** *A. latifolia* var. *latifolia* Scale bars for (b) apply to all except (a).

Pappus subplumose, tawny. **Achenes** glandular-pubescent, 4–8 mm long. Moist to wet meadows, open forest, often along streams; subalpine, rarely lower. AK to CA, NV, UT, CO.

The hemispheric heads and tawny pappus help distinguish this species. See *A. diversifolia.*

Arnica parryi A.Gray **Stems** mostly solitary, erect, 30–70 cm. **Herbage** glandular, villous. **Leaves**: basal petiolate; blades lanceolate, entire to denticulate, 3–12 cm long; cauline 2 to 4 pairs, short-petiolate, reduced upward. **Heads** 3 to 9; lateral heads nodding in flower; involucre campanulate, 10–16 mm high, glandular-pubescent. **Rays** absent. **Pappus** short-barbed, stramineous. **Achenes** glabrous, 4–6 mm long. Meadows, woodlands, open forest; montane to subalpine. YT south to CA, NV, UT, CO.

Arnica rydbergii Greene **Stems** solitary or clustered, 5–25 cm. **Herbage** glabrate to glandular-puberulent. **Leaves**: basal lanceolate to narrowly elliptic, petiolate, entire or rarely denticulate, 2–5 cm long; cauline 2 to 4 pairs, lanceolate or oblanceolate, sessile or subsessile. **Heads** 1 to 3; involucre turbinate, 6–12 mm high. **Rays** 6 to 10; ligules 8–15 mm long. **Pappus** short-barbed, white. **Achenes** hirsute, 4–7 mm long. Stony soil of meadows, turf, fellfields; upper montane to alpine. BC, AB south to UT, CO.

The smaller size and ×eric habitat help distinguish this species from *Arnica sororia* and *A. fulgens.* The type locality is in the Little Belt Mtns.

Arnica sororia Greene **Stems** solitary, 20–40 cm. **Herbage** glandular, pubescent, sometimes with tufts of wvhite hair among leaf bases. **Leaves**: basal blades short-petiolate, narrowly lanceolate, entire, 3–10 cm long; cauline 2 to 4 pairs, similar, sessile above. **Heads** 1 to 5; involucre hemispheric, 10–15 mm high. **Rays** 9 to 16; ligules 1–2 cm long. **Pappus** short-barbed, white. **Achenes** hirsute, 3–7 mm long. Grasslands, sagebrush steppe, open forest; plains, valleys, montane. BC to SK south to CA, NV, UT, WY. (p.494)

Collections suggest that *Arnica sororia* has horizontal rhizomes shallower than *A. fulgens.*

Artemisia L. Sagebrush, sagewort, mugwort, wormwood

Annual to perennial herbs, subshrubs, or shrubs. **Herbage** usually aromatic. **Leaves** alternate. **Inflorescence** leafy-bracted, spiciform, racemose or paniculate. **Heads** small, discoid, hemispheric to turbinate; phyllaries scarious-margined, green to white, in 4 to 7 series; receptacle naked to pubescent. **Disk flowers** perfect, yellow (reddish), male or female; style branches flattened. **Pappus** absent. **Achenes** usually glabrous or glandular.[41,366,414]

Members of the *Artemisia ludoviciana-tilesii-michauxiana* complex may be difficult to determine with certainty; *A. ludoviciana* ssp. *ludoviciana* approaches *A. ludoviciana* ssp. *candicans* which approaches *A. tilesii,* and *A. ludoviciana* ssp. *incompta* approaches *A. michauxiana. Artemisia vulgaris* L. is reported for MT,[366] but all specimens I have seen were cultivated plants. *Artemisia spinescens* flowers in spring; other species flower in summer or fall. Reports of *A. norvegica* Fries were based on misidentified specimens of *A. campestris* and *A. michauxiana. Artemisia rigida* (Nutt.) A.Gray was reported for MT,[366,418] but I have seen no specimens.

1. Well-developed shrubs; stems woody well above ground level 2
1. Plants herbaceous or subshrubs (woody only at the base) 9

2. Leaves deeply divided into filiform segments >5 mm long 3
2. Leaves entire to lobed into oblong to oblanceolate lobes, mostly <5 mm long .5

3. Leaves green, glabrous to sparsely villous, 2 to 3 times divided . . . . *A. abrotanum*
3. Leaves grayish, canescent to tomentose, once divided 4

4. Inflorescence spiciform; bracts divided . . . . *A. rigida*
4. Inflorescence paniculate; bracts entire . . . . *A. tripartita*

5. Some or all of the leaves entire . . . . *A. cana*
5. Leaves lobed 6

6. Undersides of leaves densely covered with conspicuous green to amber resin dots . . . . *A. nova*
6. Resin dots inconspicuous or absent 7

7. Plants >40 cm high . . . . *A. tridentata*
7. Plants <40 cm high 8

8. Bracts subtending flowers linear; herbage tomentose . . . . .A. arbuscula
8. Flowers subtended by leaf-like, lobed bracts; herbage villous . . . . A. spinescens

9. Mature foliage green, glabrous to sparsely villous 10
9. Leaves grayish, canescent to tomentose at least on the lower surface 14

10. Most or all leaves entire . . . . .A. dracunculus
10. Leaves lobed or divided 11

11. Most leaves ≤1 cm long; dwarf shrubs . . . . A. spinescens
11. Most leaves >1 cm long; herbaceous 12

12. Involucre 1–2 mm long; plants annual; leaves divided into filiform segments . . . . A. annua
12. Annuals to biennials; involucre ≥2 mm long; leaf segments linear to lanceolate 13

13. Ultimate leaf segments dentate, sharply acute. . . . . .A. biennis
13. Ultimate leaf segments entire, rounded at the tip . . . . .A. campestris

14. Heads with hairs on receptacle amongst the flowers 15
14. Receptacle without hair 17

15. Ultimate leaf segments ≥2 mm wide. . . . . .A. absinthium
15. Ultimate leaf segments ca. 1 mm wide. 16

16. Subshrub; involucre 2–3 mm high; inflorescence paniculate . . . . A. frigida
16. Herbaceous; involucre 3–4 mm high; inflorescence spiciform to racemose. . . . . A. scopulorum

17. Leaves densely sericeous . . . . .A. campestris
17. Leave tomentose at least below 18

18. Leaves entire 19
18. Leaves lobed to divided 22

19. Leaves green above. 20
19. Leaves gray above. 21

20. Involucre wider than high . . . . A. tilesii
20. Involucre higher than wide . . . . A. ludoviciana

21. Plants subshrubs, not rhizomatous; leaves whiter beneath than above . . . . A. longifolia
21. Rhizomatous; leaves about equally gray above and beneath . . . . A. ludoviciana

22. Ultimate leaf segments linear to filiform; plants mostly <15 cm high .A. pedatifida
22. Ultimate leaf segments oblanceolate to linear; plants usually >15 cm high 23

23. Involucre wider than high . . . . A. tilesii
23. Involucre higher than wide 24

24. Leaves lobed ≤1/2-way to the midvein . . . . A. ludoviciana
24. Leaves deeply divided 25

25. Leaves twice deeply divided; ultimate segments ≤1 mm wide. . . . . A. michauxiana
25. Leaves 1 to 2 times divided; most ultimate segments >1 mm wide . . . . A. ludoviciana

Artemisia abrotanum L. Shrub or subshrub 50–150 cm. **Stems** glabrate to puberulent.
Herbage glabrous to sparsely villous, aromatic. **Leaves** petiolate; blades 3–6 cm long, deeply 2
to 3 times pinnately divided into filiform lobes. **Inflorescence** open-branched paniculate; heads
nodding. **Involucre** hemispheric, 1.5–2 mm high; phyllaries green, sparsely tomentose; receptacle
glabrous. **Disk flowers** 5 to 20, perfect and female, glandular; corolla ca. 1 mm long, yellow. **Achenes**
glabrous, <1 mm long. Fields, yards; plains, valleys. Sparingly introduced to north-temperate N. America; native to
Eurasia.

Formerly cultivated for medicinal purposes, now employed as an ornamental; many of our collections appear to be
from cultivated plants.

Artemisia absinthium L. WORMWOOD, ABSINTHIUM Taprooted perennial. **Stems** erect, often
branched, 40–80 cm. **Herbage** canescent. **Leaves** basal and cauline, petiolate; blades ovate, 2–10
cm long, 2 to 3 times pinnately divided into linear-lanceolate segments, reduced and less divided
upward. **Inflorescence** leafy-paniculate; heads nodding. **Involucre** hemispheric, 2–3 mm high;
phyllaries green, tomentose; receptacle villous. **Disk flowers** perfect and female, 40 to 70, glabrous
to glandular; corolla 1–1.5 mm long. **Achenes** glabrous, ca. 1 mm long. Roadsides, fields, talus, streambanks; plains,
valleys. Introduced throughout temperate N. America; native to Europe.

Artemisia annua L. ANNUAL WORMWOOD Taprooted annual. **Stems** erect, simple or branched, 30–80 cm. **Herbage** glabrous. **Leaves** cauline, subsessile; blades lanceolate to ovate, 5–15 cm long, 1 to 2 times pinnately divided into linear, sometimes lobed segments. **Inflorescence** paniculate, leafy-bracteate, with nodding tips. **Involucre** hemispheric, ca. 1 mm high, wider than high; phyllaries green, glabrous; receptacle glabrous. **Disk flowers** perfect, 5 to 20, glabrous; corolla ca. 1 mm long, yellow. **Achenes** glabrous, <1 mm long. Fields, pastures; plains. Introduced to most of U.S.; native to Eurasia. Collected in Toole and Yellowstone cos. more than 70 years ago.

Artemisia arbuscula Nutt. LOW SAGEBRUSH Shrub 10–40 cm high. **Herbage** tomentose, sometimes with obscure resin dots beneath tomentum. **Leaves** cuneate, 5–25 mm long, 3-lobed distally. **Inflorescence** spiciform; bracts linear. **Involucre** campanulate, 3–5 mm high; phyllaries green, tomentose; receptacle glabrous. **Disk flowers** perfect, 4 to 8, glabrous or glandular; corolla ca. 2 mm long. **Achenes** resinous, <1 mm long. Sagebrush steppe; montane in Beaverhead Co. WA to MT south to CA, NV, UT, CO. (p.504)

1. Lobes of leaves rounded; stems flexuous, prostrate-spreading . . . . .ssp. *longiloba*
1. Lobes of leaves acute; stems stiffly erect to ascending . . . . ssp. *arbuscula*

Artemisia arbuscula ssp. *longiloba* (Osterh.) L.M.Shultz flowers July to mid-August in fine-textured soil of stream terraces; *A. arbuscula* ssp. *arbuscula* flowers mid-August to September in stony calcareous soil.

Artemisia biennis Willd. Annual or biennial. **Stems** erect, simple or branched, 20–90 cm. **Herbage** glabrous, inodorous. **Leaves** basal and cauline, petiolate; blades oblanceolate, 2–15 cm long, 1 to 2 times pinnately divided into dentate, lanceolate segments. **Inflorescence** compound-spiciform; bracts linear or lobed. **Involucre** hemispheric, 2–3 mm high; phyllaries green to purple, glabrous; receptacle glabrous. **Disk flowers** 20 to 60, the outer female, glandular; corolla ca. 1 mm long, yellow. **Achenes** glabrous, ≤1 mm long. Roadsides, fields, margins of wetlands; plains, valleys. Throughout temperate N. America; native to western states, adventive elsewhere.

Artemisia campestris L. [*A. caudata* Michx., *A. borealis* Pallas] Taprooted biennial or perennial with a simple or branched caudex. **Stems** ascending to erect, 5–50 cm. **Herbage** villous to glabrate. **Leaves**: basal and cauline, petiolate; blades 1–6 cm long, ovate, 1 to 2 times pinnately divided into linear segments; cauline reduced, becoming once pinnate. **Inflorescence** paniculate; bracts linear above. **Involucre** campanulate, 2–4 mm high; phyllaries green to brown, glabrous to villous; receptacle glabrous. **Disk flowers** 15 to 50, female and male, glabrous; corolla ca. 2 mm long, green, yellow, or red. **Achenes** glabrous, ≤1 mm long. Throughout N. America. (p.494)

1. Plants taprooted biennial; basal rosette withered at flowering; herbage glabrate . . . . .var. *caudata*
1. Plants perennial from a branched caudex; rosettes persistent at flowering; herbage
   sparsely to densely villous 2
2. Stems mostly ≤25 cm; inflorescence spike-like; involucres 3–4 mm high . . . . var. *borealis*
2. Stems usually >25 cm; inflorescence paniculate; involucres mostly 2–3 mm high . . . . var. *scouleriana*

Artemisia campestris var. *caudata* (Michx.) E.J.Palmer & Steyerm., with glabrous involucres ca. 2 mm high, occurs in very sandy soil of dunes and around outcrops on the plains; *A. campestris* var. *scouleriana* (Besser) Cronquist [*A. campestris* ssp. *pacifica* (Nutt.) H.M.Hall & Clem.], with involucres mostly 2–3 mm high, is found in stony or sandy soils of grasslands, fields, streambanks, roadsides; plains, valleys, montane; *A. campestris* var. *borealis* (Pall.) M.Peck, with involucres mostly 3–4 mm high, often forms small mats and occurs in fellfields, meadows; subalpine, alpine.

Artemisia cana Pursh SILVER SAGEBRUSH Shrubs 50–100 cm high, capable of resprouting after fire. **Stems** tomentose to canescent. **Herbage** sparsely to densely tomentose. **Leaves** linear to linear-oblanceolate, 2–5 cm long. **Inflorescence** densely leafy-paniculate. **Involucre** campanulate, 3–4 mm high; phyllaries green to brown, tomentose; receptacle glabrous. **Disk flowers** perfect, 4 to 20, glabrous to glandular; corolla 2–3 mm long. **Achenes** resinous, 1–2 mm long. Sagebrush steppe; plains, valleys, montane. AB to MB south to CA, NV, UT, CO, NE. (p.504)

1. Leaves entire, gray, densely tomentose, 3–5 mm wide . . . . ssp. *cana*
1. Leaves more sparsely tomentose, sometimes lobed at the tip, ≤2 mm wide . . . . ssp. *viscidula*

Artemisia cana ssp. *cana* frequently occurs in sandy soil, often on stream terraces or disturbed areas of grasslands; plains, valleys; *A. cana* ssp. *viscidula* (Osterh.) Beetle is found on loamy soil of gentle hills, slopes, stream terraces; montane in Beaverhead and Gallatin cos.

Artemisia dracunculus L. Wɪʟᴅ ᴛᴀʀʀᴀɢᴏɴ Short-rhizomatous perennial. **Stems** clustered, simple, erect, 30–90 cm. **Herbage** glabrous, aromatic. **Leaves** cauline, fascicled, linear-lanceolate, sessile, 1–8 cm long, usually entire. **Inflorescence** leafy paniculate. **Involucre** hemispheric, 1.5–3 mm long; phyllaries green, tan, or purple, glabrous or glandular; receptacle glabrous. **Disk flowers** 15 to 40, female and male, glandular; corolla 1–2 mm long, yellow to purple. **Achenes** glabrous, ≤1 mm long. Grasslands, meadows, sagebrush steppe; plains, valleys. AK to ON south to CA, AZ, NM, TX, Mexico; Eurasia.

Early-season leaves are pubescent.

Artemisia frigida Willd. Fʀɪɴɢᴇᴅ sᴀɢᴇᴡᴏʀᴛ Subshrub from a branched caudex forming mats with short, sterile stems and clustered leaves. **Stems** ascending or erect, often branched, 10–60 cm, woody at the base. **Herbage** tomentose to canescent, aromatic. **Leaves** basal and cauline, petiolate; basal blades oblong, 5–12 mm long, 2 to 3 times divided into linear segments, ca. 1 mm wide; cauline alternate, similar with dissected stipules. **Inflorescence** narrow leafy-paniculate or -racemose; bracts lobed. **Involucre** hemispheric, 2–3 mm high; phyllaries green to whitish, tomentose; receptacle hirsute. **Disk flowers** 30 to 70, perfect and female, glandular; corolla ca. 1.5 mm long, yellow. **Achenes** glabrous, ca. 1 mm long. Grasslands, sagebrush steppe, coniferous woodlands; plains, valleys, montane, rarely higher. AK to QC south to WA, AZ, NM, TX, IA, IN.

Artemisia longifolia Nutt. Woody at the base from a branched caudex. **Stems** ascending or erect, usually simple, 12–80 cm, woody at the base. **Herbage** aromatic, tomentose; leaves gray above, whitish beneath. **Leaves** cauline, sessile to short-petiolate, linear-lanceolate, 2–10 cm long, entire. **Inflorescence** paniculate; heads erect to nodding; bracts linear. **Involucre** campanulate, 3–4 mm high; phyllaries green, tomentose; receptacle glabrous. **Disk flowers** 10 to 30, perfect and female, glandular; corolla 1–2.5 mm long, yellow or purple. **Achenes** glabrous, <1 mm long. Shale-derived soil of grasslands, coniferous woodlands; plains. BC to MB south to ID, WY, NE. (p.504)

Artemisia ludoviciana Nutt. Rhizomatous perennial. **Stems** simple, erect, 15–90 cm. **Herbage** tomentose, leaves sometimes glabrate above, aromatic. **Leaves** cauline, sessile to short-petiolate, 1–9 cm long, linear-lanceolate to obovate, entire, lobed or deeply divided. **Inflorescence** open to dense, racemose to paniculate, leafy below. **Involucre** campanulate, 2–5 mm high; phyllaries green to brown, glabrate to tomentose; receptacle glabrous. **Disk flowers** 10 to 55, perfect and female, glabrous to glandular; corolla 1.5–2.5 mm long, yellow. **Achenes** glabrous, ca. 0.5 mm long. Throughout N. America. (p.504)

1. Leaves glabrate and greenish above 2
1. Leaves gray-tomentose on both surfaces 3

2. Phyllaries glabrate; inflorescence paniculate; leaves deeply lobed . . . . ssp. *incompta*
2. Phyllaries tomentose; inflorescence racemose; leaves entire to shallowly lobed . . . . .ssp. *lindleyana*

3. Leaves oblanceolate to obovate, at least the lower lobed 1/2-way to the midvein. . . . . .ssp. *candicans*
3. Leaves lanceolate, entire or shallowly lobed <1/2-way to midvein. .ssp. *ludoviciana*

Artemisia ludoviciana ssp. *ludoviciana*, with mostly entire, tomentose leaves, and tomentose phyllaries, is widespread in grasslands, sagebrush steppe, meadows; plains, valleys, montane; *A. ludoviciana* ssp. *candicans* (Rydb.) D.D.Keck [*A. ludoviciana* var. *latiloba* Nutt.], with tomentose leaves lobed 1/2-way to the midrib and tomentose phyllaries, is found in grasslands, streambanks, roadsides; valleys, montane; *A. ludoviciana* ssp. *incompta* (Nutt.) D.D.Keck, with deeply lobed leaves glabrate on the upper surface and glabrate phyllaries, occurs in stony soil of talus slopes, rock outcrops, sagebrush steppe; montane, subalpine; *A. ludoviciana* ssp. *lindleyana* (Besser) P. Lesica (ined.) [*A. lindleyana* Besser], with entire to lobed leaves, glabrate above and tomentose phyllaries, is found on stony shores along streams; valleys west of the Divide. The four subspecies have different combinations of a few variable characters and show different habitat associations. Subspecies *incompta* merges into *A. michauxiana* at higher elevations;[103] many specimens from Glacier National Park are intermediate between these two taxa.

Artemisia michauxiana Besser Perennial from a much-branched caudex. **Stems** simple or branched, ascending to erect, 20–70 cm, glabrous or glandular. **Leaves** cauline, petiolate, aromatic; blades glabrous above, tomentose beneath, 5–25 mm long, obovate, twice divided; ultimate segments linear-lanceolate, ≤1 mm wide. **Inflorescence** narrow, racemose; heads erect. **Involucre** hemispheric, 2.5–3.5 mm high; phyllaries green to brown, glabrous to glandular; receptacle glabrous. **Disk flowers** 25 to 50, perfect and female, glandular; corolla 1.5–2 mm long, yellow to red. **Achenes** glabrous, ca. 0.5

mm long. Stony soil of fellfields, outcrops, talus; upper montane to alpine. BC, AB south to CA, NV, UT, CO. See *A. ludoviciana*. (p.507)

Artemisia nova A.Nelson [*A. arbuscula* Nutt. var. *nova* (A.Nelson) Cronquist] Bʟᴀᴄᴋ sᴀɢᴇʙʀᴜsʜ Shrub 15–50 cm high, not sprouting after fire. **Herbage** tomentose with green or gold resin dots beneath, aromatic. **Leaves** cuneate, 5–15 mm long, 3-lobed distally. **Inflorescence** simple- or compound-spiciform; bracts linear. **Involucre** turbinate, 2.5–5 mm high; outer phyllaries green, tomentose; inner tan, resinous; receptacle glabrous. **Disk flowers** perfect, 2 to 6, glandular; corolla 2–3 mm long. **Achenes** glabrous or resinous, ca. 1 mm long. Stony, calcareous soil of sagebrush steppe, juniper or mountain mahogany woodlands; valleys, montane. MT south to CA, AZ, NM. (p.504)

Artemisia pedatifida Nutt. Subshrub 3–10 cm, forming mats of short, sterile shoots. **Herbage** canescent to tomentose. **Leaves** petiolate; blades 4–10 mm long, 1 to 2 times ternately divided into linear lobes. **Inflorescence** spiciform or narrowly racemose; bracts mostly lobed. **Involucre** globose, 2–3 mm high; phyllaries green, tomentose; receptacle glabrous. **Disk flowers** 8 to 15, female and male; corolla 2–3 mm long, yellow, glandular. **Achenes** glabrous, ca. 1 mm long. Shale-derived soil of terraces in sagebrush steppe; valleys. ID, MT, WY, CO. (p.504)

Artemisia scopulorum A.Gray Herbaceous perennial from a branched caudex. **Stems** ascending to erect, 5–20 cm. **Herbage** strigose to canescent. **Leaves**: basal and cauline, petiolate; blades 1–3 cm long, ovate, twice lobed into linear-oblanceolate segments; cauline few, reduced. **Inflorescence** spiciform to racemose; bracts lobed. **Involucre** hemispheric, 3–4 mm high; phyllaries black with green centers, villous; receptacle villous. **Disk flowers** 20 to 45, perfect and female; corolla ca. 2 mm long, yellow with purple, villous tips. **Achenes** glabrous, ca. 1 mm long. Stony, non-calcareous soil of turf, fellfields; alpine. MT south to NV, UT, NM.

Artemisia spinescens D.C.Eaton [*Picrothamnus desertorum* Nutt.] Bᴜᴅ sᴀɢᴇ Dwarf shrub 5–15 cm. **Stems** profusely branched; dead branches often forming thorns. **Herbage** villous. **Leaves** petiolate; blades 4–10 mm long, 1 to 2 times divided into oblanceolate lobes. **Inflorescence** leafy-bracted, short-racemose on branch tips. **Involucre** globose, ca. 2 mm high; phyllaries few, green, villous; receptacle glabrous. **Disk flowers** 7 to 20, female and male ca. 1 mm long, yellow, villous. **Achenes** villous, ca. 1 mm long. Sandy soil of sagebrush steppe; valleys of Beaverhead and Carbon cos. OR to MT south to CA, AZ, NM.

Artemisia tilesii Ledeb. Rhizomatous perennial. **Stems** simple, erect, 30–80 cm. **Herbage** tomentose; leaves glabrate above. **Leaves** cauline, petiolate; blades 3–7 cm long, lanceolate, entire to lobed; ultimate segments acute. **Inflorescence** narrow, leafy paniculate; heads erect to spreading. **Involucre** hemispheric, 3–5 mm high, wider than high; phyllaries brown, sparsely tomentose; receptacle glabrous. **Disk flowers** 30 to 70, perfect and female, glabrous; corolla 2–3 mm long, yellow. **Achenes** glabrous, 1–2 mm long. Grasslands, meadows; montane, subalpine. AK to QC south to OR, ID, MT.

Artemisia ludoviciana is similar but has more campanulate heads.

Artemisia tridentata Nutt. Bɪɢ sᴀɢᴇʙʀᴜsʜ Shrubs 30–250 cm high, not sprouting following fire. **Herbage** densely tomentose, aromatic. **Leaves** cuneate to oblong, 1–4 cm long, 3-lobed distally. **Inflorescence** paniculate, inconspicuously linear-bracteate. **Involucre** turbinate, 2.5–4 mm high; outer phyllaries green, tomentose, the inner tan; receptacle glabrous. **Disk flowers** perfect, 3 to 8, glandular; corolla 1.5–2 mm long. **Achenes** glandular, sometimes villous, ca. 1 mm long. Usually deep soils of sagebrush steppe, steep open slopes; plains, valleys to subalpine. BC, AB south to CA, AZ, NM, NE. (p.504)

1. Plants often >1 m; inflorescences arising from throughout the crown; leaves long-tapered to the base. . . . . ssp. *tridentata*
1. Plants usually <1 m; inflorescences arising from the outer edges of the crown resulting in a flat-topped appearance; leaves less tapered 2

2. Plants mostly 30–50 cm; leaves mostly ≤15 mm long, oblong, shortly tapered to the petiole. . . . . ssp. *wyomingensis*
2. Plants usually >50 cm; most leaves >15 mm long, narrowly oblong 3

3. Involucres 4–5 mm high; some leaves >3 cm long, often irregularly lobed . . . . ssp. *spiciformis*
3. Involucres ≤4 mm high; most leaves <3 cm long, lobes equal. ssp. *vaseyana*

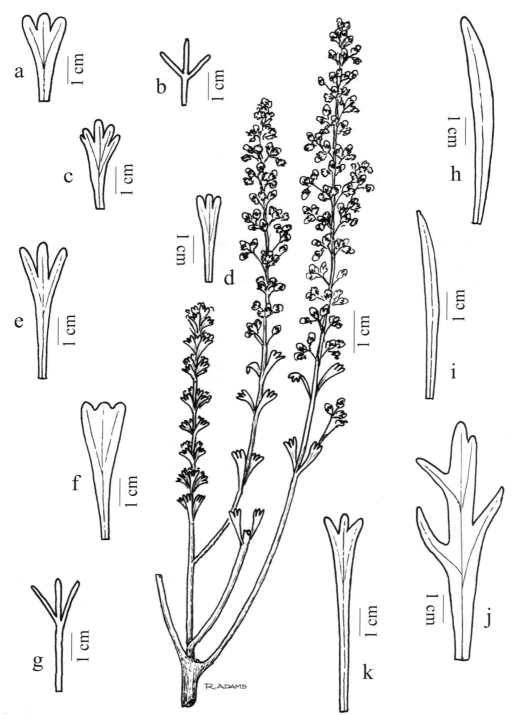

Plate 90. *Artemisia*. **a.** *A. arbuscula* var. *arbuscula*, **b.** *A. pedatifida*, **c.** *A. arbuscula* ssp. *longiloba*, **d.** *A. tridentata* ssp. *wyomingensis*, **e.** *A. nova*, **f.** *A. tridentata* ssp. *vaseyana*, **g.** *A. tripartita* ssp. *tripartita*, **h.** *A. cana* ssp. *cana*, **i.** *A. longifolia*, **j.** *A. ludoviciana* ssp. *candicans*, **k.** *A. tridentata* ssp. *tridentata*

Our four subspecies are ecologically distinct but morphologically intergradent. *Artemisia tridentata* ssp. *tridentata* occurs in often sandier soils of stream terraces or sandhills; valleys, montane, mainly in southwest cos.; *A. tridentata* ssp. *wyomingensis* Beetle & A.M.Young is found on loamy to clay soils; plains, valleys; *A. tridentata* ssp. *vaseyana* (Rydb.) Beetle occurs in loamy soils; montane, subalpine. *A. tridentata* ssp. *spiciformis* (Osterh.) Kartesz & Gandhi is found on cool slopes; subalpine in southwest cos. The latter ssp. is thought to be of hybrid origin but its actual relationship to other subspecies is unknown.

Artemisia tripartita Rydb. THREE-TIP SAGEBRUSH Shrubs 5–80 cm high, sprouting after fire. **Herbage** densely tomentose, aromatic. **Leaves** 1–3 cm long, linear or linear-oblong and deeply divided distally into 3 linear lobes. **Inflorescence** paniculate; bracts linear. **Involucre** turbinate, 2.5–3.5 mm high; outer phyllaries green, canescent; the inner brown to purple, glandular; receptacle glabrous. **Disk flowers** perfect, 3 to 11, glandular; corolla 1.5–2 mm long, yellow to reddish. **Achenes** glabrous or resinous, ca. 2 mm long. Sandy to clay soils of sagebrush steppe; valleys to subalpine. BC to NV, ID, WY. (p.504)

1. Shrubs 20–80 cm; lobes of leaves ≤0.7 wide . . . . ssp. *tripartita*
1. Plants 10–20 cm high; lobes of leaves ca. 1 mm wide . . . . ssp. *rupicola*

Artemisia tripartita ssp. *tripartita* occurs from valleys to montane; *A. tripartita* ssp. *rupicola* Beetle occurs in exposed subalpine sites.

Balsamorhiza Hook. ex Nutt. Balsamroot

Perennial herbs from a large taproot and simple or branched caudex. **Stems** unbranched. **Leaves** mainly basal, petiolate, entire to pinnately lobed; cauline leaves 1 or 2, reduced, alternate or opposite. **Inflorescence** of 1 to 3 long-pedunculate heads. **Heads** radiate; involucre turbinate to hemispheric; phyllaries in 2 to 3 overlapping, unequal series, green; receptacle with scales clasping the achenes. **Ray flowers** female, yellow. **Disk flowers** perfect, yellow; style branches with linear appendages. **Pappus** absent. **Achenes** 3- to 4-angled, glabrous.

1. Basal leaf blades sagittate, entire. . . . . *B. sagittata*
1. Basal leaf blades pinnately lobed 2

2. Leaf blades white- or silvery-tomentose . . . . *B. incana*
2. Leaves green, hispid to villous but not tomentose 3

3. Outer phyllaries surpassing the inner; leaf blades 30–60 cm long . . . .*B. macrophylla*
3. Outer phyllaries usually as long or shorter than the inner; leaf blades 15–25 cm long . . . . *B. hookeri*

Balsamorhiza hookeri Nutt. [*B. hispidula* W.M.Sharp] **Stems** ascending, 10–30 cm. **Herbage** puberulent. **Basal leaf blades** narrowly lanceolate, 8–20 cm long, pinnately divided into numerous toothed or lobed, lanceolate to ovate segments. **Involucre** hemispheric, 10–15 mm high; phyllaries lanceolate, ciliate, subequal. **Ray ligules** 15–40 mm long. Sagebrush steppe; montane. WA to MT south to CA, AZ, CO. Our plants are var. *hispidula* (Sharp) Cronquist. Known from Beaverhead and Deer Lodge cos.

Balsamorhiza incana Nutt. **Stems** ascending, 15–50 cm. **Herbage** tomentose. **Basal leaf blades** lanceolate, 5–25 cm long, pinnately divided into numerous entire to shallowly lobed, lanceolate to ovate segments. **Involucre** hemispheric, 12–25 mm high; phyllaries lanceolate, tomentose, outer shorter than the inner. **Ray ligules** 25–35 mm long. **Disk corollas** 6–9 mm long, lobes hairy to glandular. Grasslands, rarely sagebrush steppe; valleys, montane. WA, MT, OR, ID, WY.

Plants from Carbon Co. are less tomentose than typical.

Balsamorhiza macrophylla Nutt. **Stems** ascending, 30–100 cm. **Herbage** sparsely villous, sometimes sparsely glandular. **Basal leaf blades** lanceolate to ovate, 30–60 cm long, pinnately lobed into several lobed, obovate to ovate leaflets. **Involucre** hemispheric, 12–40 mm high, 20–30 mm across; phyllaries lanceolate, ciliate, ca. equal. **Ray ligules** 35–50 mm long. **Disk corollas** ca. 15 mm long. Grasslands, dry meadows; montane. ID, MT, UT, WY. Known from Beaverhead and Gallatin cos.

Balsamorhiza sagittata (Pursh) Nutt. ARROWLEAF BALSAMROOT **Stems** erect to ascending, 15–80 cm. **Herbage** strigose to tomentose. **Basal leaf blades** sagittate, 10–30 cm long, entire. **Involucre** campanulate, 15–22 mm high, 12–25 mm across; phyllaries lanceolate, tomentose, outer as long or longer than the inner. **Ray ligules** 20–40 mm long. **Disk corollas** 6–9 mm long. Grasslands, sagebrush steppe; valleys, montane. BC, AB south to CA, AZ, CO, SD. (p.507)

Stems and leaves expand after flowering. Collections from Carbon and Lewis & Clark cos. have deeply lobed sagittate leaves and may be hybrids with *Balsamorhiza incana*. The type locality is northeast of Lincoln.

Bellis L. Daisy

Bellis perennis L. ENGLISH DAISY Fibrous-rooted perennial. **Herbage** sparsely strigose. **Leaves** all basal, petiolate, spatulate, entire to serrate, 2–6 cm long. **Inflorescence** a solitary head on an erect peduncle 5–15 cm. **Heads** radiate; involucre 4–6 mm high; phyllaries in 1 series, narrowly ovate, strigose; receptacle conical, naked. **Rays** white, female; ligules 4–9 mm long. **Disk flowers** yellow, perfect; corolla 1–2 mm long; style branch appendages deltoid. **Pappus** absent. **Achenes** 1–2 mm long, glabrous, obconic. Lawns; valleys. Introduced to much of the northern U.S.; native to Eurasia; collected in Lake and Missoula cos.

Bidens L. Beggar's ticks

Annual or perennial herbs. **Leaves** cauline, opposite, simple, lobed or dissected. **Inflorescence** corymbiform, few-flowered. **Heads** radiate or discoid; involucres hemispheric; phyllaries in 2 unequal series; the outer longer, leaflike; the inner membranous, purplish with hyaline margins; receptacle flat with scales similar to the phyllaries. **Ray flowers** yellow, usually sterile, often absent. **Disk flowers** yellow to reddish, perfect; style branches short, deltoid. **Pappus** of 2 to 4 usually barbed awns. **Achenes** flattened to obconic, 3- to 4-angled.

1. Plants aquatic with flaccid stems and filiform-dissected, submerged leaves . . . . *B. beckii*
1. Stems erect; leaves not filiform-dissected 2

2. Blades of all or most leaves simple 3
2. Blades of most leaves deeply divided 4

3. Rays 2–15 mm long; pappus usually of 4 awns . . . . *B. cernua*
3. Rays absent or inconspicuous; pappus usually of 3 awns . . . . *B. comosa*

4. Outer phyllaries 10 to 16 . . . .*B. vulgata*
4. Outer phyllaries 5 to 8 . . . . *B. frondosa*

Bidens beckii Torr. ex Spreng. [*Megalodonta beckii* (Torr. ex Spreng) Greene] Glabrous aquatic perennial with flaccid stems to 2 m long. **Leaves** sessile, 1–3 cm long, dimorphic; submersed leaves 1 to 3 times divided into filiform segments; emergent leaves becoming simple, lanceolate, serrate. **Inflorescence**: heads usually solitary; peduncles 2–10 cm long. **Involucre**: outer phyllaries oblanceolate, 5–10 mm long, glabrous; inner phyllaries 7–10 mm long. **Rays** 8; ligules 10–15 mm long. **Disk corollas** 5–6 mm long. **Pappus** 2 to 6 awns, 13–25 mm long. **Achenes** cylindric, 10–15 mm long, unarmed. Water to 2 m deep in lakes, sloughs, slow streams; valleys. BC to QC south to OR, MT, MO, PA, NY.

Nearly all of our specimens are sterile, suggesting that flowering is rare or occurs late in the growing season.

Bidens cernua L. Annual. **Stems** erect, often branched, 10–80 cm, often puberulent. **Herbage** glabrate. **Leaves** short-petiolate to sessile, lanceolate, serrate, 2–15 cm long. **Inflorescence**: heads few, nodding at maturity; peduncles 1–4 cm long. **Involucre**: outer phyllaries 5 to 8, spreading to reflexed, linear-oblanceolate, 15–40 mm long, minutely ciliate; inner phyllaries narrowly ovate, 7–12 mm long. **Rays** 6 to 8 (absent); ligules 2–15 mm long. **Disk corollas** 3–4 mm long. **Pappus** of usually 4 awns 2–4 mm long. **Achenes** 4-angled, compressed, 5–6 mm long. Vernally inundated, often disturbed shores of streams, ponds, lakes, marshes; plains, valleys. Circumboreal, throughout N. America. (p.507)

Bidens comosa (A.Gray) Wiegand [*B. tripartita* L. misapplied] Annual. **Stems** erect, 10–60 cm. **Herbage** glabrous. **Leaves** short-petiolate, 3–15 cm long, simple, lanceolate, entire to serrate or shallowly lobed. **Inflorescence**: heads 1 to 3; peduncles 3–15 cm long. **Involucre**: outer phyllaries spreading, linear-lanceolate to -oblanceolate, 15–45 mm long, entire to serrate, shallowly lobed, minutely ciliate; inner phyllaries ovate, 6–10 mm long. **Rays** absent or 1 to 5, reduced. **Disk corollas**

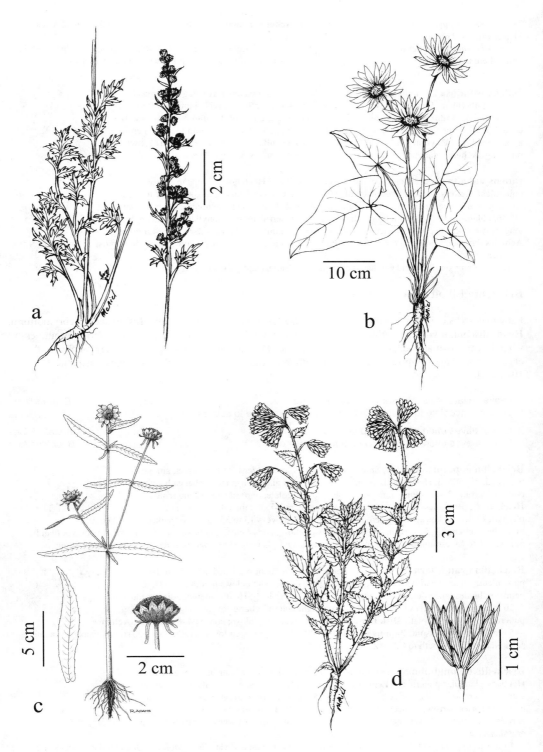

Plate 91. a. *Artemisia michauxiana*, b. *Balsamorhiza sagittata*, c. *Bidens cernua*, d. *Brickellia grandiflora*

2–4 mm long. **Pappus** usually 3 awns, 3–5 mm long. **Achenes** flattened, 6–9 mm long. Streambanks, fields, roadsides; plains, valleys. Throughout much of temperate N. America.

Bidens comosa is sometimes considered conspecific with the Eurasian *B. tripartita*. *Bidens connata* Mulh. ex Willd., with tuberculate achenes and also part of this complex, is reported for MT,[387] but I have seen no specimens.

Bidens frondosa L. Annual. **Stems** erect, 20–80 cm, often branched. **Herbage** glabrous to sparsely puberulent. **Leaves** distinctly petiolate; blades 1–8 cm long, divided into 3 lanceolate, serrate leaflets. **Inflorescence**: heads several; peduncles 1–8 cm long. **Involucre**: outer phyllaries ca. 8, linear-oblanceolate, 1–2 cm long, ciliate; inner phyllaries ovate, 4–8 mm long with erose, appendaged tips. **Rays** absent or 3 to 5, reduced. **Disk corollas** 2–3 mm long. **Pappus** 2, awns 2–4 mm long. **Achenes** flattened, 6–12 mm long. Riverbanks; plains, valleys. Throughout N. America.

Bidens vulgata Greene Annual. **Stems** erect, 30–100 cm. **Herbage** glabrous to sparsely puberulent. **Leaves** distinctly petiolate; blades divided into 3 or 5 lanceolate, serrate leaflets, 1–8 cm long. **Inflorescence**: heads usually few; peduncles 2–12 cm long. **Involucre**: outer phyllaries 10–16, oblanceolate, 8–30 mm long, ciliate, sometimes serrulate; inner phyllaries ovate, 6–8 mm long. **Rays** absent or 3 to 5, reduced. **Disk corollas** 2–3 mm long. **Pappus** 2, awns 2–4 mm long. **Achenes** flattened, 5–12 mm long. Banks of ponds, marshes, slow streams; plains. Throughout most of temperate N. America.

A specimen from Richland Co. has sharply dentate leaflets and puberulent phyllaries.

Brickellia Elliott Thoroughwort

Perennial herbs or subshrubs from a branched caudex. **Leaves** cauline, simple. **Inflorescence** corymbiform. **Heads** discoid; involucres turbinate to campanulate; phyllaries well-imbricate in 3 to 7 series, prominently veined, scarious-margined; receptacle flat, glabrous. **Disk flowers** whitish, perfect; style branches with clavate appendages. **Pappus** 1 series of smooth to plumose, capillary bristles. **Achenes** prismatic, 10-ribbed.

1. Leaves ovate to deltoid, serrate, long-petiolate . . . .*B. grandiflora*
1. Leaves lanceolate to oblanceolate, mostly entire, sessile to subsessile 2

2. Lower leaves often dentate; pappus plumose . . . .*B. eupatorioides*
2. Lower leaves entire; pappus smooth to barbellate. . . . . *B. oblongifolia*

Brickellia eupatorioides (L.) Shinners [*Kuhnia eupatorioides* L.] **Stems** erect, simple or branched, 20–60 cm. **Herbage** short-puberulent. **Leaves** glandular-punctate beneath, subsessile, entire to serrate, 2–4 cm long, linear-lanceolate to lanceolate, opposite below, alternate above. **Heads** erect; peduncles 5–20 mm long. **Involucres** narrowly campanulate, 6–12 mm high; phyllaries linear to lanceolate, green, puberulent. **Disk flowers** 15 to 35, cream to pinkish; corolla 6–9 mm long. **Achenes** 3–5 mm long, hairy. **Pappus** plumose. Grasslands, sagebrush steppe, pine or juniper woodlands; plains. MT to MA south to AZ, TX, AL, FL, Mexico. Our plants are var. *corymbulosa* (Torr. & A.Gray) Shinners.

Brickellia grandiflora (Hook.) Nutt. **Stems** erect, simple or branched, 25–80 cm. **Herbage** puberulent; leaves glandular-punctate beneath. **Leaves** opposite below, alternate above, long-petiolate; blades lance-deltoid, subcordate, serrate, 2–7 cm long. **Heads** often nodding; peduncles 4–10 mm long. **Involucres** turbinate, 5–12 mm high; outer phyllaries ovate, long-acuminate, green, puberulent; inner lanceolate. **Disk flowers** 20 to 40, cream to pinkish; corolla 4–8 mm long. **Achenes** 4–5 mm long, short-hispid. **Pappus** short-barbed. Rocky soil of open forest, eroding slopes, stream banks; montane, subalpine. BC, AB south to CA, AZ, NM, TX. (p.507)

Brickellia oblongifolia Nutt. **Stems** ascending to erect, usually branched, 15–30 cm. **Herbage** glandular-puberulent. **Leaves** mostly alternate, subsessile, lanceolate to oblanceolate, entire, 1–3 cm long. **Heads** erect; peduncles 5–15 mm long. **Involucres** campanulate, 10–15 mm high; phyllaries narrowly lanceolate, green, glandular. **Disk flowers** 25 to 50, cream to purplish; corolla 6–10 mm long. **Achenes** 3–7 mm long, glandular. **Pappus** barbed. Sparsely vegetated soil of grasslands, sagebrush steppe; valleys. BC south to CA, AZ, NM.

Both var. *linifolia* (D.C.Eaton) B.L.Rob., with setose achenes, and var. *oblongifolia,* with glandular achenes, are reported for MT.[354] Achenes of our plants are both glandular and setose.

Canadanthus G.L.Nesom New World aster

Canadanthus modestus (Lindl.) G.L.Nesom [*Aster modestus* Lindl.] Rhizomatous perennial.
Stems erect, 30–80 cm. **Herbage** glandular-puberulent. **Leaves** cauline, sessile, lanceolate,
weakly serrate to entire, 4–10 cm long, clasping above, the lowest deciduous. **Inflorescence**
leafy-corymbiform, often short with few heads; peduncles stipitate-glandular. **Heads** radiate;
involucre narrowly campanulate, 6–10 mm high; phyllaries subequal, linear-lanceolate, glandular,
green or purplish, the outer foliaceous. **Ray flowers** female, 20 to 65, purple; ligules 7–12 mm long. **Disk flowers** 40
to 65, perfect, whitish to purplish; corolla 5–7 mm long; style branches deltoid. **Pappus** of numerous capillary bristles,
brownish. **Achenes** fusiform, 4- to 9-nerved, 2–4 mm long, sparsely strigose. Moist meadows, thickets, open forests,
especially along streams, wetlands; valleys, montane. AK to ON south to OR, MT, MN.

Carduus L. Thistle

Annual or biennial. **Stems** erect, spiny-winged. **Leaves** basal and alternate, pinnately lobed, spiny on
margins. **Inflorescence**: heads solitary to few in racemose arrays. **Heads** discoid; involucres hemispheric;
phyllaries numerous in 7 to 10 series, the outer spine-tipped; receptacle flat, bristly. **Disk flowers** perfect,
purple (white), narrow, linear-lobed; style branches short, without appendages. **Pappus** minutely barbed,
capillary bristles. **Achenes** ovoid, glabrous.
   Both of our species were introduced from Eurasia.

1. Involucral bracts ≥2 mm wide; involucre 2–4 cm high; heads usually solitary . . . . *C. nutans*
1. Involucral bracts <2 mm wide; involucre ≤2 cm high; heads often several per stem . . . .*C. acanthoides*

Carduus acanthoides L. **Stems** 30–150 cm. **Herbage** sparsely villous on midveins. **Leaves**:
basal wing-petiolate; blades oblanceolate, deeply pinnately lobed, 10–20 cm long; cauline sessile,
reduced upward. **Inflorescence** with several erect heads on spiny stems; peduncles 1–2 cm long.
Involucres 14–20 mm high; phyllaries linear-lanceolate, appressed, the outer spine-tipped, inner
purple, long-attenuate. **Disk corollas** 12–20 mm long. **Achenes** 2–3 mm long. Fields, roadsides,
disturbed grasslands; valleys. Introduced to northern U.S.

Carduus nutans L. Musk thistle **Stems** 30–150 cm. **Herbage** glabrate to villous. **Leaves**:
basal wing-petiolate; blades oblanceolate, pinnately lobed, 6–20 cm long; cauline sessile, reduced
upward. **Inflorescence** of solitary, sometimes nodding heads on spiny peduncles 2–30 cm long.
Involucres 2–4 cm high; phyllaries lanceolate, outer reflexed; inner unarmed. **Disk corollas** 15–25
mm long. **Achenes** 4–5 mm long. Fields, roadsides, disturbed grasslands, riparian meadows; valleys,
montane. Introduced throughout N. America

Carthamus L.

Carthamus tinctorius L. Safflower Glabrous annual. **Stems** erect, branched, 30–100 cm.
Leaves cauline, alternate, 2–10 cm long, oblanceolate to ovate with spiny-dentate margins,
sessile. **Heads** discoid, few in a cymose inflorescence. **Involucre** ovoid, 15–25 mm high;
phyllaries lanceolate, in several well-imbricate series, the outer foliaceous, the inner lance-linear,
spine-tipped, shorter than the outer; receptacle convex, bristly. **Disk flowers** perfect, reddish-
yellow; corolla 2–3 cm long; the tube slender with an expanded limb; style branches fused. **Pappus** absent or of linear
scales. **Achenes** oblong, 4-angled, 6–7 mm long, glabrous. Fields, roadsides; valleys. Cultivated for oil and sparingly
introduced throughout much of western N. America; native to southern Europe, probably not persistent in MT.

Centaurea L. Knapweed

Annual to perennial, taprooted herbs. **Leaves** basal and alternate, simple, often lobed. **Inflorescence**: heads
solitary or in corymbiform arrays. **Heads** discoid, though sometimes appearing radiate; involucres ovoid
to hemispheric; phyllaries well-imbricate in numerous series, the tips with fringed or spiny appendages;
receptacle flat, bristly. **Disk flowers** perfect, the outer often sterile and asymmetrical; style branches short,
united with a thickened ring at the base. **Pappus** absent or deciduous, capillary bristles. **Achenes** ellipsoid,
glabrate, sometimes with an elaiosome.

All of our species are introduced from Europe or N. Africa. *Centaurea montana* is reported for MT,[212] but I am not aware of any naturalized populations. A plant that appears to be *C. dealbata* Willd. was collected once along a road in Mussellshell Co.

1. Flowers white or yellow 2
1. Flowers pink to purple 4

2. Flowers white . . . . *C. diffusa*
2. Flowers yellow 3

3. Outer involucral bracts with a spine >1 cm long. . . . . *C. solstitialis*
3. Outer involucral bracts with a fringed appendage, unarmed . . . . *C. macrocephala*

4. Pappus absent . . . . *C. jacea*
4. Pappus of short to long bristles, (sometimes easily detached) 5

5. Involucral bracts tipped with a spine 1–3 mm long. . . . . *C. virgata*
5. Tips of involucral bracts fringed or pectinate but not spine-tipped 6

6. Plants annual . . . . *C. cyanus*
6. Plants perennial 7

7. Involucres 8–13 mm high; outer bracts fringed above the middle only . . . .*C. maculosa*
7. Involucres 12–25 mm; outer bracts fringed below mid-length 8

8. Flowers rose, the outer enlarged; pappus bristles 4–5 mm long . . . . *C. scabiosa*
8. Flowers purple, outer usually not enlarged; pappus bristles ca. 1 mm long. . . . . *C. nigra*

Centaurea cyanus L. CORNFLOWER, BACHELOR'S BUTTONS Annual. **Stems** erect, branched, 20–60 cm. **Herbage** sparingly tomentose. **Leaves** petiolate below; blades linear-oblanceolate, 2–10 cm long, entire or with few linear lobes. **Inflorescence** corymbiform; heads solitary on branch tips. **Involucres** campanulate, 12–16 mm high; outer phyllaries ovate, green with a white to brown, scarious fringe at the tip; inner lanceolate with a blunt, white-fringed tip. **Disk flowers** 25 to 35, blue; the outer enlarged, resembling 5-lobed rays; inner corollas 11–15 mm long. **Achenes** 4–6 mm long. **Pappus** bristles in 2 series. Roadsides; valleys. Introduced ornamental sparingly escaped throughout most of N. America.

Centaurea diffusa Lam. DIFFUSE KNAPWEED Annual to rarely short-lived perennial. **Stems** ascending to erect, branched, 30–50 cm. **Herbage** puberulent to sparsely tomentose, glandular-punctate. **Leaves** petiolate below, the lowest deciduous; blades oblanceolate, 1–15 cm long, deeply pinnate to bipnnate into linear lobes. **Inflorescence** paniculate. **Involucres** ovoid, 10–13 mm high; outer phyllaries ovate, striate, green, tan above, spiny-margined; inner linear-lanceolate. **Disk flowers** 25 to 35, white, sometimes purple above; the outer slender, inconspicuous; inner corollas 12–13 mm long. **Achenes** 2–3 mm long. **Pappus** absent or inconspicuous. Stony soils of roadsides; valleys. Introduced to most of northern U.S. and adjacent Canada.
    Centaurea virgata has smaller heads.

Centaurea jacea L. [*C. pratensis* Thuill. in part] BROWN KNAPWEED Perennial. **Stems** erect, simple or branched, 20–80 cm. **Herbage** hispid to puberulent. **Leaves** long-petiolate below; blades oblanceolate, 4–15 cm long, sparingly dentate. **Inflorescence** corymbiform; heads few. **Involucres** hemispheric, 13–18 mm high; phyllaries ovate, brownish with scarious, white, lacerate, terminal appendages. **Disk flowers** 40 to 100, purple; the outer enlarged; inner corollas 15–18 mm long. **Achenes** 2–3 mm long. **Pappus** absent. Roadsides, fields; valleys. Introduced in western and northeast U.S. and Canada. See *C. nigra*.

Centaurea macrocephala Muss.Puschk.ex Willd. Perennial. **Stems** erect, little branched, 50–150 cm. **Herbage** short-villous, thinly arachnoid. **Leaves** petiolate below; blades oblanceolate, 5–30 cm long, entire or dentate; the upper wavy-margined, clasping to decurrent. **Heads** solitary. **Involucres** hemispheric, ca. 2–3 cm high; outer phyllaries obovate, green to tan with a brown, scarious, highly fringed appendage wider than the bract. **Disk flowers** numerous, yellow; the outer with a slightly expanded limb; the inner ca. 2 cm long. **Achenes** 7–8 mm long. **Pappus** of many flattened bristles. Grasslands; valleys, plains; Gallatin and Pondera cos. Introduced ornamental sparingly escaped in northwest and northeast N. America.

Centaurea maculosa Lam. [*C. stoebe* L. ssp. *micranthos* (S.G.Gmel. ex Gugler) Hayek, *C. biebersteinii* DC.] SPOTTED KNAPWEED Perennial. **Stems** erect, branched, 25–100 cm. **Herbage** sparsely tomentose, glandular-punctate. **Leaves** long-petiolate below; blades ovate, 3–12 cm long, deeply pinnately 1 to 2 times divided into linear-oblanceolate lobes. **Inflorescence** corymbiform with several heads. **Involucres** ovoid, 8–13 mm high; phyllaries green, striate, sparsely glandular; the outer ovate with white to brown fringed tips; inner lanceolate with swollen tips. **Disk flowers** 30 to 40, rose-purple (white); the outer enlarged, asymmetrical; inner corollas 12–15 mm long. **Achenes** 3–4 mm long. **Pappus** of deciduous, stiff bristles. Grasslands, roadsides, meadows, open forest, woodlands; plains, valleys, montane. Introduced throughout temperate N. America. (p.519)

Centaurea maculosa sensu lato consists of diploid and polyploid races; the former are biennial, while the latter (*C. stoebe*) are perennial. Differences in nomenclature result partly from this dichotomy. All populations introduced into N. America are thought to be the latter.[311]

Centaurea nigra L. [including *C. pratensis* Thuill. in part, *C.* ×*moncktonii* C.E.Britton] Perennial. **Stems** erect, branched, 25–80 cm. **Herbage** puberulent to sparsely tomentose. **Leaves** petiolate below; blades oblanceolate, 3–8 cm long, sparingly dentate. **Inflorescence** corymbiform; heads few. **Involucres** broadly campanulate, 13–18 mm high; phyllaries with black tips; the outer ovate with wiry-pectinate terminal appendages; inner oblanceolate with scarious, crenate margins. **Disk flowers** 40 to 100, purple; the outer rarely enlarged; inner corollas 15–18 mm long. **Achenes** 2–3 mm long. **Pappus** of short, black bristles. Roadsides; valleys. Introduced to western and northeastern U.S. and adjacent Canada.

Centaurea nigra and *C. jacea* form a hybrid complex, and most likely all of our specimens are some sort of hybrid between these two. These hybrids have often been called *C. pratensis*. Our material can be divided into two forms: *C. jacea* with lacerate, hylaine appendages on the outer phyllaries and *C. nigra* with wiry pectinate appendages.[212] One specimen from Glacier National Park has smaller heads and more deeply lobed leaves, possibly the result of introgression with *C. maculosa*. See *C. scabiosa*.

Centaurea scabiosa L. BROWN KNAPWEED Perennial. **Stems** erect, branched, 30–100 cm. **Herbage** glabrate. **Leaves** petiolate; blades lanceolate to ovate, 5–25 cm long, pinnately lobed, more deeply so above. **Inflorescence** with few heads. **Involucres** broadly campanulate, 12–25 mm high; phyllaries green with dark, scarious, pectinately branched appendages at the tip; the outer ovate; inner oblanceolate. **Disk flowers** numerous, rose-purple; the outer enlarged; inner corollas 20–25 mm long. **Achenes** 4–5 mm long, puberulent. **Pappus** of numerous bristles. Roadsides, streambanks; valleys; collected in Missoula and Lewis & Clark cos. Introduced in western and northeast U.S. and Canada.

Centaurea nigra and *C. jacea* have nearly entire leaves.

Centaurea solstitialis L. YELLOW STAR ROMANZOFFIA THISTLE Annual. **Stems** winged, erect, simple or   branched above, 10–80 cm. **Herbage** scabrous, tomentose. **Leaves** short-petiolate, deciduous below; blades linear-oblanceolate, 5–15 cm long, the lower pinnately lobed. **Inflorescence** solitary heads in corymbiform arrays. **Involucres** ovoid, 13–17 mm high; phyllaries pale green; the outer ovate, the tips with a long, spreading spine and several smaller bristles; inner lanceolate with swollen tips. **Disk flowers** yellow; the outer and inner similar; corollas 13–20 mm long. **Achenes** 2–3 mm long. **Pappus** of bristles or absent. Fields, grasslands, roadsides; valleys. Introduced throughout most of U.S. and southern Canada.

Centaurea virgata Lam. Biennial to short-lived perennial. **Stems** erect, branched, 20–50 cm. **Herbage** puberulent to sparsely tomentose, glandular-punctate. **Leaves** petiolate below, the lowest deciduous; blades oblanceolate, 2–10 cm long, deeply pinnate into linear lobes. **Inflorescence** paniculate with numerous heads; peduncles short. **Involucres** ovoid, 6–9 mm high; outer phyllaries ovate, striate, tan, spreading, spiny-margined; inner linear-lanceolate with scarious lacerate tips. **Disk flowers** 10 to 14, pink; the outer slender, 3-lobed; corollas 7–9 mm long, as long as the inner. **Achenes** glabrous, 2–3 mm long. **Pappus** of short bristles, deciduous. Fields, roadsides; valleys. Introduced to OR and MT south to CA, NV, UT. Known from Jefferson and Lewis & Clark cos. See *C. diffusa*.

Chaenactis DC. Dusty maiden

Chaenactis douglasii (Hook.) Hook. & Arn. [*C. alpina* (A.Gray) M.E.Jones] Biennial to perennial herbs. **Stems** ascending to erect, 1–50 cm. **Herbage** tomentose, less so with age, sometimes glandular above. **Leaves** basal and cauline, petiolate; blades oblong, 1–8 cm long, twice pinnately dissected into small oblong lobes. **Heads** discoid, 1 to several in corymbose arrays; peduncles 1–8 cm long. **Involucres** campanulate 8–15 mm high; phyllaries linear-lanceolate, green, 10 to 25, puberulent to tomentose, glandular. **Disk flowers** perfect, white; corolla 5–7 mm long, the throat longer than the tube; style branches elongate; appendages inconspicuous. **Pappus** of 4 to 20 hyaline oblong scales. **Achenes** clavate, 5–8 mm long, hirsute. Stony or sandy, poorly vegetated soil. BC, ND south to CA, AZ, NM, SD.

1. Plants with a simple caudex; stems leafy; heads usually >2 . . . . var. *douglasii*
1. Most plants mat-forming with a branched caudex; leaves all basal; heads 1 or 2 . . . . var. *alpina*

Chaenactis douglasii var. *alpina* Gray is found in alpine fellfields, open slopes; *C. douglasii* var. *douglasii* occurs in grasslands, sagebrush steppe, woodlands, open forest, talus; plains, valleys, montane. The two vars. have often been considered separate species, but the differences are most likely environmentally induced since some populations of var. *alpina* have plants with a simple caudex and/or >2 heads.

Chondrilla L.

Chondrilla juncea L. RUSH SKELETONWEED Taprooted perennial with milky sap. **Stems** ascending to erect, branched, 25–100 cm. **Herbage** hirsute to glabrate. **Leaves** basal and cauline; basal petiolate, oblong, pinnately lobed, sharply dentate, 5–13 cm long, deciduous; cauline alternate, linear-oblong, 2–10 cm long. **Heads** ligulate, short-peduncled in axils of upper reduced leaves. **Involucres** 8–12 mm high, sparsely tomentose, subtended by small bracts; phyllaries 5 to 9, linear-lanceolate in 1 series. **Ray flowers** perfect, yellow, 7 to 15; ligules 4–6 mm long. **Pappus** capillary bristles. **Achenes** cylindric, glabrous, ribbed, 3–4 mm long with a 5–6 mm beak. Roadsides, fields; valleys; collected in Sanders and Treasure cos. Introduced to northwest and northeast U.S. and adjacent Canada; native to Eurasia.

Chrysothamnus Nutt. Rabbitbrush

Chrysothamnus viscidiflorus (Hook.) Nutt. Shrubs. **Stems** spreading to erect, 10–50 cm, glabrous to puberulent; twigs brittle, white; bark gray. **Leaves** alternate, entire, linear, usually somewhat twisted, glabrous to puberulent, resinous, 1–4 cm long with 3 to 5 veins. **Inflorescence** tight-corymbiform with several heads on short peduncles. **Heads** discoid; involucres obconic, 4–9 mm high; phyllaries imbricate in 3 to 5 series, narrowly lanceolate, white to greenish; receptacle convex, glabrous. **Disk flowers** 4 to 8, perfect, yellow; corolla 4–7 mm long, tube shorter than throat; style branches linear. **Pappus** barbellate, capillary bristles. **Achenes** obconic, 2–4 mm long, 5-angled, pubescent. Sagebrush steppe, grasslands; valleys, montane. BC, MT south to CA, AZ, NM, NE.

1. Young twigs and leaves glabrate except for ciliate margins . . . . ssp. *viscidiflorus*
1. Young twigs and leaves puberulent . . . . ssp. *lanceolatus*

Chrysothamnus viscidiflorus ssp. *viscidiflorus* occurs throughout western MT; *C. viscidiflorus* ssp. *lanceolatus* (Nutt.) H.M.Hall & Clem. is found in the southwest. Plants from Lake Co. have the puberulent stems of ssp. *lanceolatus* but the glabrous leaves of ssp. *viscidiflorus*. Dwarf plants with solitary heads, densely puberulent leaves >3 mm wide were collected above treeline in the Centennial Mtns. Several species formerly placed in *Chrysothamnus* are now in *Ericameria* and *Lorandersonia*.

Cichorium L. Chicory

Cichorium intybus L. Taprooted perennial with milky sap. **Stems** erect, usually branched, 30–120 cm. **Herbage** glabrate to puberulent; stems hirsute below. **Leaves** basal and cauline; basal petiolate, oblanceolate, pinnately lobed, dentate, 8–20 cm long; cauline few, alternate, becoming auriculate-clasping. **Heads** ligulate, sessile in axils of reduced upper leaves. **Involucres** 9–15 mm high, glabrate to sparsely ciliate, cylindric; phyllaries imbricate in 2 series, lanceolate; receptacle flat, glabrous. **Ray flowers** perfect, 8 to 25, sky blue; ligules ca 15 mm long. **Pappus** of tiny scales. **Achenes** glabrous, 5-angled, 2–3 mm long. Roadsides, fields; valleys. Introduced throughout much of temperate N. America; native to Europe.

Cirsium Mill. Thistle

Biennial to perennial herbs. **Leaves** basal and alternate, dentate to pinnately lobed. **Herbage** spiny on stems and leaf margins, veins. **Inflorescence**: heads solitary to several in racemose arrays. **Heads** discoid, usually erect; involucres ovoid to hemispheric; phyllaries numerous in 5 to 20 series, the outer spine-tipped, sometimes with a darkened keel-like resin gland toward the tip; receptacle flat, bristly. **Disk flowers** perfect (1 exception), narrow, the throat little expanded, lobe linear, unequal; style swollen just below the almost completely fused branches. **Pappus** of plumose bristles. **Achenes** glabrous, compressed-ovoid sometimes with a thickened collar at the summit.

Plants described as biennial are better thought of as monocarpic; they may live for one to many years but flower only once before dying. *Cirsium* gets my vote for the most confusing genus in MT. Many of the species are connected to each other by intergradation.[210] The situation is made more difficult by the fact that recent treatments[103,115,211] are at odds on key characters. *Cirsium canovirens*, *C. pulcherrimum*, and *C. subniveum* compose one complex, and *C. eatonii*, *C. hookerianum* and *C. scariosum* are another. The presence of wings decurrent on the stem from leaf bases is variable both within and among species and does not seem like a good diagnostic character. In some species the size and even presence of the resinous keel on phyllaries seems to vary within the same population.

1. Involucres <2 cm high 2
1. Involucres of largest heads ≥2 cm high 3

2. Flowers perfect; leaves white-tomentose beneath. . . . . *C. pulcherrimum*
2. Flowers unisexual; leaves arachnoid-villous but green beneath . . . . *C. arvense*

3. Heads clustered in capitate or spike-like arrays; peduncles mostly ≤2 cm long. .4
3. Inflorescence more open; some peduncles >2 cm long. 11

4. Inner phyllaries attenuate, without dilated or erose tips 5
4. Upper part of inner phyllaries with scarious, erose margins. 9

5. Leaves densely tomentose beneath, dark green and glabrate above . . . . *C. pulcherrimum*
5. Leaves not as above 6

6. At least 1 involucre >30 mm high . . . . *C. brevistylum*
6. Involucres ≤30 mm high 7

7. Outer phyllaries glabrate or arachnoid on the margins; stem often pinkish . . . . *C. scariosum*
7. Outer phyllaries arachnoid on margins and across the face; stems not pink. 8

8. Leaves often thinly tomentose beneath; upper leaves often enclosing involucres;
   plants not very succulent . . . . *C. eatonii*
8. Leaves densely tomentose beneath; uppermost little obscuring the involucre;
   plants succulent . . . . *C. hookerianum*

9. Phyllaries lacking darkened, resinous keel. . . . . *C. scariosum*
9. Outer phyllaries usually with a raised, darkened, resinous keel. 10

10. Outer phyllaries with erose margins; upper montane, subalpine . . . . *C. longistylum*
10. Outer phyllaries with entire margins; low elevations. . . . .*C. canescens*

11. Leaves green beneath . . . . *C. vulgare*
11. Leaves tomentose beneath 12

12. At least inner phyllaries with expanded, scarious, erose tips 13
12. Inner phyllaries acuminate, not expanded or erose 14

13. Outer phyllaries with erose margins; upper montane to subalpine. . . . . *C. longistylum*
13. Outer phyllaries with entire margins; low elevations. . . . .*C. canescens*

14. Leaves white-tomentose on the upper surface making it gray. . . . . *C. undulatum*
14. Leaves glabrate or sparsely arachnoid above, the green surface not obscured .15

15. Involucral bracts (especially the inner) with minute sessile glands . . . . *C. pulcherrimum*
15. Involucral bracts without minute glands 16

16. Perennial, spreading by horizontal roots; plants of plains and valleys. *C. flodmanii*
16. Plants taprooted, monocarpic; montane to subalpine . . . .*C. canovirens*

Cirsium arvense (L.) Scop. CANADA THISTLE Strongly rhizomatous perennial; individual stems unisexual. **Stems** erect, often branched above, 30–100 cm. **Herbage** glabrate; leaves sometimes tomentose beneath. **Leaves** short-petiolate, oblanceolate to elliptic, 3–15 cm long, dentate to deeply pinnate; lowest usually deciduous. **Inflorescence** few to several heads per stem in a corymbiform array; peduncles 0–4 cm long. **Involucres** 1–2 cm high; phyllaries imbricate in 6 to 8 series; the outer ovate with a darkened, resinous keel-tip; inner linear; spines short or absent. **Disk corollas** purple; male 10–16 mm long, longer than the pappus; female 14–18 mm long, shorter than mature pappus. **Achenes** 2–4 mm long; collar absent. Moist, usually disturbed soil of fields, meadows, thickets, roadsides, woodlands, open forests, often along streams, wetlands; plains, valleys, montane. Introduced throughout temperate N. America; native to Eurasia.

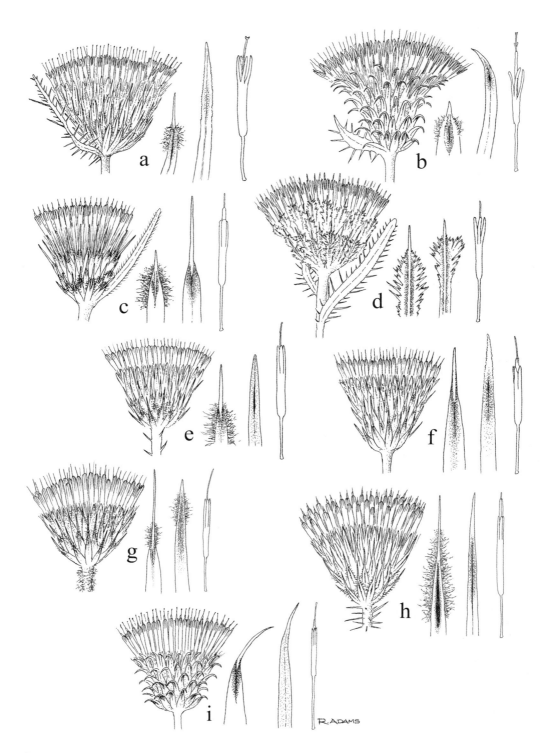

Plate 92. **Cirsium**. **a.** *C. pulcherrimum-1*, **b.** *C. flodmanii*, **c.** *C. hookerianum*, **d.** *C. longistylum*, **e.** *C. pulcherrimum-2*, **f.** *C. scariosum*, **g.** *C. canovirens*, **h.** *C. tweedyi*, **i.** *C. undulatum*

Cirsium brevistylum Cronquist Taprooted biennial. **Stems** erect, often unbranched, 30–200 cm. **Herbage** sparsely villous or tomentose. **Leaves** short-petiolate, oblanceolate to lanceolate, 8–35 cm long, pinnately lobed. **Inflorescence** several heads clustered at the stem tip; peduncles 5–10 mm long. **Involucres** hemispheric, 25–35 mm high, arachnoid; phyllaries subequal, in 5 to 10 series, linear-lanceolate, green, unkeeled; spines 2–4 mm long. **Disk corollas** white to purple, ca. 25 mm long. **Achenes** 3–5 mm long. Meadows, moist forest openings, burned or logged forests; valleys, montane. BC, MT south to CA, ID.

Heads of *Cirsium scariosum* are not as large.

Cirsium canescens Nutt. Taprooted biennial. **Stems** erect, mostly unbranched, 20–100 cm. **Herbage** tomentose. **Leaves** short-winged-petiolate, oblong to elliptic, 10–25 cm long, pinnately lobed. **Inflorescence** few to several heads on stem tips or upper leaf axils; peduncles 0–10 cm long. **Involucres** hemispheric, 3–4 cm high, sparsely arachnoid; phyllaries imbricate in 6 to 9 series, lanceolate, green, resinous-keeled; the outer appressed; the inner expanded, erose; spines 2–4 mm long. **Disk corollas** white to pinkish, 20–35 mm long. **Achenes** 5–7 mm long. Grasslands, woodlands; plains. MT and ND south to CO, NE, IA. Reported for Garfield, Sheridan and Wibaux cos.[151]

Cirsium canovirens (Rydb.) Petr. [*C. subniveum* Rydb., *C. cymosum* (Greene) J.T.Howell var. *canovirens* (Rydb.) D.J.Keil, *C. inamoenum* (Greene) D.J. Keil] Taprooted biennial. **Stems** erect, branched above or not, 30–100 cm, sometimes with spiny wings from leaf bases. **Herbage** tomentose to arachnoid; upper leaf surfaces sometimes sparsely so. **Leaves** short-petiolate; blade linear-oblanceolate, 8–25 cm long, pinnately lobed, becoming clasping or decurrent upward. **Inflorescence** heads 1 to few at tips of peduncles forming corymbiform arrays; peduncles 1–8 cm long. **Involucres** broadly campanulate, 15–30 mm high, sparsely arachnoid-tomentose; phyllaries imbricate in 6 to 10 series, green, linear-lanceolate, sometimes with a darkened, often resinous midvein; spines 2–6 mm long, spreading. **Disk corollas** white to light purple, 15–26 mm long. **Achenes** 5–8 mm long. Sagebrush steppe, grasslands, open forest; montane, subalpine. OR to MT to CA, NV, WY. (p.514)

Cirsium canovirens and *C. subniveum* have traditionally been considered distinct species based on the presence of darkened and resinous phyllary keels in the former.[103] I relocated the type locality of *Cirsium canovirens* in Madison Co. in 2011, and collected several specimens at two different times of year. The majority of the flowering heads on fresh specimens did not have darkened phyllary keels and would have been identified as *C. subniveum*. However, most of these same plants did have a darkened keel after drying. The similarity between these two species has been recognized previously,[213,265] and it seems most realistic to assume there is only one entity in our area since they cannot be reliably told apart. The older name is *C. canovirens*. See *C. pulcherrimum*.

Cirsium eatonii (A.Gray) B.L.Rob. [*C. tweedyi* (Rydb.) Petr., *C. polyphyllum* (Rydb.) Petr.] Taprooted perennial, often from a branched caudex. **Stems** erect, often unbranched, 10–70 cm. **Herbage** glabrate to arachnoid. **Leaves** short-petiolate; blade linear-lanceolate, 10–30 cm long with 10 to 20 pairs of shallow, closely-spaced lobes. **Inflorescence** capitate or spike-like with few, nearly sessile heads. **Involucres** broadly campanulate, 2–3 cm high, enclosed by uppermost leaves that obscure the phyllaries; phyllaries arachnoid especially on the margins, subequal in 4 to 5 series, green, linear-lanceolate; outer with spine tips 7–10 mm long; inner acuminate. **Disk corollas** white, 17–23 mm long. **Achenes** 5–7 mm long. Meadows, turf, fellfields; subalpine, alpine. OR to MT south to NV, UT, NM. Our plants are var. *murdockii* S.L.Welsh. (p.514)

Cirsium hookerianum is more succulent and not as spiny.

Cirsium flodmanii (Rydb.) Arthur Taprooted perennial, spreading by horizontal roots. **Stems** erect, often unbranched, 25–80 cm. **Herbage** tomentose; upper leaf surfaces becoming glabrate. **Leaves** winged-petiolate; blade oblanceolate to oblong, 4–25 cm long, deeply pinnately lobed to subentire. **Inflorescence** 1 to 2 heads at the stem tips; peduncles 5–25 mm long. **Involucres** broadly campanulate, 2–3 cm high, sparsely arachnoid; phyllaries imbricate in 7 to 12 series, green; outer lanceolate with a darkened, resinous keel-tip; inner linear; spines 2–4 mm long, spreading. **Disk corollas** white to purple, 2–3 cm long. **Achenes** 3–5 mm long. Grasslands, meadows, often along streams, wetlands; plains, valleys. AB to QC south to CO, NE, IL, MI. (p.514)

The colonial habit and more glabrate upper leaf surfaces help distinguish this species from *Cirsium undulatum*. The type locality is in Gallatin Co.

Cirsium hookerianum Nutt. ELK THISTLE Taprooted biennial. **Stems** erect, often unbranched, 25–90 cm. **Herbage** thinly to moderately arachnoid-tomentose; lower leaf surfaces usually whitish. **Leaves** short-petiolate; blades linear-oblanceolate to linear-lanceolate, 8–30 cm long, dentate to pinnately lobed. **Inflorescence** capitate or spike-like with usually 3 to 12 subsessile heads subtended by leaf-like bracts; peduncles 2 cm long. **Involucres** campanulate, 2–3 cm high, arachnoid-tomentose across the phyllaries that are imbricate in 4 to 8 series, whitish to purplish, linear-lanceolate; sometimes with a darkened resinous keel; inner acuminate, ciliate; spines 3–5 mm long, erect. **Disk corollas** white to light purple, 20–25 mm long. **Achenes** 5–7 mm long. Meadows, grasslands, woodlands, open slopes; valleys to subalpine. BC, AB south to WA, ID, WY. (p.514)

Many plants, especially from higher elevations, have a compact inflorescence and appear very similar to *Cirsium scariosum* Nutt. which has more nearly glabrous phyllaries. *Cirsium hookerianum* and *C. longistylum* intergrade where they occur together in Fergus and Judith Basin cos. See *C. eatonii.*

Cirsium longistylum R.J.Moore & Frankton Taprooted biennial. **Stems** erect, sometimes branched, 40–100 cm. **Herbage** glabrate to arachnoid, densely tomentose beneath the leaves. **Leaves** petiolate; blades oblanceolate to linear-lanceolate, 6–20 cm long, pinnately lobed, becoming auriculate-clasping upward. **Inflorescence** spike-like to racemose with 3 to 15 heads subtended by clustered leaf-like bracts; peduncles 0–15 cm long. **Involucres** campanulate, 15–30 mm high, villous; phyllaries subequal in 4 to 8 series, green, lanceolate; some of the outer with a scarious, erose upper margin, sometimes with a darkened, resinous keel; spines 2–4 mm long, erect. **Disk corollas** white, 15–25 mm long. **Achenes** 5–7 mm long. Meadows, moist forest openings, roadsides; upper montane, lower subalpine. Endemic to north-central MT. See *Cirsium hookerianum.* (p.514)

Cirsium pulcherrimum (Rydb.) K.Schum. Taprooted biennial to perennial, sometimes with a branched caudex. **Stems** tomentose to arachnoid, erect, sometimes branched, 15–60 cm. **Herbage**: stems and leaf undersides densely white-tomentose; upper leaf surfaces glabrate, dark green. **Leaves** short-petiolate; blades linear to oblanceolate, 5–15 cm long, entire to pinnately lobed; upper leaf bases decurrent as wings on the stem. **Inflorescence** racemose, capitate above with clusters of 1 to 3 subsessile heads; peduncles 5–20 mm long. **Involucres** campanulate, 15–25 mm high, sparsely arachnoid; phyllaries imbricate in 6 to 7 series, green, linear-lanceolate; outer usually with a darkened resinous keel; inner acuminate, ciliate; spines 2–9 mm long, erect. **Disk corollas** light purple, 18–20 mm long. **Achenes** 5–6 mm long. Sagebrush steppe, grasslands, juniper woodlands; plains. MT south to UT, CO, NE. (p.514)

Cirsium canovirens lacks the minute phyllary glands; *C. flodmanii* does not have clustered heads. Plants from the Pryor Mtns. have more open inflorescences and less contrast between the upper and lower leaf surfaces than plants of the Great Plains.

Cirsium scariosum Nutt. ELK THISTLE Taprooted biennial. **Stems** erect, mostly unbranched, 15–80 cm. **Herbage** sparsely arachnoid-villous to arachnoid-tomentose, often purplish. **Leaves** petiolate below; blades linear to linear-oblanceolate, 10–40 cm long, shallowly to deeply pinnately lobed. **Inflorescence** capitate with ca. 5 to 20 subsessile heads subtended by leaf-like bracts mostly longer than the flowers; peduncles 0–1 cm long. **Involucres** ovoid to hemispheric, 15–30 mm high; phyllaries imbricate in 5 to 10 series, green, not keeled; outer lanceolate, glabrate or arachnoid on the margins with a spine 1–8 mm long; inner linear-lanceolate with a hispid-ciliate to scarious-erose tip. **Disk corollas** white to purple, 20–26 mm long. **Achenes** 4–7 mm long. Moist to wet, sometimes calcareous meadows, grasslands; valleys to subalpine. BC, AB south to CA, AZ, NM. Our plants are var. *scariosum.* (p.514)

Plants with erose-appendaged inner phyllary tips occur only in the southwest counties. See *C. brevistylum, C. hookerianum.*

Cirsium undulatum (Nutt.) Spreng. Taprooted biennial. **Stems** erect, often branched, 20–100 cm. **Herbage** arachnoid-tomentose throughout; stems and leaf undersides densely white-tomentose. **Leaves** petiolate; blade lanceolate to obovate, 5–20 cm long, undulate, deeply pinnately lobed to subentire. **Inflorescence**: heads mostly solitary at stem tips, forming open corymbiform arrays; peduncles 1–10 cm long. **Involucres** ovoid to hemispheric, 25–35 mm high; phyllaries imbricate in 8 to 12 series, green; outer lanceolate, sparsely arachnoid mainly on the margins with a darkened, resinous keel and spreading spine tip 3–5 mm long; inner linear-lanceolate, acuminate. **Disk corollas** pink to purple, 25–40 mm long. **Achenes** 6–7 mm long. Grasslands, sagebrush steppe, roadsides; plains, valleys, lower montane. BC to MB south to CA, AZ, TX, MO, Mexico. See *C. flodmanii.* (p.514)

Cirsium vulgare (Savi) Ten. BULL THISTLE Taprooted biennial. **Stems** erect, often branched, 30–120 cm. **Herbage** green, spreading-hirsute to sparsely arachnoid. **Leaves** short-petiolate; blades lanceolate to obovate, 5–30 cm long, deeply pinnately lobed with prominent whitish veins beneath. **Inflorescence** heads mostly solitary at stem tips, forming open corymbiform arrays; peduncles 1–8 cm long. **Involucres** campanulate, 25–40 mm high, sparsely arachnoid; phyllaries imbricate in 10 to 12 series, green, linear-lanceolate; outer keeled but not resinous, with an erect or spreading spine tip 2–10 mm long; inner linear, acuminate. **Disk corollas** purple, 2–3 cm long. **Achenes** 3–5 mm long. Disturbed meadows, thickets, roadsides; plains, valleys, montane. Introduced throughout N. America; native to Eurasia.

Conyza Less. Horseweed

Conyza canadensis (L.) Cronquist [*Erigeron canadensis* L.] CANADA FLEABANE Taprooted annual. **Stems** erect, usually unbranched, 10–100 cm. **Herbage** glabrate to sparsely hirsute, especially on the leaf margins. **Leaves** cauline, alternate, linear-oblanceolate, 2–8 cm long, short-petiolate, dentate below. **Inflorescence** narrow to corymbiform, paniculate, leafy-bracted below. **Heads** numerous, radiate but apparently discoid; involucres campanulate, 2–5 mm high; phyllaries weakly imbricate in 2 to 4 series, green with a tan center and scarious margins, glabrate, linear-lanceolate; receptacle flat, glabrous. **Ray flowers** 20 to 30, female, white to pink; ligules ≤1 mm long. **Disk flowers** perfect, 8 to 25, yellow; corolla 2–3 mm long, tube shorter than the throat; style branch appendages deltoid. **Pappus** of capillary bristles. **Achenes** oblong, compressed, ca. 1 mm long, sparsely strigose. Disturbed meadows, grasslands, roadsides; plains, valleys. Native throughout much of temperate N. America.

Coreopsis L.

Coreopsis tinctoria Nutt. [*C. atkinsoniana* Dougl. ex Lindl.] Glabrous, taprooted annual or biennial. **Stems** erect, usually branched, 20–80 cm. **Leaves** cauline, opposite, petiolate; blades ovate, 2–10 cm long, 1 to 2 times deeply divided into 3 to 7 well-separated, long, linear to oblanceolate segments. **Inflorescence** open, corymbiform, linear-bracted; peduncles 1–10 cm long. **Heads** radiate; involucres campanulate; phyllaries ovate, dimorphic in 2 series; outer ca. 2 mm long; inner purplish, 5–9 mm long, striate; receptacle flat with scales between the flowers. **Ray flowers** sterile, 8, yellow; ligules 6–15 mm long with a dark base. **Disk flowers** perfect, purple to brown; corolla 2–3 mm long, 4-lobed, tube shorter than the throat; style branch appendages short, blunt. **Pappus** absent or inconspicuous. **Achenes** oblong, flattened, 1–3 mm long, glabrous. Moist soil along rivers, streams, ponds; plains, valleys. Throughout temperate N. America.

Varieties have been described based on the presence of an achene wing; however, this trait varies continuously in our specimens without ecological or geographic significance.

Crepis L. Hawksbeard

Annual or perennial, mostly taprooted herbs with milky sap. **Stems** usually branched above. **Leaves** petiolate, simple, entire to deeply lobed, mainly basal; cauline leaves alternate, reduced or absent. **Inflorescence** open, corymbiform arrays. **Heads** ligulate; involucre cylindric to campanulate; phyllaries in 1 to 2 unequal series; receptacle usually glabrous, flat to convex. **Ray flowers** perfect, yellow. **Pappus** white capillary bristles. **Achenes** subcylindric, 10- to 20-ribbed, sometimes beaked.

Species with pinnately lobed leaves belong to a complex that includes diploids, polyploids and their apomictic hybrids; distinctions between these species are blurred.[20] *Crepis capillaris* (L.) Wallr., an annual with dentate leaves, is reported for MT,[49] but I have seen no specimens.

1. Plants weedy annuals of fields, roadsides 2
1. Plants biennial or usually perennial 3

2. Herbage glabrate or sparsely tomentose; inner faces of phyllaries hairy. . . . . *C. tectorum*
2. Herbage yellowish-hispid; inner faces of phyllaries glabrous. . . . . *C. setosa*

3. Leaf blades entire; plants glabrous. 4
3. At least lower leaves dentate to lobed (sometimes obscure); herbage mostly
    puberulent to tomentose. 6

4. Stems naked or with few small leaves; some basal leaf blades >2 cm long; stems
    often hispid. . . . . *C. runcinata*
4. Stems with leaves; leaf blades ≤2 cm long; herbage glabrous. 5

5. Achenes with a beak ca.1 mm long; inflorescence well above basal leaves . . . . *C. elegans*
5. Achenes narrowed but not beaked; heads mostly among the leaves . . . . *C. nana*

6. Involucre glabrous or sparsely pubescent on outer bracts. . . . . *C. acuminata*
6. Involucre tomentose, setose, or glandular 7

7. Involucre (sometimes base of stem) with black, sometimes glandular setae. 8
7. Stems and involucre without black setae 10

8. Setae of involucre gland-tipped . . . . *C. occidentalis*
8. Setae not gland-tipped 9

9. Base of stem yellowish-setose . . . . *C. modocensis*
9. Stem not setose . . . . *C. atribarba*

10. Stems scapose; leaf blades oblanceolate, entire to shallowly lobed . . . . *C. runcinata*
10. Stems leafy; leaf blades lobed (rarely linear and unlobed). 11

11. Lobes of leaf blades linear . . . . *C. atribarba*
11. Leaf blade lobes lanceolate to deltoid 12

12. Lobes of leaves usually entire; involucral bracts usually 8. . . . . *C. intermedia*
12. Some leaf lobes dentate; involucral bracts often >8 . . . . *C. occidentalis*

Crepis acuminata Nutt. Taprooted perennial. **Stems** erect, 20–60 cm. **Herbage** sparsely tomentose. **Leaf blades** lanceolate, 5–20 cm long, pinnately lobed ≥1/2-way to the midvein into lanceolate, usually entire, often runcinate segments. **Heads** 30–80. **Involucre** cylindric, 7–12 mm high; phyllaries lanceolate, glabrous; the outer sometimes puberulent. **Rays** 5 to 10; ligules 6–9 mm long. **Achenes** pale brown, 6–9 mm long, not beaked. Grasslands, sagebrush steppe; plains, valleys, montane. WA to MT south to CA, AZ, NM, NE, IA. See *C. intermedia*.
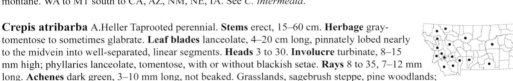

Crepis atribarba A.Heller Taprooted perennial. **Stems** erect, 15–60 cm. **Herbage** gray-tomentose to sometimes glabrate. **Leaf blades** lanceolate, 4–20 cm long, pinnately lobed nearly to the midvein into well-separated, linear segments. **Heads** 3 to 30. **Involucre** turbinate, 8–15 mm high; phyllaries lanceolate, tomentose, with or without blackish setae. **Rays** 8 to 35, 7–12 mm long. **Achenes** dark green, 3–10 mm long, not beaked. Grasslands, sagebrush steppe, pine woodlands; plains, valleys, montane. BC to MB south to NV, UT, CO, NE.

   Plants with or without setose involucres occur throughout the state. See *C. intermedia*.

Crepis elegans Hook. Taprooted biennial or perennial. **Stems** erect, 10–30 cm. **Herbage** glabrous. **Leaf blades** ovate to elliptic, 1–4 cm long, entire to obscurely dentate. **Heads** 10 to 100. **Involucre** cylindric, 5–10 mm high; phyllaries linear-lanceolate, glabrous. **Rays** 6 to 10; ligules 2–4 mm long. **Achenes** light brown, 4–5 mm long, beak ca. 1 mm long. Moist, gravelly or sandy soil of moraine and along streams, roadsides; montane, occasionally subalpine. AK to NT south to WY, ON. (p.519)

Crepis intermedia A.Gray Taprooted perennial. **Stems** erect, 20–70 cm. **Herbage** gray-tomentose, sometimes sparse. **Leaf blades** acuminate, lanceolate, 8–20 cm long, pinnately lobed 1/2- to 3/4-way to the midvein into lanceolate, mostly entire segments. **Heads** 10 to 40. **Involucre** turbinate, 10–15 mm high; phyllaries lanceolate, tomentose, usually without setae. **Rays** 7 to 12; ligules 14–30 mm long. **Achenes** tan, 6–9 mm long, not beaked. Grasslands, meadows, open pine forest; valleys, montane, rarely subalpine. BC to MB south to CA, AZ, NM.

   Crepis intermedia appears intermediate between *C. acuminata*, and *C. atribarba* in involucre vestiture, leaf incision, and number of heads.

Crepis modocensis Greene Taprooted perennial. **Stems** erect, 10–35 cm. **Herbage** gray-tomentose; stems yellowish-setose below. **Leaf blades** lanceolate to ovate, 3–10 cm long, pinnately lobed 1/2- to 3/4-way to the midvein into linear-lanceolate, dentate, forward-pointing segments; petioles setose. **Heads** 1 to 9. **Involucre** turbinate, 9–16 mm high; phyllaries linear-lanceolate, tomentose, black-setose, eglandular. **Rays** 10 to 60; ligules 7–12 mm long. **Achenes** dark green to brown, 4–7 mm long, narrowed above but not beaked. Sagebrush steppe, grasslands; plains, valleys, montane. BC to CA, NV, UT, CO. Our plants are ssp. *modocensis*.

   A form that branches from the base called ssp. *subacaulis* (Kellogg) Babc. & Stebbins is reported for MT,[49] but all specimens I have seen are best assigned to the typical ssp.

Crepis nana Richardson Taprooted perennial. **Stems** prostrate to ascending, 1–5 cm. **Herbage** glabrous, often glaucous. **Leaf blades** ovate to elliptic, 1–2 cm long, entire. **Heads** 5 to 30, often partially hidden among the leaves. **Involucre** cylindric, 6–10 mm high; phyllaries linear, glabrous, scarious-margined. **Rays** 5 to 12, often turning purplish; ligules 3–6 mm long. **Achenes** light brown, 4–9 mm long, narrowed above but not beaked. Moist, stony soil of moraine, talus slopes; alpine, rarely lower on streambanks. AK to NL south to CA, NV, UT, CO; Asia.

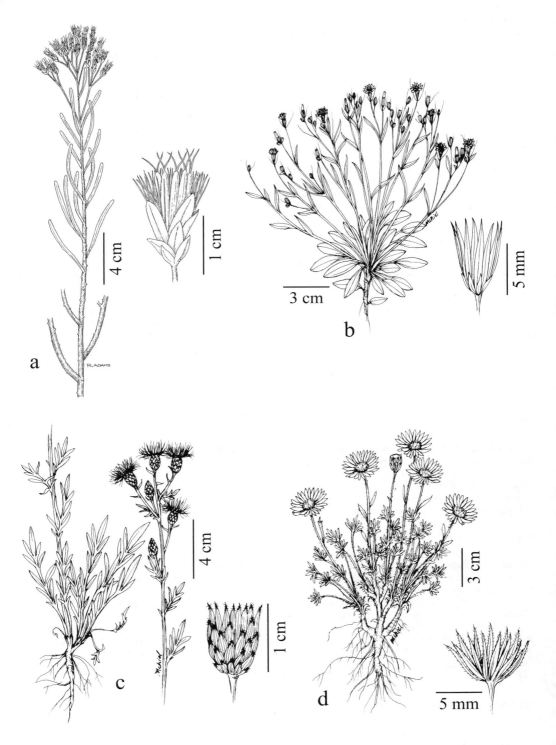

Plate 93. a. *Ericameria nauseosa* var. *oreophila*, b. *Crepis elegans*, c. *Centaurea maculosa*, d. *Erigeron compositus*

Crepis occidentalis Nutt. Taprooted perennial. **Stems** erect, 8–30 cm. **Herbage** gray-
tomentose; stems setose-glandular above. **Leaf blades** ovate to elliptic, 3–20 cm long, pinnately
lobed ca. halfway to the midvein into runcinate, lanceolate, dentate segments. **Heads** 2 to 25.
Involucre narrowly campanulate, 11–19 mm high; phyllaries linear-lanceolate, tomentose, usually
black glandular-setose. **Rays** 10 to 40; ligules 10–15 mm long. **Achenes** brown, 3–7 mm long,
narrowed above but not beaked. Grasslands, sagebrush steppe; valleys, montane. BC to SK south to CA, AZ, NM, SD.
Four subspecies are reported for MT;[49,187] however, all of our material as described above appears to be ssp. *costata*
(A.Gray) Babc. & Stebbins.

Crepis runcinata (E.James) Torr. & A.Gray Taprooted perennial. **Stems** erect or ascending,
scapose, 10–60 cm. **Herbage** glabrous to sparsely hispid. **Leaf blades** ovate, 2–12 cm long,
entire to shallowly, runcinately, pinnately lobed; cauline leaves reduced or absent. **Heads** 1 to 10.
Involucre campanulate, 8–12 mm high; phyllaries linear-lanceolate, scarious-margined, glabrous
or white stipitate-glandular. **Rays** 20 to 50; ligules 7–14 mm long. **Achenes** brown, 3–7 mm long,
narrowed above or short-beaked. Meadows, moist grasslands, often adjacent to wetlands; plains, valleys, subalpine. BC
to MB south to CA, AZ, NM, TX, MN, Mexico.

1. Involucre without stipitate glands . . . . ssp. *glauca*
1. Involucre sparsely to densely stipitate-glandular 2

2. Leaves 5–35 mm wide, 4 to 8 times as long as wide . . . . ssp. *runcinata*
2. Leaves 25–80 mm wide, 2 to 4 times as long as wide . . . . ssp. *hispidulosa*

Crepis runcinata ssp. *glauca* (Nutt.) Babc. & Stebbins is found in alkaline meadows on the plains; *C. runcinata* ssp.
runcinata occurs in wet meadows, montane, subalpine in southwest cos.; *C. runcinata* ssp. *hispidulosa* (Howell ex
Rydb.) Babc. & Stebbins is found in meadows, moist grasslands; valleys, montane. Dwarf, high-subalpine plants have
1 to 3 heads with nearly entire leaves and could be confused with *Hieracium gracile* but for the lack of involucral black
setae.

Crepis setosa Haller f. Shallowly taprooted annual. **Stems** simple, erect, 8–80 cm. **Herbage** yellowish-hispid.
Leaf blades oblanceolate, 5–30 cm long, runcinately lobed, reduced above. **Heads** 10 to 20. **Involucre** narrowly
campanulate, 6–10 mm high; phyllaries lanceolate, yellowish-setose. **Rays** 10 to 20; ligules 8–10 mm long. **Achenes**
brown, 3–5 mm long with a 1–2 mm beak. Fields, roadsides; valleys. Introduced in the U.S. to the northeast and
northwest states, CA, TX; known from Ravalli Co.; native to Europe.

Crepis tectorum L. Taprooted annual. **Stems** erect, 10–50 cm, often branched above.
Herbage glabrate to sparsely tomentose. **Leaf blades** linear-oblanceolate to linear-lanceolate,
2–10 cm long, dentate to shallowly lobed, deciduous below, becoming linear and sessile above.
Heads 10 to 40. **Involucre** campanulate, 6–9 mm high; phyllaries linear-lanceolate, tomentose,
setose, glandular. **Rays** 30 to 70; ligules 6–10 mm long. **Achenes** yellow to brown, 2–5 mm long,
narrowed above. Roadsides; plains, valleys. Introduced throughout north-temperate N. America; native to Europe.

Dyssodia Cav.

Dyssodia papposa (Vent.) Hitchc. Fibrous-rooted annual. **Stems** erect to ascending, 5–20
cm, highly branched. **Herbage** ill-smelling, glabrous or puberulent with sessile glands. **Leaves**
cauline, opposite, petiolate; the blades ovate, 1–2 cm long, deeply 1 to 2-times divided into well-
separated linear to lanceolate segments. **Inflorescence** leafy-cymose; heads terminating numerous
branches. **Heads** radiate; involucres campanulate, 5–8 mm high; phyllaries amber-glandular, 6 to 12
in 2 unequal series; outer green, linear-lanceolate; inner purplish, scarious, ovate; receptacle convex, glabrous. **Ray
flowers** 5 to 8, female, orangish; ligules 1–2 mm long. **Disk flowers** perfect, 12 to 50, yellow to reddish; corolla 2–3 mm
long; style appendages elongate, hairy. **Pappus** 15 to 20 scales in 2 series, each divided into 5 to 10 bristles. **Achenes**
obconic, 3–4 mm long, long-strigose. Often compacted soil along roads, streambanks; plains, valleys. MT to NY south
to CA, AZ, NM, TX, TN, WV, Mexico.

Echinacea Moench Purple coneflower

Echinacea angustifolia DC. [*E. pallida* (Nutt.) Nutt. var. *angustifolia* (DC.) Cronquist]
Perennial from a branched, woody taproot. **Stems** erect, 15–50 cm, mostly unbranched. **Herbage**
hirsute. **Leaves** basal and alternate on the stem, petiolate; blades linear-lanceolate to oblanceolate,
5–15 cm long, entire. **Heads** solitary, radiate; involucres hemispheric, 10–15 mm high; phyllaries

in 2 to 4 subequal series, ovate with spreading, acuminate tips; receptacle hemispheric with stiff, pointed, dark scales between the flowers. **Ray flowers** 8 to 21, sterile, light purple; ligules 1–3 cm long, reflexed. **Disk flowers** perfect, >100, purplish; corolla 5–7 mm long; style appendages acuminate, short-hairy. **Pappus** a few, short teeth. **Achenes** tan to brown, 4-angled-obconic, 4–5 mm long, glabrous. Usually sandy or stony soil of grasslands, pine woodlands; plains. SK, MB south to NM, TX, IA.

Ericameria Nutt. Goldenbush, rabbitbrush

Shrubs. Stems with brown to gray bark and greenish twigs. **Leaves** alternate, entire, sessile to subsessile. **Inflorescence** racemose, corymbiform, or heads solitary. **Heads** radiate or discoid; involucres obconic to campanulate; phyllaries in 2 to 7 series, whitish; the outer often with green tips; receptacle nearly flat, naked. **Ray flowers** absent or few, yellow, female. **Disk flowers** perfect, yellow, tube shorter than the throat; style branch appendages linear to lanceolate. **Pappus** barbellate, capillary bristles. **Achenes** 5- to 12-ribbed.[403]

Some species formerly in *Chrysothamnus*, some in *Haplopappus*.

1. Leaves linear 2
1. Leaves oblanceolate to oblong. 4

2. Leaves stipitate-glandular. . . . .*E. linearis*
2. Leaves sometimes resinous but not stipitate-glandular 3

3. Inflorescence with 1 to 5 heads; subalpine. . . . . *E. parryi*
3. Heads >5; montane or lower . . . .*E. nauseosa*

4. Twigs glabrate, sometimes resinous 5
4. Twigs white-tomentose. 6

5. Most heads solitary; phyllaries subequal . . . . *E. suffruticosa*
5. Heads >1; phyllaries imbricate . . . . *E. nana*

6. Leaves glabrate, resinous, linear-oblanceolate . . . . *E. parryi*
6. Leaves sparsely tomentose, glandular, oblanceolate to oblong. . . . .*E. discoidea*

Ericameria discoidea (Nutt.) G.L.Nesom [*Haplopappus macronema* A.Gray ssp. *macronema*] **Stems** prostrate to ascending, 10–40 cm; twigs white-tomentose; bark brown. **Leaves** oblanceolate to oblong, 1–3 cm long, glandular, sparsely tomentose, axillary fascicles absent. **Inflorescence** with 1 or few heads. **Involucres** obconic, 9–14 mm high; phyllaries subequal, lanceolate, glandular, whitish. **Rays** absent. **Disk flowers** 10 to 26; corolla 9–11 mm long. **Achenes** oblong, 5–6 mm long, villous. Stony soil of open forest, talus slopes; upper subalpine; Beaverhead Co. OR to MT south to CA, NV, UT, CO.

Ericameria linearis (Rydb.) R.P.Roberts. & Urbatsch [*Haplopappus macronema* A.Gray ssp. *linearis* (Rydb.) Hall] **Stems** ascending to erect, 10–40 cm; twigs white-tomentose. **Leaves** linear, 1–4 cm long, stipitate-glandular, sometimes tomentose, twisted; axillary fascicles sometimes present. **Inflorescence** cymose or racemose with 1 to several heads. **Involucres** obconic, 9–15 mm high; phyllaries subequal, lanceolate, glandular, inner white, outer green. **Rays** absent. **Disk flowers** 10–22; corolla 6–10 mm long. **Achenes** oblong, 5–6 mm long, pubescent. Sagebrush steppe, grasslands, often along roads; montane, rarely lower. Endemic to northwest WY, southwest MT.

Ericameria nana Nutt. [*Haplopappus nanus* (Nutt.) D.C. Eaton] **Stems** erect to ascending, 5–50 cm; twigs glabrous, often resinous. **Leaves** oblanceolate to elliptic, 10–15 mm long, glandular, resinous; fascicles in lower axils. **Inflorescence** congested-cymose. **Involucres** obconic, 5–8 mm high; phyllaries imbricate, lanceolate to elliptic, whitish, outer with green tips. **Rays** 1 to 7; ligules 3–4 mm long. **Disk flowers** 4 to 8; corolla 4–7 mm long. **Achenes** obconic, 4–6 mm long, glabrous to pubescent. Rock outcrops; montane, known from Beaverhead Co. OR to MT south to CA, NV, UT.

Ericameria nauseosa (Pall. ex Pursh) G.L.Nesom & G.I.Baird [*Chrysothamnus nauseosus* (Pall. ex Pursh) Britton] RUBBER RABBITBRUSH Plants 0.5–2 m. **Stems** ascending to erect, 1–2 m; twigs glabrate to tomentose. **Leaves** linear, 1–6 cm long, glabrous to tomentose; axillary fascicles absent. **Inflorescence** corymbiform, rounded. **Involucres** obconic, 6–12 mm high; phyllaries imbricate, lanceolate, mostly whitish, resinous, glabrate to tomentose. **Rays** absent. **Disk flowers** 4 to 6; corolla 5–10 mm long. **Achenes** tubular-obconic, 3–8 mm long, hairy. Grasslands, sagebrush steppe, badlands. BC to MB south to CA, AZ, NM, TX. (p.519)

Ericameria nauseosa var. ***nauseosa*** occurs plains, valleys, montane; *E. nauseosa* var. ***graveolens*** (Nutt.) Reveal & Schuyler occurs on the plains; *E. nauseosa* var. ***oreophila*** (A.Nelson) G.L.Nesom & G.I.Baird occurs lower montane; *E. nauseosa* var. ***speciosa*** (Nutt.) G.L.Nesom & G.I.Baird is found valleys, montane. Variety *nauseosa* and var. *graveolens* often occur together without any evidence of introgression.

Ericameria parryi (A.Gray) G.L.Nesom & G.I.Baird [*Chrysothamnus parryi* (A.Gray) Greene ssp. *montanus* L.C.Anderson] **Stems** prostrate to ascending, 10–20 cm; twigs white-tomentose. **Leaves** linear to linear-oblanceolate, 1–3 cm long, glabrous, resinous; fascicles in lower axils. **Inflorescence** cymose with 1 to 5 heads. **Involucres** obconic, 10–15 mm high; phyllaries imbricate, lanceolate, whitish, outer ciliate with green tips. **Rays** absent. **Disk flowers** 5 to 12; corolla 9–10 mm long. **Achenes** ellipsoid, ca. 8 mm long, pubescent. Stony, calcareous soil of sparsely vegetated grasslands; subalpine. OR to MT south to CA, AZ, NM. Our plants are var. ***montana*** (L.C.Anderson) G.L.Nesom & G.I.Baird which is endemic to the Centennial Mtns. of Beaverhead Co. and adjacent ID.

Ericameria suffruticosa (Nutt.) G.L.Nesom [*Haplopappus suffruticosus* (Nutt.) A.Gray] **Stems** erect to ascending, 5–25 cm; twigs glandular, brittle. **Leaves** oblanceolate to oblong, mostly with wavy margins, 1–3 cm long, stipitate-glandular. **Inflorescence** of solitary heads on branch tips. **Involucres** obconic, 8–14 mm high; phyllaries subequal, narrowly oblanceolate, green or whitish with green tips. **Rays** 0 to 8, 7–10 mm long. **Disk flowers** 15 to 40; corolla 8–11 mm long. **Achenes** ellipsoid, 5–8 mm long, pubescent. Stony soil of meadows, grasslands; subalpine. MT south to CA, NV, WY.
Plants from the Bitterroot Range have narrowly oblanceolate leaves without wavy margins.

Erigeron L. Fleabane, daisy

Annual, biennial or perennial herbs. **Leaves** basal and/or alternate, simple; lower petiolate. **Inflorescence** mostly open, coymbiform or paniculate or heads solitary. **Heads** usually radiate, sometimes apparently discoid (disciform); involucre turbinate to hemispheric; phyllaries in 2 to 5 series, mostly green, linear-lanceolate, subequal, little imbricate; receptacle flat to conic, without scales. **Ray flowers** female. **Disk flowers** perfect, yellow, slender; style branches with deltoid appendages. **Pappus** usually in 2 series; the outer short, inconspicuous; the inner of numerous, minutely barbed, capillary bristles. **Achenes** oblong, compressed, usually sparsely strigose.[98,306] See *Symphyotrichum*.
A report of *Erigeron bellidiastrum* Nutt.[113] was based on a misidentified specimen.

9. Ligules of ray flowers ≤3 mm long . . . . *E. lonchophyllus*
9. Ray ligules of most heads ≥3 mm long. 10

10. Plants with leafy stolons. . . . . *E. flagellaris*
10. Stolons absent 11

11. Ligules of ray flowers erect, 3–5 mm long . . . .*E. acris*
11. Ray ligules 4–10 mm long 12

12. Stems densely hirsute with spreading to ascending hairs; ray flowers with pappus bristles . . .*E. divergens*
12. Stems sparsely to moderately strigose with appressed to ascending hairs;
    ray flowers lacking pappus bristles. . . . . *E. strigosus*

13. Plants fibrous-rooted from a caudex or short, horizontal rhizome 14
13. Plants with a near-vertical taproot 29

14. Stem leaves linear 15
14. Stem leaves oblanceolate to ovate. 22

15. Phyllaries minutely glandular 16
15. Phyllaries eglandular (sometimes with septate hairs that may appear glandular) 20

16. Surface of involucral bracts obscured by long, tangled, often septate hairs 17
16. Involucre sparsely hairy, the surface readily visible 19

17. Disk corollas 2.5–3.5 mm long . . . . *E. simplex*
17. Disk corollas ≥4 mm long 18

18. Ligules of ray flowers 7–8 mm long, often white. . . . .*E. lanatus*
18. Ray ligules ≥8 mm long, blue . . . . *E. grandiflorus*

19. Ligules 8–15 mm long; basal leaves oblanceolate to oblong. . . . . *E. formosissimus*
19. Ligules 6–9 mm long; basal leaves linear to linear-oblanceolate. . . . .*E. ursinus*

20. Surface of involucral bracts obscured by long, tangled, often septate hairs . . . .*E. humilis*
20. Involucre sparsely hairy, the surface readily visible 21

21. At least some stems with >1 head; low elevations . . . . *E. glabellus*
21. Heads solitary; high elevations. . . . .*E. gracilis*

22. Hairs of phyllaries septate with dark crosswalls at the base; rays white or whitish 23
22. Hairs of phyllaries without dark crosswalls; rays usually blue or purple. 24

23. Stems 3–10 cm; cauline leaves linear-oblanceolate . . . .*E. lanatus*
23. Stems ≥10 cm; cauline leaves oblanceolate to ovate. . . . .*E. coulteri*

24. Ray ligules ≥2 mm wide; pappus in 1 series . . . . *E. peregrinus*
24. Ray ligules ≤1 mm wide 25

25. Leaves of upper stem linear-oblanceolate . . . . *E. formosissimus*
25. Stem leaves ovate to oblanceolate. 26

26. Stems 4–10 cm; cauline leaves reduced 27
26. Stems >10 cm; cauline leaves little reduced 28

27. Rays white . . . .*E. evermannii*
27. Rays purple . . . .*E. leiomerus*

28. Leaves glabrate with ciliate margins; stems sparsely hairy . . . . *E. speciosus*
28. Leaves and stems moderately to densely hirsute . . . . *E. subtrinervis*

29. Involucre woolly-villous with septate hairs 30
29. Involucre glabrate, strigose or hirsute, but not densely covered with tangled hairs. 33

30. Basal leaves linear to linear-oblanceolate 31
30. At least some basal leaves oblanceolate to oblong 32

31. Disk corollas mostly ≥3.5 mm long; achenes mostly ≥1.5 mm long. . . . . *E. lackschewitzii*
31. Disk corollas usually ≤3.5 mm long; achenes usually ≤1.5 mm long . . . .*E. ochroleucus*

32. Ligules of ray flowers 7–8 mm long, often white. . . . .*E. lanatus*
32. Ray ligules ≥8 mm long, blue. . . . . ≥*E. grandiflorus*

33. Basal leaf blades all filiform to long-linear 34
33. At least some basal leaf blades narrowly oblanceolate to oblong or spatulate 40

34. Rays yellow . . . .*E. linearis*
34. Rays white to purple. 35

35. Basal leaves filiform, >1 cm long and ≤2 mm wide . . . . *E. filifolius*
35. Basal leaves linear, at least some wider than 2 mm. 36

36. Some hairs of stem and petioles >1 mm long . . . . *E. pumilus*
36. Hairs of stem and petioles <1 mm long 37

37. Rays blue. 38
37. Rays white 39

38. Disk corollas mostly ≥3.5 mm long; achenes mostly ≥1.5 mm long. . . . . *E. lackschewitzii*
38. Disk corollas usually ≤3.5 mm long; achenes usually ≤1.5 mm long . . . .*E. ochroleucus*

39. Leaves and stems densely spreading-hirsute; rays often erect. *E. parryi*
39. Leaves and stems strigose; rays spreading . . . .*E. ochroleucus*

40. Rays ≥70 41
40. Rays 15–60 43

41. Mid-stem and petiole hairs strongly divergent, some >1 mm long. . . . . *E. pumilus*
41. Hairs of petiole and stem <1 mm long 42

42. Basal leaves often with 3 veins . . . .*E. caespitosus*
42. Basal leaves with only 1 obvious vein (midvein) . . . .*E. divergens*

43. Rays white 44
43. Rays pink, blue or purple 48

44. Basal leaves glabrate. 45
44. Basal leaves strigose to hirsute 46

45. Involucre 2–3 mm high with purple-septate hairs. . . . . *E. radicatus*
45. Involucre 3–4 mm high, hairs not purple-septate . . . .*E. evermannii*

46. Basal leaves with only 1 obvious vein (midvein) . . . .*E. ochroleucus*
46. Some basal leaves with 3 veins 47

47. Pappus in 1 series; stem hairs appressed . . . . *E. eatonii*
47. Pappus in 2 series; stem hairs spreading to ascending or deflexed . . . .*E. caespitosus*

48. Hairs of stem appressed. 49
48. Hairs of stem spreading to ascending or deflexed 54

49. Basal leaf blades ovate . . . . *E. tweedyi*
49. Basal leaf blades linear-oblanceolate or lanceolate. 50

50. Basal leaf blades glabrate; stem leaves 0 to 2. . . . . *E. radicatus*
50. Leaf blades strigose; stem leaves on well-developed plants more than 2 51

51. Involucre 6–10 mm high. . . . . *E. lackschewitzii*
51. Involucre 3–6 mm high. 52

52. Involucre 3–4 mm high. . . . . *E. tener*
52. Involucre 4–6 mm high. 53

53. Involucre sparsely villous as well as glandular. . . . . *E. eatonii*
53. Involucre glandular but not hairy . . . .*E. leiomerus*

54. Basal leaves with a midvein only 55
54. At least some basal leaves with 3 veins. 56

55. Basal leaf surfaces moderately to densely hirsute. . . . .*E. asperugineus*
55. Basal leaf surfaces glabrate to sparsely hairy . . . . *E. rydbergii*

56. Involucral bracts with a raised dorsal ridge; basal leaves rounded at the tip. . . . .*E. caespitosus*
56. Involucral bracts without a dorsal ridge; basal leaves pointed at the tip . . . .*E. corymbosus*

Erigeron acris L. [*E. nivalis* Nutt.] Short-lived, tap- or fibrous-rooted perennial. **Stems** ascending to erect, 5–60 cm. **Herbage** sparsely hirsute, stipitate-glandular. **Leaves** basal and cauline; blades oblanceolate, 2–10 cm long, entire, narrower and sessile above. **Heads** 1 to 30, radiate. **Involucres** hemispheric, 4–8 mm high; phyllaries in 2 to 3 series, glandular, hirsute, sometimes purple. **Rays** white to pink, 40 to 50; ligules erect, filiform, 3–5 mm long. **Disk**

corollas 4–5 mm long. **Achenes** ca. 2 mm long. Stony soil of open forest, cliffs, streambanks, trails; valleys to alpine. Circumboreal south to CA, CO, MN, ME.

1. Stems >30 cm with many heads; plants usually taprooted . . . . .var. *kamtschaticus*
1. Stems 5–30 cm with 1 to 6 heads; plants usually fibrous-rooted . . . . var. *debilis*

Erigeron acris var. *debilis* A.Gray is found in open forest, streambanks, cliffs, talus, fellfields; montane to alpine; *E. acris* var. *kamtschaticus* (DC.) Herder occurs on streambanks, roadsides, talus; valleys, montane. The similar *E. lonchophyllus* is eglandular and generally occurs in more permanently wet habitats.

Erigeron allocotus S.F.Blake Taprooted perennial with a usually branched caudex. **Stems** ascending to erect, 5–10 cm. **Herbage** hirsute, minutely glandular. **Leaves** mainly basal; blades spatulate, 1–2 cm long, mostly 3-lobed at the tip. **Heads** 1 to 4, radiate. **Involucres** campanulate, 4–6 mm high; phyllaries in 2 to 3 series, glandular, sparsely hirsute. **Rays** white to pink, 20 to 40; ligules filiform, 3–6 mm long. **Disk corollas** ca. 3 mm long. **Achenes** ca. 2 mm long. Calcareous fellfields, rock outcrops; valleys, montane. Endemic to the Bighorn and Pryor ranges of Carbon Co., MT and adjacent WY.

Erigeron asperugineus (D.C.Eaton) A.Gray Taprooted perennial with a simple or branched caudex. **Stems** prostrate to ascending, 3–8 cm. **Herbage** short-hirsute. **Leaves** basal and cauline; blades oblanceolate to obovate, 1-veined, entire, 5–20 mm long, narrower above. **Heads** radiate, solitary. **Involucres** campanulate, 5–8 mm high; phyllaries in 3 to 4 series, glandular, hirsute, often purple-tipped. **Rays** 10 to 25, violet; ligules 5–10 mm long. **Disk corollas** 4–5 mm long. **Achenes** 2–3 mm long. Stony, usually calcareous soil of open slopes, turf; subalpine, alpine. Beaverhead Co., MT, ID south to NV, UT.

Erigeron caespitosus Nutt. Taprooted perennial usually with a branched caudex. **Stems** ascending, 4–30 cm. **Herbage** short-hirsute to strigose. **Leaves** basal and cauline; blades linear-oblanceolate to obovate, entire, 1–8 cm long, the largest with 3 veins, narrower and sessile above. **Heads** radiate, 1 to 10. **Involucres** hemispheric, 4–7 mm high; phyllaries in 3 to 4 series, minutely glandular, villous-hirsute. **Rays** 30 to 100, white to blue; ligules 5–10 mm long. **Disk corollas** 2–4 mm long. **Achenes** ca. 2 mm long. Rocky soil of open slopes, grasslands, sagebrush steppe, open forest, fellfields, rock outcrops; all elevations. AK south to AZ, NM, NE.

Montana plants are extremely variable; Nesom[306] mentions four forms peculiar to MT and adjacent states. Plants from the Great Plains tend to be taller with better-developed cauline leaves, erect basal leaves, several heads and white rays. High-elevation plants from southwest MT are shorter with simple stems, solitary heads and blue to violet rays. Low-elevation plants from southwest MT have mainly 2 to 3 heads, white to pink rays and broader, often clasping cauline leaves. *Erigeron caespitosus* rarely occurs at high elevations in northwest MT.

Erigeron compositus Pursh Cut-leaf daisy Taprooted perennial with a usually branched caudex. **Stems** ascending to erect, 2–25 cm. **Herbage** hirsute, rarely glabrate, often glandular. **Leaves** mostly basal; blades obovate, 5–25 mm long, 1 to 3 times divided into crowded to well-separated, linear-oblanceolate to oblanceolate lobes. **Heads** solitary, radiate or disciform. **Involucres** campanulate, 5–9 mm high; phyllaries in 2 to 3 series, glandular, hirsute, often purple-tipped. **Rays** 20 to 60, usually white; ligules to 10 mm long. **Disk corollas** 3–4 mm long. **Achenes** 2–3 mm long. Sandy or rocky soil of grasslands, sagebrush steppe, outcrops, open slopes, fellfields; all elevations. AK to CA, AZ, CO, SD, Greenland, QC. (p.519)

There is a great deal of variation in ray length and leaf shape and vestiture, thought to be the result of hybridization, and agamospermy.[309] Varieties have been described but are thought to have little utility.[103,306]

Erigeron corymbosus Nutt. Taprooted perennial with a slender-branched caudex. **Stems** ascending to erect, 10–40 cm. **Herbage** short-hirsute, often minutely glandular. **Leaves** basal and cauline; blades linear-lanceolate to oblanceolate, entire, 6–16 cm long, the largest with 3 veins, becoming linear above. **Heads** radiate, 1 to 10. **Involucres** hemispheric, 4–7 mm high; phyllaries in 2 to 3 series, short-hirsute, minutely glandular. **Rays** 35 to 65, blue; ligules 6–11 mm long. **Disk corollas** 3–5 mm long. **Achenes** 2–3 mm long. Grasslands, sagebrush steppe; valleys, montane. BC, MT south to UT, WY.

Erigeron coulteri Porter Fibrous-rooted perennial from a short, sometimes branched rhizome. **Stems** ascending to erect, 10–50 cm. **Herbage** glabrate to sparsely strigose. **Leaves** basal and cauline; blades oblanceolate to ovate, 2–10 cm long, sometimes weakly dentate, sessile and clasping above. **Heads** 1 to 2, radiate. **Involucres** hemispheric, 6–10 mm high; phyllaries in 2 or 3 series, minutely glandular, villous with septate hairs. **Rays** 45 to 140, white; ligules 7–12 mm long.

Disk corollas 2–4 mm long. **Achenes** 1–2 mm long. Meadows, open forest; montane, subalpine. OR to MT south to CA, NM.

Erigeron divergens Torr. & A.Gray Taprooted biennial or short-lived perennial. **Stems** erect, 10–40 cm, often branched at the base of a simple caudex. **Herbage** short-hirsute, hairs spreading or ascending. **Leaves** basal and cauline; blades obovate to oblanceolate, 1–5 cm long, entire; basal usually withered at flowering; cauline becoming sessile, linear above. **Heads** at least several, radiate, nodding in bud. **Involucres** hemispheric, 3–5 mm high; phyllaries in 3 to 4 series, glandular, hirsute with a brown midvein. **Rays** 70 to 150, white, 4–8 mm long. **Disk corollas** 2–3 mm long. **Achenes** ca. 1 mm long. Grasslands, sagebrush steppe, open forest, roadsides; plains, valleys, montane. BC, AB south to CA, AZ, NM, TX, Mexico.

Erigerion pumilus has linear basal leaves present at flowering.

Erigeron eatonii A.Gray Taprooted perennial from a mostly unbranched caudex. **Stems** ascending to erect, 5–20 cm, purple at the base. **Herbage** strigose. **Leaves** basal and cauline; blades linear to oblanceolate, entire, 5–11 cm long, sessile above. **Heads** radiate, mostly 1 or 2. **Involucres** campanulate to hemispheric, 4–6 mm high; phyllaries in 2 or 3 series, sparsely villous, minutely glandular. **Rays** 15 to 42, white to blue; ligules 5–7 mm long. **Disk corollas** 3–5 mm long. **Achenes** ca. 2 mm long; pappus in 1 series. Sagebrush steppe, open forest; montane. WA to MT south to CA, AZ, CO. Our plants are var. *eatonii*. Collected in Stillwater Co. (*Hitchcock & Muhlick 13414* WTU).

Erigeron evermannii Rydb. Taprooted perennial from a slender-branched caudex. **Stems** scapose, ascending to erect, 2–6 cm, purple at the base. **Herbage** glabrate. **Leaves** mostly basal; blades oblanceolate to oblong, entire, 1–3 cm long. **Heads** solitary, radiate. **Involucres** campanulate, 5–8 mm high; phyllaries in 2 or 3 series, sparsely villous. **Rays** 15 to 40, white; ligules 5–8 mm long. **Disk corollas** 3–5 mm long. **Achenes** ca. 3 mm long. Stony granitic soil of talus, open slopes; upper subalpine, alpine. Endemic to Ravalli Co., MT and adjacent ID.

Erigeron leiomerus is similar but occurs on calcareous soils.

Erigeron filifolius Nutt. Taprooted perennial with a usually branched caudex. **Stems** erect, 7–30 cm. **Herbage** strigose; stems canescent at the base. **Leaves** basal and cauline, filiform to linear, 1–4 cm long. **Heads** 1 to 10, radiate. **Involucres** hemispheric, 2–5 mm high; phyllaries in 3 to 4 series, sparsely villous, minutely glandular. **Rays** light blue to white; ligules 7–11 mm long. **Disk corollas** 2–4 mm long. **Achenes** ca. 1.5 mm long. Grasslands, open pine forest; valleys. BC, MT south to CA, NV, UT.

Erigeron flabellifolius Rydb. Taprooted perennial with a slender-branched caudex. **Stems** erect, 2–10 cm. **Herbage** minutely glandular. **Leaves** basal and cauline; blades cuneate to oblanceolate, 1–3 cm long, shallowly divided into crowded, ovate, sometimes dentate, ultimate lobes, less divided above. **Heads** solitary, radiate. **Involucres** campanulate to hemispheric, 7–10 mm high; phyllaries in 2 to 3 series, glandular, purplish. **Rays** 50 to 70, white; ligules 5–9 mm long. **Disk corollas** 4–5 mm long. **Achenes** ca. 2 mm long; pappus in 1 series. Granitic talus, stony open slopes; alpine. Endemic to Carbon and Park cos., MT and adjacent WY.

Erigeron flagellaris A.Gray Fibrous-rooted, stoloniferous biennial or short-lived perennial. **Stems** erect, 5–15 cm, scapiform. **Herbage** strigose. **Leaves** mostly basal and on stolons; blades linear-oblanceolate to obovate, 1–4 cm long, entire, sessile, linear-oblanceolate above and on stolons. **Heads** solitary, radiate. **Involucres** hemispheric, 4–6 mm high; phyllaries in 2 to 3 series, minutely glandular, strigose. **Rays** 40 to 100, white, purplish on the back; ligules 4–9 mm long. **Disk corollas** 2–3 mm long. **Achenes** ca. 1 mm long. Usually disturbed soil of grasslands, meadows; plains, valleys to subalpine. BC, AB, ND south to CA, AZ, NM, TX, Mexico.

Erigeron formosissimus Greene Fibrous-rooted perennial from a short rhizome. **Stems** ascending, 10–40 cm. **Herbage** glabrate to sparsely villous-hirsute, glandular above. **Leaves** basal and cauline; blades oblanceolate to oblong, 2–8 cm long, entire, sessile, lanceolate above. **Heads** 1 or few, radiate. **Involucres** hemispheric, 5–8 mm high; phyllaries in 2 to 3 series, minutely stipitate-glandular. **Rays** 75 to 150, pink to purple; ligules 8–15 mm long. **Disk corollas** 3–5 mm long. **Achenes** ca. 2 mm long. Meadows; subalpine. Carbon Co., MT, SD south to AZ, NM.

Erigeron glabellus Nutt. Fibrous-rooted biennial or short-lived perennial. **Stems** erect, 8–50 cm. **Herbage** sparsely strigose to villous-hirsute. **Leaves** basal and cauline; blades linear-lanceolate to oblanceolate, 1–10 cm long, entire, sessile and narrower above. **Heads** 1 to 6, radiate. **Involucres** hemispheric, 5–10 mm high; phyllaries in 3 or 4 series, sparsely hirsute. **Rays** >125, white or blue; ligules narrow, 6–13 mm long. **Disk corollas** 3–6 mm long. **Achenes** ca. 1.5 mm long. Grasslands, meadows, woodlands, forest openings; plains, valleys, montane. AK to NT south to UT, NM, SD, WI.

Both var. *glabellus* with appressed or ascending stem hair, and var. *pubescens* Hook., with spreading stem hair, are reported for MT,[306] but too many of our specimens cannot be distinguished. Plants from the Great Plains have white rays; those from the intermountain region are generally blue.

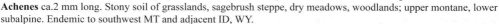

Erigeron gracilis Rydb. Fibrous-rooted perennial from slender branched caudex. **Stems** ascending, 5–20 cm. **Herbage** sparsely strigose, eglandular. **Leaves** basal and cauline; blades linear to linear-oblanceolate, entire, 1–8 cm long, becoming linear-lanceolate above. **Heads** radiate, solitary. **Involucres** hemispheric, 5–8 mm high; phyllaries in 2 or 3 series, strigose, eglandular. **Rays** 40 to 100, blue or purple; ligules 5–12 mm long. **Disk corollas** 3–5 mm long. **Achenes** ca.2 mm long. Stony soil of grasslands, sagebrush steppe, dry meadows, woodlands; upper montane, lower subalpine. Endemic to southwest MT and adjacent ID, WY.

The more common *Erigeron ursinus* has stipitate glands on the involucre.

Erigeron grandiflorus Hook. Fibrous-rooted perennial usually with a simple caudex. **Stems** ascending to erect, 3–10 cm. **Herbage** villous, often minutely glandular. **Leaves** basal and cauline; blades linear-oblanceolate to oblong, entire, 6–25 mm long, becoming linear above. **Heads** radiate, solitary. **Involucres** hemispheric, 8–10 mm high; phyllaries in 2 or 3 series, purplish, minutely glandular, tangled-villous with septate hairs. **Rays** 60 to 125, blue, 8–13 mm long. **Disk corollas** 4–5 mm long. **Achenes** 2–3 mm long. Stony soil of fellfields; alpine; Sweet Grass and Carbon cos. AK south to WY. See *E. simplex*.

Erigeron humilis Graham Fibrous-rooted perennial often with a branched caudex. **Stems** ascending to erect, 1–5 cm, purplish. **Herbage** sparsely to densely villous. **Leaves** basal and cauline; blades linear-oblanceolate to spatulate, entire, 5–20 mm long, becoming linear upward. **Heads** radiate, solitary. **Involucres** campanulate, 6–10 mm high; phyllaries in ca. 2 series, purple, woolly-villous with purple-septate hairs. **Rays** 50 to 100, light purple to white; ligules 2–4 mm long. **Disk corollas** 2–3 mm long. **Achenes** ca. 2 mm long. Sparsely vegetated soil of moraine, wet cliffs, talus; alpine. AK to Greenland south to UT, CO.

Erigeron lackschewitzii G.L.Nesom & W.A.Weber Taprooted perennial with a simple or few-branched caudex. **Stems** scapose, erect, 2–10 cm. **Herbage** villous-strigose. **Leaves** basal and cauline; blades linear-oblanceolate, entire, 1–4 cm long, linear above. **Heads** solitary, radiate. **Involucres** campanulate, 6–10 mm high; phyllaries in 3 series, minutely glandular, strongly villous, with sometimes purple-septate hairs. **Rays** 30 to 70, blue; ligules 5–8 mm long. **Disk corollas** 3–4 mm long. **Achenes** 2–3 mm long. Gravelly, calcareous soil of turf, fellfields; upper subalpine, alpine. Endemic to western MT, adjacent AB.

Blue-rayed forms of *Erigeron ochroleucus* have smaller disk flowers; the involucre of *E. grandiflorus* is densely woolly-villous. The type locality is in Flathead Co.

Erigeron lanatus Hook. Fibrous-rooted perennial from a slender-branched caudex. **Stems** ascending to erect, 3–10 cm, scapose. **Herbage** villous, minutely glandular above. **Leaves** basal and cauline; blades linear-oblanceolate to spatulate, 1–3 cm long, entire or 1- to 3-lobed at the tip. **Heads** radiate, solitary. **Involucres** hemispheric, 8–12 mm high; phyllaries in 2 or 3 series, shaggy woolly-villous, purple-tipped, minutely glandular. **Rays** 30 to 80, white to purplish; ligules 7–8 mm long. **Disk corollas** 4–5 mm long. **Achenes** 3–4 mm long. Coarse limestone talus; alpine; Flathead and Glacier cos. BC, AB sporadically south to CO.

Plants are often found sprawling between the shifting rocks. The similar *E. pallens* Cronquist, with smaller rays and involucral bracts, occurs in extreme southwest AB[225] and may occur in adjacent MT.

Erigeron leiomerus A.Gray Taprooted perennial with a slender-branched caudex. **Stems** prostrate to ascending, 4–10 cm. **Herbage** sparsely strigose. **Leaves** basal and cauline; blades oblanceolate to spatulate, 5–20 mm long, entire, narrower above. **Heads** radiate, solitary. **Involucres** hemispheric, 4–6 mm high; phyllaries in 2 or 3 series, minutely glandular, sparsely strigose, often purplish. **Rays** 15 to 60, purple; ligules 5–9 mm long. **Disk corollas** ca. 3 mm long. **Achenes** ca. 2 mm long. Stony calcareous soil of barren, eroding slopes or talus; subalpine; Beaverhead Co. MT south to NV, UT, NM. See *E. evermannii*.

Erigeron linearis (Hook.) Piper Taprooted perennial often with a branched caudex. **Stems** erect, 5–20 cm, scapose above. **Herbage** strigose. **Leaves** basal and cauline; blades linear, 1–5 cm long, entire. **Heads** solitary, radiate. **Involucres** hemispheric, 4–6 mm high; phyllaries in 2 or 3 series, minutely glandular, strigose. **Rays** 25 to 40, yellow; ligules 3–6 mm long. **Disk corollas** 2–4 mm long. **Achenes** ca. 2 mm long. Dry, stony soil of sagebrush steppe, grasslands; valleys, montane. BC, MT south to CA, NV, UT.

Erigeron lonchophyllus Hook. Short-lived, fibrous-rooted perennial. **Stems** erect, 3–35 cm, sometimes branched above. **Herbage** villous-hirsute, eglandular. **Leaves** basal and cauline; blades oblanceolate, 1–8 cm long, entire, linear above. **Heads** 1 to 12, radiate. **Involucres** turbinate, 4–10 mm high; phyllaries in 2 or 3 series, eglandular, sparsely villous, often purple-tipped. **Rays** white, 70 to 130, erect; ligules filiform, 1–3 mm long, inconspicuous. **Disk corollas** slender, 3–5 mm long. **Achenes** ca. 1.5 mm long. Wet soil along ponds, streams, wetlands; plains, valleys, montane, rarely subalpine. AK to QC south to CA, AZ, NM, NE, MI. See *E. acris*.

Erigeron ochroleucus Nutt. Taprooted perennial usually with a simple caudex. **Stems** erect, often decumbent, 2–25 cm. **Herbage** strigose to glabrate. **Leaves** basal and few or many cauline; blades linear to linear-oblanceolate, entire, 1–6 cm long. **Heads** radiate, solitary, rarely 2. **Involucres** campanulate, 5–7 mm high; phyllaries in 3 or 4 series, minutely glandular, appressed-villous with white-septate hairs. **Rays** 30 to 60, usually white, sometimes blue; ligules 4–8 mm long. **Disk corollas** 3–4 mm long. **Achenes** 2–3 mm long. Dry, stony soil of grasslands, sagebrush steppe, woodlands, fellfields, open forest; all elevations. AK to WY, NE.

Dwarf, high-elevation plants from southwest MT, sometimes with blue rays, have been called var. *scribneri* (Canby ex Rydb.) Cronquist, but there are numerous intermediate populations and some populations with both forms. *Erigeron radicatus* usually has a branched caudex and glabrate basal leaves; *E. parryi* has hirsute stems. See *E. lackschewitzii*.

Erigeron parryi Canby & Rose Taprooted perennial with a simple or branched caudex. **Stems** erect, 2–15 cm. **Herbage** hirsute with ascending to spreading hairs. **Leaves** basal and cauline, linear, entire, 1–6 cm long. **Heads** 1 to 3, radiate. **Involucres** hemispheric, 4–9 mm high; phyllaries in 1 or 2 series, densely hirsute with white-septate hairs, minutely glandular. **Rays** 20 to 40, white, usually erect; ligules 5–8 mm long. **Disk corollas** 2–4 mm long. **Achenes** ca. 2 mm long. Stony, calcareous soil of sagebrush steppe, grasslands; montane. Endemic to southwest MT. See *E. ochroleucus*.

The type locality is in Beaverhead Co.

Erigeron peregrinus (Banks ex Pursh) Greene [*E. glacialis* (Nutt.) A.Nelson] Fibrous-rooted perennial from a short rhizome. **Stems** erect, 3–50 cm. **Herbage** glabrate to sparsely strigose below the inflorescence. **Leaves** basal and cauline; blades oblanceolate to ovate, 2–15 cm long, entire, becoming clasping above. **Heads** radiate, usually solitary (2 to 4). **Involucres** hemispheric, 6–12 mm high; phyllaries in 2 or 3 series, stipitate-glandular. **Rays** 30 to 80, blue to purple; ligules 8–20 mm long. **Disk corollas** 3–6 mm long. **Achenes** ca. 2.5 mm long; pappus in 1 series. Moist meadows, turf, open forest, thickets, often where snow lies late; upper montane, alpine. AK to CA, NM. Our plants are ssp. *callianthemus* (Greene) Cronquist. (p.531)

Our subspecies is sometimes called *E. glacialis*, specifically distinct from the Pacific slope plants with eglandular involucres.[306] *Symphyotrichum foliaceum*, which is superficially similar and common in similar habitats, does not have a glandular involucre.

Erigeron philadelphicus L. Biennial or short-lived, fibrous-rooted perennial. **Stems** erect, 15–70 cm. **Herbage** villous. **Leaves** basal and cauline; blades oblanceolate to obovate, 2–12 cm long, shallowly dentate to entire, lanceolate to ovate, clasping above. **Heads** 3 to 30, radiate. **Involucres** hemispheric, 4–6 mm high; phyllaries in 2 or 3 series, villous with a tan midvein and scarious, erose margins. **Rays** narrow, 150 to 250, white to pink; ligules 3–8 mm long. **Disk corollas** 2–3 mm long. **Achenes** ca. 1 mm long. Disturbed soil of meadows, streambanks, roadsides; valleys, montane. Throughout temperate N. America.

Erigeron pumilus Nutt. [*E. concinnus* (Hook. & Arn.) Torr. & A.Gray] Taprooted perennial with a simple or branched caudex. **Stems** ascending to erect, 5–25 cm. **Herbage** long-hirsute, many spreading hairs >1 mm long. **Leaves** basal and cauline; blades linear to narrowly oblanceolate, 15–40 mm long, entire. **Heads** 1 to 15, radiate. **Involucres** hemispheric, 4–8 mm high; phyllaries in 2 to 4 series, minutely glandular, hirsute with a brownish midvein. **Rays** 50–100, white to pink; ligules 5–12 mm long. **Disk corollas** 2–5 mm long. **Achenes** 1–2 mm long. Sandy or stony soil of grasslands, sagebrush steppe, woodlands; plains, valleys, montane. BC to SK south to CA, NV, UT, CO, KS.

Three taxa have been recognized in our area;[98,306] ssp. *pumilus*, the Great Plains form with white rays, setae-like outer pappus and glabrous disk corollas; ssp. *intermedius* Cronquist has pink to blue rays with narrow scales for outer pappus and puberulent corollas; and *E. concinnus* with blue rays, outer pappus of scales and puberulent disk corollas. Our plants all have puberulent disk corollas, and plants with setiform outer pappus occur across the state. Sporadic populations west of the Continental Divide have outer pappus of scales. There is a tendency for eastern plants to have white rays; however, these trends do not seem strong enough in our area to warrant subspecific or specific segregation. A collection made in a vacant lot in Bozeman was determined to be *E. disparipilus* Cronquist by A. Cronquist based on having rays wider than 2 mm.

Erigeron radicatus Hook. Taprooted perennial with a usually branched caudex. **Stems** ascending to erect, 2–6 cm. **Herbage** leaves glabrate; stems densely strigose. **Leaves** mostly basal, linear-oblanceolate, entire, 1–5 cm long. **Heads** solitary, radiate. **Involucres** hemispheric, 4–7 mm high; phyllaries in 2 or 3 series, villous-hirsute with purple-septate hairs, sometimes minutely glandular. **Rays** 20 to 45, white; ligules 4–7 mm long. **Disk corollas** 2.5–3 mm long. **Achenes** ca. 2 mm long. Stony, often calcareous soil of fellfields, turf, open forest; subalpine, alpine. AB, SK south to UT, WY, NE. See *E. ochroleucus*.

Erigeron rydbergii Cronquist Taprooted perennial with a short-branched caudex. **Stems** erect, 2–7 cm. **Herbage** usually sparsely villous, eglandular. **Leaves** mostly basal; blades narrowly oblanceolate, entire, 1–2 cm long, becoming linear on the stem. **Heads** solitary, radiate. **Involucres** campanulate, 5–7 mm high; phyllaries in 2 or 3 series, villous with septate hairs, eglandular. **Rays** 15 to 35, blue; ligules 4–9 mm long. **Disk corollas** 3–4 mm long. **Achenes** ca. 2 mm long. Stony, usually non-calcareous soil of fellfields, turf, open forest; upper subalpine, alpine. Endemic to southwest MT and adjacent WY, ID. The type locality is in Gallatin Co.

Erigeron simplex Greene Fibrous-rooted perennial with a simple or branched caudex. **Stems** erect, 2–15 cm. **Herbage** sparsely villous. **Leaves** basal and cauline; blades linear-oblanceolate to narrowly oblong, entire, 5–40 mm long, narrower above. **Heads** radiate, solitary. **Involucres** hemispheric, 5–12 mm high; phyllaries in 2 or 3 series, purplish, minutely glandular, tangled-villous with septate hairs. **Rays** 50 to 150, white to usually blue; ligules 4–10 mm long. **Disk corollas** 3–3.5 mm long. **Achenes** 2–3 mm long. Moist soil of turf, fellfields; alpine. WA and MT south to AZ, NM.

Nesom merged this species under *Erigeron grandiflorus*;[306] the two are very similar; however, there are differences in disk corolla size and pappus bristle number, and molecular genetic analysis indicates the two are as distinct as other pairs of *Erigeron* species.[217]

Erigeron speciosus (Lindl.) DC. Fibrous-rooted from a branched rhizome. **Stems** erect, 15–60 cm. **Herbage** glabrate, villous on leaf margins and midveins. **Leaves** basal and cauline; blades lanceolate or oblanceolate to ovate, 2–10 cm long, entire, sessile above. **Heads** 1 to 10, radiate. **Involucres** hemispheric, 5–7 mm high; phyllaries in 2 or 3 series, minutely glandular. **Rays** slender, 75 to 150, blue; ligules 7–13 mm long. **Disk corollas** 3–5 mm long. **Achenes** 1–2 mm long. Grasslands, meadows, woodlands, open forest; valleys to subalpine. BC, AB south to AZ, NM, SD.

Nearly glabrous plants have been called var. *macranthus* (Nutt.) Cronquist; *E. subtrinervis* has hirsute herbage.

Erigeron strigosus Muhl. ex Willd. Fibrous-rooted annual or biennial. **Stems** erect, 20–70 cm, branched above and sometimes at the base. **Herbage** strigose. **Leaves** basal and cauline; blades ovate to oblanceolate, 3–15 cm long, entire or obscurely dentate; basal usually withered at flowering; cauline becoming linear. **Heads** 5 to 50, radiate. **Involucres** hemispheric, 3–5 mm high; phyllaries in 2 or 3 series, minutely glandular, sparsely strigose. **Rays** 80 to 125, white; ligules 3–8 mm long. **Disk corollas** 2–3 mm long. **Achenes** ca. 1 mm long; pappus absent on ray achenes. Disturbed areas of grasslands, meadows, streambanks, forest openings, roadsides; plains, valleys, montane. Throughout temperate N. America. Our plants are var. *strigosus*.

Erigeron subtrinervis Rydb. ex Porter & Britton Fibrous-rooted from a branched rhizome. **Stems** erect, 15–60 cm. **Herbage** hirsute. **Leaves** basal and cauline; blades oblanceolate to ovate, 3–7 cm long, entire, sessile above. **Heads** 1 to 6, radiate. **Involucres** hemispheric, 5–9 mm high; phyllaries in 2 or 3 series, glandular, hirsute. **Rays** 100 to 150, blue; ligules slender, 6–14 mm long. **Disk corollas** 3–4 mm long. **Achenes** ca. 2 mm long. Grasslands, sagebrush steppe, pine woodlands; plains, valleys, montane. ID, MT, ND south to NM, NE.

It seems doubtful that the difference in vesture between *Erigeron speciosus* and *E. subtrinervis* warrants recognition of two species.[187,289] Plants from the foothills of the Front Range seem to approach *E. speciosus*.

Erigeron tener (A.Gray) A.Gray Taprooted perennial with a slender-branched caudex. **Stems** ascending to erect, 3–15 cm. **Herbage** strigose. **Leaves** basal and cauline; blades oblanceolate to obovate, entire, 1–2 cm long, linear above. **Heads** radiate, solitary. **Involucres** campanulate, 3–4 mm high; phyllaries in 2 or 3 series, densely glandular, sparsely strigose. **Rays** 20 to 40, blue; ligules 3–6 mm long. **Disk corollas** 2–3 mm long. **Achenes** ca. 2 mm long. Stony soil of sagebrush steppe, grasslands; subalpine; Beaverhead Co. OR to MT south to CA, AZ, WY.

Erigeron tweedyi Canby Taprooted perennial from a simple or short-branched caudex. **Stems** ascending to erect, 3–20 cm. **Herbage** densely white-strigose. **Leaves** basal and cauline; blades obovate to oblanceolate, 7–20 mm long, entire, becoming linear-oblanceolate above. **Heads** radiate, solitary. **Involucres** hemispheric, 3–6 mm high; phyllaries in 3 to 4 series, minutely glandular, densely strigose. **Rays** 20 to 50, blue; ligules 4–7 mm long. **Disk corollas** ca. 3 mm long. **Achenes** ca. 2 mm long. Stony, usually calcareous soil of sagebrush steppe, grasslands, woodlands; valleys to subalpine. Endemic to southwest MT and adjacent ID, WY.

Erigeron ursinus D.C.Eaton Fibrous-rooted perennial from slender-branched caudex. **Stems** ascending, 4–20 cm. **Herbage** sparsely strigose, glandular above. **Leaves** basal and cauline; blades linear to narrowly oblanceolate, entire, 1–5 cm long, becoming linear above. **Heads** radiate, solitary. **Involucres** hemispheric, 5–8 mm high; phyllaries in 2 or 3 series, sparsely villous, minutely glandular, often reflexed. **Rays** 30 to 100, blue or purple; ligules 6–9 mm long. **Disk corollas** 3–4 mm long. **Achenes** 1–2 mm long. Turf, grasslands, meadows, open forest; subalpine, alpine. ID, MT south to AZ, NM. See *E. gracilis*.

Eriophyllum Lag.

Eriophyllum lanatum (Pursh) J.Forbes Taprooted perennial with a woody, branched caudex. **Stems** ascending to erect, simple, 5–40 cm. **Herbage** sparsely to densely woolly-villous. **Leaves** basal and alternate, petiolate; basal blades obovate, 5–40 mm long, lobed at the tip; cauline blades lanceolate to oblanceolate, 1–4 cm long, entire or pinnately divided into 3 to 7 linear-oblanceolate lobes. **Heads** radiate, solitary on peduncles 3–15 cm long; involucre hemispheric, 5–12 mm high; phyllaries 5 to 13, narrowly ovate in 2 series, woolly, viscid; receptacle flat, smooth. **Ray flowers** female, 11 to 13, yellow; ligules 5–16 mm long, glandular. **Disk flowers** perfect; corollas 2–5 mm long, yellow, glandular; style with short-hairy appendages. **Pappus** of 6 to 12 erose, white scales. **Achenes** 4-angled, obconic, 3–5 mm long, glabrous. Stony soil of grasslands, sagebrush steppe, open forest, open slopes. BC to CA, NV, UT, WY.

1. Involucre ≤8 mm high; stems mostly <30 cm . . . . var *integrifolium*
1. Involucre ≥9 mm high; stems often >30 cm . . . . var. *lanatum*

Eriophyllum lanatum var. *lanatum* occurs valleys, montane; *E. lanatum* var. *integrifolium* (Hook.) Smiley is found montane, subalpine.

Eucephalus Nutt. New World aster

Perennial herbs from a branched caudex or short rhizome. **Leaves** cauline, alternate, simple. **Inflorescence** racemose to corymbiform. **Heads** radiate; narrowly campanulate; phyllaries unequal in 3 to 6 overlapping series, basally thickened, often purple-tipped or -margined; receptacle pitted, glabrous. **Ray flowers** female, white to purple. **Disk flowers** perfect, yellow, often with a purple limb; style branches flattened with lanceolate appendages. **Pappus** of numerous, barbellate, capillary bristles in 2 series. **Achenes** usually 2- to 4-nerved.[2]

Formerly placed in *Aster*.

1. Some leaves >5 cm long; stems often >50 cm. . . . .*E. engelmannii*
1. Leaves usually ≤5 cm long; stems usually <50 cm . . . . *E. elegans*

Eucephalus elegans Nutt. [*Aster perelegans* A.Nelson & J.F.Macbr.] **Stems** clustered, erect, 20–50 cm. **Herbage** puberulent, glandular above. **Leaves** sessile; linear-lanceolate to linear-oblanceolate, entire, 2–6 cm long, the lowest deciduous. **Inflorescence** short, racemose to corymbiform; peduncles glandular-puberulent. **Involucre** 7–8 mm high; phyllaries imbricate, narrowly ovate, glandular-pubescent, purple-tipped and -margined. **Rays** 5 to 8, purple; ligules 7–12 mm long. **Disk corollas** 7–9 mm long. **Achenes** obconic, 1–3 mm long, strigose. Meadows, dry, open forest; montane, lower subalpine. OR to MT south to NV, UT, CO.

Eucephalus engelmannii (D.C.Eaton) Greene [*Aster engelmannii* (D.C.Eaton) A.Gray] **Stems** erect, 30–120 cm. **Herbage** glabrate to sparsely villous. **Leaves** sessile to short-petiolate; blades lanceolate to obovate, entire, 3–10 cm long, the lowest deciduous. **Inflorescence** open, leafy-corymbiform; peduncles puberulent to sparsely glandular. **Involucre** narrowly campanulate, 6–10 mm high; phyllaries imbricate, lanceolate, ciliate, glandular-pubescent; outer green; inner outlined

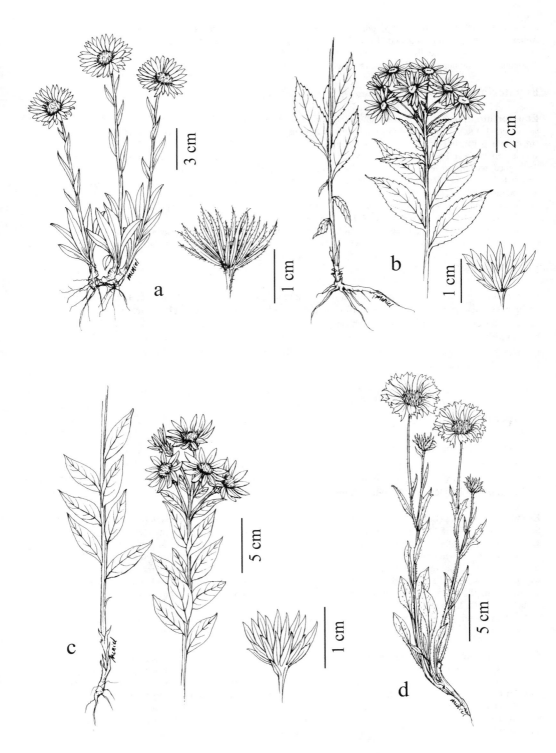

Plate 94. a. *Erigeron peregrinus*, b. *Eurybia conspicua*, c. *Eucephalus engelmannii*,
d. *Gaillardia aristata*

in purple. **Rays** 8 to 13, white to light purple; ligules 12–22 mm long. **Disk flowers** 10 to 35; corolla 7–9 mm long. **Achenes** obconic, compressed, cylindrical, 3–5 mm long, puberulent. Moist, open forest, meadows, avalanche slopes; upper montane, subalpine. BC, AB south to WA, NV, UT, CO. (p.531)

Plants from southern MT have white rays; farther north pinkish rays are more common.

Eupatorium L.

Eupatorium maculatum L. [*Eutrochium maculatum* (L.) E.E.Lamont, *Eupatoriadelphus maculatus* (L.) R.M.King & H.Rob.] Joe Pye weed Fibrous-rooted perennial. **Stems** erect, 50–150 cm, simple, purplish. **Herbage** puberulent, glandular above. **Leaves** cauline, short-petiolate, simple, in whorls of 3 or 4; blades lanceolate, 6–15 cm long, serrate. **Inflorescence** corymbiform, flat-topped, terminal or in upper leaf axils. **Heads** discoid; involucre tubular, 6–9 mm high; phyllaries imbricate in 3 to 5 series, white to purplish, ovate to oblanceolate, glabrous; receptacle convex, naked. **Disk flowers** perfect, tubular, 9 to 22, purple; corolla 4–6 mm long; style branches clavate. **Achenes** obconic, 5-angled, 3–5 mm long, glandular; pappus 1 series of capillary bristles. Riparian meadows, thickets, open forest; plains, valleys. BC to QC south to AZ, NM, TN, NC. Our plants are var. *bruneri* (A.Gray) Breitung.

Eupatorium sensu lato is a large genus that has been divided into numerous smaller genera based on microscopic characters.[219] There seems to be disagreement on how many of these genera should be circumscribed.

Eurybia (Cass.) Cass. New World aster

Rhizomatous perennial herbs. **Stems** usually unbranched. **Leaves** basal and cauline, alternate, simple. **Inflorescence** corymbiform or heads solitary. **Heads** radiate; campanulate; phyllaries in 3 to 7 overlapping, unequal series, basally thickened, often purple-tipped; receptacle pitted, glabrous. **Ray flowers** female, blue to violet. **Disk flowers** perfect, yellow, often with a purple limb; style branches flattened with lanceolate appendages. **Pappus** of numerous, barbellate, capillary bristles. **Achenes** usually 7- to 12-nerved.[63] Formerly placed in *Aster*.

1. Peduncles and involucre glandular. 2
1. Peduncles and involucre not glandular. 3

2. Leaves serrate . . . . *E. conspicua*
2. Leaves entire . . . . *E. integrifolia*

3. Peduncles glabrate to scabrous. . . . . *E. glauca*
3. Peduncles villous 4

4. Phyllaries well-imbricated, the inner outlined in purple; heads usually several . . . . *E. merita*
4. Phyllaries subimbricate, the inner purple-tipped; heads solitary to few . . . .*E. sibirica*

Eurybia conspicua (Lindl.) G.L.Nesom [*Aster conspicuus* Lindl.] Rhizomatous perennial. **Stems** erect, 30–75 cm. **Herbage** scabrous, sparsely strigose, glandular above. **Leaves** cauline, sessile or subsessile, lanceolate to obovate or ovate, strongly serrate, 5–15 cm long, the lowest deciduous. **Inflorescence** corymbiform, leafy bracteate; peduncles stipitate-glandular. **Involucre** campanulate, 9–12 mm high; phyllaries linear-lanceolate, glandular, ciliate, white below, green above with purple tips. **Rays** 12 to 35; ligules 10–15 mm long. **Disk flowers** 48 to 55; corolla 7–9 mm long. **Achenes** fusiform, 3–4 mm long, glabrous to sparsely pubescent. Dry to moist forests, woodlands; valleys to lower subalpine. BC to MB south to OR, WY, SD. (p.531)

The broad, strongly toothed leaves often cover the forest floor; however, *Eurybia conspicua* rarely flowers in the shade.

Eurybia glauca (Nutt.) G.L.Nesom [*Aster glaucodes* S.F.Blake, *Eucephalus glaucous* Nutt., *Herrickia glauca* (Nutt.) Brouillet] Rhizomatous perennial. **Stems** erect, 20–40 cm, sometimes branched. **Herbage** glabrate, leaf margins scabrous. **Leaves** cauline, sessile, narrowly lanceolate, entire, 3–7 cm long, the lowest deciduous. **Inflorescence** corymbiform, leafy bracteate with few to many heads; peduncles glabrate. **Involucre** campanulate, 5–8 mm high; phyllaries narrowly ovate, glabrous, ciliate; the inner narrower, brownish below, green above. **Rays** 10 to 20, pale blue; ligules 6–12 mm long. **Disk flowers** 12 to 32; corolla 6–8 mm long. **Achenes** fusiform, 4–5 mm long, sparsely pubescent. Stony, calcareous soil of dry, open forest; montane in Carbon Co. MT south to AZ, NM.

Eurybia integrifolia (Nutt.) G.L.Nesom [*Aster integrifolius* Nutt.] Perennial from a caudex or short rhizome. **Stems** sometimes clustered, ascending to erect, 15–60 cm. **Herbage**: leaves ciliate to pubescent; stems glandular-puberulent. **Leaves** basal and cauline, petiolate; blades lanceolate to ovate, entire, 2–15 cm long, reduced and sessile upwards. **Inflorescence** usually narrow, racemose to corymbiform with few to several heads; peduncles stipitate glandular. **Involucre** campanulate, 7–12 mm high; phyllaries oblanceolate to linear-lanceolate, glandular, ciliate, purple-tipped, white below. **Rays** 8 to 27; ligules 8–12 mm long. **Disk flowers** 20 to 50; corolla 6–8 mm long. **Achenes** fusiform, 4–5 mm long, pubescent. Grasslands, dry meadows, sagebrush steppe, talus; montane, subalpine. WA to MT south to CA, NV, UT, WY.

Eurybia merita (A.Nelson) G.L.Nesom [*Aster meritus* A.Nelson, *Aster sibiricus* L. var. *meritus* (A.Nelson) Raup] Rhizomatous perennial. **Stems** erect, 15–40 cm. **Herbage** scabrous to sparsely puberulent. **Leaves** cauline, stiff, short-petiolate; blades lanceolate to narrowly elliptic, serrulate, 1–8 cm long, sessile above. **Inflorescence** corymbiform with several heads; peduncles villous. **Involucre** campanulate, 6–9 mm high; phyllaries lanceolate, ciliate, pubescent; the inner with purple margins. **Rays** 14 to 32; ligules 7–10 mm long. **Disk flowers** 30 to 60; corolla 6–8 mm long. **Achenes** fusiform, 2–4 mm long, sparsely strigose. Sparsely vegetated, stony soil of dry forests; montane, lower subalpine. BC south to CA, UT, SD.

   Eurybia merita is sometimes considered a variety of *E. sibirica*; however, the two taxa can usually be distinguished and are ecologically distinct.

Eurybia sibirica (L.) G.L.Nesom [*Aster sibiricus* L.] Rhizomatous perennial. **Stems** ascending, 3–20 cm. **Herbage** scabrous to strigose. **Leaves** cauline, stiff, short-petiolate; blades oblanceolate, serrate to entire, 1–5 cm long, sessile above, the lowest deciduous. **Inflorescence** corymbiform with 1 to few heads; peduncles densely villous. **Involucre** campanulate, 6–9 mm high; phyllaries lanceolate, ciliate to puberulent; the inner purplish. **Rays** 12 to 50; ligules 7–12 mm long. **Disk flowers** 25 to 125; corolla 5–7 mm long. **Achenes** fusiform, 2–4 mm long, sparsely strigose. Sparsely vegetated, stony, usually calcareous soil of alpine fellfields, montane streambanks. Circumboreal south to BC, ID, MT. See *E. merita*.

Euthamia (Nutt.) Cass.

Euthamia occidentalis Nutt. [*Solidago occidentalis* (Nutt.) Torr. & A.Gray] Rhizomatous, perennial herbs. **Herbage** glabrous, glandular-punctate. **Stems** erect, simple, 50–150 cm. **Leaves** 3–10 cm long, cauline, alternate, simple, linear-lanceolate, entire, sessile, minutely ciliate. **Inflorescence** of numerous heads in small glomerules in leafy-coymbiform arrays. **Heads** radiate; involucres 3–5 mm high, obconic to campanulate, glutinous; phyllaries lanceolate, often green-tipped, imbricate in 3 to 5 series, pale, 1-veined; receptacle flat, naked. **Ray flowers** few to several, female, yellow; ligules ca. 20, 1–2 mm long. **Disk flowers** ca. 9 to 15, perfect, yellow; corolla 3–5 mm long, tubes shorter than throats; style appendages lanceolate. **Achenes** subcylindric with 2 to 4 ribs, strigose, ca. 1 mm long; pappus capillary bristles. Moist to wet meadows, often on stream terraces, around wetlands; plains, valleys. BC south to CA, AZ, NM, NE.

   Euthamia graminifolia (L.) Nutt.[*Solidago graminifolia* (L.) Salisb.], with a hemispheric inflorescence, is reported for MT,[151] but this seems unlikely. The report may be based on a specimen of *E. occidentalis* with a small inflorescence.

Evax Gaertn.

Evax prolifera Nutt. ex DC. [*Diaperia prolifera* (Nutt. ex DC.) Nutt.] Gray-woolly, taprooted annual. **Stems** erect, 2–10 cm, simple or branched at the base. **Leaves**: basal small, deciduous; cauline alternate, oblanceolate, entire, 3–15 mm long. **Inflorescence** terminal, woolly glomerules of sessile heads closely subtended by leaves. **Heads** disciform; involucre absent; receptacle conical, chaffy. **Disk flowers** tubular, minute; the outer female; inner functionally male. **Pappus** absent. **Achenes** oblong, compressed, <1 mm long. Meadows, grasslands, sagebrush steppe; plains. MT south to NM, TX, LA, AL. Known from Custer and Powder River cos.

Filago L.

Filago arvensis L. [*Logfia arvensis* (L.) Holub] Taprooted annual. **Stems** erect, mostly simple, 10–50 cm. **Herbage** white-woolly. **Leaves** alternate, linear-oblanceolate to linear-lanceolate, 1–2 cm long, entire. **Inflorescence** leafy, narrow-paniculate with heads in capitate clusters. **Heads** radiate but appearing discoid, pyriform, 3–5 mm high; true phyllaries minute; bracts of receptacle appearing like phyllaries, woolly-villous, linear-lanceolate; receptacle flat with scales enwrapping the flowers. **Flowers** slender; outer 2 to 4, female; inner 2–3 mm long, female or perfect, ca. 20. **Pappus** of inner flowers of capillary bristles; that of outer flowers absent. **Achenes** brown, cylindric, <1 mm long, glabrous. Disturbed areas of grasslands, fields, roadsides, cut-over forests; plains, valleys. Introduced BC to MB south to OR, ID, WY, NE, MI; native to Europe.

Gnaphalium is similar but has papery-tipped involucral bracts.

Gaillardia Foug.

Gaillardia aristata Pursh B<small>LANKETFLOWER</small> Taprooted perennial. Stems erect, simple or branched, 10–50 cm. **Herbage** hispid, villous. **Leaves** basal and alternate, petiolate; blades oblanceolate, 3–12 cm long, entire, dentate or pinnately lobed ca. halfway to the midvein, becoming sessile above. **Heads** radiate, solitary on peduncles 4–30 cm long; involucre hemispheric, 1–2 cm high; phyllaries ovate to lanceolate, strigose to tomentose, ciliate, spreading; receptacle convex with stiff setae. **Ray flowers** 12 to 18, sterile, yellow with purple bases; ligules 10–35 mm long, 3-lobed. **Disk flowers** perfect, 60 to 120, purple-villous at the tip; corolla 5–9 mm long; style appendages elongate, short-hairy. **Pappus** of 6 to 10 awned scales. **Achenes** obconic, hirsute at the base, 2–6 mm long. Grasslands, meadows; plains, valleys, montane. YT to MB south to OR, UT, CO, SD, MN. (p.531)

The type locality is near Lincoln.

Galinsoga Ruiz & Pav. Quickweed

Fibrous-rooted annuals. **Stems** erect, branched above. **Leaves** cauline, opposite, simple, petiolate; blades narrowly ovate, mostly serrate. **Herbage** strigose, spreading-hairy. **Inflorescence** open, leafy, cymose. **Heads** radiate; involucre campanulate; phyllaries few, subequal, narrowly ovate; receptacle convex with scarious scales. **Ray flowers** ca. 5, female, white, each subtended by a phyllary. **Disk flowers** perfect, 5 to 50, yellow; style branches acute, short-hairy. **Pappus** narrow, fringed scales. **Achenes** obconic.

1. Scales of receptacle acute, lobed ≥1/3 the length . . . . .G. parviflora
1. Scales of receptacle blunt, unlobed or lobed ≤1/3 the length. . . . . .G. quadriradiata

Galinsoga parviflora Cav. Stems 15–60 cm. Leaf blades 3–5 cm long. **Heads:** involucre 2–3 mm high, sparsely hairy. **Ray ligules** 0.5–2 mm long. **Disk flowers** 15 to 50; corolla ca. 1 mm long. **Pappus:** those of disk flowers 9 to 18 rounded; those of rays minute or absent. **Achenes** 1–2.5 mm long, short-hairy. Lawns, fields; valleys. Introduced to much of temperate N. America; native to S. America. Collected in Ravalli Co.

Galinsoga quadriradiata Ruiz & Pav. [*G. ciliata* (Raf.) S.F.Blake] **Stems** 5–50 cm. **Leaf blades** 2–7 cm long. **Heads:** involucre 3–8 mm high with glandular hairs. **Ray ligules** 1–3 mm long. **Disk flowers** 15 to 35; corolla 1–3 mm long. **Pappus** 9 to 14 scales; those of disk flowers acuminate. **Achenes** 1–2 mm long, short-hairy. Lawns, fields; valleys. Introduced to eastern U.S., southern Canada and adjacent northern states; native to tropical N. and S. America.

Gnaphalium L. Cudweed

Annual to perennial, usually taprooted herbs. **Leaves** cauline, alternate, simple, entire, mostly sessile. **Herbage** usually white woolly-tomentose. **Inflorescence** of capitate clusters (glomerules) in corymbiform or paniculate arrays. **Heads** disciform; involucres campanulate to cylindric; phyllaries imbricate in several series, partly scarious, glandular; receptacle flat, naked. **Flowers** yellowish, all disk-like; outer female; inner slender, perfect, less numerous than outer; style branches flattened. **Pappus** of barbellate, capillary bristles. **Achenes** ellipsoid to ovoid, glabrous.[103]

Some authors place the three large species in *Pseudognaphalium*.[307]

1. Pappus bristles united at the base; inflorescence cylindric, at least 3 times as long as wide . .G. purpureum
1. Pappus bristles separate; inflorescence hemispheric, ca. as long as wide 2

2. Involucre 2–4 mm high; stems mostly ≤20 cm 3
2. Involucre ≥3.5 mm high; stems usually >20 cm 4

3. Phyllaries with a rounded white tip; some leaves oblong . . . . .G. palustre
3. Outer phyllaries darkened to the pointed tip; leaves linear-oblanceolate. . . . . .G. uliginosum

4. Upper leaf surfaces glandular and pubescent but not tomentose . . . . G. macounii
4. Upper leaf surfaces white- or gray-tomentose 5

5. Leaves auriculate but not decurrent; inflorescence compact; flowers ≥100 per head . . . . *G. stramineum*
5. Leaves decurrent down the stem; inflorescence more open; flowers <60 per head . . . . . *G. microcephalum*

Gnaphalium macounii Greene [*Pseudognaphalium macounii* (Greene) Kartesz] [*G. viscosum* Kunth misapplied] Usually biennial. **Stems** erect, mostly unbranched, 30–70 cm. **Leaves** tomentose beneath, greenish above, glandular, 3–8 cm long; basal withered at flowering; cauline 4–10 cm long, narrowly lanceolate to oblanceolate, mostly sessile, often with decurrent bases. **Glomerules** terminal in a close hemispheric to pyramidal inflorescence, surpassing subtending leaves. **Involucres** 5–6 mm high; phyllaries white, acute, tomentose at the base. **Inner corollas** 3–4 mm long. **Achenes** 0.5–1 mm long; pappus deciduous separately. Disturbed soil of meadows, streambanks, burned or logged forests; valleys. Throughout most of temperate N. America. (p.538)
    Gnaphalium microcephalum has a more open inflorescence and *G. stramineum* does not have glandular herbage.

Gnaphalium microcephalum Nutt. [*Pseudognaphalium microcephalum* (Nutt.) Anderb., *P. thermale* (E.E.Nelson) G.L.Nesom] Short-lived taprooted perennial. **Stems** erect, often branched at the base, 20–70 cm. **Leaves** 3–8 cm long; basal petiolate, oblanceolate; cauline linear-oblanceolate, sessile, decurrent on the stem. **Glomerules** terminal in an open ovoid to pyramidal inflorescence, surpassing subtending leaves. **Involucres** 5–6 mm high; phyllaries white, with acute tips, sometimes tomentose at the base. **Inner corollas** 3–4 mm long. **Achenes** 0.5–1 mm long; pappus deciduous separately. Disturbed soil of roadsides, grasslands, meadows, open forest, locally common following timber harvest; valleys. BC, AB south to CA, NV, UT, WY. Our plants are var. *thermale* (E.E.Nelson) Cronquist. See *G. macounii*.

Gnaphalium palustre Nutt. Annual. **Stems** prostrate to ascending, usually branched at the base, 2–20 cm. **Leaves** basal and cauline, oblanceolate to oblong, 1–3 cm long. **Glomerules** axillary, barely surpassed by subtending leaves. **Involucres** 2–3 mm high; phyllaries brown, tomentose with rounded, white, scarious tips. **Inner corollas** ca. 1.5 mm long. **Achenes** ca. 0.5 mm long; pappus deciduous. Vernally moist soil around ponds, wetlands, roadsides; plains, valleys, lower montane. BC to SK south to CA, AZ, NM, NE.
    Gnaphalium uliginosum has more linear leaves.

Gnaphalium purpureum L. [*Gamochaeta ustulata* (Nutt.) Holub] Biennial or short-lived perennial. **Stems** erect, simple, 10–40 cm. **Leaves** 1–3 cm long; basal petiolate, oblanceolate; cauline linear, sessile. **Glomerules** terminal in a narrow, spike-like inflorescence, surpassing subtending leaves. **Involucres** 3–5 mm high; phyllaries white, with acute tips, sometimes tomentose at the base. **Inner corollas** 2–4 mm long. **Achenes** ca. 0.5 mm long; pappus deciduous in a ring. Streambanks; valleys. BC, MT south to CA. Our plants are var. *ustulata* (Nutt.) Boivin. Collected once in Missoula Co.

Gnaphalium stramineum Kunth [*G. chilense* Spreng., *Pseudognaphalium stramineum* (Kunth) Anderb.] Annual or biennial. **Stems** erect, sometimes branched at the base, 20–80 cm. **Leaves** mainly cauline, clasping, oblanceolate, 2–6 cm long. **Glomerules** terminal in a dense, hemispheric inflorescence, surpassing subtending leaves. **Involucres** 4–6 mm high; phyllaries dirty white with rounded, erose tips, tomentose at the base. **Inner corollas** 2–3 mm long. **Achenes** 0.5–1 mm long; pappus deciduous in clumps. Meadows; valleys, montane. BC south to CA, AZ, NM, TX, Mexico. See *G. microcephalum*.

Gnaphalium uliginosum L. Annual. **Stems** prostrate to erect, sometimes branched, 3–15 cm. **Leaves** mostly cauline, linear to linear-oblanceolate, 1–3 cm long. **Glomerules** axillary, surpassed by subtending leaves. **Involucres** 3–4 mm high; phyllaries greenish-brown, tomentose; outer with green to brown, pointed tips. **Inner corollas** ca. 1.5 mm long. **Achenes** ca. 0.5 mm long; pappus deciduous. Margins of wetlands, roadsides, streambanks; valleys. Introduced to most of north-temperate N. America; native to Europe.

Grindelia Willd. Gumweed

Perennial herbs. **Leaves** mainly cauline, alternate, simple, lower petiolate. **Inflorescence** paniculate or corymbiform. **Heads** radiate; involucre hemispheric to globose; phyllaries imbricate in 4 to 9 series, green, resinous, linear-lanceolate with green, punctate, recurved tips; receptacle convex, naked, pitted. **Ray flowers** female, yellow. **Disk flowers** 100 to 200, perfect, female or sterile, yellow; style branches with lanceolate appendages. **Pappus** of 2 to 8 rigid, deciduous awns. **Achenes** obovoid, compressed, sometimes 4-angled.

1. Stems short stipitate-glandular. . . . . *G. howellii*
1. Stems glabrate 2

2. Leaf blades mostly entire-margined; upper margin of achenes bumpy . . . . *G. nana*
2. Leaf margins dentate; upper margin of achenes smooth . . . . *G. squarrosa*

Grindelia howellii Steyerm. **Stems** erect, branched above, 30–90 cm. **Herbage** short stipitate-glandular. **Leaf blades** oblanceolate below to ovate and sessile above, 2–15 cm long, entire to serrate. **Involucres** 8–15 mm high. **Rays** 20 to 30; ligules 7–14 mm long. **Disk corollas** 4–6 mm long. **Achenes** 4–6 mm long, wavy at the top; pappus awns 2. Vernally moist meadows around wetlands, roadsides, disturbed grasslands, open forest; valleys, lower montane. Endemic to Missoula and Powell cos. and perhaps north ID.

The presettlement habitat of *Grindelia howellii* appears to be shallow wetlands of the Ovando Valley. However, short-lived populations are common along roads in this area. Idaho populations are also in disturbed habitats and may be adventive. A few roadside collections lack stipitate glands and may be hybridized with *G. squarrosa*.

Grindelia nana Nutt [*G. hirsutula* Hook. & Arn.] Perennial branched from the caudex. **Stems** erect, branched above, 10–40 cm. **Herbage** glabrate, glandular-punctate. **Leaf blades** oblanceolate, 1–6 cm long, mostly entire; basal sometimes present at flowering. **Involucres** 7–10 mm high. **Rays** 12 to 25; ligules 4–10 mm long. **Disk corollas** 4–5 mm long. **Achenes** 4–6 mm long, top margin knobby; pappus awns 2 or 3. Grasslands, open pine forest; valleys. Throughout most of temperate N. America.

A recent treatment has placed numerous former more geographically restricted taxa, including *Grindelia nana*, under *G. hirsutula*.[388] See *G. squarrosa*.

Grindelia squarrosa (Pursh) Dunal CURLYCUP GUMWEED Biennial to perennial, sometimes branched at the caudex. **Stems** erect, branched above, 10–70 cm. **Herbage** glabrous, glandular-punctate. **Leaf blades** linear-oblanceolate to oblong, 1–10 cm long; dentate with amber-colored, resin-tipped teeth. **Involucres** 5–12 mm high. **Rays** 24 to 40; ligules 5–12 mm long. **Disk corollas** 4–6 mm long. **Achenes** 1–4 mm long, top margin smooth; pappus awns 2 or 3. Margins of wetlands, streambanks, roadsides and other disturbed, vernally moist habitats; plains, valleys. Throughout N. America. Thought to be native to the Great Plains and Rocky Mountains and adventive elsewhere. (p.538)

"Good" *Grindelia squarrosa*, with serrate leaves and amber-tipped teeth, occurs primarily in the eastern third of the state; "good" *G. nana* is found west of the Continental Divide. The most common form of *Grindelia* in the western part of the state is intermediate, with weakly serrate leaves and white-tipped teeth. These plants can be either biennial or short-lived perennial and have been traditionally called *G. squarrosa* var. *quasiperennis* Lunell. Recently they have been combined with *G. nana* under *G. hirsutula*,[388] but I see no compelling reason to believe they are more closely related to the *G. hirsutula* complex than *G. squarrosa*.

Gutierrezia Lag. Snakeweed

Gutierrezia sarothrae (Pursh) Britton & Rusby BROOM SNAKEWEED Subshrub. **Stems** numerous, erect, 8–30 cm, unbranched below the inflorescence. **Herbage** puberulent, minutely punctate-glandular. **Leaves** cauline, linear, 1–4 cm long. **Inflorescence** flat-topped-corymbiform, composed of sessile heads in small clusters. **Heads** radiate; involucre obconic, 3–5 mm high; phyllaries 8 to 14, imbricate in 3 to 4 series, lanceolate to narrowly elliptic, glabrous, resinous with a thickened green tip; receptacle conic, naked. **Ray flowers** 3 to 7, yellow, female; ligules 1–3 mm long. **Disk flowers** perfect, yellow; corolla 2–4 mm long, funnelform; style branches with hairy, lance-linear appendages. **Pappus** 1 or 2 series of erose scales. **Achenes** clavate, 1–2 mm long, densely strigose. Sagebrush steppe, grasslands, juniper woodlands, badlands; plains, valleys to subalpine. AB to MB south to CA, AZ, NM, TX, Mexico.

Plants from the Great Plains tend to have narrower phyllaries with less pronounced green tips.

Helenium L. Sneezeweed

Helenium autumnale L. Fibrous-rooted perennial herbs. **Stems** erect, often branched, 15–60 cm, winged-decurrent from leaf bases. **Herbage** puberulent, minutely punctate. **Leaves** cauline, petiolate below, sessile above; basal withered at flowering; blades lanceolate, entire to obscurely dentate, 2–10 cm, reduced upward. **Inflorescence** corymbiform. **Heads** radiate; involucre reflexed; phyllaries subequal in 2 or 3 series, 10–15 mm long, narrowly lanceolate, green, puberulent; receptacle hemispheric, naked, pitted. **Ray flowers** 8 to 21, female, yellow, reflexed; ligules 10–15 mm long. **Disk flowers** numerous, fertile; corolla yellow, 2–4 mm long, glandular, puberulent above, tube shorter than the throat; style branches dilated. **Pappus** of 5 to 10 aristate, scarious scales. **Achenes** obconic, 4- to 5-angled, 1–2 mm long, pubescent. Moist, often gravelly soil along streams, wetlands, roadsides; plains, valleys. Throughout N. America.

Helianthella Torr. & A.Gray

Perennial herbs from a branched caudex. **Stems** usually simple, erect. **Leaves** basal and cauline, petiolate, opposite at least below, entire. **Heads** solitary or few, radiate; involucre hemispheric; phyllaries subequal in ca. 3 series, green, ciliate; receptacle convex with scales that clasp the achenes. **Ray flowers** 8 to 21, yellow, sterile. **Disk flowers** numerous, perfect, yellow; the tube shorter than the throat; style branches with a short-hairy appendage. **Pappus** of 2 awn-like scales and 2 to 4 short scales. **Achenes** obconic, compressed, strigose with ciliate margins.

1. Receptacle scales soft; phyllaries lanceolate to ovate; heads often nodding . . . . *H. quinquenervis*
1. Receptacle scales firm; phyllaries linear-lanceolate; heads erect . . . . *H. uniflora*

Helianthella quinquenervis (Hook.) A.Gray **Stems** 50–150 cm. **Herbage** glabrate to sparsely strigose. **Leaf blades** oblanceolate to elliptic, often 5-veined, 8–25 cm long. **Heads** often nodding; involucres ca. 3 cm across; phyllaries lanceolate to ovate, acuminate; receptacle scales soft. **Ray ligules** 2–4 cm long. **Disk corollas** 4–7 mm long. **Achenes** 8–10 mm long. Meadows, woodlands; montane. OR, MT, SD south to AZ, NM.

Helianthella uniflora (Nutt.) Torr. & A.Gray **Stems** 30–70 cm. **Herbage** scabrous to stiff-puberulent. **Leaf blades** narrowly lanceolate to lanceolate, 3-veined, 5–12 cm long. **Heads** erect; involucres 10–25 mm across; phyllaries linear-lanceolate; receptacle scales stiff. **Ray ligules** 15–30 mm long. **Disk corollas** 5–7 mm long. **Achenes** 3–5 mm long. Sagebrush steppe, grasslands, forest openings; montane, lower subalpine. BC to NV, AZ, CO.

Helianthus L. Sunflower

Annual or perennial herbs. **Leaves** cauline and sometimes basal, simple, petiolate, alternate or opposite. **Heads** radiate, solitary or few on stem tips; involucre nearly hemispheric; phyllaries subequal in several series; receptacle flat with scales between flowers. **Ray flowers** 5 to 30, yellow, sterile. **Disk flowers** numerous, perfect, yellow, tube shorter than the throat; style branches slender. **Pappus** of few, deciduous scales. **Achenes** obovoid, 4-angled, compressed, mostly glabrous.

Helianthus laetiflorus Pers., a hybrid between *H. tuberosus* and *H. pauciflorus*, is reported for MT;[351] *H. pumilus* Nutt., similar to but smaller than *H. pauciflorus*, is reported for MT,[152] but I have seen no specimens.

1. Plants taprooted annuals 2
1. Plants rhizomatous perennials 3

2. Phyllaries lanceolate, short-ciliate; central receptacle bracts white-villous at the tip . . . . *H. petiolaris*
2. Phyllaries nearly ovate, long-ciliate; central receptacle bracts hispid at the tip . . . . *H. annuus*

3. Phyllaries narrowly ovate . . . .*H. pauciflorus*
3. Phyllaries narrowly lanceolate, acuminate 4

4. Leaves ovate . . . . *H. tuberosus*
4. Leaves narrowly lanceolate or lanceolate 5

5. Many leaves folded up along the midvein . . . . *H. maximiliani*
5. Leaves not folded . . . .*H. nuttallii*

Helianthus annuus L. COMMON SUNFLOWER Taprooted annual. **Stems** erect, often branched, 25–100 cm. **Herbage** scabrous to sparsely hispid. **Leaves** alternate; blades weakly to strongly serrate, lanceolate to ovate, sometimes cordate, 5–20 cm long. **Involucres** 15–35 mm across; phyllaries ovate, long-acuminate, hispid, long-ciliate, 12–25 mm long. **Ray ligules** 16–45 mm long. **Disk corollas** 5–8 mm long, lobes often deep red. **Achenes** 4–5 mm long. Disturbed soil of fields, grasslands, streambanks, roadsides; plains, valleys. Throughout N. America, introduced worldwide.

Size of heads is variable, perhaps reflecting hybridization between native and cultivated forms. On average *Helianthus petiolaris* is shorter with smaller heads.

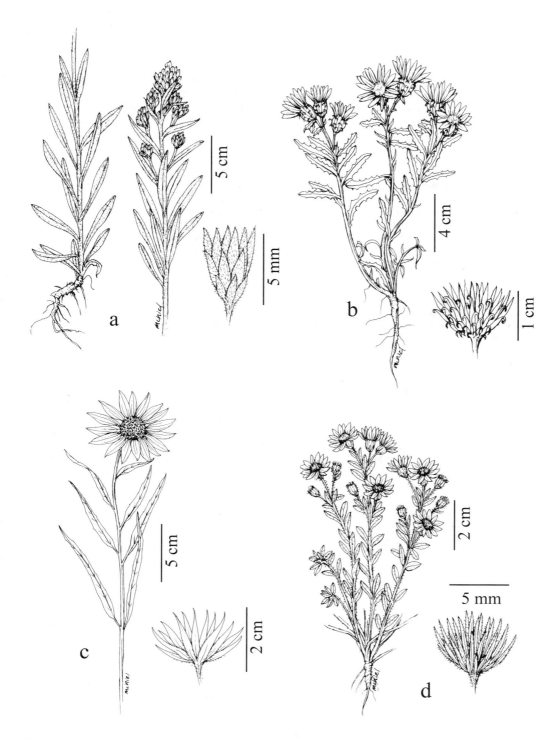

Plate 95. a. *Gnaphalium macounii*, b. *Grindelia squarrosa*, c. *Helianthus nuttallii*, d. *Heterotheca villosa*

Helianthus maximiliani Schrad. Rhizomatous perennial with thickened roots. **Stems** erect, usually simple, 45–200 cm. **Herbage** hispid-strigose, glandular. **Leaves** alternate; blades entire, narrowly lanceolate, acuminate, 7–15 cm long. **Involucres** 12–25 mm across; phyllaries lanceolate, long-acuminate, strigose, glandular, ciliate, 6–15 mm long. **Ray ligules** 10–25 mm long. **Disk corollas** 5–7 mm long. **Achenes** 3–4 mm long. Moist to wet meadows, thickets, often along streams, ponds; plains, valleys. Throughout most of U.S., southern Canada, Mexico.

The Flathead Co. record is from an old mill site.

Helianthus nuttallii Torr. & A.Gray Rhizomatous perennial with tuberous roots. **Stems** erect, simple or branched above, 60–200 cm. **Herbage** glabrate to scabrous or short-strigose, sometimes glaucous. **Leaves** opposite to mostly alternate; blades entire to serrate, narrowly lanceolate to lanceolate, 5–15 cm long. **Involucres** 12–22 mm across; phyllaries linear-lanceolate, long-acuminate, hispid to strigose, ciliate, 8–20 mm long. **Ray ligules** 15–30 mm long. **Disk corollas** 6–7 mm long. **Achenes** 3–5 mm long. Wet meadows, thickets along streams, wetlands; plains, valleys, lower montane. BC to QC south to CA, AZ, NM, AR, MO. (p.538)

Both ssp. *nuttallii*, with mainly alternate, acuminate leaves, and ssp. *rydbergii* (Britton) R.W.Long, with opposite, round-tipped leaves, are reported for MT.[351] Although some plants from northwest MT have mostly alternate leaves, the shape is the same as opposite-leaved plants.

Helianthus pauciflorus Nutt. [*H. rigidus* (Cass.) Desf., *H. subrhomboideus* Rydb., *H. laetiflorus* Pers. misapplied] Rhizomatous perennial. **Stems** erect, usually simple, 30–100 cm. **Herbage** scabrous to hispid-strigose. **Leaves** opposite at least below; blades serrate, lanceolate to ovate, 3–12 cm long. **Involucres** 14–23 mm across; phyllaries broadly lanceolate to ovate, glabrate, minutely ciliate, 6–10 mm long. **Ray ligules** 10–30 mm long. **Disk corollas** 4–8 mm long. **Achenes** 4–6 mm long. Grasslands, meadows, roadsides; plains, valleys. Throughout all of temperate N. America except the southwest. Our plants are ssp. **subrhomboideus** (Rydb.) O.Spring & E.E.Schill.

The larger *Helianthus pauciflorus* ssp. *pauciflorus* is reported for MT, but I have seen no specimens.

Helianthus petiolaris Nutt. Taprooted annual. **Stems** erect, often branched, 15–60 cm. **Herbage** appressed- hispid. **Leaves** alternate; blades entire to serrate, lanceolate, 2–8 cm long. **Involucres** 12–30 across; phyllaries lanceolate, hispid, ciliate, 7–14 mm long. **Ray ligules** 10–25 mm long. **Disk corollas** 4–7 mm long, lobes often deep red. **Achenes** 3–5 mm long. Disturbed soil of grasslands, roadsides; plains, valleys. Throughout temperate N. America. See *H. annuus*.

Helianthus tuberosus L. Jerusalem artichoke Perennial from often tuber-bearing rhizomes. **Stems** erect, usually simple, 40–120 cm. **Herbage** hispid-pubescent. **Leaves** opposite below, alternate above; blades serrate, ovate, 5–15 cm long. **Involucres** 15–25 mm across; phyllaries lanceolate, acuminate, ciliate, puberulent, ca. 1 cm long. **Ray ligules** 2–4 cm long. **Disk corollas** 6–7 mm long. **Achenes** 5–7 mm long. Fields, grasslands; plains. Throughout most of temperate N. America except the southwest U.S. Known from Richland and Sheridan cos.

Heterotheca Cass. Goldenaster

Heterotheca villosa (Pursh) Shinners [*Chrysopsis villosa* (Pursh) Nutt. ex DC.] Hairy goldenaster Taprooted perennial with a simple or branched caudex. **Stems** mostly ascending, woody at the base, 10–50 cm. **Herbage** strigose to hirsute, often stipitate-glandular. **Leaves** cauline, petiolate below; blades oblanceolate to obovate, 1–5 cm long, entire. **Heads** radiate, several  in leafy-bracted corymbiform arrays; peduncles 0–3 cm long. **Involucre** narrowly campanulate, 5–10 mm high; phyllaries imbricate in 4 to 5 series, green, narrowly lanceolate, white-margined, glabrate to strigose, sometimes glandular; receptacle nearly flat, glabrous. **Ray flowers** 10 to 20, yellow, female, fertile; ligules 5–9 mm long. **Disk flowers** 20 to 50, yellow, perfect; corolla 4–7 mm long; style branches flattened with elongate appendages. **Pappus** dimorphic, short, of outer scales and inner capillary bristles. **Achenes** obconic, 2–3 mm long, strigose with 4 to 8 ribs. Grasslands, fields, woodlands, streambanks, roadsides. BC to ON south to CA, AZ, NM, TX, IL, MI. (p.538)

1. Stems sparsely to densely stipitate-glandular, but not densely hirsute . . . . .var. *minor*
1. Stems and upper leaves densely hirsute, eglandular. 2

2. Peduncles very short; heads immediately subtended by leaves; leaves oblong to obovate . . . . . var. *foliosa*
2. Peduncles longer; heads subtended by reduced bracts; leaves more oblanceolate . . . . var. *villosa*

Heterotheca villosa var. **minor** (Hook.) Semple is our most common form; plains, valleys to rarely subalpine; *H. villosa* var. *villosa* and *H. villosa* var. *foliosa* (Nutt.) V.L.Harms occur mainly on the plains, the latter often along streams.

Plants vary greatly in leaf shape and vestiture, sometimes among sympatric populations; var. *ballardii* (Rydb.) Semple is reported for MT[357], but I have seen no specimens.

Hieracium L. Hawkweed

Fibrous-rooted perennial herbs, sometimes with rhizome-like caudex branches or stolons. **Herbage** with milky sap. **Leaves** basal and cauline, alternate, simple. **Inflorescence** paniculate, corymbiform, or heads solitary. **Heads** ligulate; involucre cylindric to campanulate, subtended by tiny bracts (calyculi); phyllaries 5 to 21 in few unequal to subequal series; receptacle flat, pitted, naked. **Ray flowers** perfect, yellow or white with several apical lobes. **Pappus** 1 or 2 series of white or tawny, capillary bristles. **Achenes** cylindric, 10-ribbed, not beaked, glabrous.[117,154]

The introduced, yellow-flowered hawkweeds are taxonomically confusing because of hybridization and agamospermy; there are more published *Hieracium* names in the European flora than for any other genus.[399]

1. Stems subscapose with 0–2 leaves near the base 2
1. Stems with >2 leaves 5

2. Flowers burnt orange (drying pinkish) . . . . *H. aurantiacum*
2. Flowers yellow 3

3. Pappus bristles in 2 series; plants mostly <30 cm tall; subalpine, alpine . . . . *H. gracile*
3. Pappus bristles in 1 series; plants mostly ≥25 cm tall; valleys, montane 4

4. Upper stem and involucre with dense glandular setae. . . . . *H. caespitosum*
4. Glandular setae scattered among setae of upper stem and involucre . . . . *H. praealtum*

5. Flowers white . . . . *H. albiflorum*
5. Flowers yellow 6

6. Upper leaves sparsely hirsute; basal leaves usually withered at flowering . . . . *H. umbellatum*
6. Leaves densely hirsute; the basal present at flowering . . . . *H. scouleri*

Hieracium albiflorum Hook. WHITE-FLOWERED HAWKWEED Plants usually from a simple caudex. **Stems** simple, erect, 20–80 cm. **Herbage** sparsely long-hirsute, glabrate above. **Leaves** basal and cauline, petiolate, sessile above; blades oblanceolate to narrowly elliptic, 2–13 cm long, entire or shallowly dentate. **Heads** 3 to 30; involucres narrowly campanulate, 6–11 mm high; phyllaries linear, sparsely setose, stipitate-glandular. **Rays** 12 to 35, white; ligules 3–5 mm long. **Achenes** 2–4 mm long; pappus tawny. Coniferous forest; valleys to lower subalpine. NT south to CA, NV, UT, CO, SD. (p.544)

Hieracium aurantiacum L. ORANGE HAWKWEED, KING DEVIL Fibrous-rooted from a rhizome; stoloniferous. **Stems** erect, simple, subscapose, 10–70 cm. **Herbage** sparsely long-hirsute, stellate-hairy, stipitate-glandular above. **Leaves** mainly basal, petiolate; blades oblanceolate, 3–15 cm long, entire. **Heads** 3 to 12, clustered; involucres campanulate, 5–9 mm high; phyllaries linear-lanceolate, stellate-hairy, stipitate-glandular, sparsely setose-hirsute. **Rays** 25 to 100, red to orange; ligules 6–8 mm long. **Achenes** ca. 2 mm long; pappus white. Disturbed soil of forest openings, rock slides, roadsides, lawns; valleys to lower subalpine. Introduced to most of non-arid, temperate N. America; native to Europe.

Hieracium caespitosum Dumort.[*H. pratense* Tausch. misapplied, *H. floribundum* Wimm. & Grab.] MEADOW HAWKWEED Fibrous-rooted from a short rhizome, sometimes short-stoloniferous. **Stems** erect, simple, subscapose, 20–80 cm. **Herbage** long-hirsute, stellate-hairy, stipitate-glandular, sparsely on leaves, densely on stems. **Leaves** mainly basal, short-petiolate; blades lanceolate to oblanceolate, 3–15 cm long, entire to obscurely denticulate. **Heads** 5 to 50; involucres campanulate, 6–8 mm high; phyllaries linear-lanceolate, scarious-margined, stellate-hairy, hirsute, densely black setose-glandular. **Rays** 25 to 50, yellow; ligules 3–10 mm long. **Achenes** ca. 1.5 mm long; pappus white. Grasslands, roadsides; valleys to montane. Introduced to eastern and northwest, temperate N. America; native to Europe.

Hieracium praealtum is less hairy and more likely to have long, slender stolons.

Hieracium gracile Hook. [*H. triste* Willd. ex Spreng. var. *gracile* (Hook.) A.Gray] ALPINE HAWKWEED Plants fibrous-rooted. **Stems** simple or branched above, erect, 3–40 cm. **Herbage** glabrate or sparsely tomentose on the stem. **Leaves** mainly basal; blades lanceolate to oblanceolate, entire to obscurely dentate, 1–6 cm long. **Heads** 1 to 25; involucre campanulate, 7–12 mm high; phyllaries linear-lanceolate, black-setose, stipitate-glandular, tomentose. **Rays** 20 to 60, pale yellow;

ligules 2–5 mm long. **Achenes** 1–3 mm long; pappus white. Meadows, turf, fellfields, open forest, often where snow lies late; subalpine, lower alpine, rarely lower. AK to NT south to CA, NM. (p.544)

Hieracium triste, which occurs north of MT, is similar but has several correlated characters that can be used to separate the two species.[154] See *Crepis runcinata*.

Hieracium praealtum Villars ex Gochnat Fibrous-rooted from a short rhizome, often stoloniferous. **Stems** erect, simple, subscapose, 30–80 cm. **Herbage** glabrate to sparsely long-hirsute. **Leaves** mainly basal, short-petiolate; blades oblanceolate, 3–10 cm long, entire. **Heads** 10 to 30; involucres narrowly campanulate, 6–9 mm high; phyllaries scarious-margined, linear-lanceolate, sparsely stellate-hairy, stipitate-glandular, setose-hirsute. **Rays** 60 to 80, yellow; ligules 4–6 mm long. **Achenes** 1–2 mm long; pappus white. Roadsides, grasslands; valleys, montane. Introduced to northeast and northwest U.S. and adjacent Canada; native to Europe.

Montana plants have been variously determined as *Hieracium floribundum* Wimm. & Grab. and *H. piloselloides* Vill.; however, our plants best fit the description of *H. praealtum* in Flora Europea.[399] Researchers in BC have come to the same conclusion.[117,154] See *H. caespitosum*.

Hieracium scouleri Hook. [*H. albertinum* Farr, *H. cynoglossoides* Arv.-Touv.] Plants from slender caudex branches. **Stems** simple, erect, 15–100 cm. **Herbage** long-hirsute, stellate-hairy, sometimes glabrate above. **Leaves** basal and cauline, sessile above; blades lanceolate to oblanceolate, entire, 2–15 cm long. **Heads** 5 to 50; involucre campanulate, 9–13 mm high; phyllaries linear-lanceolate, setose-hirsute, stellate-hairy, sometimes stipitate glandular. **Rays** 20 to 45, yellow; ligules 6–12 mm long. **Achenes** ca. 3 mm long; pappus white or tawny. BC, AB to CA, NV, UT, WY.

1. Herbage becoming glabrate and often glaucous above.....................................var. *scouleri*
1. Herbage setose-hirsute throughout 2

2. Involucre stipitate-glandular with setae <5 mm long ....var. *griseum*
2. Involucre obscurely glandular, densely hirsute, many hairs ≥5 mm long.................var. *albertinum*

Hieracium scouleri var. *scouleri* is found in open, montane coniferous forest; *H. scouleri* var. *griseum* A.Nelson [*H. cynoglossoides*] occurs in grasslands, sagebrush steppe, open forest, mostly valleys, montane; *H. scouleri* var. *albertinum* (Farr) G.W.Douglas & G.A.Allen is found in grasslands, open forest, woodlands, valleys to mostly subalpine.

Hieracium umbellatum L. [*H. canadense* Michx.] Fibrous-rooted from a short rhizome. **Stems** often branched above, erect, 30–90 cm. **Herbage** glabrate, often sparsely hirsute toward base. **Leaves** mainly cauline, basal withered at flowering, sessile above; blades oblanceolate to ovate, entire to shallowly dentate, 3–12 cm long. **Heads** 3 to 20; involucre campanulate, 8–14 mm high; phyllaries linear-lanceolate, glabrate to puberulent or sparsely short-setose. **Rays** 30 to 80, yellow; ligules 6–11 mm long. **Achenes** 2–4 mm long; pappus white or tawny. Open forest, streambanks, woodlands, grasslands, roadsides; valleys, montane. AK to Greenland south to OR, CO, IA, WV.

Two taxa are sometimes recognized based on leaf shape and vestiture; however, these characters are not strongly correlated nor associated with geography or ecology.

Hulsea Torr. & A.Gray

Hulsea algida A.Gray Taprooted perennial herbs from a branched caudex. **Stems** simple, erect, 4–30 cm. **Herbage** glandular, puberulent to villous. **Leaves** basal and cauline, wide-petiolate; blades narrowly oblanceolate, dentate to shallowly, pinnately lobed, 3–10 cm long. **Heads** solitary, radiate; involucres hemispheric, 12–23 mm high; phyllaries glandular, villous, narrowly lanceolate, subequal in 2–4 series; receptacle flat, bumpy. **Ray flowers** female, 30–60, yellow; ligules ca. 1 cm long. **Disk flowers** perfect, yellow, glandular, 5–8 mm long; tubes shorter than the throat; style branches hairy. **Pappus** of 4 lacerate, hyaline scales. **Achenes** clavate, 8–11 mm long, strigose. Stony or sandy, non-calcareous soil of fellfields, talus; alpine. MT, OR, ID, WY, CA, NV.

Hymenopappus L'Her.

Hymenopappus filifolius Hook. [*H. polycephalus* Osterh.] Taprooted perennial herbs, often with a branched caudex. **Stems** erect, often branched above, 6–40 cm. **Herbage** sparsely to densely tomentose. **Leaves** cauline and mainly basal, petiolate, 1–8 cm long; blades deeply 1–2-times pinnately divided into mostly remote, linear lobes. **Inflorescence** corymbiform with

few to numerous heads. **Heads** discoid; involucres hemispheric, 6–11 mm high; phyllaries 5–13, lanceolate to oblong, tomentose, in 2–3 unequal series, membranous-margined; receptacle flat, naked. **Disk flowers** perfect, yellow, 2–4 mm long, glandular; tubes ca. as long as the dilated throat; style branch appendages hairy. **Pappus** of several short, hyaline scales. **Achenes** obconic, 4–5-angled, 4–5 mm long, densely strigose. Grasslands, sagebrush steppe, coniferous woodlands, eroding slopes; plains, valleys to montane. AB, SK south to CA, AZ, NM, TX, Mexico.

1. Stem leaves 0–3, terminal lobes 2–6 mm long; pappus <1 mm long. . . . . var. *luteus*
1. Stem leaves 3–8, terminal lobes 10–30 mm long; pappus ca. 1 mm long . . . . .var. *polycephalus*

Hymenopappus filifolius var. ***luteus*** (Nutt.) B.L.Turner is known from Carbon Co. ***Hymenopappus filifolius*** var. ***polycephalus*** (Osterh.) B.L.Turner is more widespread.

Hymenoxys Cass.

Perennial herbs. **Leaves** basal and cauline, petiolate, alternate; blades simple or deeply divided into remote linear lobes. **Inflorescence** paniculate or heads solitary. **Heads** radiate; involucre ca. hemispheric; phyllaries in 2 to 3 series; receptacle convex, naked. **Ray flowers** female, 5 to 35, yellow, obviously 3-lobed at the tip. **Disk flowers** numerous, perfect, yellow; tube shorter than the throat; style branches flattened. **Pappus** of hyaline, pointed scales. **Achenes** obconic, 5-angled, strigose.

1. Heads solitary; involucre 15–25 mm high. . . . . *H. grandiflora*
1. Heads usually >1; involucre 5–17 mm high 2

2. Involucre 12–17 mm high; leaves entire. . . . .*H. hoopesii*
2. Involucre 5–10 mm high; leaves deeply divided. . . . . *H. richardsonii*

Hymenoxys grandiflora (Torr. & A.Gray) K.F.Parker Caudex unbranched; plants monocarpic? **Stems** erect, simple or branched at the base, 10–30 cm. **Herbage** sparsely to densely villous. **Leaves** basal and cauline, 2–7 cm long; blades pinnately 1- to 2-times divided into remote, linear segments. **Heads** solitary; involucres 15–25 mm × 2–3 cm; phyllaries subequal, linear-lanceolate, villous, united basally. **Rays** 15 to 34; ligules 12–25 mm long. **Disk corollas** 4–7 mm long. **Achenes** 3–4 mm long; pappus scales 5 to 7. Meadows; turf; upper subalpine to alpine. MT, ID, UT, WY, CO.

Hymenoxys hoopesii (A.Gray) Bierner [*Helenium hoopesii* A.Gray] Plants with a short rhizome. **Stems** erect, branched above, 20–80 cm. **Herbage** glabrate to woolly-villous. **Leaves** basal and cauline, 10–30 cm long; blades oblanceolate, entire, sessile above. **Heads** 1 to 12; involucres 12–17 mm high; phyllaries subequal, lanceolate, basally connate, tomentose. **Rays** 17 to 25, orange-yellow; ligules 15–30 mm long. **Disk corollas** 4–5 mm long. **Achenes** 3–5 mm long; pappus scales 5 to 7. Meadows; montane. OR to MT south to CA, AZ, NM. One collection from Beaverhead Co.[229]

Hymenoxys richardsonii (Hook.) Cockerell Caudex usually branched. **Stems** erect, often branched above, 5–20 cm. **Herbage** glandular-punctate, villous, sparsely so above the base. **Leaves** basal and cauline, 1–15 cm long; blades pinnately divided into 3 or 5 remote, linear segments. **Heads** 1 to 5; involucres 5–10 mm high; phyllaries villous, glandular, in 2 unequal series; outer lanceolate, united at the base; inner obovate, mucronate. **Rays** 8 to 14; ligules 7–15 mm long, glandular. **Disk corollas** 3–4 mm long, glandular. **Achenes** 2–3 mm long; pappus scales 5 to 6. Usually barren, heavy soil of grasslands, sagebrush steppe, coniferous woodlands, barren slopes; plains, valleys. AB, SK south to AZ, NM, TX. Our plants are var. *richardsonii*.

Hypochaeris L. Cat's ear

Hypochaeris radicata L. Taprooted, scapose, perennial herbs. **Stems** erect, often branched, 15–60 cm. **Herbage** hirsute below, glabrate above; sap milky. **Leaves** basal, petiolate; blades oblanceolate, runcinate, 2–15 cm long. **Inflorescence** open-paniculate, minutely bracteate with 2 to 7 heads. **Heads** ligulate; involucres turbinate, 1–2 cm high; phyllaries unequal, imbricate in 3 or 4 series, linear-lanceolate, glabrate with scarious margins; receptacle flat with scales among the flowers. **Ray flowers** 20 to 100, perfect, yellow; ligules 6–12 mm long. **Pappus** 2 series of capillary bristles; outer barbed; inner plumose. **Achenes** fusiform, 6–10 mm long, ribbed, muricate with a beak 3–5 mm long. Roadsides, lawns; valleys. Introduced to eastern and western U.S.; native to Europe.

Ionactis Greene

Perennial herbs with branched caudexs. **Leaves** cauline, alternate, simple. **Heads** solitary, radiate, campanulate; phyllaries in 2 to 6 overlapping, mostly unequal series, keeled, scarious-margined; receptacle pitted, glabrous. **Ray flowers** female, blue. **Disk flowers** perfect, yellow; style branches flattened with lanceolate appendages. **Pappus** of numerous, barbellate, capillary bristles in 2 series. **Achenes** usually 2- or 3-nerved.[305]

Species of *Ionactis* were formerly placed in *Aster*.

1. Leaves 3–12 mm long, lanceolate to oblanceolate; stems ≤15 cm . . . . *I. alpina*
1. Leaves linear, ≥10 mm long; stems often >15 cm . . . .*I. stenomeres*

Ionactis alpina (Nutt.) Greene [*Aster scopulorum* A.Gray] Perennial from a branched caudex. **Stems** sterile and fertile, erect, 4–15 cm. **Herbage** scabrous with pubescent to tomentose stems. **Leaves** cauline sessile, lanceolate to oblanceolate, entire, 3–12 mm long. **Inflorescence** a solitary head; peduncles villous. **Involucre** turbinate, 7–11 mm high; phyllaries linear, villous, glandular, keeled, scarious-margined. **Rays** 7 to 21, blue; ligules 6–15 mm long. **Disk flowers** yellow; corolla 5–7 mm long. **Pappus** in 2 series. **Achenes** linear-obovoid, 5–6 mm long, sericeous. Sagebrush steppe, dry grasslands, often in sandy soil; valleys, montane. OR to MT south to CA, NV, UT.

Ionactis stenomeres (A.Gray) Greene [*Aster stenomeres* A.Gray] Perennial from a branched caudex. **Stems** sterile and fertile, erect, 8–30 cm. **Herbage** scabrous to hispid with villous stems. **Leaves** cauline, sessile, linear, entire, 1–3 cm long; the lowest deciduous. **Inflorescence** a solitary head; peduncles puberulent to villous. **Involucre** turbinate, 8–13 mm high; phyllaries linear-lanceolate, villous, glandular, keeled, scarious-margined. **Rays** 7 to 21, blue; ligules 1–2 cm long. **Disk flowers** yellow; corolla 7–9 mm long. **Pappus** in 2 series. **Achenes** linear-obovoid, 5–6 mm long, glandular. Open forest, meadows, sagebrush steppe, often in sandy soil; upper montane, lower subalpine. Endemic to southeast BC, central ID, adjacent MT.

Iva L.

Annual or perennial herbs. **Leaves** cauline, opposite at least below. **Inflorescence**: heads pedunculate, solitary or in racemes in upper leaf axils. **Heads** disciform; involucre hemispheric; phyllaries in 1 series, connate below; receptacle convex with membranous scales between the flowers. **Outer flowers** few, female, white, tubular or absent. **Disk flowers** male, white, tubular with a non-functional style. **Pappus** absent. **Achenes** obovoid, compressed, glabrous.

1. Annuals; leaves ovate to deltoid, serrate, long-petiolate . . . . *I. xanthifolia*
1. Perennials; leaves oblanceolate to obovate, entire, short-petiolate to sessile . . . . *I. axillaris*

Iva axillaris Pursh POVERTY WEED Rhizomatous perennial. **Stems** erect, usually branched, 10–40 cm. **Herbage** strigose, sparsely to densely glandular. **Leaves** short-petiolate; blades oblanceolate to obovate, entire, 5–50 mm long. **Inflorescence** of solitary pedunculate heads in axils of reduced upper leaves. **Involucres** 2–4 mm high; phyllaries 5 to 8, obovate, ciliate, strigose. **Female flowers** ca. 5; corolla ca. 1 mm long. **Male flowers** 4 to 20, obconic; corolla 2–3 mm long. **Achenes** 2–3 mm long, gland-dotted. Gravelly or fine-textured soil of grasslands, stream terraces; plains, valleys. BC to WI south to CA, AZ, NM, TX.

Iva xanthifolia Nutt. [*Cyclachaena xanthifolia* (Nutt.) Fresen.] MARSH ELDER Taprooted annual. **Stems** erect, unbranched below, 30–100 cm. **Herbage** strigose to canescent; stem villous above. **Leaves** long-petiolate; blades ovate to deltoid, cuneate to cordate, serrate, 5–20 cm long. **Inflorescence** of numerous, branched, axillary racemes. **Involucres** 2–3 mm high; phyllaries 5 to 8, obovate, mucronate, villous, glandular. **Female flowers** 5, obscure. **Male flowers** 5 to 10; corolla ca. 2 mm long. **Achenes** 2–3 mm long, glabrate. Streambanks, roadsides, agricultural fields; plains, valleys. Throughout temperate N. America. Considered native to prairie states, adventive elsewhere. (p.544)

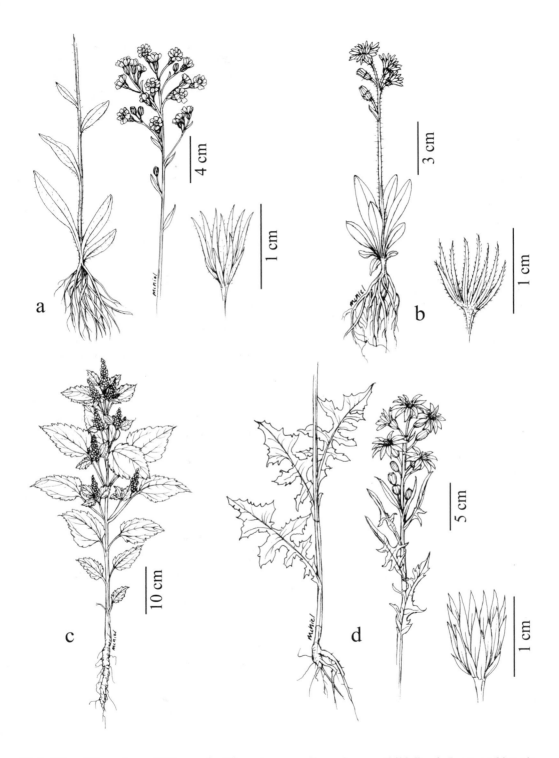

Plate 96. a. *Hieracium albiflorum* b. *Hieracium gracile*, c. *Iva xanthiifolia*, d. *Lactuca biennis*

Lactuca L. Lettuce

Annual to perennial herbs. **Stems** erect, simple below the inflorescence. **Leaves** cauline, alternate, simple with entire to lobed margins, broad-petiolate below to sessile and often clasping above. **Herbage** glabrate with milky sap, sometimes spiny. **Inflorescence** paniculate to corymbiform. **Heads** ligulate; involucre turbinate, subtended by small bracts (calyculi); phyllaries lanceolate, green, imbricate in 2 to 3 series, usually glabrous, often with orange resin dots; receptacle flat, naked. **Ray flowers** perfect, yellow or blue, 5-lobed at the tip. **Pappus** 40–80, mostly white, capillary bristles. **Achenes** obovate, compressed with few to many veins, beaked or expanded at the summit.

1. Rays 5, yellow . . . . *L. muralis*
1. Rays >10, yellow or blue 2

2. Achene beak stout, inconspicuous, ≤1 mm long 3
2. Achene beak slender, >1 mm long 4

3. Taprooted biennial; usually >60 cm tall; pappus tawny . . . . *L. biennis*
3. Plants rhizomatous; often <60 cm tall; pappus white . . . . *L. pulchella*

4. Stems prickly below; leaf blades prickly beneath; achenes 1-nerved on each face . . . . *L. serriola*
4. Stems and leaf blades little or not prickly; achenes with >1 nerve on each face 5

5. Involucre 10–15 mm high; achene beak 1–3 mm long . . . . *L. canadensis*
5. Involucre 15–22 mm high; achene beak 2–5 mm long . . . .*L. ludoviciana*

Lactuca biennis (Moench) Fernald Annual or biennial. **Stems** 60–200 cm. **Leaf blades** ovate, 10–40 cm long, deeply pinnately lobed; lobes deltoid, dentate, sparsely villous on the margins. **Involucre** 7–12 mm high; inner phyllaries reflexed in fruit. **Rays** 15 to 30, blue; ligules 1–5 mm long. **Achenes** 4–5 mm long; beaks ≤0.5 mm long, stout; pappus tawny. Moist, riparian, forest, thickets; valleys, montane. AK to NL south to CA, NM, IL, NC. (p.544)

Lactuca canadensis L. Annual or biennial. **Stems** erect, simple below, 30–200 cm. **Leaf blades** lanceolate, 10–35 cm long, sometimes pilose on the midvein below, deeply pinnately divided into lanceolate lobes. **Involucre** 10–15 mm high; phyllaries reflexed in fruit. **Rays** 15 to 20; ligules blue to yellow, ca. 4 mm long. **Achenes** 5–6 mm long; beaks 1–3 mm long. Meadows, thickets, open forest; valleys, plains; known from Park Co. Throughout N. America.

Lactuca ludoviciana (Nutt.) Riddell Biennial. **Stems** 30–150 cm. **Leaf blades** oblong, 7–30 cm long, pinnately shallowly lobed with a large terminal lobe, prickly below and on the margins, clasping above. **Involucre** 15–22 mm high; phyllaries reflexed in fruit. **Rays** 20 to 50, yellow or blue; ligules 5–6 mm long. **Achenes** 4–5 mm long; beaks 2–5 mm long. Meadows; plains; known from Big Horn and Meagher cos. SK to ON south to CA, AZ, NM, TX, LA, KY.

Lactuca muralis (L.) Gaetrn. [*Mycelis muralis* (L.) Dumort.] Glabrous and glaucous annual. **Stems** 30-90 cm. **Leaf blades** oblanceolate, 6-18 cm long, deeply pinnately lobed; lobes deltoid, shallowly lobed. **Involucre** 7-11 mm high; phyllaries reflexed in fruit. **Rays** 5, yellow; ligules ca. 7 mm long. **Achenes** ca. 3 mm long; the beak 0.5-1 mm long; pappus white. Disturbed roadsides in moist forest; valleys; one collection from Sanders Co. Introduced to northeast and northwest N. America; native to Europe.

Lactuca pulchella (Pursh) DC. [*L. oblongifolia* Nutt., *L. tatarica* (L.) C.A.Mey. var. *pulchella* (Pursh) Breitung, *Mulgedium pulchellum* (Pursh) G.Don.] Rhizomatous perennial. **Stems** erect, simple below, 20–90 cm. **Leaf blades** narrowly lanceolate to oblanceolate, 4–12 cm long; pinnately runcinately lobed to entire. **Involucre** 11–18 mm high; phyllaries erect in fruit. **Rays** 15 to 50, blue; ligules 7–12 mm long. **Achenes** 4–5 mm long; beaks ≤1 mm long. Grasslands, moist meadows, sagebrush steppe, fields; plains, valleys. Throughout all of temperate N. America except the southeast U.S.

Our only rhizomatous species. The taxonomic disposition of this species is unsettled; some authorities consider this conspecific with the European *Lactuca tatarica*, while others place it in a separate genus. Reveal considers Nuttall's specific epithet *oblongifolia* to take precedence over *pulchella*,[336] but Cronquist believes it was not validly published.[103]

Lactuca serriola L. Prickly lettuce Annual or biennial. **Stems** erect, 20–100 cm, simple and bristly below. **Leaves** clasping, oblong to oblanceolate, 5–20 cm long, prickly below and on the margins, dentate and sometimes with remote, deltoid lobes. **Involucre** 5–15 mm high; inner phyllaries purple-tipped, reflexed in fruit. **Rays** 12 to 20, yellow, drying blue; ligules 4–6 mm long.

Achenes 2–4 mm long; beaks 2–4 mm long. Disturbed soil of grasslands, fields, roadsides; plains, valleys. Introduced throughout temperate N. America; native to Europe.

Lagophylla Nutt.

Lagophylla ramosissima Nutt. Taprooted annual. **Stems** erect, branched, 10–25 cm. **Herbage** hirsute-canescent. **Leaves**: lower opposite, early deciduous; upper alternate, clustered beneath heads, short-petiolate; blades oblanceolate, entire, <1 cm long. **Inflorescence** glomerules of few heads in paniculate arrays. **Heads** radiate, inconspicuous; involucres campanulate, ca. 4 mm high; phyllaries 5 in 1 series, subequal, lanceolate, villous, stipitate-glandular, each enfolding a ray flower; receptacle hairy with scales between the disk flowers. **Ray flowers** 5, female, yellow, deeply 3-lobed at the tip; ligules 2–6 mm long. **Disk flowers** 6, functionally male; corollas yellow, tubes shorter than throat. **Pappus** absent. **Achenes** compressed-obovoid, glabrous. Disturbed soil of grasslands; valleys. WA to MT south to CA, NV. Known from Sanders Co.

Lapsana L.

Lapsana communis L. Fibrous-rooted annual. **Stems** ascending to erect, often branched above, 10–30 cm. **Herbage** villous to hirsute, stipitate-glandular. **Leaves** basal and cauline, alternate, winged-petiolate; blades oblanceolate to ovate, dentate; basal sometimes runcinately lobed, 5–10 cm long. **Inflorescence** open-paniculate. **Heads** ligulate; involucres narrowly campanulate, 5–7 mm high; phyllaries 8 to 10 in 1 series, linear-oblanceolate, subequal, glabrous; receptacle flat, naked. **Ray flowers** 8 to 15, perfect; ligules whitish, 3–5 mm long, green-tipped. **Pappus** absent. **Achenes** fusiform, glabrous, many-ribbed, 3–5 mm long. Lawns; valleys; collected in Missoula Co. Introduced to much of temperate N. America, more common in humid areas; native to Eurasia.

Leucanthemum Mill. Daisy

Leucanthemum vulgare Lam.[*Chrysanthemum leucanthemum* L.] Rhizomatous perennial. **Stems** erect, usually simple, 20–80 cm. **Herbage** glabrate. **Leaves** basal and cauline, alternate; basal long-petiolate, the blades obovate, shallowly lobed or crenate, 1–4 cm long; cauline oblanceolate, becoming sessile, 1–4 cm long. **Heads** radiate, solitary; peduncles 3–20 cm long. **Involucre** hemispheric, 1–2 cm wide; phyllaries imbricate in 2 to 4 series, lanceolate, green with brown scarious margins; receptacle nearly flat, glabrous. **Rays** 15 to 35, female, fertile; ligules white, 1–2 cm long. **Disk flowers** numerous, perfect; corolla yellow, 2–3 mm long; style branches flattened. **Pappus** absent. **Achenes** cylindric, 1–3 mm long, ribbed, glabrous. Roadsides, fields, meadows, often in forest openings or pastures converted from forest; valleys, montane. Introduced throughout temperate N. America. (p.550)

Liatris Gaertn. ex Schreb. Gay feather, blazing star

Perennial herbs from a thickened, often elongate root. **Stems** erect, unbranched. **Leaves** basal and cauline, alternate, entire, short-petiolate below. **Inflorescence** racemose to spike-like, leafy-bracteate. **Heads** discoid; involucres campanulate; phyllaries imbricate in 3 to 7 series; receptacle flat, naked. **Disk flowers** perfect, pink to purple, funnelform, long-lobed, glandular; style branch appendages purplish, petal-like. **Pappus** of barbed to plumose bristles in 1 or 2 series. **Achenes** obconic, angled, ca.10-ribbed, pubescent.

1. Disk flowers 4 to 8, ≥6 mm long; involucre campanulate . . . .*L. punctata*
1. Disk flowers 10, ≤7 mm long; involucre hemispheric . . . .*L. ligulistylis*

Liatris ligulistylis (A.Nelson) K.Schum. **Root** globose. **Stems** 20–100 cm. **Herbage** sparsely hirsute, punctate. **Leaf blades** oblanceolate, 1–15 cm long, becoming linear above. **Involucres** hemispheric, 10–15 mm high; phyllaries in 3 to 5 series, oblong, scarious-margined, purple-tipped. **Disk flowers** 30 to 70; corolla 5–7 mm long. **Achenes** 5–7 mm long; pappus barbed. Meadows, often along streams; plains. AB to MB south to NM, SD, IA.

Liatris punctata Hook. SPOTTED GAY FEATHER **Root** tuberous, often branched. **Stems** 10–70 cm. **Herbage** glabrous, punctate; leaves ciliate. **Leaf blades** linear, 2–15 cm long. **Involucres** campanulate, 7–15 mm high; phyllaries in 3 to 6 series, oblong to oblanceolate, glandular-punctate, mucronate; inner purple. **Disk flowers** 4 to 8; ligules 6–14 mm long. **Achenes** 6–9 mm long; pappus plumose. Grasslands, sagebrush steppe; plains, valleys. AB to MI south to NM, TX, LA, IL, OH. Our plants are var. *punctata*.

Lorandersonia Urbatsch, R.P.Roberts & Neubig

Lorandersonia linifolia (Greene) Urbatsch, R.P.Roberts & Neubig [*Chrysothamnus linifolius* Greene] Shrubs 0.5–2 m high with spreading roots. **Stems** glabrous; twigs whitish. **Leaves** alternate, entire, linear-lanceolate, glabrous, 2–8 cm long with 1 to 3 veins. **Inflorescence** corymbiform with numerous heads on short peduncles. **Heads** discoid; involucres obconic, 4–7 mm high; phyllaries imbricate in 3 or 4 series, narrowly lanceolate, whitish, green-tipped; receptacle convex, glabrous. **Disk flowers** 4 to 6, perfect; corolla yellow, 5–6 mm long, tube shorter than throat; style branches linear. **Pappus** barbellate, capillary bristles. **Achenes** cylindric, 2–4 mm long with 10 to 12 ribs. Fine-textured, often saline soils of stream terraces; valleys. ID, MT south to AZ, NM. Known from Carbon and Carter cos.

Lygodesmia D. Don Skeleton plant

Lygodesmia juncea (Pursh) D.Don ex Hook. Rhizomatous perennial. **Stems** erect, branched, 15–50 cm. **Herbage** glabrous. **Leaves** cauline, alternate, linear, 5–30 mm long; the upper reduced to scales. **Heads** ligulate, solitary on branch tips; involucre obconic, 9–16 mm high; phyllaries in 2 series; outer minute; inner ca. 5, linear, glabrous, scarious-margined; receptacle flat, naked. **Ray flowers** perfect, ca. 5; ligules 9–12 mm long, pinkish. **Pappus** off-white, capillary bristles. **Achenes** fusiform, 6–10 mm long, 5-ribbed, glabrous. Grasslands, sagebrush steppe, open forest, roadsides; plains, valleys. BC to MB south to AZ, NM, TX, IN, MI.

    Globose galls are common on the stems. *Lygodesmia rostrata* (A.Gray) A.Gray [*Shinnersoseris rostrata* (A.Gray) Tomb], an annual with opposite leaves, occurs in WY, ND, AB, and could occur in MT.

Machaeranthera Nees

Annual to perennial, taprooted herbs. **Leaves** cauline, alternate, petiolate; blades spiny, dentate to lobed. **Inflorescence** heads terminating branches in paniculate or corymbiform arrays. **Heads** radiate; involucre campanulate; phyllaries strongly imbricate in 3 to 8 series, lanceolate, recurved, thickened and whitish below; receptacle flat, naked, pitted. **Ray flowers** 8 to 40, female. **Disk flowers** perfect, yellow; tubes shorter than the throat; style branch appendages short-hairy. **Pappus** of unequal, capillary bristles. **Achenes** obconic, several-veined.

    A report of *Machaeranthera commixta* Greene[228] was based on a misidentified specimen of *M. canescens*.

1. Leaf blades remotely dentate to entire . . . . *M. canescens*
1. Leaf blades pinnately dissected . . . . *M. tanacetifolia*

Machaeranthera canescens (Pursh) A.Gray [*Dieteria canescens* (Pursh) Nutt., *Aster canescens* Pursh] Perennial. **Stems** ascending to erect, branched, 8–50 cm. **Herbage** canescent. **Leaf blades** linear to oblanceolate, 1–5 cm long, remotely dentate to entire. **Involucre** 6–9 mm high; phyllaries glandular, glabrate to canescent, oblanceolate. **Rays** 8 to 25, purple; ligules 5–10 mm long. **Disk corollas** 4–6 mm long. **Achenes** 3–4 mm long, strigose. Sparsely vegetated soil of grasslands, sagebrush steppe, badlands, sandhills, roadsides; plains, valleys. BC to SK south to CA, AZ, NM, TX. Our plants are var. *canescens*.

    Machaeranthera canescens is placed in *Dieteria* by some authors; however, the evidence for this change seems equivocal.[282]

Machaeranthera tanacetifolia (Kunth) Nees [*Aster tanacetifolia* Kunth] Annual or biennial. **Stems** erect, often branched, 5–20 cm. **Herbage** puberulent, stipitate-glandular. **Leaf blades** 2–10 cm long, deeply 2 to 3 times finely pinnately dissected. **Involucre** 6–11 mm high; phyllaries glandular, puberulent, linear-lanceolate, acuminate. **Rays** 8 to 40, purple; ligules 8–15 mm long. **Disk corollas** 4–6 mm long. **Achenes** 2–4 mm long, strigose. Sparsely vegetated, often sandy soil of grasslands, shrub steppe; plains, valleys. AB south to CA, AZ, NM, TX.

Madia Molina Tarweed

Taprooted, annual herbs. **Leaves** simple, opposite below, alternate above, entire. **Herbage** glandular, aromatic. **Inflorescence**: heads solitary or in glomerules in upper leaf axils. **Heads** discoid or radiate; involucre fusiform to globose; phyllaries subequal in 1 series, each subtending a ray flower; receptacle flat to convex with scales between disk flowers. **Ray flowers** 0 to 22, female, yellow. **Disk flowers** perfect or functionally male, yellow, tube shorter than throat; style branches hairy. **Pappus** absent. **Ray achenes** oblong to conical, terete to compressed, glabrate.

1. Involucre <5 mm high 2
1. Involucre ≥6 mm high 3

2. Stems branched near the base; leaves mostly opposite . . . . *M. minima*
2. Stems simple below the inflorescence; leaves alternate above . . . . *M. exigua*

3. Heads narrowly ellipsoid, 2–5 mm wide (when pressed) . . . . *M. glomerata*
3. Heads ovoid to obovoid, 6–12 mm wide 4

4. Inflorescence of glomerules at tips of short axillary branches . . . . *M. sativa*
4. Inflorescence of solitary or few, short-pedicelate heads in upper leaf axils . . . . *M. gracilis*

Madia exigua (Sm.) A.Gray **Stems** erect, simple, 8–30 cm. **Herbage** hirsute, glandular above. **Leaves** linear, 8–30 mm long. **Inflorescence**: heads solitary on axillary stem tips. **Involucre** globose, 2–4 mm high; phyllaries 4 to 8, hirsute, stipitate-glandular. **Ray ligules** ca. 1 mm long. **Disk flowers** solitary. **Achenes** 2–3 mm long, compressed with a curved beak. Disturbed soil of grasslands, open forest, rock outcrops; valleys. BC south to CA, NV, ID, MT.

Madia glomerata Hook. **Stems** erect, simple, 7–60 cm. **Herbage** villous, glandular above. **Leaves** linear to linear-lanceolate, 1–5 cm long. **Inflorescence** of glomerules at stem tips. **Involucre** fusiform, 6–9 mm high; phyllaries 4 to 6, stipitate-glandular, villous. **Rays** 0 to 3; ligules 1–3 mm long. **Disk flowers** 1–5; corolla 3–5 mm long. **Achenes** compressed, beakless, 5–6 mm long. Vernally moist, disturbed soil in open woods, grasslands, shallow ponds, roadsides; valleys, montane. AK to QC south to CA, AZ, NM, IA, MI. (p.550)

Madia gracilis (Sm.) D.D.Keck **Stems** erect, simple, 10–100 cm. **Herbage** hirsute, stipitate-glandular above. **Leaves** linear to linear-oblanceolate, 1–11 cm long. **Inflorescence** of solitary or few, short-pedicelate heads in upper leaf axils. **Involucre** obovoid, 6–11 mm high; phyllaries ca. 8, stipitate-glandular. **Rays** 3 to 10; ligules 1–8 mm long. **Disk flowers** 2–16; corolla 2–5 mm long. **Achenes** compressed, ca. beakless, ca. 3 mm long. Roadsides, other disturbed sites; BC south to CA, NV, UT; perhaps introduced in Sanders Co., MT.

Madia minima (A.Gray) D.D.Keck [*Hemizonella minima* (A.Gray) A.Gray] **Stems** ascending to erect, branched, 2–15 cm. **Herbage** villous, stipitate-glandular above. **Leaves** mostly opposite, linear to linear-lanceolate, 5–20 mm long. **Inflorescence** of small terminal glomerules. **Involucre** obovoid, 2–3 mm high; phyllaries 3 to 5, stipitate-glandular. **Rays** 3 to 5; ligules ca. 1 mm long. **Disk flowers** 1 or 2; corolla 1–3 mm long. **Achenes** compressed, ca. 2 mm long, minutely beaked. Often shallow soil of open forest, mossy rock outcrops; valleys, montane. BC south to CA, NV, ID, MT.
   Madia sensu lato is sometimes split into segregate genera, including *Hemizonella*.[73]

Madia sativa Molina **Stems** erect, simple, 20–60 cm. **Herbage** villous, stipitate-glandular. **Leaves** lanceolate, 2–8 cm long. **Inflorescence** of glomerules at tips of short axillary branches. **Involucre** obovoid, 8–14 mm high; phyllaries 5 to 13, stipitate-glandular, villous. **Rays** 5 to 13; ligules 2–4 mm long. **Disk flowers** 1 to 5; corolla 2–5 mm long. **Achenes** compressed, beakless, ca. 6 mm long. Fields, roadsides; valleys. BC to CA, sporadically introduced inland. Collected in Gallatin Co. over 100 years ago.

Malacothrix D.C. Desert dandelion

Malacothrix torreyi A.Gray Taprooted annual. **Stems** erect, branched at the base and in the inflorescence, 10–20 cm. **Herbage** with milky sap, sparsely stipitate-glandular. **Leaves** basal and cauline, broad-petiolate; blades oblong to oblanceolate, 2–8 cm long, deeply pinnately divided into deltoid lobes. **Inflorescence** leafy-bracted, paniculate. **Heads** ligulate; involucres campanulate, 8–13 mm high; phyllaries unequal; outer short, deltoid; inner subequal in 2 to 3 series, narrowly lanceolate, glabrous to glandular; receptacle mostly flat with few setae among flowers. **Ray flowers** perfect, yellow; ligules 7–10 mm long. **Pappus** of deciduous capillary bristles and 1 to 5 thickened, persistent bristles. **Achenes** cylindric, 3–4 mm long with 5 wing-like ribs, glabrous. Sandy soil of shrub steppe; valleys of Carbon Co. OR to MT south to CA, AZ, CO.

Matricaria L. Wild chamomile

Annuals or biennials. **Leaves** alternate, petiolate, deeply and finely dissected. **Herbage** glabrous. **Inflorescence** of solitary heads on stem tips. **Heads** radiate or discoid; involucre hemispheric; phyllaries scarious-margined with a green or brown midvein, subequal in 2 to 3 series; receptacle hemispheric to conic; naked. **Ray flowers** absent or white, female. **Disk flowers** perfect, yellow, throats narrow; style branches flattened. **Pappus** a low crown or absent. **Achenes** obconic, ribbed, slightly compressed, glabrous.

1. Heads radiate; rays white; herbage unscented . . . . *M. maritima*
1. Heads discoid; herbage pineapple-scented . . . . *M. matricarioides*

Matricaria maritima L. [*Tripleurospermum inodorum* (L.) Sch.Bip.] **Stems** ascending to erect, hollow, 15–40 cm. **Herbage** inodorous. **Leaves** 2–8 cm long, twice pinnately divided into filiform segments. **Heads** radiate; involucre 5–10 mm high; phyllaries linear-lanceolate; receptacle hemispheric. **Ray flowers** 10 to 25, white; ligules 6–12 mm long. **Disk corollas** 2–3 mm long. **Pappus** a low crown. Fields, roadsides; valleys. Introduced to most of N. America; native to Europe.
There is evidence for separating *Tripleurospermum* from *Matricaria*.[57,310]

Matricaria matricarioides (Less.) Porter [*M. discoidea* DC.] PINEAPPLE WEED **Stems** erect, often branched, 4–30 cm. **Herbage** pineapple-scented. **Leaves** 1–5 cm long, 1–3 times divided into crowded, short, linear-lanceolate segments. **Heads** discoid; involucre 2-5 mm high; phyllaries lanceolate to ovate. **Disk flowers** 1–1.5 mm long. **Pappus** absent. Disturbed, often compacted soil of roadsides, fields, gardens; plains, valleys. Introduced throughout N. America, native to the Pacific Northwest and Asia.
Cronquist maintains that *matricarioides* is the correct specific epithet because *discoidea* was not properly published;[103] Brouillet believes *discoidea* should be used.[62]

Microseris D.Don

Microseris nutans (Hook.) Sch.Bip. Perennial from a tuberous taproot. **Stems** erect, often branched above, 5–35 cm. **Herbage** glabrate with milky sap. **Leaves** basal and few cauline, linear, 2–20 cm long, entire or with pairs of remote, linear lobes. **Heads** ligulate, solitary on branch tips, nodding in bud; involucre campanulate, 10–18 mm high; phyllaries lanceolate in 2 unequal series; outer small; inner acuminate, black-villous, sometimes farinose; receptacle flat, naked. **Ray flowers** perfect, few to 75; ligules 7–15 mm long, yellow with purple veins. **Pappus** 15 to 30 scales, each with a plumose capillary bristle. **Achenes** fusiform, 3–8 mm long, 10- to 15-ribbed, glabrous. Moist grasslands, meadows, sagebrush steppe, open forest; valleys, montane. BC, AB south to CA, NV, UT, CO. (p.550)
Nothocalais spp. are scapose; *Crepis* spp. have pappus of capillary bristles.

Nothocalais (A.Gray) Greene

Perennials with thickened taproots and milky sap. **Stems** erect, simple, scapose. **Leaves** basal, simple, with entire or wavy margins. **Heads** solitary, ligulate; involucre campanulate; phyllaries acuminate, green, in 2 to 5 series, the inner longer and narrower than the outer; receptacle flat, naked. **Ray flowers** perfect, yellow. **Pappus** 10 to 80 white, capillary bristles or awn-like scales ca. 2 times as wide as the bristles and barbless toward the base. **Achenes** fusiform, 10-ribbed, not beaked.
Agoseris spp. have a pappus of only capillary bristles; see *Microseris*. *Nothocalais cuspidata* and *N. troximoides* are similar in appearance and habitat but tend to occur in different parts of the state; both species form hybrids with *N. nigrescens*.[79]

1. Pappus of a mixture of awn-like scales and finer capillary bristles. . . . .*N. cuspidata*
1. Pappus of awn-like scales (widened at the base) only. 2

2. Leaves narrowly lanceolate, entire; some phyllaries ovate . . . .*N. nigrescens*
2. Leaves linear-lanceolate, often wavy-margined; phyllaries lanceolate. . . . . *N. troximoides*

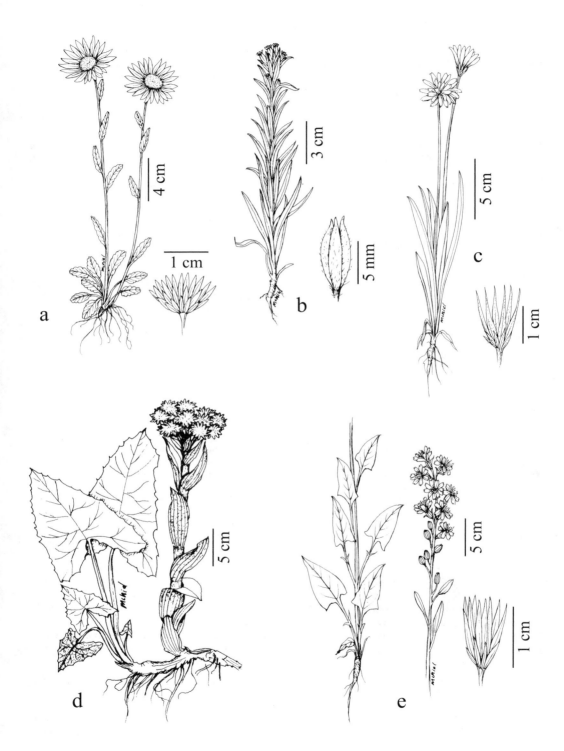

Plate 97. a. ***Leucanthemum vulgare***, b. ***Madia glomerata***, c. ***Microseris nutans***, d. ***Petasites frigidus*** var. ***sagittatus***, e. ***Prenanthes sagittata***

Nothocalais cuspidata (Pursh) Greene [*Microseris cuspidata* (Pursh) Sch.Bip.] **Stems** 5–30 cm. **Herbage** glabrate to villous, densely so on the leaf margins. **Leaves** linear-lanceolate, entire or wavy-margined, 5–20 cm long. **Involucre** 15–27 mm high; phyllaries lanceolate,13 to 34, glabrate, purple or purple-spotted. **Ray flowers** 13 to 80; ligules 1–2 cm long. **Achenes** 7–10 mm long; pappus of bristles and wider awn-like scales. Grasslands, sagebrush steppe, coniferous woodlands; plains, valleys. AB to MB south to NM, TX, MO.

Nothocalais nigrescens (L.F.Hend.) A.Heller [*Microseris nigrescens* L.F.Hend.] **Stems** 5–40 cm. **Herbage** glabrate. **Leaves** narrowly lanceolate to lanceolate, entire, 4–15 cm long. **Involucre** 15–25 mm high; phyllaries ovate, 15 to 50, glabrous, purple-spotted. **Ray flowers** 13 to 100; ligules 10–25 mm long. **Achenes** 6–10 mm long; pappus of 10 to 25 awn-like scales. Meadows, turf; upper montane to alpine. MT, ID, WY.

Nothocalais troximoides (A.Gray) Greene [*Microseris troximoides* A.Gray] **Stems** 5–25 cm. **Herbage** glabrate; leaves minutely ciliate. **Leaves** linear-lanceolate, often wavy-margined, 7–15 cm long. **Involucre** 15–24 mm high; phyllaries lanceolate, 8 to 25, glabrous, purple-spotted. **Ray flowers** 13 to 90; ligules 10–20 mm long. **Achenes** 7–13 mm long; pappus of 10 to 30 awn-like scales. Grasslands, sagebrush steppe; valleys, montane. BC south to CA, NV, UT, WY.

Onopordum L.

Onopordum acanthium L. SCOTCH THISTLE Monocarpic biennials or short-lived perennials. **Stems** erect, branched, 50–200 cm with broad, spiny wings . **Herbage** glabrate to tomentose, spiny on stems and leaf margins. **Leaves** clasping, lanceolate, 10–60 cm long, deeply dentate to shallowly pinnately divided into deltoid lobes. **Inflorescence** 1 to 3 heads at branch tips. **Heads** discoid; involucres ovoid, 20–35 mm high; phyllaries numerous in 8 to 10 series; the outer spine-tipped, glabrate to tomentose; receptacle flat, fleshy, pitted, naked. **Disk flowers** perfect, usually purple; corollas slender, 22–25 mm long; style branches fused. **Pappus** of pinkish, capillary bristles. **Achenes** obovoid, ribbed, 4–5 mm long, scabrous. Pastures, fields, roadsides; valleys. Introduced to most of U.S.; native to Eurasia.

Oreostemma Greene

Oreostemma alpigenum (Torr. & A.Gray) Greene [*Aster alpigenus* (Torr. & A.Gray) A.Gray var. *haydenii* (Porter) Cronquist] Taprooted perennial with a mostly simple caudex. **Stems** simple, lax to ascending, 2–10 cm. **Herbage** glabrous, villous above. **Leaves**: basal petiolate, linear to oblanceolate, 2–8 cm long, entire; cauline linear 5–30 mm long. **Heads** solitary, radiate; involucre turbinate to campanulate, 7–11 mm high; phyllaries in 3 or 4 series, green and purple, linear-lanceolate, ciliate, sparsely villous, little overlapping. **Ray flowers** female, 10 to 45, violet; ligules 7–12 mm long. **Disk flowers** perfect, yellow; corollas 6–9 mm long; style branches with linear-lanceolate appendages. **Pappus** of numerous, stramineous, barbellate, capillary bristles. **Achenes** usually 5- to 10-nerved, ca. 2 mm long. Stony, often sparsely vegetated fellfields, turf, meadows; upper subalpine, alpine. WA to MT south to CA, NV, WY. Our plants are var. *haydenii* (Porter) G.L.Nesom.

Petasites Mill. Coltsfoot

Petasites frigidus (L.) Fr. [*P. sagittatus* (Pursh) A.Gray] Rhizomatous, sometimes dioecious perennial. **Stems** erect, simple 15–60 cm, tomentose. **Leaves** basal and cauline; basal long-petiolate, arising from rhizome nodes after the flowering stems; blades deltoid to sagittate with dentate to lobed margins, 3–25 cm long, tomentose below, glabrate above; stem leaves (bracts) alternate, ovate, acuminate, parallel-veined, tomentose, often purplish, clasping, 2–10 cm long. **Inflorescence** corymbiform, hemispheric, becoming racemose. **Heads** radiate or disciform; involucre campanulate, 9–12 mm high; phyllaries subequal in 1 or 2 series, narrowly lanceolate, puberulent; receptacle flat, naked. **Ray flowers** absent or numerous, usually perfect, white; ligules 2–8 mm long. **Disk flowers** female (outer), perfect or functionally male (inner), white to purple; corollas 7–9 mm long, slender with an expanded throat; style ca. undivided. **Pappus** of capillary bristles. **Achenes** cylindric, 5- to 10-ribbed, glabrous, 2–3 mm long. Wet, organic soil of meadows, fens, often with willow and alder in forest openings; valleys, montane. AK to NL south to CA, CO, SD, WI, NY. (p.550)

1. Leaf margins shallowly toothed . . . . var. *sagittatus*
1. Leaves lobed nearly halfway to midvein. . . . . var. *frigidus*

Petasites frigidus var. *frigidus* is known from the northwest corner of MT; *P. frigidus* var. *sagittatus* (Pursh) Cherniawsky & R.J.Bayer is more widespread. Morphological and molecular studies indicate these taxa are better treated subspecifically;[80] the two vars. intergrade in Glacier National Park.

Picradeniopsis Rydb. ex Britt.

Picradeniopsis oppositifolia (Nutt.) Rydb. [*Bahia oppositifolia* (Nutt.) DC.] Rhizomatous perennial. **Stems** erect, 10–20 cm, usually branched. **Herbage** with minute, sessile glands, densely strigose. **Leaves** cauline, petiolate, mostly opposite; blades 1–3 cm long, deeply divided into 3 (to 5) linear lobes; the basal rarely entire. **Inflorescence** terminal with 1 to several heads. **Heads** radiate; involucre hemispheric, 4–6 mm high; phyllaries 5 to 10, ovate, subequal, green with hyaline margins, strigose; receptacle glabrous. **Ray flowers** 3 to 6, female, yellow; ligules 3–5 mm long. **Disk flowers** 30 to 60, perfect, orangish, glandular; corolla 2–4 mm long, throat much wider than tube. **Pappus** of 8 to 10 short, narrowly ovate scales. **Achenes** glandular, obconic, 3–5 mm long. Clayey, often compacted soil of grasslands, steppe, badlands, roadsides; plains. MT to ND south to AZ, NM, TX.

Platyschkuhria (A.Gray) Rydb.

Platyschkuhria integrifolia (A.Gray) Rydb. [*Bahia nudicaulis* Gray] Perennial from elongate caudex branches. **Stems** erect, unbranched below the inflorescence, 10–25 cm. **Herbage** puberulent, glandular above. **Leaves** basal and cauline, petiolate, alternate; blades narrowly ovate to elliptic, 2–8 cm long, entire. **Inflorescence** corymbiform with few heads. **Heads** radiate; involucre campanulate, 7–10 mm high; phyllaries 9 to 21, narrowly lanceolate, subequal in 2 series, green, glandular; receptacle convex, naked. **Ray flowers** 6 to 12, female, yellow; ligules 6–12 mm long. **Disk flowers** 25 to 80, perfect, orangish, glandular; corollas 5–7 mm long, throat much wider than tube. **Pappus** of 8 to 16 oblanceolate scales. **Achenes** hirsute, obconic, 4-angled, 3–5 mm long. Fine-textured soil of shrub-steppe; valleys. Carbon Co., MT south to AZ, NM.

Pleiacanthus (Nutt.) Rydb. Thorny skeletonweed

Pleiacanthus spinosus (Nutt.) Rydb. [*Lygodesmia spinosa* Nutt., *Stephanomeria spinosa* (Nutt.) Tomb] Taprooted perennial with a brown-woolly branched crown. **Stems** erect, spine-tipped, 15–40 cm, branched. **Herbage** glabrate to weakly tomentose with milky sap. **Leaves** alternate, cauline, linear, entire, 5–30 mm long, inconspicuous. **Inflorescence** heads solitary on upper spine-like branches. **Heads** ligulate; involucre cylindric, 7–13 mm high; phyllaries linear-oblanceolate in 2 series, the outer short; receptacle flat, naked. **Ray flowers** 3 to 5, perfect, pinkish. **Pappus** 2 series of minutely barbed bristles in 2 series, tan. **Achenes** tubular, 6–8 mm long, 5-ribbed, glabrous. Stony, calcareous soil of grasslands; valleys. OR to MT south to CA, AZ. Known from Madison Co.

Neither of our two specimens have flower buds even though they were collected in late July.

Prenanthes L. White lettuce

Rhizomatous, perennial herbs with milky sap. **Leaves** basal and cauline, alternate, simple, winged-petiolate to sessile. **Inflorescence** racemose to narrowly paniculate. **Heads** ligulate; involucre narrowly campanulate; phyllaries in 2 series; the outer short; the inner equal, linear-lanceolate, scarious-margined; receptacle flat, naked. **Ray flowers** perfect, white. **Pappus** of barbed, capillary bristles. **Achenes** cylindric, 5- to 12-ribbed, glabrous.

Prenanthes alata (Hook.) D. Dietr., with a corymbiform inflorescence and tomentulose phyllaries, is reported for MT,[50] but I have seen no specimens.

1. Leaves of mid-stem sessile to cordate-clasping. . . . . *P. racemosa*
1. Most stem leaves petiolate. . . . . *P. sagittata*

Prenanthes racemosa Michx. **Stems** erect, usually simple, 50–120 cm. **Herbage** glabrous; the stem hispid above. **Leaf blades** oblong below, lanceolate and sessile above, entire to dentate, 7–40 cm long. **Inflorescence** racemose to narrowly paniculate. **Involucre** 11–12 mm high; phyllaries setose. **Ray flowers** 9 to 29, white to pink; ligules 7–13 mm long. **Achenes** 5–6 mm long; pappus yellowish. Moist grassland, meadows; valleys, montane. BC to NL south to WA, CO, MO, NY. Reported from an undetermined location in MT.[187] Our plants are ssp. *multiflora* Cronquist.

Prenanthes sagittata (A.Gray) A.Nelson **Stems** erect, usually simple, 15–70 cm. **Herbage** glabrous. **Leaf blades** sagittate, dentate to lobed on the lower half, 3–15 cm long. **Inflorescence** racemose; heads nodding in fruit. **Involucre** 7–13 mm high; phyllaries glabrous. **Rays** 7 to 20, white; ligules 7–9 mm long. **Achenes** ca. 5 mm long; pappus tawny. Moist soil of thickets, wet meadows, open forest, especially along streams; montane, subalpine. BC, AB, ID, MT. (p.550)

The type locality is in Lake Co.

Psilocarphus Nutt.

Taprooted annuals. **Herbage** white- or gray-woolly. **Leaves** cauline, opposite, simple, entire. **Inflorescence**: heads solitary or in glomerules in forks of branches. **Heads** disciform; involucre and phyllaries absent; receptacle hemispheric with tomentose, sac-like scales enwrapping the flowers. **Flowers** slender, dimorphic; outer 8 to 80, female; inner 2 to 10, functionally male. **Pappus** absent. **Achenes** glabrous.

1. Receptacle bracts ca. 2 mm long . . . . *P. tenellus*
1. Receptacle bracts ca. 3 mm long 2

2. Achenes cylindric, terete. . . . .*P. elatior*
2. Achenes narrowly obovoid, compressed . . . . *P. brevissimus*

Psilocarphus brevissimus Nutt. **Stems** prostrate to erect, usually branched, 1–6 cm. **Leaves** lanceolate, 5–20 mm long. **Heads** 5–8 mm high; receptacle scales 2–3 mm long. **Achenes** obovoid, compressed, ca.1 mm long. Vernally moist to wet soil of shallow ponds, wetland margins; plains, valleys. AB, SK south to CA, NV, UT.

Psilocarphus elatior (A.Gray) A.Gray Similar to *P. brevissimus*. **Leaves** linear to oblanceolate, 10–35 mm long. **Heads** 6–8 mm high; receptacle scales 2–4 mm long. **Achenes** cylindric, terete, 1–2 mm long. Vernally moist ground of forest openings; valleys. BC south to CA, ID, MT. Collected in Lincoln Co.

Psilocarphus tenellus Nutt. **Stems** usually prostrate, branched. **Leaves** oblanceolate, 4–10 mm long. **Heads** 2–4 mm high; receptacle scales ca. 2 mm long. **Achenes** narrowly obovoid, compressed, ca.1 mm long. Vernally moist to wet soil of shallow ponds, wetland margins; plains, valleys. BC to CA east to ID and Sanders Co., MT.

Pyrrocoma Hook. Goldenweed

Taprooted perennial herbs usually with a branched caudex. **Stems** simple or branched in the inflorescence. **Leaves** basal and cauline, alternate, simple, petiolate; blades usually 1-veined. **Heads** 1 to 5, radiate to discoid; involucre turbinate to hemispheric; phyllaries imbricate in 2 to 6 series, basally firm; receptacle convex, naked, pitted. **Ray flowers** 0 to 80, yellow, female. **Disk flowers** 20 to 100, perfect, yellow, the tube ca. as long as the throat; style branch appendages deltoid. **Pappus** 1 series of tawny capillary bristles. **Achenes** cylindric, 3- to 4-angled.

Previously placed in *Haplopappus*.

1. Involucres 6–10 mm high; disk corollas 5–7 mm long 2
1. Involucres ≥10 mm high; disk corollas 6–14 mm long 3

2. Herbage glabrate to sparsely villous; heads usually ≥3; phyllaries unequal . . . . *P. lanceolata*
2. Herbage tomentose; heads mostly solitary; phyllaries subequal . . . .*P. uniflora*

3. Phyllaries oblong to ovate or spiny-serrate; rays absent . . . .*P. carthamoides*
3. Phyllaries oblanceolate, entire; heads radiate . . . . *P. integrifolia*

Pyrrocoma carthamoides Hook. [*Haplopappus carthamoides* (Hook.) A.Gray] **Stems** ascending, 8–40 cm, villous. **Leaf blades** linear-oblanceolate to oblong, entire to spiny-serrate, 3–15 cm long, pubescent. **Heads** disciform to discoid, solitary or few; involucre narrowly campanulate to hemispheric, 12–20 mm high; phyllaries in 3 to 5 series, ovate to oblong, mucronate, sometimes spiny-serrate, puberulent, white below, green above. **Rays** 0 to 30, inconspicuous. **Disk** corollas 8–12 mm long. **Achenes** 3–5 mm long, glabrous. Grasslands; valleys, montane. BC south to CA, NV, WY. (p.556)

1. Phyllaries spiny-serrate, oblanceolate . . . . var. *subsquarrosa*
1. Phyllaries entire, oblong to ovate . . . . var. *carthamoides*

Pyrrocoma carthamoides var. **subsquarrosa** (Greene) G.K.Brown & D.D.Keil occurs in calcareous soil, montane; endemic to Carbon Co. and adjacent WY; *P. carthamoides* var. **carthamoides** is found valleys, montane.

Pyrrocoma integrifolia (Porter ex A.Gray) Greene [*Haplopappus integrifolius* Porter ex A.Gray] **Stems** ascending, 7–50 cm, glabrate to sparsely villous. **Leaf blades** oblanceolate to narrowly elliptic, entire, 3–15 cm long, glabrous, becoming sessile above. **Heads** radiate, solitary or few, reduced in size below; involucre hemispheric, 1–2 cm high; phyllaries in 2 to 3 series, oblanceolate, ciliate, green above. **Rays** 18 to 45; ligules 7–18 mm long. **Disk corollas** 6–10 mm long. **Achenes** 5–7 mm long, glabrous. Moist, often calcareous soil of meadows, grasslands; valleys, montane. Endemic to southwest MT, adjacent ID, WY.

   Pyrrocoma lanceolata has smaller heads.

Pyrrocoma lanceolata (Hook.) Greene [*Haplopappus lanceolatus* (Hook.) Torr. & A.Gray] **Stems** ascending, 10–40 cm, glabrate to sparsely villous. **Leaf blades** linear-oblanceolate to oblanceolate, spiny-dentate to entire, 2–10 cm long, glabrate. **Heads** 3 to 8, rarely solitary, radiate; involucre hemispheric, 7–10 mm high; phyllaries in 3 to 4 series, lanceolate, pointed, white below, green above, ciliate to villous. **Rays** 18–45; ligules 5–11 mm long. **Disk corollas** 4–7 mm long. **Achenes** 3–5 mm long, sericeous. Moist to wet meadows; plains, valleys, montane. AB, SK south to CA, UT, CO, TX. Our plants are var. **lanceolata**.

   Pyrrocoma uniflora has fewer heads and more tomentose herbage; plants from Beaverhead Co. approach *P. uniflora*. See *P. integrifolia*.

Pyrrocoma uniflora (Hook.) Greene [*Haplopappus uniflorus* (Hook.) Torr. & A.Gray] **Stems** ascending, 5–30 cm, sparsely villous to tomentose. **Leaf blades** linear-oblanceolate to lanceolate, entire to serrate, 3–15 cm long, sparsely tomentose, sessile on the stem. **Heads** mostly solitary, radiate; involucre hemispheric, 7–10 mm high; phyllaries in 2 series, subequal, lanceolate, green, villous. **Rays** 18 to 50; ligules 6–10 mm long. **Disk corollas** 5–7 mm long. **Achenes** 2–4 mm long, sericeous. Moist to wet, usually calcareous meadows, turf; valleys to alpine. NT south to CA, NV, UT, CO. Our plants are var. **uniflora**.

   High-elevation plants have relatively broad leaf blades. See *P. lanceolata*.

Ratibida Raf. Coneflower

Ratibida columnifera (Nutt.) Wooten & Standl. PRAIRIE CONEFLOWER Taprooted perennial herb. **Stems** erect, often branched, 25–90 cm. **Herbage** strigose, punctate. **Leaves** basal and cauline, alternate, petiolate; blades 2–10 cm long, deeply pinnately divided into 3 to 13 oblanceolate, well-separated lobes. **Heads** radiate, solitary on branch tips; involucres reflexed; phyllaries in 2 series; outer linear-lanceolate, 5–10 mm long; inner short, ovate; receptacle cylindric, 15–35 mm high with small, hirsute scales clasping the flowers. **Ray flowers** 4 to 12, sterile; ligules yellow (purple), 10–25 mm long. **Disk flowers** numerous, perfect, greenish; tube shorter than the throat; style branches with a hairy appendage. **Pappus** of 1 or 2 short teeth. **Achenes** obconic, 1–3 mm long, compressed, glabrate. Grasslands, streambanks, fields, roadsides; plains, valleys, montane. BC to QC south to AZ, TX, TN, NC. (p.556)

Rudbeckia L. Coneflower

Perennial herbs from a rhizome or long caudex branches. **Stems** erect, simple. **Leaves** mainly cauline, alternate, petiolate. **Inflorescence**: heads solitary or few in leafy-paniculate arrays. **Heads** radiate or discoid; involucre reflexed; phyllaries green, subequal, 5 to 20 in 1 or 2 series; receptacle hemispheric to conical with a scale clasping each flower. **Ray flowers** sterile, 0 to 16, yellow. **Disk flowers** perfect, yellow to purple; tube shorter than the throat; style appendages hairy. **Pappus** a low crown. **Achenes** 4-sided, columnar, glabrate.

1. Rays absent; receptacle 15–45 mm high . . . . *R. occidentalis*
1. Rays present; receptacle 8–25 mm high 2

2. Leaves divided into lobed leaflets . . . .*R. laciniata*
2. Leaves simple . . . . *R. hirta*

Rudbeckia hirta L. Biennial or short-lived perennial. **Stems** erect, often branched, 40–80 cm. **Herbage** hirsute. **Leaf blades** lanceolate, 3–15 cm long, serrate. **Heads** radiate; phyllaries lanceolate, hirsute, 1–2 cm long; receptacle hemispheric, 8–12 mm high. **Rays** 8 to 16, hirsute; ligules 15–35 mm long. **Disk corollas** purple, 3–4 mm long, ca. as long as the scales. **Achenes** 1.5–3 mm long. Wetlands, streambanks, roadsides; valleys. Throughout temperate N. America.
Cronquist believes *Rudbeckia hirta* is introduced in our area.[187]

Rudbeckia laciniata L. **Stems** 50–200 cm. **Herbage** glabrate, glaucous. **Leaf blades** ovate, 10–25 cm long, deeply divided into 3 to 5 lobes or serrate leaflets; ultimate segments lanceolate; uppermost leaves ovate to lanceolate. **Heads** radiate; phyllaries ovate to lanceolate, minutely ciliate, 5–15 mm long; receptacle hemispheric to conic, 10–25 mm high. **Rays** 8–12; ligules pubescent, 2–4 cm long. **Disk corollas** yellowish, 4–5 mm long. **Achenes** 3–5 mm long. Moist meadows, often along streams; valleys, montane. Throughout temperate N. America. Our plants are var. *ampla* (A.Nelson) Cronquist.

Rudbeckia occidentalis Nutt. **Stems** 50–200 cm. **Herbage** glabrate; leaves minutely ciliate. **Leaf blades** ovate, acuminate, 5–20 cm long, entire to serrate. **Heads** discoid; phyllaries lanceolate, sparsely puberulent, 10–25 mm long; receptacle conic, 15–45 mm high. **Disk corollas** yellow below, purple above, 3–5 mm long. **Achenes** 3–5 mm long. Moist meadows, forest openings, often along streams; valleys, montane. WA to MT south to CA, NV, UT.

Saussurea DC. Sawwort

Perennial herbs with often branched caudexs. **Leaves** basal and cauline, alternate, simple, winged-petiolate. **Inflorescence** compact, corymbiform arrays. **Heads** discoid; involucres campanulate; phyllaries unequal to subequal in 3 to 5 series; receptacle flat or convex, naked or with tiny scales. **Disk flowers** perfect, blue to purple; tube longer than the expanded throat; style branches fused. **Pappus** 2 series of white capillary bristles; outer short, deciduous; inner plumose. **Achenes** oblong, glabrous 4- to 5-nerved or angled.

1. Stems >40 cm . . . . S. americana
1. Stems <25 cm 2

2. Phyllaries subequal, acuminate; receptacle naked . . . . S. nuda
2. Phyllaries well-imbricate, acute to rounded at the tip; receptacle scaly . . . . S. weberi

Saussurea americana D.C.Eaton **Stems** erect, simple, 40–100 cm. **Herbage** loosely tomentose to glabrate. **Leaf blades** lanceolate to ovate, serrate, 4–12 cm long. **Involucres** 10–15 mm high; phyllaries ovate to lanceolate, strongly imbricate in ca. 5 series, villous, tomentose and purple on the margins. **Disk flowers** 8 to 21; corollas 10–13 mm long. **Achenes** 4–6 mm long. Moist soil of tall herb meadows, open forests, avalanche slopes; montane, lower subalpine. AK south to CA, ID, MT.

Saussurea nuda Ledeb. [*S. densa* (Hook.) Rydb.] **Stems** erect, simple, 4–20 cm. **Herbage** villous to patchy-tomentose. **Leaf blades** narrowly lanceolate, tapered to the petiole, serrate, 2–7 cm long, often superceding the inflorescence. **Involucres** 10–15 mm high; phyllaries linear-lanceolate, acuminate, subequal in 3 or 4 series, villous to tomentose, purplish. **Disk flowers** 15 to 20; corollas 8–13 mm long. **Achenes** 6–7 mm long. Calcareous talus, scree; alpine. AK to MT; Asia. Our plants are var. *densa* (Hook.) Hultén.

Saussurea weberi Hultén **Stems** erect, simple, 5–15 cm. **Herbage** weakly tomentose, becoming glabrate. **Leaf blades** lanceolate to narrowly ovate, entire to weakly serrate, 2–6 cm long. **Involucres** 10–12 mm high; phyllaries lanceolate, rounded at the tip, imbricate in 3 or 4 series, villous, purplish. **Disk flowers** 9 to 10; corollas 10–12 mm long. **Achenes** 3–5 mm long. Alpine turf. MT to CO. Known from Deer Lodge and Granite cos.

Scorzonera L.

Scorzonera laciniata L. Taprooted biennial. **Stems** erect or ascending, usually branched, 6–30 cm. **Herbage** glabrous to tomentose with milky sap. **Leaves** basal and cauline, petiolate to sessile above; blades 2–20 cm long, deeply pinnately divided into remote, linear-lanceolate lobes.

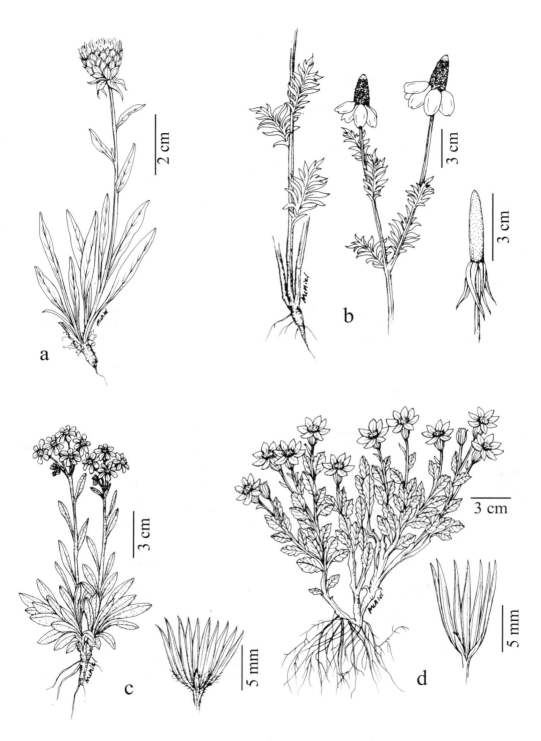

Plate 98. a. *Pyrrocoma carthamoides* var. *carthamoides*, b. *Ratibida columnifera*, c. *Senecio canus*, d. *Senecio fremontii*

Heads ligulate, solitary on branch tips, expanding in fruit; involucres campanulate 12–25 mm high in fruit; phyllaries lanceolate in 3 to 5 unequal series, sparsely tomentose, scarious-margined; receptacle flat, pitted, naked. **Ray flowers** perfect, 30 to 100, yellow; ligules 2–4 mm long. **Pappus** tawny, plumose, capillary bristles. **Achenes** fusiform, 8–17 mm long, ca. 10-nerved, glabrous. Vernally moist soil of wetland margins, roadsides; plains, valleys. Introduced from MT to NM, TX; native to Europe.

Scorzonera hispanica L. was once collected in a garden but was never naturalized.

Senecio L. Groundsel, ragwort, butterweed

Mostly perennial, rarely annual herbs. **Stems** usually simple. **Leaves** cauline and often basal, alternate; the basal petiolate, often becoming sessile above. **Inflorescence**: corymbiform or cymose arrays or heads solitary. **Heads** radiate or discoid; involucres mostly campanulate; phyllaries narrowly lanceolate, scarious-margined, equal or subequal in 1 (to 2) series, often with a few short bracts at the base; receptacle flat to convex, naked. **Ray flowers** 0 to 31, female, yellow. **Disk flowers** perfect, yellow to orange; tubes shorter than the throats; style branches truncate. **Pappus** of white, capillary bristles. **Achenes** cylindric, 5-ribbed or -angled.[26,27,187,396]

One of the largest genera in MT and the world. Several species have been segregated into *Packera* and *Tephroseris* based on incompletely congruent characters: chromosome number, pollen wall architecture and morphology.[22,258] However, there is no evidence that *Senecio* as traditionally circumscribed is not monophyletic, and more comprehensive studies are needed before such a large cosmopolitan group can be reconfigured with confidence. Leaf shape and size of the flower heads are often useful characters for determination. *S. pauperculus*, *S. indecorus*, *S. streptanthifolius*, and *S. pseudaureus* are closely related species and difficult to distinguish, as are *S. cymbalarioides* and *S. cymbalaria*; habitat may serve to distinguish them as much as morphology. *Senecio hyperborealis* Greenm. was reported for MT,[26] but it is a high-arctic species.[396] *Senecio pauciflorus* Pursh [*Packera pauciflora* (Pursh) A.Löve & D.Löve] is reported for MT,[113] but I have seen no specimens.

1. Stem leaves only gradually reduced upward; the basal often not the largest 2
1. Upper stem leaves greatly reduced compared to the basal. 10

2. Most leaf blades serrate to dentate but not divided >1/2-way to the midvein. .3
2. Most leaf blades divided >1/2-way to the midvein 8

3. Basal leaf blades suborbicular; stem leaves lanceolate to oblanceolate 4
3. Basal and stem leaves more similar in shape 5

4. Involucres 5–8 mm high; forest to tundra habitats . . . . *S. dimorphophyllus*
4. Involucres 7–12 mm high . . . . *S. elmeri*

5. Most stems with 1 to 4 heads, often lax . . . . *S. fremontii*
5. Most stems with >4 heads, usually erect to ascending 6

6. Plant annual to biennial; inflorescence with some tomentum. . . . . *S. congestus*
6. Plant a strong perennial; inflorescence glabrate without tomentum. 7

7. Leaves narrowly lanceolate, cuneate at the base . . . . *S. serra*
7. Leaves broadly lanceolate to deltoid, truncate to cordate at the base . . . .*S. triangularis*

8. Plants annual; heads discoid . . . . *S. vulgaris*
8. Plants perennial; heads radiate 9

9. Leaves rounded at the tip, at least 2 times divided . . . . *S. jacobaea*
9. Leaves with pointed tips, once lobed . . . .*S. eremophilus*

10. Heads discoid. 11
10. Heads radiate. 14

11. Stem leaves entire to serrate . . . . *S. hydrophilus*
11. Stem leaves deeply divided 12

12. Heads tubular at anthesis; plants weedy annuals of roadsides and gardens . . . . *S. vulgaris*
12. Heads campanulate; plants native perennials 13

13. Leaves thick in life, with rounded or crenate lobes. . . . . *S. debilis*
13. Leaves not noticeably thick, with sharp-pointed lobes . . . .*S. indecorus*

14. At least some stem leaves lobed 15
14. Leaves entire, dentate or serrate, not lobed. 25

15. Leaves and stems arachnoid to tomentose at flowering 16
15. Leaves and stems glabrate (sparse tomentum in axils). 18

16. Heads 1 or 2; plants subalpine to alpine . . . . *S. cymbalaria*
16. Heads usually >2; plants often below subalpine 17

17. Leaves and stems closely tomentose. . . . .*S. canus*
17. Leaves and stems sparsely to densely, loosely arachnoid. . . . . *S. plattensis*

18. Heads 1 to 3 per stem on most plants 19
18. Most stems with >4 heads 21

19. Plants of well-drained, usually stony soil; lower stem leaves clasping. . . . . *S. dimorphophyllus*
19. Plants of mesic to wet sites; stem leaves not clasping. 20

20. Involucre 9–15 mm high, bracts purplish; plants of well-drained soil . . . . *S. cymbalaria*
20. Involucre 5–9 mm high, green; plants of wetland sites. . . . . *S. cymbalarioides*

21. Basal leaves truncate to cordate at the base. 22
21. Basal leaves rounded to cuneate at the base 23

22. Basal leaves ovate to lanceolate, sharply dentate . . . . *S. pseudaureus*
22. Basal leaves suborbicular, entire to shallowly crenate. . . . . *S. dimorphophyllus*

23. Basal leaves broadly ovate to suborbicular, rounded at the base . . . . *S. dimorphophyllus*
23. Basal leaves narrowly ovate to oblanceolate, cuneate. 24

24. Leaf blades thick; plants of well-drained, stony, upland soil. . . . . *S. streptanthifolius*
24. Leaf blades thin; plants of moist to wet habitats. . . . . *S. pauperculus*

25. Some leaf surfaces tomentose or arachnoid (sometimes sparse) at flowering 26
25. Tangled hairs restricted to leaf axils, stems, or base of involucres or absent. 32

26. Plants fibrous-rooted from a button-like caudex. . . . . *S. integerrimus*
26. Plants with a short to long, sometimes branched rhizome. 27

27. Largest involucre >12 mm high . . . . *S. megacephalus*
27. Involucres ≤12 mm high. 28

28. Leaves sparsely arachnoid, green 29
28. At least underside of leaves white-hairy. 30

29. Plants ≤15 cm tall. . . . . *S. werneriifolius*
29. Plants ≥20 cm tall. . . . . *S. sphaerocephalus*

30. Plants ≤15 cm tall; stems scapose. . . . . *S. werneriifolius*
30. Plants usually >15 cm; stem leaves apparent 31

31. Leaves arachnoid and strigose beneath the longer hairs; disk flowers orangish. . . . . *S. fuscatus*
31. Leaves tomentose but not strigose beneath; disk flowers yellow. . . . . *S. canus*

32. Most stems with 1 or 2 heads. 33
32. Most stems with >2 heads 35

33. Upper cauline leaves short-petiolate; plants of talus . . . . *S. amplectens*
33. Upper cauline leaves becoming sessile; plants growing in soil 34

34. Herbage glabrous; plants of moist habitats . . . . *S. cymbalarioides*
34. Plants with arachnoid tomentum in leaf axils; plants of well-drained soil. . . . . *S. cymbalaria*

35. Most stems with >15 heads 36
35. Most stems with 3 to 15 heads. 37

36. Involucres usually sparsely arachnoid at the base. . . . . *S. integerrimus*
36. Involucres glabrous . . . . *S. hydrophiloides*

37. Plants fibrous-rooted from a button-like caudex 38
37. Plants from a more elongate, often horizontal, simple or branched caudex or rhizome 39

38. Plants glabrous. . . . . *S. crassulus*
38. Plants usually with some tomentum in leaf axils or at base of involucre . . . . *S. integerrimus*

39. Plants growing in talus with a branched caudex. 40
39. Plants usually single-stemmed; growing in soil 41

40. Heads >2 on most plants; inflorescence somewhat villous . . . . *S. elmeri*
40. Heads 1 or 2 on many stems; inflorescence glabrate . . . . *S. amplectens*

41. Phyllaries not or barely black-tipped. . . . . *S. streptanthifolius*
41. Phyllaries with conspicuous black tips 42

42. Plants usually with some tomentum in leaf axils or at base of involucre . . . . *S. lugens*
42. Plants glabrous. . . . . *S. crassulus*

Senecio amplectens A.Gray Fibrous-rooted with a branched caudex. **Stems** ascending, sometimes branched, 10–20 cm. **Herbage** glabrate. **Leaves**: basal conspicuously petiolate; blades lanceolate to ovate, 3–15 cm long, dentate or serrate; cauline reduced. **Inflorescence** of 1 to 4 heads. **Heads** radiate, nodding; involucres 8–15 mm high; phyllaries 13 to 21, glabrous, often black-tipped. **Rays** ca.13; ligules 6–15 mm long. **Disk corollas** 5–9 mm long. **Achenes** glabrous, 2–4 mm long. Talus slopes, rock outcrops; upper subalpine, alpine; known from Carbon Co. MT to NV, UT, NM. Our plants are var. *holmii* (Greene) H.D.Harr.

Senecio canus Hook. [*Packera cana* (Hook.) W.A.Weber & A.Löve] Fibrous-rooted from a simple or branched caudex or simple rhizome. **Stems** erect to ascending, sometimes branched above, 3–40 cm. **Herbage** tomentose, sometimes becoming glabrate on upper leaf surfaces. **Leaves** basal and cauline; basal blades linear-oblanceolate to ovate, entire to dentate, 1–8 cm long; cauline blades lanceolate, often with lobes or teeth, becoming sessile. **Inflorescence** corymbiform of 2 to 20 heads. **Heads** radiate; involucres 4–12 mm high; phyllaries 13 to 21, sparsely tomentose, sometimes purplish. **Rays** 8 to 13; ligules 3–12 mm long. **Disk corollas** 4–7 mm long. **Achenes** 2–4 mm, glabrous. Often stony soil of grasslands, sagebrush steppe, meadows, open forest, fellfields, talus, roadsides at all elevations; plains, valleys to alpine. BC to MB south to CA, AZ, NM, KS. (p.556)

A highly variable species; some plants from southeast MT have nearly linear leaves. Many populations from deep-soil habitats of central MT have solitary stems from a short rhizome and tend to be taller (25–40 cm) with larger involucres (8–12 mm).

Senecio congestus (R.Br.) DC. [*Tephroseris palustris* (L.) Reichb.] Fibrous-rooted annual to short-lived perennial. **Stems** hollow, erect, 15–80 cm. **Herbage** villous to sparsely tomentose. **Leaves** short-petiolate below, mainly cauline at flowering; blades oblanceolate to spatulate, dentate to shallowly lobed, 5–15 cm long, not much reduced but becoming sessile above. **Inflorescence** corymbiform with 6 to 20 heads, somewhat congested. **Heads** radiate; involucres 5–10 mm high; phyllaries ca. 21, pale, sometimes pink-tipped. **Rays** ca. 21; ligules 5–9 mm long. **Disk corollas** 6–10 mm long. **Achenes** glabrous. Moist meadows, streambanks; plains. AK to NL south to BC, AB, MT, SD, IA, MI. One collection from Roosevelt Co.

Senecio crassulus A.Gray Fibrous-rooted from a short rhizome. **Stems** ascending to erect, 20–40 cm. **Herbage** glabrous. **Leaves** thick, basal and cauline, becoming sessile above; blades oblanceolate to oblong, 2–12 cm long, serrate. **Inflorescence** corymbiform of 3 to 12 heads. **Heads** radiate; involucres 8–12 mm high; phyllaries 8 to 21, glabrous, conspicuously black-tipped. **Rays** 5 to 13; ligules 8–18 mm long. **Disk corollas** 7–10 mm long. **Achenes** 2–4 mm long, glabrous. Moist meadows, open forest, often near late-melting snow; upper montane, subalpine. OR to MT south to NV, UT, NM.

Senecio integrifolius has some patches of tomentum at least on the stem base and leaf axils. See *S. sphaerocephalus*.

Senecio cymbalaria Pursh [*S. conterminus* Greenm., *S. resedifolius* Less., *Packera contermina* (Greenm.) J.F.Bain, *P. cymbalaria* (Pursh) W.A.Weber & A.Löve] Fibrous-rooted from a short rhizome. **Stems** ascending to erect, 4–10 cm. **Herbage** glabrate to sparsely arachnoid, especially at basal leaf bases. **Leaves** thick; basal blades suborbicular to obovate, dentate, 5–20 mm long; cauline, sessile, lanceolate to oblanceolate, entire to pinnately lobed. **Inflorescence** 1 (2) heads. **Heads** radiate; involucres 9–15 mm high; phyllaries 21, glabrous, purplish. **Rays** 10 to 13; ligules 7–11 mm long. **Disk corollas** 5–8 mm long. **Achenes** 2–4 mm long, glabrous. Moist to dry, stony soil of fellfields, moraine, rock outcrops; alpine. AK to NL south to WA, ID, MT.

Senecio cymbalarioides and *S. cymbalaria* are similar, but the former is usually in soil with ample organic matter, while the latter is mostly found in poorly developed soil. Recent treatments have separated *S. conterminus* from *S. cymbalaria*; only the former is reported for MT.[396] However, our plants have characters of both, so it seems advisable to use the more inclusive name at this time. See *S. werneriifolius*.

Senecio cymbalarioides H.Buek, [*S. subnudus* DC., *Packera subnuda* (DC.) Trock & T.M.Barkley] Fibrous-rooted from a rhizome or short-branched caudex. **Stems** erect, 2–25 cm. **Herbage** glabrous. **Leaves**: basal blades suborbicular to ovate, crenate to shallowly lobed, 1–3 cm long; cauline becoming sessile, narrower, sometimes pinnately lobed. **Inflorescence** mostly 1(2) head. **Heads** radiate; involucres 5–9 mm high; phyllaries 21, glabrous, green or purplish. **Rays** ca. 13; ligules 6–12 mm long. **Disk corollas** 5–7 mm long. **Achenes** 1–3 mm long, glabrous. Wet, often organic soil of meadows, turf, streambanks; subalpine, alpine. YT south to CA, NV, UT, NM.

The epithet "*subnudus*" cannot be legitimately used for this plant in *Senecio* but can in *Packera*.[397] See *S. pauperculus*, *S. cymbalaria*.

Senecio debilis Nutt. [*Packera debilis* (Nutt.) W.A.Weber & A.Löve] Fibrous-rooted. **Stems** erect, 10–50 cm. **Herbage** glabrous or sparsely arachnoid in leaf axils. **Leaves** thick, long-petiolate below; basal blades ovate or obovate, crenate, 1–6 cm long; cauline narrower, pinnately divided into several pairs of rounded lobes. **Inflorescence** corymbiform with 3 to 15 heads. **Heads** discoid; involucres 5–8 mm high; phyllaries 13 or 21, glabrous, green. **Disk corollas** 3–7 mm long. **Achenes** 1–2 mm long, glabrous. Moist to wet, alkaline meadows; montane. MT, ID, WY, CO.

Senecio indecorus is usually not in strongly alkaline areas and has more sharply lobed leaves.

Senecio dimorphophyllus Greene [*Packera dimorphophylla* (Greene) W.A.Weber & A.Löve] Fibrous-rooted from a short rhizome or branched caudex. **Stems** erect, 10–30 cm. **Herbage** glabrous. **Leaves** thick, long-petiolate below; basal blades obovate to suborbicular, sometimes cordate, crenate, 1–4 cm long; cauline lanceolate to oblanceolate, pinnately lobed or toothed, the lower clasping. **Inflorescence** corymbiform with 1 to 12 heads. **Heads** radiate; involucres 5–8 mm high; phyllaries 13 or 21, glabrous, green with ciliate tips. **Rays** 8 or 13; ligules 3–8 mm long. **Disk corollas** 4–6 mm long. **Achenes** 2–3 mm long, glabrous. Stony soil; subalpine to alpine. MT to NV, UT, NM.

1. Basal leaf blades truncate at the base . . . . .var. *paysonii*
1. Basal leaf blades rounded at the base . . . . var. *dimorphophyllus*

Senecio dimorphophyllus var. *paysonii* T.M.Barkley occurs in open forest, fellfields, turf, often below snowbanks; endemic to Beaverhead Co. and adjacent WY, ID; *S. dimorphophyllus* var. *dimorphophyllus* is found in turf and fellfields; more widespread.

Senecio elmeri Piper [*S. spribillei* W.A.Weber] Fibrous-rooted with a branched caudex. **Stems** erect to ascending, simple, 10–20 cm. **Herbage** glabrate to weakly villous. **Leaves**: basal conspicuously petiolate; blades oblanceolate to ovate, 2–4 cm long, dentate or serrate; lower cauline often larger than basal. **Inflorescence** cymose, of 2 to 9 nodding to erect heads. **Heads** radiate; involucres 7–12 mm high; phyllaries ca. 13, glabrous, often black-tipped. **Rays** ca. 8; ligules 7–12 mm long. **Disk corollas** 8–9 mm long. **Achenes** glabrous. Talus slopes; alpine. Endemic to southern BC and WA, disjunct in Lincoln Co., MT.

The recently described *Senecio spribillei*[426] can easily be accommodated here.

Senecio eremophilus Richardson Fibrous-rooted from a short-branched rhizome. **Stems** ascending, 30–100 cm. **Herbage** glabrous. **Leaves** cauline, only gradually reduced above; blades lanceolate to oblanceolate, 4–10 cm long, pinnately divided into lanceolate lobes or teeth. **Inflorescence** leafy-bracted corymbiform of 10 to 30 heads. **Heads** radiate; involucres 7–10 mm high; phyllaries ca. 13, glabrous, barely black-tipped. **Rays** ca. 8; ligules 5–10 mm long. **Disk corollas** 6–7 mm long. **Achenes** glabrous, 2–3 mm long. Meadows, forest openings, thickets, often along streams; montane. AK to QC south to AZ, NM, SD. Our plants are var. *eremophilus*.

Senecio fremontii Torr. & A.Gray Plants from a branched caudex. **Stems** prostrate to ascending, 4–25 cm. **Herbage** glabrous, sometimes purplish. **Leaves** cauline, little reduced upward; blades oblong to obovate, mostly crenate to dentate, 5–30 mm long. **Inflorescence** of 1(2) head. **Heads** radiate; involucres 6–12 mm high; phyllaries ca. 13, glabrous. **Rays** ca. 8; ligules 5–12 mm long. **Disk corollas** 6–9 mm long, slender. **Achenes** glabrous to strigose, 2–4 mm long. Talus, moraine, rock crevices; subalpine, alpine. BC, AB, south to CA, UT, CO. Our plants are var. *fremontii*. (p.556).

Senecio fuscatus Hayek [*S. lindstroemii* (Ostenf.) A.E.Porsild, *Tephroseris lindstroemii* (Ostenf.) A.Löve & D.Löve] Fibrous-rooted from a short rhizome. **Stems** erect, 8–25 cm. **Herbage** arachnoid, sparsely strigose. **Leaves** basal and cauline; blades lanceolate to obovate, 2–6 cm long, entire to crenate, winged-petiolate below. **Inflorescence** of 2 to 8 heads. **Heads** radiate; involucres 5–10 mm high; phyllaries ca. 21, glabrate, purplish. **Rays** 13 or 21; ligules 8–14 mm long.

Disk corollas 5–7 mm long, orangish. **Achenes** 2–4 mm long, pubescent. Moist turf; alpine. Circumpolar south to WY.

Senecio fuscatus was originally described from Europe. In the broad sense it is a circumpolar species which has been divided into several geographic taxa.[28]

Senecio hydrophiloides Rydb. [*S. foetidus* Howell] Fibrous-rooted. **Stems** erect, 30–90 cm, sometimes clustered. **Herbage** glabrate to obscurely arachnoid on the stem. **Leaves** basal and cauline, narrower upward; blades lanceolate to elliptic, 4–15 cm long, usually serrate. **Inflorescence** corymbiform of 15 to 30 heads. **Heads** radiate; involucres 5–8 mm high; phyllaries 8–21, glabrous, black-tipped. **Rays** ca. 8; ligules 4–9 mm long. **Disk corollas** 5–7 mm long. **Achenes** ca. 3 mm long, glabrous. Wet meadows, moist open forest, thickets; valleys, montane. BC, AB south to CA, NV, UT, WY.

Senecio hydrophilus has a glaucous, hollow stem, mostly entire leaves and occurs in alkaline habitats.

Senecio hydrophilus Nutt. Fibrous-rooted. **Stems** erect, hollow, 40–100 cm. **Herbage** glabrate, glaucous, especially on the stem. **Leaves** thick, basal and cauline, narrower upward; blades lanceolate to oblanceolate, 2–20 cm long, entire or obscurely serrate. **Inflorescence** congested-corymbiform of 15 to 50 heads. **Heads** usually discoid (radiate); involucres 7–10 mm high; phyllaries 8 to 13, glabrous, sometimes black-tipped. **Rays** ca. 5 or absent; ligules ca. 5 mm long. **Disk corollas** 5–6 mm long. **Achenes** 2–4 mm long, glabrous. Wet alkaline meadows; valleys, montane. BC south to CA, NV, UT, CO, SD. See *S. hydrophiloides*.

Senecio indecorus Greene [*Packera indecora* (Greene) A.Löve & D.Löve] Fibrous-rooted. **Stems** erect, 20–70 cm. **Herbage** glabrate to sparsely arachnoid. **Leaves** long-petiolate below, clasping above; basal blades ovate, dentate, 1–5 cm long; cauline lanceolate to oblanceolate, pinnately divided into acute lobes. **Inflorescence** corymbiform with 5 to 20 heads. **Heads** discoid; involucres 6–10 mm high; phyllaries 13 or 21, glabrous, purplish. **Disk corollas** 5–8 mm long. **Achenes** 2–3 mm long, glabrous. Fens, wet meadows, open forest, thickets, woodlands, often along streams; valleys, montane. AK to NL south to CA, ID, WY, MN.

Similar to *Senecio pauperculus* but lacking rays; *S. pauciflorus* has fewer heads and occurs in drier habitats; see *S. debilis*.

Senecio integerrimus Nutt. Fibrous-rooted from a small caudex. **Stems** erect, 10–90 cm. **Herbage** sparsely arachnoid with tufts of tomentum at stem and leaf bases. **Leaves** basal and cauline; blades linear-lanceolate to ovate or obovate, 2–15 cm long, entire to usually serrate, narrower and sessile above. **Inflorescence** corymbiform of 4 to 25 heads. **Heads** radiate; involucres 5–13 mm high; phyllaries mostly 13 to 21, usually sparsely arachnoid, sometimes black-tipped. **Rays** 8–13; ligules 5–11 mm long. **Disk corollas** 6–8 mm long. **Achenes** 2–4 mm long, short-hairy on the ribs. BC to MB south to CA, NV, NM, NE, IA.

1. Phyllaries black-tipped . . . . var. *exaltatus*
1. Phyllaries not black-tipped 2

2. Basal leaves linear-lanceolate . . . . var. *scribneri*
2. Basal leaves lanceolate to ovate . . . . var. *integerrimus*

Senecio integerrimus var. *exaltatus* (Nutt.) Cronquist is found in mesic grasslands, sagebrush steppe, dry forest openings; valleys to lower alpine; *S. integerrimus* var *integerrimus* occurs in grasslands, meadows, woodlands; plains; *S. integerrimus* var. *scribneri* (Rydb.) T.M. Barkley is found in grasslands, sagebrush steppe on the plains and is endemic to central MT. Variety *ochroleucus* (Gray) Cronquist, with white rays, is reported for MT,[27] but I have seen no specimens. See *S. crassulus*.

Senecio jacobaea L. Tansy ragwort Short-lived, taprooted perennial. **Stems** erect, 20–80 cm. **Herbage** sparsely arachnoid. **Leaves** mainly cauline at flowering; blades narrowly oblanceolate, deeply divided into dentate lobes, 4–20 cm long, becoming sessile above. **Inflorescence** corymbiform with 10 to 60 heads. **Heads** radiate; involucres 4–6 mm high; phyllaries ca. 13, glabrous. **Rays** ca. 13; ligules 6–10 mm long. **Disk corollas** 4–5 mm long. **Achenes** 1–2 mm long, puberulent. Disturbed soil of open forest, meadows, often associated with timber harvest or fire; valleys. Introduced to northeast and northwest N. America; native to Europe.

Senecio lugens Richardson Plants from horizontal or inclined rhizomes. **Stems** erect, 7–40 cm. **Herbage** glabrate, sparsely arachnoid in leaf axils. **Leaves** basal and cauline; blades ovate to oblanceolate, 1–12 cm long, entire to minutely serrate, becoming linear-lanceolate and sessile above. **Inflorescence** corymbiform, of 3 to 12 heads. **Heads** radiate; involucres 6–9 mm high; phyllaries ca. 13, glabrous to sparsely arachnoid, conspicuously black-tipped. **Rays** 5 to 13; ligules 6–12 mm long. **Disk corollas** 6–8 mm long. **Achenes** 1–3 mm long, glabrous. Turf, meadows, open forest; subalpine, alpine. AK to NT south to WA and WY.

Senecio integerrimus is similar but has a button-like caudex instead of bending into a horizontal rhizome. See *S. sphaerocephalus*.

Senecio megacephalus Nutt. Fibrous-rooted from a short or branched caudex. **Stems** erect, 20–60 cm. **Herbage** arachnoid. **Leaves** basal and cauline; blades linear-lanceolate to oblanceolate, entire, 3–15 cm long, greatly reduced above. **Inflorescence** a solitary (2 to 4) head. **Heads** radiate; involucres 10–25 mm high; phyllaries ca. 21, arachnoid. **Rays** ca. 13; ligules 12–22 mm long. **Disk flowers** 9–14 mm long. **Achenes** 2–6 mm long, glabrous. Stony soil of meadows, grasslands, open forest, fellfields; upper montane to alpine. Endemic to northwest MT, adjacent ID, Canada.

Senecio pauperculus Michx. [*Packera paupercula* (Michx.) A.Löve & D.Löve] Fibrous-rooted. **Stems** erect, 20–60 cm. **Herbage** glabrous or tomentose at leaf bases. **Leaves**: basal petiolate; blades ovate to lanceolate, cuneate, dentate, 2–6 cm long; cauline becoming sessile or clasping, lanceolate to oblanceolate, deeply pinnately divided into narrow lobes. **Inflorescence** corymbiform with 3 to 10 heads. **Heads** radiate; involucres 5–10 mm high; phyllaries 13 or 21, glabrous. **Rays** 8 or 13; ligules 4–12 mm long. **Disk corollas** 5–6 mm long. **Achenes** 1–3 mm long, glabrous. Moist or wet meadows, fens, thickets, forest openings, often along streams; valleys, montane. Throughout all of N. America except the desert southwest.

Similar to *Senecio cymbalarioides* which has thicker leaves, is smaller, has fewer heads, and occurs at higher elevations; see *S. pseudaureus*.

Senecio plattensis Nutt. [*Packera plattensis* (Nutt.) W.A.Weber & A.Löve] Fibrous-rooted from a rhizome or short caudex. **Stems** erect, 10–40 cm. **Herbage** arachnoid-tomentose, sometimes sparsely so on leaves. **Leaves**: basal long-petiolate; blades ovate to lanceolate, crenate to lobed, 2–5 cm long; cauline becoming sessile, narrower, pinnately lobed. **Inflorescence** corymbiform with 4 to 20 heads. **Heads** radiate; involucres 5–9 mm high; phyllaries 13 or 21, tomentose. **Rays** 8 to 10; ligules 5–10 mm long. **Disk corollas** 5–7 mm long. **Achenes** 1–3 mm long, puberulent on the ribs. Mesic grasslands, pine woodlands; plains. SK to QC south to NM, KS, LA, AL.

Where they occur together, *Senecio pauperculus* generally occurs in moister habitats than *S. plattensis*.

Senecio pseudaureus Rydb. [*Packera pseudaurea* (Rydb.) W.A.Weber & A.Löve] Plants from a short, branched rhizome. **Stems** erect, 20–80 cm. **Herbage** glabrate or arachnoid at leaf bases. **Leaves**: basal long-petiolate; blades ovate, truncate to cordate, dentate, 2–8 cm long; cauline becoming sessile, lanceolate to oblanceolate, deeply dentate to acutely, pinnately lobed. **Inflorescence** corymbiform with 5 to 20 heads. **Heads** radiate; involucres 5–10 mm high; phyllaries 13 or 21, glabrous. **Rays** 8 or 13; ligules 4–10 mm long. **Disk corollas** 4–8 mm long. **Achenes** 1–3 mm long, glabrous. Moist forest, thickets, meadows, woodlands; valleys to lower subalpine. BC to SK south to CA, NM, KS, MO.

Senecio pauperculus is similar but rarely in the shade. The type locality is in the Little Belt Mtns.

Senecio serra Hook. Plants from a branched caudex. **Stems** erect, 40–150 cm. **Herbage** glabrate. **Leaves** cauline, short-petiolate, gradually reduced upward; blades linear-lanceolate to lanceolate, cuneate-based, 5–15 cm long, finely serrate. **Inflorescence** paniculate with 30 to 90 heads. **Heads** radiate; involucres 6–7 mm high; phyllaries 8 to 13, glabrous, dark-tipped. **Rays** 5 to 8; ligules 4–8 mm long. **Disk flowers** 6–8 mm long. **Achenes** 2–4 mm long, glabrous. Tall herb meadows, open forest, woodlands, thickets, often along streams; valleys to montane. WA to MT south to CA, NV, UT, CO. Our plants are var. *serra*.

Senecio triangularis has more truncate leaves and occupies somewhat wetter habitats.

Senecio sphaerocephalus Greene [*S. lugens* Richardson var. *hookeri* D.C.Eaton] Plants from a stout horizontal rhizome. **Stems** erect, 25–80 cm. **Herbage** sparsely arachnoid. **Leaves** basal and cauline; blades ovate to oblanceolate, 4–15 cm long, entire to minutely dentate, becoming linear-lanceolate and sessile above. **Inflorescence** corymbiform of 7 to 30 heads. **Heads** radiate; involucres 5–9 mm high; phyllaries ca. 21, sparsely arachnoid, black-tipped. **Rays** ca. 13; ligules 5–9 mm long. **Disk corollas** 5–8 mm long. **Achenes** 1–3 mm long, sparsely puberulent. Moist to wet meadows, thickets; upper montane, subalpine. OR, ID, MT, NV, UT, WY.

Similar morphologically and ecologically to *Senecio lugens* and sometimes understandably considered a variety of it.

Senecio streptanthifolius Greene [*Packera streptanthifolia* (Greene) W.A.Weber & A.Löve] Fibrous-rooted from a simple or branched short rhizome. **Stems** erect to ascending, 10–50 cm. **Herbage** glabrate, sparsely arachnoid in leaf axils. **Leaves** basal and cauline; basal blades oblanceolate to ovate, cuneate, entire to dentate, 2–4 cm long; cauline blades oblanceolate, entire or with lobes or teeth, becoming sessile above. **Inflorescence** corymbiform of 3 to 20 heads. **Heads** radiate; involucres 4–7 mm high; phyllaries 13 to 21, glabrous. **Rays** 8 to 13; ligules 4–10 mm long. **Disk corollas** 4–7 mm long. **Achenes** 1–3 mm long, glabrous. Stony soil of open forest, grasslands, sagebrush steppe, dry meadows; montane, subalpine. YT, NT south to CA, NM.

Leaf shape in *Senecio streptanthifolius* is highly variable; *S. dimorphophyllus* is similar but has ovate to suborbicular basal leaves that contrast sharply with the stem leaves.

Senecio triangularis Hook. Plants from a branched caudex. **Stems** erect, 10–100 cm. **Herbage** glabrate. **Leaves** cauline, short-petiolate, only gradually reduced upward; blades lanceolate to deltoid, trunctae or cordate, 2–15 cm long, serrate. **Inflorescence** corymbiform with 1 to 30 heads. **Heads** radiate; involucres 7–10 mm high; phyllaries ca. 13, glabrous. **Rays** ca. 8; ligules 5–12 mm long. **Disk corollas** 5–9 mm long. **Achenes** 3–4 mm long, glabrous. Moist soil of forests, meadows, thickets, avalanche slopes, often along streams; montane, subalpine, rarely alpine. AK to NT south to CA, AZ, NM. See *S. serra*. (p.565)

Senecio vulgaris L. Taprooted annual. **Stems** erect, sometimes branched, 10–50 cm. **Herbage** sparsely villous. **Leaves** basal and cauline; blades oblong to oblanceolate; dentate to pinnately lobed, sessile upward, 1–5 cm long. **Inflorescence** leafy-paniculate with 5 to 15 heads. **Heads** discoid; involucres narrow, 5–9 mm high; phyllaries ca. 21, glabrous. **Disk corollas** 4–6 mm long. **Achenes** ca. 2 mm, sparsely strigose. Roadsides, gardens, yards; plains, valleys. Introduced throughout N. America; native to Eurasia.

Senecio werneriifolius (A.Gray) A.Gray [*Packera werneriifolia* (Gray) W.A.Weber & A.Löve] Fibrous-rooted from a branched rhizome. **Stems** ascending to erect, 3–10 cm, scapiform. **Herbage** sparsely arachnoid. **Leaves** mostly basal; blades oblong to ovate, entire to dentate at the tip, 5–15 mm long, greatly reduced on the stem. **Inflorescence** of 1 to 3 heads. **Heads** radiate (discoid); involucres 7–9 mm high; phyllaries 13 or 21, glabrous or sparsely arachnoid. **Rays** (0)8 to 13; ligules 5–9 mm long. **Disk flowers** 5 to 7 mm long. **Achenes** 2–3 mm long, glabrous. Fellfields, talus slopes; alpine. ID, MT south to CA, AZ, NM.

Senecio cymbalaria is similar but has more prominent, lobed stem leaves; plants from the Front Range west of Choteau appear intermediate.

Solidago L. Goldenrod

Perennial herbs. **Leaves** cauline and sometimes basal, alternate, simple, petiolate below, becoming sessile above. **Inflorescence** racemose, paniculate or coymbiform. **Heads** radiate; involucres campanulate to cylindric; phyllaries imbricate in 3 to 5 series, green with a prominent midvein; receptacle barely convex, usually naked. **Ray flowers** few to several, female, yellow. **Disk flowers** numerous, perfect, yellow; tubes shorter than throats; style appendages lanceolate, flattened. **Pappus** of capillary bristles. **Achenes** subcylindric with 8 to 10 ribs.[188,152,359]

Solidago speciosa Nutt. occurs in adjacent SD and could be in MT.

1. Plants strongly rhizomatous; basal leaves lacking 2
1. Plants with a branched caudex or short rhizome; basal leaves present or withered 6

2. Stem and leaves strongly puberulent . . . . *S. mollis*
2. Leaves glabrate or sparsely strigose to minutely ciliate but not very hairy on the surface 3

3. Stem pubescent below the inflorescence 4
3. Stem glabrous below the inflorescence 5

4. Leaf blades narrowly lanceolate, largest usually on lower stem; rays 10–16 . . . . *S. canadensis*
4. Leaf blades oblanceolate, largest at mid-stem; rays 6–12 . . . . *S. velutina*

Solidago canadensis L. [*S. lepida* DC. var. *salebrosa* (Piper) Semple, *S. altissima* L. var.
gilvocanescens (Rydb.) Semple] CANADA GOLDENROD Strongly rhizomatous. **Stems** erect, simple,
25–90 cm. **Herbage** glabrate to puberulent, at least at mid-stem and above. **Leaves** cauline, sessile
or subsessile; blades narrowly lanceolate, serrate, 4–10 cm long, only gradually reduced upward.
Inflorescence narrowly to broadly pyramidal; heads secund on spreading branches. **Involucres**
narrowly campanulate, 3–5 mm high; phyllaries linear-lanceolate, whitish, puberulent. **Rays** 10 to 16; ligules 1–3 mm
long. **Disk flowers** 5 to 9; corollas 3–4 mm long. **Achenes** ca. 1 mm long, strigose. Moist soil of meadows, thickets,
streambanks, open forest, fields, roadsides. AK to NL south to CA, TX, FL. (p.565)

1. Leaves glabrate on the upper surfaces . . . . var. *salebrosa*
1. Leaves noticeably hairy on the upper surfaces . . . . var. *gilvocanescens*

Solidago canadensis var. *salebrosa* (Piper) M.E.Jones occurs in the valleys to lower subalpine; *S. canadensis* var.
gilvocanescens Rydb. occurs on the plains. Specimens of *S. gigantea* with poorly developed inflorescences approach *S.
canadensis*.

Solidago gigantea Aiton Strongly rhizomatous. **Stems** erect, simple, 40–100 cm, sometimes
clustered. **Herbage** glabrous; the inflorescence puberulent. **Leaves** cauline, sessile, short-
petiolate; blades lanceolate, serrate, 4–12 cm long, gradually reduced upward. **Inflorescence**
pyramidal; heads secund on arching branches. **Involucres** campanulate, 3–4 mm high; phyllaries
linear-lanceolate, ciliate. **Rays** 9 to 15; ligules 2–3 mm long. **Disk flowers** 7 to 12; corollas 3–5 mm
long. **Achenes** 1–1.5 mm long, sparsely strigose. Meadows, thickets, roadsides, often along streams; plains, valleys. BC
to NS south to TX, FL, Mexico.
  Inflorescences of *Solidago canadensis* generally have straighter branches. Montane plants often have stipitate glands
in the inflorescence; some authors refer such plants to *S. canadensis*.[359]

Solidago missouriensis Nutt. MISSOURI GOLDENROD Rhizomatous. **Stems** ascending to
erect, simple, 10–60 cm, sometimes clustered. **Herbage**: leaves glabrous, minutely ciliate; the
inflorescence glabrate to puberulent. **Leaves** cauline, often also basal; basal petiolate; blades
oblanceolate, entire to serrate, 3–12 cm long, usually 3-veined; cauline gradually reduced and
becoming sessile upward. **Inflorescence** pyramidal to ovoid with spreading branches. **Involucres**
campanulate, 2–5 mm high; phyllaries lanceolate to ovate, sometimes ciliate. **Rays** 7 to 14; ligules 1–3 mm long. **Disk
flowers** 8 to 20; corollas 3–4 mm long. **Achenes** 1–2 mm long, glabrous to strigose. Grasslands, meadows, fields, open
forest; plains, valleys, montane, rarely higher. BC to ON south to AZ, NM, TX, TN.
  There is a tendency for plants from west of and near the Continental Divide to have larger involucres; these plants
have been called var. *extraria* A.Gray. Other plants from the western part of the state lack basal leaves at flowering
and are referred to var. *fasciculata* Holz. However, recent treatments disavow varietal distinctions.[103,359] *Solidago
multiradiata* generally has larger and fewer heads and occurs at higher elevations; *S. nemoralis* has puberulent stems.

Solidago mollis Bartl. Rhizomatous. **Stems** erect, simple, 15–60 cm. **Herbage** densely
puberulent. **Leaves** all cauline, short-petiolate to sessile, oblanceolate to elliptic, serrate, 3–10
cm long; the lowest 3-veined. **Inflorescence** ovoid to conical with secund branches. **Involucres**
campanulate, 3–5 mm high; phyllaries lanceolate to oblanceolate, ciliate. **Rays** 6 to 10; ligules 1–3

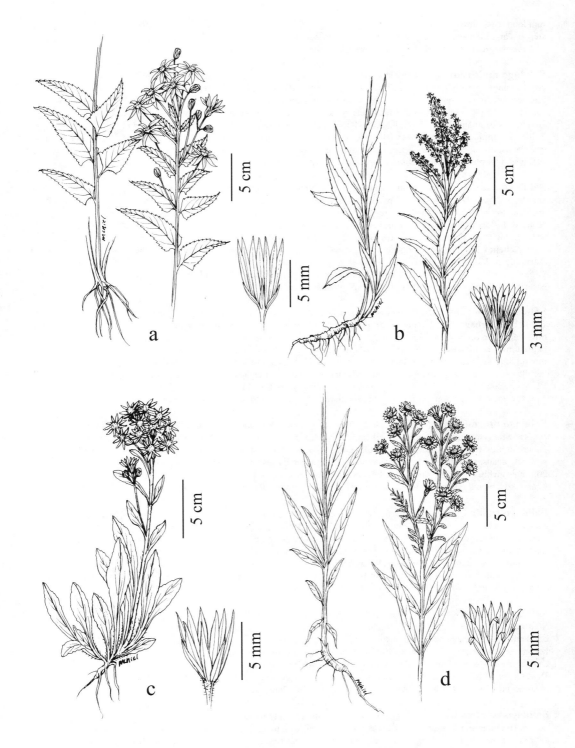

Plate 99. a. *Senecio triangularis*, b. *Solidago canadensis*, c. *Solidago multiradiata*, d. *Symphyotrichum eatonii*

mm long. **Disk flowers** 3 to 8; corollas 2–5 mm long. **Achenes** 1–2 mm long, strigose. Grasslands, meadows, sagebrush steppe, coniferous woodlands; plains, valleys. AB to MB south to NM, TX, IA, MI.

   Solidago rigida has basal leaves and a branched caudex rather than rhizomes.

Solidago multiradiata Aiton Plants with a branched caudex. **Stems** ascending, simple, 5–40 cm. **Herbage** sparsely villous to glabrous. **Leaves** basal and cauline; basal blades oblong to oblanceolate, mostly serrate, 1–7 cm long, with 1 prominent vein; basal leaf petioles broad, ciliate; cauline becoming sessile upward. **Inflorescence** hemispheric- to oblong-paniculate with few heads. **Involucres** campanulate, 5–7 mm high; phyllaries lanceolate to oblanceolate, ciliate. **Rays** 12 to 18; ligules 3–5 mm long. **Disk flowers** 10 to 35; corollas 4–6 mm long. **Achenes** 1–3 mm long, strigose. Often stony soil of grasslands, turf, meadows, dry open forest, talus, fellfields; subalpine, alpine, lower along rivers. AK to QC south to CA, AZ, NM. Our plants are var. *scopulorum* A.Gray. See *S. missouriensis*. (p.565)

Solidago nana Nutt. Plants from a short rhizome or branched caudex. **Stems** ascending, simple, 5–30 cm. **Herbage** densely puberulent. **Leaves** basal and cauline; basal petiolate; blades spatulate to oblanceolate, mostly entire, 15–35 mm long, usually 1-veined; cauline oblanceolate, reduced. **Inflorescence** corymbiform. **Involucres** campanulate, 4–6 mm high; phyllaries narrowly oblong, minutely ciliate. **Rays** 6 to 10; ligules 2–4 mm long. **Disk flowers** 8 to 20; corollas 4–5 mm long. **Achenes** 1–2 mm long, strigose. Often stony soil of grasslands, sagebrush steppe, meadows, open forest; valleys to subalpine. MT south to NV, AZ, NM.

   Solidago nemoralis has fewer disk flowers and is more likely to have a secund and curved inflorescence. Plants at the eastern margin of the range in MT approach *S. missouriensis*.

Solidago nemoralis Aiton Plants from a branched caudex. **Stems** erect to ascending, simple, 15–40 cm. **Herbage** densely puberulent. **Leaves** basal and cauline; basal petiolate; blades oblanceolate to oblong, entire to serrate, 2–8 cm long, 1-veined; cauline becoming narrower and entire upward. **Inflorescence** linear-secund, often curved. **Involucres** narrowly campanulate, 4–6 mm high; phyllaries narrowly oblanceolate. **Rays** 5 to 11; ligules 1–3 mm long. **Disk flowers** 3 to 10; corollas 4–5 mm long. **Achenes** 1–2 mm long, strigose. Moist grasslands, meadows, sagebrush steppe; plains, valleys. BC to QC south to NM, TX, FL. Our plants are var. *decemflora* (DC.) Fernald.

   A specimen from Teton Co. appears intermediate to *Solidago missouriensis*. See *S. nana, S. missouriensis*.

Solidago ptarmicoides (Torr. & A.Gray) B.Boivin [*Aster ptarmicoides* Torr. & A.Gray] Plants with a branched caudex. **Stems** ascending, simple, 10–40 cm with old leaf bases attached. **Herbage** glabrate; leaves ciliate; stems puberulent. **Leaves** basal and cauline; blades narrowly lanceolate, entire, 2–10 cm long. **Inflorescence** corymbiform, with few heads. **Involucres** campanulate, 5–7 mm high; phyllaries linear, glabrous, often glutinous. **Rays** 10 to 20; ligules white, 6–8 mm long. **Disk flowers** 30 to 36; corollas ca. 4 mm long. **Achenes** 1–2 mm long, glabrous. Grasslands; plains. SK to QC south to OK, OH, SC. Known from Richland and Wibaux cos.

Solidago rigida L. [*Oligoneuron rigidum* (L.) Small] STIFF GOLDENROD Plants from a branched caudex or short rhizome. **Stems** erect, simple, 20–70 cm. **Herbage** densely puberulent. **Leaves** basal and cauline; basal long-petiolate with lanceolate to ovate, entire to serrate blades 5–12 cm long; cauline becoming sessile upward. **Inflorescence** corymbiform, hemispheric. **Involucres** campanulate, 5–8 mm high; phyllaries linear-oblong, striate, canescent. **Rays** 6 to 13; ligules 2–3 mm long. **Disk flowers** 14 to 35; corollas 4–5 mm long. **Achenes** ca. 2 mm long, glabrate. Grasslands, pine woodlands, fields; plains. BC to ON south to NM, TX, LA, GA. Our plants are var. *humilis* Porter. See *S. mollis*.

Solidago simplex Kunth [*S. spathulata* DC.] Plants with a branched caudex or short rhizome. **Stems** ascending, simple, 5–30 cm. **Herbage** glabrous, glutinous in the inflorescence. **Leaves** basal and cauline; basal blades oblong to spatulate, serrate to entire, 1–10 cm long; cauline becoming narrower, sessile upward. **Inflorescence** paniculate, linear to oblong. **Involucres** campanulate, 3–8 mm high; phyllaries oblong to linear-oblong. **Rays** 7 to 16; ligules 2–5 mm long. **Disk flowers** 6 to 31; corollas 3–6 mm long. **Achenes** 2–3 mm long, strigose. Grasslands, meadows, open forest; valleys to subalpine. AK to QC south to AZ, NM, SC, OH, PA. Our plants are ssp. *simplex*.

Solidago velutina DC. [*S. sparsiflora* A.Gray] Plants from a branched rhizome. **Stems** erect to ascending, simple, 15–60 cm. **Herbage** sparsely strigose. **Leaves** short-petiolate, mainly cauline, the largest at mid-stem; blades oblanceolate, entire to dentate, 2–10 cm long, 1- or 3-veined. **Inflorescence** branched-secund, often curved. **Involucres** narrowly campanulate, 3–5 mm high; phyllaries narrowly lanceolate. **Rays** 6 to 12; ligules 3–6 mm long. **Disk flowers** 5 to 17; corollas 3–6 mm long. **Achenes** 1–3 mm long, strigose. Sagebrush steppe; valleys of Stillwater Co. OR to MT south to CA, AZ, NM, TX, Mexico. Our plants are ssp. *sparsiflora* (A.Gray) Semple.

Sonchus L. Sow thistle

Annual or perennial herbs with milky sap. **Herbage** glabrous. **Leaves** basal and cauline, alternate, simple, entire to lobed, prickly-margined. **Inflorescence** paniculate. **Heads** ligulate; involucre campanulate; phyllaries linear-lanceolate, unequal, imbricate in 3 to 5 series; receptacle flat to convex, naked. **Ray flowers** perfect, yellow, >80 per head. **Pappus** 10 to 80, white, capillary, smooth or barbed bristles. **Achenes** ellipsoid to oblong, 4- to 10-ribbed, glabrous, not beaked.

Our species are all native to Europe and introduced throughout N. America.

1. Rhizomatous perennials; ray ligules 7–13 mm long; involucre 1–2 cm high . . . .*S. arvensis*
1. Taprooted annuals; rays 3–6 mm long; involucre 8–15 mm high 2

2. Achenes bumpy between the ribs. . . . .*S. oleraceus*
2. Achenes smooth between the ribs . . . . *S. asper*

Sonchus arvensis L. [*S. uliginosus* M.Bieb.] Rhizomatous perennial. **Stems** erect, 30–140 cm, sometimes branched above. **Leaves** linear-oblanceolate, pinnately lobed, 4–30 cm long, winged-petiolate below to sessile and auriculate above. **Involucres** 1–2 cm high; phyllaries glabrous or rarely stipitate-glandular. **Ray ligules** 7–13 mm long. **Achenes** brown, 2–4 mm long, minutely bumpy. Wet meadows, marshes, ditch banks, irrigated fields; plains, valleys.

1. Involucre glabrous or nearly so . . . . .ssp. *uliginosus*
1. Involucre stipitate-glandular . . . . ssp. *arvensis*

Sonchus arvensis ssp. *arvensis* is less common and occurs mainly in the western cos.; *S. arvensis* ssp. *uliginosus* (M.Bieb.) Nyman is more common and widespread. They are sometimes considered separate species, but most recent treatments consider them conspecific;[103,199,399] they have no ecological distinction in MT.

Sonchus asper (L.) Hill Taprooted annual. **Stems** erect, 25–90 cm tall, simple or branched. **Leaves** oblong to oblanceolate, pinnately toothed or lobed less than halfway to the midvein, 5–20 cm long, winged-petiolate below, auriculate-clasping above. **Involucres** 8–14 mm high; phyllaries glabrous. **Ray ligules** 3–5 mm long, inconspicuous. **Achenes** tan to red-brown, 2–3 mm long, smooth between the ribs. Disturbed soil of roadsides, pastures, lawns, streambanks; plains, valleys.

Sonchus oleraceus L. Taprooted annual. **Stems** erect, 30–100 cm tall, mostly simple. **Leaves** lanceolate to oblanceolate, pinnately lobed to mostly entire, 5–25 cm long, winged-petiolate below, becoming sessile and auriculate above. **Involucres** 8–15 mm high; phyllaries glabrous or with patches of tomentum. **Ray ligules** 3–6 mm long, inconspicuous. **Achenes** brown, 2–4 mm long, bumpy between the ribs. Disturbed soil of roadsides, fields, lawns; plains, valleys.

Sphaeromeria Nutt.

Subshrubs forming loose mats of short, sterile stems. **Herbage** glandular-punctate, aromatic, tomentose. **Leaves** alternate, fan-shaped, 3-lobed at the tip, cuneate at the base, petiolate. **Heads** discoid; involucre hemispheric; phyllaries in 2 to 3 subequal series, scarious-margined; receptacle conic, naked. **Disk flowers** perfect, yellow; style branches flattened. **Pappus** absent. **Achenes** obconic, ribbed, glabrous or glandular, 3- to 5-ribbed.

1. Leaves shallowly lobed at the tip . . . . *S. argentea*
1. Leaves deeply lobed into linear segments . . . . *S. capitata*

Sphaeromeria argentea Nutt. [*Tanacetum nuttallii* Torr. & A.Gray] Chicken sage **Flower stems** erect, 4–15 cm. **Leaves** 5–15 mm long, shallowly 3-lobed at the tip. **Inflorescence** tightly corymbiform to capitate. **Involucre** 2–4 mm high; phyllaries 12 to 16, sparsely tomentose. **Disk corollas** 2–3 mm long. **Achenes** 2–3 mm long. Sparsely vegetated, often calcareous soil of grasslands, sagebrush steppe; valleys, montane in Beaverhead Co. MT, ID, NV, WY, CO.

The leaves are similar to those of *Artemisia tridentata*.

Sphaeromeria capitata Nutt. [*Tanacetum capitatum* (Nutt.) Torr. & A.Gray] **Stems** erect, 2–15 cm. **Leaves** 8–25 mm long, deeply lobed into linear segments. **Inflorescence** capitate. **Involucre** 2–3 mm high; phyllaries 5 to 8, tomentose. **Disk corollas** 1–3 mm long. **Achenes** 1–2 mm long. Shallow, calcareous soil of grasslands, sagebrush steppe, coniferous woodlands; valleys, montane. MT to UT, CO.

Stenotus Nutt., Goldenweed

Taprooted perennial herbs from a branched caudex. **Leaves** basal and cauline, alternate, simple, entire, short-petiolate; blades usually 3-veined. **Heads** solitary or few, radiate; involucre campanulate to hemispheric; phyllaries imbricate in 2 to 4 series, basally firm, scarious-margined; receptacle convex, naked, pitted. **Ray flowers** 5 to 17, yellow, female. **Disk flowers** 14 to 45, perfect, yellow, the tube shorter than the throat; style branch appendages linear-lanceolate. **Pappus** 1 series of capillary bristles. **Achenes** obconic, 6- to 12-nerved, usually sericeous.[284]

Previously placed in *Haplopappus*.

1. Stems villous to tomentose. . . . .*S. lanuginosus*
1. Stems glabrate to sparsely villous 2

2. Phyllaries blunt. . . . .*S. armerioides*
2. Phyllaries acute to acuminate. 3

3. Heads solitary; stems well surpassing the leaves . . . .*S. acaulis*
3. Heads >1; stems barely surpassing the leaves . . . . *S. multicaulis*

Stenotus acaulis (Nutt.) Nutt. [*Haplopappus acaulis* (Nutt.) A.Gray] Mat-forming. **Stems** simple, erect, subscapose, 2–12 cm, mostly glabrate. **Leaves** mainly basal; blades oblanceolate, 1–4 cm long, mostly glabrate, sometimes minutely ciliate, becoming linear on the stem. **Involucres** 7–12 mm high; phyllaries lanceolate, acute, white below with green tips, glabrous, viscid. **Rays** 5 to 13; ligules 5–11 mm long. **Disk corollas** 5–8 mm long, glabrous. **Achenes** 2–5 mm long, sometimes glabrous. Sparsely vegetated, often shallow soil of grasslands, sagebrush steppe, coniferous woodlands, rock outcrops, more common in calcareous soil; plains, valleys to rarely alpine. OR to MT south to CA, AZ, CO.

Some plants from Beaverhead Co. have puberulent or glandular herbage.

Stenotus armerioides Nutt. [*Haplopappus armerioides* (Nutt.) A.Gray] Cespitose. **Stems** mostly simple, erect, 4–15 cm, glabrate, viscid. **Leaves** basal and cauline, ascending, rigid; blades linear-oblanceolate to oblong, 2–7 cm long, glabrate, viscid, sometimes minutely ciliate. **Involucres** 7–13 mm high; phyllaries oblong, white below with green tips, glabrous, viscid. **Rays** 5 to 15; ligules 7–14 mm long. **Disk corollas** 4–8 mm long, glabrate. **Achenes** 2–6 mm long. Grasslands, sagebrush steppe, open slopes, confined to calcareous soils west of the Continental Divide; plains, valleys. SK south to AZ, NM, KS.

Stenotus lanuginosus (A.Gray) Greene [*Haplopappus lanuginosus* A.Gray] Mat-forming. **Stems** simple, ascending to erect, subscapose, 4–30 cm, villous to tomentose, glandular. **Leaf blades** linear to oblanceolate, 1–5 cm long, sparsely villous to tomentose. **Involucres** 7–15 mm high; phyllaries lanceolate, green, ciliate, glandular. **Rays** 9 to 17; ligules 6–14 mm long. **Disk corollas** 5–7 mm long, sericeous. **Achenes** 2–5 mm long. Stony soil of sagebrush steppe, grasslands, fellfields, coniferous woodlands; montane to alpine. WA, MT, ID, CA, NV.

Our plants have been assigned to var. *andersonii* (Rydb.) C.A.Morse which is endemic to southwest MT and adjacent ID, but some populations have characters of the more widespread var. *lanuginosus*.[284]

Stenotus multicaulis Nutt. [*Haplopappus multicaulis* (Nutt.) A.Gray, *Oönopsis multicaulis* (Nutt.) Greene] Mat-forming from a branched caudex. **Stems** simple, ascending to erect, villous, becoming glabrate, 3–6 cm, barely exceeding the leaves. **Leaf blades** linear, 1–5 cm long, glabrate to sparsely villous. **Involucres** 7–9 mm high; phyllaries oblanceolate to obovate, acuminate, green above, ciliate. **Rays** 6 to 8; ligules 7–10 mm long. **Disk flowers** 3–6 mm long, puberulent. **Achenes** 2–4 mm long, sericeous. Barren clay soil of badlands; plains. MT, WY, SD, NE. Known from Carter Co.

There seems to be no compelling evidence to segregate this species into *Oönopsis*.

Stephanomeria Nutt. Skeletonweed, wire lettuce

Perennial herbs. **Herbage** glabrate with milky sap. **Leaves** alternate, narrow, inconspicuous; the basal withered at flowering. **Inflorescence** paniculiform with solitary heads on bracteate branch tips. **Heads** ligulate; involucre cylindric; phyllaries green, in 2 series, the outer (calyculi) much shorter, inner 5–6, glabrous; receptacle flat, naked. **Ray flowers** perfect, mostly 5 to 13, pinkish. **Pappus** white, plumose bristles. **Achenes** cylindric, 5-angled, glabrous.

1. Lower leaves pinnate with backward-pointing lobes; achenes bumpy. . . . . *S. runcinata*
1. Leaves entire or with remote teeth; achenes smooth. . . . . *S. tenuifolia*

Stephanomeria runcinata Nutt. Rhizomatous. **Stems** erect, 5–30 cm, branched above. **Leaves** 2–9 cm long, linear-lanceolate with pinnate, recurved lobes; the basal present at flowering. **Involucre** 9–14 mm high. **Rays** 5 or 6; ligules 5–12 mm long. **Achenes** 3–5 mm long, bumpy. Grasslands, sagebrush steppe, pine woodlands; plains, valleys. AB, SK south to UT, CO, NE.

Stephanomeria tenuifolia (Raf.) H.M.Hall Taprooted or rhizomatous. **Stems** erect, 30–75 cm, branched above. **Leaves** 5–8 cm long, linear, entire or with remote teeth; basal absent at flowering. **Involucre** 7–10 mm high. **Rays** 4 to 6; ligules 3–9 mm long. **Achenes** 3–6 mm long, smooth. Sandy or rocky soil of grasslands, sagebrush steppe; valleys. BC to SK south to CA, AZ, NM, TX.

Symphyotrichum Nees American aster, New World aster

Perennial or annual herbs. **Stems** usually unbranched below. **Leaves** basal and alternate, simple; **Inflorescence** paniculate, racemose, or heads solitary. **Heads** radiate; turbinate to hemispheric; phyllaries in several imbricate, mostly unequal series; receptacle pitted, glabrous. **Rays** female, white, pink, or blue. **Disk flowers** perfect, yellow to reddish; style branches flattened with lanceolate appendages. **Pappus** of numerous barbellate, capillary bristles. **Achenes** usually 3- to 5-nerved.[66]

Symphyotrichum and *Erigeron* are often confused. Phyllaries of the former are mostly of several different lengths and overlap each other like shingles (imbricate); phyllaries of *Erigeron* also overlap but are more nearly the same length. This difference is not absolute, and determinations can sometimes be difficult. *Symphyotrichum novae-angliae* (L.) Nesom, native to eastern N. America and often grown ornamentally, is reported for MT,[66] but I have seen no specimens representing naturalized populations; *S. frondosum* (Nutt.) Nesom [*Aster frondosus* (Nutt.) T. & G.] is reported for Sheridan Co.,[113] but I have seen no specimens.

1. Rays absent or barely exceeding the pappus; plants annual. 2
1. Rays conspicuous; plants perennial. 3

2. Ray ligules absent . . . . *S. ciliatum*
2. Ray ligules ca. 2 mm long . . . . *S. frondosum*

3. Involucre glandular. 4
3. Involucre glabrous to hairy but not glandular 5

4. Tips of main leaves rounded or bluntly acute; mainly Great Plains . . . . *S. oblongifolium*
4. Leaf tips acute, minutely mucronate; mainly mountain valleys. . . . . *S. campestre*

5. Outer phyllaries mucronate, spreading or recurved; rays white. 6
5. Phyllaries not mucronate, usually appressed; rays often pink to violet 7

6. Involucre ≤4.5 mm high; stems clustered on a caudex or short rhizome; inflorescence
   branches secund . . . . *S. ericoides*
6. Involucre ≥5 mm high; stems from extensive rhizomes; heads not strongly secund . . . .*S. falcatum*

7. Leaf surfaces pubescent 8
7. Leaf surfaces glabrous to sparsely strigose. 9

8. Involucre 5–7 mm high; reticulations of leaf veins longer than wide . . . .*S. ascendens*
8. Involucre 7–9 mm high; leaf vein reticulations ca. as long as wide . . . . *S. molle*

9. Middle and upper cauline leaves ovate, auriculate clasping 10
9. Cauline leaves petiolate or sessile but not auriculate. 11

10. Phyllaries distinctly imbricate; peduncles often glaucous; mature achenes glabrous . . . . *S. laeve*
10. Phyllaries subequal; peduncles not glaucous; mature achenes hairy . . . . *S. foliaceum*

11. Basal leaf blades ovate, cordate, truncate, or rounded at the base, often serrate . . . . *S. ciliolatum*
11. Basal leaf blades oblanceolate and entire or absent 12

12. Stem pubescence in vertical lines from below leaf bases 13
12. Stem glabrous or pubescence evenly or patchily distributed but not in lines 16

13. Heads many; inflorescence ca. 1/2 the length of the stem 14
13. Heads relatively few; inflorescence ≤1/3 the length of the stem 15

14. Outer involucral bracts oblanceolate . . . . *S. eatonii*
14. Involucral bracts linear-lanceolate . . . . *S. lanceolatum*

15. Lower cauline leaves deciduous at flowering time . . . . *S. boreale*
15. Lower leaves usually present at flowering . . . . *S. subspicatum*

16. Phyllaries not or weakly imbricate; the outer as long or longer than the inner 17
16. Phyllaries well-imbricate; the outer definitely shorter than the inner 18

17. Leaves linear to lanceolate, little reduced upward; inflorescence nearly 1/2 the
    length of the stem . . . . *S. eatonii*
17. Leaves lanceolate to oblong, strongly reduced above; inflorescence ca. 1/3 the
    length of the stem . . . . *S. foliaceum*

18. Leaves minutely hispid-ciliate; leaf vein reticulations ca. twice as long as wide . . . . *S. ascendens*
18. Leaves not consistently ciliate; leaf vein reticulations ca. as long as wide 19

19. Lower cauline leaves deciduous at flowering time . . . . *S. boreale*
19. Lower leaves usually present at flowering 20

20. Base of phyllaries brownish; some leaves often serrate . . . . *S. subspicatum*
20. Base of phyllaries white; leaves entire . . . . *S. spathulatum*

Symphyotrichum ascendens (Lindl.) G.L.Nesom [*Aster ascendens* Lindl., *A. chilensis* Nees ssp. *ascendens* (Lindl.) Cronquist] Rhizomatous perennial. **Stems** ascending to erect, 15–60 cm. **Herbage** glabrate to strigose. **Leaves** cauline, sessile or subsessile, glabrate to scabrous, ciliate, linear to linear-oblanceolate, entire, 2–10 cm long, the lowest deciduous. **Inflorescence** paniculate; peduncles strigose. **Involucre** campanulate, 5–7 mm high; phyllaries well-imbricate, oblanceolate, strigose, white below, green-tipped. **Rays** 15 to 40, blue to pink; ligules 5–10 mm long. **Disk flowers** 15 to 40, yellow, often purple-tipped; corolla 4–7 mm long. **Achenes** obconic, 2–3 mm long, pubescent. Grasslands, dry meadows, sagebrush steppe, streambanks, roadsides; plains, valleys, montane. BC to SK south to CA, AZ, NM, ND. (p.572)
   Symphyotrichum ascendens seems to thrive with moderate disturbance. The minutely hispid-ciliate leaves and bracts help identify this species.

Symphyotrichum boreale (Torr. & A.Gray) A.Löve & D.Löve [*Aster junciformis* Rydb.] Perennial from thin rhizomes. **Stems** erect, 15–40 cm. **Herbage** glabrous; leaves minutely ciliate; stems puberulent in lines above. **Leaves** cauline, sessile, linear, 3–8 cm long, entire; the lower withering. **Inflorescence** leafy paniculate with few heads; peduncles strigose. **Involucre** narrowly campanulate, 5–8 mm high; phyllaries erect, linear to lanceolate, glabrous, green above, white below, sometimes purplish. **Rays** 20 to 50, white or pale blue; ligules 6–12 mm long. **Disk flowers** 20–30, pale yellow; corollas 3–7 mm long. **Achenes** obovoid, ca. 1 mm long, sparsely hairy. Wet, organic, often calcareous soil of fens, wet meadows, thickets; valleys, montane. Across northern N. America to WA, CO, IA, PA.
   Reports[66] of *Symphyotrichum welshii* (Cronquist) G.L.Nesom were based on this species and *S. eatonii*. The type locality is in Gallatin Co.

Symphyotrichum campestre (Nutt.) G.L.Nesom [*Aster campestris* Nutt.] Rhizomatous perennial. **Stems** erect, 15–40 cm, often with short, sterile stems from leaf axils. **Herbage** glabrate to puberulent, glandular above. **Leaves** cauline, linear to lanceolate, sessile or subsessile, entire, 2–5 cm long, the lowest deciduous. **Inflorescence** open, leafy-paniculate; peduncles strigose, glandular. **Involucre** campanulate, 5–7 mm high; phyllaries linear-lanceolate, acuminate, stipitate-glandular, white below, green above, scarious-margined. **Rays** 15 to 30, violet; ligules 6–10 mm long. **Disk flowers** 25–40, yellow; corollas 4–6 mm long. **Achenes** obconic, 1–2 mm long, pubescent. Grasslands, moist to dry meadows, sagebrush steppe; plains, valleys, montane. BC, AB south to CA, NV, CO. (p.572)
   Symphyotrichum oblongifolium is very similar but occurs to the east of *S. campestre*.

Symphyotrichum ciliatum (Ledeb.) G.L.Nesom [*Aster brachyactis* S.F.Blake] Taprooted annual. **Stems** erect, usually branched, 5–70 cm. **Herbage** glabrate; leaf margins sparsely retrorse-scabrous. **Leaves** cauline, linear, entire, petiolate below, 1–12 cm long. **Inflorescence** open leafy-paniculate with suberect branches. **Involucre** hemispheric, 5–10 mm high; phyllaries glabrous, linear-lanceolate, nearly equal, green, white-margined below, mucronate. **Rays** absent. **Disk flowers** ca. 14, filiform, white to pink; corollas 2–4 mm long; style well exserted. **Achenes** obovoid, 2- to 4-nerved, purplish, ca. 2 mm long. Vernally wet margins of ponds, lakes, streams, tolerant of saline conditions; plains, valleys. North-temperate N. America; Eurasia. (p.572)

Symphyotrichum ciliolatum (Lindl.) A.Löve & D.Löve [*Aster ciliolatus* Lindl.] Rhizomatous perennial. **Stems** erect, 20–100 cm. **Herbage** glabrate, ciliate-petiolate. **Leaves**: basal and cauline, petiolate; blades lanceolate to narrowly ovate, serrate, cordate to rounded at the base, 3–12 cm long, wing-petiolate, gradually reduced upward. **Inflorescence** paniculate with few leaf-like bracts. **Involucre** campanulate, 5–7 mm high; phyllaries linear-lanceolate, acuminate, unequal, glabrous, green, scarious. **Rays** 14 to 20, blue; ligules 10–15 mm long. **Disk flowers** 14–25, yellow; corollas 4–6 mm long. **Achenes** obovoid, 1–2 mm long, glabrous. Open forest, meadows; valleys, montane; collected in Cascade and Meagher cos. over 100 years ago. YT to QC south to WY, SD, IL, NY.

Symphyotrichum eatonii (A.Gray) G.L.Nesom [*Aster eatonii* (A.Gray) Howell, *S. bracteolatum* (Nutt.) G.L.Nesom] Rhizomatous perennial. **Stems** erect, 20–80 cm. **Herbage** glabrous; stems puberulent. **Leaves** cauline, sessile, linear to lanceolate, entire, 4–10 cm long, the lowest deciduous. **Inflorescence** leafy-paniculate, usually open with many heads; peduncles puberulent. **Involucre** campanulate, 5–8 mm high; phyllaries weakly imbricate to subequal; the outer oblanceolate, glabrate to ciliate, green with white bases. **Rays** 20 to 40, pink; ligules 5–12 mm long. **Disk flowers** 35 to 60, yellow to pinkish; corollas 4–7 mm long. **Achenes** cylindric, ca. 3 mm long, pubescent. Moist soil of meadows, thickets, often along streams, wetlands; valleys, montane. BC to SK south to CA, AZ, NM. (p.565, 572)

   Symphyotrichum lanceolatum is similar in habit and habitat but has narrower outer phyllaries.

Symphyotrichum ericoides (L.) G.L.Nesom [*Aster pansus* (S.F.Blake) Cronquist, *A. ericoides* L. var. *pansus* (S.F.Blake) B.Boivin] Perennial from a branched caudex. **Stems** ascending to erect, sometimes branched, 20–70 cm, often clustered. **Herbage** strigose to pubescent. **Leaves** cauline, linear, sessile, entire, 1–5 cm long, often with fascicles or short branches in the axils, the lowest deciduous. **Inflorescence** open, leafy-paniculate with numerous secund heads; peduncles pubescent to hispid. **Involucre** campanulate, 3–5 mm high; phyllaries imbricate, spreading, oblanceolate, mucronate, sparsely hispid, ciliate, green above, white below. **Rays** 10 to 18, white; ligules 3–6 mm long. **Disk flowers** 6 to 12, yellow; corolla 2–3 mm long. **Achenes** oblong, 1–2 mm long, sparsely hairy. Moist meadows, grasslands, often around streams, wetlands; plains, valleys, montane. BC to QC south to AZ, NM, TX, AL, VA, Mexico. Our plants are var. *pansum* (S.F.Blake) G.L.Nesom. See *S. falcatum*.

Symphyotrichum falcatum (Lindl.) G.L.Nesom [*Aster falcatus* Lindl.] Perennial from rhizomes or a branched caudex. **Stems** ascending to erect, 15–70 cm. **Herbage** curly-puberulent. **Leaves** cauline, linear, sessile or subsessile, entire, 2–5 cm long, the lowest deciduous. **Inflorescence** open, leafy-paniculate with numerous heads; peduncles puberulent. **Involucre** campanulate, 5–8 mm high; phyllaries spreading, linear-oblanceolate, mucronate, short-hispid, green and recurved above, white below. **Rays** 20 to 35, white; ligules 4–7 mm long. **Disk flowers** 25–40, yellow; corollas 3–5 mm long. **Achenes** obconic, 2–3 mm long, brown, strigose. Grasslands, dry meadows; plains, valleys, montane. AK to MB south to AZ, TX, IA. (p.572)

   The similar *Symphyotrichum ericoides* occurs in moister sites and has smaller heads.

Symphyotrichum foliaceum (Lindl. ex DC.) G.L.Nesom [*Aster foliaceus* Lindl. ex DC.] Rhizomatous perennial. **Stems** ascending to erect, 6–80 cm. **Herbage** glabrate, sparsely hairy on stems. **Leaves** basal (often deciduous) and cauline; blades ovate to lanceolate, 2–15 cm long, entire; the lower with broad petioles, sometimes sessile and clasping above. **Inflorescence** leafy paniculate with 1 to several heads; peduncles glabrous, puberulent or villous. **Involucre** campanulate, 6–15 mm high; phyllaries, outer lanceolate to oblanceolate, green, white-margined, sometimes foliaceous, inner narrower. **Rays** 15 to 60, blue or violet; ligules 6–18 mm long. **Disk flowers** 50 to 150, yellow; corollas 4–7 mm long. **Achenes** cylindric, 2–4 mm long, hairy. BC, AB south to CA, AZ, NM. (p.572)

1. Outer phyllaries oblanceolate, longer than the inner . . . . var. *foliaceum*
1. Outer phyllaries lanceolate, equal to or slightly shorter than the inner. 2

2. Stems ≤ 25 cm; heads usually solitary; outer phyllaries often purplish . . . . var. *apricum*
2. Stems usually >25 cm; heads often ≥2; phyllaries green. 3

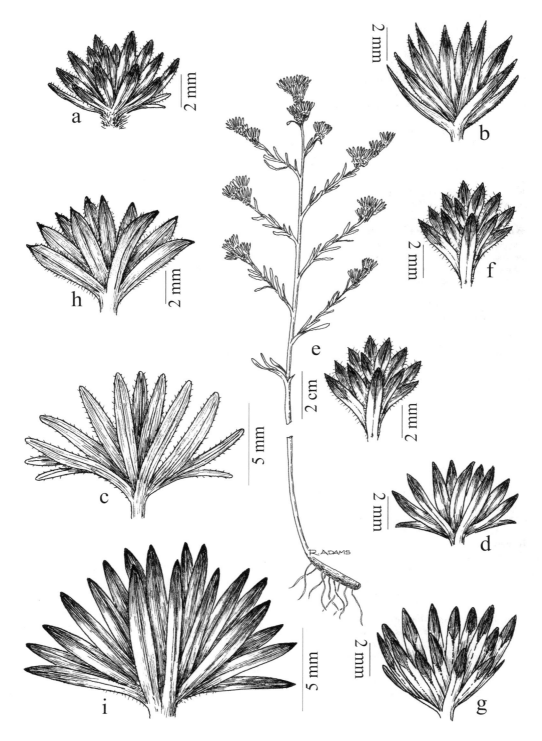

Plate 100. **Symphyotrichum**. **a.** *S. ascendens*, **b.** *S. campestris*, **c.** *S. ciliatum*, **d.** *S. eatonii*, **e.** *S. falcatum*, **f.** *S. laeve*, **g.** *S. lanceolatum*, **h.** *S. foliaceum* var. *foliaceum*, **i.** *S. spathulatum*

3. Upper stem leaves strongly auriculate-clasping; phyllaries relatively broad . . . . var. *cusickii*
3. Upper stem leaves not or weakly clasping; phyllaries relatively narrow. .var. *parryi*

Symphyotrichum foliaceum var. *apricum* (A.Gray) G.L.Nesom is found in subalpine to alpine meadows, turf in southwest cos; *S. foliaceum* var. *parryi* (D.C.Eaton) G.L.Nesom is found in montane to lower subalpine moist meadows, fens; *S. foliaceum* var. *foliaceum* occurs in subalpine to alpine open forest, dry meadows; *S. foliaceum* var. *cusickii* (A.Gray) P. Lesica (ined.) is found in moist to wet, montane forest in northwest cos.Variety *canbyi* (A.Gray) G.L.Nesom differs from var. *parryi* by having more rounded phyllaries; however, the distinction seems obscure in our area so I believe it is best to combine the two under the latter name. *Symphyotrichum hendersonii* (Fernald) Nesom [*Aster foliaceus* (Lindl.) var. *lyallii* (Gray) Cronq] is reported for MT,[66,187] but I have seen no specimens that could not be accommodated under *S. foliaceum* var. *cusickii*. The *S. foliaceum* complex consists of several diploid species and their polyploid hybrid derivatives.[1,3] The most recent treatment[66] designates the diploid *S. cusickii* and *S. hendersonii* as species even though morphological lines between them and *S. foliaceum* are blurred by hybrids. I have chosen to follow Cronquist's treatment where all taxa are treated as varieties of a single species. Placement of many specimens will be difficult. White-rayed populations of var. *apricum* are documented for the Gravelly and Tobacco Root ranges.

Symphyotrichum laeve (L.) A.Löve & D.Löve [*Aster laevis* L.] Perennial from a branched caudex or short rhizomes. **Stems** erect, 20–80 cm. **Herbage** glabrous. **Leaves** cauline, rarely basal, petiolate below, auriculate-clasping above; blades lanceolate to ovate, entire to serrate, 2–15 cm long. **Inflorescence** leafy-paniculate, open with many heads; peduncles often glaucous, glabrous to puberulent in lines, the bracts clasping. **Involucre** campanulate, 5–9 mm high; phyllaries linear-lanceolate, imbricate, glabrous, green above with white bases and hyaline margins. **Rays** 15 to 30, blue; ligules 7–12 mm long. **Disk flowers** 20 to 30, yellow, turning purplish; corollas 5–7 mm long. **Achenes** oblong, 2–4 mm long, glabrous. Moist grasslands, meadows, forest openings, especially along streams, roads or other disturbed areas; plains, valleys, montane. Throughout N. America. Our plants are *Symphyotrichum laeve* var. *geyeri* (A.Gray) G.L.Nesom. (p.572)

The clasping bracts just below the involucre and especially the glabrous achenes are the best characters to separate this from *Symphyotrichum foliaceum* var. *parryi*.

Symphyotrichum lanceolatum (Willd.) G.L.Nesom [*Aster hesperius* A.Gray] Rhizomatous perennial. **Stems** erect, 40–100 cm. **Herbage** glabrate; stems puberulent in vertical lines from leaf bases. **Leaves** cauline, sessile, linear to lanceolate, entire, 5–15 cm long, often fascicled, the lowest deciduous. **Inflorescence** open, leafy-paniculate, usually with many heads; peduncles villous. **Involucre** narrowly campanulate, 5–9 mm high; phyllaries linear-lanceolate, weakly imbricate, glabrate to ciliate, green with white bases. **Rays** 18 to 45, white to blue; ligules 4–12 mm long. **Disk flowers** 18 to 52, yellow; corollas 4–7 mm long. **Achenes** obovoid, 1–3 mm long, sparsely hairy. Moist, sometimes saline soil of meadows, usually around streams, ponds; plains, valleys. BC to QC south to CA, AZ, NM, TX, IA, WI. Our plants are var. *hesperium* (A.Gray) G.L.Nesom.[358] See *S. eatonii, S. spathulatum.* (p.572)

Symphyotrichum molle (Rydb.) G.L.Nesom [*Aster mollis* Rydb.] Rhizomatous perennial. **Stems** ascending to erect, 30–60 cm. **Herbage** pubescent to sericeous. **Leaves** basal and cauline; blades oblanceolate, 1–5 cm long, entire; the lower petiolate. **Inflorescence** leafy-paniculate with several heads; peduncles pubescent. **Involucre** campanulate, 7–9 mm high; phyllaries spreading, oblanceolate, pubescent, green above, white below; inner narrower. **Rays** 20 to 35, violet; ligules 12–20 mm long. **Disk flowers** 35 to 70, yellow; corollas 5–7 mm long. **Achenes** cylindric, 2–4 mm long, hairy. Grasslands, meadows; montane. Endemic to Big Horn Co., MT and adjacent WY.[246]

Symphyotrichum oblongifolium (Nutt.) G.L.Nesom [*Aster oblongifolius* Nutt.] Perennial from rhizomes or a branched caudex. **Stems** ascending, 10–30 cm, often with short, sterile branches with small leaves. **Herbage** hispid and glandular above. **Leaves** cauline, oblanceolate, sessile, clasping above, entire, 1–4 cm long, the lowest deciduous. **Inflorescence** open, paniculate with heads at tips of small-leaved branches; peduncles glandular. **Involucre** campanulate, 5–7 mm high; phyllaries imbricate, oblanceolate, acuminate, crystaline-glandular, white below, green above, scarious-margined. **Rays** 25 to 35, violet; ligules 6–8 mm long. **Disk flowers** 30 to 40, yellow becoming purple; corollas 4–5 mm long. **Achenes** obovoid, 2–3 mm long, pubescent. Grasslands, pine woodlands; plains. MT to PA south to NM, TX, AL, Mexico. See *S. campestre.*

Symphyotrichum spathulatum (Lindl.) G.L.Nesom [*Aster occidentalis* (Nutt.) Torr. & A.Gray] Rhizomatous perennial. **Stems** erect to ascending, 15–60 cm. **Herbage** glabrate. **Leaves** basal and cauline, petiolate below; blades linear to lanceolate, 3–12 cm long, entire; the upper sessile and narrower. **Inflorescence** open-paniculate with few heads; peduncles glabrate to densely strigose. **Involucre** campanulate, 6–9 mm high; phyllaries linear-oblanceolate, imbricate, green with white bases, glabrous to ciliate. **Rays** 15 to 40, blue or violet; ligules 7–15 mm long. **Disk flowers** 30 to 80, yellow becoming purple; corollas 4–8 mm long. **Achenes** cylindric, 2–4 mm long, hairy. Wet meadows, thickets, fens, open forest; valleys to subalpine. BC, AB south to CA, UT, NM. Our plants are var. *spathulatum*. (p.572)

 Symphyotrichum boreale has thinner stems with lines of pubescence below the leaves; *S. lanceolatum* has lines of pubescence on the stems and many heads. See *S. foliaceum.*

Symphyotrichum subspicatum (Nees) G.L.Nesom [*Aster subspicatus* Nees] Rhizomatous perennial. **Stems** erect to ascending, 15–50 cm. **Herbage** glabrate to sparsely puberulent. **Leaves** cauline, petiolate below; blades narrowly lanceolate to oblanceolate, 3–10 cm long, entire to serrate; the upper sessile. **Inflorescence** open-paniculate with few to several heads; peduncles sparsely strigose. **Involucre** campanulate, 5–10 mm high; phyllaries oblanceolate, imbricate, green with yellowish bases, glabrous. **Rays** 15 to 45, violet; ligules 8–14 mm long. **Disk flowers** 50 to 75, yellow; corollas 4–7 mm long. **Achenes** cylindric, 2–4 mm long, hairy. Streambanks, wetlands; valleys, montane. BC, AB south to CA, ID, MT.

 Probably derived from hybridization between *Symphyotrichum chilense, S. foliaceum,* and other species of *Symphyotrichum.*[1,66]

Tanacetum L. Tansy

Rhizomatous perennials. **Herbage** glandular-aromatic. **Leaves** alternate, petiolate, basal and cauline. **Inflorescence** corymbiform, flat-topped. **Heads** radiate or discoid; involucres hemispheric; phyllaries scarious-margined in 2 to 5 unequal series; receptacle flat, naked. **Ray flowers** female or sterile, white, yellow, or absent. **Disk flowers** perfect, yellow. **Pappus** an inconspicuous crown. **Achenes** obconic to cylindric, ribbed, usually glandular.[421]

 All our species were introduced from Eurasia for horticulture.

1. Leaves simple, not lobed . . . . *T. balsamita*
1. Leaves deeply lobed to divided 2

2. Heads radiate; herbage puberulent . . . . *T. parthenium*
2. Heads discoid; herbage glabrate . . . . *T. vulgare*

Tanacetum balsamita L. [*Chrysanthemum balsamita* L.] Costmary **Stems** erect, 50–150 cm. **Herbage** strigose to canescent. **Leaf blades** simple, ovate, 3–10 cm long, reduced upward. **Involucre** 5–8 mm across; phyllaries in 3 or 4 series, dilated at the tip. **Rays** absent. **Disk corollas** ca. 2 mm long. **Achenes** 1–2 mm long, 4- to 5-angled. Roadsides, fields, irrigation ditches; valleys. Introduced in northern U.S. and adjacent Canada.

Tanacetum parthenium (L.) Sch.-Bip. [*Chrysanthemum parthenium* (L.) Bernh.] Rhizomatous or taprooted. **Stems** erect, 25–80 cm. **Herbage** puberulent. **Leaf blades** 1–10 cm long, pinnately divided; segments oblanceolate, deeply serrate. **Involucre** 5–7 mm across; phyllaries in 2 or 3 series. **Ray flowers** female, white; ligules 2–8 mm long. **Disk corollas** ca. 2 mm long. **Achenes** 1–2 mm long, 5- to 10-angled. Roadsides and around habitations; valleys. Introduced in eastern U.S. and BC to MT south to NV, UT, CO.

Tanacetum vulgare L. Common tansy **Stems** erect, 40–120 cm, unbranched below the inflorescence. **Herbage** glabrate. **Leaf blades** 5–15 cm long, pinnately divided into lanceolate, serrate to sharply lobed leaflets. **Involucre** 5–10 mm across; phyllaries in 2 to 3 subequal series. **Rays** absent. **Disk corollas** 1–3 mm long. **Achenes** 1–2 mm long, cylindric, 4- to 5-angled. Moist, disturbed meadows, often along streams; plains, valleys. Introduced throughout U.S., Canada.

Taraxacum F.H.Wigg. Dandelion

Taprooted, perennial, scapose herbs. **Herbage** glabrate with milky sap. **Leaves** all basal, oblong with recurved lobes on the lower half at least, broad-petiolate. **Heads** solitary on naked peduncles, ligulate; involucre campanulate; phyllaries green, in 2 series, the outer (calyculi) shorter; receptacle flat, naked. **Ray flowers** perfect, ≥15, yellow. **Pappus** fine, white bristles. **Achenes** obconic, 4- to 5-angled, long-beaked, glabrous with small, pointed bumps below the beak.[61,103]

 The introduced species are told by the sharply reflexed outer involucral bracts. The genus is a famous agamic complex. Asexual races are common making circumscription of species difficult and open to personal opinion. It is clear that the taxonomy is far from being settled. Until a global monograph is completed, I prefer to err on the conservative side.

1. Outer phyllaries reflexed. 2
1. Outer phyllaries ascending to erect, appressed to the inner 3

2. Achenes olive-green; terminal lobe of leaves usually ≥twice as long as the lateral . . . .*T. officinale*
2. Achenes reddish; terminal leaf lobe usually ca. as large as the lateral . . . . *T. laevigatum*

3. Peduncles 1–5 cm long; involucre ≤12 mm high; phyllaries without lobed tips . . . . *T. lyratum*
3. Peduncles 2–20 cm long; involucre ≥10 mm high; phyllaries often with lobed tips . . . .*T. ceratophorum*

Taraxacum ceratophorum (Ledeb.) DC. [*T. eriophorum* Rydb.] **Leaves** 2–15 cm long, margins shallowly to deeply lobed. **Peduncles** prostrate to erect, 2–10 cm. **Involucre** 10–16 mm high; inner phyllaries often with swollen, lobed, scarious tips; outer phyllaries erect to ascending. **Ray ligules** 4–10 mm long. **Achenes** 3–4 mm long, olive-green; beaks 6–10 mm long. Moist, often sparsely vegetated soil of meadows, turf, spruce forest, sagebrush steppe; montane to alpine, more common higher. Circumboreal south to CA, NV, NM. (p.580)
    Taraxacum ceratophorum and *T. eriophorum* were separated based on achene color and lobing of phyllary tips. However, all of our plants have some inner phyllaries with lobed tips. Although some plants have red-brown achenes and others have olive-green achenes, this distinction does not correlate with other morphological or ecological characters. Plants from montane or subalpine spruce forests are notably larger with more divided leaves than typical specimens and resemble *T. officinale* except for the erect phyllaries.

Taraxacum laevigatum (Willd.) DC. [*T. erythrospermum* Andrz.] **Leaves** 3–15 cm long, deeply lobed; terminal lobe usually ≤twice as large as the lateral ones. **Peduncles** ascending to erect, 5–25 cm. **Involucre** 8–25 cm high; inner phyllaries with slightly swollen tips; outer phyllaries reflexed. **Ray ligules** 5–10 mm long. **Achenes** 2–3 mm long, reddish-brown; beaks 6–10 mm long. Grasslands, sagebrush steppe, dry forest, woodlands, roadsides; plains, valleys to montane. Introduced throughout N. America; native to Eurasia.
    Taraxacum erythrospermum may be the correct name for this species (Brouilllet 2006a). Similar to *T. officinale* and often confused with it, *T. laevigatum* occurs in somewhat drier habitats, although the distinction is far from perfect. The majority of our specimens identified as *T. officinale* were actually *T. laevigatum*.

Taraxacum lyratum Ledeb. [*T. scopulorum* (A.Gray) Rydb.] **Leaves** 2–5 cm long, deeply and regularly lobed. **Peduncles** prostrate to ascending, 1–5 cm. **Involucre** 6–12 mm high; inner phyllaries pointed; outer phyllaries erect to ascending. **Ray ligules** 4–7 mm long. **Achenes** green, black, or brown ca. 3 mm long; beaks 3–5 mm long. Meadows, turf, often among large rocks; alpine, rarely lower. AK to NV, AZ; Asia.

Taraxacum officinale F.H.Wigg. **Leaves** 4–30 cm long, deeply lobed with a large, terminal lobe. **Peduncles** erect, 5–40 cm, sometimes villous. **Involucre** 1–2 cm high; inner phyllaries pointed; outer phyllaries reflexed. **Ray ligules** 7–12 mm long. **Achenes** 3–4 mm long, olive-green; beaks 7–12 mm long. Vernally moist soil of meadows, thickets, woodlands, lawns, roadsides; plains, valleys, montane, occasionally as high as alpine. Introduced throughout N. America; native to Eurasia. See *T. laevigatum*.

Tetradymia DC. Horsebrush

Shrubs. Leaves alternate, simple, entire, often in secondary fascicles. **Heads** discoid; involucre cylindric to obconic; phyllaries few in 1 series; receptacle flat, naked. **Disk flowers** 4 to 6, perfect, yellow; tube longer than the throat; style branches flattened. **Pappus** white, capillary bristles. **Achenes** obconic, 5-ribbed, hairy.

1. Stems unarmed; heads in a corymbiform inflorescence. . . . . *T. canescens*
1. Stems spiny; heads axillary . . . . *T. spinosa*

Tetradymia canescens DC. Plants to 80 cm tall. **Stems** unarmed; twigs tomentose. **Primary leaves** narrowly oblanceolate, 1–2 cm long, tomentose; fascicled leaves smaller. **Inflorescence** heads terminating branches in a corymbiform array. **Involucre** 7–10 mm high; phyllaries 4,

tomentose. **Disk corollas** 7–12 mm long. **Achenes** 3–5 mm long, glabrous or sericeous. Grasslands, sagebrush steppe; valleys, montane. BC south to CA, AZ, NM.

Tetradymia spinosa Hook. & Arn. SPINY HORSEBRUSH Plants to 100 cm tall. **Stems** spiny, tomentose, highly branched. **Primary leaves** converted to spines ca. 1 cm long; fascicled leaves linear, glabrous, 7–15 mm long. **Inflorescence** of axillary heads. **Involucre** 8–12 mm high; phyllaries 4 to 6, tomentose. **Disk corollas** 7–12 mm long. **Achenes** 6–8 mm long, densely puberulent. Sagebrush steppe; plains, valleys. OR to MT south to CA, UT, NM.

Tetraneuris Greene

Perennial herbs from a branched caudex. **Leaves** mostly basal, obscurely petiolate, simple, entire. **Heads** solitary, radiate; involucre nearly hemispheric; phyllaries in 3 series; receptacle convex, naked. **Ray flowers** female, 7-15, yellow, glandular, deeply lobed. **Disk flowers** numerous, perfect, yellow; tube shorter than the throat; style branches flattened. **Pappus** of 4 to 8 hyaline, awned scales. **Achenes** obconic, strigose.

1. Involucral bracts often purplish, scarious-margined, sparsely villous. . . . . *T. torreyana*
1. Involucral bracts not thin-margined, densely villous. . . . . *T. acaulis*

Tetraneuris acaulis (Pursh) Greene [*Hymenoxys acaulis* (Pursh) K.F.Parker] Caudex usually branched, often mat-forming. **Stems** simple, erect, scapose, 4–15 cm. **Herbage** sparsely to densely villous. **Leaf blades** linear-oblanceolate to oblong, entire, 1–5 cm long. **Heads** solitary; involucres 5–10 mm high; phyllaries subequal, lanceolate, densely villous, sometimes purple-tipped. **Ray ligules** 6–10 mm long. **Disk corollas** 3–5 mm long, puberulent. **Achenes** 2–4 mm long; pappus scales 5 to 8. Stony soil of grasslands, stream terraces; plains, valleys. AB, SK south to NM, TX. Our plants are var. *acaulis*.

Tetraneuris torreyana (Nutt.) Greene [*Hymenoxys torreyana* (Nutt.) K.F.Parker] Caudex branched, mat-forming. **Stems** simple, erect, scapose, 3–10 cm. **Herbage** punctate-glandular, villous on stem and leaf bases. **Leaf blades** narrowly oblanceolate, entire, 1–5 cm long. **Heads** solitary; involucres 7–11 mm high; phyllaries subequal, lanceolate to ovate, scarious-margined, glandular, villous. **Ray ligules** 7–11 mm long. **Disk corollas** 4–5 mm long, glandular. **Achenes** 3–4 mm long; pappus scales 5–7. Stony, calcareous soil of grasslands, fellfields; montane, lower subalpine; Carbon Co. MT, WY, UT, CO.

Thelesperma Less.

Thelesperma subnudum A.Gray [*T. marginatum* Rydb.] Glabrous perennial from a branched caudex. **Stems** erect, simple, 10–25 cm. **Leaves** cauline, opposite; lower petiolate, deeply 3-lobed; lobes linear-lanceolate, 1–3 cm long; upper reduced, simple. **Inflorescence** of 1 to 3 terminal heads; peduncles 2–15 cm long. **Heads** discoid; involucre hemispheric, 6–9 mm high; phyllaries in 2 very unequal series; the inner narrowly ovate, striate, scarious-margined, united below; outer short, often reflexed; receptacle flat with a scale subtending each flower. **Disk flowers** perfect, yellow; corollas 4–6 mm long, tube shorter than the throat; style branches flattened. **Pappus** of 2 minute awns or absent. **Achenes** cylindric, compressed, glabrous, 5–7 mm long, enclosed in the scale. Grasslands, sagebrush steppe, coniferous woodlands; plains, valleys. AB south to NV, AZ, NM, SD. Our plants are var. *marginatum* (Rydb.) Cronquist.

Tonestus A.Nelson

Taprooted perennial herbs. **Stems** simple. **Leaves** basal and cauline, alternate, simple, short-petiolate, entire. **Heads** solitary, radiate or discoid; involucre campanulate; phyllaries subequal in 3 or 4 series, green, basally firm; receptacle convex, naked, pitted. **Ray flowers** 10 to 23, yellow, female. **Disk flowers** 25 to 65, perfect, yellow; the tube shorter than the throat; style branch appendages linear-lanceolate. **Pappus** 1 series of capillary bristles. **Achenes** cylindric.

Molecular studies suggest that this is an artificial group composed of only distantly related species.[285]

1. Leaves and phyllaries ciliate, sometimes glandular . . . . *T. pygmaeus*
1. Leaves and phyllaries stipitate-glandular but not ciliate . . . . *T. lyallii*

Tonestus lyallii (A.Gray) Nelson [*Haplopappus lyallii* A.Gray] Plants with slender caudex branches that sometimes root. **Stems** prostrate to ascending, 2–10 cm, stipitate-glandular. **Leaf blades** lanceolate to oblanceolate, 1–6 cm long, stipitate-glandular, sessile on the stem. **Involucres** campanulate, 7–11 mm high; phyllaries lanceolate, stipitate-glandular, often purple. **Ray ligules** 7–12 mm long. **Disk corollas** 6–8 mm long. **Achenes** 2–5 mm long, villous. Fellfields, talus, rock outcrops, eroding slopes; subalpine, alpine. BC, AB south to CA, NV, CO.

Tonestus pygmaeus (Torr. & A.Gray) A.Nelson [*Haplopappus pygmaeus* (Torr. & A.Gray) A.Gray] Plants with a branched caudex. **Stems** erect, 1–9 cm, villous. **Leaf blades** linear to oblanceolate, 1–5 cm long, ciliate. **Involucres** hemispheric, 8–10 mm high; phyllaries oblong, ciliate. **Ray ligules** 6–8 mm long. **Disk corollas** 4–7 mm long. **Achenes** 2–5 mm long, villous. Fellfields, alpine. MT to NM. Known from Missoula Co.

   Tonestus pygmaeus has not been collected in MT for 90 years; it is reported to be glandular, but the one MT specimen I have seen is not.

Townsendia Hook. Easter daisy

Taprooted biennial or perennial herbs. **Leaves** alternate, simple, entire, tapered to the petiole. **Heads** solitary on stem tips, radiate; involucre campanulate to hemispheric; phyllaries narrow, in 3 to 6 unequal, imbricate series; receptacle flat to hemispheric, naked. **Ray flowers** female, white to blue. **Disk flowers** perfect, yellow, numerous; tube shorter than the narrow throat; style with short-hairy appendages. **Pappus** of white bristle-like scales. **Achenes** obconic, flattened, 2-ribbed.[40]

1. Plants biennial, without a woody, branched caudex; leafy stems apparent 2
1. Stems obscure; plants perennial; often with a woody branched caudex 3

2. Involucre 7–10 mm high, the bracts green; rays pink, drying purplish . . . . *T. florifer*
2. Involucre 9–20 mm high, the bracts scarious; rays usually blue to violet. . . . . *T. parryi*

3. Herbage villous with tangled hairs 4
3. Herbage glabrate to densely strigose. 5

4. Involucre 6–10 mm high. . . . . *T. spathulata*
4. Involucre 10–25 mm high. . . . . *T. condensata*

5. Heads on leafy stems 3–40 mm long. 6
5. Heads sessile among basal leaves 7

6. Rays blue; achenes usually glabrous; herbage strigose; high elevations . . . .*T. montana*
6. Rays pinkish; achenes hairy; herbage densely strigose; low elevations . . . . *T. incana*

7. Leaf blades glabrate. . . . . *T. leptotes*
7. Leaf blades moderately to densely strigose. 8

8. Involucre 7–12 mm high; phyllaries with a fringed tip. . . . .*T. hookeri*
8. Involucre 12–20 mm high; phyllaries ciliate but without an apical tuft . . . . *T. exscapa*

Townsendia condensata Parry Perennial with a branched caudex, forming small mats. **Stems** ca. 1 cm, inconspicuous. **Herbage** villous. **Leaves** spatulate, 5–20 mm long. **Involucre** broadly campanulate, 10–25 mm high; phyllaries acuminate, villous. **Rays** pink, drying lavender; ligules 6–15 mm long. **Disk corollas** 6–7 mm long. **Achenes** 3–5 mm long, pubescent. Fellfields on exposed ridges, slopes; alpine; Glacier Co., reported for Park Co.[40] AB south to CA, UT, WY.

   Cronquist confused this with the smaller-headed *T. spathulata*.[187]

Townsendia exscapa (Richardson) Porter Perennial with a branched caudex, forming small mats. **Stems** ca.1 cm, inconspicuous. **Herbage** strigose. **Leaves** linear-oblanceolate, 1–4 cm long. **Involucre** broadly campanulate, 12–20 mm high; phyllaries ciliate, acute to acuminate, purple-tipped. **Rays** white to pink; ligules 10–13 mm long. **Disk corollas** 8–10 mm long, shorter than the pappus. **Achenes** 3–5 mm long, pubescent with hairs branched at the tip. Grasslands, sagebrush steppe; plains, valleys. AB to MB south to NV, AZ, NM, TX.

Townsendia florifer (Hook.) A.Gray Biennial or short-lived perennial. **Stems** erect, 3–15 cm. **Herbage** appressed-villous. **Leaves** narrowly spatulate, 15–30 mm long. **Involucre** hemispheric, 7–10 mm high; phyllaries strigose, acute. **Rays** pinkish, drying purplish; ligules 6–10 mm long. **Disk corollas** 4–6 mm long. **Achenes** 3–5 mm long, pubescent. Sagebrush steppe; montane. WA to MT to NV, UT, WY. Known from Beaverhead Co. See *T. parryi*.

Townsendia hookeri Beaman [*T. nuttallii* Dorn] Perennial with a branched caudex, forming small mats. **Stems** nearly obsolete. **Herbage** densely strigose. **Leaves** linear or linear-oblanceolate, 5–30 mm long. **Involucre** campanulate, 7–12 mm high; phyllaries sparsely strigose, acuminate, ciliate, scarious-margined with a small tuft of cilia at the tip. **Rays** white to pink; ligules 7–10 mm long. **Disk corollas** 4–6 mm long, shorter than the pappus. **Achenes** 3–5 mm long, pubescent with hairs branched at the tip. Grasslands, sagebrush steppe, coniferous woodlands; plains, valleys, montane. YT to UT, CO, NE.

Some plants from Beaverhead Co. have linear-oblanceolate leaves and pappus of ray flowers much shorter than that of the disk flowers; these plants have been called *Townsendia nuttallii*. I have chosen not to recognize this form as distinct because plants with intermediate leaves may have either long or short ray pappus, and both forms may occur in the same population.

Townsendia incana Nutt. Short-lived perennial with a branched caudex, forming small mats. **Stems** ascending, 1–3 cm. **Herbage** densely strigose. **Leaves** linear, 1–2 cm long. **Involucre** broadly campanulate, 6–9 mm high; phyllaries strigose, ciliate, scarious-margined, acute. **Rays** pinkish; ligules 4–9 mm long. **Disk corollas** 4–5 mm long, often smaller than the pappus. **Achenes** 3–5 mm long, pubescent; hairs branched at the tip. Sandy, usually calcareous soil of sagebrush steppe, grasslands, coniferous woodlands; valleys, montane; known from Carbon Co. MT to NV, AZ, NM.

Townsendia leptotes (A.Gray) Osterh. Perennial with a branched caudex, forming small mats. **Stems** ascending, 1–3 cm. **Herbage** strigose. **Leaves** linear-oblong, 12–25 mm long. **Involucre** campanulate, 8–12 mm high; phyllaries strigose. **Rays** white to pinkish; ligules 5–7 mm long. **Disk corollas** 4–6 mm long. **Achenes** 3–4 mm long, pubescent with hairs branched at the tip. Fellfields, sparsely vegetated soil of meadows, grasslands; montane, subalpine. MT to CA, NV, UT, NM. Collected once in Madison Co.[40]

Townsendia montana M.E.Jones [*T. alpigena* Piper] Perennial with a branched caudex, forming small mats. **Stems** 1 cm; heads on peduncles 3–40 mm long. **Herbage** strigose. **Leaves** oblong to spatulate, 1–3 cm long. **Involucre** campanulate, 6–11 mm high; phyllaries strigose, ciliate, often purplish, acute or rounded. **Rays** blue; ligules 4–7 mm long, sometimes glandular-puberulent. **Disk corollas** 3–6 mm long. **Achenes** 4–5 mm long, mostly glabrous. Stony, usually calcareous soil of fellfields, meadows; subalpine, alpine. OR, ID, MT, UT, WY, CO.

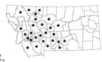

Townsendia parryi D.C.Eaton Biennial or short-lived perennial. **Stems** erect, 2–25 cm. **Herbage** villous. **Leaves** spatulate, 1–4 cm long. **Involucre** hemispheric, 9–20 mm high; phyllaries ciliate, sparsely strigose, scarious-margined. **Rays** blue to violet; ligules 8–17 mm long. **Disk corollas** 4–6 mm long. **Achenes** 3–5 mm long, pubescent. Sparsely vegetated soil of grasslands, sagebrush steppe, dry forest, woodlands; valleys to rarely alpine. BC, AB south to OR, NV, UT, WY. (p.580)

The size of the flower heads varies greatly. Many plants from the Bozeman area have small heads and could be mistaken for *Townsendia florifer*. One population from Jefferson Co. has short, decumbent stems and white to pink, minutely glandular rays; a similar plant from WY was described as *T. parryi* var. *alpina* A.Gray.

Townsendia spathulata Nutt. Perennial with a branched caudex, forming small mats. **Stems** <1 cm. **Herbage** woolly-villous. **Leaves** narrowly oblong to spatulate, 5–10 mm long. **Involucre** broadly campanulate, 6–10 mm high; phyllaries villous, acute. **Rays** white to pinkish; ligules 4–10 mm long, glandular-puberulent. **Disk corollas** 4–7 mm long, often shorter than the pappus. **Achenes** 2–4 mm long, pubescent with hairs branched at the tip. Stony, calcareous soil of fellfields, grasslands; montane to alpine. MT, WY.

Plants from Broadwater Co. have nearly linear leaves and sparsely strigose phyllaries. See *T. condensata*.

Tragopogon L. Goatsbeard

Taprooted biennial or monocarpic perennial herbs. **Stems** erect, simple or branched above. **Herbage** patchily tomentose, becoming glabrate with milky sap. **Leaves** in a basal rosette and alternate-clasping on the stem, simple, entire, linear, acuminate. **Heads** solitary, pedunculate on branch tips, ligulate; involucre campanulate; phyllaries linear-lanceolate, long-acuminate in 1 series; receptacle convex, naked. **Ray flowers** perfect, numerous, yellow or purple. **Pappus** of plumose bristles united at the base, off-white to tan, borne on a beak. **Achenes** fusiform, 5- to 10-ribbed, long-beaked, minutely bumpy.

All our species have been introduced throughout N. America from Europe. *Tragopogon miscellus* G.B.Owenby, a hybrid between *T. dubius* and *T. pratensis*, was reported for Park Co., but it has not been relocated there.[376]

1. Flowers purple . . . . *T. porrifolius*
1. Flowers yellow 2

2. Peduncle swollen, tapered to the involucre; flowering involucre ≥20 mm high . . . . *T. dubius*
2. Peduncle not swollen below the head; involucre ≤20 mm high at flowering. . . . . *T. pratensis*

Tragopogon dubius Scop. **Stems** 30–110 cm. **Cauline leaves** 4–20 cm long. **Involucre** 2–4 cm high, to 6 cm in fruit; peduncle inflated, nearly as wide as the involucre. **Rays** yellow; ligules 10–15 mm long; the outer shorter than the phyllaries. **Achene** 20–35 mm long including the beak. Grasslands, sagebrush steppe, open forest, woodlands, fields, roadsides, lawns; plains, valleys, montane. (p.580)

Tragopogon porrifolius L. Salsify, oyster-root, vegetable oyster **Stems** 30–80 cm. **Cauline leaves** 2–25 cm long, recurved at the tip. **Involucre** 2–3 cm high, to 5 cm in fruit; peduncle inflated and tapering to the involucre. **Rays** purple; ligules 15–20 mm long; the outer shorter or equal to the phyllaries. **Achene** 2–3 cm long including the beak. Roadsides, fields; valleys. Cultivated.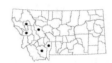

Tragopogon pratensis L. **Stems** 15–100 cm. **Cauline leaves** 2–25 cm long, recurved at the tip. **Involucre** 15–20 mm high, to 35 mm in fruit; peduncle much narrower than the involucre. **Rays** yellow, ca. 15 mm long; the outer longer than the phyllaries. **Achene** 15–20 mm long including the beak. Grasslands, meadows, roadsides; plains, valleys.

Triniteurybia Brouillet, Urbatsch & R.P.Roberts

Triniteurybia aberrans (A.Nelson) Brouillet, Urbatsch & R.P.Roberts [*Haplopappus aberrans* (A.Nelson) H.M.Hall, *Tonestus aberrans* (A.Nelson) G.L.Nesom & D.R.Morgan] Perennial herb with slender caudex branches. **Stems** ascending to erect, 5–20 cm, simple. **Herbage** glandular-pubescent. **Leaves** basal and cauline, alternate, petiolate; blades oblanceolate to oblong, spiny-serrate, 1–6 cm long. **Inflorescence** of few sessile heads in upper leaf axils. **Heads** discoid; involucre campanulate, 8–11 mm high; phyllaries imbricate in 3 or 4 series, lanceolate, glandular, green-tipped; receptacle convex, naked, pitted. **Disk flowers** perfect, yellow; corollas 6–7 mm long, cylindric; style branch appendages lanceolate. **Pappus** of 2 series of tawny, capillary bristles. **Achenes** fusiform, ca. 3 mm long, sericeous. Rock crevices in open forest; montane. Endemic to central ID and adjacent Ravalli, Co., MT.

Verbesina L. Crownbeard

Verbesina encelioides (Cav.) Benth. & Hook. f. ex A.Gray Taprooted annual. **Stems** erect, usually branched, 20–100 cm. **Herbage** strigose to puberulent. **Leaves** cauline, alternate above, simple, petiolate; blades lanceolate, serrate, 2–9 cm long. **Inflorescence**: heads on branch tips in a leafy, corymbiform array. **Heads** radiate; involucre hemispheric, 6–12 mm high; phyllaries green, in 1 or 2 subequal series, lanceolate, strigose, villous; receptacle hemispheric with scales clasping the achenes. **Disk flowers** perfect, yellow; tube shorter than the throat. **Ray flowers** pistillate or sterile; ligules yellow, 8–10 mm long, 3-lobed at the tip. **Pappus** of 2 awns; absent on ray achenes. **Achenes** winged, 3–5 mm long, strigose. Grasslands, meadows; plains, valleys. Throughout U.S., Mexico, probably introduced in northern states. Known from Beaverhead and Blaine cos.

Viguiera Kunth

Viguiera multiflora (Nutt.) S.F.Blake [*Heliomeris multiflora* Nutt.] Perennial herb with a branched, woody crown. **Stems** ascending to erect, simple, 20–120 cm. **Herbage** strigose. **Leaves** cauline, opposite, simple, short-petiolate, entire, narrowly lanceolate to oblanceolate, 3–8 cm long. **Inflorescence** a corymbiform array. **Heads** radiate; involucre hemispheric, 5–10 mm high; phyllaries subequal in 2 or 3 series, linear-lanceolate, green, puberulent, outer longest; receptacle convex with scales clasping the achenes. **Disk flowers** perfect, yellow; corollas 3–4 mm long, tubes shorter than throat. **Ray flowers** sterile, yellow; ligules 7–20 mm long, puberulent. **Pappus** absent. **Achenes** obconic, 4-angled, glabrous, 1–3 mm long, black. Open forest, sagebrush steppe; montane. MT to CA, AZ, NM, TX. Our plants are var. *multiflora*.
 There is conflicting evidence for placing this species in *Heliomeris*.[103,352]

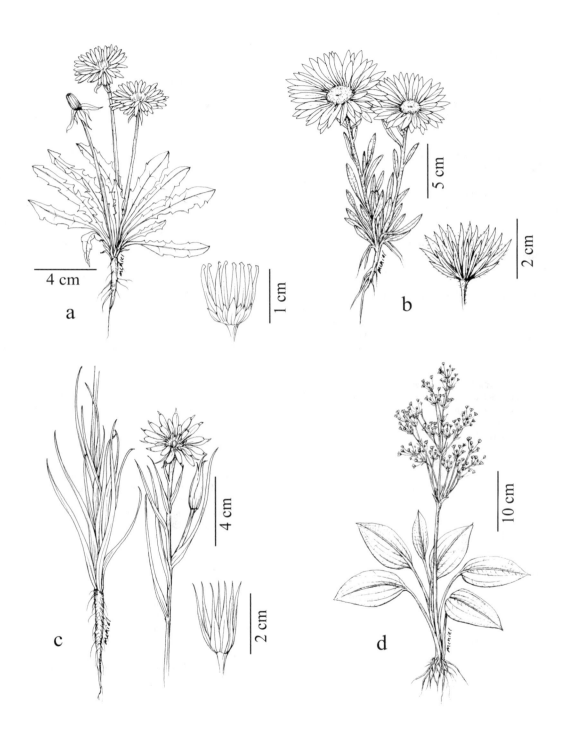

Plate 101. a. *Taraxacum ceratophorum*, b. *Townsendia parryi*, c. *Tragopogon dubius*, d. *Alisma plantago-aquatica*

Wyethia Nutt. Mule's ears

Taprooted perennials. **Stems** ascending to erect, usually simple below the inflorescence. **Leaves** basal and/ or alternate on the stem, simple, entire, petiolate. **Inflorescence** of 1 to 3 heads. **Heads** radiate; involucre hemispheric to turbinate; phyllaries green, in 2 or 3 series, outer often longest; receptacle low-convex with scales clasping the achenes. **Disk flowers** perfect, yellow; tubes shorter than the throat. **Ray flowers** female, white to yellow. **Pappus** an obscure crown. **Achenes** obconic, 4-angled, compressed.

Type specimens of *Wyethia amplexicaulis* and *W. helianthoides* were collected in 1833 by Nathaniel Wyeth. Labels give the type localities as "About the Flathead River" and "Camas Plains, Flathead River" respectively. Weber[424] believes that these locations are incorrect because Wyeth was in northwest MT in late April and May, too early for the plants to be in flower. *Wyethia helianthoides* has not otherwise been collected in the Flathead drainage; however, there are collections of *W. amplexicaulis* from north of Missoula taken in late May.

1. Basal leaves absent; herbage scabrous to hispid . . . .*W. scabra*
1. Basal leaves conspicuous; herbage glabrous to villous 2

2. Rays white; stem leaves subsessile but not clasping . . . . *W. helianthoides*
2. Rays yellow; stem leaves clasping . . . . *W. amplexicaulis*

Wyethia amplexicaulis (Nutt.) Nutt. **Stems** 30–80 cm. **Herbage** glabrous, aromatic. **Leaves**: basal long-petiolate; blades lanceolate to elliptic, 20–40 cm long; stem leaves clasping, smaller. **Involucres** 2–3 cm high, the central hemispheric, others campanulate, smaller; phyllaries glabrous, resinous. **Disk corollas** 9–10 mm long. **Rays** 13 to 21, yellow; ligules 25–40 mm long. **Achenes** 8–9 mm long, glabrous. Grasslands, sagebrush steppe, riparian meadows; valleys, montane. WA to MT south to NV, UT, CO.

Wyethia helianthoides Nutt. **Stems** 10–40 cm. **Herbage** villous. **Leaves**: basal petiolate; blades lanceolate to elliptic, 6–20 cm long; stem leaves petiolate, smaller. **Involucres** 15–20 mm high, hemispheric; phyllaries ciliate. **Disk corollas** 8–10 mm long. **Rays** 13 to 25, white; ligules 2–4 cm long. **Achenes** 9–11 mm long, strigose. Meadows, sagebrush steppe, often along streams; montane, lower subalpine. OR, ID, MT, NV, WY.

Wyethia scabra Hook. [*Scabrethia scabra* (Hook.) W.Weber] **Stems** 20–60 cm. **Herbage** short-hispid. **Leaves**: basal absent; stem leaves narrowly lanceolate, sessile or subsessile, 8–18 cm long. **Involucres** 15–40 mm high, hemispheric; phyllaries hispid, ciliate. **Disk corollas** 7-9 mm long. **Rays** 10 to 18, yellow; ligules 15–50 mm long. **Achenes** 7–9 mm long, glabrous. Calcareous soil of juniper woodlands; valleys, montane. MT to AZ, NM. Collected in Big Horn and Carbon cos.

Xanthisma DC.

Taprooted perennial herbs with woody, branched caudexs. **Leaves** mainly cauline, alternate, petiolate; blades dentate to lobed, spiny-margined, 1-veined. **Heads** radiate or discoid, solitary or few in corymbiform arrays; phyllaries imbricate in 3 to 6 series, lanceolate, thickened, whitish below, mucronate; receptacle convex, bristly, pitted. **Ray flowers** absent or 15 to 60, female. **Disk flowers** perfect, yellow; the tube shorter than the throat; style branch appendages lanceolate. **Pappus** of 2 to 4 series of slender, attenuate, tawny bristles. **Achenes** narrowly obovoid, sericeous, dimorphic; disk achenes compressed, ribbed; ray achenes 3-sided.

1. Heads discoid; leaves dentate . . . . *X. grindelioides*
1. Heads radiate; leaves lobed . . . . *X. spinulosum*

Xanthisma grindelioides (Nutt.) D.R.Morgan & R.L.Hartman [*Machaeranthera grindelioides* (Nutt.) Shinners, *Haplopappus nuttallii* Torr. & A.Gray] **Stems** ascending to erect, branched above, 5–20 cm. **Herbage** villous to sparsely tomentose, sometimes minutely glandular. **Leaf blades** lanceolate to oblanceolate, 1–3 cm long, spiny-dentate. **Heads** discoid, solitary on branch tips; involucre hemispheric, 6–9 mm high; phyllaries in 3 to 5 series, stipitate-glandular. **Disk flowers** 15 to 50; corollas 4–6 mm long. **Achenes** 2–4 mm long. Stony or sandy soil of grasslands, badlands; plains, valleys. AB, SK south to AZ, NM, NE. Our plants are var. *grindelioides*.

Xanthisma spinulosum (Pursh) D.R.Morgan & R.L.Hartman [*Haplopappus spinulosus* (Pursh) DC., *Machaeranthera pinnatifida* (Hook.) Shinners] **Stems** ascending to erect, branched, 5–40 cm. **Herbage** tomentose, viscid. **Leaf blades** 1–4 cm long, deeply pinnately lobed; lobes remote, linear, spine-tipped, sometimes dentate. **Inflorescence** corymbiform with heads solitary on branch tips. **Heads** radiate; involucre hemispheric, 6–9 mm high; phyllaries in 5 or 6 series, tomentose, viscid; the outer green; the inner whitish. **Rays** 14 to 60, yellow; ligules 4–10 mm long. **Disk flowers** 30 to 150; corollas 4–6 mm long. **Achenes** ca. 2 mm long. Stony or sandy soil of grasslands, sagebrush steppe, roadsides; plains, valleys. AB to MB south to CA, NM, TX, IA, Mexico. Our plants are var. ***spinulosum***.

Variety *glaberrimum* (Rydb.) D.R.Morgan & R.L.Hartman, with glabrate foliage, occurs on the Great Plains and could be found in MT.

Xanthium L. Cocklebur

Annuals. **Leaves** cauline, alternate, petiolate, simple. **Inflorescence** heads axillary or terminal, solitary to clustered. **Heads** discoid, unisexual; male involucres flat-hemispheric; phyllaries few in 1 or 2 series; receptacle conical with scales between flowers; female involucres ellipsoid; phyllaries with hooked spines in several series, enclosing the flowers in a bur. **Male flowers** 20 to 50, white; a tube of anthers and an aborted style. **Female flowers** 2 without a corolla. **Pappus** absent. **Achenes** fusiform, enclosed in the bur.

1. Bur 10–12 mm long, puberulent; stems spiny at the nodes; leaves lanceolate . . . .*X. spinosum*
1. Bur ≥20 mm long, stipitate-glandular; stems not spiny; leaves deltate-cordate . . . . *X. strumarium*

Xanthium spinosum L. **Stems** erect, strigose, 30–120 cm with spines at the nodes. **Leaf blades** lanceolate, 2–8 cm long, deeply pinnately lobed, becoming more entire upward, strigose above, canescent beneath. **Burs** 10–12 mm long, 1-beaked, puberulent, prickly. Roadsides; valleys. Introduced to most of U.S.; native to S. America. Collected 100 years ago in Ravalli Co.

Xanthium strumarium L. **Stems** erect, strigose, 10–100 cm. **Leaf blades** cordate-deltate, 2–12 cm long, serrate, sometimes shallowly few-lobed, sparsely short-hispid. **Burs** 2–4 cm long, 2-beaked, stipitate-glandular. Streambanks, roadsides, fields; plains, valleys. A native weed occurring throughout N. America. Our plants are var. ***canadense*** (Mill.) Torr. & A.Gray.

Xylorhiza Nutt. Woody aster

Xylorhiza glabriuscula Nutt. [*Machaeranthera glabriuscula* (Nutt.) Cronquist & D.D.Keck] Taprooted perennial with a branched, woody caudex. **Stems** ascending to erect, 10–20 cm, sometimes branched near the base. **Herbage** finely villous. **Leaves** cauline, alternate, sessile, linear to oblanceolate, entire, 2–6 cm long. **Heads** radiate, 1 or 2 on naked peduncles at stem tips; involucres campanulate, 7–14 mm high; phyllaries scarious-margined, linear-lanceolate, villous, acuminate, imbricate in ca. 3 series; receptacle flat, naked. **Disk flowers** perfect, yellow; corollas 5–6 mm long, the tube and throat ca. equal. **Ray flowers** female, 12 to 25, white; ligules 10–15 mm long. **Pappus** of barbellate bristles. **Achenes** obconic, 4–7 mm long, 4-ribbed, sericeous. Clay or bentonitic soil of steppe, badlands; valleys. MT, SD, WY, UT, CO. Known from Carbon Co.

MONOCOTYLEDONS

BUTOMACEAE: Flowering Rush Family

Butomus L. Flowering rush

Butomus umbellatus L. Perennial scapose herb from a fleshy rhizome. **Herbage** glabrous. **Leaves** mainly basal, 2-ranked, ascending to erect, linear, 4–8 mm wide. **Inflorescence** an umbel enclosed by 3 ovate, purplish bracts, reflexed at anthesis; pedicels unequal, 1–9 cm long; scape 3-angled, 50–100 cm. **Flowers** perfect, regular; sepals 3, green; petals 3, pinkish, 8–12 mm long; stamens 9; ovaries superior; pistils 6. **Fruit** many-seeded, beaked follicles, 8–10 mm long. Perennially inundated marshes; valleys. Introduced to much of northern U.S. and adjacent Canada; native to Eurasia. Known from Lake and Flathead cos.

ALISMATACEAE: Water-plantain Family

Aquatic or emergent, scapose, perennial herbs. **Herbage** glabrate with milky sap. **Leaves** basal, long-petiolate, simple, often floating. **Inflorescence** terminal, scapose, bracteate, racemose or paniculate with opposite or whorled pedicels or branches. **Flowers** perfect or unisexual, regular; sepals 3, green, separate; petals 3, white, separate; stamens ≥6; ovaries superior; pistils numerous. **Fruits** achenes (follicles).

Emergent leaf blades are generally broader than submersed leaves which may be linear.

1. Leaves arrow-shaped (submersed ones linear); achenes in a globose cluster . . . . *Sagittaria*
1. Floating leaf blades elliptic; achenes in a wheel-like cluster 2

2. Achenes beaked; petals with erose margins; rare . . . . *Damasonium*
2. Achenes rounded at the tip; petals entire-margined; common . . . . *Alisma*

Alisma L. Water plantain

Leaf blades elliptic. **Inflorescence** a compound, bracteate, whorled panicle. **Flowers** perfect; sepals ovate; petals obovate, erose, white; stamens 6 to 9; pistils 10 to 25, whorled. **Achenes** compressed-ovate with a short, curved beak, arranged in a whorl like sections of an orange.

1. Leaf blades linear to narrowly elliptic; lower panicle branches with 1 whorl of flowers . . . . . . *A. gramineum*
1. Leaf blades elliptic; lower panicle branches often with >1 whorl of flowers . . . . *A. plantago-aquatica*

Alisma gramineum Lej. **Scapes** 2–25 cm. **Leaf blades** narrowly elliptic or linear when submersed, 3–30 cm long. **Flowers**: sepals 2–3 mm long; petals 2–4 mm long, quickly deciduous. **Achenes** 2–3 mm long; clusters 3–6 mm across. Marshes, margins of sloughs, slow-moving streams; plains, valleys. Across northern U.S. and adjacent Canada; Eurasia.

Alisma plantago-aquatica L. [*A. triviale* Pursh] **Scapes** 6–50 cm. **Leaf blades** elliptic, 4–20 cm long. **Flowers**: sepals 2–4 mm long; petals 1–3 mm long. **Achenes** ca. 2 mm long; clusters 4–7 mm across. Marshes, margins of sloughs, slow-moving streams; plains, valleys. Throughout temperate N. America; Eurasia. (p.580)

Our plants are part of a circumboreal complex that has often been split into numerous species with much taxonomic confusion.[101]

Damasonium Mill.

Damasonium californicum Torr. [*Machaerocarpus californicus* (Torr.) Small] Cormose. **Scapes** 20–40 cm. **Leaf blades** lanceolate to narrowly elliptic, 3–12 cm long. **Inflorescence** a raceme of 1 to 4 whorls; bracts ovate, 4–10 mm long. **Flowers** unisexual, female below the male; sepals oblong to ovate, reflexed, 4–5 mm long; petals rhombic, white with a yellow, clawed base, 7–10 mm long, erose-margined; stamens 6; pistils 6 to 10. **Achenes** compressed-obovoid, 3–6 mm long with a beak 3–6 mm long, spreading at maturity, tightly packed in a wheel-like head. Mossy riverbanks; valleys. MT, OR, ID, NV, CA. Collected once in Lincoln Co. at a site inundated by Lake Koocanusa.

Sagittaria L. Arrowhead, wapato

Roots with tuber-bearing rhizomes. **Emergent leaf blades** sagittate, lobes lanceolate to deltoid; submersed linear. **Inflorescence** a bracteate raceme with whorled pedicels. **Flowers** unisexual or perfect, female below the male; sepals ovate, reflexed; petals obovate, white, clawed at the base; stamens numerous; pistils numerous. **Achenes** winged, compressed-obovate with a short curved beak, tightly packed in a globose head.

1. Bracts of inflorescence lanceolate, ≥10 mm long; basal leaf lobes as long or shorter
   than the terminal lobe. . . . . *S. cuneata*
1. Inflorescence bracts ovate, often <10 mm long; basal leaf lobes ca. as long as
   the terminal lobe. . . . . *S. latifolia*

Sagittaria cuneata E.Sheld. **Scapes** 6–50 cm. **Leaf blades** 4–15 cm long, lateral lobes shorter or as long as the terminal. **Inflorescence** of 2–10 whorls; bracts lanceolate, 10–35 mm long. **Flowers**: sepals 4–7 mm long; petals 7–12 mm long. **Achenes** 2–3 mm long; heads 5–13 mm across. Shallow water of ponds, lakes, sloughs; plains, valleys. Throughout all temperate N. America but the southeast U.S.

Plants in deeper water remain vegetative with only linear, submersed leaves.

Sagittaria latifolia Willd. **Scapes** 4–50 cm. **Leaf blades** 2–15 cm long, lateral lobes ca. as long as the terminal. **Inflorescence** of 2–7 whorls. **Flowers**: sepals 5–10 mm long; petals 1–2 cm long. **Achenes** 2–4 mm long; heads 6–20 mm across. Shallow water of ponds, lakes, sloughs; plains, valleys. Throughout most of N. America.

Completely submersed leaves are rare.

HYDROCHARITACEAE: Waterweed Family

Elodea Michx. Waterweed

Dioecious, herbaceous, submersed aquatic perennials. **Stems** branched, rooting at the nodes. **Leaves** sessile, opposite to whorled, finely serrate. **Inflorescence** 1 to 3 axillary flowers borne on a long, pedicel-like perianth with a tubular sheath (spathe) at the base. **Flowers** usually unisexual; sepals 3, linear, separate; petals 3 or absent, oblong, separate; stamens 3 to 9; ovary inferior; stigmas 3, 2-lobed. **Fruit** a many-seeded, indehiscent capsule.

Flowers often float on the water's surface; male flowers may be free-floating.

1. Upper leaves opposite, 10–20 mm long . . . . *E. bifoliata*
1. Upper leaves in whorls of 3 or 4, 4–15 mm long 2

2. Leaves 1.5–2.5 mm wide, ovate. . . . . *E. canadensis*
2. Leaves 1.0–1.5 mm wide, nearly linear . . . .*E. nuttallii*

Elodea bifoliata H.St.John [*E. longivaginata* H.St.John] **Leaves** opposite, 10–20 × 1.5–2.5 mm, minutely serrate, narrowly lanceolate to linear. **Male flowers**: spathe 2–5 cm long, inflated above; sepals 3–5 mm long; petals 4–5 mm long; stamens 7 to 9. **Female flowers**: spathe 3–7 cm long, slightly inflated above; sepals 2–3 mm long; petals 3–4 mm long. **Capsule** 8–10 mm long. Shallow, often saline water of ponds; plains, valleys. BC to MN south to OR, UT, NM.

Elodea canadensis Michx. **Leaves** 5–12 × 1.5–2.5 mm, minutely serrate, narrowly lanceolate and opposite below to more linear and 3-whorled above. **Male flowers**: spathe to 15 mm long, inflated above; sepals 3–5 mm long; petals 4–5 mm long; stamens 7 to 9. **Female flowers**: spathe 1–2 cm long, not inflated; sepals ca. 2 mm long, purplish; petals 2–3 mm long. **Capsule** 5–6 mm long. Shallow to deep water of lakes, ponds, sloughs, slow streams; valleys. Throughout most of temperate N. America. (p.588)

Elodea nuttallii (Planch.) H.St.John **Leaves** 4–15 × 1–1.5 mm, linear, 3- to 4-whorled. **Male flowers**: sessile; sepals ca. 2 mm long; petals minute or absent; stamens 9. **Female flowers**: spathe 9–15 mm long; sepals 1–2 mm long; petals 1–2 mm long. **Capsule** 4–8 mm long. Shallow to deep water of ponds, lakes, sloughs; valleys, montane. Throughout most of temperate N. America.

SCHEUCHZERIACEAE

Scheuchzeria L.

Scheuchzeria palustris L. Perennial, rhizomatous herb. **Stem** ascending, 15–40 cm. **Leaves** alternate, terete, 5–35 cm long, erect with broad, sheathing bases. **Inflorescence** a several-flowered raceme with leaf-like bracts. **Flowers** perfect; tepals 6, greenish-white, 2–3 mm long; stamens 6; ovaries superior; pistils 3. **Fruits** 3, spreading, ovoid follicles, 5–8 mm long with 1 or 2 seeds each. *Sphagnum* fens; valleys, montane. Circumboreal south to CA, ID, MT, IA, NJ. Our plants are var. *americana* Fernald (p.588)

JUNCAGINACEAE: Arrow-grass Family

Herbaceous perennials. **Leaves** mainly at the base of the stem, terete, linear, sheathing the stem at the base; sheaths hyaline-ligulate. **Inflorescence** a spike or raceme. **Flowers** perfect or unisexual, regular; tepals 0, 1, or 6, separate; stamens 1 to 6; ovary superior; pistils 1, 3, or 6, connate. **Fruit** a nutlet or schizocarp.

1. Tepals 6; inflorescence racemose . . . . *Triglochin*
1. Tepals 0 or 1; inflorescence a spike . . . . *Lilaea*

Lilaea Bonpl. Flowering quilwort

Lilaea scilloides (Poir.) Hauman Scapose. **Leaves** 5–25 cm long. **Inflorescence** a pedunculate, bracteate spike, 5–40 mm long. **Perfect and male flowers** with tepals 1–2 mm long; stamens 1; pistils 1. **Female flowers** borne at the base of the spike; tepals 0 or 1; pistil 1. **Fruit** a ribbed, glabrous, beaked or beakless nutlet 2–5 mm long. Shallow water of ponds; valleys; Lake and Phillips cos. BC to SK south to CA, NV, MT.

Triglochin L. Arrow grass

Short-rhizomatous. **Leaves** mostly basal, 2-ranked. **Inflorescence** an ebracteate, spike-like raceme. **Flowers** perfect; tepals 6, separate, elliptic, yellowish or purplish; stamens 6, pistils 3 or 6, connate. **Fruit** a schizocarp, 1 seed per follicle.
  Plants produce cyanide and are poisonous.

1. Fruit obconic with 3 carpels and 3 stigmas. . . . . *T. palustre*
1. Fruit ovoid with 6 carpels and stigmas . . . . *T. maritima*

Triglochin maritima L. [*T. concinna* Burtt Davy] **Stems** ascending to erect, 15–70 cm. **Leaves** 4–40 cm long. **Flowers**: tepals 1–2 mm long; pistils 6 (sometimes 3 aborted). **Fruit** erect, narrowly ovoid, 3–6 mm long on pedicels 2–4 mm long. Moist to wet, usually saline or alkaline soil of pond margins, wet meadows, shallow marshes; plains, valleys, montane. Circumboreal south to CA, NM, OH, NY, S. America.
  Triglochin concinna has been separated on plant size and a bilobed ligule, but these characters are only weakly correlated and without geographic or ecological association in MT.[168]

Triglochin palustre L. **Stems** ascending to erect, 10–40 cm. **Leaves** 5–30 cm long. **Flowers**: tepals 1–2 mm long; pistils 3. **Fruit** erect, narrowly obconic, 5–10 mm long on pedicels 2–6 mm long. Wet, usually alkaline, often organic soils of fens, shallow wetlands; plains, valleys, montane. Circumboreal south to CA, NM, IA, NY, S. America. (p.588)

POTAMOGETONACEAE: Pondweed Family

Potamogeton L. Pondweed

Submergent or floating-leaved, rhizomatous, aquatic perennials. **Stems** flaccid. **Leaves** mostly with entire or wavy margins, alternate or opposite above; submersed leaves and floating leaves often different; stipules membranous, sheathing. **Inflorescence** continuous or interrupted, cylindric, pedunculate, axillary spikes. **Flowers** perfect, small; tepals 4, separate, minute; stamens 4; pistils 4; stigmas sessile. **Fruit** an ovoid, short-beaked achene.
  Some species produce free-floating turions (large buds) that sink and overwinter in mud until forming new plants the following summer. Some of the narrow-leaved species with leaves attached to the stipule have been segregated into *Stuckenia*.[169]

1. Leaves sessile and clasping the stem at the base 2
1. Leaves not clasping the stem 4

2. Stems zig-zag; some leaves >10 cm long . . . . *P. praelongus*
2. Stems mostly straight; leaves usually <10 cm long with wavy margins 3

3. Leaves linear-oblong, minutely serrate. . . . . *P. crispus*
3. Leaves lanceolate to ovate, entire . . . . *P. richardsonii*

4. Submersed leaves attached to the stipule not directly to the stem 5
4. Submersed leaves attached directly to the stem at the base of the stipule 9

5. Submersed leaves >2 mm wide; margins minutely serrate toward the tip . . . . *P. robbinsii*
5. Submersed leaves ≤2 mm wide; margins entire. 6

6. Plants with some narrowly elliptic floating leaves. . . . . *P. diversifolius*
6. Plants with linear to filiform submersed leaves only. 7

7. Basal portion of stipules of lower leaves swollen, brownish. . . . . *P. vaginatus*
7. Stipule tubular, tight to the stem and green at the base. 8

8. Leaves long-attenuate, tapering for at least 2 mm to the tip . . . . *P. pectinatus*
8. Leaf tips rounded to acute but not tapering for more than 2 mm . . . . *P. filiformis*

9. Leaves all submersed and ≤5 mm wide. 10
9. At least the floating leaves ≥5 mm wide 16

10. Leaves with >10 veins; stems flattened with winged edges. . . . . *P. zosteriformis*
10. Leaves with <10 veins; stems terete or flattened but not winged. 11

11. Stems with small globose glands at leaf nodes (use 10X). 12
11. Stems lacking globose glands 14

12. Some leaves with 5 or more veins (use 10X). . . . . *P. friesii*
12. Leaves with 3 veins 13

13. Achene 1.5–2.5 mm long; leaves 1–2 mm wide. . . . .*P. pusillus*
13. Achene 2.5–3.6 mm long; leaves 2–4 mm wide. . . . .*P. obtusifolius*

14. Tepals 2–3 mm long; some leaves with >3 veins . . . . *P. gramineus*
14. Tepals 1–2 mm long; leaves with 1 to 3 veins 15

15. Spikes congested, subcapitate; leaves flaccid. . . . .*P. foliosus*
15. Spikes more elongate with separate whorls of fruits; leaves stiff. . . . . *P. strictifolius*

16. Some submersed leaves ≥20 mm wide 17
16. Submersed leaves ≤20 mm wide 18

17. Floating leaves with ≥25 veins; submersed leaves folded and falcate. . . . . *P. amplifolius*
17. Floating leaves with <25 veins; nearly flat . . . .*P. illinoensis*

18. Submersed leaves ≤2 mm wide. 19
18. Submersed leaves >2 mm wide. 20

19. Floating leaves ≤3 cm long. . . . . *P. diversifolius*
19. Floating leaves ≥4 cm long. . . . . *P. natans*

20. Stems somewhat flattened; achenes strongly compressed . . . .*P. epihydrus*
20. Stems terete; achenes with rounded margins 21

21. Submersed leaves with petioles 2–10 cm long . . . .*P. nodosus*
21. Submersed leaves sessile to short-petiolate 22

22. Floating leaves reddish, short-petiolate . . . . *P. alpinus*
22. Floating leaves green, some with petioles ca. as long as the blade . . . . *P. gramineus*

Potamogeton alpinus Balb. **Stems** terete, little branched. **Submersed leaves** linear-lanceolate, 4–12 cm × 5–12 mm; stipules 15–35 mm long, not closely sheathing. **Floating leaves** (often lacking) short-petiolate; blades oblanceolate to narrowly elliptic, reddish, 4–7 cm long. **Spikes** to 13 cm long, continuous. **Tepals** green becoming reddish, 2–3 mm long. **Fruits** 3–4 mm long, keeled with a curved beak. Shallow, fresh or alkaline water of ponds, lakes; valleys, montane. Circumboreal south to CA, NV, UT, CO.

Potamogeton gramineus has long-petiolate floating leaves that are not reddish.

Potamogeton amplifolius Tuck. **Stems** terete, little branched. **Submersed leaves** petiolate; blades narrowly ovate, often falcate, 8–20 × 2–5 cm; stipules 4–10 mm long, not sheathing. **Floating leaves** long-petiolate; blades elliptic, 5–10 cm long. **Spikes** 3–6 cm long, continuous.

Tepals green or purplish, 2–3 mm long. **Fruits** 4–6 mm long, barely keeled with a short, curved beak. Shallow to deep, fresh water of lakes, ponds; valleys, montane. BC to NL south to CA, WY, OK, TN, VA.

Potamogeton crispus L. Plants producing turions. **Stems** flattened, branched. **Submersed leaves** linear-oblong with serrate, wavy margins, sessile with a clasping base, 3–8 cm × 4–10 mm; stipules 3–7 mm long, shredding early. **Floating leaves** absent. **Spikes** 1–2 cm long, continuous. Tepals yellowish, ca. 1 mm long. **Fruits** 5–6 mm long, keeled with a straight beak. Shallow to deep, fresh or saline water of ponds, lakes, slow streams; plains, valleys. Introduced throughout most of N. America; native to Europe.

Our few specimens are sterile. See *Potamogeton richardsonii*.

Potamogeton diversifolius Raf. **Stems** terete, branched. **Submersed leaves** linear, 1–3 cm × ≤1 mm, attached to the stipule; stipules 2–6 mm long, sheathing. **Floating leaves** petiolate; blades elliptic, 1–3 cm long. **Spikes** 2–10 mm long, continuous with a peduncle 3–15 mm long, much shorter below than above, continuous. Tepals green, ≤1 mm long. **Fruits** 1–2 mm long, keeled with a minute beak. Shallow, fresh or saline water of ponds, lakes; plains, valleys, montane. OR to VT south to CA, AZ, TX, LA, FL, Mexico.

Potamogeton epihydrus Raf. **Stems** flattened, little branched. **Submersed leaves** linear, 5–20 cm × 2–8 mm; stipules 8–20 mm long, not sheathing, deciduous. **Floating leaves** petiolate; blades elliptic, 2–6 cm long. **Spikes** 15–40 mm long, continuous. Tepals green, 1–3 mm long. **Fruits** 2–4 mm long, strongly keeled with a minute beak. Shallow, fresh water of ponds, sloughs, lakes; valleys. AK to NL south to CA, ID, CO, AL, GA.

Potamogeton filiformis Pers. [*Stuckenia filiformis* (Pers.) Börner] Plants tuberous. **Stems** terete, branched. **Submersed leaves** filiform, only the ultimate 1 mm tapered, 1–8 cm × 1–2 mm, attached to the stipule; lower stipules 5–15 mm long, sheathing. **Floating leaves** absent. **Spikes** 1–5 cm long with distinct whorls of flowers. Tepals brownish, 1–2 mm long. **Fruits** 2–3 mm long, obscurely keeled with a minute beak. Shallow, fresh water of lakes, ponds, slow streams; valleys, montane. Circumboreal south to CA, AZ, NM, NE, NY.

Subspecies have been recognized based on stipule size and durability, but many specimens are difficult to place, and the characters seem unrelated to geography or ecology in our material. *Potamogeton pectinatus* has larger fruits, larger stipules, more tapered leaf tips and often occurs in saline or alkaline water.

Potamogeton foliosus Raf. **Stems** slightly compressed, branched. **Submersed leaves** linear 1–4 cm × ca. 1 mm; stipules 5–10 mm long, sheathing but fraying above. **Floating leaves** absent. **Spikes** capitate, 1–6 mm long. Tepals green, ≤1 mm long. **Fruits** 1.5–2 mm long, keeled with a minute, erect beak. Shallow, fresh water of ponds, ditches, sloughs; valleys. Throughout temperate N. America. Our plants are var. *foliosus*.

Potamogeton friesii Rupr. **Stems** branched, slightly compressed with globose glands at the nodes. **Submersed leaves** linear, 1–8 cm × 1–3 mm; stipules 6–18 mm long, sheathing but fraying with age. **Floating leaves** absent. **Spikes** ca. 1 cm long with separate whorls of flowers. Tepals green, 1–2 mm long. **Fruits** ca. 2 mm long, barely keeled with a curved beak. Fresh, usually shallow water of lakes, ponds; plains, valleys. Circumboreal south to WA, UT, NE, IN, VA.

Potamogeton gramineus L. **Stems** terete, branched. **Submersed leaves** linear to linear-lanceolate, 2–10 cm × 3–15 mm; stipules 1–3 cm long, not sheathing. **Floating leaves** long-petiolate; blades elliptic, 2–7 cm long. **Spikes** 15–40 mm long, continuous. Tepals green, ca. 2 mm long. **Fruits** 2–3 mm long, keeled with a short, curved beak. Shallow or deep, sometimes alkaline water of ponds, lakes; plains, valleys, montane. Circumboreal south to most of U.S. (p.588)

Plants often occur in ephemerally aquatic habitats. See *Potamogeton alpinus*.

Potamogeton illinoensis Morong **Stems** terete, sometimes branched. **Submersed leaves** short-petiolate; blades linear-lanceolate to oblanceolate, 6–20 cm × 15–35 mm; stipules 2–6 cm long, not sheathing. **Floating leaves** (often absent) petiolate; blades elliptic, 3–12 cm long. **Spikes** stout, 2–7 cm long. Tepals green, 2–3 mm long. **Fruits** 3–4 mm long, keeled with a rounded beak ca. 1 mm long. Moderate to deep, fresh water of lakes, sloughs, rivers; valleys. Throughout N. America.

Potamogeton praelongus has clasping leaves.

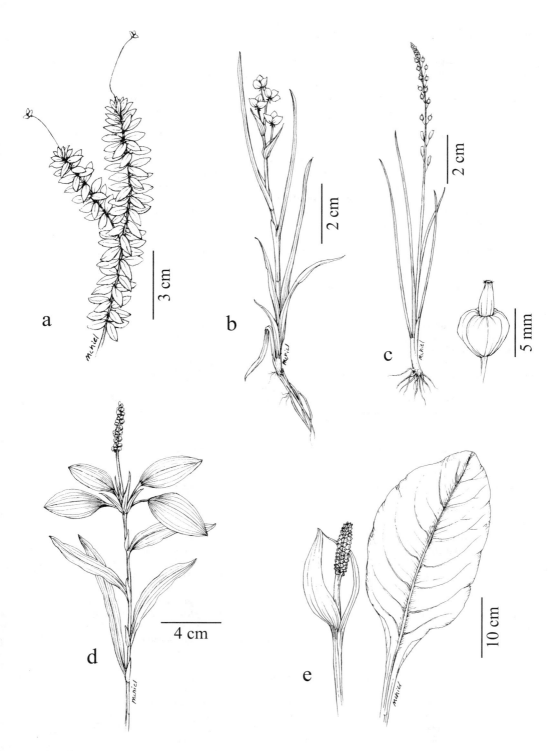

Plate 102. a. *Elodea canadensis*, b. *Scheuchzeria palustris*, c. *Triglochin palustris*, d. *Potamogeton gramineus*, e. *Lysichiton americanus*

Potamogeton natans L. Rhizomatous. **Stems** terete, little branched. **Submersed leaves** linear, terete, 8–15 cm × ca. 1 mm; stipules 2–12 cm long, not sheathing. **Floating leaves** long-petiolate; blades elliptic to ovate, 3–9 cm long. **Spikes** 3–5 cm long, continuous. **Tepals** green, 1–2 mm long. **Fruits** 4–5 mm long, rounded on back with a curved beak ca. 1 mm long. Shallow to deep water of ponds, lakes, sloughs; valleys. Circumboreal south to most of U.S.

Potamogeton nodosus Poir. **Stems** terete, little branched. **Submersed leaves** petiolate; blades lanceolate, 10–20 cm × 8–20 mm; stipules 2–12 cm long, not sheathing, deciduous. **Floating leaves** long-petiolate; blades narrowly elliptic, 5–12 cm long. **Spikes** stout, 3–6 cm long, continuous. **Tepals** green, 1–3 mm long. **Fruits** 3–4 mm long, rounded on back with a short beak. Shallow water of ponds; plains, valleys. Circumboreal south through N. America to S. America.

Potamogeton obtusifolius Mert. & W.D.J.Koch Plants producing turions. **Stems** slightly compressed, branched with globose glands at the nodes. **Submersed leaves** linear, 3–8 cm × 2–4 mm; stipules 1–2 cm long, membranous, sheathing. **Floating leaves** absent. **Spikes** ca. 1 cm long. **Tepals** green, ca. 1 mm long. **Fruits** barely keeled, 3–3.5 mm long including the erect beak. Shallow, fresh water of ponds, sloughs; valleys, montane. Circumboreal south to WA, WY, NY.

Potamogeton pectinatus L. [*Stuckenia pectinata* (L.) Börner] Plants tuberous. **Stems** terete, branched. **Submersed leaves** filiform, 1–10 cm × <1 mm, tapering at the tip for >3 mm, attached to the stipule; lower stipules 16–30 mm long, sheathing. **Floating leaves** absent. **Spikes** 1–5 cm long with distinct whorls of flowers. **Tepals** brownish-green, 1–1.5 mm long. **Fruits** 3–4 mm long, obscurely keeled with a slender, curved beak. Shallow to deep, often saline or alkaline water of ponds, lakes, streams, ditches; plains, valleys, montane. Circumboreal south to Mexico. See *P. filiformis*.

Potamogeton praelongus Wulfen **Stems** terete, little branched, somewhat zigzag. **Submersed leaves** sessile, clasping, lanceolate, minutely wavy, 6–20 × ca. 1 cm; stipules 2–7 cm long, not sheathing. **Floating leaves** absent. **Spikes** 2–4 cm long, continuous. **Tepals** green, 2–3 mm long. **Fruits** 4–5 mm long, keeled with an erect beak. Deep water of large lakes; valleys. AK to NL south to CA, UT, CO, IN, NY.

Potamogeton amplifolius has falcate leaves; see *Potamogeton illinoensis*, *P. richardsonii*.

Potamogeton pusillus L. [*P. berchtoldii* Fieber] **Stems** terete, branched, often with globose glands at some nodes. **Submersed leaves** linear, 1–6 cm × 1–2 mm; stipules 5–30 mm long, sheathing or not but quickly fraying. **Floating leaves** absent. **Spikes** 5–12 mm long, continuous or with distinct whorls of flowers. **Tepals** green, 1–1.5 mm long. **Fruits** 1–2 mm long, obscurely keeled with a short, curved beak. Shallow to deep, fresh to saline water of ponds, lakes, marshes, slow streams; plains, valleys, montane. Circumboreal south to most of U.S., Mexico, S. America.

Both var. *pusillus* and var. *tenuissimus* (Mert. & W.D.J.Koch) Fernald, with capitate spikes and broad gold or white stripes along the midvein, are reported for MT. Many of our plants can be referred to the former, while others have some characters of both vars. There appears to be little ecological or geographic difference associated with the two forms in our area.

Potamogeton richardsonii (A.Benn.) Rydb. **Stems** terete, little branched. **Submersed leaves** sessile, clasping, ovate to lanceolate with wavy margins, 2–10 cm × 6–20 mm; stipules 1–2 cm long, not sheathing, quickly fraying. **Floating leaves** absent. **Spikes** 1–3 cm long, continuous. **Tepals** green, ca. 2 mm long. **Fruits** 2–4 mm long, obscurely keeled with an erect beak. Shallow to deep, sometimes alkaline water of ponds, lakes, slow rivers; valleys to montane. AK to NL south to CA, CO, IA, NY.

Potamogeton praelongus has zigzag stems and longer leaves. The leaves of *P. crispus* are also wavy but are oblanceolate with serrate margins.

Potamogeton robbinsii Oakes **Stems** terete, branched. **Submersed leaves** linear, 3–8 cm × 2–5 mm, attached to the stipule; stipules 1–3 cm long, shredding above the leaf attachment. **Floating leaves** absent. **Spikes** 7–15 mm long with separate flowers. **Tepals** yellowish, 1–1.5 mm long. **Fruits** ca. 2 mm long, keeled with a curved beak. Shallow, fresh water of lakes; valleys. AB to NL south to CA, UT, WY, MI, NY.

Potamogeton strictifolius A.Benn. **Stems** simple or branched, slightly compressed without globose glands at the nodes. **Submersed leaves** linear, rigid, 2–7 cm × 0.5–2 mm; stipules 8–20 mm long, sheathing but fraying with age. **Floating leaves** absent. **Spikes** 8–15 mm long with separate whorls of flowers. **Tepals** green, 1–2 mm long. **Fruits** 2–3 mm long, barely keeled with a beak ca. 0.5 mm long. Fresh or somewhat alkaline, shallow water of lakes; montane. YT to NB south to UT, SD, WI, VA. Known from Park Co.

Potamogeton vaginatus Turcz.[*Stuckenia vaginata* (Turcz.) Holub] **Stems** terete, branched.  **Submersed leaves** linear, 2–10 cm × 1–2 mm, attached to the stipule; stipules 1–5 cm long, sheathing, brown, swollen at the base. **Floating leaves** absent. **Spikes** 1–5 cm long with well-separated whorls of flowers. **Tepals** brownish-green, 1–2 mm long. **Fruits** 2–3 mm long, obscurely keeled and beaked. Mostly shallow, often alkaline water of lakes, streams; valleys, montane. Circumboreal south to OR, UT, CO, NY.

The sheaths of *Potamogeton pectinatus* are not swollen but clasp the stem tightly.

Potamogeton zosteriformis Fernald **Stems** flattened, winged, branched. **Submersed leaves**  linear, 5–20 cm × 2–5 mm; stipules 15–30 mm long, not sheathing. **Floating leaves** absent. **Spikes** 15–25 mm long, continuous. **Tepals** green, 1–2 mm long. **Fruits** 3–5 mm long, keeled with a curved beak. Shallow to deep water of lakes, ponds, sloughs; valleys to montane. AK to NL south to CA, UT, KS, NY.

NAJADACEAE: Water-nymph Family

Najas L. Water nymph

Monoecious or dioecious, submersed, aquatic annuals. **Stems** branched, lax, floating. **Leaves** serrate, linear with a dilated base, opposite or verticillate, appearing whorled. **Inflorescence** axillary, solitary, sessile or short-pedicellate. **Flowers** unisexual; male flowers with a lobed perianth segment enclosing a hyaline bract and 1 stamen; female flowers with or without a perianth; pistil 1; stigmas 2 to 4, sessile. **Fruit** a fusiform to narrowly ovoid achene.

1. Seeds smooth; leaves often >15 mm long . . . . *N. flexilis*
1. Seeds with a honeycombed surface; leaves ≤15 mm long . . . .*N. guadalupensis*

Najas flexilis (Willd.) Rostk. & W.L.E.Schmidt **Stems** 5–40 cm. **Leaves** 10–35 mm long; the  dilated base 2–3 mm wide. **Female flowers** 2–5 mm long. **Achenes** ca. 3 mm long; seeds smooth, shiny, sometimes minutely spotted. Usually shallow, often alkaline water of lakes, sloughs; valleys. BC to NL south to CA, NV, KY, VA.

Najas guadalupensis (Spreng.) Magnus **Stems** 5–30 cm. **Leaves** 5–15 mm long; the  dilated base 1–2 mm wide. **Female flowers** 1–4 mm long. **Achenes** ca. 2 mm long; seeds coat honeycomb-pitted. Shallow, fresh water of lakes, sloughs; valleys. BC to NL south to CA, NV, KY, VA. Our plants are ssp. ***guadalupensis***.

RUPPIACEAE: Ditch-grass Family

Ruppia L. Ditch grass

Ruppia maritima L. [*R. cirrhosa* (Petagna) Grande] Submersed, aquatic, rhizomatous, perennial herb. **Stems** filiform, branched, terete, 6–50 cm. **Leaves** alternate, filiform, 3–20 cm long, attached at the base to a membranous, sheathing stipule 5–20 mm long with or without a lamina to 3 cm long. **Inflorescence** axillary umbels of few to several 2-flowered spikes; peduncles to 30 cm long and coiled in fruit. **Flowers** perfect, regular; perianth absent; stamens 2; pistils 4 to 6; ovaries superior. **Fruit** ovoid, drupe-like, 2–3 mm long, beaked, 1-seeded, borne on a pedicel. Shallow, saline or alkaline water of ponds, sloughs, lakes; plains, valleys, montane. Cosmopolitan.

Plants of inland North America have highly coiled peduncles and are considered a separate species.[167]

ZANNICHELLIACEAE: Horned-pondweed Family

Zannichellia L. Horned pondweed

Zannichellia palustris L. Monoecious, rhizomatous, submersed, aquatic perennials.
Stems filiform, branched 15–50 cm. **Leaves** mostly opposite, filiform, 2–10 cm long; stipules
membranous, deciduous, 1–4 mm long. **Inflorescence** axillary cymes. **Flowers** unisexual, without
a perianth; male flowers 1 stamen; female flowers of 4 or 5 pistils, distinct; ovaries superior.
Fruits a curved-oblong, compressed, pedicellate achene 2–3 mm long with a curved beak 1–2
mm long. Shallow, fresh to alkaline water of ponds, lakes, ditches; plains, valleys, montane. Cosmopolitan.

ARACEAE: Arum Family

Lysichiton Schott Skunk cabbage

Lysichiton americanus Hultén & H.St.John Scapose, rhizomatous perennials with a
skunky odor. **Leaves** basal, petiolate; blades elliptic to obovate, entire, 20–100 cm long, greatly
expanding in fruit. **Inflorescence**: scape 15–80 cm, expanding in fruit; flowers densely clustered
on a cylindrical spike (spadix) 3–14 cm long, mostly enclosed by a deciduous, yellow, hood-like
bract (spathe) 10–20 cm long that sheaths the scape. **Flowers** perfect, yellowish; tepals 4, ca. 2
mm long, separate; stamens 4; ovary superior, 1- or 2-celled; style 1. **Fruit** a green to red, ovoid 1- or 2-seeded berry,
3–10 mm long embedded in the fleshy spadix. Vernally inundated, wet, organic soil of spruce swamps, alder thickets;
valleys. AK to CA, ID, MT. (p.588)

ACORACEAE: Sweet-flag Family

Acorus L. Sweet flag

Acorus americanus (Raf.) Raf. [*A. calamus* L. misapplied] Scapose, rhizomatous perennial with a pleasant odor.
Leaves basal, folded, flattened, linear, erect, sheathing below, 50–100 cm long. **Inflorescence** 20–60 cm; flowers
densely clustered on a narrowly cylindrical spike (spadix), 4–7 cm long, subtended by an erect, linear, leaf-like spathe
15–70 cm long, that appears to be a continuation of the stem. **Flowers** perfect; tepals 6, separate, brown, ca. 1 mm
long; stamens 6; ovary superior, 1- to 3-celled. **Fruit** a leathery, oblong, glabrous, several-seeded berry 3–5 mm long
embedded in the spadix. Shallow water of marshes along lakes, streams; valleys. AK to NL south to WA, MT, IA, NY.
Known from Flathead and Lake cos.
   This species was formerly placed in the Araceae.

LEMNACEAE: Duckweed Family

Perennial, free-floating, aquatic herbs. **Stems** (called fronds) flattened, green, discoid to paddle-shaped,
sometimes with a few dangling roots. **Inflorescence** inconspicuous, solitary, a small sac containing 1 or 2
male and 1 female flowers. **Flowers** unisexual; perianth absent; male flowers 1 stamen; female flowers of 1
pistil; ovary superior, 1-celled. **Fruit** a follicle with 1 to 5 seeds.
   Many fronds are sterile; flowers are rarely observed, and reproduction is mainly vegetative. Some
authors consider the inflorescence to be a single flower with 1 pistil and 1 or 2 stamens.[233]

1. Fronds without veins on the upper surface; roots absent. . . . . *Wolffia*
1. Fronds with 1 to several veins on the upper surface and 1 to several roots 2

2. Fronds with 1 root per frond. . . . . *Lemna*
2. Fronds with ≥2 roots per frond . . . . *Spirodela*

Lemna L. Duckweed

Fronds solitary or attached together, flattened, each with a solitary root. **Inflorescence** an open or closed
pouch on the margin of the frond. **Seeds** several, ribbed.

1. Fronds paddle-shaped, attached together by elongated stipes . . . .*L. trisulca*
1. Fronds oblong to ovate, not attached by slender stipes. 2

2. Fronds oblong, 1–2 mm long . . . .*L. valdiviana*
2. Fronds ovate, some >2 mm long . . . . *L. minor*

Lemna minor L. [*L. turionifera* Landolt] **Fronds** ovate, 1–8 mm long; solitary or in small clusters, often purplish on the lower surface. **Roots** 1–15 mm long. Permanent fresh water of ponds, lakes, slow streams; plains, valleys, montane. Circumboreal south to most of temperate N. America. (p.602)

    Lemna turionifera has been separated from *L. minor* by forming turions, differences in frond color and disposition of veins,[233] but these characters are difficult to discern from fresh or pressed material without special preparation, so I have combined them under the older name. Most of our material would be considered *L. turionifera* if the two were differentiated.

Lemna trisulca L. **Fronds** shaped like canoe paddles, 6–12 mm long including the stipe, connected together, mainly green. **Roots** solitary or absent, 5–30 mm long. Fresh to somewhat saline or alkaline water of ponds, lakes, marshes; plains, valleys. Circumboreal south to most of temperate N. America.

Lemna valdiviana Phil. **Fronds** oblong, 1–2 mm long; usually in clusters, green. **Roots** 1–15 mm long. Our 1 location is from a warm spring in Granite Co. MT to CA, AZ, NM, TX and eastern U.S., Mexico, S. America.

Spirodela Schleid. Duck meal

Spirodela polyrhiza (L.) Schleid. **Fronds** obovate to orbicular, flattened, 3–8 mm long, attached together in groups of 3 to 10, green. **Roots** 1–3 cm long, several from each frond. **Inflorescence** a flower-bearing pouch on each side of the frond that also serves as point of attachment to other fronds. **Seeds** 1 or 2, ribbed or smooth. Fresh water of ponds, lakes, often with *Lemna minor*; valleys. Circumboreal through all of N. America.

Wolffia Horkel ex Schleid. Water meal

Fronds solitary, green to brown, globular or ovoid, without roots. **Inflorescence** a dorsal cavity with 1 male and 1 female flower. **Seed** solitary, smooth.

    Wolffia borealis (Engelm.) Landolt is reported for MT,[233] but I have seen no specimens. This genus contains the world's smallest vascular plants.

1. Fronds subglobose, upper surface green. . . . .*W. columbiana*
1. Fronds oblong to narrowly ellipsoid, upper surface with brown pigment cells . . . . *W. brasiliensis*

Wolffia brasiliensis Wedd. [*W. punctata* Griseb.] **Fronds** elongate-globular, 0.5–1 mm long, green with minute brown spots. Open fresh water of ponds; valleys in Lake and Ravalli cos. Eastern and northwest U.S. to S. America.

Wolffia columbiana H.Karst. **Fronds** globular, ca. 1 mm long, green. Open fresh water of ponds, sloughs, marshes; valleys. AB south to CA, MT, eastern U.S., S. America.

COMMELINACEAE: Spiderwort Family

Tradescantia L. Spiderwort

Perennial herbs with fleshy roots. **Leaves** alternate, folded, linear with a swollen, sheathing base. **Inflorescence** a compact, few- or several-flowered, umbelliform, terminal cyme subtended by 2 leaf-like bracts. **Flowers** perfect, regular; sepals 3, ovate, separate; petals 3, separate, broadly ovate; stamens 6; pistil 1; ovary superior, 3-celled. **Fruit** an oblong, few-seeded capsule.

1. Sepals with glandular hairs only; leaf sheath 2 to 4 times blade width. . . . . *T. occidentalis*
1. Sepals with a mixture of glandular and eglandular hairs; leaf sheath 1 to 2 times blade width . . *T. bracteata*

Tradescantia occidentalis (Britton) Smyth **Stems** erect, 5–50 cm, usually branched. **Herbage** glabrous. **Leaves** long-acuminate, 9–40 cm × 4–15 mm. **Bracts** 6–15 cm long. **Flowers** campanulate; pedicels 1–2 cm long; sepals 6–13 mm long with glandular hairs; petals blue, 7–16 mm long; filaments pilose. **Capsule** 4–7 mm long. Sandy soil of grasslands; plains. MT to MN south to AZ, NM, TX, LA. Our plants are var. *occidentalis*.

Tradescantia bracteata Small **Stems** erect to ascending, 5–45 cm, sometimes branched. **Herbage** glabrate. **Leaves** long-acuminate, 15–29 cm × 9–20 mm. **Bracts** 6–30 cm long. **Flowers**: pedicels 2–3 cm long; sepals 10–13 mm long, with glandular and eglandular hairs; petals rose, ca. 18 mm long; filaments pilose. **Capsule** 5–6 mm long. Grasslands, thickets; plains. MT to MN south to OK, KS. Known from Carter and Custer cos.

Our specimens approach *Tradescantia bracteata* by having larger flowers and narrower leaf blades relative to the sheaths, but they do not seem like typical *T. bracteata* from central U.S.[152]

JUNCACEAE: Rush Family

Mostly perennial, grass-like herbs. **Leaves** (sometimes absent) linear, with a sheath enclosing the stem. **Flowers** perfect, regular; tepals 6 in 2 similar series, lanceolate, often scarious; stamens 6 (3); pistil 1; ovary superior, 3-celled; stigmas 3. **Fruit** a globose to narrowly ovoid, many-seeded capsule.[59]

1. Leaves glabrous; capsule with many seeds . . . . *Juncus*
1. Leaves usually with at least some long hairs near the collar; capsule 3-seeded . . . . *Luzula*

Juncus L. Rush

Plants caespitose or rhizomatous, glabrate. **Stems** terete or flattened. **Leaves** with open sheaths, sometimes projecting above into auricles or a ligule; blades flat, channeled, or terete with membranous crosswalls. **Inflorescence** terminal, cymose, open to capitate, subtended by a leaf-like bract. **Flowers** sometimes subtended by a small bract and/or 2 closely subtending, scarious bracteoles (prophylls); tepals lanceolate, usually with scarious margins. **Capsules** short-beaked, sometimes 3-sided; seeds sometimes with tail-like appendages.[174]

Mature capsules are usually needed for positive identification. *Juncus confusus, J. tenuis, J. dudleyi,* and *J. interior* are similar; the last three have similar habitats and have broad geographic overlap, differing only in auricle characters. They might better be treated as a single species.[101] Plants of *J. nodosus* and *J. torreyi* occasionally produce galls that look like a slender, purplish, Brussel sprout rosette at a node of the rhizome or stem; these galls are caused by a small aphid-like insect. *Juncus tweedyi* Rydb. is known from Yellowstone National Park in WY and is reported for MT,[113,174] but I have seen no specimens.

1. Plants fibrous-rooted annual with flowers in the axils of leaves as well as in terminal clusters . . . *J. bufonius*
1. Plants perennial without flowers in leaf axils 2

2. Leaves consisting of papery sheaths only; conspicuous green blades lacking 3
2. Stem with at least 1 leaf with a conspicuous blade 6

3. Inflorescence of 1 to 4 flowers . . . . *J. drummondii*
3. Inflorescence with >4 flowers 4

4. Stems densely clustered from short rhizomes . . . . *J. effusus*
4. Stems sometimes tufted but mostly spread along creeping rhizomes 5

5. Stem-like main involucral bract ca. as long as the stem (inflorescence appearing
    to be borne at mid-stem . . . . *J. filiformis*
5. Involucral bract much shorter than the stem (inflorescence appearing to be on
    upper third of the stem) . . . .*J. balticus*

6. Leaves iris-like, folded in half and enfolding those above at the base (equitant). . . . . *J. ensifolius*
6. Leaves not equitant 7

7. Leaf blades terete (round in cross-section), usually divided cross-wise into sections (septate) 8
7. Leaf blades flat or inrolled 22

8. Stems with a single head-like cluster of flowers; high montane to alpine. 9
8. Flowers not strongly clustered or most stems with >1 head; habitat usually lower 12

9. Each flower with a pair of scarious bracts (prophylls) closely attached at the base of the tepals . . . . *J. hallii*
9. Prophylls lacking (bracteoles may be present but not enveloping the base of the flower). 10

10. Flowers >5 . . . . *J. mertensianus*
10. Flowers mostly 1 to 4 11

11. Flowers 1 or 2, exceeded by subtending, leaf-like bract; capsules indented on top. . . . . *J. biglumis*
11. Flowers 2 to 4, as long as subtending, papery bracts; capsules rounded to the tip . . . . *J. triglumis*

12. Flowers in clusters (heads) of 3 to many flowers 13
12. Inflorescence paniculate; flowers all pedicellate, not clustered 20

13. Flowers in clusters of 3 to 12 flowers 14
13. Most flower heads with >12 flowers each. 17

14. Tepals mostly >3 mm long 15
14. Tepals 2–3 mm long 16

15. Stamens 3; seeds with appendages as long as the body . . . . *J. tweedyi*
15. Stamens 6; seeds apiculate but not appendaged . . . .*J. nevadensis*

16. Capsule tapered to the tip, 3–4 mm long, exserted . . . . *J. articulatus*
16. Seed capsule rounded to the tip, 2–3 mm long, barely exserted . . . .*J. alpinoarticulatus*

17. Heads 10–15 mm wide, sometimes clustered together . . . .*J. torreyi*
17. Heads smaller, mostly well-separated 18

18. Plants caespitose; stamens 3 in most flowers . . . . *J. acuminatus*
18. Plants rhizomatous; stems arising singly or tufted; stamens 6. 19

19. Capsules rounded to the tip; tepals dark brown or purple . . . .*J. nevadensis*
19. Capsules tapered to a long point; tepals green to tan . . . . *J. nodosus*

20. Tepals 5–7 mm long . . . . *J. parryi*
20. Tepals 3–5 mm long 21

21. Inflorescence with 2 to 7 flowers; capsules indented on top . . . . *J. hallii*
21. Inflorescence of 5 to 20 flowers; capsules rounded to the tip. . . . .*J. vaseyi*

22. Each flower with a pair of scarious bracts (prophylls) closely attached at the base of the tepals 23
22. Prophylls lacking (bracteoles may be present but not enveloping the base of the flower). 28

23. Outer tepals curved inward (hooded). 24
23. Outer tepals not hooded. 25

24. Anthers and filaments equal; capsules longer than the tepals; riverbanks. . . . . *J. compressus*
24. Anthers ca. 3 times as long as filaments; capsules ca. as long as the tepals; meadows. . . . .*J. gerardii*

25. Auricles at the top of leaf sheaths >1 mm long, membranous . . . .*J. tenuis*
25. Auricles ≤1 mm long or absent, leathery or membranous 26

26. Auricles leathery, yellowish. . . . . *J. dudleyi*
26. Auricles white to purplish, membranous. 27

27. Capsule indented on top, completely divided into 3 cells. . . . . *J. confusus*
27. Capsule rounded on top, dividing walls not going all the way to the center. . . . . *J. interior*

28. Tepals brown, without a greenish midstripe . . . . *J. castaneus*
28. Tepals with a green, fading to brown, midstripe 29

29. Seeds with white appendages as long as the body . . . . *J. regelii*
29. Seeds apiculate but not appendaged. 30

30. Tepals 4–5 mm long, silvery-margined; capsule rounded to the tip . . . . *J. longistylis*
30. Tepals 3–4 mm long, brown-margined; capsule truncate on top . . . . *J. covillei*

Juncus acuminatus Michx. Short-rhizomatous, caespitose. **Stems** erect, terete, 40–80 cm, tufted. **Leaves** basal and cauline; blades subterete, 1–2 mm wide; auricles rounded. **Inflorescence** open with erect branches and clusters of 10 to 20+ flowers ca. 1 cm across; main bract shorter than the inflorescence. **Flowers**: prophylls absent; tepals green to light brown, 2–4 mm long, acuminate; stamens 3. **Capsules** 2–4 mm long, slightly exserted, tapering to the beak; seeds apiculate. Pond margins; valleys. BC to NS south to most of U.S., Mexico, S. America; rare on the Great Plains. One collection from Teton Co.

Juncus alpinoarticulatus Chaix [*J. alpinus* Vill.] Rhizomatous. **Stems** erect, terete, 10–40 cm, often tufted. **Leaves** basal and cauline; blades terete, 0.5–2 mm wide; auricles rounded. **Inflorescence** open with clusters of 3 to 6 flowers at tips and forks of erect branches; main bract smaller than the inflorescence. **Flowers**: prophylls absent; tepals mostly brown, 2–3 mm long; outer acute; inner rounded, slightly shorter; stamens 6. **Capsules** 2–3 mm long, barely exserted, rounded to the beak; seeds apiculate. Wet mineral or organic soil of streambanks and margins of ponds, fens; plains, valleys to lower subalpine. AK to Greenland south to WA, UT, CO, MN, VT. (p.598)

Some specimens appear intermediate between *Juncus alpinoarticulatus* and *J. articulatus*; the former is thought to be a hybrid derivative of the latter.[59]

Juncus articulatus L. Rhizomatous. **Stems** erect, terete, 10–70 cm, often tufted. **Leaves** basal and cauline; blades terete, 1–2 mm wide; auricles acute to rounded. **Inflorescence** open with erect to spreading branches and clusters of 3 to 10 flowers; main bract smaller than the inflorescence. **Flowers**: prophylls absent; tepals green to brown, 2–3 mm long, acute to acuminate, ca. equal; stamens 6. **Capsules** 3–4 mm long, exserted, acuminate, tapering to the beak; seeds with minute appendages. Wet mineral soil of stream or ditch banks, margins of lakes, wetlands; plains, valleys. BC to NL south to OR, AZ, NE, KY, VA. See *J. alpinoarticulatus*. (p.598)

Juncus balticus Willd. [*J. arcticus* Willd. var. *balticus* (Willd.) Trautv.] Baltic rush, wiregrass Strongly rhizomatous. **Stems** erect, terete, 15–90 cm, sometimes tufted. **Leaves** mainly basal; blades and auricles absent; sheaths light to dark brown. **Inflorescence** congested to open with sessile to pedicellate flowers; main bract terete, erect, appearing like a continuation of the stem. **Flowers** with prophylls; tepals brown to purple, 4–5 mm long, acuminate; stamens 6. **Capsules** 3–4 mm long, rounded to the beak; seeds apiculate. Wet mineral soil of meadows, low areas in grasslands, thickets, often along streams, ditches, tolerant of saline and alkaline conditions; plains, valleys to lower subalpine. Circumboreal south to S. America, absent from southeast U.S. (p.598)

Perhaps this species is better considered as part of the circumpolar *Juncus arcticus* complex; but I prefer to maintain the name that is most widely used in MT. Plants with stout stems >5 mm wide have been called var. *vallicola* Rydb.

Juncus biglumis L. Fibrous-rooted perennial forming small tufts. **Stems** erect, 2–10 cm. **Leaves** basal; blades channeled, 0.5–1.5 mm wide; auricles absent. **Inflorescence** a small head with 1 to 3 sessile flowers; main bract brown, greater than the inflorescence. **Flowers**: prophylls absent; tepals dark brown, 2–3 mm long, acute; stamens 6. **Capsules** 3–4 mm long, indented on top; seeds with a small appendage. Wet, cold, shallow soil of alpine seeps, rock ledges, along streams; alpine. Circumpolar south to CO.

Juncus bufonius L. Toad rush Fibrous-rooted annual. **Stems** ascending to erect, terete, 2–15 cm, branched at the base. **Leaves** basal and cauline; blades channeled, <1 mm wide; auricles absent. **Inflorescence** open with ascending branches and clusters of 1 to 3 sessile flowers in axils of upper leaves (nodal bracts). **Flowers** with prophylls; tepals narrow, green with scarious margins, 4–7 mm long, outer acuminate, inner acute, shorter; stamens 6. **Capsules** 3–4 mm long, rounded at the tip; seeds apiculate. Vernally moist, often disturbed soil of meadows, streambanks, margins of ponds, wetlands; plains, valleys, montane. Circumboreal south to most of N. America.

Juncus castaneus Sm. Rhizomatous. **Stems** erect, terete, 5–30 cm, not tufted. **Leaves** basal and cauline; blades channeled or folded, 1–3 mm wide; auricles absent. **Inflorescence** 1 to 3 clusters of 2 to 8 flowers each; main bract usually greater than the inflorescence. **Flowers**: prophylls absent; tepals purple or brown, 4–6 mm long, narrow, acute; stamens 6. **Capsules** 5–8 mm long, exserted, acutely tapered to the beak; seed appendages 0.5–1 mm long. Seeps, wet rock ledges, along streams; upper subalpine, alpine. Circumpolar south to NV, UT, NM.

Juncus compressus Jacq. Rhizomatous. **Stems** erect, terete, 10–40 cm, sometimes tufted. **Leaves** basal and cauline; blades flat or channeled, ≤1 mm wide; auricles rounded. **Inflorescence** open with 5–60 pedicellate flowers; main bract shorter or longer than the inflorescence. **Flowers** with prophylls; tepals purplish, 2–2.5 mm long, rounded, hood-like, inner shorter and wider; stamens 6. **Capsules** 2–3 mm long, exserted, rounded to the beak; seeds without appendages. Gravelly or sandy riverbanks, roadside ditches; plains. Introduced sparingly to northern U.S. and adjacent Canada; native to Eurasia.

The similar *Juncus gerardii* occurs in alkaline meadows.

Juncus confusus Coville Caespitose. **Stems** erect, terete, 10–60 cm. **Leaves** basal; blades flat or inrolled, <1 mm wide; auricles rounded, <1 mm long. **Inflorescence** congested with 3 to 25 subsessile flowers; main bract greater than the inflorescence. **Flowers** with prophylls; tepals green to brown, 2–5 mm long with acute tips; stamens 6. **Capsules** 3–4 mm long, included, rounded and indented on top; seeds apiculate. Moist, often disturbed soil of meadows, open forest, rock outcrops, along trails, roads, streams; valleys to lower subalpine. BC to SK south to CA, AZ, UT, CO. (p.598)

   Many reports of the similar *Juncus tenuis* are referable here.

Juncus covillei Piper Rhizomatous. **Stems** erect, compressed, 6–20 cm, sometimes clumped. **Leaves** basal and cauline; blades flat, 1–2 mm wide; auricles acute or absent. **Inflorescence** congested with 3 to 7 sessile flowers in each of 1 to 3 clusters; main bract shorter or longer than the inflorescence. **Flowers**: prophylls absent; tepals brown to purplish with a green midstripe, 3–4 mm long, acute; stamens 6. **Capsules** 3–5 mm long, truncate on top; seeds without appendages. BC to CA, ID, MT.

1. Inner tepals rounded at the tip, ca.1 mm shorter than the mature capsule . . . . .var. *covillei*
1. All tepals acute or acuminate, ca. as long as the capsule . . . . var. *obtusatus*

Juncus covillei var. ***covillei*** occurs on riverbanks; Mineral and Missoula cos.; *J. covillei* var. ***obtusatus*** C.L.Hitchc is found in montane to subalpine meadows; Ravalli Co.

Juncus drummondii E.Mey. Caespitose. **Stems** erect, terete, 6–35 cm. **Leaves** basally disposed; blades reduced to a bristle, sheaths brown; auricles absent. **Inflorescence** congested with 1 to 3 short-pedicellate flowers; main bract terete, erect, ca. as long as the inflorescence. **Flowers** with prophylls; tepals brown to purplish, 4–7 mm long, acuminate; stamens 6. **Capsules** 3–7 mm long, indented on top; seeds with long appendages. Moist meadows, open forest, turf, often along streams, around boulders, or where snow accumulates; subalpine to alpine. AK to AB south to CA, UT, CO. (p.598)

   Juncus parryi is similar but has at least one stem leaf blade. High-elevation plants have shorter tepals and have been called var. *subtriflorus* (E.Mey.) C.L.Hitchc.

Juncus dudleyi Wiegand [*J. tenuis* Willd. var. *dudleyi* (Wiegand) F.J.Hermann] Caespitose. **Stems** erect, terete, 15–80 cm. **Leaves** basal; blades flat or inrolled, ca. 1 mm wide; auricles yellowish, thickened, <1 mm long. **Inflorescence** open to congested with 6 to 30 sessile flowers; main bract usually much greater than the inflorescence. **Flowers** with prophylls; tepals green, 3–5 mm long, subequal with acuminate tips; stamens 6. **Capsules** 2–4 mm long, rounded on top; seeds apiculate. Meadows, margins of streams, ponds, lakes, ditches; plains, valleys. Throughout most of boreal and temperate N. America.

Juncus effusus L. [*J. pylaei* Laharpe] Short-rhizomatous, caespitose. **Stems** erect, terete, hollow, 40–100 cm. **Leaves** mainly basal; blades reduced to a bristle; auricles absent; sheaths purple to brown. **Inflorescence** open or congested with numerous, pedicellate flowers; bract terete, erect, appearing like a continuation of the stem. **Flowers** with prophylls; tepals brown with green midvein, 2–3 mm long, acuminate; stamens 3. **Capsules** ca. 2 mm long, truncate on top; seeds apiculate. Streambanks, wet meadows, marshes; valleys. BC to MT south to CA, northeast N. America; Europe. Our plants are ssp. ***effusus*** which is introduced from Europe.[454]

Juncus ensifolius Wikstr. [*J. tracyi* Rydb.] Rhizomatous. **Stems** erect, terete, 10–60 cm. **Leaves** basal and cauline; blades equitant (folded flat in half longitudinally), 1–5 mm wide; auricles absent. **Inflorescence** of capitate clusters, 3–12 mm across, each with 3 to many sessile flowers; main bract smaller than the inflorescence. **Flowers**: prophylls absent; tepals tan, black, or purplish, 3–4 mm long, acuminate; stamens 3 or 6. **Capsules** rounded on top, 3–4 mm long; seeds apiculate or with a short appendage. Moist to wet soil around seeps, ponds, lakes, along streams, ditches; plains, valleys to subalpine. AK to QC south to CA, AZ, NM, TX, Mexico.

   This species is usually divided into 2 to 4 taxa based on number of stamens, tepal color and leaf width. Plants with 3 stamens (var. *ensifolius*) tend to have wider leaves than those with 6 stamens (var. *montanus* (Engelm.) C.L.Hitchc.), but there is no ecological or geographic associations with these traits in our area. *Juncus tracyi* is now considered a form of var. *montanus* having seeds with long appendages.[59]

Juncus filiformis L. Spreading-rhizomatous. **Stems** erect, terete, 4–30 cm. **Leaves** mainly basal; blades and auricles absent; sheaths green to light brown. **Inflorescence** congested with 5 to 20 short-pedicellate flowers; main bract terete, erect, appearing like a continuation of and often longer than the stem. **Flowers** with prophylls; tepals green to tan, 2–4 mm long, acuminate; stamens 6. **Capsules** 2–3 mm long, rounded on top; seeds unappendaged. Fens, lake margins; montane. Circumboreal south to OR, ID, UT, CO, MI, PA.

Juncus gerardii Loisel. Rhizomatous. **Stems** erect, terete, 10–40 cm. **Leaves** basal and cauline; blades flat or channeled, ca. 1 mm wide; auricles obscure. **Inflorescence** open with 10 to 30 pedicellate flowers; main bract smaller than the inflorescence. **Flowers** with prophylls; tepals green to tan, 2.5–3 mm long, rounded, hood-like, subequal; stamens 6. **Capsules** ca. 2.5 mm long, truncate on top; seeds without appendages. Wet, saline or alkaline meadows; valleys. Sporadic across northern U.S. and adjacent Canada; Eurasia. Possibly introduced in at least part of the N. American range. Known from Madison Co.

Juncus hallii Engelm. Short-rhizomatous, caespitose. **Stems** erect, terete, 10–40 cm, tufted. **Leaves** mainly basal, bladeless; the 1 or 2 cauline leaves have terete blades ≤1 mm wide; auricles minute. **Inflorescence** congested with 2 to 7 subsessile flowers; lowest bract terete, erect or ascending, shorter or longer than the inflorescence. **Flowers** with prophylls; tepals brown, 3–5 mm long, acute; stamens 6. **Capsules** 4–6 mm long, indented on top; seeds appendaged. Moist to wet meadows; upper montane, subalpine. WA to MT south to UT, WY.

The retuse capsule separates *Juncus hallii* from *J. parryi*.

Juncus interior Wiegand Caespitose. **Stems** erect, terete, 6–70 cm. **Leaves** basal; blades flat or inrolled, ≤1 mm wide; auricles white or purplish, <1 mm long. **Inflorescence** congested with 3–30 sessile or short-pedicellate flowers; main bract usually greater than the inflorescence. **Flowers** with prophylls; tepals green, 3–5 mm long with acuminate tips; stamens 6. **Capsules** 3–4 mm long, rounded on top; seeds apiculate. Moist, disturbed soil on margins of ditches, streams, ponds; plains, valleys. SK to ON south to NM, TX, TN.

Juncus occidentalis (Coville) Wiegand occurs in adjacent ID and differs from *J. interior* mainly by having slightly larger auricles.

Juncus longistylis Torr. [*J. orthophyllus* Coville] Rhizomatous. **Stems** erect, terete, 20–80 cm. **Leaves** basal and cauline; blades flat, 1–2 mm wide; auricles truncate, 1–3 mm long. **Inflorescence** of 1 to 5 subcapitate clusters of 3 to 12 sessile flowers each, 6–15 mm across; main bract much smaller than the inflorescence; bracteoles numerous, silvery-membranous. **Flowers**: prophylls absent; tepals silvery with a green and brown mid-stripe, 4–6 mm long, acute; stamens 6. **Capsules** 3–5 mm long, rounded on top; seeds apiculate. Moist to wet meadows on margins of streams, fens, seeps, ponds, tolerant of alkali; valleys, montane, rarely higher. BC to NL south to CA, AZ, NM, NE. (p.598)

The silvery-margined tepals and bracteoles help distinguish this rush. *Juncus orthophyllus* is supposed to differ from *J. longistylis* by having brown tepals;[59] however, this seems like a weak character because the tepals of *J. longistylis* become brownish with age. See *J. regelii*.

Juncus mertensianus Bong. Rhizomatous. **Stems** erect, terete, 5–35 cm, sometimes tufted. **Leaves** basal and cauline; blades terete, 1–2 mm wide; auricles acute to rounded, ca. 1 mm long. **Inflorescence** capitate with 1 cluster of 6 to 60 flowers, 8–15 mm across; main bract usually greater than the inflorescence. **Flowers**: prophylls absent; tepals purple to black, 3–5 mm long, acuminate, inner slightly shorter and broader; stamens 6. **Capsules** 2–4 mm long, rounded on top; seeds without appendages. Wet soil along streams, lakes, seeps; subalpine to alpine, rarely lower. AK to SK south to CA, AZ, NM.

Juncus nevadensis S.Watson Rhizomatous. **Stems** erect, terete, 20–60 cm, sometimes tufted. **Leaves** basal and cauline; blades terete, septate, 1–2 mm wide; auricles acute to rounded, 1–3 mm long. **Inflorescence** open with erect to ascending branches and several clusters, 5–10 mm across, of 5 to 11 flowers each; main bract shorter than the inflorescence. **Flowers**: prophylls absent; tepals purple or brown, 3–4 mm long, acuminate, inner slightly smaller; stamens 6. **Capsules** 2.5–4 mm long, rounded on top; seeds apiculate. Moist to wet, sometimes gravelly soil around ponds, lakes, streams, rarely fens; valleys to lower subalpine. BC to SK south to CA, AZ, NM. Our plants are var. **badius** (Suksd.) C.L.Hitchc. (p.598)

Juncus nodosus has narrower, lighter tepals and more tapered capsules.

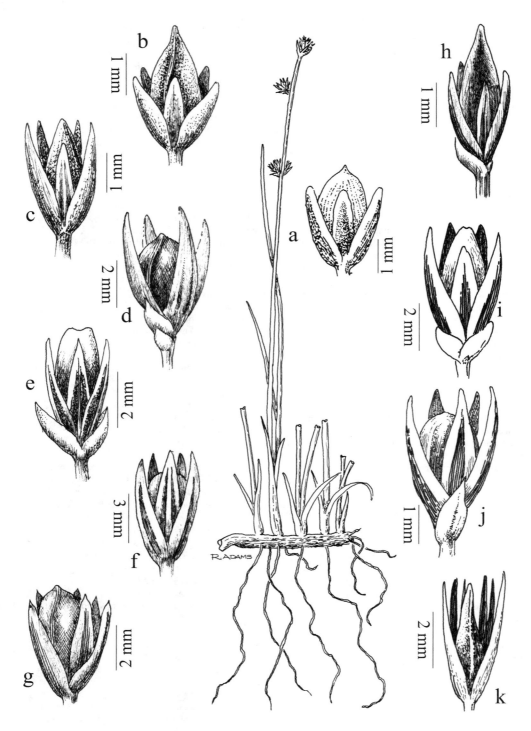

Plate 103. *Juncus*. **a.** *J. alpinoarticulatus*, **b.** *J. articulatus*, **c.** *J. balticus*, **d.** *J. confusus*, **e.** *J. drummondii*, **f.** *J. longistylis*, **g.** *J. nevadensis*, **h.** *J. nodosus*, **i.** *J. parryi*, **j.** *J. tenuis*, **k.** *J. torreyi*

Juncus nodosus L. Rhizomatous. **Stems** erect, terete, 15–50 cm. **Leaves** basal and cauline; blades terete, septate, 0.5–2 mm wide; auricles rounded, ca. 1 mm long. **Inflorescence** open with erect to spreading branches and several clusters of 6 to 30 spreading flowers; main bract greater than the inflorescence. **Flowers**: prophylls absent; tepals green to light brown, 2.5–4 mm long, acuminate, subequal; stamens usually 6. **Capsules** 4–5 mm long, long-tapered to the beak; seeds apiculate. Wet meadows, wet gravelly soil along streams, ponds, lakes; plains, valleys, rarely montane. AK to NL south to AZ, NM, NE, OH, MD. See *J. torreyi*, *J. nevadensis*. (p.598)

Juncus parryi Engelm. Caespitose. **Stems** erect, terete, 4–25 cm. **Leaves** mainly basal, bladeless; the 1 or 2 cauline leaves have terete blades <1 mm wide; auricles obscure. **Inflorescence** somewhat congested with 1 to 3 pedicellate flowers; main bract terete, erect, greater than the inflorescence. **Flowers** with prophylls; tepals green to brown, 5–7 mm long, acuminate; stamens 6. **Capsules** 6–8 mm long, acute; seeds appendaged. Stony soil of dry to moist meadows, turf, outcrops, often near streams or shores; subalpine or alpine. BC, AB south to CA, NV, UT, CO. (p.598)

Juncus drummondii and *J. hallii* have no stem leaf blades.

Juncus regelii Buchenau Rhizomatous. **Stems** erect, compressed, 10–60 cm, often clustered. **Leaves** basal and cauline; blades flat, 1–3 mm wide; auricles small, pointed. **Inflorescence** open to congested with 1 to 5 clusters of 10 to 30 flowers; main bract usually shorter than the inflorescence. **Flowers**: prophylls absent; tepals brown with a broad green midstripe and silvery margins, 4.5–6 mm long, acute, apiculate, subequal; stamens 6. **Capsules** 4–5 mm long, truncate on top; seeds with appendages as long as the body. Marshes, fens, wet meadows, often along streams; montane, subalpine. BC to CA, NV, UT, WY.

Juncus longistylis has lighter tepals and is more common at lower elevations.

Juncus tenuis Willd. Caespitose. **Stems** erect, terete, 10–60 cm. **Leaves** basal; blades flat or inrolled, <1 mm wide; auricles membranous, 2–5 mm long. **Inflorescence** open to congested with 5 to 40 sessile flowers; main bract greater than the inflorescence. **Flowers** with prophylls; tepals green, 3–4.5 mm long, subequal with acute tips; stamens 6. **Capsules** 2–4 mm long, rounded on top; seeds apiculate. Usually disturbed soil of meadows, open forest, roadsides; plains, valleys. Throughout N. America. (p.598)

Juncus torreyi Coville Rhizomatous. **Stems** erect, terete, 15–60 cm from tuberous-thickened rhizomes. **Leaves** basal and cauline; blades terete, septate, 1–3 mm wide; auricles rounded, 1–2 mm long. **Inflorescence** of 1 to 10 heads each with 25 to 100 sessile flowers, 10–15 mm across, some heads clustered; main bract terete, ascending, greater than the inflorescence. **Flowers**: prophylls absent; tepals green to tan, 3–4 mm long, long-acuminate, inner shorter; stamens 6. **Capsules** 4–6 mm long, long-tapered to the beak; seeds apiculate. Moist to wet meadows, margins of streams, ditches, ponds, often where saline or alkaline; plains, valleys, rarely higher. Throughout most of temperate N. America. (p.598)

Juncus nodosus has smaller heads.

Juncus triglumis L. [*J. albescens* (Lange) Fernald] Fibrous-rooted, caespitose. **Stems** erect, 3–20 cm. **Leaves** basal; blades terete, septate, <1 mm wide; auricles minute, rounded. **Inflorescence** a small head with 2 to 4 sessile flowers subtended by 2 scarious bracts shorter than the inflorescence. **Flowers**: prophylls absent; tepals brown to whitish, 3–5 mm long, acute; stamens 6. **Capsules** 3–6 mm long, rounded, shorter or longer than the tepals; seeds with appendages as long as the body. Wet, shallow, gravelly soil along alpine meltwater streams, seeps. Circumpolar south to UT, CO.

1. Tepals whitish, higher than the mature capsule . . . . var. *albescens*
1. Tepals brown, smaller than the mature capsule . . . . var. *triglumis*

Juncus triglumis var. *albescens* Lange occurs in the northwest and south-central mountains; *J. triglumis* var *triglumis* is only in south-central cos.

Juncus vaseyi Engelm. Short-rhizomatous, caespitose. **Stems** erect, terete, 20–100 cm. **Leaves** basal; blades terete, ≤1 mm wide; auricles scarious, whitish, minute. **Inflorescence** open to congested with 5 to 20 pedicellate flowers; bract shorter or greater than the inflorescence. **Flowers** with prophylls; tepals green to tan, 3.5–4.5 mm long, subequal with acuminate tips; stamens 6. **Capsules** 4–5 mm long, rounded on top; seeds short-appendaged. NT to NL south to ID, CO, IA, NY. Reported for northwest MT;[59] a report for Cascade Co.[113] was based on a misidentified specimen of *J. hallii*.

Luzula DC. Wood rush

Perennial, caespitose, often short-rhizomatous. **Stems** terete. **Leaves** basal and cauline with a closed sheath; blades flat; auricles absent. **Inflorescence** terminal, subtended by a leaf-like bract, open-paniculate or of 1 to several spike-like clusters. **Flowers** closely subtended by 2 scarious, lacerate bracts (prophylls) as well as a more distant bracteole; tepals scarious-margined, subequal; stamens 6. **Capsules** globose, 3-seeded.[101,390]

1. Flowers usually ≥3 in cylindical or globose clusters 2
1. Flowers 1 or 2 on tips of branches in an open inflorescence 4

2. Inflorescence a single cylindrical drooping spike . . . . *L. spicata*
2. Inflorescence of ≥2 globose or cylindrical clusters 3

3. Inflorescence with clusters of 3 to 6 flowers; alpine . . . . *L. arcuata*
3. Clusters with 8 to 15 flowers; montane. . . . . *L. campestris*

4. Anthers mostly 0.8–1.2 mm long, much longer than the filaments; capsules and
   tepals 2.5–3.5 mm long . . . . *L. hitchcockii*
4. Anthers ≤0.8 mm long, shorter or equal to the filaments; capsules and tepals ≤2.5 mm long 5

5. Inflorescence branches divaricately spreading to reflexed. . . . . *L. divaricata*
5. Inflorescence branches arched-spreading 6

6. Bracteoles and prophylls tangled-ciliate; stems <30 cm tall. . . . . *L. piperi*
6. Bracteoles and prophylls entire or lacerate, barely ciliate; stems often >30 cm. . . . . *L. parviflora*

Luzula arcuata (Wahlenb.) Sw. **Stems** ascending, 2–10 cm. **Leaves** glabrate; blades 1–3 mm wide. **Inflorescence** arching, of 2 to 12 clusters of 2 to 6 flowers; bracteoles ciliate. **Flowers**: tepals brown to purple, 1.5–2.5 mm long, outer mucronate; prophylls purplish. **Capsules** ellipsoid, ca. 2 mm long. Moist rock ledges, turf; alpine. Circumpolar south to WA, MT. Our plants are ssp. *unalaschkensis* (Buchenau) Hultén.

Luzula campestris (L.) DC. [*L. multiflora* (Ehrh.) Lej., *L. comosa* E.Mey.] **Stems** erect, 10–40 cm. **Leaves** sparsely long-hairy; blades 2–6 mm wide. **Inflorescence** of 1 to few erect or ascending, globose to cylindric clusters of 5 to 20 flowers, 5–10 mm long. **Flowers**: tepals purplish to tan, white-margined, 2.5–3.5 mm long; prophylls whitish. **Capsules** globose, ca. 2 mm long. Moist meadows, thickets, moist to dry forests, often along streams; valleys to subalpine. AK to NL south to OR, ID, WY, KY, GA.

The taxonomy of the *L. campestris* complex awaits a global monograph before taxa can be assigned names with confidence.[101,390]

Luzula divaricata S.Watson **Stems** erect to ascending, 20–40 cm. **Leaves** glabrate; blades 3–7 mm wide. **Inflorescence** open with 1 to 2 flowers on tips of divaricately spreading to reflexed branches. **Flowers**: tepals acuminate, mucronate, green to tan, ca. 2 mm long; prophylls whitish to tan, sparsely ciliate. **Capsules** globose, ca.2 mm long. Cedar-hemlock forests; valleys. WA to MT south to CA. Known from Lincoln Co.

Luzula hitchcockii Hämet-Ahti [*L. glabrata* (Hoppe) Desv. misapplied] Sometimes long-rhizomatous. **Stems** erect to ascending, 10–30 cm. **Leaves** glabrate; blades 4–10 mm wide, often turning golden-brown with age. **Inflorescence** open with 1 or 2 flowers on tips of spreading branches. **Flowers**: purple to brown, 2.5–3.5 mm long; prophylls purplish. **Capsules** ovoid, 2.5–4 mm long with a beak ≥0.5 mm long. Moist meadows, open forest, often in late snowmelt sites; subalpine, rarely lower or higher. BC, AB, WA, OR, ID, MT. See *L. piperi*. (p.602)

Luzula parviflora (Ehrh.) Desv. Sometimes long-rhizomatous. **Stems** erect to ascending, 20–50 cm. **Leaves** glabrate; blades 3–8 mm wide. **Inflorescence** open with 1 or 2 flowers on tips of arching branches. **Flowers**: tepals tan to purplish, 1.5–2 mm long; prophylls whitish to tan, lacerate or entire. **Capsules** globose, ca. 2 mm long. Meadows, thickets, conifer forest, often around streams, wetlands; valleys to subalpine, more common lower. Circumboreal south to CA, AZ, NM, SD. See *L. piperi*.

Luzula piperi (Coville) M.E.Jones [*L. wahlenbergii* Rupr. misapplied] **Stems** erect to ascending, 6–30 cm. **Leaves** glabrate; blades 2–7 mm wide. **Inflorescence** open with solitary flowers on tips of arching branches; bracteoles ciliate. **Flowers**: tepals purplish-brown with white margins, ca. 2 mm long; prophylls whitish, often tangled-ciliate. **Capsules** ovoid, ca. 2 mm long. Usually stony soil of rock ledges, talus, meadows, turf; subalpine, alpine. AK to QC south to WA, MT.

Luzula hitchcockii has larger flowers; several specimens appear intermediate between the two species; *L. parviflora* has lighter-colored tepals.

Luzula spicata (L.) DC. **Stems** erect, 3–30 cm. **Leaves** sparsely hairy; blades 1–2 mm wide. **Inflorescence** a nodding, solitary, terminal, cylindric, often compound spike 5–25 mm long with numerous flowers. **Flowers**: tepals purplish to brown, mucronate, 2–3 mm long; prophylls whitish, tangled-ciliate. **Capsules** globose, ca. 2 mm long. Grasslands, meadows, open forest, talus, rock ledges, turf; montane to alpine, more common higher. Circumboreal south to CA, UT, NM.

TYPHACEAE: Cattail Family

Rhizomatous emergent perennials. **Leaves** basal and cauline, alternate, grass-like, sheathing at the base. **Inflorescence** of few to several globose to cylindrical, unisexual heads, male above. **Flowers** unisexual, sessile, each subtended by small bracts or hairs; sepals and petals absent or undifferentiated; male flowers of 1 to 7 stamens; female flowers with 1 ovary and 1 or 2 styles. **Fruit** an obovoid to ellipsoid, beaked achene.

Sparganium is often placed in its own family, the Sparganiaceae.

1. Flowers in cylindrical heads >10 cm long. . . . .*Typha*
1. Flowers in globose heads <3 cm long . . . .*Sparganium*

Sparganium L. Bur reed

Stems erect or floating. **Leaves** with enlarged, sheathing bases. **Inflorescence** of globose heads, sessile or pedunculate, continuous to well-separated on a simple or branched rachis; the lower subtended by leaf-like bracts. **Flowers** unisexual, sessile, densely clustered, each subtended by scales; stamens 2 to 5; styles 1. **Achene** narrowed below, beaked above, clustered in a bur-like head.

A report of *Sparganium androcladum* (Engelm.) Morong for Gallatin Co.[113] was based on a misidentified specimen.

1. Stigmas 2; inflorescence usually branched . . . .*S. eurycarpum*
1. Stigma 1; inflorescence usually simple. 2

2. Staminate heads 1(2); mature achene beak ca. 1 mm long. . . . .*S. natans*
2. Staminate heads usually ≥2; mature achene beak 2–4 mm long. . . . .*S. angustifolium*

Sparganium angustifolium Michx. [*S. multipedunculatum* (Morong) Rydb., *S. emersum* Rehmann, *S. simplex* Huds.] **Stems** 5–50 cm, often floating. **Leaves** 2–12 mm wide, sometimes keeled, often exceeding the inflorescence. **Inflorescence** unbranched, of 2 to 6 sessile, contiguous or separate male heads, and 2 to 4, sessile or pedunculate female heads, 8–25 mm wide at maturity. **Styles** unbranched, 0.6–2 mm long. **Achenes** fusiform; body 3–6 mm long, constricted in the middle, tapering gradually to the beak, 2–4 mm long. Shallow, fresh water of lakes, ponds, sloughs, ditches, fens; plains, valleys to subalpine. Circumboreal south to CA, NM, SD, PA. (p.602)

Often considered two taxa: *S. angustifolium* sensu stricto has narrower, unkeeled leaves, contiguous and smaller female heads, and shorter achene beaks than *S. emersum*.[113,207] Some collections can easily be assigned to one or the other of these taxa, but half of our specimens have some characters of both, and the traits are mostly continuous in our area, so I have chosen to recognize just one species.[55] Plants referred to *S. multipedunculatum*, the common form in our area, have been placed under both *S. emersum* and *S. angustifolium*.[187,207]

Sparganium eurycarpum Engelm. **Stems** erect, 30–100 cm. **Leaves** 5–15 mm wide, keeled at least toward the base, flattened toward the tip. **Inflorescence** usually branched, each branch with 5 to 14 separate sessile male heads, and 1 or 2, sessile or pedunculate female heads, 15–25 mm wide at maturity. **Styles** 2-lobed. **Achenes** obovoid, 4–8 mm long, tapering gradually to abruptly to the 2–4 mm long beak. Shallow to deep, fresh water of lakes, ponds, sloughs; valleys. NT to NL south to CA, NM, KY, VA, Mexico; Asia.

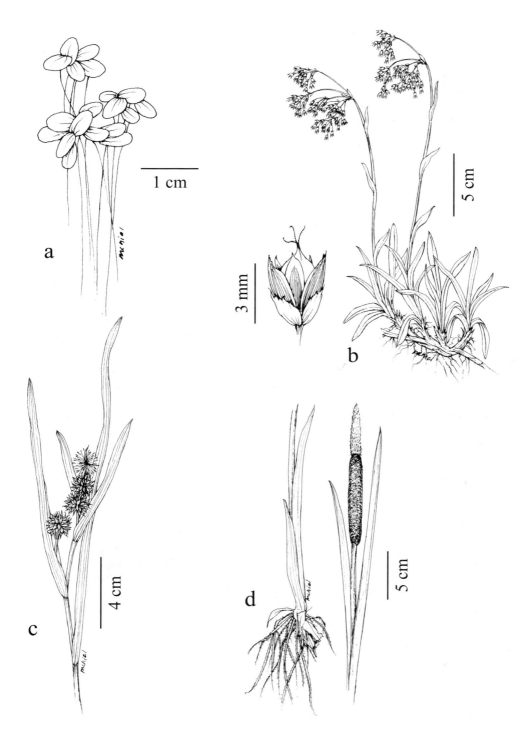

Plate 104. a. *Lemna minor*, b. *Luzula hitchcockii*, c. *Sparganium angustifolium*, d. *Typha latifolia*

Sparganium natans L. [*S. minimum* (Hartm.) Fries] **Stems** floating, 10–60 cm. **Leaves** 2–5 mm wide, not keeled. **Inflorescence** unbranched with usually 1 sessile male head, and 1 to 4 sessile or pedunculate female heads, 7–12 mm wide at maturity. **Styles** unbranched. **Achenes** fusiform, 2–4 mm long, abruptly tapered to the ca. 1 mm long beak. Shallow to deep water of lakes, ponds, fens; valleys, montane. Circumboreal south to CA, UT, WY, ND, IL, NY.

Typha L. Cattail

Stems erect. **Leaves** erect with scarious margins, sheathing at the base, filled with spongy tissue. **Inflorescence** a terminal, ebracteate, spike-like, cylindrical head; the upper portion male, light brown, separate or contiguous with the lower, dark brown, female portion. **Flowers** unisexual, sessile, densely clustered, each subtended by scales and hairs; stamens 1 to 5; style 1. **Achene** stipitate, fusiform, ca. 1 mm long hidden by hairs below, with the persistent style above.

Plants with separate male and female spikes and narrowly lanceolate, brown stigmas are thought to be hybrids between the two species below and have been called *Typha ×glauca* Godr. Some specimens identified as *T. angustifolia* may be this hybrid.

1. Stigmas oblanceolate or wider; male and female spikes contiguous . . . . *T. latifolia*
1. Stigmas linear; male and female spikes usually separated by at least 5 mm . . . . *T. angustifolia*

Typha angustifolia L. **Stems** 80–150 cm. **Leaves** 3 to 12 mm wide. **Inflorescence**: male spike 8–12 cm × 8–10 mm; separated from the female spike 7–15 cm × 6–9 mm. **Flowers**: stigma linear, brown; pollen grains shed separately. Shallow water of marshes, ponds, lakes, often where saline; plains, valleys. Eastern and central N. America; Eurasia; introduced to the western states.

Typha latifolia L. **Stems** 1–2 m. **Leaves** glaucous, 6–18 mm wide. **Inflorescence**: male spike 7–15 × 1–2 cm; contiguous or separated from the female spike 7–15 × 1–3 cm. **Flowers**: stigma lanceolate; black at the tip; pollen grains shed in tetrads. Shallow to moderately deep water of marshes, ponds, lakes, ditches; plains, valleys, montane. Circumboreal south to all of temperate N. America. (p.602)

CYPERACEAE: Sedge Family

Annual or perennial, usually monoecious herbs. **Stems** 3-angled or terete. **Leaves** basal and usually cauline, alternate, linear, grass-like, entire, composed of a lower sheath enclosing the stem and a flat to V-shaped blade. **Inflorescence** terminal, consisting of spikes arranged in spikes, racemes or panicles. **Flowers** often unisexual, each subtended by a scale; petals and sepals lacking; perianth reduced to a papery sac, bristles, a scale or absent; stamens 3; ovary superior with 1–3 stigmas. **Fruit** a 2–3-sided achene.

Carex and *Kobresia* are distinctive by the presence of a sac (perigynium) surrounding the seed. Species of *Amphiscirpus, Bolboschoenus, Schoenoplectus* and *Trichophorum* are placed in *Scirpus* by some authorities.

1. Achene surrounded in a membranous wrapper or sac (perigynium); bristles lacking; flowers unisexual . . .2
1. Perigynium lacking; achene with bristles or scales at the base; flowers bisexual .3

2. Perigynium a closed sac fully enclosing the achene . . . . *Carex*
2. Perigynium partly open, exposing the achene . . . . *Kobresia*

3. Leaf sheaths on lower 1/3 of stem without blades, but upper leaf blades well-developed. . . . . *Dulichium*
3. Leaf blades lacking entirely or leaves well-developed at the base of the plant 4

4. Leaves appear to be lacking, all reduced to bladeless sheaths. 5
4. At least 1 leaf with a distinct blade (sometimes short) present. 6

5. Spikelets solitary. . . . . *Eleocharis*
5. Spikelets numerous . . . . *Schoenoplectus*

6. Achene bristles >8, elongate, forming a cottony tuft. . . . . *Eriophorum*
6. Achene bristles 0–8, elongate in only 1 species. 7

7. Achene subtended by 1 or more bristles in addition to a scale 8
7. Bristles absent; achene subtended only by scales. 14

Amphiscirpus Oteng-Yeb.

Amphiscirpus nevadensis (S.Watson) Oteng-Yeb. [*Scirpus nevadensis* S.Watson] Strongly
rhizomatous perennial. **Stems** erect, terete, 10–40 cm. **Leaves** basal and lower-cauline; blades
inrolled, 0.5–2 mm wide. **Inflorescence** terminal, of 1 to 6 sessile spikelets; involucral bracts 1
to 3; the lowest leaf-like, spreading to erect; the others scale-like. **Spikelets** ovoid to lanceoloid,
6–12 mm long, of numerous overlapping, spirally arranged scales; scales 4–5 mm long, glabrous,
hyaline-brown, with a pale midrib, acute or minutely apiculate. **Flowers** perfect, each at the base of a scale; perianth of
1 to 6 spinulose bristles; stamens 3; style 2-branched. **Achene** obovoid, 2-sided, brown with a reticulate surface, 2–2.5
mm long. Muddy shores of saline ponds, calcareous fens; plains, valleys. BC to SK south to CA, NV, UT, CO, NE, S.
America.

Bolboschoenus (Asch.) Palla Tuberous bulrush

Rhizomatous and tuberous perennial herbs. **Stems** erect, 3-angled, swollen at the base. **Leaves** mainly
cauline; blades flat. **Inflorescence** terminal, congested or open with sessile or clustered-pedunculate
spikelets; involucral bracts 2 to 5, leaf-like, erect or spreading. **Spikelets** of several overlapping, spirally
arranged scales; scales puberulent with an awn arising from the notch at the tip. **Flowers** perfect, each
at the base of a scale; perianth of 3 to 6 spinulose bristles; stamens 3; style 2- or 3-branched. **Achene**
obovoid, 2- or 3-sided, short-beaked.

1. Achene 2-sided; style 2-branched . . . .*B. maritimus*
1. Achene 3-sided; style 3-branched . . . .*B. fluviatilis*

Bolboschoenus fluviatilis (Torr.) Soják [*Scirpus fluviatilis* Torr.] **Stems** 20–150 cm. **Leaf**
blades 6–15 mm wide. **Inflorescence** congested; spikelets sessile on the culm or clustered on
short peduncles; involucral bracts 3 to 6, leaf-like. **Spikelets** lanceoloid, 1–2 cm long; scales
6–9 mm long, hyaline-brown; awn 2–3 mm long. **Flowers**: bristles 3 to 6; stamens 3; style
3-branched. **Achene** obovoid, 3-sided, gray to brown, 3–5 mm long. Marshes, streambanks;
plains. BC to QC south to CA, AZ, AL, VA.

Bolboschoenus maritimus (L.) Palla [*Scirpus maritimus* L.] **Stems** 20–80 cm. **Leaf blades**
2–10 mm wide. **Inflorescence** congested; spikelets sessile on the culm or rarely clustered on
short peduncles; involucral bracts 2 or 3, leaf-like. **Spikelets** ovoid, 10–25 mm long; scales 7–10
mm long, hyaline-brown; awn 1–2 mm long. **Flowers**: bristles 3 to 6, deciduous; stamens 3; style
2-branched. **Achene** obovoid, 2-sided, gray to brown, 2.5–4 mm long. Muddy shores of usually saline

or alkaline ponds, streams; plains, valleys, montane. Circumboreal south through most of U.S. except southeastern states. Our plants are ssp. *paludosus* (A.Nelson) T.Koyama.

Bulbostylis Kunth

Bulbostylis capillaris (L.) C.B.Clarke Tufted, glabrate annuals. **Stems** erect, 5–15 cm. **Leaves** basal and cauline, ca. 0.5 mm wide; sheaths ciliate at the summit; ligules absent. **Inflorescence** terminal, several spikelets borne in an umbel-like inflorescence subtended by few leaf-like bracts. **Spikelets** red-brown, narrowly ovoid, 3–5 mm long with spirally arranged flowers. **Flowers** perfect, consisting of an ovate scale ca. 2 mm long; 1(2) anther; and an ovary with a 3-parted style. **Achenes** yellowish, 3-sided, 1 mm long, transversely ribbed. Sandy, disturbed soils, often along streams; valleys. Eastern and southern N. America; known from Carbon Co. where it was possibly introduced.

Carex L. Sedge

Perennial herbs. **Leaves** basal and/or cauline; ligule present. **Herbage** mostly glabrous. **Inflorescence** of 1 to many unisexual or bisexual spikes arranged in spikes, racemes or panicles; bisexual spikes with male flowers segregated into either the top or bottom portion; at least the lowest spikelet subtended by a leaf-like bract **Flowers** unisexual; each subtended by a papery scale; male flowers of 3 stamens; female flower an ovary enclosed in a papery, sac-like perigynium; style 1. **Fruit** a 2-sided (with 2 styles) or 3-sided (with 3 styles) achene.[24,173,381]

This is the largest genus in MT. Descriptions of perigynia are for those with mature achenes. Measurements and descriptions of perigynia and female scales were taken at mid-spike. Reports of *Carex bigelowii* Torr. ex Schwein. and *C. tincta* (Fernald) Fernald[173] were based on misidentifications.[446] *Carex ebenea* Rydb.,[113,170] *C. microglochin* Wahlenb.,[51,173] *C. luzulifolia* Boott.,[24] *C. occidentalis*,[24,113] *C. multicostata* Mack.,[24,173] and *C. straminiformis*[113,173] are reported for MT, but I have seen no specimens. One immature specimen from Missoula Co. could be *C. angustata* Boott [*C. eurycarpa* Holm] (L. Standley annotation on *Barkley 2401* MONTU); otherwise I have seen no MT vouchers for this species.

1. Plants unisexual (dioecious). . . . . Group A
1. Both male and female flowers on the same plant (monoecious; white, thread-like
 filaments indicate male flowers) 2

2. Stems with a solitary, bisexual, terminal spike (plants with densely clustered spikes
 may appear like a solitary spike, but they will key below) . . . . Group B
2. Stems with 2 or more spikes 3

3. All or most spikes unisexual; male spikes usually narrower and above the female spikes 4
3. All spikes similar in appearance; at least one spike bisexual. 6

4. Stigmas 2; achenes 2-sided. . . . . Group C
4. Stigmas 3; achenes 3-sided. 5

5. Beaks of perigynia >0.5 mm long. . . . . Group D
5. Perigynia beaks ≤0.5 mm long. . . . . Group E

6. Stigmas 3; achenes 3-sided. . . . . Group F
6. Stigmas 2; achenes 2-sided. 7

7. One or more spikes with male flowers above female flowers . . . . Group G
7. One or more spikes with male flowers at the base. 8

8. Perigynium with rounded edges, barely thin-edged, the body filled or nearly filled by the achene . . Group H
8. Margins of perigynia wing-margined or obviusly thin-edged, not filled by the achene . . . . .Group I

Group A. Plants unisexual (dioecious)

1. Spike solitary 2
1. Spikes >1, sometimes densely clustered 3

2. Spike 6–15 mm long; perigynia glabrous; conspicuously veined. . . . .*C. dioica*
2. Spike 1–4 cm long; perigynia pilose; veins inconspicuous. . . . . *C. scirpoidea*

3. Perigynia ca. 2 mm long; the beak <0.5 mm long . . . . *C. simulata*
3. Perigynia 3–4.5 mm long; the beak 1–2 mm long 4

4. Inflorescence ≤1 cm wide . . . . *C. praegracilis*
4. Inflorescence >1 cm wide . . . . *C. douglasii*

Group B. Stems with a solitary, terminal spike

1. Largest leaf blades ≥1 mm wide 2
1. Leaf blades <1 mm wide 7

2. Perigynia 5–7 mm long, usually 1 or 2 per spike . . . . *C. geyeri*
2. Perigynia 2–4 mm long, usually >3 per spike (except *C. obtusata*) 3

3. Perigynia pilose . . . . *C. scirpoidea*
3. Perigynia glabrous 4

4. Lower perigynia spreading or reflexed at maturity . . . . *C. nigricans*
4. Perigynia erect to ascending 5

5. Perigynium suborbicular, noticeably larger than the achene . . . . *C. capitata*
5. Perigynium ovoid to ellipsoid, nearly filled by the achene 6

6. Perigynia green below, brown above, not strongly ribbed; leaves often curled . . . . *C. rupestris*
6. Perigynia purplish, strongly ribbed; leaves not curled . . . . *C. obtusata*

7. Plants densely ceaspitose; stems arising among old leaf bases 8
7. Plants rhizomatous; sometimes forming tufts but not dense clumps with old leaf bases 11

8. Perigynia 2–2.5 mm long, pedicellate; the lower ones spreading to reflexed at maturity . . . . . . *C. pyrenaica*
8. Perigynia 2.5–4 mm long, erect to ascending, not noticeably pedicellate .9

9. Perigynia thin-edged, not filled by the achene; male portion of spike <5 mm long . . . . *C. nardina*
9. Perigynia with rounded margins, nearly filled by the achene; male portion of spike >5 mm long 10

10. Perigynia puberulent above with a beak ≤0.5 mm long; plants usually of lower elevations . . . . . . *C. filifolia*
10. Perigynia glabrate with a beak ≥0.5 mm long; plants usually alpine . . . . *C. elynoides*

11. Perigynia green, rounded on top without a beak . . . . *C. leptalea*
11. Perigynia brown to purple with a beak ca. 0.5 mm long 12

12. Perigynia suborbicular . . . . *C. capitata*
12. Perigynia elliptic to ovoid 13

13. Stigmas 2; achene 2-sided . . . .*C. dioica*
13. Stigmas 3; achene 3-sided 14

14. Perigynia 3–4.5 mm long, not filled by the achene; the beak ≤0.5 mm long . . . .*C. engelmannii*
14. Perigynia 2–3 mm long, filled by the achene; the beak ≥0.5 mm long . . . . *C. obtusata*

Group C. Terminal spike male, narrower than female spikes; achene 2-sided

1. Perigynia 3–4.5 mm long 2
1. Perigynia 2–3 mm long 3

2. Female scales ovate, unawned; perigynia 3–4.5 mm long . . . . *C. saxatilis*
2. Female scales lanceolate, awned; perigynia 2.5–3.5 mm long . . . .*C. nebrascensis*

3. Perigynia yellow to green; male spike 3–10 mm long . . . .*C. aurea*
3. Perigynia green to brown; longest male spike >10 mm long 4

4. Perigynia veinless 5
4. Perigynia with few to many raised veins on each face 7

5. Perigynia inflated, plump, coppery; lower female scales just longer than the perigynia . . . . *C. aperta*
5. Perigynia flattened; female scales shorter or equal to the perigynia 6

6. Lowest bract as long or longer than the inflorescence . . . .*C. aquatilis*
6. Lowest bract much shorter than the inflorescence . . . . *C. scopulorum*

7. Plants rhizomatous; leaf blades 3–10 mm wide; female scales awn-tipped . . . .*C. nebrascensis*
7. Plants strongly caespitose; leaf blades 1–2 mm wide; scales not awned . . . .*C. kelloggii*

Group D. Terminal spike male, narrower than female; achene 3-sided; beak >0.5 mm long

1. Body of perigynia pubescent 2
1. Body of perigynia glabrous (rarely ciliate on upper half) 5

2. Uppermost male spike ≥2 cm long; female spikes 1–4 cm long. 3
2. Uppermost male spike ≤2 cm long; female spikes <15 mm long 4

3. Leaves 1–2 mm wide, inrolled; organic soils . . . . *C. lasiocarpa*
3. Leaves 2–5 mm wide, flat; inorganic soils . . . .*C. pellita*

4. Bract subtending lowest spike shorter than inflorescence; spikes not borne at stem base . . . . *C. inops*
4. Lowest bract longer than the inflorescence; spikes often borne at stem base. . . . . *C. rossii*

5. Perigynia beaks ≤1 mm long 6
5. Perigynia beaks >1 mm long 14

6. Perigynia (including beak) 2–3.5 mm long 7
6. Perigynia >3.5 mm long 10

7. Bract subtending the lowest spike shorter than the inflorescence . . . . *C. capillaris*
7. Lowest bract mostly longer than the terminal male spike. 8

8. Uppermost female spike overlapping the terminal male spike . . . .*C. viridula*
8. Uppermost female spike well-separated from terminal male spike 9

9. Plants strongly rhizomatous, turf-forming; montane. . . . . *C. saxatilis*
9. Plants short-rhizomatous, caespitose; plains, valleys . . . . *C. diluta*

10. Upper margins of perigynia ciliate . . . . *C. petricosa*
10. Perigynia completely glabrous 11

11. Male spike(s) much narrower than the female spikes . . . . *C. saxatilis*
11. Male spike(s) similar to the female or biexual spikes below. 12

12. Most female scales with a prominent pale midvein . . . . *C. spectabilis*
12. Female scales solid blackish or with an indistinct midvein. 13

13. Female spikes ascending; basal leaf blades present; perigynia broadly ovate . . . . *C. paysonis*
13. Lowest female spike spreading or arching; basal blades absent; perigynia narrowly ovate . . *C. podocarpa*

14. Beak of perigynia ≥3 mm long 15
14. Beak of perigynia <3 mm long 20

15. Beak of perigynia bidentate with spreading tips ca. as long as the tubular portion
   (sometimes broken off). 16
15. Perigynia beak tips shorter than the tubular portion and little spreading 18

16. Upper leaf sheaths pilose. . . . . *C. atherodes*
16. Upper leaf sheaths glabrous or nearly so. 17

17. Female spikes on long spreading or drooping peduncles . . . . *C. comosa*
17. Female spikes ascending. . . . . *C. laeviconica*

18. Perigynia abruptly narrowed to the beak which is longer than the body . . . .*C. sprengelii*
18. Perigynia gradually narrowed to a beak as long or shorter than the body 19

19. Perigynia ascending to erect . . . . *C. vesicaria*
19. Lower perigynia spreading to reflexed . . . . *C. retrorsa*

20. Leaf-like bract subtending lowest spike shorter than the inflorescence . . . . *C. luzulina*
20. Lowest bract exceeding the inflorescence 21

21. Terminal male spike <2 cm long. . . . . *C. flava*
21. Terminal male spike on most plants >2 cm long. 22

22. Lowest female spike spreading to nodding . . . .*C. hystericina*
22. Lowest female spike ascending 23

23. Body of perigynia (excluding beak) ≤3 mm long. . . . .*C. amplifolia*
23. Body of perigynia >3 mm long 24

24. Middle perigynia spreading nearly perpedicular to axis of the spike 25
24. Perigynia ascending to nearly erect 26

25. Leaf blades 4–10 mm wide, yellow-green, flat or V-shaped in cross-section. . . . . *C. utriculata*
25. Leaf blades rolled, 2–4 mm wide, glaucous, bluish-green . . . . *C. rostrata*

26. Beaks of perigynia ca. 1 mm long . . . .*C. lacustris*
26. Perigynia beaks 2–3 mm long . . . . *C. vesicaria*

Group E. Terminal spike male, narrower than female; achene 3-sided; beak ≤0.5 mm long

1. Perigynia pubescent. 2
1. Perigynia glabrous 7

2. Longest female spike >15 mm long 3
2. Longest spike ≤15 mm long 4

3. Leaves 1–2 mm wide, inrolled; organic soils . . . . *C. lasiocarpa*
3. Leaves 2–5 mm wide, flat; inorganic soils . . . .*C. pellita*

4. Bract subtending lowest spike inconspicuous, sheathing 5
4. Lowest bract obvious but not sheathing. 6

5. Terminal male spike 3–6 mm long . . . . *C. concinna*
5. Terminal male spike 10–18 mm long . . . .*C. concinnoides*

6. Bract subtending lowest spike shorter than inflorescence; spikes not borne at stem base . . . . *C. inops*
6. Lowest bract longer than the inflorescence; spikes often borne at stem base. . . . . *C. rossii*

7. Perigynia (including beak) <3.5 mm long. 8
7. Perigynia >3.5 mm long 22

8. Lowest female spike ascending to erect. 9
8. Lowest female spike spreading or arched 19

9. Perigynia, leaves and stems bluish-green (glaucous) . . . . *C. livida*
9. Perigynia and foliage not glaucous. 10

10. Female scales awned. 11
10. Female scales without awns. 12

11. Perigynia <1.5 mm wide; the beak absent . . . . *C. pallescens*
11. Perigynia 1.5–2 mm wide; beak 0.2–0.5 mm long . . . . *C. torreyi*

12. Bract subtending lowest spike exceeding the inflorescence 13
12. Lowest bract not longer than the inflorescence 14

13. Uppermost female spike(s) overlapping the terminal male spike. . . . .*C. viridula*
13. Uppermost female spike separate from the terminal male spike . . . . *C. diluta*

14. Perigynia green to tan 15
14. Perigynia purplish to black, at least above. 17

15. Perigynia ca. 2 mm long; spikes few-flowered . . . .*C. eburnea*
15. Perigynia 2.5–3.5 mm long; spikes many-flowered 16

16. Female scales tan; spikes well-separated . . . . *C. crawei*
16. Female scales blackish; spikes overlapping. . . . .*C. raynoldsii*

17. Perigynia 2.5–4 mm long . . . . *C. paysonis*
17. Perigynia 1.5–2.5 mm long. 18

18. Male and female spikes 3–5 mm long . . . . *C. glacialis*
18. Male spikes 1–2 cm long; female spikes 5–20 mm long . . . . *C. parryana*

19. Leaves ≤0.5 mm wide; stems weak . . . .*C. eburnea*
19. Leaves ≥1 mm wide 20

20. Plants caespitose without long rhizomes; terminal male spike ca. 5 mm long. . . . . *C. capillaris*
20. Plants rhizomatous; terminal male spike usually >5 mm long 21

21. Terminal male spike 10–35 mm long; female spikes sometimes with male flowers at the tip . . . . *C. limosa*
21. Terminal male spike 4–12 mm long; female spikes sometimes with basal male flowers . . . .*C. magellanica*

22. Lowest female spike spreading or arched 23
22. Lowest female spike ascending 25

23. Terminal male spike <3 mm wide . . . . *C. limosa*
23. Terminal male spike ≥3 mm wide 24

24. Female scales with a prominent pale midvein; leaves 3–6 mm wide . . . . *C. spectabilis*
24. Female scales dark; leaves 1–3 mm wide . . . . *C. podocarpa*

25. Leaves ca. 1 mm wide . . . . *C. petricosa*
25. Leaves 2–8 mm wide 26

26. Perigynia inflated, plump . . . . *C. raynoldsii*
26. Perigynia wrinkled or partly flattened 27

27. Lower female scales with a prominent pale midvein; basal leaves often bladeless;
     perigynia often papillose above . . . . *C. spectabilis*
27. Midvein of female scales obscure or absent; basal leaf blades present; perigynia not
     papillose . . . . *C. paysonis*

Group F. Spikes bisexual or female, all similar; achene 3-sided

1. Some female scales leaf-like and as long as the spike 2
1. Female scales not leaf-like and not much longer than the spike 4

2. Beak of perigynia 2–3 mm long . . . . *C. backii*
2. Beak of perigynia 0.5–1.5 mm long 3

3. Leaves dark green with narrow, white margins . . . . *C. saximontana*
3. Leaves yellow-green without white margins . . . . *C. cordillerana*

4. Beak of perigynia <0.5 mm long 5
4. Beak of perigynia 0.5–1.5 mm long 15

5. Female scales awned . . . . *C. buxbaumii*
5. Female scales without awns 6

6. Spikes sessile, congested in a globose to ovoid cluster . . . . *C. albonigra*
6. Spikes separated or in a loose cluster; at least the lowest spike pedunculate 7

7. Largest spike ≥10 mm long 8
7. Spikes ≤10 mm long 13

8. Lowest spikes nodding, >15 mm long . . . . *C. mertensii*
8. Lowest spikes ascensing, often <15 mm long 9

9. Perigynia 2–3 mm long 10
9. Perigynia 3–4.5 mm long 11

10. Terminal spike usually 1/3 longer and wider than the lower spikes . . . . *C. idahoa*
10. Terminal spike ca. as long and wide as the lower ones . . . . *C. parryana*

11. Perigynia smooth . . . . *C. epapillosa*
11. Perigynia papillose (use 20X to see papillae) 12

12. Female scales as long or longer than the perigynia . . . . *C. chalciolepis*
12. Female scales as long or shorter than but nearly as broad as the perigynia . . . . *C. atrosquama*

13. Perigynia green, sometimes bronze tinged; scales barely hyaline-margined . . . . *C. media*
13. Perigynia bronze to dark brown; scales with apparent white margins 14

14. Spikes clustered; perigynia solid-colored . . . . *C. norvegica*
14. Lowest spike with a long peduncle compared to the upper; perigynia spotted below . . . . *C. stevenii*

15. Perigynia, including beak, ≥4.5 mm long, ciliate above 16
15. Perigynia <4.5 mm long, glabrous 17

16. Leaves ca. 1 mm wide . . . . *C. petricosa*
16. Leaves 3–8 mm wide . . . . *C. luzulina*

17. Perigynia ≤3 mm long . . . . *C. idahoa*
17. Perigynia ≥3 mm long 18

18. Beaks of perigynia 1–1.5 mm long 19
18. Beaks of perigynia 0.5–1 mm long 20

19. Leaves 1–4 mm wide; stems 5–20 cm . . . . .C. fuliginosa
19. Leaves 3–8 mm wide; stems 15–70 cm . . . . C. luzulina

20. Perigynia smooth 21
20. Perigynia papillose (use 20X to see papillae). 22

21. Spikes loosely clustered; perigynia broadly elliptic to obovate, coppery with green margins . C. epapillosa
21. Spikes tightly clustered; perigynia circular, purplish with green edges. . . . . C. pelocarpa

22. Spikes tightly clustered. . . . . C. nelsonii
22. Lowest spike obviously pedunculate 23

23. Female scales as long or longer than the perigynia. . . . . C. chalciolepis
23. Female scales as long or shorter than but nearly as broad as the perigynia . . . . .C. atrosquama

Group G. Achene 2-sided; spikes all similar, bisexual or female; male flowers above

1. Spikes with 1 to 3 perigynia each. . . . . C. disperma
1. Some spikes with more than 3 perigynia 2

2. Inflorescence short-branched below (branches may look like long spikes) 3
2. Spikes all attached directly to the main inflorescence axis. 7

3. Hyaline upper portion of leaf sheaths cross-corrugate. 4
3. Leaf sheaths not cross-corrugate. 5

4. Perigynia 4–5 mm long; beak ca. as long as the body. . . . . C. stipata
4. Perigynia 2.5–3.5 mm long; beak shorter than the body . . . . C. vulpinoidea

5. Hyaline upper portion of leaf sheaths with inconspicuous red dots only . . . . C. diandra
5. Upper leaf sheaths coppery blotched as well as red-dotted. 6

6. Leaves 3–5 mm wide . . . . C. cusickii
6. Leaves 1–3 mm wide . . . . .C. prairea

7. Plants caespitose; stems strongly tufted 8
7. Plants rhizomatous or stoloniferous; stems usually 1 or few arising together 15

8. Largest leaves > 3 mm wide; female scales whitish. . . . . C. gravida
8. Leaves mostly ≤3 mm wide; female scales mostly brownish 9

9. Stems <15 cm; plants alpine . . . . C. maritima
9. Stems ≥50 cm; lower elevations. 10

10. Hyaline upper portion of leaf sheaths cross-corrugate. . . . . C. neurophora
10. Leaf sheaths not cross-corrugate. 11

11. Perigynia beak 0.5–1 mm long, <1/3 as long as body . . . . C. vallicola
11. Perigynia beak 1–1.5 mm long, >1/3 as long as body 12

12. Plants of upland habitats (grasslands). . . . . C. hoodii
12. Plants of wet meadows, streamsides, fens 13

13. Inflorescence 1 to 3 times as long as wide. . . . . C. jonesii
13. Inflorescence >3 times as long as wide 14

14. Hyaline upper portion of leaf sheaths with small red dots . . . . C. diandra
14. Upper leaf sheaths coppery blotched on ventral surface . . . . .C. prairea

15. Beaks of perigynia 1–3 mm long 16
15. Beaks of perigynia mostly ≤1 mm long. 18

16. Perigynia mostly ≥4.5 mm long, somewhat longer than the scales .C. siccata
16. Perigynia 3–4.5 mm long, concealed by the scales 17

17. Some spikes usually bisexual; inflorescence ≤1 cm wide . . . . C. praegracilis
17. Spikes usually all unisexual; inflorescence >1 cm wide . . . . C. douglasii

18. Perigynia ca. 2 mm long. . . . . .C. simulata
18. Perigynia 2.5–4 mm long 19

19. Stems ≥40 cm; leaves 2–5 mm wide . . . . C. sartwellii
19. Stems 2–25 cm; leaves 1–3 mm wide 20

20. Plants of *Sphagnum* fens with prostrate stems (stolons) on the surface . . . . *C. chordorrhiza*
20. Rhizomatous plants of uplands 21

21. Perigynium beak weakly serrulate; plants of low elevations. . . . . *C. eleocharis*
21. Perigynium beak entire; plants alpine. . . . . *C. maritima*

Group H. Achene 2-sided; spikes all similar, female or bisexual; male flowers below; perigynium with rounded edges, usually filled by the achene (not in *C. plectocarpa*)

1. Beak of perigynia >1 mm long 2
1. Beak of perigynia ≤1 mm long 4

2. Perigynia spreading at maturity; spikes appearing star-like . . . . *C. echinata*
2. Perigynia ascending to erect; spikes not star-like 3

3. Beak of perigynia ca. 1 mm long; spikes strongly overlapping. . . . . *C. arcta*
3. Beak of perigynia 1.5–2 mm long; lower spikes well-separated. . . . . *C. deweyana*

4. Beak of perigynia <0.5 mm long. 5
4. Beak of perigynia 0.5–1 mm long. 10

5. Perigynia dark brown . . . .*C. illota*
5. Perigynia green to tan 6

6. Achene ca. 1/2 as large as the perigynium; upper subalpine, alpine . . . . *C. plectocarpa*
6. Perigynium filled by the achene or nearly so; lower elevations 7

7. Female scales medium brown; subalpine to alpine . . . . *C. praeceptorum*
7. Female scales whitish to pale tan; mostly montane 8

8. Perigynia 3–3.5 mm long . . . . *C. tenuiflora*
8. Perigynia 2–2.5 mm long 9

9. Most spikes with >10 perigynia; perigynia 2–3 mm long . . . .*C. canescens*
9. Most spikes with 5–10 perigynia; perigynia 2–2.5 mm long. . . . . *C. brunnescens*

10. Spikes congested in a globose to conical head 11
10. Spikes well-separated in the inflorescence, at least below 14

11. Beak tip of perigynia serrulate . . . . *C. arcta*
11. Beak tips entire-margined. 12

12. Perigynia <1 mm wide; the beak serrulate below. . . . . *C. leporinella*
12. Perigynia often ≥1 mm wide. 13

13. Perigynia narrowly elliptic to oblanceolate, erect . . . .*C. lachenalii*
13. Perigynia lanceolate, ascending to spreading . . . .*C. illota*

14. Perigynia spreading at maturity; spikes appearing star-like . . . .*C. interior*
14. Perigynia ascending to erect; spikes not star-like 15

15. Perigynia 2–2.5 mm long; the beak indistinct ca. 0.5 mm long . . . . *C. brunnescens*
15. Perigynia 2.5–3.5 with a distinct beak >0.5 mm long . . . . *C. laeviculmis*

Group I. Achene 2-sided; spikes all similar, bisexual or female; male flowers below; perigynium with winged or thin-edged margins

1. Bracts subtending lowest spike(s) leaf-like, greatly exceeding the inflorescences 2
1. Bract subtending the lowest spike shorter than the inflorescence, often inconspicuous 3

2. Beaks of perigynia 1–2 mm long; bract subtending lowest spike 1 to 5 times as long
as the inflorescence . . . . *C. athrostachya*
2. Perigynia beaks 3–5 mm long; lowest bracts >5 times as long as the inflorescence. . . . . *C. sychnocephala*

3. Margin of perigynia entire, lacking a winged or thin-edged margin . . . .*C. illota*
3. Perigynia margin serrulate above, margin winged or thin-edged. 4

4. Perigynia 2–3 mm wide (measure several) 5
4. Perigynia 0.5–2 mm wide. 12

5. Perigynia 5.5–8 mm long . . . . *C. petasata*
5. Most perigynia <6 mm long 6

6. Body of perigynia circular or nearly so 7
6. Body of perigynia lanceolate to ovate. 8

7. Perigynia wing margins crinkled; plants upper montane to subalpine . . . . *C. straminiformis*
7. Perigynia not crinkle-margined; plants mainly of plains and valleys . . . . *C. brevior*

8. Terminal 0.5 mm of the bidentate perigynium beak with entire margins (sometimes broken off) 9
8. Perigynium beak serrulate to the tip or nearly so. 10

9. Beaks of perigynia ca. 1 mm long . . . . *C. macloviana*
9. Perigynia beaks 1.5–2.5 mm long . . . . *C. haydeniana*

10. Inflorescence elongate, often bent; lower spikes well-separated. . . . .*C. foenea*
10. Inflorescence congested in a capitate to ovoid head 11

11. Beaks of perigynia 2–3 mm long . . . .*C. multicostata*
11. Perigynium beaks <2 mm long . . . .*C. tahoensis*

12. Perigynia 0.5-1.5 mm wide. 13
12. Perigynia ≥1.5 mm wide. 21

13. Perigynium body narrowly lanceolate. 14
13. Perigynium body lanceolate to ovate 17

14. Perigynia beaks ca. 1 mm long, only the upper half exposed above the female scales . . . . . *C. leporinella*
14. Perigynia beaks 1.5–2 mm long, most of the beaks exposed 15

15. Perigynia 4–5 mm long. . . . .*C. stenoptila*
15. Perigynia 3–4 mm long. 16

16. Bract of lowest spike inconspicuous; tip of perigynia beak weakly serrulate . . . .*C. crawfordii*
16. Lowest bract leaf-like; beak tip entire. . . . . *C. athrostachya*

17. Beak of perigynia serrulate to the tip . . . . *C. bebbii*
17. At least terminal 0.4 mm of beak entire 18

18. Outer surface of perigynia convex, the margin thin-edged but not conspicuously
    wing-margined . . . . *C. pachystachya*
18. Perigynia flat except where distended by the achene; wing margin almost 1/2 the width of the achene. .19

19. Beaks of perigynia 1.5–2.5 mm long . . . . *C. haydeniana*
19. Beaks of perigynia 1–1.5 mm long. 20

20. Perigynia at least partly a shiny bronze color; female scales bronze often
    with a conspicuous hyaline margin . . . . *C. macloviana*
20. Female scales brown with an inconspicuous hyaline margin; perigynia green to brown,
    not shiny. . . . .*C. microptera*

21. Terminal 0.4 mm of beak entire (sometimes broken off) 22
21. Beak of perigynia serrulate to the tip 25

22. Perigynia beaks indistinct, <1 mm long . . . . *C. phaeocephala*
22. Perigynia beaks distinct, ≥1 mm long. 23

23. Perigynia beaks 1.5–2.5 mm long; spikes overlapping in an elongate, often
    bent inflorescence . . . . *C. praticola*
23. Perigynia beaks 1–1.5 mm long; spikes usually congested in an ovoid head 24

24. Outer surface of perigynia convex, the margin thin-edged but not conspicuously
    wing-margined . . . . *C. pachystachya*
24. Perigynia flat except where distended by the achene; wing margin almost
    1/2 the width of the achene . . . . *C. macloviana*

25. Female scales concealing the perigynia or nearly so. 26
25. Female scales shorter than the perigynia and exposing ca. the upper half of the beak 29

26. Inflorescence somewhat nodding; lower spikes well-separated. . . . .*C. foenea*
26. Inflorescence congested into a globose or ovoid head 27

27. Female scales bronze with a wide hyaline margin. . . . .*C. tahoensis*
27. Female scales tan or lighter. 28

28. Lower surfaces of perigynia with conspicuous veins longer than the achene; male
    scales brown. . . . .*C. multicostata*
28. Veins on lower perigynia surfaces inconspicuous or absent; female scales whitish . . . . *C. xerantica*

29. Inflorescence open; spikes well-separated . . . . *C. tenera*
29. Spikes congested, at least above. 30

30. Perigynia lanceolate, 4–5 mm long, >3 times as long as wide. . . . . *C. scoparia*
30. Perigynia ovate, 3–4 mm long; ≤3 times as long as wide. . . . . *C. bebbii*

Carex albonigra Mack. Caespitose. **Stems** erect, 8–25 cm. **Leaves** basal and lower-cauline; blades 2–5 mm wide. **Inflorescence** of 2 to 4 short-pedunculate, erect spikes forming a tight cluster; lowest bract inconspicuous. **Spikes** 8–15 mm long, terminal largest; male flowers basal. **Perigynia** erect, purplish-black, elliptic, 2–4 mm long, including the short beak, scabrous on the margins; stigmas 3. **Female scales** dark brown to black with light margins, as long as but narrower than the perigynia. **Achene** 3-sided, partly filling the perigynium. Dry to moist turf, fellfields; alpine. AK to NT south to CA, AZ, NM. See *C. atrosquama*. (p.616)

Carex amplifolia Boott Rhizomatous. **Stems** erect, 50–100 cm, reddish at the base, tufted. **Leaves** basal and cauline; sheaths hispidulous; blades 1–2 cm wide. **Inflorescence** of 4 to 8 overlapping, pedunculate spikes; the lowest bract greater than the inflorescence. **Spikes** ascending, 3–10 cm long, linear; the uppermost male, otherwise female. **Perigynia** ascending, green to brown, ovoid, glabrous, 3–4 mm long including the 1–2 mm long beak. **Female scales** acute to mucronate, reddish-brown with a hyaline tip and a light midstripe, longer or shorter than the perigynia. **Achene** 3-sided, filling the perigynium. Cedar forests along rivers; valleys. BC south to CA, ID, MT; collected in Sanders Co.
This species superficially resembles *Carex scopulorum* var. *prionophylla*, but the perigynia are distinctive.

Carex aperta Boott Short-rhizomatous. **Stems** erect, 25–60 cm, loosely tufted. **Leaves** basal and cauline; blades 2–5 mm wide. **Inflorescence** of 3 to 5 well-separated spikes; the lowest bract equal to or smaller than the inflorescence. **Spikes** ascending; the uppermost 1 or 2 male; the lower female, 1–3 cm long, short-pedunculate, cylindrical. **Perigynia** ascending, coppery, suborbicular to obovate, veinless, glabrous, plump, 2–3 mm long with a tiny beak. **Female scales** acute to acuminate, dark brown with a pale midstripe and margins, longer but narrower than the perigynia. **Achene** 2-sided, partly filling the perigynium. Marshes, wet meadows, shores of ponds, lakes, streams; valleys, montane. BC, AB, WA, OR, ID. MT.
The inflated, coppery perigynia separate this from *Carex aquatilis, C. nebrascensis* and *C. kelloggii.*

Carex aquatilis Wahlenb. Rhizomatous. **Stems** erect, 30–100 cm, often tufted. **Leaves** basal and cauline; blades 2–6 mm wide. **Inflorescence** of 3 to 10 well-separated spikes; the lowest bract greater than the inflorescence. **Spikes** erect; the uppermost 1 to 3 male; the lower female or androgynous, subsessile to pedunculate below, 12–60 mm long, cylindric. **Perigynia** ascending, green to tan, obovate, veinless, glabrous, 2–3 mm long with a minute beak. **Female scales** lanceolate, acute, brown with a pale midstripe, equal to or smaller and narrower than the perigynia. **Achene** 2-sided, smaller than the perigynium. Marshes, wet meadows, fens, streambanks; plains, valleys, montane. Circumboreal south through temperate N. America. Our plants are var. *aquatilis.*
There are a few specimens with arching lower spikes that would be identified as var. *dives* (T.Holm) Kük., a Pacific slope form, but their occurrence appears random and not worthy of distinction. See *Carex nebrascensis, C. kelloggii.*

Carex arcta Boott Caespitose. **Stems** erect, 15–50 cm. **Leaves** basal and lower-cauline; blades 1–4 mm wide. **Inflorescence** of 7 to 15 sessile spikes in a tight conical cluster; lowest bracts short. **Spikes** 5–10 mm long, all similar; female flowers above; male below. **Perigynia** ascending, green, lanceolate, 2–3 mm long, including the broad, serrulate beak, ca. 1 mm long; stigmas 2. **Female scales** tan with hyaline margins and a green midvein, shorter and narrower than the perigynia. **Achene** 2-sided, filling the perigynium. Wetlands along streams, ponds, lakes, often with *C. athrostachya*; valleys, montane. YT to QC south to CA, ID, MT, MI, NY. (p.617)
Carex canescens has fewer spikes and perigynia with shorter beaks.

Carex atherodes Spreng. Rhizomes deep. **Stems** erect, 40–100 cm, tufted. **Leaves:** basal and cauline; blades 3–10 mm wide; sheaths pilose. **Inflorescence** of 4 to 10 well-separated spikes; the lowest bract equal to or greater than the inflorescence. **Spikes** ascending to erect; the uppermost 2 to 4 male; the lower female, 2–7 cm long, pedunculate, cylindric. **Perigynia** ascending, green to tan, lanceolate, veiny, glabrous, 7–10 mm long including an apically divided beak, 3–5 mm long; stigmas 3. **Female scales** brown, lanceolate, long-awned, narrower and as long as the perigynia. **Achene** 3-sided, smaller than the perigynium. Moist to wet meadows, marshes, riparian forests, often around lakes, ponds where inundated early but eventually drying; plains, valleys. Circumboreal south to CA, AZ, NM, MO, VA.

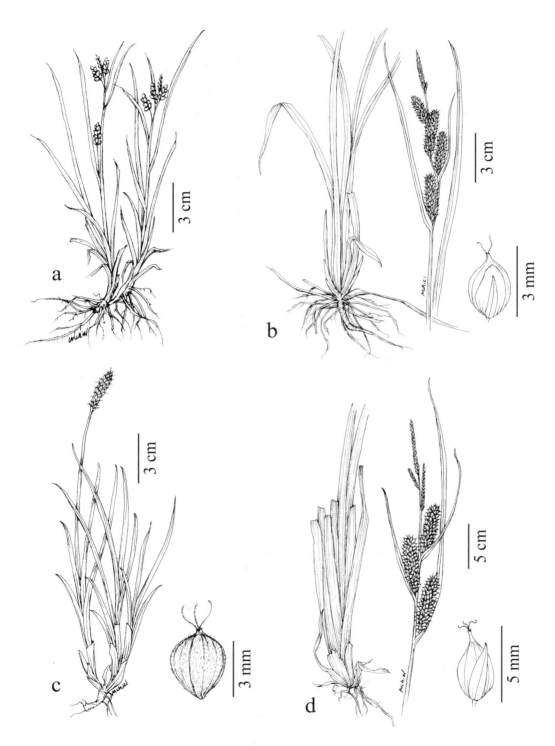

Plate 105. a. **Carex aurea**, b. **Carex kelloggii**, c. **Carex scirpoidea** ssp. **scirpoidea**, d. **Carex utriculata**

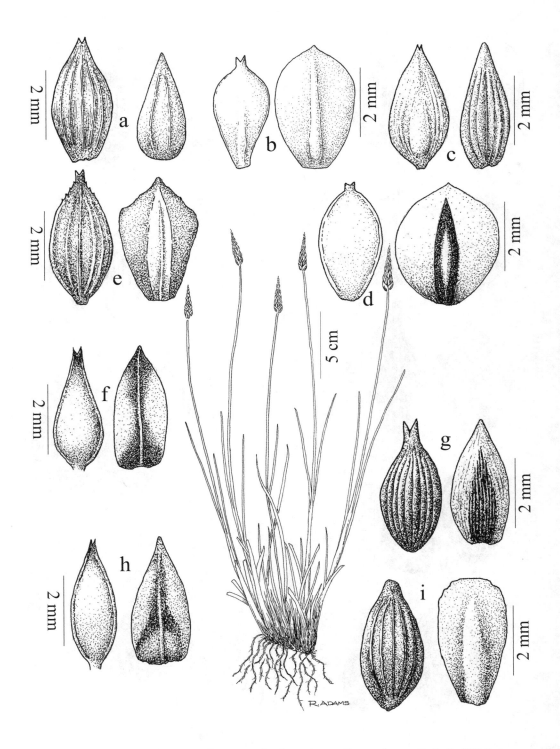

Plate 106. *Carex* **Group B. a.** *C. dioica*, **b.** *C. elynoides*, **c.** *C. engelmannii*, **d.** *C. filifolia*,
e. *C. nardina*, **f.** *C. nigricans*, **g.** *C. obtusata*, **h.** *C. pyrenaica*, **i.** *C. rupestris*

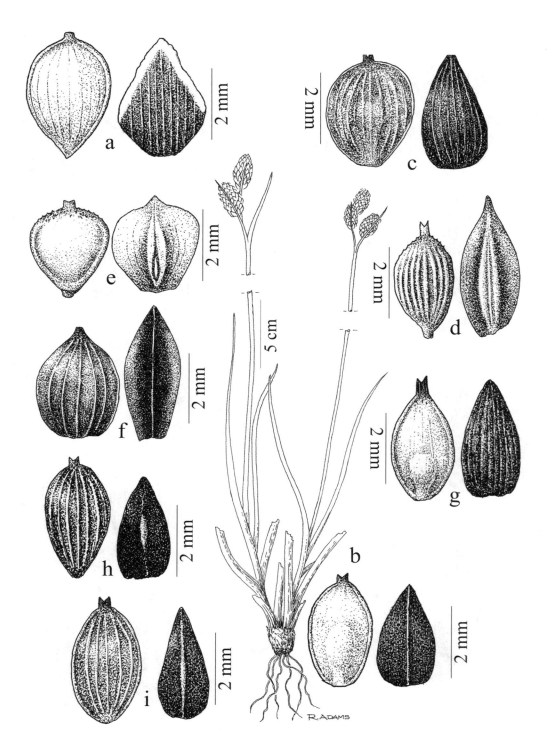

Plate 107.*Carex* Group E. **a.** *C. albonigra*, **b.** *C. atrosquamma*, **c.** *C. epapillosa*, **d.** *C. idahoa*, **e.** *C. parryana*, **f.** *C. paysonis*, **g.** *C. podocarpa*, **h.** *C. raynoldsii*, **i.** *C. spectabilis*

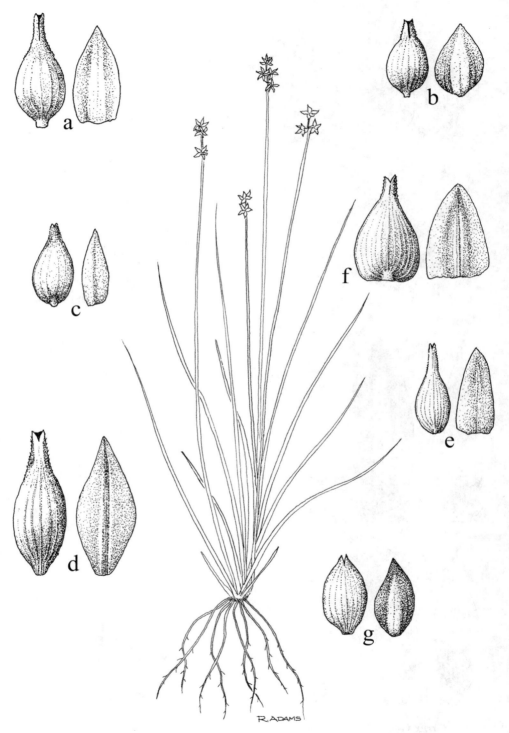

Plate 108. *Carex* **Group H**. **a.** *C. arcta*, **b.** *C. brunnescens*, **c.** *C. canescens*, **d.** *C. deweyana*,
e. *C. echinata*, **f.** *C. interior*, **g.** *C. praeceptorum*

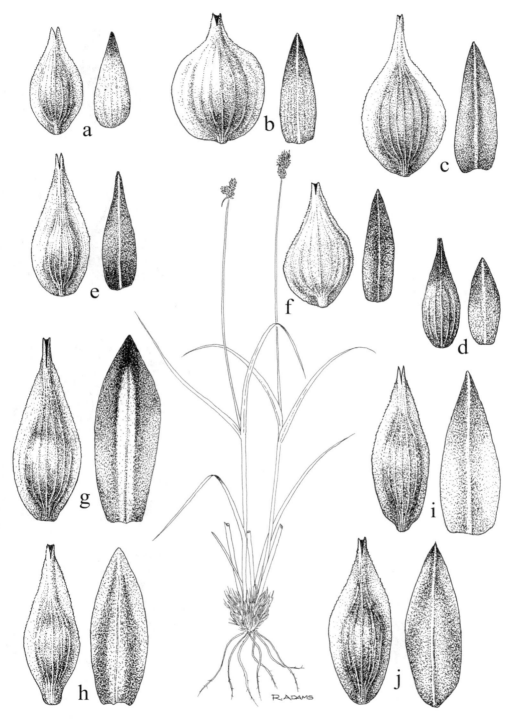

Plate 109. **Carex Group I**. **a.** *C. bebbii*, **b.** *C. brevior*, **c.** *C. haydeniana*, **d.** *C. illota*, **e.** *C. microptera*, **f.** *C. pachystachya*, **g.** *C. petasata*, **h.** *C. phaeocephala*, **i.** *C. praticola*, **j.** *C. tahoensis*

Carex athrostachya Olney Caespitose. **Stems** erect, 15–80 cm. **Leaves** basal and lower-cauline blades 1–4 mm wide. **Inflorescence** of 3 to 8 sessile spikes in a dense head; lowest bract spreading, linear, usually longer than the inflorescence. **Spikes** 5–9 mm long, all similar; female flowers above; male below. **Perigynia** erect, (narrowly) lanceolate, green to tan, wing- and serrulate-margined, 3–4 mm long including the indistinct 1–2 mm long beak, serrulate below, entire above; stigmas 2. **Female scales** lanceolate, acuminate, light brown with a pale midvein, as wide and just shorter than the perigynia. **Achene** 2-sided, much smaller than the perigynium. Wet or vernally wet soil of meadows, sparsely vegetated shores of streams, ponds; plains, valleys, montane. AK to MB south to CA, AZ, TX, SD.

The narrow perigynia with serrate edges and long bract of the inflorescence help distinguish this sedge. Most but not all plants have the long lower bract; those without resemble *C. crawfordii*. Plants may sometimes flower the first year.

Carex atrosquama Mack. [*C. atrata* L. var. *atrosquama* (Mack.) Cronquist] Caespitose with old leaves at the often reddish base. **Stems** erect, 10–45 cm. **Leaves** basal and lower-cauline; blades 1–4 mm wide. **Inflorescence** of 3 to 5 short-pedunculate, ascending to erect, loosely clustered spikes. **Spikes** 10–18 mm long, all similar; male flowers basal on uppermost spike; otherwise all female. **Perigynia** green to purplish, elliptic, 3–4 mm long, scabrous on the margins; beaks 0.3–0.5 mm long; stigmas 3. **Female scales** blackish throughout, as long or barely shorter than the perigynia but nearly as broad. **Achene** 3-sided, partly filling the perigynium. Moist to dry meadows, turf, fellfields; subalpine, lower alpine. AK to NT south to UT, CO. (p.616)

Carex atrosquama, *C. epapillosa* and *C.chalciolepis* are similar and have all been included within the Eurasian *C. atrata* L.[302]

Carex aurea Nutt. [*C. garberi* Fernald, *C. hassei* L.H.Bailey] Rhizomatous. **Stems** erect, 3–30 cm. **Leaves** basal and cauline; blades 1–4 mm wide. **Inflorescence** of 2 to 4 well-separated spikes; the lowest bract greater than the inflorescence. **Spikes** ascending or spreading; the uppermost male; the lower female, pedunculate, 5–20 mm long, narrowly cylindric. **Perigynia** spreading, white to green, becoming golden, obovoid, fleshy, papillose, 2–3 mm long, loosely aggregated; stigmas 2. **Female scales** ovate, sometimes awn-tipped, light brown with a pale midvein and hyaline margins, shorter but nearly as wide as the perigynia. **Achene** 2-sided, smaller than the perigynium. Moist meadows, turf, usually around streams, ponds, seeps, wetlands; plains, valleys to lower alpine. AK to NL south to CA, NM, TX, OH, NY. (p.614)

Long-stalked female spikes may arise from the base. *Carex hassei* and *C. garberi* are sometimes segregated from *C. aurea* based on the spike internode distance, perigynia color, presence of perigynia in the terminal spike and prominence of perigynium ribs (Hermann 1970, Ball and Reznicek 2002). However, these characters do not covary and are not correlated with habitat or geographic range in our material.[187]

Carex backii Boott Caespitose. **Stems** erect, 14–25 cm. **Leaves** basal and lower-cauline, some greater than the inflorescence; blades 1–5 mm wide, yellowish. **Inflorescence** of 1 to 4 well-separated spikes; lowest bract greater than the inflorescence. **Spikes** bisexual, similar, erect to ascending, pedunculate, 5–10 mm long not including the basal female scales; male flowers above, few, inconspicuous. **Perigynia** 2 to 5, ascending, green, obovoid, glabrous, loosely aggregated, 5–6 mm long, tapered to the smooth 2–3 mm long beak; stigmas 3. **Female scales** green, elongate and bract-like below, ovate, shorter above. **Achene** 3-sided, filling the perigynium. Grasslands, thickets; valleys, montane; known from Flathead and Stillwater cos. BC to NL south to CO, SD, IL, NY.

Carex backii, *C. cordillerana* and *C. saximontana* all have the distinctive leaf-like female scales and have been treated as a single species, but they differ by several morphological characters[349] and have geographic integrity in MT; the latter two species have smaller perigynia than *C. backii*.

Carex bebbii (L.H.Bailey) Olney ex Fernald Caespitose. **Stems** erect, 20–70 cm. **Leaves**: basal blades absent; cauline blades 1–3 mm wide. **Inflorescence** of 4 to 10 sessile spikes, aggregated into an ovoid to elongate head. **Spikes** 4–9 mm long, all similar; female flowers above; male below; lowest bract inconspicuous. **Perigynia** erect, glabrous, green to tan, ovate, flattened, wing-margined, 3–4 × 1–2 mm, gradually tapering to the indistinct beak, 1–2 mm long, serrulate to the tip; stigmas 2. **Female scales** tan-hyaline, sometimes with a green midvein, shorter and narrower than the perigynia. **Achene** 2-sided, smaller than the perigynium. Wet meadows, marshes, streambanks, shores of lakes, ponds; valleys. AK to NL south to OR, UT, CO, IA, IN, OH, NY. (p.618)

Carex crawfordii has narrower perigynia. See *C. brevior*, *C. scoparia*.

Carex brevior (Dewey) Mack. ex Lunnel Caespitose. **Stems** erect, 20–80 cm. **Leaves**: basal blades absent; cauline blades 2–3 mm wide. **Inflorescence** of 3–6 sessile spikes, overlapping or aggregated into a capitate to elongate head; lowest bract inconspicuous. **Spikes** 5–10 mm long, all similar; female flowers above; male below. **Perigynia** erect, glabrous, green to tan, orbicular, flattened, wing-margined, 3.5–5 × 2–3 mm including the distinct beak, 1.5–2.5 mm long, serrulate to the tip; stigmas 2. **Female scales** tan with a white to green midvein and hyaline margins, narrower and shorter than the perigynia. **Achene** 2-sided, smaller than the perigynium. Moist grasslands, meadows, woodlands; plains, valleys. BC to QC south to WA, AZ, NM, TX, MS, GA. (p.618)

The nearly orbicular perigynia helps separate this species from *Carex bebbii*.

Carex brunnescens (Pers.) Poir. Caespitose **Stems** erect, 20–50 cm. **Leaves** basal and cauline; blades 1–2 mm wide. **Inflorescence** of 4 to 8 sessile, well-separated spikes; lowest bract shorter than the inflorescence. **Spikes** 3–7 mm long, all similar; female flowers above; male below. **Perigynia** erect to ascending, green to tan, ovate, 2–2.5 mm long with a serrulate beak, ca. 0.5 mm long; stigmas 2. **Female scales** ovate, green to tan with a dark midvein and hyaline margins, shorter but nearly as wide as the perigynia. **Achene** 2-sided, filling the perigynium. Wet meadows, seeps, often in cedar forests; montane. Circumboreal south to OR, UT, WI, AL. Our plants are ssp. **brunnescens**. (p.617)

Carex canescens has more perigynia in larger spikes, and *C. laeviculmis* has perigynia with more distinct beaks. See *C. disperma*.

Carex buxbaumii Wahlenb. Rhizomatous. **Stems** erect, 25–80 cm. **Leaves:** basal bladeless; cauline blades 1–3 mm wide. **Inflorescence** of 2 to 5 short-pedunculate, erect, barely overlapping spikes; lowest bract ca. as long as the inflorescence. **Spikes** 1–2 cm long; uppermost bisexual with male flowers below; otherwise all female. **Perigynia** ascending to spreading, green, pappilose, elliptic, 3–4.5 mm long, beakless, somewhat inflated; stigmas 3. **Female scales** narrowly lanceolate, brown with a pale midvein, awned, as long but narrower than the perigynia. **Achene** 3-sided, smaller than the perigynium. Fens, wet meadows; valleys, montane. Circumboreal south to UT, CO, AR, SC.

Carex canescens L. Caespitose. **Stems** erect, 10–40 cm. **Leaves** basal and cauline; blades 1–3 mm wide. **Inflorescence** of 4 to 8 sessile, overlapping to clustered spikes; lowest bract short. **Spikes** 4–9 mm long, all similar; female flowers above, male below. **Perigynia** ascending, pale green to light tan, ovate, 2–3 mm long with an indistinct, serrulate beak <0.5 mm long; stigmas 2. **Female scales** ovate, whitish to green with a green midvein, nearly as long and wide as the perigynia. **Achene** 2-sided, filling the perigynium. Wet, usually organic soil of meadows, marshes, fens; montane, lower subalpine. Circumboreal south to CA, AZ, NM, IN, SC. Our plants are ssp. **canescens**. See *C. brunnescens*, *C. arcta*. (p.617)

Carex capillaris L. Caespitose. **Stems** erect, 2–40 cm. **Leaves** mainly basal; blades 1–3 mm wide. **Inflorescence** of 2 to 5 arching-pedunculate spikes; the lowest bract shorter than the inflorescence. **Spikes:** the uppermost male, ca. 5 mm long; the lower female, 4–12 mm long, cylindric. **Perigynia** spreading to ascending, green to tan, narrowly ovoid, glabrous, 2.5–3.5 mm long with a beak ca. 0.5 mm long; stigmas 3. **Female scales** pale with a darker midstripe, shorter and ca. as wide as the perigynia. **Achene** 3-sided, nearly filling the perigynium. Wet, usually organic soil of streambanks, wet meadows, fens, turf; montane to alpine. Circumboreal south to CA, UT, NM, SD, NY.

Alpine (var. *capillaris*) and the low-elevation (var. *major* Drejer) forms of *Carex capillaris* are sometimes recognized.[173]

Carex capitata L. Short-rhizomatous. **Stems** erect, 5–15 cm, tufted. **Leaves** mainly basal; blades ca. 1 mm wide, inrolled. **Inflorescence** a solitary spike; lowest bract absent. **Spike** ovoid, 5–10 mm long; male flowers above female. **Perigynia** ascending, glabrous, suborbicular, 3–3.5 mm long, tan to brown, with a distinct, entire beak ca. 0.5 mm long; stigmas 2. **Female scales** brown with hyaline margins, exposing only the tip of the perigynia. **Achene** 2-sided, smaller than the perigynium. Turf, rock ledges; alpine. Circumpolar south to CA, NV, CO, NH. Known from Carbon Co.

Carex chalciolepis T.Holm [*C. heteroneura* W.Boott var. *chalciolepis* (T.Holm) F.J.Herm., *C. atrata* L. var. *chalciolepis* (T.Holm) Kük.] Caespitose. **Stems** erect, 20–50 cm. **Leaves** basal and lower-cauline; blades 1–4 mm wide. **Inflorescence** of 2 to 6 short-pedunculate, erect spikes forming a loose cluster; lowest spike sometimes arching. **Spikes** 10–25 mm long, all similar; male flowers basal on uppermost spike; otherwise all female. **Perigynia** erect, purplish to green, elliptic, 4–4.5 mm long, scabrous on the upper margin, filled by the achene; beaks 0.3–0.5 mm long; stigmas 3. **Female scales** narrowly lanceolate, blackish with thin, hyaline margins, as long or longer but much narrower than the perigynia. **Achene** 3-sided, much smaller than the perigynium. Moist meadows; subalpine. MT, NV, UT, CO, AZ, NM. Known from Carbon and Ravalli cos.

Carex chordorrhiza Ehrh. ex L.f. Stoloniferous. **Stems** usually prostrate, clothed in old leaf bases, 10–25 cm. **Leaves** cauline; blades 1–2 mm wide, channeled. **Inflorescence** of 3 to 5 sessile spikes, densely aggregated, appearing like 1 ovoid head; lowest bract scale-like. **Spikes** 4–8 mm long, all similar; male flowers above, few female below. **Perigynia** spreading to ascending, brown, ovoid, 2.5–3.5 mm, with an entire beak 0.5 mm long; stigmas 2. **Female scales** brown with a pale midvein and margins, ca. as long as the perigynia. **Achene** 2-sided, filling the perigynium. *Sphagnum* fens; montane. Circumboreal south to OR, MT, IN and NY.

Carex comosa Boott Short-rhizomatous. **Stems** erect, 50–100 cm, tufted. **Leaves**: basal and cauline; blades 5–12 mm wide. **Inflorescence** of 5 to 9 overlapping, pedunculate spikes; the lowest bract equal to or longer than the inflorescence. **Spikes** unisexual; the uppermost 1 to 4 male, linear; the lower female, 3–7 cm long, cylindric, spreading to nodding. **Perigynia** spreading, pale green, lanceolate, veiny, glabrous, 6–9 mm long including a deeply divided beak, 3–5 mm long with long spreading tips; stigmas 3. **Female scales** tan, lanceolate, awned, shorter and narrower than the perigynia. **Achene** 3-sided, smaller than the inflated perigynium. Marshes; valleys. ON, QC south to TX, MS, SC, disjunct in WA to MT south to CA. Known from Flathead Co.

Carex concinna R.Br. Rhizomatous. **Stems** ascending to erect, 5–20 cm, often tufted. **Leaves** mainly basal; blades 1–3 mm wide. **Inflorescence** of 3 to 4 overlapping spikes; the lowest bract inconspicuous. **Spikes** short-pedunculate, ascending; the uppermost male, 3–6 mm long; the lower female, 4–8 mm long, few-flowered. **Perigynia** ascending to spreading, yellow-green, pubescent, obovoid, 2–3 mm long with a minute beak; stigmas 3. **Female scales** ovate, brown with a pale midvein and hyaline margins, shorter and narrower than the perigynia. **Achene** 3-sided, filling the perigynium. Wet, often calcareous soil of seeps, forests (often spruce), streambanks; montane. YT to NL south to OR, CO, MI.
   The more common *Carex concinnoides* has a larger male spike, sessile female spikes and occurs in drier habitats.

Carex concinnoides Mack. Rhizomatous. **Stems** erect, 6–40 cm, often tufted. **Leaves** basal and cauline; blades 2–5 mm wide. **Inflorescence** of 2 to 4 sessile, overlapping spikes; the lowest bract inconspicuous. **Spikes** ascending; the uppermost male, 10–18 mm long; the lower female, 5–10 mm long, few-flowered. **Perigynia** ascending, pale green, pubescent, obovoid, 2.5–3 mm long with a short beak; stigmas 4 (3–5). **Female scales** ovate, purplish-black with hyaline margins, ca. as large as the perigynia. **Achene** obovoid, filling the perigynium. Moist to dry, coniferous forest; valleys to lower subalpine. BC, AB south to CA, ID, MT.
   The type locality is in Flathead Co. See *C. concinna*.

Carex cordillerana Saarela & B.A.Ford **Stems** erect, 7–40 cm. **Leaves** basal and lower-cauline, some greater than the inflorescence, yellowish; blades 1–5 mm wide. **Inflorescence** of 1 to 4 well-separated spikes; lowest bract greater than the inflorescence. **Spikes** erect to ascending; similar, pedunculate, 5–10 mm long, not including the basal female scales; male flowers above, few, inconspicuous. **Perigynia** 3 to 5, ascending, green, obovoid, glabrous, loosely aggregated, 3–4.5 mm long, with a smooth beak 0.5–1.5 mm long; stigmas 3. **Female scales** green, elongate, bract-like below, ovate and shorter above. **Achene** 3-sided, filling the perigynium. Open forest along streams; valleys. BC, AB south to OR, UT, CO. See *C. backii*.

Carex crawei Dewey Rhizomatous. **Stems** erect, 5–20 cm. **Leaves** basal and cauline; blades 1–4 mm wide. **Inflorescence** of 2 to 5 well-separated spikes; the lowest bract shorter than the inflorescence. **Spikes** ascending; the uppermost male, 1–2 cm long; the lower female, 8–20 mm long, pedunculate, sometimes arising from the base. **Perigynia** spreading to ascending, green to tan, glabrous, ovoid, 2.5–3.5 mm long with an obscure beak; stigmas 3. **Female scales** ovate, tan with pale margins and a green midvein, shorter and narrower than the perigynia. **Achene** 3-sided, partly filling the perigynium. Wet, gravelly, calcareous soil of streambanks, shores; plains, valleys. BC to NL south to WA, UT, WY, OK, AL, GA.

Carex crawfordii Fernald Densely caespitose. **Stems** erect, 20–50 cm. **Leaves** basal and cauline; blades 1–4 mm wide. **Inflorescence** of 3 to 8 sessile spikes, aggregated into an elongate head; lowest bract inconspicuous. **Spikes** 5–9 mm long, all similar; female flowers above male. **Perigynia** erect, light brown with green margins, glabrous, narrowly lanceolate, narrowly wing-margined, 3–4 × 0.6–1.2 mm, tapered to the 1.5–2 mm long beak, weakly serrulate to the tip; stigmas 2. **Female scales** golden-brown with a green midvein, shorter and as wide as the perigynia. **Achene** 2-sided, nearly filling the perigynium. Moist meadows, forest openings, shores of streams, lakes; valleys, montane. AK to NL south to OR, ID, MT, MO, NY.
   Perigynia of *Carex petasata* and *C. scoparia* are larger; those of *C. bebbii* are more ovate; See *C. athrostachya*.

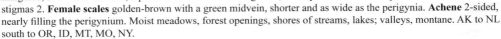

Carex cusickii Mack. ex Piper & Beattie Caespitose. **Stems** erect, 30–90 cm. **Leaves**: basal bladeless; cauline sheaths coppery opposite the blades that are 3–5 mm wide. **Inflorescence** of numerous, sessile spikes, congested on cylindric, ascending, overlapping branches; lowest bract usually inconspicuous. **Spikes** 5–8 mm long, all similar; male flowers above female or all female. **Perigynia** ascending to spreading, lanceolate, brown to green, glabrous, serrulate-margined, 2.5–3.5

mm, tapered to the flattened, grooved beak, 1–1.5 mm long; stigmas 2. **Female scales** hyaline-brown, lanceolate, hiding the perigynia. **Achene** 2-sided, filling the perigynium. Wet, organic soil of fens, marshes; valleys. BC south to CA, ID, WY.

Perigynia of *Carex cusickii* are greener, veinier, and more hidden by the scales than those of *C. diandra*.

Carex deweyana Schwein. [*C. bolanderi* Olney, *C. leptopoda* Mack.] Caespitose. **Stems**
ascending to erect, 15–70 cm. **Leaves** basal and lower-cauline; blades 2–5 mm wide.
Inflorescence of 2 to 6 sessile, well-separated to overlapping spikes; lowest bract usually shorter
than the inflorescence. **Spikes** 5–10 mm long, all similar; female flowers above male. **Perigynia**
ascending to spreading, green to tan, narrowly elliptic, 3.5–5 mm long including the 1.5–2 mm
long, serrulate beak; stigmas 2. **Female scales** ovate, hyaline with a green midvein, shorter and as wide as the perigynia.
Achene 2-sided, filling the perigynium. Moist to wet forest openings, thickets; valleys to lower subalpine. AK to NL
south to WA, CO, IL, NY. (p.617)

Carex laeviculmis has smaller perigynia. *Carex deweyana* is sometimes divided into three ecologically similar species
in our area, based mainly on continuous, overlapping characters.[187]

Carex diandra Schrank **Stems** erect, 15–75 cm. **Leaves**: basal bladeless; cauline sheaths white,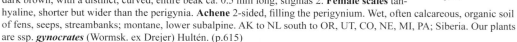
minutely red-dotted opposite the blades that are 1–3 mm wide. **Inflorescence**: spikes numerous,
sessile, congested, ovoid, clustered on overlapping branches; lowest bract inconspicuous. **Spikes**
3–7 mm long, all similar; male flowers above female. **Perigynia** ascending, tan to
brown, glabrous, serrulate-margined, 2.5–3 mm tapered to the ca. 1 mm long beak; stigmas 2.
Female scales tan, hyaline-margined, lanceolate, barely shorter and narrower than the perigynia. **Achene** 2-sided, filling
the perigynium. Often calcareous fens, marshes; valleys to montane. Circumboreal south to CA, UT, CO, IL, PA. See *C. cusickii*.

Carex diluta M.Bieb. Short-rhizomatous. **Stems** erect, 15–50 cm, tufted. **Leaves** mainly basal, yellowish; blades 2–5
mm wide. **Inflorescence** of 3 to 5 well-separated, pedunculate spikes; the lowest bract greater than the inflorescence.
Spikes: 1–2 cm long; the uppermost male, erect, 8–25 mm long; the lower female, ascending to spreading, cylindric.
Perigynia ascending to spreading, yellowish, narrowly obovoid, glabrous, veiny, 2.5–3.5 mm long; beak 0.4–0.7 mm
long. **Female scales** ovate, acute, scarious-margined, shorter than the perigynium. **Achene** 3-sided, smaller than the
perigynium. Moist alkaline meadows; valleys. Introduced to Madison Co.; native to Europe.

The exact determination of the MT specimen is not certain.[24]

Carex dioica L. [*C. gynocrates* Wormsk. ex Drejer] Long-rhizomatous. **Stems** erect to
ascending, 3–20 cm. **Leaves** mainly basal; blades ca. 0.5 mm wide. **Inflorescence** a solitary
bractless spike. **Spike** ovoid, 6–15 mm long; male flowers above female (sometimes all female).
Perigynia spreading to reflexed at maturity, glabrous, narrowly ovoid, ribbed, 2.5–4 mm long,
dark brown, with a distinct, curved, entire beak ca. 0.5 mm long; stigmas 2. **Female scales** tan-
hyaline, shorter but wider than the perigynia. **Achene** 2-sided, filling the perigynium. Wet, often calcareous, organic soil
of fens, seeps, streambanks; montane, lower subalpine. AK to NL south to OR, UT, CO, NE, MI, PA; Siberia. Our plants
are ssp. **gynocrates** (Wormsk. ex Drejer) Hultén. (p.615)

The female scales of *Carex nigricans* and *C. pyrenaica* fall before maturity.

Carex disperma Dewey Rhizomatous. **Stems** erect, 10–50 cm, somewhat tufted. **Leaves** basal
and lower-cauline; blades 1–3 mm wide. **Inflorescence** of 2 to 6 sessile, remote to overlapping
spikes, bractless. **Spikes** 2–5 mm long, all similar; few male flowers above the female. **Perigynia**
1 to 3 per spike, spreading to ascending, green to tan, red-veined, glabrous, elliptic, 2–3 mm long
with an obscure beak; stigmas 2. **Female scales** ovate, hyaline with a tan midvein, shorter and
narrower than the perigynia. **Achene** 2-sided, filling the perigynium. Wet, organic soil, often beneath shrubs, in swamps,
fens, wet coniferous forest; valleys, montane. Circumboreal south to CA, NM, IN, NJ.

The few-flowered, sessile spikes along with the moist, shady habitat help distinguish this sedge. *Carex leptalea* has a
similar habitat but only one spike; *C. brunnescens* has a beak 0.5-1 mm long.

Carex douglasii Boott Dioecious, rhizomatous. **Stems** erect, 5–30 cm. **Leaves** lower-cauline;
blades 1–2 mm wide. **Inflorescence** of 3 to 15 spikes; densely clustered in an ovoid head; lowest
bract usually shorter than the inflorescence. **Spikes** unisexual, sessile, ascending; female ovoid,
8–15 mm long; male lanceolate. **Perigynia** erect, lanceolate, green to tan, glabrous, 3.5–4.5 mm
long including the slender beak 1–2 mm long; stigmas 2. **Female scales** ovate, tan with pale
margins and midvein, hiding the perigynia. **Achene** 2-sided, partly filling the perigynium. Moist, often compacted or
disturbed soil of grasslands, meadows, stream or lake terraces; plains, valleys, montane. BC to SK south to CA, NM, KS.

Carex eburnea Boott Rhizomatous. **Stems** erect, 10–25 cm, tufted. **Leaves** basal and lower-cauline; blades <0.5 mm wide. **Inflorescence** of 3 to 5 remote spikes; lower bract bladeless. **Spikes** ascending or spreading, unisexual; the male terminal, ca. 5 mm long; the lower female, long-pedunculate, 3–5 mm long, few-flowered. **Perigynia** ascending to spreading, green becoming tan, narrowly obovoid, glabrous, ca. 2 mm long with a beak <0.5 mm long, loosely aggregated; stigmas 3. **Female scales** ovate, hyaline with a dark midstripe, shorter but as wide as the perigynia. **Achene** 3-sided, filling the perigynium. Moist to wet forest, woodlands; plains, valleys. AK to NL south to BC, MT, NE, AR, AL.

Carex echinata Murray [*C. muricata* L. misapplied] Caespitose. **Stems** erect, 12–40 cm. **Leaves** basal and cauline; blades 1–2 mm wide. **Inflorescence** of 2 to 5 sessile, well-separated spikes; lowest bract inconspicuous. **Spikes** ca. 5 mm long; the uppermost bisexual with a few male flowers below; the lower female. **Perigynia** spreading, brown, ovate, thin-edged, 2–3 mm long, tapered to the serrulate beak, 1–1.5 mm long; stigmas 2. **Female scales** ovate, hyaline-tan with a green midvein, shorter than the perigynia. **Achene** 2-sided, filling the perigynium. Fens, wet meadows; montane. BC to QC south to CA, NV, UT, CO, TN, NC. Our plants are ssp. *echinata*. (p.617)
    Carex interior has shorter beaks, but intermediate specimens do occur.

Carex eleocharis L.H.Bailey [*C. stenophylla* Wahlenb. var. *eleocharis* (L.H.Bailey) Breitung, *C. duriuscula* C.A.Mey.] Rhizomatous. **Stems** erect, 4–25 cm, loosely tufted. **Leaves:** basal bladeless; cauline blades 1–2 mm wide. **Inflorescence** of 3 to 8 spikes in a dense ovoid cluster; lowest bract inconspicuous. **Spikes** sessile, similar, ascending, ovoid; 4–7 mm long; male flowers above the few female. **Perigynia** ascending, ovate, brown, glabrous, 2.5–3.5 mm long tapered to an indistinct serrulate beak 0.5–1 mm long; stigmas 2. **Female scales** ovate, apiculate, dark brown with hyaline margins, hiding the perigynia. **Achene** 2-sided, filling the perigynium. Dry, often sandy soil of sparsely vegetated or disturbed grasslands, sagebrush steppe; plains, valleys, montane. AK south to ID, NM, CO, KS, IL.
    This sedge is sometimes considered to be a form of the Eurasian *Carex stenophylla*;[187] or a form of the Eurasian *C. duriuscula*.[24] I have chosen to retain the oldest, widely used name. *Carex inops* has more widely spaced spikes.

Carex elynoides T.Holm Densely caespitose. **Stems** erect, 3–12 cm with old leaf sheaths at the base. **Leaves** basal and lower-cauline; blades ca. 0.5 mm wide. **Inflorescence** a solitary bractless spike. **Spike** narrow, 8–18 mm long; male flowered portion usually >5 mm long; female flowers below male. **Perigynia** obovoid, 2.5–4 mm long, brown, with an entire beak ≥0.5 mm long and sometimes a few hairs above; stigmas 3. **Female scales** brown with hyaline margins, mostly larger than the perigynia. **Achene** 3-sided, filling the perigynium. Fellfields, dry turf; alpine. MT south to NV, UT, NM. (p.615)
    Carex nardina has fewer male flowers; *Kobresia myosuroides* is also similar, but the perigynium is split down 1 side; see *C. filifolia*.

Carex engelmannii L.H.Bailey [*C. breweri* Boott var. *paddoensis* (Suksd.) Cronquist] Rhizomatous. **Stems** erect, 1–15 cm, densely tufted. **Leaves** basal and lower-cauline; blades ≤0.5 mm wide. **Inflorescence** a solitary spike; lowest bract scale-like. **Spike** ovoid, 8–14 mm long; male flowers above; female below. **Perigynia** elliptic, 3–4.5 mm long, brown, tapered to a short, indistinct, entire beak; stigmas 3. **Female scales** brown, as wide but shorter than the perigynia; upper sometimes awn-tipped. **Achene** 3-sided, smaller than the wrinkled perigynium. Fellfields, turf; alpine. BC to NV, UT, CO. (p.615)

Carex epapillosa Mack. [*C. atrata* L. var. *erecta* W.Boott, *C. heteroneura* W.Boott var. *epapillosa* (Mack.) F.J.Herm.] Caespitose. **Stems** erect, 20–70 cm. **Leaves** basal and lower-cauline; blades 3–6 mm wide. **Inflorescence** of 3 to 6 short-pedunculate, erect to ascending spikes forming a loose cluster. **Spikes** 1–2 cm long, all similar; male flowers basal on uppermost spike; otherwise all female. **Perigynia** erect, coppery with green glabrous margins, elliptic to obovate, 3–4 mm long; the beak 0.3–0.5 mm long; stigmas 3. **Female scales** blackish, often with thin, hyaline margins, lanceolate, much narrower than the perigynia. **Achene** 3-sided, much smaller than the perigynium. Moist meadows, turf; subalpine, alpine. BC, AB south to CA, NV, UT, CO. See *C. atrosquama*. (p.616)

Carex filifolia Nutt. Densely caespitose. **Stems** erect, 4–30 cm with old leaf sheaths at the base. **Leaves** basal and lower-cauline; blades ca. 0.5 mm wide. **Inflorescence** a solitary bractless spike. **Spike** narrow, 1–2 cm long; male flowered portion 5–10 mm long; female flowers below. **Perigynia** obovoid, 3–4 mm long, brown and white, puberulent with a beak ≤0.5 mm long; stigmas 3. **Female scales** suborbicular, brown with pale margins, hiding the perigynia. **Achene** 3-sided, filling the perigynium. Often sandy or gravelly soil of grasslands, sagebrush steppe, pine woodlands; plains, valleys. AK to MB south to CA, AZ, NM, KS. Our plants are var. *filifolia*. (p.615)

The similar *Carex elynoides* occurs near or above treeline and has nearly glabrous perigynia.

Carex flava L. Caespitose. **Stems** erect, 10–60 cm. **Leaves** basal and cauline; blades 2–4 mm wide. **Inflorescence** of 3 to 6 ascending, separate spikes; the lowest bract longer than the inflorescence. **Spikes**: the uppermost male, 8–17 mm long; the lower female, 8–15 mm long, ellipsoid, short-pedunculate. **Perigynia** spreading to reflexed, green to yellow, ovoid, glabrous, 4–5 mm long with a distinct, gently curved beak 1.5–2.5 mm long; stigmas 3. **Female scales** lanceolate, coppery with a green midvein, smaller than the perigynium. **Achene** 3-sided, nearly filling the perigynium. Wet, often gravelly or peaty soil of meadows, lake shores, streambanks, fens; valleys, montane. Circumboreal south to BC, ID, MT, IN, NY.

Carex viridula has smaller perigynia with straight beaks.

Carex foenea Willd. [*C. aenea* Fernald] Caespitose. **Stems** erect, 30–60 cm. **Leaves**: basal blades absent; cauline blades 1–4 mm wide. **Inflorescence** of 3 to 8 sessile spikes, well-spaced below. **Spikes** 8–10 mm long, all similar; all female or male flowers basal. **Perigynia** ascending, green to pale brown, ovate, thin-margined, 4–5 mm long including the ca. 1 mm long serrulate beak; stigmas 2. **Female scales** green to light brown with lighter margins, ca. as long as the perigynium. **Achene** 2-sided, smaller than the perigynium. Moist to dry soil of grasslands, meadows, open forest; montane; collected in Petroleum Co. AK to NL south to MT, SD, MI, PA.

Carex fuliginosa Schkuhr [*C. misandra* R.Br.] Caespitose. **Stems** erect, 5–20 cm. **Leaves** basal and cauline; blades 1–4 mm wide. **Inflorescence** of 3 to 5 pedunculate spikes; the lowest bract shorter or longer than the inflorescence. **Spikes** 5–17 mm long, lanceolate, similar; the uppermost bisexual with male flowers below; the lower female, ascending to arching. **Perigynia** ascending, green to black, narrowly lanceolate, glabrous, 3.5 × 0.5–1 mm, tapered to a serrulate beak 1.5 mm long; stigmas 3. **Female scales** brown to blackish with a pale midstripe, exposing the periginia beak. **Achene** 3-sided, nearly filling the perigynium. Meadows, turf; upper subalpine,to alpine. Circumpolar south to UT, CO.

Perigynia of *Carex podocarpa* have short beaks.

Carex geyeri Boott ELK SEDGE Branched-rhizomatous, densely caespitose. **Stems** erect, 10–35 cm, clustered. **Leaves**: basal bladeless; cauline blades 1–2 mm wide. **Inflorescence** a solitary, bractless spike. **Spike** narrow, 10–35 mm long; male-flowered portion above, 8–25 mm long; female flowers below, 1 to 3, well-separated. **Perigynia** ellipsoid with a thick, basal stalk, 5–7 mm long, green to brown, glabrous; stigmas 3. **Female scales** acuminate, light brown with hyaline margins, hiding the perigynia. **Achene** 3-sided, filling the perigynium. Open forests, grasslands, meadows; valleys to alpine. BC, AB south to NV, UT, CO.

Leaves remain green during winter; stems flower near the ground in early spring then elongate in fruit.

Carex glacialis Mack. Caespitose. **Stems** erect, 2–6 cm. **Leaves** mainly basal; blades <1 mm wide. **Inflorescence** of 2 to 4 overlapping spikes; lower bract inconspicuous. **Spikes** erect, unisexual; the male terminal, 3–5 mm long; the lower female, short-pedunculate, 3–5 mm long. **Perigynia** spreading, green to purplish-black, obovoid, glabrous, 2–2.5 mm long with a distinct beak ca. 0.5 mm long, loosely aggregated; stigmas 3. **Female scales** ovate, purplish-black with a hyaline margin and green midvein, shorter than the perigynium. **Achene** 3-sided, nearly filling the perigynium. Calcareous fellfields on cool upper slopes; alpine. Circumpolar south to BC, AB, MT. Known from Flathead and Pondera cos.[254]

Carex gravida L.H.Bailey Caespitose. **Stems** erect, 25–75 cm. **Leaves** basal and cauline; blades 3–7 mm wide; sheaths whitish. **Inflorescence** of 5 to 12 sessile spikes, clustered into an elongate head; lowest bract shorter than the inflorescence. **Spikes** 6–9 mm long, all similar; male flowers above female. **Perigynia** ascending to spreading, green, broadly ovate, narrowly wing-margined, 3–4.5 × 2–3 mm with a serrulate, bidentate beak 1–1.5 mm long; stigmas 2. **Female scales** hyaline to brown with a green midstripe, awned, ca. as long and wide as the perigynium. **Achene** 2-sided, smaller than the perigynium. Deciduous or pine woodlands; plains. SK to ON south to NM, TX, TN, NC.

The similar *Carex hoodii* occurs in the mountainous portion of MT and usually has narrower leaves.

Carex haydeniana Olney Densely caespitose. **Stems** erect, 8–25 cm. **Leaves** basal and lower-cauline; blades 1–3 mm wide. **Inflorescence** of 4 to 7 sessile spikes, densely clustered into a globose to ovoid head; lowest bract inconspicuous. **Spikes** 6–10 mm long, all similar; female flowers above male. **Perigynia** ascending, brown, green-edged, ovate, flattened, serrulate, wing-margined, 4.5–6 × 2–3 mm, the beak 1.5–2.5 mm long, serrulate below, entire above; stigmas 2.

Female scales brown, sometimes with a pale midstripe, exposing the tip of the perigynia. **Achene** 2-sided, much smaller than the perigynium. Moist to wet, often stony soil of meadows, turf, often where snow lies late; subalpine, alpine. BC, AB south to CA, UT, CO. (p.618)

Characters that separate this species from *Carex microptera* are continuous, although most specimens can be easily placed; see *C. microptera*.

Carex hoodii Boott Densely caespitose. **Stems** erect, 20–70 cm. **Leaves**: basal bladeless; cauline blades 1–3 mm wide. **Inflorescence** of 4 to 8 sessile spikes, densely clustered into a globose to ovoid head; lowest bract short. **Spikes** 5–8 mm long, all similar; male flowers above female. **Perigynia** ascending to spreading, brown with green margins, ovate, narrowly wing-margined, 3.5–5 × 1.5–2 mm; the serrulate beak bidentate, 1–1.5 mm long; stigmas 2. **Female scales** brown with a green midstripe and hyaline margins, just shorter and narrower than the perigynia. **Achene** 2-sided, smaller than the perigynium. Grasslands, meadows, open forests; plains, valleys to lower subalpine. BC to SK south to CA, NV, UT, CO, SD.

Carex microptera and *C. pachystachya* have shorter heads with female flowers above the male; see *C. gravida*.

Carex hystericina Muhl. ex Willd. Short-rhizomatous. **Stems** erect, 20–80 cm, loosely tufted. **Leaves** basal and cauline; blades 3–8 mm wide. **Inflorescence** of 2 to 5 well-separated spikes; the lowest bract equal to or greater than the inflorescence. **Spikes** pedunculate, arching to ascending; the terminal male, linear, 1–3 cm long; the lower female, 1–4 cm long, cylindric. **Perigynia** ascending to reflexed, green to tan, narrowly ovate, veiny, glabrous, 4–6 mm long including a distinct beak 2–2.5 mm long; stigmas 3. **Female scales** brown, awn-like, serrulate, narrower and shorter than the perigynia. **Achene** 3-sided, much smaller than the perigynium. Wet meadows, marshes, thickets, often around streams, springs; plains, valleys. BC to QC south to UT, NM, TX, AR, VA.

Carex retrorsa has wider female scales.

Carex idahoa L.H.Bailey [*C. parryana* Dewey ssp. *idahoa* (L.H.Bailey) D.F.Murray] Short-rhizomatous. **Stems** erect, 12–60 cm, tufted. **Leaves**: basal and lower-cauline; blades 2–4 mm wide. **Inflorescence** of usually 3 subsessile, erect, overlapping spikes; lowest bract short. **Spikes** 6–25 mm long; uppermost largest, female or bisexual with male flowers below; lower all female. **Perigynia** ascending, green, pappilose, obovoid, 2.5–3 mm long, with a brown, serrulate beak ca. 0.3 mm long; stigmas 3. **Female scales** lanceolate, black with a pale midstripe, concealing the perigynia, becoming shorter and more rounded above. **Achene** 3-sided, nearly filling the perigynium. Moist meadows, often on stream terraces; montane, lower subalpine. MT, OR, ID, CA, UT. (p.616)

All of our specimens appear to be completely female. The terminal spike of *Carex parryana* is not wider than the lower ones. *Carex hallii* Olney has lighter and shorter female scales than *C. idahoa*. In our plants scales become shorter and lighter in the upper portion of the spike.

Carex illota L.H.Bailey Caespitose. **Stems** erect, 5–35 cm. **Leaves** basal and cauline; blades 1–2 mm wide. **Inflorescence** of 3 to 6 sessile spikes congested in a dense globose head; lowest bract short. **Spikes** 4–7 mm long, all similar; female flowers above male. **Perigynia** ascending to spreading, dark brown, lanceolate, thin-margined, 2.5–3.5 × ca. 1 mm including the 0.5 mm long, entire beak; stigmas 2. **Female scales** ovate, purplish-brown, shorter and narrower than the perigynia. **Achene** 2-sided, nearly filling the perigynium. Wet meadows, turf, often along streams or around ponds, lakes; subalpine, alpine. BC, AB south to CA, NV, UT, CO. See *C. lachenalii*. (p.618)

Carex inops L.H.Bailey [*C. heliophila* Mack., *C. pensylvanica* Lam. var. *digyna* Boeck.] Rhizomatous. **Stems** erect, 8–30 cm, tufted. **Leaves** basal and cauline; blades 1–4 mm wide. **Inflorescence** of 2 to 3 well-separated sessile spikes; the lowest bract shorter than the inflorescence. **Spikes** unisexual; the uppermost male, linear, 5–20 mm long; the lower female, ascending, 5–12 mm long, ovoid. **Perigynia** ascending, greenish, obovoid, short-stalked, pilose, 3–4 mm; the bidentate beak 0.5–1 mm long; stigmas 3. **Female scales** ovate, awned, tan to brown with a green midstripe and hyaline margins, not completely concealing the perigynia. **Achene** 3-sided, filling the perigynium. Sandy or stony soil of grasslands, sagebrush steppe, woodlands, forest openings; plains, valleys, montane. AB to MB south to CA, NM, IA, IN. Our plants are var. **heliophila** (Mack.) Crins.

Plants may have long-pedunculate, female spikes at the base.

Carex interior L.H.Bailey Caespitose. **Stems** erect, 15–50 cm. **Leaves** basal and cauline; blades 1–2 mm wide. **Inflorescence** of 3 to 6 sessile, separate spikes; lowest bract inconspicuous. **Spikes** 3–5 mm long, the uppermost bisexual with male flowers below; lower female spikes sometimes with a few male flowers. **Perigynia** spreading, green, ovate, 2–3 mm long tapered to the indistinct, serrulate beak, 0.5–1 mm long; stigmas 2. **Female scales** ovate, hyaline-tan with a green midvein, shorter than the perigynia. **Achene** 2-sided, filling the perigynium. Wet, organic soil of fens, meadows, swamps; valleys, montane. AK to NL south to CA, AZ, AR, VA, Mexico. (p.617)

Carex jonesii L.H.Bailey Caespitose. **Stems** erect, 15–30 cm. **Leaves** clustered near the base; cauline blades 1–3 mm wide. **Inflorescence** of 5 to 10, sessile spikes, congested in an ovoid head; lowest bract inconspicuous. **Spikes** ca. 5 mm long, all similar, male flowers above, inconspicuous; female below. **Perigynia** ascending, lanceolate, convex, veiny, light brown, 3–4 × ca. 1.5 mm; the beak 1–1.5 mm long, serrulate below, entire above; stigmas 2. **Female scales** brown with paler margins, as wide but shorter than the perigynia. **Achene** 2-sided, filling the perigynium. Wet meadows often along streams; montane; known from Missoula Co. (*Hart 426*, RM). WA to MT south to CA, CO.

Carex kelloggii W.Boott [*C. lenticularis* Michx. var. *lipocarpa* (T.Holm) L.A.Standl.] Caespitose. **Stems** prostrate to erect, 10–60 cm. **Leaves** basal and cauline; blades 1–2 mm wide. **Inflorescence** of 3 to 6 overlapping to clustered spikes; the lowest bract equal to or greater than the inflorescence. **Spikes** ascending; the uppermost male or bisexual; the lower female, 1–3 cm long, short-pedunculate, cylindric. **Perigynia** ascending, green, ovate, few-veined, glabrous, 2–3 mm long with an indistinct beak. **Female scales** rounded, blackish (rarely light tan) with a pale green midstripe, shorter and narrower than the perigynia. **Achene** 2-sided, nearly filling the perigynium. Wet or vernally wet, often sparsely vegetated soil around streams, lakes, ponds, fens; valleys to subalpine. AK to CA, AZ, CO. (p.614)

    Carex aquatilis and *C. nebrascensis* are not as strongly caespitose. A recent study indicates that *C. kelloggii* and the more eastern *C. lenticularis* are separate species.[118]

Carex lachenalii Schkuhr [*C. bipartita* All.] Caespitose in small tufts. **Stems** erect, 3–20 cm. **Leaves** mainly basal; blades 1–2 mm wide. **Inflorescence** of 2 to 4 sessile, clustered spikes; lower bract short. **Spikes** 5–10 mm long, all similar; female flowers above; male below. **Perigynia** erect, brown, narrowly oblanceolate, 2.5–3 mm long with an entire beak 0.5–1 mm long; stigmas 2. **Female scales** ovate, brown with a lighter midvein and hyaline margins, exposing only the beak tip. **Achene** 2-sided, filling the perigynium. Wet, organic, sometimes shallow soil of turf, rock ledges, streambanks; alpine. Circumpolar south to CO, UT.

    The few, clustered spikes and perigynia that are not thin-edged help identify this sedge. *Carex illota* has more spreading perigynia and an inflorescence as wide as long; achene beaks of *C. leporinella* are longer.

Carex lacustris Willd. Rhizomatous. **Stems** erect, 50–80 cm. **Leaves** basal and cauline; blades 3–12 mm wide. **Inflorescence** of 5 to 8 short-pedunculate spikes; the lowest bract greater than the inflorescence. **Spikes** unisexual, erect ascending; the uppermost 1 to 4 male, narrow, 2–3 cm long; the lower female, 2–6 cm long, cylindric. **Perigynia** ascending, green, broadly lanceoloid, inflated, glabrous, 5–7 mm long tapered to an indistinct beak ca. 1 mm long; stigmas 3. **Female scales** acute to acuminate, tan with a pale midstripe and margins, awned, smaller than the perigynia. **Achene** 3-sided, much smaller than the perigynium. Marshes; valleys of Lake Co. AB to QC south to ID, MT, MO, VA.

Carex laeviconica Dewey Rhizomes deep. **Stems** erect, 30–100 cm, tufted. **Leaves**: basal and cauline; blades 3–6 mm wide; sheaths glabrous. **Inflorescence** of 4 to 9 well-separated spikes; the lowest bract equal to or greater than the inflorescence. **Spikes** ascending; the uppermost 2 to 4 male, linear; the lower female, 3–7 cm long, pedunculate, cylindric. **Perigynia** ascending, yellowish-green, lanceolate, veiny, glabrous, 5–8 mm long including a deeply divided beak 3–5 mm long; stigmas 3. **Female scales** tan, lanceolate, awn-tipped, narrower and ca. as long as the perigynia. **Achene** 3-sided, smaller than the perigynium. Wet meadows, often around ponds; plains. SK to ON south to MT, KS, IN.

    Carex comosa has nodding spikes; *C. atherodes* has hairy leaf sheaths.

Carex laeviculmis Meinsh. Densely caespitose. **Stems** erect, 15–70 cm. **Leaves** basal and cauline; blades 1–2 mm wide. **Inflorescence** of 3 to 6 sessile, well-separated spikes; lowest bract shorter than the inflorescence. **Spikes** 4–6 mm long, all similar; female flowers above; male below. **Perigynia** ascending, green, ovate, 2.5–3.5 mm long, tapered to the weakly serrulate beak 0.5–1 mm long; stigmas 2. **Female scales** ovate, white to tan with a green midvein, shorter but as wide as the perigynia. **Achene** 2-sided, filling the perigynium. Wet meadows, thickets, forest openings; montane, lower subalpine. AK south to CA, ID, WY. See *C. brunnescens*, *C. deweyana*.

Carex lasiocarpa Ehrh. Rhizomatous. **Stems** erect, 30–90 cm, loosely tufted. **Leaves**: basal bladeless; cauline blades 1–2 mm wide, inrolled. **Inflorescence** of 3 to 4 well-separated sessile spikes; the lowest bract leaf-like, ca. as long as the inflorescence. **Spikes** unisexual, ascending; the uppermost 1 or 2 male, linear, 2–4 cm long; the lower female, 1–3 cm long, cylindrical. **Perigynia** ascending, purplish, ovoid, pilose, 3–4 mm; the beak 0.5–1 mm long with spreading teeth; stigmas

3. **Female scales** lanceolate, short-awned, purple with a pale midstripe, longer or shorter than the perigynia. **Achene** 3-sided, smaller than the perigynium. Saturated, organic soil, often forming near monocultures in fens; valleys to lower subalpine. Circumboreal south to CA, ID, MT, IL, VA. Our plants are var. *americana* Fernald. See *Carex pellita*.

Carex leporinella Mack. Densely caespitose. **Stems** erect, 5–30 cm. **Leaves**: basal and cauline; blades 1–2 mm wide. **Inflorescence** of 3 to 8 sessile spikes, loosely aggregated in an ovoid head; lowest bract shorter than the inflorescence. **Spikes** 4–8 mm long, all similar; female flowers above male. **Perigynia** ascending, brown, lanceolate, barely wing-margined, 2.5–4 × <1 mm, the beak, ca. 1 mm long, serrulate below, entire above; stigmas 2. **Female scales** medium brown with a pale midstripe and hyaline margins, hiding the perigynia. **Achene** 2-sided, barely smaller than the perigynium. Moist meadows, turf, often on cool slopes; subalpine, alpine. WA to MT south to CA, NV, UT, WY.

    Carex phaeocephala has larger perigynia with more conspicuously winged margins.

Carex leptalea Wahlenb. Finely rhizomatous. **Stems** prostrate to ascending, 5–40 cm, tufted. **Leaves** basal and lower-cauline; blades <1 mm wide. **Inflorescence** a solitary bractless spike. **Spike** narrow, 4–7 mm long; male flowered portion above, 1–2 mm long; female flowers below, few. **Perigynia** narrowly elliptic, veiny, thick-stalked, 1.5–2.5 mm long, green, beakless; stigmas 3. **Female scales** green to tan, awn-tipped, shorter than the perigynia. **Achene** 3-sided, nearly filling the perigynium. Often calcareous fens, thickets, wet spruce or cedar forests, often on hummocks; valleys, montane. Throughout N. America.

    Often occurring with *Carex disperma*, which has several small spikes.

Carex limosa L. Long-rhizomatous with yellow roots. **Stems** erect, 12–40 cm. **Leaves**: basal bladeless; cauline blades 0.5–2 mm wide. **Inflorescence** of 2 or 3 well-separated spikes; the lowest bract shorter than the inflorescence. **Spikes** mostly unisexual; the uppermost male, linear, 10–35 mm long; the lower female, arching-pedunculate, 8–25 mm long, cylindric. **Perigynia** ascending, green to tan, elliptic, glabrous, 2–4 mm with a tiny beak; stigmas 3. **Female scales** ovate, coppery, acute to apiculate, ca. as long as the perigynia. **Achene** 3-sided, nearly filling the perigynium. Fens; valleys, montane. Circumboreal south to CA, NV, UT, WY, IA, NY.

    Some lateral spikes may have male flowers at the tip, while in *Carex magellanica* these male flowers are basal.

Carex livida (Wahlenb.) Willd. Long-rhizomatous. **Stems** erect, 10–40 cm. **Leaves** pale green, basal and lower-cauline; blades 1–4 mm wide. **Inflorescence** of 2 or 3 well-separated spikes; the lowest bract longer or shorter than the inflorescence. **Spikes** unisexual; the uppermost male, narrow, 7–20 mm long; the lower female, ascending, short-pedunculate, 1–2 cm long, cylindric. **Perigynia** ascending, pale green, narrowly elliptic, glabrous, glaucous, 2–3.5 mm long, beakless, tapered at both ends; stigmas 3. **Female scales** tan to brown with a green midstripe, ca. as long as the perigynia. **Achene** 3-sided, nearly filling the perigynium. Calcareous fens; valleys, montane. Circumboreal south to CA, WY, IN, NY.

Carex luzulina Olney [*C. fissuricola* Mack.] Caespitose. **Stems** erect, 15–70 cm. **Leaves** basal and cauline; blades 3–8 mm wide. **Inflorescence** of 2 to 5 ascending spikes; the lowest bract shorter than the inflorescence. **Spikes**: the uppermost male or bisexual with male flowers below, 1–3 cm long; the lower female, 10–25 mm long, pedunculate, cylindric. **Perigynia** ascending, green to purple, narrowly lanceolate, glabrous or sparsely setose above, 3.5–4.5 mm long with an entire, indistinct beak ca. 1 mm long; stigmas 3. **Female scales** purplish to black, sometimes with a pale midstripe, shorter and ca. as wide as the perigynia. **Achene** 3-sided, smaller than the perigynium. Moist to wet meadows; montane, subalpine. BC south to CA, NV, UT, WY.

1. Female scales brownish with the pale midvein obvious; perigynia with some green . . . . var. *ablata*
1. Female scales blackish with an obscure midvein; perigynia purple . . . . var. *atropurpurea*

The terminal spike of ***Carex luzulina*** var. ***ablata*** (L.H.Bailey) F.J.Herm. is usually male; while that of **C.** *luzulina* var. ***atropurpurea*** Dorn is usually bisexual. The two forms have similar habitats and MT distributions. Scales of var. *ablata* have been described as purplish-black with a distinct midvein not extending to the tip (Hermann 1970, Ball and Reznicek 2002). However, the type of var. *ablata* has brown scales with a distinct midvein. I have chosen to follow Dorn who recognizes our black-scaled plants as var. *atropurpurea* (Dorn 2001). Perigynia of many of these specimens have sparsely setose beaks and could be referred to *C. fissuricloa*, but the distinction seems slight and is not correlated with any ecological or geographic differences in our area.[187]

Carex macloviana d'Urv. Densely caespitose. **Stems** erect, 10–25 cm. **Leaves** basal and cauline; blades 2–4 mm wide. **Inflorescence** of 3 to 9 sessile spikes, densely clustered into a globose to ovoid head; lowest bract inconspicuous. **Spikes** 5–10 mm long, all similar; female flowers above; male below. **Perigynia** erect to ascending, green to shiny bronze, ovate, flattened, wing-margined, 3–4 × ca. 1.5 mm, the beak ca. 1 mm long, serrulate below, entire above; stigmas 2. **Female scales** gold or coppery to shiny brown sometimes with broad, hyaline margins and a pale midstripe, as wide and shorter than the perigynia. **Achene** 2-sided, smaller than the perigynium. Wet meadows, turf; alpine. AK to Greenland south to CO.

Carex macloviana seems like a high-elevation form of C. microptera; the coppery perigynia and female scales with broad hyaline margins are the subtle characters that distinguish this species.

Carex magellanica Lam. [C. paupercula Michx.] Long-rhizomatous with yellow roots. **Stems** erect, 15–60 cm. **Leaves**: basal and cauline; blades 2–4 mm wide. **Inflorescence** of 3 to 5 well-separated spikes; the lowest bract shorter than the inflorescence. **Spikes** mostly unisexual; the uppermost male, linear, 4–12 mm long; the lower female, arching-pedunculate, 6–15 mm long, cylindric. **Perigynia** ascending, green to brown, elliptic, glabrous, glaucous, 2–3 mm long, beakless; stigmas 3. **Female scales** ovate, brown with a green midstripe, acuminate, ca. as long or longer than the perigynia, deciduous. **Achene** 3-sided, nearly filling the perigynium. Organic soil of Sphagnum fens, wet forest and around ponds; valleys. Circumboreal south to UT, CO, OH, NY, S. America. See C. limosa.

Carex maritima Gunnerus [C. incurviformis Mack.var. danaensis (Stacey) F.J.Herm.] Short-rhizomatous. **Stems** erect to ascending, 2–15 cm, loosely tufted. **Leaves** mainly basal; blades ca. 1 mm wide. **Inflorescence** 3 to 5 bisexual spikes, densely clustered in a hemispheric head; lowest bract inconspicuous. **Spikes** ca. 5 mm long; male flowers above, inconspicuous; female below. **Perigynia** ascending, glabrous, narrowly ellipsoid, ribbed, 3.5–4 × 1–2 mm, green to brown, with an indistinct, mostly entire beak 0.5–1 mm long; stigmas 2. **Female scales** brown with hyaline margins, shorter than the perigynia. **Achene** 2-sided, slightly smaller than the perigynium. Wet turf often along small streams; alpine. Circumpolar south to CA, CO.

Most of our plants fit the description of C. maritima (sensu stricto) better than C. incurviformis, and the former is the older name.

Carex media R.Br. [C. norvegica Retz. ssp. inferalpina (Wahlenb.) Hultén] Caespitose. **Stems** erect, 15–50 cm. **Leaves** basal and lower-cauline; blades 2–3 mm wide. **Inflorescence** of 3 to 4 short-pedunculate, ascending spikes forming a loose cluster; lowest bract ca. as long as the inflorescence. **Spikes** 3–10 mm long, all similar; uppermost bisexual with male flowers basal; lower all female. **Perigynia** erect, green to bronze-tinged, elliptic, 2–2.5 mm long, including the serrulate beak <0.5 mm long; stigmas 3. **Female scales** blackish, sometimes with thin, pale margins, shorter than the perigynia. **Achene** 3-sided, nearly filling the perigynium. Shores of steams, lakes, wet spruce forest; montane, subalpine. Circumboreal south to OR, MT, IA, MI.

The similar Carex stevenii has darker and somewhat smaller perigynia.

Carex mertensii Prescott ex Bong. Caespitose. **Stems** erect to ascending, 40–80 cm. **Leaves** basal and cauline; blades 3–8 mm wide. **Inflorescence** of 5 to 10 overlapping, pedunculate, arching spikes; lowest bract longer than the inflorescence. **Spikes** cylindric, 1–4 cm long, bisexual, all similar; male flowers few, basal; uppermost spike sometimes mostly male. **Perigynia** erect, green to tan, elliptic, 2.5–4.5 mm long, glabrous, flattened with a tiny beak; stigmas 3. **Female scales** lanceolate, blackish with a light midvein, shorter and narrower than the perigynia. **Achene** 3-sided, much smaller than the perigynium. Thickets, moist forest, often along streams, lakes, roads; montane, subalpine. AK to CA, ID, MT; Asia.

Carex microptera Mack. [C. festivella Mack., C. limnophila F.J.Herm.] Densely caespitose. **Stems** erect, 15–90 cm. **Leaves** basal and cauline; blades 1–5 mm wide. **Inflorescence** of 4 to 10 sessile spikes densely clustered into a globose to ovoid head; lowest bract scale-like. **Spikes** 4–10 mm long, all similar; female flowers above male. **Perigynia** ascending, green to brown, narrowly ovate, flattened, wing-margined, 3–4.5 × 1–1.5 mm, the beak 1–1.5 mm long, serrulate below, entire above. **Female scales** brown sometimes with a pale midstripe and margins, shorter but nearly as wide as the perigynia. **Achene** 2-sided, much smaller than the perigynium. Moist to wet meadows, grasslands, shores of streams, lakes, occasionally open forest, pastures; valleys to alpine. YT south to CA, AZ, NM, SD. (p.618)

Carex microptera is likely our most common, caespitose, montane sedge. It and C. haydeniana, C. pachystachya, C. macloviana, and C. praticola are closely related and difficult to distinguish.

Carex nardina Fr. Densely caespitose. **Stems** erect, 5–15 cm with old leaf sheaths at the base. **Leaves** mostly basal; blades ca. 0.5 mm wide. **Inflorescence** a solitary bractless spike. **Spike** lanceolate, 5–15 mm long; male-flowered portion 2–4 mm long; female flowers below. **Perigynia** obovate, 3–4 mm long, green to brown, thin-edged, serrulate or ciliate on the margins with a short, entire beak; stigmas 2 or 3. **Female scales** brown with hyaline margins and a pale midstripe, as wide and long as the perigynia. **Achene** usually 3-sided, partly to nearly filling the perigynium. Stony soil of fellfields, turf; alpine. Circumpolar south to WA, NV, UT, CO. See *C. elynoides*. (p.615)

Carex nebrascensis Dewey NEBRASKA SEDGE Rhizomatous. **Stems** erect, 8–80 cm. **Leaves** basal and cauline; blades 3–10 mm wide. **Inflorescence** of 3 to 6 well-separated spikes; the lowest bract equal to or shorter than the inflorescence. **Spikes** ascending; the uppermost 1 or 2 male; the lower female, 1–5 cm long, pedunculate, cylindric. **Perigynia** ascending, green to tan, obovate, 5- to 9-veined, glabrous, 2.5–3.5 mm long with a short divided beak. **Female scales** acuminate, awn-tipped, dark brown with a pale midstripe, longer and narrower than the perigynia. **Achene** 2-sided, filling the perigynia. Wet meadows, marshes, pastures, banks of streams, ponds, tolerant of alkaline and saline soil; plains, valleys, montane. AB, SK south to CA, AZ, NM, KS, IL.

Carex nebrascensis has nerves on the perigynia, while *C. aquatilis* does not.

Carex nelsonii Mack. Caespitose. **Stems** erect, 7–20 cm. **Leaves** basal and lower-cauline; blades 2–3 mm wide. **Inflorescence** of (1)2 to 4 sessile spikes forming a tight globose cluster; lowest bract inconspicuous. **Spikes** 7–10 mm long, bisexual, all similar; male flowers basal. **Perigynia** ascending, white to green, purple above, papillose, 3–4 × ca. 1.5 mm with a serrulate beak 0.8–1 mm long; stigmas 3. **Female scales** blackish, ca. as long and wide as the perigynia. **Achene** 3-sided, smaller than the perigynium. Moist meadows, cliffs; alpine. MT, WY, UT, CO; collected in Park and Carbon cos.

Carex pelocarpa has perigynia wider than the scales.

Carex neurophora Mack. Caespitose. **Stems** erect, 30–60 cm. **Leaves:** basal bladeless; cauline blades 2–3 mm wide; upper sheaths cross-corrugate ventrally. **Inflorescence** of 5 to 10 sessile spikes, congested in an ovoid head; lowest bract inconspicuous. **Spikes** ca. 5 mm long, all similar, male flowers above, inconspicuous; female below. **Perigynia** spreading to ascending, lanceolate, convex, veiny, light brown, 2.5–4 × 1–1.5 mm; the beak 1–1.5 mm long, serrulate below, entire above; stigmas 2. **Female scales** brown with a paler midvein, as wide but shorter than the perigynia. **Achene** 2-sided, filling the perigynium. Wet meadows, thickets, forest openings, often along streams; montane, lower subalpine. WA, OR, ID, MT, WY, CO.

Carex nigricans C.A.Mey. Rhizomatous. **Stems** erect, 10–40 cm. **Leaves:** basal and cauline; blades 1–3 mm wide. **Inflorescence** a solitary bractless spike. **Spike** ovoid, 8–20 mm long; male flowered portion 3–8 mm long; female flowers below. **Perigynia** ascending, spreading to reflexed at maturity, narrowly lanceolate, brown to purplish above, 2–3 mm long not including the distinct pedicel, tapered to an indistinct, entire beak; stigmas 3. **Female scales** blackish, concealing the perigynia at first but soon deciduous. **Achene** 3-sided, smaller than the perigynium. Moist meadows, along streams, in depressions; alpine, rarely lower. AK south to CA, UT, CO. (p.615)

Carex nigricans forms solid colonies in depressions that cache snow. Occasionally spikes are all male. *Carex pyrenaica* forms small, dense tussocks and usually occurs in drier or better-drained soils.

Carex norvegica Retz. Caespitose. **Stems** erect, 5–15 cm. **Leaves** basal and lower-cauline; blades ca. 2 mm wide. **Inflorescence** of 2 to 4 short-pedunculate, ascending spikes forming a loose cluster; lowest bract as long or shorter than the inflorescence. **Spikes** 4 to 10 mm long, all similar; uppermost bisexual with male flowers basal; lower all female. **Perigynia** erect, bronze below, blackish above, elliptic, ca. 2 mm long, including the serrulate beak <0.5 mm long; stigmas 3. **Female scales** blackish with pale margins, shorter or equal to the perigynia. **Achene** 3-sided, nearly filling the perigynium. Wet soil along streams, seeps; alpine. Greenland to ON; Europe; disjunct in Carbon and Park cos., MT.

Some previous treatments[113,188] have included *Carex media* and *C. stevenii* in *C. norvegica* sensu lato.

Carex obtusata Lilj. Rhizomatous. **Stems** erect, 3–15 cm, solitary. **Leaves** basal and lower-cauline; blades 0.5–2 mm wide. **Inflorescence** a solitary bractless spike. **Spike** narrow, 5–12 mm long; male flowered portion 3–8 mm long; female flowers below, few. **Perigynia** ovoid, ribbed, 2–3 mm long, purplish, with a distinct, entire, bidentate beak 0.5–1 mm long; stigmas 3. **Female scales** hyaline with a tan center, apiculate, nearly concealing the perigynia. **Achene** 3-sided, filling the perigynium. Mesic grasslands, meadows, turf, fellfields; valleys to alpine, more common lower. Circumboreal south to UT, NM, SD, MN. (p.615)

Carex pachystachya Cham. ex Steud. [*C. preslii* Steud.] Densely caespitose. **Stems** erect, 15–70 cm. **Leaves**: basal blades absent; cauline blades 2–4 mm wide. **Inflorescence** of 4 to 12 sessile spikes, aggregated into an ovoid head. **Spikes** 4–8 mm long, all similar; female flowers above male. **Perigynia** ascending, green to brown, ovate, convex, wing-margined, 3–4.5 × 1–2 mm, the beak 1–1.5 mm long, serrulate below, entire above; stigmas 2. **Female scales** medium brown with a pale midstripe and hyaline margins, nearly as wide but shorter than the perigynia. **Achene** 2-sided, a little smaller than the perigynium. Coniferous forest openings, moist meadows, often along streams, lakes; valleys to subalpine. AK to CA, NV, UT, CO. (p.618)

Plants called *Carex preslii* have green perigynia that contrast sharply with the dark scales; while *C. pachystachya* (sensu stricto) has brown perigynia. However, the majority of our specimens have brown perigynia with green edges; plants with green perigynia occur only in the northwest portion of the state, but there is no ecological difference among the forms. See *C. microptera*.

Carex pallescens L. Short-rhizomatous. **Stems** erect, 20–60 cm, tufted. **Leaves** basal and cauline; pubescent; blades 2–4 mm wide. **Inflorescence** of 3 to 5 overlapping, pedunculate spikes; the lowest bract greater than the inflorescence. **Spikes** ascending; 1–2 cm long; the uppermost male; otherwise female, cylindric. **Perigynia** ascending to spreading, green, narrowly ellipsoid, glabrous, veiny, 2–2.5 mm long without a beak. **Female scales** narrowly ovate, short-awned, hyaline with a green midvein, shorter than the perigynia. **Achene** 3-sided, filling the perigynium. Wet meadows; valleys. AB to MB south to CO, SD, WI and QC to NL south to OH, NY; Eurasia; known from Ravalli Co.

Carex torreyi is similar but has a short beak.

Carex parryana Dewey Rhizomatous. **Stems** erect, 15–40 cm, tufted. **Leaves**: basal and lower-cauline; blades 2–4 mm wide. **Inflorescence** of 2 to 4 subsessile, erect, overlapping spikes; lowest bract shorter than the inflorescence. **Spikes** cylindric; uppermost male or bisexual with male flowers below, 1–2 cm long; lower all female, 5–20 mm long. **Perigynia** erect, green to purplish, papillose, ellipsoid, 1.5–2.5 mm long; beak serrulate, <0.5 mm long; stigmas 3. **Female scales** ovate, brown with a pale midstripe and hyaline margins, concealing the perigynia. **Achene** 3-sided, filling the perigynium. Moist to wet, often alkaline or saline meadows, often along streams; plains, valleys, montane. AK to ON south to NV, UT, CO. See *C. idahoa*. (p.616)

Carex paysonis Clokey [*C. tolmiei* Boott misapplied] Short-rhizomatous. **Stems** erect, 10–50 cm with dried leaves at the base. **Leaves** basal and lower-cauline; blades 2–5 mm wide. **Inflorescence** of 3 to 6, ascending, overlapping spikes; lowest bract shorter than the inflorescence. **Spikes**: upper 1 or 2 male or bisexual with male flowers above, 10–25 mm long; lower female, short-pedunculate, 5–25 mm long, ovoid to cylindric. **Perigynia** ascending, whitish below, purplish above, glabrous, flattened, broadly ovate, 2.5–4 × 1.5–2.5 mm with a minute beak; stigmas 3. **Female scales** lanceolate, blackish with an indistinct midstripe, narrower but as long as the perigynia. **Achene** 3-sided, much smaller than the perigynium. Moist to wet meadows, turf, often where snow lies late; subalpine, alpine, more common higher. BC, AB south to OR, UT, WY. (p.616)

Carex paysonis, *C. spectabilis* and *C. podocarpa* appear to form a continuum in northwest MT; there are intermediates between the first and the second pairs. *Carex podocarpa* has more nodding female spikes and fewer leaves; *C. spectabilis* has narrower perigynia and scales with a more prominent midvein.

Carex pellita Willd. [*C. lanuginosa* Michx. misapplied] Rhizomatous. **Stems** erect, 20–80 cm, tufted. **Leaves**: basal bladeless; cauline blades 2–5 mm wide, flat. **Inflorescence** of 3 to 5 well-separated, sessile spikes; the lowest bract greater than the inflorescence. **Spikes** unisexual, erect to ascending; the uppermost 1 or 2 male, linear; the lower female, 1–4 cm long, cylindric. **Perigynia** spreading to ascending, green to purplish, ovoid or obovoid, pilose, 3–4 mm long with a beak ca. 1 mm long with spreading teeth ca. 0.5 mm long; stigmas 3. **Female scales** ovate, purplish with a pale midstripe, acuminate or awned, nearly as long as the perigynia. **Achene** 3-sided, filling the perigynium. Wet mineral soil of wet meadows, marshes, often around streams, ponds; plains, valleys. Throughout N. America.

Carex lasiocarpa has narrower leaves and is usually found in organic soils.

Carex pelocarpa F.J.Herm. [*C. nova* L.H.Bailey var. *pelocarpa* (F.J.Herm.) Dorn] Caespitose. **Stems** erect, 5–45 cm. **Leaves** basal and lower-cauline; blades 1–4 mm wide. **Inflorescence** of 3 to 4 sessile spikes forming a tight globose cluster; lowest bract inconspicuous. **Spikes** 5–12 mm long, similar; male flowers basal. **Perigynia** ascending, shiny, purplish with green edges, circular, 2.5–3.5 × 1.5–2 mm with a short, distinct, entire beak; stigmas 3. **Female scales** blackish, narrower and ca. as long as the perigynia. **Achene** 3-sided, much smaller than the perigynium. Meadows, turf, often along streams; alpine. OR, ID, MT, NV, UT, CO. See *C. nelsonii*.

Carex petasata Dewey Densely caespitose. **Stems** erect, 15–60 cm. **Leaves**: basal blades absent; cauline blades 2–3 mm wide. **Inflorescence** of 3 to 6 sessile, overlapping spikes; lowest bract inconspicuous. **Spikes** 8–18 mm long, all similar, bisexual; female flowers above male. **Perigynia** erect, green to tan, narrowly ovate, flattened, wing-margined, 5.5–7 × 2–2.5 mm, tapered to the indistinct beak, 1.5–2 mm long, serrulate below, entire above; stigmas 2. **Female scales** light brown with a pale midvein, concealing the perigynia. **Achene** 2-sided, smaller than the perigynium. Moist grasslands, meadows; valleys to subalpine. AK south to CA, AZ, CO. (p.618)
    The similar *Carex tahoensis* has perigynia with flattened, deeply divided tips and slightly smaller perigynia.

Carex petricosa Dewey Rhizomatous. **Stems** erect, 5–20 cm, tufted. **Leaves** basal and cauline; blades ca. 1 mm wide. **Inflorescence** of 2 or 4 pedunculate spikes; the lowest bract shorter than the inflorescence. **Spikes** 5–17 mm long, lanceolate, similar, the uppermost 1 or 2 male, 5–15 mm long; the lower ascending, 8–10 mm long, female or bisexual with male flowers above. **Perigynia** ascending, green below, brown above, lanceolate, minutely ciliate on the upper margins, 4.5–5 × 1.5–2 mm, tapered to an indistinct serrulate beak 0.5–1 mm long; stigmas 3. **Female scales** purplish-brown with pale margins, smaller than the perigynium. **Achene** 3-sided, smaller than the perigynium. Stony turf; alpine; collected in Glacier Co. AK south to MT, QC; Asia.

Carex phaeocephala Piper Densely caespitose. **Stems** erect, 5–40 cm. **Leaves**: basal and cauline; blades 1–2 mm wide. **Inflorescence** of 2 to 5 sessile spikes, aggregated in an ovoid to elongate head; the lower just overlapping; lowest bract inconspicuous. **Spikes** 5–12 mm long, all similar; female flowers above, male below. **Perigynia** ascending, brown, usually with green margins, narrowly elliptic, wing-margined, 3–5.5 × 1.5–2 mm, the indistinct beak ca. 1 mm long, serrulate below, entire above; stigmas 2. **Female scales** tan to brown with a pale midstripe and hyaline margins, concealing the perigynia. **Achene** 2-sided, smaller than the perigynium. Dry to moist, stony soil of meadows, turf, fellfields, open forest, cliffs; subalpine, alpine. AK south to CA, NV, UT, CO. (p.618)
    Carex leporinella has smaller perigynia and darker spikes.

Carex plectocarpa F.J.Herm. [*C. lenticularis* Michx. var. *dolia* (M.E.Jones) L.A.Standl., *C. eleusinoides* Turcz. ex Kunth misapplied] Caespitose. **Stems** ascending to prostrate, 5–30 cm. **Leaves** basal and cauline; blades 1–3 mm wide. **Inflorescence** of 3 or 4 short-pedunculate, overlapping spikes; lowest bract shorter or longer than the inflorescence. **Spikes** 8–20 mm long; uppermost bisexual with male flowers below; lower spikes female. **Perigynia** ovate, ascending, green- and purple-blotched, 2–3 mm long with a beak ca. 0.2 mm long; stigmas 2. **Female scales** narrowly ovate, blackish with a pale midvein, shorter than the perigynia. **Achene** 2-sided, smaller than the perigynium. Wet, stony soil along shallow meltwater streams, ponds, lakes; upper subalpine, lower alpine. Known from Glacier National Park, Park Co., MT and adjacent WY.[118]

Carex podocarpa R.Br. [*C. spectabilis* Dewey misapplied] Short-rhizomatous. **Stems** erect, 15–60 cm tufted. **Leaves**: basal leaves bladeless; cauline blades 1–3 mm wide. **Inflorescence** of 3 to 5, well-separated spikes; lowest bract ca. as long as the inflorescence. **Spikes**: uppermost male, 7–25 mm long; lower female, pedunculate, ovoid to cylindric, ascending to arching, 1–3 cm long. **Perigynia** ascending, purplish-brown, glabrous, flattened, narrowly ovate, 4–5 × 1.5–2 mm with a beak 0.5–1 mm long; stigmas 3. **Female scales** lanceolate, acute, mostly solid purplish-black, as long but narrower than the perigynia. **Achene** 3-sided, much smaller than the perigynium. Wet meadows, rock ledges, especially near streams; subalpine, lower alpine. AK south to OR, ID, MT. See *C. paysonis, C. fuliginosa*. (p.616)

Carex praeceptorum Mack. Caespitose in small tufts. **Stems** erect, 7–30 cm. **Leaves** basal and cauline; blades 1–3 mm wide. **Inflorescence** of 3 to 5 sessile, overlapping to clustered spikes; lower bract short. **Spikes** 4–8 mm long, all similar; female flowers above male. **Perigynia** ascending, green, ovate, 1.5–2 mm long with an indistinct, slightly serrate beak <0.5 mm long; stigmas 2. **Female scales** medium brown with a lighter midvein and hyaline margins, shorter but as wide as the perigynia. **Achene** 2-sided, filling the perigynium. Moist to wet meadows, fens; subalpine, alpine. BC to CA, NV, UT, CO. (p.617)
    Carex canescens has lighter scales and occurs at lower elevations.

Carex praegracilis W.Boott Rhizomatous. **Stems** erect, 10–70 cm. **Leaves**: basal bladeless; cauline blades 1–2 mm wide. **Inflorescence** of 5 to 15 overlapping spikes in an elongate cluster; lowest bract shorter than the inflorescence. **Spikes** sessile, similar, ascending, ovoid; 5–8 mm long; male flowers above, few; female flowers below; sometimes spikes or inflorescences unisexual. **Perigynia** erect, narrowly ovate, green to brown, glabrous, 3–4 mm long, tapered to a serrulate beak 1–2 mm long; stigmas 2. **Female scales** lanceolate, tan with pale margins and midvein, hiding the perigynia. **Achene**

2-sided, nearly filling the perigynium. Moist soil of meadows, woodlands, often around ponds, streams; plains, valleys, montane. YT to QC south to CA, AZ, NM, MO, VA, Mexico.

The similar *Carex simulata* has perigynia with smaller beaks and is often found in organic soils, while *C. praegracilis* is usually found in mineral, sometimes alkaline soil.

Carex prairea Dewey Densely caespitose. **Stems** erect, 30–60 cm. **Leaves**: basal and cauline; cauline sheaths coppery opposite the blades that are 1–3 mm wide. **Inflorescence** elongate, of several sessile, ascending spikes, overlapping above, separate and sometimes branched below; lowest bract inconspicuous. **Spikes** 3–6 mm long, all similar; male flowers above; female below or all female. **Perigynia** ascending, lanceolate, tan to brown, glabrous, serrulate-margined, 3–4 mm tapered to the flattened, grooved beak 1.5–2 mm long; stigmas 2. **Female scales** brown lanceolate, concealing the perigynia. **Achene** 2-sided, filling the perigynium. Wet, organic soil of calcareous fens; valleys. YT to NL south to MT, IL, VA. Known from Lincoln Co.

Inflorescences of *Carex diandra* and *C. cusickii* are usually more crowded with more spikes.

Carex praticola Rydb. [*C. platylepis* Mack.] Densely caespitose. **Stems** erect, 20–60 cm. **Leaves**: basal blades absent; cauline blades 1–4 mm wide. **Inflorescence** of 3 to 7 sessile, overlapping spikes, aggregated above, separate below; lowest bract inconspicuous. **Spikes** ovoid, 5–12 mm long, ascending, all similar; female flowers above; male below. **Perigynia** ascending, green to tan, lanceolate, flattened, wing-margined, 3.5–5.5 × 1.5–2 mm, tapered to the beak 1.5–2.5 mm long, serrulate below, entire above; stigmas 2. **Female scales** bronze with a pale midvein and margins, exposing only the tip of the perigynia. **Achene** 2-sided, a little smaller than the perigynium. Moist to wet meadows, moist forest, woodlands; plains, valleys to lower subalpine. AK to Greenland south to CA, NV, UT, CO, SD, IL. (p.618)

Carex pachystachya has more congested inflorescences; *C. petasata* has larger perigynia.

Carex pyrenaica Wahlelnb. [*C. micropoda* C.A.Mey.] Caespitose. **Stems** erect, 3–20 cm with old leaf sheaths at the base. **Leaves**: basal blades absent; cauline crowded near the base; blades <1 mm wide. **Inflorescence** a solitary bractless spike. **Spike** lanceolate, 6–20 mm long; male flowered portion 2–4 mm long; female flowers below. **Perigynia** ascending, spreading at maturity, lanceolate, green to bronze, 2–2.5 mm long not including the distinct pedicel, tapered to an indistinct, entire beak; stigmas 3. **Female scales** brown to purple with hyaline margins and an indistinct midstripe, nearly concealing the perigynia at first but eventually deciduous. **Achene** 3-sided, nearly filling the perigynium. Moist to dry, well-drained soil of meadows, fellfields, rocky slopes; alpine, rarely lower. Circumpolar south to OR, UT, CO. (p.615)

Carex nardina has sessile obovate perigynia. North American plants are sometimes treated as separate from Eurasian *C. pyrenaica*, but a global study is needed for certainty.[303] See *C. nigricans*.

Carex raynoldsii Dewey Caespitose with old leaves at the base. **Stems** erect, 15–80 cm. **Leaves** basal and lower-cauline; blades 2–8 mm wide. **Inflorescence** of 3 to 6 pedunculate to subsessile, ascending, separate to overlapping, oblong to ovoid spikes. **Spikes** unisexual; terminal spike male, 6–15 mm long; lower female, 1–2 cm long. **Perigynia** green to yellowish, obovoid, inflated, 3–4 × 1.5–2.5 mm, with a distinct, bidentate beak <0.5 mm long; stigmas 3. **Female scales** purplish-black, shorter, narrower than the perigynia. **Achene** 3-sided, smaller than the perigynium. Moist grasslands, meadows; montane, subalpine. BC to SK south to CA, NV, UT, CO. (p.616)

Carex retrorsa Schwein. Rhizomatous. **Stems** erect, 20–70 cm. **Leaves** basal and cauline; blades 3–8 mm wide. **Inflorescence** of 4 to 7 subsessile to pedunculate spikes; the lowest bract much greater than the inflorescence. **Spikes** unisexual, ascending; the uppermost 1 to 3 male, linear, 1–5 cm long; the lower female, 1–5 cm long, cylindrical, subsessile to short-pedunculate. **Perigynia** spreading, the lowest reflexed, green to tan, broadly ovate, inflated, glabrous, 6–10 mm long, tapered to a beak 3–5 mm long; stigmas 3. **Female scales** acuminate, tan with hyaline margins, smaller than the perigynia. **Achene** 3-sided, much smaller than the perigynium. Marshes, margins of streams, ponds; valleys. BC to QC south to OR, UT, WY, IL, NY.

Carex utriculata and *C. vesicaria* have smaller perigynia.

Carex rossii Boott [*C. brevipes* W.Boott ex B.D.Jackson, *C. deflexa* Hornem. var. *boottii* L.H.Bailey] Rhizomatous. **Stems** erect, 5–40 cm, tufted. **Leaves** basal and cauline; blades 1–3 mm wide. **Inflorescence** of 3 to 6 well-separated spikes; the lowest bract longer than the inflorescence. **Spikes** unisexual; the uppermost male, linear, 3–12 mm long; the lower female, sessile, 3–15 mm long, with well-separated flowers, pedunculate below, sometimes arising from near the base. **Perigynia** ascending, green, ellipsoid, pilose, 2.5–4 mm; the bidentate beak 0.7–2 mm long; stigmas 3. **Female scales** often aristate, hyaline-bronze with a green midvein, shorter than the perigynia. **Achene** 3-sided, filling the perigynium. Open forest, grasslands, turf at all elevations. AK to Greenland south to CA, AZ, NM, WI, MI, WV.

The long-stalked spikes arising from the base help distinguish this inconspicuous sedge; it lacks the dark female scales of *Carex concinna* and *C. concinnoides*. Plants of exposed, often high-elevation sites are more caespitose and have shorter beaks; they have been segregated as *C. deflexa* or *C. brevipes*, but the variation appears continuous and does not seem to warrant recognition at the species level.

Carex rostrata Stokes Rhizomatous. **Stems** erect, 40–80 cm. **Leaves** basal and cauline; blades 2–4 mm wide, glaucous-green, inrolled. **Inflorescence** of 4 to 6 well-separated spikes; the lowest bract longer than the inflorescence. **Spikes** unisexual, ascending; the uppermost 1 to 3 male, linear; the lower female, 1–5 cm long, subsessile to pedunculate, cylindrical. **Perigynia** spreading, green to tan, broadly ovate, inflated, glabrous, 3.5–5.5 mm long; the distinct beak bidentate, 1–2 mm long; stigmas 3. **Female scales** acuminate, brown with a pale midstripe, smaller than the perigynia. **Achene** 3-sided, much smaller than the perigynium. *Sphagnum* fens; montane; Flathead and Missoula cos. Circumboreal south to WA, ID, MT, WI.
Carex utriculata has yellow-green leaf blades that are wider and not inrolled.

Carex rupestris All. Curly sedge Rhizomatous. **Stems** erect, 2–15 cm, tufted. **Leaves** mainly basal; blades 1–3 mm wide, curled at the tip. **Inflorescence** a solitary bractless spike. **Spike** narrow, 1–2 cm long; male flowered portion usually 5–10 mm long; female flowers below, few. **Perigynia** narrowly ellipsoid, 2–3.5 mm long, green below, brown above, with a beak ≤0.5 mm long; stigmas 3. **Female scales** brown with hyaline margins, mostly concealing the perigynia. **Achene** 3-sided, nearly filling the perigynium. Gravelly soil of grasslands, fellfields, turf, often on exposed slopes, ridges; valleys to alpine, more common higher. Circumboreal south to UT, CO. (p.615)
The curled, relatively broad leaves help distinguish this sedge from other small, single-spike species.

Carex sartwellii Dewey Rhizomatous. **Stems** erect, 40–100 cm. **Leaves:** basal bladeless; cauline sheaths green-stripped opposite the blades that are 2–5 mm wide. **Inflorescence** of numerous, sessile spikes, congested in an elongate cluster; lowest bract shorter than the inflorescence. **Spikes** 5–10 mm long, all similar; unisexual or male flowers above female. **Perigynia** ascending, ovate, brown to green, glabrous, 3–4 mm tapered to the indistinct, serrulate beak 0.5–1 mm long; stigmas 2. **Female scales** tan with hyaline margins, shorter than the perigynia. **Achene** 2-sided, nearly filling the perigynium. Marshes, along streams; valleys to montane. NT south to ID, CO, MO, NY.
Carex diandra forms tussocks and has white leaf sheaths.

Carex saxatilis L. Rhizomatous. **Stems** erect, 20–50 cm. **Leaves** basal and cauline; blades 1–4 mm wide. **Inflorescence** of 2 to 4 well-separated spikes; the lowest bract greater than the inflorescence. **Spikes** unisexual; the upper 1 or 2 male, linear, 8–25 mm long; the lower female, ascending to arching, 1–2 cm long, pedunculate, short-cylindric. **Perigynia** ascending, tan to brown, elliptic, glabrous, 3–4.5 mm long with a distinct beak ≤0.5 mm long; stigmas 2 or rarely 3. **Female scales** ovate, brown to purplish with a pale midvein and margins, shorter than the perigynia. **Achene** 2- or rarely 3-sided, smaller than the perigynium. Wet, gravelly or shallow soil on margins of lakes, streams; valleys to subalpine, more common higher. Circumboreal south to WA, UT, CO. Our plants are var. *major* Olney.
Carex vesicaria, *C. utriculata* and other members of this group all have longer beaks and 3-sided achenes.

Carex saximontana Mack. [*C. backii* Boott misapplied] **Stems** erect, 25–35 cm. **Leaves** basal and lower-cauline, white-margined, often greater than the inflorescence; blades 2–5 mm wide. **Inflorescence** of 1 to 4 well-separated spikes; lowest bract greater than the inflorescence. **Spikes** erect to ascending, all similar, pedunculate, 5–10 mm long, not including the basal female scales; male flowers above, few, inconspicuous. **Perigynia** 2 to 6, ascending, green, obovoid, papillose, loosely aggregated, 3–4.5 mm long, with a serrulate beak 0.5–1 mm long; stigmas 3. **Female scales** green, elongate, bract-like below, ovate and shorter above. **Achene** 3-sided, filling the perigynium. Pine forest, deciduous woodlands, thickets; plains. SK, MB south to CO, NE, MN. See *C. backii*.

Carex scirpoidea Michx. Dioecious, rhizomatous. **Stems** erect, 5–40 cm, tufted. **Leaves** basal and cauline or basal bladeless; blades 1–3 mm wide. **Inflorescence** a solitary spike; bract inconspicuous. **Spikes** unisexual, cylindric, 1–4 cm long. **Perigynia** ovate to elliptic, sometimes flattened, brown to purple above, pilose, 2.5–4 mm long, with a distinct beak ≤0.5 mm long; stigmas 3. **Female scales** brown to purple with hyaline margins and a pale midvein, sometimes ciliate or puberulent, concealing the perigynia or nearly so. **Achene** 3-sided, smaller or nearly filling the perigynium. Circumboreal south to CA, NV, UT, CO, MI, NY. (p.614)

1. Dried, old basal leaves present; mainly high-elevations. . . . . .ssp. *pseudoscirpoidea*
1. Basal leaves bladeless; old basal leaves absent 2

2. Perigynia narrowly elliptic to lanceolate, flattened, 3–4 mm long; high-elevations. . . . . . . ssp. *stenochlaena*
2. Perigynia obovoid, 2–3 mm long; low- to mid-elevations . . . . .ssp. *scirpoidea*

Carex scirpoidea ssp. *scirpoidea* occurs in moist to wet, usually calcareous meadows, fens, valleys to alpine; *C. scirpoidea* ssp. *pseudoscirpoidea* (Rydb.) D.A.Dunlop and *C. scirpoidea* ssp. *stenochlaena* (T.Holm) A.Löve & D.Löve are found in alpine meadows, turf.

Carex scoparia Schkuhr ex Willd. Densely caespitose. **Stems** erect, 30–70 cm. **Leaves:** basal blades absent; cauline blades 1–3 mm wide. **Inflorescence** of 3 to 10 sessile overlapping spikes, aggregated above. **Spikes** 6–12 mm long, all similar; female flowers above male. **Perigynia** erect, green to pale brown, lanceolate, flattened, wing-margined, 4–5 × 1.5–2 mm, tapered to the serrulate beak 1.5–2 mm long; stigmas 2. **Female scales** light brown with a green midvein and hyaline margins, shorter and narrower than the perigynia. **Achene** 2-sided, much smaller than the perigynium. Wet meadows, marshes, along rivers; valleys; Missoula and Ravalli cos. BC to NL south to CA, ID, MT, KS, MS, GA.
Perigynia of *Carex bebbii* are not as narrow. See *C. crawfordii*.

Carex scopulorum T.Holm [*C. prionophylla* T.Holm] Rhizomatous. **Stems** erect, 8–70 cm. **Leaves** basal and/or cauline; blades 2–6 mm wide. **Inflorescence** of 3 to 5 overlapping spikes; the lowest bract shorter than the inflorescence. **Spikes** ascending to erect; the uppermost male; the lower female, subsessile to short-pedunculate, 8–25 mm long, cylindric. **Perigynia** erect, green to tan, brown above, lanceolate to suborbicular, veinless, glabrous, papillose, 2–3 mm long with a minute beak. **Female scales** acute to rounded, almost completely blackish, narrower and subequal to the perigynia. **Achene** 2-sided, smaller than the perigynium. Wet meadows, thickets, fens. YT south to CA, UT, CO.

1. Basal leaf sheaths <6 cm long; perigynium rounded on top. . . . . var. *bracteosa*
1. Basal leaf sheaths 5–15 cm long; perigynium acute . . . . var. *prionophylla*

Carex scopulorum var. *bracteosa* (L.H.Bailey) F.J.Herm. is found in the subalpine zone, rarely lower; *C. scopulorum* var. *prionophylla* (T.Holm) L.A.Standl. occurs montane, lower subalpine.

Carex siccata Dewey [*C. foenea* Willd. misapplied] Rhizomatous. **Stems** erect, 15–40 cm. **Leaves:** basal bladeless; cauline blades 1–3 mm wide. **Inflorescence** of 6 to 12 sessile, loosely aggregated spikes; lowest bract inconspicuous. **Spikes** all similar, 5–10 mm long, unisexual or bisexual with male flowers above. **Perigynia** ascending, brown, ovate, 4–6 mm long, tapered to the distinct, serrulate, flattened beak 1–3 mm long; stigmas 2. **Female scales** brown with a pale midvein and silvery margins, shorter than the perigynia. **Achene** 2-sided, filling the perigynium. Grasslands, open pine forest; valleys, montane, known from Carbon Co. YT to QC south to WA, AZ, NM, IL, PA.
There is a great deal of confusion on the identity of this species.

Carex simulata Mack. Rhizomatous. **Stems** erect, 15–60 cm. **Leaves:** basal bladeless; cauline blades 1–2 mm wide. **Inflorescence** of 4 to 12 overlapping spikes in an oblong to ovoid cluster; lowest bract inconspicuous. **Spikes** sessile, similar, ascending, ovoid, 5–10 mm long; sometimes unisexual or male flowers above few female. **Perigynia** erect, broadly ovate, brown, shiny, glabrous, ca. 2 mm long with a beak ≤0.5 mm long; stigmas 2. **Female scales** ovate, tan with pale margins and midvein, hiding the perigynia. **Achene** 2-sided, filling the perigynium. Wet, often organic soil of fens, wet meadows; valleys, montane. AB, SK south to CA, AZ, NM. See *C. praegracilis*.

Carex spectabilis Dewey Short-rhizomatous. **Stems** erect, 20–60 cm, loosely tufted. **Leaves** basal and cauline; basal leaves sometimes bladeless; cauline blades 3–6 mm wide. **Inflorescence** of 3 to 6 overlapping spikes; lowest bract shorter than the inflorescence. **Spikes:** uppermost male 1–2 cm long; lower female, pedunculate, ascending, 1–3 cm long, ovoid to cylindric, lowest sometimes arching. **Perigynia** ascending, green and purplish, glabrous, flattened, narrowly elliptic, 3.5–4.5 × 1.5–2 mm with a distinct beak ca. 0.5 mm long; stigmas 3. **Female scales** lanceolate, acuminate, blackish with a pale midvein sometime prolonged into an awn tip, as wide and ca. as long as the perigynia. **Achene** 3-sided, smaller than the perigynium. Moist to wet meadows, turf, forest openings; subalpine, rarely alpine. AK south to CA, UT, WY. See *C. paysonis*. (p.616)

Carex sprengelii Dewey ex Spreng. Caespitose from short rhizomes. **Stems** erect, 20–70 cm. **Leaves** basal and cauline; blades 2–4 mm wide. **Inflorescence** of 3 to 6 pedunculate spikes; the lowest bract ca. as long as the inflorescence. **Spikes:** the uppermost male, 10–25 mm long; the lower female, cylindric, arching, 10–35 mm long. **Perigynia** ascending, green to tan, glabrous, elliptic, 4–6.5 mm long including the distinct beak 1.5–3.5 mm long; stigmas 3. **Female scales**

hyaline-tan with a green or tan midstripe, nearly as long as the perigynia. **Achene** 3-sided, filling the perigynium. Woodlands, open forest, thickets, moist meadows; plains, valleys, montane. BC to QC south to CO, MO, NY.

Carex stenoptila F.J.Herm. Densely caespitose. **Stems** erect, 15–75 cm. **Leaves** basal and cauline; blades 1–2 mm wide. **Inflorescence** of 7–10 sessile spikes congested in a dense ovoid to suborbicular head; lowest bract short. **Spikes** 5–8 mm long, all similar; female flowers above; male below. **Perigynia** ascending, green to tan, narrowly lanceolate, barely wing-margined, 4–5 × 1–1.5 mm including the ca. 1.5 mm long beak, serrulate below, entire above; stigmas 2. **Female scales** narrowly ovate, brown with hyaline margins, shorter than the perigynia. **Achene** 2-sided, smaller than the perigynium. Rocky soil of open forest; montane. MT, WY, UT, CO.

    Carex microptera is similar but has more winged perigynia, though some plants appear to have perigynia of both kinds.

Carex stevenii (T.Holm) Kalela [*C. norvegica* Retz. ssp. *stevenii* (T.Holm) D.F.Murray] Caespitose. **Stems** erect, 10–50 cm. **Leaves** basal and lower-cauline; blades 2–3 mm wide. **Inflorescence** of 2 or 3 pedunculate, ascending spikes; the lowest separate; lowest bract shorter than the inflorescence. **Spikes** 4–9 mm long, all similar; uppermost bisexual with male flowers basal; lower all female. **Perigynia** erect, bronze to dark brown, obovate, 1.5–2.5 mm long including the pappilose beak <0.5 mm long; stigmas 3. **Female scales** blackish with pale margins, shorter than the perigynia. **Achene** 3-sided, smaller than the perigynium. Wet meadows, along streams; subalpine. ID, MT, WY, UT, CO, NM.

    Carex stevenii seems like a cordilleran form of *C. media*.

Carex stipata Muhl. ex Willd. Densely caespitose. **Stems** erect, 20–80 cm. **Leaves** basal and cauline; blades 5–11 mm wide; upper sheaths cross-corrugate ventrally. **Inflorescence** of numerous sessile spikes arranged on branches of a dense, narrowly lanceolate head; lowest bract inconspicuous. **Spikes** 3–9 mm long, all similar; male flowers above, inconspicuous; female below. **Perigynia** spreading to ascending, lanceolate, convex, veiny, green to brown, 4–5 × 1–1.5 mm; the serrulate beak 2–3 mm long; stigmas 2. **Female scales** hyaline-tan with a green midvein, shorter than the perigynia. **Achene** 2-sided, filling the base of the perigynium. Thickets, swamps, marshes, often around lakes, ponds, streams; valleys, montane. Throughout Canada and U.S.; Asia. Our plants are var. *stipata*.

    Carex cusickii has smaller, more ovate perigynia and lacks cross-corrugated leaf sheaths.

Carex sychnocephala J.Carey Caespitose. **Stems** erect, 5–50 cm. **Leaves** basal and cauline; blades 1–3 mm wide. **Inflorescence** of 4 to 12 sessile spikes in a dense head; lower bracts erect, much longer than the inflorescence. **Spikes** 4–8 mm long, all similar; female flowers above few male. **Perigynia** erect, linear-lanceolate, green to tan, wing- and serrulate-margined, 3–6.5 × 0.8–1 mm, tapered to the serrulate beak 3–5 mm long; stigmas 2. **Female scales** lanceolate, acuminate, tan-hyaline with a green midvein, shorter or longer than the perigynia. **Achene** 2-sided, smaller than the perigynium. Vernally wet soil around lakes, ponds; plains, valleys. AK to QC south to WA, CO, MO, NY.

    Some plants appear to flower their first year.

Carex tahoensis Smiley [*C. xerantica* L.H.Bailey misapplied] Densely caespitose. **Stems** erect, 15–30 cm. **Leaves** basal and cauline; blades 1–3 mm wide. **Inflorescence** of 4 to 6 sessile, overlapping spikes; lowest bract inconspicuous. **Spikes** 9–13 mm long, all similar; female flowers above male. **Perigynia** erect, green to brown, narrowly ovate, convex, wing-margined, ca. 5.5 × ca. 2 mm, tapered to a serrulate beak 1–2 mm long; stigmas 2. **Female scales** bronze with wide hyaline margins, concealing the perigynia. **Achene** 2-sided, smaller than the perigynium. Grasslands, meadows; valleys to lower subalpine. YT south to CA, UT, CO. (p.618)

    Material from the Rocky Mtns. formerly referred to *Carex xerantica* was determined to be *C. tahoensis*.[269]

Carex tenera Dewey Caespitose. **Stems** erect, 20–80 cm. **Leaves**: basal blades absent; cauline blades 2–3 mm wide. **Inflorescence** somewhat nodding, of 4 to 8 sessile, well-separated spikes. **Spikes** 4–8 mm long, all similar; female flowers above male. **Perigynia** erect, green to brown, ovate, narrowly wing-margined, 3.5–4 × 1.5–2 mm long, gradually tapering to the indistinct, serrulate beak 1–1.5 mm long; stigmas 2. **Female scales** pale tan with hyaline margins, shorter than the perigynia. **Achene** 2-sided, just smaller than the perigynium. Riparian meadows; valleys. BC to QC south to OR, WY, MO, KY, SC.

    Carex bebbii has more clustered spikes.

Carex tenuiflora Wahlenb. Caespitose. **Stems** weak, often prostrate, 20–50 cm. **Leaves** basal and cauline; blades ca. 1 mm wide. **Inflorescence** of 2 to 4 sessile, aggregated spikes; lowest bract inconspicuous. **Spikes** 4–6 mm long, all similar, bisexual; female flowers above. **Perigynia** spreading to ascending, pale green, narrowly ovoid, 3–3.5 mm long with an inconspicuous beak; stigmas 2. **Female scales** hyaline to tan with a green midvein, shorter than the perigynia. **Achene** 2-sided, filling the perigynium. Hummocks in *Sphagnum* fens; montane. Circumboreal south to BC, MT, CO; collected once in Flathead Co.

Carex disperma and *C. leptalea* also have lax stems, beakless perigynia and occur in fens, but the former has widely separated spikes, while the latter has only one spike.

Carex torreyi Tuck. Short-rhizomatous. **Stems** erect, 15–50 cm, tufted. **Leaves** basal and cauline, pubescent; blades 1–3 cm wide, pubescent. **Inflorescence** of 3 to 7 overlapping, short-pedunculate spikes; the lowest bract shorter or longer than the inflorescence. **Spikes** ascending; the uppermost male, 3–10 mm long; lower female, 6–12 mm long. **Perigynia** ascending to spreading, green, obovoid, glabrous, veiny, ca. 3 mm long with a beak <0.5 mm long. **Female scales** ovate, short-awned, tan with hyaline margins, shorter than the perigynia. **Achene** 3-sided, filling the perigynium. Woodlands, open forest; plains, valleys. AB to MB south to CO, SD, WI.

Carex utriculata Boott [*C. rostrata* Stokes misapplied] BEAKED SEDGE Rhizomatous. **Stems** erect, 50–100 cm. **Leaves** basal and cauline; blades 4–10 mm wide, yellowish-green, not inrolled. **Inflorescence** of 4 to 8 well-separated spikes; the lowest bract greater than the inflorescence. **Spikes** unisexual; the uppermost 1 to 3 male, linear; the lower female, ascending, 2–10 cm long, subsessile to pedunculate, cylindric. **Perigynia** perpendicular-spreading, green to tan, elliptic, inflated, glabrous, 4–6 mm long; the distinct narrow bidentate beak ca. 1.5 mm long; stigmas 3. **Female scales** acuminate to awned, brown with hyaline margins and pale midvein, smaller than the perigynia. **Achene** 3-sided, much smaller than the perigynium. Abundant in wet soil to standing water of marshes, fens, swamps, around streams, ponds; plains, valleys to subalpine. Circumboreal south to CA, AZ, NM, SD, TN, VA. (p.614)

This is our most common, coarse, wetland sedge. It has been called *Carex rostrata*,[113,187,188] but this name correctly applies to a similar but much less common species confined to northern fens and bogs.

Carex vaginata Tausch. Long-rhizomatous. **Stems** erect, 10–40 cm. **Leaves** glabrous, basal and lower-cauline; blades 2–4 mm wide. **Inflorescence** of 3 to 4 well-separated spikes; the lowest bract shorter than the inflorescence, sheathing. **Spikes** unisexual; the uppermost male, narrow, 1–2 cm long; the lower female, ascending, pedunculate, 1–3 cm long, cylindric. **Perigynia** ascending, green to brown, narrowly elliptic, glabrous, 3–5 mm long with a beak 0.5–1.2 mm long; stigmas 3. **Female scales** with a green midstripe and hyaline margins, shorter than the perigynia. **Achene** 3-sided, nearly filling the perigynium. Fens, spruce swamps; valleys. Circumboreal south to MT, WI, NY. Known only from Lincoln Co.[380]

Carex vallicola Dewey Caespitose. **Stems** erect, 20–40 cm. **Leaves**: basal and cauline; blades 1–2 mm wide. **Inflorescence** of 5 to 10 sessile spikes crowded in a cylindric head; lowest bract inconspicuous. **Spikes** 4–7 mm long, all similar; male flowers above the few female. **Perigynia** ascending to spreading, green to tan, ellipsoid, ca. 3–4 × 1.5–2 mm, with a short, distinct, entire beak; stigmas 2. **Female scales** hyaline with a dark midstripe, shorter but as wide as the perigynia. **Achene** 2-sided to orbicular, tightly filling the perigynium. Grasslands, sagebrush steppe, open forest, woodlands; plains, valleys to lower subalpine. BC to CA, AZ, NM, SD.

Carex vesicaria L. [*C. exsiccata* L.H.Bailey] Caespitose from short rhizomes. **Stems** erect, 20–60 cm. **Leaves** basal and cauline; blades 3–7 mm wide. **Inflorescence** of 3 to 6 separate to overlapping spikes; the lowest bract greater than the inflorescence. **Spikes** unisexual, ascending; the uppermost 1 to 3 male, linear; the lower female, 2–5 cm long, subsessile to short-pedunculate, cylindric. **Perigynia** ascending to erect, green to purplish, lanceoloid to ovoid, often partly inflated, glabrous, 6–8 mm long, tapered to a beak 2–3 mm long; stigmas 3. **Female scales** acuminate, brown with a pale midstripe and margins, awn-tipped, smaller than the perigynia. **Achene** 3-sided, much smaller than the perigynium. Marshes, wet meadows, shores, wetlands that dry later in the summer; valleys, montane. Circumboreal south to CA, UT, MO, VA.

Our plants are often classified into two varieties or species: var. *major* Boott [*C. exsiccata*] has a longer, narrower perigynia with an indistinct beak compared to var. *vesicaria*. Characters overlap, and both perigynia shapes can occur on the same plant. Perigynia of *C. utriculata* have a more distinct beak and are usually nearly horizontal; those of *C. lacustris* have short, stout beaks.

Carex viridula Michx. [*C. oederi* Retz. var. *viridula* (Michx.) Kük.] Caespitose. **Stems** erect, 3–50 cm. **Leaves** basal and cauline; blades 1–2 mm wide. **Inflorescence** of 3 to 5 ascending, overlapping (rarely remote) spikes; the lowest bract longer than the inflorescence. **Spikes**: the uppermost male, 5–15 mm long; the lower female, 3–12 mm long, ellipsoid, subsessile (pedunculate). **Perigynia** spreading, green to tan, obovoid, glabrous, 2–3.5 mm long with a distinct, entire beak 0.5–1 mm long; stigmas 3. **Female scales** hyaline-bronze with a green or pale midvein, smaller than the perigynia. **Achene** 3-sided, nearly filling the perigynium. Fens, shores of lakes, streams, often where calcareous; plains, valleys, montane. Circumboreal south to CA, UT, NM, IL, PA. See *C. flava*.

Carex vulpinoidea Michx. Caespitose. **Stems** erect, 40–80 cm. **Leaves**: basal bladeless; cauline sheaths cross-wrinkled opposite the blades that are 2–6 mm wide. **Inflorescence** elongate-congetsed, of numerous, sessile spikes, clustered on cylindric, spreading to ascending, overlapping branches subtended by conspicuous linear bracts. **Spikes** 4–7 mm long, all similar; male flowers above female or all female. **Perigynia** ascending to spreading, ovate, green to gray, glabrous, 2.5–3.5 mm tapered to the flattened, serrulate beak, 1–1.5 mm long; stigmas 2. **Female scales** hyaline-tan with a green midvein, awned, as long as but narrower than the perigynia. **Achene** 2-sided, nearly filling the perigynium. Wet meadows, marshes; plains, valleys. Throughout temperate and boreal N. America.

Carex xerantica L.H.Bailey Densely caespitose. **Stems** erect, 15–70 cm. **Leaves** basal and cauline; blades 1–3 mm wide. **Inflorescence** of 3–5 sessile spikes, separate below, overlapping above; lowest bract inconspicuous. **Spikes** 7–14 mm long, all similar; female flowers above; male below. **Perigynia** erect, tan, ovate, convex, wing-margined, 4–5 × 1.5–2 mm, tapered to a serrulate beak 1–2 mm long; stigmas 2. **Female scales** white to silvery with a darker midvein, hiding the perigynia. **Achene** 2-sided, smaller than the perigynium. Grasslands, woodlands; plains. BC to ON south to WY, NE, MN. Reported for MT,[269] but I have seen no specimens; see *C. tahoensis*.

Cyperus L. Flatsedge

Annual or perennial herbs. **Stems** triangular. **Leaves** basal and cauline with a closed sheath. **Inflorescence** of few to numerous, sessile or pedunculate spikelets arranged in capitate clusters (spikes), the clusters sometimes also arranged in clusters, each subtended by a leaf-like, sheathless bract longer than the spike. **Spikelets** of several to numerous, stiff, overlapping scales alternating on opposite sides of the axis. **Scales** with a stiff, green midvein. **Flowers** perfect, each at the base of a scale; perianth and perigynium absent; stamens 3; style 3- (sometimes 2-) branched. **Achene** 2- or 3-sided.

1. Plants rhizomatous perennials with bulbous-based stems. . . . . *C. schweinitzii*
1. Plants fibrous-rooted annuals or short-lived perennials 2

2. Spikelets borne in elongate spikes, longer than wide. 3
2. Spikelets borne in capitate spikes, ca. as long as wide 4

3. Scales 1.5–2 mm long; spikelets somewhat flattened . . . . *C. erythrorhizos*
3. Scales 3–4 mm long; spikelets cylindric. . . . . *C. strigosus*

4. Scales prominently many-veined with a long, spreading to reflexed tip. . . . . *C. squarrosus*
4. Scales with 3 or fewer prominent veins and short tips 5

5. Scales acuminate with 3 prominent veins . . . . *C. acuminatus*
5. Scales rounded and apiculate at the tip with only the midvein prominent . . . . *C. bipartitus*

Cyperus acuminatus Torr. & Hook. Caespitose annual. **Stems** ascending to erect, 2–5 cm. **Leaves** basal, 0.5–1 mm wide, U-shaped in cross-section. **Inflorescence** of hemispheric spikes of sessile spikelets. **Spikelets** flattened, 3–7 mm long. **Scales** acuminate, yellowish, 1–2.5 mm long, 3-veined with short ascending tips. **Flowers**: stamen 1; style 3-branched. **Achene** narrowly obovate, 3-sided, brown, ca. 1 mm long. Riverbanks, gravel bars; valleys. Throughout the U.S., Mexico. Known from Cascade and Sanders cos.

Cyperus bipartitus Torr. [*C. rivularis* Kunth] Caespitose annual. **Stems** ascending to erect, 2–8 cm. **Leaves** mainly basal, 0.5–2 mm wide, U-shaped in cross-section. **Inflorescence** of hemispheric spikes of subsessile spikelets. **Spikelets** compressed, 3–15 mm long. **Scales** purplish, 2–2.5 mm long with rounded, minutely apiculate tips and obscure veins. **Flowers**: stamens 2 or 3; style 2-branched. **Achene** obovate, black, ca. 1 mm long, 2-sided, apiculate. Moist, annually-flooded river gravels; valleys. Throughout temperate N. America to S. America. Known from Ravalli Co.

Cyperus erythrorhizos Muhl. Caespitose annual. **Stems** ascending to erect, 10–70 cm. **Leaves** mainly basal, 2–9 mm wide, V- or W-shaped in cross-section. **Inflorescence** of hemispheric clusters of ovoid spikes of sessile spikelets. **Spikelets** cylindric, 3–10 mm long. **Scales** gold, apiculate, 1.5–2 mm long with 3 obscure, median veins. **Flowers**: stamens 3; style 3-branched. **Achene** elliptic, brown, 3-sided, 0.5–1 mm long, apiculate. Riverbanks; plains, valleys. Throughout most of temperate U.S., Mexico. Known from Prairie Co.

Cyperus schweinitzii Torr. Rhizomatous perennial. **Stems** erect, 10–40 cm, bulbous-based. **Leaves** basal and lower-cauline, 1–2 mm wide, U-shaped in cross-section. **Inflorescence** of ovoid spikes of sessile spikelets. **Spikelets** somewhat compressed, 6–12 mm long. **Scales** green to tan, apiculate, 2.5–3.5 mm long, prominently veined. **Flowers**: stamens 3; style 3-branched. **Achene** elliptic, brown, 3-sided, 2–2.5 mm long, apiculate. Sandhills, dunes, beneath sandstone outcrops; plains. AB to QC south to WA, NM, TX, KY, NY, Mexico.

Cyperus squarrosus L. [*C. aristatus* Rottb.] Caespitose annual. **Stems** ascending to erect, 1–5 cm. **Leaves** basal, 0.5–2 mm wide, mostly flat. **Inflorescence** of hemispheric to obovoid spikes of sessile or subsessile spikelets. **Spikelets** flattened, 3–8 mm long. **Scales** long-acuminate, light brown, ca. 1.5 mm long, 7- to 9-veined with long spreading to reflexed tips. **Flowers**: stamen 1; style 3-branched. **Achene** narrowly obovate, 3-sided, brown, 0.5–1 mm long. Sandy or gravelly, vernally-flooded soil along rivers; plains, valleys. Throughout temperate and tropical N. America, cosmopolitan. (p.641)

Cyperus strigosus L. Caespitose annual to perennial. **Stems** ascending to erect, 10–60 cm. **Leaves** mainly basal, 2–10 mm wide, flat. **Inflorescence** of ovoid to hemispheric clusters of spikes of sessile spikelets. **Spikelets** cylindric, 4–10 mm long. **Scales** pale brown with a green midvein, 3–4 mm long, several-veined. **Flowers**: stamens 3; style 3-branched. **Achene** narrowly elliptic, dark brown, 3-sided, ca. 2 mm long. Streambanks; valleys; Flathead and Missoula cos. Throughout U.S. and adjacent Canada.

Dulichium Per.

Dulichium arundinaceum (L.) Britton Rhizomatous, perennial herb. **Stems** terete, solitary, erect, hollow, 25–70 cm. **Leaves** cauline; the lower sheaths brown, bladeless; upper blades leathery, 1–12 cm × 2–8 mm. **Inflorescence** of axillary spikes 15–35 mm long, each composed of several spikelets. **Spikelets** cylindric, 10–25 mm long, of 2-ranked, overlapping, green to tan, acuminate scales 5–10 mm long. **Flowers** perfect, each at the base of a scale; perianth of 6 to 9 retrorsely barbed bristles 5–8 mm long; stamens 3; style 2-branched. **Achene** 2-sided, narrowly ellipsoid, short-stalked, 2–4 mm long. Shallow water or saturated peat of fens; valleys, lower montane. Boreal N. America south to CA, WY, TX, AL, FL.

Eleocharis R.Br. Spike rush

Annual or perennial herbs. **Stems** terete, appearing leafless. **Leaves** 2, basal, bladeless. **Inflorescence** a solitary, terminal, bractless spikelet. **Spikelet** ovoid to conical, of numerous spirally-arranged scales; all but the very lowest subtending a flower. **Flowers** perfect; perianth of 0 to 10 barbed bristles; stamens usually 3; styles 2- or 3-branched, thickened at the base. **Achene** obovoid, or obovate and 2-sided (lenticular) with a crown (tubercle) on top.

Eleocharis atropurpurea (Retz.) J.Presl & C.Presl has been reported for MT,[113] but I have seen no specimens. *Eleocharis flavescens* (Poir.) Urb. is known from Yellowstone National Park and reported for MT,[187] but I have seen no specimens.

1. Stigmas 3; achenes obovoid, nearly round in cross-section 2
1. Stigmas 2; achenes mostly 2-sided (lenticular) 6

2. Achene 0.5–1.5 mm long including the tubercle. 3
2. Achene 2–3 mm long 5

3. Scales <1.5 mm long; plants without rhizomes . . . .*E. bella*
3. Scales 1.5–3 mm long 4

4. Mature achenes yellow; lowest scale sterile, rounded, hyaline, clasping the stem;
   leaf sheaths purplish. . . . .*E. tenuis*
4. Mature achene gray; lowest scales subtending a flower, similar to the ones above;
   leaf sheaths usually not purplish . . . .*E. acicularis*

5. Lowest scale fertile, not enlarged or completely hyaline; flowers 3 to 9. . . . .*E. pauciflora*
5. Lowest scale sterile, enlarged, hyaline, often red-spotted; some stems with >9 flowers . . . . *E. rostellata*

6. Achenes 2–3 mm long; tubercle ≥0.5 mm long, set off from achene by a short stalk . . . .*E. palustris*
6. Achenes <2 mm long; tubercle smaller and confluent with the body 7

7. Scales <20 per spikelet; tubercle conical . . . . *E. flavescens*
7. Scales >20 per spikelet; tubercle flattened-triangular . . . . *E. ovata*

Eleocharis acicularis (L.) Roem. & Schult. Slenderly rhizomatous perennial. **Stems**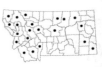
ascending to erect, 1–25 cm, tufted. **Spikelet** 2–7 mm long with 3 to 15 flowers. **Scales** purplish
with a green midrib and hyaline margins, 1.5–2.5 mm long. **Flowers**: bristles 0 to 4, exceeding
the achene; stigmas 3. **Achenes** gray, honeycombed, obovoid, ca. 1 mm long with a flattened,
triangular tubercle. Muddy shores or shallow water on margins of ponds, lakes, streams, ditches;
plains, valleys, montane. Circumboreal to S. America.
    The fine rhizomes that distinguish this from the much less common *Eleocharis bella* can be difficult to detect.

Eleocharis bella (Piper) Svenson Fibrous-rooted annual. **Stems** ascending to erect, 1–8 cm,
tufted. **Spikelet** 1–3 mm long with 3 to 15 flowers. **Scales** ca. 1 mm long, purplish with a green
midrib and hyaline margins. **Flowers**: bristles absent; stigmas 3. **Achenes** whitish, honeycombed,
obovoid, 0.5–1 mm long with a flattened, apiculate tubercle. Vernally wet soil along ponds,
streams; valleys to montane. WA to MT south to CA, AZ, NM. See *E. acicularis*.

Eleocharis ovata (Roth) Roem. & Schult. [*E. engelmannii* Steud., *E. obtusa* (Willd.) Schult.]
Fibrous-rooted annual. **Stems** ascending to erect, 4–20 cm, tufted. **Spikelet** 4–10 mm long with
numerous flowers. **Scales** brown with hyaline margins and a lighter midrib, 2–2.5 mm long.
Flowers: bristles 0 to 7, equaling or exceeding the achene; stigmas usually 2. **Achenes** gray to
brown, smooth, obovate, lenticular, ca. 1.5 mm long with a broad, low tubercle 0.1–0.2 mm high.
Vernally wet to shallowly submersed soil on the margins of ponds, lakes, sloughs; valleys. Circumboreal south to AZ,
OK, TN, VA.
    Alternative treatments recognize 3 species with overlapping ranges in western N. America[371] based on overlapping
characters; further study is needed.[371]

Eleocharis palustris (L.) Roem. & Schult. (*E. macrostachya* Britton, *E. erythropoda* Steud.)
Rhizomatous perennial. **Stems** erect with purple bases, 7–90 cm solitary or few together.
Spikelet 5–25 mm long with numerous flowers. **Scales** brown to purple with a green or tan
midrib, 2.5–4 mm long, the lowest 1 or 2 empty, bract-like. **Flowers**: bristles 4 to 6, exceeding the
achene; stigmas 2. **Achenes** yellow to brown, smooth, obovate, lenticular, 2–3 mm long including
the short-stemmed triangular tubercle 0.5–1 mm long. Vernally wet to shallow water of wet meadows, margins of lakes,
ponds, marshes, fens; plains, valleys, montane. Circumboreal south through U.S., Mexico. (p.641)
    This is our common, conspicuous, low-elevation spike rush. Alternative treatments[371] recognize at least 3 species with
broadly overlapping ranges and ecology. Distinctions are based on overlapping or weak vegetative characters that have
no ecological or geographic integrity in our area.

Eleocharis pauciflora (Lightf.) Link [*E. quinqueflora* (Hartmann) O.Schwarz, *E. suksdorfiana*
Beauverd] Short-rhizomatous perennial. **Stems** erect, 5–25 cm, tufted. **Spikelet** 4–8 mm long
with ca. 5 flowers. **Scales** brown to purple, 3–6 mm long, the lowest larger. **Flowers**: bristles
absent or 3 to 6, longer or shorter than the achene; stigmas 3. **Achenes** brown, smooth, narrowly
obovate, 3-sided, 2–3 mm long including the conical tubercle confluent with the body. Moist to
wet, often peaty, often calcareous soil of fens, wet meadows, seeps; montane to rarely alpine. Circumboreal south to CA,
AZ, NM, NE, IL, NY.
    Two species are sometimes recognized in our area for what is here considered *E. pauciflora*.[371] These segregates are
separated primarily on vegetative characters that could be environmentally induced.

Eleocharis rostellata (Torr.) Torr. Caespitose perennial. **Stems** 10–80 cm, tufted, erect or
arching, rooting at the tip and forming new plants. **Spikelet** 4–10 mm long with few to many
flowers. **Scales** 2–6 mm long, brown to purple with broad hyaline margins, rounded, the lowest
empty. **Flowers**: bristles ca. 6, mostly equaling the achene; stigmas 3. **Achenes** green-grayish,
smooth, obovoid, ca. 2 mm long with a conical tubercle confluent with the body. Spring-fed,
marly marshes, calcareous fens; valleys, montane. BC to NS south through most of U.S. to Mexico.

Eleocharis tenuis (Willd.) Schult. [*E. elliptica* Kunth] Rhizomatous perennial. **Stems** erect, 10–40 cm with purple sheaths, tufted. **Spikelets** 3–8 mm long with 10 to 30 flowers. **Scales** 2–3 mm long, brown to purple with hyaline margins, the lowest larger, sterile, rounded. **Flowers**: bristles absent or reduced; stigmas 3. **Achenes** yellow, honeycombed, obovate, 1–1.5 mm long; tubercle triangular, 0.1–0.2 mm long. Fens, moist alkaline meadows around ponds, streams; valleys, montane. BC to QC south to MT. Our plants are var. ***borealis*** (Svenson) Gleason.

This variety has been segregated as *Eleocharis elliptica* based on continuous or overlapping characters.[371]

Eriophorum L. Cotton grass

Perennial herbs. **Stems** erect, terete or 3-angled. **Leaves** basal and cauline. **Inflorescence** terminal, of 1 to several spikelets subtended by 1 to few bracts. **Spikelets** appearing like a cottony tuft in fruit, obconic, of numerous spirally-arranged scales, the lowest not subtending a flower. **Flowers** perfect; perianth of 8 to 25 smooth, long bristles; stamens 1 to 3; styles 3-branched. **Achene** oblong, 3-sided.

Eriophorum brachyantherum Trautv. & C.A.Mey. has been reported for MT[187] based on a misinterpretation of a map.[198] Reports of *E. scheuchzeri*[187] are probably based on misidentified specimens of *E. callitrix*. See *Trichophorum alpinum*.

1. Spikelet solitary, erect; leaf-like involucral bracts absent 2
1. Spikelets >1, pedunculate, nodding; subtended by leaf-like involucral bracts 4

2. Bristles reddish-brown at maturity; valleys to lower subalpine . . . . *E. chamissonis*
2. Bristles white; alpine. 3

3. Plants caespitose; lowest scale nearly as long as the spikelet . . . . *E. callitrix*
3. Plants rhizomatous; lowest scale much shorter than the spikelet . . . .*E. scheuchzeri*

4. Leaf-like involucral bract solitary, shorter than the inflorescence. . . . . *E. gracile*
4. Involucral bracts >1, at least 1 as long or longer than the inflorescence 5

5. Midvein of scales disappearing well before the tip. . . . .*E. angustifolium*
5. Midvein prominent, extending to tip of scale . . . .*E. viridicarinatum*

Eriophorum angustifolium Honck. [*E. polystachion* L.] Rhizomatous. **Stems** 15–50 cm. **Leaves**: flat; blades 2–5 mm wide. **Inflorescence** 2 to 8 nodding, pedunculate spikelets; bracts ≥2, leaf-like, as long as the inflorescence. **Spikelets** 1–2 cm long in fruit; lowest scale smaller than the spikelet. **Fertile scales** brown to greenish-black with a pale midvein not reaching the tip, 5–10 mm long. **Bristles** white. **Achene** 2–5 mm long, minutely apiculate. Organic soil of fens, wet meadows; valleys to lower subalpine. Circumboreal south to OR, UT, NM.

Eriophorum callitrix Cham. Caespitose. **Stems** 7–25 cm. **Leaves**: upper leaf blades rudimentary; lower blades V-shaped or triangular in cross-section, ca.1 wide. **Inflorescence** a solitary, erect spikelet; bracts absent. **Spikelets** 1–2 cm long in fruit; lowest scale ovate, ca. as long as the spikelet. **Fertile scales** blackish with hyaline margins, 7–12 mm long. **Bristles** white. **Achene** ca. 2 mm long. Wet, organic turf; alpine; Carbon Co. Circumpolar south to BC, WY.

Eriophorum chamissonis C.A.Mey. Rhizomatous. **Stems** 15–60 cm. **Leaves**: upper leaf blades rudimentary; lower blades V-shaped or trinagular in cross-section, 1–2 mm wide. **Inflorescence** a solitary, erect spikelet; bracts absent. **Spikelets** 15–35 mm long in fruit; lowest scale lanceolate, shorter than the spikelet. **Fertile scales** brown to black with pale margins, 4–7 mm long. **Bristles** becoming tawny-brown. **Achene** 2–3 mm long, apiculate. Organic soil of fens, wet meadows; valleys to lower subalpine. Circumboreal south to OR, UT, CO, MN. (p.641)

Eriophorum gracile W.D.J.Koch ex Roth Rhizomatous. **Stems** 25–60 cm. **Leaves**: V-shaped in cross-section; blades 1–2 mm wide. **Inflorescence** 2 to 5 ascending to nodding, pedunculate spikelets; bract 1, leaf-like, shorter than the inflorescence. **Spikelets** 15–25 mm long in fruit; lowest scale ca. as long as the spikelet. **Fertile scales** ovate, greenish-black with a pale midvein, not reaching the tip, 3–4 mm long. **Bristles** white. **Achene** 1.5–3 mm long. Fens; valleys to montane. Circumboreal south to CA, NV, UT, CO, IL, PA.

Eriophorum viridicarinatum (Engelm.) Fernald Rhizomatous. **Stems** 25–60 cm. **Leaves**: flat; blades 2–5 mm wide. **Inflorescence** 2 to 10 nodding, pedunculate spikelets; bracts ≥2, leaf-like, as long as the inflorescence. **Spikelets** 15–30 mm long in fruit; lowest scale ca. as long as the

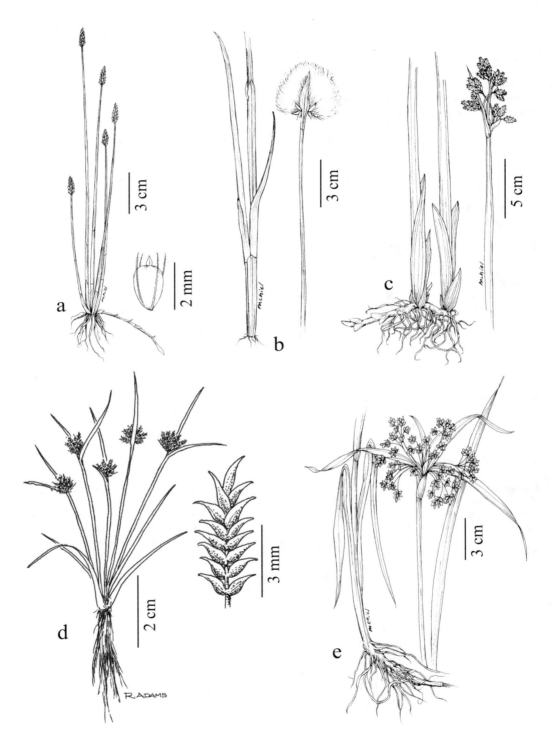

Plate 110. a. *Eleocharis palustris*, b. *Eriophorum chamissonis*, c. *Schoenoplectus acutus*, d. *Cyperus squarrosus*, e. *Scirpus microcarpus*

spikelet. **Fertile scales** 4–6 mm long, greenish-black with a pale midvein that reaches the tip. **Bristles** whitish. **Achene** 2–4 mm long, minutely apiculate. Calcareous fens; valleys, montane. AK to NL south to BC, ID, CO, IL, PA.

Kobresia Willd.

Caespitose, perennial herbs. **Stems** erect, obscurely 3-angled. **Leaves** basal and cauline with persistant sheaths. **Inflorescence** terminal, of small spikelets in simple or compound spikes, each subtended by a hyaline, scale-like bract. **Spikelets** 1- to 3-flowered, unisexual or bisexual; male flowers mostly above. **Flowers** unisexual, each wrapped in a scale (perigynium); male flowers of 3 anthers; female flowers of an ovary with 3 stigmas. **Fruit** a 3-sided, narrowly ovoid, short-beaked achene.

Very similar in general appearance to *Carex* spp.; the open perigynium is diagnostic.

1. Inflorescence compound, of few to several spikes. . . . . *K. simpliciuscula*
1. Inflorescence a simple terminal spike. 2

2. Perigynia ≤4 mm long; spikes 2–3 mm wide; subtending bracts 3–5 mm long . . . . *K. myosuroides*
2. Perigynia ≥3.5 mm long; spikes 4–5 mm wide; subtending bracts 5–6 mm long. . . . . *K. sibirica*

Kobresia myosuroides (Vill.) Fiori [*K. bellardi* (All.) Degl. ex Loisel.] **Stems** 5–15 cm. **Leaves** inrolled, erect, ca. 0.5 mm wide; basal bladeless. **Inflorescence** 6–17 mm long, a linear spike; bracts 3–5 mm long. **Spikelets:** upper male, 1-flowered; lower female, 1- or 2-flowered. **Perigynia** light brown, 2–4 mm long. **Achenes** 2–3 mm long. Moist to dry, often exposed turf; alpine. Circumpolar south to CA, NM.

Very similar in habitat and general appearance to *Carex nardina* and *C. elynoides*.

Kobresia sibirica (Turcz. ex Ledeb.) Boeck. [*K. macrocarpa* Clokey ex Mack.] **Stems** 5–15 cm. **Leaves** inrolled, ca. 0.5 mm wide; basal bladeless. **Inflorescence** 1–2 cm long, a solitary spike; bracts 5–6 mm long. **Spikelets:** upper male, 1-flowered; lower bisexual, 1- to 3-flowered. **Perigynia** light brown, 3.5–5 mm long. **Achenes** 2.5–4 mm long. Wet turf around seep areas; alpine. Circumpolar south to CO. Known from Carbon Co.

Kobresia simpliciuscula (Wahlenb.) Mack. **Stems** 5–20 cm. **Leaves** flat, 1–2 mm wide. **Inflorescence** of several aggregated spikes 3–6 mm long; bracts 3–4 mm long. **Spikelets:** upper male, 1-flowered; lower female or bisexual, 1- or 2-flowered. **Perigynia** brown, 2.5–3 mm long. **Achenes** ca. 3 mm long. Calcareous fens, alkaline meadows, moist organic turf; montane to lower alpine; Glacier and Teton cos. Circumboreal south to UT, CO.

Lipocarpha R.Br.

Lipocarpha micrantha (Vahl) G.C.Tucker [*Hemicarpha micrantha* Vahl] Tufted, glabrous annual. **Stems** erect, 5–15 cm. **Leaves** basal and lower-cauline, ≤0.5 mm wide; sheaths closed. **Inflorescence** terminal (appearing lateral) of 1 to 3 sessile spikes subtended by 2 or 3 leaf-like bracts; the longest erect and appearing to be a continuation of the stem. **Spikes** ovoid, 1–3 mm long, composed of spirally-arranged, 1-flowered spikelets; each spikelet composed of 1 or 2 scales, the outer largest, the inner subtending the ovary or absent. **Flowers** perfect, consisting of 1 stamen and an ovary with a 2-parted style. **Achenes** obovoid, dark brown, ca. 0.5 mm long. Moist, sandy soil; valleys. BC to QC south to CA, AZ, NM, TX, AL, FL. Collected in Carbon Co.

Reports of *Hemicarpha drummondii* Nees for MT[113] are referable here.

Schoenoplectus (Rchb.) Palla Bullrush

Rhizomatous perennial herbs. **Stems** erect, terete or 3-angled. **Leaves** mainly basal or lower-cauline; blades sometimes absent. **Inflorescence** terminal, of few to numerous sessile or pedunculate, capitate to open-paniculate spikelets; involucral bracts 1 to few; lowest erect, longer than the inflorescence, appearing to be an elongation of the stem. **Spikelets** terete, of several overlapping, spirally-arranged scales; scales usually with a short awn arising from the notch at the tip. **Flowers** perfect, each at the base of a scale; perianth of 4 to 8 spinulose bristles; stamens 3; style 2- or 3-branched. **Achene** obovoid, 2- or 3-sided, short-beaked.[386]

Often included in *Scirpus*.

1. Plants aquatic; stems submerged or floating on the surface . . . . *S. subterminalis*
1. Plants emergent or terrestrial. 2

2. Stems terete 3
2. Stems 3-angled 5

3. Nearly all spikelets on individual peduncles . . . . *S. heterochaetus*
3. Peduncles with clusters of sessile spikelets at the tip 4

4. Scales white to gray, contrasting with the minute red lines; awns twisted, 1–2 mm long . . . . *S. acutus*
4. Scales reddish-brown at least above, the red lines obscure; awns straight or bent,
   <1 mm long . . . . *S. tabernaemontani*

5. Main bract mostly (2)3–10 cm long; sterile, scale-like bracts present at the base
   of lowest spikelet . . . . *S. pungens*
5. Main bracts usually 1–3 cm long; scale-like bracts absent . . . . *S. americanus*

Schoenoplectus acutus (Muhl. ex Bigelow) A.Löve & D.Löve [*Scirpus acutus* Muhl. ex
Bigelow] HARDSTEM BULRUSH, TULE **Stems** terete, 50–200 cm, filled with spongy air cavities.
Leaf blades reduced to scales. **Inflorescence** open or congested with small clusters of spikelets
on tips of spreading to erect peduncles. **Spikelets** 5–10 mm long. **Scales** 3–4.5 mm long, ciliate,
sparsely puberulent, gray to tan-hyaline with short, dark red, parallel lines; awn twisted, 1–2 mm
long. **Flowers:** bristles ca. 6, ca. as long as achene; stigmas 2 or 3. **Achene** usually 2-sided, gray to nearly black, 2–3
mm long. Fresh or saline marshes around ponds, lakes, sloughs; plains, valleys, montane, rarely higher. Throughout
temperate N. America. (p.641)

Recent treatments recognize two varieties:[370] var. *acutus* has 2 stigmas and small air cavities, while var. *occidentalis*
(S.Watson) S.G.Sm. has 3 stigmas and larger cavities. However, these characters are often not correlated in our
specimens, and there is no ecological or geographic distinctions in our area.

Schoenoplectus americanus (Pers.) Volkart ex Schinz & R.Keller [*Scirpus americanus* Pers. *Scirpus olneyi* A.Gray]
Stems 40–150 cm, 3-sided with concave sides. **Leaf blades** short, flat or folded, 2–8 mm wide. **Inflorescence** compact,
of 2 to 15 sessile spikelets. **Spikelets** 6–15 mm long. **Scales** 3–4 mm long, glabrous, reddish to purplish-brown; awn
ca. 0.5 mm long. **Flowers:** bristles 5 or 6, ca. as long as achene; stigmas 2 or 3. **Achene** 2- or 3-sided with rounded
edges, brown, ca. 3 mm long. Usually saline or alkaline, wet meadows, marshes; valleys; known from Beaverhead and
Madison cos. BC to NS south to CA, AZ, NM, TX, MS, FL, Mexico, S. America.

Schoenoplectus heterochaetus (Chase) Soják [*Scirpus heterochaetus* Chase] **Stems** terete, 50–200 cm, filled
with spongy air cavities. **Leaf blades** reduced to scales. **Inflorescence** usually open with solitary spikelets on tips of
spreading to erect peduncles. **Spikelets** 5–15 mm long. **Scales** 2.5–4 mm long, lacerate to ciliate, light brown with short,
dark red, parallel lines; awn ca. 0.5 mm long. **Flowers:** bristles 4, ca. as long as achene; stigmas 3. **Achene** 3-sided, dark
brown, 2–3 mm long. Marshes around lakes, reservoirs; plains; Phillips Co. AB to QC south to CA, ID, WY, OK, NY.

Schoenoplectus pungens (Vahl) Palla [*Scirpus pungens* Vahl] [*S. americanus* Pers.
misapplied] **Stems** 3-sided, 10–200 cm. **Leaf blades** flat or folded, 1–4 mm wide. **Inflorescence**
compact, of 1 to 6 sessile spikelets. **Spikelets** 7–20 mm long. **Scales** 3–6 mm long, glabrous,
light brown to blackish; awn bent, 0.5–2 mm long. **Flowers:** bristles 4 to 8, shorter to as long
as the achene; stigmas 2 or 3. **Achene** 2- or 3-sided, brown, ca. 3 mm long. Usually saline
or alkaline, wet meadows, marshes; plains, valleys. Throughout temperate N. America to S.
America; Europe.

Schoenoplectus subterminalis (Torr.) Soják [*Scirpus subterminalis* Torr.] Aquatic. **Stems**
terete, 20–80 cm, submergent or floating, not emergent. **Leaf blades** folded, ≤1 mm wide.
Inflorescence a solitary spikelet. **Spikelet** 7–12 mm long. **Scales** 4–6 mm long, glabrous,
membranous, light brown; the tip not notched or awned. **Flowers:** bristles 6, ca. as long as
achene; stigmas 3. **Achene** brown, 3-sided, 2–4 mm long. Shallow water of lakes, ponds; valleys.
AK to NL south to OR, ID, MT.

Schoenoplectus tabernaemontani (C.C.Gmel.) Palla [*Scirpus validus* Vahl] SOFTSTEM
BULRUSH **Stems** terete, 80–150 cm, filled with spongy air cavities. **Leaf blades** reduced to scales.
Inflorescence usually open with small clusters of spikelets on tips of spreading to erect peduncles.
Spikelets 5–12 mm long. **Scales** 2–3 mm long, glabrous, ciliate, tan to reddish with short, dark
red, parallel lines; awn ca. 0.5 mm long. **Flowers:** bristles 6, ca. as long as achene; stigmas 2.
Achene 2-sided, dark brown, 1.5–3 mm long. Fresh or saline marshes around ponds, lakes, slow streams; plains, valleys.
Throughout temperate N. America.

Scirpus L. Bulrush, clubrush

Perennial herbs. **Stems** ascending to erect, 3-angled. **Leaves** mainly cauline; blades flat or folded; basal often absent. **Inflorescence** mostly terminal, umbelliform, of solitary or clustered spikelets on ascending to spreading peduncles; involucral bracts 1 to 4, leaf-like. **Spikelets** terete, of numerous overlapping, spirally-arranged, glabrous scales. **Flowers** perfect, each at the base of a scale; perianth of 3 to 6 bristles; stamens 1 to 3; style 3-branched. **Achene** obovoid, 2–3-sided, short-beaked.[386,441]

1. All or nearly all spikelets individually pedunculate 2
1. Spikelets sessile in pedunculate clusters 3

2. Basal leaf sheaths whitish; flower bristles brownish. . . . . *S. pendulus*
2. Basal leaf sheaths brown; flower bristles white . . . .*S. atrocinctus*

3. Flower bristles strongly contorted, longer than the achenes or scales. . . . .*S. cyperinus*
3. Flower bristles straight or curved, ca. as long or shorter than the scales. 4

4. Scales with an awn tip ca. 0.5 mm long; bristles ca. as long as scales . . . . *S. pallidus*
4. Scales acute to apiculate; bristles ca. as long as the achenes 5

5. Achene 2-sided; scales dark green . . . .*S. microcarpus*
5. Achene 3-sided; scales blackish. . . . . *S. atrovirens*

Scirpus atrocinctus Fernald Densely caespitose. **Stems** 30–80 cm, clustered on short rhizomes. **Leaves**: blades 2–4 mm wide; basal sheaths brown. **Inflorescence** open-cymose; nearly all spikelets pedunculate. **Spikelets** 3–5 mm long. **Scales** 1.5–2 mm long, reddish-brown to black with a green midvein, tip rounded, sometimes apiculate. **Flowers**: bristles 6, contorted, smooth, white, longer than the scales; stigmas 3. **Achene** 2- or 3-sided, whitish, 0.5–1 mm long. Marshes around ponds, reservoirs; valleys. NT south to WA, WY, SD, IL, WV. Known from Lincoln and Sanders cos.

Scirpus atrovirens Willd. **Stems** 30–80 cm, clustered on short rhizomes. **Leaves**: blades 4–8 mm wide; basal sheaths green to tan. **Inflorescence** of few or several globose heads on short peduncles. **Spikelets** 2–5 mm long. **Scales** 1–1.5 mm long, blackish with pale midribs; tip acute to apiculate. **Flowers**: bristles 6, straight or curved, white, spinulose, ca. as long as the achene; stigmas 3. **Achene** obscurely 3-sided, white to light brown, ca. 1 mm long. Fresh water marshes; valleys. MT to NL south to AZ, TX, GA. Collected in Flathead Co.

Scirpus cyperinus (L.) Kunth. Caespitose. **Stems** 80–120 cm, clustered on short rhizomes. **Leaves**: blades 2–6 mm wide; basal sheaths brown. **Inflorescence** open-cymose with spikelets clustered at tips of short or long peduncles. **Spikelets** 3–5 mm long. **Scales** ca. 1.5 mm long, brown; tip rounded to acute. **Flowers**: bristles 6, contorted, smooth, white, longer than the scales; stigmas 3. **Achene** 2- or 3-sided, whitish, 0.5–1 mm long. Shores of reservoirs; valleys. MB to NS south to TX, AL, FL, Mexico; disjunct and perhaps introduced in WA, OR, and Sanders CO., MT.

Scirpus microcarpus J.Presl & C.Presl Rhizomatous **Stems** 30–100 cm, rarely clustered. **Leaves**: blades 5–12 mm wide; basal sheaths reddish. **Inflorescence** open, compound-umbelliform. **Spikelets** 3–6 mm long. **Scales** 1–2.5 mm long, dark green; tip rounded to apiculate. **Flowers**: bristles 3–6, straight or curved, spinulose, ca. as long as the achene, white turning brown; stigmas 2. **Achene** 2-sided, whitish, 1–1.5 mm long. Marshes, beaver ponds, streambanks; valleys, montane. AK to NL south to CA, AZ, NM, KY, WV. (p.641)
Scales of *Scirpus pallidus* have awns ca. 0.5 mm long.

Scirpus pallidus (Britton) Fernald Short-rhizomatous. **Stems** 50–150 cm. **Leaves**: blades 6–15 mm wide; basal sheaths white to green. **Inflorescence** unequally umbellate, of several to numerous, pedunculate, capitate clusters. **Spikelets** 3–5 mm long. **Scales** 2–2.5 mm long, blackish with a pale midvein and an awn ca. 0.5 mm long. **Flowers**: bristles 6, straight or curved, spinulose, ca. as long as the scales; stigmas 3. **Achene** 2- or 3-sided, whitish, ca. 1 mm long. Fresh water or alkaline marshes, streambanks; plains, valleys. BC to ON south to OR, AZ, NM, TX, MO. See *S. microcarpus*.

Scirpus pendulus Muhl. **Stems** 30–80 cm, clustered on short rhizomes. **Leaves**: blades 2–6 mm wide; basal sheaths whitish. **Inflorescence** open, cymose; spikelets on individual short peduncles. **Spikelets** 3–8 mm long. **Scales** 1.5–2 mm long, brown with prominent green midribs; tip apiculate. **Flowers**: bristles 6, strongly contorted, smooth, brown, longer than the achene; stigmas 3. **Achene** 3-sided, gray-brown, ca. 1 mm long. Wet meadows, streambanks; valleys. OR to QC south to CA, AZ, NM, TX, AL, FL. Known from Missoula Co.

Trichophorum Pers.

Perennial herbs. **Stems** terete or 3-angled with old stems and leaf sheaths at the base. **Leaves** mainly basal or lower-cauline; blades sometimes absent. **Inflorescence** a solitary, terminal spikelet; involucral bracts 1 to few, scale-like, the lowest short-awned. **Spikelets** terete, of several overlapping, spirally-arranged, scales. **Flowers** perfect, each at the base of a scale; perianth of 0 to 6 smooth or scabrous bristles; stamens 3; style 3-branched. **Achene** obovoid, ca. 3-sided, short-beaked.[198]

Plants are similar to *Eleocharis* but for the small bract subtending the spikelet.

1. Bristles at least 2 times as long as the spikelet; stems 3-sided . . . . *T. alpinum*
1. Bristles absent or ca. as long as the spikelet; stems terete 2

2. Bristles absent; stems <15 cm . . . . *T. pumilum*
2. Bristles 6, stems usually >15 cm . . . . *T. cespitosum*

Trichophorum alpinum (L.) Pers. [*Eriophorum alpinum* L., *Scirpus hudsonianus* (Michx.) Fernald] **Stems** 3-angled, 10–40 cm, loosely clustered on short rhizomes. **Leaf blades** ca. 0.5 mm wide. **Spikelet** 5–7 mm long, many-flowered; awn of lowest bract inconspicuous, shorter than the spikelet. **Scales** 4–5 mm long, glabrous, brown with a pale midstripe; tips acute. **Flowers**: bristles 6, white, greatly exceeding the spikelet. **Achene** ca. 1.5 mm long. Wet organic soil of fens, seeps; montane, subalpine. Circumboreal south to BC, MT. Known from Flathead and Glacier cos.

Similar to *Eriophorum* spp. which have >6 bristles per flower.

Trichophorum cespitosum (L.) Hartm. [*Scirpus cespitosus* L.] Plants forming large hemispheric tussocks. **Stems** terete, 10–40 cm, densely clustered. **Leaf blades** reduced to scales at the base, often with 1 short blade on the lower stem, ≤1 mm wide. **Spikelet** 3–6 mm long with 2 to 4 flowers; awn of lowest bract barely longer than the spikelet. **Scales** 3–4 mm long, glabrous, brown, apiculate. **Flowers**: bristles 6, brown, barely exceeding the scale awn. **Achene** ca. 1.5 mm long. Wet, organic soil of fens, turf; montane to alpine. Circumboreal south to OR, ID, UT, MT. (p.653)

Trichophorum pumilum (Vahl.) Schinz & Thell. [*Scirpus pumilus* Vahl, *S. rollandii* Fernald] Caespitose. **Stems** terete, 5–25 cm, clustered on thin rhizomes. **Leaf blades** ca. 0.5 mm wide. **Spikelet** 3–4 mm long with 3 to 6 flowers; awn of lowest bract shorter than the spikelet. **Scales** 2.5–3.5 mm long, glabrous, brown with a pale midstripe; tip rounded. **Flowers**: bristles absent. **Achene** 1.5–2 mm long. Hummocks in calcareous fens; montane. Circumboreal south to CA, WY. Known from Glacier and Teton cos.

POACEAE (Graminae) Grass Family
Contributed by Matt Lavin

Annual or perennial, usually monoecious herbs. **Stems** terete, rarely flattened. **Leaves** basal and often cauline, alternate, linear, parallel-veined, entire, composed of a lower sheath enclosing the stem and a flat, inrolled, or folded blade. **Inflorescence** terminal, consisting of spikelets arranged in usually open to spike-like panicles, less commonly racemes or solitary spikes. **Flowers** often bisexual; each enclosed by a lemma and palea; petals and sepals reduced to 1 to 6 small translucent scales (lodicules); stamens 1 to 6; ovary superior with 1 to 3 stigmas. **Fruit** a caryopsis, rarely an achene (as in *Eragrostis* and *Sporobolus*).

The Poaceae includes about 10,035 species and 668 genera, making it one of the largest flowering plant families.[384] Poaceae is distributed world-wide from deserts to tropical rain forests and from coastal saline areas to alpine and arctic tundra. It is one of the most economically important plant families and includes rice, wheat, barley, rye, corn, millet, and sorghum, as well as many forage, ornamental, and turf species. Poaceae is distinguished from other monocots by small flowers that are each enclosed by a lemma and palea. The flower, lemma, and palea comprise the grass floret. Florets alternate along opposite sides of a rachilla to form the grass spikelet, which is delimited by a basal pair of glumes (lemma-like but sterile bracts). Grasses show outstanding morphological diversity in the architecture of seed dispersal. Seeds may be dispersed by the fruit only (*Eragrostis* and *Sporobolus*), the floret, the entire spikelet, a section of the inflorescence, or the entire inflorescence. The structures aiding dispersal of the grass fruit, whether fruits or entire inflorescences, are taxonomically useful for identifying genera. This treatment is derived from local and regional floras.[30,31,51,101,113,144,145,182,183] Grass genera and tribes are distinguished in part by whether they are cool or warm season.[89,145,202,216,372,383,401] We seldom use the taxonomic ranks of subspecies and varieties because many of these infraspecific taxa have neither geographical nor ecological integrity. If a phenotype is pronounced or aligns with ecology or geogeography, we tend to distinguish it as a species in this treatment.

Key to the traditional groups of grass genera

1. Spikelets dorsally compressed so that the dorsal surface of the glumes and lemmas is flat, rounded and rarely keeled, the glumes or lowest lemma are more or less flat and sandwich the innermost (uppermost) floret and obscure it in side view even when under a lens; disarticulation below the glumes, at least the spikelet is the unit of dispersal; each spikelet with one fruit-producing (plump) terminal floret above one sterile (reduced) or staminate (hollow) floret 2
1. Spikelets laterally compressed so that the glumes and lemmas are often keeled, the glumes and all the lemmas are more or less folded in half (keeled) along the midrib, the florets are splayed out and easily seen in side-view with a naked eye or lens; disarticulation often above the glumes; each spikelet with one or more fruit-producing (plump) floret, sometimes with sterile or staminate florets mostly above but sometimes below the fruit-producing floret. 3

2. Inflorescences of paired spikelets, one sessile and the second pedicellate (the pedicellate spikelet often reduced to a sterile floret or a small glume, or rarely entirely absent); glumes hard or bony in texture; fertile lemmas membranous, much thinner in texture than the glumes; awns usually present (sometimes early deciduous) and borne from the fertile lemmas. . . . . .Andropogoneae (p.647)
2. Inflorescences usually without paired spikelets; glumes leafy or papery in texture, the first glume often reduced to absent; fertile lemmas hard and bony in texture, much thicker in texture than the glumes; awns when present borne from glumes or sterile lemmas . . . . . . Paniceae (p.650)

3. Growth habit of stout reeds, mostly 2–3 m tall; leaf blades often 2 cm wide or wider; inflorescences dense plume-like panicles; ligule mostly of hairs . . . . Phragmites
3. Growth habit of slender herbs, usually <2 m tall; leaf blades much <2 cm wide; inflorescences not dense plume-like panicles; ligules various 4

4. Inflorescences of single, terminal bilateral (2-sided or distichous) spikes; both glumes present and often awn-like; spikelets attached laterally to main rachis (sides of lemmas face the main inflorescence rachis) [*Elymus condensatus* and *Elymus flavescens* of this tribe commonly develop a panicle rather than a spike, but each panicle branch is spike-like in having sessile spikelets arranged in a two-sided fashion] . . . . Triticeae (p.652)
4. Inflorescences of racemes or open to spike-like panicles, or of laterally or digitately-arranged spikes; if a single terminal bilateral spike, then only the second glume present (excepting the terminal-most spikelet) and never awn-like and the spikelets attached edge-wise (dorsally) to main rachis 5

5. Spikelets arranged in secund lateral or digitate spikes; ligule of hairs (but *Schedonnardus* with a membranous one); mostly one fertile floret per spikelet . . . . .Chlorideae (p.649)
5. Spikelets arranged in spikes, racemes, or panicles, if secund (e.g., *Beckmannia* in tribe Aveneae or *Dactylis* in tribe Poeae) then panicle branches not in discrete linear arrays of spikelets. 6

6. 1 floret per spikelet. 7
6. >1 floret per spikelet (including sterile or rudimentary florets) 12

7. Glumes absent, vestigial, or forming a small cup; riparian or emergent aquatic grasses . . .Oryzeae (p.650)
7. Glumes well-developed; usually other than emergent aquatic grasses 8

8. Floret bearing a tripartite awn; ligule of hairs . . . . .Aristida
8. Floret unawned or bearing a single undivided awn; ligule various. 9

9. Lemma tightly rolled around palea and enclosed fruit, hardened or bony at maturity and much thicker in texture than glumes, the lemma thus appearing as the hard seed coat; lemma awned from the tip although the awn is early deciduous in some species. . . . . Stipeae (p.652)
9. Lemma not tightly rolled around palea and enclosed fruit, membranous at maturity and of the same texture as the glumes, the lemma thus not appearing as the hard seed coat; lemma unawned, awned from the back, or awned from the tip 10

10. Ligule partially or entirely a fringe of hairs, the hairs as long or longer than a sometimes-present basal membranous portion . . . . .Eragrosteae (p.649)
10. Ligule entirely membranous, but often toothed or split, or with a very short rim of hairs along the larger basal membranous portion. 11

11. Lemmas distinctly 3-nerved, or 1-nerved (as in some species of *Muhlenbergia* and all species of *Sporobolus*); if present, the long-awn is borne from the tip of lemma; tips of glumes surpassing the tip(s) of the floret(s) or not; spikelets disarticulating above glumes; ligule membranous or hairy; inhabiting dry or saline   sites (some *Muhlenbergia* inhabit wet sites) and flowering during mid- to late-summer (C4 grasses); stems and leaves wiry or tough textured, some species occasionally more fragile and easily broken . . . . .Eragrosteae (p.649)
11. Lemmas indistinctly many-nerved (3-nerved in some species of *Alopecurus*); if present, the long awn is borne from the back of lemma (*Beckmannia*, *Cinna*, and *Hierochloe* are very short-awned from near the tip; *Apera* only is with long awn from tip); tips of glumes surpassing the tip(s) of the floret(s); spikelets disarticulating below or above glumes; ligule always membranous; inhabiting usually wet sites or flowering during the earlier part of summer (C3 grasses); stems and leaves never wiry or tough. . . .Aveneae (p.647)

12. Plants dioecious, the staminate spikelets exerted well above basal leaves, the pistillate spikelets not; collar often with conspicuous long straight hairs; ligule fringed with fine teeth; rhizomes well-developed and woody; older leaves usually with stiff, pungent tips, conspicuously distichous especially in fresh condition . . . . *Distichlis*
12. Plants mostly with bisexual spikelets but if rarely dioecious, then not with the arrangement above; collar hairless or with soft curly hairs; ligule various; rhizomes various but not woody; older leaves soft or wiry but not pungent at tips and not distichously arranged 13

13. Ligule with a conspicuous fringe of hairs or a basal membrane fringed with hairs as long as the membrane 14
13. Ligule entirely membranous or with a very short fringe of hairs on the larger membranous portion 15

14. Glumes as long as the lowest (1st) floret and usually as long or longer than all florets; lemmas indistinctly many-nerved; lemma-awn somewhat flattened dorsally and distinctly twisted, and arising from just below the distinctly bifid apex of the lemma. . . . . *Danthonia*
14. Glumes shorter than the lowest (1st) floret, if longer, then the lemmas distinctly 3-nerved; lemma-awn, if present, slender, round in cross-section, not distinctly twisted, and not arising from just below a distinctly bifid apex of the lemma . . . . .Eragrosteae (p.649)

15. Lemmas awned from at least the lower two-thirds of the back; if awnless, then one or both glumes longer than the florets; disarticulation either above or below the glumes . . . . Aveneae (p.647)
15. Lemmas awned from the tip or nearly so, often from between two small teeth; if lemmas awnless, then both glumes shorter than the florets; disarticulation always above the glumes 16

16. Venation of lemmas prominent (rib-like and parallel); tips of lemmas often (but not always) broadly rounded to truncate (*Melica subulata* and *Schizachne* have pointed lemma tips); leaf sheaths closed to the throat or nearly so at least on young leaves; unit of dispersal is the floret (*Torreyochloa* of the tribe Poeae could key here but the margins of the leaf sheath overlap – they are not fused into a solid cylinder) . . . . Meliceae (p.650)
16. Venation of lemmas rarely prominent; tips of lemmas commonly pointed (*Puccinellia*, *Torreyochloa*, and some *Poa* species have blunt lemma tips); margins of leaf sheath overlapping, except in the genus *Bromus* (*Bromus* differs from Meliceae in having hairy lemmas and a unit of dispersal that includes the fruit attached to the palea). . . . . Poeae (p.651)

Andropogoneae

1. Plants monoecious (staminate and pistillate spikelets on the same plant); pistillate spikelets on a lateral woody cob and staminate spikelets in a terminal panicle; cultivated corn or maize . . . . *Zea*
1. Plants with bisexual spikelets. 2

2. Inflorescences apparently of racemes or spikes (spikelet pairs are arranged as spikes) positioned either digitately or laterally at the end of the main stem and/or lateral branch . . . . . . *Andropogon*
2. Spikelet pairs arranged in diffuse panicles. 3

3. Pedicellate spikelets completely absent; leaf sheath with distinct flaps of tissue (auricles) projecting upward from the top of the leaf sheath at the junction with the leaf blade . . . . *Sorghastrum*
3. Pedicellate spikelets well-developed and staminate; sheath not bearing auricles of any kind. . . . . *Sorghum*

Aveneae

1. Spikelets with one floret (if glumes are strongly ribbed, see *Ventenata*) 2
1. Spikelets with two or more florets. 10

2. Spikelets oval, flat, arranged in a contracted panicle of numerous secund spikes . . . .*Beckmannia*
2. Spikelets longer than broad; panicles diffuse or contracted, but if contracted and spike-like,
   then spikelets not oval in outline. 3

3. Glumes long-awned, the awn longer than body of glume . . . . *Polypogon*
3. Glumes awnless or short-awned 4

4. Sterile lemmas appearing as one or two scales subtending and usually adhering to fertile lemma . *Phalaris*
4. Sterile lemmas not present, no scale-like structures subtending and adhering to fertile lemma 5

5. Inflorescence a compact cylindrical spike, the main rachis not visible without removal of spikelets 6
5. Inflorescence open to contracted but not a compact cylindrical spike 7

6. Glumes united at base, unawned or minutely awned and then not abruptly so; lemmas awned;
   disarticulation below glumes. . . . .*Alopecurus*
6. Glumes separate to base, abruptly short-awned; lemmas awnless; disarticulation above glumes . . *Phleum*

7. Florets with a tuft of hairs from base, the hairs 1/4 to as long as the lemma . . . . *Calamagrostis*
7. Florets not hairy at base or hairs short. 8

8. Glumes about 4 mm long and equaling lemma; lemma usually short-awned from tip; disarticulation
   below glumes . . . . *Cinna*
8. Glumes usually <4 mm long and exceeding the lemma; lemma awnless or awned from back;
   disarticulation above glumes 9

9. Lemmas long-awned, the awn 5–10 mm long and arising from the upper third of the lemma;
   palea firm, evidently 2-nerved . . . .*Apera*
9. Lemmas awnless or short-awned, the awn <4.5 mm long and arising from below the upper
   third of lemma; palea thin membranous, not evidently 2-nerved . . . . *Agrostis*

10. Lemmas awnless or awn-tipped. 11
10. Lemmas awned or at least one lemma in each spikelet bearing an awn. 17

11. Spikelets with one well-developed floret above 1 or 2 scale-like reduced lemmas . . . . *Phalaris*
11. Spikelets with >1 well-developed floret, whether staminate or bisexual. .12

12. Second glume widest above the middle, about twice as wide as the first glume;
    disarticulation below glumes . . . .*Sphenopholis*
12. Second glume widest at about midpoint, not greatly different from first glume; disarticulation
    above glumes (in the genus *Holcus* disarticulation occurs both above and below the glumes). 13

13. Leaf sheaths velvety hairy; disarticulation above and below the glumes (though the glumes
    usually fall after the florets); glumes hairy on veins; lemmas hairless and shiny . . . . *Holcus*
13. Leaf sheaths not velvety hairy; disarticulation above glumes; glumes and lemmas various 14

14. Spikelets about as broad as long and with 3 florets; panicle pyramidal, the lower branches
    much longer than those above. . . . .*Anthoxanthum*
14. Spikelets longer than broad, with 2 or more florets; panicle open or contracted and spike-like. 15

15. Spikelets 1 cm long or longer; panicle open and diffuse . . . .*Avena*
15. Spikelets much <1 cm long; panicle contracted and spike-like 16

16. Rachilla prolonged above the upper floret and appearing as a tuft of hairs. . . . . *Trisetum*
16. Rachilla prolonged but lacking the tuft of hairs. . . . . *Koeleria*

17. Glumes 1.5–2 cm long or longer; panicle open and diffuse; annuals. . . . .*Avena*
17. Glumes much <1.5 cm long, if >1 cm long, then panicle contracted; annuals and perennials 18

18. Lemmas uniformly awned within each spikelet 19
18. Lemmas not uniformly awned within each spikelet, often one or more of the lemmas unawned,
    or else one lemma with a bent awn and another lemma bearing a straight awn 23

19. Lemmas awned from near base or up to midpoint. 20
19. Lemmas awned from above the middle 22

20. Spikelets 11–15 mm long, usually with 3 to 5 florets . . . .*Helictotrichon*
20. Spikelets 4–5 mm long, usually with 2 florets; occasional specimens of *Calamagrostis purpurascens*
    with two florets per spikelet will key out here, but they differ from *Deschampsia* by having
    lemmas with conspicuous callus hairs 21

21. Spikelets ovoid to oblong, 3.5–5 mm; ligules blunt, never acute, 2.5–5 mm; plants perennial . . . *Vahlodea*
21. Spikelets lanceolate, 2–4 mm; ligules acute, 5–10(12) mm; plants annual or perennial . . . . .*Deschampsia*

22. Ligule hairy; awn of lemma flat-twisted and emanating from between two narrow and widely spaced teeth of the bifid apex. . . . . *Danthonia*
22. Ligule membranous; awn of lemma hair-like and bent and emanating from just above the middle of the lemma back. . . . . *Trisetum*

23. The second glume 2× as long and as wide as the first glume and completely enfolding and hiding the much shorter cluster of 3 florets; of the three florets per spikelet, the lower two are awned from the back and enfold the hairless and awnless uppermost floret. . . . . *Anthoxanthum*
23. Second glume not 2× as long and as wide as first glume or not completely enclosing and hiding the cluster of three or fewer florets 24

24. Spikelets about as broad as long, bronze-colored, with 3 florets; lower two florets distinctly hairy, awned from a deeply bifid apex; uppermost (fertile) floret shiny or hairless and unawned . . *Anthoxanthum*
24. Spikelets distinctly longer than broad, not bronze-colored, with 2 or more florets; upper-most florets bearing an awn. 25

25. Spikelets with 2 florets, the lowermost unawned, the uppermost with a short hooked awn; herbage velvety-hairy; disarticulation below glumes . . . . *Holcus*
25. Spikelets with 2 to 4 florets, all of which bear an awn, none of which is hooked; herbage not velvety hairy; disarticulation above glumes (or tardily below) 26

26. Glumes prominently many veined; callus at base of spikelet and the upper floret well-developed; first floret awned from tip with a straight awn; second and possibly third florets awned from back with long bent awns; the lemma tips bifid with each of the two tips bearing a hair-like projection; spikelet with lowest persistent floret ultimately disarticulating and rendering naked wiry pedicels and inflorescence branches . . . . *Ventenata*
26. Glumes smooth to weakly veined; callus lacking at base of spikelet and florets; first floret awned from middle or below with a long bent awn; second floret unawned or usually with a shorter or straighter awn; the lemma tip single-pointed and not bearing a hair-like projection; glumes remaining attached to the soft and pliable pedicels and inflorescence branches . . . . *Arrhenatherum*

Chlorideae

1. Lateral spikes 5–15 cm long, about 1 mm thick, not appressed to the main inflorescence rachis; spikelets closely appressed to the rachis of the lateral spike; unit of dispersal is the entire inflorescence . . . . *Schedonnardus*
1. Lateral or digitate spikes <8 cm long, >4 mm thick, appressed or not to the main rachis; spikelets divergent from the lateral spike rachis; unit of dispersal is a lateral spike, a spikelet, or a floret 2

2. Stoloniferous sodgrass; plants dioecious; pistillate spikelets fascicled among leaves; staminate spikelets arranged on lateral secund spikes; the unit of dispersal a cluster of several pistillate spikelets that coalesce into a single burr formed from the fused hardened glumes. . . . . *Buchloe*
2. Bunchgrass, if with stolons then the inflorescence comprises digitately-arranged secund spikes; plants not dioecious; all spikelets arranged along lateral (or digitate) spikes; unit of dispersal other than a burr 3

3. Spikelets with 4 or more bisexual (well-developed) florets. 4
3. Spikelets with 1 bisexual (well-developed) floret 5

4. Lemmas usually with hairs along veins, at least basally, the tips often with toothed margins; secund spikes (in the sole MT species) arranged mostly laterally . . . . *Leptochloa*
4. Lemma veins glabrous, the tips with smooth margins (entire); secund spikes (in the sole MT species) arranged mostly digitately . . . . *Eleusine*

5. Stem height 40 cm or more tall; lateral spikes ascending to upwardly appressed to the main inflorescence rachis; rhizomes well-developed . . . . *Spartina*
5. Stem height under 40 cm tall; lateral spikes divergent from the main inflorescence rachis; rhizomes absent to weakly developed (bunched or mat/sod-forming grasses) 6

6. Inflorescences of lateral spikes; plants not bearing stolons . . . . *Bouteloua*
6. Inflorescences of digitate spikes; plants bearing stolons . . . . *Cynodon*

Eragrosteae

1. Spikelets with only 1 floret 2
1. Spikelets with 2 or more florets 6

2. Callus hairs about one-half as long as lemma; stems 60–150 cm tall; rhizomes well-developed; a grass of sandy soils. . . . . *Calamovilfa*
2. Callus hairs, if any, much <one-half as long as lemma; stems often <60 cm tall; rhizomes, if any, short and slender; grasses of various substrates. 3

3. Lemmas awned, the awn >1 mm long . . . . *Muhlenbergia*
3. Lemmas awnless or only awn-tipped, ≤0.5 mm long 4

4. Ligule membranous or short-fringed; fruit enclosed in lemma and palea; commonly other than bunchgrasses (except *M. cuspidata* and *M. montana*). . . . . *Muhlenbergia*
4. Ligule long-hairy; fruit falling free of lemma and palea or enclosed within them; bunchgrasses 5

5. Inflorescence a dense cylindrical spike-like panicle; plants with a mat-forming habit; unit of dispersal is the spikelet . . . .*Crypsis*
5. Inflorescence an open to somewhat contracted panicle, if densely contracted then because it is partially or entirely enclosed in a leaf sheath; plants erect, not with a mat-forming habit; unit of dispersal is an individual fruit (technically an achene) . . . . *Sporobolus*

6. Spikelets in open panicles, these raised well above the leaves; stems distinctly bunched . . . . . . *Eragrostis*
6. Spikelets in contracted panicles, mostly concealed in leafy clusters; stems bunched to more commonly mat-forming. . . . . *Munroa*

Meliceae

1. Callus prominently and densely long-hairy; lemma awns 10–15 mm long. . . . . *Schizachne*
1. Callus hairless or with very fine, short hairs; lemma awnless or with an awn <5 mm long 2

2. Lemmas with 3 prominent veins (not including the marginal ones); florets 2–3 and widely spaced so that the rachilla is readily visible . . . . *Catabrosa*
2. Lemmas with 5 or more prominent veins (not including the marginal ones); florets 4 or more and closely spaced so that the rachilla is hidden. 3

3. Lemma veins prominent and equally spaced; second glume 1-veined; rhizomes well-developed; florets <4 mm long; riparian sites . . . . *Glyceria*
3. Lemma veins conspicuous but not equally spaced; second glume 5- to 7-veined; rhizome not well-developed, but stem base bulbous (except for *Melica smithii*); florets greater than 6 mm long; dry sites   . . . . *Melica*

Oryzeae

1. Monoecious annuals; pistillate spikelets positioned above the staminate spikelets; lemmas of pistillate spikelets awned; staminate spikelets with 6 stamens; ligules mostly >5 mm long. . . . .*Zizania*
1. Perennials with bisexual florets; lemmas unawned; stamens 3; ligules ≤1 mm long . . . .*Leersia*

Paniceae

1. Spikelets enclosed in a hard spiny burr of fused bristles . . . . *Cenchrus*
1. Spikelets not enclosed in a hard spiny burr, bristles absent or not united and then only subtending spikelets. 2

2. Inflorescences of slender digitate secund spikes; ligules completely membranous. . . . . *Digitaria*
2. Inflorescences not of secund spikes; ligules mostly of hairs 3

3. Spikelets subtended by several to many bristles (sterile branches) that often exceed the spikelet in length. . . . . *Setaria*
3. Spikelets not subtended by bristles or sterile branches 4

4. Ligule absent; veins of second glumes and sterile lemmas bearing stiff coarse hairs . . . .*Echinochloa*
4. Ligule a conspicuous fringe of hairs; veins of second glumes and sterile lemmas hairless or bearing soft hairs 5

5. Plants annual or perennial, not developing a rosette of short, broad basal leaves during the cool season; plants flowering in the warm season only . . . .*Panicum*
5. Plants perennial, developing a rosette of short, broad basal leaves during the cool season; plants flowering during the cool season and warm season; cleistogamous spikelets commonly produced on small axillary inflorescences during the late summer and fall . . . . *Dichanthelium*

Poeae

1. Inflorescences of terminal spikes or contracted secund panicles, all or most spikelets sessile or
nearly so 2
1. Inflorescences of racemes or open panicles, all or most spikelets distinctly pedicellate, if spikelets
are mostly sessile (as in *Puccinellia* or some *Poa*), then they are not arranged in a secund fashion 6

2. Inflorescence a two-ranked (bilateral or distichous) spike; spikelets oriented edgewise to rachis
(back or front of lemma faces the main inflorescence rachis); first glume absent on all but the very
terminal spikelet (this is because the main inflorescence rachis assumes the protective cover
otherwise provided by the first glume) . . . . *Lolium*
2. Inflorescence a secund spike-like or contracted panicle; spikelets not oriented edgewise to rachis
(side of lemma faces the main inflorescence rachis); both glumes present. 3

3. Spikelets dimorphic (of two distinct morphologies), one sterile and of empty flattened lemmas,
the other fertile and with 1 or 2 florets . . . . *Cynosurus*
3. Spikelets not dimorphic (all appearing nearly identical); sterile lemmas, if any, in the same
spikelets as the fertile lemmas 4

4. Leaf sheaths strongly keeled; disarticulation above glumes; perennials forming large tussocks. . . . *Dactylis*
4. Leaf sheaths not keeled; small annuals (*Poa secunda* and related species might be keyed out
here because of the subtly secund aspect of the inflorescence, but they are perennial bunchgrasses,
not small annuals; if so, go to couplet 6) 5

5. Disarticulation below glumes; lemmas prominently nerved, with a blunt tip. . . . . *Sclerochloa*
5. Disarticulation above glumes; lemmas inconspicuously nerved, with an awn tip. . . . . *Vulpia*

6. Tips of lemmas are transformed into juvenile leaf blades; florets irregular, not bilaterally symmetrical, the
florets replaced by small vegetative bulbs whereby each floret vegetatively propagates a new plant. 7
6. Tips of lemmas not transformed into juvenile leaf blades; florets bilaterally symmetrical, the florets
not vegetatively propagating new plants 8

7. Plants of low elevation mostly disturbed settings; common throughout MT. . . . .*Poa bulbosa*
7. Plants of high elevation alpine settings; known only from Glacier National Park . . . . . *Festuca viviparoidea*

8. Tip of lemma blunt or truncate or nearly so, the edges ragged; veins of lemmas parallel;
spikelets very often sessile or nearly so, closely appressed to the subtending rachis. 9
8. Tip of lemma pointed, with smooth edges, or if with ragged-edged tips (as in *Poa*), then spikelets
conspicuously pedicellate; veins of lemmas converging toward a pointed apex; spikelets sessile
to distinctly pedicellate, but in any case diverging away from the subtending rachis. 10

9. Stems bunched; veins of lemmas mostly 3 to 5 and obscure; ligules mostly ≤3 mm long; leaf
blades folded or inrolled, or if flat then usually 1–3 mm wide; plants of alkaline habitats . . . . *Puccinellia*
9. Stems strongly rhizomatous; veins of lemmas mostly 7 to 9 and prominent, raised from the
surface; ligules ≥3 mm long; leaf blades flat, mostly 4–15 mm wide; plants of mostly fresh
water habitats . . . . *Torreyochloa*

10. Callus of floret with conspicuously long hairs, the hairs about 1/5 as long as the lemma; lemmas
awnless; stems arising from standing water and about 1 cm in diameter at water level . . . . *Scolochloa*
10. Callus of floret hairless, short-hairy, or with cobwebby hairs; lemmas with or without awns; plants of
dry sites or at least not growing in standing water; stems much <5 mm in diameter towards the base. . .11

11. Leaves prominently veined particularly on the inner surface; plants dioecious and forming thick
coarse bunches, each stem bearing numerous old leaf sheaths from previous seasons; dry
mountain habitats. . . . .*Leucopoa*
11. Leaves not prominently and deeply veined on the inner surface; plants not dioecious, or if so (as
in some species of *Poa*), then leaves with at most two median lines on the upper surface;
individual stems not bearing numerous old leaf sheaths from previous seasons; habitats various 12

12. Lemmas awnless; leaves usually flat or folded, sometimes inrolled, but in any case the leaf tips
prow-like or abruptly contracted to a pointed apex; leaves with a prominent median venation
on upper surface. 13
12. Lemmas awned or awn-tipped; leaves mostly flat or rolled on drying and the tips not prow-like,
gradually tapering to a pointed apex; leaves without a prominent whitish, parallel venation 14

13. Floret 1; glumes minute; lemmas 3-nerved . . . .*Phippsia*
13. Florets 2 or more per spikelet; glumes the same size as the individual lemmas; lemmas 5-nerved . . . *Poa*

14. Lemmas keeled and tightly overlapping, nerves prominent on back; lemma awn arising from slightly below the very tip of lemma; leaf sheaths closed, at least initially; ligule not bilobed, the ligule not conspicuous from a side view of leaf; the palea and attached fruit are the unit of dispersal . . . *Bromus*
14. Lemmas rounded on back and not tightly overlapping (rachilla often slightly visible from side view), not prominently nerved; lemma awn arising from the very tip of lemma; leaf sheaths open; ligule often bilobed, cleft in the middle, the individual ligule lobes conspicuous from a side view of leaf (if not bilobed, as in *Festuca pratensis* and *F. arundinacea*, then auricles of leaf blade are well-developed); the floret is the unit of dispersal. 15

15. Plants annual; spikelets in a secund arrangement. . . . .*Vulpia*
15. Plants perennial; spikelets not in a distinctly secund arrangement 16

16. Auricles at junction of leaf sheath and blade conspicuous, well-developed; leaf blades flat, mostly >3 mm wide; stems mostly >1 m tall; lemmas awnless or awn-tipped . . . . *Schedonorus*
16. Auricles at junction of sheath absent; leaf blades mostly inrolled or flat but then typically <3 mm wide; stems of various height; lemmas usually with an awn >2 mm long, less commonly awnless. *Festuca*

Stipeae

1. Floret oval in outline (short and squat); awn deciduous, usually 1–4 times longer than lemma; callus short and blunt . . . .*Oryzopsis*
1. Floret linear in outline (long and cylindrical); awn persistent, usually several to many times longer than lemma; callus sharp-pointed . . . . *Stipa*

Triticeae

1. Spikelets mostly 1 per node, if 2 per node, then glumes and lemmas aligned in the same plane and the midribs of each in alignment 2
1. Spikelets mostly >1 per node, if 1 per node then lemmas out of alignment with glume, which is most evident by the midribs of glumes being out of alignment with those of lemmas 6

2. Plants perennial; native or introduced forage species (if glumes are small, very narrow, and tapered to a sharp point, and the backs of the lemmas are out of alignment with the backs of the glumes) . . . . *Agropyron*
2. Plants annual; introduced crops or grasses of disturbed settings 3

3. Glume and lemmas with an asymmetrical midrib that curves to the side; the tip of the glumes blunt and often notched to one side of the midrib. 4
3. Glume and lemmas with a symmetrical midrib; the tip of the glumes gradually tapering into an awn tip. . . .5

4. Inflorescence nearly cylindrical; the single spikelet per node embedded into the thickened inflorescence rachis; rachis disarticulating such that the unit of dispersal is a spikelet embedded in the inflorescence rachis . . . .*Aegilops*
4. Inflorescence not cylindrical; the single spikelet per node not embedded into the inflorescence rachis; rachis not disarticulating and the unit of dispersal is typically the grain (i.e., fruiting heads don't shatter). . . . . *Triticum*

5. Spikelets appressed to the main rachis and diverging at an angle of much <45°; plants typically >50 cm tall; glumes and lemmas soft or papery in texture at maturity; lemmas with conspicuously ciliate margins. . . . . *Secale*
5. Spikelets diverging from the main rachis at an angle much greater than 45°; plants usually <40 cm tall; glumes and lemmas hardened and burr-like at maturity; lemmas lacking ciliate margins. . *Eremopyrum*

6. Spikelets 3 per node and with 1 floret per spikelet; the lateral 2 spikelets at a node pedicellate and sterile or staminate, the central spikelet sessile and fertile (except 6-row *Hordeum vulgare* where all three spikelets are sessile and produce a grain). . . . . *Hordeum*
6. Spikelets mostly in groups of 2 per node, but if 3 or more, then all spikelets fertile and with often >1 floret per spikelet . . . .*Elymus*

Aegilops L. Goatgrass

Aegilops cylindrica Host [*Triticum cylindricum* Ces., Pass. & Gibbelli] JOINTED GOATGRASS Cool season, annual bunchgrass. **Stems** 25–50 cm. **Leaves**: blades 2–5 mm wide, flat and lax; sheaths with overlapping margins; ligule membranous. **Inflorescence** a 2-sided but cylindrical spike 5–15 cm long. **Spikelets** 8–14 mm long, one per node, partially to completely embedded within the

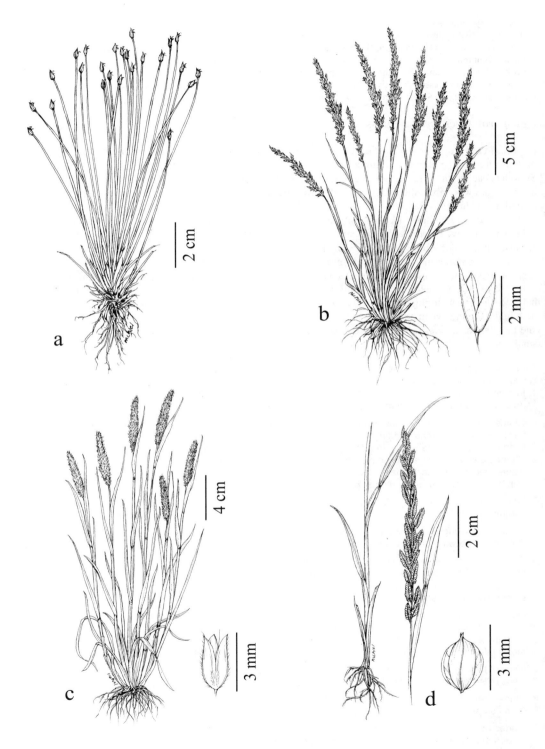

Plate 111. a. *Trichophorum cespitosum*, b. *Agrostis exarata*, c. *Alopecurus aequalis*, d. *Beckmannia syzigachne*

rachis, usually comprising several florets; glumes broad and asymmetrical, usually awned by extension of the eccentric midrib, especially on the distal-most spikelets. **Lemmas** similar to glumes; awnless to awned especially on the distal-most spikelets; palea well-developed. **Disarticulation** below the glumes; unit of dispersal the spikelet and internode of the adjacent inflorescence rachis. Disturbed settings including in and around crop fields where wheat is commonly cultivated. Introduced throughout the U.S. except in the southeast.

Aegilops can introgress with cultivated wheat, *Triticum aestivum*, to form a hybrid referred to as *Aegilotriticum sancti-andreae* (Degen) Soó; collected in Chouteau Co.

Agropyron Gaertn. Wheatgrass

Cool season, rhizomatous to bunched perennials. **Leaf blades** flat and lax; sheaths with overlapping margins; ligule membranous. **Inflorescence** a terminal 2-sided spike with mostly one spikelet per node. **Spikelets** each with >1 fertile floret; glumes often broad and aligned with subtending floret. **Lemmas** unawned or awned from the tip. **Disarticulation** above or below glumes; dispersal unit the floret or sometimes a cluster of florets.

Agropyron here includes all perennial wheatgrasses native and introduced on the North American continent that have mostly 1 spikelet per node, several florets per spikelet, and broad glumes that are aligned with the subtending lemmas.[101] The key below is provided to assist in the identification of species that have been traditionally referred to *Agropyron* in North America. The hybrid *Agrositanion saxicola* (Scribn. & J.G.Sm.) Barkworth & D.R.Dewey, a hybrid of *Elymus elymoides* and *Pseudoroegneria spicata*, collected in McCone and Sanders Counties, is similar to *Agropyron scribneri*. This hybrid has an erect habit and a non-shattering or tardily disarticulating spike, and a habitat in disturbance-prone settings rather than in the mountains, in contrast to *A. scribneri*. The divergent awns from the glume tips are distinctive and these are as long as those of the lemma. The spikelets are mostly one per node and have glumes with a small second awn tip in a manner similar to the glumes of *Elymus elymoides*.

1. Internodes of the main rachis <1/4 the length of the spikelets; the spikelets diverging at a
   wide angle from the main rachis (40–80°) . . . . *A. cristatum*
1. Internodes of main rachis >1/2 the length of the spikelets; the spikelets diverging at very
   acute angles from the main rachis (<30°). 2

2. Glumes and often the lemmas blunt or flat at the tip, or if pointed, then the glumes and lemmas
   thick and rigid, resisting bending; plants usually glaucous; margins of at least lower leaf sheaths
   lined with hairs (ciliate); cultivated forage grasses with heavy rigid stems. 3
2. Glumes and lemmas usually pointed and often awned; lemmas pliable and easily bent;
   plants greenish or occasionally glaucous; margins of at least lower leaf sheaths not lined
   with hairs; stems flexible and often hollow 4

3. Lemmas distinctly hairy . . . . *A. intermedium* var. *trichophorum*
3. Lemmas not hairy. . . . .*A. intermedium* var. *intermedium*

4. Rhizomes lacking or very short; bunchgrasses 5
4. Rhizomes present; grasses usually not bunched but creeping 7

5. Glumes long-awned (12 mm or more), the awns of the glumes widely divergent or outwardly curving;
   stems prostrate to strongly decumbent; inflorescence rachis disarticulating at maturity . . . .*A. scribneri*
5. Glumes awnless, or only awn-pointed, or glumes with a short straight awn (<10 mm);
   stems erect; inflorescence rachis continuous, not disarticulating at maturity. 6

6. Most inflorescence internodes >10 mm long; glume length ≤length of internode of adjacent
   rachis; lemmas with widely divergent awns, or awnless. . . . . *A. spicatum*
6. Most inflorescence internodes <10 mm long; glume length ≥length of internode of adjacent
   rachis internodes; lemmas awnless or with straight awns (rarely with divergent awns). . . . .*A. trachycaulum*

7. Awns of lemmas 5–12 mm long, strongly divergent or curved outward . . . .*A. albicans*
7. Awns of lemmas lacking or rarely >5 mm long and then straight 8

8. Glumes and lemmas hairy with long white hairs; awnless or nearly so . . . *A. intermedium* var. *trichophorum*
8. Glumes hairless or nearly so; lemmas hairless or hairy; awned or awnless .9

9. Blades mostly flat, lax, (3.5)4–11 mm wide, green, not rigid . . . . *A. repens*
9. Blades mostly inrolled, rigid, up to 5 mm wide, green or glaucous, often stiff and upwardly directed 10

10. Lemmas prominently hairy, hairs easily visible without magnification . . . . *A. dasystachyum*
10. Lemmas hairless or short-hairy 11

11. Glumes rigid, narrowly long-tapering to a narrow sharp-pointed tip, the distal half of the glume usually with just the single mid-vein. . . . . *A. smithii*
11. Glumes flexible, widest at or above the base and shortly tapering from the mid-portion to an acute tip, the distal half of the glume usually with >1 evident vein. . . . . *A. dasystachyum* var. *riparium*

Agropyron albicans Scribn. & J.G.Sm. [*Agropyron griffithsii* Scribn. & J.G.Sm., *Elymus albicans* (Scribn. & J.G. Sm.) A. Löve] MONTANA WHEATGRASS Rhizomatous, sometimes bunched. **Stems** often glaucous, 35–80 cm. **Leaves**: blades 2–4 mm wide, often stiff and inrolled; sheaths and collar often hairy. **Inflorescence** 3–15 cm long; rachis continuous; internodes <10 mm long. **Spikelets** 10–29 mm long, appressed to the main rachis, with 4 to 9 florets; glumes shorter than spikelet, with a broad base, narrowly tapering to a point or short awn. **Lemmas** hairy, with a divergent awn 5–15 mm. Open, dry shrub steppe vegetation. Mainly the Pacific Northwest and northern Rocky Mountain regions.

An ephemeral hybrid between *Agropyron spicatum* and *A. dasystachyum*. This species is much like thickspike wheatgrass but with awned glumes and lemmas and perhaps with more of a bunchgrass growth habit.

Agropyron cristatum (L.) Gaertn. [*Agropyron desertorum* (Fisch. ex Link) Schult.] CRESTED WHEATGRASS Bunchgrass. **Stems** mostly 30–80 cm. **Leaves**: blades 2–9 mm wide, often open but becoming inrolled upon drying; leaf sheaths and collar hairless. **Inflorescence** 2.5–7(–11) cm long; rachis continuous; the internodes 0.5–2.5(–5) mm long. **Spikelets** 6–11(–16) mm long, widely divergent from the main rachis; florets mostly 2 to 8 per spikelet; glumes shorter than florets, narrow and tapering to an awn tip. **Lemmas** hairless, awn-tipped or with awns to 5 mm long. Open dry sites at middle to low elevations. Introduced and cultivated throughout much of N. America except in the southeast.

Standard and Fairway crested wheatgrass are two common cultivars; Standard is less leafy and has narrower (lanceolate) flowering spikes and seed heads compared to Fairway. (p.656)

Agropyron dasystachyum (Hook.) Scribn. & J.G.Sm. [*Elymus lanceolatus* (Scribn. & J.G.Sm.) Gould] THICKSPIKE WHEATGRASS Rhizomatous. **Stems** usually glaucous, 35–80 cm. **Leaves**: blades 2–5 mm wide, often stiff and somewhat inrolled; leaf sheaths and collar often hairy. **Inflorescence** 3–15 cm long; rachis continuous; internodes <10 mm long. **Spikelets** 10–29 mm long, appressed to the main rachis; florets mostly 4 to 9; glumes shorter than florets, broadest above the base and shortly tapering to an acute tip. **Lemmas** sparsely to densely hairy, awn-tipped or with an awn up to 5(–10) mm. Open dry sites at lower elevations. Throughout the western half of N. America and eastward in southern Canada. (p.656)

Distinguished from *A. smithii* by little other than glumes rendered short because of the lack a long-tapering tip. *Agropyron dasystachyum* var. *riparium* (Scribn. & J. G. Sm.) Bowden, streambank wheatgrass, differs from thickspike by hairless to lightly scabrous lemmas and growing in more mesic habitats, usually in more clayey soils. This variety is widely seeded in MT. Thickspike wheatgrass varies in the degree of hairiness and also inhabits wet sites, including the edges of stock ponds and ephemeral streams. The distinction of streambank wheatgrass is thus questionable.

Agropyron intermedium (Host.) Beauv. [*Elymus hispidus* (Opiz) Melderis, *Thinopyrum intermedium* (Host.) Barkworth & Dewey, *Agropyron elongatum* Host., *Thinopyrum ponticum* (Podp.) Barkworth & Dewey] INTERMEDIATE WHEATGRASS Rhizomatous or bunched. **Stems** few to extensively bunched, 0.5–1.5 m. **Leaves**: blades 4–12 mm wide, firm, flat or rolled upon drying; leaf sheaths often with ciliate margins. **Inflorescence** (6–)12–30 cm long; rachis continuous, thick and somewhat concave on the side toward the spikelet; the internodes 10 or more mm long. **Spikelets** (10–)14–32 mm long, appressed to the main rachis; florets mostly 6 to 12 per spikelet; glumes shorter than florets, thick and blunt or notched, rarely pointed. **Lemmas** blunt to pointed, sometimes with a minute awn. Open, dry disturbed sites especially in pastures and along roads and trails. A forage grass and soil stabilizer that has been introduced throughout much of western N. America and sporadically eastward. (p.656)

Populations of *A. intermedium* that are well over 1 m tall, strongly bunched, non-rhizomatous, and perhaps flowers during late summer are referred to *Agropyron elongatum* (tall wheatgrass). Populations of *A. intermedium* with hirsute lemmas are referred to as pubescent wheatgrass, *Agropyron intermedium* var. *trichophorum* (Link) Halac. (or *Thinopyrum intermedium* (Host.) Barkworth & Dewey ssp. *barbulatum* (Schur.) Barkworth & Dewey). These distinctions may refer to selectively bred cultivars and don't really hold up over broad geographical scales. All of these grasses require frequently disturbed ecologies and thus founder events may play a role in local differentiation.

Agropyron repens (L.) Beauv. [*Elymus repens* (L.) Gould; *Elytrigia repens* (L.) Nevski] QUACKGRASS Rhizomatous. **Stems** slender, 0.7–1.1 m. **Leaves**: blades 4–11 mm wide, flat, lax to ascending; sheaths often with retrorse soft hairs. **Inflorescence** 5–13 cm long; rachis continuous, slender; the internodes <10 mm long. **Spikelets** 9–16 mm long, appressed to the rachis; florets 3 to 7 per spikelet; glumes shorter than florets, tapering to an awn tip. **Lemmas** tapering to an awn

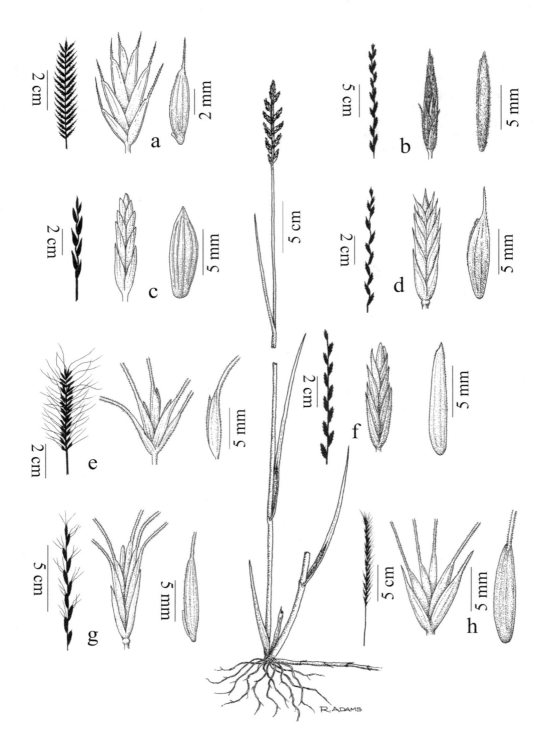

Plate 112. **Agropyron**. **a.** *A. cristatum*, **b.** *A. dasystachyum*, **c.** *A. intermedium*, **d.** *A. repens*,
e. *A. scribneri*, **f.** *A. smithii*, **g.** *A. spicatum*, **h.** *A. trachycaulum*

tip (rarely unawned) or with an awn usually about 5(–10) mm long. Moist, disturbed settings especially around lawns, in gardens, along roadsides, and riparian areas from lower to middle elevations. Introduced throughout N. America excepting the very southeast. (p.656)

Agropyron repens occasionally has large glumes like those of *A. trachycaulum* but broad lax leaf blades and extensive rhizomatous stands don't suggest the bunchgrass *A. trachycaulum*, which has ascending leaf blades.

Agropyron scribneri Vasey [*Elymus scribneri* (Vasey) M.E. Jones] SCRIBNER'S WHEATGRASS Bunchgrass. **Stems** prostrate to strongly decumbent, 20–40 cm. **Leaves:** blades 2–4 mm wide, flat to commonly inrolled; sheaths and collar usually hairless. **Inflorescence** 3.5–8 cm long; rachis slender and disarticulating at maturity; the internodes much <10 mm long. **Spikelets** 11–15 mm long, appressed to the main rachis; florets 3 to 6 per spikelet; glumes shorter than florets, tapering to an awn 12–18 mm long. **Lemmas** tapering to an awn 15–25 mm long, the awns strongly divergent at maturity. On exposed slopes especially in the subalpine to alpine. Mostly throughout the western U.S. (p.656)

Some specimens from Deerlodge and Beaverhead cos. appear to be hybrids between *A. scribneri* (disarticulating rachis, decumbent stems) and *A. trachycaulum* (large glumes).

Agropyron smithii Rydb. [*Elymus smithii* (Rydb.) Gould; *Pascopyrum smithii* (Rydb.) Löve] WESTERN WHEATGRASS Rhizomatous. **Stems** mostly hairless and usually glaucous, 35–80 cm; blades 2–5 mm wide, often stiff and inrolled; sheaths and collar often hairless. **Inflorescence** 3–15 cm long; rachis continuous; the internodes <10 mm long. **Spikelets** (sometimes 2 per node) 10–29 mm long, appressed to the main rachis; florets mostly 4 to 9 per spikelet; glumes shorter than to as long as florets, rigid and gradually long-tapering from the base to a narrow sharp tip (the acuminate tip is often slightly curved). **Lemmas** awn-tipped or with an awn up to 5(–10) mm. Open dry settings, often where historically disturbed. Throughout much of N. America except in the southeast. (p.656)

Resembling *Agropyron dasystachyum* in morphology and ecology but differing in long-tapered glume tips.

Agropyron spicatum (Pursh) Scribn. & Smith [*Elymus spicatus* (Pursh) Gould; *Pseudoroegneria spicata* (Pursh) A.Löve; also known as *Agropyron inerme* Scribn. & Smith] BLUEBUNCH WHEATGRASS (state grass of MT) Bunchgrass. **Stems** 3–9 dm. **Leaves:** blades 1–3 mm wide, flat to rolled; sheath and collar often hairless. **Inflorescence** 4–20 cm long; rachis continuous; internodes >10 mm long. **Spikelets** 10–15(–26) mm long, appressed to the main rachis; florets mostly 3 to 7 per spikelet; glumes shorter than florets, awnless or short-awned. **Lemmas** hairless to inconspicuously hairy, with a divergent awn up to 2 cm long, rarely awnless. Open dry vegetation to open understory. Throughout much of western N. America and rarely elsewhere. (p.656)

Specimens lacking lemma awns ("inerme") are likely reclamation cultivars. *Agropyron arizonicum* Scribn. & Smith, native to the southwestern U.S., has been cultivated in Gallatin Co. and has a flexuous rather than straight inflorescence rachis, as well as leaf blades 4–6 mm wide. A hybrid population of *Agropyron spicatum* and *Elymus elymoides* is referred to as *Agrositanion saxicola* (see notes under *Agropyron*). A robust form of *A. spicatum* called *Elymus wawawaiensis* J.R.Carlson & Barkworth might have been introduced into west-central MT as the cultivar 'Secar'.[31] It grows on shallow rocky soils and has a "vigorous" growth habit, and "more imbricate spikelets and narrower, stiff glumes" compared to *Agropyron spicatum*.

Agropyron trachycaulum (Link) Malte [*Agropyron subsecundum* (Link.) A.S. Hitchcock; *Elymus trachycaulus* (Link) Gould; *Agropyron caninum* (L.) P.Beauv.] SLENDER WHEATGRASS Bunchgrass. **Stems** erect to sometimes decumbent, 3–10 dm. **Leaf blades** 3–8 mm wide, flat to rolled, ascending; leaf sheath and collar often hairless. **Inflorescence** 5–20 cm long; rachis continuous; the internodes <10 mm long. **Spikelets** (9–)12–18 mm long, appressed to the main rachis; florets mostly 3 to 7 per spikelet; glumes often nearly as long or longer than florets, glumes awn-tipped to long-awned. **Lemmas** awn-tipped or sometimes with a straight awn up to 20 mm long. Open dry sites at lower elevations to the subalpine, but ecologically confined to disturbance-prone settings. Throughout much of N. America except in the southeast. (p.656)

Agropyron trachycaulum var. *unilaterale* is a tall form of slender wheatgrass in which most lemmas have awns up to 40 mm. Agropyron trachycaulum var. latiglume is common in alpine and subalpine habitats in the mountains of western MT and is distinguished by its short inflorescences with closely overlapping spikelets (caused by relatively short internodes). Occasional specimens of *A. trachycaulum* from Glacier National Park bear a panicle of spikelets, reminiscent of species of *Elymus*, most notably *Elymus flavescens*. In the Tendoy Mts. of Beaverhead Co., *Agropyron trachycaulum* inhabits alkaline meadows. These specimens are similar to *Agropyron albicans*, which is also common in the Tendoys, but the glumes are clearly those of slender wheatgrass.

Agrostis L. Bentgrass

Cool season, bunched or rhizomatous annuals and perennials. **Leaf blades** flat and lax; sheaths with overlapping margins; ligule membranous. **Inflorescences** of diffuse to contracted panicles. **Spikelets** with 1 floret; glumes commonly <2.5 mm, nearly equal or the first slightly shorter, both with mostly a distinctive scabrous midvein. **Lemma** translucent and enclosed by glumes, awnless or with a slender awn; palea weakly developed or at least mostly enclosed in the floret. **Disarticulation** above the glumes; unit of dispersal the floret.

Populations and specimens of *Agrostis oregonensis* and *A. stolonifera* have been observed with spikelets morphing into vegetative bulblets possibly in response to infection. Such specimens might not be readily distinguished as *Agrostis*, but the small glumes with scabrous margins remain diagnostic.

1. Palea evident, nearly half to as long as the lemma 2
1. Palea absent or <a third as long as the lemma 4

2. Rhizomes or stolons not present . . . . *A. humilis*
2. Rhizomes or stolons present 3

3. Panicle branches bearing spikelets only on the outer half, not along the basal one-half of
   the rachis; inflorescence rachises remaining open after anthesis; ligules 0.5–2 mm long . . . . . . *A. capillaris*
3. Panicle branches bearing spikelets for the entire length, from base to tip; inflorescence
   rachises usually closing or distinctly ascending after anthesis; ligules 2–7 mm long . . . . *A. stolonifera*

4. Rhizomes long and slender; anthers 1–1.2 mm long . . . .*A. pallens*
4. Rhizomes absent; anthers shorter than 1 mm long 5

5. Lemma awned, the awn abruptly bent at about the middle . . . . *A. mertensii*
5. Lemma unawned or with a short straight awn 6

6. Panicle branches contracted, at least some branches bearing spikelets for the entire length 7
6. Panicle branches open, bearing spikelets only on the outer half of the rachis (the lower
   one-half of rachis near the stem not bearing spikelets) 8

7. Plants slender; alpine; blades 5 cm or shorter and 1–2 mm wide . . . . *A. variabilis*
7. Plants stout; low to medium altitudes; blades 8–10 cm long and 1–6 mm wide. . . . . *A. exarata*

8. Panicle branches long and slender, spreading horizontally at maturity, the branches again
   branched above the middle; leaves stiff, scabrous. . . . . *A. scabra*
8. Panicle branches usually somewhat upright, the branches again branched below the middle;
   leaves lax, not scabrous. 9

9. Plants of high mountain meadows; 10–30 cm tall; spikelets about 1.5–2 mm long . . . .*A. idahoensis*
9. Plants of medium to low altitude wet areas; usually >30 cm tall; spikelets 2–3 mm long. . . . . .*A. oregonensis*

Agrostis capillaris L. [*Agrostis tenuis* Sibth] Colonial bentgrass Rhizomatous perennial.
Stems decumbent at base, 30–50 cm. **Leaves**: blades 1.5–4 mm wide; ligules 0.5–2 mm long.
Inflorescence a diffuse panicle 5–15 cm long. **Spikelets** 2–2.1 mm long. **Lemmas** 1.8–2 mm
long, awnless; palea about one-half as long as the lemma. Disturbance-prone settings from mostly
northwestern MT. Introduced throughout much of N. America except in most central states and
provinces.

Agrostis exarata Trin. Spike bentgrass Perennial bunchgrass with short rhizomes. **Stems** 20–
100 cm. **Leaves**: blades mostly 2–8 mm broad; ligules 0.5–2 mm long. **Inflorescence** a contracted
panicle 5–20 cm long. **Spikelets** 2–3 mm long. **Lemmas** 1.2–2 mm long, awnless or with a short
delicate awn; palea minute to absent. Moist sites including streamsides. Throughout much of N.
America. (p.653)

Similar to *Agrostis pallens* but differing by anthers 0.2–0.3 mm long and leaves clustered basally.

Agrostis humilis Vasey [*Agrostis thurberiana* A.S. Hitchcock, *Podagrostis humilis* (Vasey)
Björkam, *Podagrostis thurberiana* (A.S. Hitchcock) Hultén] Low bentgrass, alpine bentgrass
Perennial bunchgrass occasionally with short rhizomes. **Stems** 7–15(–50) cm. **Leaves**: blades
mostly basal, 0.5–1 mm wide; ligules 0.5–4 mm long. **Inflorescence** a contracted panicle 1–3(–5)
cm long. **Spikelets** 1.7–2 mm long. **Lemmas** 1.7–1.9 mm long, awnless to short-awned; palea about
two-thirds as long as lemma. Alpine and subalpine meadows, or dry mountain meadows at lower elevations. Throughout
western N. America and sporadically elsewhere.

Agrostis humilis is placed in *Podagrostis* because of a prolonged rachilla, short awn from the lemma base, and conspicuous palea.[31] These traits are all found in species of *Agrostis*, however. The first two of these are poorly developed in MT specimens. The florets of *A. humilis* have consistently large and conspicuous paleas, which differs in the ecologically and morphologically similar *A. mertensii* and *A. variabilis*. *Agrostis thurberiana* is a taller-statured form of *Agrostis humilis* (up to 0.5 m tall) that grows in the mountains below the subalpine zone.

Agrostis idahoensis Nash IDAHO BENTGRASS Perennial bunchgrass. **Stems** 20–30 cm. **Leaves:** blades 1–2 mm wide; ligules 1–4 mm long. **Inflorescence** an open panicle 4–6 cm long. **Spikelets** 1.6–2 mm long. **Lemmas** 1.4–1.9 mm long, awnless; palea absent. Moist meadows and other grassy sites in open to closed montane understory and subalpine to alpine zones. Throughout western N. America.

Agrostis mertensii Trin. [*Agrostis borealis* Hartm.] ARCTIC BENTGRASS, NORTHERN BENTGRASS Perennial bunchgrass. **Stems** 10–40 cm. **Leaves:** blades 1–3 mm wide; ligules 1–3 mm long. **Inflorescence** 5–15 cm long, of narrow but open panicles. **Spikelets** 2.5–3 mm long. **Lemmas** 2.4–2.8 mm long, bearing a straight or bent awn 2.5–3 mm long, from above the middle; palea absent or minute. Alpine rocky slopes. Throughout all of Canada, New England, and AK, south along the Rocky Mountains, and disjunct in central Appalachia.

Agrostis mertensii is most similar to *A. variabilis* but readily distinguished by its lemma bearing an awn.

Agrostis oregonensis Vasey OREGON BENTGRASS Perennial bunchgrass. **Stems** 30–50 cm. **Leaves:** blades mostly basal, 2–4 mm wide; ligules 1–6 mm long. **Inflorescence** an open panicle 6–12 cm long. **Spikelets** 2.5–3 mm long. **Lemmas** 2.5–3 mm long, awnless; palea absent to minute. Moist meadows at middle elevations. Very western N. America extending east to ID, MT, WY.

Agrostis pallens Trin. [*Agrostis diegoensis* Vasey] DUNE BENTGRASS Perennial bunchgrass with long and slender rhizomes. **Stems** 60–80 cm. **Leaves:** blades 2–6 mm wide; ligules 1–6 mm long. **Inflorescence** a contracted to somewhat open panicle, 7–11 cm long. **Spikelets** 3–3.5 mm long. **Lemmas** 2.2–2.5 mm long, awnless or with an apiculate awn; palea lacking or minute. Meadows; collected in Beaverhead Co. Very western N. America extending east to ID, MT.

Similar to *A. exarata* but differing by anthers 1–1.3 mm long, leaves positioned more on the stem than at the base, and an inflorescence that ranges from spicate to narrowly open.

Agrostis scabra Willd. ROUGH BENTGRASS, TICKLEGRASS Perennial bunchgrass. **Stems** 20–70 cm. **Leaves:** blades 1–2 mm wide, scabrous; ligules 1–5 mm long. **Inflorescence** an open diffuse panicle 1–3 dm long. **Spikelets** 2–2.2 mm long. **Lemmas** about 1.5–2 mm long, awnless or with an awn up to 4 mm long; palea minute or absent. Open moderately disturbed sites including roadsides at all elevations. Throughout nearly all of N. America.

Agrostis rossiae Vasey, Ross bentgrass, a WY endemic from around hot springs in Yellowstone National Park, differs in its annual growth form, contracted panicles, and shorter stature (<2 dm tall).

Agrostis stolonifera L. [*A. gigantea* Roth, *A. palustris* Hudson, *A. alba* L. var. *palustris* (Hudson) Persoon, *Agrostis alba* L. misapplied, *A. stolonifera* L. var. *major* (Gaudin) Farw.] CREEPING BENTGRASS, REDTOP Rhizomatous perennial or stoloniferous sod grass, often forming dense mats or patches. **Stems** decumbent at base, 20–120 cm. **Leaves:** blades 2–6 mm wide; ligules 2–7 mm long. **Inflorescence** a mostly diffuse panicle 5–30 cm long, sometimes contracting after anthesis. **Spikelets** 2–2.5 mm long (up to 10 mm long if infected). **Lemmas** 2–2.2 mm long, awnless or rarely minutely awned from the back; palea up to two-thirds as long as lemma. Wet meadows, streamsides, roadsides, pastures, and lawns from low to middle elevations. Introduced throughout all of N. America.

Agrostis stolonifera is considered a lawn pest when it forms dense colonies capable of competing with Kentucky bluegrass. *Agrostis gigantea* is an introduced perennial that differs by a strongly rhizomatous growth habit lacking stolons, culms 20–120 cm tall, panicles 8–30 cm long with the longest lower panicle branches 4–9 cm long, and spikelets and panicle branches tending to be reddish-tinted. *Agrostis stolonifera* produces stolons but not rhizomes, has culms 8–60 cm tall, panicles 3–20 cm long with the longest lower panicle branches 2–6 cm long, and spikelets and panicle branches tending to be straw-colored. *Agrostis gigantea* supposedly has a greater ecological predilection to more extreme climates including hot summers, cold winters, and drought. Both species are tentatively treated under *Agrostis stolonifera* because the presence of stolons or rhizomes covaries with no other convincing morphologies.

Agrostis variabilis Rydb. Mountain bentgrass Perennial bunchgrass. **Stems** 10–20 cm.
Leaves: blades 1–2 mm wide; ligules 1–3 mm long. **Inflorescence** a contracted panicle 2–6 cm
long. **Spikelets** 2–2.3 mm long. **Lemmas** 1.8–2.1 mm long, awnless; palea absent. Alpine and
subalpine meadows. Throughout much of western N. America.

    Agrostis variabilis is similar to *A. mertensii* and *A. humilis* in having a small-statured bunchgrass
habit in subalpine to alpine settings. *Agrostis variabilis* is readily distinguished from *A. humilis* by its florets that lack
paleas and from *A. mertensii* by lacking lemma awns.

Alopecurus L. Foxtail

Cool season, bunched to rhizomatous perennials and annuals. **Leaf blades** flat and lax; sheaths with
overlapping margins; ligule membranous. **Inflorescences** a cylindrical compact spike-like panicle.
Spikelets with 1 floret; glumes fused at the base. **Lemma** short to long-awned; palea well-developed.
Disarticulation below glumes; unit of dispersal the spikelet.

1. Panicle ovoid; glumes densely woolly over entire surface . . . . *A. magellanicus*
1. Panicle cylindrical; glumes hairy mostly on the veins. 2

2. Spikelets 4–6 mm long; anthers 2–4 mm long; leaf blades often >4 mm wide; perennials 3
2. Spikelets 2–4 mm long; anthers up to 1 mm long; leaf blades often <4 mm wide; annuals and perennials .4

3. Awns straight, included in or barely extending (up to 2 mm) beyond glumes; lemma obliquely
   truncate; strongly rhizomatous grasses forming large diffuse patches; leaf blades scabrous on
   upper surface . . . .*A. arundinaceus*
3. Awns bent, extending >2 mm beyond glumes; lemma acute; well-developed bunchgrass not producing
   rhizomes (some stems may root at lower nodes); leaf blades smooth on upper surface . . . . . . *A. pratensis*

4. Awns from about the middle of lemma, extending slightly to 1 mm beyond glumes . . . .*A. aequalis*
4. Awns form the lower third of lemma, extending well beyond 1 mm past the glumes 5

5. Anthers 0.3–0.5 mm long; glumes 2–2.5(–3) mm long; annuals . . . .*A. carolinianus*
5. Anthers 1.2–2.2 mm long; glumes 2–3.5 mm long; perennials. . . . . *A. geniculatus*

Alopecurus aequalis Sobol. Stream foxtail, shortawn foxtail Perennial bunchgrass,
occasionally a winter annual. **Stems** 20–60 cm. **Leaves:** blades 2–5 mm broad. **Inflorescence** a
cylindrical panicle 2–6 cm long, 4–5 mm in diameter. **Spikelets** 1.9–2.1 mm long; glumes silky
along the veins. **Lemmas** awnless or awn extending up to 1 mm beyond lemma tip, arising from or
just below the middle of the lemma. Stagnant water, streamsides and flood banks, and wet meadows at
most elevations. Throughout most of N. America excepting much of the southeast. (p.653)

Alopecurus arundinaceus Poir. Creeping meadow foxtail Rhizomatous perennial. **Stems**
1–2 m tall, 1 to several from nodes of rhizomes. **Leaves:** blades 6–14 mm wide; ligule 2–5 mm
long. **Inflorescence** a cylindrical panicle 4–10 cm long, 7–10 mm in diameter. **Spikelets** 4–5.5 mm
long; glumes silky mostly from along marginal and midrib veins. **Lemmas** with awn from base but
extending only to tip of lemma and rarely beyond. Hay meadows, along ditches, alkaline seeps, and
wet meadow at low to middle elevations. Introduced throughout the northern Great Plains and Rocky Mountain regions.

    Similar to *Alopecurus pratensis* but *A. arundinaceus* has broader and more scabrous (silica-rich) leaf blades. Cultivar
releases of this species (e.g., "Garrison creeping foxtail") may have enhanced the widespread distribution of this no-
longer-desirable hay grass.

Alopecurus carolinianus Walter Carolina foxtail Annual (and winter annual) bunchgrass.
Stems 10–30 cm. **Leaves:** blades 2–4 mm wide. **Inflorescence** a cylindrical panicle 1–3 cm long,
<5 mm in diameter. **Spikelets** 2.5–3 mm long; glumes silky along the keel and marginal veins.
Lemmas with an awn from the base and extending 2–3 mm past tip of lemma. Moist disturbed
areas, wet meadows and seep areas at low to moderate elevations. Throughout most of the U.S. and
adjacent southwestern Canada.

    A co-occurrence with *Psilocarphus brevissimus* (Asteraceae) suggests this species is common to low swales in dry
settings.

Alopecurus geniculatus L. Water foxtail Perennial bunchgrass. **Stems** 10–30 cm. **Leaves:**
blades 2–4 mm wide. **Inflorescence** a cylindrical panicle 1–3 cm long, <5 mm in diameter.
Spikelets 1.8–2.1 mm long; glumes silky along the keel and marginal veins. **Lemmas** with an
awn from base and extending 2–3 mm past tip of lemma. Moist disturbed areas, wet meadows,

and muddy and gravelly banks at low to moderate elevations. Throughout most of N. America excepting much of the southeastern U.S. and north-central Canada.

The difference between *Alopecurus geniculatus* and *A. carolinianus* may involve little other than the robust perennial growth form of the former. *Alopecurus geniculatus* may have a much more distinct and larger bunched habit compared to *A. carolinianus*.

Alopecurus magellanicus Lam. [*Alopecurus alpinus* J.E. Smith, *Alopecurus borealis* Trin.] ALPINE FOXTAIL Rhizomatous perennial. **Stems** few-bunched, 20–100 cm. **Leaves**: blades 2–6 mm broad. **Inflorescence** a dense ovoid panicle 1.5–4 cm long, >1 cm in diameter. **Spikelets** 3–4 mm long; glumes silky over entire surface. **Lemmas** with an awn extending 2.5–3 mm beyond lemma tip, emanating from the middle. Moist montane to subalpine meadows and in hot spring areas. Throughout all of Canada and the northern Rocky Mountain region of the U.S.

Dense stands of this species are distinctive because the ovoid inflorescences are subtended by a flag leaf that is angled and oriented uniformly in the same direction throughout a population.

Alopecurus pratensis L. MEADOW FOXTAIL Perennial bunchgrass. **Stems** 30–100 cm. **Leaves**: blades 3–8 mm wide. **Inflorescence** a cylindrical panicle 3.5–8 cm long, 5–10 mm in diameter. **Spikelets** 5–6 mm long; glumes silky along the keel and marginal veins. **Lemmas** awned from near the base, awn extending 3–4 mm beyond lemma tip. Disturbed moist sites like roadsides and often in seasonally inundated settings (e.g., margins of reservoirs). Throughout most of N. America excepting much of the very south.

Andropogon L. Bluestem

Warm season, bunched or rhizomatous perennials. **Leaf blades** flat and lax; sheaths with overlapping margins; ligule membranous. **Inflorescence** terminating lateral stem branches with one to several spikes (of spikelet pairs). **Sessile spikelet** bisexual; pedicellate spikelet rudimentary to well-developed and staminate. **Lemmas** hardened and dorsally compressed; palea well-developed. **Disarticulation** below the glumes; unit of dispersal the floret cluster.

Typically in grasslands and open understory but in MT *Andropogon* is most common and diverse along roadsides and similar disturbed settings. *Andropogon* here includes the genera *Bothriochloa* Kuntze and *Schizachyrium* Nees.[152]

1. Hairs of rachis and pedicels mostly longer than 3 mm; glumes mostly yellowish; awn of fertile lemma mostly 4–7 mm long, straight; rhizomes prominent . . . . *A. hallii*
1. Hairs of rachis and pedicels 3 mm or less long; glumes mostly purplish or greenish; awn of fertile lemma bent, mostly 9–18 mm long; rhizomes short or absent 2

2. Racemes or spikes of spikelet pairs clustered at the tip of the stem, lateral inflorescences absent; pedicellate spikelet borne on a pedicel that is flattened and grooved on both sides, membranous in the groove and easily punctured with a dissecting needle . . . . *A. ischaemum*
2. Racemes or spikes of spikelet pairs clustered at the tips of stem and lateral branches; pedicellate spikelet borne on a pedicel that is not flattened or strongly grooved, pedicel hard or tough textured throughout. . . .3

3. Stem branches terminating in a single spike of spikelet pairs; pedicellate spikelet much smaller than the sessile spikelet, mostly 1–3 mm long; leaf sheath keeled . . . .*A. scoparius*
3. Stem branches terminating in at least several mostly digitately-arranged spikes of spikelet pairs; pedicellate spikelet similar in size to the sessile spikelet, usually >5.5 mm long; leaf sheath not keeled . . . . *A. gerardii*

Andropogon gerardii Vitman. BIG BLUESTEM Bunchgrass. **Stems** 1–2 m, often glaucous, sometimes with short rhizomes. **Leaves**: blades 5–10 mm wide; ligule 1–2.5 mm long with a fringe of hairs. **Inflorescence** 4–16 cm long; internodes 3–6 mm long, whitish hairy at nodes and along margins of internodes, the hairs to 3 mm long. **Spikelets**: sessile and pedicellate spikelets 6–11 mm long; glumes mostly purplish, acute. **Lemmas**: fertile lemma 4–6 mm long, awn (6)10–18 mm long, bent. Along roadsides and similarly disturbed settings in eastern MT. Throughout most of N. America except in the very western tier of states and provinces.

Andropogon gerardii, along with *Andropogon hallii*, *Bouteloua hirsuta*, *Buchloe dactyloides*, *Dichanthelium wilcoxianum*, *Muhlenbergia cuspidata*, and *Stipa spartea*, are the only grass species occurring in Montana that could be of Great Plains origin.

Andropogon hallii Hack. Sand bluestem Bunchgrass bearing extensive rhizomes. **Stems** 1–2 m, often glaucous. **Leaves**: blades 3–10 mm wide; ligule 2–4.5 mm long with a fringe of hairs. **Inflorescence** 5–7 cm long; internodes 4–6 mm, densely whitish-hairy at nodes and along margins of internodes, the hairs 3–5 mm long. **Spikelets**: sessile and pedicellate spikelets 7–10 mm long; glumes mostly yellowish, acute. **Lemmas**: fertile lemma 4–5 mm long, awn 4–7 mm long, straight. Along roadsides and similarly disturbed settings in eastern MT. Throughout most of N. America in the central tier of provinces and states.

   Plants intergrading between *Andropogon gerardii* and *A. hallii* may have well-developed rhizomes but with hairs on inflorescence rachis <3 mm long, glumes mostly purplish, and lemma awns 10 mm long or longer.

Andropogon ischaemum L. [*Bothriochloa ischaemum* (L.) Keng] Yellow bluestem Bunchgrass. **Stems** 6–10 dm; leaf sheaths not keeled. **Leaves**: blades 1–3 mm wide; ligule 0.5–1.5 mm long with a fringe of hairs. **Inflorescence** with numerous (2 to 8) racemes or spikes, each about 3–7 cm long, terminating the main stem in an apparently digitate arrangement. **Spikelets**: rachis and pedicels silky-hairy along margins; sessile spikelets 3–4.5 mm long; glumes purplish to greenish, acute. **Lemmas**: fertile lemma with an awn 9–17 mm long; pedicellate spikelet about as large as the sessile spikelet. Roadsides; Introduced in southern U.S. and in in Lake Co., MT.

Andropogon scoparius Michx. [*Schizachyrium scoparium* (Michx.) Nash] Little bluestem Bunchgrass occasionally with short rhizomes. **Stems** 30–100 cm, often red or rust colored when mature; sheaths flattened and keeled. **Leaves**: blades 2–5 mm wide; ligule 1–3 mm long. **Inflorescence** with numerous lateral branches each bearing a single spike 3–6 cm long; rachis and pedicels long-hairy along margins. **Spikelets**: sessile spikelets 6–8 mm long; glumes purplish to greenish, acute. **Lemmas**: fertile lemma 4–6.5 mm long, with an awn 5–13 mm long; pedicellate spikelet reduced to a small glume 1–3 mm long. On gravelly, rocky, and sandy soils especially in overgrazed rangelands and roadside. Throughout most of N. America except at extreme northern latitudes.

Anthoxanthum L. Hornwort

Cool season, bunched to sometimes single-stemmed perennials. **Leaf blades** flat and lax; sheaths with overlapping margins; ligule membranous. **Inflorescence** an open to contracted panicle. **Spikelets** with 3 florets; at least the first glume as long as the floret cluster. **Lemmas** of first and second floret staminate and awnless or awned from the back, third floret fruit-bearing; palea enclosed in the floret. **Disarticulation** above glumes; unit of dispersal a floret cluster.

   The well-known sweetgrass genus *Hierochloe* R.Br. is included in *Anthoxanthum*.

1. Inflorescence an open pyramidal panicle; glumes subequal in size and shape; the lower two staminate lemmas awnless or essentially so . . . . *A. hirtum*
1. Inflorescence a contracted spicate panicle; first glume distinctly shorter and narrower than second; the lower two staminate lemmas with an awn 3–5 mm long . . . . *A. odoratum*

Anthoxanthum hirtum (Schrank) Schouten & Valdkamp [*Hierochloe odorata* (L.) P.Beauv., *Hierochloe hirta* (Schrank) Borbas ssp. *arctica* (Presl) G.Weim.] Northern sweetgrass Bunchgrass with creeping rhizomes. **Stems** 1–few, bunched, 30–100 cm. **Leaves**: blades 2–6 mm wide. **Inflorescence** a pyramidal panicle 4–12 cm long. **Spikelets** 3–6 mm long. **Lemmas**: the first two larger and more hairy that the glabrous uppermost floret (which is concealed by the lower two), unawned or with an awn up to 0.5 mm long. Wet meadows from low to high elevations especially in western MT. Throughout the U.S., including AK, but not in the southeast. (p.684)

Anthoxanthum odoratum L. Sweet vernalgrass Perennial bunchgrass. **Stems** 30–60 cm. **Leaves**: blades 3–7 mm wide. **Inflorescence** a contracted panicle 2–9 cm long, about 10 mm in diameter. **Spikelets** 8–10 mm long. **Lemmas** the first two awned, hairy, and concealing the third or uppermost glabrous and unawned lemma. Roadsides and similarly disturbed sites. Introduced throughout N. America excepting many of the central states and provinces.

Apera Adans. Silkybent

Apera interrupta (L.) P.Beauv. [*Agrostis interrupta* L.] Dense silkybent, longawn bentgrass Cool season, annual bunchgrass. **Stems** few-bunched, 15–40 cm. **Leaves**: blades 0.5–1 mm wide, flat and ascending; sheaths with overlapping margins; ligule membranous. **Inflorescence** a contracted panicle 2–8 cm long. **Spikelets** with one floret 1.7–2 mm long; glume with a scabrous

midvein. **Lemmas** 1.5–1.9 mm long, awned from the upper two-thirds with a slender awn 5–9 mm long; palea about as long as the lemma. Disturbed moist settings including depressions where water collects and lake margins at moderate elevations. Sporadic but introduced throughout N. America except in the central tier of provinces and states and in the southeast.

Sometimes included in the genus *Agrostis*.

Aristida L. Threeawn

Aristida purpurea Nutt. [*Aristida longiseta* Steud., *Aristida purpurea* var. *longiseta* (Steud.) Vasey, *A. fendleriana* Steud., *Aristida purpurea* var. *fenderliana* (Steud.) Vasey] PURPLE THREEAWN Warm season, perennial bunchgrass. **Stems** 15–40 cm. **Leaves:** blades curly to straight, 1–2 mm wide; ligules hairy, 0.3–0.5 mm long; sheaths with overlapping margins. **Inflorescence** a contracted panicle 6–25 mm long. **Spikelets** 10–18 mm long excluding awns, with one floret; first glume 1/2 the length of the second. **Lemma** 9–15 mm long (including awn column), gradually tapering into a tripartite awn, awn segments 2–5 cm long; palea enclosed in floret. **Disarticulation** above the glumes; dispersal unit the floret. Open dry often shrub-steppe settings and on various soils. Throughout much of the western two-thirds of N. America.

Aristida is classified into Aristideae, a tribe superficially similar to Stipeae except for the ligule hairy, tripartite awn, and warm season physiology. *Aristida purpurea* var. *fendleriana* is distinguished by lemma awns about 20–40 mm long, whereas *Aristida purpurea* var. *longiseta* has lemma awns 40–100 mm long. Leaf blades vary from congested basally to distributed along the stem, but these forms don't covary with awn length except that short curly basal leaf blades co-occur with shorter lemma awns, but not always. Both forms can be found at a single site, however.

Arrhenatherum P.Beauv. Oatgrass

Arrhenatherum elatius (L.) Presl TALL OATGRASS Cool season, perennial bunchgrass. **Stems** 40–120 cm, base of stem not enlarged or bulbous. **Leaves:** blades 5–9 mm wide, flat and lax; sheaths with overlapping margins; ligule membranous. **Inflorescence** a contracted panicle, 12–30 cm long, the branches short and verticillate. **Spikelets** 8–10 mm long, with 2 florets; glumes 1-nerved. **Lemmas** 8–9 mm long; staminate (first) floret with a twisted bent awn 1–1.6 cm long, occasionally lemma of fertile (second) floret short- to long-awned; palea well developed. **Disarticulation** above glumes; dispersal unit the floret cluster. Mountain meadows, rangeland, pastures, and roadside ditches, and sometimes cultivated. Introduced throughout much of N. America. Our plants are ssp. *elatius.*

Arrhenatherum elatius ssp. *elatius* lacks densely hairy nodes and the swollen basal internodes, in contrast to the ssp. *bulbosum* (Willd.) Schubl. & G. Martens. "Weed-free" hay may be spreading this grass because it seems abundant where horse use is common, even in some backcountry mountain meadows.

Avena L. Oat

Cool season, bunched annuals and perennials. **Leaf blades** flat and lax; sheaths with overlapping margins; ligule membranous. **Inflorescence** an open panicle. **Spikelets** with 2–4 florets; glumes 9- to 11-nerved. **Lemmas** long-awned from back; palea well-developed; disarticulation above glumes. **Dispersal unit** the individual florets.

Avena sterilis L., animated oat, is occasionally cultivated near Billings for its large spikelets used in dry arrangements. *Avena barbata* Pott ex Link, slender oat, may occur in MT and would be distinguished from *Avena fatua* by long lemma teeth tapering into a small hair-point and capillary-curving pedicels that would render a very open arrangement of the spikelets along each branch. Although many oatgrasses used to be classified in the tribe Aveneae, they are no longer recognized as a group.[30,31] Some phylogenetic evidence suggests that Aveneae characters, such as large glumes combined with awned lemma backs or dimorphic florets, often mark groups of closely related species.[377]

1. Lemmas hairy to essentially hairless, with a twisted bent awn; callus of lemma ring-like and conspicuously hairy; spikelets usually with 3 florets. . . . . *A. fatua*
1. Lemmas hairless and awnless or awn straight; callus of lemma inconspicuous; spikelets usually with 2 florets . . . .*A. sativa*

Avena fatua L. WILD OAT Annual bunchgrass. **Stems** 50–100 cm. **Leaves:** blades 5–10 mm wide. **Inflorescence** an open panicle. **Spikelets** 20–25 mm long, usually with at least 3 florets. **Lemmas** hairless to most often long-hairy, with a twisted bent blackish awn 3–4 cm long and

a conspicuously hairy callus. In and around cultivated fields, roadsides, and pastures at low to moderate elevations. Introduced throughout most of N. America.

Avena sativa L. [*Avena fatua* var. *sativa* (L.) Hausskn.] COMMON OAT Cultivated annual bunchgrass. **Stems** 50–100 cm. **Leaves:** blades 5–10 mm wide. **Inflorescence** an open panicle 10–25 cm long. **Spikelets** 17–22 mm long, usually with 2 to 3 florets. **Lemmas** hairless, awnless or with a straight whitish awn 0.1–3 cm long, with an inconspicuous hairless callus. Escaping cultivation at margins of fields and along roadsides and similarly disturbed settings. Introduced throughout most of N. America.

Beckmannia Host Sloughgrass

Beckmannia syzigachne (Steudel) Fernald AMERICAN SLOUGHGRASS Cool season, annual bunchgrass. **Stems** few to many, bunched, 10–80 cm. **Leaves:** blades 4–10 mm wide, flat and lax; sheaths with overlapping margins; ligule membranous. **Inflorescence** a contracted secund panicle 5–30 cm long. **Spikelets** with 1 floret, circular, strongly laterally flattened, 2.5–3 mm long; glumes 1 nerved, strongly laterally compressed. **Lemmas** with an apiculate tip (approaching a very short awn in some spikelets); palea well-developed. **Disarticulation** below glumes; dispersal unit the spikelet. Wet meadows, along ditches, streams, and lake banks, often rooted below water level. Throughout most of N. America except in the southeast. (p.653)

Bouteloua Lag. Grama

Warm season bunched or matted perennials or annuals. **Leaf blades** flat and ascending, sheaths with overlapping margins; ligule hairy. **Inflorescence** a panicle with 1 to many short lateral secund spikes. **Spikelets** with 1 fruit-bearing floret and 1 or 2 rudimentary spikelets. **Lemma** distinctly 3-nerved, awned; palea mostly enclosed in floret. **Disarticulation** above glumes or below; unit of dispersal the floret or a lateral spike of spikelets.

Bouteloua hirsuta Lag. (hairy grama) is expected in southeastern MT;[151] it occurs throughout much of the U.S. except in the extreme west and some of the most eastern states.

1. Plants annual; stems several in a bunch, sprawling to erect . . . . *B. barbata*
1. Plants perennial; stems many in a bunch, nearly always erect 2

2. Lateral spikes numbering 20 to 50 or more on a single stem, mostly oriented in a downward direction; disarticulation below glumes; the unit of dispersal is a lateral spike of spikelets . . .*B. curtipendula*
2. Lateral spikes numbering 1 to 10 on a single stem, divergent and not oriented in a downward direction; disarticulation above glumes; the unit of dispersal is a floret 3

3. Rachis of a lateral spike projecting beyond the terminal spikelets; black gland-based hairs present on at least the second glume; callus of sterile lemma not hairy . . . . *B. hirsuta*
3. Rachis of a lateral spike not projecting beyond the terminal spikelets; clear gland-based hairs on the second glume or absent; callus of sterile lemma densely hairy . . . .*B. gracilis*

Bouteloua barbata Lag. SIXWEEKS GRAMA Annual bunchgrass. **Stems** 10–40 cm long. **Leaves:** blades 1–2 mm wide, often curled. **Inflorescence** of 4 to 7 secund, lateral divergent spikes, each 1–2 cm long with 25 to 45 spikelets. **Spikelets** 3–4 mm long. **Lemmas** hairy and with an awn to 2 mm long; disarticulating from the spikelet; rudimentary florets 1 or 2. Open, dry disturbed sites, probably ephemeral; collected in Gallatin and Carbon cos. Throughout southwest N. America.

Bouteloua curtipendula (Michx.) Torr. SIDEOATS GRAMA Perennial bunchgrass with scaly rhizomes. **Stems** 4–7 dm long. **Leaves:** blades 2–6 mm wide, somewhat curled. **Inflorescence** of 20–50 secund, lateral downward-oriented spikes, each 1–2 cm long with 3–6 spikelets. **Spikelets** 4–5 mm long. **Lemmas** essentially hairless, awn-tipped, these falling with the lateral spike of spikelets; rudimentary florets 1. Open, dry rocky slopes and open understory often where *Yucca glauca* is common. Throughout much of N. America.

Bouteloua gracilis (Willd. ex Kunth) Lag. ex Griffiths BLUE GRAMA Perennial bunchgrass or sodgrass. **Stems** 20–50 cm long. **Leaves:** blades 1–2 mm wide, often curled. **Inflorescence** of 1 to 4 secund, lateral divergent spikes, each 2–4.5 cm long with 50 to 80 spikelets, each lateral rachis terminated by a spikelet. **Spikelets** 4–5.5 mm long, second glume essentially hairless or with clear gland-based hairs. **Lemmas** essentially hairless, short-awned, these disarticulating from the spikelet; rudimentary florets 1, densely hairy. Open, dry shrub-steppe vegetation and open understory. Throughout much of N. America except at extreme northern latitudes and in much of the southeast.

Bromus L. Brome

Cool season, single-stemmed to bunched perennials and annuals. **Leaf blades** flat and lax; sheath margins mostly fused; ligule membranous. **Inflorescence** an open to contracted panicle, rarely a raceme. **Spikelets** with several to many florets; glumes weakly to strongly keeled. **Lemmas** often hairy and strongly keeled, prominently veined, awned from a bifid apex; palea fused to the fruit. **Disarticulation** above glumes; dispersal unit the floret or caryopsis and palea.

Bromus is most closely related to the tribe Triticeae and thus has been segregated as tribe Bromeae.[31] *Bromus latiglumis* (Shear) A.S. Hitchcock [*Bromus altissimus* Pursh], earlyleaf brome, might be found in northeastern MT where it would be distinguished from other perennial bunched native bromegrasses by its culms with 9 to 20 nodes, leaf sheath densely pilose in the region of the collar and throat, and auricles 1–2.5 mm long on most lower leaves.

1. Plants annual, including winter annuals 2
1. Plants perennial 8

2. Lemmas gradually tapered into 2 narrow teeth; awn >10 mm long; first glume 1-veined 3
2. Lemmas somewhat rounded at tip, awnless or with an awn usually <10 mm; first glume 3- to 5-veined . . .4

3. Lemma body (without awn) 9–12 mm long. . . . . *B. tectorum*
3. Lemma body (without awn) 20–35 mm long. . . . . *B. diandrus*

4. Spikelets awnless or with awn up to 1 mm long; florets inflated and appearing like the rattles
   in the tail of a rattlesnake . . . . *B. briziformis*
4. Spikelets awned, the awn 2 mm long or longer; not inflated 5

5. Lemma usually completely wrapping around the fruit, at least at the base, often rounded and well-
   separated from others (rachilla of spikelet often visible); palea equal in length to the lemma . . . *B. secalinus*
5. Lemma usually not wrapping around the fruit, often flattened laterally and tightly overlapping the
   adjacent lemmas; palea distinctly shorter in length than the lemma (except in *B. hordeaceus*) 6

6. Panicle narrow, the branches short and ascending; pedicels mostly shorter than spikelets . . *B. hordeaceus*
6. Panicle more open; pedicels much longer than the spikelets. 7

7. Spikelets usually 5–10 mm wide; lemmas 5–7 mm wide, diamond-shaped in outline when flattened;
   spikelets borne mostly singly from branch ends; awns curved outward >90 degrees . . . . *B. squarrosus*
7. Spikelets <6 mm wide; lemmas 4–5 mm wide, elliptic in outline when flattened; awns straight
   or curved outward <90 degrees; spikelets often several from branch ends. . . . . *B. japonicus*

8. Rhizomes prominent; panicle branches ascending to erect. . . . . *B. inermis*
8. Rhizomes not well-developed; panicle branches spreading or nodding to drooping (except *B. carinatus*) . .9

9. Spikelets strongly laterally compressed or flattened; lemmas sharply folded along the
   prominent keel; 1st glume 3- to 5-veined, 2nd glume 5- to 9-veined . . . . *B. carinatus*
9. Spikelets weakly compressed or not laterally flattened; lemmas broadly rounded on back;
   1st glume 1- to 3-veined, 2nd glume 3- to 5-veined 10

10. Inflorescences spreading and ascending; ligules 2–6 mm long; lemma awns mostly
    >5 mm long. . . . . *B. vulgaris*
10. Inflorescences distinctly drooping; ligules 0.5–2 mm long; lemma awns mostly <5 mm long 11

11. Lemmas hairy along margins and sometimes over the lower back; glumes not hairy except
    sometimes at midvein; first glume usually 1-veined; leaf blades usually >5 mm wide. . . . . *B. ciliatus*
11. Lemmas usually evenly hairy over the entire back, more densely so near the base towards
    the margins; glumes usually hairy; first glume mainly 3-veined; leaf blades usually ≤5 mm wide . *B. porteri*

Bromus briziformis Fisch. & C.A.Mey. Rattlesnake brome Annual bunchgrass. **Stems** 15–50 cm. **Leaves:** blades 2–5 mm wide. **Inflorescence** an open panicle or raceme 4–12 cm long. **Spikelets** 9–14 mm long, appearing inflated, with 5 to 9 florets. **Lemmas** awnless or awn-tipped. Roadsides, edges of crop fields, parking lots, and pastures. Introduced throughout most of N. America except in the southeast and some central provinces and states. (p.667)

Bromus carinatus Hook. & Arn. [*Bromus marginatus* Nees ex Steud., *B. polyanthus* Scribn. ex Shear] MOUNTAIN BROME Perennial bunchgrass. **Stems** 50–100 cm. **Leaves**: blades 6–12 mm wide. **Inflorescence** an open panicle at anthesis, up to 10 cm long. **Spikelets** 17–22 mm long with 5 to 10 florets. **Lemmas** with an awn 8–17 mm long. Open understory and adjacent meadows in the mountains. Throughout much of the western half of N. America and disjunct in the New England region. (p.667)

Bromus carinatus is distinguished from other perennial bromes by its strongly keeled lemmas that are gray-green distally and often straw-colored basally with no purplish pigment. *Bromus marginatus* indistinctly differs from *B. carinatus* by having lemma awns 4–7 mm long; *B. polyanthus* has glabrous lemmas and leaf sheaths (that latter at least in the vicinity of the throat). *Bromus aleutensis* Trin. ex Griseb. (Aleut brome) may occur in northwestern MT and would be distinguished from *B. carinatus* by its lower inflorescence branches that measure up to 20 cm long and bear 1 to 3 spikelets distally and stems 3–7 mm thick. *Bromus carinatus* in contrast has lower inflorescence branches <10 cm long that bear 1 to 5 spikelets variously positioned along the branch and stems that are <4 mm thick. *Bromus catharticus*, rescue grass, may occur in MT; it has lemma awns <3.5 mm long or lemmas with 9 to 13 prominent veins.

Bromus ciliatus L. FRINGED BROME Perennial bunchgrass. **Stems** 70–100 cm. **Leaves**: blades 7–10 mm wide. **Inflorescence** an open panicle 15–25 cm long. **Spikelets** 9–13 mm long with 5 to 8 florets. **Lemmas** with a straight awn 3–4 mm long. Open to closed montane understory. Throughout most of N. America excepting much of the southeast. (p.667)

Very similar to *B. porteri* in the drooping panicle and montane understory habitat. *Bromus richardsonii* Link. is reported from south-central MT and is very similar to *B. ciliatus* except that the backs of the upper lemma surfaces are hairy, the 2nd glumes are 9–11 mm long, and the anthers are ca. 2 mm long, while *B. ciliatus* has glabrous upper lemma surfaces, 2nd glumes ca. 7–8.5 mm long, and anthers ca. 1–1.4 mm long.

Bromus diandrus Roth [*Bromus rigidus* Roth] RIPGUT BROME Annual bunchgrass. **Stems** 40–90 cm. **Leaves**: sheaths sparsely to densely long-soft-hairy; blades 2–5 mm wide. **Inflorescence** 8–30 cm, open, drooping. **Spikelets** strongly compressed with 3 to 9 florets; glumes keeled, glabrous; lower 8–16 mm long, 1-veined; upper 10–18 mm long, 3-veined. **Lemma** body 20–25 mm long, back rounded, densely silky-hairy, teeth 3–5 mm long, thread-like; awn 14–22 mm, bent, twisted below middle. Disturbed sites, partial shade, open understory, edges of buildings, and rocky sites; collected in Gallatin and Missoula cos. Introduced throughout most of western and southern N. America.

Bromus hordeaceus L. [*Bromus mollis* L.] SOFT BROME Annual bunchgrass. **Stems** 20–80 cm. **Leaves**: blades 2–3 mm wide. **Inflorescence** a narrow panicle 3–10 cm long. **Spikelets** 8–13 mm long with 5 to 7 florets. **Lemmas** with a straight awn 6–10 mm long. Disturbed sites, especially gravelly roadsides, lots, and loose steep rocky sites. Introduced throughout much of N. America. (p.667)

Bromus inermis Leyss. [*Bromus inermis* Leyss. ssp. *pumpellianus* (Scribn.) Wagnon, *B. inermis* var. *pumpellianus* (Scribn.) C.L. Hitchc., *B. inermis* var. *purpurascens* (Hook.) Wagnon, *B. pumpellianus* Scrib.] SMOOTH BROME Rhizomatous perennial. **Stems** 30–130 cm. **Leaves**: blades 4–12 mm wide, often with a "W" wrinkle about halfway along the leaf blade. **Inflorescence** a narrow panicle 7–18 cm long. **Spikelets** 15–30 mm long, with 8 to 15 florets. **Lemmas** awnless, ≤2 mm long when lemmas glabrous, or 3–10 mm long when lemmas hairy. Roadsides, pastures, meadows, and grasslands from low to middle montane elevations. Introduced throughout most of N. America, whereas native populations (*Bromus pumpellianus*) occur at higher latitudes and elevations through much of Canada and the Pacific Northwest region of the U.S.

Native populations are referred to *Bromus pumpellianus* or *Bromus inermis* var. *purpurascens*, Pumpelly's brome, which occupies montane rather than low elevation settings. They differ from introduced populations (*B. inermis* in the strict sense) by having hairs on the leaf sheaths, leaf blades, and lemmas, ligules 1.2–5 mm long, and lemma awns 3–10 mm long. *Bromus inermis* in the strict sense has hairless leaf sheaths and blades, ligules 0.5–1 mm long, lemmas often awnless or awn-tipped. The covariation of these morphologies and ecology is not evident in MT.

Bromus japonicus Thunb. JAPANESE BROME Annual bunchgrass. **Stems** 20–60 cm. **Leaves**: blades 2–4 mm wide. **Inflorescence** an open panicle 10–20 cm long, the branches drooping to one side and at least some flexuous, each branch bearing several spikelets. **Spikelets** 8–13 mm long, with 5 to 11 florets. **Lemmas** broad; awn 2–6 mm long, straight to curved downward at maturity. Roadsides, pastures, overgrazed sagebrush steppe, sites with strongly fluctuating water levels, and other open disturbed dry sites. Introduced throughout nearly all of N. America. (p.667)

This is perhaps the most common annual brome found throughout MT. *Bromus berteroanus* Colla [*B. trinii* Desv.], Chilean chess, is reported to grow in MT.[31] It would be identified as *B. japonicus* (or *Bromus squarrosus*) but the lemma

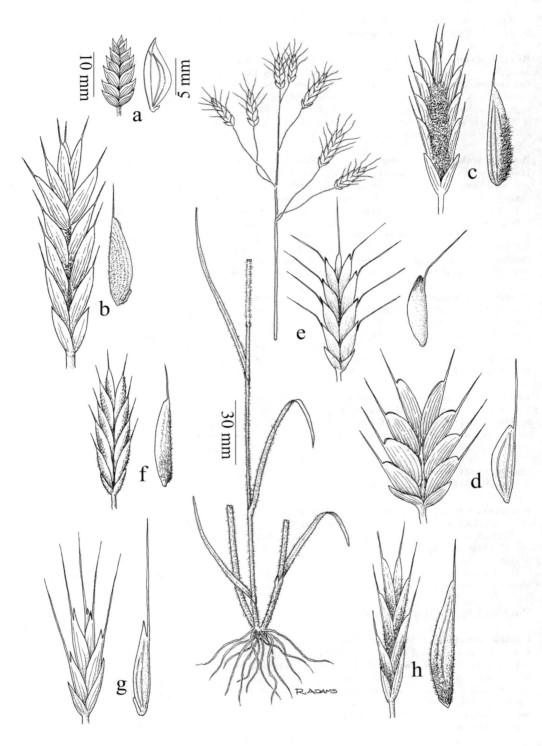

Plate 113. **Bromus**. **a.** *B. brizaeformis*, **b.** *B. carinatus*, **c.** *B. ciliatus*, **d.** *B. hordaceus*, **e.** *B. japonicus*, **f.** *B. porteri*, **g.** *B. tectorum*, **h.** *B. vulgaris*

awns in *B. berteroanus* are geniculate and strongly twisted at the base. *Bromus japonicus* is sometimes included under *B. arvensis*, but see notes under *B. secalinus*.

Bromus porteri (J.M.Coulter) Nash [*Bromus anomalus* Rupr. ex Fourn. misapplied] PORTER
BROME, NODDING BROME Perennial bunchgrass. **Stems** 30–90 cm. **Leaves**: blades 3–5 mm wide.
Inflorescence an open panicle 6–10 cm long. **Spikelets** 15–20 mm long, with 5 to 11 florets.
Lemmas hairy with an awn (1)2–3 mm long. Open to closed montane understory, sometimes at
edges of forested sites. Throughout much of the western one-half of N. America. (p.667)
   Bromus porteri is a northern form of the Texas endemic *B. anomalus* and is very similar to *B. ciliatus* but differs by its distinctly hairy glumes and lemmas.

Bromus secalinus L. CHESS BROME, RYE BROME. Annual bunchgrass. **Stems** 30–60 cm. **Leaves**:
blades 4–8 mm wide. **Inflorescence** a nodding panicle 7–12 cm long. **Spikelets** 10–15 mm long,
with 5 to 8 florets. **Lemmas** with a straight or down-curved awn 10–20 mm long. Disturbed sites,
roadsides, and overgrazed rangelands. Introduced throughout most of N. America.
   Bromus secalinus can be distinguished from the more common *B. japonicus* by having spikelets
with lemmas distinctly separated enough to render the rachilla visible, distinctly twisted and erect awns, hairless to
nearly hairless leaf sheaths, and stem bases ca. 2–3 mm in diameter. *Bromus arvensis* L., field brome, is reported from
MT and is distinguished from *B. secalinus* by its purplish-tinged spikelets, anthers 2.5–5 mm long, and lower leaf
sheaths with soft appressed hairs. *Bromus commutatus* Schrader is reported from MT.[113] It is distinguished by lemmas
with straight awns and spikelets borne from straight spreading pedicels.

Bromus squarrosus L. CORN BROME Annual bunchgrass. **Stems** loosely bunched, 40–60 cm.
Leaves: blades 2–4 mm wide. **Inflorescence** an open panicle 10–20 cm long; the branches short,
not secund, each bearing one spikelet, at least some flexuous. **Spikelets** 10–16 mm long, with
7 to 12 florets. **Lemmas** broad with a flattened awn 2–6 mm long, bending outward at maturity.
Disturbed sites, especially roadsides. Introduced across southern Canada and northern U.S.
   This species is very similar to *B. japonicus* but is distinguished by lemmas with distinctly hyaline margins (up to 0.9
mm broad), which render a miniature rattlesnake-tail appearance to the spikelet.

Bromus tectorum L. CHEATGRASS, DOWNY BROME Annual bunchgrass. **Stems** 20–50 cm.
Leaves: blades 2–4 mm wide. **Inflorescence** an open, often nodding panicle 2–15 cm long.
Spikelets 10–17 mm long, with 3 to 6 florets. **Lemmas** 9–12 mm long, gradually tapered into
two narrow teeth; awn straight or twisted, 12–20 mm long. Roadsides, overgrazed rangeland and
sagebrush steppe, and open dry understory. Throughout most of N. America. (p.667)
   Poorly developed plants may be as small as 4 cm and have a single spikelet. *Bromus rubens* L., red brome, has been
reported from Montana (Invaders Database: http://invader.dbs.umt.edu/) and is similar to *B. tectorum* except for the
short erect pedicels that render an erect congested panicle.

Bromus vulgaris (Hook.) COMMON BROME, COLUMBIA BROME Perennial bunchgrass. **Stems**
80–100 cm. **Leaves**: blades 5–10 mm wide. **Inflorescence** an open, nodding or drooping panicle
7–15 cm long. **Spikelets** 2–2.5 mm long, with 5 to 7 florets. **Lemmas** with a straight to twisted
awn 12–18 mm long. Montane grasslands and open understory. In N. America mainly along the
western tier of provinces and states. (p.667)
   Distinguished by its conspicuous ligule >2 mm long and inflorescences of ascending to erect spikelets, which are
glabrate with tightly clasping lemmas (i.e., the florets are not conspicuously splayed out laterally).

Buchloe Engelm. Buffalograss

Buchloe dactyloides (Nutt.) Engelm. [*Bouteloua dactyloides* (Nutt.) J.T.Columbus] Warm
season, dioecious, perennial sodgrass with creeping stolons. **Stems** 4–10 cm. **Leaves**: leaf blades
1–1.5 mm wide, curly; sheaths with overlapping margins; ligule hairy. **Inflorescence**: staminate
of 1 to 3 secund, lateral, divergent spikes, each 8–15 mm long with 10 to 15 spikelets; pistillate of
4 to 5 spikelets clustered into a short burr-like head or enclosed in sheathing leaf bases, this falling
entire. **Spikelets** 4–6 mm long with 1 floret, the second glume becoming hardened and bony in texture and forming the
burr-like covering. **Lemmas** distinctly 3-nerved, awnless, fused 2 to 5 together into a burr-like structure; palea enclosed
in floret. **Disarticulation** above glumes in staminate inflorescence or below glumes in pistillate inflorescence; unit
of dispersal is 3 to 5 spikelets enclosed in a burr-like covering. Disturbance-prone, open dry sites including pastures,
roadsides, and lawns. Throughout much of central N. America.
   The genus *Buchloe* has been included within *Bouteloua* but limited taxon sampling and weakly resolved phylogenetic
groupings suggest more evidence is needed.[92]

Calamagrostis Adans. Reedgrass

Cool season, bunched to rhizomatous perennials. **Leaf blades** flat and lax or inrolled and ascending; sheaths with overlapping margins; ligule membranous. **Inflorescence** an open to contracted panicle. **Spikelets** with 1 floret (sometimes 2 in *C. purpurascens*); glumes with a scabrous keel. **Lemmas** long- to short-awned, with a conspicuously bearded callus; palea enclosed in floret. **Disarticulation** above glumes; unit of dispersal the floret.

   Calamagrostis epigeios (L.) Roth, bushgrass or chee reedgrass, a tall introduced rhizomatous perennial with callus hairs much longer than the subtended floret, might be found in southwestern MT because it is known, albeit rarely, from reseeded rangeland and roadsides in adjacent areas of ID and WY.

1. Mature panicle open and diffuse, the branches >2 cm long, the main inflorescence rachis plainly
   visible through young spikelets (before callus hairs of lemmas begin to unfold at fruit maturity);
   ligule with a deeply cut jagged margin; glume midrib scabrous . . . . *C. canadensis*
1. Mature panicle contracted, the branches mostly <2 cm long, the main inflorescence rachis
   hidden or nearly so even by young spikelets with callus hairs of lemmas still concealed;
   ligule not deeply cut; glume midrib smooth 2

2. Awns bent when mature, usually protruding outside of the glumes 3
2. Awns straight or nearly so, not bent, usually hidden inside glumes 7

3. Awns longer than the glumes and exserted beyond them 4
3. Awns not longer than glumes, but exposed by projecting between them. .5

4. Leaf blades mostly basal, usually inrolled, 2–5 mm wide, prominent veins of inner (upper)
   surface distinctively lined with short straight hairs . . . . *C. purpurascens*
4. Leaf blades basal and from the stem, flat, 6–10 mm wide, inner (upper) surface glabrous. . . . . . . *C. tweedyi*

5. Collar of leaf sheath or at least some of them covered with dense short hairs; stems loosely
   bunched . . . . *C. rubescens*
5. Collar of leaf sheath hairless; stems often solitary or tightly bunched 6

6. Callus and rachilla hairs abundant and about 1/2 as long as the lemma; stems often
   solitary to few-bunched and connected by long rhizomes . . . . *C. montanensis*
6. Callus and rachilla hairs scant and about 1/3 as long as lemma; stems tightly bunched
   and bearing short and thick rhizomes. . . . .*C. koelerioides*

7. Rhizomes very short; stems densely bunched. . . . . *C. scopulorum*
7. Rhizomes long; stems scattered to loosely bunched . . . . *C. stricta*

Calamagrostis canadensis (Michx.) P.Beauv. [*C. canadensis* (Michx.) P. Beauv. var. *langsdorffii* (Link) Inman] BLUEJOINT REEDGRASS Rhizomatous bunchgrass. **Stems** 60–110 cm. **Leaves:** blades mostly 3–10 mm wide. **Inflorescence** an open panicle 7–35 cm long. **Spikelets** 4–5 mm long. **Lemmas** with an awn extending at most <1 mm beyond lemma tip; callus hairs nearly as long as the lemmas. Middle to high elevations in mountain meadows and open understory, often along streams or lake margins. Throughout much of N. America except in much of the southeast. (p.675)
   The broad green leaf blades combined with the diffuse inflorescences comprising soft green to purplish-tinged spikelets (with glumes diverging to expose the floret inside) are characteristics of this species. Regardless, *Calamagrostis canadensis* still might be confused with *C. rubescens* or *C. stricta*. The mostly greenish spikelets, deeply cut ligule, and scabrous glume midribs should distinguish *C. canadensis*. *Calamagrostis canadensis* var. *macouniana* (Vasey) Stebbins is known from a Ravalli Co. collection and is characterized by its diffuse inflorescence with sparse and diminutive spikelets. Given that *C. canadensis* intergrades with some of the species below, the formal recognition of such morphological variants seems unjustified.

Calamagrostis koelerioides Vasey FIRE REEDGRASS Rhizomatous bunchgrass. **Stems** 70–100 cm. **Leaves:** blades 2–4 mm wide. **Inflorescence** a contracted panicle 8–12 cm long. **Spikelets** 4.5–5.5 mm long. **Lemma** with a bent awn barely surpassing the glumes; callus hairs mostly <1/2 the length of lemma. Montane understory. The very western states of the U.S. and extending east to MT, WY.

Calamagrostis montanensis Scribn. ex Vasey PLAINS REEDGRASS Rhizomatous. **Stems** single to few-bunched, 20–50 cm. **Leaves:** blades 2–3 mm broad. **Inflorescence** a contracted panicle 5–10 cm long. **Spikelets** 3.5–6 mm long. **Lemma** with a bent awn barely if at all surpassing the glume tips; callus hairs one-half to two-thirds as long as lemma. Mostly moderately disturbed settings in open, dry shrub steppe vegetation from low to middle elevations, in the north-central states and provinces of N. America.

Without inflorescences this species might be mistaken for *Agropyron dasystachyum* or *A. smithii* but the conspicuous ligule >2 mm long and lack of wheatgrass clasping auricles readily distinguish plains reedgrass from the wheatgrasses. This grass is typically <0.5 m tall in dry settings, whereas the similar *Calamagrostis stricta* is >0.5 m tall in wet settings (e.g., roadside ditches, meadows, etc.).

Calamagrostis purpurascens R.Br. PURPLE REEDGRASS Rhizomatous bunchgrass. **Stems** 20–70 cm. **Leaves**: blades 2–5 mm wide. **Inflorescence** a contracted panicle 3–8 dm long. **Spikelets** mostly 6–7 mm long. **Lemmas** with a bent awn surpassing glumes by 2–4 mm; callus hairs <one-third the length of lemma. Dry mountain meadows to subalpine and alpine slopes. Throughout much of N. America but mostly absent from the eastern half of the U.S.

Calamagrostis rubescens Buckley. PINE REEDGRASS Rhizomatous. **Stems** 60–100 cm. **Leaves**: blades 2–5 mm wide. **Inflorescence** a contracted panicle 4–15 cm long. **Spikelets** 4–4.5 mm long. **Lemmas** with a bent awn barely surpassing if at all the glume tips; callus hairs >1/2 the lemma length, sometimes less. Montane understory, often of spruce and fir forests. Throughout much of western N. America.

Calamagrostis scopulorum M.E.Jones DITCH REEDGRASS Rhizomatous bunchgrass. **Stems** 50–80 cm. **Leaves**: blades 5–7 mm wide. **Inflorescence** a contracted panicle 6–12 cm long. **Spikelets** 6–7 mm long. **Lemma** with a slender straight awn not surpassing the lemma tip; callus hairs >1/2 as long as the lemma. Dry understory at middle elevations. Throughout the Rocky Mountain states in the U.S.

This species is similar to *Calamagrostis purpurascens* in having a robust bunched growth habit but it occurs at lower elevations, has short and mostly straight lemma awns, and scabrous (not hairy) veins along the upper surface of the leaf blade.

Calamagrostis stricta (Timm) Koeler [*C. inexpansa* A. Gray and *C. neglecta* (Ehrh.) Gaertn.] NORTHERN REEDGRASS Rhizomatous bunchgrass. **Stems** loosely bunched, 50–100 cm. **Leaves**: blades mostly 2–5 mm wide. **Inflorescence** a contracted panicle 6–15 cm long. **Spikelets** 3.5–5 mm long. **Lemma** with a slender straight awn barely surpassing lemma tip; callus hairs >1/2 the length of lemma. Streamsides, gravel benches, wet meadows, and seasonally inundated settings such as roadside ditches. Throughout much of N. America except in the southeast.

Calamagrostis tweedyi (Scribn.) Scribn. ex Vasey TWEEDY'S REEDGRASS Rhizomatous bunchgrass. **Stems** 40–80 cm. **Leaves**: blades 6–10 mm wide. **Inflorescence** a contracted panicle 8–15 cm long. **Spikelets** mostly 6–7 mm long. **Lemma** with a bent awn surpassing glume tips by 4–7 mm; callus hairs extending <1/3 the length of the lemma. Montane understory; collected in Mineral and Ravalli cos. WA, ID, MT.

The broad green leaf blades combined with the bunched habit are characteristic of this species.

Calamovilfa (A.Gray) Hack. ex Scribn. & Southworth Sandreed

Calamovilfa longifolia (Hook.) Scribn. PRAIRIE SANDREED Warm season, rhizomatous perennial, often bunched or in patches. **Stems** 60–150 cm. **Leaves**: blades 3–8 mm wide, wiry; sheaths with overlapping margins; ligule hairy. **Inflorescence** a narrow to open panicle 15–40 cm long. **Spikelets** 4–6 mm long, with 1 floret; glumes enclosing the floret. **Lemmas** with 1 distinct vein, awnless with a long hairy callus. **Disarticulation** above glumes; unit of dispersal the floret. Mostly in open dry, moderately disturbed settings at lower elevations. Throughout southern Canada and northern U.S. and south to NM.

Catabrosa P.Beauv. Whorlgrass

Catabrosa aquatica (L.) P.Beauv. BROOKGRASS, WATER WHORLGRASS Cool season, stoloniferous perennial. **Stems** creeping and rooting at basal nodes, 10–110 cm. **Leaves**: blades flat and lax, 5–14 mm wide, often wrinkled; ligule 1–8 mm long. **Inflorescence** an open panicle 7–28 cm long. **Spikelets** 2–3.5 mm long, with 2 to 3 well-spaced florets; glumes weakly 1-veined. **Lemmas** blunt-tipped and prominently 3-veined, awnless; palea well-developed. **Disarticulation** above glumes; unit of dispersal the floret. Wet meadows, edges of lakes, springs, and slow running streams from low to middle elevations. Throughout much of N. America but absent from the eastern half of the U.S.

Catabrosa is more closely related to *Puccinellia* than to *Melica*.[377] It is keyed out with *Melica* because of the leaf and spikelet morphology mentioned in the keys to groups of grasses.

Cenchrus L. Sandbur

Cenchrus longispinus (Hack.) Fernald MAT SANDBUR Warm season, annual bunchgrass. **Stems** decumbent, 18–60 cm. **Leaves**: blades 3–6 mm wide, flat and lax; ligule a fringe of hairs 1–2 mm long; sheaths with overlapping margins. **Inflorescence** a spike of burrs, with each burr enclosing 1 to 3 spikelets (the burrs are derived from fused sterile branches). **Spikelets** 6–8 mm long; glumes acute, the first 2–3 mm long, the second 4–6 mm long. **Lemmas**: fertile lemmas with acute apices, awnless, hard-textured if fruit-brearing; palea enclosed by lemma. **Disarticulation** below the glumes; unit of dispersal the burr or cluster of spikelets. Open dry to moist sites in disturbance-prone settings. Throughout nearly all of the U.S. and sporadically in southern Canada.

Cinna L. Woodreed

Cool season, bunched to single-stemmed perennials. **Leaf blades** flat and lax; sheaths with overlapping margins; ligule membranous. **Inflorescence** an open drooping panicle. **Spikelets** with 1 floret; glumes with a scabrous and keeled midrib. **Lemmas** with a well-developed hairless callus, usually long awned; palea enclosed by lemma. **Disarticulation** below glumes; unit of dispersal the spikelet.

Both species inhabit moist understory.

1. Inflorescence a dense panicle, the branches ascending to erect at maturity; stems commonly bulbous-based; second glume 3-veined, usually 4.5–6 mm long . . . . *C. arundinacea*
1. Inflorescence an open panicle, the branches lax to drooping at maturity; stems not bulbous-based; second glume usually 1-veined, mostly 2.5–4 mm long . . . . *C. latifolia*

Cinna arundinacea L. STOUT WOODREED Rhizomatous. **Stems** 80–170 cm. **Leaves**: blades 6–15 mm wide; ligules 3–9 mm long. **Inflorescence** a contracted panicle 15–45 cm long. **Spikelets** 4.5–5.5 mm long. **Lemma** with an awn 0.5–1.5 mm long borne from the upper half. Reported from Sheridan Co.[151] Sporadic but only locally common throughout the eastern two-thirds of the N. America.

Cinna latifolia (Trevis. ex Goepp.) Griseb. DROOPING WOODREED Rhizomatous. **Stems** 60–150 cm. **Leaves**: blades 7–17 mm wide; ligules 3–7 mm long. **Inflorescence** an open but drooping panicle, 12–40 cm long. **Spikelets** 3–4 mm long. **Lemmas** with an awn about 1 mm borne from the upper half. Montane understory. Throughout much of N. America but absent from the south-central and southeastern states. (p.675)

Crypsis Aiton Pricklegrass

Crypsis alopecuroides (Piller & Mitterp.) Schrad. [*Heleochloa alopecuroides* (Piller & Mitterp.) Host ex Roemer] FOXTAIL PRICKLEGRASS Warm season, annual bunchgrass or in large patches. **Stems** 10–30 cm. **Leaves**: blades 1–3 mm wide, flat and ascending; ligule hairy, ca. 1 mm long. **Inflorescence** a cylindrical spike-like panicle 1–6 cm long, 2.5–4 mm wide. **Spikelets** 2–2.5 mm long with 1 floret; glumes enclosing the floret. **Lemmas** 1.5–2.5 mm long. **Disarticulation** below glumes; unit of dispersal the spikelet. Sandbars and banks of large rivers. Introduced primarily in the western U.S.

Crypsis schoenoides (L.) Lamark [*Heleochloa schoenoides* (L.) Host ex Roemer], swamp pricklegrass, an annual exotic bunchgrass, may eventually be found in eastern MT. It would have an ovoid inflorescence 5–9 mm wide that remains partially enclosed by the leaf sheath and a lemma 2.5–3 mm long. This is in contrast to *Crypsis alopecuroides* with its narrower inflorescence usually extending beyond the uppermost leaf sheath.

Cynodon Rich. Bermudagrass

Cynodon dactylon (L.) Pers. Warm season, stoloniferous, mat- or sod-forming perennial. **Stems** 10–40 cm. **Leaves**: blades flat and ascending, 2–4 mm wide; ligule hairy, <1 mm long; sheaths with overlapping margins. **Inflorescence** of 3 to 6 digitally-arranged secund spikes, the branches 3–5 cm long. **Spikelets** closely appressed, 2–3 mm long with 1 floret, sometimes with vestigial florets. **Lemmas** 3-nerved, awnless, 2–3 mm long. **Disarticulation** above glumes; unit of dispersal the floret. Frequently disturbed sites, roadsides, lawns, and reclaimed areas; collected in Missoula Co. Introduced in North America, mainly in the U.S. excepting some north-central states.

Cynosurus L. Dogstail grass

Cool season, single-stemmed to bunched perennials and annuals. **Leaf blades** flat and ascending; sheaths with overlapping margins; ligule membranous. **Inflorescence** a contracted secund panicle. **Spikelets** dimorphic, a sterile spikelet paired with a fertile spikelet; sterile spikelets with numerous flat lemmas; the fertile with fewer florets. **Lemmas** rounded on back, inconspicuously veined, awned or awn-tipped; palea well-developed. **Disarticulation** above glumes; unit of dispersal the floret.

1. Plants perennial; inflorescence <1 cm thick; lemmas awnless or very short-awned . . . .*C. cristatus*
1. Plants annual; inflorescence >1 cm thick; lemmas with an awn 5–15 mm long. . . . .*C. echinatus*

Cynosurus cristatus L. Crested dogstail Perennial bunchgrass. **Stems** 20–60 cm. **Leaves**: blades 1–2 mm wide. **Inflorescence** a contracted secund panicle 4–11 cm long, usually <1 cm wide. **Spikelets** 4–7 mm long; the fertile ones with 2 to 5 florets. **Lemmas** 3–5 mm long, mostly awn-tipped. Creeksides, draining hot springs, expected along roadsides and similarly disturbed settings, occasionally a pasture grass; collected in Ravalli Co. Introduced mainly in eastern and western N. America.

Cynosurus echinatus L. Spiny dogstail Annual bunchgrass. **Stems** 30–90 cm. **Leaves**: blades 3–10 mm wide. **Inflorescence** a contracted secund panicle 2–4 cm long, usually >1 cm wide. **Spikelets** 7–12 mm long; the fertile ones with 1 to 4 florets. **Lemmas** with an awn 5–15 mm long. Forest edges, along roads, in grain fields and similarly disturbed open sites. Introduced mainly in the very western states and provinces of N. America and throughout much of the eastern U.S.; reported for Mineral Co.[29]

Dactylis L. Orchardgrass

Dactylis glomerata L. Cool season, perennial bunchgrass, commonly forms tussocks. **Stems** 80–200 cm. **Leaves**: blades flat and lax to ascending, 4–8 mm wide; ligules membranous, 3–9 mm long; sheaths with overlapping margins and dorsally keeled. **Inflorescence** a contracted secund panicle 4–20 cm long. **Spikelets** 5–8 mm long with 3 to 6 florets. **Lemmas** 4–6 mm long, keeled and inconspicuously veined, awnless or awn-tipped; palea enclosed in lemma. **Disarticulation** above glumes; unit of dispersal the floret. Open meadows and open understory where moderately disturbed, including road and trailsides, as well as in lawns. Introduced throughout most of N. America.

Danthonia DC. Oatgrass

Cool season, bunched perennials. **Leaf blades** flat and ascending; sheaths with overlapping margins; ligule hairy. **Inflorescence** a contracted to open but sparsely flowered panicle. **Spikelets** with several to many florets; glumes enclosing the florets. **Lemma** tapering to a bifid tip and bearing a flattened, twisted awn from between the two teeth; palea well-developed. **Disarticulation** above glumes; unit of dispersal the floret.

Danthonia is classified into Danthonieae, a cool season grass tribe superficially similar to the Aveneae group except for the hairy ligules.

1. Spikelets 1 per stem, rarely 2 or 3; plants 30 cm tall or less; sheaths at least near the throat conspicuously long-hairy with spreading or downward-oriented gland-based hairs 2–4 mm long. . . . . *D. unispicata*
1. Spikelets usually 3 to many, if as few as 2–3 then the plants taller than 30 cm; sheaths hairless to hairy and then the hairs usually <2 mm long 2

2. Raceme open, lateral branches wide-spreading; glumes widely open and exposing the florets inside (similar to *D. unispicata* in this last trait) . . . . *D. californica*
2. Panicle or raceme contracted, lateral branches appressed-ascending; glumes tightly enclosing the florets 3

3. Lemmas hairless on back, hairy only along the margins and callus. . . . .*D. intermedia*
3. Lemmas hairy over the back and margins 4

4. Glumes 17–22 mm long; lemmas 10–15 mm long, not tightly clasping each other because the callus of each is longer than it is wide . . . .*D. parryi*
4. Glumes 12 mm long or shorter; lemmas <5 mm long, tightly clasping each other because the callus of each is shorter than it is long . . . .*D. spicata*

Danthonia californica Bol. CALIFORNIA OATGRASS **Stems** 30–100 cm. **Leaves**: blades 2–4 mm
wide; sheaths and collar hairless to hairy; ligule 1–1.5 mm long. **Inflorescence** a raceme 3–6
cm long, with usually 2 to 4 spikelets. **Spikelets** 15–25 mm long, with 5 to 7 florets. **Lemmas**
not tightly enfolded by the glumes, 7–10 mm long, hairless except along margins, with awns 8–11
mm long. Sporadic but locally common in open dry meadows, shrub steppe, and open understory.
Throughout most of western N. America.

Danthonia californica is a taller-statured version of D. unispicata with slightly more and larger spikelets in each
inflorescence.

Danthonia intermedia Vasey TIMBER OATGRASS. **Stems** 10–50 cm. **Leaves**: blades 1–4 mm
wide; sheaths essentially hairless; collar hairy; ligule 0.5–0.7 mm long. **Inflorescence** a
contracted panicle or raceme 3–6 cm long, with 4 to 10 spikelets. **Spikelets** 11–15 mm long, with
4 to 7 florets. **Lemmas** tightly enfolded by the glumes, 3–6 mm long, hairless except along margins,
with awns 6–8 mm long. Dry mountain meadows at middle to high elevations. Throughout all of
western North America and across southern Canada. (p.675)

This is the most common Danthonia in MT and is readily distinguished from the similar D. parryi by its leaves
folded along the midrib and glumes ≤15 mm long.

Danthonia parryi Scribn. PARRY'S OATGRASS **Stems** 30–70 cm. **Leaves**: blades 2–3 mm wide;
sheaths essentially hairless; the collar usually hairy; ligule 0.5–1 mm long. **Inflorescence** a
contracted panicle or raceme 3–7 cm long, with 4 to 10 spikelets. **Spikelets** 17–22 mm long, with
4 to 6 florets. **Lemmas** tightly enfolded by glumes, 7–10 mm long, long-hairy over the entire back,
with awns 12–14 mm long. Dry mountain meadows. AK south to SK, NM.

Aside from the larger glumes and more densely hairy lemmas, the flat leaves distinguish this species from the more
common D. intermedia.

Danthonia spicata (L.) P.Beauv. ex Roem. & Schult. POVERTY OATGRASS **Stems** 30–60 cm.
Leaves 1–3 mm wide; sheaths essentially hairless to hairy; the collar usually hairy; ligule 0.5–1
mm long. **Inflorescence** a contracted panicle or raceme 3–5 cm long, with 5 to 10 spikelets.
Spikelets 10–15 mm long with 4 to 7 florets. **Lemmas** tightly enfolded by glumes, 3–5 mm long,
sparsely but uniformly hairy on back except at the tip, with awns 5–8 mm long. Dry mountain
meadows. Throughout nearly all of N. America.

The florets of Danthonia spicata tightly clasp each other, are usually <5 mm long, and have a short callus and
conspicuous but sparse hairs over the backs of the lemmas. This in contrast to those of D. intermedia, which are not
strongly clasping, are usually >5 mm long, have a long callus (longer than wide), and conspicuous hairs concentrated
along the margins (usually lower margins) and not on the backs of the lemmas. The tightly clasping nature of the floret
cluster of Danthonia spicata is probably due to the small basal callus of each floret. In the other species of Danthonia,
the callus is longer than wide and effectively separates the florets enough so that they don't tightly clasp each other.

Danthonia unispicata (Thurb.) Munro ex Macoun ONESPIKE OATGRASS **Stems** 15–30 cm.
Leaves: blades 1–2 mm wide; sheath at least near the throat densely long-hairy, the hairs wide-
spreading and 2–4 mm long; collar long-hairy; ligule 0.2–0.5 mm long. **Inflorescence** with mostly
1 spikelet, rarely 2 or 3. **Spikelets** 14–24 mm long, with 4 to 9 florets. **Lemmas** 7–11 mm long,
hairless on the back, sparsely hairy along margins, with awns 7–12 mm long. Open understory and
shrub steppe from middle to subalpine elevations. Throughout much of the western half of N. America.

Deschampsia P.Beauv. Hairgrass

Cool season, bunched perennials and annuals. **Leaf blades** flat or inrolled and lax to ascending; sheaths
with overlapping margins; ligule membranous. **Inflorescence** an open to contracted panicle. **Spikelets**
with mostly 2 florets, the rachilla hairy; glumes nearly equal and enclosing the florets. **Lemmas** uniformly
awned from below the middle; palea enclosed in lemma. **Disarticulation** above glumes; unit of dispersal
the floret.

Deschampsia atropurpurea is now treated under the genus Vahlodea.

1. Plants annual; spikelets 7–9 mm long; the glumes narrowly tapering to a point; the
   lemma awns distinctly bent; leaves very few, short, and withering by anthesis . . . . D. danthonioides
1. Plants perennial; spikelets <7 mm long; the glumes broadly tapering to a point; the
   lemma awns straight; leaves ample, forming a basal tuft, and persisting past anthesis 2

2. Panicle narrow, the branches appressed to the main rachis; glume mostly green, purplish
distally . . . . *D. elongata*
2. Panicle open, the branches spreading; glume mostly purplish, whitish to golden distally . . . . . . *D. cespitosa*

Deschampsia cespitosa (L.) P.Beauv. BUNCHED HAIRGRASS Perennial. **Stems** 30–100 cm.
Leaves: blades mostly 1.5–3 mm wide; ligules 2–10 mm long. **Inflorescence** an open panicle
8–25 cm long. **Spikelets** 3.5–6 mm long; glumes purplish but distally straw- or silvery-colored.
Lemmas 2–5 mm long. Montane to alpine meadows, in open dry understory to moist alpine
meadows. Throughout N. America except in the southeast.

Deschampsia danthonioides (Trin.) Munro ANNUAL HAIRGRASS Annual. **Stems** 15–40 cm.
Leaves: blades 1–1.5 mm wide, withering by anthesis; ligules 2–3 mm long. **Inflorescence** a
contracted to open panicle, 5–20 cm long. **Spikelets** 6.5–9 mm long; glumes gradually tapering
to a sharp-pointed tip. **Lemmas** 1.5–3 mm long. Regularly disturbed settings in open dry montane
understory. Throughout all of western N. America and sporadically in the eastern U.S. (Appalachia).

Deschampsia elongata (Hook.) Munro SLENDER HAIRGRASS Perennial. **Stems** 20–80 cm.
Leaves: blades 1–2 mm wide; ligules 3–8 mm long. **Inflorescence** a contracted panicle 5–25 cm
long, the branches appressed to the main rachis. **Spikelets** 4–7 mm long; glumes pale green but
distally purplish-colored. **Lemmas** 2–3 mm long. Open to closed montane understory. Throughout
all of western N. America and sporadically in the eastern U.S. (Appalachia).

Dichanthelium (Hitchc. & Chase) Gould Rosette grass

Cool season, bunched perennials. **Leaf blades** flat and ascending (vegetative leaves forming rosettes);
sheaths with overlapping margins; ligule hairy. **Inflorescence** a diffuse open panicle; panicle branches
occasionally clustered among leaves during the fall bear self-pollinating spikelets. **Spikelets** with 1 fertile
floret; glumes unequal in length, the first shorter than half the length of the spikelet. **Lemma** with a blunt
apex, awnless, hard-textured if fruit-brearing; palea enclosed in floret. **Disarticulation** above the glumes;
unit of dispersal the floret.

1. Spikelets 1.5–2 mm long; the larger second glume often purplish at maturity (the spikelet
thus rendered mostly dark in color); ligule 2–6 mm long . . . . *D. acuminatum*
1. Spikelets 2–4 mm long; the larger second glume usually greenish at maturity (the spikelet
thus predominantly green in color); ligule 0.5–3 mm long 2

2. Stem leaves mostly 5–15 mm wide, upper surface glabrous, lower surface hairy;
spikelets mostly 2.5–4 mm long, 1.5–2.5 mm wide; ligule 1–3 mm long . . . . *D. oligosanthes*
2. Stem leaves mostly 2–5 mm wide, upper and lower surfaces hairy; spikelets mostly
2–2.5 mm long, 0.5–1.2 mm wide; ligule 0.5–1 mm long . . . . *D. wilcoxianum*

Dichanthelium acuminatum (Sw.) Gould & C.A. Clark [*Dichanthelium lanuginosum*
(Elliot) Gould var. *sericeum* (Schmoll) Spellenb.] PACIFIC PANICGRASS **Stems** 10–30 cm. **Leaves**:
blades 5–10 mm wide, usually <8 times longer than wide; ligule 2–6 mm long. **Inflorescence** an
open panicle 3–8 cm long. **Spikelets** 1.5–2 mm long; glumes hairy, first 0.5–0.75 mm long, second
1.5–2 mm long. **Lemmas**: fertile lemma blunt, globe-like. Often forming dense stands on wet soils
around edges of hot springs. Our plants are ssp. *sericeum* (Schmoll) Freckmann & Lelong. Throughout most of N.
America; ssp. *sericeum* is more narrowly confined to the U.S. in the Rocky Mountain states, UT, CO, WY, ID, MT.
    Many subspecies of this species are distinguished. *Dichanthelium acuminatum* ssp. *fasciculatum* (Torr.) Gould
potentially occurs along the southern and eastern tier of MT cos. and is distinguished by longer stems that measure
30–100 cm and longer leaf blades usually >8 times longer than wide.

Dichanthelium oligosanthes (Schult.) Gould [*Panicum scribnerianum* Nash] SCRIBNER'S
ROSETTE GRASS **Stems** 2–5 dm, usually spreading. **Leaves**: blades 5–15 mm wide, mostly hairless;
ligule 1–3 mm long. **Inflorescence** an open panicle 3–9 cm long. **Spikelets** 2.7–4.2 mm long;
glumes hairless to sparsely hairy, first 1–1.2 mm long, second 2.5–3 mm long. **Lemmas**: fertile
lemma blunt, spherical. Disturbed sites and open understory. Throughout most of North America
except at extreme northern latitudes. Our plants are ssp. *scribnerianum* (Nash) Freckmann & Lelong.
    Subspecies *oligosanthes*, which is distributed more in the eastern and southern part of N. America, differs from ssp.
scribnerianum by having ligules mostly >2 mm long and an abaxial leaf blade surface that is uniformly hairy.

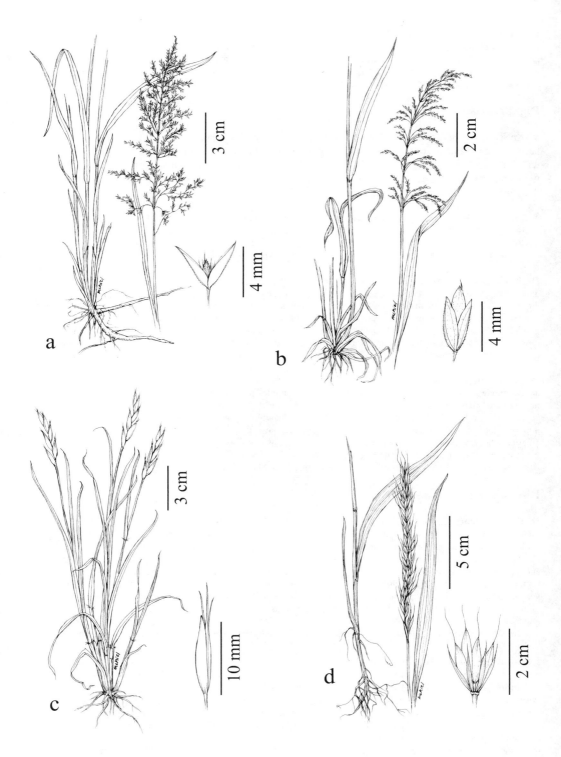

Plate 114. a. *Cinna latifolia*, b. *Calamagrostis canadensis*, c. *Danthonia intermedia*, d. *Elymus glaucus*

Dichanthelium wilcoxianum (Vasey) Freckman [*Panicum wilcoxianum* Vasey] FALL ROSETTE GRASS **Stems** 10–20 cm, usually spreading. **Leaves**: blades 2–5 mm wide, distinctly hairy; ligule to 1 mm long. **Inflorescence** an open panicle 2–5 cm long. **Spikelets** 2.4–3.2 mm long; glumes conspicuously hairy, first 1–1.2 mm long, second 2.5–3 mm long. **Lemmas**: fertile lemma blunt, spherical. Open dry disturbed settings; collected in Carter and Custer cos. Throughout most central states and provinces in N. America.

Digitaria Haller Crabgrass

Warm season, bunched annuals. **Leaf blades** flat and ascending; sheaths with overlapping margins; ligule hairy or membranous. **Inflorescence** of digitately or almost digitately-arranged, secund spikes. **Spikelets** with 1 fertile floret; glumes unequal in length, the first shorter than half the length of the spikelet. **Lemma** with a blunt apex, awnless, hard-textured if fruit-brearing; palea enclosed in floret. **Disarticulation** above the glumes; unit of dispersal the floret.

1. Spikelets 1.5–2 mm long; first lemma and second glume with glandular hairs; fertile lemma dark brown at maturity; stem nodes hairless . . . . *D. ischaemum*
1. Spikelets 2.8–3 mm long; first lemma and second glume hairless to occasionally hairy but then lacking glandular hairs; fertile lemma light brown at maturity; stem nodes hairy . . . . *D. sanguinalis*

Digitaria ischaemum (Schreb.) Schreb. ex Muhl. SMOOTH CRABGRASS **Stems** 18–45 cm, mostly decumbent; nodes hairless. **Leaves**: blades 3–5 mm wide; ligules 1–2 mm long, membranous, throat hairy. **Inflorescence** of 2 to 7 digitately or almost digitately-arranged secund spikes, 3–9 cm long. **Spikelets** 1.5–2 mm long, ovate; first glume very short, about 0.2 mm long; the second glume 1.5–2 mm long, with glandular hairs. **Lemmas**: fertile lemma ovate, dark brown at maturity. Disturbance-prone sites such as roadsides and in lawns. Introduced throughout nearly all of N. America except at the extreme northern latitudes.

Digitaria sanguinalis (L.) Scop. HAIRY CRABGRASS **Stems** 30–60 cm, decumbent and rooting at lower nodes; nodes hairy. **Leaves**: blades 4–7 mm wide; ligules 1–2 mm long, membranous, throat sparsely hairy. **Inflorescence** of 7 to 10 digitately or almost digitately-arranged secund spikes, 8–12 cm long. **Spikelets** 2.8–3 mm long, narrowly ovate; first glume very short, about 0.3 mm long; the second glume 1–2 mm long, without glandular hairs. **Lemmas**: fertile lemma narrowly elliptic, light brown at maturity. Disturbance-prone sites such as roadsides and in lawns. Throughout nearly all of N. America except at the extreme northern latitudes.

    Digitaria ciliaris (Retz.) Koeler, reported from south-central MT,[30] is very difficult to distinguish from *D. sanguinalis*.

Distichlis Raf. Saltgrass

Distichlis spicata (L.) Greene [*Distichlis stricta* (Torr.) Rydb.] Warm season, strongly rhizomatous perennial. **Stems** 10–40 cm. **Leaves**: blades 2–5 mm wide, flat and ascending, distichously-arranged, stiff and pungent-tipped. **Inflorescence** a contracted panicle or raceme; the staminate inflorescence well exerted above leaves and the pistillate inflorescence 1–6 cm long and typically clustered among leaves, the two types of spikelets superficially similar, although the staminate averaging more florets than the pistillate. **Spikelets** 10–17 mm long, with 5 to 16 florets; glumes faintly veined. **Lemmas**: 4–6 mm long, many-veined, awnless, thin-textured in the staminate spikelet. **Disarticulation** above glumes; unit of dispersal is the floret. Alkali flats and less commonly in shrub steppe at lower elevations. Throughout much of N. America.

    Distichlis is classified in tribe Aeluropodeae, a group of strongly rhizomatous, dioecious, and warm season grasses.

Echinochloa P.Beauv. Cockspur grass

Echinochloa crus-galli (L.) P.Beauv. [*E. muricata* (P.Beauv.) Fernald] BARNYARD GRASS Warm season, annual bunchgrass. **Stems** 10–120 cm long, few- to many-bunched, usually sprawling to ascending. **Leaves**: blades 5–25 mm wide, flat and lax; ligule absent or represented by a pigmented rim; sheaths with overlapping margins. **Inflorescence** a contracted secund panicle 5–22 cm long. **Spikelets** 2.5–4.5 mm long (excluding awns) with one fertile floret; first glume 1–1.5 mm long; second glume 2.5–4.5 mm long, gradually tapering to a sharp-pointed tip, with stiff coarse hairs along veins. **Lemmas**: sterile lemma like second glume and with an awn 0.5–20 mm long; fertile lemma awnless with an acute apex, globe-like. **Disarticulation** above the glumes; unit of dispersal the floret. Regularly disturbed settings such as cracks in sidewalks and parking lots, in lawns, along roadsides, and around cultivated fields. Introduced and native throughout most of N. America.

Echinochloa muricata (rough barnyard grass) is distinguished from *E. crus-galli* by having a slender acuminate hairless tip on the fertile lemma, which happens to be the common morphology among MT specimens of *Echinochloa*. The abruptly contracted hairy tip of the fertile lemma, which supposedly distinguishes *Echinochloa crus-galli*, is found in relatively few specimens and mostly from Custer, Missoula, and Toole cos. These lemma tip morphologies don't covary with other traits (e.g., length or abundance of hairs on the inflorescence branches). *Echinochloa frumentacea* Link (billion-dollar grass) has been cultivated in Gallatin Co. and is distinguished from *E. crus-galli* by having pale green spikelets lacking awns on the sterile lemma and inflorescence rachises that are at most sparsely hairy.

Eleusine Gaertn. Goosegrass

Eleusine indica (L.) Gaertn. INDIAN GOOSEGRASS. Warm season, annual bunchgrass. **Stems** flat and ascending, 30–90 cm. **Leaves**: blades 3–7 mm wide; sheaths with overlapping margins (dorsally keeled); ligule membranous. **Inflorescence** a panicle of 4 to 15 secund spikes, most of them digitately-arranged, the branches 5–16 cm long. **Spikelets** 4–7 mm long, with 5 to 7 florets; glumes similar to lemmas. **Lemmas** 3–4 mm long, distinctly 3-nerved, awnless; palea well-developed. **Disarticulation** above glumes; unit of dispersal the floret. Regularly disturbed settings such as roadsides, vacant lots, and lawns. Introduced throughout much of the southern portion of North America. One collection from Dawson Co., but reported to be sporadic throughout MT.[31]

Elymus L.

Cool season, rhizomatous or bunched perennials. **Leaf blades** flat and lax to ascending; sheaths with overlapping margins; ligule membranous. **Inflorescence** a terminal bilateral spike with >1 spikelet per node. **Spikelets** each with >1 fertile floret; glumes often narrow and not aligned with subtending floret. **Lemmas** unawned or awned from the tip. **Disarticulation** above or below glumes; dispersal unit the floret or sometimes a cluster of florets.

Elymus traditionally includes all perennial wheatgrasses native and introduced in N. America that have 2 or mostly 2 spikelets per node. Those species of *Elymus* with 3 (or more) spikelets per node are distinguished from *Hordeum* by the presence of >1 fertile floret per spikelet. Species of *Elymus* with mostly 1 spikelet per node can be distinguished from *Agropyron* by having very narrow sharp-pointed glumes and florets that are turned 90° relative to the glumes. Such a definition conflicts with some genetic data that are useful to plant breeders. The key below is provided to identify MT species that have been traditionally referred to *Elymus* in N. America. The tendency to lump the perennials into a "catch-all" *Elymus*[113] does not reflect species relationships any better than does the traditional wheatgrass classification.[101] Species of *Elymus* with the "SH" genome arose via hybridization from *Pseudoroegneria* ("S" genome) and *Hordeum* ("H" genome) and are distinct from *Elymus* species containing other genomes.[29,172,214,256] Much more evidence is needed, however, before a comprehensive taxonomy can be settled involving the species traditionally assigned to *Agropyron* and *Elymus*.

1. Plants commonly in large clumps to 1 m in diameter; stems usually well >1 m tall; spikelets
   mostly three or more per node; blades usually >10 mm wide; ligule usually >2 mm long 2
1. Plants not clump-forming or in clumps much <1 m in diameter; stems usually <1 m tall;
   spikelets mostly two per node (sometimes 2–3 spikelets per node in *Elymus junceus*);
   blades usually <10 mm wide; ligule usually <2 mm long 3

2. Lemmas hairless; stems hairless; spikes >1.5 cm thick, often with secondary branches;
   leaf blades 15–35 mm wide . . . . *E. condensatus*
2. Lemmas sparsely to distinctly hairy; stems lightly hairy at least near the nodes; spikes
   <1.5 cm thick, with no secondary branches; leaf blades 5–15 mm wide, mostly <15 mm wide. . .*E. cinereus*

3. Stems not bunched or only very loosely so; rhizomes well-developed, usually long,
   slender, and creeping (rarely short as in *E. ambiguus*) 4
3. Stems usually bunched (occasionally single-stemmed in *Elymus glaucus*); rhizomes
   absent or very short and not conspicuous 7

4. Lemmas conspicuously long-hairy, the hairs dense and long 5
4. Lemmas hairless to lightly hairy, the hairs very short, often sparse 6

5. Lemmas awnless or awn to 1 mm long; auricles absent or poorly developed; lemma hairs
about 2 mm long, very dense and obscuring the lemma surface . . . . *E. flavescens*
5. Lemma awns up to 6 mm long; auricles usually present; lemma hairs about 1 mm long,
not so dense as to obscure the surface of the lemma . . . . *E. innovatus*

6. Stems 2.5–12 mm thick, 1–3 m tall; spikelets 2–6 per node; lemmas hairy . . . . *E. angustus*
6. Stems 1–3 mm thick, about 1 m or less tall; spikelets mostly 2 per node; lemmas glabrous . . . . *E. triticoides*

7. Rachis fragile and breaking at maturity under very slight pressure, the florets and spikelets remaining
attached to the rachis joints (compare also *E. junceus* with its tardily disarticulating rachis) 8
7. Rachis continuous or disarticulating with force; the florets disarticulating above the glumes 9

8. Awns of glumes and lemmas 20–100 mm long, the glumes often becoming divided distally
into two or three segments; fertile florets 1–4 per spikelet . . . . *E. elymoides*
8. Awns of glumes and lemmas 4–20 mm long, the glumes never becoming divided
distally into segments; fertile florets not developing (plants sterile) . . . . *E. macounii*

9. Awns of lemmas 20–35 mm long, strongly curved outward; the inflorescence spike
drooping at maturity, seldom erect . . . . *E. canadensis*
9. Awns of lemmas much <20 mm, also straight or only slightly curved outward; the
inflorescence spike always erect at maturity 10

10. Glumes narrowly tapering imperceptibly from the base into an awn-tip, with only 1 distinct vein; the
inflorescence rachis disarticulating at nodes but requiring mechanical disturbance to do so . . . *E. junceus*
10. Glumes broadest above the base, with 3–5 distinct veins, awn distinct from body of glume; the
inflorescence rachis remaining solid and never shattering with age even with mechanical disturbance . . 11

11. Basal 1 mm of glume bony, round in cross-section, not green, outwardly curving;
spikelets 10–13 mm long; lemmas 6–9 mm long . . . . *E. virginicus*
11. Basal 1 mm of glume not bony, flattened, greenish, and not outwardly curved;
spikelets 12–20 mm long; lemmas 9–14 mm long . . . . *E. glaucus*

Elymus angustus (Trin.) Pilg. ALTAI WILDRYE Rhizomatous bunchgrass, mostly cultivated. **Stems** 1–3 m. **Leaves:** blades 6–15 mm wide, usually flat. **Inflorescence** 15–35 mm long, the rachis continuous and erect. **Spikelets** 2 to 6 per node, 17–25 mm long; glumes narrow, gradually tapering from the base to a narrow, pointed tip, about as long as 1st lemma, veins not evident or with a single vein. **Lemmas** mostly 2 or 3 per spikelet, densely hairy, awnless or with a short awn to 2.5 mm long. Cultivated and escaped around crop fields (Gallatin and Fallon cos.). Introduced as a cultivated plant and escaping mainly in southwestern Canada.

This species is very similar to *E. cinereus* except that its growth habit is mostly single- to few-stemmed and the inflorescence is less compact with longer internodes such that spikelet clusters are separate from each other.

Elymus canadensis L. CANADA WILDRYE Bunchgrass. **Stems** 80–150 cm. **Leaves:** blades 4–12 mm wide, usually flat. **Inflorescence** 7–19 cm long, the rachis continuous and usually drooping. **Spikelets** mostly 2 (rarely more) per node, 12–16 mm long (excluding awns); glumes 3- to 5-veined, tapering to an awn up to 12 mm long. **Lemmas** mostly 3 to 4 per spikelet, short-hairy, tapering gradually into an awn up to 3.5 cm long, the awns divergent and curved. Most common along road cuts where likely introduced for soil stabilization. Throughout much of N. America.

Elymus cinereus Scribn. & Merr. [*Leymus cinereus* (Scribn. & Merr.) A.Löve] BASIN WILDRYE Bunchgrass. **Stems** 1–2 m tall, forming large bunches or tussocks. **Leaves:** blades 7–12 mm wide, usually flat. **Inflorescence** 8–22 cm long, the rachis continuous and erect. **Spikelets** mostly 3 (sometimes more) per node, 11–15 mm long; glumes narrow, gradually tapering from the base to a narrow, pointed tip, about as long as 1st lemma, veins not evident or with a single vein. **Lemmas** mostly 3 to 5 per spikelet, sparsely short-hairy, awnless or with a short awn to 5 mm long. Open shrub steppe to open montane understory, in theory near seeps, as well as moderately disturbed sites such as along rural roads subject to little disturbance (e.g., grading, spraying, etc.). Mainly throughout much of the western half of N. America.

Elymus condensatus J.Presl [*Leymus condensatus* (J.Presl) A.Löve] GIANT WILDRYE Bunchgrass. **Stems** 1–3 m tall, forming large patches in which may be found short rhizomes. **Leaves:** blades up to 30 mm wide, usually flat. **Inflorescence** 15–50 cm long, the rachis continuous and erect. **Spikelets** mostly >3 per node, about 15 mm long; glumes narrow, gradually tapering from the base to a narrow, pointed tip, about as long or longer than the first lemma, veins not evident or with a single vein. **Lemmas** usually about 5 per spikelet, usually hairless, awnless to short-awned. Frequently disturbed settings, mine tailings; collected in Stillwater Co. Western N. America in BC, AB, CA.

Elymus elymoides (Raf.) Swezey [*Sitanion hystrix* (Nutt.) Smith] SQUIRRELTAIL Bunchgrass. Stems 20–60 cm. **Leaves**: blades 1–5 mm wide, flat to folded. **Inflorescence** 4–14 cm long, the erect rachis readily disarticulating at maturity. **Spikelets** usually 2 per node (1 to 3), 11–14 mm long; glumes gradually tapering from the base to a narrow acute tip, often bifid, rarely trifid, distinctly 1- to 2-veined, imperceptibly tapering into 1 or 2 awns that are 2–11 mm long. **Lemmas** mostly 2 to 6 per spikelet, hairless to hairy, tapering into an awn 2–10 cm long. Open, dry shrub steppe as well as in moderately disturbed sides such as along gravel back roads. Throughout most of the western half of N. America.

This species often hybridizes readily with other species of *Elymus* or *Agropyron*. See notes under *Agropyron spicatum* where *Agrositanion saxicola*, a hybrid between *Agropyron spicatum* and *Elymus elymoides*, is diagnosed.

Elymus flavescens Scribn. & J.G.Sm. [*Leymus flavescens* (Scribn. & J.G.Sm.) Pilg.] YELLOW WILDRYE Rhizomatous. Stems solitary to few-bunched, 60–100 cm. **Leaves**: blades 3–6 mm wide, inrolled. **Inflorescence** 10–22 cm long, the rachis continuous and erect. **Spikelets** mostly 2 per node, 10–20 mm long; glumes narrow, faintly 3-nerved, tapering to an awn tip. **Lemmas** mostly 2 to 5 per spikelet, copiously long-hairy, awnless or awn-tipped. Sand dunes and adjacent sandy soils, collected in Beaverhead and Madison cos. Pacific Northwest region in AB, WA, OR, ID, MT, UT.

Elymus glaucus Buckley BLUE WILDRYE Bunchgrass. **Stems** few-bunched, 60–130 cm. **Leaves**: blades 5–12 mm wide, flat. **Inflorescence** 6–16 cm long, the rachis continuous and erect. **Spikelets** usually 2 per node, 12–20 mm long; glumes distinctly broadest in the lower half, with 3 to 5 veins, the base mostly greenish, not indurate, distally tapering to a short awn or awn tip. **Lemmas** mostly 2 to 4 per spikelet, essentially hairless, with a straight awn 10–20 mm long. Dry montane meadows and open understory. Throughout much of N. America but absent from some central states and eastern provinces and states. (p.675)

Elymus innovatus Beal [*Leymus innovatus* (Beal) Pilg.] DOWNY RYEGRASS Rhizomatous. Stems 50–100 cm. **Leaves**: blades 3–6 mm wide, inrolled. **Inflorescence** 6–14 cm long, the rachis continuous and erect. **Spikelets** mostly 2 per node, 10–20 mm long; glumes gradually tapering from the base to a very narrow, pointed tip, mostly 1-nerved, tapering to an awn tip. **Lemmas** mostly 2 to 4 per spikelet, distinctly hairy, bearing an awn 1–6 mm long. Moderately disturbed settings such as along streambanks of the North Fork Sun River (Teton and Lewis & Clark cos.). Throughout southern and western Canada, as well as AK, south along the Rocky Mountain states to CO.

Resembling *E. flavescens* but the lemmas are less conspicuously hairy and bear a distinct awn.

Elymus junceus Fisch. [*Psathyrostachys juncea* (Fisch.) Nevski] RUSSIAN WILDRYE Bunchgrass. Stems few to many-bunched, 50–100 cm. **Leaves**: blades 2–7 mm wide, flat to inrolled. **Inflorescence** erect, 6–12 cm long; the inflorescence rachis disarticulating at maturity. **Spikelets** 2 or 3 per node, 8–11 mm long; glumes gradually tapering from the base to a very narrow pointed tip, 1-veined. **Lemmas** 2 or 3 per spikelet, distinctly short-hairy, tapering to an awn tip or awn that is <4 mm long. Introduced for forage and reclamation, sometimes escaping, possibly not long-persisting. Introduced throughout many of the central states and provinces, as well as OR, AK.

Elymus macounii Vasey [*Agrohordeum macounii* (Vasey) Lepage, *Elyhordeum macounii* (Vasey) Barkworth & D.R.Dewey] MACOUN'S WILDRYE, MACOUN'S BARLEY. Bunchgrass. **Stems** 20–60 cm. **Leaves**: blades 2–5 mm wide, flat to inrolled. **Inflorescence** 5–10 cm long, the erect rachis disarticulating at maturity. **Spikelets** 2 per node (sometimes evident by just 3 glumes at a node), 12–15 mm long; glumes narrow, 1- to 3-veined, tapering to a short to long awn. **Lemmas** 1 or 2 per spikelet, hairless, tapering to an awn 4–10 mm long. Open, dry moderately disturbed sites at lower elevations. Throughout most of northern and western N. America and mostly absent from eastern U.S.

This species is a hybrid formed by the crossing of *Agropyron trachycaulum* and *Hordeum jubatum*. Similar to *Elymus glaucus* but the glumes are very narrow and usually 1(3)-nerved, the inflorescence rachis disarticulates, and the species occupies typically non-montane disturbance-prone settings.

Elymus triticoides Buckley [*Leymus triticoides* (Buckley) Pilg.] BEARDLESS WILDRYE Rhizomatous. Stems 50-100 cm. **Leaves**: blades 3–8 mm (–10 mm in cultivation) wide, flat to inrolled, somewhat stiff. **Inflorescence** 5–14 cm long, the rachis continuous and erect. **Spikelets** mostly 2 per node, 10–18 mm long; glumes linear and with a very narrow pointed tip, 1-veined, tapered to mostly an awn tip. **Lemmas** 3 to 8 per spikelet, mostly hairless, tapered to an awn tip or short awn that is <3 mm long. Dry montane meadows where disturbance is somewhat regular (overgrazing). Throughout much of the western half of N. America.

Elymus triticoides is superficially very similar to *Agropyron smithii* but has mostly 2 spikelets per node, each with very narrow linear glumes that are characteristic of *Elymus* as traditionally characterized. In addition, the florets are twisted about 90° out of alignment with the glumes, which further distinguishes *Elymus* from *Agropyron* in the traditional sense.

Elymus virginicus L. Virginia wildrye Bunchgrass. **Stems** 8–13 dm. **Leaves**: blades 5–12 mm wide, flat. **Inflorescence** 6–12 cm long, the rachis continuous and erect. **Spikelets** usually 2 per node, 10–13 mm long; glumes distinctly broadest in the lower half, with 3 to 5 veins, the base yellowish, distinctly indurate, distally tapering to a short awn or awn tip. **Lemmas** mostly 2 to 4 per spikelet, essentially hairless, bearing a straight awn 10–20 mm long. Shady moist understory settings at lower elevations. Across southern Canada and over most of the eastern two-thirds of the U.S.

Eragrostis Wolf Lovegrass

Warm season, bunched annuals. **Leaf blades** flat and ascending; sheaths with overlapping margins; ligule hairy. **Inflorescence** usually an open panicle. **Spikelets** with 3 to many florets; glumes shorter than lemmas. **Lemma** with 3 distinct veins, awnless; palea well-developed and sometimes persistent. **Disarticulation** above glumes; unit of dispersal the achene.

Redfieldia flexuosa (Thurb.) Vasey, blowout grass, occurs in adjacent WY and the Dakotas and is superficially similar to *Eragrostis* except that it is a perennial rhizomatous, native of sand dunes and sites prone to wind-scouring, the stems are 50–70 cm, and the inflorescence is a very diffuse panicle mostly 30–35 cm long.

1. Spikelets <1.5 mm wide; lower glumes 0.3–0.6 mm long; lemmas with inconspicuous lateral veins . . . .*E. pilosa*
1. Spikelets >1.5 mm wide; lower glumes >0.5 mm long; lemmas with conspicuous lateral veins 2

2. Nodes of stems, keel of lemmas and glumes, or panicle branches bearing raised glands with central depressions; pedicels mostly <4 mm long, which render spikelets congested within the panicle 3
2. Nodes of stems, keel of lemmas and glumes, or panicle branches not bearing glands of any sort; pedicels mostly 4–10 mm long, which render spikelets diffused throughout an open panicle (except in *E. hypnoides*). 4

3. Spikelets mostly 2.5 mm or more wide; glands prominent on keel of most lemmas . . . . *E. cilianensis*
3. Spikelets mostly 2 mm wide or less; glandular depressions mostly on panicle branches and leaves . . . .*E. minor*

4. Stems spreading horizontally and rooting at nodes, forming mats; panicles 1–3 cm long. . . . . *E. hypnoides*
4. Stems upright or at least not rooting at nodes or forming mats 5

5. Panicles 5–25 cm long; lemmas 1–2.2 mm long; fruits brownish; native. . . . .*E. pectinacea*
5. Panicles 10–40 cm long; lemmas 1.6–3 mm long; fruits light brown to white; cultivated. . . . . *E. tef*

Eragrostis cilianensis (All.) Vign. ex Janchen Stinkgrass **Stems** spreading to ascending, 10–40 cm long. **Leaves**: blades 1–3 mm wide. **Inflorescence** a contracted to open panicle 7–15 cm long. **Spikelets** 6–10 mm long. **Lemmas**: about 10 to 30 per spikelet. Inhabiting disturbed areas along roadsides, pastures, and fields on well-drained soils. Introduced; throughout all of N. America.

The disagreeable odor is emitted from the glandular bumps on the leaves and inflorescences.

Eragrostis hypnoides (Lam.) Britton, Sterns & Poggenb. Teal lovegrass **Stems** spreading, 10–30 cm long. **Leaves**: blades 1–3 mm wide. **Inflorescence** a contracted to open panicle 2–5 cm long. **Spikelets** 5–8 mm long. **Lemmas**: about 7 to 20 florets. Mud flats, gravel and sand bars, and sandy banks mostly along the Yellowstone and Missouri Rivers in MT. Throughout most of N. America.

Eragrostis minor Host [*Eragrostis poacoides* Beauv. ex Roem. & Schult., *E. eragrostis* P.Beauv.] Little lovegrass **Stems** spreading, 5–30 cm long. **Leaves**: blades 1–3.5 mm wide. **Inflorescence** an open panicle 5–12 cm long. **Spikelets** 4–8 mm long. **Lemmas**: 10 to 19 per spikelet. Frequently disturbed settings such as cracks in or along sidewalks, driveways, and parking lots, and along roads and railroad tracks. Introduced; throughout most of N. America.

Superficially like *Eragrostis pectinacea* except that in addition to bearing some raised glands, *E. minor* also has short pedicels (<4 mm) that render a congested inflorescence, which contrasts to the open diffuse inflorescence of *E. pectinacea*. The disagreeable odor is emitted from the glandular bumps on the leaves and inflorescences.

Eragrostis pectinacea (Michx.) Nees ex Steud. [*E. diffusa* Buckley] Tufted lovegrass
Stems 1–5 dm long. **Leaves**: blades 2–3 mm wide. **Inflorescence** an open panicle 5–20 cm long.
Spikelets 5–8 mm long. **Lemmas**: 6 to 12 per spikelet. Sporadic in MT and known from disturbed
sites including sandy or gravelly banks of the Yellowstone and Big Horn Rivers. Throughout most
of N. America.

Eragrostis pilosa (L.) P. Beauv. India lovegrass **Stems** 1–5 dm. **Leaves**: blades 1–4.5 mm wide. **Inflorescence**
an open panicle 5–25 cm long. **Spikelets** 3.5–10 mm long. **Lemmas**: 5 to 17 per spikelet. Roadsides, railroad
embankments, edges of gardens and cultivated fields; collected in Yellowstone Co. Throughout most of N. America.

Eragrostis tef (Zuccagni) Trotter Teff **Stems** 25–60 cm. **Leaves**: blades 2–5.5 mm wide. **Inflorescence** an open
panicle 10–40 cm long. **Spikelets** 4–11 mm long. **Lemmas**: 7 to 16 per spikelet. Rarely escaping cultivation in vicinity
of crop fields; collected in Gallatin Co. Introduced in N. America and known otherwise from SC.

Eremopyrum (Ledeb.) Jaubert & Spach False wheatgrass

Eremopyrum triticeum (Gaertn.) Nevski [*Agropyron triticeum* Gaertn.] Annual wheatgrass
Cool season annual bunchgrass. **Stems** erect to somewhat spreading, 15–30 cm. **Leaves**: blades
flat to inrolled, ascending, 2–6 mm wide; sheaths with overlapping margins; ligule membranous.
Inflorescence 1–2 cm long, a bilateral terminal spike with a continuous rachis. **Spikelets** one per
node, 8–12 mm long with several florets very crowded and diverging at wide angles (40–80°); the
internodes 1/10–1/8 the length of the spikelets; glumes smaller than lemmas; glumes and lemmas becoming hardened
or bony in texture at maturity. **Lemmas** tapering to an awn tip; palea enclosed by lemma. **Disarticulation** below the
glumes; unit of dispersal the burr-like spike of spikelets. Roadsides, gravel pits, margins of garbage dumps, and other
regularly disturbed settings with sandy or gravelly soils where competition with other grasses is minimal. Introduced
throughout much of western N. America except at the extreme northern latitudes.

This grass resembles a diminutive form of crested wheatgrass.

Festuca L. Fescue

Cool season, bunched perennials. **Leaf blades** inrolled and ascending; sheaths with overlapping margins;
ligule membranous and bilobed. **Inflorescence** a contracted to open panicle. **Spikelets** with several to
many florets, these often not tightly overlapping; glumes smaller than lemmas. **Lemma** rounded on back
and with an inconspicuous venation, awned or awn-tipped; palea mostly enclosed in floret. **Disarticulation**
above glumes; dispersal unit the floret.

The broad- and fine-leaved *Festuca* species have been resolved into sister lineages.[394] The broad-leaved
lineage includes species of *Leucopoa* and *Lolium*, as well as *Festuca arundinacea* and *F. pratensis*, for
example. The fine-leaved lineages include the species of *Vulpia* and the remaining species of *Festuca*
(e.g., *F. rubra*, *F. idahoensis*, etc.). The segregation of *Leucopoa*, *Lolium*, *Schedonorus* (including *Festuca
arundinacea* and *F. pratensis*), and *Vulpia* is now well-supported by phylogenetic evidence. Distinguishing
among *Festuca* species relies to a large degree on micromorphologies that are not readily useable and thus
not adopted in this treatment.[107]

1. Spikelets with an irregular symmetry, comprising small vegetative bulb-like plants. . . . . *F. viviparoidea*
1. Spikelets with bilaterally symmetrical. 2

2. Awns of lemma 0–2 mm long (if awns <2 mm long and plants <35 cm tall, go to couplet 7). 3
2. Lemmas with awns often much longer than 1.5 mm 5

3. Auricles of sheath conspicuous and well-developed; leaf blades flat, mostly >3 mm
    wide; stems mostly about 1 m tall or taller . . . . see genus *Schedonorus*
3. Auricles of sheath absent; leaf blades mostly inrolled or flat and <3 mm wide; stems of various height 4

4. Leaf margins smooth or minutely scabrous; basal leaf blades distinctly tapering distally; new
    basal sheaths green, old basal sheaths with white veins contrasting against dark inter-veins
    and bearing a hard, persistent, short blade; lemmas averaging 5–6 mm long. . . . . *F. viridula*
4. Leaf margins strongly scabrous; basal leaf blades not distinctly tapering distally; new
    basal sheaths reddish or purplish, old basal sheaths light colored with a deciduous blade
    appearing to be grazed off; lemmas averaging 8–9 mm long . . . . *F. campestris*

5. Stems averaging <35 cm; panicles mostly <10 cm long, narrow; spikelets with mostly 2 to 4 florets 6
5. Stems averaging 40–100 cm; panicles mostly 10–20 cm long, open; spikelets with mostly 5 to 7 florets. . .7

6. Stems densely short-hairy below the panicle, the epidermis dark purplish; exposed alpine
   tundra . . . . *F. baffinensis*
6. Stems not hairy below the panicle, epidermis greenish; montane to alpine. . . . . *F. idahoensis*

7. Rhizomes well-developed; stems not bunched or bunched with decumbent bases; basal leaf
   sheaths dark brown or reddish, prominently nerved, ultimately shredding into persistent fibers . . . . .*F. rubra*
7. Non-rhizome bunchgrasses with erect stems; basal leaf sheaths green, or if reddish or
   purplish, then not shredding into persistent fibers 8

8. Leaf blades flat, lax, 3–10 mm wide; ligules not distinctly bilobed . . . .*F. subulata*
8. Leaf blades inrolled, stiff, about 1 mm wide; ligules distinctly bilobed 9

9. Awns as long or longer than the lemmas; pedicels and inflorescence branches spreading or
   drooping; lemmas scabrous on the upper half . . . . *F. occidentalis*
9. Awns shorter than the lemmas; pedicels and inflorescence branches stiff and ascending;
   lemmas scabrous at most in the upper third. 10

10. New basal leaf sheaths green; old leaf blades remaining on old sheaths; leaf blades
    smooth-edged or weakly scabrous; stems usually averaging <1 m. . . . . *F. idahoensis*
10. New basal leaf sheaths reddish or purplish; old leaf blades disarticulating from old sheaths;
    leaf blades with strongly scabrous edges; stems averaging close to 1 m . . . . *F. campestris*

Festuca baffinensis Polunin Baffin fescue Perennial bunchgrass. **Stems** 10–20 cm. **Leaves:**
blades ca. 1 mm wide, inrolled. **Inflorescence** a narrow panicle 1–3 cm long; the peduncle hairy
and usually pigmented with a dark purplish color. **Spikelets** 4–7 mm long, with 3 to 4 florets.
Lemmas with an awn tip or awn ≤2.5 mm long. Alpine meadows. Circumpolar south to UT, CO.

   This species is barely distinct from the short-statured, high elevation forms of *Festuca idahoensis*
and is arbitrarily distinguished by the hairy and usually dark pigmented peduncle.

Festuca campestris Rydb. [*Festuca hallii* (Vasey) Piper, *F. scabrella* Torr. ex Hook] Rough
fescue Perennial bunchgrass. **Stems** 60–100 cm. **Leaves:** blades 1–2 mm wide, inrolled to
partially flattened, distinctly scabrous; old leaf sheaths persistent in tuft. **Inflorescence** a narrow
panicle 5–15 cm long. **Spikelets** 8–12 mm long, with 4 to 6 florets. **Lemmas** awnless or awn-tipped.
Grasslands; valleys to montane. BC to SK, ON south to OR, ID, MT, CO. (p.684)

   Differing from *Festuca hallii*, plains rough fescue, which occurs mostly to the north in Canada, by spikelets with few
florets and glumes that overtop the distal-most florets. Such "oatgrass" spikelets in MT rough fescue do not segregate
geographically. Most MT specimens lack the "oatgrass" spikelet morphology. The few specimens that have them come
from western MT, and these don't otherwise differ from the more common forms of *Festuca campestris*.

Festuca idahoensis Elmer Idaho fescue Perennial bunchgrass. **Stems** 10–80 cm. **Leaves:**
blades 1 mm wide, inrolled. **Inflorescence** an open panicle 2–20 cm long; the peduncle
glabrous and usually greenish in color. **Spikelets** 5–10 mm long, with 3 to 7 florets. **Lemmas**
with an awn tip or awn ≤5 mm long. Grasslands, meadows at all elevations. Throughout western
N. America except at extreme northern latitudes.

   On drier sites and at high elevations this species is short-statured and distinguished as *Festuca
brachyphylla* Schult. & Schult. f. [*Festuca ovina* L. var. *brevifolia* (R. Br.) S. Watson], alpine fescue, *Festuca minutiflora*
Rydb., small-flower fescue, or *Festuca saximontana* Rydb., Rocky Mountain fescue. Such segregates are distinguished
by highly overlapping traits mostly involving sizes and patterns of coloration on spikelets, florets, lemma awns, and
leaf-blade. *Festuca rubra* differs by a loosely bunched rhizomatous habit, moist loamy habitats, and reddish-tinged leaf
sheaths with prominent whitish nerves. *Festuca idahoensis* is non-rhizomatous, strongly bunched, and grows in open,
rocky dry sites. Occasional specimens might best be assigned to either *F. idahoensis* or *F. rubra* using morphology and
ecology. *Festuca ovina*, sheep fescue, is introduced from Europe and rare in N. America because it is no longer used in
seed trade. It differs from *F. idahoensis* by technical characters involving leaf sclerenchyma patterns, leaf blades with 1
to 3 indistinct ribs (vs. 1 to 5 distinct ribs), and anthers 1.4–2.6 mm long (vs. 2.4–4.5 mm).

Festuca occidentalis Hook. Western fescue Perennial bunchgrass. **Stems** 40–100 cm.
Leaves: blades 1 mm wide, smooth to slightly scabrous. **Inflorescence** an open, often drooping
panicle 5–12 cm long. **Spikelets** 6–9 mm long, with 3 to 5 florets. **Lemmas** with a slender,
flexuous awn 4–12 mm long. Moist soils of forest understory, along streams, lakes; valleys to
subalpine. Very western N. America and extending east into MT, WY, SD, to MI, ON.

   The long awns and drooping panicle are distinctive, the latter often inferred from herbarium specimens by the arching
inflorescence rachis and the arrangement of spikelets that are all swept to one side.

Festuca rubra L. Red fescue Perennial bunchgrass or not with well-developed, slender rhizomes. **Stems** 20–80 cm. **Leaves**: blades 1–1.5 mm wide, inrolled; lower sheaths reddish or brown, becoming fibrous in age. **Inflorescence** a loosely contracted panicle 7–15 cm long. **Spikelets** 6–8 mm long, with 4 to 7 florets. **Lemmas** with an awn 1–3 mm long. Grasslands, meadows; valleys, montane. Throughout nearly all of N. America, both native and introduced from Eurasia.

Indigenous to northern MT [*F. rubra* ssp. *arctica* (Hack.) Govor., *F. rubra* ssp. *vallicola* (Rydb.) Pavlick] but naturalized elsewhere from lawn plantings. The predilection of *F. rubra* for moist loamy soils may help distinguish this species from *F. idahoensis*. *Festuca rubra* has reddish stem bases >1 mm in diameter and distinctly whitish-veined. The veins persist as strands after the sheath lamina has disintegrated. Stem bases are decumbent and occur as single or loosely bunched stems interconnected by rhizomes. Distinctly bunched stems, however, occur sporadically and *F. rubra* then differs from *F. idahoensis* only by stem thickness and ecological setting.

Festuca subulata Trin. Bearded fescue Perennial bunchgrass. **Stems** loosely bunched, 60–100 cm. **Leaves**: blades 4–10 mm wide, flat. **Inflorescence** an open panicle with drooping branches, 8–25 cm long. **Spikelets** 7–10 mm long, with 3 to 5 florets. **Lemmas** with an awn 5–15 mm long. Open forest, riparian woodlands; valleys, montane. Throughout most of western N. America and extending east into SD.

The broad leaf blades and open panicles are very characteristic of this species.

Festuca viridula Vasey Greenleaf fescue Perennial bunchgrass. **Stems** 40–80 cm. **Leaves**: blades 1.5–2 mm wide, flat. **Inflorescence** an open panicle 6–15 cm long. **Spikelets** 8–11 mm long, with 3 to 7 florets. **Lemmas** unawned or awned to 1.5 mm. Grasslands, meadows; subalpine, alpine. BC to CA, NV, MT.

This species might be characterized as a slender form of *Festuca campestris* except that the persistent, older leaf sheaths do not have abscised blades but rather taper to highly reduced blades that are commonly hardened into a claw-shape, a feature unique to this species.

Festuca viviparoidea Krajina ex Pavlick [*Festuca ovina* L. var. *vivipara* L.; *Festuca vivipara* (L.) Sm. misapplied] Northern fescue, viviparous fescue. Perennial bunchgrass. **Stems** 10–25 cm. **Leaves**: blades ca. 1 mm wide. **Inflorescence** a narrow panicle 3–5 cm long. **Spikelets** of variable length depending on number and size of the vegetative bulblets. **Lemmas** uncommon, sometimes awn-tipped. Alpine turf; collected in Glacier Co. AK to Greenland south to BC, MT, WY; Asia.

Glyceria R. Br. Mannagrass

Cool season, rhizomatous perennials. **Leaf blades** flat and lax; sheaths with fused margins; ligule membranous. **Inflorescence** an open and diffuse or contracted panicle. **Spikelets** with mostly 3 to many florets; glumes 1-veined. **Lemma** with 5 to 9 prominent, evenly spaced veins; palea well-developed. **Disarticulation** above glumes; unit of dispersal the floret.

1. Panicle narrow, the branches ascending; spikelets round in cross-section, 8 mm long or longer, usually with 8–12 florets (rarely as few as 5 florets); lemmas 3–4 mm long; ligules mostly 5–13 mm long . . . . *G. borealis*
1. Panicle open, the branches spreading; spikelets laterally flattened, <8 mm long, usually with 3–7 florets (rarely as many as 8 florets); lemmas <3 mm long; ligules mostly 1–7 mm long 2

2. Spikelets 4–7 mm long (rarely as short as 3 mm long); glumes with mostly acute tips and the midrib prominent to the tip, first glume mostly 1.5–2.5 mm long; lemma tips somewhat flat; ligules often 5–7 mm long (sometimes as short as 3 mm long); at least some leaf blades 6 mm wide or wider (if margins of leaf sheaths overlap, check *Torreyochloa pallida*) . . . . *G. grandis*
2. Spikelets 2–4 mm long (rarely as long as 5 mm long); glumes with mostly rounded ends and the midrib not extending to the tip, first glume <1.5 mm long (and often no >1 mm long); lemma tips prow-shaped; ligules mostly 1–5 mm long (rare as long as 6 mm long); leaf blades 2 mm wide or wider 3

3. Leaf blades, at least some if not most, 6 mm wide or wider; ligules 2–6 mm long; stems mostly about 1 m tall or more . . . . *G. elata*
3. Leaf blades 2–5 mm wide; ligules 1–4 mm long; stems 30–100 cm tall. . . . . *G. striata*

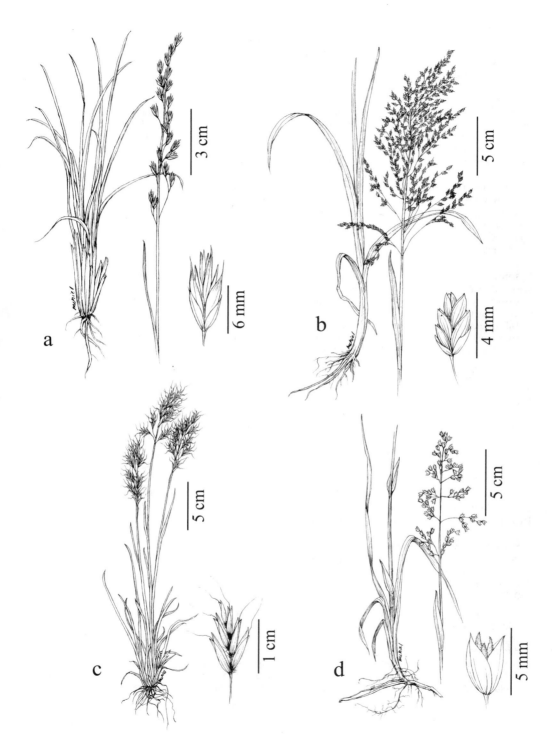

Plate 115 a. *Festuca campestris*, b. *Glyceria elata*, c. *Helictotrichon hookeri*, d. *Anthoxanthum hirtum*

Glyceria borealis (Nash) Batch. NORTHERN MANNAGRASS, SMALL FLOATING MANNAGRASS **Stems** 80–150 cm. **Leaves**: leaf blades 2–5 (7) mm wide; ligules 5–11 mm long. **Inflorescence** a narrow panicle 20–45 cm long, branches ascending. **Spikelets** 8–15 mm long, round in cross-section. **Lemmas** 8 to 12 per spikelet. Wet meadows and along stream and lake margins. Sporadic throughout much of N. America, although absent in the southeast.

The many long florets per spikelet render a very distinctive appearance to the cylindrical spikelet. In some specimens, however, the florets can be splayed out laterally, slightly obscuring the cylindrical spikelet construction.

Glyceria elata (Nash ex Rydb.) M.E.Jones TALL MANNAGRASS **Stems** 1–2 m. **Leaves**: blades 6– 10 mm wide; ligules 2–6 mm long. **Inflorescence** an open panicle 15–30 cm long, the branches spreading and often drooping. **Spikelets** 3–5 mm long, laterally compressed. **Lemmas** 2 to 7 per spikelet. Wet areas along streams and lake margins. Throughout N. America. (p.684)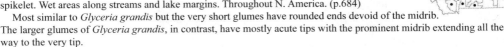

Most similar to *Glyceria grandis* but the very short glumes have rounded ends devoid of the midrib. The larger glumes of *Glyceria grandis*, in contrast, have mostly acute tips with the prominent midrib extending all the way to the very tip.

Glyceria grandis S. Watson AMERICAN MANNAGRASS **Stems** 1–2 m. **Leaves**: blades 2–12 mm wide; ligules mostly 5–7 mm long. **Inflorescence** an open diffuse panicle, 20–40 cm long, the branches spreading and often drooping. **Spikelets** 3–7 mm long, laterally compressed. **Lemmas** 2 to 7 (rarely 8) per spikelet. Banks of steams and along lake margins. Throughout most of N. America excepting many southeastern states.

Torreyochloa pallida var. *pauciflora* (pale false mannagrass) may be confused with *Glyceria grandis*. *Torreyochloa pallida* has sheath margins that overlap (they are fused in *Glyceria*) and the ligules are confined mostly to the front of the leaf blade (in *Glyceria* the ligules often wrap and fuse around the stem). The spikelets of *Torreyochloa pallida* are distinctly broad and the lemmas have a black band that delimits the hyaline tip from the greenish lemma body. In addition, the lemma tip of *T. pallida* is broad and in the plane of the main lemma body. The glumes of *Torreyochloa pallida* are rounded, devoid of a midrib towards the tip, and much shorter than the first lemma. In *Glyceria grandis*, the lemma tip is greenish or with an inconspicuous narrow hyaline tip, and the entire lemma tip is contracted from the main lemma body to give the entire lemma body, including the tip, a scoop-shape. The glumes of *Glyceria grandis* are mostly acute, with a midrib extending to the tip, and are about half as long as the first lemma.

Glyceria striata (Lam.) Hitchc. FOWL MANNAGRASS **Stems** often bunched, 30-100 cm. **Leaves**: blades 2–5 mm wide; ligules 1–4 mm long. **Inflorescence** an open panicle, 10–20 cm long, branches spreading but if ascending then the tips drooping. **Spikelets** 2–4 mm long. **Lemmas** 2 to 7 florets per spikelet. Common along lake shores and in wet meadows. Throughout most of N. America.

Very similar to *Glyceria elata* except in having a shorter stature and narrower leaves.

Helictotrichon Besser ex Schult. & Schult. f. Alpine spikeoat

Helictotrichon hookeri (Scribn.) Henr. [*Avenua hookeri* (Scribn.) Holub] SPIKEOAT Cool season, perennial bunchgrass. **Stems** 20–40 cm. **Leaves**: ascending, the larger blades (2)3–4 mm wide, flat to folded along midrib; sheaths with overlapping margins; ligule membranous. **Inflorescence** a contracted panicle 2–11 cm long. **Spikelets** 11–15 mm long with mostly 3 to 6 florets; glumes nearly enclosing the florets. **Lemmas** 3–5 mm long, uniformly awned from the middle, the awns twisted and bent; palea well-developed. **Disarticulation** above glumes; unit of dispersal the floret. Northern plains and middle to high elevation montane meadows and open understory. NM north to northwestern Canada, east to QC. (p.684)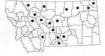

High elevations populations of *H. hookeri* approach *H. mortonianum* from the southern Rocky Mountains in having short inflorescences and spikelets with 2 florets but they differ in having wider leaves. *Helictotrichon* is often confused with *Danthonia* species but the membranous ligule and awn originating from the middle of the back of the lemma of *Helictotrichon* differ from the hairy ligule and nearly terminal awn of *Danthonia*.

Holcus L. Velvetgrass

Holcus lanatus L. Cool season, perennial bunchgrass. **Stems** 30–80 cm, covered with dense soft hairs, often decumbent and rooting at nodes. **Leaves**: blades 4–8 mm wide, flat and lax to ascending; sheaths with overlapping margins; ligule membranous. **Inflorescence** a contracted panicle 8–15 cm long, opening up at or after anthesis. **Spikelets** with 2 florets 3.5–4.5 mm long; glumes enclosing the florets. **Lemmas** 2 per spikelet; lower floret unawned and fruit bearing; upper lemma staminate and with a hooked awn; palea well-developed. **Disarticulation** below glumes and tardily above the glumes; unit of dispersal the spikelet or floret. Frequently disturbed sites, pastures, roadsides, croplands; collected in Ravalli Co. Throughout much of N. America excepting the very north-central provinces.

Hordeum L. Barley

Cool season, bunched perennials and annuals. **Leaf blades** flat and lax to ascending; sheaths with overlapping margins; ligule membranous. **Inflorescence** a two-sided terminal spike with short internodes. **Spikelets** exactly 3 per node and with one floret each; the lateral two pedicellate and sterile and the central one sessile and fertile (all three fertile at a node in six-row barley, *Hordeum vulgare*); glumes awn-like. **Lemma** unawned or awned from the tip; palea enclosed in floret. **Disarticulation** above glumes; dispersal unit the floret (or fruit)

Mostly inhabiting frequently disturbed settings. *Hordeum leporinum* Link [*H. murinum* L. ssp. *leporinum* (Link) Arcang.], hare barley, is very common in adjacent states and is expected to occur in MT. It differs from *H. geniculatum* by having lateral spikelets with well-developed staminate florets, glumes of the central spikelet with ciliate margins, and well-developed auricles mostly >1 mm long.

1. Plants perennial; awns slender. 2
1. Plants annual; awns stout. 3

2. Glume awns 30–80 mm long; inflorescence spike lax, 5–12 cm long, nearly as broad as
   long when mature and fresh; common to dry often disturbed sites . . . .*H. jubatum*
2. Glume awns <30 mm long; inflorescence spike erect, 2–6 cm long, cylindrical; plants
   mostly of meadow or riparian sites. . . . . *H. brachyantherum*

3. Leaf blades mostly 6–15 mm wide; rachis not disarticulating at maturity; awns of lemmas
   >4 cm long or lacking; cultivated barley; escaping roadside or near crop fields . . . . *H. vulgare*
3. Leaf blades <6 mm wide; rachis readily disarticulating at maturity; awns of lemmas ≤4 cm
   long; not cultivated and common to disturbed settings, often on alkaline soils 4

4. Glumes of central spikelet 2 times wider at the middle than at the base . . . .*H. pusillum*
4. Glumes of central spikelet awn-like, not distinctly wider above the base. . . . .*H. geniculatum*

Hordeum brachyantherum Nevski. [*Hordeum jubatum* L. ssp. *intermedium* Bowden, *Hordeum caespitosum* Scribn. ex Pammel] MEADOW BARLEY Perennial. **Stems** 25–60 cm. **Leaves:** lower blades and sheaths usually hairless, blades flat to folded, 1–3 mm wide. **Inflorescence** erect, 2–6 cm long; the inflorescence rachis disarticulating. **Spikelets:** the 2 laterals sterile; the pedicel 0.5–1 mm long; glumes awn-like, 7–15 mm long (rarely to 30 mm). **Lemmas:** fertile lemma 7–8 mm long. Montane meadows and along streams. Throughout much of western N. America and sporadically in eastern states and provinces.

Hordeum caespitosum (*H. jubatum* ssp. *intermedium*), intermediate barley, is an alleged hybrid between *H. brachyantherum* and *H. jubatum*. This hybrid is similar to meadow barley in its erect inflorescence and predilection to wet meadow settings. With awns 20–30 mm long (instead of <20 mm long), this hybrid is more similar to foxtail barley (*H. jubatum*). The ecological and inflorescence similarity of intermediate and meadow barley suggests these two might be the same.

Hordeum geniculatum All. [*Hordeum marinum* Huds. ssp. *gussonianum* (Parl.) Thell.] MEDITERRANEAN BARLEY Annual. **Stems** few, bunched, often bent at nodes on dried specimens, 10–40 cm. **Leaves:** lower blades and sheaths usually hairy, blades flat, 2–4 mm wide. **Inflorescence** erect, 1–3 cm long; rachis disarticulating. **Spikelets:** the 2 laterals sterile; the pedicel 0.5–1 mm long; glumes awn-like, 8–15 mm long. **Lemmas:** fertile lemma 5–6.5 mm long. Frequently disturbed settings, alkaline soils; collected in Carter and Custer cos. Introduced in western N. America and sporadic in some eastern states.

Very similar to *H. glaucum* Steudel, which may occur in MT and would be distinguished from *H. geniculatum* by having hairless lower leaf sheaths, a fertile floret in all three spikelets, pedicels of the lateral spikelets that measure 1.5–3 mm long, and a rachilla projection that is stout rather than bristle-like.

Hordeum jubatum L. FOXTAIL BARLEY Perennial. **Stems** 20–70 cm tall; lower blades and sheaths usually hairless, less commonly hairy, blades flat, lax, 2–5 mm wide. **Inflorescence** lax, 5–12 cm long; rachis disarticulating at maturity. **Spikelets** the 2 laterals sterile; the pedicel 0.5–1.5 mm long; glumes awn-like, 25–75 mm long. **Lemmas:** fertile lemma 6–8 mm long. Common in frequently disturbed, open dry sites (e.g., roadsides, trailsides, pastures, vacant lots, and overgrazed rangeland), including areas of high salinity. Throughout nearly all of N. America. (p.689)

Hordeum pusillum Nutt. Little barley Annual. **Stems** 10–30 cm tall; lower sheaths and blades hairless to hairy, blades flat, 1–4 mm wide. **Inflorescence** erect, 2–7 cm long; rachis disarticulating at maturity. **Spikelets:** the 2 laterals staminate; the pedicel 0.5–1 mm long; glumes awn-like but distinctly broad above the base, 8–14 mm long. **Lemmas:** fertile lemma 5–7 mm long. Open, dry disturbed sites especially where water intermittently stands; e.g., depressions in pastures, rangeland, and roadsides. Throughout the U.S., BC, AB.

Little barely was cultivated as a grain crop by some of the early N. American civilizations from 800–100 BC.[264]

Hordeum vulgare L. [*Hordeum distichon* L.] Common barley, six-row barley, two-row barley Annual (winter or spring). **Stems** 50–120 cm. **Leaves:** lower sheaths and blades usually hairless, blades flat, 6–15 mm wide. **Inflorescence** erect, 6–12 cm long; rachis not readily disarticulating at maturity. **Spikelets** all fertile in six-row barley, the laterals sterile in two-row barley; pedicel about 0.5 mm long; glumes awn-like though slightly broadened above the base, 8–15 mm long. **Lemmas:** fertile lemma 8–12 mm long, the awns of some varieties of barley are >4 cm long, while other varieties lack such an awn. Cultivated grain crop, escapes not persisting. Introduced and escaping throughout nearly all of N. America.

Six-row barley is derived from two-row barley, the result of a single gene mutation. Two-row barley has uniform grain size and is preferred for malt. Six-row barely produces grains of unequal size and varying in germination rates such that malt from such grain is of lower quality. Two-row barley is used for brewing Michelob whereas six-row barley is used for Budweiser. Beardless or hooded barley (a variety often cultivated for hay because it lacks lemma awns) is distinctive in that the awn is transformed into a trifurcate lamina.

Koeleria Pers. Junegrass

Koeleria macrantha (Ledeb.) Schult. [*K. cristata* Pers. misapplied, *K. nitida* Nutt., *K. pyramidata* (Lam.) P.Beauv. misapplied] Prairie junegrass Cool season, perennial bunchgrass. **Stems** 20–60 cm. **Leaves:** blades 1–3 mm wide, flat and ascending; sheaths with overlapping margins; ligule membranous. **Inflorescence** a contracted to occasionally open pyramidal panicle 4–11 cm long. **Spikelets** 4–5 mm long, with 2 to 4 florets; glumes somewhat enclosing the florets. **Lemmas** awnless or with a very short awn <1 mm long; palea very well-developed. **Disarticulation** above glumes; unit of dispersal the floret or fruit. Low to middle elevations in sagebrush steppe, dry montane meadows, and other open, dry sites with high native cover (indicative of infrequent disturbance). Throughout nearly all of N. America excepting some very southeastern states and very northwestern provinces. (p.689)

Leersia Sw. Cutgrass

Leersia oryzoides (L.) Sw. Rice cutgrass Cool season, rhizomatous perennial. **Stems** 80–160 cm, solitary to bunched, often rooting at nodes. **Leaves:** blades 7–14 mm wide, flat and lax; sheaths retrorsely scabrous with overlapping margins; ligules membranous, truncate, 0.8–1 mm long. **Inflorescence** an open panicle 10–25 cm long. **Spikelets** 4–5 mm long, with one floret; glumes lacking. **Lemma** bristly, awnless; palea equaling lemma and tightly clasped within it. **Disarticulation** above the glumes; dispersal unit the floret. Sporadic along fresh water lake and stream margins. Widespread, but only locally common throughout most of N. America except at extreme northern latitudes.

Leptochloa P.Beauv. Sprangletop

Leptochloa fusca (L.) Kunth Bearded sprangletop Warm season annual bunchgrass. **Stems** 10–40 cm. **Leaves:** blades 2–4 mm wide, flat and ascending; sheaths with overlapping margins; ligule membranous. **Inflorescence** a panicle of secund lateral spikes, but these often congested in leaf sheaths and adjacent leaf blades. **Spikelets** 4–7 mm long, with 5 to 7 florets; glumes similar in size to lemmas. **Lemmas** hairy, awn-tipped, distinctly 3-nerved; palea enfolded by floret. **Disarticulation** above glumes; unit of dispersal the floret. Open, dry and regularly disturbed settings including roadsides. Throughout nearly all of the U.S. and sporadically across southern Canada. Our plants are ssp. *fascicularis* (Lam.) N.Snow.

MT specimens have inflorescences overtopped by a rosette of leaves, and thus appear stunted.

Leucopoa Griseb. Spike fescue

Leucopoa kingii (S.Watson) W.A.Weber. [*Festuca kingii* (S.Watson) Cassidy, *Hesperochloa kingii* (S.Watson) Rydb.] Cool season perennial bunchgrass. **Stems** 30–50 cm. **Leaves**: blades 3–7 mm wide, lax to ascending; sheaths with overlapping margins, those of older leaves very persistent; ligule membranous. **Inflorescence** a spike-like panicle 9–19 cm long, the branches appressed to the main rachis. **Spikelets** 6–10 mm long; glumes similar in size to lemmas. **Lemmas** 3 to 7 per spikelet, unawned to slightly awn-tipped. **Disarticulation** above glumes; unit of dispersal the floret. On well-drained rocky slopes in the mountains, particularly in open dry understory, steppe, or sparsely vegetated slopes; throughout the western half of the U.S.

The bunched and robust stem bases are reminiscent of *Festuca campestris*. *Leucopoa* is distinguished by its broader leaf blades that do not abscise from the old leaf sheaths, deeply furrowed inside surface of leaf blades, and new leaf sheaths that are straw-colored.

Lolium L. Ryegrass

Cool season, bunched to single-stemmed annuals to short-lived perennials. **Leaf blades** flat and lax to ascending; sheaths with overlapping margins; ligule membranous. **Inflorescence** a bilateral or two-sided terminal spike. **Spikelets** turned edgewise to the rachis, with several to many florets; glumes both present on the terminal spikelet, all lateral spikelets with one glume. **Lemma** often keeled and inconspicuously veined, unawned to short-awned; palea enclosed in floret. **Disarticulation** above glumes; unit of dispersal the floret.

All of these have been introduced into N. America, often in seed mixes intended for lawn establishment. *Lolium* is most closely related to *Schedonorus*, which justifies the narrow circumscription of *Festuca* to include only the narrow-leaved perennial species.[106]

1. Glumes ≥length of floret-cluster; spikelets with <10 florets; annuals 2
1. Glumes <length of floret-cluster; spikelets with <10 or >10 florets; annuals to perennials. 3

2. Lemmas usually much <8 mm long and without any keel, circular in cross-section such that each lemma does not nest within the next lower one on the same side of the rachilla . . . . . . *L. temulentum*
2. Lemmas usually 8 or more mm long and laterally flattened, at least distally keeled, elliptic in cross-section and each lemma is nested within the next lower one on the same side of the rachilla . . *L. persicum*

3. Lemmas often more than 9 per spikelet, most with awns >1 mm long; most of the large leaves tending to be 3–8 mm broad. . . . .*L. multiflorum*
3. Lemmas fewer than 9 per spikelet, awnless or barely awn-tipped; most of the larger leaves tending to be <4 mm wide (but sometimes up to 6 mm broad). . . . .*L. perenne*

Lolium multiflorum Lam. [*Lolium perenne* L. var. *multiflorum* (Lam.) Husnot] ITALIAN RYEGRASS, ANNUAL RYEGRASS Biennial to short-lived perennial bunchgrass. **Stems** 20–150 cm, often single-stemmed in lawn settings. **Leaves**: blades 3–6 mm wide. **Inflorescence** 15–40 cm long. **Spikelets** 10–30 mm long; glumes about as long as an individual floret. **Lemmas** often 10 to 20 per spikelet, keeled, at least the upper ones with an awn 5–15 mm long. Regularly disturbed sites, roadsides, and newly planted lawns. Introduced throughout much of N. America.

Very similar to *Lolium perenne* and intermediate forms exist.

Lolium perenne L. PERENNIAL RYEGRASS Perennial bunchgrass. **Stems** 30–150 cm. **Leaves**: blades 2–5 mm wide. **Inflorescence** 5–25 cm long. **Spikelets** 7–19 mm long; glumes about as long as an individual floret. **Lemmas** mostly 5 to 9 per spikelet, keeled, awnless or awn-tipped. Regularly disturbed sites, roadsides, trailsides, and newly planted lawns. Introduced throughout much of N. America.

Lolium persicum Boiss. & Hohen. ex Boiss. PERSIAN RYEGRASS Annual bunchgrass. **Stems** 10–40 cm. **Leaves**: blades 2–6 mm wide. **Inflorescence** 10–25 cm long. **Spikelets** 10–26 mm long; glumes about as long as the spikelet length. **Lemmas**: 4 to 9 per spikelet, with a slender flexuous awn 5–15 mm long. Regularly disturbed settings. Introduced along the southern tier of Canadian provinces and sporadically south into mainly NY, MT, ND, WY, CO.

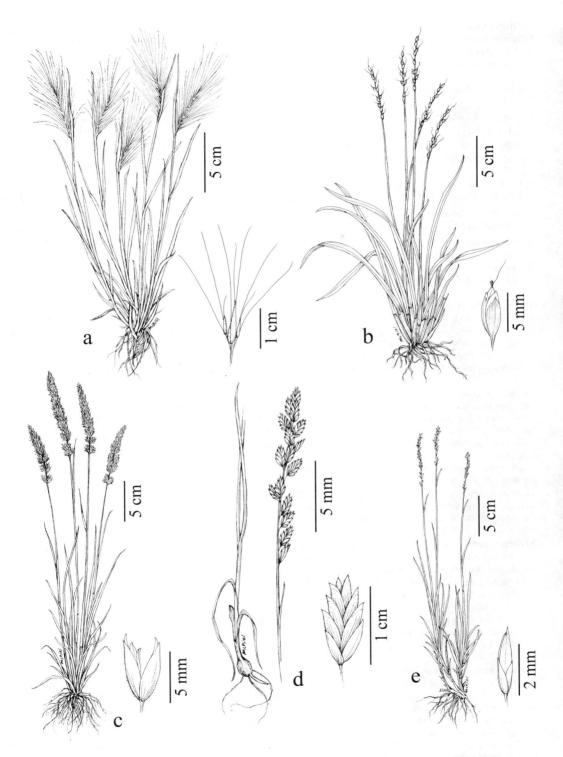

Plate 116. a. *Hordeum jubatum*, b. *Oryzopsis asperifolia*, c. *Koeleria macrantha*, d. *Melica spectabilis*, e. *Muhlenbergia richardsonis*

Lolium temulentum L. DARNELL RYEGRASS Annual bunchgrass. **Stems** 60–90 cm. **Leaves**: blades 2–10 mm wide. **Inflorescence** 10–30 cm long. **Spikelets** 10–25 mm long; glumes about as long as the spikelet length. **Lemmas**: 5 to 10 per spikelet, awnless or awned to 15 mm long. Frequently disturbed sites; collected in Gallatin Co. Introduced throughout much of N. America.

Melica L. Melicgrass

Cool season, bunched to single-stemmed perennials. **Leaf blades** flat and lax to ascending; sheaths with fused margins; ligule membranous. **Inflorescence** a raceme or contracted panicle. **Spikelets** with mostly 2 to 4 florets; glumes mostly 3- to 5-veined. **Lemma** with prominent parallel veins, unawned to awned; palea well-developed but enclosed in floret. **Disarticulation** above glumes; unit of dispersal the floret.

1. Lemmas awned; leaf blades 6–12 mm wide; stems not having corms at base . . . . *M. smithii*
1. Lemmas awnless or only awn-pointed; leaf blades 1.5–10 mm wide; corms present 2

2. Lemmas gradually tapering to a narrowly pointed tip; leaf blades 2–10 mm wide . . . . *M. subulata*
2. Lemmas blunt or abruptly tapered to a narrow tip; leaf blades 1.5–5 mm wide 3

3. Pedicels short, stiff, straight; glumes from 1/2–2/3 as long to equaling the spikelets;
   ligules 2–6 mm long; corms closely clustered . . . . *M. bulbosa*
3. Pedicels, or at least some, slender, twisted or curved; glumes usually <1/2 as long as the
   spikelets; ligules 0.1–2 mm long; corms separated by short rhizomes . . . . *M. spectabilis*

Melica bulbosa Geyer ex Porter & J.M.Coult. ONIONGRASS Bunchgrass. **Stems** 30–80 cm tall, often few-bunched, each from a corm base. **Leaves**: leaf blades 2–5 mm wide; ligules 2–6 mm long. **Inflorescence** a narrow panicle or raceme 10–30 cm long; the pedicels stiff, erect, mostly bearing 1 or 2 spikelets. **Spikelets** 12–20 mm long. **Lemmas**: 4 to 7 per spikelet, awnless. Montane dry meadows, rocky slopes, and open understory. Throughout much of the western half of N. America.
  Melica bulbosa is similar to *M. spectabilis* in sometimes having clustered sterile florets at the end of the spikelet.

Melica smithii (Porter ex A.Gray) Vasey SMITH'S MELICGRASS Bunchgrass. **Stems** 60–100 cm tall, from a slender base. **Leaves**: blades 6–12 mm wide; ligules 2–4 mm long. **Inflorescence** a narrow panicle or raceme 10–30 cm long, branches solitary, spreading or bent downwards. **Spikelets** 12–18 mm long. **Lemmas**: 3 to 5 per spikelet, with a 2–8 mm awn borne from the tip. Open to closed montane understory. Throughout many of the states and provinces along the Canada-U.S. border.
  The leaves of this species are on average about 1 cm broad. *Melica smithii* and *M. subulata* might be confused with the genus *Bromus* (e.g., *Bromus porteri*) because all concerned occur in montane meadow and understory settings, have a similar stature, and also have closed leaf sheaths. The lemmas of *Melica* are either glabrous, prominently veined, or have an awn originating from very near the lemma tip. *Bromus* lemmas are typically hairy, less prominently veined, and the awn originates from the back of a bifid lemma tip.

Melica spectabilis Scribn. [*Melica subulata* (Griseb.) Scribn. var. *pammelii* (Scribn.) C.L.Hitchc.] PURPLE ONIONGRASS Bunchgrass. **Stems** 60–100 cm tall, loosely bunched each from a corm base. **Leaves**: blades 2–5 mm wide; ligules 1–2 mm long. **Inflorescence** a narrow panicle 5–25 cm long, the branches slender and twisted. **Spikelets** 10–18 mm long. **Lemmas**: 4 to 8 per spikelet, awnless. Open montane slopes and meadows. Throughout the western U.S. and southwestern Canada. (p.689)
  In addition to the twisted and sometimes abruptly curved pedicels, the purplish bands on the lemmas of *M. spectabilis* cover much more surface area from base to tip, whereas these purplish bands in *M. bulbosa* are narrow and more confined to the lemma apex.

Melica subulata (Griseb.) Scribn. [*Melica subulata* (Griseb.) Scribn. var. *pammelii* (Scribn.) C.L. Hitchc.] ALASKA ONIONGRASS Bunchgrass. **Stems** 60–120 cm tall, loosely bunched or in patches, most stems from a corm base. **Leaves**: blades 2–10 mm wide; ligules 1–4 mm long. **Inflorescence** a narrow panicle or raceme 10–25 cm long, with few branches. **Spikelets** 10–20 mm long. **Lemmas**: 4 to 8 per spikelet, mostly awn-tipped. Open montane slopes and understory. Sporadic throughout western N. America.
  This species may be mistaken for small individuals of *Bromus inermis*. However, the onion-like base of the stem readily distinguishes *Melica* from *Bromus*. Also, *Bromus inermis* rarely has a spikelet as small as those of this species.

Muhlenbergia Schreb. Muhly

Warm season, bunched perennials and annuals, often with rhizomes. **Leaf blades** flat and ascending; sheaths with overlapping margins; ligule membranous. **Inflorescence** an open to sometimes contracted or spike-like panicle. **Spikelets** with 1 floret, rarely with 2; glumes shorter than lemma or enclosing the floret. **Lemma** with 3 (rarely 1 or 2) distinct veins, awnless or awned, sometimes with a long-hairy callus; palea enclosed in floret. **Disarticulation** above glumes; unit of dispersal the floret.

Muhlenbergia mexicana (L.) Trin., wirestem muhly, is common just west of Lolo Pass in ID and might be eventually found in MT. It is a native rhizomatous perennial similar to *Muhlenbergia glomerata* or *M. racemosa* but with glumes measuring only 3–4 mm long including the awn tip. The similar MT *Muhlenbergia* species have glumes mostly 4–8 mm, including the awns that are about as long as the lemma body. A specimen of *Muhlenbergia montana* (Nutt.) Hitchc. is deposited in the MONT herbarium with the note "Collected in Montana but location unknown. G.F.Payne 1951" This perennial bunchgrass is marked by loose spicate panicles where each spikelet bears a lemma that is long-awned (10–20 mm) and longer than the subtending glumes, and the upper glume is distinctly three-toothed.

1. Plants annual; mostly weakly rooted and completing life cycle in six weeks or so. 2
1. Plants perennial; robust bunchgrasses or strongly rhizomatous, roots not attaching to pulled-up stems . . .3

2. Panicles narrow, <4 cm long; glumes not hairy; lemmas 1.5–3 mm long. . . . . *M. filiformis*
2. Panicles open, mostly 10–20 cm long; glumes lightly hairy; lemmas 1–1.5 mm long . . . .*M. minutissima*

3. Bunchgrasses; stem bases with knotty bulbous nodes, the basal most covered with scaly
   sheaths . . . . *M. cuspidata*
3. Rhizomatous grasses with single to bunched stems often forming extensive patches;
   rhizomes extensive, scaly, and branching 4

4. Panicles open when fully extended from leaf sheath; the spikelets on long slender pedicels . . *M. asperifolia*
4. Panicles narrow, contracted, or spike-like when fully extended; the spikelets on short pedicels 5

5. Leaf blades 2 mm wide or less, usually short and inrolled; lemmas about twice as long as
   glumes; anthers 1–1.5 mm long; stems below nodes roughened with undulations, bumps,
   or cross-corrugations . . . . *M. richardsonis*
5. Leaf blades mostly >3 mm wide, flat; lemmas barely as long or shorter than the glumes;
   anthers mostly <1 mm long; stems below nodes smooth. 6

6. Hairs of the lemma callus almost or fully as long as the lemma, the hairs 2–3 mm long;
   lemmas with an awn 3–8 mm long . . . .*M. andina*
6. Hairs of the lemma callus <half the length of lemma, the hairs 0.2–1.5 mm long; lemmas
   awn-tipped or (as in *M. mexicana*) with an awn up to 6 mm long. 7

7. Lower stem internodes terete (round in cross-section) and tending to be covered by leaf
   sheaths, when exposed they are lightly hairy with a dull texture; stems unbranched or
   branching only at base; ligules 0.2–0.6 mm long; anthers 0.8–1.5 mm long; lemma lightly
   hairy at the base and up the margins. . . . . *M. glomerata*
7. Lower stem internodes often compressed (elliptical in cross-section) and tending to be
   exposed, often glabrous and shiny in texture; stems commonly branching from above ground
   nodes; ligules 0.6–1.5 mm long; anthers 0.4–0.6 mm long; lemma lightly hairy at base only . . *M. racemosa*

Muhlenbergia andina (Nutt.) Hitchc. FOXTAIL MUHLY Perennial with well-developed rhizomes. **Stems** 30–80 cm tall, often loosely bunched. **Leaves**: blades 2–4 mm wide; ligule about 1 mm long. **Inflorescence** a spike-like panicle 5–13 cm long. **Spikelets** 2.5–3.5 mm long. **Lemmas** with an awn 3–9 mm long; callus hairs as long as the lemma. Damp but often well-drained soils. Throughout much of the western half of the U.S. and southwestern Canada.

Muhlenbergia asperifolia (Nees & Meyen ex Trin.) Parodi. SCRATCHGRASS Perennial with slender rhizomes. **Stems** 10–50 cm tall, spreading to erect and sometimes aggregated into patches. **Leaves**: blades 1–3 mm wide; ligule about 1 mm long. **Inflorescence** an open panicle 5–20 cm long. **Spikelets** 1–2 mm long. **Lemmas** usually 1, rarely 2, per spikelet, awn-tipped; callus hairs very short to lacking. Damp and often alkaline soils along streams and low areas in pastures. Throughout most of southern Canada and the U.S., except in the southeast.

Muhlenbergia cuspidata (Torr. ex Hook) Rydb. Plains muhly Perennial. **Stems** 20–40 cm long, the area below the stem node smooth and without small bumps, stem bases hard, scaly, and bulb-like. **Leaves**: blades mostly 2–3 mm wide; ligule about 1 mm long. **Inflorescence** a narrow panicle 5–12 cm long. **Spikelets** 2.5–3.5 mm long. **Lemmas** slightly hairy often towards the base only, awnless or awn-tipped; callus hairs very short or lacking. Short grass prairie and occasionally sagebrush steppe on moderately disturbed settings including road cuts. Throughout the eastern flanks of the Rocky Mountains from NM to AL and east through south-central Canada and sporadically to the very eastern U.S.

Differs from *Muhlenbergia richardsonis* mainly by its bunched growth form that includes erect stems that remain straight even in dried herbarium specimens. Isolated stems in herbarium collections of *M. cuspidata* have greenish leaf blades 2–3 mm wide, blunt ligules <1 mm long, stems that are straight in line above and below nodes, and stem regions below the nodes that are hairy and with uniform parallel veins. Isolated stems in herbarium collections of *M. richardsonis* will have narrower grayish-green leaf blades, acute ligules >1 mm long, stems that are often distinctly bent at nodes, and stem regions below the nodes that are hairless or nearly so and with veins rendered wavy by transverse connections, undulations, or bumps.

Muhlenbergia filiformis (Thurb. ex S. Watson) Rydb. Pullup muhly Annual. **Stems** 5–18 cm. **Leaves**: blades 0.5–1.5 mm wide; ligule 1–2 mm long. **Inflorescence** a spike-like panicle 2–6 cm long. **Spikelets** 2–2.5 mm long. **Lemmas** awnless or awn-tipped; callus hairs very short or lacking. Moist well-drained soils, around hot springs, and edges of montane meadows. Throughout most of the western half of the U.S. and into BC.

Muhlenbergia glomerata (Willd.) Trin. Spiked muhly Perennial with well-developed woody rhizomes. **Stems** loosely bunched and often in patches, 30–90 cm tall, branching mostly at ground level; internodes lightly hairy and with a dull texture. **Leaves**: blades 2–6 mm wide; ligule <1 mm long. **Inflorescence** a contracted panicle 3–11 cm long. **Spikelets** 4–6 mm long. **Lemmas** lightly hairy at the base and up the margins, awn-tipped or rarely short-awned; callus hairs short to absent. On regularly disturbed well-drained soils including road cuts. Throughout the northern half of the U.S. and across southern Canada and north to NT.

Muhlenbergia minutissima (Steud.) Swallen. Annual muhly Annual bunchgrass. **Stems** 5–39 cm. **Leaves**: blades 1–2 mm wide; ligule about 1 mm long. **Inflorescence** an open panicle 5–15 cm long. **Spikelets** 1–1.5 mm long. **Lemmas** unawned to awn-tipped; callus hairs few to absent. Open, dry to moist sites on well-drained soils; collected in Gallatin Co. Sporadic throughout western U.S.

Muhlenbergia racemosa (Michx.) Britton, Sterns & Poggenb. Marsh muhly Perennial with well-developed rhizomes. **Stems** loosely bunched and often in patches, 30–80 cm tall, often branching above ground level; internodes with a polished and shiny texture. **Leaves**: blades 2–6 mm wide; ligule about 0.5–1.5 mm long. **Inflorescence** a contracted panicle 3–15 cm long. **Spikelets** 4–6 mm long. **Lemmas** lightly hairy at base only, awn-tipped or rarely short-awned; callus hairs short to absent. Open, dry to moist or alkaline sites on well-drained soils. Across southern Canada and throughout most of the U.S., excepting many east and southeast states.

Muhlenbergia richardsonis (Trin.) Rydb. Mat muhly Perennial with well-developed rhizomes. **Stems** decumbent and often mat-forming, 5–25 cm tall, rough with small bumps just below the nodes, bases slender just like the more distal portions of the stem. **Leaves**: blades mostly 1–2 mm wide; ligule about 1–2 mm long. **Inflorescence** a narrow contracted panicle 2–12 cm long. **Spikelets** 2–3 mm long. **Lemmas** not hairy, awn-tipped; callus hairs short to absent. Open dry to moist sites often on well-drained and somewhat disturbed soils from low to montane elevations. Throughout western N. America and extending eastward across southern Canada and the northern U.S. (p.689)

Differing from *M. cuspidata* by lemmas that are slightly hairy, perhaps only at the base (see *M. cuspidata* for further notes).

Munroa Torr. False buffalograss

Munroa squarrosa (Nutt.) Torr. Warm season, annual bunchgrass. **Stems** often mat-forming, 3–10 cm. **Leaves**: blades 1.5–2.5 mm wide, flat and ascending, clustered at nodes, the throat with stiff hairs to 2 mm long; ligule a fringe of hairs 0.5–1 mm long; sheaths with overlapping margins. **Inflorescence** a cluster of spikelets concealed among leaf sheaths. **Spikelets** 6–8 mm long with 3 to 4 florets; glumes as long as florets. **Lemmas** 3 to 5 per spikelet, awn-tipped with 3 distinct veins; palea enclosed in floret. **Disarticulation** below glumes; unit of dispersal the inflorescence and associated leaf cluster.

Sporadic in open, dry and regularly disturbed sites such as long roadsides and anthills. Throughout much of the western half of the U.S. and south-central Canada.

Oryzopsis Michx. Ricegrass

Cool season, bunched perennials. **Leaf blades** flat and lax or ascending; sheaths with overlapping margins; ligule membranous. **Inflorescence** a raceme to an open or contracted panicle. **Spikelets** with one plump floret; glumes enclosing the floret. **Lemma** hardened at maturity and enveloping the floret, with a deciduous awn, callus not sharp but usually distinctly hairy; palea enclosed in floret. **Disarticulation** above glumes; unit of dispersal the floret.

Oryzopsis should be included in *Stipa* except for *Oryzopsis asperifolia* and *O. micrantha* have not been formally transferred to that genus (see notes under *Stipa*).

1. Lemmas covered with long silky hairs that extend beyond the tip of the lemma, with a
   straight to weakly bent awn; panicle branching dichotomous (or not) . . . . *O. hymenoides*
1. Lemma glabrous or covered with short-appressed hairs that don't extend beyond the
   lemma tip, with a straight or bent awn; panicle branching, not dichotomous 2

2. Panicle open, branches wide-spreading to bent downward; lemmas glabrous; awn straight . . *O. micrantha*
2. Panicle narrow, contracted, the branches short and erect or appressed; lemmas hairy; awn bent 3

3. Glumes 6–8 mm long; awn mostly 5–10 mm long, sinuous or gradually bent; stems
   spreading or prostrate; leaf blades flat, mostly 5–9 mm wide . . . . *O. asperifolia*
3. Glumes 3.5–5 mm long; awn mostly 3–5 mm long, sharply bent; stems erect; leaf blades
   inrolled, about 1.5 mm wide . . . . *O. exigua*

Oryzopsis asperifolia Michx. [*Urachne asperifolia* (Michx.) Trin.] ROUGHLEAF RICEGRASS
Stems 20–60 cm tall, erect to widely spreading. **Leaves:** blades mostly 5–9 mm wide, scabrous; ligules 0.5–1 mm long. **Inflorescence** a raceme or narrow panicle 3–11 cm long. **Spikelets** 6–8 mm long. **Lemmas** sparsely hairy, with a deciduous awn 6–12 mm long. Dry meadows and open understory. Across northern U.S., throughout the Rocky Mountain states, and across much of Canada, excepting the extreme northern latitudes. (p.689)

Spreading stems from a leaf bunch that includes broad green leaf blades are distinctive to this species, as are the large spikelets that superficially approximate those of cultivated rice. This species is the sole representative of *Oryzopsis* in the Flora of North America.[31]

Oryzopsis exigua Thurb. [*Piptatherum exiguum* (Thurb.) Dorn; *Stipa exigua* (Thurb.) Columbus & J.P. Smith] LITTLE RICEGRASS **Stems** 15–35 cm. **Leaves:** blades 0.5–1.5 mm wide, scabrous; ligules 1–3 mm long. **Inflorescence** a raceme or narrow panicle 3–7 cm long. **Spikelets** 3.5–4.5 mm long. **Lemmas** mostly hairless, with a deciduous bent awn 4–6 mm long. Dry, open montane understory. Throughout much of western N. America.

This species is distinguished from the superficially similar *Oryzopsis micrantha* by its shorter and contracted panicles that bear larger spikelets (4–5 mm) with more persistent and bent awns.

Oryzopsis hymenoides (Roem. & Schult.) Ricker ex Piper [*Achnatherum hymenoides* (Roem. & Schult.) Barkworth; *Stipa hymenoides* Roem. & Schult.] INDIAN RICEGRASS **Stems** 30–60 cm. **Leaves:** blades about 1 mm wide (because they are typically inrolled), usually smooth; ligules mostly 1–2 mm long. **Inflorescence** an open panicle 10–20 cm long. **Spikelets** 5–6 mm long. **Lemmas** densely long-hairy, with a deciduous awn 4–6(9) mm long. Open, dry exposed sites on a diversity of substrates including sandy and gravelly soils. Throughout western and central U.S. and the southwestern portion of Canada.

A form of *Oryzopsis hymenoides* having a contracted panicle with non-dichotomous branching is referred to as *Oryzopsis bloomeri* (Bolander) Ricker [*Stipa bloomeri* Bolander; *Achnatherum ×bloomeri* (Bolander) Barkworth], which was collected in a frequently disturbed site in Yellowstone Co. coexisting with *Stipa viridula* and *O. hymenoides* (presumably a hybrid between these two). It is a large bunchgrass 1–1.3 m tall with a narrow or spicate panicle but has spikelets like those of *O. hymenoides*. *Oryzopsis contracta* (B.L.Johnson) Schltr. [*Achnatherum contractum* (B.L.Johnson) Barkworth; *Stipa contracta* (B.L.Johnson) W.A.Weber], contracted ricegrass, is a perennial bunchgrass very similar to *O. hymenoides* and differs primarily in having dichotomous inflorescence branches where one branch is usually about half the length of the other. It has perhaps shorter lemmas (up to 3.5 mm versus up to 4.5 mm long) with longer awns (up to 9 mm versus up to 6 mm long). *Oryzopsis contracta* is of possible hybrid origin between

O. hymenoides and *O. micrantha* and was collected in Beaverhead Co. and otherwise in western and central WY and northern CO.[360,361]

Oryzopsis micrantha (Trin. & Rup.) Thurb [*Piptatherum micranthum* (Trin. & Rupr.) Barkworth] LITTLESEED RICEGRASS **Stems** 30–80 cm. **Leaves**: blades 1–2.5 mm wide, smooth; ligules 1–2 mm long. **Inflorescence** a narrow to open panicle 5–20 cm long. **Spikelets** 3–3.5 mm long. **Lemmas** hairless to sparsely hairy, with a deciduous straight awn 5–8 mm long. Moist meadows, streambanks, and understory. Throughout much of western and central U.S. and southwestern Canada.

    Oryzopsis micrantha can have a contracted panicle during early stages of growth, and thus appear similar to *O. exigua*. However, the lemmas of *O. micrantha* are often glabrous and bear a straight awn.

Panicum L. Panicgrass

Warm season, bunched or rhizomatous perennials and annuals. **Leaf blades** flat and lax to ascending; sheaths with overlapping margins; ligule hairy. **Inflorescence** of diffuse open panicles. **Spikelets** with one fertile floret; glumes unequal, with the first much shorter than the second glume, which is a long as the lemmas. **Lemmas** usually rounded on back, awnless, hard-textured if fruit-brearing; palea enclosed in floret. **Disarticulation** above the glumes; unit of disarticulation the floret.

1. Plants perennial, rhizomatous; stems erect, and typically >1 m tall; sheaths hairless; glumes and lemmas keeled; ligules 3–8 mm long . . . . *P. virgatum*
1. Plants annual, without rhizomes; stems sprawling to ascending and stems usually <1 m long; sheaths glabrous or covered with glandular-based hairs; glumes and lemmas usually not keeled; ligules 0.5–3 mm long 2

2. Spikelets 5–5.5 mm long; first glume about 2/3 as long as spikelet. . . . . *P. miliaceum*
2. Spikelets 2–3 mm long; first glume <1/2 as long as spikelet 3

3. Stems and leaves conspicuously hairy with long-spreading hairs; first glume acute, >1/3 as long as the length of the floret; leaf sheaths not compressed . . . .*P. capillare*
3. Stems and leaves glabrous or nearly so; first glume blunt, rounded, sometimes acute, <1/3 as long as the length of the floret; leaf sheaths more or less compressed . . . . *P. dichotomiflorum*

Panicum capillare L. WITCHGRASS Annual bunchgrass. **Stems** sprawling or spreading, 10–60 cm long. **Leaves**: blades 5–14 mm wide; sheath densely to lightly covered with glandular-based hairs; ligules a short fringe of hairs, 0.5–1.3 mm long. **Inflorescence** an open panicle 15–25 cm long. **Spikelets** 2–3 mm long; first glume 1–1.8 mm long; second glume 2–3 mm long. **Lemmas**: sterile lemma not keeled; fertile lemma blunt, yellowish at maturity. Along roadsides, railroad tracks, sidewalks, and in lawns. Widespread in N. America except at extreme northern latitudes. (p.696)

Panicum dichotomiflorum Michx. FALL PANICGRASS Annual bunchgrass. **Stems** spreading to ascending, 30–100 cm long. **Leaves**: blades 4–15 mm wide; sheath glabrous; ligules a short fringe of hairs, 1–1.5 mm long. **Inflorescence** an open panicle 15–25 cm long. **Spikelets** 2.5–3.5 mm long; first glume 0.5–1 mm long; second glume 2–3 mm long. **Lemmas**: sterile lemma not keeled; fertile lemma blunt, yellowish at maturity. Frequently disturbed sites (railroad yard); collected in Missoula Co. Throughout most of N. America.

Panicum miliaceum L. BROOMCORN MILLET Annual bunchgrass. **Stems** robust and ascending, 30–100 cm. **Leaves**: blades 7–14 mm wide; sheath densely to lightly covered with glandular-based hairs; ligule a fringed membrane 2–3 mm long. **Inflorescence** an open panicle 12–22 cm long. **Spikelets** 5–5.5 mm long; first glume 2.5–3.5 mm long; second glume 5–5.5 mm long. **Lemmas**: sterile lemma not keeled; fertile lemma blunt, straw-colored to burnt-orange at maturity. Recently disturbed settings such as graded roadsides and sometimes cultivated and escaped, but probably not long-persisting. Introduced throughout most of North America except at extreme northern latitudes.

Panicum virgatum L. SWITCHGRASS Perennial bunchgrass with well-developed rhizomes. **Stems** 60–150 cm. **Leaves**: blades 2–10 mm wide; sheath essentially hairless; ligule a fringed membrane 3–5 mm long. **Inflorescence** an open panicle 12–40 cm long. **Spikelets** 3.5–4 mm long; first glume 3–3.5 mm long; second glume 3.5–4.5 mm long, distally keeled. **Lemmas**: sterile lemma distally keeled; fertile lemma blunt, light greenish at maturity. From disturbance-prone settings, especially roadsides and similar sites. Throughout much of N. America except for the very western and northern tier of states and provinces.

Phalaris L. Canarygrass

Cool season, bunched to single-stemmed perennials. **Leaf blades** flat and lax to ascending; sheaths with overlapping margins; ligule membranous. **Inflorescence** a contracted panicle. **Spikelets** with 3 florets, the lowest two scale-like and sterile; glumes enclosing the florets. **Lemma** awnless; palea enclosed in floret. **Disarticulation** above glumes; unit of dispersal the floret.

1. Glume keels not winged; inflorescence a contracted panicle mostly 10–40 cm long; plants perennial with rhizomes; stems commonly >1 m; spikelets 4–5 mm long, the lowest two florets comprising minute hairy scales . . . . *P. arundinacea*
1. Glume keels winged, the wings 0.2–1 mm wide; inflorescence a dense ovoid spike mostly 1.5–5 cm long; plants annual; stems <60 cm tall; spikelets 6–8 mm long, the lowest two florets comprising well-developed hairless scales . . . . *P. canariensis*

Phalaris arundinacea L. Reed canarygrass Rhizomatous perennial. **Stems** mostly 90–200 cm tall, forming dense stands. **Leaves:** blades 6–18 mm wide; ligules 4–10 mm long. **Inflorescence** a contracted panicle 8–35 cm long. **Spikelets** 4–7 mm long. **Lemmas** with one hairless seed-bearing floret and two lower sterile florets that are each a minute scale covered with long hairs. Riparian and wetland habitats mostly at lower to middle elevations throughout the state. Common throughout N. America, excepting some of the southeast. (p.696)

    Although a widespread and abundant colonizer of wetlands, this species was collected in MT by Lewis & Clark.[322]

Phalaris canariensis L. Annual canarygrass Annual bunchgrass. **Stems** 30–80 cm. **Inflorescence** a dense ovoid panicle 2–5 cm long, 1.5–2 cm wide. **Spikelets** 7–9 mm long. **Lemmas** with one lightly-hairy seed-bearing floret subtended by two hairless sterile florets or scales. An ephemeral that sporadically germinates from seed that falls from bird feeders and such, sometimes in lawns. Introduced throughout much of North America.

    Phalaris aquatica L. is reported from Flathead and Ravalli cos.[31] It is very similar to *P. canariensis* except for a perennial habit, an inflorescence that is a cylindrical spike 5–15 cm long, spikelets with one sterile floret or two unequal-sized sterile florets, and predilection for disturbed soils subject to seasonal flooding.

Phippsia (Trin.) R.Br. Icegrass

Phippsia algida (Sol.) R.Br. Cool season, perennial bunchgrass. **Stems** 3–12 cm. **Leaves:** blades 0.5–2.5 mm wide, flat and spreading to ascending; sheaths with overlapping margins; ligule membranous. **Inflorescence** a contracted panicle 1–2 cm long. **Spikelets** 1–2 mm long with one floret; glumes minute. **Lemmas** usually keeled and inconspicuously veined, hairless, awnless; palea enclosed in floret. **Disarticulation** above glumes; unit of dispersal the floret. On wet, alpine gravelly soils; collected in the Beartooth Mountains from Carbon and Stillwater cos. Throughout much of Canada and the Northern Rockies in the U.S.

Phleum L. Timothy

Cool season, bunched perennials. **Leaf blades** flat and lax to ascending; sheaths with overlapping margins; ligule membranous. **Inflorescence** a cylindrical spicate panicle. **Spikelets** with 1 floret; glumes nearly equal, awn-tipped. **Lemmas** awnless; palea well-developed. **Disarticulation** above glumes; unit of dispersal the floret.

1. Panicle ovate-cylindrical, up to 3 times longer than wide; upper leaf sheath inflated, very loosely surrounding the stem; plants usually under 5 dm tall; stem base not bulbous. . . . . *P. alpinum*
1. Panicle long-cylindrical; much >3 time longer than wide; upper leaf sheath not inflated, tightly surrounding the stem; plants usually much >5 dm tall; stem base bulbous. . . . . *P. pratense*

Phleum alpinum L. Alpine timothy **Stems** 20–50 cm tall, not bulbous at base. **Leaves:** blades 4–7 mm wide; sheath subtending the inflorescence inflated; ligules 1–4 mm long. **Inflorescence** 1–6 cm long and 5–13 mm wide. **Spikelets** 3–4.5 mm long. **Lemmas** 2–2.5 mm long. Common in mountain meadows, open understory; subalpine to alpine. Throughout most of Canada and the western U.S. (p.696)

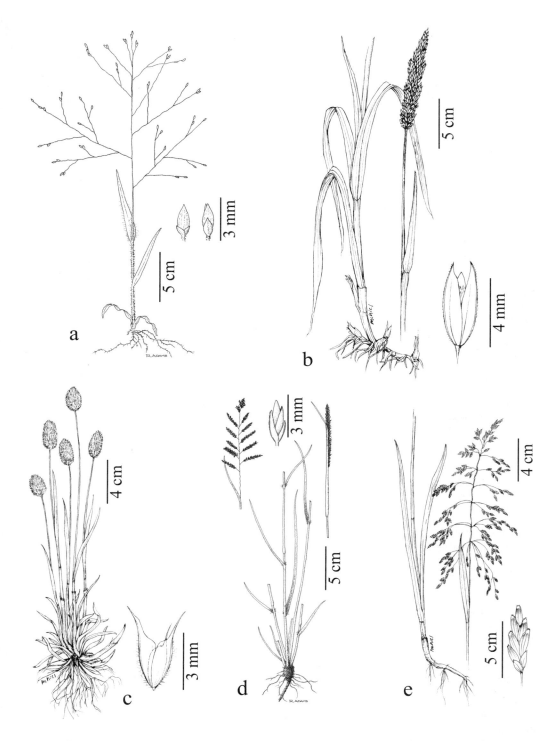

Plate 117. a. ***Panicum capillare***, b. ***Phalaris arundinacea***, c. ***Phleum alpinum***, d. ***Sporobolus cryptandrus***, e. ***Torreyochloa pallida***

Phleum pratense L. Timothy **Stems** 50–140 cm tall, bulbous at the base. **Leaves**: blades 5–9 mm wide; sheath subtending inflorescence not inflated. **Inflorescence** 6–12 cm long and 5–8 mm wide. **Spikelets** 3–4 mm long. **Lemmas** 2.5–2 mm long. Mountain meadows, pastures, rangeland, and along roadsides and trails at middle to higher elevations. Introduced throughout most of North America.

Phragmites Adans. Reed

Phragmites australis (Cav.) Trin. ex Steud. [*Phragmites communis* Trin.] COMMON REED Cool season, rhizomatous perennial. **Stems** 1.5–3.5 m, forming dense stands. **Leaves**: blades 2–4 cm wide, flat and lax to ascending; sheaths with overlapping margins; ligules hairy, 3–6 mm long. **Inflorescence** a plumose panicle 15–32 cm long. **Spikelets** 11–14 mm long, with 3 to 8 florets, the florets covered by silky hairs from the rachilla; glumes shorter than the florets. **Lemmas** hairless, with an awn-like tip; the rachilla with long silky hairs; palea well-developed. **Disarticulation** above the glumes; unit of dispersal the floret. Margins of ponds, marshes, and river flood plains. Throughout nearly all of North America. Our plants are **ssp. *americanus*** Saltonstall, P.M. Peterson & Soreng.[350]

This is the largest native MT grass and the plume-like panicle can persist into the winter and is valued for winter bouquets. MT populations comprise only the N. American native ssp. *americanus*. This subspecies is distinguished from an introduced subspecies (ssp. *australis*) by ligules 1–1.7 mm (versus 0.4–0.9 mm in the introduced subspecies), loose leaf sheaths that are shed in the fall (versus tightly adherent persistent leaf sheaths), green stems with maroon nodes (versus yellow nodes), and larger glumes, the second of which ranges from 5.5–11 mm (versus 4.5–7.5 mm). *Phragmites* is classified into Arundinae, a cool season grass tribe that includes tall, robust perennial reed-like grasses (over 2 m tall) with hairy ligules, stout rhizomes, and inflorescences of dense terminal plumose panicles (e.g., *Arundo* and *Cortaderia* are included in this tribe).

Poa L. Bluegrass

Cool season, rhizomatous, sod-forming, or bunched perennials and annuals. **Leaf blades** flat or folded and ascending; sheaths with overlapping margins; ligule membranous. **Inflorescence** an open to contracted panicle. **Spikelets** with several to many florets; glumes about the dimensions of the lemma. **Lemma** keeled or rounded on back, inconspicuously veined, blunt-tipped and awnless. **Disarticulation** above glumes; unit of dispersal the floret.

Morphologically divergent species of *Poa* can be very similar to one another. Some species can be so morphologically variable as to blur species distinctions (e.g., high elevation forms of *Poa secunda* converging with *Poa pattersonii*).

1. Second glume broadest just above the middle because of a broadening of the translucent margin in this region; annuals with no remains of old leaf sheaths; stems mostly <20 cm tall; inflorescences of open panicles usually much <6 cm long. . . . .*P. annua*
1. Second glume broadest at or towards base with no expansion of the width of any translucent margin; perennials with the remains of old leaf sheaths evident; stems most commonly longer than 20 cm tall (except *Poa lettermanii*); inflorescences of open to closed panicles but usually not short (<6 cm long) and open (except *Poa alpina*) 2

2. Spikelets becoming small vegetative bulbs at maturity, the bulbs dark purple at base; individual glumes and lemmas are asymmetrical because they are actually minute leaf blades with growth involving twisting and turning; plants not setting seed but rather reproducing from these vegetative spikelet-bulbs; base of stem bulbous . . . . *P. bulbosa*
2. Spikelets not becoming small vegetative bulbs at maturity, spikelets not distinctly dark purple at base; individual glumes and lemmas are bilaterally symmetrical and not appearing as minute leaf blades; plants setting seed and not reproducing from vegetative spikelet-bulbs; base of stem usually not bulbous 3

3. Stems single to loosely bunched or sod-forming; rhizomes present 4
3. Stems tightly bunched; rhizomes absent 10

4. Stems flattened, 2-edged, often basally decumbent, nodes of stem usually distinctly but subtly bent and banded with a conspicuously dark band set off from the green stem by a straw-colored band; leaf blades mostly from stem and few if any basal; the pedicels and inflorescence branches are conspicuously and often densely scabrous . . . .*P. compressa*
4. Stems cylindrical or nearly so, usually erect and straight, nodes not consistently bent and banded; leaf blades mostly basal and few from the stem; the pedicels and inflorescence branches not conspicuously or densely scabrous. 5

5. Lemma base or callus (adjacent rachilla) with tangled cobwebby hairs. 6
5. Lemma and callus (adjacent rachilla) lacking cobwebby hairs. 7

6. Lemmas 2–4 mm long, often hairless or scabrous between veins; dry to moist sites often associated
   with lawns and other cultivated settings but also escaping cultivation and inhabiting open dry
   sites (e.g., sagebrush steppe) as well as sometimes the edges of riparian sites or meadows . . *P. pratensis*
6. Lemmas 4–6 mm long, hairy between veins at least on lower half; mountain meadows
   mostly at or above timberline . . . . *P. arctica*

7. Lower and older leaf sheaths hairy to densely scabrous; lemmas and glumes strongly keeled
   and often well-separated on long rachilla internodes; in flower the plumose stigma branches
   may be common but no stamens (anthers or filaments) will be evident (plants are pistillate);
   ligules mostly <2 mm long; plants of mostly dry understory settings in western MT . . . . *P. wheeleri*
7. Lower and older leaf sheaths without hairs; lemmas and glumes often tightly overlapping but if
   well-separated on long rachilla internodes then they are not keeled (they are rounded on the back); in
   flower the plumose stigmas co-occur with evident stamens (anthers or filaments; plants are bisexual);
   ligules usually well >2 mm long; plants of alpine or subalpine settings or of open dry or saline sites 8

8. Lemmas 4–6 mm long, conspicuously hairy on the sides and evident without having to remove
   the subtending glumes or other lemmas; panicle open and the spikelets laterally compressed
   and the glumes and lemmas distinctly keeled; plants of alpine meadows . . . . *P. arctica*
8. Lemmas 3–4 mm long, not conspicuously hairy on the sides, any lemma hairs are visible only after
   removing subtending glumes and other lemmas; panicle mostly contracted, but sometimes an open
   panicle in high mountain settings, spikelets dorsally compressed, but if the glumes and lemmas are
   splayed out laterally they are always with rounded backs (not keeled); plants of mostly low
   elevation open dry or saline sites, sometimes up to high subalpine and rarely alpine sites 9

9. Stems usually single or loosely bunched, rhizomatous; plants often glaucous; panicle contracted;
   basal leaves not tightly bunched into a small tuft but diffuse and borne from base to the lower stem
   regions; low elevation dry or saline settings particularly in the eastern two-thirds of the state . . . . . .*P. arida*
9. Stems tightly bunched, rarely with rhizomes (but if so, then tightly bunched); plants mostly
   greenish; basal leaves diffuse or most commonly tightly bunched into a small tuft; panicle
   contracted to open; highly variable habitats but mostly dry sites from low to high elevations . . . .*P. secunda*

10. Spikelets dorsally compressed, rounded in cross-section or the keels of lemmas and glumes lacking or
    faint; florets tightly overlapping each other when the spikelet is closed, or florets well-separated from each
    other on long rachilla internodes when the spikelet is open; the terminal and subterminal florets project
    well beyond the lower florets, a feature that is evident when the spikelet is closed or open . . . .*P. secunda*
10. Spikelets laterally compressed, narrowly elliptic in cross-section, lemmas and glumes folded
    along the midrib and the keels very distinct; florets splayed out laterally such that individual
    lemmas are exposed and the lemma keel is readily visible at least in the upper half; the terminal
    and perhaps subterminal florets are often reduced in size and mostly concealed or rendered
    inconspicuous by the overlapping lower florets 11

11. Lemmas with long cobwebby hairs at base or on callus (sometimes just distinguishable in *P. interior*). . .12
11. Lemmas lacking cobwebby hairs 18

12. The lower glume distinctly narrower than the second glume and characteristically arched or sickle-
    shaped, diverging at an angle distinct from that of the second glume; lemma with cobwebby hairs
    only on keel base or perhaps to the very base of the marginal veins; spikelets <4 mm long . . . . . *P. trivialis*
12. The lower glume not distinctly narrower than the second glume, very similar to the second
    glume especially in width and the angle it diverges from the main axis of the spikelet; lemma
    with cobwebby hairs on at least the lower keel and marginal veins; spikelets often >4 mm long 13

13. Lower panicle branches 1–2 at a node and very slender or capillary and bearing spikelets only
    towards the very distal ends; anthers <1 mm long. 14
13. Lower panicle branches often >2 at a node or not distinctly slender or not necessary bearing
    spikelet towards the distal ends; anthers often >1 mm long. 15

14. Lemmas 2–3 mm long, marginal veins distinctly hairy; panicle branches smooth, the lower
    ones often oriented downward at maturity; glumes nearly equal in length; dry and often
    disturbed sites (e.g., steep slopes) in open montane forest settings up to subalpine . . . . *P. reflexa*
14. Lemmas 3–4 mm long, marginal veins lacking dense hairs; panicle branches usually
    scabrous and ascending at maturity; glumes distinctly unequal in length; moist meadows
    and riparian sites and usually in subalpine and alpine settings . . . . *P. leptocoma*

15. Stems mostly (3)4–12 dm tall, loosely bunched, frequently producing stolons; ligule 1.5–6 mm long; panicle mostly 13–30 cm long . . . . *P. palustris*
15. Stems 2–3(4) dm tall, tightly bunched, not producing stolons; ligule 0.5–5.5 mm long; panicle mostly 5–15 cm long 16

16. Second glume usually 2–3 mm long; panicles mostly 3–15 cm long; ligules 0.5–1.5 mm long; at and below alpine regions . . . . *P. interior*
16. Second glume usually 3–4.5 mm long; panicles mostly 2–5 cm long; ligule 1–5.5 mm long; alpine regions. 17

17. Flowering stems arising from a mat of leaves, branching (tillering) occurring at or below ground level and contained within the subtending leaf sheath; anthers mostly 0.6–1.2 mm . . . . *P. pattersonii*
17. Flowering stems arising from bunched stems with few if any basal leaves, branching (tillering) occurring just above ground and rupturing subtending leaf sheaths; anthers mostly 1.2–2.5 mm *P. glauca*

18. Lemmas lacking hairs or sometimes sparsely hairy on lower surface or with a scabrous midvein 19
18. Lemmas conspicuously hairy on keel, marginal veins, or both, shorter hairs sometimes between veins .20

19. Lemmas 2–3 mm long; glumes as long as the floret cluster (oat-like spikelets); anthers about 0.5 mm long; stems mostly <1 dm tall; leaf blades persistent on leaf sheath even on older leaves; on rocky alpine slopes and ridges at the highest elevations. . . . . *P. lettermanii*
19. Lemmas 4–6 mm long; glumes shorter than subtended florets; anthers 2–3 mm long; stems mostly >1 dm tall; leaf blades commonly deciduous from persistent sheaths especially on old leaves; common from open subalpine forests to low elevations in open dry, shrub-steppe settings. . . . .*P. fendleriana*

20. Leaf blades mostly 2–5 mm wide; panicle open to loosely contracted, the lower branches often divergent or drooping. 21
20. Leaf blades mostly 1–2 mm wide; panicle mostly contracted to rarely somewhat open 22

21. Panicle mostly 2–7 cm long, pyramidal, usually about as long as broad, erect; leaves mostly basal; plants of high montane settings. . . . .*P. alpina*
21. Panicles mostly 5–20 cm long, loosely contracted to open, usually longer than broad and drooping and secund; leaves distributed along the lower stem regions; plants of low elevation to low montane settings . . . .*P. stenantha*

22. Inflorescence of predominantly greenish spikelets congested into thick (1–2 cm wide) and short (2–7 cm long) spikes; bunchgrasses bearing many basal leaves >5 cm in length. . . . .*P. fendleriana*
22. Inflorescence of predominantly purplish spikelets loosely spaced in slender (<1 cm wide) inflorescences (sometimes >7 cm long); bunchgrasses bearing few if any basal leaves mostly <5 cm in length 23

23. Flowering stems arising from a mat of leaves, branching (tillering) occurring at or below ground level and contained within the subtending leaf sheath; anthers mostly 0.6–1.2 mm . . . . *P. pattersonii*
23. Flowering stems arising from bunched stems with few if any basal leaves, branching (tillering) occurring just above ground and rupturing subtending leaf sheaths; anthers mostly 1.2–2.5 mm . . . . *P. glauca*

Poa alpina L. ALPINE BLUEGRASS Perennial bunchgrass. **Stems** 10–30 cm. **Leaves**: blades 2–4.5 mm wide; ligules 2–4 mm long. **Inflorescence** a pyramidal panicle 3–7 cm long. **Spikelets** 4–6.5 mm long. **Lemmas** strongly keeled and uniformly hairy. Mostly subalpine to alpine or montane open understory. Higher latitudes in N. America, and throughout the Intermountain, Rocky Mountain, and Pacific Northwest regions. (p.701)

 Poa alpina is distinguished by basal leaf blades that are usually <5 cm long and up to 5 mm wide combined with a pyramidal panicle mostly <7 cm, and strongly compressed spikelets up to 6.5 mm long. Most likely confused with montane forms of *Poa secunda* with open panicles, but *Poa alpina* always has lemmas that are conspicuously hairy and distinctly folded (keeled) along the midrib. Potentially confused with high elevation forms of *Poa fendleriana* but the conspicuously hairy lemmas and relatively few basal leaf blades up to 4.5 mm wide (without the many persistent old leaf sheaths) will distinguish *Poa alpina* from *Poa fendleriana*.

Poa annua L. ANNUAL BLUEGRASS Perennial bunchgrass. **Stems** usually decumbent, 5–20 cm long. **Leaves**: blades 1–3 mm wide; ligules 0.5–3 mm long. **Inflorescence** usually a pyramidal panicle 2–6 cm long. **Spikelets** 3–5 mm long. **Lemmas** distinctly keeled, conspicuously hairy at least at base along the midrib. Lawns, pastures, roadsides, and moist disturbed sites. Introduced throughout all of N. America.

 The second glume that is distinctly broad at or above the middle due to the hyaline (translucent) margins is distinctive of *Poa annua*. The annual *Poa bolanderi* occurs in ID just over the west-central border of MT and differs by having larger inflorescences 4–16 cm long, glumes that are tapered from the base into a narrow tip, and lemma bases bearing cobwebby hairs.

Poa arctica R.Br. [*Poa grayana* Vasey, *P. arctica* R.Br. ssp. *grayana* (Vasey) A.Löve & D.Löve & Kapoor] ARCTIC BLUEGRASS Rhizomatous perennial. **Stems** 20–60 cm. **Leaves**: blades 2–5 mm wide, basal; ligules 2–7 mm long. **Inflorescence** an open pyramidal panicle 4–12 cm long. **Spikelets** 5–8 mm long. **Lemmas** with cobwebby hairs at base, conspicuously hairy on keel and margin veins, at least on the lower half. Alpine to subalpine. At higher latitudes in N. America, and throughout the Intermountain and Rocky Mountain regions, and much of the Pacific Northwest. (p.701)

Similar to an alpine form of *Poa pratensis* but not sod-forming and with larger lemmas bearing many more hairs on either side of the midrib.

Poa arida Vasey [*Poa glaucifolia* Scribn. & T.A.Williams] PLAINS BLUEGRASS Rhizomatous perennial sometimes loosely bunched. **Stems** 20–60 cm. **Leaves**: blades 2–4 mm wide, mostly on stem towards stem base; ligules 2–4 mm long. **Inflorescence** usually a narrow panicle, less commonly open, 5–11 cm long. **Spikelets** 4–7 mm long. **Lemmas** lightly hairy on the keel and marginal veins. Shrub steppe and similar open, dry sites. Primarily of the Great Lakes and Great Plains regions, and east slope of the Rocky Mountains.

Poa arida is similar to and co-occurs with *Poa secunda*. Both have a bunched habit (albeit loosely in *Poa arida*), long narrow inflorescences, and a predilection to open, dry settings. The spikelets of both *Poa arida* and *Poa secunda* are dorsally compressed or at least the lemmas have rounded backs (see notes under *Poa secunda*). *Poa arida*, however, lacks the dense basal leaf bunch characteristic of most forms of *Poa secunda*, has a rhizomatous growth habit, and thicker stems (about 2 mm in diameter). Forms of *Poa arida* with open panicles have been referred to as *Poa glaucifolia*.

Poa bulbosa L. [*Poa bulbosa* ssp. *vivipara* (Koel.) Arcang.] BULBOUS BLUEGRASS Perennial bunchgrass. **Stems** 20–60 cm, with a bulbous base. **Leaves**: blades 1–2.5 mm wide, mostly basal; ligules 1–3 mm long. **Inflorescence** a contracted panicle, 3–10 cm long. **Spikelets** 4–15 mm long. **Lemmas** usually transformed into small vegetative bulbs, hairy along veins. Roadsides, trailsides, open dry sites. Introduced throughout most of the U.S. Sometimes seeded in open range to compete with less desirable colonizing grasses.

Bulbous bluegrass has distinctive spikelets that are transformed into small vegetative bulbs. When these are few or lacking, bulbous bluegrass could be confused with *Poa fendleriana*. The bulbous stem bases and growth in disturbed settings distinguish such *Poa bulbosa* from *Poa fendleriana*, the latter of which is a native bunchgrass of open, dry undisturbed vegetation.

Poa compressa L. CANADA BLUEGRASS Rhizomatous perennial. **Stems** 20–50 dm, wiry and flattened, nodes often distinctly bent and banded with contrasting lighter and darker bands. **Leaves**: blades 1.5–4 mm wide, from stem, rarely basal; ligules 1–3 mm long. **Inflorescence** a narrow panicle 2–8 cm long. **Spikelets** 3–6 mm long. **Lemmas** with cobwebby hairs at very base. Roadsides, road cuts, trailsides, along ditches, and on open slopes and other open sites with a combination of moisture and regular moderate disturbance. Introduced throughout most of North America. (p.701)

This species is distinguished by its wiry stems with banded and bent nodes, the lower of which are distinctly exposed from the leaf sheath rendering this node character all the more distinctive. In age, the leaf sheaths become straw-colored, which contrast with the greenish stems immediately below. The scattered stems typically lack basal leaves and all leaf blades along the stem are usually similar in length. The usually narrow panicle has branches that are ridged longitudinally and these ridges are strongly lined with scabrous hairs.

Poa fendleriana (Steud.) Vasey [*Poa cusickii* Vasey, *P. epilis* Scribn., *P.* ×*nematophylla* Rydb.] MUTTONGRASS Perennial bunchgrass. **Stems** 20–60 cm. **Leaves**: blades 1–3 mm wide; ligule 1–6 mm long. **Inflorescence** a contracted panicle 2–11 cm long. **Spikelets** 4–8 mm long. **Lemmas** mostly without hairs in MT except at base, keeled. Shrub-steppe to montane understory to open subalpine and alpine settings. Throughout the western half of N. America. (p.701)

A common early flowering bunchgrass in the sagebrush steppe but later flowering at higher elevations. The persistent old leaf sheaths from which the leaf blades uniformly abscise and from which the relatively slender flowering stems arise is a fairly consistent feature of this species and is reminiscent of the persistent leaf sheaths of *Festuca campestris*. *Poa fendleriana* is tentatively circumscribed here to include *Poa cusickii* (Cusick's bluegrass). These two species are very similar in ecology, geography, and morphology, including the form of apomixis, the short dense inflorescences, and persistent leaf sheaths. A high elevation form referred to as *Poa epilis* is distinguished by lacking hairs on the lower lemma surfaces and having flat leaves scattered along the stem. Most MT populations of *Poa fendleriana* have lemmas with few hairs regardless of elevation. The basal leaf bunch is sometimes lacking at low elevations just as it is in alpine settings. The existence of *Poa* ×*nematophylla*, of hybrid origin between *Poa fendleriana* and *Poa cusickii*, suggests that *Poa cusickii* may be the same as *Poa fendleriana*.

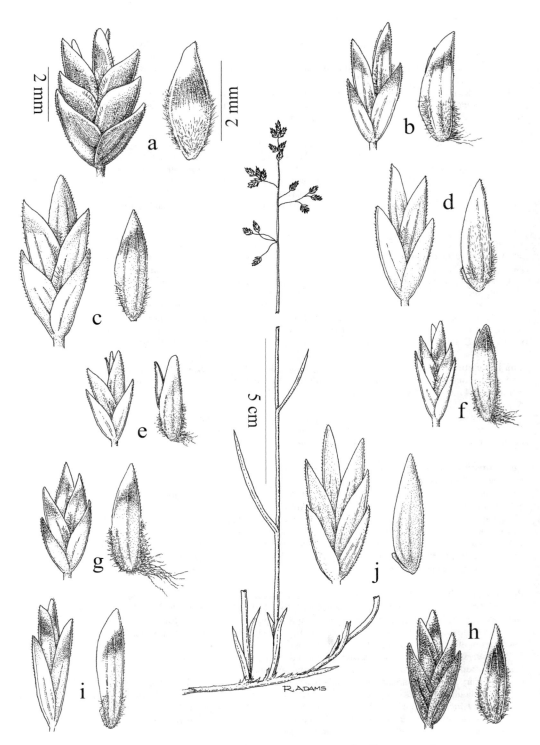

Plate 118. **Poa**. **a.** *P. alpina*, **b.** *P. arctica*, **c.** *P. compressa*, **d.** *P. fendleriana*, **e.** *P. interior*, **f.** *P. palustris*, **g.** *P. pratensis*, **h.** *P. glauca*, **i.** *P. secunda*, **j.** *P. wheeleri*

Poa glauca Vahl [*P. rupicola* Nash ex Rydb., *Poa glauca* Vahl ssp. *rupicola* (Nash ex Rydb.) W.A.Weber] TIMBERLINE BLUEGRASS Perennial bunchgrass. **Stems** 10–20 cm. **Leaves**: blades mostly 1–2.5 mm wide; ligules 1–4 mm long. **Inflorescence** a narrow panicle 2–8 cm long. **Spikelets** 4–6 mm long. **Lemmas** covered with few hairs, or hairy to sometimes cobwebby on keel and marginal veins. Alpine and subalpine settings. Throughout the Rocky Mountain regions and at higher latitudes in N. America. (p.701)

This species is difficult to distinguish from *Poa pattersonii*, but in addition to the characters given in the key the lemmas of some specimens of *P. glauca* tend to be covered with few if any hairs (the variant referred to as *P. glauca* ssp. *rupicola*), in contrast to the more hairy lemmas of *P. pattersonii*.

Poa interior Rydb. [*Poa nemoralis* L. ssp. *interior* (Rydb.) W.A.Weber] INLAND BLUEGRASS Perennial bunchgrass. **Stems** 10–50 cm. **Leaves**: blades 1–3 mm wide; ligules 1–2 mm long. **Inflorescence** a narrow panicle 4–15 cm long. **Spikelets** 3–5 mm long. **Lemmas** mostly glabrous but with cobwebby hairs at very base. Montane grassy slopes and open forest to occasionally alpine. Throughout much of N. America especially at higher latitudes or elevations. (p.701)

Poa interior represents a smaller-statured higher-elevation form of *Poa palustris* and perhaps is adapted to drier settings.

Poa leptocoma Trin. BOG BLUEGRASS, MARSH BLUEGRASS Perennial bunchgrass. Stems 20–100 cm. **Leaves**: blades 1–4 mm wide; ligules 2–4 mm long. **Inflorescence** a loosely contracted panicle, 5–15 cm long, the branches slender and capillary. **Spikelets** 4–7 mm long. **Lemmas** hairy, mostly along midrib and to some degree along the marginal veins, keeled. Moist alpine and subalpine meadows. Sporadic throughout western N. America.

This species is very similar to *Poa reflexa* in having delicate inflorescence branches and lemmas with cobwebby hairs. The characters listed in the key to species, especially including the scabrous inflorescence branches and moister habitat, should help to distinguish *Poa leptocoma* from *Poa reflexa*. *Poa leptocoma* may have a more robust and taller stature than *P. reflexa*, perhaps because it is a hexaploid, in contrast to the tetraploid *P. reflexa*.

Poa lettermanii Vasey LETTERMAN'S BLUEGRASS Perennial bunchgrass. **Stems** 2–10 cm. **Leaves**: blades 0.5–2 mm wide; ligules 1–3 mm long. **Inflorescence** a contracted panicle 1–3 cm long. **Spikelets** 3–4 mm long. **Lemmas** hairless, keeled. Alpine ridges and ledges. Throughout western N. America.

This very small-statured ground-hugging species is a high elevation wind-swept form of *Poa fendleriana* (e.g., the short spicate panicle and essentially hairless lemmas). The glumes tend to be as large as the floret cluster in a spikelet, which renders an "oatgrass" spikelet.

Poa palustris L. FOWL BLUEGRASS Perennial bunchgrass, often stoloniferous. **Stems** loosely bunched, 40–100 cm. **Leaves**: blades 2–8 mm wide; ligules 2–5 mm long. **Inflorescence** an open panicle 13–30 cm long. **Spikelets** mostly 3–4 mm long. **Lemmas** mostly hairless but with cobwebby hairs at very base, strongly keeled. Meadows and open ground at low to middle elevations. Native and introduced throughout N. America, excepting parts of the southeast. (p.701)

This species may sometimes be confused with *Poa pratensis*, but its inflorescence has many more and smaller spikelets that are much more diffusely arranged on the open panicle, and the lemmas are not so hairy excepting the cobwebby hairs just at the very lemma base. Also, basal leaves are few and more on the stem in a manner similar to *Poa compressa* and in contrast to *Poa pratensis*.

Poa pattersonii Vasey [*Poa abbreviata* R.Br. ssp. *pattersonii* (Vasey) A.Löve & D.Löve & Kapoor] PATTERSON'S BLUEGRASS Perennial bunchgrass. **Stems** 10–20 cm. **Leaves**: blades 0.5–2 mm wide; ligules 1–5 mm long. **Inflorescence** a contracted panicle 2–5 cm long. **Spikelets** 4–6 mm long. **Lemmas** keeled, densely hairy to cobwebby on keel and marginal veins. Alpine meadows and ridges or other open, exposed sites. Throughout most of western N. America.

Poa pattersonii may have spikelets with glumes and lemmas lacking a keel, similar to those of *Poa secunda* and *Poa arida*, but this is occasional. *Poa pattersonii* would then be distinguished from *Poa secunda* by its lemmas with distinctly hairy keels and marginal nerves and often cobwebby lemma base and callus (close inspection of such specimens should reveal distinctly keeled lemmas that contrast to those of *Poa secunda* with rounded backs). *Poa laxa* Haenke ssp. *banffiana* Soreng, Banff bluegrass, is recorded for MT in Fergus and Flathead counties, but these specimens lie intermediate between *Poa pattersonii* (small anthers) and *Poa glauca* (tillers rupturing the subtending leaf sheath) and thus further weaken the distinction between *P. pattersonii* and *P. glauca*.

Poa pratensis L. [*Poa pratensis* L. ssp. *agassizensis* (B.Boivin & D.Löve) R.L.Taylor & MacBryde, *Poa pratensis* L. ssp. *alpigena* (Fr. ex Blytt) Hiitonen] KENTUCKY BLUEGRASS Perennial rhizomatous sodgrass. **Stems** 10–80 cm. **Leaves:** blades 2–4 mm wide, mostly basal; ligules 1–2 mm long. **Inflorescence** an open pyramidal panicle, 3–13 cm long. **Spikelets** 4–6 mm long. **Lemmas** with a cobwebby base and hairy along the mid and margin veins. Low to high elevations in open vegetation, roadsides, dry meadows, lawns, and riparian habitats. Introduced and native and a fairly aggressive colonizer that is common throughout N. America, especially at more northern latitudes or higher elevations. (p.701)

    Poa pratensis is one introduced species that can most readily immigrate into fairly undisturbed native dry-site vegetation (e.g., sagebrush steppe) in high abundance. The spikelets of *Poa pratensis*, though arranged in a small open panicle, are relatively large and congested and in this regard are very different from smaller spikelets that are more diffusely arranged in *Poa palustris*, with which *Poa pratensis* may be confused. Specimens having smooth rather than scabrous panicle branches are referred to ssp. *alpigena* if of alpine regions, and ssp. *agassizensis* if not of alpine regions.

Poa reflexa Vasey & Scribn. ex Vasey NODDING BLUEGRASS Perennial bunchgrass. **Stems** few-bunched, 20–40 cm. **Leaves:** blades 2–4 mm wide; ligules 1.5–3.5 mm long. **Inflorescence** a nodding open panicle 5–13 cm long, the branches delicate and capillary, the lowermost often oriented downward. **Spikelets** 4–6 mm long. **Lemmas** with a cobwebby base, hairy on keel and especially along the marginal veins. Moderately disturbed settings in the mountains, including road cuts, steep slopes with loose soil, and open understory. Western N. America in the Rocky Mountain and Intermountain regions.

    Poa reflexa is very similar to *Poa leptocoma*, and in addition to the key characters *Poa reflexa* tends to inhabit drier settings than *Poa leptocoma*. *Poa paucispicula* Scribn. & Merr. would key out here because of the small spikelets with usually hairy marginal veins of the lemmas and smooth panicle branches. *Poa paucispicula* (recorded from Carbon, Deerlodge, Park, and Ravalli cos.) differs only in having slightly less hairy lemma veins and shorter (1–2 cm) lower panicle branches that may not reflex at anthesis. This morphology suggests *Poa paucispicula* is intermediate between *Poa reflexa* and *Poa leptocoma*. Given that the last two are distinguished often with difficulty, segregation of a third species seems unwarranted.

Poa secunda J.Presl [*Poa ampla* Merr., *P. canbyi* (Scribn.) Howell, *P. gracillima* Vasey, *P. juncifolia* Scribn., *P. nevadensis* Vasey ex. Scribn., *P. sandbergii* Vasey, *P. scabrella* (Thurb.) Benth. ex Vasey, and *P. secunda* ssp. *juncifolia* (Scribn.) Soreng] SANDBERG BLUEGRASS Perennial bunchgrass. **Stems** 20–120 cm. **Leaves:** blades 1–3 mm wide; ligules 1–5 mm long. **Inflorescence** a contracted to open panicle mostly 7–24 cm long. **Spikelets** 6–10 mm long. **Lemmas** hairless or scabrous to short-hairy, rounded on back. Shrub steppe vegetation and in open dry and saline settings throughout the state from low to high elevations. Western and northern N. America. (p.701)

    Poa secunda includes *Poa juncifolia*, which traditionally was distinguished by a growth form involving branching above ground level, thus rupturing leaf sheaths and spikelets with glabrous lemmas. *Poa ampla* is included here and is traditionally distinguished by glaucous stems and leaves, flat leaf blades scattered along the stem instead of concentrated at the base, and a predilection to gravelly well-drained soils along or near riparian settings. *Poa gracillima* (slender bluegrass) is an ecotype of mountain meadows and forest understory and is distinguished by an open panicle. At montane elevations, *Poa secunda* produces open panicles and rhizomes with higher frequency, as if *Poa arida* was being reconstituted perhaps by hybridization of *Poa secunda* with sympatric *Poa arctica* (which produces rhizomes). These rhizomatous *Poa secunda* specimens are not *Poa arida* because they retain a tightly bunched habit. Herbarium collections of rhizomatous *Poa secunda* come from mountain meadows and explicitly indicate a rarity that suggests hybrid progeny of limited local abundance. In spite of the high morphological variation of *Poa secunda*, the spikelet morphology remains distinctly uniform and is otherwise found only in *Poa arida* and rarely in *Poa pattersonii* (see comments under that species). The lemmas and glumes of *Poa secunda* have rounded backs. When not receptive to pollen, the spikelet is tightly closed in a dorsally compressed manner. In that case, the spikelet is nearly cylindrical in cross-section, and the long narrow terminal floret with its acute apex protrudes distinctively through the clasping lower lemmas, which are typically broader and blunter. When the lemmas are splayed out laterally during pollen reception, the dorsal compression of the spikelet is evident only in the rounded backs of the lemmas. *Poa secunda* can have persistent leaf sheaths from which the blades have uniformly abscised, as in *Poa fendleriana*. The non-keeled spikelets arranged in long slender spicate panicles should distinguish *Poa secunda* from *Poa fendleriana*, which often grow in sympatry and flower together during the early summer in shrub-steppe settings. The spikelet morphology of *Poa secunda* is somewhat similar to that of *Puccinellia nuttalliana*, but the spikelets of the latter are firmly appressed to very scabrous pedicels and inflorescence branches. *Poa secunda* has few if any scabrous hairs.

Poa stenantha Trin. [*Poa macroclada* Rydb.] NORTHERN BLUEGRASS Perennial bunchgrass. **Stems** 30–60 cm. **Leaves:** blades 2–4 mm wide; ligules 2–5 mm long. **Inflorescence** a nodding panicle 5–16 cm long. **Spikelets** 7–9 mm long. **Lemmas** hairy on veins, keeled. Wet mountain meadows; collected in Deer Lodge Co. Pacific Northwest region, CO, and UT.

Poa stenantha is apparently similar to *P. secunda*, from which it is distinguished by its leaf blades that tend to be flat (not folded) and >2 mm wide, lemmas that are strongly folded along the keel and are glabrous between the main veins, and callus hairs when present that are longer than 0.2 mm. A form of *Poa stenantha* with open panicles and glabrous lemma calluses is called *Poa macroclada*, which occurs in the U.S. Rocky Mountain portion of the distribution of *P. stenantha*.

Poa trivialis L. Rough bluegrass Perennial, rhizomatous. **Stems** loosely bunched, 40–100 cm tall, often decumbent and rooting at nodes. **Leaves**: blades 2–5 mm wide, lax, somewhat scabrous; ligules 3–10 mm long. **Inflorescence** an oblong open panicle 8–25 cm long. **Spikelets** 3–5 mm long. **Lemmas** 2.5–3.5 mm long, with a cobwebby base, keeled. Moist disturbed settings. Introduced throughout much of N. America.

    This species is very similar to *Poa palustris* in habitat, habit, and inflorescence and spikelet morphology. The sickle-shaped or curved first glume that is conspicuously narrower than the second glume is distinctive to *Poa trivialis* and distinguishes it from the similar *Poa palustris*.

Poa wheeleri Vasey [*Poa nervosa* (Hook.) Vasey var. *wheeleri* (Vasey) C.L. Hitchc.] Wheeler's bluegrass Perennial bunchgrass sometimes with short rhizomes (i.e., widely diverging tillers). **Stems** 40–70 cm tall, erect; sheaths glabrous to more commonly short-hairy or scabrous. **Leaves**: blades 2–3.5 mm wide; ligules 0.5–2 mm long. **Inflorescence** an open panicle, often nodding, 5–11 cm long. **Spikelets** 6–9 mm long. **Lemmas** 3–6 mm long, hairless or scabrous. Open to closed montane understory. Throughout western N. America. (p.701)

    Poa wheeleri could be confused with *Poa pratensis* but the lemmas lack cobwebby hairs and indeed the lemmas of *Poa wheeleri* are commonly hairless or nearly so. This feature is conspicuous because the lemmas are often not tightly overlapping at anthesis. This renders the lemma readily visible from base to tip, and their large size, being strongly folded along the midrib keel, and the often very greenish coloration additionally readily distinguishes *Poa wheeleri*. That the populations of this species are mostly pistillate means that at anthesis many plumose stigma branches will be visible with no evidence of stamens (filaments or anthers).

Polypogon Desf. Rabbitsfoot grass

Polypogon monspeliensis (L.) Desf. Annual rabbitsfoot grass Cool season, annual bunchgrass. **Stems** erect to decumbent, 10–50 cm. **Leaves**: blades 4–6 mm wide, flat and ascending; sheaths with overlapping margins; ligule membranous. **Inflorescence** a dense spike-like panicle, 2–15 cm long, tawny-yellowish to whitish when mature. **Spikelets** with 1 floret, 2–2.5 mm long excluding the awns; glumes with awns 5–10 mm long, enclosing the floret. **Lemmas** 1–1.5 mm long, awned from near the tip with a delicate awn no longer than the lemma; palea well-developed. **Disarticulation** below glumes; unit of dispersal the spikelet. Disturbed places with moist soil including shady stream and ditchbanks. Introduced throughout N. America, excluding a very east-central portion.

Puccinellia Parl. Alkaligrass

Cool season, bunched to single-stemmed perennials. **Leaf blades** flat to inrolled and ascending; sheaths with overlapping margins; ligule membranous. **Inflorescence** an open or contracted panicle. **Spikelets** with several to many florets; glumes similar in dimensions to lemma. **Lemma** usually keeled and distinctly parallel-veined, tip blunt or truncate with usually ragged edges, unawned; palea enclosed in floret. **Disarticulation** above glumes; unit of dispersal the floret.

    Mostly in open settings on alkaline and gravelly substrates, often coastal or interior along gravelly banks and beaches.

1. Lemmas ≤2 mm, broadly rounded or at least blunt-tipped; anthers 0.5–0.8 mm long;
   lower panicle branches oriented downward at maturity . . . . *P. distans*
1. Lemmas mostly >2 mm long, blunt-tipped to acute; anthers 0.7–1.8 mm long; lower panicle
   branches upright or ascending 2

2. Inflorescence 5–10 cm long; lemmas acute, midvein extending to the tip through the otherwise
   hyaline margin; leaves mostly in a short basal bunch; blades tending to be ≤2 mm wide . . . . . . *P. lemmonii*
2. Inflorescence 10–20 cm long; lemmas blunt-tipped to acute, midvein ending before the hyaline
   marginal end of the lemma; leaves distributed along stem, not in a basal bunch; blades
   tending to be >2 mm wide . . . . *P. nuttalliana*

Puccinellia distans (Jacq.) Parl. Weeping alkaligrass Perennial bunchgrass. **Stems** 10–50 cm.
Leaves: blades 2–4 mm wide, flat or inrolled. **Inflorescence** a pyramidal panicle 5–18 cm long.
Spikelets 3–6 mm long. **Lemmas** 1.5–2 mm long, with a blunt tip. Alkaline and saline substrates.
Introduced throughout N. America, but absent from the southeast region.

The small florets that measure mostly <2 mm long also tend to have a conspicuous hyaline lemma
tip compared to the other two species in MT. This exotic is commonly misidentified as *Puccinellia nuttalliana* but the
open panicle of spikelets with small (ca. 2 mm long) blunt florets is distinctive of *P. distans*.

Puccinellia lemmonii (Vasey) Scribn. Lemmon's alkaligrass Perennial bunchgrass. **Stems** 15–30 cm. **Leaves**:
blades 1–2 mm wide, mostly inrolled. **Inflorescence** a pyramidal panicle 3–17 cm long. **Spikelets** 4–7 mm long.
Lemmas 2.5–4 mm long, with an acute tip. Moist meadows, alkaline soils; collected in Beaverhead Co. Along the
western corridor of states within the U.S.

The hyaline (translucent) lemma tip is bisected by a conspicuously darker pigmented midrib. This color contrast
renders a superficially awn-tipped appearance to the lemma.

Puccinellia nuttalliana (Schult.) Hitchc. [*Puccinellia airoides* (Schult.) S.Watson &
J.M.Coult., *P. cusickii* Weath.] Nuttall's alkaligrass Perennial bunchgrass. **Stems** 30–70 cm.
Leaves: blades 1–3 mm wide, inrolled at maturity. **Inflorescence** an open pyramidal panicle 10–30
cm long. **Spikelets** 4–8 mm long. **Lemmas** 2.2–3 mm long, with a blunt tip. Moist alkaline soils.
Throughout N. America, but not in the southeast region.

If specimens identified as *Puccinellia nuttalliana* lack conspicuously scabrous pedicels and inflorescence branches,
then *Poa secunda* might be the correct identity of the specimen in hand.

Schedonnardus Steud. Tumblegrass

Schedonnardus paniculatus (Nutt.) Trel. Warm season, perennial bunchgrass. **Stems**
ascending, 10–40 cm long. **Leaves**: blades 1–2 mm wide, flat and ascending; ligule membranous,
1–1.5 mm long; sheaths with overlapping margins. **Inflorescences** of 5 to 10 secund lateral
divergent spikes, each 5–15 cm long with 10 to 20 spikelets. **Spikelets** 3–4 mm long, closely
appressed to the rachis, with 1 floret; glumes narrow. **Lemmas** 2–4 mm long, distinctly 3-nerved,
awnless or awn-tipped; palea enclosed in floret. **Disarticulation** below glumes; unit of dispersal is the entire
inflorescence. Disturbed areas including overgrazed rangeland, prairie dog towns, lawns, and shoulders of roads and
railroad tracks. Throughout central and southwestern N. America.

Schedonorus P.Beauv. Broadleaf fescue

Cool season, bunched perennials. **Leaf blades** flat and lax to ascending; sheaths with overlapping margins;
ligule membranous. **Inflorescence** an open panicle. **Spikelets** with 4 to 20 florets; glumes similar in
dimension to lemmas. **Lemma** rounded on back and weakly veined, unawned. **Disarticulation** above the
glumes; unit of dispersal the floret.

Introduced mainly as forage grasses. Although these species have traditionally been included within
Festuca, phylogenetic evidence suggests they are most closely related to *Lolium*.[106]

1. Lemmas mostly awn-tipped to awn 4 mm long, lemma margins, midrib, or back usually scabrous
   especially distally; auricles ciliate, with at least 1 or 2 hairs along the margins of the auricle;
   panicle branches at the lowest node usually paired, the shorter with 1 to 13 spikelets, the
   longer with 3 to 19 spikelets . . . .*S. arundinaceus*
1. Lemmas unawned or with an awn tip to 0.2 mm long, lemma margins, midrib, and back usually
   smooth, sometimes slightly scabrous distally; auricles glabrous; panicle branches at the lowest
   node 1 or 2, if paired the shorter with 1 or 2(3) spikelets, the longer with 2 to 6(9) spikelets . . . *S. pratensis*

Schedonorus arundinaceus (Schreb.) Dumort. [*Festuca arundinacea* Schreb., *Lolium*
arundinaceum (Schreb.) S.J.Darbyshire, *Schedonorus phoenix* (Scop.) Holub] Tall fescue
Sometimes with short rhizomes. **Stems** mostly 90–150 cm. **Leaves**: blades 5–12 mm wide, flat.
Inflorescence a contracted panicle 10–30 cm long. **Spikelets** 9–15 mm long, with mostly 4 to 8
florets. **Lemmas** awn-tipped or awn up to 4 mm long. Moist meadows and along ditch banks and
irrigated fields. Introduced throughout most of N. America.

Schedonorus pratensis (Huds.) P.Beauv. [*Festuca pratensis* Huds., *Lolium pratense* (Huds.)
S.J.Darbyshire] MEADOW FESCUE **Stems** mostly 50–100 cm. **Leaves**: blades 3–7 mm wide, flat.
Inflorescence a contracted, erect to nodding panicle, 10–23 cm long. **Spikelets** 6–9 mm long,
with 4 to 9 florets. **Lemmas** awnless. Moist meadows and pastures at lower elevations. Introduced
throughout all of N. America.

Schizachne Hack. False melic

Schizachne purpurascens (Torr.) Swallen Cool season, perennial bunchgrass. **Stems** loosely
bunched and connected by short rhizomes, 30–80 cm. **Leaves**: blades 2–5 mm wide, often flat
and lax; ligule 1–1.5 mm long; sheaths with fused margins. **Inflorescence** an open panicle 7–13
cm long, branches in pairs or single, drooping, bearing 1 or 2 spikelets. **Spikelets** 11–16 mm long,
with 4 to 5 florets; glumes with mostly 3 to 5 faint veins. **Lemmas** 8–12 mm long with 7 prominent
veins; hairs on callus about one-quarter the length of lemma; awned from between two apical teeth, the awn 9–15 mm
long; palea enclosed in floret. **Disarticulation** above glumes; unit of dispersal the floret. Open understory settings.
Throughout the Rocky Mountain region and the northern latitudes of N. America.

   Because of the long awn arising from the backside of a bifid lemma tip, *Schizachne purpurascens* is potentially
misidentified as a perennial *Bromus* (e.g., *Bromus ciliatus*). The lemma callus with a tuft of hairs 3–4 mm long, the main
lemma surface that is completely hairless, and a ligule fused into a solid cylinder around the stem readily distinguish
Schizachne from *Bromus*. *Schizachne* has a conspicuous palea that separates from the tight clasp of the lemma during
fruit maturation to expose the long-ciliate raised veins of the palea.

Sclerochloa P.Beauv. Hardgrass

Sclerochloa dura (L.) P. Beauv. COMMON HARDGRASS Cool season, annual prostrate bunchgrass.
Stems ca. 10 cm, spreading. **Leaves**: blades flat and spreading with prow-like tips, 1–2 mm
wide; sheaths with overlapping margins; ligule membranous. **Inflorescence** a contracted secund
panicle or raceme 1–5 cm long. **Spikelets** 6–10 mm long, with usually 3 florets; glumes similar to
lemmas. **Lemmas** with prominent parallel venation and a blunt tip, unawned; palea well-developed.
Disarticulation below glumes; unit of dispersal the spikelet. Sporadic and possibly ephemeral on disturbed soils along
roadsides, parking lots, and driveways. Sporadically introduced throughout most of N. America.

Scolochloa Link. Rivergrass

Scolochloa festucacea (Willd.) Link COMMON RIVERGRASS Cool season, perennial, rhizomatous, aquatic. **Stems** 1–2
m. **Leaves**: blades 5–11 mm wide, flat and ascending; ligule membranous, 4–8 mm long; sheaths with overlapping
margins. **Inflorescence** an open panicle 15–26 cm long. **Spikelets** 7–11 mm long, with 5 to 9 florets; glumes similar
to lemmas. **Lemmas** 4–9 mm long, each with a distinctly hairy callus, usually keeled and distinctly parallel-veined,
unawned but with an acute tip; palea enclosed in floret. **Disarticulation** above glumes; unit of dispersal is the floret.
Peatlands, swamps, and near lake outlets; collected in Flathead Co. Throughout the northwest region of N. America.

   Superficially similar to *Schedonorus arundinaceus* (*Festuca arundinacea*) but in aquatic settings and with thick stems
at water level 6–8 mm in diameter and borne from stout rhizomes, and with conspicuous hairs borne from the lemma
callus. Like *Schizachne*, the two raised palea veins of *Scolochloa* are distinctively lined with ciliate hairs.

Secale L. Rye

Secale cereale L. CEREAL RYE Cool season, annual bunchgrass. **Stems** 50–150 cm. **Leaves**:
blades flat and ascending, 6–12 mm wide; sheaths with overlapping margins; ligule membranous.
Inflorescence a two-sided terminal spike with short internodes, 6–12 cm long; rachis readily
disarticulating at maturity. **Spikelets** 1 per node, 14–18 mm long, mostly with 2 florets; glumes
narrow and linear. **Lemmas** with distinctively scabrous veins, distinctly keeled, the margins
conspicuously pectinate-ciliate, awn 20–40 mm long; palea enclosed in floret. **Disarticulation** below the glumes; unit of
dispersal the spikelet and subtending internode of the inflorescence rachis. Occasionally planted or colonizing roadsides
and fields. Introduced throughout most of N. America.

   Secale is known to increase in abundance in wheat producing areas after successive years of drought. *Triticale* is
a hybrid between *Secale cereale* and *Triticum aestivum* that may occasionally be found. This hybrid is like *Secale* in
having lemma margins that are distinctly pectinate-ciliate, but like *Triticum* in its wide asymmetrically keeled glumes.

Setaria P.Beauv. Bristlegrass

Warm season, bunched annuals. **Leaf blades** flat and ascending; sheaths with overlapping margins; ligule hairy. **Inflorescence** a terminal spicate panicle with sterile green to yellowish bristles subtending the spikelets. **Spikelets** with one fertile floret; glumes unequal, the first much shorter than the second. **Lemma** rounded on back, awnless, hard-textured if fruit-brearing; palea enclosed in floret. **Disarticulation** below the glumes; unit of dispersal the spikelet.

1. Bristles retrorsely barbed (angled backward), inflorescence easily clinging to clothes or fur. . . *S. verticillata*
1. Bristles antrorsely barbed (angled forward), inflorescence not clinging to clothes or fur 2

2. Fruiting lemma coarsely corrugated; margins of sheaths hairless except at throat; longest bristles
   5–9 mm long, often golden-yellow to straw-colored especially distally; ligule 1 mm long or less. . . *S. pumila*
2. Fruiting lemma smooth or very finely corrugated; margins of sheaths conspicuously hairy;
   longest bristles 9–12 mm long, often green- to purplish-pigmented, not often straw-colored;
   ligule usually >2 mm long 3

3. Inflorescence a cylindrical spike-like panicle 1–9 cm long; disarticulation below glumes,
   the fertile lemma mostly enclosed inside the spikelet. . . . . *S. viridis*
3. Inflorescence a lobed spike-like panicle 20–30 cm long; disarticulation above glumes,
   the fertile lemma falling free from the spikelet . . . . *S. italica*

Setaria italica (L.) P.Beauv. FOXTAIL BRISTLEGRASS **Stems** 80–100 cm long. **Leaves:** blades 8–28 mm wide; sheaths ciliate along margins; ligule 1–2 mm long. **Inflorescence** a usually nodding spike-like panicle 15–30 cm long; longest sterile bristles 9–12 mm long, antrorsely barbed. **Spikelets** 2.8–3.2 mm long. **Lemmas** reticulately veined but not conspicuously transversely corrugated. Frequently disturbed sites; collected in Big Horn and Power River cos. Cultivated, introduced sporadically throughout all of N. America, excepting the Intermountain Region.

Setaria pumila (Poir.) Roem. & Schult. [*Setaria glauca* (L.) P.Beauv., *S. lutescens* (Weigel) F.T.Hubbard] YELLOW FOXTAIL **Stems** 15–60 cm long. **Leaves:** blades 5–9 mm wide; sheaths hairless; ligule <1 mm long. **Inflorescence** an erect spike-like panicle 5–15 cm long; sterile bristles 4–7 mm long, antrorsely barbed. **Spikelets** 2.5–3 mm long. **Lemmas** strongly corrugated at maturity. Disturbed sites including in and around cultivated fields. Introduced throughout N. America.

Setaria verticillata (L.) P.Beauv. HOOKED BRISTLEGRASS **Stems** 10–50 cm. **Leaves:** blades 5–12 mm wide; sheaths usually hairy at throat; ligule 1–1.5 mm long. **Inflorescence** an erect to nodding spike-like panicle 5–13 cm long; sterile bristles 4–7 mm long, retrorsely barbed. **Spikelets** 2–2.5 mm long. **Lemmas** reticulately veined but not conspicuously transversely corrugated. Disturbed sites along roads and pastures and in lawns. Introduced throughout N. America, excepting the very southeast.

Setaria viridis (L.) P.Beauv. GREEN BRISTLEGRASS **Stems** 10–40 cm long. **Leaves:** blades 3–10 mm wide; sheaths ciliate along margins; ligule 1–1.5 mm long. **Inflorescence** an erect to nodding cylindrical panicle 3–15 cm long; longest sterile bristles 8–12 mm long, antrorsely barbed. **Spikelets** 1.5–2.2 mm long. **Lemmas** reticulately veined but not conspicuously transversely corrugated. Regularly disturbed settings such as along roadsides, pastures, and ditches. Introduced throughout nearly all of N. America.

Sorghastrum Nash Indiangrass

Sorghastrum nutans (L.) Nash Warm season, perennial bunchgrass with short rhizomes. **Stems** 1–2 m. **Leaves:** blades 5–10 mm wide, flat and lax to ascending; ligule 2–7 mm long, membranous; sheaths ciliate along margins. **Inflorescence** a contracted panicle, 15–30 cm long; rachis and pedicels long-hairy along margins. **Spikelets** paired, one sessile, the other pedicellate; the sessile seed-bearing, 5–8 mm long; the pedicellate mostly absent and represented by only the naked pedicel; glumes straw-colored, acute, rounded on back and completely enveloping the florets. **Lemmas** with a bent awn 10–18 mm long; palea inconspicuous. **Disarticulation** below the glumes; unit of dispersal the spikelet pair. Escaping from cultivation in Yellowstone Co. and reported from reclamation areas in Rosebud Co. (fide J. Rumely). Throughout central and eastern N. America.

Sorghum Moench Sorghum

Warm season, rhizomatous perennials or robust annuals. **Leaf blades** flat and lax to ascending; sheaths with overlapping margins; ligule hairy. **Inflorescence** a dense panicle. **Spikelets** paired, one sessile the other pedicellate; the sessile seed-bearing; the pedicellate at least staminate; glumes rounded on back and completely enveloping the florets. **Lemma** thin-textured, with a deciduous awn; palea inconspicuous. **Disarticulation** below the glumes; unit of dispersal the spikelet pair.

Two species have been introduced into N. America. Young growth often generates cyanide.

1. Plants perennial from stout rhizomes; panicles open and diffuse; disturbed sites . . . . *S. halepense*
1. Plants annual, non-rhizomatous; panicles dense; cultivated . . . . *S. bicolor*

Sorghum bicolor (L.) Moench [*S. vulgare* Pers.] Grain sorghum Annual, cultivated bunchgrass. **Stems** 50–100 cm. **Leaves:** blades 3–5 cm wide; ligules 1–5 mm long, membranous but densely covered with hairs. **Inflorescence** a dense panicle 10–30 (60) cm long; rachis and pedicels short-hairy. **Spikelets** 4–9 mm long; glumes purplish at maturity. **Lemmas** unawned or with an awn 5–20 mm long, bent and early deciduous. Cultivated mostly for forage (e.g., milo and sorgo), occasionally as an ornamental, rarely escaping; collected in Broadwater and Missoula cos. Escaping cultivation throughout much of the U.S. and eastern Canada.

Sorghum halepense (L.) Pers. Johnsongrass Rhizomatous perennial. **Stems** 1–1.5 m. **Leaves:** blades 10–15 mm wide; ligule 2–4 mm long, membranous with a short fringe of hairs. **Inflorescence** an open and diffuse panicle 15–35 cm long; rachis short-hairy; pedicels long-hairy along margins. **Spikelets** 4–6.5 mm long; glumes purplish to greenish-yellow, acute, hairless or nearly so. **Lemma** unawned or with an awn 8–13 mm long, twisted and bent, readily deciduous. Frequently disturbed sites; collected in Gallatin Co. Introduced throughout much of N. America, excepting most of Canada.

Spartina Schreb. Cordgrass

Warm season, rhizomatous perennials. **Leaf blades** flat and ascending; sheaths with overlapping margins; ligule hairy. **Inflorescence** a panicle with several lateral secund spikes, the spikes often closely appressed to the main inflorescence rachis. **Spikelets** with 1 floret; glumes unequal, the first much shorter than the second. **Lemma** distinctly 3-nerved, awnless or awn-tipped; palea enclosed in floret. **Disarticulation** above glumes; unit of dispersal the floret.

1. Second glume 10–20 mm long including the awn, greatly exceeding the lemma in length;
   lateral spikes 5–15 cm long; plants robust, 1–3 m tall; leaf blades mostly >5 mm wide . . . . *S. pectinata*
1. Second glume 7–9 mm long, awnless or awn-tipped, only barely exceeding the lemma in
   length; lateral spikes 2–5 cm long; plants <1 m tall; leaf blades mostly ≤5 mm wide. . . . . *S. gracilis*

Spartina gracilis Trin. Alkali cordgrass **Stems** 40–80 cm. **Leaves:** blades 3–5 mm wide; ligule 1–1.2 mm long. **Inflorescence** of 4 to 8 secund lateral spikes often appressed to the main rachis, each 3–4 cm long with 20 to 30 spikelets. **Spikelets** 7–9 mm long; glumes awnless or awn-tipped. **Lemmas** distally smooth. Alkaline seeps and meadows. Throughout western and north-central N. America.

Spartina pectinata Bosc ex Link. Prairie cordgrass **Stems** 1–2 m. **Leaves:** blades 3–5 mm wide; ligule 2–3 mm long. **Inflorescence** of 8 to 15 secund lateral spikes often appressed to the main rachis, each 3–4 cm long with 20 to 30 spikelets. **Spikelets** 10–20 mm long; glumes short-awned. **Lemmas** distally scabrous. Along streams, marshes and sloughs, often where alkaline. Throughout much of N. America, excepting the very southwest and southeast.

Sphenopholis Scribn. Wedgescale

Cool season, bunched perennials. **Leaf blades** flat and ascending; sheaths with overlapping margins; ligule membranous. **Inflorescence** a contracted panicle. **Spikelets** with 2 or 3 florets; glumes unequal, the second broadly obovate or wedge-shaped. **Lemma** awnless; palea enclosed in floret. **Disarticulation** below glumes; unit of dispersal mainly the spikelet.

1. Panicle branches loose and not spike-like; spikelets commonly borne from pedicels >1 mm long, thus not strongly fascicled or aggregated; 2nd glume broadest just below the apex . . . . . *S. intermedia*
1. Panicle branches contracted, often spike-like; spikelets commonly borne from pedicels <1 mm long, thus strongly fascicled and aggregated; 2nd glume distinctly broadest near the apex . . . . . *S. obtusata*

Sphenopholis intermedia (Rydb.) Rydb. SLENDER WEDGESCALE, WEDGEGRASS **Stems** 30–100 cm. **Leaves:** blades flat, 2–6 mm wide. **Inflorescence** nodding open panicle, 7–20 cm long. **Spikelets** 2.5–4 mm long; second glume typically broadest below the tip. **Lemmas** 2–3 mm long. Wet sites, often in disturbance-prone settings. Sporadic throughout most of N. America.

Sphenopholis obtusata (Michx.) Scribn. PRAIRIE WEDGESCALE, PRAIRIE WEDGEGRASS **Stems** 30–100 cm. **Leaves:** blades flat, 2–6 mm wide. **Inflorescence** an erect, dense, panicle with spicate branches, 7–20 cm long. **Spikelets** 2.5–3.5 mm long; second glume broadest right near the tip. **Lemmas** 2–3 mm long. Wet to dry sites, but usually where water is at least intermittently abundant. Throughout most of N. America.

Sporobolus R.Br. Dropseed

Warm season, bunched perennials and annuals. **Leaf blades** flat and ascending to lax; sheaths with overlapping margins; ligule hairy. **Inflorescence** an open to contracted panicle, often partially or entirely enclosed in the leaf sheath. **Spikelets** with usually 1 floret; glumes unequal, the first 1/2 as long as the second. **Lemma** with 1 distinct vein, awnless or awned; callus sometimes long-hairy. **Disarticulation** above the glumes; unit of dispersal the fruit, sometimes the floret.

Sporobolus flexuosus (Thurb. ex Vasey) Rydb. is reported from west-central MT.[31] This species is distinguished from *S. cryptandrus* by having spreading pedicels and reflexed secondary branches of the inflorescence, rendering a very open panicle. The attachment point of the secondary branches is hairy and the entire panicle has a subovate outline because the lower branches are no longer than the branches in the middle of the inflorescence.

1. Plants annual; panicle up to 4 cm long, usually partially enclosed in the leaf sheath or spathe . *S. neglectus*
1. Plants perennial; panicle >4 cm long, exposed or variously enclosed in the leaf sheath or spathe 2

2. Panicle diffuse and usually fully exposed (not much enclosed by the leaf sheath), panicle dimensions about as long as wide; stems bases decumbent, hardened and woody . . . . *S. airoides*
2. Panicle contracted or open but then much longer than wide, usually partially enclosed in the leaf sheath; bases of stems erect, but if decumbent, then not hardened and woody 3

3. Spikelets 2.5 mm long or shorter . . . . *S. cryptandrus*
3. Spikelets 3 mm long or longer . . . . *S. compositus*

Sporobolus airoides (Torr.) Torr. ALKALI SACATON Perennial, often forming broad tussocks. **Stems** 40–80 cm. **Leaves:** blades 2–4 mm wide; throat with long hairs. **Inflorescence** a diffuse panicle 10–40 cm long, not much enclosed in leaf sheath. **Spikelets** 2–2.5 mm long. **Lemmas** 1.5–2.5 mm long. On moderately alkaline soils and moderately disturbed open dry vegetation, including road and trailsides. Throughout much of central and western N. America.

Sporobolus compositus (Poir.) Merr. [*Sporobolus asper* (P.Beauv.) Kunth] COMPOSITE DROPSEED Perennial. **Stems** 30–80 cm. **Leaves:** blades 2–4 mm wide; throat with a tuft of hairs. **Inflorescence** a contracted panicle 5–15 cm long, mostly enclosed in leaf sheath. **Spikelets** 4–6 mm long. **Lemmas** 3–6 mm long. Disturbance-prone settings on a diversity of soils; collected in Carter Co. Sporadic throughout most of N. America.

Sporobolus heterolepis (A.Gray) A.Gray, prairie dropseed, occurs in the adjacent Dakotas and is distinguished from *S. compositus* by spikelets having at least one glume as long or longer than the lemma and an open panicle usually entirely exerted from the leaf sheath.

Sporobolus cryptandrus (Torr.) A.Gray SAND DROPSEED Perennial. **Stems** 20–80 cm. **Leaves:** blades 2–5 mm wide; throat with a conspicuous tuft of whitish hairs. **Inflorescence** a contracted to open panicle 8–20 cm long, partially enclosed in leaf sheath. **Spikelets** 2–2.5 mm long. **Lemmas** 1.4–2.5 mm long. Occurring in disturbed settings and rarely in sagebrush steppe, common along roadsides and there conspicuous because of inflorescences each enclosed in an arcuate leaf sheath. Throughout most of N. America. (p.696)

Sporobolus neglectus Nash PUFFSHEATHED DROPSEED Annual. **Stems** 10–40 cm. **Leaves:** blades 1–2 mm wide; throat with a tuft of hairs. **Inflorescence** a narrow panicle 1–4 cm long, partially enclosed in leaf sheath. **Spikelets** 2–3 mm long. **Lemmas** 1.5–2.9 mm long. Sporadic in open dry disturbed sites. Throughout much of N. America, excepting the Intermountain region and the very southeast.

The annual *Sporobolus vaginiflorus* (Torrey) Wood, poverty dropseed, is known from adjacent ID and differs from *S. neglectus* by spikelets usually >3 mm long and hairy lemmas.

Stipa L. Needlegrass

Cool season, bunched perennials. **Leaf blades** flat and ascending to lax; sheaths with overlapping margins; ligule membranous. **Inflorescence** a raceme to an open or contracted panicle. **Spikelets** with one plump to cylindrical floret; glumes enveloping the floret. **Lemmas** hardened at maturity and enveloping the floret, with a deciduous to persistent awn; callus blunt to sharp but usually distinctly hairy; palea enclosed in floret. **Disarticulation** above the glumes; unit of dispersal the floret.

For the tribe Stipeae (including *Oryzopsis* and *Stipa*), genetic and other evidence strongly suggests that morphology is a poor predictor of phylogenetic relationships.[32,203,344,457,458,459] It is clear that the many species of Stipeae represent a recent widespread diversification in mostly open, dry settings where most species retain the ability to interbreed with at least one close relative. Species of tribe Stipeae are best treated as belonging to a single genus, *Stipa*. However, this has not been done here because two species of *Oryzopsis* in Montana (*O. asperifolia* and *O. micrantha*) have not formally been placed into *Stipa*.

1. Awn plumose (feathery), with divergent dense whitish hairs in lower portion. 2
1. Awn smooth, scabrous, or with appressed hairs in lower portion. 3

2. Ligules 2–6 mm long (especially of upper leaves), conspicuous . . . . *S. thurberiana*
2. Ligules 1 mm or less long, inconspicuous . . . . *S. occidentalis*

3. Lemmas mostly >8 mm long; glumes 1.5–4 cm long. 4
3. Lemmas <8 mm long; glumes 1.5 cm or shorter 5

4. Lemmas 8.5–14(17) mm long, with white to brownish hairs evenly or patchily distributed, sometimes glabrous immediately above the callus; awns mostly 6–10.5 cm long; glumes mostly 15–25 mm long . . . . *S. comata*
4. Lemmas 15–25 mm long, with brownish hairs usually patchily distributed; awns 9–19 cm long; glumes mostly 25–45 mm long. . . . . *S. spartea*

5. Panicle open, branches spreading and spikelets borne only toward the branch tips. . . . . *S. richardsonii*
5. Panicle narrow or contracted, branches short and appressed, bearing spikelets toward the base 6

6. Lemmas densely hairy, the hairs spreading and 1–3 mm long; awn arising between two terminal lemma teeth; leaves <1 mm wide. . . . . *S. pinetorum*
6. Lemmas hairless or with short appressed hairs ≤1 mm long or lemmas not toothed; leaves mostly 1–5 mm wide 7

7. Tufts of hair usually present at junction of leaf sheath and blade and at lower panicle nodes; mostly plains and foothills; florets at maturity mostly >1 mm thick, the tip differentiated from the awn in color and texture; palea glabrous and <1/2 as long as lemma . . . . *S. viridula*
7. Tufts of hair lacking at junction of leaf sheath and blade and at lower panicle nodes; mostly in montane meadows; florets at maturity <1 mm thick, the tip not differentiated from the awn in color and texture; palea either hairy or >1/2 as long as lemma or both .8

8. Lower leaf sheaths conspicuously hairy. . . . . *S. nelsonii*
8. Lower leaf sheaths hairless or nearly so 9

9. Awns mostly ≤2 cm long; lemmas 4–5 mm; palea well over half as long as lemma; glumes mostly <9 mm long; leaf blades 0.5–2 mm wide . . . . *S. lettermanii*
9. Awn mostly >2 cm long; lemmas 5.5–7 mm; palea mostly half as long as lemma or less; glumes 9–10 mm long; leaf blades 2–5 mm wide . . . . *S. nelsonii*

Stipa comata Trin. & Rupr. [*Hesperostipa comata* (Trin. & Rupr.) Barkworth; *Stipa curtiseta* (A. S. Hitchcock) Barkworth, *Hesperostipa curtiseta* (Hitchc.) Barkworth, *Stipa spartea* Trin. var. *curtiseta* A.S. Hitchcock, *Stipa comata* var. *intermedia* Scribn. & Tweedy, *Hesperostipa comata* ssp. *intermedia* (Scribn. & Tweedy) Barkworth] NEEDLE-AND-THREAD **Stems** 30–70 cm. **Leaves:** blades 2–4 mm wide; hairy or hairless at throat region; ligule 3–6 mm long. **Inflorescence** a

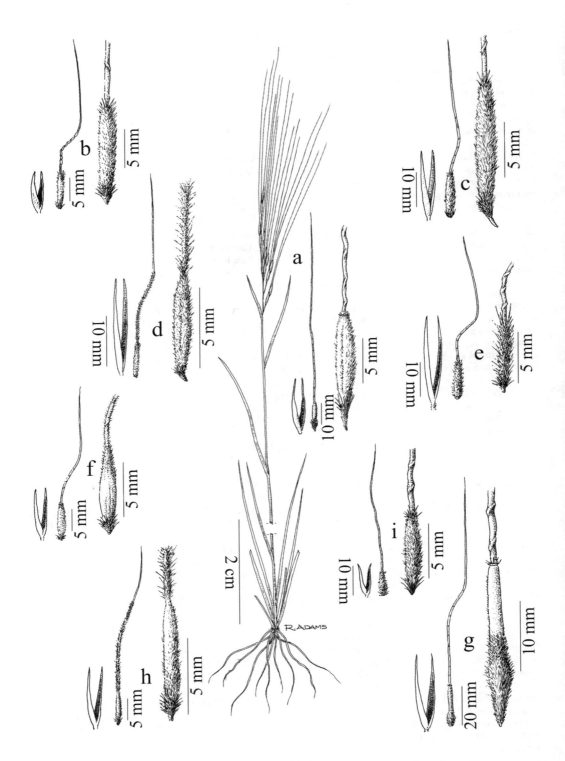

Plate 119. **Stipa**. **a.** *S. comata*, **b.** *S. lettermanii*, **c.** *S. nelsonii*, **d.** *S. occidentalis*, **e.** *S. pinetorum*, **f.** *S. richardsonii*, **g.** *S. spartea*, **h.** *S. thurberiana*, **i.** *S. viridula*

contracted to sometimes open panicle 15–32 cm long. **Spikelets** 16–35 mm long. **Lemmas** evenly to unevenly whitish-hairy; awn 6.5–22 cm long. Open, dry settings including sagebrush steppe with high native plant cover as well as well-grazed rangelands and along secondary roads. Throughout N. America, excepting the very southeast. (p.711)

The density, cover, and degree of whiteness of the lemma hairs is so variable as to render *Stipa comata* indistinct from *Stipa curtiseta*. Intermediate needle-and-thread, *Stipa comata* ssp. *intermedia*, is known from a few specimens in Gallatin, Madison, and Silver Bow cos. and is marked by a distinctively straight and relatively short terminal awn segment (<8 cm). Few, if any characters covary with such an awn, however. For example, MT specimens of intermediate needle-and-thread taken by mid-summer flowering have inflorescences partly included within the subtending leaf sheath, even though this entity is diagnosed as having completely exerted panicles.

Stipa lettermanii Vasey [*Achnatherum lettermanii* (Vasey) Barkworth] LETTERMAN'S
NEEDLEGRASS Perennial bunchgrass. **Stems** 30–60 cm. **Leaves**: blade 0.5–2 mm wide; essentially
hairless at throat region; ligule 3–4 mm long. **Inflorescence** a contracted panicle 10–15 cm long.
Spikelets 7–9 mm long. **Lemmas** with dense appressed hairs; awn 1–2 cm long. Open understory
at middle to high elevations. Sporadic over much of western N. America. (p.711)

Stipa nelsonii Scribn. [*Stipa columbiana* Macoun; *Achnatherum nelsonii* (Scribn.) Barkworth;
Stipa williamsii Scribn.] COLUMBIA NEEDLEGRASS **Stems** 30–80 cm. **Leaves**: blades 2–5 mm
wide; hairless at throat region; sheaths conspicuously hairy or glabrate; ligule 1–2 mm long.
Inflorescence a contracted panicle 10–30 cm long. **Spikelets** 8–12 mm long. **Lemmas** appressed-hairy; awn 2–4 cm long. Open dry settings, including mountain big sagebrush steppe, at middle
elevations in the mountains. Throughout the Rocky Mountain and Intermountain regions. (p.711)

Stipa nelsonii may have hairless or hairy stems and leaf sheaths, although hairless forms are most common in MT. *Stipa nelsonii* can be very difficult to distinguish from *Stipa viridula*. *Stipa nelsonii* grows in the mountains, whereas *Stipa viridula* grows at lower elevation often in open shrub steppe settings. This general ecological difference, along with morphology, may help distinguish these two. The bony-textured tip of the floret of *Stipa viridula*, supposedly characteristic of the genus *Nassella*, is distinct from that of *Stipa nelsonii* (an *Achnatherum*), where the beak of the floret is not set off in coloration and texture from the main body of the floret.

Stipa occidentalis Thurb. [*Achnatherum occidentale* (Thurb.) Barkworth ssp. *pubescens*
(Vasey) Barkworth] WESTERN NEEDLEGRASS, PUBESCENT WESTERN NEEDLEGRASS **Stems** 20–40 cm.
Leaves: blade 2–3 mm wide; hairless to somewhat hairy near the throat; ligule 0.5–1 mm long.
Inflorescence a narrow contracted panicle 10–15 cm long. **Spikelets** 5–8 mm long. **Lemmas**
appressed-hairy; with an awn 3–4 cm long that is plumose in the lower portion. Sagebrush steppe and
other open dry settings in the mountains. Throughout much of western N. America. (p.711)

Stipa pinetorum M.E.Jones [*Achnatherum pinetorum* (M.E.Jones) Barkworth] PINE NEEDLEGRASS **Stems** 10–30 cm.
Leaves: blades 0.5–0.8 mm wide; essentially hairless at the throat; ligule <0.5 mm long. **Inflorescence** a contracted
panicle 10–14 cm long. **Spikelets** 8–10 mm long. **Lemmas** with dense, spreading long hairs; awn 15–20 mm long. Open
to closed dry understory at middle elevations; collected in Gallatin Co. Throughout most of western N. America. (p.711)

Stipa richardsonii Link [*Achnatherum richardsonii* (Link) Barkworth] RICHARDSON'S
NEEDLEGRASS **Stems** 40–80 cm. **Leaves**: blades 1–2 mm wide; scabrous near the throat; ligule <0.5
mm. **Inflorescence** an open panicle 10–15 cm long. **Spikelets** 7–8 mm long, pendent. **Lemmas**
appressed-hairy; awn 1.5–3 cm long. Open dry settings and open montane understory, including
mountain big sagebrush steppe. Throughout the northern Rocky Mountain region. (p.711)

Stipa spartea Trin. [*Hesperostipa spartea* (Trin.) Barkworth] PORCUPINEGRASS **Stems**
70–110 cm. **Leaves**: blade 3–5 mm wide; essentially hairless at throat; ligule 2–6 mm long.
Inflorescence a contracted panicle 15–25 cm long. **Spikelet** 25–45 mm long. **Lemmas** unevenly
brownish-hairy; with a stout awn 10–20 cm long. Sporadic in open dry settings at low elevations.
Throughout most of central and northern N. America. (p.711)

Stipa thurberiana Piper [*Achnatherum thurberianum* (Piper) Barkworth] THURBER'S NEEDLEGRASS **Stems** 40–70 cm.
Leaves: blades 1–2 mm wide; essentially hairless at the throat; ligule 3–6 mm long. **Inflorescence** a contracted panicle
5–16 cm long. **Spikelets** 10–15 mm long. **Lemmas** with appressed hairs; awn 3–5 cm with plumose hairs on the lower
portion. Sagebrush steppe and understory of pine forests; collected in Gallatin and Park cos. Throughout much of
western N. America. (p.711)

Known to hybridize with *Oryzopsis hymenoides*. The purplish glumes and long ligule may be the only characters
distinguishing *Stipa thurberiana* from *S. occidentalis*.

Stipa viridula Trin. [*Nassella viridula* (Trin.) Barkworth] GREEN NEEDLEGRASS **Stems** 50–100 cm. **Leaves**: blades 1–2 mm wide; hairy at the throat; ligules 1–2 mm long. **Inflorescence** a contracted panicle 7–14 cm long. **Spikelets** 9–12 mm long. **Lemmas** appressed-hairy, with an awn 2–3.5 cm long. Sagebrush steppe and other open dry sites at lower elevations, especially east and south of the Continental Divide. Throughout N. America except in the eastern portion of the continent.

The whitish-tipped floret with a bony texture contrasts to the color and texture of the adjacent awn, and a zone of abscission between the two renders a tardily disarticulating awn. Such a floret tip and awn weakly distinguishes *Stipa viridula* from the morphologically similar *S. nelsonii*. (p.711)

Torreyochloa Church False mannagrass

Torreyochloa pallida (Torr.) Church var. *pauciflora* (J. Presl) J.I.Davis [*Puccinellia pauciflora* (J.Presl) Munz] PALE FALSE MANNAGRASS Cool season, rhizomatous, perennial aquatic. **Stems** 40–70 cm. **Leaves**: blades 1–3 mm wide, mostly flat and lax to ascending; sheaths with overlapping margins; ligule membranous. **Inflorescence** an open panicle 6–12 cm long. **Spikelets** 4–7 mm long, with 4 to 7 florets; glumes similar to lemmas. **Lemmas** hairless or scabrous with a blunt tip, rounded on back and with prominent parallel venation; palea well-developed. **Disarticulation** above glumes; unit of dispersal the floret. Alkaline and fresh water aquatic, often at higher elevations, primarily in western MT. Throughout western N. America. (p.696)

Very similar to *Glyceria grandis* except *Glyceria* has prominent ligules that wrap and fuse around the stem to form a solid cylinder, like the subtending leaf sheath. In *Torreyochloa*, the ligule is limited largely to the front (inside) of the leaf blade. The margins of the leaf sheath are overlapping rather than fused. Also, the lemma tips of *Torreyochloa* are broad and the hyaline distal margin is delimited from the green lemma body by a black band. See *Glyceria grandis*.

Trisetum Pers. Oatgrass

Cool season, bunched perennials. **Leaves**: blades flat and ascending; sheaths with overlapping margins; ligule membranous. **Inflorescence** a contracted to sometimes open panicle. **Spikelets** with 2 (to 4) florets; glumes enveloping the florets. **Lemma** with a bent awn from the middle, rarely awnless; palea well-developed. **Disarticulation** above glumes; unit of dispersal the floret.

1. Lemmas awnless, merely awn-tipped, or with a short straight awn 2
1. Lemmas with a well-developed awn that is curved or bent 3

2. Awn absent or up to 2 mm long but then not surpassing the tip of lemma . . . . *T. wolfii*
2. Awn 4–6 mm long, well-surpassing the tip of lemma . . . . *T. orthochaetum*

3. Both glumes about equal in length and width; panicle spike-like, the individual branches usually obscure . . . . *T. spicatum*
3. First glume distinctly narrower than second glume and often shorter; panicle open to contracted, but in any case at least the lower inflorescence branches are distinct 4

4. Most panicle branches bearing spikelets their entire length; panicles narrow and nodding only towards the tip; upper glume widest at or below the middle and tapering to the tip; first glume 3–5 mm long . . . . *T. canescens*
4. Most panicle branches bearing spikelets only towards the branch ends; panicles open and completely nodding or drooping; upper glume widest above the middle and rounded to the tip; first glume 1–3 mm long . . . . *T. cernuum*

Trisetum canescens Buckley TALL TRISETUM **Stems** loosely bunched, 40–100 cm. **Leaves**: blades mostly 5–9 mm wide. **Inflorescence** a loosely contracted to open panicle 10–25 cm long; the lower branches with clusters of spikelets distinctly set off from rest of inflorescence. **Spikelets** 7–9 mm long. **Lemmas** 2 to 4 per spikelet; awns 7–13 mm long, bent, from upper one-half of lemma. Middle to high elevation forest understory. Common throughout western N. America.

Trisetum flavescens (L.) P. Beauv. has been cultivated in Gallatin Co. and is very similar to *T. canescens* except that the spikelets are mostly with 4 florets and arranged in dense, yellowish panicles.

Trisetum cernuum Trin. NODDING TRISETUM **Stems** 60–100 cm. **Leaves**: blades 6–12 mm wide. **Inflorescence** an open, lax panicle 10–30 cm long, the branches often drooping. **Spikelets** 8–11 mm long. **Lemmas** 2–3 per spikelet; awns 8–13 mm long, bent, from upper half of lemma. Open to closed mountain understory. Common throughout western N. America.

Trisetum orthochaetum Hitchc. Hybrid false oat, Bitterroot trisetum **Stems** few to a bunch, 80–110 cm tall. **Leaves**: blades 2–7 mm wide. **Inflorescence** a contracted and somewhat nodding panicle 13–20 cm long. **Spikelets** 7–9 mm long. **Lemmas** mostly 2 or 3; awns 4–6 mm long, straight, from upper half of lemma. Boggy meadows near the edge of forests; collected in Missoula and Glacier cos. Endemic to MT.

Trisetum orthochaetum is possibly a hybrid between *T. canescens* and *T. wolfii*.

Trisetum spicatum (L.) K.Richt. Spike trisetum **Stems** 15–50 cm. **Leaves**: blades 1.5–3 mm wide. **Inflorescence** a spicate panicle 10–25 cm long. **Spikelets** 7–9 mm long. **Lemmas** 2 to 4 per spikelet; awn 5–7 mm, bent, from upper half of lemma. Mostly montane to subalpine understory. Common throughout western N. America. (p.716)

Trisetum wolfii Vasey. Wolf's trisetum **Stems** loosely bunched, 50–90 cm. **Leaves**: blades 4–6 mm wide. **Inflorescence** a contracted panicle 10–40 cm long. **Spikelets** 5–7 mm long. **Lemmas** 2 per spikelet; awns absent. Open grasslands, wet meadows, and edges of forests in the western MT. Common throughout western N. America.

Because of the absence of lemma awns, this species is superficially similar to *Koeleria macrantha* (prairie junegrass), except that it has other *Trisetum* characteristics, including the well-developed hairs along the prolonged rachillas. Both *Koeleria* and *Trisetum* have a characteristic silvery-greenish-purplish patterning to the coloration of the glumes and lemmas, thus rendering further similarity between *T. wolfii* and *K. macrantha*. *Trisetum wolfii* otherwise is similar to *Trisetum spicatum* excepting its awnless lemmas.

Triticum L. Wheat

Triticum aestivum L. Common wheat Cool season, annual bunchgrass. **Stems** 50–120 cm. **Leaves**: blades 5–15 mm wide, flat and lax to ascending; sheaths with overlapping margins; ligule membranous. **Inflorescence** 5–12 cm long, a two-sided terminal spike with short internodes; rachis continuous. **Spikelets** 1 per node, 10–14 mm long excluding awns, with 3 to 5 florets; glumes asymmetrically keeled, broad and blunt, often notched at the tip. **Lemmas** broad, also asymmetrically keeled, awned or awnless; palea well-developed. **Disarticulation** above the glumes; unit of dispersal the fruit. Introduced and cultivated. Escaping or planted for soil stabilization throughout all of N. America, but then not long persisting.

Vahlodea Fr. Mountain hairgrass

Vahlodea atropurpurea (Wahlenb.) Fr. ex Hartm. [*Deschampsia atropurpurea* (Wahl.) Scheele] Cool season, perennial bunchgrass. **Stems** few-bunched, 20–60 cm. **Leaves**: blades 4–6 mm wide, flat and ascending; sheaths with overlapping margins; ligule membranous. **Inflorescence** a contracted to open panicle, drooping to erect, 4–8 cm long. **Spikelets** ovoid, 2-flowered, 5.5–6 mm long; glumes enveloping the florets, distal portion purplish. **Lemmas** 2 per spikelet, awned from near the middle; rachilla hairs long; palea enclosed in floret. **Disarticulation** above the glumes; unit of dispersal the floret. Alpine to subalpine meadows. Throughout most of Canada and the Pacific Northwest region of the U.S.

Vahlodea atropurpurea, formerly *Deschampsia atropurpurea*, represents a distinct lineage apart from *Deschampsia* and close relatives.[81,82]

Ventenata Koeler North Africa grass

Ventenata dubia (Leers) Coss. & Durieu. Ventenata, wiregrass Cool season, annual bunchgrass. **Stems** few-bunched, 50–70 cm. **Leaves**: blades 1–3 mm broad, inrolled and ascending; sheaths with overlapping margins; ligule membranous. **Inflorescence** an open panicle, 2–4 dm long, only the very distal branch ends bearing spikelets. **Spikelets** 6–10 mm long, mostly with 3 florets; callus of spikelet well-developed; glumes distinctly ribbed or veined, enveloping the florets. **Lemmas** dimorphic; the lowermost staminate and with a straight awn; the upper lemmas fertile and with a bent and twisted awn, readily disarticulating; callus bearded; palea well-developed. **Disarticulation** above and below glumes; unit of dispersal the floret. Sporadic in MT along roadsides, pastures, and range with a significant disturbance history. Introduced and of sporadic or ephemeral occurrence in the Pacific Northwest region and also in east-central N. America.

Locally abundant populations can be ephemeral and not detected in subsequent years. The prominently ribbed glumes of *Ventenata* are distinctive. During the late season, this grass will have spikelets bearing only the single basal persistent floret with a terminal straight awn. This partial spikelet obscures its oatgrass affinities. Regardless, the many-ribbed glumes, the stiff and wiry inflorescence branches, and pedicels that persist after the spikelets have disarticulated are diagnostic of *Ventenata*.

Vulpia C.C.Gmel. Annual fescue

Cool season, bunched annuals. **Leaf blades** inrolled to flat and ascending; sheaths with overlapping margins; ligule membranous. **Inflorescence** an open to usually contracted secund panicle with appressed branches, rarely a raceme. **Spikelets** with several florets; glumes similar in dimension to lemmas, sometimes distinctly shorter. **Lemma** rounded and smooth on back, awn-tipped to long-awned; palea enclosed in floret. **Disarticulation** above glumes; unit of dispersal the floret.

The treatment below follows Cronquist et al.[101], which lists the many names of *Vulpia* that are now considered synonyms of the species listed below.

1. Lower glumes <one-half the length of the upper glumes . . . .*V. myuros*
1. Lower glumes one-half or more the length of the upper glumes 2

2. Panicle branches 1–2 per node; spikelets with 4–17 florets; rachilla internodes 0.5–0.7 mm long; awn of the lowermost lemma in each spikelet 0.3–9 mm long . . . .*V. octoflora*
2. Panicle branches 1 per node; spikelets with 1–8 florets; rachilla internodes 0.6–1.2 mm long; awn of the lowermost lemma in each spikelet 2–20 mm long 3

3. Panicles narrow, branches oriented upward or ascending, the branches and pedicels all lacking axillary callus thickenings . . . . *V. bromoides*
3. Panicles open, branches (at least the lower ones) and sometimes the spikelets oriented outward or downward; branches with callus thickenings in the axil (base of branch) . . . . . . *V. microstachys*

Vulpia bromoides (L.) Gray [*Festuca bromoides* L.] BROME FESCUE **Stems** 10–30 cm. **Leaves:** blades 1 mm wide. **Inflorescence** a narrow panicle 2–8 cm long. **Spikelets** 6–8 mm long. **Lemmas** with an awn 6–11 mm long. Disturbed dry sites in the Cabinet Mts. (Sanders Co.) and the National Bison Range (Lake Co.). Sporadic throughout much of N. America, excluding the north-central region.

Distinguished from *Vupia myuros* by a longer first glume.

Vulpia microstachys (Nutt.) Munro [*Festuca microstachys* Nutt.] SMALL FESCUE **Stems** 20–40 cm. **Leaves:** blades 1 mm wide. **Inflorescence** an open panicle 5–10 cm long. **Spikelets** 5–7 mm long. **Lemmas** with an awn 5–8 mm long. Shrub steppe vegetation as well as disturbed sites; collected in Missoula Co. Introduced throughout western N. America.

Vulpia myuros (L.) C.C.Gmel. [*Vulpia megalura* (Nutt.) Rydb., *Festuca megalura* Nutt., *Vulpia myuros* (L.) K. C. Gmelin var. *hirsuta* Hack.] RAT-TAIL FESCUE **Stems** 20–60 cm. **Leaves:** blades 1 mm wide. **Inflorescence** a narrow panicle 7–20 cm long. **Spikelets** 5–8 mm long. **Lemmas** with an awn 5–12 mm long. Frequently disturbed sites in open, dry shrub steppe vegetation. Introduced in open disturbed sites throughout most of N. America, excepting the north-central region.

The very short first glume is distinctive of this species.

Vulpia octoflora (Walter) Rydb. [*Festuca octoflora* Walter, *Vulpia octoflora* var. *glauca* (Nutt.) Fernald, *Vulpia octoflora* var. *hirtella* (Piper) Henr.] SIXWEEKS FESCUE **Stems** 8–30 cm. **Leaves:** blades 1–2 mm wide. **Inflorescence** a narrow panicle 2–8 cm long. **Spikelets** 6–10 mm long. **Lemmas** awn-tipped or with an awn 1–7 mm long. Open, dry disturbed sites usually with some history of overgrazing. Throughout all of N. America.

Spikelets may be paired, one subsessile and the other pedicellate. Three varieties are weakly distinguished and in MT. Perhaps the most distinctive is var. *hirtella* with its scabrous lemmas, which contrast to glabrous lemmas in var. *octoflora*; var. *glauca* has spikelets 4–6.5 mm (vs. 5.5–13) and first lemma awns 0.3–3 mm (vs. 2.5–9).

Zea L. Maize

Zea mays L. CORN Warm season, annual, monoecious bunchgrass. **Stems** few-bunched, 1–3 m. **Leaves:** blades up to 1 dm wide, flat and lax to ascending; sheaths with overlapping margins; ligule hairy, membranous, 5–10 mm long. **Inflorescence** dimorphic; staminate in a terminal panicle of digitately-arranged spikes; pistillate a woody cob enclosed in leafy bracts and located in the axils of the main stem. **Spikelets** with one fertile floret; glumes enveloping the florets, at least in the staminate spikelets; staminate spikelets distinctly paired as one sessile and one pedicellate; pistillate bearing long styles (known as silk). **Lemmas** thin-textured to inconspicuous, rounded on back, awnless; palea enclosed in floret. **Disarticulation** below the glumes; unit of dispersal the inflorescence and spikelet. Rarely found outside of cultivation and then near crop fields. Different varieties and races are grown for grain and forage throughout most of N. America, especially in more southern latitudes.

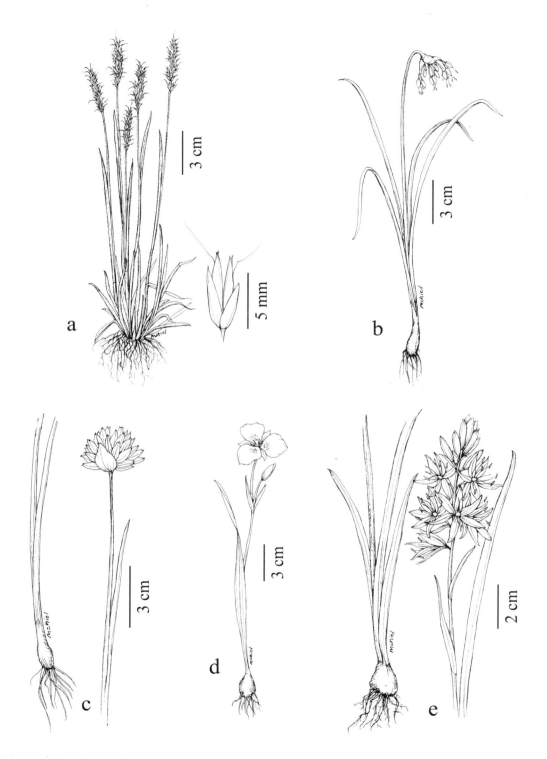

Plate 120. a. *Trisetum spicatum*, b. *Allium cernuum*, c. *Allium schoenoprasum*, d. *Calochortus apiculatus*, e. *Camassia quamash*

Zizania L. Wild rice

Zizania palustris L. [*Z. aquatica* L. misapplied] NORTHERN WILD RICE Cool season, annual reeds. **Stems** scattered or in dense stands, rooting in mud, 1–3 m; sheath essentially hairless with overlapping margins; collar hairy; ligule membranous, 10–15 mm long. **Leaves**: blades 1–4 cm broad, flat and ascending. **Inflorescence** large open panicles, 4–6 dm in length; the lower branches bearing pendulous staminate spikelets; the upper bearing appressed-ascending pistillate spikelets. **Spikelets** with one floret; staminate 6–8 mm long; pistillate 10–15 mm long (to 20 mm long in fruit), nearly round in cross-section; glumes of both spikelet types minute and consisting of a small cup just below point of disarticulation. **Lemmas** of pistillate spikelet hispid toward tip and tapered to an awn 6–40 mm long; lemmas of staminate spikelets thin-textured and ephemeral. **Disarticulation** above glumes; unit of disarticulation the floret and fruit. Marshes, lakes, and streams in fresh, flowing water at lower elevations; often planted for waterfowl use. Throughout much of N. America, especially at northern latitudes.

PONTEDERIACEAE: Pickerel Weed Family

Heteranthera Ruiz & Pav. Mud plantain

Heteranthera dubia (Jacq.) MacMill. Perennial, glabrous, aquatic herbs. **Stems** flaccid, submersed or floating, ca. 30 cm, sometimes rooting at the nodes. **Leaves** alternate, sessile, linear, 1–4 mm wide; stipules membranous, tubular, basally clasping the stem. **Inflorescence** axillary, a solitary flower; the tube sheathed at the base by a membranous, tubular spathe 12–45 mm long. **Flowers** perfect, radially symmetrical, ephemeral; perianth united below into an elongate tube 2–4 cm long; perianth 6-lobed; lobes ca. 5 mm long, blue; stamens 3; ovary superior; style 3-lobed. **Fruit** a many-seeded, 3-celled capsule. Shallow water of riverine sloughs, backwaters; valleys. BC to QC south to CA, MT, ID, MT, TX, VA. Collected in Flathead and Sanders cos.

Our specimens are in preflowering condition, although they were collected in August.

ASPARAGACEAE: Asparagus Family

Asparagus L.

Asparagus officinalis L. Rhizomatous, glabrous perennial. **Stems** erect, branched, 60–150 cm. **Leaves** scale-like, each subtending 4–15 linear branchlets 8–25 mm long. **Inflorescence** of 1 or 2 pendent, pedicellate flowers in leaf axils. **Flowers** perfect or some unisexual, regular, campanulate; tepals 2–6 mm long, yellow-green, united below; stamens 6; ovary superior, 3-celled; style 1, 3-branched. **Fruit** a globose, red berry 5–8 mm wide with 2 to 4 seeds. Disturbed, moist grasslands, meadows, fields, roadsides; plains, valleys. A garden escape introduced from Europe throughout N. America.

LILIACEAE: Lily Family

Herbaceous perennials. **Leaves** alternate (whorled), entire, usually parallel-veined. **Flowers** bisexual, radially symmetrical; sepals and petals separate, 3 each, or 6 identical tepals; stamens usually 6; ovary superior, 3-celled; style 1, 3-lobed. **Fruit** a berry or 3-celled capsule.

Many recent studies suggest that the Liliaceae be decomposed into numerous segregate families,[339] but there does not seem to be agreement on how these families should be circumscribed. I have adopted a broad concept of the Liliaceae, similar to that of Cronquist[98] with the exception that *Asparagus* is placed in a separate family, Asparagaceae, and *Smilax* in the Smilacaceae. *Hemerocallis fulva* (L.) L., the common introduced daylily, is a sterile triploid that was collected on roadsides of Ravalli and Carbon cos., but it does not appear to be truly naturalized.

1. Leaves lance-shaped, elliptic or ovate, <8 times as long as wide 2
1. Leaves linear, grass-like, >8 times as long as wide 11

2. Tepals >15 mm long; flowers mostly ≤5 per stem 3
2. Tepals <15 mm long; flowers usually >3 per stem 6

3. Leaves 3 in a whorl at the top of the stem . . . . *Trillium*
3. Leaves >3 or not whorled, often all basal. 4

4. Tepals ca. 20 mm long; fruit a blue berry . . . .*Clintonia*
4. Tepals >25 mm long; fruit a capsule. 5

5. Plants scapose; ≤30 cm tall . . . . *Erythronium*
5. Plants with cauline leaves; stems usually >30 cm . . . . *Lilium*

6. Flowers stalked, 1 to 3 in the axils of leaves or on branch tips 7
6. Flowers ≥3 in a terminal inflorescence. 8

7. Flowers at the tips of branches. . . . . *Prosartes*
7. Flowers in leaf axils . . . . *Streptopus*

8. Tepals united; flowers urn-shaped, nodding. . . . .*Convallaria*
8. Tepals separate most of their length, rotate, not nodding. 9

9. Plants >80 cm high; tepals green, ≥8 mm long; fruit a capsule . . . . *Veratrum*
9. Plants <60 cm high; tepals 2–5 mm long; fruit a berry. 10

10. Leaves cordate-based; tepals 4 . . . . *Maianthemum*
10. Leaves sessile but not cordate; tepals 6 . . . . *Smilacina*

11. Flowers 1 to 4 12
11. Flowers >4 16

12. Plants scapose; peduncle and pedicels below ground. . . . . *Leucocrinum*
12. Plants with leafy, above ground stems. 13

13. Perianth of well-differentiated sepals and petals . . . . *Calochortus*
13. Perianth parts all similar in size 14

14. Tepals white, ca. 1 cm long; leaves ≤1 mm wide . . . . *Lloydia*
14. Tepals yellow, reddish or purple-mottled; leaves >1 mm wide 15

15. Tepals 1–2 cm long. . . . . *Fritillaria*
15. Tepals ≥3 cm long. . . . . *Lilium*

16. Flowers stalked, all arising from the stem tip, forming a flat-topped or hemispheric
  inflorescence (umbel). 17
16. Inflorescence elongate 18

17. Tepals united below into a tube . . . . *Triteleia*
17. Tepals essentially separate; perianth campanulate . . . .*Allium*

18. Leaves folded lengthwise (equitant); flower bracts funnel-like around the pedicel . . . . *Tofieldia*
18. Leaves not equitant; bracts linear-lanceolate, not funnel-like. 19

19. Basal leaves wiry, numerous, forming tussocks. . . . .*Xerophyllum*
19. Basal leaves few, not tussock-forming 20

20. Tepals white with a basal spot; capsule with a pointed beak . . . . *Zigadenus*
20. Tepals red-green to blue; capsule rounded on top. 21

21. Tepals blue, 2–3 cm long . . . .*Camassia*
21. Tepals reddish-green, 10–15 mm long. . . . . *Stenanthium*

Allium L. Onion

Scapose perennials from bulbs, onion-scented. **Bulbs** solitary or clustered, reforming annually; inner coats pale, membranous; outer coats brown or gray, membranous or fibrous. **Leaves** linear or linear-lanceolate, flat or terete, basal, sometimes persisting to anthesis. **Inflorescence** an erect scape terminated by an umbel subtended by 2 to 3 sheathing, membranous bracts. **Flowers** regular, mostly campanulate; tepals 6 in 2 usually similar whorls, separate, lanceolate to ovate; stamens 6; ovary sometimes crested; stigma usually narrowly capitate. **Fruit** an ovoid, 3- or 6-seeded capsule.

1. Leaves terete, hollow . . . .*A. schoenoprasum*
1. Leaves flat or channeled (U-shaped in cross-section) 2

2. Sessile bulblets present in the umbel; flowers few. . . . . *A. geyeri*
2. Umbel without bulblets 3

3. Umbel nodding in flower; erect in fruit . . . . *A. cernuum*
3. Umbel erect in flower and fruit 4

Allium acuminatum Hook. **Bulbs** sometimes clustered, globose; outer coat dingy white, membranous, honeycombed. **Scapes** terete, 20–35 cm. **Leaves** 2 to 3, subterete to channeled, 0.5–2 mm wide, withering. **Umbel** hemispheric with 10 to 30 flowers; pedicels 5–25 mm long; bracts 2, lanceolate to ovate, acuminate. **Flowers** pink to magenta; outer tepals 7–14 mm long; inner tepals smaller; ovary obscurely crested; stamens included. **Seed surface** minutely roughened. Grasslands, open forest, rock outcrops; valleys, montane. BC, MT south to CA, AZ, NM. Known from Ravalli and Sanders cos. (p.720)

Allium brandegeei S.Watson **Bulbs** ovoid, solitary or clustered; outer coat membranous with hexagonal reticulations. **Scapes** terete, 3–10 cm. **Leaves** 2, channeled, sometimes falcate, 1–3 mm wide, persistent. **Umbel** compact, hemispheric with 8 to 25 flowers; pedicels 5–15 mm long; bracts 2, ovate, acuminate. **Flowers** white to rose with green midveins; tepals 5–8 mm long; ovary nearly crestless; stamens included. **Seed surface** smooth. Shallow soil of cliffs and outcrops; montane. OR to MT south to NV, UT, CO. Known from Park Co. (*Evert 30190* RM); reported for southwest MT.[313] (p.720)

Allium brevistylum S.Watson **Bulbs** elongate-ovoid on a thick rhizome; outer coat dirty-white, membranous with elongate cells, often wrapped in old gray, parallel fibers. **Scapes** flattened, 12–50 cm. **Leaves** 2–5, linear, 2–6 mm wide, persistent. **Umbel** narrow-hemispheric with 7–15 flowers; pedicels 5–20 mm long; bracts 2, ovate, acute to acuminate. **Flowers** magenta; tepals 6–12 mm long; ovary crestless; stigma 3-lobed; stamens included. **Seed surface** minutely roughened. Moist grasslands, meadows, woodlands, open forest, thickets, especially aspen groves, riparian areas; valleys to lower subalpine. MT south to UT, NM. (p.720)

Allium cernuum Roth NODDING ONION **Bulbs** elongate-ovoid, solitary or clustered; outer coat membranous with elongate cells, pink beneath. **Scapes** terete, 10–40 cm, sometimes >1 per bulb. **Leaves** 3 to 7, flat, 2–6 mm wide, persistent. **Umbel** nodding, hemispheric with 5–40 flowers; pedicels 8–15 mm long; bracts 2, lanceolate, acute. **Flowers** white to pink; tepals 4–6 mm long; ovary with prominent triangular crests; stamens exserted. **Seed surface** smooth to minutely roughened. Meadows, grasslands, woodlands, open forest; plains, valleys to subalpine. BC to ON south to AZ, NM, TN, WV. (p.716)

The distinctive nodding umbel becomes nearly erect in fruit.

Allium columbianum (Owenby & Mingrone) P.M.Peterson, Annable & Rieseberg [*A. douglasii* Hook. var. *columbianum* Owenby & Mingrone] **Bulbs** globose, solitary or clustered; outer coat membranous, lacking reticulations. **Scapes** terete, 20–40 cm. **Leaves** 2, falcate, flat, 3–12 mm wide, persistent. **Umbel** hemispheric to globose with 20 to 50 flowers; pedicels 6–10 mm long; bracts 3, ovate, apiculate. **Flowers** rose to magenta; tepals 6–9 mm long; ovary with 6 low, rounded crests; stamens ca. as long as the tepals. **Seed surface** smooth. Vernally moist or wet soil of grasslands, meadows; valleys. Endemic to eastern WA and adjacent ID, disjunct in Ravalli and Sanders cos., MT. (p.720)

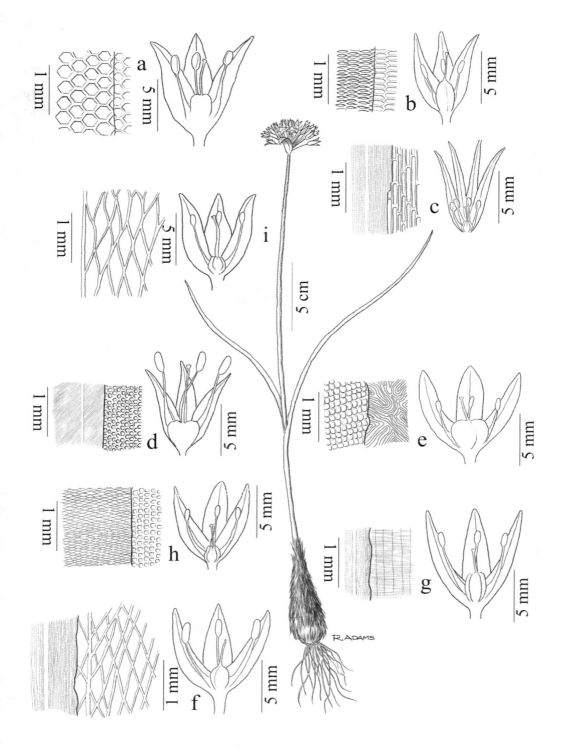

Plate 121. ***Allium***. **a.** *A. acuminatum*, **b.** *A. brandegei*, **c.** *A. brevistylum*, **d.** *A. columbianum*, **e.** *A. fibrillum*, **f.** *A. geyeri* var. *tenerum*, **g.** *A. parvum*, **h.** *A. simillimum*, **i.** *A. textile*

Allium fibrillum M.E.Jones ex Abrams **Bulbs** globose to ovoid, usually solitary; outer coat membranous with obscure, contorted reticulations. **Scapes** terete, 3–15 cm. **Leaves** 2(–4), nearly flat, 1–3 mm wide, barely persistent. **Umbel** hemispheric with 10 to 20 flowers; pedicels 5–12 mm long; bracts 2, ovate, apiculate. **Flowers** white with green midveins; tepals 5–10 mm long; ovary with 3 low crests; stamens included. **Seed surface** smooth. Shallow, vernally moist soil of grasslands, rock outcrops; montane, lower subalpine. Endemic to eastern WA, adjacent OR, ID, MT. (p.720)

Allium geyeri S.Watson **Bulbs** narrowly ovoid, often clustered; outer coat fibrous, net-like. **Scapes** terete, 10–50 cm. **Leaves** 2 or 3, channeled, 1–3 mm wide, persistent. **Umbel** hemispheric or narrower with 0 to 25 flowers and 0 to 25 sessile ovoid bulbils 3–6 mm long; pedicels 5–12 mm long; bracts 2 or 3, ovate, acute to apiculate. **Flowers** white to pink; tepals 5–7 mm long; ovary with small crests adjacent to the style; stamens included. **Seed surface** minutely roughened. BC to SK south to AZ, NM, TX. (p.720)

1. Many flowers replaced by sessile bulbils . . . . .var. *tenerum*
1. Bulbils absent in the inflorescence . . . . var. *geyeri*

Allium geyeri var. *geyeri* is diploid and found in vernally moist, often stony soil of grasslands, valleys in Flathead Co. *Allium geyeri* var. *tenerum* M.E.Jones is polyploid and reproduces primarily by bulbils; it occurs in moist grasslands, meadows, woodlands and riparian thickets; plains, valleys, montane. The deep pink tepals of var. *geyeri* do not contrast with the midveins as do the whitish tepals of *A. textile*.

Allium parvum Kellogg **Bulbs** usually ovoid; outer coat membranous mostly without reticulations. **Scapes** flattened, 3–6 cm. **Leaves** 2, flat, falcate, 1–4 mm wide, persistent. **Umbel** congested-hemispheric with 5 to 30 flowers; pedicels 5–10 mm long; bracts 2, ovate, acute to acuminate. **Flowers**: tepals white with a purple midvein; 8–12 mm long; ovary with 3 crests adjacent to the style; stamens included. **Seed surface** smooth. Open soil of sagebrush steppe, open forest; montane. OR to MT south to CA, NV, UT. Known from Beaverhead and Ravalli cos. (p.720)

Allium schoenoprasum L. **Bulbs** lanceoloid, clustered on a short rhizome; outer coat membranous with persistent parallel veins. **Scapes** terete, hollow, 15–70 cm. **Leaves** 2 to 5, terete, hollow, 1–7 mm wide, persistent. **Umbel** congested with 15 to 50 flowers; pedicels 2–10 mm long; bracts 2, ovate to suborbicular, apiculate. **Flowers** pink to rose; tepals 8–12 mm long; ovary crestless; stamens included. **Seed surface** minutely roughened. Wet meadows, especially along streams, lakes; valleys to lower alpine. Circumboreal south to OR, CO, MN, NY. (p.716)

Allium simillimum L.F.Hend. **Bulbs** ovoid, usually solitary; outer coat membranous with hexagonal reticulations. **Scapes** slightly flattened, partly subterranean, 2–6 cm. **Leaves** 2, channeled, ca. 1 mm wide, persistent. **Umbel** compact, campanulate to hemispheric with 5 to 15 flowers; pedicels 2–5 mm long; bracts 2, ovate, apiculate. **Flowers**: tepals white with green to purple midveins, 6–8 mm long; ovary with 3 crests adjacent to the style; stamens included. **Seed surface** smooth. Open, often shallow soil of meadows, outcrops, scree slopes; upper montane, lower subalpine. Endemic to central ID and Gallatin and Ravalli cos., MT. (p.720)

Allium textile A.Nelson & J.F.Macbr. **Bulbs** ovoid, usually clustered; outer coat fibrous, net-like. **Scapes** terete, 5–25 cm. **Leaves** 2, channeled, 1–2 mm wide, barely persistent. **Umbel** hemispheric with 10 to 30 flowers; pedicels 5–20 mm long; bracts 3, ovate, acuminate. **Flowers**: tepals white sometimes tinged with pink with brown to pink midveins, 6–7 mm long; ovary with 6 crests adjacent to the style; stamens included. **Seed surface** smooth. Grasslands, sagebrush steppe; plains, valleys, montane. AB to MN south to NV, UT, NM. (p.720)

Tepals are more likely to be pink-tinged near or west of the Continental Divide; see *Allium geyeri*.

Calochortus Pursh Mariposa, sego lily

Glabrous perennials from bulbs. **Stems** usually unbranched. **Bulbs** solitary, fibrous- or membranous-coated. **Leaves** usually solitary, basal, sheathing below; blade linear-lanceolate, flat or channeled. **Inflorescence** terminal, of 1 to few pedicellate flowers subtended by 1 to few leaf-like bracts. **Flowers** regular, campanulate; sepals 3, distinct; petals 3, distinct, obovate, narrowed to a short claw with a gland near the base, often ornamented with hair or color above and below it; stamens 6, included; style 3-lobed. **Fruit** a fusiform to ovoid, 3-angled, many-seeded capsule.

The size of petals can vary greatly among years, possibly depending on soil moisture.

1. Basal leaf flat, much broader than stem leaves, usually ≥½ as long as the stem .2
1. Basal leaf channeled, barely broader than stem leaves, often <½ as long as the stem 4

2. Anthers blunt-tipped . . . . *C. eurycarpus*
2. Anthers apiculate 3

3. Gland at base of petal circular, ca. 1.5 mm across. . . . . *C. apiculatus*
3. Gland at base of petal hemispheric, >2 mm across . . . . *C. elegans*

4. Petals purple; sepals longer than petals. . . . . *C. macrocarpus*
4. Petals mainly white or greenish; sepals as long or shorter than petals .5

5. Anthers shortly-apiculate; hairs on petal branched at the tip . . . . *C. gunnisonii*
5. Anthers blunt-tipped; petal hairs simple 6

6. Petals with a green, longitudinal stripe; glabrate on the inside. . . . . *C. bruneaunis*
6. Petals not green-striped; sparsely hairy on the inside . . . .*C. nuttallii*

Calochortus apiculatus Baker **Stem** 10–35 cm. **Leaf blade** 5–18 mm wide. **Bracts** 2 or more, 1–5 cm long. **Flowers** 1 to 3; sepals 15–25 mm long; petals white, drying yellowish with serrulate margins; inner surface often yellow basally, hairy with a small, dark circular gland near the base; anthers lanceolate, apiculate. **Capsule** nodding, ovoid 1–3 mm long. Grasslands, drier meadows, forest openings; valleys to lower subalpine. Endemic to northeast WA and adjacent BC, AB, ID, MT. (p.716)

Calochortus bruneaunis A.Nelson & J.F.Macbr. [*C. nuttallii* Torr. var. *bruneaunis* (A.Nelson & J.F.Macbr.) Owenby] **Stem** 40–60 cm. **Leaf blade** linear, ca. 5 mm wide. **Bracts** 2–4 cm long. **Flowers** 1 to 4; sepals 1–4 cm long, basally purplish; petals white, 2–4 cm long, apiculate; outer surface with a green midvein; inner surface glabrous with a dark, crescent-shaped mark above the small, circular gland; anthers oblong. **Capsule** erect, spindle-shaped, 3–7 cm long. Sagebrush steppe; montane. OR to MT south to CA, UT, NV. Known from Beaverhead Co.

Calochortus elegans Pursh **Stem** 5–25 cm. **Leaf blade** 2–10 mm wide. **Bracts** usually 2, 5–30 mm long. **Flowers** 1 to 5; sepals 10–15 mm long, purplish; petals white; inner surface hairy, purplish above and below the fringed, hemispheric gland; anthers lanceolate, apiculate. **Capsule** ellipsoid, nodding, 1–2 cm long. Meadows, moist grasslands, open forests; valleys, montane. WA to MT south to CA, ID. Our plants are var. **selwayensis** (H.St.John) Owenby.

Variety *elegans* is found in east-central ID and could occur in Beaverhead Co.

Calochortus eurycarpus S.Watson **Stem** 15–60 cm. **Leaf blade** linear to linear-lanceolate, 2–10 mm wide. **Bracts** 1–4 cm long. **Flowers** 1 to 3; sepals yellowish, 15–30 mm long; petals white or purplish, 3–4 cm long; inner surface sparsely hairy with a purple, flattened-hemispheric mark 4–6 mm above the fringed bullet-shaped gland; anthers oblong. **Capsule** ellipsoid, erect, 2–3 cm long. Grasslands, sagebrush steppe, open forest; valleys, montane. WA to MT south to NV, WY.

Calochortus gunnisonii S.Watson **Stem** 20–40 cm. **Leaf blades** 1–2 mm wide. **Bracts** 1–7 cm long. **Flowers** 1 or 2; sepals 15–45 mm long, marked with purple; petals greenish to purplish, 3–5 cm long; inner surface with a yellow-fringed, crescent-shaped gland, purple above and below, extending the width of the petal; anthers apiculate. **Capsule** spindle-shaped, erect, 2–5 cm long. Usually calcareous soil of grasslands, sagebrush steppe, woodlands, open forest; valleys, montane. MT south to AZ, NM. Our plants are var. **gunnisonii**.

Calochortus macrocarpus Douglas **Stem** 20–50 cm. **Leaf blade** 1–2 mm wide, becoming inrolled. **Bracts** 3–10 cm long. **Flowers** 1 to 4; sepals 3–6 cm long, purple; petals purple, 3–5 cm long, apiculate; outer surface with a green midvein; inner surface with an ovate, yellow-fringed gland and a crescent-shaped purple mark above; anthers lanceoloid. **Capsule** spindle-shaped, erect, 4–5 cm long. Grasslands, pine woodlands; valleys. BC, MT south to CA, NV.

Calochortus nuttallii Torr. **Stem** 10–40 cm. **Leaf blade** linear, 1–3 mm wide. **Bracts** 2–8 cm long. **Flowers** 1 to 4; sepals 2–3 cm long, greenish; petals white, purple-tinged, yellow at the base, 3–5 cm long; inner surface nearly glabrous, with a purple, crescent-shaped mark above the circular, fringed gland; anthers oblong. **Capsule** spindle-shaped, erect, 3–6 cm long. Grasslands, sagebrush steppe, open forest, woodlands; plains, valleys. MT, ND south to CA, AZ, CO.

Camassia Lindl. Camas

Camassia quamash (Pursh) Greene Glabrous perennials from globose bulbs. **Stems** erect, 20–60 cm. **Leaves** basal, linear-lanceolate, 2–15 mm wide. **Inflorescence** a terminal raceme; each flower subtended by a green, linear bract; pedicels 5–10 mm long. **Flowers** nearly regular, star-shaped, blue; tepals spreading, separate, linear-lanceolate, 15–25 mm long; stamens shorter than tepals; anthers yellow; style 3-lobed. **Fruit** an ellipsoid, 3-angled capsule 6–12 mm long. Deep soil of moist to wet meadows, grasslands; valleys to lower subalpine. BC, AB south to CA, UT, WY. Our plants are var. *quamash*. (p.716)

Clintonia Raf. Bead lily

Clintonia uniflora (Menzies ex Schult.) Kunth Scapose, rhizomatous perennials. **Scape** ascending to erect, 7–12 cm, long-hairy with 1 small bract. **Leaves** 2 to 3, basal; the blade 8–20 cm long, narrowly obovate, appressed, pubescent beneath. **Inflorescence** 1(2) terminal flower. **Flowers** regular, campanulate, opening to star-shaped, white; tepals separate, lanceolate, 15–25 mm long; stamens shorter than tepals; anthers yellow; style barely 2-lobed. **Fruit** an ellipsoid, blue berry 6–13 mm long with 10 to 18 seeds. Moist coniferous forest; montane, lower subalpine. AK to CA, ID, MT. (p.724)

Convallaria L. Lily of the valley

Convallaria majalis L. Scapose, glabrous, rhizomatous perennials. **Scape** erect, 5–25 cm, minutely bracteate. **Leaves** 2–3, basal, sheathing, petiolate; the blade 10–20 cm long, elliptic. **Inflorescence** a 1-sided raceme with nodding flowers. **Flowers** white, regular, urn-shaped, 5–10 mm long; tepals united; lobes 2–3 mm long; stamens shorter than tepals; ovary superior, 3-celled; style barely 3-lobed. **Fruit** a globose, reddish berry 6–12 mm long with 1 to 5 seeds. Rarely escaped into moist, shady sites near habitations; valleys. Introduced sparingly throughout N. America. Collected in Lake Co.

Erythronium L. Glacier lily

Erythronium grandiflorum Pursh Scapose, glabrous perennials from slender bulbs. **Scape** erect, 7–30 cm, ebracteate. **Leaves** 2, basal, short-petiolate; the blade 5–20 cm long, narrowly elliptic, fleshy. **Inflorescence** solitary or few, terminal, nodding flowers. **Flowers** regular, narrowly campanulate, opening to star-shaped in full sun; tepals yellow, rarely white, separate, narrowly lanceolate, 15–40 mm long, reflexed during the day; stamens shorter than tepals; anthers yellow or red; stigma capitate or 3-lobed. **Fruit** an erect, oblong-ovoid, 3-lobed, many-seeded capsule 25–50 mm long. BC, AB south to CA, UT, CO. (p.724)

1. Tepals white, basally yellow . . . . var. *candidum*
1. Tepals yellow . . . . var. *grandiflorum*

Erythronium grandiflorum var. *grandiflorum* occurs in open forest, moist meadows, montane, subalpine; *E. grandiflorum* var. *candidum* (Piper) Abrams is found in open forest; montane; endemic to northwest MT and adjacent ID and northeast WA. There is evidence to consider these taxa separate species.[136] Tepals extend forward over the anthers and stigmas at night and on rainy days, but reflex backward when it's sunny.

Fritillaria L. Fritillary

Glabrous perennials from often clustered bulbs. **Leaves** basal or cauline, alternate, sessile, fleshy. **Inflorescence** terminal, of 1 to 4 nodding, pedicellate flowers. **Flowers** regular, campanulate; tepals separate; stamens shorter than the tepals; ovary superior; stigma unblobed. **Fruit** an erect 3-celled, many-seeded capsule.

1. Tepals yellow; capsule rounded, not winged . . . . *F. pudica*
1. Tepals mottled with purple and brown; capsule winged . . . . *F. atropurpurea*

Fritillaria atropurpurea Nutt. Chocolate lily, leopard lily **Stems** 10–60 cm. **Leaves** numerous, some nearly whorled, linear-lanceolate, 4–12 cm × 2–10 mm. **Flowers** 1 to 4, broadly campanulate; tepals mottled purple, brown, and yellow, lanceolate, 1–2 cm long. **Capsule** erect, obovoid, 6-winged, 15–20 mm long. Grasslands, sagebrush steppe, badlands, drier open forests, thickets; plains, valleys, montane. WA to ND south to CA, AZ, NM, NE.

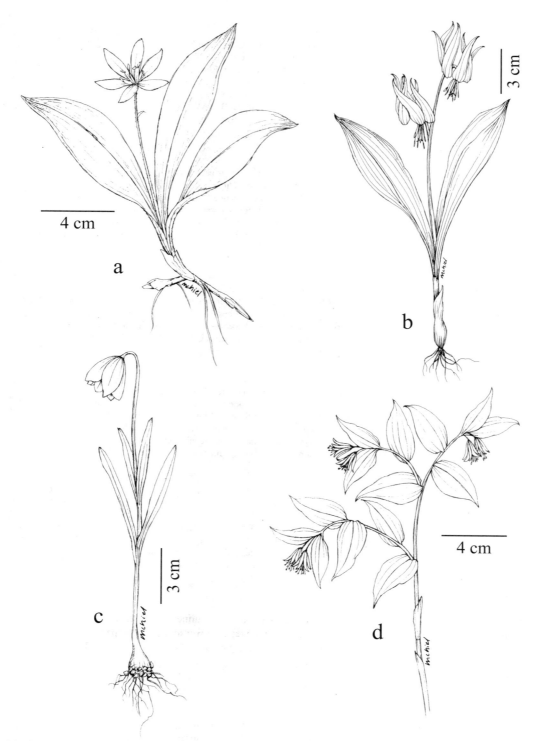

Plate 122. a. *Clintonia uniflora*, b. *Erythronium grandiflorum*, c. *Fritillaria pudica*, d. *Prosartes trachycarpa*

Fritillaria pudica (Pursh) Spreng. Yellow bell **Stems** 8–30 cm. **Leaves** 2–4, linear-oblong, 3–12 cm long, clustered near midstem. **Flowers** solitary, campanulate; tepals yellow, oblong, 1–2 cm long. **Capsule** erect, obovoid, 15–35 mm long, rounded on top. Grasslands, sagebrush steppe, open forest; valleys to montane. BC, AB south to CA, NV, UT, WY. (p.724)

Tepals turn brick-red with age. One collection from the Pryor Mtns. has >4 leaves and tepals 3 cm long.

Leucocrinum Nutt. ex Gray Star lily

Leucocrinum montanum Nutt. ex A.Gray Stemless, rhizomatous, fibrous-rooted perennial. **Leaves** basal, sheathing, linear, 5–15 cm × 2–5 mm, leathery. **Inflorescence** an umbel of several pedicellate flowers; peduncle and pedicels below ground. **Flowers** regular, white, tubular; tube 4–8 cm long; lobes spreading, lanceolate, 1–2 cm long; stamens included, inserted in the tube; ovary superior; style 3-lobed. **Fruit** an obovoid capsule, 5–7 mm long, subterranean. Grasslands; plains, valleys. OR to ND south to CA, AZ, NM, NE.

Lilium L. Lily

Glabrous perennials from scaly bulbs. **Stems** erect, unbranched. **Leaves** cauline, alternate, whorled above, sessile, fleshy. **Inflorescence** a terminal, bracteate raceme, of 1 to few pedicellate flowers; bracts leaf-like. **Flowers** large, regular, campanulate with spreading to reflexed, separate tepals; stamens ca. as long as the tepals; ovary superior; stigma barely lobed. **Fruit** an erect 3-celled capsule; seeds numerous, flattened.

1. Flowers nodding; tepals yellowish, reflexed upward . . . . *L. columbianum*
1. Flowers erect; tepals red-orange, spreading . . . .*L. philadelphicum*

Lilium columbianum Hanson Tiger lily **Stems** 40–100 cm. **Leaves** lanceolate, 3–8 cm long. **Flowers** nodding, 2 to 5; tepals yellowish with black-spotted bases, lanceolate, 35–45 mm long, reflexed; stamens exserted. **Capsule** cylindric, 3–4 cm long. Moist forest openings; valleys. BC to MT south to CA. Known from Lincoln Co.

Lilium philadelphicum L. Philadelphia lily, wood lily **Stems** 20–60 cm. **Leaves** linear-lanceolate, 3–7 cm long. **Flowers** 1 to 3; tepals reddish with yellowish and black-spotted bases, elliptic, long-clawed, 5–7 cm long, spreading; stamens barely included. **Capsule** narrowly ovoid, 2–4 cm long. Moist, usually calcareous meadows, grasslands, fens, woodlands; valleys. BC to QC south to ID, NM, IL, GA.

Lloydia Salisb. ex Rchb. Alp lily

Lloydia serotina (L.) Salisb. ex Rchb. Glabrous perennials from a short, vertical, bulb-like rhizome wrapped in old leaf bases. **Stems** ascending to erect, 4–12 cm. **Leaves** few, alternate, linear, 0.5–1 mm wide. **Inflorescence** usually a solitary, terminal flower. **Flowers** regular, campanulate; tepals spreading, separate, oblanceolate, 6–10 mm long, white with purplish veins; stamens shorter than tepals; style with 3 short lobes. **Fruit** an obovoid capsule 6–8 mm long. Moist turf, meadows, rarely fellfields; upper subalpine, alpine, rarely lower in calcareous soil. AK to NV, UT, NM. Our plants are var. *serotina*.

Maianthemum F.H.Wigg. May lily

Maianthemum canadense Desf. Rhizomatous perennials. **Stems** erect, 5–20 cm. **Leaves** cauline, 2 or 3 (solitary on sterile shoots), short-petiolate; blade cordate, ovate, 4–7 cm long, seemingly clasping the stem. **Inflorescence** an ebracteate raceme with 12 to 25 flowers. **Flowers** regular, star-shaped; tepals 4, white, separate, obovate, ca. 2 mm long, spreading; stamens ca. as long as the tepals; stigma 2-lobed. **Fruit** a globose berry, 4–6 mm long, red, mottled when immature. Moist forest, woodlands; plains. BC to NL south to WY, SD, IA, GA. One collection from Carter Co.

Maianthemum dilatatum (Alph.Wood) A.Nelson & J.F.Macbr. with distinctly petiolate leaves, occurs in north ID and could be found in adjacent MT.

Prosartes D.Don Fairy bell

Rhizomatous perennials. **Stems** branched, puberulent. **Leaves** cauline, sessile, lanceolate to ovate, acuminate. **Inflorescence** terminal, of 1–3 pendent, pedicellate flowers. **Flowers** narrowly campanulate; tepals 6, white, separate; stamens exserted; style slightly 3-lobed. **Fruit** a reddish, few-seeded, ellipsoid berry.

Molecular evidence indicates that the genus *Disporum* is confined to Asia.[365]

1. Top of berry and ovary bumpy; leaf margins glabrous or spreading-hairy . . . . *P. trachycarpa*
1. Top of berry and ovary smooth or hairy; leaf margins with uniformly forward-pointing short hairs. .*P. hookeri*

Prosartes hookeri Torr. [*Disporum hookeri* (Torr.) G.Nicholson] **Stems** 25–90 cm. **Leaves** 5–15 cm long, short-hairy below, nearly glabrous above with uniformly forward-pointing short hairs on the margins. **Flowers:** tepals 8–15 mm long, linear-elliptic; ovary and style base hairy. **Berry** 7–10 mm long, apiculate. Moist coniferous forest; valleys; montane. BC, AB south to CA, ID, MT.

Prosartes trachycarpa S.Watson [*Disporum trachycarpum* (S.Watson) Benth. & Hook.f.] **Stems** 15–60 cm. **Leaves** 3–10 cm long, short-hairy below, glabrous above; the margins glabrous or with spreading hairs. **Flowers:** tepals 8–12 mm long, linear-oblanceolate; ovary papillose. **Berry** 8–14 mm long, rounded on top, bumpy. Moist or dry forest, woodlands, thickets; plains, valleys, montane. BC to ON south to OR, AZ, NM, NE, MN. (p.724)

Smilacina Desf. False Solomon's seal, Solomon's plume

Rhizomatous perennials. **Stems** simple. **Herbage** glabrate to puberulent. **Leaves** cauline, sessile, lanceolate to ovate, acute. **Inflorescence** a terminal raceme or panicle. **Flowers** small, star-shaped; tepals white, narrowly lanceolate, separate, spreading; style slightly 3-lobed. **Fruit** a few-seeded, globose berry.

Smilacina and *Maianthemum* are sometimes united under the latter name,[230] but there is little evidence to support the change.

1. Inflorescence a branched panicle; berries red; tepals 1–3 mm long . . . . *S. racemosa*
1. Inflorescence a narrow raceme; berries green and dark-striped; tepals 3–5 mm long. . . . .*S. stellata*

Smilacina racemosa (L.) Desf. [*Maianthemum racemosum* (L.) Link] **Stems** 20–80 cm. **Leaves** 7–15 cm long. **Inflorescence** a branched, pyramidal panicle with numerous flowers. **Flowers:** tepals 1–1.5 mm long; filaments petal-like, longer than the tepals. **Berry** red, 4–8 mm long. Open forest, thickets, woodlands, avalanche slopes; plains, valleys to lower subalpine. NT to NL south to CA, AZ, NM, LA, MO, GA.

Our plants are sometimes considered a taxon separate from plants in eastern N. America.[230,231]

Smilacina stellata (L.) Desf. [*Maianthemum stellatum* (L.) Link] **Stems** 15–60 cm. **Leaves** 4–12 cm long, often folded. **Inflorescence** a simple, often zig-zag raceme with 3 to 12 flowers. **Flowers:** tepals 3–5 mm long; stamens shorter and narrower than tepals. **Berry** green with dark longitudinal stripes, 5–12 mm long. Grasslands, moist meadows, thickets, moist, often riparian forest, woodlands; plains, valleys, montane. AK to NL south to CA, AZ, NM, OK, VA. (p.727)

Stenanthium (A.Gray) Kunth

Stenanthium occidentale A.Gray Bronze bells Glabrous perennial from an elongate bulb. **Stems** ascending to erect, 15–55 cm. **Leaves** petiolate, basal; blades linear-oblanceolate, 5–25 cm × 2–20 mm. **Inflorescence** a simple or branched, linear-bracteate raceme with 3 to 20 well-separated, nodding flowers. **Flowers** regular, campanulate, yellow-green to purplish; tepals united at the base, 8–15 mm long, linear-lanceolate with spreading, acuminate tips; stamens barely included; ovary superior or slightly inferior; style deeply 3-branched. **Fruit** a lanceoloid, 3-beaked capsule ca. 15 mm long. Moist, often shallow soil of coniferous forest, often along streams, cool shaded slopes, cliffs; valleys to lower alpine. BC, AB south to CA, ID, MT. (p.727)

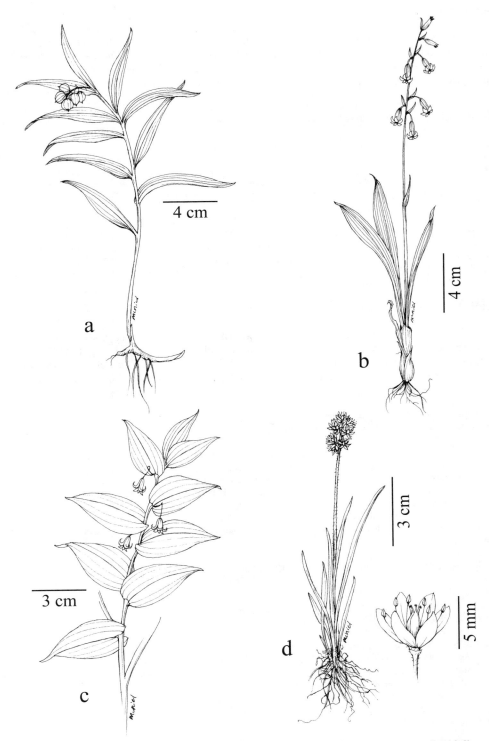

Plate 123. a. *Smilacina stellata*, b. *Stenanthium occidentale*, c. *Streptopus amplexifolius*, d. *Tofieldia glutinosa*

Streptopus Michx. Twisted stalk

Rhizomatous perennials. **Leaves** cauline, alternate, ovate, acuminate. **Inflorescence**: 1(2) pendent, pedunculate flower from leaf axils. **Flowers** campanulate, regular; tepals separate, linear-lanceolate with spreading tips; stamens 6, shorter than tepals. **Fruit** a globose to ellipsoid, yellow or red berry.
Streptopus streptopoides (Ledeb.) Frye & Rigg, with saucer-shaped flowers, occurs in adjacent BC.

1. Stamens unequal; tepals widely spreading; peduncles bent . . . . *S. amplexifolius*
1. Stamens subequal; tepals barely spreading; peduncles curved . . . . *S. lanceolatus*

Streptopus amplexifolius (L.) DC. **Stems** erect, usually branched above, 30–100 cm. **Herbage** glabrate. **Leaves** clasping, 4–10 cm long. **Inflorescence**: solitary pendent flowers on jointed peduncles. **Flowers** yellow-green; tepals 6–12 mm long with spreading to recurved tips; stamens unequal; style barely 3-lobed. **Berry** yellow or red, 8–12 mm long. Moist to wet forest, thickets, especially along streams, lakes; valleys to lower subalpine. Circumboreal south to CA, NM, MN, VA. (p.727)

Streptopus lanceolatus (Aiton) Reveal [*S. roseus* Michx.] **Stems** erect, usually simple, 15–30 cm. **Herbage** sparsely pubescent. **Leaves** not clasping, 5–9 cm long. **Inflorescence**: 1 or 2 pendent flowers on short-hairy pedicels. **Flowers** yellow-green, streaked reddish; tepals 6–10 mm long with barely spreading tips; stamens ca. equal; style 3-branched. **Berry** red, 5–6 mm long. Wet forest, valleys. BC, AB south to OR, MT and ON to NL south to WI, WV. One collection from Flathead Co. (Boivin 13,576, DAO).

Tofieldia Huds. False asphodel

Short-rhizomatous perennials. **Stems** simple, erect. **Leaves** basal and lower-cauline, sessile, linear, glabrous, folded and enfolding each other basally (equitant). **Inflorescence** a terminal, globose, bracteate raceme; each flower with a bract united around the base. **Flowers** small, regular, star-shaped; tepals white, linear-oblong, separate, spreading; stamens ca. as long as the tepals; styles 3, spreading. **Fruit** a many-seeded, globose, 3-beaked capsule.

1. Stems glandular, >10 cm tall; capsules 5–8 mm long . . . . *T. glutinosa*
1. Stems glabrous, <10 cm tall; capsules 2–3 mm long . . . . *T. pusilla*

Tofieldia glutinosa (Michx.) Pers. [*T. occidentalis* S.Watson, *Triantha occidentalis* (S.Watson) R.R.Gates ssp. *montana* (C.L.Hitchc.) Packer] **Stems** stipitate-glandular, 6–45 cm. **Leaves** 3–25 cm × 2–6 mm. **Flowers** in groups of 3; tepals 4–7 mm long. **Capsule** 5–8 mm long. Wet meadows, fens, especially around lakes, streams, wet rock ledges; subalpine, alpine, uncommon lower. AK to CA, ID, WY. Our plants are ssp. *montana* C.L.Hitchc. (p.727)

Some treatments distinguish our plants as a separate species, *Tofieldia occidentalis*, but Hitchcock (1944) presented a convincing argument against this split. Packer[314] places these plants in *Triantha* but provides no evidence for the recircumscription.

Tofieldia pusilla (Michx.) Pers. **Stems** glabrous, 3–12 cm. **Leaves** 15–25 × 1–2 mm. **Flowers**: tepals 1–1.5 mm long. **Capsule** 2–3 mm long. Permanently moist, often shallow soil of turf; alpine. Circumpolar south to BC and Glacier Co., MT.

Trillium L. Wake-robin

Trillium ovatum Pursh Short-rhizomatous, fibrous-rooted, glabrous perennials. **Stems** erect, 10–40 cm, naked below. **Leaves** (bracts) 3, sessile, whorled at the stem tip, 3–13 cm long, broadly ovate, acuminate. **Inflorescence** a solitary, terminal flower; pedicel 15–60 mm long. **Flower** regular, star-shaped; sepals green, narrowly lanceolate, 15–35 mm long; petals white, fading pink, separate, 25–40 mm; stamens shorter than petals; anthers yellow; style 3-lobed. **Fruit** a broadly ovoid, 3-winged, many-seeded capsule, 12–25 mm long. Moist to wet forest; valleys, lower montane. BC, AB south to CA, ID, CO. (p.730)

Triteleia Douglas ex Lindl.

Triteleia grandiflora Lindl. [*Brodiaea douglasii* S.Watson] WILD HYACINTH Scapose, glabrous
perennials from fibrous-coated corms. **Scapes** erect, 30–70 cm. **Leaves** 1 or 2, basal, linear, 2–15
mm wide. **Inflorescence** a terminal umbel subtended by few scarious, lanceolate bracts; pedicels
8–25 mm long. **Flowers** funnelform, blue; tepals united below into a tube 9–15 mm long, the mouth
nearly closed; lobes spreading 7–10 mm long; stamens attached to corolla at 2 levels; anthers blue;
ovary stipitate; stigmas 3. **Fruit** an ovoid capsule 6–12 mm long. Grasslands, open forest; valleys, lower montane. BC,
MT south to CA, UT, WY.

Veratrum L. False hellebore, corn lily

Rhizomatous perennials. **Stems** erect, unbranched, hollow. **Leaves** cauline, sessile, ovate to elliptic,
pleated between veins, twisted. **Herbage** sparsely tomentose above. **Inflorescence** terminal, bracteate
panicles or compound racemes. **Flowers** short-pedicellate, regular, saucer-shaped; the upper bisexual; the
lower functionally male; tepals separate, fused to the ovary at the base (hypanthium); stamens included;
ovary 3-celled, slightly inferior; styles 3. **Fruit** a glabrous, ovoid, 3-lobed capsule.

1. Tepals green, 5–9 mm; lower panicle branches spreading to drooping at anthesis . . . . *V. viride*
1. Tepals white, 10–17 mm; panicle branches ascending . . . . *V. californicum*

Veratrum californicum Durand **Stems** 1.5–2 m. **Leaves** 15–25 cm long, reduced upward. **Inflorescence** with
numerous flowers in ascending branches. **Flowers** off-white; tepals elliptic, 10–17 mm long with dark green glands at
the base. **Capsule** glabrous, 2–3 cm long. Wet riparian meadows, thickets; montane, lower subalpine in Granite Co. WA
to MT south to CA, AZ, NM.

Veratrum viride Aiton **Stems** 1–2 m. **Leaves** 12–35 cm long, reduced upward. **Inflorescence**
with numerous flowers in spreading to drooping branches at anthesis. **Flowers** greenish;
tepals oblanceolate, 5–9 mm long, tomentose on the outside, short-fringed at the base. **Capsule**
glabrous,1–2 cm long. Moist forest openings, meadows, avalanche slopes; montane, subalpine. AK
to CA, ID, MT and NL to GA. Our plants are var. *eschscholtzianum* (Roem. & Schult.) Breitung.
(p.730)

Xerophyllum Michx.

Xerophyllum tenax (Pursh) Nutt. BEARGRASS Glabrate, rhizomatous perennial. **Stems** leafy,
erect, 50–120 cm. **Leaves**: basal, linear, wiry, scabrous on the margins, 15–60 cm × 2–4 mm,
forming large tussocks; cauline leaves alternate, similar. **Inflorescence** a terminal raceme,
expanding from hemispheric to linear-oblong and up to 40 cm long; pedicels ascending, 2–4 cm
long. **Flowers** regular, star-shaped; tepals white, separate, narrowly oblong, 5–10 mm, spreading;
stamens longer than tepals; styles 3. **Fruit** a globose, 3-lobed, several-seeded capsule, 5–8 mm long. Coniferous forest,
often where fire has reduced tree cover, avalanche slopes, meadows; montane, subalpine. BC, AB south to CA, ID, MT.
(p.730)
  Plants often have several rosettes; well known for periodic mass flowering.

Zigadenus Michx. Death camas

Mostly glabrous perennials from bulbs. **Stems** erect, unbranched. **Leaves** mainly basal, linear, grass-like,
reduced on the stem. **Inflorescence** a terminal, bracteate raceme or panicle; bracts small, linear, scarious
upward. **Flowers** regular, rotate, greenish- or yellowish-white; tepals obovate, sometime clawed with a
yellow or green basal gland; stamens as long or longer than the tepals; ovary superior to partly inferior;
styles 3. **Fruit** an erect 3-celled, 3-beaked, ovoid, many-seeded capsule.
  The bulbs, capsules and seeds of these plants are poisonous, although the leaves may be grazed by elk.
Although *Zigadenus paniculatus* is reported for MT,[113,187,353] the distinction between it and *Z. venenosus* is
not obvious here. There are a few collections with 2 to 4 inflorescence branches with male flowers that are
herein treated as *Z. paniculatus*, but they are mixed with typical *Z. venenosus* and display no ecological or
geographic distinction. It seems to me that these specimens could be accommodated in *Z. venenosus*.

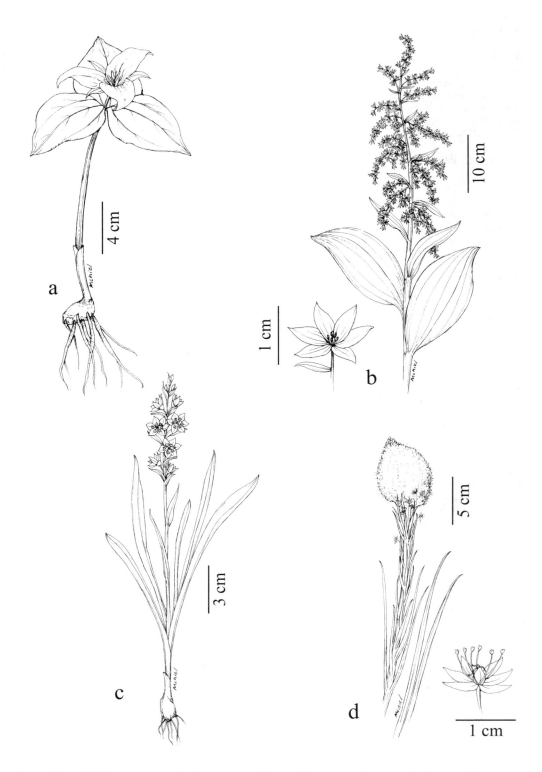

Plate 124 a. *Trillium ovatum*, b. *Veratrum viride*, c. *Zigadenus elegans*, d. *Xerophyllum tenax*

1. Tepals 7–10 mm long . . . . *Z. elegans*
1. Tepals 3–6 mm long 2

2. Inflorescence paniculate with several branches; flowers of lowest branches male . . . . *Z. paniculatus*
2. Inflorescence a raceme with 0 to 2 branches; lowest flowers perfect . . . .*Z. venenosus*

Zigadenus elegans Pursh **Stems** 10–60 cm. **Leaves** glaucous, 8–30 cm × 2–10 mm.
Inflorescence a raceme or branched panicle; flowers well-separated below; pedicels 5–40 mm
long. **Flowers**: tepals 6–11 mm long, the inner longer with a claw 0.5–1 mm long; stamens as long
as the tepals. **Capsule** 12–20 mm long. Moist meadows, open forest, often along streams, stony,
calcareous soil of exposed slopes, ridges; montane to lower alpine. AK to MB south to OR, AZ, NM,
Mexico. (p.730)

    One of few plants able to thrive in both wet meadows and limestone fellfields where it generally has fewer flowers
and shorter leaves. The type locality is northeast of Lincoln.

Zigadenus paniculatus (Nutt.) S.Watson **Stems** 25–50 cm. **Leaves** 5–25 cm × 2–6 mm.
Inflorescence a crowded panicle; pedicels 5–15 mm long; some flowers on lowest branches
functionally male. **Flowers**: tepals 4–6 mm long; the inner longer; stamens longer than the tepals.
Capsule 10–18 mm long. Grasslands; valleys to subalpine. WA to MT south to CA, NM. See genus
discussion.

Zigadenus venenosus S.Watson **Stems** 15–40 cm. **Leaves** 10–25 cm × 2–4 mm.
Inflorescence a crowded, puberulent raceme; pedicels 5–15 mm long. **Flowers**: tepals 3–5.5
mm long; the inner longer; stamens longer than the tepals. **Capsule** 8–13 mm long. Grassland,
sagebrush steppe, dry meadows, rock outcrops; plains, valleys to lower subalpine. BC to SK south
to CA, NV, UT, CO, NE.

IRIDACEAE: Iris Family

Rhizomatous perennial herbs. **Stems** simple. **Leaves** equitant (folded in half lengthwise and enfolding next
higher leaf at the base). **Inflorescence** a bracteate raceme or cyme; bracts (spathes) leaf-like, enfolding the
stem. **Flowers** perfect, regular, pedicellate; perianth of 6 similar tepals or 3 sepals and 3 petals; stamens 3,
united below; ovary inferior; style 3-branched. **Fruit** a 3-celled, many-seeded capsule.

1. Leaves ≥5 mm wide; flowers ca. 6 cm long; sepals and petals unlike . . . . *Iris*
1. Leaves <5 mm wide; flowers ca. 1 cm long; sepals and petals similar . . . . *Sisyrinchium*

Iris L. Iris, flag

Glabrous perennials. **Stems** erect. **Leaves** equitant, basal or cauline, linear-lanceolate. **Inflorescence** a few-
flowered raceme; each pedicel subtended by paired bracts (spathes). **Flowers** erect; perianth united at the
base; sepals petal-like, spreading, oblong with dark veins; petals smaller, erect; style branches petal-like,
2-lobed, spreading over the sepals; stigma on the underside of the style branch; stamens between the styles
and sepals. **Capsule** with 2 columns of seeds in each of the 3 cells.

1. Sepals predominantly yellow . . . .*I. pseudacorus*
1. Sepals mainly blue and white. . . . . *I. missouriensis*

Iris missouriensis Nutt. Blue flag **Stems** 10–50 cm. **Leaves** basal, 10–40 cm × 3–8 mm wide.
Inflorescence of 1 to 3 flowers; spathes inflated, united at the base, sometimes membranous-
margined, 4–8 cm long; pedicels 2–5 cm long. **Flowers**: sepals 4–6 cm long, whitish with a yellow
midstripe and blue veins and margins; petals blue, 3–4 cm long; lobes of style branch white, 3–6
mm long. **Capsule** ellipsoid, 2–4 cm long. Moist to wet meadows, thickets, woodland margins; plains,
valleys, montane. BC, AB south to CA, AZ, NM, NE. (p.733)

Iris pseudacorus L. Yellow flag **Stems** clumped, 35–90 cm. **Leaves** basal and cauline, 5–60 cm × 6–15 mm.
Inflorescence of 3 to 10 flowers, often branched above; spathes 4–7 cm long, the outer keeled; pedicels 2–4 cm long.
Flowers: sepals 5–6 cm long, yellow with brown penciling; petals yellow, ca. 3 cm long; lobes of style branches yellow,
fringed, ca. 5 mm long. **Capsule** obovoid, 3-angled, 3–7 cm long. Marshes, wet meadows, irrigation ditches; valleys.
Native to Europe; introduced throughout temperate N. America; collected in Lake and Sanders cos.

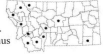

Sisyrinchium L. Blue-eyed grass

Glabrous, short-rhizomatous perennials. **Stems** flattened, scape-like. **Leaves** equitant, basal and lower-cauline, linear. **Inflorescence** umbel-like, subtended by paired bracts (spathes); pedicels bracteate at the base. **Flowers** rotate, ascending to spreading; perianth barely united at the base; sepals and petals (tepals) similar, spreading, narrowly elliptic, awned; ovary stipitate-glandular; style branches slender, erect. **Capsule** globose to ovoid, 3-celled.

1. Flowers white or slightly blue-tinged. . . . . *S. septentrionale*
1. Flowers blue. 2

2. Outer spathe more than twice as long as the inner . . . . *S. montanum*
2. Outer spathe ca. twice as long as the inner . . . . *S. idahoense*

Sisyrinchium idahoense E.P.Bicknell **Stems** clumped, 10–30 cm. **Leaves** 1–2 mm wide. **Inflorescence** of 1 to 4 flowers; inner spathe 12–25 mm long; outer spathe 25–45 mm long. **Flowers**: tepals 8–13 mm long, blue with yellow bases. **Capsule** globose 3–6 mm high. Moist to wet, often saline or alkaline meadows; valleys, montane. BC, MT south to CA, AZ, NM.

Two forms are sometimes recognized, var. *idahoense* and var. *occidentale* (E.P.Bicknell) Douglass M.Hend., based on continuous characters of the spathe; however, there is no geographic or ecological distinction between the two in MT.

Sisyrinchium montanum Greene [*S. angustifolium* Mill. misapplied] **Stems** clumped, 12–40 cm. **Leaves** 1–3 mm wide. **Inflorescence** of 2 to 4 flowers; inner spathe 22–28 mm long; outer spathe 3–7 cm long. **Flowers**: tepals 7–11 mm long, blue with yellow bases. **Capsule** globose, 4–6 mm high. Moist grasslands, moist to wet meadows; plains, valleys, montane. YT to NL south to NM, OK, IN, NY. Our plants are var. *montanum*. (p.733)

Sisyrinchium septentrionale Bicknell **Stems** 10–40 cm. **Leaves** 1–2 mm wide. **Inflorescence** of 2 to 4 flowers; inner spathe 1–4 cm long; outer spathe 2–6 cm long. **Flowers**: tepals 8–9 mm long, blue with yellow bases. **Capsule** globose, 3–5 mm high. Moist meadows, often adjacent to wetlands; plains. BC to SK south to WA, MT. Known from Sheridan Co.

AGAVACEAE: Century-plant Family

Yucca L.

Yucca glauca Nutt. Semi-woody, acaulescent perennials, sometimes with multiple rosettes. **Leaves** simple, linear-lanceolate, spine-tipped, concave in cross-section, 15–50 cm × 3–6 mm, glaucous; margins white with curled, spreading fibers. **Scape** 20–40 cm. **Inflorescence** a raceme, 40–100 cm long with leaf-like bracts. **Flowers** perfect, regular, pendent, broadly overlapping; tepals 6, separate, elliptic, greenish-white, 30–45 mm long; stamens 6, included; ovary superior; stigma 3-lobed. **Fruit** a 6-celled obovoid capsule, 3–6 cm long; seeds numerous, black, flattened. Stony or sandy soil of grasslands, badlands; plains, valleys. AB south to NM, TX, KS.

SMILACACEAE: Catbrier Family

Smilax L. Greenbrier, catbrier

Smilax herbacea L. [*S. lasioneura* Hook.] CARRION FLOWER Dioecious, rhizomatous, perennial, herbaceous vine. **Stems** glabrate, ascending or climbing, 30–120 cm; tendrils from the leaf axils. **Leaves** simple, petiolate, alternate; blades ovate, cordate to truncate, 5–9 cm long, net-veined, glaucous and puberulent beneath. **Inflorescence** axillary umbels with numerous flowers; peduncles 3–8 cm. **Flowers** unisexual, star-shaped, green; tepals 6, separate, spreading, oblong, 3–4 mm long; stamens 6, shorter than the tepals, reduced in female flowers; ovary 3-celled, superior; stigmas 3. Fruit a 3- or 6-seeded, blue berry 6–10 mm long. Woodlands, riparian forests, thickets; plains. SK to ON south to WY, TX, MS, FL. Our plants are var. *lasioneura* (Hook.) A.DC.

One collection from Carter Co. lacks tendrils.

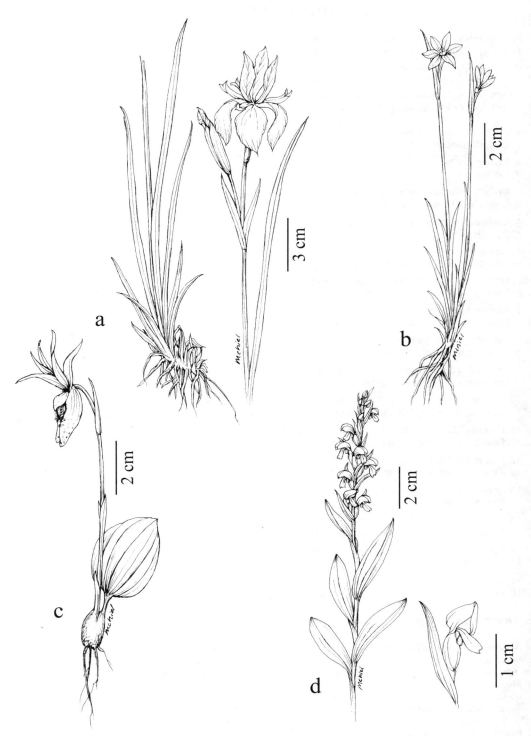

Plate 125. a. *Iris missouriensis*, b. *Sisyrinchium montanum* c. *Calypso bulbosa*,
d. *Coeloglossum viride*

ORCHIDACEAE: Orchid Family

Herbaceous perennials often with fleshy roots. **Leaves** simple, entire, often sheathing, basal or alternate. **Inflorescence** a spike, raceme, panicle, or flowers solitary. **Flowers** perfect, irregular; perianth segments 6, unlike, in 2 whorls; the lower petal (lip) often enlarged and modified to form a sac or spur; fertile stamens 1 or 2, united with the style, forming a column; ovary inferior, 3-celled. **Fruit** a capsule with numerous minute seeds, opening by longitudinal slits.[260,345]

Species of *Coeloglossum*, *Piperia* and *Platanthera* were formerly placed in *Habenaria*.

1. Green leaves all basal or absent entirely 2
1. Green leaves on at least the upper 2/3 of the stem 9

2. Flowers solitary. . . . . *Calypso*
2. Flowers >1 3

3. Green leaves completely absent; stems yellow to purple. . . . . *Corallorhiza*
3. Green basal leaves present 4

4. Basal leaf solitary (rarely 2) 5
4. Basal leaves more than 1. 6

5. Lip lobed and spotted. . . . . *Amerorchis*
5. Lip with entire margins, not spotted . . . . *Platanthera*

6. Upper stem and inflorescence glandular-hairy. . . . . *Goodyera*
6. Stem and inflorescence glabrous. 7

7. Flowers lacking a spur from the lip petal . . . . *Liparis*
7. Flowers with a backward- or downward-pointing spur 8

8. Leaves opposite, nearly orbicular, appressed to the ground . . . . *Platanthera*
8. Leaves not opposite, not strictly appressed to the ground, narrowly elliptic. . . . . *Piperia*

9. Leaves 2, opposite each other at mid-stem 10
9. Leaves usually more than 2, alternate 11

10. Leaves >7 cm long; sepals purplish . . . . *Cypripedium*
10. Leaves <7 cm long; sepals green or whitish . . . . *Listera*

11. Flowers 1 to 3; lip petal >12 mm long, pouch-shaped . . . . *Cypripedium*
11. Flowers mostly more than 3; lip petal usually <12 mm long. 12

12. Flowers purplish . . . . *Epipactis*
12. Flowers white or greenish-white. 13

13. Upper stem glandular-hairy . . . . *Goodyera*
13. Stems glabrous. 14

14. Flowers spirally-arranged in the spike . . . . *Spiranthes*
14. Flowers in a spike-like inflorescence but not spirally-arranged 15

15. Lip petal 3-lobed at the tip (central lobe small). . . . . *Coeloglossum*
15. Lip rounded on the tip. 16

16. Leaves on lower third of stem only. . . . . *Piperia*
16. Leaves present above mid-stem . . . . *Platanthera*

Amerorchis Hultén

Amerorchis rotundifolia (Banks ex Pursh) Hultén [*Orchis rotundifolia* Banks ex Pursh]
Stems scapose, 8–25 cm. **Leaf** basal, solitary, sessile, glabrous, ovate to elliptic, 3–7 cm long. **Inflorescence** a terminal spike of 2 to 8 flowers; each subtended by a small, leaf-like bract. **Flowers**: sepals narrowly ovate, 6–9 mm long, the lower 2 spreading; upper petals narrowly oblong, white to rose, 6–8 mm long, forming a hood with the upper sepal; lip petal 6–10 mm long, purple-spotted, oblong, 3-lobed; the terminal lobe wedge-shaped, emarginate with a curved, tubular, basal spur nearly as long as the lip. **Capsule** ellipsoid, 10–14 mm long. Wet spruce forest, often near calcareous seeps, fens; montane. AK to Greenland south to WY, MI, NY.

Calypso Salisb. Fairy slipper

Calypso bulbosa (L.) Oakes **Stems** 5–20 cm, scapose with 2 sheathing bracts. **Leaf** basal, solitary, petiolate; blade glabrous, ovate, 3–6 cm long. **Inflorescence** a solitary flower subtended by a purplish, linear bract, 8–20 mm long. **Flowers**: sepals purple, lanceolate, 14–22 mm long, erect; upper 2 petals similar; lip petal forming a white- and purple-spotted pouch, ca. 2 cm long, open at the top, yellow within; margin of the orifice (lamina) bearded with paired horns near the front; column obovate, yellowish, hood-like over the lip. **Capsule** lanceoloid, 20–35 mm long. Dry to wet coniferous forest; valleys, montane. Circumboreal south to CA, AZ, CO, MI, NY. (p.733)

1. Hairs on lip yellow; lamina white, sometimes with a few small spots . . . . var. *americana*
1. Hairs on lip white; lamina purple-mottled . . . . var. *occidentalis*

Calypso bulbosa var. *occidentalis* (Holz.) B.Boivin is confined to west-central counties; *C. bulbosa* var. *americana* (R.Br.) Luer is more widespread; intermediates occur near Missoula.

Coeloglossum Hartm.

Coeloglossum viride (L.) Hartm. [*Habenaria viridis* (L.) R.Br., *Dactylorhiza viridis* (L.) R.M.Bateman., Pridgeon & M.W.Chase] **Stems** 10–35 cm, leafy. **Leaves** shiny green, cauline, 4–10 cm long, oblong to oblanceolate with sheathing bases; basal leaves bladeless. **Inflorescence** a raceme of 8 to 40 flowers, each subtended by a leaf-like bract, 5–55 mm long. **Flowers** green; sepals broadly obovate, 4–7 mm long, forming a hood over the column; upper petals beneath the hood, ca. 4 mm long; lip petal 6–11 mm long, strap-shaped, 3-lobed at the tip; the base with a pair of small lateral lobes and forming a sac-like spur behind. **Capsule** ellipsoid, 7–12 mm long. Moist grasslands, open forest, woodlands; plains, valleys, montane. Our plants are var. *virescens* (Muhl.) Luer. Circumboreal south to WA, NM, IL, NC. (p.733)

Plants from western MT favor calcareous soil; those from the Great Plains tend to have more deeply lobed lips.

Corallorhiza Gagnebin Coral root

Plants lacking chlorophyll. **Stems** purple or yellow, clustered on a coral-like rhizome. **Leaves** bladeless sheaths. **Inflorescence** a narrow, terminal, several-flowered, minutely-bracteate raceme. **Flowers**: sepals nearly equal, yellow to purplish, cupped forward; upper petals similar to the sepals but shorter; lip petal oblong to obovate, often laterally lobed. **Capsules** ellipsoid, 3-ribbed, pendent.

The sole source of nutrition for these orchids is soil-inhabiting fungi that invade the coral-like rhizome and are digested; MT has five of the seven N. American species.

1. Stems yellow 2
1. Stems red to purple 4

2. Lip petal with a fringed margin, lacking small basal lobes; sepals purplish . . . . *C. wisteriana*
2. Lip petal with undulating margin and small basal lobes; sepals pale 3

3. Flowers 4 to 12; upper petals 4–7 mm long; capsule ca. 10 mm long . . . . *C. trifida*
3. Flowers 10 to 30; upper petals 6–10 mm long; capsule 15–20 mm long . . . . *C. maculata*

4. Petals striped; lip petal without small lobes at the base . . . . *C. striata*
4. Petals spotted or clear; lip sometimes with tiny basal lobes 5

5. Lip petal pinkish to purplish . . . . *C. mertensiana*
5. Lip petal white often with small purple spots 6

6. Lip petal with small, paired lobes at the base . . . . *C. maculata*
6. Lip petal entire or fringed, not lobed . . . . *C. wisteriana*

Corallorhiza maculata (Raf.) Raf. Spotted coral root **Stems** purple, occasionally yellowish, 15–50 cm. **Inflorescence** with 10 to 30 flowers; bracts 1–5 mm long. **Flowers**: sepals and upper petals 6–10 mm long, purple (yellowish), somewhat spreading; lip petal 4–9 mm long, white with purple spots and a pair of small lanceolate basal lobes. **Capsule** 15–22 mm long. Moist to wet coniferous forest; valleys, montane. BC to NL south to CA, AZ, NM, IN, SC. (p.738)

This species parasitizes root rot fungi. A yellow form of this species is sometimes mistaken for *Corallorhiza trifida*. Plants with unspotted lips occur on the east shore of Flathead Lake.

1. Floral bracts mostly ≤1 mm long; lip not expanded toward the tip . . . . var. *maculata*
1. Floral bracts mostly ≥1.5 mm long; lip with dilated tip . . . . var. *occidentalis*

Corallorhiza maculata var. *maculata* and *C. maculata* var. *occidentalis* (Lindl.) Ames have similar distributions and ecology in MT.

Corallorhiza mertensiana Bong. **Stems** reddish, 15–40 cm. **Inflorescence** with 10 to 30 flowers; bracts 1–3 mm long. **Flowers** pink; sepals 6–10 mm long, forming a small sac-like spur at the base; upper petals clustered with the upper sepal, arching over the column; lip petal 6–9 mm long, with a pair of small lanceolate basal lobes. **Capsule** 10–25 mm long. Moist, coniferous forest, often spruce/fir/lodgepole; valleys, montane. AK to CA, ID, WY.

Corallorhiza striata Lindl. STRIPED CORAL ROOT **Stems** purple, 10–45 cm. **Inflorescence** with 7 to 25 flowers; bracts 2.5–5 mm long. **Flowers**: sepals and upper petals 9–16 mm long, cupped forward, white with purple longitudinal stripes; lip petal 8–12 mm long, white, purple-stripped, unlobed, with 2 bumps on the inner surface at the base. **Capsule** 12–25 mm long. Coniferous forest, deciduous woodlands; plains, valleys, montane. BC to NL south to CA, AZ, NM, TX, Mexico. Our plants are var. *striata*.

Corallorhiza trifida Châtel YELLOW CORAL ROOT **Stems** pale yellow, 5–25 cm. **Inflorescence** with 4 to 12 flowers; bracts 0.5–1.5 mm long. **Flowers**: sepals and upper petals 4–7 mm long, cupped forward, yellow to white; lip petal 3–5 mm long, white with obscure basal lobes. **Capsule** 6–11 mm long. Moist forest, thickets, under shrubs in fens, often with moss or along streams; valleys, montane. Circumboreal south to CA, NV, ID, UT, NM, IL, WV.
    Corallorhiza wisteriana has purple and white flowers. See *C. maculata*.

Corallorhiza wisteriana Conrad **Stems** yellow to purplish, 10–30 cm. **Inflorescence** with 5 to 20 flowers; bracts 0.5–4 mm long. **Flowers**: sepals slender, purplish, 6–9 mm long; upper petals clustered with the upper sepal, arching over the column; lip petal 5–7 mm long, unlobed, with a fringed margin and sometimes a few tiny spots. **Capsule** 8–10 mm long. Moist to dry coniferous forest, often under Douglas fir on calcareous soils; valleys, montane. ID to NY south to AZ, NM, TX, AL, FL, Mexico. See *C. trifida*.

Cypripedium L. Lady's slipper

Leaves cauline, lanceolate to elliptic, pleated, sheathing the stem, the lowest bladeless. **Herbage** sparsely to densely glandular-pubescent. **Inflorescence** a terminal raceme of 1 to few flowers, each subtended by a leaf-like bract. **Flowers** large; sepals and upper petals similar, spreading, linear to lanceolate, dark-colored; lower sepals united under the lip (except *C. passerinum*); lip petal pouch-like; style column arched over the top of the lip opening. **Fruit** an ellipsoid capsule.

1. Leaves 2, opposite . . . . *C. fasciculatum*
1. Leaves usually >2, alternate 2

2. Lip petal yellow, sometimes with purple spots at the base . . . .*C. calceolus*
2. Lip petal white, sometimes tinged purple 3

3. Sepals twisted, >20 mm long . . . .*C. montanum*
3. Sepals not twisted, 10–15 mm long . . . .*C. passerinum*

Cypripedium calceolus [*C. parviflorum* Salisb., *C. pubescens* Willd., *C. calceolus* L. var. *parviflorum* (Salisb.) Fernald misapplied] YELLOW LADY'S SLIPPER **Stems** 12–45 cm. **Leaves** alternate, broadly lanceolate to elliptic, 5–17 cm long. **Inflorescence** of 1 or 2 flowers; bracts 3–12 cm long. **Flowers**: sepals lanceolate 25–35 mm long, purple-striped; upper petals linear, twisted, 3–5 cm long; lip petal yellow, often with small reddish spots near the base, 20–35 mm long. **Capsule** ca. 2 cm long. Moist to wet, often calcareous meadows, fens; plains, valley, montane. AK to NL south to NV, AZ, NM, MO, GA; Europe. Our plants are var. *pubescens* (Willd.) Correll.
    North American plants have been segregated as a separate species based on a smaller lip petal, but the taxon is highly variable phenotypically.[364] I prefer to emphasize the similarities with European material.

Cypripedium fasciculatum Kellogg ex S.Watson CLUSTERED LADY'S SLIPPER **Stems** ascending to erect, densely pubescent, 15–35 cm, lengthening in fruit. **Leaves** 2, broadly elliptic, 7–12 cm long, opposite on upper half of stem. **Inflorescence** of 1 to 5 flowers, nodding, erect in fruit; bracts 15–35 mm long. **Flowers**: sepals lanceolate 1–2 cm long, purplish; lip petal dull yellow to purplish, 8–12 mm long. **Capsule** 1–2 cm long. Moist to dry forest, thickets; montane; Lake and Mineral cos. WA to MT south to CA, UT, CO.

Cypripedium montanum Douglas ex Lindl. WHITE LADY'S-SLIPPER **Stems** 20–50 cm. **Leaves** alternate, lanceolate to elliptic, 5–15 cm long. **Inflorescence** of 1 to 3 (mostly 2) flowers; bracts 3–12 cm long. **Flowers**: sepals lanceolate, 3–5 cm long, brownish-purple; upper petals linear, 4–6 cm long, twisted; lip petal white, 2–3 cm long. **Capsule** 2–3 cm long. Moist to dry forest, usually in deep, moist soil but sometimes shallow, stony calcareous soil; montane. AK to CA, OR, ID, WY. (p.738)

Cypripedium passerinum Richardson **Stems** 20–40 cm. **Leaves** alternate, lanceolate to narrowly elliptic, 8–18 cm long. **Inflorescence** of usually 1 flower; bracts 25–50 mm long. **Flowers**: upper sepals ovate, 15–20 mm long, green; lower separate, 10–13 mm long; upper petals white, 12–15 mm long; lip petal white with purple mottling, 12–18 mm long. **Capsule** ca. 2 cm long. Wet, calcareous, organic soil of coniferous forests, seeps, often beneath spruce at the forest-fen ecotone; montane. AK to QC south to MT.

Epipactis Zinn Helleborine

Epipactis gigantea Douglas ex Hook. GIANT HELLEBORINE Rhizomatous. **Stems** erect, 30–80 cm. **Leaves** cauline, lanceolate, pleated, sheathing the stem, bladeless below, narrower above, 6–20 cm long, glabrate. **Inflorescence** a terminal raceme of 4 to 10 flowers subtended by narrowly lanceolate, leaf-like bracts 1–12 cm long. **Flowers**: sepals lanceolate, spreading, green with brown veins, 1–2 cm long; upper petals smaller than uppermost sepal and just beneath it, ovate, acute, pinkish, 8–13 mm long; lip petal ovate, sac-like at the base, trough-like in front, 3-lobed, green and reddish with reddish veins, 12–20 mm long with a red callous in the bottom of the base; style column short, just beneath the upper petals. **Fruit** an obovoid to ellipsoid capsule 2–3 cm long, glabrate. Saturated, calcareous soil of often warm seeps and springs where the ground doesn't freeze hard; valleys. BC south to CA, AZ, NM, TX, Mexico.

Epipactis helleborine (L.) Crantz, a garden escape with smaller flowers than *E. gigantea*, is reported for Lewis and Clark Co.[187] but does not appear to be naturalized.

Goodyera R. Br. Rattlesnake plantain, rattlesnake orchid

Rhizomatous. **Stems** glandular-pubescent. **Leaves** glabrate, basal and lower-cauline, petiolate; blades lanceolate to ovate, reduced to bracts above. **Inflorescence** a terminal, bracteate spike of several to numerous flowers. **Flowers** glandular; sepals greenish, narrowly ovate, separate, equal; upper petals white, similar to and forming a hood with the uppermost sepal; lip petal sac-like at the base with a tongue-like tip, hidden by the lateral sepals; style column short. **Fruit** a glandular-pubescent, erect capsule.

1. Inflorescence ≥6 cm long; hood ≥5 mm long; leaf midribs white . . . . *G. oblongifolia*
1. Inflorescence 3–6 cm long; hood <4 mm long; midrib not white . . . . *G. repens*

Goodyera oblongifolia Raf. **Stems** 15–40 cm. **Leaf blades** elliptic, 3–7 cm long, usually white-variegated with a white midvein. **Inflorescence** 5–13 cm long, spiraled or 1-sided; bracts 3–10 mm long. **Flowers**: lateral sepals 4–7 mm long; hood 5–8 mm long; lip petal 4–6 mm long. **Capsule** 7–10 mm long. Moist to wet coniferous forest; valleys to lower subalpine. BC to NL south to CA, AZ, NM.

Only a small fraction of the plants bloom in any given year.

Goodyera repens (L.) R.Br. **Stems** 10–20 cm. **Leaf blades** ovate, 15–30 mm long, lacking a white midvein. **Inflorescence** 3–6 cm long, 1-sided; bracts 3–10 mm long. **Flowers**: lateral sepals 3–4 mm long; hood 3–5 mm long; lip petal 2–4 mm long. **Capsule** ca. 5 mm long. Open, drier, coniferous forest; montane. Circumboreal south to AZ, NM, SD, TN, NC. Known from Flathead and Judith Basin cos. (p.738)

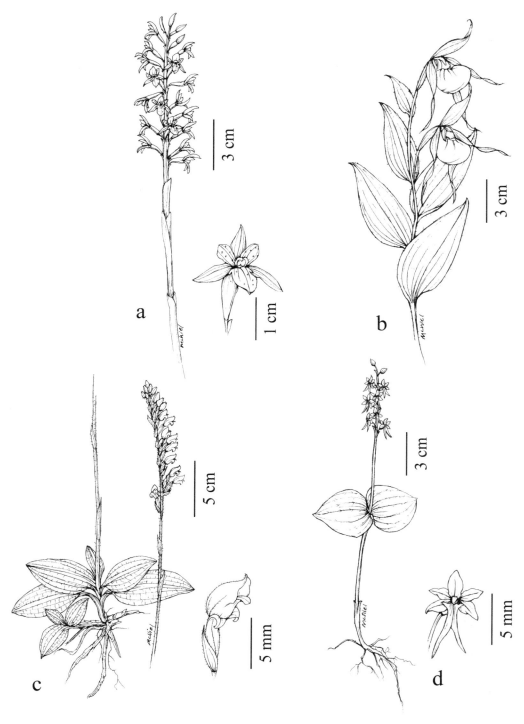

Plate 126. a. *Corallorhiza maculata*, b. *Cypripedium montanum*, c. *Goodyera repens*, d. *Listera cordata*

Liparis Rich. Twayblade

Liparis loeselii (L.) Rich. FEN ORCHID **Stems** 7–20 cm, scapose, bulbous-based. **Leaves** 2, basal, petiolate, erect; blade glabrous, lanceolate, 4–10 cm long. **Inflorescence** a terminal, bracteate raceme of few, barely overlapping flowers; bracts 2–4 mm long. **Flowers** yellowish-green; sepals linear-lanceolate, 5–7 mm long, spreading; upper 2 petals shorter, narrower; lip petal oblong, 3–5 mm long, pendent; column broad at the base. **Capsule** pedicellate, narrowly ellipsoid, 10–14 mm long. Calcareous fens; montane in Lake Co. BC to QC south to WA, MT, LA, VA.

Listera R.Br. Twayblade

Fibrous-rooted. **Leaves** 2, opposite near mid-stem, sessile, glabrous. **Inflorescence** a terminal bracteate raceme of few to several, little-overlapping flowers. **Flowers** greenish; sepals and upper petals similar, linear to lanceolate, spreading to erect; lip petal pendent to nearly horizontal, oblong to obovate, often notched or lobed at the tip. **Fruit** an ascending to spreading capsule.

1. Lip petal divided at least halfway into pointed, divergent lobes . . . .*L. cordata*
1. Lip shallowly lobed or indented 2

2. Lip petal blunt at the tip with barely a shallow indentation; common . . . .*L. caurina*
2. Lip with distinct lobes at least 1/5 length of the petal 3

3. Lip minutely puberulent, abruptly narrowed to a basal claw. . . . .*L. convallarioides*
3. Lip glabrous, slightly tapered to the base. . . . . *L. borealis*

Listera borealis Morong **Stems** 5–15 cm, glandular-pubescent above. **Leaves** ovate, 2–4 cm long. **Inflorescence** of 3 to 12 flowers; bracts 1–3 mm long. **Flowers**: sepals 4–5 mm long, linear-lanceolate; lip petal declined but not pendent, 7–11 mm long, broadly oblong with 2 rounded lobes at the tip. **Capsule** ellipsoid, ca. 8 mm long. Moist to wet coniferous forest; montane, lower subalpine. Often beneath spruce at the margin of wetlands; known from Granite and Madison cos. AK to NL south to WA, UT, CO.

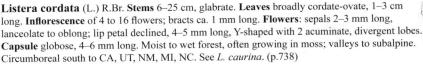

Listera caurina Piper **Stems** 8–30 cm, glandular-pubescent above. **Leaves** broadly elliptic to ovate, 3–7 cm long. **Inflorescence** of 5 to 35 flowers; bracts 3–8 mm long. **Flowers** sepals 2–4 mm long, narrowly lanceolate; lip petal declined but not pendent, 4–7 mm long, pear-shaped, blunt-tipped, gradually tapered to the base. **Capsule** obovoid, 4–7 mm long. Dry to wet forest; valleys to lower subalpine. AK to CA, OR, ID, MT.
   Listera cordata, the other common twayblade, has a smaller, deeply divided lip.

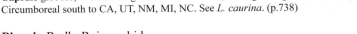

Listera convallarioides (Sw.) Nutt. ex Elliott **Stems** 8–25 cm, glandular-pubescent above. **Leaves** broadly elliptic, 3–7 cm long. **Inflorescence** of 5–25 flowers; bracts 3–6 mm long. **Flowers** sepals 4–7 mm long, linear-lanceolate, reflexed backward with the upper petals; lip petal horizontal, 8–12 mm long, oblong-triangulsr with 2 rounded lobes at the tip and a distinct green claw from 2 small basal lobes. **Capsule** ovoid ca. 8 mm long. Moist to wet forest, thickets; valleys, montane. AK to NL south to CA, UT, CO, MI, NY; Asia.

Listera cordata (L.) R.Br. **Stems** 6–25 cm, glabrate. **Leaves** broadly cordate-ovate, 1–3 cm long. **Inflorescence** of 4 to 16 flowers; bracts ca. 1 mm long. **Flowers**: sepals 2–3 mm long, lanceolate to oblong; lip petal declined, 4–5 mm long, Y-shaped with 2 acuminate, divergent lobes. **Capsule** globose, 4–6 mm long. Moist to wet forest, often growing in moss; valleys to subalpine. Circumboreal south to CA, UT, NM, MI, NC. See *L. caurina*. (p.738)

Piperia Rydb. Rein orchid

Glabrous with tuberous roots. **Leaves** basal and lower-cauline, sheathing the stem, becoming withered at anthesis. **Inflorescence** a terminal, bracteate, spike-like raceme of numerous flowers. **Flowers** greenish to whitish; lateral sepals lanceolate, spreading; upper sepal shorter and ovate; upper petals similar to the upper sepal; lip petal ovate, blunt-tipped, similar to the upper sepal, forming a backward- or downward-pointing spur at the back; style column short, just beneath the upper petals. **Fruit** an ellipsoid to cylindric capsule.
   Formerly placed in *Habenaria*; found in drier habitats than species of *Platanthera*.

1. Spur 3–5 mm long, ca. as long as the lip . . . . *P. unalascensis*
1. Spur >7 mm long, >2 times as long as the lip 2

2. Flowers white, crowded . . . . *P. elegans*
2. Flowers yellow-green, overlapping but not crowded . . . .*P. elongata*

Piperia elegans (Lindl.) Rydb. [*Platanthera elegans* Lindl., *Habenaria elegans* (Lindl.) Bol.]
Stems 25–70 cm. **Leaves** oblong, 7–20 cm long. **Inflorescence** of crowded flowers, 9–30 cm
long; bracts 5–8 mm long. **Flowers** white; sepals 4–5 mm long; lip petal 3–5 mm long; spur
weakly curved, 8–12 mm long. **Capsule** 6–8 mm long. Drier, coniferous forest; valleys, montane.
BC to CA, ID, MT.

Piperia elongata Rydb. [*Piperia elegans* (Lindl.) Rydb. var. *elata* (Jeps.) Luer, *Habenaria elegans* (Lindl.) Bol. var.
elata Jeps.] **Stems** 30–40 cm. **Leaves** oblong, 7–15 cm long. **Inflorescence** of overlapping flowers, 10–20 cm long;
bracts 9–14 mm long. **Flowers** green; sepals 4–6 mm long; lip petal 4–6 mm long; spur weakly curved, 6–15 mm long.
Capsule 5–10 mm long. Moist to wet meadows; valleys. BC, MT south to CA. One collection from Lincoln Co.

Piperia unalascensis (Spreng.) Rydb. [*Habenaria unalascensis* (Spreng.) S.Watson] **Stems**
20–60 cm. **Leaves** oblong to oblanceolate, 6–20 cm long. **Inflorescence** of well-separated
flowers, 6–25 cm long; bracts 2–6 mm long. **Flowers** yellowish-green; sepals 2–4 mm long; lip
petal 2–3 mm long; spur clavate, 3–4 mm long. **Capsule** 5–8 mm long. Moist to more often dry
forest, thickets, streambanks; valleys, montane. AK to QC south to CA, UT, NM, SD. (p.741)

Platanthera Rich. Bog orchid

Glabrous with tuberous or fleshy roots. **Leaves** basal and/or cauline, opposite or alternate, sheathing the
stem. **Inflorescence** a terminal bracteate, spike-like raceme of few to numerous flowers. **Flowers** greenish
to whitish; lateral sepals lanceolate, spreading; upper petals similar to the upper sepal and forming a hood
with it; lip petal lanceolate, blunt-tipped, forming a backward- or downward-pointing spur at the back;
style column short, just beneath the upper petals. **Fruit** an ellipsoid to cylindric capsule.[260]

Formerly placed in *Habenaria*. *Platanthera hyperborea*, *P. dilatata* and *P. stricta* compose a complex
with many intermediates.

1. Leaves all basal, usually 1 or 2; stem leaves absent 2
1. Stems leafy. 3

2. Leaves 2, opposite, appressed to the ground; lip petal 7–13 mm long . . . .*P. orbiculata*
2. Leaves 1 to 3, alternate; lip petal 3–5 mm long . . . .*P. obtusata*

3. Flowers bright white; lip distinctly broadened at the base . . . .*P. dilatata*
3. Flowers yellowish or greenish; lip narrow, gradually tapered 4

4. Spur swollen at the tip, ≤2.5 times as long as wide, ca. half as long as lip petal . . . .*P. stricta*
4. Spur cylindric, >3 times as long as wide, ca. as long as the lip . . . .*P. hyperborea*

Platanthera dilatata (Pursh) Lindl. ex L.C.Beck [*Habenaria dilatata* (Pursh) Hook.] **Stems**
15–70 cm. **Leaves** cauline, 3–20 cm long, lanceolate. **Inflorescence** dense, 4–25 cm long with
10 to 50+ flowers; bracts 5–40 mm long. **Flowers** white; lateral sepals 4–8 mm long; hood 4–6
mm long; lip petal 4–9 mm long, broadened at the base, pendent; spur sac-like to slender, curved,
2–12 mm long. **Capsule** 8–15 mm long. Wet soil of meadows, fens, thickets, open forest, often along
streams, ditches; valleys to lower subalpine. AK to NL south to CA, NV, UT, CO, IL, NY. (p.741)

Our plants have been placed in 3 varieties without geographic or ecological distinction in our area: var. *albiflora*
(Cham.) Ledeb., var. *dilatata* and var. *leucostachya* (Lind.) Ames have spurs smaller than, equal to, or longer than the
lip respectively. Varietal designations based on a single continuous variable seem arbitrary. Hybrids with *Platanthera
hyperborea* [*P. huronensis* (Nutt.) Lindl.] are reported for MT.[363]

Platanthera hyperborea (L.) Lindl. [*P. aquilonis* Sheviak, *Habenaria hyperborea* (L.) R.Br.]
Stems 15–60 cm. **Leaves** cauline, 3–15 cm long, lanceolate. **Inflorescence** dense, 3–20 cm long
with 8 to 50+ flowers; bracts 1–4 cm long. **Flowers** yellowish-green; lateral sepals 3–6 mm long;
hood 2–5 mm long; lip petal 3–5 mm long, narrowly lanceolate, spreading forward; spur cylindric,
curved, 3–6 mm long. **Capsule** 10–15 mm long. Often calcareous fens, wet meadows, seeps; montane.
AK to Greenland south to CA, NM, NE, IN, NY.

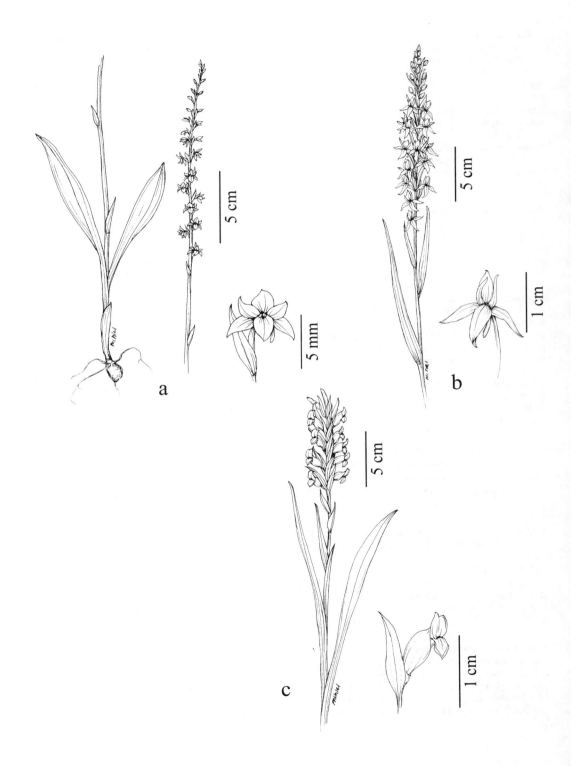

Plate 127. a. *Piperia unalascensis*, b. *Platanthera dilatata*, c. *Spiranthes romanzoffiana*

Sheviak recognized 3 species in what is here considered *Platanthera hyperborea*:[363] *P. hyperborea* sensu stricto with a broadly dilated lip base, is restricted to Greenland and Iceland; *P. huronensis* has a more linear lip and white flowers; and *P. aquilonis* is autogamous with the stamen and stigma touching. The latter 2 species are reported for MT, but are not easy to distinguish from herbarium material;[355] all of the material I have seen from MT would be referable to *P. aquilonis*. *Platanthera stricta* has more elliptic leaves and a shorter, more sac-like spur. See *P. dilatata*.

Platanthera obtusata (Banks ex Pursh) Lindl. [*Habenaria obtusata* (Banks ex Pursh) Richardson] **Stems** scapose, 6–20 cm. **Leaves** basal, petiolate, solitary; blade 3–10 cm long, oblong to elliptic. **Inflorescence** open, 2–8 cm long with 2 to 12 flowers; bracts 5–15 mm long. **Flowers** greenish-white; lateral sepals 3.5–5 mm long; hood 2–4 mm long; lip petal 3–5 mm long, nearly linear, reflexed parallel to the spur; spur slender and curved, 4–7 mm long. **Capsule** 6–8 mm long. Coniferous forest-fen ecotones, wet meadows, often beneath spruce; montane. Circumboreal south to OR, UT, CO, WI, NY. Our plants are ssp. **obtusata**.

Platanthera orbiculata (Pursh) Lindl. [*Habenaria orbiculata* (Pursh) Hook.] **Stems** scapose, 25–60 cm. **Leaves** basal, 2, opposite, appressed to the ground, broadly elliptic, 7–15 cm long, glossy. **Inflorescence** open, 5–20 cm long with 5 to 30 flowers; bracts 5–15 mm long. **Flowers** white; lateral sepals 6–10 mm long; hood 4–6 mm long; lip petal 7–13 mm long, linear-lanceolate, pendent; spur slender, curved, 10–20 mm long. **Capsule** 10–12 mm long, cylindric. Moist coniferous forest; valleys to lower subalpine. AK to NL south to OR, MT, IN, GA. Our plants are var. **orbiculata**.

Platanthera stricta Lindl. [*P. saccata* (Greene) Hultén, *Habenaria saccata* Greene, *H. stricta* A.Rich. & Galeotti,] **Stems** 20–80 cm. **Leaves** cauline, 3–12 cm long, oblong to lanceolate. **Inflorescence** 6–25 cm long with 5 to 40 flowers; bracts 6–40 mm long. **Flowers** yellowish-green; lateral sepals 3–7 mm long; hood 2–5 mm long; lip petal 4–6 mm long, narrowly lanceolate, pendent; spur sac-like, shallowly 2-lobed, 2–4 mm long. **Capsule** 8–11 mm long. Wet meadows, fens, thickets, moist forest openings, more common in non-calcareous soils; valleys to lower subalpine. AK to AB south to CA, NM.

Several specimens from the northwest part of the state have spurs that appear intermediate to *Platanthera hyperborea*, but most of these plants have an open inflorescence and are best referred here.

Spiranthes Rich. Lady's tresses

Glabrous with fleshy or tuberous roots. **Leaves** basal and lower-cauline, sheathing the stem, reduced to mere scales above. **Inflorescence** a terminal bracteate spike of numerous flowers arranged in spiraling columns. **Flowers** white to yellowish; sepals greenish, lanceolate; lower forward-pointing; upper sepal and upper petals crowded together, united below and forming a forward-pointing hood; lip petal tongue-shaped, wavy-margined, concave at the base and enfolding the short column. **Fruit** an ovoid capsule.

1. Lip petal constricted in the middle with an acute tip . . . . *S. romanzoffiana*
1. Lip petal blunt-tipped, not constricted in the middle . . . .*S. diluvialis*

Spiranthes diluvialis Sheviak **Stem** erect, 15–40 cm. **Leaves** linear-lanceolate, 3–20 cm long. **Inflorescence** glandular-puberulent, 2–8 cm long, not as densely-flowered as in *S. romanzoffiana*; bracts 6–12 mm long. **Flowers**: lower sepals spreading-erect; hood 7–12 mm long; lip petal strap-shaped, blunt-tipped, 6–9 mm long with a declined tip. **Capsule** 10–15 mm long. Wet, calcareous, riparian meadows; valleys of Beaverhead and Jefferson cos. WA to MT south to NV, CO, NE. Listed as threatened under the Federal Endangered Species Act.

Spiranthes romanzoffiana Cham. **Stem** 5–40 cm. **Leaves** narrowly oblanceolate, 3–15 cm long. **Inflorescence** viscid, 2–10 cm long with numerous tightly contiguous flowers in 3 or 4 columns; bracts 7–12 mm long. **Flowers**: lower sepals with spreading tips; hood 5–10 mm long; lip petal 7–11 mm long with a dilated, reflexed tip. **Capsule** 6–10 mm long. Wet meadows, fens, occasionally moist grasslands; valleys to subalpine. AK to NL south to CA, NM, NE, IA, NY. (p.741)

Plants in grasslands produce above ground stems only in wet years.

Index

The accepted scientific names of families are in upper-case letters. The accepted scientific names of genera, species, subspecies and varieties are bolded as are the page numbers on which they are described. When a species has only one subspecies or variety in Montana, these subspecific epithets are not listed in the index.

Errata (Second Printing, 2014)

The following errors could lead to a misidentification. I suggest circling these page numbers in the book to alert yourself that a change on that page is needed.

P. 5   2nd paragraph. There are 56 counties in Montana.

P. 41   Group A; second line of couplet 2 should read: Leaves needle-like…..3.

P. 41   Group A; second line of couplet 3 should read Fruit berry-like, dry or juicy….4.

P. 86   *Anemone multifida* var. *tetonensis* may have ultimate leaf segments >3 mm wide.

P. 95   *Ranunculus gelidus* is in the text but not in the key. The short beak and deeply lobed leaves would cause it to key to *R. sceleratus* which occurs below subalpine.

P. 120   *Chenopodium botrys* has glandular but not farinose leaves. It will key to *C. rubrum* in the key.

P. 141   Couplet 13 should read: (13) Petal appendages 2...*S. parryi;* (13) Petal appendages 4–6 ...*S. oregana.*

P. 218   *Rorippa* key couplet 8 should read: (8) Silicles 1–2 mm long…*R. austriaca;* (8) Silicles >2 mm long...*R. palustris.*

P. 220   Figure 34a is *Rorippa sylvestris.*

P. 221   *Rorippa sylvestris* can have a basal rosette.

P. 255   Figure 42m is *Saxifraga mertensiana.*

P. 293   *Astragalus leptaleus* is in the text but not the key. It would fall in Group C, lead 14 or 15 but differs from those species by having creeping underground stems.

P. 374   Flower color of *Centaurium erythraea* is yellowish to pink.

P. 394   Flower color of *Phlox alyssifolia* is white to lilac.

P. 394   Couplet 9 in the key; leaves of *Phlox pulvinata* may or may not be >1 mm wide.

P. 397   Flower corollas of *Polemonium* are blue or occasionally white.

P. 483   Group H, couplet 2; *Xanthisma* has white or tawny bristles.

P. 569   Couplet 9; upper cauline leaves are auriculate clasping but not always ovate.

P. 581   *Xanthisma* has pappus of off-white to tawny spinulose bristles.

P. 634   *Carex scirpoidea* ssp. *scirpoidea* occurs at low to high elevations.